ALLE ZEIT WACH
1842

Julien L. Van Lancker

Molecular and Cellular Mechanisms in Disease

1

Bioenergetics · Cell Specificity · Inborn Errors of Metabolism · Malnutrition · Calcium and Phosphorus Iron and Bile Pigments · Coagulopathies · Hormones Body Fluids and Electrolytes

With 266 Figures

Springer-Verlag
Berlin Heidelberg New York 1976

Professor JULIEN L. VAN LANCKER, M.D., Department of Pathology, U.C.L.A. School of Medicine, Los Angeles, California 90024/U.S.A.

ISBN 3-540-06932-1 Springer-Verlag Berlin Heidelberg New York
ISBN 0-387-06932-1 Springer-Verlag New York Heidelberg Berlin

Library of Congress Cataloging in Publication Data. Van Lancker, Julien L. 1924– Molecular and cellular mechanisms in disease. Includes bibliographical references and indexes. 1. Pathology, Molecular. 2. Pathology, Cellular. I. Title. [DNLM: 1. Pathology. 2. Molecular biology. QZ4 V259m] RB27.V33 611'.0181 76-484

Printed in Germany.

Typesetting, printing, and bookbinding by Universitätsdruckerei H. Stürtz AG, Würzburg

The Sick Rose

O Rose, thou art sick!
The invisible worm
That flies in the night,
In the howling storm,

Has found out thy bed
Of crimson joy,
And his dark secret love
Does thy life destroy.

William Blake (1757–1827)

Preface

In spite of ingenious experiments, imaginative theories, and unshakable faith in supreme forces, there is no way to know how life began. What is certain is that in the course of the development of the universe existing sources of energy fused to generate atoms, and atoms mingled to become small molecules. At some point by chance or design—according to one's belief, but no one's evidence—small molecules such as hydrogen, oxygen, carbon dioxide, water, and ammonia reacted to yield larger molecules with the property most essential to life: self-replication. Such molecules had to achieve a proper balance between the stability needed for their survival in the environment and the mutability for the generation of many forms of life. How amino acids were created or how DNA, RNA, and proteins developed remains a mystery. But we know that a simple core of nucleic acid embedded in a protein coat made the simplest unit of life (except for viroids). Whether viruses are a primitive or degenerated form of life is not known.

Once proteins appeared, their great structural plasticity allowed them to react with other elements such as sulfur, iron, copper, and zinc. After an incalculable number of years, some of the proteins became capable of catalyzing the synthesis of new nucleic acids, new proteins, and other compounds such as polysaccharides and lipids. How macromolecules came in contact and blended in an organized molecular community capable of overcoming the threats of the environment (hyposmolarity, ultraviolet light, heat, and cold) is unknown. Yet, at some time the first cell was born and was able to replicate, generate its own source of energy, and respond to stimuli. Later, as multicellular organisms appeared in the ocean and on the earth's crust, they grew more complex and became functionally and morphologically specialized. Specialized cells were grouped into organs, and as the complexity of the total organism increased, patterns of communication between the various organs—blood circulation, hormones, and the nervous system—became indispensable.

Evolution tells us how man developed in the biosphere ultimately to dominate it. Because of the complexity and the beauty of life, one often forgets that life was threatened as soon as it began. DNA, a child of the sun, was probably one of the sun's first victims. Survival meant adaptation to the environment. What did not adapt did not live.

When the human species appeared, it found itself like other living things surrounded by an environment not entirely hostile, but certainly not always friendly. Men and woman could not escape the cosmic rays, and often they were the victims of the poisons in the plants they ate. The ecological balance of nature is cruel. Life thrives on life, men hunt and slaughter, but they too are the prey of other living organisms—viruses, bacteria, parasites, and fungi.

Like other mammals, men had to develop mechanisms of adaptation and defenses against the environment. When these adaptation and defense mechanisms are overwhelmed, humans lose their health.

In a broad sense, health is a state in which the individual functions optimally—physically and mentally—without the help of drugs and without overt or incipient disease. Disease is a departure from health.

Pathology is the study of mechanisms of disease production. Although since the Renaissance a great deal has been learned about the morphological distortions of disease, no rational explanation of disease mechanisms was available until Virchow developed the cellular theory. Cells are derived from other cells and are the targets of all forms of injury: trauma, radiation, chemicals, nutritional deficiencies, viruses, bacteria, parasites, and fungi. When cells are injured they may adapt and sometimes acquire survival advantages over the other cells of the body (cancer), they may be multilated, or they may die.

Defense mechanisms in response to injury include metabolic alterations, immunological responses, inflammation, blood coagulation, regeneration, and wound healing. In these events cellular and humoral factors interact, but even humoral factors are of cellular origin.

Thus, the cellular theory is as well founded today as it was in the time of Virchow. However, because of the unprecedented deployment of our understanding of molecular and cellular biology during the last four decades, the pathologist can focus more accurately his study and understanding of disease mechanisms. Many injuries and reactions to injuries can be described at the level of integration much simpler than the cellular level.

The primary targets of injury are molecules, molecular complexes, and subcellular organelles. Although in his book on cholera, Snow titled one chapter, "Cholera, a Molecular Disease," molecular pathology originated with Pauling's studies on sickle cell anemia. Pauling's studies demonstrated that the alteration of a single base in DNA was responsible for the complex symptoms of the disease. Since then many molecules, molecular complexes, and subcellular organelles have been found to be the target of cellular injury. Moreover, those molecules involved in reactions to injuries, such as antibodies and components of complement, have been purified and identified. Some reactions to injuries were found to be much more subtle than those observable microscopically. Damage to DNA can be repaired by a sequence of enzymic steps; toxins are detoxified in many molecular steps involving, for example, mixed-function oxidases and conjugation.

An understanding of disease mechanisms is indispensable for accurate diagnosis and appropriate therapy. Therefore, some of our present knowledge about the molecular and cellular mechanisms of disease is presented here. The book cannot be comprehensive because too little is known about the mechanisms of disease, and no one could hope to master all available knowledge.

The book is addressed to students of disease whether they are clinicians or researchers. Hopefully, it will contribute to making medicine more of a science, guide the investigator in unexplored territory, and help the practitioner respond to the request of Macbeth, "...find her disease/And purge it to a sound pristine health" (*Macbeth* act 5, scene 3, line 52).

Even the most clairvoyant among us cannot predict the future of medicine, but it seems likely that as the root of modern pathology grows deeper in the fertile soil of the new biological sciences, the fruit to be harvested will be beneficial to mankind. One can hope that in medicine the prophecy of the poet Andre Van Hasselt will materialize, "...and the fruits will forego the promise of the flowers."

Los Angeles, Spring 1976 J.L. VAN LANCKER

Acknowledgments

Many contributed to the completion of this book. I am especially grateful to Joyce Glover and Ruth Sauber, the typists, Diana Van Lancker, who edited the first drafts, and Mary Lou Motl and Eleanor Dunham for editorial assistance.

Discussions with graduate students, medical students, and colleagues helped me greatly in finding the sources and organizing the material. Special credit is due to Dr. Gloria Hepner who, in 1969, wrote a first draft of the chapter on immunology, and Dr. Vijaya Melnick who read the manuscript of the first two chapters and provided constructive criticism.

Many of the illustrations were assembled during a period of 20 years. I am grateful to my colleagues at the University of Wisconsin, Brown University, and UCLA who helped me collect the material.

The work done in this laboratory, described in some of the chapters, was supported by a Damon Runyon Fellowship, a Career Development Award from the National Institutes of Health, grants from the American Cancer Society, and grants from the United States Public Health Service.

Notes on the References. I am aware that due to the necessity of being selective, I have omitted many important contributions in the text. Moreover, it was not practical to quote every reference. Instead, I have, whenever possible, provided the reader with a list of reviews that should lead him to the information mentioned. Individual references were mentioned only when they were recent or were judged to be especially critical to the development of our knowledge in the field.

Contents · Volume 1

Contents · Volume 2

Chapter 1

Cellular Sources of Energy

When in the nineteenth century pioneering pathologists discovered histopathology, it became obvious that the cell, the structural unit of living tissue, is also the structural target of injury. At first, injury could be described only at the level of those structures observable with the light microscope. Today, with the advances of biochemistry and electron microscopy, the causes of disease can often be traced to molecular and ultrastructural levels. The goal of this first chapter is to describe those molecular mechanisms that govern the sources of energy and specificity in the cell.

Tissue Fractionation Techniques

Before studying the intracellular distribution of various biochemical compounds and multiple enzyme systems, the preparative methods of cellular organelles should be described and evaluated [1–11].

Homogenization of Tissues

The discovery of the homogenate method by Potter and Elvehjem [12] may well be one of the most significant steps in cellular biochemistry. For a long time, investigators had sought to break each individual cell while keeping the intracellular components intact. This aim was at least partially achieved with a simple and ingenious instrument called the Potter-Elvehjem homogenizer, a glass tube in which a glass or plastic piston moves freely. The tissue is suspended in an appropriate solution and compressed between the pestle and the wall of the tube. The area is too small for an intact cell, but permits nuclei, mitochondria, fragmented endoplasmic reticula, and soluble fractions to pass freely between the piston and tube.

A perfect homogenate should contain no intact cells, and the cellular fractions (nuclei, mitochondria, and microsomes) should be undamaged and distributed uniformly in the suspension without aggregation. This, however, is never achieved. Instead, a small percentage of cells remains intact, some of the granules are destroyed, and one fraction is always contaminated by another.

Despite these limitations, studies on homogenates reflect the events in the intact cell and present advantages over histochemical methods, because reactions can be better quantitated and the medium is better controlled. However, in contrast to histochemical techniques, which provide information on individual cells, homogenate studies must deal with information gathered from a nonhomogeneous cellular population. This disadvantage can seldom be overcome. (There is an exception in liver, where reticuloendothelial cells can be separated from parenchymal cells.)

Such factors as the nature of the tissue, pH, temperature, ionic strength, and osmotic pressure of the medium greatly influence the quality of a homogenate. Whereas some tissues such as liver, brain, and pancreas, are easy to obtain in uniform suspensions, others (skin, for example) are practically impossible to homogenize with standard methods. Efficient homogenization is achieved only after freezing with liquid air, a procedure that cannot be followed by differential centrifugation of the cell fraction.

Inasmuch as the pH of liver homogenates usually remains between 6 and 7, its variation is seldom a problem. In contrast, in kidney homogenates, the pH drops rapidly and the cytoplasmic granules are destroyed if buffer is not added.

To prevent bursting of cytoplasmic organelles, it is necessary to homogenize in saline or sucrose solutions. Bensley and Claude [13] developed the electrolyte medium, Schneider [14] was responsible for using 0.25 M sucrose, and investigators at the Rockefeller Institute later introduced the 0.88 M sucrose medium. In general, although the 0.88 M sucrose preparations are preferred for electron microscopic examination, the 0.25 M solutions are usually more appropriate for biochemical studies. The high viscosity of the 0.88 M homogenate causes considerable friction, leading to the rupture of some of the cytoplasmic organelles during homogenization. In preparing mouse pancreas homogenate, for example, a high proportion of granules containing acid phosphatase and zymogen granules are ruptured when the tissue is homogenized in 0.88 M sucrose.

When homogenates other than liver are prepared, the relative advantages of various homogenizing media should be studied. Remember that 0.25 M sucrose is hypotonic, and 0.88 M sucrose is hypertonic. Neither medium provides intact mitochondrial preparations. The mitochondria swell in the first instance, and a large portion are lost in 0.88 M sucrose. To obviate these disadvantages, the use of 0.44 M sucrose solutions containing traces of citric acid (pH 6.2) has been proposed. Raffinose solution supplemented with dextran has also been used as a homogenizing medium. However, the value of these methods remains uncertain, since few morphological and biochemical studies on mitochondria prepared in such media are available.

Dextran and polyvinylpyrrolidone (PVP) solutions have also been used to prepare cell fractions for electron microscopy or for biochemical analysis. These media have few advantages over the classical sucrose solution, and the high nitrogen content of PVP makes nitrogen determination of the cell fraction difficult. Finally, organic solvents have also been used to prepare tissue homogenates for differential centrifugation with the hope of preventing leakage of water soluble constituents from their respective organelles. Homogenates prepared in such solvents usually provide the investigator with cell preparations that are difficult to identify morphologically, and hence the biochemical information obtained from such preparations is often difficult to interpret.

Differential Centrifugation Technique

When particulate matter is suspended in a liquid medium, it settles at the bottom of the container under the influence of gravity (see Fig. 1-1). Centrifugation

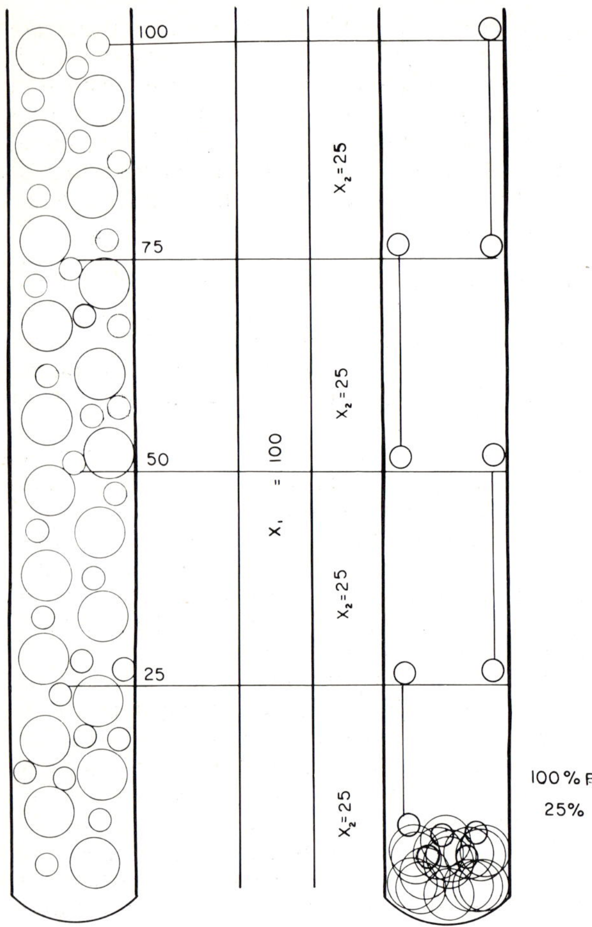

Fig. 1-1. Distribution of cellular components in homogenizing tube

considerably increases the gravitational field because a new force, the centrifugal force, is exerted on that particle. Since the sedimentation rate of spherical particles depends on their size and density, the various types of cellular organelles can be separated by refined centrifugation techniques. To convey a better understanding of these technical procedures, it is worthwhile to outline the theory on which they are based.

The force acting on a body sedimenting in the gravitation field is a function of the mass of the body: $f=mg$ where m is the effective mass and g is gravity $=980$ cm/s. This formula must be modified to calculate the forces acting on particles in a centrifugal field. The reader will readily understand the conversion if he keeps in mind the following: (1) The mass of a particle is equal to the product of the volume (V) and the density ρ, or $m=V\rho$; (2) the volume of a sphere of radius a equals $4\pi\, \alpha^3/3$; and (3) in keeping with Archimedes' principle, the effective density of a body plunged in a liquid medium is equal to the difference between the density of that body and the density of the medium.

Thus, when a body falls in a medium of density:

$$f_1 = \frac{4\pi\, a^3(\rho-\rho_1)}{3} \times g.$$

If the medium is a liquid, the viscosity of the medium exerts friction on the particle, interfering with its free motion. Consequently, in order to cause further downward movement, a force must be applied on the particle according to Stokes's law:

$$f_2 = 6\pi\, a\, \eta \frac{dx}{dt}$$

where a is the radius of the particle, η the viscosity of the medium, and dx/dt the velocity through the medium.

When the falling sphere has reached a constant speed equal to the gravitational field, $f_1 = f_2$:

$$\frac{4/3\pi\, a^3(\rho-\rho_1)}{3} \times g = 6\pi\, a\, \eta \frac{dx}{dt}$$

solving the equation for dx/dt:

$$\frac{dx}{dt} = \frac{2g\, r^2(\rho-\rho_1)}{9\eta}. \tag{1}$$

A free-moving body tends to move in a straight line. When it is directed along a curved path, as is the case in the centrifuge, it still tends to progress tangentially. This continuous effort to pursue a tangential path exerts a new force on the particle that acts against the force maintaining the particle on the curved track. Where ω is the angular velocity, and x a point distant from the axis of rotation, this centrifugal force can be stated $\omega\, 2x/g$. Thus it follows from (1):

$$\frac{dx}{dt} = \frac{2g\, a^2(\rho-\rho_1)}{9\eta} \times \frac{\omega^2 x}{g} = \frac{2a^2\, \omega^2\, x(\rho-\rho_1)}{9\eta}.$$

If this equation is solved for dx/x and integrated, it can be rewritten:

$$1 \frac{x}{x_0} = \frac{2\omega^2\, a^2(\rho-\rho_1)}{g} (t-t_0).$$

Now the value $(1\, x/x_0)/\omega^2(t-t_0) = S$ is a constant for a given particle moving in a given medium at a specific temperature. It can be derived from (2):

$$S = \frac{2a^2(\rho-\rho_1)}{9}. \tag{2}$$

When the physical chemist studies sedimentation characteristics of a protein preparation, he determines these values as a function of time. For that purpose, a light beam is passed through the sample and with the appropriate optical devices, the rate at which the particle moves can be recorded on a photographic plate. If the system is monodispersed (containing protein molecules of the same size), a single boundary is obtained. If the system is polydispersed, several boundaries appear.

Once the values of S are known, it is simple to calculate the size of the particle and deduce its molecular weight. Unfortunately, such precise physicochemical methodology cannot be applied to the separation of cellular organelles. In such experimentation the centrifugal forces used are determined empirically. Furthermore, because of the intricate structural relationships between granules, the adsorption of one type of granule onto another, and the great variety of sizes within each group of major cellular organelles, a pellet containing only a single type of cellular organelle has never been prepared.

Despite these limitations, pellets containing the bulk of zymogen granules, nuclei, mitochondria, and microsomes can be prepared easily by differential centrifugation. Such relative success is due precisely to the difference in the sedimentation characteristics of the various types of cellular organelles.

Let us for a moment make the following assumptions: two spherical particles of identical densities but different sizes are uniformly suspended in a homogeneous medium. The radius of one of these particles equals the arbitrary value of 10, and the radius of the other is 5. This suspension of particles is centrifuged at centrifugal forces just sufficient to sediment all the larger particles. Under these conditions, the maximum distances traveled by both particles equal x_1 and x_2, and the ratio of these distances expresses the degree of separation of the two particles. In view of the above assumptions, the ratio x_1/x_2 is only a function of the ratio of the square of the radius of the large particle over the square of the radius of the small particle (see Fig. 1-1). Then x_1/x_2 is equal to 4.

$$\frac{x_1}{x_2} = \frac{A^2}{a^2} = \frac{100}{25} = 4.$$

Under these conditions, the pellet of large particles will be contaminated by one-quarter of the small particles in the tube. But it is also clear that a single washing will reduce the contamination of the pellet to one-quarter of what it was in the unwashed pellet.

According to the conditions of our assumptions, the ratio of A/a equals only 2, and the separating capacity is 4. The latter value theoretically should be much greater when nuclei are separated from mitochondria or mitochondria from microsomes. The respective average diameter of these cell structures is 6 μ, 1 μ, and 0.15 μ, so that the separation capacity between nuclei and mitochondria should be at least 1/10, and one or two washings should reduce the contamination of the larger particles by the smaller to practically zero.

Such a degree of separation is never achieved for at least three reasons: (1) the size of mitochondria and microsomes varies widely; (2) small particles tend to aggregate or be absorbed by larger cellular structures; (3) some cellular organelles are readily fragmented, and the fragments are either absorbed by other cellular components or are sedimented with smaller particles.

Separation of Cellular Components

In classical tissue fractionations, homogenate studies are divided by differential centrifugation into four parts: the nuclei, the mitochondria, the microsomes, and the supernatant fluid (see Fig. 1-2). In the case of 0.25 M homogenates, the nuclear pellet can be sedimented with 600 *g* for 10 minutes, the mitochondria with 10,000 *g* for 5 minutes, and the microsomes at 100,000 *g* for 15 minutes. In liver, the mitochondrial pellet can be further divided into a fraction containing the large mitochondria and another containing the smaller mitochondria.

When the variation in size of cellular organelles is to be studied, two methods are available—centrifugation in gradient concentrations of sucrose and preparation of many pellets. In the first method, the sucrose concentration may be gradually increased by adding, with the aid of a special device, a concentrated solution (for example, a 33% solution) to a tube already containing a sucrose solution. Another method consists of layering, by careful pipetting, various concentrations of sucrose within the tube.

After centrifugation of the tubes containing a constant gradient of sucrose (usually in the swinging bucket rotor of the Spinco model L of the ultracentrifuge), the fluid is forced to flow out of the tube by injecting a concentrated sucrose solution (for example, 50%) with a needle inserted at the bottom of the plastic tube. If the investigator is interested only in biochemical determinations unaffected by freezing or thawing, the entire tube may be frozen and cut at various levels. Each fraction can then be tested for enzyme activity.

Kuff and Hogeboom [15] were responsible for initiating the gradient concentration methods. They first studied the separation of polystyrene particles of known diameter. By plotting the fraction of the original concentration on the ordinate and the distance from the meniscus in the tube on the abscissa, they obtained an S-shaped curve. Such a plot gives the concentration in each portion of the tube but does not give a good representation of the variation in particle size. Size variation can be obtained by plotting the increment of the concentration in each fraction versus the calculated particle size. With a suspension of uniform size particles, a sharp peak is expected.

Of course, size variation is not the sole factor determining the particles' sedimentation rates—the density and shape of the particles are also involved. Density equilibration offers only approximate information about the former, and it is impossible to define the latter of these parameters. Therefore, calculations based on the assumption that the particles are spheres and that the density is 1.2 g/cm express only a relative truth.

The method of fractionation in a concentration gradient proved useful in studying the heterogeneity of the mitochondrial fraction and the origin of some enzymes associated with the nucleus. By preparing a large number of cell fractions, Novikoff [16] was the first to demonstrate that acid phosphatase is found in a particle of a different size from the particle containing cytochrome oxidase. These findings were then extended in Kuff's laboratory by using ultracentrifugation methods in gradient concentrations of sucrose. The De Duve group extended

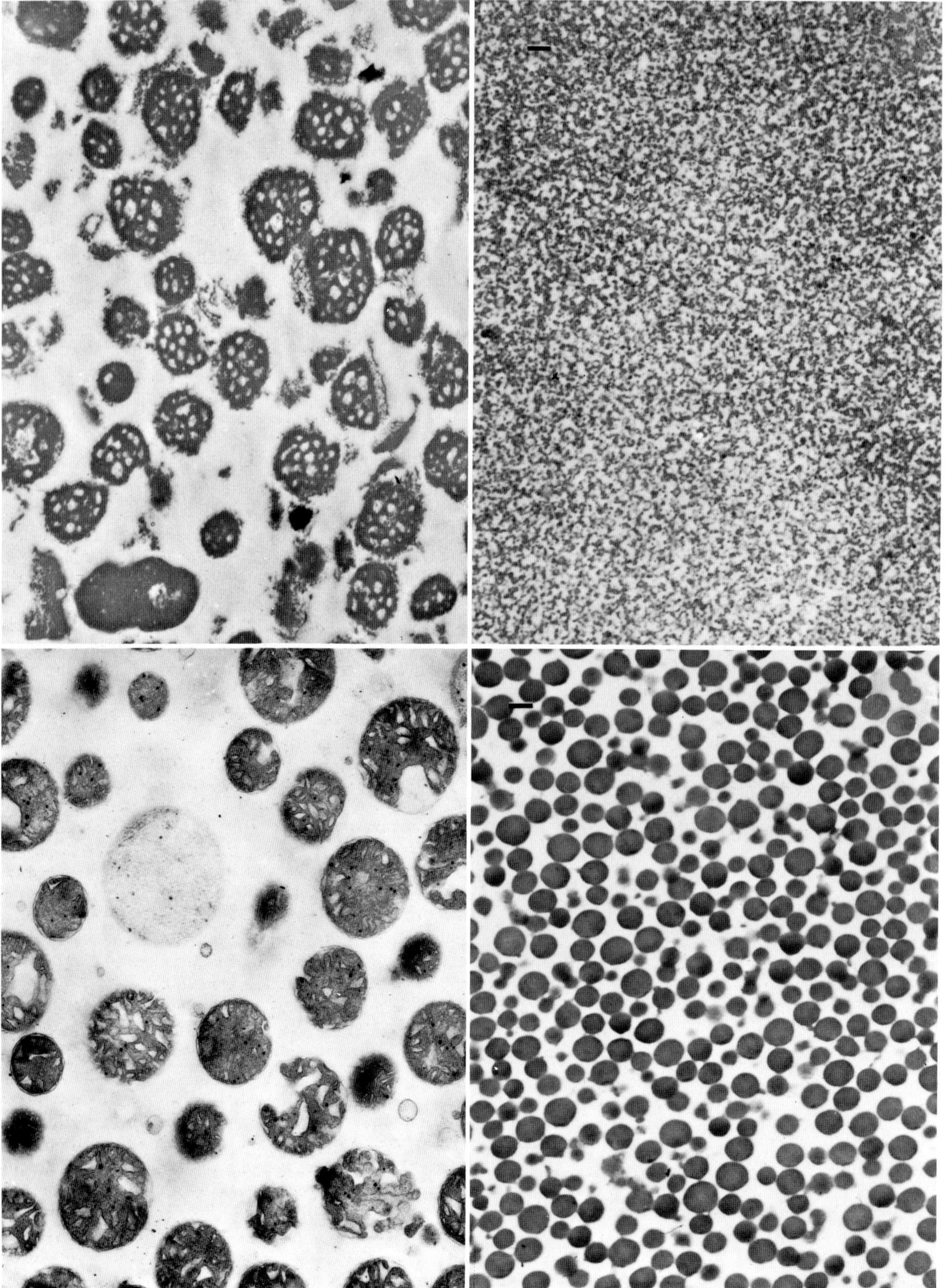

Fig. 1-2. Cellular organelles obtained by tissue fractionation. Electron micrograph of the nuclei (upper left), mitochondria (lower right), microsomes (upper right), and zymogen granules (lower right) prepared by differential centrifugation. The nuclear preparation picture was provided by Palade, the others were prepared in collaboration by Van Lancker and Swift

this research on acid phosphatase to other hydrolases, and their investigations culminated in the concept of "lysosomes", which is discussed in the chapter concerned with cellular necrosis.

Roodyn [17] placed a suspension of liver nuclei in a tube containing several layers, each with a different sucrose concentration. With this method he demonstrated that after centrifugation the succinoxidase activity associated with the nuclear pellet is found in a layer different from the one containing DNA. In contrast, after centrifugation in layered sucrose solution, a large portion of the aldolase associated with nuclear suspension remained with the layer that contained DNA. These findings led Roodyn to conclude that some aldolase activity is native to the nucleus.

Intracellular Distribution of Biochemical Components

Some cellular biochemical constituents, such as DNA and the glycolytic enzymes, are found in every cell, while others (pepsin) are limited to a few. The particular mosaic formed is an expression of differentiation. Among cellular components, those common to all cells have been most extensively investigated. The biochemical cytology of only liver tissue has been studied on a large scale, and much of our knowledge of cellular components is based on information gathered from it.

This review of the biochemical composition of the cytoplasmic fractions is divided into three parts: (1) the distribution of basic cellular constituents (nucleic acids, protein, nitrogen, and phospholipids); (2) the study of those constituents in a specific cytoplasmic organelle; and (3) the intracellular distribution of multiple-enzyme systems.

Distribution of Basic Cellular Constituents

The amount of nitrogen in each fraction obtained by ultracentrifugation is an excellent indication of the size of that fraction in the total homogenate. In rat liver, 18% of the nitrogen is associated with the nuclear fraction, 30% is in the supernatant, and the remaining 52% is distributed between the mitochondrial and microsomal fractions (17% and 37%, respectively). In mouse pancreas, there is three times as much microsomal as mitochondrial nitrogen. This is in keeping with the high rate of protein synthesis and low cytochrome oxidase and succinic cytochrome c reductase activity in the pancreas.

In contrast to DNA, which is exclusively nuclear, RNA is found in both nucleus and cytoplasm. The bulk of RNA is associated with the microsomes and the supernatant. The association of RNA with other cytoplasmic granules remains controversial. While Novikoff [18] has prepared a mitochondrial fraction completely free of RNA, most other investigators have found a small portion (from 1–5%) of RNA in the mitochondrial pellet. If mitochondria were capable of autoduplication, the absence of nucleic acids would be puzzling. Except for the zymogen granules of the pancreas, few cytoplasmic organelles other than mitochondria or microsomes have been analyzed for RNA. It is unanimously concluded that a small portion of RNA is associated with the zymogen granules prepared from mouse, rat, or guinea pig pancreas; however, this also might be a contaminant.

Although a small fraction of the phospholipids is associated with the nuclear fraction, 27% is mitochondrial. The exact intraorganelle localization and role of phospholipids remain unknown, but their presence in mitochondria is thought to be related to the electron transport mechanism. Microsomes are the intracellular granules richest in phospholipids.

Reference Compounds

Cell fractions presumably could best be identified by electron microscopic examination. This method, however, has its limitations; it is not quantitative, and preparations optimal for morphological examination are often unsuitable for biochemical analysis. Furthermore, electron microscopy is often impractical in the routine biochemistry laboratory.

Sometimes biochemists must identify cell fractions by using reference biochemicals or "markers"—compounds found in high proportion in a specific organelle. Obviously, all fractions are best characterized when submitted to correlated biochemical and morphological analysis.

Ideally, a large part of a reference compound should be associated with a specific organelle; it should be stable and not leak out of the granules during the homogenization and fractionation procedure; it should not be artificially adsorbed; there should be no interference with the quantitative determination of the reference element by a similar substance; and, finally, if an enzyme is used as a marker, the method of assay should measure all the potential activity associated with the specific organelle.

DNA is probably the most typical of all reference compounds; more than 90% of DNA is in the nuclear pellet, and the small amount in other fractions results from contamination. Eighty per cent of cytochrome oxidase or succinic cytochrome c reductase is associated rith the mitochondrial fraction. By counting the mitochondrria in the nuclear fraction, Schneider and Hogeboom [19] demonstrated that the cytochrome oxidase activity measured in this pellet could be accounted for by contaminating mitochondria.

No reference compound existed to identify the microsomal fraction until Hers and coworkers demonstrated that 88% of glucose-6-phosphatase is recovered in the microsomal pellet. Unfortunately, since that enzyme is present only in a few tissues, several constituents usually must be analyzed to characterize the microsomal fraction. In the absence of a specific enzyme, the safest marker is the ratio of RNA to

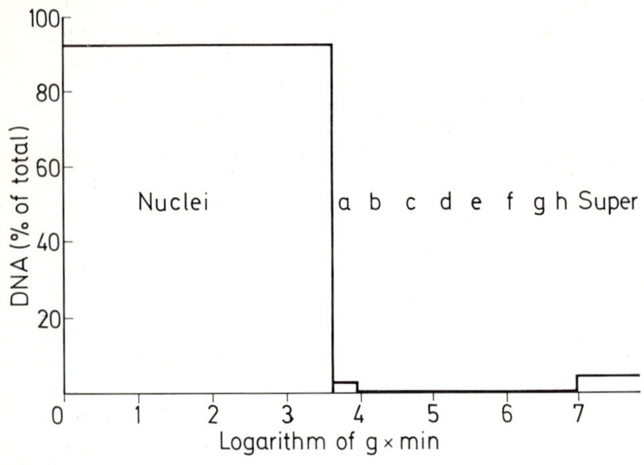

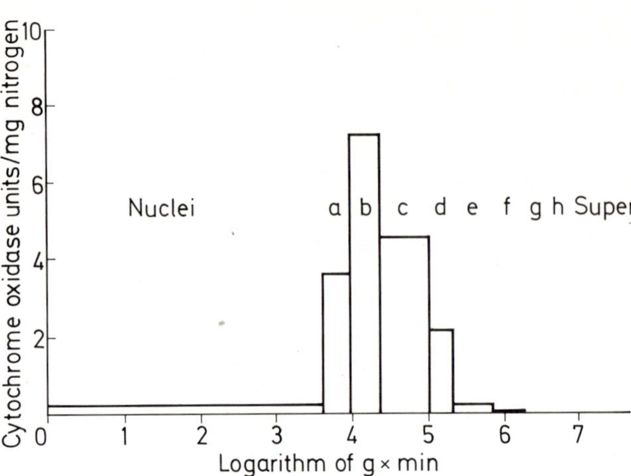

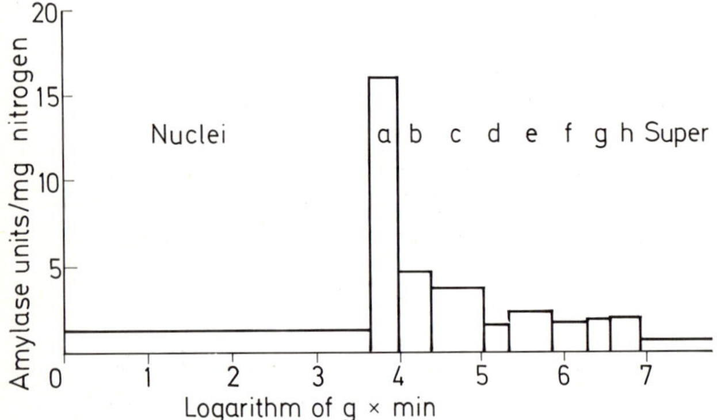

Fig. 1-3. The distribution of DNA, cytochrome oxidase, and amylase in per cent of total amounts in the homogenate or in units of activity per milligram of nitrogen in nuclei and 9 cytoplasmic fractions obtained by increasing the centrifugal forces logarithmically. 95 per cent of the cytochromal oxidase and 15 per cent of the amylase are associated with mitochondrial or zymogen granular fractions, respectively (Ref.: Van Lancker, J.L. and Holtzer, R.L.: Tissue fractionation studies of mouse pancreas. J. Biol. Chem. **237**, 2359–2363, 1959)

nitrogen. On the basis of specific sedimentation characteristics associated with typical enzyme composition, De Duve and coworkers have defined a new type of granule called lysosomes. Although only 40% of acid phosphatase is associated with these granules, the authors consider it an excellent marker.

In the pancreas, amylase might have been a marker for the zymogen granules if a high portion had not been associated with the microsomes. Fortunately, since amylase binding in the microsome is different from that in the zymogen granules, the contamination of each cell fraction by either zymogen granules or microsomes can be estimated by measuring the amylase activity in various media (*e.g.*, 0.25 M sucrose or distilled water, see Fig. 1-3).

In conclusion, the morphological identification of each cell fraction should be completed by its biochemical characterization. This should include the determination of basic compounds (nitrogen, nucleic acids, and phospholipids) and the assay of specific reference enzymes (cytochrome oxidase, glucose-6-phosphatase, and acid phosphatase).

Distribution of Multiple-Enzyme Systems

In the next part of this book, the intracellular localization of some of the most important metabolic functions is discussed. These functions can be divided into three major groups: first, the bioenergetic pathways that provide the chemical energy necessary for cell maintenance, repair, response to stimuli, and reproduction; second, the biosynthetic pathways that provide the building blocks and the macromolecules of the cell's genetic material, biological catalysts, and structural elements; and third, the catabolic pathways, many of which remain poorly understood, although some are known to yield metabolites eliminated through the excretory organs.

Injury affects one or more of the individual steps of these pathways in ways that will become apparent later. Even if an injury primarily affects a single step in a single pathway, the biochemical lesion is soon reflected in several more steps and in other pathways. Proper interpretation of the various biochemical alterations that result from a primary injury requires detailed knowledge of all main and alternative pathways and their enzymes and coenzymes.

Bioenergetic Pathways

Glucose is an important source of energy and building blocks. In mammalian tissue, it is degraded essentially in four metabolic pathways: glycolysis, the hexose monophosphate shunt, the tricarboxylic acid cycle, and the uronic cycle.

Glycolysis

Pasteur first recognized the ability of living cells to derive energy from glucose in the absence of oxygen by a process similar to fermentation. After Meyerhof discovered the transformation of glycogen to lactic acid by a muscle extract, it took 50 years of research to demonstrate each step of the glycolytic pathway as it now appears on metabolic maps. Glycolysis is now thought to exist in all bacteria, plant, and animal cells [20–29].

Inasmuch as insulin facilitates the cell's utilization of glucose, it seems logical to introduce the study of glycolysis by analyzing the mode of action of insulin on glucose utilization. However, insulin's mode of action is so closely connected with studies on diabetes that the description of the hormone's mechanism of action is deferred to the section on diabetes.

In liver, glucose enters the glycolytic cycle after phosphorylation to glucose-6-phosphate, a reaction catalyzed by hexokinase in the presence of ATP. Glucose-6-phosphate is then transformed into its isomer, fructose-6-phosphate (see Fig. 1–4), which after the fixation of a second molecule of phosphorus is split into dihydroxyacetone phosphate and glyceraldehyde-3-phosphate. In a single step glyceraldehyde-3-phosphate is oxidized and again phosphorylated to form 1,3-diphosphoglycerate. Because this oxidation reaction occurs in the absence of oxygen, it must be linked to a reduction—the reduction of NAD to NADH. NAD, in turn, is regenerated when lactic acid is formed from pyruvic acid in the presence of lactic dehydrogenase (see Fig. 1-6).

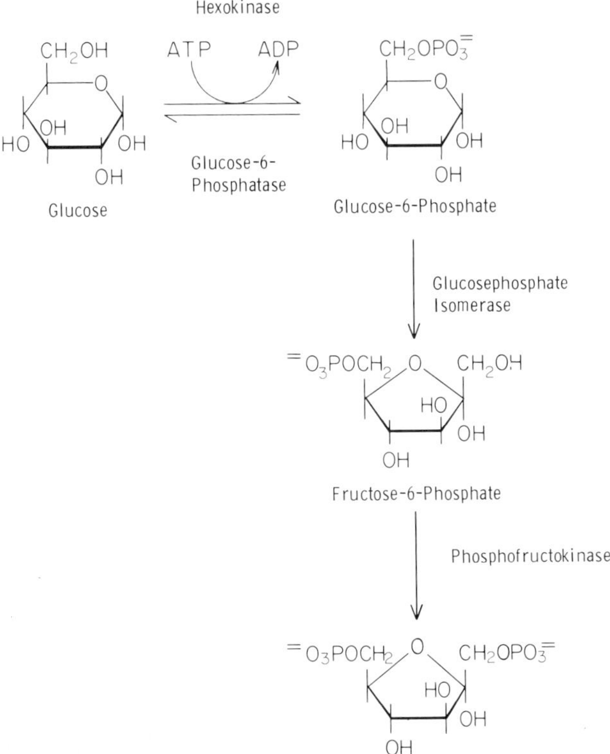

Fig. 1-4. Glycolytic pathway conversion of glucose to fructose-1,6-diphosphate

During glycolysis, four molecules of ATP are generated, but because two ATP molecules are involved in the formation of fructose-1,6-diphosphate, the net gain in ATP is only two molecules per molecule of glucose. Furthermore, the high-energy phosphate of glycolysis is available only when 1,3-diphosphoglycerate or phosphoenolpyruvic acid is formed, because the other phosphorylated intermediates contain low-energy phosphate bonds. In muscle, the effective use of this chemical energy is associated with the transfer of phosphorus from ATP to creatine by the catalytic action of creatine kinase to form phosphocreatine (an extremely labile organic compound).

Hexokinase and Glucokinase

Glucose can enter metabolism only through its conversion to glucose-6-phosphate. Hexokinase is the enzyme catalyzing that reaction at the expense of ATP [30–32].

$$\text{Glucose} + \text{ATP} \leftrightharpoons \text{Glucose-6-phosphate} + \text{ADP}$$

In discussing hexokinase, liver and all other sources of the enzyme, must be distinguished. Hexokinase is present in all cells. It has been purified from various mammalian tissues, and has been crystallized from yeast (molecular weight 94,000). The specificity of hexokinases prepared from various tissues has been investigated. Crystallized yeast hexokinase has a rather broad specificity for the phosphate acceptor. It will phosphorylate several sugars—glucose, mannose, fructose, and galactosamine. Its affinity for glucose, fructose, and galactosamine is equal; its affinity for mannose is half that for glucose. D-Galactose, D-xylose, L-arabinose, L-rhamnose, sucrose, lactose, maltose, trehalose, and raffinose are not attacked by the enzyme. The attack of at least four different hexoses by a crystalline enzyme could be explained by cocrystallization of four different enzymes with similar physicochemical properties. This is unlikely because the amount of sugar hydrolyzed is considerably reduced if the four substrates are added to the incubation mixture simultaneously. This indicates that the substrates compete for a single active center.

Mammalian hexokinase exists in four major isoenzymic forms that can be separated by electrophoresis. The distribution pattern is the same in different tissues of rat, hamster, rabbit, monkey, and cow for types I, II, and III. Type IV has been found only in liver [33].

Tissues vary, however, in the relative amounts of the various isoenzymes present, and isoenzymes have been named after the tissues in which they predominate. Thus, type I is called brain hexokinase, type II muscle, and type IV liver. Type III seems to be present in approximately similar amounts in all investigated tissues.

Sols and Crane [34, 35] have investigated the specificities of various mammalian hexokinases. The specificities of heart, muscle, intestine, and brain hexokinase seem to be similar. Brain hexokinase attacks pentoses and heptoses in addition to glucose. Changes in the length of the carbon chain of the substrate do not abolish activity; however, the affinity of hexokinase for pentose and heptose is much lower than that for glucose. While the K_m (in moles per liter) is 8×10^{-6} for glucose, it is 2×10^{-4} for glucoheptulose, and 2 for arabinose. Steric alterations or substitutions in carbons 1, 2, and 3 also reduce the affinity of the substrate for the enzyme. Mannose is a relatively good substrate, but hexokinase has a lower affinity for fructose and is practically inactive on galactose.

These studies on the specificity of hexokinase are not without physiological significance. As already pointed out, all sugars used for metabolism must be phosphorylated. When hexokinase is the only enzyme capable of phosphorylating sugars, the relative affinity of hexokinase for these various cell sugars determines their use in metabolism. On the other hand, if a specific kinase exists for each sugar metabolized, the relative proportion of sugar to enter metabolism depends upon the relative enzyme activities. In this respect, it is interesting that *Schistosoma mansoni* have four different hexokinases for glucose, fructose, mannose, and glucosamine. In yeast, a single enzyme acts on these four substrates. In adult mammals, the situation rather resembles that of *Schistosoma*.

The exact role of the isoenzymic form of hexokinase and the variability in tissue specificity need to be better understood. Several models for hexokinase's mode of action have been proposed, but too little is known of the molecular structure of the enzyme to permit a detailed description of the molecular contacts between enzyme and substrate. The reader is referred to specialized review articles on the subject [36]. The role of hexokinase in the regulation of glycolysis is discussed in a separate chapter. Suffice to point out here that the enzyme is inhibited by glucose-6-phosphate and that the inhibition is relieved by P_i.

An addition to the classical hexokinases I, II, and III, the liver contains another enzyme capable of phosphorylating D-glucose, namely, glucokinase. Glucokinase differs from the other hexokinases in at least three ways: (1) the K_m of glucokinase (10^{-2} M) for D-glucose is 3,000 times greater than the K_m of hexokinase (10^{-5}); (2) in contrast to the other hexokinases, which are not altered by dietary or hormonal changes, glucokinase activity is reduced in diabetes or after administration of a low-carbohydrate diet. Insulin or carbohydrate administration restores glucokinase activity to normal by triggering the *de novo* synthesis of the enzyme; (3) glucokinase is absent in fetal liver and appears only one week after birth.

The high K_m of glucokinase is of considerable physiological interest. The amount of glucose that the enzyme can phosphorylate in a unit of time is greater than the level of glucose in the blood circulation. Consequently, the level of circulating glucose determines the amount of glucose phosphorylated by the liver [37–42].

Already in 1929, it was observed that glucose utilization in liver depended upon glucose concentration in blood. When the glucose level is below 151 mg/100 ml of blood, the liver pours glucose in the blood (gluconeogenesis). Only when the concentration of glucose is higher will the liver take in glucose (for glycogen synthesis or glucose dissimilation). Consequently, the affinity for glucose of the molecule that regulates the entry of glucose in the liver cell must be lower than that of similar molecules in other tissues. Furthermore, in view of the effect of insulin on carbohydrate metabolism in liver, the glucose-regulating molecule should be insulin sensitive. Liver glucokinase fulfills these requirements.

The hexokinase activity has also been measured in the course of development of the fetal liver and showed that only one type of hexokinase existed in the newborn liver, whereas two were found in adult liver. The adult liver contains a hexokinase and a glucokinase with high affinity for glucose; the former is inhibited by glucose-6-phosphate, and the latter is not. Only the low-K_m hexokinase is found in fetal liver (high affinity for glucose). In contrast, the high-K_m glucokinase develops only after birth. The significance of these two enzymes in controlling carbohydrate metabolism in health and disease will become evident later.

The existence in the same tissue of two different enzymes involved in phosphorylating glucose raises intriguing problems. For example, is glucokinase present in all hepatic cells or only in cells devoid of hexokinase? Does a new hepatic cell containing glucokinase, but not hexokinase, appear during the first week after birth?

Phosphoglucose Isomerase

Glucose-6-phosphate is converted to fructose-6-phosphate in the presence of a phosphoglucose isomerase. The enzyme was discovered and purified from muscle where it was separated from a phosphomannose isomerase, which catalyzes the conversion of mannose-6-

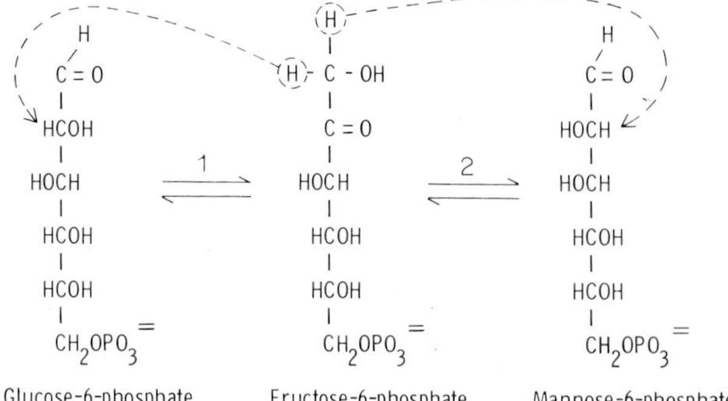

Fig. 1-5. Mechanism of action of phosphohexose isomerases. *1* Phosphoglucose isomerase, *2* Phosphomannose isomerase

phosphate into fructose-6-phosphate. These two phosphohexose isomerases differ in at least two ways—ion requirement and specificity. Phosphomannose isomerase requires metallic ions for activity, but no such requirement is known for phosphoglucose isomerase. Each of these enzymes attacks a different hydrogen on the carbon 1 of fructose. The mechanism of action of the two hexose isomerases is demonstrated in Fig. 1-5.

Sorbitol-6-phosphate is a competitive inhibitor of phosphoglucose isomerase. In contrast, 1,5-sorbitan-6-phosphate has no effect on the enzyme activity, which suggests that the phosphohexose isomerase substrate must be present in the open form. Thus, before a sugar such as glucose-6-phosphate can be acted upon by phosphoglucose isomerase, an opening of the ring is imperative. The mechanism by which such a molecular ransformation occurs, however, remains obscure [43].

Phosphofructokinase

Fructose-6-phosphate is further phosphorylated to yield fructose-1,6-diphosphate, and then hexose diphosphate is split to yield two triose phosphate sugars—dihydroxyacetone phosphate and D-phosphoglyceraldehyde. The phosphofructokinase and aldolase reactions are discussed when fructose metabolism is reviewed. We are concerned here only with the fate of the triose phosphates (see Fig. 1-6).

Triose Phosphate Isomerase

An enzyme catalyzes the interconversion of dihydroxyacetone phosphate into D-phosphoglyceraldehyde in presence of NAD. Thus, triose phosphate isomerase breaks a carbon-hydrogen bond in the hydroxymethyl group of the D-phosphoglyceraldehyde to yield dihydroxyacetone phosphate. The equilibrium of that reaction favors the formation of the dihydroxyacetone phosphate. From the description of the glycolytic pathway, it is evident that dihydroxyacetone phosphate is produced in two different enzymic reactions, catalyzed by aldolase or triose phosphate isomerase. The exact mechanism of the reaction is not known, but it has ben suggested that it involves the formation of an enolate anion that is bound to the enzyme.

Glycerophosphate Dehydrogenase

Dihydroxyacetone phosphate is converted catalytically to L-α-glycerophosphate under the influence of glycerophosphate dehydrogenases, of which there are at least two. One is found in the soluble fraction of muscle homogenates; it has been crystallized and experiments have shown that the enzyme molecule contains NAD. The role of this enzyme has been clarified. Most of the original studies were done on insect

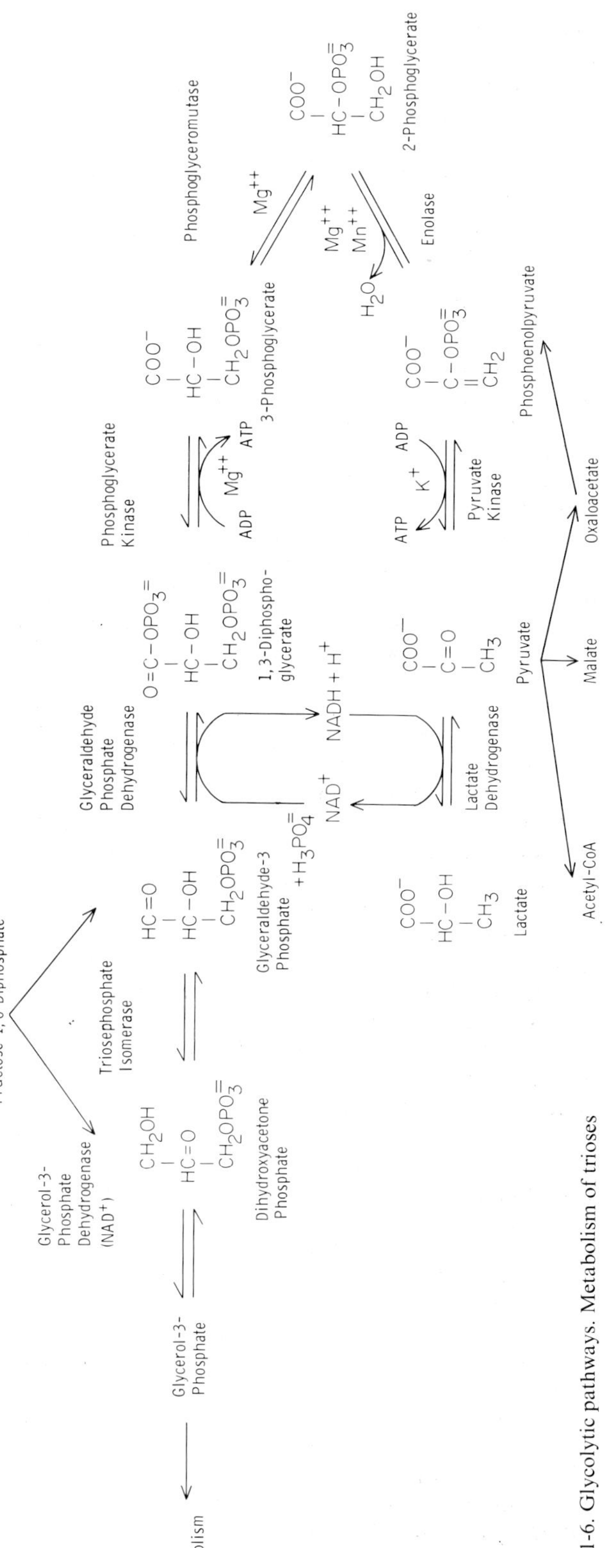

Fig. 1-6. Glycolytic pathways. Metabolism of trioses

muscle, where the enzyme couples the reduction of dihydroxyacetone with the oxidation of NADH. Under stress, the rate of glycolysis is accelerated in muscle, and the NAD required for the performance of the cycle is regenerated through the conversion of pyruvate to lactic acid. In insect muscle, however, as much as 50% of the NADH oxidation is coupled with the reduction of the dihydroxyacetone phosphate.

Glycerophosphate dehydrogenase has also been found in many mammalian tissues, but it is not known how much the dihydroxyacetone phosphate-L-α-glycerophosphate oxidation-reduction system contributes to the regeneration of NAD in mammals. It is clear, however, that in mammals the conversion of pyruvic to lactic acid is mainly responsible for the oxidation of NADH. In view of the high rates of aerobic glycolysis in tumors, it is interesting that glycerophosphate dehydrogenase is absent in some tumors.

α-Glycerophosphate dehydrogenase was also found in mitochondria and purified from pig brain after it had been released from the mitochondria with phospholipase A. Preliminary studies of its molecular structure suggest that it is a ferroflavoprotein. Although the enzyme migrated in a single band in the electrophoresis apparatus, it had the properties of a polydispersed system in the ultracentrifuge.

The incubation of 3-phosphoglyceraldehyde with ADP, NAD, and two enzymes (D-triose phosphate dehydrogenase and phosphoglycerate kinase) leads to the appearance of D-3-phosphoglyceric acid: ATP and $NADH_2$. These coupled reactions lead to the oxidation of glyceraldehyde and the phosphorylation of ADP to yield ATP. This is the key reaction of the glycolytic pathway, because it converts the energy made available by the oxidation of glucose into utilizable chemical energy. The names of Warburg, Cori, Krebs, Wieland, and Racker have all been associated with the elucidation of the molecular mechanism of this coupling of oxidation and phosphorylation. Warburg [44] proposed that phosphorus reacts with the aldehyde carbon, which is later oxidized to yield the acyl phosphate. However, evidence suggests that the phosphorylation of ADP is secondary to the oxidation of the aldehyde.

Triose Phosphate Dehydrogenase

Triose phosphate dehydrogenase has been crystallized from rabbit muscle, and even after repeated crystallization, it contains two moles of bound NAD per mole of protein. The nucleotide can be removed by passing the enzyme through charcoal, but this treatment renders the enzyme preparation unstable. Both the stability and the activity of the enzyme can be restored by adding NAD to the medium. The presence of NAD in the enzyme molecule can readily be demonstrated by ultraviolet spectrophotometry. The exact mode of attachment of NAD to the enzyme molecule is not known, but it has been established that the enzyme contains SH groups belonging to cysteine residues—cysteine probably is part of the tripeptide glutathione, which may constitute the prosthetic group of the glyceraldehyde dehydrogenase. Indeed, the SH groups appear to play a central role in the reaction and are believed to bind to NAD. In presence of the aldehyde, the bond is ruptured, the NAD is reduced, and an S—C bond is formed between the sulfur atom of the cysteine and the carbon atom of the aldehyde.

Phosphoglycerate Kinase

Phosphoglycerate kinase is the enzyme involved in transferring the acyl phosphate of the 1,3-diphosphoglyceric acid to ADP. ATP and 3-phosphoglycerate are the products of the reaction. As is the case for most kinases, Mg^{++} is an absolute requirement. 3-Phosphoglyceric acid is converted to 2-phosphoglyceric acid under the catalytic influence of phosphoglyceromutase.

Phosphoglyceromutase

The enzyme has been purified from rabbit muscle, and its molecular weight was found to be 64,000. The mechanism of action of phosphoglyceromutase is reminiscent of the mechanism of action of phosphoglucomutase. A diphospho intermediate compound—3,2-diphosphoglyceric acid—is formed. It is not known, however, whether the intermediate is strongly complexed with the enzyme. Phosphoglyceromutase is found in relatively large amounts in the erythrocytes, where 2,3-diphosphoglyceric acid represents a considerable amount of the organic phosphate present.

Enolase

In the next reaction of the glycolytic pathway, 2-phosphoglycerate loses a molecule of water to yield phosphoenolpyruvate. The enzyme catalyzing that reaction is enolase, which has been crystallized from yeast and from muscle. The purified enzyme requires magnesium for activity; in fact, the metal is an intrinsic part of the molecule. Fluoride inhibits enolase, probably by forming an Mg-F1-PO_4 complex. Fluoride does not inhibit the enzyme if phosphate is absent from the incubation medium or if magnesium is replaced by manganese as an activator. The affinity of enolase for magnesium has been used for the purification of the enzyme on magnesium sulfomethyl-cellulose columns. The amino acid composition of the crystallized yeast enzyme has been established as a long polypeptide chain composed of 600 amino acids with a single N-terminus (alanine). Enolase contains methionine but not cysteine.

The effect of various proteolytic enzymes on enolase activity has been investigated. In some experiments, the enzyme was incubated with aminopeptidase, in others with carboxypeptidase, and in a third group of experiments enolase was treated with aminopeptidase first and with carboxypeptidase later. The activity of the enzyme is maintained even when 90 amino acid residues are eliminated at the N-terminus using aminopeptidase, or when carboxypeptidase eliminates 80 residues from the C-terminus. Such a finding indicates that the active center of the enzyme is somewhere on the remaining polypeptide chain. The enzyme activity is lost if the enzyme is submitted to the consecutive actions of aminopeptidase and carboxypeptidase. The molecule requires at least one of its end tails for catalytic activity.

In liver, most of the phosphoenolpyruvate is formed by the continued action of pyruvate carboxykinase and pyruvate carboxylase. In fact, the carboxykinase reaction determines whether the original substrate, oxaloacetate, is converted to glucose or oxidized in the Krebs cycle. The conversion of pyruvic acid to phosphoenolpyruvate is a rate-limiting step in the conversion of 3-carbon compounds (alanine and lactic acid, for example) to glucose. When glucogenesis is increased (*e.g.*, in diabetes and after the administration of glucocorticoids), carboxylase and carboxykinase activities are also increased.

Various isozymic forms of enolase exist. In rats, electrophoretic and chromatographic procedures have separated an enolase I found in liver, spleen, and brain; an enolase II found in heart; and an enolase III found in muscle.

Pyruvic Kinase

A reversible reaction catalyzes the conversion of pyruvate to phosphopyruvate, and the enzyme involved is pyruvic kinase. The equilibrium of that reaction is on the side of the formation of ATP. Thus, pyruvate kinase is the enzyme responsible for the conversion of phosphoenolpyruvate to pyruvate. The enzyme has been crystallized from muscle; it requires ADP, potassium, and magnesium and is noncompetitively inhibited by some estrogenic steroids. Steroids alter the enzyme's viscosity and electrophoretic properties. From this observation, it was assumed that steroids act by modifying the protein molecule.

Although pyruvic kinase is found in many tissues, two main forms have been purified extensively—the L-type from liver and the M-type from skeletal muscle. Both isoenzymes are found in mature liver, but, of course, the L-type predominates. Variations in the isoenzymic pattern take place during cell maturation, differentiation, and cancerization. The significance of these observations will be reviewed in the chapter on cancer.

Lactic Dehydrogenase

Pyruvic acid may either be oxidized to water and CO_2 in the Krebs cycle or reduced to lactic acid by various lactic dehydrogenases. Thus, under stress, when the need for ATP is great and oxygen uptake is not rapid enough to secure the aerobic oxidation of the terminal products of glycolysis, it is necessary to regenerate NAD, the hydrogen acceptor in the anaerobic oxidation of glyceraldehyde. This is done by reducing pyruvic acid to lactic acid. The enzyme involved in that reaction is lactic dehydrogenase, and NADH serves as coenzyme.

This enzyme activity has been investigated extensively, and a review of all the known properties of lactic dehydrogenase is beyond the scope of this book; only a few relevant findings will be discussed here. Recent reviews on lactic dehydrogenase are available [45–47]. Zinc has repeatedly been found in association with the enzyme molecule, but when compounds that complex with zinc are added to the enzyme preparation, the preparation loses zinc but nevertheless maintains activity. Therefore, zinc may not play a vital role in the enzyme activity. However, at least under some circumstances, some lactic dehydrogenases are known to function with a flavoprotein as a coenzyme. Some lactic dehydrogenases contain as many as six SH groups, two of which can be substituted without altering the enzyme's activity. The remaining SH groups are essential to activity and apparently are not involved in the binding of NAD. In contrast, the tyrosyl group of the enzyme appears to be involved in NADH binding, since NADH protects lactic dehydrogenase from inactivation by iodine.

The most striking observation with respect to lactic dehydrogenase is the multiplicity of enzyme molecules with similar catalytic properties. A number of molecules having lactic dehydrogenase activity separate upon electrophoresis. This observation was made with preparations obtained from different tissues. Thus, by electrophoresis on agar, polyacrylamide, or other suitable media, five lactic dehydrogenases have been separated from tissues such as heart and kidney, and three from liver and plasma.

All lactic dehydrogenases are tetrameric. Electrophoresis separates two parent types—the H predominant in heart muscle and the M predominant in skeletal muscle. With appropriate treatment, the polymeric enzyme can be split to monomers with a molecular weight of approximately 35,000. Although LDH-H and LDH-M each contain four identical subunits, the other bands obtained by electrophoresis result from hybridization of H subunits with M subunits in the following fashion: H_3M, H_2M_2, and HM_3. The H and M subunits are clearly different polypeptide chains as indicated by their amino acid composition, helical structure, and other factors. Thus, at least two different genes must be involved in the coding of the lactic dehydrogenases, and these genes are believed to be located on two different chromosomes.

This molecular complexity is intriguing and must have a physiological meaning. Kaplan and his associates have proposed the following explanation. In skeletal muscle the main function of LDH is to convert pyruvate to lactate for the purpose of oxidizing the NADH generated by glyceraldehyde phosphate oxidation. The reverse reaction, namely, the oxidation of lactate, almost never occurs in muscle.

The lactate produced mainly in muscle diffuses in blood and reaches the heart where the principal function of LDH-H is oxidizing lactate to pyruvate, which can then be utilized through the Krebs cycle. In fact, if blood concentrations of lactate are increased, the uptake of that metabolite by the heart is also increased. Thus, the H-type is the true lactic dehydrogenase, while the M-type is really a pyruvic reductase.

By what mechanism is the oxidation of lactate favored in the heart? In the presence of high pyruvate concentration, LDH-H forms with pyruvate and NAD an inactive ternary complex, thus blocking the conversion of pyruvate to lactate. The addition of NADH to the complex activates the enzyme, and pyruvate is converted to lactate with the formation of NAD. These findings have led to the conclusion that the activity of the enzyme is regulated by the ratio of NADH to NAD in the medium for any given concentration of pyruvate.

In the functionally normal heart, the ratio of NADH to NAD is believed to be such that LDL-H prevents the conversion of pyruvate to lactate but does not interfere with the conversion of lactate to pyruvate, thus making lactate produced in other tissues a suitable, aerobically oxidizable fuel. The human heart does not seem to thrive on glucose alone—it uses approximately 10 g a day of glucose, but it also consumes 10 g of lactate.

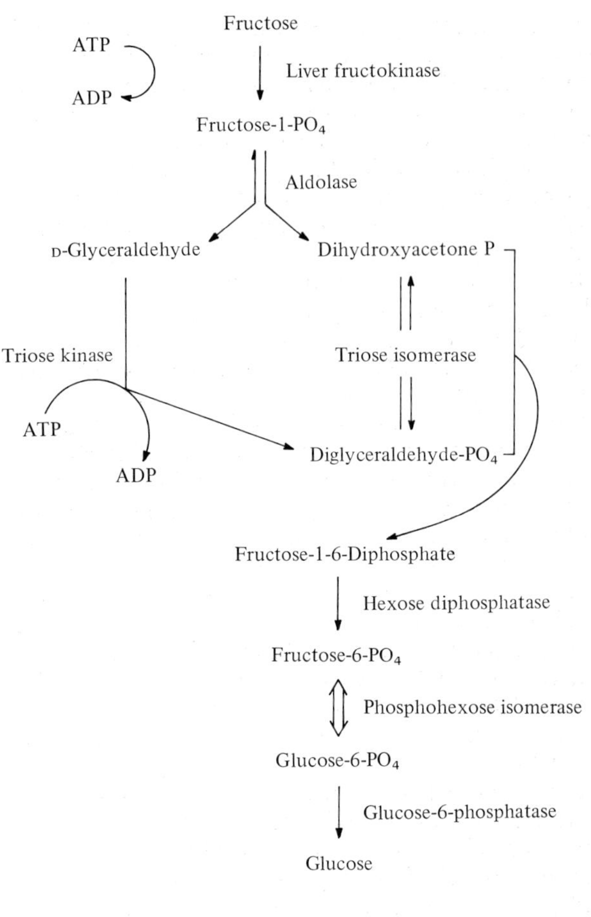

Fig. 1-7. Fructose metabolism in liver

Fructose Phosphorylation

Glucose is not the only hexose used for glycolysis—fructose, mannose, and galactose can also enter the glycolytic cycle after phosphorylation. Like glucose, fructose can be used only after phosphorylation in one of three ways [33]: (1) phosphorylation to fructose-6-phosphate by hexokinase, (2) phosphorylation to fructose-6-phosphate by a specific fructokinase, and (3) phosphorylation to fructose-1-phosphate by fructokinase (Fig. 1-7). It is well established that the glucokinase of liver and muscle can also phosphorylate fructose. Fructose can enter muscle metabolism only in the form of fructose-6-phosphate. This is strikingly different from liver metabolism in which fructose is converted to fructose-1-phosphate by a specific fructokinase.

These conclusions were based on advanced isotopic studies showing that if [^{14}C] fructose-1 is converted to fructose-6-phosphate, it can go directly to glycogen through the isomerase and phosphoglucomutase reaction without first being split into trioses. Under those conditions, all the ^{14}C of the hexoses entering the glycogen molecule will be recovered in carbon 1, and this is precisely what happens in muscle. In liver, the label of the [^{14}C] fructose-1 is randomized; why this happens will become obvious when the pathway for fructose-1-phosphate has been outlined.

A kinase catalyzing the formation of fructose-1-phosphate in muscle has been described. It appears that the mechanism of fructose utilization in muscle is not entirely resolved. Neither the affinity of the hexokinase for fructose nor the marginal activity of fructokinase explains the efficient utilization of fructose by muscle.

The fate of fructose-1-phosphate is varied—it may be phosphorylated again in the presence of 1-phosphofructokinase, magnesium, and ATP to yield fructose-1,6-diphosphate and ADP. 1-Phosphofructokinase is found in liver and muscle. Fructose-1,6-diphosphate may then be used by the glycolytic cycle.

Aldolase

Aldolase (see Fig. 1-7) is the enzyme that catalyzes the splitting of hexose diphosphate to yield triose phosphates [49–51]. This reaction is reversible, and, in fact, the equilibrium favors hexose diphosphate formation. Thus, aldolase condenses dihydroxyacetone phosphate and D-glyceraldehyde phosphate in a typical aldol condensation this type of reaction explains the origin of the name "aldolase." The enzyme has been crystallized from a variety of sources, among which are muscle (mol wt 150,000) and yeast (mol wt 122,000). Metal ions (zinc, iron, or cobalt) are required for activity. The specificity of the enzyme is absolute for dihydroxyacetone phosphate; it is broad for the aldehyde. Thus, glyceraldehyde, formaldehyde, acetoaldehyde, and D-erythrose may be used in the aldol condensation reaction.

In view of the rigid specificity requirements for dihydroxyacetone phosphate, the enzyme can be expected to form a complex with that triose.

Rabbit muscle aldolase can attack three different substrates, fructose-1,6-diphosphate, sedoheptulose-1,7-diphosphate, and fructose-1-phosphate. Although the cleavage rates of the first two substrates are almost equal, that of fructose-1-phosphate is slow. The phosphate in position 1 is thought to be indispensable for activity, and that in position 6—although dispensable—activates the enzyme activity.

Aldolase catalyzes both a stereospecific exchange of a hydrogen of dihydroxyacetone phosphate and an exchange reaction between carbons 4, 5, and 6 of the fructose and the three carbons of glyceraldehyde. Since glyceraldehyde can also be derived from dihydroxyacetone, the carbons of fructose will randomize (see Fig. 1-8).

Like lactic dehydrogenase, aldolase exists in multiple forms, the existence of which were not immediately obvious on electrophoresis. Electrophoresis of crystalline aldolase yields a single but rather broad band, and only through electrofocusing were five different forms of the enzyme discovered.

Further investigation indicated that like LDH, aldolase is a tetramer composed of two different types of subunits referred to as α and β, thus the five different forms are tetramers in which the α- and β-units are combined in different proportions.

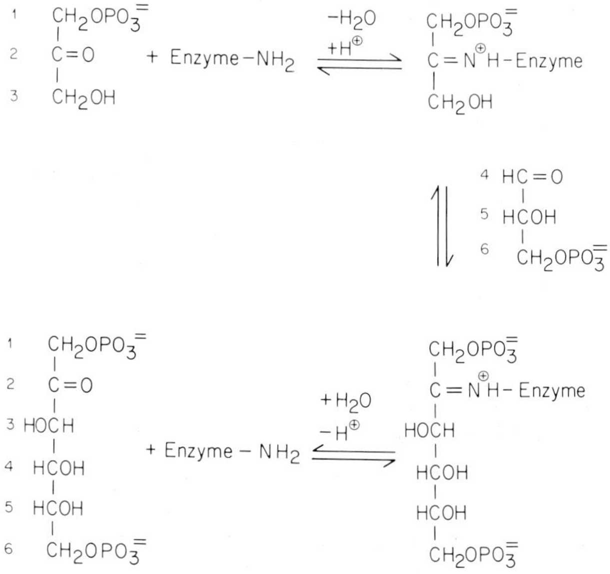

$$\underset{\substack{1\\2\\3}}{}\begin{array}{l} CH_2OPO_3^{=} \\ | \\ C{=}O \\ | \\ CH_2OH \end{array} + \text{Enzyme}{-}NH_2 \underset{}{\overset{-H_2O,\ +H^{\oplus}}{\rightleftharpoons}} \begin{array}{l} CH_2OPO_3^{=} \\ | \\ C{=}\overset{\oplus}{N}H{-}\text{Enzyme} \\ | \\ CH_2OH \end{array}$$

$$\Updownarrow \quad \begin{array}{ll} 4 & HC{=}O \\ & | \\ 5 & HCOH \\ & | \\ 6 & CH_2OPO_3^{=} \end{array}$$

$$\begin{array}{ll} 1 & CH_2OPO_3^{=} \\ & | \\ 2 & C{=}O \\ & | \\ 3 & HOCH \\ & | \\ 4 & HCOH \\ & | \\ 5 & HCOH \\ & | \\ 6 & CH_2OPO_3^{=} \end{array} + \text{Enzyme}-NH_2 \overset{+H_2O,\ -H^{\oplus}}{\rightleftharpoons} \begin{array}{l} CH_2OPO_3^{=} \\ | \\ C{=}\overset{\oplus}{N}H{-}\text{Enzyme} \\ | \\ HOCH \\ | \\ HCOH \\ | \\ HCOH \\ | \\ CH_2OPO_3^{=} \end{array}$$

Fig. 1-8. Mechanism of aldolase action

In contrast to the subunits that form the various types of LDH, those of aldolase are coded for by a single gene; therefore, the conversion of α to β is a posttranscriptional event involving the deamination of a single asparagine residue in the α-subunits.

Although not completely known, the amino acid sequence of the rabbit muscle aldolase subunit is largely elucidated. (A more detailed description on protein structure appears in the chapter on inborn errors of metabolism.) The molecule contains 364 amino acids with a proline NH_2-terminal, a tyrosine in the COOH-terminal position, and 8 cysteine residues. A critical residue is the lysine 221 which is believed to form a Schiff base with dihydroxyacetone phosphate. (The role of Schiff bases in enzymic reactions is discussed in more detail in the section devoted to transaminases.) In the model proposed by Lai und Horecker [51], this lysine is near the center of the molecule. If one follows the contour of the molecule from this critical residue 221, which must be at the active center, toward the tyrosine carboxy terminal, three SH groups are well exposed on the surface of the molecule. They occupy positions 193, 171, and 143.

Three other SH groups are believed to be located between the center of the molecule and the proline NH_2-terminal, occupying positions 231, 280, and 339, but these are buried in the infolds of the molecule. Further evidence has shown that other lysine, histidine, and tyrosine residues in various positions in the sequence modulate the activity of the enzyme. On the basis of this information, models for the mode of action of aldolase have been proposed. These are presented here only schematically in the direction of dealdolization. First, the carbon 1 of fructose-1,6-diphosphate binds to an unidentified position of the active site through electrostatic bonding; carbon 6 is believed to interact with the second lysine residue. Second, the 6-amino group of lysine 221 of the enzyme interacts with the carbamyl group of the substrate, forming a "Schiff base" through the elimination of a molecule of water. The protonation of the Schiff base nitrogen draws electrons, which help to labilize the bond between carbons 3 and 4 of the substrate [51].

Triose Metabolism

Glyceraldehyde can be used by the glycolytic pathway after phosphorylation. This reaction is apparently catalyzed by triokinase (see Fig. 1-9), an enzyme found in liver [53]. The affinity of triokinase is not restricted to glyceraldehyde; it will also phosphorylate dihydroxyacetone, although the rate of phosphorylation of that compound is much slower than that of glyceraldehyde. Glyceraldehyde is also oxidized to glyceric acid, or reduced to the alcohol glycerol. The first of these reactions is catalyzed by an enzyme isolated from rat liver, which seems to be the same as acetoaldehyde dehydrogenase. The reduction to glycerol is catalyzed by alcohol dehydrogenase (see chapter on alcoholism). However, glyceric acid cannot be used by the glycolytic

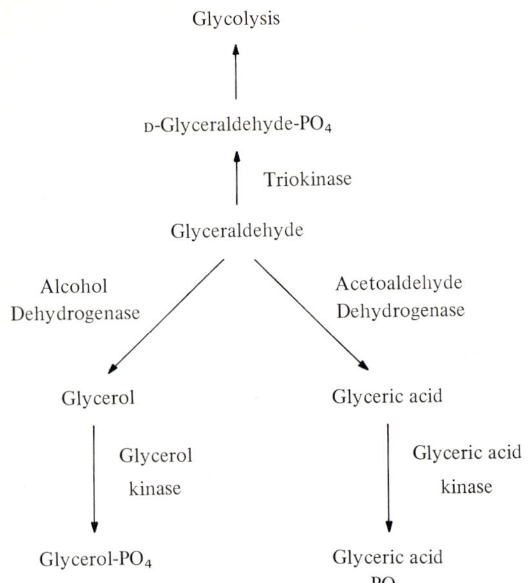

Fig. 1-9. Glyceraldehyde metabolism

cycle until it has been phosphorylated. A glyceric acid kinase has been found in rat and horse liver. A glycerokinase that phosphorylates the alcohol in the presence of ATP and α-glycerophosphate has also been isolated from pigeon liver. Triose phosphate can then be utilized by the Krebs cycle.

The fate of dihydroxyacetone phosphate is also varied—it may be transformed to α-glycerophosphate, or through the reaction catalyzed by the triose isomerase, it may yield D-glyceraldehyde-3-phosphate, which in the presence of triose phosphate dehydrogenase and phosphoglycerate kinase yields 3-phosphoglycerate. This degradation pathway is complete only in liver, and it is not known to what extent it operates in muscle.

Glycogen Breakdown

Primarily the heart, brain, and skeletal muscles, but all other organs as well, constantly require glucose to maintain anaerobic and aerobic bioenergetic pathways. Therefore, most organs have devised mechanisms to store glucose in the form of a large polymer—glycogen—which is made available when needed in the process of glycogenolysis. In muscle and liver, glycogen is rapidly synthesized and degraded. Sugars (glucose, fructose, and galactose) are not the only substances used for glycogen synthesis, since glycogen is also formed from amino, pyruvic (of aerobic or anaerobic origin), and lactic acids. Glycogen synthesis is thus partly due to the reversion of glycolysis.

Glycogen is a highly branched polysaccharide made of glucose units bound together by 1,4-glycosidic linkages in the main chain (11–18 glucose units) and 1,6-glycosidic linkages in the cross-links between the main chain and the branches. Breakdown occurs by hydrolysis and phosphorolysis (see Fig. 1-10). The elements of a molecule of water or phosphoric acid are thus incorporated into the glycosidic linkage. Phosphorylase splits the 1,4-glycosidic linkages, and amylo-1,6-glucosidase splits the 1,6-linkages [54–58].

Glycogen breakdown requires an enzyme capable of splitting α-1,4-linkages—phosphorylase (see Fig. 1-11), and one that splits 1,6-linkages—amylo-1,6-glucosidase (see Fig. 1-10). Moreover, since the product of glycogen breakdown by phosphorylase is glucose-1-PO_4, which is inaccessible to glycolysis, a phosphoglucomutase must convert glucose 1 into glucose-6-phosphate.

Muscle Phosphorylase

Muscle phosphorylase exists in (a) and (b) forms. The latter is half the molecular weight of the former and requires AMP as an activator, but since the AMP concentration (5×10^{-5} M per liter) required for the activation is not reached in muscle, the (b) form is practically inactive. In passing from one form to another, two enzymes are involved—a proteolytic enzyme catalyzes the breakdown of phosphorylase (a), and another catalyzes the formation of (a) from (b) in the presence of ATP, magnesium, and manganese. The stoichiometry of the second reaction has been studied; four molecules of ATP for each molecule of phosphorylase (b) appear to be involved. It has been shown that both phosphorylase (a) and (b) contain pyridoxal phosphate; the (b) form contains two such molecules, and the (a) form contains four. The role of the coenzyme in the mechanism of action of phosphorylase is unknown, but the enzyme loses its activity when the coenzyme is dissociated from the main protein molecule by ammonium sulfate; activity is restored by the addition of pyridoxal phosphate.

Pyridoxal phosphate binds to the apoenzyme by means of a 4-formyl group. One of the covalent bonds of the formyl group involves the ε-amino group of one of the protein's lysine residues. Isonicotinic acid, which combines with the 4-formyl groups, prevents the combination of enzyme with coenzyme.

The presence of pyridoxal phosphate confers specific absorbance properties onto the holoenzyme. The absorption maximum is at 330 mμ, a wavelength far removed from that of the Schiff base. Therefore, the combination of apoenzyme and coenzyme in the active phosphorylase molecule does not involve the formation of a Schiff base. Krebs and others studied the site of attachment of the phosphorus in the phosphorylase molecule by preparing ^{32}P-labeled phosphorylase. After the enzyme was treated with trypsin, it became clear that the phosphorus was present in a basic hexapeptide containing two basic amino acids. The hexapeptide has the following amino acid sequence: lysine, glutamine, isoleucine, serine phosphate, valine, and arginine.

It seems that there is only one phosphorylated serine residue in the long phosphorylase molecule. The position of the phosphoserine probably coincides with the

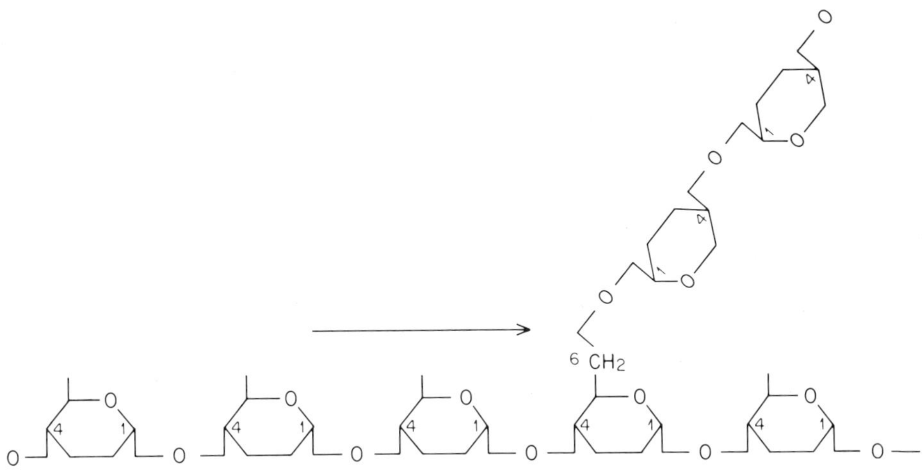

Fig. 1-10. Glycogen metabolism. Site of action of glycogen branching enzyme and of amylo-1,6-glucosidase

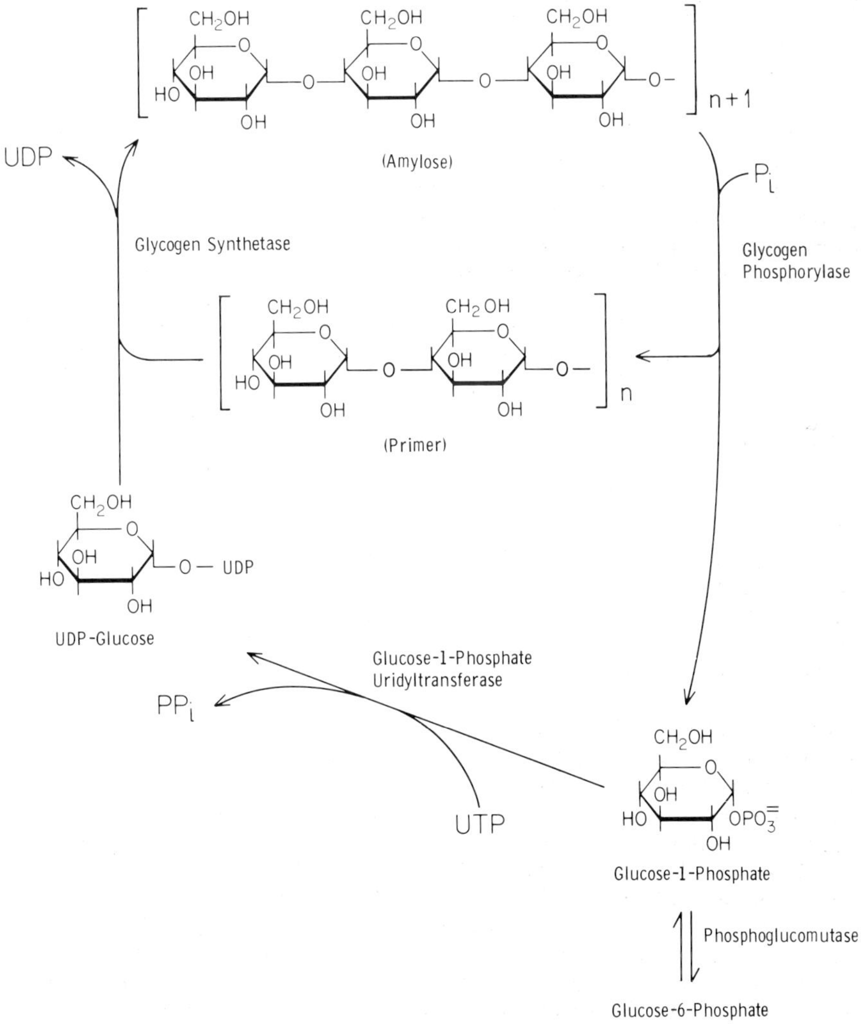

Fig. 1-11. Glycogen metabolism

active site of the enzyme. The fact that the polypeptide is basic also explains the dimerization of the phosphorylase molecule that occurs simultaneously with the phosphorylation of phosphorylase (b). The heavy positive charge of phosphorylase (b) is responsible for sufficient electrostatic repulsion to maintain the molecule in the monomeric state. As soon as the phosphate neutralizes these charges, dimerization occurs, and phosphorylase (b) yields phosphorylase (a).

The activation of phosphorylase is probably due to phosphorylation of the protein molecule rather than to dimerization. Phosphorylase (a) and (b) were isolated and crystallized from lobster muscle, and their sedimentation characteristics were found to be identical. Thus, in lobster muscle, the only difference between the active and inactive forms probably resides in the presence of phosphorus in the former. It was further observed that phosphorylase (a) contains one molecule of pyridoxal phosphate per 100,000 g of protein, and that, in contrast to (b), phosphorylase (a) did not require AMP for activity. A kinase that converts the (b) to the (a) form and a phosphatase that catalyzes the opposite reaction have also been purified from lobster muscle. Phosphorylase kinase has been purified from rabbit muscle. The enzyme is much more active if it is preincubated with ATP and magnesium prior to exposure to the substrate. The activity of the preincubated enzyme can be further stimulated by heparin, glycogen, and cyclic AMP.

It is well known that glycogen disappears from muscle during contraction, presumably from the activation of phosphorylase kinase through a mechanism different from that put into gear by epinephrine. Indeed, during muscle contraction, kinase appears to be stimulated by the effect of Ca^{++} ions rather than by that of cyclic AMP.

Liver Phosphorylase

The properties of liver phosphorylase were studied primarily in Sutherland's [59] laboratory. After devising methods for purifying the enzyme, the investigators demonstrated that liver contains an inactivating enzyme with phosphatase properties which dephosphorylates the enzyme to yield inactive phosphorylase and inorganic phosphate. In contrast to muscle phosphorylase, inactive and active liver phosphorylase had the same molecular weight. Sutherland and associates further succeeded in purifying "phosphorylase kinase," which catalyzes the passage from inactive to active phosphorylase. The activity of the kinase depends upon the presence of 3′,5′-cyclic adenosine phosphate and magnesium ions.

Thus, in liver, as in muscle, phosphorylase exists in both inactive and active forms. Enzymes in both tissues catalyze the passage from one form to another. But whereas in muscle the passage from inactive to active form is accompanied by a dimerization of the molecule and an increase in molecular weight (at least in mammals), in liver the enzyme's molecular weight is not altered.

Although it is apparently not necessarily related to phosphorylase activation, dimerization occurs near the active center of the molecule. The fact that it can occur in muscle phosphorylase and not in liver phosphorylase suggests that the molecular structures of the two enzymes differ.

In muscle phosphorylase, the passage from the inactive to the active form is stimulated only by epinephrine. (It has been suggested that in the liver fluke, *Fasciola hepatica,* serotonin stimulates phosphorylase. In this fluke, serotonin also stimulates phosphofructokinase, probably through the intermediate of 3′,5′-cyclic adenylic acid.) In liver, both epinephrine and glucagon stimulate the activity of the phosphorylase. In liver, the activation has been demonstrated *in vivo* in cell-free preparations; epinephrine does not act directly on phosphorylase, but it stimulates the formation of 3′,5′-cyclic adenylic acid, which in turn catalyzes the conversion of inactive phosphorylase to active phosphorylase in the supernatant. The cyclic nucleotide is formed from ATP in the presence of a particulate enzyme (adenine cyclase), which is found in heart, muscle, liver, brain, and other organs. The reaction requires magnesium, and, as can be expected, 3′,5′-cyclic adenosine accumulates when the medium contains fluoride.

Caffeine and puromycin also stimulate the conversion of phosphorylase (a) to phosphorylase (b), probably by interfering with the breakdown of the 3′,5′-cyclic adenosine nucleotide. A diesterase capable of splitting the cyclic phosphate of the 3′,5′-cyclic adenosine phosphate has been found in beef heart. The effect of epinephrine on phosphorylase is important because it constituted the first explanation of a hormone's mechanism of action in molecular terms.

Phosphoglucomutase

Glycogen breakdown yields glucose-1-phosphate, which must be converted into glucose-6-phosphate to enter the glycolytic pathway. The enzyme responsible for that reaction is phosphoglucomutase. The phosphate in position 1 is transferred to the carbon 6 of glucose without exchange with inorganic phosphate. It is glucose-1,6-diphosphate that acts as the phosphate donor in this reaction, and diphosphohexose is strongly complexed with the enzyme. Thus, the following steps occur in the conversion of glucose-1-phosphate to glucose-6-phosphate. The phosphorylated enzyme reacts with glucose-1-phosphate to yield a glucose-1,6-diphosphate and the dephosphorylated enzyme. The hexose diphosphate is dephosphorylated to yield glucose-6-phosphate and the phosphorylated enzyme. The phosphate's position in the enzyme molecules has been determined—it exists in the form of a phosphoserine—the part of a polypeptide sequence resembling that which includes the serine residue of thrombin and chymotrypsin.

The exact specificity of phosphoglucomutase is not known; however, galactose, mannose, glucosamine, N-acetylglucosamine, and ribose phosphate probably are substrates for the enzyme. The importance of the phosphoglucomutase reaction in relation to pathology will become more obvious when the pathogenesis of galactosemia is discussed.

Amylo-1,6-glucosidase

The specificity of phosphorylase is restricted to the α-1,4-linkage of the glycogen molecules. Phosphorylase attacks the reducing ends and splits the glucose molecules of the main and side chains until a 1,4-linkage near a branching point (the 1,6-linkage) is reached. Phosphorylase cannot split or bypass 1,6-linkages. The product of phosphorolysis, called the limit dextrin, can be further hydrolyzed by adding phosphorylase and amylo-1,6-glucosidase simultaneously. The latter enzyme is capable of splitting the 1,6-glucosidic linkages. Glycogen is also attacked by amylase, which is reviewed in another chapter.

Intracellular Distribution of the Glycolytic and Glycogenolytic Enzymes

Only cytoplasmic glycolysis is considered here; nuclear glycolysis is reviewed in another section.

The intracellular distribution of a few enzymes involved in glycolysis has been investigated by measuring the formation of the specific product and determining the enzyme activity of the cell fractions as well as the total homogenate. Nearly 75% of all aldolase, phosphorylase, phosphoglucomutase, and hexose diphosphatase activity in the homogenate is found in the soluble fraction. However, since the formation of lactic acid is highest in the soluble fraction, it is classically stated that all enzymes involved in glycolysis are present in the supernatant. This is obviously an oversimplification. The findings indicate only that all enzymes necessary for glycolysis are in the soluble fraction, but they do not exclude the possibility that several noninvestigated enzymes exist in high concentration in other cell fractions.

Furthermore, the soluble fraction is probably the cell fraction that differs most from actual intracellular distribution. It contains numerous constituents, which in the intact cells are probably associated with other cellular organelles. The soluble fraction is arbitrarily defined as the supernatant of a pellet centrifuged either at 1,500,000 $g \times$ minutes or at 3,000,000 $g \times$ minutes. If this preparation is centrifuged longer (6,000,000 $g \times$ minutes), smaller particles are sedimented, and some of the glycolytic enzymes possibly could be associated with these small cell particles.

Although the supernatant is essentially structureless at the resolution of the electron microscope, there is an intimate intramolecular association between enzymes and coenzymes; for instance, two molecules of NAD are solidly attached to the triosedehydrogenase, probably through the mediation of two glutathione molecules. The NADH formed in the reaction can either transfer electrons to an acceptor molecule or exchange its hydrogen with free NAD.

The metabolic activity of the soluble fraction is intimately connected with that of other cellular organelles; final breakdown products of pyruvic acid are utilized in the Krebs cycle by mitochondria, and the maintenance of glycolysis depends upon the presence of NAD, a coenzyme synthesized at least partly in the nucleus.

Interference between mitochondrial oxidation and glycolysis have been repeatedly demonstrated. Such interferences are not surprising; mitochondrial oxidation and glycolysis use several metabolites that are identical in structure—namely, ATP, NAD, and inorganic phosphate. These interferences are discussed in the review of the Pasteur effect. Since triose phosphate dehydrogenation and ATP synthesis are coupled, glycolysis depends upon the availability of ADP and phosphate. The amount of ADP in the medium is rate limiting, and the rate of glycolysis is regulated by the concentration of these substances, which varies with the cell's consumption of ATP.

Effects of isolated membranous components of the endoplasmic reticulum on the performance of the glycolytic pathway in the supernatant have been suggested. However, there is no evidence that the membranes of the endoplasmic reticulum and glycolysis interact in the intact cell; therefore, these effects *in vitro* probably result from the presence of an ATPase in the membranes of the endoplasmic reticulum.

The supernatant is not a stagnant fluid trapped in the cytoplasmic structures; on the contrary, it circulates throughout the tubules and vesicles of the endoplasmic reticulum, which may form compartments separating different types of soluble fractions. All these fractions are pooled during homogenation procedures. Regulation of compartmentalization possibly through hormonal controls has been invoked to explain the fate of glucose-6-phosphate. This metabolite is indeed at the crossroads of four main biochemical pathways: gluconeogenesis, glycogen synthesis, the hexose monophosphate shunt, and glycolysis. Glucose-6-phosphatase and hexokinase are abundant in liver, and each catalyzes an irreversible reaction. For this reason, the glucose utilization for glycogen synthesis would be difficult to explain in the absence of compartmentalization.

Among those enzymes involved in or related to glycolysis, it is precisely glucose-6-phosphatase and hexokinase that have been found in particulate form. Thus, glucose-6-phosphate is constantly associated with microsomes. Ninety per cent of the hexokinase activity, at least in brain, can be sedimented at relatively low centrifugal forces (18,000 $g \times$ 30 minutes). Smaller amounts of that enzyme are also found in a particulate form in other tissues.

Although it seems logical to conclude a discussion on glycolysis by considering the Pasteur and Crabtree effects, it is more appropriate to review these phenomena in the chapter devoted to cancer metabolism.

Glycogen Synthesis

Glycogen is a branched polysaccharide of glucose. The hexose units are linked together by 1,4-glycosidic linkages, except at the branching points where the linkage involves the carbon 1 of one hexose and the carbon 6 of the other hexose. Because of glycogen's structure, its biosynthesis requires at least two different types of enzymes—those catalyzing the biosynthesis of the α-1,4-linkage and those catalyzing the biosynthesis of the 1,6-linkage.

Role of Phosphorylase

In view of the specificity of the phosphorylase that disrupts the α-1,4-glucosidic linkages at the nonreducing end of glycogen, one might readily conceive that reversal of the phosphorylase reaction is responsible for glycogen synthesis. The thermodynamics of the phosphorylase reaction are not in conflict with such a hypothesis.

The equation for the phosphorylase reaction at equilibrium can be written as follows:

$$\text{At equilibrium, } K = \frac{(\text{glucose-1-phosphate})}{(\text{polysaccharide}) \times P_i} = 0.088$$

Since there is no significant change in the molecular concentration of polysaccharides during the breakdown process, the equilibrium formula can also be written:

$$\text{At equilibrium, } K = \frac{(\text{glucose-1-phosphate})}{P_i} = 0.088$$

Thus, for all practical purposes, the equilibrium favors glycogen synthesis, but the free energy of the reaction is obviously small. The equilibrium is considerably modified by pH; for example, a slight change toward acidic pH favors glycogen biosynthesis.

Inasmuch as phosphorylase is specific for the formation of a 1,4-glycosidic linkage, it will stimulate either the synthesis of a straight glycosidic chain with 1,4-glycosidic bonds or the interaction of glucose-1-phosphate with the nonreducing ends of the branches of an activating polysaccharide. Hence, phosphorylase catalyzes glycogen synthesis only if a primer containing 1,6-glycosidic linkages is available. It follows that although both glycogen and amylopectin are good primers, the straight amylose chain synthesized by the action of phosphorylase on glucose-1-phosphate is not active.

Other Pathways for Glycogen Synthesis

Although theoretically one could involve phosphorylase in glycogen biosynthesis, the evidence suggests that another mechanism of glycogen synthesis is operating in the living cell. In fact, phosphorylase *in vivo* probably is involved only in glycogen breakdown. In early studies, phosphorylase activation by epinephrine always led to glycogen breakdown but never to synthesis. Furthermore, experiments with breast muscle homogenate suggest that phosphorylase is not involved in glycogen synthesis. The problem gained new impetus when Leloir [60] discovered a liver enzyme that forms glycogen from UDP-glucose. Lipmann and coworkers soon demonstrated that skeletal muscles also synthesize glycogen from UDP-glucose, and that the addition of phosphorylase kinase increases the concentration of phosphorylase but inhibits the incorporation of [^{14}C]glucose into glycogen. But, as will be seen later, one of the most convincing arguments in favor of the operation of a new pathway for glycogen synthesis came from observation of patients with a special type of glycogen storage disease characterized by low phosphorylase levels. Also, some tumors low in glycogen have abnormally low levels of glycogen synthetase but normal levels of phosphorylase.

The anabolic and catabolic glycogen pathways are now well accepted. The central metabolite is, of course, glucose-1-phosphate. In the presence of UTP and an enzyme called UDP-glucose pyrophosphorylase, the glucose-1-phosphate is converted into UDP-glucose and pyrophosphate is formed. The enzyme has been found in various tissues and has been purified from skeletal muscle. The enzyme is specific for α-glucose-1-phosphate. Its affinities for Mg^+ pyrophosphate and UDP-glucose are great.

The reaction catalyzed by UDP-glucose pyrophosphorylase is intriguing in terms of thermodynamics. An energy-rich pyrophosphate bond is involved. The pyrophosphate probably is reutilized, but it is not known to what purpose. In the next steps, the enzyme UDP-glucose-glycogen synthetase or UDP-glucose-glycogen transglucosylase catalyzes the polymerization of glucose-1-phosphate. This reaction is irreversible; therefore, glycogen breakdown can occur only through phosphorylase. Glycogen synthetase, a widely distributed enzyme, has been purified from yeast and muscle. The purified enzyme is free of phosphorylase. Although its specificity is not rigidly restricted to UDP-glucose, its affinity is greatest for that compound; however, TDP-glucose, ADP-glucose, and UDP-galactose can also function as glucose donors.

Glycogen synthetase is activated by glucose-6-phosphate as a function of enzyme phosphorylation and the pH of the medium. The phosphorylated enzyme is inactive at all pH values in the absence of glucose-6-phosphate. The unphosphorylated enzyme is active at optimum pH in the absence of glucose-6-phosphate, but it requires glucose-6-phosphate for activity at alkaline pH. The physiological significance of the activation by glucose-6-phosphate remains to be elucidated.

Experiments in rats made diabetic with alloxan are significant. It has been known for a long time that glycogen synthesis is impaired in liver slices of alloxan-diabetic rats. Treatment of the animals with insulin restores to normal the liver's ability to synthesize glycogen. Further studies demonstrated that the activity of the glycogen synthetase was low in liver preparations from alloxan-diabetic rats, but could be restored by adding glucose-6-phosphate to the incubation medium. Such experiments support the proposition which postulated that the rate of glycogen synthesis was regulated by the concentration of glycogen synthetase or of its activator. Insulin may also increase glycogen synthesis by stimulating the formation of glucose-6-phosphate. Differential centrifugation of liver homogenate demonstrated that the glycogen-synthesizing enzyme is associated with particles sedimenting with the microsomal preparation; however, further investigation, including electron microscopic examination of the preparation containing the enzyme, demonstrated that the enzyme is associated with glycogen granules rather than with the membranes of the endoplasmic reticulum.

Branching Enzyme

Neither phosphorylase nor the glycogen synthetase is capable of synthesizing the 1,6-glycosidic bonds. A "branching enzyme," also referred to as the amylose-amylopectine *trans*-α-glucosidase, is necessary. This enzyme converts the straight chain of amylose into a branched chain by catalyzing the formation of 1,6-glycosidic linkages. Chains varying in length from 6–11 glucose units are good substrates for the enzyme. The branching enzyme of liver has been purified and prepared free of amylase activity.

Pentose Phosphate Pathway

In the previous section we considered three possible pathways for glucose-6-phosphate: dephosphorylation by glucose-6-phosphatase, conversion to fructose-6-phosphate followed by oxidation through the Embden-Meyerhof pathway, and conversion to glucose-1-phosphate followed by glycogen synthesis. A fourth pathway exists—the hexose monophosphate shunt [61–63].

Glucose-6-Phosphate Dehydrogenase

This is an important enzyme in pathology and pharmacology; it is the enzyme missing in the erythrocytes of patients with primaquine sensitivity. The enzyme, known for a long time and called Zwischenferment, or glucose-6-phosphate dehydrogenase, transfers two electrons from glucose-6-phosphate to NADP to yield 6-phospho-D-gluconic acid and NADPH. The reaction is often assumed to go directly from glucose-6-phosphate to phosphogluconic acid. In reality, the overall transformation occurs in two steps: first, the pyranose ring is dehydrogenated to yield 6-phosphogluconic lactone, and then that compound is hydrolyzed to the acid. The hydrolysis is either spontaneous or catalyzed by a specific lactonase, an enzyme which has been found in yeast, bacteria, and mammalian tissues. Although the first step in the transformation of glucose-6-phosphate to phosphogluconic acid is reversible, the second is irreversible.

Glucose-6-phosphate dehydrogenase was crystallized from bovine mammary glands and from yeast. It is inhibited by certain steroids (dehydroisoandrosterone, pregnenolone), but the physiological significance of such inhibitory effects is not known. One could imagine that the steroids inhibit their own production by a feedback mechanism that interferes with the biosynthesis of NADPH.

6-Phosphogluconate Dehydrogenase

After oxidation, the phosphogluconic acid undergoes decarboxylation. In this reaction catalyzed by 6-phosphogluconate dehydrogenase, as in that catalyzed by the malic enzyme, no free intermediate has been detected, and, again, the reaction presumably occurs on a single protein. The oxidative carboxylation is NADP-dependent and yields D-ribulose-5-phosphate, CO_2, and NADPH as final products. This is the only step in the cycle in which CO_2 is formed.

In the pentose pathway, the CO_2 formed is derived from the carbon 1 of glucose, in contrast to the Embden-Meyerhof pathway, which yields equal amounts of CO_2 from carbons 1 and 6 of glucose. Therefore, if glucose labeled in carbon 1 or carbon 6 is oxidized by a system in which the Embden-Meyerhof pathway operates alone, equal amounts of radioactive CO_2 are derived from glucose labeled in carbon 1 and carbon 6. But if glucose is also oxidized by the hexose monophosphate pathway, more CO_2 is derived from carbon 1 than from carbon 6. Hence, the production of CO_2 derived from glucose labeled in carbon 1 or carbon 6 is often used to evaluate the relative activity rates of each pathway.

Metabolism of Ribulose-5-Phosphate

Ribulose-5-phosphate may undergo two different types of isomerization. The first involves carbon 1 and carbon 2 of the pentose to yield D-ribose-5-phosphate, and the second concerns carbon 3 and yields D-xylulose-5-phosphate. The first reaction is catalyzed by an isomerase, the second by an epimerase. Both enzymes are widely distributed in nature and have been found in most tissues; purified isomerase was first prepared from alfalfa, and spleen epimerase has also been extensively purified.

In the next reaction, an enzyme called transketolase splits the xylose phosphate to 2-glyceraldehyde phosphate and a 2-carbon compound that combines with the enzyme to form an active glycolaldehyde. A glycolaldehyde enzyme complex has been isolated as well as an enzyme-substrate intermediate. The active glycolaldehyde condenses with the carbon 1 of 2-ribose-5-phosphate to yield a compound with seven carbons: 2-sedoheptulose-7-phosphate. This reaction illustrates a more general biochemical process in which two carbons are split from a monosaccharide to yield a ketol that remains attached to the protein molecules, thus forming the activated glycolaldehyde. These reactions are catalyzed by transketolases, enzymes widely distributed in plants and animals, which have been purified and crystallized from yeast. Transketolases require thiamine pyrophosphate and magnesium for activity and involve both a donor and an acceptor. The specificity of the enzyme for the donor is not well understood, but sedoheptulose-7-phosphate, fructose-6-phosphate, xylose-5-phosphate, and hydroxypyruvate can all act as donors. All these compounds have the hydroxyl group in position 3 at the L-site of the substrate.

In the next reaction, which is catalyzed by transaldolase, three carbons of the sedoheptulose—C_1–C_3—are transferred to glyceraldehyde, yielding 2 erythrose-4-phosphate and 2 fructose-6-phosphate.

A molecule of dihydroxyacetone is split from sedoheptulose-7-phosphate and is condensed with glyceraldehyde in an aldol condensation reaction. Since no free dihydroxyacetone accumulates in the medium, the enzyme has been called transaldolase rather than aldolase, and the formation of a dihydroxyacetone enzyme intermediate has been demonstrated. The enzyme was purified 700 times from yeast and was found to catalyze the reaction without the help of a prosthetic group or cofactor.

The cycle is completed when the erythrose-4-phosphate has been converted to hexoses. The 4-carbon compounds can serve as acceptors of two carbons of xylose-4-phosphate in a reaction catalyzed by transketolase and yield fructose-6-phosphate and glyceraldehyde phosphate.

Fructose-6-phosphate and glyceraldehyde constitute the channels through which the metabolites of the pentose phosphate cycle are reconverted to glucose-6-phosphate. Indeed, phosphoglucose isomerase converts fructose-6-phosphate to glucose-6-phosphate, and in the presence of triose isomerase and aldolase, glyceraldehyde phosphate may be converted to fructose-1,6-diphosphate. The metabolic steps involved in the pentose pathway are summarized in Fig. 1-12.

Role of the Pentose Phosphate Pathway in Metabolism

Now that the individual steps of the pathway have been outlined, it will be easier to understand some aspects of its role in metabolism and its distribution in the various organs. The pentose phosphate pathway offers a means by which the carbon 1 of glucose is oxidized and pentoses are formed. Pentoses are either used in the synthesis of nucleotides or go through the cycle and regenerate glucose-6-phosphate. NADPH is an important product of the pathway, as will be clear later. As a matter of fact, NADPH production through the glucose-6-phosphate dehydrogenase and the lactonase reactions is often used as a source of NADPH in biochemical experimentation.

The cycle is driven to completion by several irreversible steps. Although the transaldolase, transketolase, and isomerase reactions are reversible, the hydrolysis of phosphogluconolactone and the oxidation reaction producing ribulose-5-phosphate are irreversible.

A number of intermediates common to both the hexose monophosphate shunt and the glycolytic pathway are glucose-6-phosphate, fructose-6-phosphate, fructose-6,1-diphosphate, and triose phosphate. Thus, the two pathways can be expected to compete for intermediates, and, indeed, when a reconstituted glycolytic system made of purified enzymes is added to the reconstituted hexose monophosphate shunt, glucose oxidation by the shunt is inhibited by glycolysis.

Like glycolysis and the Krebs cycle, the pentose pathway constitutes an effective source of ATP, but in the presence of these two main sources, it is unlikely that the pentose pathway plays an essential role in the production of high-energy phosphates. Therefore, it is tempting to conclude that the hexose monophosphate shunt provides metabolites that are not extensively produced by other pathways. At least two components produced during glucose oxidation through the shunt pathway play an important role in metabolism—ribose-5-phosphate and NADPH. The coenzyme could be used as a source of ATP, but is more likely used in reductive synthesis. Such an interpretation of the role of the hexose monophosphate pathway implies that various degrees of pathway activity exist, depending upon the cell type. The hexose monophosphate pathway will not necessarily be ubiquitous like glycolysis or the Krebs cycle. Therefore, the presence of the pentose pathway constitutes an important aspect of biochemical differentiation of the cell.

Many have attempted to quantitate the relative activity of the glycolytic and the pentose pathways usually by comparing the amounts of CO_2 formed from [^{14}C]1-glucose and [^{14}C]6-glucose. Again, identical amounts of radioactive CO_2 are formed from carbons 1 and 6 of glucose when it is oxidized through the Embden-Meyerhof pathway exclusively. But if the hexose monophosphate shunt is also functioning, the amount of radioactive CO_2 formed from carbon 1 exceeds that formed from carbon 6. However, quantitation of the activity of the hexose monophosphate shunt is difficult. The most important problem is that after several turns of the cycle, the label of the carbon 1 of glucose is randomized. All equations used to calculate the relative activities of the glycolytic and the hexose monophosphate pathways must take this randomization into account.

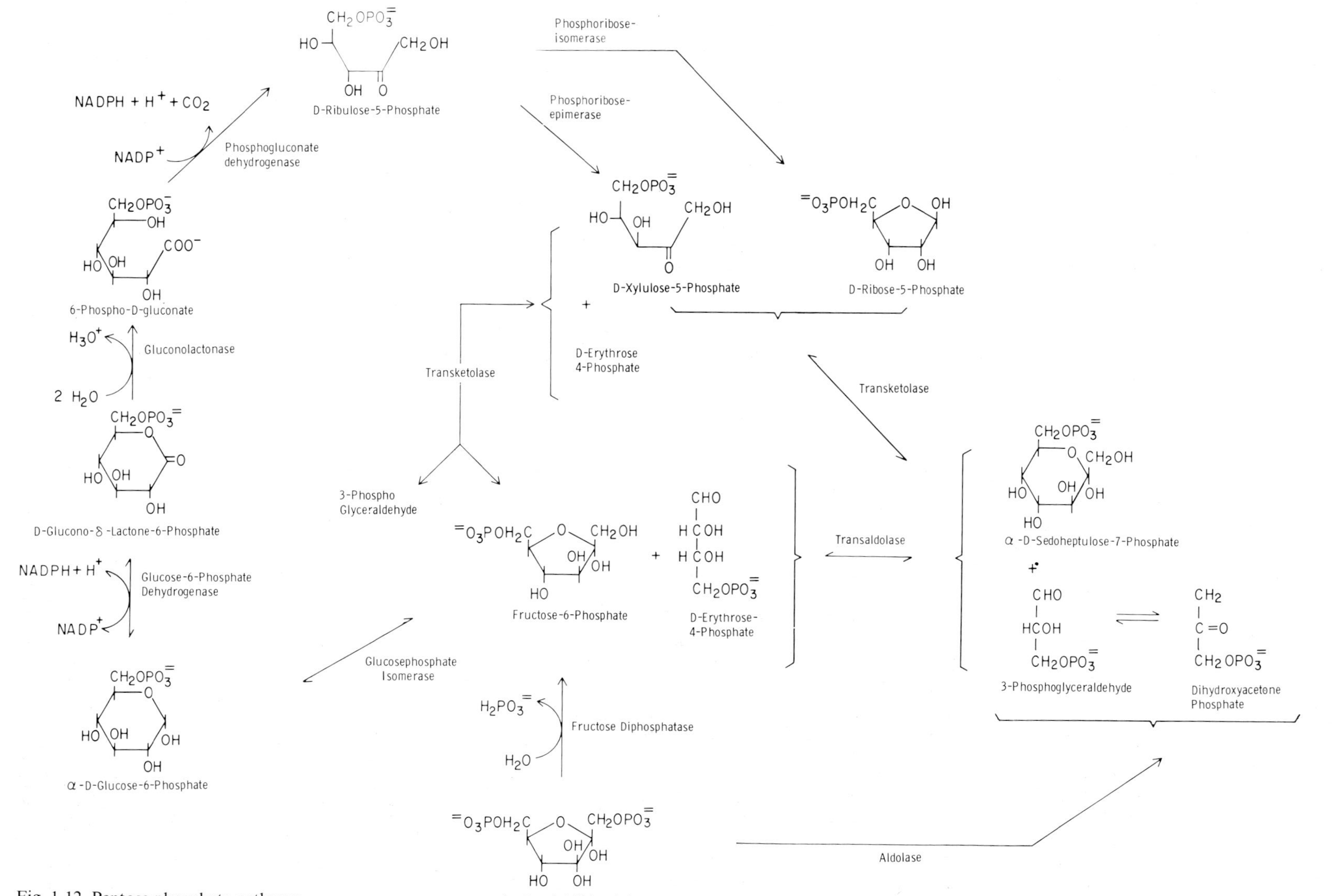

Fig. 1-12. Pentose phosphate pathway

Distribution of the Pentose Pathway in the Organism

Mammalian tissues can be divided into three groups with regard to the relative distribution of the glycolytic pathway and the hexose monophosphate shunt. First, there are tissues in which the glycolytic pathway is the only oxidative pathway; second, tissues in which the hexose monophosphate shunt is present but incomplete; and third, those in which the hexose monophosphate shunt is present and complete. In the third group, the amount of glucose oxidized through the hexose monophosphate shunt may exceed, be equal to, or be less than the amount oxidized through glycolysis. For example, brain slices metabolize glucose only through the Embden-Meyerhof pathway, while the amounts of glucose dissimilated through the Embden-Meyerhof pathway and through the hexose monophosphate shunt are equal in liver slices. In testes and bone marrow, the amount of glucose dissimilated through the hexose monophosphate shunt exceeds the amount of glucose dissimilated through glycolysis. In mammary glands, it is estimated that during the period of lactation 60% of the glucose is used through the shunt pathway, and although the enzymes of the hexose monophosphate shunt are abundant in fetal pig heart, little enzyme activity is present in adult pig heart. The tissues in which the hexose monophosphate shunt is active are precisely those in which either steroids (requiring NADPH for biosynthesis) or nucleotides (requiring ribose-5-phosphate) are also actively synthesized.

It should be kept in mind that calculations of the relative activities of the hexose monophosphate shunt and the classical Embden-Meyerhof pathway are based on the assumption that the triose phosphates, dihydroxyacetone phosphate and D-glyceraldehyde-3-phosphate, are constantly equilibriated. However, isotopic studies suggest that this is not the case. In view of the complexities of glucose metabolism, it is not surprising that the values for the relative activities of the glycolytic pathway and the hexose monophosphate shunt vary considerably from one laboratory to another.

The significance of the relative contributions of these two pathways for glucose oxidation appears in leukocyte metabolism. During phagocytosis, the rate of glucose oxidation is considerably increased, and most of the increase results from stimulation of the hexose monophosphate pathway.

Glucuronic Pathway

The carbon 6 of glucose may be oxidized to CO_2 in a series of reactions different from the Embden-Meyerhof pathway. The formation of glucuronic acid constitutes the first step of that pathway. Except for the formation of UDP-glucose, there is no phosphorylation involved in these reactions, and most of the steps are catalyzed by dehydrogenases which are NAD- or NADP-dependent. The reaction thus leads to the formation of NADH or NADPH. This pathway is called the glucuronic acid cycle because it starts with the formation of glucuronic acid. It ends with D-xylose, which is reconverted through the hexose monophosphate shunt to glucose-6-phosphate. While investigating the various steps of the pathway, researchers found intermediates important to the pathologist—namely, glucuronic acid (which combined with aglycone forms glucuronides) and L-xylose, the compound abundant in the urine of pentosuric patients. As a matter of fact, the urge to elucidate the mechanism of pentosuria was largely responsible for the discovery of many of the intermediate steps of the glucuronic pathway. The biosynthesis of ascorbic acid is also closely associated with the glucuronic pathway.

While UDP-glucose is a good precursor of glucuronic acid, glucose is not efficiently converted to glucuronic acid. These findings exclude the possibility of direct oxidation of glucose to form glucuronic acid. When UDP-glucuronic acid was found in liver, it was soon assumed that glucose oxidizes to glucuronide at the level of the UDP-linked hexose. Later, a NAD-dependent dehydrogenase that catalyzes the oxidation of UDP-glucose to UDP-glucuronic acid was found in the supernatant of liver homogenates. In that reaction, the alcohol function of carbon 6 of glucose is oxidized to yield a carboxyl group.

One could anticipate that such a reaction might occur in two successive steps: (1) the oxidation of the alcohol to yield the aldehyde, and (2) the oxidation of the aldehyde to yield the acid. If such a pathway occurred, two different dehydrogenases might be involved. In reality, an aldehyde intermediate was not found, and apparently a single dehydrogenase (uridine diphosphoglucose dehydrogenase) catalyzes the double oxidation. However, two molecules of NAD are required and two of NADH are formed. The further oxidation of the glucuronic acid involves nonphosphorylated intermediates [64–66].

After hydrolysis of UDP-glucuronide to glucuronide-1-phosphate and uridylic acid by UDP-glucuronide pyrophosphatase, the glucuronide-1-phosphate is further hydrolyzed in most tissue to glucuronide by a specific phosphatase. A number of observations made *in vivo* and *in vitro* suggest that in liver, the conversion of glucuronic acid-1-phosphate to glucuronate involves an alternative pathway. The activity of the glucuronic acid-1-phosphate is low in liver. If the enzyme activity is further inhibited *in vivo* by the administration of barbiturates, or *in vitro* by other inhibitors of phosphatases, glucuronic acid-1-phosphate does not accumulate; on the contrary, dephosphorylated glucuronate increases. Furthermore, compounds that stimulate the formation of ascorbic acid, a metabolite derived from glucuronic acid, also stimulate the formation of UDP-transglucuronylase. For these reasons, it has been postulated that the formation of glucuronate in liver involves UDP-transglucuronylase (glucuronyl transferase).

In the next step, the D-glucuronic acid is reduced to gulonate, and NADPH acts as a hydrogen donor. This reaction is catalyzed by a dehydrogenase, NADP-L-hexonate dehydrogenase, which has been purified from hog kidney and rat liver. The enzymes prepared from both sources are active on a variety of substrates, including iduronic acid and glucuronic acid. The liver enzyme also acts on D-glucuronolactone yielding L-gulonolactone [67].

Gulonate is a key metabolite. It is at the crossroads of two important pathways—the biosynthesis of ascorbic acid and the metabolic chain that links the glucuronic pathway to the hexose monophosphate shunt. Thus, gulonate may either be oxidized to yield gulonolactone, which is further converted to ascorbic acid, or it may be reduced to yield L-xylose. The oxidation of L-gulonate to L-gulonolactone is catalyzed by a lactonase, of which there are several in liver. Aldonolactonase, lactonase I, is found in the soluble fraction of rat liver; it has a broad specificity and hydrolyzes a wide variety of 5-, 6-, and 7-carbon aldonolactones; gulonolactone-1-phosphate; and gluconolactone [68].

Uronolactonase (lactonase II) hydrolyzes D-glucuronolactone. It is a microsomal enzyme with a more restricted specificity. The exact specificity and the exact number of lactonases remain unsettled. The enzymes have not yet been purified extensively. It was already mentioned that D-glucuronolactone can be converted to gulonolactone under the influence of the L-NADP hexonate dehydrogenase. This provides an alternative pathway for ascorbic acid, whereby gulonate is bypassed. It has also been suggested that specific lactonases act on glucuronic acid to yield gulonolactone, but the physiological significance of glucuronic acid lactonase remains unsettled.

L-Gulonolactone is further oxidized under the catalytic influence of a gulonolactone oxidase; 2-L-ketogulonolactone is then assumed to be an intermediate. The oxidase is found in microsomes and has been solubilized with the aid of bile salts, snake venom, and sonic oscillation. It has not been purified extensively. The conversion of gulonic acid to xylose involves the oxidation of the alcohol group of carbon 3 to the ketone and decarboxylation of carbon 1. 1,3-Keto-L-gulonic acid* is an intermediate, and, apparently, two different enzymes are involved in that reaction. The first enzyme, an NAD-dependent dehydrogenase with broad specificity, is a β-L-hydroxy acid dehydrogenase. In addition to catalyzing the conversion of gulonic acid to 3-keto-L-gulonic acid, it converts D-gluconate into D-ribose and acetoacetate into L(+)-β- hydroxybutyrate. The second enzyme is responsible for the decarboxylation.

* It was also postulated that ascorbic acid and L-xylose have a common intermediate—namely, 3-keto-L-gulonic acid. Although the compound has definitely been established to be an intermediate for L-xylose, it is not likely to be involved in the biosynthesis of ascorbic acid.

The conversion of L-xylose to D-xylose involves the reduction of the alcohols of carbon 2 to yield xylitol, which is then further oxidized to yield the D-isomers. The first of these two reactions is catalyzed by xylitol reductase and is NADP-dependent. The second is catalyzed by D-xylose reductase, an enzyme that requires NADH as a cofactor. The enzymes have been found in pig kidney mitochondria. Thus, the conversion of L-xylose to D-xylose occurs in two coupled reactions in which the hydrogen of NADPH is passed to NADH:

$$\begin{aligned} &\text{Xylose} + \text{NADPH} \rightleftharpoons \text{Xylitol} + \text{NADPH} \\ &\text{Xylitol} + \text{NAD} \rightleftharpoons \text{D-Xylose} + \text{NADH} \end{aligned}$$

D-xylose is readily converted to xylose-5-phosphate by a D-xylose kinase in the presence of ATP.

Intracellular Distribution of Glucuronic and Pentose Pathways

All the enzymes of the hexose monophosphate shunt have been found in the supernatant fraction of the cell. The enzymes of the glucuronic pathway, with the exception of NADP-L-hexonate dehydrogenase and aldonolactonase, which are found in the supernatant fraction of the cell, have a complex intracellular distribution. All the enzymes involved in glucuronic acid-1-phosphate synthesis and in the conversion of glucuronic phosphate or glucuronic acid to L-xylose or ascorbic acid are associated with the endoplasmic reticulum. The enzymes involved in the conversion of L-xylose to D-xylose are found in mitochondria. Thus, the complete glucuronic pathway involves three different cell fractions.

Since the soluble fraction and the endoplasmic reticulum seem to be closely related, the mechanism of transfer of the product of a reaction catalyzed in one cell fraction to the other cell fraction may be a simple one. It is also possible that the endoplasmic reticulum separates pools of "soluble protein," each containing its own battery of enzymes. Under such conditions, the products of the reaction catalyzed in the endoplasmic reticulum could conceivably be directed into alternative metabolic pathways. A mechanism by which the intracellular distribution of the enzyme controls cellular metabolism may be the selection of a given pathway on the basis of cell need. The transfer of the product of enzyme reactions associated with the endoplasmic reticulum to mitochondria (L-xylose) must be followed by extrusion from the mitochondria of the final product of xylose metabolism, D-xylose-5-phosphate, if the glucuronic pathway is to be complete. The mechanism that controls the passage of these metabolites through the mitochondrial membrane is most intriguing. Knowledge of this mechanism would probably provide new clues on metabolic controls.

Tricarboxylic Acid Cycle

Glycolysis, an anaerobic process, oxidizes glucose to yield a 3-carbon compound, pyruvic acid. Obviously, a large part of the chemical energy stored in the glucose molecule remains unavailable for cellular metabolism. Fortunately, there exists a biochemical device capable of oxidizing pyruvic acid to CO_2 and water in the presence of oxygen. It is known as the Krebs cycle, the tricarboxylic acid cycle, or the citric acid cycle [69–73] (see Fig. 1-13).

Although the Krebs cycle is usually considered as part of carbohydrate metabolism, it should be emphasized that amino acids (final products of protein catabolism), acetyl coenzyme A, and acetoacetic acid (final product of fat catabolism) are all oxidized through the citric acid cycle. The Krebs cycle is thus a system capable of oxidizing the final products of carbohydrates, proteins, and fats.

The energy released during the oxidation process is stored in high-energy bonds. Because a large number of metabolites are oxidized by the Krebs cycle, this pathway obviously constitutes the principal source of the cell's chemical energy. However, in addition to this main role, the Krebs cycle, by the reversal of some of its catabolic reactions, also produces numerous substrates for various biosynthetic processes.

The reactions in which pyruvic acid is oxidized to form CO_2 and acetic acid are of the greatest significance since they constitute the link between glycolysis and the Krebs cycle. These reactions involve a considerable number of coenzymes: thiamine pyrophosphate, lipoic acid, CoA, and NAD. Much of our knowledge of pyruvic acid oxidation depended on the discovery of CoA and lipoic acid, and it might be useful to review the biochemistry of lipoic acid before we enter into more detail. Refer to the chapter on vitamins for a review of the metabolism and catabolism of thiamine, CoA, and NAD.

Lipoic Acid

The coenzyme lipoic acid was first found in bacteria where it could replace acetate as a growth factor; later it was identified in the soluble fraction of liver. It has been crystallized and was demonstrated to be a fatty acid that exists in either a reduced or an oxidized form. Lipoic acid undoubtedly stimulates the growth of many bacteria. It has also been claimed that it stimulates growth in chickens and rats, but the results in animals are still controversial. The effect of lipoic acid on mammalian growth is not clear, and little is known of its biosynthesis.

The thioester group of lipoic acid can be reduced in various ways. In the laboratory, the oxidized lipoic acid can be reduced by strong reducing agents (zinc). The longer wavelength of ultraviolet radiation also reduces the oxidized form. In the cell, enzymes catalyze the reduction of lipoic acid.

In as much as reduced lipoic acid is an antioxidant, it is difficult to distinguish a specific biological action from a nonspecific antioxidant effect. The protective effects of lipoic acid against X-irradiation and methemoglobin intoxication are thus difficult to interpret. Lipoic acid has often been used in the therapy of chronic and acute liver disease. Although these observations are not yet conclusive, lipoic acid appears to decrease the abnormal accumulation of pyruvic and lactic acid in the blood occurring in some liver injuries, but the reader has undoubtedly noticed that, like cysteine, lipoic acid contains SH groups. Cysteine is known to prevent fatty infiltration in the liver. It is impossible at present to decide if the effect of lipoic acid is linked to its specific metabolic role or results from its general molecular composition [74, 75].

Pyruvic Acid Decarboxylation

The decarboxylation of pyruvic acid is an example of a more general type of biochemical reaction: the decarboxylation of α-keto acids. The reaction is complex and occurs in several consecutive steps. The intermediates have been identified, but little is known of the enzymes involved. The reaction starts with the complexion of pyruvic acid with one molecule of enzyme-bound thiamine pyrophosphate. This is followed by decarboxylation of pyruvic acid and the formation of an intermediate, 2-acetylthiamine pyrophosphate, in which the aldehyde carbon of the acetyl is bound to the carbon 2 of the thiozole ring of the thiamine pyrophosphate. In the second step, the aldehyde is oxidized, the disulfide bond of enzyme-bound lipoic acid is reduced, and the free enzyme-bound thiamine pyrophosphate is restored. The third step of the reaction involves the transacylation from reduced lipoic acid to CoA. Finally, lipoic acid is reoxidized by the catalytic activity of an NAD-dependent flavoprotein, lipoic dehydrogenase (see Fig. 1-14).

In the last two steps of this sequence, the activities of two enzymes are involved, a transacylase and a dehydrogenase. It is interesting that while the first offers a specificity for the (−) isomer of lipoic acid, the second is active on both the (−) and (+) isomers. It has been proposed that pyruvic acid decarboxylation occurs on a single protein complex, and that thiamine pyrophosphate and lipoic acid are both enzyme bound. However, the bond between thiamine pyrophosphate and the enzyme is not so tight that it cannot be split. The coenzyme can be removed, but it must be restored to regenerate enzyme activity. Magnesium, ATP, and inorganic phosphate are necessary for the binding of lipoic acid to the enzyme.

The metabolic map (see Fig. 1-13) illustrates the reactions involved in the citric acid cycle. After the condensation of acetyl-CoA with oxaloacetate to form citric acid, this carbon compound is transformed into isocitrate, which is itself oxidized and decarboxylated by the same enzyme to yield α-ketoglutarate and CO_2. Oxalosuccinate is an intermediate in this reaction.

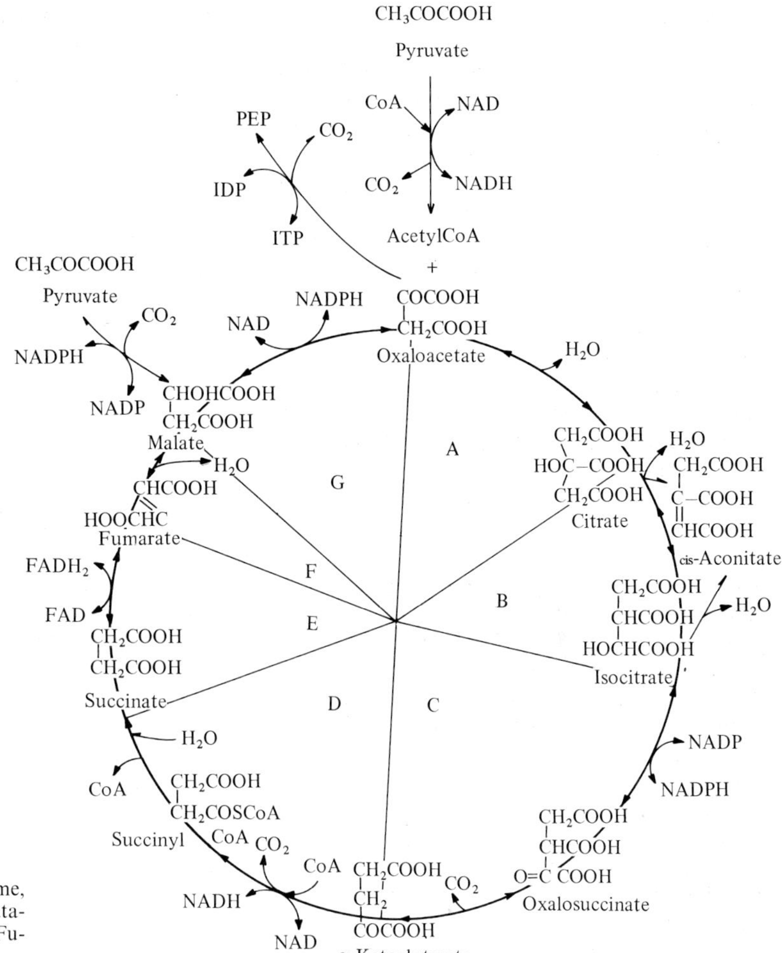

Fig. 1-13. Krebs cycle. *A* condensing enzyme, *B* Aconitase, *C* Isocitric dehydrogenase, α-ketoglutarate decarboxylase, *E* Succinic dehydrogenase, *F* Fumarase, *G* Malic dehydrogenase

I - Pyruvic Decarboxylase

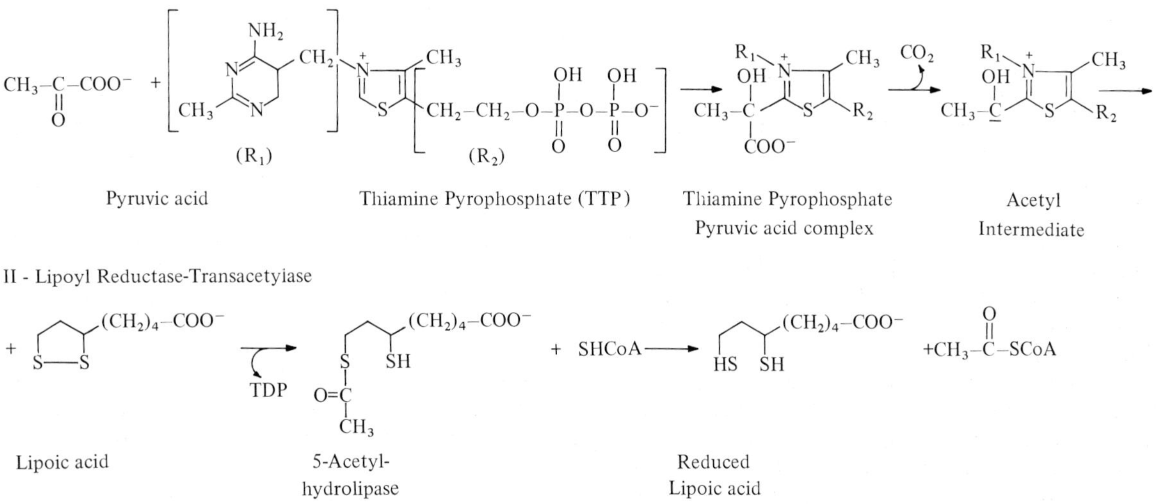

$CH_3-C(=O)-COO^-$ + [Thiamine Pyrophosphate: (R_1) CH_2 – thiazolium ring (CH_3) – (R_2) $CH_2-CH_2-O-P(O)(OH)-O-P(O)(OH)-O^-$] → $CH_3-C(OH)(COO^-)$–thiazolium (R_1, CH_3, R_2) $\xrightarrow{-CO_2}$ $CH_3-C(OH)$=thiazolium (R_1, CH_3, R_2) →

Pyruvic acid; Thiamine Pyrophosphate (TTP); Thiamine Pyrophosphate Pyruvic acid complex; Acetyl Intermediate

II - Lipoyl Reductase-Transacetyiase

+ Lipoic acid (S—S ring, $(CH_2)_4-COO^-$) $\xrightarrow{TDP}$ 5-Acetyl-hydrolipase ($S-C(=O)-CH_3$, SH, $(CH_2)_4-COO^-$) + SHCoA → Reduced Lipoic acid (HS, SH, $(CH_2)_4-COO^-$) $+CH_3-C(=O)-SCoA$

Lipoic acid; 5-Acetyl-hydrolipase; Reduced Lipoic acid

Fig. 1-14. Pyruvic acid decarboxylation

α-Ketoglutarate is oxidized and decarboxylated to yield succinyl-CoA in the presence of a multiple-enzyme system and a variety of coenzymes. The decarboxylation of α-ketoglutarate is, like that of pyruvic acid, another example of α-keto acid decarboxylation.

The Fate of Succinyl Coenzyme A

Succinyl CoA may yield succinate by at least three different pathways (see Fig. 1-15). It may be hydrolyzed by the action of succinyl CoA deacylase; the cleavage of the coenzyme may be linked to the phosphorylation of IDP or GDP; or the coenzyme may be transferred from succinate to a keto acid.

The breakdown of succinyl coenzyme A may also be linked to the phosphorylation of GDP and IDP to yield GTP and ITP in the presence of inorganic phosphate. The enzyme catalyzing that reaction has been named the phosphorylating enzyme, and has been prepared in a crude form from heart muscle. This preparation also contains another enzyme, nucleoside diphosphate kinase, which catalyzes the transfer of phosphorus from GTP or ITP to ADP to yield ATP. The phosphorylation of ADP coupled to the oxidation of α-ketoglutarate is the only substrate level phosphorylation in the Krebs cycle, and, as can be expected, it is not inhibited by dinitrophenol. When ^{18}O-labeled phosphate is used in the reaction, the label appears in the carboxyl of succinate, and therefore, it has been assumed that a phosphorylated intermediate is formed, but such an intermediate has never been identified.

Succinyl CoA can also yield succinate by a third type of reaction, one in which the coenzyme is transferred to a keto acid. The CoA transferase catalyzing this reaction has also been found in heart muscle. It is specific for succinyl CoA, but the coenzyme may be transferred to any keto acid the chain length of which varies from that of acetoacetic acid to that of ketocaproate. It is interesting that the capacity of the enzyme to catalyze the transfer decreases when the length of the CoA acceptor is increased from ketovalerate to ketocaproate.

In the presence of oxygen and the cytochrome system, succinate is finally transformed to fumarate by a dehydrogenation process, and in turn, fumarate is converted into malate, which itself is dehydrogenated to form oxaloacetate.

The enzymes involved in all these reactions have been extensively studied. The effects of pH, ionic strength, inhibitors, and the specificity of the enzymes have all been investigated, and it would be impossible to review here all the literature concerned with these properties. However, we shall describe some of these properties because they illustrate the complexity of multiple-enzyme reactions and emphasize the danger of jumping to conclusions about the intracellular distribution of enzymes.

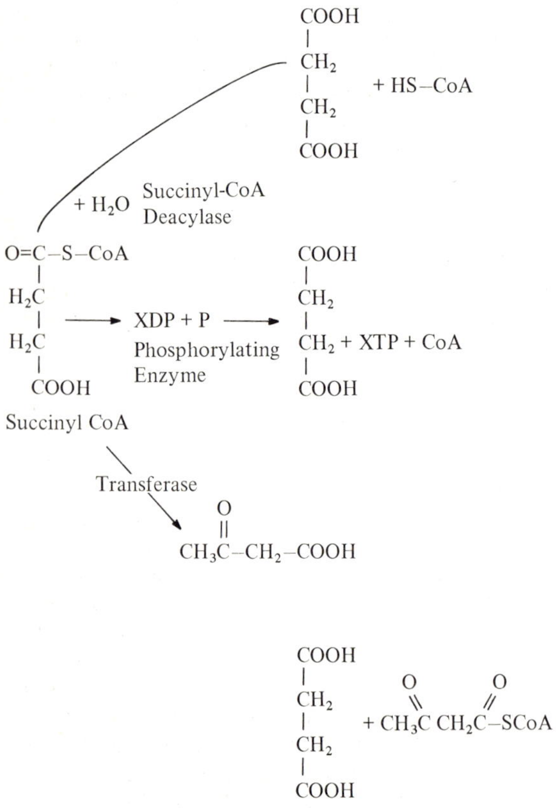

Succinyl CoA ($O{=}C{-}S{-}CoA$ / H_2C / H_2C / $COOH$):

- $+ H_2O$, Succinyl-CoA Deacylase → $COOH{-}CH_2{-}CH_2{-}COOH + HS{-}CoA$
- → $XDP + P$ → (Phosphorylating Enzyme) → $COOH{-}CH_2{-}CH_2{-}COOH + XTP + CoA$
- Transferase → $CH_3\overset{O}{\overset{\|}{C}}{-}CH_2{-}COOH$ → $COOH{-}CH_2{-}CH_2{-}COOH + CH_3\overset{O}{C}\,CH_2\overset{O}{C}{-}SCoA$

Fig. 1-15. Fate of succinyl-CoA

Citric Acid Synthesis

The reaction that uses most of the acetyl-CoA formed by pyruvic acid is the condensation of acetic acid and the keto form of oxaloacetic acid to yield citric acid by forming a carbon-to-carbon bond between the methyl carbon of acetyl-CoA and the carbonyl carbon of oxaloacetate. This reaction is interesting in more than one respect. In addition to the fact that the reaction introduces the oxidation product of fatty acids, carbohydrates, and proteins into the tricarboxylic cycle by converting the 4-carbon chain of oxaloacetic acid into a 6-carbon chain (citric acid), the condensing enzyme presents a fascinating stereospecificity.

The reaction is catalyzed by an enzyme called the condensing enzyme, which was isolated and crystallized from pig heart. The general specificity of the enzyme is not definitely established. However, it is known that the enzyme catalyzes the condensation of acetate derivatives. For example, fluoroacetate can be condensed to form fluorocitrate.

Three-Point Attachment Theory

The enzymes involved in the biosynthesis and metabolic conversion of citric acid manifest an interesting type of stereospecificity. Since citric acid is a symmetrical molecule, the two acetate groups of citric acid are equally apt to participate in organic reactions. If citric acid acted as a symmetrical molecule in the reversible reaction involving acetyl-CoA and oxaloacetate, and in the conversion of citric acid to α-ketoglutarate (Fig. 1-16), in which carboxyl-labeled acetic acid is used as a precursor, one might anticipate that the two carboxyl groups of α-ketoglutaric acid would be equally labeled. Isotopic experiments of Potter and Heidelberger [76] clearly demonstrate that this was not the case, and that all the label could be recovered in the α-carboxyl of α-ketoglutarate. These investigators first prepared biologically labeled citric acid by incubating reinforced liver homogenates with $[^{14}C]O_2$. The labeled precursor was recovered and then added to another liver homogenate supplemented with arsenite, a compound that blocks α-ketoglutarate decarboxylation and causes labeled α-ketoglutarate to accumulate. The product was then isolated and the site of labeling was traced. All the label was found to be associated with the α-carboxyl of α-ketoglutarate. These results were later confirmed in several other laboratories.

The findings supported the "three point attachment theory" that Ogston [77] had proposed precisely to explain the biological stereospecificity of symmetrical organic compounds. This concept will be best understood if one takes a three-dimensional view of the citric acid molecule. The reader should first be reminded that the carbon atom has four sp^3 orbitals, and the angles between them are 109.5°. Thus, the forces represented by the valence bonds are symmetrically distributed around the center of the atom, and the angle that separates them is 109.5°. Consequently, if the center of a carbon atom is placed in the center of a tetrahedron, the valences will be represented by lines directed from the center to the four corners of the tetrahedron.

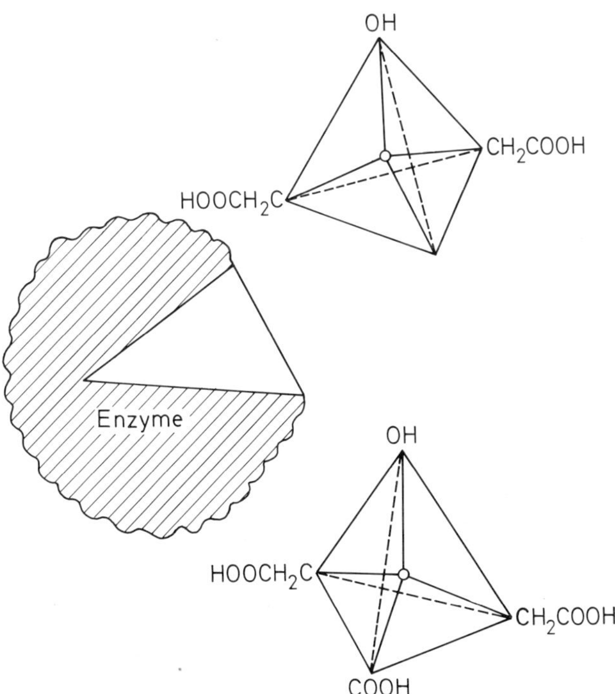

Fig. 1-16. Schematic representation of three-point attachment theory

Let us now place the carbonyl atom of citric acid in the center of such a tetrahedron and draw the line of valences. The hydroxyl is placed above, the carboxyl below, and the two acetic groups to the left and right of a plane passing through the center. Let us now represent the enzyme as a large protein molecule with an active center shaped to accept the citric acid molecule, and by way of an example, consider its effect on aconitase.

To make the double bond of the *cis*-aconitate, aconitase must be presented with the carbonyl carbon and with at least one of the methylene carbons. If the enzyme and substrate contact at only two points, the molecule can rotate to fit the active center of the enzyme and both methylene carbons are equivalent; but if there are three points of attachment, the two methylene carbons are no longer equivalent. Thus, when the molecule is presented to the enzyme in configuration A, the carboxyl group will be at the right of the central plane. It will be at the left when the positions of the molecules have been reversed.

To explain the selective reactivity of only one of the two identical groups, the Ogston theory further postulates that the reactivity is different at each of the points of attachment of the enzyme with the substrate. The three-point attachment theory thus explains how symmetrical organic molecules can be acted upon by an enzyme as if they presented a specific type of stereoisomerism. However, the absolute enzyme-substrate relationship remains unknown. In addition to citric acid, several other symmetrical substrates are known to be specifically oriented by their enzymes.

Aconitase

Little was known of the mechanisms of action of aconitase because all attempts at purification failed until it was demonstrated that the enzyme requires iron and cysteine for activity. It has since become evident that the mechanism of the reaction catalyzed by aconitase is extremely complex. *Cis*-aconitate was thought to be an intermediate, but the compound could not be isolated. This led Speyer and Dickman [78] to suggest that a common intermediate exists between citrate on the one hand, and aconitate and isocitrate on the other (see Fig. 1-17). These authors found that when citric acid labeled with heavy water was used as a substrate, the isocitrate was extensively labeled, while only traces of deuterium were found in *cis*-aconitate. This suggested that *cis*-aconitate is not on the pathway leading from isocitric to citric acid. To explain these results, the authors postulated an intermediate common to *cis*-aconitate and isocitrate consisting of a tricarboxylic acid forming a complex with iron and cysteine. Such a complex would then be capable of intramolecular hydrogen transfer between the carbonium

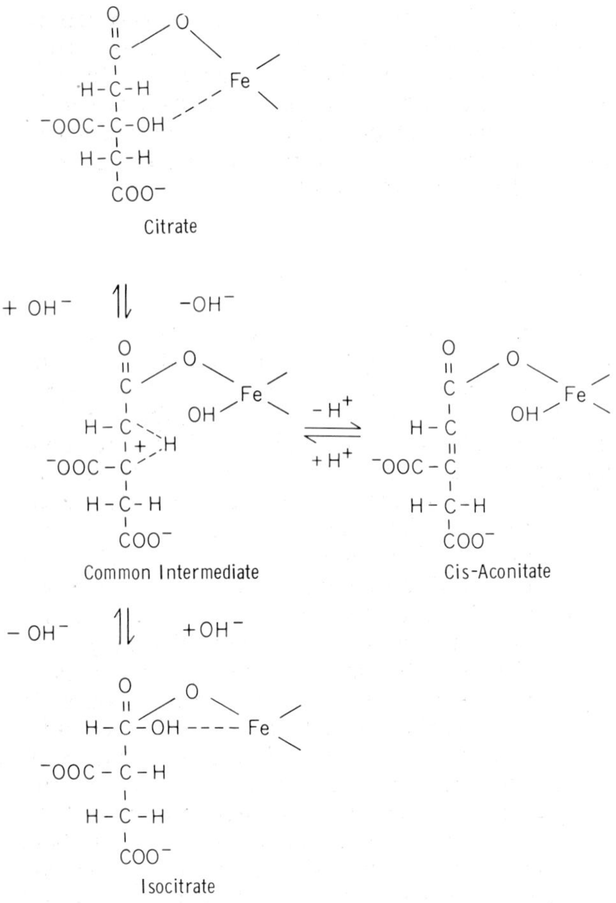

Citrate

$+OH^-$ ⇅ $-OH^-$

Common Intermediate ⇌ ($-H^+$ / $+H^+$) Cis-Aconitate

$-OH^-$ ⇅ $+OH^-$

Isocitrate

Fig. 1-17. Mechanism of action of aconitase

ion and the oxygen atom. This transfer converts citrate to isocitrate. As Dickman [79] proposed, such a complex might offer a new mechanism of regulating the rate at which the Krebs cycle operates. The amounts of isocitrate formed would depend on the ratio of ferric and ferrous ions available in the medium, and thus be a function of the medium's oxido-reduction potential.

The ability of aconitase to form both isocitrate and *cis*-aconitate is not well understood. It should be pointed out that, here again, the enzyme is capable of two different reactions: an intramolecular conversion in the process of transforming citrate to isocitrate and a dehydration in the reaction transforming citrate into *cis*-aconitate. All attempts to separate these two catalytic properties have failed except in *Aspergillus niger*, in which two separate aconitases have been found.

Isocitric Dehydrogenase

The next reaction, the transformation of isocitrate into α-ketoglutarate, is also catalyzed by a single enzyme, although the process involves both dehydrogenation and decarboxylation. The overall reaction of isocitric dehydrogenation requires a pyridine nucleotide and magnesium ions. The pyridine nucleotide is necessary for dehydrogenation, whereas the magnesium ion is required for decarboxylation. Both NAD- and NADP-specific isocitric dehydrogenases have been found in mammalian tissue. The former differ from the latter by their intracellular distribution, their requirements for an adenosine-5′-monophosphate, and by the fact that the NAD reaction is irreversible under conditions in which the NADP reaction is readily reversible. The significance of the presence of the two types of enzymes will appear more clearly when the intracellular distribution of isocitric dehydrogenase is examined.

α-Ketoglutarate Decarboxylase

As pyruvic acid decarboxylation constitutes the link between glycolysis and the Krebs cycle, α-ketoglutaric decarboxylation divides the reactions involving 6-carbon acids (citrate, isocitrate, and oxalosuccinate) and those involving 4-carbon acids (succinate, fumarate, and malate). The analogy between the two reactions is not restricted to their role in intermediate metabolism, but extends also to the mechanism of action of the two multiple-enzyme systems. In α-ketoglutaric decarboxylation, the overall reaction leads to the formation of CO_2 and succinate. CoA, NAD, thiamine, lipoic acid, and magnesium are requirements for this multiple-enzyme system activity.

As is so often the case with multiple-enzyme systems, in beginning studies investigators needed a particulate enzyme preparation to demonstrate the reaction. Later, however, a soluble enzyme preparation became available, and it was demonstrated that the various steps of the reaction are performed by a high molecular weight compound to which thiamine pyrophosphate and lipoic acid are attached. The detailed mechanism of the reaction is far from understood; it is not even known how many enzymes are involved. In bacteria, the complex has been resolved into two separate fractions named A and B, and neither fraction was found to be active by itself. In other words, α-ketoglutarate is oxidized and decarboxylated only if the A and B fractions are recombined.

Succinic Dehydrogenase

Succinic dehydrogenase catalyzes the transfer of electrons from succinate to fumarate. Although in the cell the enzyme is tightly bound to mitochondria, it has been prepared in a soluble form from heart muscle and yeast. The enzyme probably contains SH groups and one flavin molecule for every protein molecule. The property of that enzyme is reviewed in more detail in the section on electron transport.

Fumarase

Like aconitase, fumarase catalyzes a reversible hydration-dehydration reaction. The fumarase manifests stereospecificity for both substrate and product. The substrate must be fumarate and not L-maleate. The product is only L-maleate. (Fumarate is *trans*: L-maleate is *cis*-ethylene dicarboxylic acid.) The kinetics of the fumarase reaction are most peculiar, but this is not the place to discuss their mechanism. Suffice it to point out that the enzyme kinetics are greatly influenced by the pH, and that fumarate may react with the enzyme not only as a substrate by combining with the active site, but also as an ion by combining with some part of the molecule other than the active site. This dual mode of attachment of substrate to the enzyme considerably complicates the kinetics of the reaction.

Pyruvate Metabolism

Pyruvate may enter the Krebs cycle by a pathway different from that involving the formation of acetyl-CoA. Indeed, the keto acid can be converted to a dicarboxylic acid by at least two different biochemical processes. One of them involves the carboxylation of pyruvic acid leading to the formation of oxaloacetic acid. The actual substrate in that reaction is not pyruvic acid, but phosphoenolpyruvic acid. The enzyme involved requires manganese, and IDP acts as the phosphorus acceptor. The reaction is coupled with the pyrophosphate transfer (catalyzed by nucleoside diphosphokinase) from ITP to ADP.

$$\begin{array}{l} CH_3COOH \\ | \\ C{=}O \\ | \\ COOH \end{array} + CO_2 \xrightleftharpoons[\text{Phosphoenolpyruvate kinase}]{IDP \;\; ITP} \begin{array}{l} CH_2 \\ | \\ C{-}OPO_3H_2 \\ | \\ COOH \end{array}$$

Oxaloacetic acid — Phosphoenolpyruvate

$$ITP + ADP \xrightleftharpoons[\text{Nucleoside Diphosphokinase}]{} IDP + ATP$$

Avidin blocks the reaction, an observation which suggests that biotin is required for the decarboxylation. The malic enzyme catalyzes the decarboxylation of malic acid to pyruvic acid. The reverse of the reaction leads to the formation of the dicarboxylic acid at the expense of a pyruvic acid. The malic enzyme has many properties in common with isocitric dehydrogenase. It catalyzes an oxidative decarboxylation. Its coenzyme, NADP, acts as a hydrogen acceptor, and no intermediate has been isolated in the reaction. The molecular mechanism of the reaction is not known.

$$\begin{array}{l} COOH \\ | \\ CH_2 \\ | \\ HCOH \\ | \\ COOH \end{array} \xrightleftharpoons[\text{Malic enzyme}]{NADP \;\; NADPH} CO_2 + \begin{array}{l} CH_3 \\ | \\ C{=}O \\ | \\ COOH \end{array}$$

Malic acid — Pyruvic acid

We have considered two types of carboxylation; in the first case it was the substrate and in the second the NADPH that provided the energy for forming a C—C bond. Other types of carboxylations will be discussed when the biological role of biotin is reviewed. Both these carboxylations may be affected by hormones, diabetes, and starvation.

Krebs Cycle Distribution in Tissue

It is highly probable that the Krebs cycle functions in all cells. Several reports have implied that it does not exist in the skin either because some enzymes essential to its performance could not be found in skin homogenate or because oxygen uptake was not stimulated by the addition of various substrates such as citrate or α-ketoglutarate to the skin slices. However, these first conclusions were followed by contradictory observations; α-ketoglutarate and other substrates were later found to stimulate oxygen uptake, and two enzymes of the citric acid cycle—malic isocitrate and succinic dehydrogenase—were detected in rat skin.

It has been clearly demonstrated that the Krebs cycle functions in bone. Thus, in normal bone, oxaloacetate and pyruvate are converted to citrate, which is rapidly oxidized. Citrate accumulates under the influence of the parathyroid hormone without increasing its rate of oxidation. In contrast, estrogens accelerate both citric production and citrate oxidation. However, these phenomena may be several steps removed from the primary mechanism triggering phosphorus and calcium fixation on bone tissue, in view of the fact that no correlation between Krebs cycle activity and the uptake of phosphorus and calcium from the medium has been established. Thus, this effect of the hormones on the Krebs cycle may simply reflect hormonal stimulation of growth and protein synthesis in those tissues.

Intracellular Distribution of the Krebs Cycle Enzymes

The relationship between the enzymes of the Krebs cycle and cellular structure is surely one of the most challenging chapters in cellular biochemistry. For years investigators sought to determine two things: the intracellular distribution of the enzymes of the Krebs cycle, and the relationship between enzymes and the mitochondrial structures. We shall now review these two lines of investigation [80].

Among the cell fractions obtained by differential centrifugation, only the mitochondria are capable of oxygen uptake when various metabolites of the Krebs cycle are added to the medium. It was concluded that the enzymes of the Krebs cycle are present in the mitochondria and that each of these enzymes was found exclusively in the mitochondria. It should be noted that such experiments measure overall oxygen uptake, and that they are therefore inadequate to demonstrate the presence of Krebs cycle enzymes in fractions other than the mitochondria, since cytochrome oxidase is found only in mitochondria. The assertion that all enzymes of the Krebs cycle are present exclusively in mitochondria was proved wrong when the intracellular distribution of the enzymes of the Krebs cycle was tested by measuring the formation of the specific product of each reaction.

The first breakthrough came when Schneider and Hogeboom [81] directly measured NADP isocitric dehydrogenase activity and demonstrated that only 10% of the total enzyme activity is present in the mitochondrial fraction. According to their estimate, this amount is not sufficient to explain the normal oxidation rate of the Krebs cycle metabolites. Assuming that the NADP isocitric dehydrogenase in the mitochondria is a contaminant, they concluded that at least some of the enzymes involved in the Krebs cycle are extramitochondrial.

Two subsequent observations supported the hypothesis that the enzymes are external to the mitochondria. Speyer and Dickman [78] demonstrated that a large portion of aconitase activity is found outside of the mitochondria. This suggests that the transformation of citric acid into isocitric acid also occurs outside the mitochondria. However, the work of Speyer and Dickman leaves no doubt that the aconitase associated with the mitochondria is genuine to that granule. In addition, it was observed in Schneider's laboratory that most of the hepatic cell citrate can be recovered within the normal or the fluorocitrate-poisoned mitochondria. These findings suggest that citrate dissimilates outside of the mitochondria. At this point, it seemed logical to assume that the conversion of citrate to α-ketoglutarate takes place outside of the mitochondria. Such a concept, however, could no longer prevail when Plaut and Sung [82] demonstrated the presence of NAD isocitric dehydrogenase in mammalian mitochondria. This enzyme was then suspected to be the one involved in the oxidation of the Krebs cycle intermediate.

It was found that in brain, for example, the activity of the NAD-dependent dehydrogenase is much greater than that of the NADP-dependent dehydrogenase.

A new series of observations seems to reconcile all these points of view. It was confirmed that most of the NADP isocitric dehydrogenase is extramitochondrial, except for 10% of the enzyme that is truly mitochondrial and not due to contamination by the supernatant NADP isocitric dehydrogenase. But it is apparently the NAD isocitric dehydrogenase which is active in the Krebs cycle, since in mitochondria that have been deprived of their nucleotides, the Krebs cycle activity can be restored by adding NAD but not by adding NADP. NAD isocitric dehydrogenase has been extracted from beef mitochondria. The enzyme is activated by ADP and inhibited by ADP and NADH [83].

A third possible pathway by which the isocitrate is oxidized has been suggested, in which NADP isocitric dehydrogenase reduces NADP, the hydrogen of which is then transferred to NAD through the catalytic action of a transhydrogenase.

Although these findings add to our understanding of the performance of the Krebs cycle in the mitochondria, they do not explain the role of the NADPH isocitric dehydrogenase or of aconitase outside of the mitochondria. Is this presence due to leakage from the organelles during the preparation procedure, or are the enzymes truly supernatant enzymes? The fact that mitochondria can retain large concentrations of citrate makes leakage of large enzyme molecules improbable. Furthermore, the findings of Speyer and Dickman [78] suggest that the intra- and extramitochondrial aconitase might be two different enzymes. They found that the optimum pH for the intramitochondrial aconitase was 5.4, and 7.3 for the supernatant aconitase. However, when the mitochondria are frozen and thawed, the aconitase released from its organelles has a pH optimum of 7.3, and, therefore, the experiments of Speyer and Dickman [78] do not allow us to decide whether the aconitase of the supernatant leaked out of the mitochondria, or whether it was present before the cellular disruptions.

Various relationships between enzymes of the Krebs cycle and mitochondria are possible. For instance, all enzymes could be enclosed within mitochondrial structures; or the enzymes could take part in the structural build-up of the cell. There is no evidence demonstrating that all enzymes of the Krebs cycle are part of the mitochondria. The existence of enzymes with multiple catalytic properties (isocitric dehydrogenase, aconitase, and malic dehydrogenase) and the failure to separate the multiple steps of an overall reaction (pyruvic and α-ketoglutarate oxidation) are sometimes taken as evidence for the participation of the enzyme in the building-up of the mitochondrial structure, but these arguments do not take into account the limitations of the actual biochemical methods, and, therefore, conclusions based upon them are premature.

The various catalytic properties of aconitase, for example, might be due to the presence of several proteins so similar in many respects that separation is difficult. Nevertheless, as we have already pointed out, the two catalytic properties of aconitase have been separated in *A. niger*. Furthermore, the claim that these various catalytic properties are due to a single protein because the preparation is declared pure on the basis of ultracentrifugal and electrophoretic analysis is unjustified, because these methods demonstrate only that the protein components of the material under study are similar in size and mobility. Most protein chemists would require more severe criteria of purity, such as end-group analyses. This line of reasoning applies *a fortiori* to the multiple-enzymes systems.

The Respiratory Chain

The secrets of life have fascinated scientists and wizards of all times. Leonardo da Vinci the universal genius, and Paracelsus, the courageous and clairvoyant physician of the early Renaissance, both remarked that men and fire die alike in absence of air. The first constructive step toward the solution to the problem of respiration came when Priestley, an English clergyman, demonstrated that air was a mixture of several gases, of which only one was absolutely necessary for life. In fact, Priestley discovered what was later called oxygen, and predicted by mere deduction most of the medical applications of oxygen.

The mechanism by which oxygen maintains life was not understood until Lavoisier completed his fascinating investigations. This great French scientist re-isolated oxygen as a pure gas, gave it its name ("oxygen" means "acid-former" because Lavoisier thought that oxygen participated in the formation of all acids), and developed a theory of combustion. (It is amazing that Lavoisier, aware of the English clergyman's work, never acknowledged it, despite the fact that his devoted wife translated Priestley's work into French.)

In one of the most remarkable scientific collaborations, Lavoisier, the chemist, and Laplace, the physicist, carefully measured oxygen intake and CO_2 output in laboratory animals, and were thus able to calculate the respiratory quotient:

$$\mathrm{Rq} = \frac{\text{Volume of } CO_2 \text{ expired}}{\text{Volume of } O_2 \text{ inspired}}.$$

By correlating these results with the formation of heat in the living organism, they came to the conclusion that during respiration, carbon is oxidized to CO_2. Today we know, of course, that the CO_2 expired comes essentially from decarboxylation processes. In contrast, biological oxidations involve the oxidation of hydrogen to form water. This Lavoisier could not know until Cavendish, the noble misanthrope, demonstrated that oxygen and hydrogen combine to form water; but as soon as this revolutionary finding was made available to him, Lavoisier predicted that part of the oxygen inspired would be used to oxidize hydrogen and lead to the formation of water. Much had to be done yet before the site of animal respiration (the cell) was discovered, and before the nature of the catalyst (enzymes rather than metals) could be demonstrated.

A great step in the understanding of biological oxidation was made when the concept of oxidation was itself broadened. At present, oxidation is defined as the reaction that involves a loss of electrons. This, of course, leads to a change in valence, and may occur in various ways: by direct reaction with an oxidizing compound (chlorine or oxygen); by a loss of electrons (as in the passage from ferrous to ferric iron); and by dehydrogenation. Direct oxidation in the presence of oxygen, though common in inorganic chemistry, is rather rare in organic chemistry in which the dehydrogenation mechanism is more frequent. A classical example of dehydrogenation is the oxidation of an aldehyde to the corresponding carboxylic acid through the hydrate.

It was Wieland who pointed out, in 1912, that dehydrogenation was the pattern of oxidation in living tissues. Insofar as any loss of electrons in an element produces a corresponding gain in electrons in another element, oxidation must always be accompanied by a coreduction, and the hydrogen liberated in a biological reaction must be transferred to a hydrogen acceptor. This transfer is catalyzed by a specific enzyme. In biological oxidation, the hydrogen is transferred from the substrate to oxygen through a chain of hydrogen acceptors (pyridine nucleotides, flavin nucleotides, quinones, or cytochromes) in the presence of specific dehydrogenases.

The hydrogen is finally transferred to molecular oxygen in a reaction catalyzed by a specific oxidase (cytochrome oxidase). Biological oxidations are, therefore, a function of the dehydrogenases and of the hydrogen acceptors. These various factors will be considered in the following paragraphs [84–92].

The hydrogen that enters the electron transport chain is generated in the Krebs cycle and is derived from pyruvate and water. The water is introduced in the substrates during the hydration of fumarate and use of acetyl and succinyl CoA.

Pyridine Nucleotides

Two pyridine nucleotides, diphosphopyridine nucleotide (NAD) and triphosphopyridine nucleotide (NADP), are involved in electron transport. In NADP, a third phosphoric group combines with the carbon 2 of the nucleotide's ribose moiety.

NAD is the active form of a vitamin, nicotinamide. The evidence available indicates that nicotinic acid is the intermediate in NAD synthesis. Thus, if the vitamin nicotinamide is used for NAD synthesis, it is used only after deamination. Nicotinic acid reacts with PRPP to yield the mononucleotide of nicotinic acid which forms deamido NAD in the presence of ATP and an enzyme present in liver nuclei. The amidation of the deamido NAD occurs in the presence of ATP, glutamine, and an enzyme present in the supernatant of liver homogenates.

Although nicotinamide is a vitamin, it can also be synthesized from tryptophan. Therefore, the NAD levels of the organism depend on both nicotinamide and tryptophan intake. The pathway of NAD biosynthesis starting from tryptophan is discussed in detail later. NADP is much less abundant in the cell than NAD. The biosynthesis of NADP probably involves the reaction of NAD with ATP in the presence of a nucleotide pyrophosphorylase.

It is interesting to comment on some *in vivo* experiments in which nicotinamide administration increased NAD amounts in mouse liver. One could expect nicotinamide to act as precursor, and adenine to be mobilized for NAD synthesis. Instead, the stimulation of NAD synthesis after the administration of nicotinamide results from an acceleration of the *de novo* synthesis of both adenine and ribose. The stimulation of NAD synthesis after nicotinamide administration is inhibited in hyperthyroid animals or after azaserine administration. The effect of azaserine can readily be explained on the basis of the role played by glutamine in purine biosynthesis. In contrast, the mechanism by which the thyroid acts on NAD synthesis is not clear. The effect of the thyroid hormone on NADP concentration is discussed in another part of the book.

A variety of enzymes attack the NAD molecule, splitting either the amino group, the glycosidic linkage, or the pyrophosphate bridge. Taka-diastase attacks NAD to yield deamino NAD by splitting the amino group that was attached to the carbon 2 of adenine. Taka-diastase is an adenosine deaminase that will also deaminate ADP, ATP, NAD, and NADP. NADase splits the glycosidic bond liberating free nicotinamide and a molecule containing one ribose attached to one adenosine by a pyrophosphate bond. NADase is inhibited by nicotinamide; therefore, adding the vitamin to the *in vitro* incubation mixture inhibits NAD breakdown. Thus, it is obvious that in those systems where NAD synthesis is studied, nicotinamide may stimulate the biosynthetic process in two different ways: either by acting as a precursor or by blocking NAD. Nucleotide pyrophosphatase attacks the pyrophosphate bridge and splits the dinucleotide into two mononucleotides, namely, nicotinamide mononucleotide and adenylic acid. The enzyme will also split NADH, NADP, and NADPH.

Thus, in the cell, the newly synthesized NAD is faced with several foes, and the cell must either provide NAD at a rate more rapid than that at which the coenzyme is hydrolyzed, or the new NAD must be kept separate from the catabolic enzymes. The intracellular distribution of the enzymes involved in NAD metabolism, therefore, is of capital interest. The enzyme that catalyzes the formation of deamido NAD from ATP and nicotinic acid mononucleotide is in the nucleus, as was first shown by Hogeboom and Schneider [93]. Later, Elyane Baltus also demonstrated that, at least in the starfish, this reaction is restricted to the nucleolus [94].

The formation of deamino NAD in the nucleus provides us with a neat hypothesis for the control of cytoplasmic anabolism by the nucleus, and it has been proposed that the rate of NAD synthesis in the nucleus controls the transfer of genetic information to the cytoplasm. However, this was postulated when the pathway of NAD synthesis was thought to use nicotinamide as a precursor. In such a pathway, the pyrophosphorylase reaction would be the last step in NAD synthesis. Since NAD is involved in a chain of reactions supplying energy to numerous biochemical processes in the cytoplasm, the nucleus might regulate cytoplasmic anabolism by increasing or reducing the rate of deamino NAD synthesis. The mechanism by which deamido NAD is transferred to the cytoplasm is not clear. Little is known of the intracellular distribution of the nucleotide pyrophosphatase and NADase.

The metabolism of the pyrimidine nucleotides is further complicated by the existence of transhydrogenases [95]. The difference between NAD and NADP resides in the number of phosphates that are part of the molecule. The difference in the number of phosphates probably does not affect the oxidation-reduction potential of the coenzymes, since the same fractions of the molecules of NAD and NADP are involved in the oxidation reaction. Thus, it is conceivable that the hydrogen may be readily transferred from NADPH to NAD, and *vice versa*. This transhydrogenation reaction is of more than theoretical interest because NADPH could enter the electron transport chain by passing its electron to NAD, and no separate pathway for NADPH and NADH would be necessary. Such transhydrogenation reactions were first observed in bacteria, but soon their presence in mammalian tissue was discovered.

At first, the enzymes involved were found to be flavoproteins, capable of transferring an electron from NADH or NADPH to an electron acceptor. NADH and NADPH cytochrome c reductase, as well as NADH diaphorase, were found to have transhydrogenase properties. The flavin nucleotide is not involved in the reaction, because it can be extracted from the protein without interfering with the transhydrogenation. Later, true transhydrogenases were found. A transhydrogenase has been purified from the ammonium sulfate fraction of rat liver mitochondria. Kaplan and his associates purified a transhydrogenase from heart mitochondria.

Theoretically, transhydrogenase should be capable of catalyzing a variety of reactions, namely:

1. NADPH + NAD $\rightleftharpoons$ NADP + NADH
2. NADH + NADP $\rightleftharpoons$ NAD + NADPH
3. NADPH + NADP $\rightleftharpoons$ NADP + NADPH
4. NADH + NAD $\rightleftharpoons$ NAD + NADH

It is difficult to study the hydrogen transfer from the reduced molecule to an identical oxidized molecule as in reactions 3 and 4, unless the acceptor can be recognized and easily separated chemically from the hydrogen donor.

The pyridine analogs, such as deamino or acetyl pyridine, constitute remarkable tools for the study of these reactions. According to Kaplan, a number of transhydrogenations may occur:

1. NADPH + NAD $\rightleftharpoons$ NADP + NADH
2. NADH + deamino NAD $\rightleftharpoons$ NAD + deamino NADH
3. NADH + acetyl pyridine NADP $\rightleftharpoons$ NAD + acetyl pyridine NADPH
4. NADPH + acetyl pyridine NAD $\rightleftharpoons$ NADP + acetyl pyridine NADH
5. NADPH + acetyl pyridine NADP $\rightleftharpoons$ NADP + acetyl pyridine NADPH

If it is assumed, as mentioned above, that this reaction is significant in controlling the availability of the two reduced coenzymes, it is important to investigate: the equilibrium of each of these reactions; the enzyme involved in each reaction; the enzyme's intracellular distribution; and enzyme activity in the intact cell. The solutions to these various problems are still fragmentary, but some of the available evidence is worthy of mention.

The various types of transhydrogenase reactions are not necessarily present at equal rates in every tissue. In beef heart, there exists an enzyme capable of the reversible transhydrogenation between NAD and NADPH or NADP and NADPH, but not of the transfer hydrogen from NADPH to deamino NADP. In brain the transhydrogenation from NADH to deamino NAD is possible, whereas the transhydrogenation from NADPH to NAD does not occur.

In Kaplan's laboratory a transhydrogenase has been partially purified from pig heart that can catalyze the heteromolecular, but not the isomolecular reaction. It is a macromolecule probably lipoproteic in nature found in the mitochondrial membrane. Thus, the enzyme catalyzes the reversible conversion of NADPH to NADH. The role of the transhydrogenase in metabolism is still unclear. Although it almost certainly permits NADPH oxidation via NADH, NADPH is not oxidized by mitochondria in the absence of small amounts of NAD.

Also, under some circumstances when ATP is generated in the mitochondria, NADH may be converted to NADPH. In fact, the presence of ATP drives the enzyme to produce more NADPH than can be expected from calculating the equilibrium in the absence of ATP.

Although the switch in equilibrium might suggest that the enzyme that catalyzes the conversion of NADH to NADPH is different from that which catalyzes the reverse reaction, all evidence suggests that the same protein catalyzes both reactions. Therefore, it is assumed that some conformation change in the enzyme has taken place, possibly as a result of the binding of the transhydrogenase molecule with a high-energy intermediate.

Finally, it is of interest that the level of conversion of NADH to NADPH is somewhat related to the levels of activity of NADP isocitric dehydrogenase. When the NADP isocitric dehydrogenase level is high compared to that of NAD dehydrogenase, as in heart, the transhydrogenase is also high. The reverse is true in brain mitochondria, which have low levels of NADPH isocitric hydrogenase and transhydrogenase.

The importance of understanding the regulation of NAD synthesis could not escape the reader. One can postulate various mechanisms of regulation: (1) the existence of anabolic enzyme inhibitors; (2) the adjustment of a delicate balance between anabolic and catabolic enzyme activities; and (3) a regulation of the enzyme biosynthesis by repression of the gene responsible for their elaboration.

The third hypothesis has found some support from the work of Pardee [96] and his collaborators. These investigators studied in *Escherichia coli* the activity of nicotinic acid, mononucleotide pyrophosphorylase, nicotinic acid dinucleotide pyrophosphorylase, NAD synthetase, and NAD kinase—four enzymes required for NADP elaboration. Pyrophosphorylase activity was found to be relatively low; and pyrophosphorylase was the only enzyme affected by changes in the NAD concentration added to a culture of *E. coli* mutant requiring nicotinic acid for growth. The enzyme activities increased when the nicotinic acid concentrations were low and decreased when large amounts of nicotinic acid or NAD were added to the medium, and the changes in the enzyme activities did not result from the appearance of inhibitors.

On the basis of these findings, the investigators concluded that pyrophosphorylase activity controls the rate of NAD biosynthesis, and, therefore, its concentration in the bacterial cell; and that the activity of that enzyme is regulated by repressive genes.

Flavin Nucleotides

In flavin nucleotides, riboflavin (vitamin B_2) is the nucleotide that combines with phosphoric acid. In riboflavin, the nitrogen in position 9 of a 6,7-dimethylisoalloxazine forms a linkage with the carbon 1 of a pentose sugar, namely, the D-ribitol. Hence, the flavin nucleotide differs from the nucleotides derived from nucleic acids in at least three ways: the nature of the base, the nature of the sugar, and the nature of the linkage between sugar and base. Riboflavin is a derivative of the sugar alcohol ribitol, and the nature of the linkage between base and sugar is nonglycosidic.

The role of riboflavin* in mammalian tissue is reviewed in detail in the section on vitamins. We need only discuss here those steps that catalyze the phosphorylation of riboflavin to the mononucleotide and those that catalyze the condensation of two molecules of the mononucleotide to form the flavin dinucleotide.

A specific kinase, flavokinase, yields flavin mononucleotide in the presence of riboflavin and ATP, and probably magnesium. Schrecker and Kornberg [97] described an enzyme that catalyzes the synthesis of flavin adenine dinucleotide from flavin mononucleotide and ATP. The enzyme was isolated from yeast, and similar enzymes have been found in animal tissues. The enzyme is called flavin adenine dinucleotide pyrophosphorylase.

The flavin adenine nucleotides act as hydrogen acceptors in a large number of reactions, several of which are listed below:

* The biosynthesis of riboflavin was studied in bacteria. By using glucose labeled in carbon 1, 4, 5, or 6. Plaut has demonstrated that ribitol is partly derived from glucose metabolized through the hexose monophosphate shunt. However, all these experiments were done in one type of bacteria, and the relative participation of the various glucose pathways can be expected to vary with the relative importance of the pathways in the bacterial cell. Little is known of the biosynthesis of the isoalloxazine ring. However, adenine labeled in the pyrimidine ring was shown to be incorporated in the isoalloxazine ring, which itself is then incorporated entirely. This, of course, explains why adenine labeled only in carbon 8 is not incorporated in isoalloxazine, whereas glycine (incorporated in carbon 6 of adenine) and formate (incorporated in carbon 2 of adenine) are partly recovered in the isoalloxazine ring.

amino acid + $^1/_2$—O_2 $\xrightarrow[FAD]{}$ keto acid + NH_3 + H_2O

D-amino acid oxidase (sheep liver)

amino acid + $^1/_2$—O_2 $\xrightarrow[FMN]{}$ keto acid + NH_3 + H_2O

L-amino acid oxidase (rat kidney)

glycine + $^1/_2$—O $\xrightarrow[FAD]{}$ glyoxylate + NH_3 + H_2O

Glycine oxidase (kidney or liver – swine)

D-aspartic acid + $^1/_2$—O_2 $\xrightarrow[FAD]{}$ oxaloacetate + NH_3 + H_2O

Aspartic acid oxidase (rabbit kidney)

N-methyl-L-amino acid + $^1/_2$—O_2 $\xrightarrow[FAD]{}$ L-amino acid + H CHO + H_2O_2

NADH + dye $\xrightarrow[FAD]{}$ NAD = reduced dye

Heart diaphorase

NADH + cytochrome c $\xrightarrow[\text{FAD or FMN}]{}$ NAD + reduced cytochrome c

NADH cytochrome c reductase (animal tissues)

NADPH + cytochrome c $\xrightarrow{\text{FAD or FMN}}$ NADP + reduced cytochrome c

NADPH cytochrome c reductase

NADH + cytochrome b_5 $\xrightarrow{\text{FAD or FMN}}$ NAD + reduced cytochrome b_5

NADH cytochrome b_5 reductase (liver microsomes)

NADH + 2-methyl-1,4- $\xrightarrow[FAD]{}$ NAD + naphthoquinone

2-methylhydronaphthoquinone

NAD naphthoquinone reductase (animal tissues)

succinate + cytochrome c $\xrightarrow[FAD]{}$ fumarate + reduced cytochrome c

Succinic dehydrogenase (animal mitochondria)

In all these reactions, the flavin nucleotide combines with a protein to form a flavoprotein. In the reactions involved in electron transport, the flavoprotein acts as a hydrogen carrier and plays an important role in the transfer of electrons from pyridine nucleotides to cytochromes. Among the flavoproteins of interest in the electron transport chain are: (1) coenzyme 1-diaphorase, (2) the Warburg flavoprotein, (3) NADH cytochrome c reductase, (4) NADPH cytochrome c reductase, (5) NADH cytochrome b_5 reductase, (6) NADH 2-methyl-1,4-naphthoquinone reductase, and (7) succinic cytochrome c reductase. The most important among these are succinic cytochrome c reductase and NADH cytochrome c reductase. The following discussion of these two flavoproteins may serve as an example for the others.

The prosthetic group of succinic dehydrogenase is a ferroflavoprotein containing one flavin molecule and four iron atoms. The overall molecule is rather large and has a molecular weight of 200,000. As early as 1939, Keilin and Hartree observed that the addition of succinate to a succinic dehydrogenase preparation from heart muscle reduced the absorption at 455 mμ. In addition, riboflavin deficiency was found to decrease succinic dehydrogenase activity in liver. These observations suggested that flavins are important constituents of the succinoxidase system. However, all attempts to demonstrate the presence of flavin in the protein after treatment with strong acid or denaturation by heat failed, and only when the enzyme was submitted to tryptic digestion was the presence of flavin in the succinic dehydrogenase molecule conclusively demonstrated. The flavin component so obtained is similar to FAD but differs from it in a variety of ways; namely, by its fluorescent intensity, its RF values, its efficiency as a growth factor, and its coenzyme activity in the amino oxidase system. These differences are now partially explained. The riboflavin moiety is attached to the protein moiety of the flavoprotein through covalent bonds; indeed, trypsin digestion of the proteins releases a flavin dinucleotide attached to a small peptide fraction (see Fig. 1-18). The flavin dinucleotide moiety of the flavopeptide can be further catabolized by splitting the pyrophosphate linkage. This yields the classical AMP and another nucleotide resembling riboflavin-5-phosphate, the exact nature of which remains unsettled.

Iron is necessary for enzyme activity; the activity decreases with the amount of iron lost in the preparation. Therefore, it is tempting to assume that iron acts as an electron carrier. The successive reduction and oxidation of the iron in the course of succinic oxidation have never been demonstrated, but the action of many inhibitors on the succinic dehydrogenase system can be explained by assuming the formation of metal complexes.

In spite of the fact that succinic dehydrogenase has been known to exist for half a century and that many laboratories have attempted to study its mode of action, little is known of its mode of regulation. The 23 amino acids adjacent to the flavin have been sequenced.

A gamut of physicochemical and biochemical techniques have demonstrated that succinate, coenzyme Q_{10}, and ATP activate succinate dehydrogenase, and these modulations are much more effective on enzymes embedded in intact mitochondria than on particulate of partially solubilized preparations [98]. On the basis of these observations, a tentative hypothesis for succinate dehydrogenase regulation has been proposed. The main purpose of the enzyme is to remove the

Flavin
(NH_2)Ser-His-Thr-Val-Ala
Ala-Glx-Gly-Gly-Ile
Asx = Leu-Ala-Ala-Gly
Asx-Met-Asp-Glu-Asx
Glx-Typ-Arg (COOH)

Fig. 1-18. Amino acid sequence of a flavin containing tryptic peptide from succinate dehydrogenase

succinate that accumulates. The cycle is believed to be triggered by the ATP/ADP ratio; whenever that ratio is low, the enzyme is deactivated. During the deactivation period, succinate accumulates and NADH remains in the reduced state. The accumulation of succinate activates the enzyme, and as a result the reduction of coenzyme Q_{10} is accelerated and ATP accumulates. Since the two latter factors are positive modulators, the enzyme is further activated and the succinate levels decrease with reduction in the ATP/ADP ratio. Of course, it is assumed, if not convincingly established, that the changes in enzyme activity result from modification in enzyme conformation.

NADH cytochrome c reductase was isolated from pigeon breast and pig heart muscle. The enzyme was shown to contain four atoms of iron per flavin molecule. NADH cytochrome c reductase, like succinic dehydrogenase, is a ferroflavoprotein. The ratio of iron to flavin is four. The enzyme contains sulfhydryl groups that can be titrated by classical methods, but their oxidation has no effect on the enzymatic activity. In contrast, the removal of the metal leads to a decrease in the ability of the enzyme to reduce cytochrome c. As for succinic dehydrogenase, the structure of the flavin in NADH cytochrome c reductase is not clear. It was demonstrated that it is not flavin mononucleotide, but the identity of the flavin component with flavin adenine dinucleotide is not established; in fact, the flavin component differs from the classical FAD by its chromatographic properties and its behavior in enzymic assays. It is not known if it is a structural variation of the flavin nucleotide or if the nucleotide is conjugated to a peptide.

In 1929, Straub prepared from the Keilin-Hartree heart muscle preparation an enzyme that contained flavin adenine dinucleotide in its prosthetic group, and that catalyzed the oxidation of NADH by dyes, not by ferrous cytochrome c. This enzyme was called diaphorase. Besides the fact that diaphorase reduces only dyes, the enzyme differs from the NADH cytochrome c reductase in a variety of ways; among them are its solubility and its iron content.

The Cytochromes

The discovery of the cytochromes illustrates how a simple and ingenious experiment may open new fields. In 1925, Keilin examined with a microspectroscope the honey bee's thoracic muscles after pressing them between a slide and cover slip. He found several characteristic absorption bands in the spectrum. Since the compounds responsible for the absorption bands were in the suspension of most animal tissue examined, Keilin assumed that they were essential to all cells and called them "cytochromes."

Cytochromes are constituents of many living cells. The extensiveness of the development of the cytochrome system in the cell is related to its respiratory activity. Cytochromes are all conjugated proteins with a porphyrin as prosthetic group. The cytochromes are part of a large number of intracellular hemoproteins—such as hemoglobin, myoglobin, peroxidases, and catalase—that can pass from a reduced to an oxidized form and have a spectrum specific for either form. Since porphyrins constitute such an important component of the cytochromes, some of their properties will be reviewed here [99–104].

Porphyrins are widely spread natural compounds. They are found in feces (coproporphyrin), urine (uroporphyrin), amniotic fluid, meconium, hemoglobin, and cytochromes. They are all formed by the condensation of four pyrrole rings, but differ from each other in the nature of their side chains. As it appears from the formula, the pyrrole ring that constitutes the unit structure of the porphyrin is the same ring that appears in proline, hydroxyproline, and indole.

The chemistry of porphyrin was most confusing until Hans Fischer clarified this chapter of organic chemistry by synthesizing practically all known porphyrins starting from the pyrrole ring. It is obvious from studying the structure of the porphyrin that a variety of isomers can be synthesized; however, only a restricted number are found in nature. The biosynthesis of porphyrins will be discussed in the section devoted to porphyrias. In the following paragraphs, the structure and the role of cytochromes in electron transport are reviewed.

Many cytochromes have been discovered since Keilin's first observation. Their nomenclature has developed empirically and has become most confusing. The important cytochromes can be grouped in three main families: cytochrome a, cytochrome b, and cytochrome c. Each member of a family is characterized by specific absorption bands—alpha, beta, and gamma. The first occurs between 600 and 500 mμ, the second around 500 and 525 mμ, and the third in the neighborhood of 430 mμ (Soret band). Upon oxidation, the alpha and beta bands become faint and only weak absorption is observed. During this oxidation process, the iron passes from the ferrous to the ferric form.

Cytochrome c. Among cytochromes, cytochrome c is best known because of its solubility and extractibility. In spite of those properties, only recently has cytochrome c been obtained in a crystalline form from penguin pectoral muscle and from a variety of other sources ranging from yeast to ox and pig heart.

The porphyrin of the prosthetic group of cytochrome c is protoporphyrin IX, the same as in hemoglobin, peroxidase, and catalase. The mode of attachment of the porphyrin to the protein moiety of the cytochrome is of great interest, but it can only be understood in relation to the amino acid sequence.

The fact that the heme prosthetic group is deeply buried within the protein has important implications as to the movement of the electron from the donor to the cytochrome iron and *vice versa* from the iron to the electron receptor. It seems inescapable that the amino acid residues are involved in the transfer of such electrons [105–108]. The role of the amino acid

For instance, the concentration of cytochrome in muscle increases after thyroxine is administered but decreases in hypothyroid animals. However, this constitutes a nonspecific effect of thyroxine, and other calorigenic hormones have similar effects on the concentration of cytochrome c.

Cytochrome c can easily be oxidized to ferric cytochrome c by ferric cyanide or other oxidizing agents. The reduced form can be regenerated by the addition of reducing agents, such as cysteine, ascorbic acid, or hydrosulfide.

Cytochrome c_1. Cytochromes in the c family have been found in bacteria and have been named cytochrome c_1, c_2, c_3, c_4, c_5, etc; c_1 is the only one also present in mammalian tissue. Cytochrome c_1 has spectral characteristics similar to those of cytochrome c. It was found in the heart muscle preparation of Keilin and Hartree. Cytochrome c_1 differs from cytochrome c in two ways: it is not easily extractible, and it is thermolabile. It resembles cytochrome c in that it does not react with oxygen, carbon monoxide, or cyanide.

Table 1-1 gives the alpha, beta, and gamma absorption bands of some cytochromes, including cytochrome c. Whereas the alpha band of purified cytochrome c has its maximum exactly at 550 mμ, the spectrum of heart mitochondria has its maximum at 551 mμ. However, when the mitochondria are heated, the band position moves to 550 mμ. These results suggested to Keilin and Hartree that the spectrum of heart mitochondria is the resultant of two spectra. Indeed, repeated washings of the particles led to the discovery of a thermolabile, insoluble cytochrome with a maximum alpha band at 553–554 mμ.

Table 1-1. Absorption bands of some cytochromes

Reduced	α	β	γ
a (heart)	605	—	445
a_1 *(Acetobacter pasteurianus)*	590	—	435–440
b (yeast)	564	525	430
c (heart)	554	521	415
c_1 (heart)	554	524	418
b_5 (liver)	556	526	423
CO complexes			
a	590	—	430
a_1	590	—	427

Cytochrome c_1 can be reduced by succinate, and its reduction is inhibited by antimycin A. Its oxidation potential is probably of a value intermediate between the oxidation potential of cytochrome c and cytochrome b. It can now be prepared from submitochondrial fragments and purified by treatments of the preparation with butanol and deoxycholate. Such a preparation yields a polymer, the monomer of which contains one hemin. The exact molecular weight of the unit is not clear, but the proposed values range between 37,000 and 70,000. The structure of the prosthetic group of cytochrome c_1 has not been established. However, it appears to be very similar, if not identical to that of cytochrome c.

Cytochrome b. Like cytochrome c_1, cytochromes a and b are not readily extractible. They can be solubilized only after rather drastic treatments, such as the addition of cholate or deoxycholate, salt extractions, and treatment with hydrolytic enzymes. Thus, except for their spectra and some basic properties such as their capacity to react with carbon monoxide, oxygen, and cyanide, little is known of the structure of cytochromes a and b. Mammalian tissues contain both mitochondrial and microsomal cytochromes of the b type. They differ in their intracellular localization, but also in their spectra and their response to antimycin A.

Mitochondrial cytochrome b was solubilized for the first time by Cooperstein and his associates, and attempts to purify cytochrome b have been made. The enzyme has now been prepared free of cytochrome oxidase, succinic dehydrogenase, and cytochromes c and a. However, the criteria of purity are not always reliable. The best purification procedures have yielded preparations without enzymic activity. In the absence of cationic detergents, the protein tends to aggregate, but with such detergents a minimum molecular weight of 28,000 has been obtained.

Little is known of the catalytic properties of cytochrome b; investigators must rely on the spectrum characterization of cytochrome b rather than on its enzymic activity to evaluate the purity of the preparation. Little is known of the heme group of mammalian cytochrome b. It is significant that cytochrome b extracted from *Micrococcus denitrificans* and *Pseudomonas denitrificans* contains protoheme, the same heme as is in cytochrome c. The nature of the binding between the prosthetic group and the protein is not known. The firm thioester bond in cytochrome c probably is not present in cytochrome b.

Spectrophotometric and potentiometric techniques have shown that more than one species of cytochrome b is now known to function in mitochondria and participate in electron transport. They are referred to as cytochrome bK or b565 and cytochrome bt or b562 [197].

Until recently, all cytochromes were found in the mitochondria, but Strittmatter and Ball [117] discovered a cytochrome strongly attached to the microsome particles of the cell. It was named cytochrome m. In view of its spectral characteristics, it was called cytochrome b_5 by other investigators. The role of cytochrome b_5 was at first not clear. However, it was certainly not involved in the oxidation of succinate; cytochrome b_5 is found mainly in microsomes, and succinate oxidation is restricted to mitochondria. Furthermore, succinate does not reduce cytochrome b_5 and cytochrome b_5 oxidation is insensitive to cynaide. These observations indicated that cytochrome oxidase does not act as the final hydrogen acceptor.

It is now clear that cytochrome b_5 is part of a complex enzyme machinery involved in detoxification and referred to as the mixed-function oxidase. In the sequence, electrons are transferred from NADH to cytochrome b_5, and under the catalytic influence of a cytochrome b_5 reductase, the electrons are further transferred from cytochrome b_5 to cytochrome p_{450}. Cytochrome b_5 has been prepared in two different fashions. In the first, pancreatic lipase is used to detach the cytochrome from the membrane, and under those circumstances a protein with a molecular weight of 10,000–11,000 daltons is obtained. The protein must contain the catalytic site because the heme remains attached to it. However, a much larger protein is obtained if cytochrome b_5 is further extracted. The cytochrome has a molecular weight of 120,000 daltons. The large molecule is a polymer that can be resolved into smaller monomers of 26,000 daltons. The exact mode of binding of the heme to the protein is not known, although tyrosine, lysine, and sulfhydryl groups have been excluded. It is, however, established that the heme stabilizes the protein. A study of the amino acid sequence of cytochrome b_5 has revealed that it is an amphipatic molecule composed of a hydrophilic heme-binding segment and a hydrophobic portion that promotes polymerization in aqueous solution but facilitates protein solubilization in lipid. The significance of this special molecular arrangement will become clearer when we discuss the structure of the cell membrane. The amino acid sequence of cytochrome b_5 from rabbit, bovine, pig, human, chicken, and *E. coli* has been established.

The liver mixed-function oxidase system catalyzes the hydroxylation of fatty acids, hydrocarbons (see section on carcinogenesis in Chapter 16), and many drugs. Because of their broad involvement in metabolism, mixed-function oxidases have been researched so extensively that even a superficial review of the published work is impossible. For further information, the reader is referred to specialized texts [193].

It is, however, of particular significance that the liver system has been solubilized and found to be made of four essential fractions: cytochrome p_{450}, NADPH cytochrome p_{450} reductase, and a heat-labile phospholipid fraction essential for electron transfer with phosphatidylcholine as the active component [194].

An intriguing observation was made when it was attempted to determine the stoichiometry of the reaction. Studies of aminopyrine oxidation yielded different ratios of O_2 to NADPH consumed in the reaction. In one case, the molar ratio was 1; in the other, 2. Measurement of the reactants during benzphetamine hydroxylation revealed that the stoichiometry was as expected, provided that small amounts of catalase were added to the incubation mixture. In the absence of catalase the oxygen uptake is higher than expected. The exact role of catalase in the reaction or the exact chemical structure of the oxygen-containing intermediate on which catalase acts remains unknown. The formation of H_2O_2 during the hydroxylation reaction has, however, been excluded. If one keeps in mind the close relationship of the peroxisomes, the catalase-containing bodies, with the endoplasmic reticulum one cannot help wondering about their role in facilitating the reaction of the mixed-function oxidases.

Five different steps have been distinguished in the hydroxylation reaction: (1) combination of the substrate (RH) with the oxidized form of cytochrome p_{450}, (2) reduction of the ferric state of the cytochrome's iron atom to the ferrous state by the transfer of one electron from NADPH through the catalytic action of the reductase (the reaction requires phosphatidylcholine), (3) formation of a ternary complex through the binding of oxygen, (4) formation of a superoxide anion bound to the oxidized cytochrome p_{450} after an intramolecular electron transfer, and (5) finally in the molecular complex (oxidized cytochrome, superoxide, and substrate), the superoxide attacks the substrate which becomes hydroxylated, and at the same time the reductase provides another electron and a molecule of water is formed as one of the products of the reaction.

In conclusion, a second cytochrome-containing respiratory chain, found in the membranes of the endoplasmic reticulum, functions in the reduction of NAD or NADP. In this sequence the electron is transferred from NADH to cytochrome b_5 to cytochrome p_{450}, and finally to oxygen, which can then be used in the mixed oxidation reaction.

According to Estabrook and his associates, as much as 25% of the oxygen that enters the liver is used through that extramitochondrial pathway. The regulation of the interaction between these two pathways remains to be discovered [118].

Cytochrome oxidase. The reaction catalyzed by cytochrome oxidase consists of adding four electrons to an oxygen molecule obtained from the atmosphere. The oxygen reductase system, the terminal link of the electron transport chain, is assumed to be composed of two cytochromes of type a, namely, cytochrome a and cytochrome a_3. The latter differs from the former by its reactivity with CO, oxygen, and cyanide. Cytochrome a_3 forms with CO complexes with specific absorption maxima at 589 and 430 mμ. The present evidence suggests that cytochrome a_3 is actually cytochrome oxidase. The CO derivative of cytochrome a_3 exhibits photochemical properties very similar to those of the CO combination product of the "Atmung" enzyme.

Despite the differences in catalytic and spectroscopic properties, cytochrome a and a_3 are similar in many of their chemical properties. For instance, it is impossible to separate the two compounds by ultracentrifugation, electrophoresis, chromatography, ammonium sulfate fractionation, or serial addition of deoxycholate, and, therefore, the proportion of cytochrome a or cytochrome a_3 present in cytochrome $a+a_3$ preparations is not well known. However, measurements of the absorption spectrum obtained after reaction with CO and cyanide has established that 50% of the Soret band and 28% of the 605 mμ peak can be accounted for by the presence of cytochrome a in the cytochrome $a+a_3$ preparation.

Studies of the mammalian cytochromes of the a type have raised at least three problems: (1) the identity of cytochromes a and a_3; (2) the presence of copper in the cytochrome molecule; and (3) the mechanism of oxygen reduction. Again, the preferred theory postulates the existence of two different cytochromes, but this concept is not accepted by all biochemists, and some still believe in a unitarian theory which postulates that cytochromes a and a_3 are actually a single molecule. For example, it was demonstrated that cytochrome a, provided it is in the ferrous form, can also react with carbon monoxide; furthermore, the presence of cytochrome c renders cytochrome a autoxidizable. Finally, the oxidation of reduced cytochrome c_1 requires cytochromes a and c, and the rate of the reaction is optimal if the ratio of a/c is equal to 1.

Perhaps the strongest evidence in favor of the theory that the cytochrome oxidase system is composed of a single protein comes from the physicochemical studies of Takemori. Purified cytochrome oxidase appears in the ultracentrifuge as a nondispersed protein (sedimentation constant=21.9; diffusion constant=3.58×10^{-7}; partial specific volume=0.72). The molecular weight of the protein was calculated on the basis of these studies and was found to equal 530,000.

The heme determination indicates that the cytochrome oxidase system contains five molecules of heme per molecule of protein. This evidence is not conclusive for at least two reasons. First, the purified preparation of cytochrome oxidase has very low catalytic activities, and therefore its true relationship with the native mitochondrial enzyme remains unsettled. Second, the data do not exclude the possibility that cytochrome oxidase is formed by the intimate complexion of two proteins, cytochromes a and a_3.

There is no doubt that some copper is always associated with cytochrome $a+a_3$ preparations. However, it has not been satisfactorily established that copper is essential to the activity of the protein; in fact, conflicting results have been obtained in that respect. Copper associated with cytochrome oxidase is difficult to eliminate by dialysis, but prolonged dialysis against chelating agents finally releases a high proportion of it. After dialysis, the catalytic properties of cytochrome $a+a_3$ are proportional to the amount of copper that remains associated with the cytochrome molecules, according to some investigators. Some compounds, such as bathocuproine sulfonate, chelate copper preferentially. Some investigators claim that the addition of an amount of a chelating agent that would chelate about 60% of the copper associated with the cytochrome destroys its catalytic properties.

Perhaps the most convincing argument in favor of a direct role of copper in the cytochrome oxidase reaction comes from studies done with the electron spin resonance spectrometer. Such studies demonstrate that some of the copper ions undergo valency changes during the oxidation-reduction process. The site of action of the copper in the electron transport chain would then be as follows:

$$\text{Reduced cytochrome c} \rightarrow \text{Cu} \rightarrow \text{cytochrome a} \rightarrow \\ \rightarrow \text{cytochrome } a_3{}^* \rightarrow O_2.$$

It is not known how the two components of cytochrome oxidase interact in the course of electron transfer. The fact that the two polypeptide chains are so closely linked suggests that the two proteins must collaborate in the oxidation process either through protein-protein, heme-heme, copper-heme, or other interactions. Heme-heme interactions have been excluded by measuring optical rotary disperson or circular dichroism. Yong and King [119] have partially purified cytochrome oxidase from beef heart, and results of their experiments using electron spin resonance techniques suggest copper-heme interaction. Thus, the detection of copper by ESR (electron spin resonance) depends upon the valence of both cytochrome a and cytochrome a_3. Yong and King believe that one of the copper ions, the one detectable by ESR, is sandwiched between cytochrome a and cytochrome a_3. The other copper ion is assumed to function as an electron sink. Whether such changes will prove useful to the interpretation of injuries that develop in copper deficiency and excess only time can prove.

By investigating the effect of disulfide bond reagents on the catalytic properties of purified cytochrome oxidase, it was demonstrated that those agents which react with disulfide bonds (cysteine, British antilewisite [BAL], and glutathione) inhibit cytochrome oxidase activity. The effect can be reversed by adding oxidized glutathione to the medium. These results suggest the existence of disulfide bonds essential to the activity, probably located within the peptide chain rather than between the two chains binding them together, as is the case for ribonuclease and insulin.

The reader is referred to King and Yong's text [195] for further information on the interdependence of the components of cytochrome oxidase.

Vitamin K, Vitamin E, and Ubiquinone

In addition to the pyridine and flavin nucleotides and cytochromes, several other components are suspected of being involved in electron transport. Among them are vitamin K, vitamin E, and ubiquinone; these three compounds are quinones with a long fatty acid side chain. Their role in electron transport is as yet unclear, and it will be reviewed only briefly here.

The evidence in favor of the role of vitamin K in electron transport is based on a variety of facts. Among them are the observations that vitamin K analogs**

* This transfer of electrons from cytochrome a to cytochrome a_3 would be the step in this pathway that is inhibited by cyanide.

** It should be kept in mind that Dicumarol not only uncouples phosphorylation and oxidation by acting in the electron transport chain at the site of phosphorylation corresponding to the position of vitamin K in the electron transport chain, but Dicumarol is also active at the site of phosphorylation at the level of cytochrome oxidase.

uncouple oxidative phosphorylation and that vitamin K deficiency in chickens leads to a reduction of the oxidation rate when α-ketoglutarate is used as a substrate. Extraction of vitamin K with isooctane leads to inhibition of succinoxidase activity; the enzyme activity can be restored only by adding vitamin K to the system. After ultraviolet irradiation of isolated liver mitochondria or bacteria extract, oxidative phosphorylation is impaired and can be reactivated by adding vitamin K to the system.

On the basis of these observations, Martius has proposed a scheme in which vitamin K influences electron transfer between NAD, flavoprotein, and cytochrome b. This view was further supported when a reductase was discovered that catalyzes the oxidation of NADH in the presence of vitamin K_1. It has not yet been established whether the reduced vitamin K can be oxidized in turn by the respiratory chain. The mechanism by which vitamin K affects oxidative phosphorylation is not clear.

1. $p_{450}\,(Fe^{3+}) + RH \rightarrow p_{450}\,(Fe^{3+})RH$

2. $p_{450}\,(Fe^{3+})RH \rightarrow p_{450}\,Fe^{2+}RH$

3. $p_{450}\,(Fe^{2+})RH \rightarrow p_{450}\,(Fe^{2+})\,(O_2)\,(RH)$

4. $p_{450}\,(Fe^{2+})\,(O_2)\,(RH) \rightarrow p_{450}\,(Fe^{3+})\,(O_2)^-\,(RH)$

5. $p_{450}\,(Fe^{3+})\,(O_2)^-\,(RH) \rightarrow p_{450}\,(Fe^{3}) + ROH + H_2O$

Vitamin E is abundant in the heart; it is thought to be concentrated entirely in heart mitochondria. The concentration of vitamin E in the heart is in keeping with the concentration of the other components involved in the electron transport chain. Vitamin E-deficient animals have increased respiration rates. These findings have been thought ot indicate that vitamin E plays a role in the control of oxidative phosphorylation. It was also observed that isooctane extraction of various enzyme preparations leads to inactivation of NADH oxidase, succinoxidase, and cytochrome oxidase systems. The inactivation can be restored to normal by adding vitamin E to the system; however, the reactivation is completely nonspecific, and a variety of lipids may have the same effect [120, 121].

In the course of studies on the succinoxidase system, a new compound was discovered which appeared to be involved in electron transport. It was first thought to be a steroid, but later was demonstrated to have the structure of a quinone. Since the compound has been found in all tissues investigated, it was called ubiquinone. At about the same time, another group of investigators found that a new coenzyme, coenzyme Q, is involved in the succinoxidase system. When coenzyme Q was later crystallized, it was observed to resemble the ubiquinone of Morton in its ultraviolet and visible spectral characteristics. However, the two compounds differ in chromatographic behavior, infrared spectra, and melting points. Although it is now generally accepted that ubiquinone and coenzyme Q are similar compounds, the observed differences are unexplained.

The solubility properties of coenzyme Q constitute the basis for its purification. The coenzyme can be extracted with heptane or isooctane, then treated with an alkaline solution containing pure alcohol, and after chromatographic analysis on aluminum columns, it can be crystallized from ethanol or methanol. Samples obtained by such procedures all have the same characteristics on silicon-treated paper and present identical visible and infrared spectra as well as identical coenzyme activities.

The structure of coenzyme Q has now been elucidated. It is an orthoquinone with three short side chains and a long one. Among the short side chains are two hydroxymethyls and a methyl group. The long side chain is made of a varied number of isoprenoid groups. Ubiquinones prepared from a number of sources differ in the length of their side chain (*e.g.*, coenzyme Q isolated from animal tissues contains ten isoprenoid groups, but yeast ubiquinone contains only six).

At least two main sources of experimental evidence suggest that ubiquinone plays a role in the electron transport chain. First, extracting the electron transport particle with isooctane leads to a loss of activity that can be restored by adding ubiquinone. Second, ubiquinone can be alternately reduced and oxidized in the course of electron transport. Inasmuch as coenzyme Q is a neutral lipid completely insoluble in water but very soluble in hydrocarbons, extracting the electron transport particle with isooctane deprives the electron transport system of its coenzyme Q contingent. The activity of the particle can be restored by adding coenzyme Q, although this effect is not restricted to ubiquinone; such compounds as vitamin E, vitamin K, or analogs of coenzyme Q may also restore the activity. Only when the particle is extracted with both isooctane and deoxycholate is restoration specifically dependent on the addition of coenzyme Q, but then cytochrome c must also be added to the system.

Coenzyme Q undergoes rapid reduction in the presence of succinate or an NADH-linked substrate. The reduced form of the coenzyme can be oxidized if the mitochondria are incubated under aerobic conditions. The oxidation rate is largely dependent upon the presence of ADP and inorganic phosphate. The omission of P_i from the incubation mixture leads to a considerable reduction in the oxidation rate. If ADP is added to the system, the reduced coenzyme is rapidly oxidized.

On the basis of inhibition studies with antimycin, it has been postulated that coenzyme Q occupies, in the electron transport chain, a position intermediate between the succinate dehydrogenase flavoprotein and cytochrome c_1 in the succinoxidase system, and between cytochrome b and cytochrome c_1 in the NADH oxidase system.

A protein complex capable of rapidly reducing coenzyme Q in the presence of succinate (succinic coenzyme Q reductase), NADH (NADH coenzyme Q reductase), and cytochrome c (cytochrome c coenzyme Q reductase) has been purified. If these enzymes are recombined with purified cytochrome oxidase, succinate and NADH are oxidized in ways similar to their oxidation in the respiratory chain.

Electron Transport Chain

As early as in 1925, Keilin's work indicated that the individual compounds involved in the electron transport chain act in sequence. He observed that urethane blocks electron transport from cytochrome b to cytochrome c. Thus, cytochrome b precedes cytochrome c in the sequence of electron transfer. It was also known that carbon monoxide competes with oxygen for cytochrome a_3, and it was therefore concluded that cytochrome a_3 and cytochrome oxidase are identical compounds. From this early knowledge, the following sequence emerged: cytochrome b, cytochrome c, cytochrome a.

Numerous investigators, among them Warburg, Chance, Slater and Green, further contributed to the solution of the sequence of events of the respiratory chain. Obviously, the most logical approach to establishing the succession of steps in a multiple-enzyme system is to reconstitute the entire system after isolating and purifying each of its components. For a long time, this approach was impractical for studying the sequence of steps involved in respiration. The enzymes of the electron transport chain are strongly attached to structural elements of the cell and are therefore insoluble. Furthermore, the structure itself seems to play an important role in maintaining the efficiency of the respiratory chain, and the electron transport chain proved difficult to reconstruct in soluble systems. However, some components of the chain can be prepared in a purified form. Among these are the pyridine nucleotides, some flavoproteins, and cytochrome c. The study of these purified systems has shown that NADH is directly oxidized by a flavoprotein, that cytochrome c is readily reduced by NADH under anaerobic conditions, and that cytochrome c is oxidized by molecular oxygen in the absence of NADH. Such studies on the purified compounds have established the sequence NADH, flavoprotein, cytochromes, oxygen.

Because cytochromes a and a_3 cannot be separated and cytochrome oxidase is necessarily the last step in the chain, the following sequence becomes obvious: cytochrome b, cytochrome c, cytochrome a.

A more detailed description of the electron transport chain was obtained by using three other methods: selective inhibition, spectrophotometric kinetics, and determination of the respective oxidation-reduction potentials. The first two methods are based on the spectral properties of the various components of the electron transport chain. It is indeed fortunate that the pyridine nucleotides, the cytochromes, and the flavoproteins have specific absorption spectra, and that the spectra undergo considerable changes when they pass from the reduced to the oxidized form.

In selective inhibition, the electron transport chain is blocked with specific inhibitors, and the oxidation-reduction states of the components below and above the block are studied. Those below the block are reduced, and those above the block are oxidized. The kinetic studies are based on the assumption that the sequence of the spectrophotometric changes resulting from the oxidation and reduction of each component of the electron transport chain reflects the sequence of the chemical reactions. On the basis of such spectrophotometric studies on phosphorylating particles, intact mitochondria, and intact cells, the following sequence has been obtained: NADH, cytochrome b, cytochrome c_1, cytochrome c, cytochrome a, cytochrome a_3, O_2.

Some of these steps can be bypassed in nonphosphorylating particles, and the spectrophotometric methods do not exclude the possibility that some of the components that seem to participate in the electron transport chain are only parasites. Definite information on these mechanisms will be obtained only when *in vitro* reconstruction with purified enzyme systems is possible.

The following can be concluded:

Flavoproteins, NADH, cytochrome c reductase, and succinate cytochrome c reductase must occupy the first step of the electron transport chain because they are reacting with their specific substrates in isolated form (see Fig. 1-20).

Cytochrome b is known to transfer electrons from the substrate to cytochrome c because (1) narcotics allow the oxidation of cytochrome b but leave cytochrome c reduced; (2) the oxidation-reduction potential of cytochrome b is below that of cytochrome c; and (3) kinetic studies have shown that cytochrome c is oxidized before cytochrome b, and cytochrome b is oxidized before the flavoprotein.

Cytochrome c_1 acts before cytochrome c because (1) cytochrome c is oxidized more rapidly than cytochrome c_1; (2) the oxidation-reduction potential of cytochrome c_1 falls between the oxidation-reduction potential of cytochrome c and that of cytochrome b; (3) antimycin prevents the oxidation of cytochrome c_1; and (4) NADH or succinate cytochrome c reductase reduces cytochrome c_1 in the absence of cytochrome c.

That cytochromes a and a_3 must act after cytochrome c is indicated by the fact that purified cytochrome a oxidizes cytochrome c. Cytochrome a_3 is by definition at the end of the chain since it has been demonstrated to be identical with cytochrome oxidase. Cytochrome a is located between cytochromes c and a_3. In fact, the two cytochromes are practically inseparable, and since a_3 is terminal, a must be between a_3 and c. Furthermore, the oxidation-reduction potential of cytochrome a is somewhat higher than that of cytochrome c but slightly below that of cytochrome a_3.

Many of these conclusions found direct confirmation when mitochondria were systematically dissected in Green's laboratory [122]. It is clear that most of the biological oxidations occur in the mitochondria.

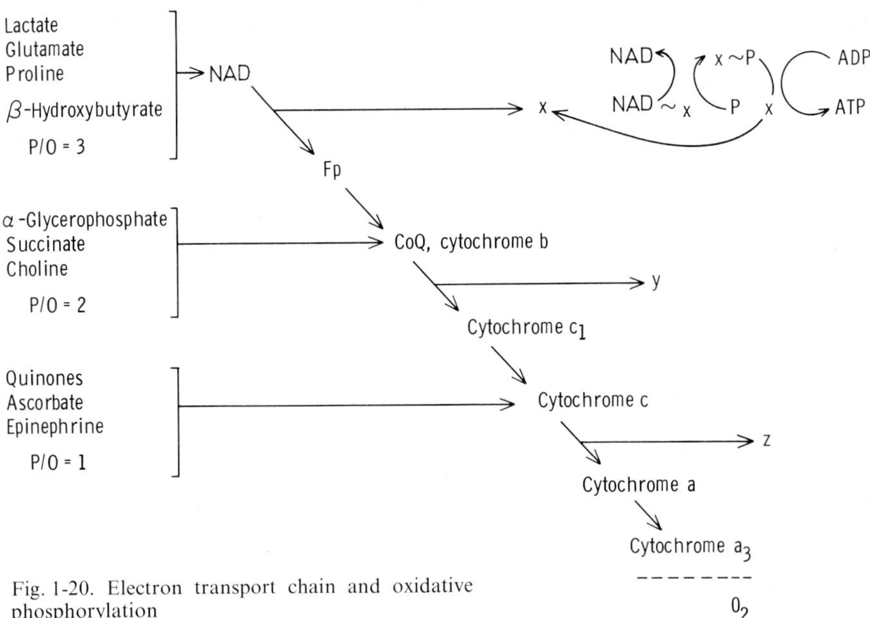

Fig. 1-20. Electron transport chain and oxidative phosphorylation

The mitochondrion is a device capable of converting the energy liberated during the oxidation of pyruvic acid into the usable chemical energy stored in the molecule, ATP. The electrons liberated in the course of pyruvic acid oxidation are transferred through a chain of electron carriers to molecular oxygen. In the course of that transfer, the chemical energy is liberated in small packages to phosphorylate ADP.

Green and his coworkers were intrigued by the nature of the elementary unit which, within the mitochondria, can perform all or part of these energy conversions. Heart mitochondria were chosen for study because they are sturdy and can be prepared on a large scale. When submitted to sonic radiation, these mitochondria, lose their Krebs cycle enzymes and their ability to couple oxidation and phosphorylation, but particles can be recovered that are perfectly capable of carrying out electron transport. These particles are small compared to the original mitochondria (1:50,000). However, the electron transport chain that they contain probably is unaltered. Their coupling ability can be restored by adding specific proteins. Using cholate, deoxycholate, butanol, petroleum ether, and cyclohexane, it is possible to fractionate the electron transport particle (ETP) into a number of catalytic complexes.

The stepwise degradation of the ETP yielded four different catalytic complexes the molecular composition of which was further investigated. The position occupied by each complex within the complete chain could then be determined and the integral catalytic properties of the ETP reconstructed. The composition of these four complexes is given in Table 1-2. Two observations recorded in the table—the presence of a nonheme metal group in each of the four catalytic complexes and the existence of dissociable groups—deserve further comment.

Table 1-2

Complex	Electron transfer reaction
I	NADH → f_D → $(Fe—Fe)_4$ → CoQ NADH-coenzyme Q reductase
II	Succinate → f_s → $(Fe—Fe)_4$ → CoQ Succinic-coenzyme Q reductase
III	QH_2 → $(b)_2$ → (Fe—Fe) → c_1 → c QH_2-cytochrome c reductase
IV	Reduced cytochrome c → (a—Cu) → (a_3—Cu) → O_2 Reduced cytochrome c-oxygen reductase (cytochrome oxidase)

The sequence of electron flow within the complexes of the electron transfer chain.

The nonheme metal complexes contain iron in three of the molecular complexes and copper in the cytochrome oxidase system. The role of copper has already been discussed. The role of the nonheme iron remains unknown, but electron spin resonance studies and investigations of the effects of various inhibitors all suggest that nonheme iron is a genuine participant in the electron transport chain. Furthermore, Green has proposed that the dissociable groups are mobile elements in the electron transport chain capable of moving from one complex to another, thereby holding the complexes together.

The ETP can be further divided by appropriate chemical treatment into even smaller particles that contain all the components of the electron transport chain (except cytochrome c) in concentrations greater than that found in the intact mitochondrion. These particles are referred to as elementary particles.

While the elementary particles were being prepared from mitochondria, two major groups of mitochondrial components were separated out: a colorless structural protein and the enzymes of the Krebs cycle. The exact composition of the structural protein is not known. The evidence suggests that only one structural protein has a molecular weight of about 25,000 in the monomeric form. Native proteins probably exist in the form of polymers. The portion of structural protein is considerable, representing between 50 and 60% of the total mitochondrial protein. The biochemical findings of Green and his associates acquire particular significance in the light of the morphological observations of Fernández-Morán [123]. This skillful electron microscopist demonstrated that mitochondria contain repeating units of a diameter of about 125 A. These units, present in the outer membrane of the cristae, tend to form groups of six particles composed of two opposite pairs of three particles linearly arranged.

On the basis of the biochemical studies and the morphological findings of Morán, Green proposed a schematic representation of the interrelationships between structural protein, elementary particles, and Krebs cycle dehydrogenases. However, it remains to be established that the particles Morán demonstrated electron microscopically and the elementary particles Green described are identical; this view has been challenged by Chance and his associates who found that after stripping the particles from the membrane, the components of the respiratory chain are associated with the membrane and not with the particles [124].

Attempts to resolve and reconstitute the electron transport chain have also been undertaken in Racker's laboratory. The investigators succeeded in reconstructing from purified succinate dehydrogenase, cytochromes b, c, c_1, and oxidase, coenzyme Q_{10}, and phospholipids a molecular complex capable of catalyzing the oxidation of succinate by molecular oxygen at a rate comparable to that of phosphorylating submitochondrial particles sensitive to antimycin. Cytochrome b_1 inactivation was apparently responsible for the difficulty encountered in reconstituting this segment of the electron transport chain. This problem was overcome by preparing the respiratory protein with guanidine in the presence of succinate and glycerol. Surprisingly, reconstitution of this functional segment of the electron transport chain did not require an additional nonheme iron protein [125].

Oxidative Phosphorylation

We have already remarked that the anaerobic oxidation of glucose releases only a small fraction of the available chemical energy ($C_6H_{12}O_6 \rightarrow 2\ CH_3CHOH\text{-}COOH + 16{,}000$ calories) and that, in contrast, the aerobic oxidation of glucose through the Krebs cycle makes available larger amounts of usable energy ($C_6H_{12}O_6 \rightarrow 6\ CO_2 + CH_2O + 288{,}000$ calories). An appropriate understanding of the molecular mechanism involved in converting the energy released during oxidation into a form of energy usable for cell maintenance, function, and reproduction is vital to the understanding of life itself.

In reviewing the metabolism of the glycolytic process, we described a reaction in which the exergonic dehydrogenation of glyceraldehyde-3-phosphate is coupled to the endergonic phosphorylation of ADP to ATP. This was referred to as substrate-linked phosphorylation. It was demonstrated in numerous laboratories that the oxidation of glucose in various respiring biological preparations (intact cells, tissue slices, homogenates, etc.) was invariably associated with the uptake of inorganic phosphorus, which can be recovered from various organic compounds, such as creatine phosphate and phosphorylated sugars. These observations raised two questions: (1) What is the nature of the phosphorylated intermediates? and (2) How many molecules of inorganic phosphorus disappear in the phosphorylating process?

It is now well established that ATP is the phosphate donor in many of the reactions cited above, that ATP is formed in the course of aerobic respiration, and that the ratio of the number of atoms of phosphorus esterified per atom of oxygen used is referred to as the P/O ratio. The P/O ratio varies depending on the nature of the oxidizable substrate used in the experiment. Thus when ketoglutarate is used, the P/O ratio is 4 because there is an additional substrate-level oxidation. When succinate is used, the P/O ratio is 2 because succinate oxidation does not involve the pyridine nucleotide.

Belitzer and Tsibakowa [126] showed that the P/O ratio is greater than one. Careful analysis of the oxygen uptake versus the amount of inorganic phosphorus present in the system led Ochoa to conclude that three molecules of inorganic phosphorus are converted to organic phosphorus for each molecule of oxygen used in the respiring system. A P/O ratio of three is the generally accepted figure, but is difficult to reproduce experimentally. For example, in liver the P/O ratio may vary between 2.6 and 2.8 when β-hydroxybutyrate is used. When it was established that ADP is the acceptor for the inorganic phosphorus and that ATP is the product of the reaction, a definition of oxidative phosphorylation could be enounced. Oxidative phosphorylation is a biochemical process coupling the oxidation of foodstuffs (carbohydrates, lipids, proteins) to the phosphorylation of ADP to yield ATP [126–131].

The chemical energy liberated during oxidation of the Krebs cycle substrates is captured in the electron transport chain and delivered in small packages to yield the energy-rich phosphate bond of ATP.

The process of oxidative phosphorylation involves a dual mechanism: a conservation of energy within the electron transport chain, and the transfer of that energy from the electron transport chain to ADP.

From the description of the Krebs cycle and the electron transport chain, it should be clear that succinate and NADH are two substances that link the Krebs cycle to the electron transport chain, and that the exergonic transfer of the electrons released by each of these compounds triggers the alternating reduction and oxidation of each element of the electron transport chain*. Now, since all the enzymes involved in the electron transport chain are found in the mitochondria, oxidation and phosphorylation must also be coupled in mitochondria. Thus the two processes of electron transport and phosphorylation of ADP are intimately interconnected. As a matter of fact, a detailed understanding of the molecular mechanism of oxidative phosphorylation has long been delayed because of this rigid structural integration of the multiple-enzyme system.

Although few fields of biochemistry have been tackled by such a great concentration of talent, much of our information on the mechanisms coupling oxidation to phosphorylation is incomplete and often controversial. Among the problems under investigation are the sites of phosphorylation on the electron transport chain, the nature of the carriers that transfer the energy from the chain to the inorganic phosphate, and the mechanism of action of uncouplers.

Sites of Phosphorylation

Using inhibitors and spectrophotometric analysis, several laboratories concluded that the sites of phosphorylation on the electron transport chain are located between NAD and cytochrome c reductase, between cytochrome b and cytochrome c, and between cytochrome c and cytochrome oxidase. The location of these sites has later been confirmed by P/O ratio measurements with fragments of mitochondria containing all the elements of the electron transport chain necessary to phosphorylation at one of these sites.

Particles prepared by Lehninger were able to couple electron transport to the phosphorylation of ADP. This investigator first prepared a classical mitochondrial pellet in 0.25 M sucrose. The mitochondria were then suspended in a solution containing 0.8% digitonin, a treatment that dissolves mitochondria and yields a clear supernatant from which membranes can be separated by centrifuging the preparation at 50,000 *g* for 25 minutes. If the supernatant is further centrifuged at 100,000 *g* for 25 minutes, a new pellet is sedimented which contains the phosphorylating particles. When measured in terms of the amount of nitrogen present, the particles show phosphorylating activities four times greater than that found in intact mitochondria. They also contain all cytochromes, each at a molar concentration close to that reported to exist in the intact mitochondria by Chance and Williams. NAD is present in a bound form, but in ratios lower than one mole of NAD per mole of cytochrome. These low levels of NAD strikingly contrast with the rather high values that have been reported for the intact mitochondrion, in which the NAD concentration on a molar basis is about 40–60 times that of cytochrome. Thus, Lehninger's observations suggest that the effective mitochondrial NAD is tightly bound to the organelle's structure.

The phosphorylating particles have many of the properties of intact mitochondria. They are sensitive to inhibitors of the electron transport chain, such as Amytal and cyanide. In addition, such uncoupling agents as 2,4-dinitrophenol, Dicumarol, pentachlorophenol, gramicidin, arsenic, and methylene blue can uncouple the transport of electrons from the phosphorylation of ADP in the phosphorylating particle. However, the phosphorylating particle differs from the intact mitochondria. For instance, oxidative phosphorylation in intact mitochondria is readily uncoupled by thyroxine and calcium; in contrast, in the phosphorylating particle, oxidative phosphorylation is unaffected by adding the hormone or the metal to the incubation mixture.

The three sites of phosphorylation suspected to exist in intact mitochondria have also been recognized in the phosphorylating particle where these sites do not function at full capacity. Although the first of these sites (between NAD and the flavoprotein) operates at 90% of its full capacity, the second (between cytochrome c and cytochrome c_1) is active at only 20% of its normal phosphorylating ability, and the third (between cytochrome c and oxygen) reaches 70% of the normal phosphorylating rates.

Some of the chemical properties of the phosphorylating particle have also been elucidated. The particle consists mainly of two basic components, protein (65%) and phospholipids (20%). In addition to the various components of the electron transport chain, the phosphorylating particle also contains ATPase, the activity of which is so high that it may well explain the particle's low phosphorylating ability. Like that associated with mitochondria, the ATPase found in the particle is a highly specific enzyme that splits ATP to yield ADP, and is without effect on inosine, guanosine, or cytidine triphosphate. There is more bound copper and iron in these particles than can be accounted for by the ferroproteins and cuproproteins of the electron transport chain.

Intermediates and Coupling Mechanisms

The net result of the oxidative phosphorylation reaction is a transfer of hydrogen in the presence of inorganic phosphate and ADP to molecular oxygen to yield a molecule of water. ATP is formed in the process.

Thus in oxidative phosphorylation, the free energy associated with the reduced electron transport chain

* Oxidative phosphorylation can be studied with a preparation of liver mitochondria incubated at 30° C in the presence of a Krebs cycle substrate. In liver, hydroxybutyrate is most appropriate as a substrate because after oxidation to acetoacetate, it is not further metabolized. Obviously, the presence of ADP and a phosphorus acceptor such as glucose is also necessary in the system. ATP is very unstable; it may break down spontaneously or it may be catabolized by ATPase. These deleterious effects can be obviated by adding fluoride (0.40 M), which inhibits ATPase, or by trapping the phosphorus high-energy bond in glucose in the presence of hexokinase. NADP is also required. Its breakdown by NADPase in the system can be obviated by adding nicotinamide.

is in some way transferred to ADP and inorganic phosphate to make ATP. The notion of a P/O ratio equal to three is so strongly anchored in the mind of the biologist that he fails to be aware of the relative inefficiency of this energy transfer. The free-energy change in the conversion of NADH and NAD under aerobic conditions at pH 7 is −50 kilocalories per mole. This oxidation involves one atom of oxygen, and three molecules of ATP are formed. The synthesis of each of them corresponds to +8 kilocalories. Thus it is obvious that the system is only 50% efficient. The biochemical properties of the reduced carriers and of the final product are known, but one of the most intriguing aspects in oxidative phosphorylation remains unresolved—namely, the molecular properties of the intermediates involved in the transfer of free energy from the carrier to ATP.

The existence of such intermediates was first postulated by Lardy and Wellman [133], who obtained indirect evidence supporting the concept by spectrophotometric methods and by following the fate of ^{18}O in the incubation system. The characteristics of these intermediates remain unknown, but various models have been proposed.

Most models postulate the existence of an intermediate capable of conserving and transferring energy. Thus, in a reaction coupled with an oxidation reduction, the hydrogen carrier (A) is energized, and a complex is formed with another substance (X) or with inorganic phosphate itself.

This complex is important because it includes the formation of a new bond with a high free energy of hydrolysis. The first model includes a reduced hydrogen donor (A), an oxidized hydrogen acceptor (B), and an intermediate (X). In the first step of the reaction, the oxidation of the hydrogen donor is coupled with the formation of a high-energy complex, A∼X, which then undergoes phosphorolysis in the presence of inorganic phosphorus and ADP to yield ATP(I).*

In a second model (II), the intermediate is postulated to react directly with P_i, and the reduced hydrogen donor (A) is thus assumed to form a complex with the inorganic phosphorus. The energized complex reacts with ADP to yield ATP.

$$\text{I.}\quad AH_2 + B + X \rightarrow A{\sim}X + BH_2$$
$$A{\sim}X + ADP + P_i \rightarrow A + X + ATP$$
$$\text{II.}\quad AH_2 + P_i \rightarrow AH_2P_i$$
$$AH_2P_i + B \rightarrow AP_i + BH_2$$
$$AP_i + ADP \rightarrow A + ATP$$

The molecular description of the chain of intermediates and coupling factors in the process of oxidative phosphorylation would untie one of the most intriguing knots in our understanding of the process of life, and it is thus understandable that biochemists have eagerly attacked the problem.

Research in the field of intermediates of oxidative phosphorylation has developed in two main directions: (1) Some investigators attacked the problem directly and attempted either to reconstruct *in vitro* the enzyme system responsible for the coupling of electron transport and phosphorylation, or to isolate intermediates; (2) others have approached the problem by studying elementary reactions suspected to participate in oxidative phosphorylation. So much data has been gathered on all the aspects of oxidative phosphorylation that it would be unrealistic to attempt to cover the subject comprehensively. We will consider only a few representative experiments in the hope of illustrating the amplitude and complexity of the problem.

A logical starting point for such studies is to look for phosphorylated derivatives of any of the components that might serve as intermediates in oxidative phosphorylation, such as phosphorylated derivatives of the elements of the electron transport chain, whether they be proteins or low molecular weight cofactors.

Pinchot [134] tried to dissect the molecular mechanism that couples oxidation to phosphorylation at the first site of phosphorylation in the bacterium *Alcaligenes faecalis* (see Fig. 1-21). His system is composed of an NADH oxidase system, a coupling enzyme, magnesium, and a polynucleotide. The NADH oxidase system is particulate-bound, contains cytochromes, and is capable of oxidizing NADH even in the absence of phosphorylation. The polyribonucleotide is an RNase-resistant tetranucleotide composed of two molecules of adenylic acid, one of guanylic acid, and one of a terminal uridylic acid. If the NADH oxidase system is incubated with the coupling enzyme, inorganic phosphorus, and ADP, but without magnesium and the polynucleotide, the coenzyme is oxidized, although the oxidation of NADH is not coupled with

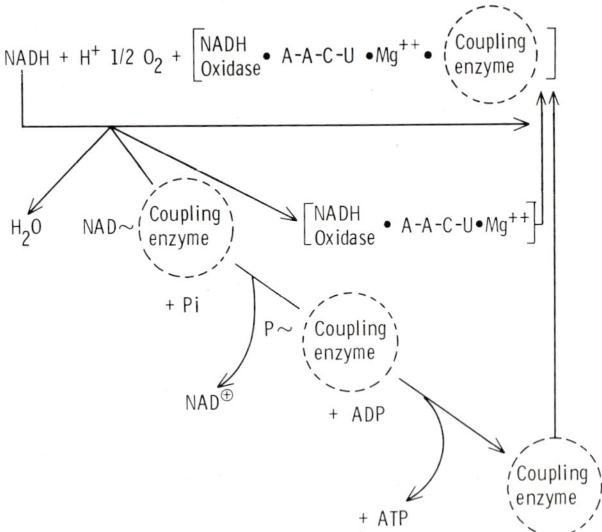

Fig. 1-21. First site of phosphorylation. (According to Pinchot)

* It is assumed in the above scheme that the high-energy complex exists in the oxidized form, but some investigators have proposed that the energy transfer occurs while the complex is in the reduced form.

Inasmuch as respiration drives the energy-linked reduction in the absence of inorganic phosphate and in the presence of oligomycin, energy coupling does not appear to involve formation of phosphorylated high-energy intermediates.

the phosphorylation of ADP. When magnesium and the polynucleotide are added to the incubation mixture, the enzyme becomes part of the particles that contain the NADH oxidase system, and the oxidation of NADH is coupled to the phosphorylation of ADP.

Further investigation led Pinchot to outline the individual steps of oxidative phosphorylation at Site I as they are presented in Fig. 1-21. In this sequence of reactions, the energy of oxidation, which has been transferred to NADH, is subsequently transferred to the NADH-enzyme complex, from there to the P_i-enzyme complex, and finally to ATP. The mode of attachment of the NADH or the phosphorus to the coupling enzyme is not known, but a histidine residue may be involved in the phosphorylation of the enzyme.

Again, a phosphorylated protein may be an intermediate; therefore, Boyer's [135] discovery of a phosphohistidine compound that may act as an intermediate in oxidative phosphorylation was most intriguing. The compound is either 1- or 3-phosphohistidine. It is not known which of the two isomers is the natural compound, but only one of them is active. The phosphorylated histidine residue was found to be part of a polypeptide sequence, but little information is available on the protein that contains the phosphorylated histidine, although that protein has been obtained in a pure form.

The phosphorylated histidine could play a role in oxidative phosphorylation. Boyer and his associates found an enzyme in the mitochondria of beef heart that catalyzes the formation of ^{32}P-labeled histidine at the expense of [^{32}P]ATP. Such a reaction could be considered part of the reversal of oxidative phosphorylation. It is interesting that the same enzyme preparation also catalyzes the exchange between inorganic phosphate and oxygen—a reaction discussed in more detail later.

Phosphohistidine has more recently been shown to be part of the succinate thiokinase, but this does not exclude a potential role of phosphohistidine in oxidative phosphorylation. The only mitochondrial phosphoprotein in which maximal ^{32}P incorporation coincides with maximal ATP synthesis is bound phosphohistidine.

Whatever part phosphohistidine plays in oxidative phosphorylation, the reason behind this research is still valid and can serve as a model for future experiments. Therefore, it may be worthwhile to consider on theoretical grounds what possible connection could exist between the phosphorylated histidine and coupling of oxidation (for example, of NADH) and phosphorylation of ADP. Such a question will be more easily answered if we first outline ADP's role in controling respiration. In passing from the resting to the working state, the oxygen consumption of muscle increases up to 65 times. It follows that a mechanism controls the oxidation rate in the muscle cell. Oxygen uptake probably is regulated by a number of different control mechanisms, for instance, glucose penetration in the cell. Lardy and Wellman [133] demonstrated that a molecular mechanism controlling oxygen uptake also operates at the level of oxidative phosphorylation. These investigators prepared mitochondria free of ATPase activity and observed that in this system the oxygen uptake was reduced when all the ADP was converted to ATP, but the uptake was considerably increased if ADP and P_i were added to the incubation medium.

Chance and his associates [136] and Slater's [120] group took advantage of Lardy and Wellman's important experiments to investigate which step in the sequence of the coupling reaction is inhibited by ADP deprivation. Chance studied the extent to which ADP and phosphate control electron transport in mitochondria isolated from various sources. This was done by measuring the oxygen consumption in the suspension with the aid of a vibrating platinum electrode connected to the appropriate electronic recording equipment. In agreement with the observations of Lardy and Wellman, the oxygen consumption was found to be low in the absence of ADP, to increase four to ten times after adding the nucleotide to the suspension, and to drop as ADP is converted to ATP. Three different substrates were used in these experiments—glutamate, hydroxybutyrate, and succinate. The ratio of the disappearing ADP and of the oxygen consumed was in all cases that expected from the classical values of the P/O ratio. The Pennsylvania group further observed that the pyridine nucleotides in mitochondria were mainly in the reduced form, and that the reduced nucleotide accumulates in the cell in ATP deprivation. However, the concentration of the reduced coenzymes decreases stoichiometrically with the concentration of ADP in the medium.

As these studies developed, the evidence accumulated to suggest that the reduced pyridine nucleotides in the mitochondria were protein-bound. The most suggestive observation in that respect was the emission of fluorescent light by the mitochondrial preparation when illuminated with a beam of light ranging from 340–360 mμ wavelength. When the concentration of the phosphate acceptor is limiting, the fluorescence increases and exceeds by as much as six times that expected from the determination of the concentration of the pyrimidine nucleotide in the mitochondrial preparation. That this enhancement of fluorescence results from the formation of a protein-bound pyrimidine nucleotide is suggested by Boyer and Theorell [137], whose studies on alcohol dehydrogenase showed that the fluorescence of NADH bound to the purified enzyme was greater than that observed for the free pyrimidine nucleotide. It is thus remarkable that the maximum of the emission spectrum in the mitochondria occurs at 443 mμ a peak very close to that observed for the NADH-alcohol dehydrogenase complex. The fluorescence emitted by the NADH-alcohol dehydrogenase complex can be further enhanced by adding histidine to the medium. It is assumed that under these conditions a ternary complex, NADH-alcohol dehydrogenase-histidine, is formed.

These observations bring us back to our original concern: the relationship between NADH and the phosphorylated histidine. In the scheme outlined by Pinchot, protein-bound NADH is the initial reactant in the sequence of energy-conserving reactions. This finding is in agreement with the observations made by Chance in intact mitochondria.

As mentioned above, Boyer and his associates discovered a phosphorylated histidine and obtained some indirect evidence that it might participate in oxidative phosphorylation. On the basis of further studies of the fluorescent spectrum emitted by mitochondria, Estabrook [138] has proposed a reaction mechanism that might explain the relationship of these two compounds in the coupling process.

Models of the NADH and histidine molecules can be built and placed in such a position that they satisfy a specific hypothesis describing the mechanism of their interaction. It then becomes obvious that the pyridine ring of the coenzyme and the imidazole ring of histidine can be placed in two adjacent parallel planes, a situation that favors a charge transfer from the nitrogen of the pyridine ring to that of the imidazole ring. Such a charge transfer labilizes the hydrogen of the imidazole, which is then readily removed by a hydrogen acceptor. This makes an electrophilic compound out of the imidazole ring. Consequently, the lone electrons of the pyridine ring can be transferred to the imidazole ring, which is therefore apt to attack the phosphoric acid nucleophilically, leading to the formation of a PN bond and liberation of the phosphate hydroxyl.

(This might be an appropriate time to remind the reader of important definitions in chemistry—electrophilic and nucleophilic compounds. An electrophilic reagent is a compound that in the course of its reaction with another compound acquires electrons or seeks to share electrons belonging to some other atom or ion. The nucleophilic reagent is one that transfers electrons or shares its electrons with another atom. Hydroxyl groups are thus nucleophilic.)

Griffiths [139] isolated an intermediate that might satisfy many of the requirements of a high-energy intermediate in oxidative phosphorylation. The new compound, first discovered by paper chromatography, seems to be a phosphorylated derivative of reduced NAD (Fig. 1-22). When the paper was developed in the isobutyric acid/ammonia solvent of Krebs and Hems, a new ultraviolet quenching spot was observed. It had moved far ahead of the spots made by NADH or radioactive P_i. After the compound was eluted with 0.1 M tris buffer at pH 7.2, it became obvious that the new compound contained both NADH and P_i. The compound could not be NADP, because its spectral, chromatographic, and stability characteristics were different from those of NADP. As might have been expected, the new compound turned out to be extremely unstable, and only by resorting to paper electrophoresis could Griffiths elute enough material for further chemical and biochemical investigations.

The new compound contains approximately three moles of total phosphate (after washing) and one mole of labile phosphate per mole of NADH. The spectrum of the new compound has been identified and suggests that the phosphorus group occupies the position 2, 4, or 6 of the NADH molecule. (Available evidence that cannot be reviewed here favors position 6.)

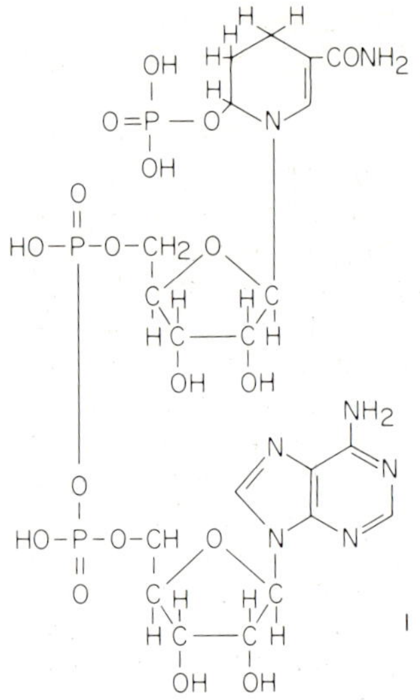

I With closed ring

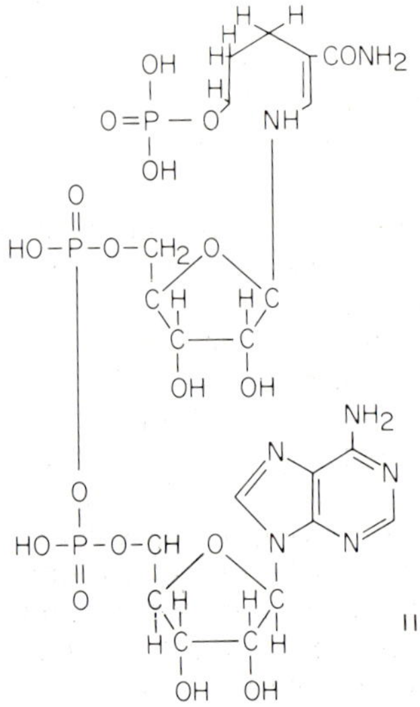

II With open ring

Fig. 1-22. Coupling of the first site of phosphorylation in *Alcaligenes faecalis*

It is highly probable that the new compound plays a role in oxidative phosphorylation. It is not formed in the absence of succinate, and it can yield ATP from ADP in the presence of magnesium. Furthermore, the formation of the intermediate is inhibited by dinitrophenol. Oligomycin has no effect on the formation of the intermediate but blocks the formation of ATP from the phosphorylated derivative. Amytal has an opposite effect; it blocks the formation of the phosphorylated derivative but does not affect the transphosphorylation of ADP.

Griffiths' study is only one example of the overlapping of biochemistry and organic chemistry, a healthy marriage of disciplines. The sooner the molecular mechanisms of the reactions that govern life are solidly established, the easier it will be to interpret the chemistry of disease.

The logic of these working hypotheses of the mechanism of oxidative phosphorylation seemed so compelling that it was impossible to resist the temptation of presenting the original unchallenged interpretations before seeding doubt in the reader's mind. Yet, it must be added in all fairness that several aspects of this partial molecular interpretation of some of the steps of oxidative phosphorylation are under severe criticism either because of lack of direct evidence or the existence of conflicting evidence. For example, the role of the enzyme-NAD complex in the scheme of Pinchot has been challenged, and, as already mentioned, the phosphohistidine formed in mitochondria may not be directly involved in oxidative phosphorylation. One must reluctantly conclude that direct evidence for the existence of phosphorylated or nonphosphorylated ("energy carrier" intermediates) does not exist.

Coupling Factors

If the scheme outlined in Fig. 1-21 were applicable to each of the coupling stages, elucidation of the coupling mechanism would be well on its way. Unfortunately, all investigators do not share that optimism, and Racker, who studied the coupling mechanism of mitochondria at the second and third sites of phosphorylation, has emphasized the complexity of the sequence of energy-conserving reactions.

Racker and his associates [140–143] have disrupted mitochondria by various means, including sonication, mechanical disruption in a shaker, and trypsin and urea treatment. By combining one or more of these procedures, the investigators solubilized a number of factors referred to as F_1, F_2, F_3, F_4, etc., which restore phosphorylation to the particulate submitochondrial residues. Many of these factors have been partially purified. For a complete review of these elaborate studies, refer to Racker's original paper; only one simple illustration of the basic principle guiding these studies is presented here.

Among the many systems Racker used, one is a particulate system prepared by exposing beef mitochondria to sonic oscillation. The system can oxidize succinate and NADH but cannot couple the oxidation of the substrate to the phosphorylation of ADP. Coupling can be achieved by adding to the system two soluble proteins referred to by Racker as factor 1 (F_1) and factor 4 (F_4). F_1 is a protein with ATPase activity obtained by sonic disintegration of mitochondria. F_4 is prepared by alkaline extraction of the mitochondria. F_1, which has been extensively purified, is a decamer with a molecular weight of 284,000, each monomer weighing 26,000.

The nature and exact role of the protein is still obscure, but the results clearly demonstrate the multiplicity of the coupling factors. In fact, further investigation by Racker suggests that the efficiency in the coupling of oxidation and phosphorylation is closely related to the integrity of the structural relationship between the proteins involved in the reaction. It was observed that oligomycin can uncouple oxidative phosphorylation and inhibit ATPase activity in intact mitochondria, but it does not affect the purified F_1 protein. When a mitochondrial membrane preparation that is free of F_1 but contains the respiratory chain, structural proteins, and phospholipids is added to F_1, sensitivity to oligomycin is restored. The presence of the respiratory chain clearly is not necessary to confer intramitochondrial properties of F_1. The addition of a preparation called CF_0, devoid of the respiratory chain components and phospholipids, masks the ATPase activity of F_1. However, if phospholipids are also added to the CF_0-F_1 combination, ATPase activity reappears and the ATPase activity is sensitive to oligomycin. Observations of this type led to the conclusion that the coupling mechanism involves several coupling factors rigidly arranged structurally. At each site of phosphorylation is a structurally organized chain of coupling factors, which is much like the rigid chain of compounds transferring the electrons.

Other coupling factors have been described and sometimes partially purified [145, 146]. The exact intracellular role of these coupling factors is not known. Although it is clear that they restore oxidative phosphorylation in submitochondrial particles, it is not certain whether this effect is specific or direct. It is indeed distressing that albumin stimulates oxidative phosphorylation, probably by binding natural uncouplers; and CoA restores oxidative phosphorylation in uncoupled submitochondrial fractions by facilitating the removal of free fatty acids, which act as uncouplers.

Therefore, much more work is needed before a clear understanding of the role of coupling factors will be available. Yet, one thing is certain, such investigations have permitted the systematic reconstruction of oxidative phosphorylating sites, a fact undreamed of a decade ago.

Thus, by combining the NADH-ubiquinone reductase complex with a heterogeneous fraction solubilized from bovine heart submitochondrial fractions (referred to as "hydrophobic protein") and phospholipids (phosphatidylethanolamine, phosphatidylcholine, cardiolipid), Ragan and Racker and his associates [146] were able to reconstitute vesicles. These vesicles, which can be isolated on sucrose gradients, carry on the phosphorylation of ADP to ATP, and the oxidative phosphorylation is sensitive to uncouplers.

Other Reactions

Now that we have recognized the complexity of the problem of coupling mechanisms, we can consider some apparently unrelated observations. A series of reactions associated with the intact mitochondria is likely to be involved in oxidative phosphorylation. The significance of these reactions resides in the fact that they could be examples of the reversal of the steps involved in oxidative phosphorylation. Among them are ATPase release, the ATP-$^{32}P_i$ exchange, the $H_2{}^{18}O$-P_i and the $H_2{}^{18}O$-ATP exchange reactions, and the ATP-ADP exchange reaction.

ATPase hydrolyzes ATP to yield ADP. In phosphorylating mitochondria with a P/O a ratio of about 2.6, ATPase activity is low, or practically nonexistent, but ATPase is released as soon as oxidative phosphorylation is uncoupled in the mitochondria, for example, under the influence of dinitrophenol. Under those conditions the ATPase activity may rise to five or fifteen times the normal levels. The degree of uncoupling is correlated with ATPase release, but correlation is far from perfect, and therefore it is not clear how much of a role ATPase plays in oxidative phosphorylation.

The mechanism of enzyme release is interesting *per se*. An enzyme can be maintained in an inactive form by various mechanisms—compartmentalization, intermolecular reactions, and intramolecular reactions.

The organization of the cellular structure may be such that the enzyme is separated from the substrate. The zymogens in the zymogen granules constitute a classical example of this mode of latency. It has also been proposed that many of the acid hydrolases are shielded from their intracellular substrates by their enclosure in a specific cytoplasmic organelle.

Some enzymes are rendered inactive by specific or unspecific inhibitors, as in the case of ribonuclease and deoxyribonuclease. Other enzymes have latent activity because their active center is in some way inactivated by some polypeptidic chain in the molecule. Once the inhibiting chain has been removed by the action of proteolytic enzymes, the enzyme becomes fully active, as is the case for example, for thrombin and chymotrypsin. Finally, the activity of some enzymes is kept latent because of the absence of a cofactor (*e.g.*, phenylpyruvic hydroxylase in the developing liver).

ATPase is latent not because it is secluded in an osmotic bag or trapped to a cellular structure. The latency seems to be intramolecular in nature; the uncoupling agent activates the enzymes even after extensive purification. A new outlook on the mechanism of ATPase activity has been brought about by experiments in which the SH groups present in submitochondrial fragments were titrated and related to ATPase activity. Two phases were observed in the fragments' response to the titration of the SH groups. When half of the SH groups are titrated, all the ATPase activity is released, but if all the SH groups are titrated, ATPase activity is inhibited. Such findings suggest that there might be two classes of SH groups in the ATPase molecule—a first that protects the active site from reacting with water, a second directly involved in the reaction.

Considering the role that ATPase might play in oxidative phosphorylation, it is not surprising that the enzyme has been the object of much research. Several soluble forms of ATPase have now been obtained. One of these was prepared from heart mitochondria and has been purified to a practically homogeneous preparation. This soluble form of ATPase resembles the particulate enzyme; it is stimulated by magnesium and dinitrophenol and inhibited by ATP and azides. It restores oxidative phosphorylation when added to heart particles that have all the other components necessary for oxidative phosphorylation but cannot couple without ATPase.

Another interesting aspect of the ATPase activity is the multiplicity of the protein molecules responsible for that catalytic property. The possible role of ATPase in oxidative phosphorylation raises new problems: Is it the same enzyme that is active at the three coupling sites, or do three different ATPases exist, each of which is active at a different site? Studies of Lehninger suggest that the second possibility cannot be excluded. If mitochondrial ATPase is stimulated by NAD when the pH of the medium is changed, three different stages of activations can be detected, each of which is characterized by a different pH optimum. Thus, it has been proposed that the ATPase of Site I has a pH optimum of 8.5, that of Site II of 7.4, and Site III 6.3.

Studies with ATPase inhibitors also suggest that different proteins might be involved in catalyzing the breakdown of ATP. Oligomycin, chlorpromazine, and azide inhibit the ATPase activity elicited by both magnesium and NAD. In contrast, amytal interferes only with the ATPase activity revealed by NAD. These observations have been interpreted to suggest that the ATPase stimulated by NAD is different from that stimulated by magnesium.

Many people try to reconstruct the steps of oxidative phosphorylation by studying the exchange reactions outlined above. These reactions were described first in intact mitochondria, and later in mitochondrial fragments obtained by treating the organelle with digitonin or ultrasound. The enzymes involved have occasionally been prepared in a soluble form.

The role played by exchange reactions in oxidative phosphorylation is not known; these reactions could be the steps of the coupling mechanism, but in reverse. Some indirect evidence suggests that they are indeed involved in oxidative phosphorylation. First, as already mentioned, these exchange reactions are affected by agents uncoupling oxidative phosphorylation. Their response to the uncouplers, however, does not correlate with the effect of these compounds on the efficiency of oxidative phosphorylation itself. Secondly, these reactions are influenced by the carriers' state of oxidation.

If these exchange reactions are involved in the coupling mechanism, the oxidation or reduction of the carrier could be expected to influence their process. In the light of this assumption, it is interesting that the ATP-$^{32}P_i$ exchange is strongly inhibited when the carriers are maintained in the reduced state. The oxidation or reduction of the carriers does not affect the ATP-ADP exchange reaction of particles obtained by digitonin treatment of the mitochondria, but it inhibits the reaction in intact mitochondria. And maintaining the carriers in the reduced forms also inhibits the ATPase activity.

The ATP-$^{32}P_i$ exchange reaction has been studied in digitonin and sonic particles of mitochondria. The reaction requires ADP, and the rate of the reaction is a function of the relative concentration of ATP, ADP, and inorganic phosphorus. The reaction can be inhibited by NAD, pentachlorophenol, gramicidin, and other agents.

The $H_2{}^{18}O$-P_i exchange reaction has been observed in mitochondria and in digitonin fragments. This reaction is activated by ATP in the absence of respiration. It is inhibited by oligomycin, azide, and sucrose.

The ATP-ADP exchange reaction has been observed in intact mitochondria and in mitochondrial particles obtained by sonication or treatment with digitonin. Among its inhibitors are NAD, oligomycin, dihydroxycoumarin, and gramicidin. Inorganic phosphate stimulates the exchange. The response of the ATP-ADP exchange reaction to either stimulants or inhibitors varies with the tightness of the coupling in the preparation.

Chemiosmotic and Conformational Coupling

Throughout this discussion of oxidative phosphorylation, we have assumed that the coupling mechanisms involve the formation of high-energy intermediates. This chemical hypothesis is not accepted by all; the chemiosmotic hypothesis of oxidative phosphorylation was proposed by Mitchell in 1961, and in 1966 Boyer [148] proposed a new hypothesis involving conformation coupling.

The chemiosmotic theory is based on two fundamental assumptions. First, the enzymes of the electron transport chain and ATPase are vectorially distributed on a membrane. The position of the enzymes' active sites determines on which side of the membrane the product of the reaction is delivered. Second, the membrane is impermeable to ions. Uncouplers are thought to abolish the permeability barrier to protons presented by the membrane. It is difficult to evaluate the validity of the hypothesis, because all arguments that have been marshalled in its support can also be used to support the other proposed mechanism of coupling.

In Mitchell's model, the active sites of succinate dehydrogenase and cytochrome oxidase are in opposite directions. As a result, all protons (H+) are released at one side of the membrane, and all OH are released on the opposite side. Since the membrane is impermeable to ions, the concentration of H+ on one side and of OH− on the other side generates an electrical potential. The free energy of the electrical potential is thought to drive the reverse reaction of an anisotropic ATPase, which catalyzes the condensation of ADP and P_i with the formation of a molecule of water. The molecular orientation of the active site of the ATPase is such that it releases the OH formed in the reverse reaction in the pool of protons, and its H+ in the OH pools. In the absence of ADP and inorganic phosphate, protons accumulate in the coupling membrane and as a result block respiration.

If valid, the chemiosmotic theory indicates that the ultimate step of one of the most essential reactions of live coupling of oxidation and phosphorylation is not simply molecular, but is also vectorial because of the special orientation of the molecules. The separate distribution of the products generates a physical force capable of driving the phosphorylation of ATP, and there is no need to postulate the existence of high-energy intermediates. In conclusion, it seems likely that an adequate and comprehensive interpretation of the mechanism of coupling may well require cooperation of biochemists and physicists.

Boyer proposes that there is no compelling evidence for the existence of intermediates with a high-energy bond and challenges the notion that the energy generated by the oxidation forms a high-energy precursor of ATP or is used in some way for the formation of the covalent anhydride bond of ATP. Recently Boyer and his associates have provided evidence suggestive of a mechanism by which changes in protein conformation could yield oxidative phosphorylation [148].

The alternative assumes that the free energy of the oxidation-reduction process is stored in a protein (a respiratory carrier), which undergoes conformational changes, either as a result of covalent bond formation or rupture. Efficient energy transfer through conformational changes in protein assumes that the coupling of oxidation and phosphorylation takes place on a complex and rigidly organized macromolecular structure. If this is the case, then it is unlikely that the process will be easily reconstructed *in vitro*.

These investigators remarked that the P_i-HOH exchange is not blocked by uncouplers of oxidative phosphorylation but is inhibited by oligomycin. On the basis of these findings, they predicted that ATP is formed at the catalytic site of the mitochondrial membrane even when an amount of energy sufficient for ATP synthesis is not available. On the basis of experiments too complex to be described here, but in which the effect of uncouplers and oligomycin on the P_i-HOH, P_i-ATP, and ATP-HOH exchanges were compared, they conclude that the P_i-HOH exchange results from the release of noncovalently bound ATP at the site of oxidative phosphorylation.

Thus, they propose that ATP is formed at the catalytic site from ADP and P_i through reversal of hydrolysis, but with minimal or no expense of energy. The energy generated by oxidative phosphorylation is then used to release the tightly but not covalently bound ATP from the catalytic site. Energy requiring conformational changes in the proteins at the catalytic site may be involved in the release.

talline enzyme, although the velocity of the reaction decreases when the length of the carbon chain increases.

Even in the crystalline state, crotonase catalyzes different types of reactions (see Fig. 1-24). The discussion of these properties is simpler if some basic concepts in organic chemistry are reviewed. Many carboxylic acids have at least two different names—one is based on molecular structure and has been assigned by the International Union of Chemists (IUC); another, referred to as the trivial name, usually is derived from the name of the source in which the carboxylic acid was first discovered (butyric, oleic, etc.). The position of the substitutes on the carbon chain is denoted by either Greek letters or Arabic numerals. The numerals should be used with the IUC nomenclature, and the Greek letters correspond in numerical order with the trivial name. It is important to remember that the letter or number does not refer to the first carbon but to that carbon immediately adjacent to the carboxyl.

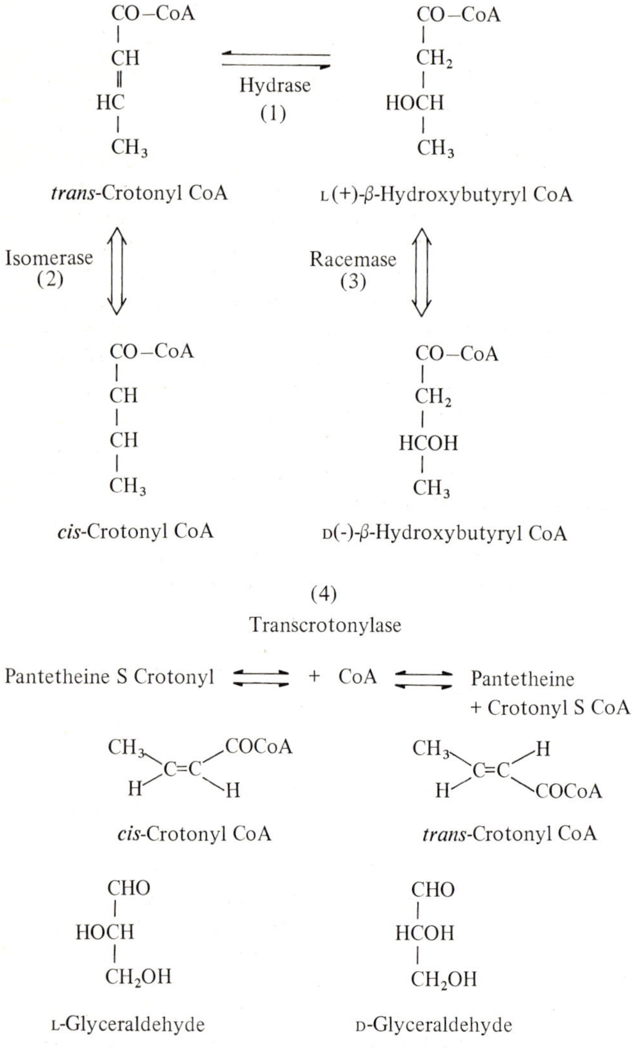

CO–CoA | CH ‖ HC | CH_3 ⇌ Hydrase (1) ⇌ CO–CoA | CH_2 | HOCH | CH_3

trans-Crotonyl CoA   L(+)-β-Hydroxybutyryl CoA

Isomerase (2) ⇕   Racemase (3) ⇕

CO–CoA | CH | CH | CH_3 CO–CoA | CH_2 | HCOH | CH_3

cis-Crotonyl CoA   D(-)-β-Hydroxybutyryl CoA

(4)
Transcrotonylase

Pantetheine S Crotonyl ⇌ + CoA ⇌ Pantetheine + Crotonyl S CoA

CH_3(H)C=C(H)COCoA CH_3(H)C=C(COCoA)H

cis-Crotonyl CoA   *trans*-Crotonyl CoA

CHO | HOCH | CH_2OH CHO | HCOH | CH_2OH

L-Glyceraldehyde   D-Glyceraldehyde

Fig. 1-24. Reactions catalyzed by crotonase

The second concept essential to an appreciation of the specificity of crotonase is that of geometrical isomerism. Remember that when single carbon-to-carbon bonds are formed, the four valence bonds of the carbon atoms rotate freely, but when a double bond is introduced the free rotation is lost. Because of these spatial restrictions, two types of isomers (*cis* and *trans*) may exist if each carbon contains two or more different substitutes. The *cis* and *trans* forms have been presented in Fig. 1-24. This form of stereoisomerism of crotonyl CoA should not be confused with optical isomerism, which is conferred upon the molecule by the addition of a hydroxyl group to the carbon. Such a substitution makes the carbon asymmetrical, and the compound containing it deviates polarized light to the right (dextrorotatory) or left (levorotatory) depending on the spatial position of the hydroxyl in the molecules. The symbol (−) is used to indicate that the compound is dextrorotatory, and (+) indicates levorotatory.

The absolute structures of such molecules can be related to the structure of glyceraldehyde, and they are accordingly referred to by the symbols (see Fig. 1-24). There is no constant correlation between the direction of the optical rotation and the absolute structure of the molecule. Thus, a compound may be dextrorotatory and have an absolute formula related to that of the glyceraldehyde. With these concepts in mind, the specificity of crotonase can be easily understood.

Crotonase hydrates both the *cis* and the *trans* forms of crotonyl CoA. The enzyme, however, can differentiate the two substrates because the rate of the reaction differs depending on the geometrical isomerism of the substrate. Thus, the hydration of the *trans* form is much more rapid than that of the *cis* form. When the substrate is the *cis* form of crotonyl CoA, the product is the D(−) hydroxyacyl-CoA, while the L(+) hydroxyacyl-CoA is formed when the *trans* form is hydrated.

In addition to "hydrase" properties, crystalline crotonase is believed to possess three different catalytic properties: it acts as an isomerase that converts the *cis* to the *trans* form; it catalyzes the racemization of the D(−) hydroxybutyryl CoA to the L(+) hydroxybutyryl CoA (racemase activity); and it catalyzes the transfer of crotonyl-S-pantetheine to CoA to yield crotonyl CoA (transcrotonylase activity). This association of several catalytic properties with a single crystalline protein suggests either that crystallization is no absolute criterion for purity or that pure protein may present several rather different catalytic properties, each of which may be elicited at different active sites.

Hydroxyacyl Dehydrogenase

The oxidation of the L(−) hydroxy fatty acyl-CoA of chain lengths ranging from four to twelve carbons is catalyzed by an enzyme first discovered in mitochondria and later crystallized. The enzyme requires NAD as a coenzyme, and small amounts of NAD are associated with the crystalline molecule. The reaction catalyzed by hydroxyacyl dehydrogenase is reversible, and the direction of the reaction changes with the pH of the medium. The enzyme preferentially reduces the ketoacyl at a low pH, whereas it favors the oxidation of the hydroxyacyl at a high pH. This effect of pH on the direction of the enzyme reaction suggests that changes in pH may influence the regulation of fatty acid metabolism. Unfortunately, it is not known to what extent intracellular pH is changed or to what extent such changes affect the native enzyme.

The enzyme hydroxyacyl dehydrogenase described above is specific for the L-isomer. Apparently, some mammalian tissue can also oxidize the D-isomer, but it is not clear what enzymic mechanism is responsible for that reaction. Although the presence of an enzyme that specifically catalyzes the oxidation of the D-hydroxyacyl ester to yield the keto acid has been proposed by some, others believe that the D-hydroxyacyl is transformed to the L-hydroxy acid by enzymes with racemase activity—namely, crotonase and another racemase. The equilibrium of the enzyme reaction is modified by the presence of magnesium in the medium. The modification of the equilibrium probably results from the complexion of magnesium with the keto acid. Eliminating the product favors the formation of the hydroxyacyl. Hydroxybutyrate can also be oxidized by an enzyme found in the mitochondria of many tissues, such as brain, kidney, heart, and liver. Hydroxybutyrate dehydrogenase has been isolated, solubilized, purified from beef heart, and demonstrated to require lecithin for activity.

Ketoacyl Thiolase

The oxidation of the L(−) hydroxy fatty acyl-CoA is catalyzed by ketoacyl thiolases. An enzyme has been isolated from beef and sheep liver mitochondria that catalyzes such a reaction on fatty ketoacyl with chain lengths of four to eighteen carbons. The enzyme molecule contains SH groups that are essential for activity. The cell probably contains more than one thiolase. A thiolase specific for acetoacetic acid has been found and is discussed in the section on ketosis.

Intracellular Distribution of Fatty Acid Oxidation

Now that we have completed this analytical examination of the enzymes involved in fatty acid oxidation, it is possible to take a synthetic view of the pathway as it is outlined in Fig. 1-23. On examination of that figure, it is obvious that two metabolites—ATP and oxaloacetate—provided by the Krebs cycle, are needed for proper functioning of the fatty acid oxidation pathway. ATP is required in the thiokinase reaction where the free energy of the pyrophosphate is used to activate the fatty acid, and oxaloacetate escorts acetyl-CoA through the Krebs cycle.

In addition to acetyl-CoA, the products of fatty acid oxidation include four electrons that must be transferred to the electron transport chain. The links between that chain and the oxidation pathway are the flavoprotein ETF for the electrons liberated during the acyl dehydrogenase reaction and NAD for the electrons liberated in the hydroxyacyl dehydrogenase reaction.

The electrons freed during the oxidation of fatty acids or of the Krebs cycle metabolites are ultimately transferred to the electron transport chain. This transfer can best be achieved if the oxidizing system and the electron transport chain are maintained in close contact in their cellular structures. It is therefore not surprising that, with few exceptions, all the enzymes of the fatty acid oxidation pathway are found in mitochondria. It has not yet been possible to reconstruct the exact pattern of the integration of each of these enzymes within the mitochondrial structure. Available information suggests only that these enzymes are not freely soluble within the mitochondrion. However, they are not as tightly bound to the mitochondrial structure as the enzymes of the electron transport chain.

Branched Fatty Acid Oxidation

The preceding discussion on fatty acid oxidation was concerned with the degradation of straight-chain, saturated fatty acids of carbon chain lengths ranging from four to eighteen carbons. However, living organisms also contain straight-chain fatty acids up to C_{22}, branched fatty acids, and unsaturated fatty acids. Other mechanisms different from classical oxidation have been described, namely α and ω oxidations.

The branched fatty acids the oxidation of which will be considered now are carbon compounds derived from amino acid metabolism. Isovalerate, methylbutyrate, and isobutyrate are related to the metabolism of leucine, isoleucine, and valine, respectively. The interest in the metabolism of branched fatty acids stems partly from the fact that the branched amino acids from which they are derived are precisely those that accumulate in maple syrup disease.

The pathway for the oxidation of isovaleryl CoA is outlined in Fig. 1-25. Isovaleryl CoA borrows several of the enzymes involved in the oxidation of straight-chain fatty acids. The novelty in the pathway involves the two last steps of the sequence of reactions: the CO_2 fixation and the cleavage of the molecule of β-hydroxymethyl glutaryl CoA. Lynen challenged this scheme because from the standpoint of organic chemistry it seemed difficult to explain the participation of hydroxyvaleryl CoA as a CO_2 acceptor. Lynen discovered another pathway in which methylcrotonyl CoA acts as the CO_2 acceptor in the presence of ATP and a carboxylase that requires biotin for activity. Methylglutaconyl CoA is the product of the carboxylase reaction. A specific hydrase, methylglutaconase, was found in mycobacteria and in chicken liver. β-Hydroxymethylglutaryl CoA is the product of the hydrase reaction. A cleavage enzyme found in liver requiring magnesium and glutathione for activity splits β-hydroxymethylglutaryl CoA to yield acetoacetic acid

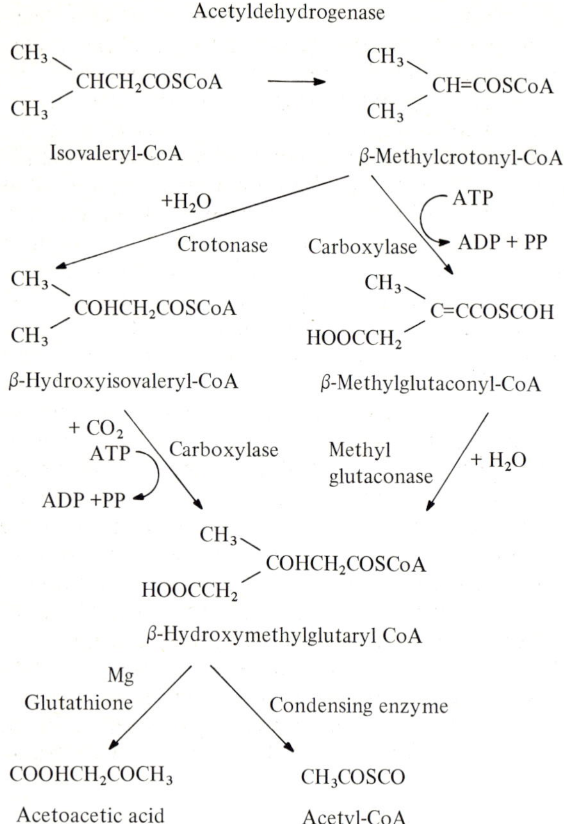

Fig. 1-25. Pathway of oxidation of isovaleric CoA

and acetyl-CoA. The reaction involved in the cleavage is reversible. Since free-energy changes favor the condensation of acetoacetic acid and acetyl-CoA to form hydroxymethylglutaryl CoA, the enzyme is also referred to as a condensing enzyme.

The production of acetoacetic acid in this pathway will undoubtedly awaken the pathology student's interest because acetoacetic acid is a key metabolite in ketosis and a precursor of mevalonic acid, and therefore of cholesterol.

The degradation reactions of methylbutyryl CoA are again catalyzed by the enzymes that are involved in the oxidation of the straight-chain fatty acids: acyl dehydrogenase, crotonase, hydroxyacyl dehydrogenase. Only the last step of the sequence of reactions requires a specialized enzyme, ketoacyl thiolase. The products of that reaction are propionic acid and acetyl-CoA.

At least two steps in the degradation of isobutyryl CoA—a deacylation and a dehydrogenation—involve enzymes that have not yet been discussed (see Fig. 1-26). The deacylase has been purified from pig heart. The enzyme's specificity is not restricted to hydroxybutyryl CoA; it also attacks propionyl CoA. The dehydrogenase has been obtained from various sources including mammalian tissues. The reaction's equilibrium favors reduction, but the enzyme's exact specificity remains unsettled.

The breakdown of the branched fatty acids brings us to the metabolism of propionyl CoA, the terminal product of the oxidation of the odd-number fatty acids. There are at least two pathways for the use of propionyl CoA—one is analogous to the oxidation of isobutyric acid, and the other involves CO_2 fixation. Although both pathways seem to exist in mammalian tissues, all available evidence indicates that most of the propionic acid is carboxylated to methylmalonic acid (see Fig. 1-27).

The arguments in favor of the existence of the malonyl semialdehyde pathway outlined in the upper part of figure include the following observations: the discovery of acrylyl CoA in bacteria; the fact that acrylyl CoA is a good substrate for crotonase; and the existence in bacteria and mammalian tissues of a specific hydroxypropionyl dehydrogenase. The enzyme is NAD-dependent for activity and has been purified 200 times. The product of the reaction is malonyl semialdehyde which can be either oxidized to malonate or split to yield acetyl-CoA and CO_2.

Propionyl CoA is also converted to methylmalonyl CoA by CO_2 fixation, and methylmalonyl CoA is in turn transformed to succinyl CoA. Two important

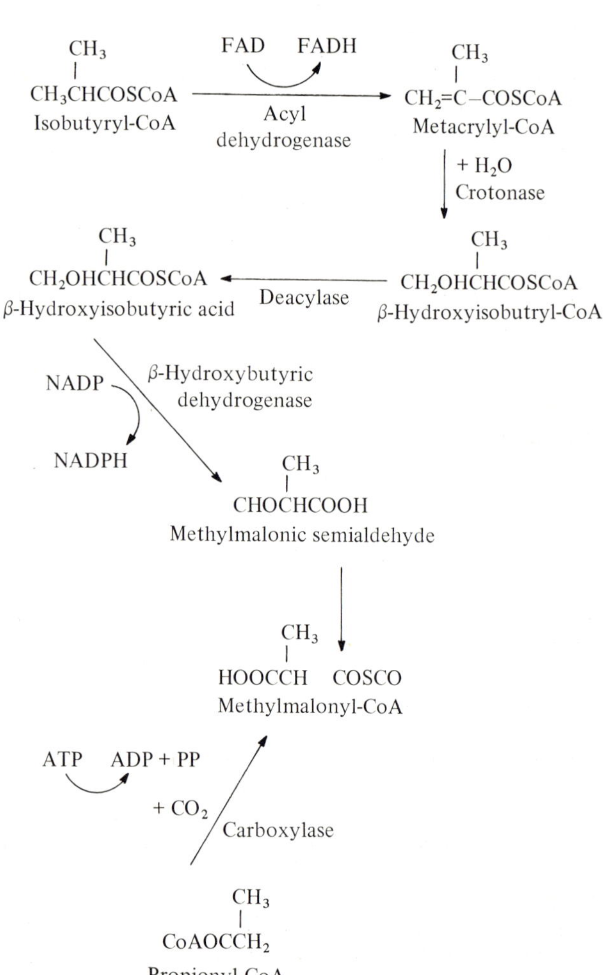

Fig. 1-26. Pathway of oxidation of isobutyryl CoA

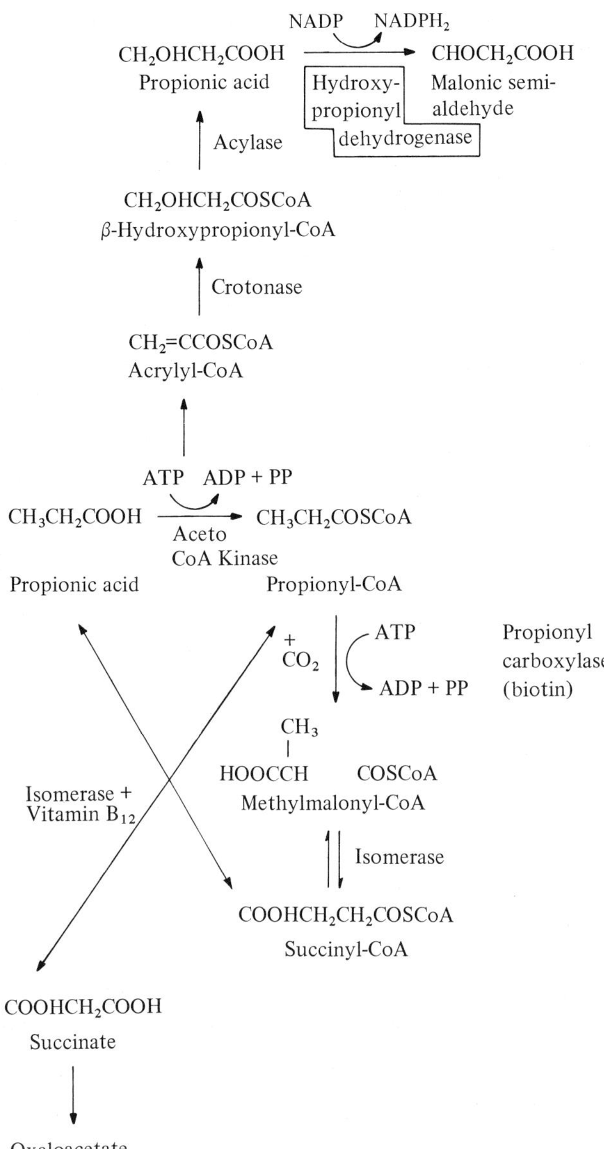

Fig. 1-27. Propionic acid metabolism

enzymes are involved in these conversions: propionyl carboxylase and methylmalonyl CoA isomerase. The carboxylation of propionic acid was first demonstrated with the aid of a mitochondrial extract. The enzyme responsible for the reaction, later crystallized from pig heart, has a molecular weight of 700,000. Both crude and crystalline enzyme preparations were found to be specific for propionyl CoA. The affinities of the enzyme for butyryl- and acetyl-CoA are 4 and 1%, respectively, of the affinity measured for propionyl CoA. Early experiments demonstrated that the ability of a chicken liver mitochondrial extract to form propionic acid is markedly decreased when the animals are given a biotin-deficient diet. The reaction catalyzed by mitochondrial extracts is inhibited by avidin. Biotin was later found to form a covalent bond with the crystallized enzyme, which contains four moles of biotin per enzyme molecule.

The propionyl CoA carboxylation is another reaction in which the energy required to synthesize a carbon-to-carbon bond is provided by ATP. Thus, the free energy of the pyrophosphate bond is transferred to CO_2, which forms an N-carboxy bond with the biotin enzyme complex. Some of the molecular details of the propionyl carboxylation reaction are presented in Fig. 1-27. The isomerization of methylmalonyl CoA to succinyl CoA also involves vitamin B_{12}. At first it was demonstrated that the activity of the isomerase was considerably decreased in the liver of vitamin B_{12}-deficient rats. Later the stimulating effect of 5,6-dimethylbenzimidazole carbamide coenzyme was demonstrated with partially purified preparations of the liver enzyme. The methylmalonate CoA-isomerase complex purified from sheep liver has been divided into two different protein fractions, one with isomerase and the other racemase activity. The isomerase acts on only one of the enantiomorphs of methylmalonyl CoA, but the absolute formula of the enantiomorph that serves as substrate for the isomerase is unknown. A racemase catalyzes the conversion of the alternate enantiomorph to the substrate of the isomerase.

A description of fatty acid oxidation is incomplete unless it includes an outline of ketone body metabolism. The fate of acetoacetic acid, however, is more appropriately discussed in the chapter on diabetes.

α- and ω-Oxidations

Mead and Lewis [160] have reported results that indicate that oxidation of some fatty acid may start with the α-carbon rather than with the β-carbon. Middle-chain and to a lesser degree long-chain fatty acids can be oxidized in this manner. The sequence of reactions that leads to complete oxidation of the α-carbon of the fatty acid is assumed to involve the intermediate formation of α-hydroxyl intermediates. But the enzymes catalyzing each intermediate step of the reaction have not been identified. In addition to an unknown protein fraction, a monamine oxidase, ferrous iron, and molecular oxygen appear to be needed.

$$\text{RCHOH COOH} + O_2 \xrightarrow[\text{Enzyme}]{Fe^{2+}} \text{R CO COOH}$$

$$\text{RCO COOH} \xrightarrow[\text{Enzyme}]{Fe^{2+}} \text{RCOOH} + CO_2 + H_2O + \text{Enzyme}$$

Refsum disease, also called heredopathia atactica polyneuritiformis, is a rare hereditary neurological disease that is transmitted as an autosomal recessive. It results from the absence of an enzyme (phytanic acid oxidase) which catalyzes the α-oxidation of phytanic acid (3,7,11,15-tetramethylhexadecanoic acid). Because of the methyl side chain in position 3 the β-carbon oxidation of phytanic acid is blocked and the breakdown of phytanic acid can only start through α-oxidation. The product of the α-oxidation is pristanic acid which can be further degraded by β-oxidation [154–196].

A third form of fatty acid oxidation has been described which involves the terminal methyl carbon. The reaction has been studied in brain, kidney, and plants. Although the details of the reaction are still being investigated, it appears that three steps are involved: (1) hydroxylation of the ω-carbon, which is catalyzed by the microsomal mixed-function oxidases and requires O_2 and reduced pyridine nucleotides, (2) formation of a complex between a specific NAD dehydrogenase and the ketoacid, and (3) decarboxylation of the terminal carboxyl in the presence of decarboxylase [196].

Fatty Acid Synthesis

As glycogen is an important reserve of chemical energy because it stores the molecules of glucose to be oxidized aerobically or anaerobically, fats are stores of fatty acids that will be converted to acetyl-CoA, propionic acid, or acetoacetic acid.

When biochemists started to untangle complex metabolic chains, the steps of the breakdown pathways were often elucidated before the piecemeal reconstruction of the biosynthetic puzzle. In most cases it was optimistically assumed that the synthetic pathways simply followed in reverse the steps of the breakdown pathways. But within a few decades, the young students of physiological chemistry observed to their dismay that the biosynthetic pathways are different from the breakdown pathways.

Many observations made in intact animals or *in vitro* suggested the existence of independent oxidative and biosynthetic pathways for fatty acid metabolism. One of the most interesting findings was made on liver slices of pancreatectomized cats. Two metabolic alterations have been observed in these preparations: (1) a disturbance of fatty acid oxidation leading to ketosis, and (2) retarded incorporation of labeled acetate into fatty acids. If, in addition to pancreatectomy, a hypophysectomy is performed on the cat (Houssay preparation), ketone bodies still accumulate in the liver slices, but incorporation of acetate into fatty acids is normal.

These results indicate that it is possible to design experimental conditions under which fatty acid oxidation is inhibited but fatty acid synthesis proceeds normally. Thus, these two pathways differ from each other in one or more steps of the sequence of reactions.

Further evidence obtained from intact animals suggested that more than one pathway for fatty acid synthesis might exist. After injection of 1-[^{14}C]acetate, palmitate is labeled throughout the molecule, but only the last carboxyl of stearate is labeled. These differences in labeling can now be explained: the random labeling results from the repeated condensation of 2-carbon compounds, and the terminal labeling results from a process known to occur in mitochondria and referred to as "elongation."

Elongation

At least two pathways are now known to exist in mitocheondria for fatty acid synthesis: one is biotin-insensitive and is responsible for the elongation of the fatty acid molecules by a 2-carbon unit, and another is biotin-sensitive and resembles the pathway discovered in liver supernatant fractions. The enzymes involved in fatty acid elongation have not yet been purified extensively. It appears that with the exception of the condensation of two carbons on palmitate to yield stearate, the elongation involves two reactions occurring at the same time—an elongation and a desaturation. Elongation has been demonstrated with intact mitochondria, acetone powders of mitochondria, and mitochondrial subparticles. It catalyzes the condensation of a 2-carbon chain with a medium- or long-chain fatty acid. In spite of the enzyme's broad specificity (C_8 to C_{20}), butyryl CoA and hexanoyl CoA are poor substrates. Not all the requirements of the reactions are known, but ATP, NADH, and NADPH are necessary for elongation. There are also indications using *in vitro* systems that pyridoxal phosphate may be involved in the reaction. After the enzyme system is passed through a charcoal column, part but not all the activity is lost, and the remaining activity can be stimulated 2.5 times by adding pyridoxal phosphate *in vitro*. This finding suggests that the elongation reaction differs from the classical thiokinase reaction. The results obtained after investigation of the elongation reaction in the pyridoxal-deficient animal remain inconclusive [158–162].

Fatty Acid Synthesis in Mitochondria

The exact contribution made by mitochondria to the overall rate of fatty acid synthesis in the cell is not known. Studies on heart homogenates suggested that heart sarcosomes are responsible for a great deal of fatty acid synthesis, and that the enzyme activity associated with the supernatant may well result from leakage out of the mitochondria in the course of their isolation.

Acetyl-CoA Carboxylase

All the enzymes involved and all the intermediates formed in each individual step of fatty acid synthesis are not known, but the following observations are now well established: the first intermediate in the biosynthetic process is malonyl CoA, which in the presence of liver supernatant is rapidly converted to higher fatty acids.

Acetyl-CoA carboxylase is the enzyme that catalyzes the condensation of acetyl-CoA and CO_2. Acetyl-CoA carboxylase is found in the cell supernatant. However, if the supernatant fluid is centrifuged with very high centrifugal forces, a pellet that contains the CoA carboxylase is sedimented. In the discussion of the intracellular distribution of the enzymes involved in glycolysis, it was stated that the concept of a "structureless supernatant" was a relative one, and that centrifugation at higher forces might indicate that many of the multiple enzymes discussed are in reality associated with small particles.

Acetyl-CoA carboxylase has now been purified from a number of tissues, including mammalian, and the enzyme exists in the form of inactive subunits and active polymers. For example, in avian liver the smallest subunit has a molecular weight of 410,000, the largest 8,000,000. Electrophoresis of purified rat liver enzymes has yielded 3 bands whose molecular weights are 118,000, 125,000, and 215,000, respectively. The role of the subunit is not clear, but the discovery that each mole of rat enzyme contains 2 moles of phosphate suggests that polymerization may be associated with phosphorylation, as is the case for phosphorylase.

The specificity of the acetyl-CoA carboxylase is not strictly restricted to acetyl-CoA; propionyl CoA can also function as substrate. The enzyme has rigid requirements for manganese that cannot be replaced by any other metal. ATP is also necessary for the reaction, and the amount of ATP converted to ADP and the amount of malonyl CoA formed are stoichiometrically related. UTP can substitute for ATP, provided that high concentrations of the uridine nucleotide are added to the incubation mixture. The carboxylase reaction is blocked by avidin, and the enzyme preparation extensively purified from chicken livers (mol wt 500,000–750,000) contains 1 mole of biotin per mole of enzyme.

Because the activity of the acetyl-CoA carboxylase was found to be much lower than that of other enzymes involved in fatty acid synthesis, the conversion of acetyl-CoA to malonyl CoA was believed to be rate limiting for fatty acid synthesis. Whether this is so is now in doubt because the earliest experiments were not done under optimal conditions for activity. When such conditions were achieved, the level of activity of the carboxylase was not different from that of the fatty acid synthetase multiple-enzyme complex.

However, there is evidence that the carboxylase activity can be regulated in two ways: allosterically and through *de novo* synthesis. Although the requirements for the modulation of activity depend somewhat upon the enzyme source, activity is stimulated by citrate (probably by causing polymerization of the enzyme). In contrast, palmityl CoA inhibits the enzyme, and the inhibition is competitive with citrate. The details of the allosteric regulation of this complex molecule are not known. The enzyme activity increases markedly in rats fed fat-free diets and in weaning rats. (The enzyme increase does not occur in minimal deviation hepatomas.) Immunological techniques and studies with labeled amino acids have shown that such an increase results from *de novo* synthesis rather than accelerated breakdown or allosteric activation. Acetyl-CoA carboxylase activity as well as fatty acid synthetase and ATP lyase activities are consistently high in genetically obese, hyperglycemic mice, but the enzyme activity is depressed in alloxan-diabetic rats or rats fed a high-fat diet. The role of biotin as a prosthetic group is discussed in the chapter on vitamins.

Conversion of Malonyl CoA to Fatty Acid in Liver

A second multiple-enzyme system present in liver cytosol catalyzes the conversion of malonyl CoA to fatty acids. Among the requirements for the reactions are acetyl-CoA and NADPH; the products are palmitic acid, CO_2, CoASH, and NADP.

In vitro incorporation studies with labeled acetyl-CoA and malonyl CoA suggested the following stoichiometry:

$$CH_2COSCoA + 7HOOCCH_2COSCoA + 14NADPH + 14H^+$$

$$CH_3(CH_2)14COOH + 7CO_2 + 8CoASH + 6H_2O + 14NADPH$$

On the basis of such stoichiometry, it is possible to outline a scheme for the biosynthesis of palmitic acid by repeated condensation of a 2-carbon chain. Malonyl CoA provides the 2-carbon chain needed in each step of the elongation of the fatty acid chain, except the last. Acetyl-CoA provides the two last carbons of the chain.

$$CH_3COOH + COOHCH_2COCOSCoA \rightarrow CO_2 + C_4$$
$$C_4 + COOHCH_2COCOSCoA \rightarrow CO_2 + C_6$$
$$C_6 + COOHCH_2COCOSCoA \rightarrow CO_2 + C_8$$
$$C_8 + COOHCH_2COCOSCoA \rightarrow CO_2 + C_{10}$$
$$C_{10} + COOHCH_2COCOSCoA \rightarrow CO_2 + C_{12}$$
$$C_{12} + COOHCH_2COCOSCoA \rightarrow CO_2 + C_{14}$$
$$C_{14} + COOHCH_2COCOSCoA \rightarrow CO_2 + C_{16}$$

In this sequence of reactions, malonyl CoA provides fourteen, and acetyl-CoA two of the carbons of palmitic acid.

The conversion of malonyl CoA to palmitic acid is a complex reaction that involves a multiple-enzyme system in the course of which several intermediates are formed. Much of our knowledge of the intricacies of that reaction stems from work by Wakil and Lynen and their associates. The available evidence suggests that the formation of palmitic acid involves the intermediate condensation of acetyl-CoA and malonyl CoA. This reaction is accompanied by the release of CO_2 and leads to the formation of a β-keto acid, which is in turn reduced to the β-hydroxy acid. The elimination of one molecule of water from the hydroxy acid yields the α-β-unsaturated fatty acid, which is saturated by further reduction.

The molecular mechanism of fatty acid synthesis can best be understood if the anabolism of fatty acid is first examined in bacteria. The fatty acid synthetase complex of bacteria presents two major advantages: it is readily solubilized, and it can be resolved into separate protein components. Three major stages of fatty acid synthesis can be described. First, acetyl-CoA reacts with an SH protein to yield bound acetyl derivatives and free CoA. This reaction is catalyzed by a fatty acid transacylase. Second, the acetyl *S*-protein complex condenses with the malonyl moiety of the malonyl CoA to yield a protein-bound acetoacetyl derivative. Malonyl transacylase catalyzes the formation of the acetoacetyl-protein complex. Third, the acetoacetyl S-protein complex is reduced and degraded in the presence of NADPH to yield the protein-bound butyrate derivative.

Because the bacterial enzymes are much more readily soluble than the mammalian, the steps of fatty acid synthesis have been elucidated in *E. coli*. The multiple-enzyme complex has been resolved into at least 7 protein components, 6 of which have specific enzyme activities. The other is referred to as the acyl carrier protein. The bacterial carrier protein (mol wt approximately 16,000) is rich in acidic residues and contains a 4′-phosphopantothenic prosthetic group, which, as we shall see, plays a key role in the binding of the acyl group. The prosthetic group is bound to a serine residue of the carrier protein. The carrier protein acts as an acceptor of the acetyl group attached to the SH group of CoA in a reaction catalyzed by an enzyme, acetyl-CoA-ACP transacylase:

$$CH_3CO\text{-}S\text{-}CoA + HS\text{-}ACP \leftrightharpoons CH_3CO\text{-}S\text{-}ACP + CoASH$$

Thus, in the reaction the acetyl group is transferred from the SH group of acetyl-CoA to a thioester linkage with the SH group of the 4′-phosphopantothenic moiety. The acetyl-CoA-ACP transacylase has been partially purified; although not absolutely specific for acetyl-CoA, it reacts poorly with other CoA derivatives, such as propionyl CoA. An acetyl enzyme intermediate is formed in the course of the reaction.

$$\text{acetyl-CoA} + E \leftrightharpoons \text{acetyl } E + CoA$$

$$\text{acetyl } E + ACP \rightleftharpoons \text{acetyl-ACP} + E$$

The transfer of the malonyl group from the SH group of malonyl CoA to ACP is catalyzed by malonyl CoA-ACP transacylase.

$$HOOCCH_2CO\text{-}S\text{-}CoA + HSACP \rightleftharpoons HOOCCH_2CO\text{-}S\text{-}ACP + CoA\text{-}SH$$

The malonyl-CoA-ACP transacylase has been purified to homogeneity from *E. coli* and has a molecular weight of 36,500. A malonyl enzyme intermediate is formed, and the malonyl moiety is known to be attached to a serine residue of the enzyme.

$$\text{malonyl CoA} + E \rightarrow \text{malonyl } E + CoA$$

$$\text{malonyl } E + ACP \rightarrow \text{malonyl ACP} + E$$

Although their exact mechanism of action is not known, compounds containing SH groups (e.g., dithiothreitol) stimulate enzyme activity.

A β-ketoacyl-ACP synthetase condenses acetyl ACP and malonyl ACP to form acetoacetyl ACP, CO_2, and ACP. The enzyme has been crystallized and has a molecular weight of 66,000.

$$HOOCCH_2CO\text{-}S\text{-}ACP + CH_3CO\text{-}SACP \rightleftharpoons CH_3COCH_2CO\text{-}S\text{-}ACP + CO_2 + ACP$$

The reaction occurs in two steps:

$$\text{acetyl S-ACP} + HS\text{-}E \rightarrow \text{acetyl-S-E} + ACP\text{-}SH$$

$$\text{acetyl-S-E} + \text{malonyl-S-ACP} \rightleftharpoons \text{acetoacetyl-S-ACP} + HS\text{-}E + CO_2$$

Cysteine residues seem to be essential to enzyme activity and are involved in acetyl binding.

The NADPH-dependent reduction of the β-ketoacyl-ACP to the D(−) β-hydroxyacyl-ACP is catalyzed by β-ketoacyl-ACP reductase.

$$CH_3COCH_2CO\text{-}S\text{-}ACP + NADPH \rightleftharpoons CH_3CHOHCH_2\text{-}CO\text{-}S\text{-}ACP + NADP$$

Saturated and unsaturated fatty acids of various lengths are reduced by the enzyme.

The dehydration is catalyzed by the β-hydroxyacyl-ACP dehydrase, an enzyme which again has been extensively purified in bacteria.

$$CH_2CHOHCH_2CO\text{-}S\text{-}ACP \rightleftharpoons H_2O + CH_3CH = CHCO\text{-}S\text{-}ACP$$

$$\text{acetyl S-ACP} + HS\text{-}E \rightarrow \text{acetyl-S-E} + ACP\text{-}SH$$

$$\text{acetyl-S-E} + \text{malonyl-S-ACP} \rightleftharpoons \text{acetoacetyl-S-ACP} + HS\text{-}E + CO_2$$

$$CH_3COCH_2CO\text{-}S\text{-}ACP + NADPH \rightleftharpoons CH_3CHOHCH_2\text{-}CO\text{-}S\text{-}ACP + NADP$$

$$CH_2CHOHCH_2CO\text{-}S\text{-}ACP \rightleftharpoons H_2O + CH_3CH = CHCO\text{-}S\text{-}ACP$$

The reduction of the *trans*-2-unsaturated acyl-ACP thioester is catalyzed by the enoyl-ACP-reductase, which has again been partially purified from *E. coli.*

$$CH_3\text{-}CH = CHCO\text{-}S\text{-}ACP + NADPH \rightleftharpoons CH_3CH_2CH_2CO\text{-}S\text{-}ACP + NADP$$

Fatty acid synthesis is believed to take place in mammalian cells according to a scheme similar to that described in bacteria, except for minor differences.

Fatty Acid Synthetase in Yeast and Avian and Mammalian Tissues

Spectral analysis of the incubation mixture that catalyzes fatty acid synthesis *in vitro* has demonstrated the presence of a typical β-ketothiolester bond that probably appears during the formation of protein-bound acetoacetate. As soon as NADPH is added to the system, the spectral characteristics of the β-ketothiolester disappear. This change in the spectrum suggests that the protein-bound keto acid is reduced to the β-hydroxy acid. Reductases catalyzing such a reaction have been found in yeast by Lynen's group and in bird livers by Walker's group.

Mammalian multiple-enzyme complexes have been purified from rat liver (540,000 dalton), rat mammary tissue (478,000 dalton), and rabbit mammary gland (910,000 dalton). No active subunits have as yet been prepared.

As will become obvious later, fatty acid synthesis is constantly modulated by dietary factors and hormones. If fatty acid synthesis is not regulated at the level of acetyl-CoA carboxylase, regulation must affect the synthetase; allosteric and adaptive regulation of the synthetase has been described, but these findings will remain difficult to interpret until more is known about the protein involved in the individual steps of fatty acid synthesis in mammalian cells.

The Mitochondrion

Function and Structure

Although the cell contains several enzymic systems generating energy, most cellular energy is provided by the intramitochondrial oxidation of the Krebs cycle substrates linked to the phosphorylation of ATP in the electron transport chain. In addition to their role in oxidative phosphorylation, mitochondria are actively involved in ion transport. The ion exchange is discussed in other parts of this book [163–174].

Early histologists interpreted the role of cytoplasmic structure on the basis of observations made on fixed material. They described fine reticular, granular, or fibrillar structures and elaborated complex theories, most of which were rejected when *in vivo* observation became available. Such observations led to the discovery of an important cytoplasmic constituent—the chondriome.

With proper staining, mitochondria appear as granules or rodlike structures, both types existing in the same cell. Under direct illumination and by examination of living cells with an oil immersion objective, the mitochondria appear as refractile filaments parallel to the long axis of the cell. Similar structures can be observed on examination of living cells with the phase contrast microscope or after supravital staining with methylene blue or Janus green.

Electron micrographs of the mitochondria demonstrate three main components: a mitochondrial membrane surrounding a clear mitochondrial space containing internal membranes or septa (see Figs. 1-28 and 1-29). The concept of individual mitochondria has recently been challenged by Hoffmann and Avers, who through three-dimensional reconstruction of the ultrastructure of yeast mitochondria have shown that what was believed to be separate organelles are in reality fragments of a single branching tubular structure 50–60 μm in length and 200–600 nm in diameter. If such findings are confirmed, they will necessitate review of accepted notions of mitochondrial function and biogenesis [175].

In view of the osmotic properties of mitochondria, the structure and biochemical composition of the outer membrane is important. After proper fixation, the membrane is seen to be composed of three layers: two electron-dense layers enclosing an electron-transparent layer. The dense layers are thinner (50 A) than the transparent (60–80 A). This triple layering is analogous to what was observed for the cell membrane. Sjöstrand has suggested that the clear central space is made of a lipid bimolecular layer, and the two outer layers are made of proteins (see Cell Membranes).

Careful examination of each denser layer of this triple-layer membrane reveals that each dense layer can be resolved into two thinner layers enclosing a clear internal space.

Within the mitochondrial cavity, which is delimited by the mitochondrial outer membrane, are triple-layered septa. The total thickness of these septa is approximately 180 A. They are composed of two dense layers 60 A thick enclosing a clear interspace 60 A thick.

Various relationships between the inner membrane in the mitochondrial cavity and the outer membrane delimiting the mitochondria have been described. Inside the mitochondria, the inner membrane may form independent septa or lamellae in a cavity that remains continuous. Another possibility is that the outer and inner membranes are in fact intimately connected. In that case, the septa may either form infolds of the three layers of the mitochondrial membrane or may just be projections of the inner electron-dense layer of the outer mitochondrial membrane. It becomes important to determine whether these projections end abruptly at some distance from the outer membrane or if they extend throughout the entire width of the mitochondrial cavity to form complete septa separating the main mitochondrial cavity into numerous smaller cavities.

The relationship between inner and outer membranes has been investigated by various methods, including osmotic change and electron microscopic studies. The osmotic behavior of mitochondria can best be explained by the existence of a bimolecular lipid layer, the internal monomolecular layer of which forms infolds. This interpretation implies that the cristae are infolds of the inner, denser layer of the mitochondrial membrane and could explain the shortening and disappearance of the cristae during swelling of isolated mitochondria. To increase its surface, the mitochondrial membrane would have to reduce the density of its infolds. This view is not supported by the electron microscopic studies of Andersson-Cedergren [176], who studied serial sections of mitochondria and reconstructed one mitochondrion in three dimensions. He concluded that the inner membranes are discs with irregular outlines because there is no continuity between the middle clear space of the central discs and the middle clear space of the outer membrane. Furthermore, he believes that the continuity between the two clear spaces observed in a few minute areas is due to fixation artifacts. The discs are not completely autonomous. Small round or ovoid structures were observed within the mitochondria and were thought

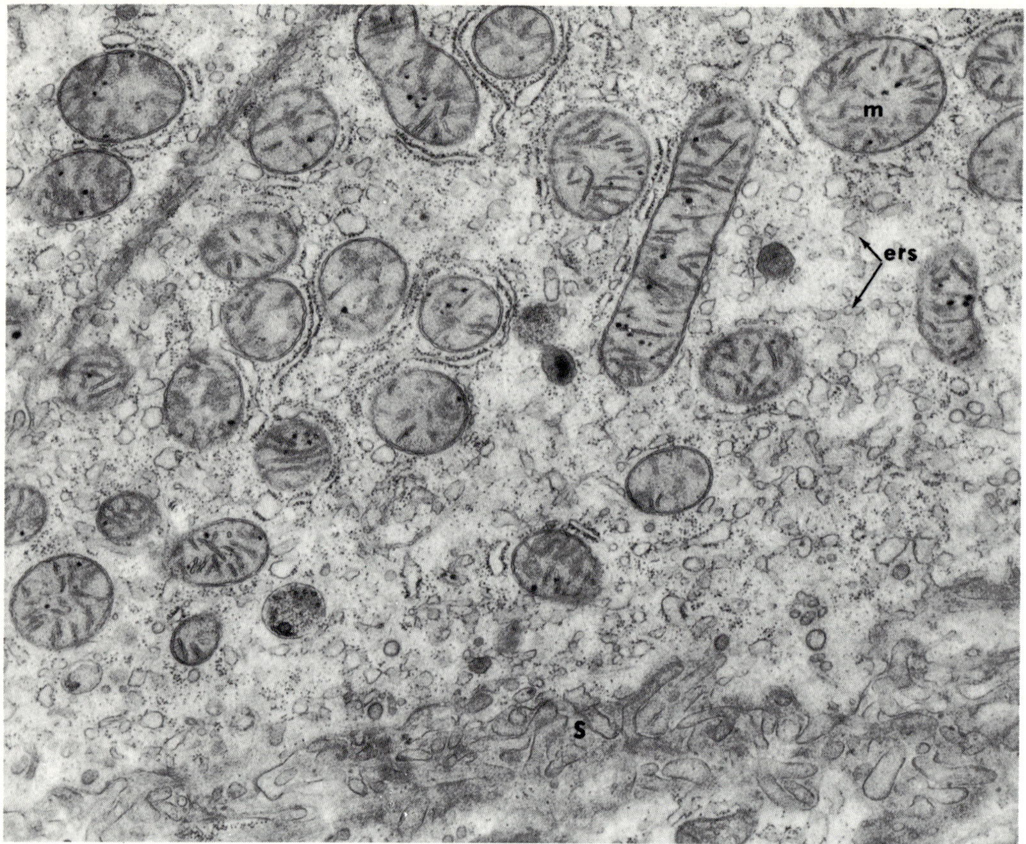

Fig. 1-28. Mitochondria in liver cell (low resolution). *m* mitochondria; *ers* smooth endoplasmic reticulum; *S* space of Dische. (From Trump)

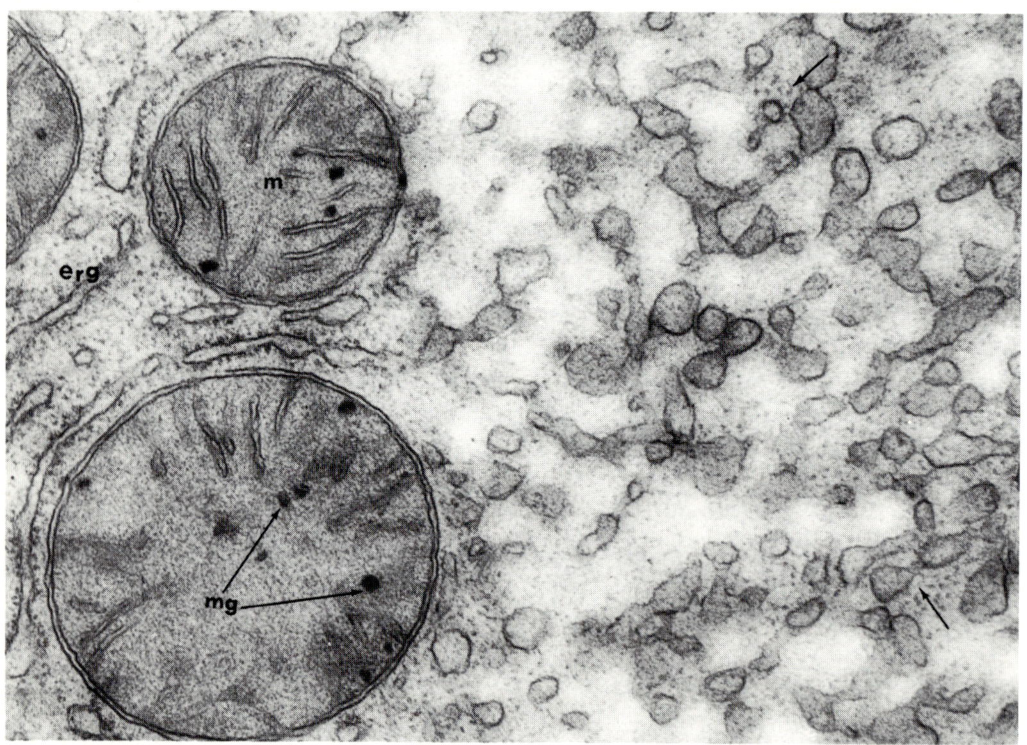

Fig. 1-29. Mitochondria (high resolution). *m* mitochondria; *mg* mitochondrial granules. (From Trump)

to be cross sections through narrow processes linking the inner discs to the outer membrane. In view of these disagreements, it seems safe to conclude that the relationship between the outer and inner membrane remains unresolved.

All mitochondria have internal membranes. Independent of the important roles that these internal membranes may play in mitochondrial physiology, either by subdividing the mitochondria into numerous small cavities or by increasing the reaction surface of the enzyme, they also constitute the morphological criteria for mitochondrial identification.

The disposition of the internal membrane is not alike in all mitochondria—they vary in direction, number, and branching. Although they are frequently perpendicular to the long axis of the mitochondria, they are occasionally parallel to the long axis. Some mitochondria have few internal membranes; others, such as muscle mitochondria, have large and dense internal membranous components. Finally, the internal lamellae of some mitochondria, particularly some muscle mitochondria, form arborizations that give rise to reticular structures inside the mitochondrial cavities.

There seems to be no doubt that the cristae play an important functional role, for the density of the intramitochondrial lamellation correlates with physiological function, the lamellation being more abundant in tissues (skeletal muscle) with the greatest energy expenditures. Furthermore, there is a correlation between the number of cristae and the amount of cytochromes found in the mitochondria.

The picture obtained by electron microscopy is static, whereas the real mitochondrion is a dynamic body that bends or straightens its structure and can be displaced throughout the cytoplasmic field. Motion pictures of mitochondria have even demonstrated mitochondrial fragmentation in both long and short axes and fusion of small mitochondria to form larger mitochondria. Mitochondria can travel from the cellular to the nuclear membrane where they appear to come in close contact with the nucleolus. The cause and the significance of these movements remain unknown.

Inner and outer mitochondrial membranes are specialized modifications of a large family of membranes that includes the nuclear and the plasma membranes and the membranes of the endoplasmic reticulum. Mitochondrial membranes function as selective barriers that are also capable of active transport and provide a framework on which catalytic proteins (*e.g.*, those of the respiratory chain) are tightly inserted in a pattern compatible with maximum efficiency. Like other membranes, mitochondrial membranes are composed of lipids and protein molecules—the fundamental building blocks. However, some features are unique to the mitochondria. For example, the mitochondrial membranes contain a lipid cardiolipin not found in the plasma membrane.

Mitochondria have not been excluded from the great membrane controversy in which the notion of a unit membrane is opposed to that of a membrane composed of subunits (see section on cell membranes). In fact, it is from studies on mitochondria that the concept of the subunit membrane emerged. Clearly the respiratory chain constitutes a functional unit, but whether it is an identifiable morphological unit is not certain. Separation and reaggregation of some portion of the respiratory chain as it has been described previously in Green's laboratory would seem to support the notion of a functional subunit, provided of course that none of the components used in the reaggregation are themselves membranous in nature. But it has been argued that the preparations used for the reaggregation do, in fact, contain membrane factors.

The inner particles, which at one point were believed to contain cytochromes, are now known to be devoid of respiratory enzymes and to contain a single enzyme—ATPase—which can be separated from and rebound to the membrane.

Although a great deal is known about the chemical composition of the mitochondrial membrane and it is established that the membrane contains a number of catalytic proteins (*e.g.*, the ATPase synthetase system, an ion transport molecular machinery and electron transport chain), the topological distribution of these proteins in the membrane is not known. All topological models proposed are at present hypothetical [177]. However, it is accepted that the mitochondrial membrane, like most if not all biological membranes, is of the "fluid mosaic model" and is composed of a lipid bilayer traversed by proteins (see plasma membrane in Chapter 16). Electron microscopic studies of the freeze-edge fractured faces of the outer and the inner membrane [178] indicate that the proteins are asymmetrically distributed not only when the inner is compared to the outer membrane, but also when the inner and outer faces of each of the fractured membranes are compared (Table 1-3).

Table 1-3. Number of particles per μ^2 seen on freeze-fractured faces

	Outer Face	Inner Face
Outer membrane	2,806	770
Inner membrane	2,120	4,208

Origin

After cell division, both daughter cells inherit mitochondria, and new mitochondria are formed before cell division occurs [179–183].

So much time was devoted to the study of the bioenergetic properties of mitochondria that investigation of their biosynthetic properties was neglected. But between 1960 and 1970 much evidence was assembled, which indicated that mitochondria possess an independent replicative system. Early morphological observation of live mitochondria with the phase microscope suggested that mitochondria divide. The presence of small amounts of RNA in association with mitochondrial fractions was considered to be in agreement with the autoduplication theory. Interest in autoduplication grew further when it was observed that mitochondria incorporate amino acids because this suggested that mitochondria were capable of synthesizing proteins. Work in the last decade has established that mitochondria possess the complete machinery for protein biosynthesis including the templates DNA and RNA.

The suspicion that mitochondria are equipped with the enzymic arsenal necessary for protein synthesis came from the observation that labeled amino acids are incorporated into proteins by mitochondria. Because of the difficulties involved in obtaining mitochondrial preparations devoid of nuclear and endoplasmic reticulum contaminants, the objection was at first raised that amino acid incorporation in mitochondrial proteins results from the presence of contaminating ribosomes and polyribosomes. Simpson claims that this is unlikely, at least in the case of muscle mitochondria, for the following reasons. First, these mitochondria were washed several times before incubation. Second, electron microscopic examination and reference-enzyme analysis indicated that the mitochondria were virtually free of microsomal contaminant. Third, the incorporation of amino acid in mitochondria resembles that observed in microsomes inasmuch as it is energy-dependent, but differs in that it is inhibited by chloramphenicol but not by ribonuclease. Mitochondria resemble bacteria in that respect. However, one should be cautious in evaluating the significance of a difference in sensitivity to protein inhibition of cytoplasmic and mitochondrial protein-synthesizing systems. For example, although chloramphenicol inhibits protein synthesis in liver mitochondria, it does not affect protein synthesis in brain mitochondria. Cyclohexamide, which inhibits protein synthesis in cytoplasm, does not affect protein synthesis in yeast mitochondria. Protein synthesis of rat brain mitochondria is, however, inhibited by cyclohexamide. And, finally, in muscle preparations, mitochondria are the cell fractions most active in incorporating amino acids *in vitro*. For these reasons it can be concluded that if amino acid incorporation results from contamination by microsomes, it must be by a fraction that can incorporate amino acids into proteins much more readily than the ordinary muscle microsomal preparation. It is significant that the contamination of mitochondria by polyribosomes has not been excluded.

Bacteria contaminating and growing in the preparation during incubation could well be responsible for the amino acid incorporation observed with mitochondrial pellets. Liver homogenates contain a surprisingly large amount of bacteria, and attempts to maintain aseptic conditions seem to be ineffective in preventing contamination; yet it is generally agreed that the bacteria in the preparation cannot account for the amino acid incorporation that has been observed. One argument is that succinate or glucose is required for amino acid incorporation into the proteins of the bacteria. Maybe a more compelling argument in support of mitochondrial rather than bacterial protein synthesis is the fact that protein synthesis in mitochondrial preparation can be modified by administration of hormones or by stimuli to proliferation (partial hepatectomy), treatments that are not expected to affect protein synthesis in contaminating bacteria.

Since the discovery of amino acid uptake in mitochondria, several investigators have attempted to demonstrate the presence of nucleic acids in mitochondria.

Both DNA and RNA are necessary for protein synthesis in mitochondria. Electron microscopists had long ago claimed that mitochondria contain DNA, but biochemical confirmation was slow because of the difficulties in eliminating contamination with nuclear DNA.

Electron microscopists had seen long strands with a double helical structure, sensitive to DNase but resistant to RNase, inside mitochondria. Biochemists have isolated mitochondrial DNA. Enough differences have been shown between nuclear and mitochondrial DNA that the two types of DNA can no longer be confused. Mitochondrial DNA differs from nuclear DNA in its chemical properties, base composition, and turnover.

Mitochondrial DNA has a molecular weight of 9×10^6 to 10×10^6 daltons.

The exact amount of DNA in mitochondria from various sources is not known because of the many pitfalls in measuring DNA content accurately. What is certain is that the amount in mitochondria is always small compared to that in nuclei. The amount of DNA in mitochondria varies depending upon the source of the organelle. For example, although the DNA content of liver mitochondria is around 1.0–1.8 µg per mg of protein, in rapidly growing tissues (mouse cell fibroblast) it ranges from 2.5 (ascites tumor) to 5.5 (hepatoma) or even 8.2 (Walker carcinoma).

The base composition of mitochondrial and nuclear DNA differs only slightly, but in some cases the differences are large enough that the two types of DNA can be separated by their bouyant densities. (Bouyant density is related to the guanosine cytosine content of the DNA.) However, this is not the case for mammalian mitochondrial DNA, the buoyant density of which resembles that of nuclear DNA (nDNA) because of the small difference in base composition.

Although heat-denatured nDNA does not readily reanneal, heat-denatured mitochondrial DNA (mtDNA) does; therefore, the two types of DNA can be separated by sedimentation on cesium chloride gradients after denaturation.

A precursor is incorporated in mtDNA at rates different from its incorporation in nDNA. A typical example is liver. [^{3}H]thymidine is not markedly incorporated in nDNA of normal liver, but it is in mtDNA. Thus, the specific activity of mtDNA is more than 40 times that of nDNA. In regenerating liver, the specific activity of mtDNA increases before the increase in specific activity of nDNA. Therefore, the metabolism of mtDNA seems to be independent of that of nDNA, and Simpson and his colleagues were able to show that mitochondria contain a DNA polymerase with chromatographic properties different from the polymerase found in the cytoplasm. Moreover, this group was also able to establish that incorporation of precursors into mtDNA resulted from *de novo* synthesis rather than from DNA repair.

Gene Expression

Mitochondrial DNA of higher organisms is devoid of histone and occurs in the form of a closed circular duplex 4.7–5.5 microns in circumference [184].

A DNA polymerase different from that found in the nucleus is present in mitochondria. It binds to a single scission of the strand, and single nucleotides are one by one bound covalently to the template strand [185].

Thus, there are many similarities between mitochondrial and bacterial DNA polymerase. However, the mitochondrial polymerase does not exhibit 5′-exonucleolytic effects, and no hairpin structures are formed in the process of replication. As for replication *in vivo,* the template strand is displaced ahead of the growing point in the first stage of replication (see chapter on cellular specificity).

Little is known of the regulation of gene expression in mitochondria, but it is significant that mitochondrial DNA of higher organisms is devoid of histones.

RNA containing methylated guanine has been extracted from mitochondria assumed not to be contaminated by supernatant RNA. When HeLa cells are grown in presence of chloramphenicol, the rate of cell growth is reduced, but the mitochondria have a normal size with normal number and appearance of cristae except for a small percentage (4–5%) that have an unusual looking configuration [186]. Similar observations were made when mice embryos were grown in the presence of chloramphenicol [187]. The biochemical alterations of these mitochondria is not known, but the number of cytochrome oxidase-positive organelles were not decreased in mitochondria of HeLa cells grown in the presence of chloramphenicol [188].

In addition to the templates, mitochondria also contain RNA polymerase and ribosomes, suggesting that transcription and translation can occur inside of mitochondria. Similarly, transfer RNA (tRNA) and activating enzymes have been isolated from mitochondrial preparations. Mitochondrial ribosomes are somewhat different (70 S instead of 80 S) from those usually found in the cytoplasm.

Although there seems to be little doubt that mitochondrial DNA is coding for protein synthesis, it is also clear that it does not code for all proteins found in mitochondria. It is not certain what protein mitochondrial DNA codes for. It has been proposed that it might dictate the biosynthesis of a structural protein, but not all mitochondrial proteins are coded by mDNA.

A number of mitochondrial proteins, cytochrome c and glutamic dehydrogenase, malic dehydrogenase and leucyl tRNA synthetase, and all 53 structural proteins of mitochondrial ribosomes are known to be synthesized in the endoplasmic reticulum and transferred to mitochondria.

On the other hand, the mitochondrial genome is believed to code for the largest polypeptide of cytochrome oxidase, an enzyme that has been highly purified and has been found to be composed of seven subunits. When the enzyme is placed on polyacrylamide gel, the exact molecular weight of the subunits is not known, but it is believed to range between 8,600 and 35,000 [189–198]. (Mitochondrial DNA also codes for cytochrome b subunits.)

Similarly, there is evidence that the ATPase activity of *Saccharomyces cervisiae* is under the control of the mitochondrial genome [190].

RNA-DNA hybridization experiments have established that mitochondrial ribosomal RNA is coded for by mitochondrial DNA.

Studies in HeLa cell and rat liver cells have shown that various kinds of mitochondrial RNA are synthesized on mitochondrial DNA. Attardi and Aloni [191] have shown that the mitochondrial DNA molecules contain one gene for a 12-S RNA, 16-S RNA, respectively, and 11 genes for 4-S and 8-S RNA.

Even if the coding properties of mtDNA are limited, the biological implications are enormous. Not only does it shake the belief that all genetic control rests with nDNA, but it also allows for unexpected cellular metabolic individuality. Moreover, the finding of mitochondrial DNA raises the question of the existence of DNA with special coding properties in other parts of the cell, such as the cell membrane or the endoplasmic reticulum.

Thus, important questions emerge from these studies on mtDNA: Is the phenotypic appearance of a cell determined only by nuclear or by a combination of nuclear and cytoplasmic templates? Are some of the abnormal genetic traits transmitted from cell to cell by cytoplasmic genes? Both histochemists and biochemists have shown that liver and brain contain populations of mitochondria that differ from each other by their enzyme mosaic. As pointed out by Wagner [192], mitochondrial heterogeneity may simply result from the fact that at all times liver and brain contain mitochondria at various stages of development. Yet is has not been excluded that heterogeneity is the consequence of "genetic" differences. If so, does the mother cell make sure that after cell division the two daughter cells receive an identical population of mitochondria, or is the cytoplasm of the two daughter cells heterogeneous? When female Mexican armadillos deliver their litter, they almost always deliver quadruplets. In spite of the fact that the four embryos are derived from a single fertilized egg, marked individual differences are found in the biochemical mosaic of the offspring. Wagner asks whether these individual differences result from modification of the regulation of nuclear genes, or from distribution of cytoplasmic templates among the dividing cells.

References

1. De Duve, C., Berthet, J.: The use of differential centrifugation in the study of tissue enzymes. Int. Rev. Cytol. **3**, 225–275 (1954)
2. Novikoff, A.B., Podber, E.: The contributions of differential centrifugation to the intracellular localization of enzymes. J. Histochem. Cytochem. **5**, 552–564 (1957)
3. Schneider, W.C., Hogeboom, G.H.: Biochemistry of cellular particles. Ann. Rev. Biochem. **25**, 201–224 (1956)
4. Schneider, W.C., Hogeboom, G.H.: Cytochemical studies of mammalian tissues: The isolation of cell components by differential centrifugation: A review. Cancer Res. **11**, 1–22 (1951)
5. Cori, C.F.: Problems of cellular biochemistry. In: Currents in biochemical research (Green, D.E., ed.), p. 198–214. New York: Interscience Publishers 1956
6. Albright, J.F., Anderson, N.G.: A method for rapid fractionation of particulate systems by gradient differential centrifugation. Exp. Cell Res. **15**, 271–281 (1958)
7. Allfrey, V.: The isolation of subcellular components. In: The cell (Brachet, J., and Mirsky, A.E., eds.), vol. 1, p. 193–290. New York: Academic Press 1959
8. Green, D.E., Järnefelt, J.: Enzymes and biological organization. Perspect. Biol. Med. **2**, 163–184 (1959)
9. Hogeboom, G.H., Schneider, W.C., Striebich, M.J.: Localization and integration of cellular function. Cancer Res. **13**, 617–632 (1953)
10. De Duve, C.: Tissue fractionation, past and present. J. Cell Biol. **50**, 20–55 (1971)
11. Coleman, R.: Membrane-bound enzymes and membrane ultrastructure. Biochim. biophys. Acta (Amst.) **300**, 1–30 (1973)
12. Potter, V.R., Elvehjem, C.A.: A modified method for the study of tissue oxidations. J. biol. Chem. **114**, 495–504 (1936)
13. Claude, A.: Fractionation of mammalian liver cells by differential centrifugation. I. Problems, methods, and preparation of extract. J. exp. Med. **84**, 51–59 (1946)
14. Schneider, R.M., Petermann, M.L.: Nuclei from normal and leukemic mouse spleen. I. The isolation of nuclei in neutral medium. Cancer Res. **10**, 751–754 (1950)
15. Hogeboom, G.H., Kuff, E.L.: Sedimentation behavior of proteins and other materials in a horizontal preparative rotor. J. biol. Chem. **210**, 733–751 (1954)
16. Novikoff, A.B., Podber, E., Ryan, J., Noe, E.: Biochemical heterogeneity of the cytoplasmic particles isolated from rat liver homogenate. J. Histochem. Cytochem. **1**, 27–46 (1953)
17. Roodyn, D.B.: The enzymic properties of rat-liver nuclei. I. Estimation of the extent to which contaminant material contributes to the activity observed in the nuclear fraction. Biochem. J.**64**, 361–368 (1956)
18. Novikoff, A.B.: Biochemical heterogeneity of the cytoplasmic particles of rat liver. Symp. Soc. exp. Biol., No. X, 93–109 (1957)
19. Schneider, W.C., Hogeboom, G.H.: Intracellular distribution of enzymes. V. Further studies on the distribution of cytochrome C in rat liver homogenates. J. biol. Chem. **183**, 123–128 (1950)
20. Cori, C.F.: Enzymatic reactions in carbohydrate metabolism. Harvey Lect. Series **41**, 253–272 (1945–1946)
21. De Duve, C., Hers, H.G.: Carbohydrate metabolism. Ann. Rev. Biochem. **26**, 149–180 (1957)
22. Korkes, S.: Carbohydrate metabolism. Ann. Rev. Biochem. **25**, 685–734 (1956)
23. Racker, E., Gatt, S.: Interreactions of glycolysis and oxidative pathways; metabolic factors in cardiac contractibility. Ann. N.Y. Acad. Sci. **72**, 427–438 (1959)
24. Stumpf, P.: Glycolysis. In: Chemical pathways of metabolism (Greenberg, D.M., ed.), 1st ed., vol. I, p. 66–108. New York: Academic Press 1954
25. Leloir, L.F.: The interconversion of sugar in nature. In: Currents in biochemical research, p. 585–608. New York: Interscience Publishers 1960
26. Mehler, A.H.: Glycolysis. In: Introduction to enzymology, p. 37–69. New York: Academic Press 1957
27. Axelrod, B.: Glycolysis. In: Chemical pathways of metabolism (Greenberg, D.M., ed.), 2nd ed., vol. 1, p. 97–128. New York: Academic Press 1960
28. Scrutton, M.C., Utter, M.F.: The regulation of glycolysis and gluconeogenesis in animal tissues. Ann. Rev. Biochem. **37**, 249–302 (1968)
29. Beloff-Chain, A., Pocchiari, F.: Carbohydrate metabolism. Ann. Rev. Biochem. **29**, 295–346 (1960)
30. Crane, R.K.: Hexokinases and pentokinases. In: The enzymes (Boyer, P., Lardy, H., Myrback, K., eds.), vol. 6, p. 47–66. New York: Academic Press 1962
31. Walker, D.G.: The nature and function of hexokinases in animal tissues. In: Essays in biochemistry (Campbell, P.N., and Greville, G.D., eds.), vol. 2, p. 33–67. New York: Academic Press 1966
32. Lardy, H.: Phosphohexokinases. In: The enzymes (Boyer, P., Lardy, H., and Myrback, K., eds.), vol. 6, p. 67–74. New York: Academic Press 1962
33. Grossbard, L., Weksler, M., Schimke, R.T.: Electrophoretic properties and tissue distribution of multiple forms of hexokinase in various mammalian species. Biochem. biophys. Res. Commun. **24**, 32–38 (1966)
34. Sols, A., Crane, R.K.: Substrate specificity of brain hexokinase. J. biol. Chem. **210**, 581–595 (1954)
35. Crane, R.K., Sols, A.: The non-competitive inhibition of brain hexokinase by glucose-6-phosphate and related compounds. J. biol. Chem. **210**, 597–606 (1954)
36. Purich, D.L., Rudolph, F.B., Fromm, H.J.: The hexokinases: Kinetic, physical and regulatory properties. Advanc. Enzymol. **39**, 249–326 (1973)
37. Viñuela, E., Salas, M., Sols, A.: Preliminary communication: Glucokinase and hexokinase in liver in relation to glycogen synthesis. J. biol. Chem. **238**, 1175–1177 (1963)
38. Salas, M., Viñuela, E., Sols, A.: Insulin-dependent synthesis of liver glucokinase in the rat. J. biol. Chem. **238**, 3535–3538 (1963)
39. Sharma, C., Manjeshwar, R., Weinhouse, S.: Effects of diet and insulin on glucose adenosine triphosphate phosphotransferases of rat liver. J. Biochem. **238**, 3840–3845 (1963)
40. Walker, D.G., Rao, S.: The role of glucokinase in phosphorylation of glucose by rat liver. Biochem. J. **90**, 360–368 (1964)
41. Gonzáles, C., Ureta, T., Sánchez, R., Niemeyer, H.: Multiple molecular forms of ATP: Hexose 6-phosphotransferase from rat liver. Biol. Biochem. Res. Commun. **16**, 347–357 (1964)
42. Oliver, I.T., Cooke, J.S.: Rat liver-glucokinase activities in starvation. Biochim. biophys. Acta (Amst.) **81**, 402–404 (1964)
43. Najjar, V.A.: Phosphoglucomutase. In: The enzymes (Boyer, P., Lardy, H., and Myrback, K., eds.), vol. 6, p. 161–178. New York: Academic Press 1962
44. Warburg, O., Christian, W.: Isolierung und Kristallisation des Proteins des oxydierenden Gärungsferments. Biochem. Z. **303**, 40–68 (1939)
45. Schwert, G.W., Winer, A.D.: Lactate dehydrogenase. In: The enzymes (Boyer, P., Lardy, H., and Myrback, K., eds.), vol. 7, p. 127–148. New York: Academic Press 1963
46. Everse, J., Kaplan, N.O.: Lactate dehydrogenase: Structure and function. Advanc. Enzymol. **37**, 61–133 (1973)
47. Kaplan, N.O., Everse, J.: Regulatory characteristics of lactate dehydrogenases. Advanc. Enzyme Regulat. **10**, 323–336 (1972)
48. Hers, H.G.: Le metabolisme du fructose. Arscia, Bruxelles: Université Catholique de Louvain 1957
49. Horecker, B.L.: Aldol and ketol condensations. J. cell. comp. Physiol., Suppl. 110, **54**, 89–108 (1960)
50. Topper, Y.J.: Aldose-ketose transformation. In: The enzymes (Boyer, P., Lardy, H., and Myrback, K., eds.), vol. 5, p. 429–441. New York: Academic Press 1961
51. Lai, C.Y., Horecker, B.L.: Aldolase: A model for enzyme structure-function relationships. In: Essays in biochemistry (Campbell, P.N., and Dickens, F., eds.), vol. 8, p. 149–178. New York: Academic Press 1972
52. Holzer, H.: Carbohydrate metabolism. Ann. Rev. Biochem. **28**, 171–222 (1959)
53. Hers, H.G.: Triokinases. In: The enzymes (Boyer, P., Lardy, H., and Myrback, K., eds.), vol. 6, p. 75–83. New York: Academic Press 1962
54. Stetten, D., Jr., Stetten, M.R.: Glycogen turnover. In: Essays in biochemistry (Graff, S., ed.), p. 291–307. New York: John Wiley & Sons 1956
55. Robbins, P.W., Lipman, F.: Recent observations on the mechanism of glycogen synthesis in muscle. Ciba Symposium on Regulation of Cell Metabolism, p. 188–193 (1959)
56. Stetten, D., Jr., Stetten, M.R.: Glycogen metabolism. Physiol. Rev. **40**, 505–537 (1960)
57. Krebs, E.G., Fischer, E.H.: Molecular properties and transformations of glycogen phosphorylase in animal tissues. Advanc. Enzymol. **24**, 263–290 (1962)
58. Krebs, H.A.: Enzyme activity and cellular structure. In: Horizons in biochemistry (Kasha, M., and Pullman, B., eds.), p. 285–294. New York: Academic Press 1962
59. Sutherland, E.W., Wosilait, W.D., Rall, T.W.: The action of glucagon on liver phosphorylase. Ciba Found. Coll. Endocrin. **9**, 179–191 (1956)
60. Leloir, L.G., Goldemberg, S.H.: Synthesis of glycogen from uridine diphosphate glucose in liver. J. biol. Chem. **235**, 919–923 (1960)
61. Neilands, J.B., Stumpf, P.K.: The hexose monophosphate shunt. In: Outlines of enzyme chemistry, p. 258–264. New York: John Wiley & Sons 1955
62. Mehler, A.H.: Alternative pathways of carbohydrate metabolism. In: Introduction to enzymology, p. 116–137. New York: Academic Press 1957
63. Hollman, S.: Non-glycolytic pathways of metabolism of glucose (transl. and rev. by Touster, O.). New York: Academic Press 1964
64. Touster, O.: Carbohydrate metabolism. Ann. Rev. Biochem. **31**, 407–450 (1962)
65. Burns, J.J., Conney, A.H.: Water-soluble vitamins, part I. Ann. Rev. Biochem. **29**, 413–436 (1960)
66. Hollmann, S.: Metabolism of glucuronic acid and ascorbic acid. In: Non-glycolytic pathways of metabolism of glucose (transl. and rev. by Touster, O.), p. 83–114. New York: Academic Press 1964
67. Cabib, E.: Carbohydrate metabolism. Ann. Rev. Biochem. **32**, 321–354 (1963)
68. Bublitz, C., York, J.L.: Some observations on pig-kidney DPN-L-gulonate dehydrogenase. Biophys. biochim. Acta (Amst.) **48**, 56–60 (1961)
69. Krebs, H.A., Lowenstein, J.M.: The tricarboxylic acid cycle. In: Metabolic pathways (Greenberg, D.M., ed.), 2nd ed. of Chemical pathways of metabolism, vol. I, p. 129–203. New York: Academic Press 1960

70. Wiame, J.M.: Le role biosynthétique du cycle des acides tricorlxyliques. Advanc. Enzymol. **18**, 241–280 (1957)
72. Martius, C., Lynen, F.: Probleme des Citronensäurecyklus (Problems of citric acid cycle). Advanc. Enzymol. **10**, 167–222 (1950)
73. Ochoa, S.: Enzymic mechanisms in the citric acid cycle. Advanc. Enzymol. **15**, 183–270 (1954)
74. Reed, L.J.: The chemistry and function of lipoic acid. Advanc. Enzymol. **18**, 319–347 (1957)
75. Schmidt, U., Altland, K., Goedde, H.W.: Biochemistry and chemistry of lipoic acids. Advanc. Enzymol. **32**, 423–469 (1969)
76. Potter, V.R., Heidelberger, C.: Biosynthesis of "asymmetric" citric acid: a substantiation of the Ogston concept. Nature (Lond.) **164**, 180–181 (1949)
77. Ogston, A.G.: Interpretation of experiments on metabolic processes, using isotopic tracer elements. Nature (Lond.) **162**, 963 (1948)
78. Speyer, J.F., Dickman, S.R.: On the mechanism of action of aconitase. J. biol. Chem. **220**, 193–208 (1956)
79. Dickman, S.R.: Aconitase. In: The enzymes (Boyer, P., Lardy, H., and Myrback, K., eds.), vol. 5, p. 495–510. New York: Academic Press 1961
80. Schneider, W.C., Hogeboom, G.H.: Biochemistry of cellular particles. Ann. Rev. Biochem. **25**, 201–224 (1956)
81. Schneider, W.C., Hogeboom, G.H.: Intracellular distribution of enzymes IX: Certain purine-metabolizing enzymes. J. biol. Chem. **195**, 161–166 (1952)
82. Plaut, G.W.E., Sung, S.C.: Diphosphopyridine nucleotide isocitric dehydrogenase from animal tissue. J. biol. Chem. **207**, 305–314 (1954)
83. Chen, R.F., Plaut, G.W.E.: Activation and inhibition of DPN-linked isocitrate dehydrogenase of heart by certain nucleotides. Biochemistry **2**, 1023–1032 (1963)
84. Chance, B. (ed.): Energy linked function of mitochondria. New York: Academic Press 1963
85. Ciba Foundation Symposium: Quinones in electron transport (Wolstenholme, G.E.W., and O'Connor, M., eds.). Boston: Little, Brown and Company 1961
86. Kleiber, M., Rogers, T.A.: Energy metabolism. Ann. Rev. Physiol. **23**, 15–36 (1961)
87. Szent-Györgyi, A.: Introduction to a submolecular biology. New York: Academic Press 1960
88. Lehninger, A.L.: Respiration-linked mechanochemical changes in mitochondria. In: Horizons in biochemistry (Kasha, M., and Pullman, B., eds.), p. 421–436. New York: Academic Press 1962
89. Green, D.E., Fleischer, S.: On the molecular organization of biological transducing systems. In: Horizons in biochemistry (Kasha, M., and Pullman, B., eds.), p. 381–420. New York: Academic Press 1962
90. Lehninger, A.L.: Bioenergetics: The molecular basis of biological energy transformations. New York: W.A. Benjamin 1965
91. King, T.E.: Reconstitution of the respiratory chain. Advanc. Enzymol. **28**, 155–236 (1966)
92. Van Dam, K., Meyer, A.J.: Oxidation and energy conservation by mitocondria. Ann. Rev. Biochem. **40**, 115–160 (1971)
93. Hogeboom, G.H., Schneider, W.C.: The synthesis of diphosphopyridine nucleotide by liver cell nuclei. J. biol. Chem. **197**, 611–620 (1952)
94. Baltus, E.: Etude de la ribonucleàse des nucléoles isolés à partir d'oocytes d'asteries (Asterias Rubens). Biophys. biochim. Acta (Amst.) **55**, 82–87 (1962)
95. Kaplan, N.O.: Pyridine nucleotide transhydrogenases. Harvey Lect. Series **66**, 105–133 (1972)
96. Imsande, J., Pardee, A.B.: Regulation of pyridine nucleotide biosynthesis in *E. coli*. J. biol. Chem. **237**, 1305–1308 (1962)
97. Schrecker, A.W., Kornberg, A.: Reversible enzymatic synthesis of flavin-adenine dinucleotide. J. Biochem. **182**, 795–803 (1950)
98. Singer, T.P., Kearney, E.B., Kenney, W.C.: Succinate dehydrogenase. Advanc. Enzymol. **37**, 189–272 (1973)
99. Goldberger, R., Green, D.E.: Properties and function of mammalian cytochromes b and c_1. In: The enzymes (Boyer, P., Lardy, H., and Myrback, K., eds.), vol. 8, p. 81–95. New York: Academic Press 1963
100. Strittmatter, P.: Microsomal cytochrome b_5 and cytochrome b_5 reductase. In: The enzymes (Boyer, P., Lardy, H., and Myrback, K., eds.), vol. 8, p. 113–145. New York: Academic Press 1963
101. Paléus, S., Paul, K.G.: Mammalian cytochrome c. In: The enzymes (Boyer, P., Lardy, H., and Myrback, K., eds.), vol. 8, p. 97–112. New York: Academic Press 1963
102. Nicholls, P.: Cytochromes—A survey. In: The enzymes (Boyer, P., Lardy, H., and Myrback, K., eds.), vol. 8, p. 3–40. New York: Academic Press 1963
103. Waino, W.W., Cooperstien, S.J.: Some controversial aspects of the mammalian cytochromes. Advanc. Enzymol. **17**, 319–448 (1956)
104. Lemberg, R.: Cytochromes of group A and their prosthetic groups. Advanc. Enzymol. **23**, 265–321 (1961)
105. Lambeth, D.O., Campbell, K.L., Zand, R., Palmer, G.: The appearance of transient species of cytochrome c upon rapid oxidation or reduction at alkaline pH. J. biol. Chem. **248**, 8130–8136 (1973)
106. Salemme, F.R., Kraut, J., Kamen, M.D.: Structural basis for function in cytochromes c—An interpretation of comparative X-ray and biochemical data. J. biol. Chem. **248**, 7701–7716 (1973)
107. Cusanovich, M.A., Gibson, Q.H.: Anomalous ligand binding by a class of high spin c-type cytochromes. J. biol. Chem. **248**, 822–834 (1973)
108. Wilson, M.T., Brunori, M., Rotilio, G., Antonini, E.: Properties of modified chromosomes. II. Ligand binding to reduced carboxymethyl cytochrome c. J. biol. Chem. **248**, 8162–8169 (1973)
109. Margoliash, E., Smith, E.L.: Isolation and amino acid composition of chymotryptic peptides from horse heart cytochrome c. J. biol. Chem. **237**, 2151–2160 (1962)
110. Margoliash, E., Smith, E.L., Kreil, G., Tuppy, H.: The complete amino acid sequence. Nature (Lond.) **192**, 1125–1127 (1961)
111. Anfissen, C.B.: Principles that govern the folding of protein chains. Science **181**, 223–230 (1973)
112. Hess, G.P., Rupley, J.A.: Structure and function of proteins. Ann. Rev. Biochem. **40**, 1013–1044 (1971)
113. Dickerson, R.E.: X-ray studies of protein mechanisms. Ann. Rev. Biochem. **41**, 815–842 (1972)
114. Dickerson, R.E.: The structure and history of an ancient protein. Sci. Amer. **266**, 58–70 (1972)
115. Margoliash, E.: The molecular variations of cytochrome c as a function of the evolution of species. Harvey Lect. Series **66**, 177–247 (1971–1972)
116. Marsh, J.B., Drabkin, D.L.: The biosynthesis of cytochrome C *in vivo* and *in vitro*. J. biol. Chem. **224**, 909–920 (1957)
117. Strittmatter, C.F., Ball, E.G.: A hemochromogen component of liver microsomes. Proc. nat. Acad. Sci. (Wash.) **38**, 19–25 (1952)
118. Estabrook, R.W., Shigematsu, A., Schenkman, J.B.: The contribution New York: Pergamon Press 1969
of the microsomal electron transport pathway to the oxidative metabolism of liver. In: Advances in enzyme regulation (Weber, G., ed.), vol. 8.
119. Yong, F.C., King, T.E.: Studies on cytochrome oxidase. J. biol. Chem. **247**, 6384–6388 (1972)
120. Slater, E.C.: The constitution of the respiratory chain in animal tissues. Advanc. Enzymol. **20**, 147–199 (1958)
121. Morrison, M.: Biological oxidation. Ann. Rev. Biochem. **30**, 11–44 (1961)
122. Green, D.E.: The mitochondrion. Sci. Amer. **210**, 67–74 (1964)
123. Fernández-Morán, H., Oda, T., Blair, P.V., Green, D.E.: A macromolecular repeating unit of mitochondrial structure and function. J. Cell Biol. **22**, 63–100 (1964)
124. Chance, B., Parsons, D.F., Williams, G.R.: Cytochrome content of mitochondria stripped of inner membrane structure. Science **143**, 136–169 (1964)
125. Yamashita, S., Racker, E.: Resolution and reconstitution of the mitochondrial electron transport system. II. Reconstitution of succinoxidase from individual components. J. biol. Chem. **244**, 1220–1227 (1969)
126. Belitzer, V.A., Tsibakowa, E.T.: The mechanism of phosphorylation as related to respiration. Biokhimiya **4**, 516–535 (1939)
127. Green, D.E.: Electron transport and oxidative phosphorylation. Advanc. Enzymol. **21**, 73–129 (1959)
128. Lehninger, A.L.: Oxidative phosphorylation in submitochondrial systems. Fed. Proc. **19**, 952–962 (1960)
129. Lehninger, A.L., Wadkins, C.L.: Oxidative phosphorylation. Ann. Rev. Biochem. **31**, 47–78 (1962)
130. Boyer, P.D.: Oxidative phosphorylation. In: Biological oxidations (Singer, T.P., ed.), p. 193–235. New York: John Wiley & Sons 1968
131. Lardy, H.A., Ferguson, S.M.: Oxidative phosphorylation in mitochondria. Ann. Rev. Biochem. **38**, 991–1034 (1969)
132. Racker, E.: Mechanisms of synthesis of adenosine triphosphate. Advanc. Enzymol. **23**, 323–399 (1961)
133. Lardy, H.A., Wellman, H.: Oxidative phosphorylations: Role of inorganic phosphate and acceptor systems in control of metabolic rates. J. biol. Chem. **195**, 215–244 (1952)
134. Pinchot, G.B.: The first phosphorylation site—Observations on mechanism. Fed. Proc. **22**, 1076–1079 (1963)
135. Boyer, P.D., Hultquist, D.E., Peter, J.B., Kreil, G., Mitchell, R.A., DeLuca, M., Hinkson, J.W., Butler, L.G., Moyer, R.W.: Role of the phosphorylated imidazole group in phosphorylation and energy transfer reactions. Fed. Proc. **22**, 1080–1087 (1963)
136. Chance, B., Williams, G.R.: The respiratory chain and oxidative phosphorylation. Advanc. Enzymol. **17**, 65–134 (1956)
137. Boyer, P.D., Theorell, H.: The change in reduced diphosphopyridine nucleotide (DPNH) fluorescence upon combination with liver alcohol dehydrogenase (ADH). Acta chem. scand. **10**, 447–450 (1956)
138. Estabrook, R.W., Gonze, J., Nissley, S.P.: A possible role for pyridine nucleotide in coupling mechanism of oxidative phosphorylation. Fed. Proc. **22**, 1071–1075 (1963)
139. Griffiths, D.E.: A new phosphorylated derivative of NAD, an intermediate in oxidative phosphorylation. Fed. Proc. **22**, 1064–1070 (1963)
140. Pullman, M.E., Penefsky, H.S., Datta, A., Racker, E.: Partial resolution of the enzymes catalyzing oxidative phosphorylation I. Purification and properties of soluble dinitrophenol-stimulated adenosine triphosphatase. J. biol. Chem. **235**, 3322–3329 (1960)
141. Pullman, M.E., Penefsky, H.S., Racker, E.: The role of ATPase in oxidative phosphorylation in biological structure and function (Goodwin, T.W., and Lindberg, O., eds.). Proc. IUB/IUBS Int. Symp. (Stockholm, 1960), vol. 2, p. 241–251. New York: Academic Press 1961
142. Penefsky, H.S., Pullman, M.E., Datta, A., Racker, E.: Partial resolution of the enzymes catalyzing oxidative phosphorylation. II. Participation of a soluble adenosine triphosphatase in oxidative phosphorylation. J. Biochem. **235**, 330–333 (1960)

143. Racker, E.: Studies of factors involved in oxidative phosphorylation. Proc. nat. Acad. Sci. (Wash.) **48**, 1659–1670 (1962)
144. Ernster, L., Lee, C.P.: Biological oxidoreductions. Ann. Rev. Biochem. **33**, 729–788 (1964)
145. Lardy, H.A., Ferguson, S.M.: Oxidative phosphorylation in mitochondria. Ann. Rev. Biochem. **38**, 991–1034 (1969)
146. Kagawa, Y., Kandrach, A., Racker, E.: Partial resolution of the enzymes catalyzing oxidative phosphorylation XXVI. Specificity of phospholipids required for energy transfer reactions. J. biol. Chem. **248**, 676–684 (1973)
147. Mitchell, P.: Chemiosmotic coupling and energy transduction. Bodmin: Glynn Research Ltd. 1968
148. Boyer, P.D., Cross, R.L., Momsen, W.: A new concept for energy coupling in oxidative phosphorylation based on a molecular explanation of the oxygen exchange reactions. Proc. nat. Acad. Sci. (Wash.) **70**, 2837–2839 (1973)
149. Erecinska, M., Wilson, D.F., Dutton, P.L., Chance, B.: Kinetic interactions at Site II during energy coupling reactions. Fed. Proc. **32**, 1981–1987 (1973)
150. Hinkle, P.C.: Electron transfer across membranes and energy coupling. Fed. Proc. **32**, 1988–1992 (1973)
151. Wakil, S.J.: Lipid metabolism. Ann. Rev. Biochem. **31**, 369–406 (1962)
152. Ball, E.G.: Regulation of fatty acid synthesis in adipose tissue. Advanc. Enzyme Regulat. **4**, 3–18 (1966)
153. The Nutrition Foundation, Inc.: Metabolism of polyunsaturated fatty acids (Stare, F.J., ed.). Nutr. Rev. **25**, 90–91 (1967)
154. Stumpf, P.K.: Metabolism of fatty acids. Ann. Rev. Biochem. **38**, 159–212 (1969)
155. King, H.K.: The chemistry of lipids in health and disease. Springfield, Illinois: Charles C. Thomas Publisher, 1960
156. Vagelos, P.R.: Lipid metabolism. Ann. Rev. Biochem. **33**, 139–172 (1964)
157. Breusch, F.L.: The biochemistry of fatty acid catabolism. Advanc. Enzymol. **8**, 343–423 (1948)
158. Bressler, R., Wakil, S.G.: Studies on the mechanism of fatty acid synthesis XI. The product of the reaction and the role of sulfhydryl groups in the synthesis of fatty acids. J. biol. Chem. **237**, 1441–1448 (1962)
159. Volpe, J.J., Vagelos, P.R.: Saturated fatty acid biosynthesis and its regulation. Ann. Rev. Biochem. **42**, 21–60 (1973)
160. Mead, J.F., Levis, G.M.: Alpha oxidation of the brain fatty acids. Biochem. biophys. Res. Commun. **9**, 231–234 (1962)
161. Prescott, D.J., Vagelos, P.R.: Acyl carrier protein. Advanc. Enzymol. **36**, 269–311 (1972)
162. Rodwell, V.W., McNamara, D.J., Shapiro, D.J.: Regulation of hepatic 3-hydroxy-3-methylglutaryl-coenzyme A reductase. Advanc. Enzymol. **38**, 373–412 (1973)
163. Sjöstrand, F.S.: Electron microscopy of mitochondria and cytoplasmic double membranes. Nature (Lond.) **171**, 30–32 (1953)
164. Palade, G.E.: Electron microscopy of mitochondria and other cytoplasmic structures. In: Enzymes: Units of biological structure and function (Gaebler, O.H., ed.). Henry Ford Hosp. Symp., p. 185–215. New York: Academic Press 1956
165. Ernster, L., Lindberg, O.: Animal mitochondria. Ann. Rev. Physiol. **20**, 13–42 (1958)
166. Lehninger, A.L.: The mitochondrion—molecular basis of structure and function. New York: W.A. Benjamin 1965
167. Green, D.E.: Mitochondrial structure and function. In: Subcellular particles (Hayashi, T., ed.), p. 84–103. New York: The Ronald Press Company 1959
168. Ball, E.G., Joel, C.D.: The composition of the mitochondrial membrane in relation to its structure and function. Int. Rev. Cytol. **13**, 99–133 (1962)
169. Green, D.E., Fleischer, S.: Mitochondrial system of enzymes. In: Metabolic pathways, p. 41–96. New York: Academic Press 1960
170. Karnovsky, M.J.: The fine structure of mitochondria in the frog nephron correlated with cytochrome oxidase activity. Exp. molec. Path. **2**, 347–366 (1963)
171. Parsons, D.F.: Mitochondrial structure: Two types of subunits on negativey stained mitochondrial membranes. Science **140**, 985–987 (1963)
172. Tapley, D.F., Kimberg, D.V., Buchanan, J.L.: The mitochondrion. New Engl. J. Med. **276**, 1124–1132 (1967)
173. Allmann, D.W., Bachmann, E., Orme-Johnson, N., Tan, W.C., Green, D.E.: Membrane systems of mitochondria. VI. Membranes of liver mitochondria. Arch. Biochem. Biophys. **125**, 981–1012 (1968)
174. Racker, E. (ed.): Membranes of mitochondria and chloroplasts. New York: Van Nostrand Reinhold Company 1970
175. Hoffmann, H.P., Avers, C.J.: Mitochondrion of yeast: Ultrastructural evidence for one giant, branched organelle per cell. Science **181**, 749–751 (1973)
176. Andersson-Cedergen, E.: Ultrastructure of motor end plate and sarcoplasmic components of mouse skeletal muscle fiber as revealed by three-dimensional reconstructions from serial sections. J. Ultrastruct. Res., Suppl. **1**, 1–191 (1959)
177. Green, D.E.: A framework of principles for the unification of bioenergetics. Ann. N.Y. Acad. Sci. **227**, 6–45 (1974)
178. Packer, L.: Membrane structure in relation to function of energy-transducing organelles. Ann. N.Y. Acad. Sci. **227**, 166–174 (1974)
179. Raff, R.A., Mahler, H.R.: The nonsymbiotic origin of mitochondria. Science **177**, 575–582 (1972)
180. Borst, P.: Mitochondrial nucleic acids. Ann. Rev. Biochem. **41**, 333–376 (1972)
181. Dawidowicz, K., Mahler, H.R.: Synthesis of mitochondrial proteins. In: Gene expression and its regulation (Kenney, F.T., Hamkalo, B.A., Favellikes, G., and August, Y.T., eds.), vol. I, p. 503–522. New York: Plenum Publishing Corp. 1973
182. Baxter, R.: Origin and continuity of mitochondria. In: Origin and continuity of cell organelles; results and problems in cell differentiation (Reinert, J., and Ursprung, H., eds.), vol. II, p. 46–64. New York: Springer 1971
183. Tzagoloff, A., Rubin, M.S., Sierra, M.F.: Biosynthesis of mitochondrial enzymes. Biochim. biophys. Acta (Amst.) **301**, 71–104 (1973)
184. Clayton, D.A., Smith, C.A., Jordan, J.M., Teplitz, M., Vinograd, J.: Occurrence of complex mitochondrial DNA in normal tissues. Nature (Lond.) **220**, 976–979 (1968)
185. Tibbetts, C., Vinograd, J.: Properties and mode of action of a partially purified deoxyribonucleic acid polymerase from the mitochondria of HeLa cells. J. biol. Chem. **248**, 3367–3379 (1973)
186. Storrie, B., Attardi, G.: Expression of the mitochondrial genome in HeLa cells. XV. Effect of inhibition of mitochondrial protein synthesis in mitochondrial formation. J. Cell Biol. **56**, 819–831 (1973)
187. Storrie, B., Attardi, G.: Mode of mitochondrial formation in HeLa cells. J. Cell Biol. **56**, 833–838 (1973)
188. Piko, L., Chase, D.: Role of the mitochondrial genome during early development in mice. J. Cell Biol. **58**, 357–378 (1973)
189. Rubin, M.S., Tzagoloff, A.: Assembly of the mitochondrial membrane system, X. Mitochondrial synthesis of three of the subunit proteins of yeast cytochrome oxidase. J. biol. Chem. **248**, 4275–4279 (1973)
190. Shannon, C., Enns, R., Wheelis, L., Burchiel, K., Criddle, R.S.: Alterations in mitochondrial adenosine triphosphatase activity resulting from mutation of mitochondrial deoxyribonucleic acid. J. biol. Chem. **248**, 3004–3011 (1973)
191. Aloni, Y., Attardi, G.: Expression of the mitochondrial genome in HeLa cells. II. Evidence for complete transcription of mitochondrial DNA. J. molec. Biol. **55**, 251–270 (1971)
192. Wagner, R.P.: Genetics and phenogenetics of mitochondria. Science **163**, 1026–1031 (1969)
193. King, T.E., Mason, H.S., Morrison, M. (eds.): Oxidases and related redox systems, vol. 2, from Proceedings of the Second Int'l Symp. on Oxidases and Related Redox Systems, Memphis, Tennessee, June 8–12, 1971. Baltimore, Maryland: University Park Press 1973
194. Coon, M.J., Autor, A.P., Boyer, R.F., Lode, E.T., Strobel, H.W.: On the mechanism of fatty acid, hydrocarbon, and drug hydroxylation in liver microsomal and bacterial enzyme systems. In: Oxidases and related redox systems, vol. 2, p. 529–553. Baltimore, Maryland: University Park Press 1973
195. King, T.E., Yong, F.-C.: Interdependence of the components of cytochrome oxidase and their spectral behavior. In: Oxidases and related redox systems, vol. 2, p. 677–700. Baltimore, Maryland: University Park Press 1973
196. Fulco, A.J.: Metabolic alterations of fatty acids. In: Annual review of biochemistry (Snell, E.E., Boyer, P.D., Meister, A., and Richardson, C.C., eds.), vol. 43, p. 215–241. Palo Alto, California: Annual Reviews Inc. 1974
197. Baltscheffsky, H., Baltscheffsky, M.: Electron transport phosphorylation. In: Annual review of biochemistry (Snell, E.E., Boyer, P.D., Meister, A., and Richardson, C.C., eds.), vol. 43, p. 871–897. Palo Alto, California: Annual Reviews Inc. 1974
198. Schatz, G., Mason, T.L.: The biosynthesis of mitochondrial proteins. In: Annual review of biochemistry (Snell, E.E., Boyer, P.D., Meister, A., and Richardson, C.C., eds.), vol. 43, p. 51–87. Palo Alto, California: Annual Reviews Inc. 1974

Chapter 2

Determination of Cellular Specificity

The first section of this book describes those pathways generating the chemical energy necessary for cell function. The following section describes the biosynthetic pathways involved in determining cellular specificity. The determination of cellular specificity primarily involves two major cellular structures—the nucleus and the endoplasmic reticulum, and three major biosynthetic pathways—DNA synthesis, RNA synthesis, and protein synthesis. Specialized pathways that are more relevant to the understanding of a given disease, for example, an inborn error of metabolism or a nutritional deficiency, are reviewed in other chapters.

The Nucleus

The cell nucleus was named after the pit or the kernel of fruit because like that structure the nucleus often appears as a well-delimited, round mass in the middle of the cell (see Fig. 2-1). It forms a central core, denser and more coarsely granular than the surrounding cytoplasm. The Greek word for nucleus is *καρυον*, and such terms as karyokinesis, karyolysis, and karyorrhexis have a Greek origin.

Morphology of the Nucleus

Early in the history of cytology and histology, several structures of the nucleus were identified. Among them are the membrane, the nucleoli, and the chromosomes. In the early days of histology, the existence of a nuclear membrane was often contested because its presence under the microscope depended upon the type of fixative used in the preparation. These doubts, however, did not survive after nuclear membranes were clearly demonstrated in photographs of cells grown on tissue culture. Further information on the structure of the nuclear membrane was gained with the advent of the electron microscope.

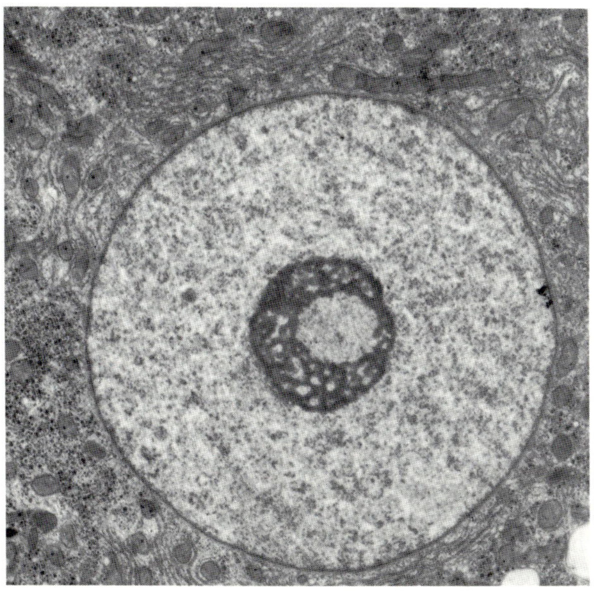

Fig. 2-1. Low-power electron micrograph of hepatocytes from a "normal" liver biopsy. The cells have a spheroidal nucleus with a prominent nucleolus, and chromatin is uniformly distributed. In the cytoplasm, mitochondria, cisternae of the ergastoplasmic reticulum in parallel arrays, glycogen, and a few lipid droplets are easily identifiable (×54,000). (From L. Zamboni)

On electron micrographs prepared from tissues fixed in osmic acid, the nuclear membrane appears as an envelope 250 A thick composed of three different layers: an inner and an outer dense lining (each measuring 60 A in thickness) and a less dense median space (120 A thick). Unlike the cell or the mitochondrial membranes, the nuclear envelope (see Fig. 2-2) is not continuous, but is frequently interrupted by circular pores 1000 A in diameter [1–9].

High-resolution examination of the pores demonstrates that the openings are formed by interruptions of the inner and outer membranes. Instead of forming an envelope that surrounds the entire nuclear mass at the level of the pores, the inner and outer membranes are curved toward each other and become confluent. The fact that the nucleus is not surrounded by a continuous membrane but rather by a perforated envelope is probably largely responsible for the difficulties the biochemist has in preparing intact cell nuclei. The pores might play an important role in the relationship between nucleus and cytoplasm. However, the pores are not free passages connecting the cytoplasmic and nuclear spaces. Careful fixation of the nuclei demonstrates a fine, fibrillar low-contrast substance which may either restrain the flow of components from cytoplasm to nucleus or may directly participate in the exchange. Although the nucleus of the metazoan cell is bounded by a membrane, no such structure has been observed in bacteria, in which the chromosomes are not separated from the cytoplasmic structures.

Available evidence suggests that chromosomes are attached to the nuclear membrane close to nuclear pores. Moreover, DNA replication appears to be initiated at these points of attachment.

The Nuclear Membrane

Although the nuclear membrane clearly serves to separate the nucleoplasm from the cytoplasm, the membrane interacts with a number of cytoplasmic and nuclear structures.

The outer membrane, like that of the endoplasmic reticulum, is studded with ribosomes in some areas. In fact, connections between the nuclear membrane and those of the endoplasmic reticulum have been observed in many cells. It is debated, however, whether the continuity involves both the inner and outer membranes, or only the outer one.

In many cells the inner membrane shows a close association with chromosomes. The attachment is not random, but occurs at preferential sites for distinct chromosomes. Reasons for the link between chromo-

other consisted of the elaboration of a completely new RNA molecule.

Obviously, two types of RNA synthesis must be distinguished in the course of incorporation studies. Although the formation of a new molecule of RNA may be involved in the coding of protein synthesis, merely adding a nucleotide to a preexisting chain probably has little to do with this process. Vincent's data suggest that there are in the nucleolus, as in the cytoplasm, two types of RNA—one which could be analogous to a soluble RNA, and the other to the messenger RNA. The significance of these studies will become clearer when we discuss protein synthesis.

Although the preceding discussion may have given the impression that all nuclear RNA is concentrated in the nucleoli, with the aid of histochemical methods, nuclear RNA was shown to be in at least three different sites: the nucleolus, the chromosome, and the spindle.

One of the most intriguing aspects of RNA metabolism in the nucleoli is the shift in RNA concentration that occurs under various physiological conditions. Nuclear RNA is very labile. It is present in varying amounts from one tissue to another. Within tissue, it often changes from cell to cell; within a cell, it may increase or decrease with the physiological conditions. Such changes have been shown by several investigators using various methods. We refer here only to some of Swift's work on starved HeLa cells and regenerating liver nucleoli.

In the HeLa cell, the RNA content per nucleolus may fall to 10% of its original value in the course of starvation. During that period a smaller decrease in the protein content of the nucleolus is observed (50% of its original value). The RNA content of the nucleolus also fluctuates during cellular division, as was demonstrated by Swift's experiments [10] with regenerating liver. During the first 24 hours preceding the high mitotic rate, the RNA in the nucleolus increases. Then, when the mitotic rate reaches its maximum 24 hours after hepatectomy, the RNA decreases in the nucleolus and returns to normal values 29–48 hours after hepatectomy. It is interesting that the high rate of nuclear RNA precedes active protein synthesis and the increase in cytoplasmic RNA. It may thus be concluded that the nucleolus is an intranuclear structure composed of RNA and protein in ratios that change under such conditions as starvation or mitosis. Furthermore, nucleolar RNA is composed of at least two different types of metabolically active molecules.

The major components of the nucleolus are RNA and proteins. The presence of both these biochemical components in that structure has definitely been established by histochemical and biochemical methods. The existence of specific catalytic proteins, such as alkaline phosphatase and NAD-synthesizing enzymes, remains to be established with more precision.

Experiments in which synchronized liver cells were cultured and incubated in presence of actinomycin either before or after mitosis established that the RNA needed for the reconstruction of nucleoli after mitosis is synthesized prior to the mitotic process [232].

This observation is in keeping with studies on liver regeneration which established that some of the RNA synthesized in the early hours after partial hepatectomy is not transferred from nucleus to cytoplasm.

Litt, Monty, and Dounce [17] prepared nucleoli from liver nuclei, analyzed the enzyme contents, and found those nucleoli to contain aldolase, arginase, catalase, acid phosphatase, and several other enzymes. The data, however, are difficult to interpret because although the amount of DNA in their preparation is extremely high (12–18%), that of RNA is surprisingly low (2–4%). Such data are not at all in accordance with the findings on the starfish nucleoli or with the histochemical studies on mammalian nucleoli. Therefore, Dounce's preparation probably is not composed of clean nucleoli, but includes contaminating cytoplasmic and other nuclear fragments.

Busch and his associates [18–21] attempted to isolate nucleoli from either Walker tumors or liver treated with thioacetamide. They identified the nucleoli with the aid of azure and claim to have obtained, after sonication of the nuclei followed by differential centrifugation, a nucleolar preparation 90–95% free of contamination. The preparation is rich in DNA, relatively poor in RNA; but because the yield in nucleoli is low, their nucleolar preparation can hardly be considered representative of the original nuclear preparation.

A more promising technique was that described by workers of the Rockefeller Institute [3], who prepared nucleoli from a crop of clean mammalian nuclei obtained by centrifuging them in sucrose gradients. Their nucleolar preparation was examined with the electron microscope and found to be a clean preparation of cellular organelles presenting all the morphological characteristics of native nucleoli.

Perhaps one of the most remarkable accomplishments in molecular biology during the 1960's is the demonstration that nucleoli play an essential role in the formation of ribosomal RNA. Evidence for this emanated from three different types of experiments: studies of enucleolated mutants, DNA-RNA hybridization, and actual isolation of the rRNA cistron.

A mutant—*Xenopus laevis*—has embryos containing nuclei devoid of nucleoli. Brown and Gurdon [22] observed that these mutants were also incapable of synthesizing ribosomal RNA. *Drosophila melanogaster* have an X chromosome with a short piece of heterochromatin that contains the nucleolar organizer. Mutants can be obtained which are either deficient in that piece of heterochromatin or contain two or three such organizers. Thus, labeled ribosomal RNA can be prepared and hybridized with the DNA, and thereby the amount of DNA that is complementary to ribosomal RNA is measured. Such experiments reveal that the maximum amount of RNA that hybridizes with a given DNA is directly proportional to the number of nuclear organizers found in the organism used as a source of DNA. Similar hybridization experiments have been done with embryos of homozygous, heterozygous, nucleolated, and enucleolated Xenopus embryos.

Studies in bacterial systems suggested that the cistrons responsible for ribosomal RNA synthesis were arranged in repetitive linear sequence. A similar situation is likely to occur in multicellular organisms. When DNA extracted from a Xenopus embryo is placed on a cesium chloride density gradient, a satellite band appears that contains approximately 0.15–0.2% of the total DNA. This satellite band does not develop when the DNA is extracted from enucleolated mutants. Furthermore, the satellite band preferentially hybridizes with ribosomal RNA. It is not certain whether this RNA is multicistronic, each cistron coding for the biosynthesis of a specific ribosomal RNA, or whether the RNA is composed of redundant identical cistrons. The weight of the evidence accumulated in studies from heterozygous organizers, partial organizers, and organizers located on different chromosomes favors redundancy [23].

A Xenopus embryo carrying a single heterozygous organizer can support the development of perfect nucleoli. X-Rays induce breaks in the organizer that may be followed by rearrangement, and even these partial organizers are capable of supporting the development of normal nucleoli. In *Chironomus thummi*, the organizers are located on two different points of the genome, and therefore the chromosomes carrying the organizer can be separated by micromanipulation. The RNA of the individual chromosome can be hydrolyzed and its base composition studied by electrophoresis. Edström [24, 25] found that the content of adenine, guanine, cytosine, and uracil was the same for RNA extracted from chromosome II of the Chironomus, which contains one of the organizers, and chromosome III, which contains the other organizer.

It seems now well established that the major if not the sole function of the nucleolus is to produce ribosomal RNA. Thus, the two large molecular weight RNA's found in all eukaryotic cells derive from a 45-S* precursor. Certain base and certain ribose residues of the precursor are then methylated, and a series of cleavages ultimately yields the 28-S and 18-S RNA.

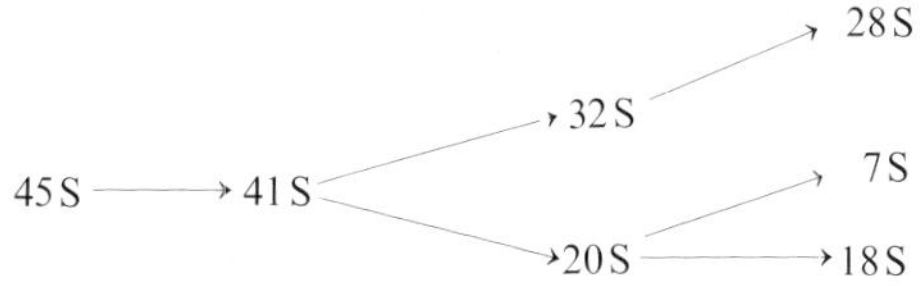

While they were in Busch's laboratory, Siebert and his associates [27] studied the intranuclear distribution of a number of enzymes known to be associated with the nuclear fraction (see below). Siebert concluded that RNA polymerase and ribonuclease were two nuclear enzymes primarily found in the nucleolus. Nucleotide phosphorylase and ATPase A were only in part associated with the nucleolus. Other nuclear enzymes were not associated with the nucleolus. The significance of these results is difficult to evaluate. Enzyme activities of individual preparations vary considerably, and no attempts were made to include a statistical analysis of the data. Furthermore, the methodology does not prevent redistribution of the enzyme under study as as result of adsorption or leakage.

* 45-S RNA is in reality a mixture of 45-, 46-, and 47-S RNA. 47S is the primary precursor which by cleavage yields 46S, which in turn yields 45S [26].

Nucleolus and Chromosomes

Two types of relationships between nucleolus and chromosomes have been considered; an indirect relationship in which the size and number of nucleoli are assumed to be dependent on the chromosomal stock of the cells, and a direct relationship in which the nucleolus is thought to be derived directly from chromosomal material. Studies on plants seem to have established conclusively that the number and size of nucleoli are characteristic of a species and, therefore, under genetic control. The number of nucleoli was found to be directly related to the number of chromosomes in the heteroploid nuclei of amphibians. Although no such conclusive relationships have been satisfactorily established in mammalian tissue, mammalian chromosomes possibly control nucleolar composition in a more subtle manner than by affecting their number and size.

An intimate relationship of nucleolus and chromosomes is so well established in the minds of some that White [28] concluded: "The bodies known as nucleoli are attached to certain chromosomes (always the same ones, for a particular organism) and are hence to be looked upon as part of those chromosomes, rather than as independent constituents of the nucleus." This concept of nucleoli originating from chromosomes is not without experimental support. Already in 1928, Heitz [29] demonstrated that the nucleolus is formed from a specific part of the chromosome, a chromosomal structure later called the organizing element by McClintock [30]. It was soon observed that the nucleolus is not formed by one or a few chromosomes, but that all chromosomes contribute material that is later incorporated into the nucleolus.

Electron microscopic studies confirmed the concept of the chromosomal origin of the nucleolus. There are two main schools of thought: one maintains that the nucleolus is formed from widely dispersed material assembled at the site of the chromosomes, and a second believes that the nucleolus is directly formed from chromosomal material. A brief discussion of the second concept illustrates the problem.

Histochemical studies by Swift [31] have indicated that within the nucleus there are three main sites that contain RNA: the spindle, the chromosomes, and the nucleolus. Spindle RNA can be demonstrated with histochemical techniques during telophase and early interphase. This type of RNA is usually close to the chromosome, but has no immediate connection with that structure. Swift has developed a staining technique that allows the experimenter to distinguish between nucleolar RNA and chromosomal RNA. This method is based on the fact that after specific fixation, the nucleolus forms a chromate lipid complex that resists dehydration and can be stained with Sudan black. With the aid of such techniques, investigators were able to recognize nucleolar material during telophase in very close contact with the nucleolus.

In addition, Swift demonstrated that this juxtachromosomal nucleolar material contains RNA. However, in contrast to the spindle RNA, which merely lies between the chromosomes, this nucleolar RNA seems to be intimately attached to the chromosomal structure.

Sources of Chemical Energy in the Nucleus

Studies with radioactive precursors have demonstrated that even nondividing nuclei undergo active metabolic changes. The turnover of RNA, protein, and lipids is very high in the interphase nucleus, and during mitosis at least the last steps of DNA synthesis are achieved in the nucleus. Such an active metabolism indicates that a chemical source of energy and a pool of enzymes are available in the nucleus. This raises a very important question: Does the nucleus parasitize the cytoplasm for its supply of enzymes and energy, or does it stock its own catalytic proteins? We have seen that the nucleus contains a high proportion of proteins, all of which—except for the basic proteins—could act as enzymes.

Among all particulate fractions of the cells, the nucleus is most likely to suffer leakage during tissue fractionation. As pointed out before, in contrast to mitochondria, the nucleus is not surrounded by a continuous membrane, but rather by a perforated envelope. As with any other particulate preparation, it is difficult to evaluate the amount of material that has been adsorbed onto the nucleus. Deoxyribonucleic acid and ribonucleic acid are both apt to adsorb proteins; therefore, it would not be surprising if proteins that are sometimes considered genuine constituents of the nucleus remain associated with the nuclear preparation only because they are artifacts of the preparative procedure.

Biochemists have demonstrated at least 80 different enzymes in preparations of nuclei. Among them are enzymes of the main bioenergetic pathways and a large number of hydrolytic enzymes. On the basis of their relationship to the nucleus, the enzymes in nuclear preparations can be divided into three different groups: (1) those likely to be contaminants; (2) those clearly related to the nucleus; and (3) those present in the nucleus.

The succinoxidase and cytochrome oxidase found in the nuclear pellet obtained by tissue fractionation are probably not normal constituents of the nucleus. The activities associated with nuclei obtained by tissue fractionation are proportional to the number of mitochondria contaminating the preparation. No appreciable amount of cytochrome c or flavoprotein is found in the nuclear preparation. Therefore, it seems that no complete electron transport chain is present in the nucleus, and the fragments of the chain that are present are thus likely to be contaminants.

Enzymes of the Krebs cycle are not found in significant amounts in the nucleus. In the absence of a Krebs cycle, glycolysis may constitute an important source of energy in the nucleus. Although the evidence favors this multiple-enzyme system in the nucleus, the existence of nuclear glycolysis remains to be established with certainty. Different groups of investigators using different methods to prepare nuclei found glycolytic enzymes in the nucleus. Although appreciable glycolytic activity was found in both liver and plant nuclei in two different laboratories, Siebert [32, 33] claims that the rate of glycolysis in their nuclear preparations is not in keeping with the metabolic activity of that cellular organelle.

The problem of nuclear glycolysis is twofold: What part of the glycolytic cycle is present in the nucleus, and is its presence compatible with an appropriate energy supply? Although for years authors have eagerly debated this matter in the literature, few new facts have been added to answer either of these two questions. The situation can be summarized as follows: hexokinase has not been found in the nucleus in appreciable amounts, but several enzymes (aldolase, triose phosphate dehydrogenase, enolase, and lactic dehydrogenase) are present in appreciable quantities. Such an enzyme mosaic suggests that nuclear glycolysis would have to start with fructose-1,6-diphosphate, a substrate that the cytoplasm would provide. If this were true, none of the high-energy phosphates made available during the oxidation of glucose would be needed to trigger the glycolytic process by the phosphorylation of glucose. All the ATP generated could be used for other metabolic processes. If the nucleus is so permeable to fructose-1,6-diphosphate, what prevents its permeability to ADP or to ATP; and if ATP could travel freely from cytoplasm to nucleus, why would the nucleus need its own bioenergetic system?

The presence of glycolytic enzymes in the nuclear preparation does not prove that those enzymes are normal constituents of the nucleus in the intact cell. Much more sophisticated and up-to-date investigations are required to demonstrate such an association beyond a reasonable doubt. To our knowledge, such investigations have been attempted only in the case of aldolase.

Roodyn [34] studied the distribution of aldolase in liver nuclear preparations centrifuged in a gradient concentration of sucrose. He demonstrated that although the distributions of aldolase and DNA were similar, cytochrome oxidase and succinoxidase were found in particles of different sizes. His results suggested that aldolase and DNA are closely associated. The possibility of contamination was eliminated in experiments in which aldolase was added to the nuclear preparation because the added aldolase, in contrast to the aldolase originally associated with the nuclear pellet, could be easily washed off.

Even if it is accepted that glycolysis is present in

the nucleus, we may not conclude that glycolysis supplies the nucleus with all the energy required for its metabolic activity. Siebert [33] does not believe that the glycolytic activity of the nucleus provides all its energy requirement; however, these conclusions were based on studies on nuclei prepared in aqueous media, in which nuclei are prone to leakage. Siebert has since reversed his position on that matter on the basis of studies on nuclei prepared in nonaqueous media, which, unfortunately, are not free of cytoplasmic contamination.

Assuming that neither the Krebs cycle nor glycolysis could generate enough ATP to satisfy normal nuclear functions, some investigators looked for other mechanisms of ATP generation. Allfrey and his associates [35, 36] claimed to have found a pathway of oxidative phosphorylation specific for the nucleus. Their arguments for a specific oxidative phosphorylating system in the nucleus are based on both indirect and direct evidence. As indirect evidence, they submit the fact that protein synthesis, a process that requires energy, is extremely active in the nucleus; from this they conclude that the nucleus in some way must provide the chemical energy necessary for this metabolic activity.

When they further observed that the normal nucleus contains a high proportion of mono-, di-, and trinucleotides of adenine, they claimed to have provided direct proof of their theory by demonstrating that the mono- or dinucleotides in the nucleus may be converted to ATP when oxygen is present. (The nucleotides can be extracted from the nucleus with acetate buffer at pH 5.1.) This conversion certainly suggested the existence of an intranuclear process of oxidative phosphorylation. As in mitochondria, oxidative phosphorylation in the nucleus is inhibited by uncouplers or agents blocking the electron transport chain. Nuclear oxidative phosphorylation is blocked by cyanide, azide, and antimycin A, or by dinitrophenol; but, in contrast to mitochondria, it is resistant to Janus green, methylene blue, carbon monoxide, Dicumarol, and calcium.

Nuclear oxidative phosphorylation is difficult to quantify. Although oxygen uptake in the nucleus can be measured, no exact P/O ratio is available. This is because only the AMP already present in the nuclear preparation can be converted to ATP; any AMP added to the nuclei remains unaltered. An intriguing observation is the effect of DNase on the phosphorylation of AMP. (Allfrey has proposed that DNase blocks ATP synthesis in the nucleus indirectly; namely, by inhibition of the nuclei by the histones, which after DNA extraction are no longer associated with DNA by salt linkages [35].) The enzymic extraction of 55% of the DNA in the nucleus leads to the loss of nuclear phosphorylation properties, which can be restored by adding DNA to the system. The effect of DNA is not specific because DNA can be replaced by RNA, polyadenylic acid, heparin, chondroitin sulfate, and polyethylene sulfate. Oxidative phosphorylation in thymus nuclear preparation has been confirmed in two laboratories. Whole body doses of ionizing radiation inhibited oxidative phosphorylation in thymus nuclei.

It remains to be established, however, that this peculiar phosphorylating system is native to the nucleus and does not represent contamination of the nuclear fraction by intact cells or by intact or fragmented mitochondria. The arguments against mitochondrial contamination rest on several observations. Among them are the absence of cytochrome oxidase in nuclei and the difference between nuclear and mitochondrial oxidative phosphorylation with respect to their responses to inhibitors.

Stern and Timonen [37] demonstrated that thymus nuclei contain little cytochrome oxidase, but this finding is not in agreement with observations made on liver, spleen, or pancreas nuclei prepared in 0.25 M sucrose. Although the difference between the effect of inhibitors on nuclear and mitochondrial oxidative phosphorylation excludes the possibility of contamination of nuclei by intact mitochondria, it does not eliminate the possibility of contamination by fragmented mitochondria. Inhibitors do not affect the phosphorylating particle and the intact mitochondria the same way.

Electron microscopic examination of nuclei prepared in 0.25 M sucrose in the presence of calcium demonstrates clumping of the intranuclear structures. Clumping should warn investigators of the possibility of agglomeration of cytoplasmic material. It is also believed that the effect of DNase on nuclear oxidative phosphorylation constitutes a reliable method of distinguishing nuclear and mitochondrial oxidative phosphorylation. Although this conclusion is probably justified, it can be substantiated only by adequate control studies of the effect of DNase on the oxidative phosphorylation of cytoplasmic particles.

Whole cells cannot be removed from the nuclear pellet by ordinary tissue fractionation procedures, and only by differential centrifugation in a 2.20 M sucrose solution can the normal nuclei and the intact cells be separated. Unfortunately, such a technique destroys both the nuclear phosphorylating properties and the nuclear capacity to take up radioactive amino acids or orotic acid.

Klouwen and his associates [38, 39] have further investigated the occurrence of oxidative phosphorylation in thymus nuclei; they acknowledge contamination of the nuclear pellet with as much as 10% intact cell. Yet they believe that the small levels of oxidative phosphorylation measured in nuclei (P/O ratio 0.6–1.0) do not result from contamination by intact or fragmented mitochondria. Further, they claim that the nuclear oxidative phosphorylation is sensitive to various inhibitors of the electron transport chain. Using low-temperature spectophotometry, it was established that a nuclear preparation free of cytoplasmic contamination contained cytochrome b, c, and aa_3 and that as much as 40% of cytochrome aa_3 may be found in association with calf thymus nuclei. Other researchers have also found cytochrome b and c, but not cytochrome aa_3, in calf thymus nuclear preparations.

In conclusion, in spite of the conscientious efforts of the investigator to eliminate the possibility of contamination of the nuclei with nonnuclear components and to avoid redistribution of cytochrome by absorption in the nucleus, these endeavors were only partly successful. Although not compelling, the results suggest that at least in thymus, part of the cellular oxidative phosphorylation takes place in the nucleus.

Proteins and Protein Synthesis in the Nucleus

The problem of protein synthesis in the nucleus cannot be approached without briefly considering important concepts related to the use of isotopes; namely, the meaning of the terms uptake, turnover, and synthesis. When a simple labeled compound is incorporated in a more complex biochemical molecule, the labeled molecule either changes place with an identical nonlabeled molecule already present in the complex compound, or it combines with other nonlabeled molecules to build a new complex molecule. There has been isotopic uptake in either case—through exchange in the former and through *de novo* synthesis in the latter. Turnover studies measure the rate at which an unlabeled simple compound, such as an amino acid, is incorporated into a more complex molecule such as protein. This incorporation may result from either exchange or synthesis. The isotopic uptake is expressed in terms of specific activity; that is, in counts per minute per unit weight of a purified compound. The various curves obtained by plotting specific activities versus time in the absence of breakdown of the original molecules express the results of either exchange or synthesis.

In many cases, the problem is complicated by the fact that the newly synthesized molecule may undergo rapid and complete breakdown under the influence of hydrolytic or other enzymes. The broken down molecules need to be replaced by new molecules, thus involving *de novo* synthesis. Consequently, synthesis may have occurred when there is no detectable net increase of the compounds under investigation. In such a case, only the use of radioisotopes permits measurement of the rate of synthesis. The product of molecular breakdown is often reused; it enters the pool of labeled precursors, which is continuously altered. The shape of the specific activity curves of the compound under investigation is determined by these alterations of the pool. Of course, most investigators who have used isotopes for turnover or synthesis studies have been aware of these problems and accordingly have analyzed their results critically. It is hoped that this brief outline of some of the problems brought up by the use of isotopes *in vivo* and *in vitro* will encourage the young pathologist engaging in this type of research to view his results with a critical eye.

Protein synthesis in the nucleus has been studied *in vivo* and *in vitro* in a variety of organs. Two main features have emerged from measuring the incorporation of amino acids in cellular proteins *in vivo*. First, the incorporation is high in the proteins of actively metabolizing tissue, such as pancreas and liver, but it is low in tissue with a lesser metabolic activity, such as the chicken erythrocyte. Secondly, the specific activities of nuclear and cytoplasmic proteins are equal, suggesting that the turnover of these proteins is similar in both cellular fractions. Such investigations deal with a population of proteins, and the specific activity obtained in each case is the mean of a large number of individual specific activities. Only in a few instances has it been possible to determine how the specific activity of an individual protein in the nucleus quantitatively deviates from the mean value. Such deviations were suggested in the radioautographic studies of Errera [40] and his associates. These authors found that phenylalanine is incorporated at higher rates in the nuclear proteins than in the cytoplasmic proteins. However, it remains to be established that the observations made on mammalian liver can be extended to other tissues.

More precise studies were carried out in Leblond's [41, 42] laboratory with the aid of a variety of tritium-labeled amino acids (leucine, methionine, and glycine). Although Leblond's group investigated protein synthesis in a large number of organs, their original effort was concentrated on the pyramidal and Purkinje cells of the brain, the liver cell, and the Sertoli cell and spermatocyte in the testicle. In the Purkinje cell, both nuclear and cytoplasmic proteins incorporated the labeled amino acids. The incorporation occurred early and was maintained for as long as 45 days after the amino acid was injected. There was no indication of a relationship between nuclear and cytoplasmic protein synthesis. On the contrary, the data suggested that although protein synthesis occurred simultaneously in both cell fractions, the nuclear and cytoplasmic synthetic processes were independent. A surprising observation was that nuclear protein synthesis did not seem to occur in the nucleolus, but rather in the vicinity of the chromosomes.

Leblond's radioautographic observations contradict the work of Busch and his associates [19]. After devising methods to extract and separate nuclear proteins, this latter group of investigators made a commendable effort to prepare nucleoli. They extracted protein from nucleoli obtained from animals injected with labeled amino acids and demonstrated that the nucleolar protein had a specific activity much higher than that of the protein obtained from the entire nucleus.

The discussion of the ability of isolated nuclei to synthesize protein can be centered mainly on the elegant work of Allfrey and his coworkers [43–45]. They have described their work in numerous review articles; therefore, it is summarized only briefly here. A highly purified calf nuclei preparation was used in these experiments. When incubated in the presence of labeled amino acid (lysine, alanine, glycine, etc.), the nuclei rapidly incorporate the labeled amino acid. A curve of protein-specific activity was obtained for each type of amino acid. The curves expressing the counts per minute per milligram of protein versus time presented similar shapes. A lag period for the first 30 minutes was followed by a rapid, almost linear uptake during the next 90 minutes. The possibility of contamination of the nuclear preparation by labeled amino acids was eliminated on the basis of several observations: (1) the uptake is energy-dependent, which suggests an active synthetic process; (2) the incorporation is stable; the incorporated, labeled amino acid cannot readily be replaced by cold amino acid; (3) the incorporation seems to involve peptide bonds; and (4) according to the investigators, the incorporation is irreversible. In fact, the authors have calculated the amounts of protein that might be synthesized by such a nuclear preparation. They believe that 22 protein molecules of 50,000 molecular weight are synthesized every second by each nucleus.

Apparently, all nuclear proteins participate in the uptake of the labeled amino acids as indicated by specific activities studies of nonhistone and histone-type protein. However, although the activity of nonhistone proteins was high, the amino acid uptake by histones was relatively low. The specific activity of the histones is only one-third the specific activity of the total protein. These basic experiments of the *in vitro* amino acid incorporation suggested that proteins are actively synthesized in the nuclear preparation. The cytoplasm is believed to provide the amino acids for the nucleus.

The existence of an enzymically controlled amino acid transport from cytoplasm to nucleus has been suggested. Such a mechanism has not been solidly established; however, it is interesting that the nuclear uptake of various amino acids is influenced by the concentration of sodium ions in the medium. Amino acid uptake in the nucleus seems to involve all the steps described for protein synthesis in the cytoplasm. The amino acid is first activated, and the presence of activating enzymes of various amino acids has been demonstrated in nuclei prepared in nonaqueous media. The activated amino acids are then transferred to a nuclear RNA, the exact location of which is not clear. The so-called nucleolar RNA, which is insoluble in 1 M sodium chloride, does not act as transfer RNA. The accepting nucleotide of the nuclear RNA for the amino acid is a terminal adenosine, precisely what it is in cytoplasmic RNA.

The presence of ribonuclear protein particles active in amino acid uptake has been demonstrated in nuclear preparations. The so-called nuclear ribosomes require ATP and GTP for amino acid uptake, exactly as cytoplasmic ribosomes require the same triphosphonucleotides for activity. There seems to be one major difference between nuclear and cytoplasmic ribosomes. RNase, which effectively blocks protein synthesis in the cytoplasm, is without effect on protein synthesis in the nucleus. The fact that DNA is required for amino acid incorporation into nuclear protein cannot be invoked as an argument in favor of the existence of such a metabolic pathway in the nucleus, because the effect of DNA is indirect and DNase interferes with the generation of ATP in the nucleus and therefore, abolishes the only source of energy in the system.

This remarkable similarity between the enzymes that synthesize proteins in the nucleus and those synthesizing proteins in the cytoplasm will probably raise doubts in the mind of the skeptical reader. In view of such similarities, it seems difficult to establish that the protein-synthesizing system of the nucleus is genuine to that structure and does not result from cytoplasmic contamination.

Nuclear Enzymes

We may put too much emphasis on nuclear integrity. If the nuclear envelope is as freely permeable as electron microscopy suggests, only the chromosome, nucleolus, and nuclear membrane can be considered genuine nuclear structures. Much of the nuclear sap might be a fraction of circulating cytoplasm trapped in the nuclear structure during the isolation procedure. The most representative nuclear preparation may very well be nuclei prepared in aqueous media and generously washed with the suspending medium. Such a concept is supported by the fact that the enzyme mosaic of the nucleus often mimics that of the cytoplasm, and that up to the present time, except for some enzymes involved in RNA synthesis, no nuclear enzyme has been found that is not also present in large amounts in the cytoplasm. It is quite possible, however, that the enzymes in the cytoplasm and in the nucleus, although identical in catalytic properties, differ in other respects; but this remains to be demonstrated. In his preparation made with nonaqueous media, Siebert found enzymes quite unlikely to be associated with the nuclei, such as acid phosphatase and β-glucuronidase.

Smellie [46] has claimed that the specific activity of DNA polymerase in nuclei is greater than that in cytoplasm, but the published data are incomplete and do not take into account the fact that the cytoplasmic DNase greatly interferes with polymerase activity. Furthermore, it does not exclude the possibility of coprecipitation of polymerase and DNA.

A discussion of all the enzymes known to be associated with the nucleus is unnecessary for the purposes of this text; however, it might be valuable to review the nuclear association of some hydrolytic enzymes. The presence of adenosine triphosphatase in the nucleus remains controversial. Although some investigators found it in large amounts, none was present in the nuclear preparations of others; but since dinitrophenol, which activates mitochondrial adenosine triphosphatase, has no effect on the nuclear enzyme, it seems that nuclear adenosine triphosphatase is different from the mitochondrial enzyme. The association of phosphorylase with the nucleus is also puzzling; it is not clear what part the enzyme plays in the synthesis or breakdown of nuclear glycogen.

Forty to fifty per cent of the cell's 5′-adenylic phosphatase activity was found in the nucleus, but most of the remaining activity was associated with the microsomes. This observation is interesting in several respects. First, here is a phosphatase active at an acid pH yet not associated with lysosomes. Second, inasmuch as the enzyme is present in both nuclei and microsomes, nuclear 5′-adenylic phosphatase could be attached to those nuclear granules that resemble microsomes. Numerous enzymes, such as lactic, glutamic, malic, glucose-6-phosphate, isocitric, hydroxybutyryl CoA dehydrogenase, ATPase, and β-glucuronidase have been found in association with nuclear ribosomes. Among this group of enzymes, lactic dehydrogenase and malic dehydrogenase seem to be present in the highest concentrations. Of course, it remains to be established that nuclear ribosomes are genuine and not the result of contamination by microsomes.

A number of investigators have claimed that cytochrome oxidase is associated with the nuclear membrane. Careful studies of Jarasch and Franke seem, however, to demonstrate that the presence of the enzyme in the nuclear membrane results from contamination by enzymes of mitochondrial origin [268].

Arginase, probably a manganese protein with a pH optimum between 9 and 10, is an enzyme activated by manganese and cobalt. It splits one molecule of arginine in the presence of water to form a molecule of ornithine, which is itself an intermediate in the formation of citrulline. This enzyme occupies a key position in the steps connecting the Krebs cycle to the urea-forming system. All the other steps involved in the formation of urea take place in cell fractions other than the nucleus. Therefore the presence of arginase in the nucleus remains a mystery. However, arginase, like catalase—another enzyme often found in the nuclear preparation—is not present in all nuclei, and was found only in mammalian liver nuclei.

The description of the enzymic equipment of the nucleus is not solely of academic interest. If each nucleus has a specific enzyme mosaic, this pattern may play a considerable role in determining cellular differentiation. In that case, nuclei from different tissue would have an identical stock of chromosomes surrounded by different enzymic patterns, which may affect the screening of the genetic information transferred to the cytoplasm.

The Chromosomes

Morphology of the Chromosomes

With ordinary staining techniques, chromosomes can be seen and identified only during mitosis. They cannot be recognized in the interphase cell, although a thread-like structure resembling the chromosome has been described. New cytochemical techniques have permitted visualization of the chromosomes of mammalian nuclei. In these techniques, biopsy material of bone marrow cells, for example, are trypsinized and cultured for a period of time to provide a number of cells sufficient for investigation. The proliferating cells are then blocked in mitosis by the addition of colchicine to the medium. The cells are taken up in a hypotonic medium, squashed, fixed, and stained with the Feulgen method so that the chromosome pattern of a given cell can be counted and mapped out onto a graph called an idiogram.

One of the most surprising observations made with the aid of this new technique involved a change in the accepted number of chromosomes in humans. For years, the number of chromosomes in human cells had been assumed to be 48; now it is established that they contain only 46 chromosomes. The human diploid cell contains 22 pairs of autosomal chromosomes and one pair of sex chromosomes. In females, the sex chromosomes are identical; in males, there is one sex chromosome of the X type and a sex chromosome of the Y type. Thus, the male is the heterogametic sex in humans as Wilson [47] postulated previously.

Careful examination of chromosomes from many different living sources had led Heitz [29] to propose a model for the chromosome. Under the light microscope, the mitotic chromosome looks like a rod, a V, or a comma. The basic structural element is the chromatid. Two chromatids are suspended in the fundamental substance or matrix, and these chromatids separate during anaphase. Under appropriate conditions, in each chromatid two or more individual fibers—the chromonemata—can be identified. Various foci of differentiation are present within this basic structure: the centromere, the secondary constriction, the heterochromatin, and the nucleolar organizer. The centromeere has already been described as a specific area of the chromosome coinciding with a constriction in its structure and resistant to Feulgen staining. The heterochromatin corresponds to a dense portion of the chromosome that consistently stains with basic dyes and is therefore referred to as heteropyknotic. The fine structural differences between hetero- and euchromatin are not clear, but it has been suggested that excessive spiralization of the chromonemata is responsible for the heteropyknosis. The matrix is also not well understood; it has been defined in different ways by different investigators. "Matrix" usually refers to nongenic material accumulating around the chromosome thread during mitosis, either in the form of an "achromatic matrix," free of nucleic acid, or in the form of a "chromatic matrix," containing nucleic acids.

The Role of the Chromosomes in Genetics

The notion that the chromosome constitutes the morphological entity storing genetic information stems in part from experiments in which the cytological traits of parents and offspring were carefully correlated with controlled-breeding experiments. Some genes could be directly associated with a specific chromosome recognizable under the microscope. For example, sex and chromosome patterns are closely associated. In grasshoppers and in many other orthopteran insects, the male possesses one less chromosome than the female. *Drosophila melanogaster*, both male and female, have four pairs of chromosomes, but whereas the chromosomes of each pair are alike in the female, one of the pairs is asymmetrical in the male. This pair is composed of a rod-shaped X and a hook-shaped Y chromosome (see Table 2-1).

Table 2-1

	Autosomes	Sex chromosomes	
		♀	♂
Grasshopper	2 × (6A)	XX (14)	X (13)
Drosophila	2 × (3A)	XX (8)	XX[a] (8)
Man	2 × (22A)	XX (46)	XY (46)

[a] X is rodlike, Y is hook-shaped.

Around 1910, Morgan [48, 49] was breeding Drosophila for the purpose of studying the transmission of hereditary characteristics. Most of these fruit flies have red eyes, but one in a thousand has white eyes. Naturally, Morgan attempted to mate red-eyed flies with white-eyed flies. The results were very different, depending on the sex of the red-eyed fly. If a white-eyed male was mated with a red-eyed female, all the flies born in the first generation were red-eyed. Now, if the flies of that first generation were allowed to breed together, the distribution in the color of the eyes in the next generation was as follows: one-quarter of the flies were white-eyed, and all the white-eyed flies were male. All female flies were red-eyed, but half of the males were white-eyed.

Quite different results were obtained if the reciprocal experiments were carried out: mating a white-eyed female with a red-eyed male. In the first generation, all the females were red-eyed, and all the males were white-eyed. The next generation would be divided into four groups: $^1/_4$ white-eyed females, $^1/_4$ red-eyed females, $^1/_4$ white-eyed males, and $^1/_4$ red-eyed males (see Fig. 2-4). The following conclusions could be derived from such observations: (1) the color of the eye—red or white—is carried by allelic genes; (2) when two alleles are present, the red eye is dominant; and (3) the white-eyed gene must be carried by the X chromosome because the transmission of the white-eyed character corresponds to the transmission of the X chromosome in the second generation. Thus, the male Drosophila transmits the sex-linked character to his grandson through his daughter. The daughter is the only one to inherit the X chromosome of the father, since the sons inherit the hook-shaped Y chromosome.

The mode of inheritance of numerous human diseases is similar to that just described for the sex-linked trait in Drosophila. Two diseases are most typical in this respect—color blindness and hemophilia. Color blindness is a rather common disease, but the distribution among male and female varies considerably. The

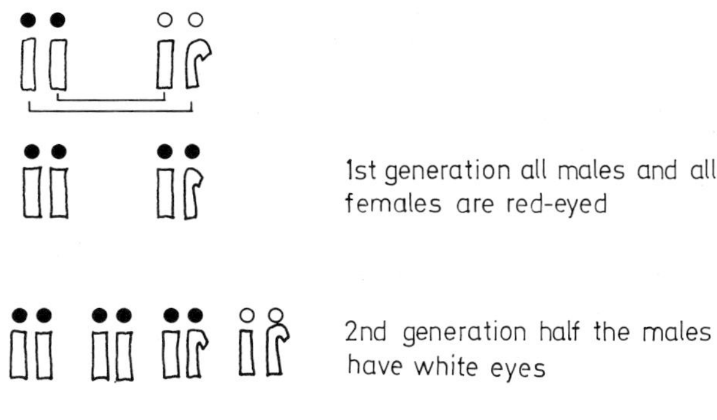

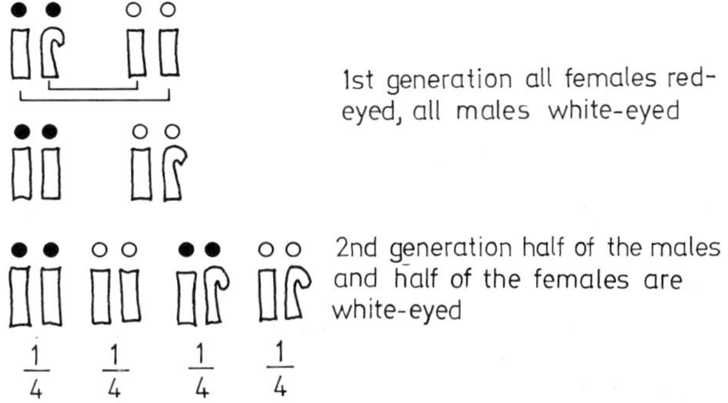

* Red eyes are represented by black circles
White eyes by open circles

Fig. 2-4. Results of controlled-breeding experiments in *Drosophila melanogaster*

incidence is around 8% among males, but it is only 0.5% among females. This marked difference in the distribution of color blindness among sexes led Wilson [47] to assume that the disease is transferred by a recessive gene contained on the X chromosome; and that in man, the male is the heterogametic sex. On the basis of these assumptions, one can expect the transmission of the color blindness gene to resemble that of the white-eye characteristic in Drosophila. Color blindness is transmitted by a color-blind father to his grandsons by the intermediary of his daughter, who alone can inherit the father's X chromosome. The color-blind mother, however, can transfer the disease to her sons and, through them, to both her granddaughters and grandsons (see Figs. 2-5 and 2-6).

The transmission of hemophilia is very similar to that of color blindness. Hemophilia is transmitted through a recessive gene linked to the X chromosome. Again, the father carries the pathogenic gene and transfers it to his grandsons through the intermediary of his daughter. Queen Victoria, who, of course, had no sign of hemophilia, is responsible for transmitting the disease among the royal families of Europe.

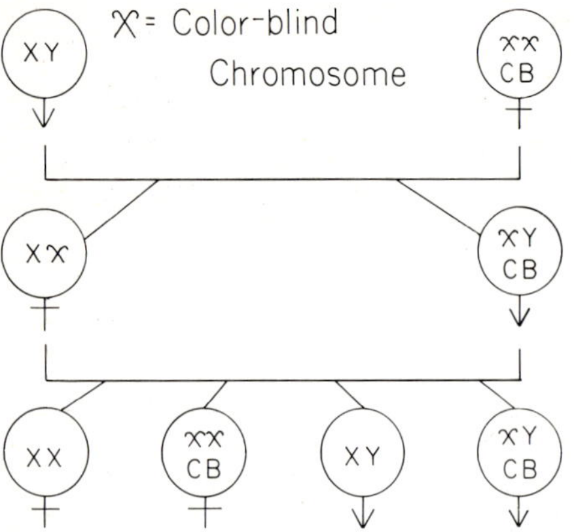

Fig. 2-5. Mode of inheritance of colorblind chromosome (mother)

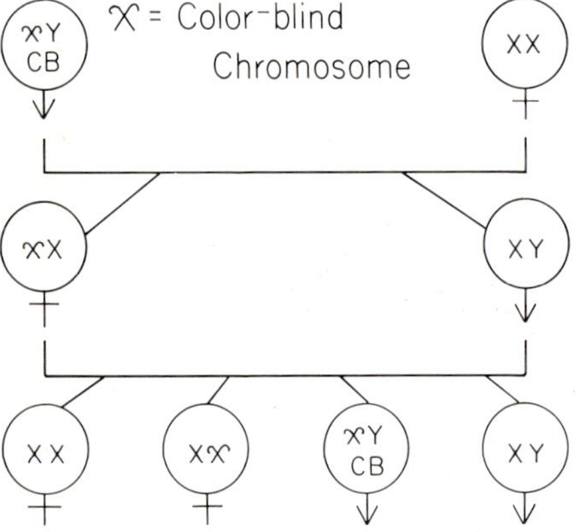

Fig. 2-6. Mode of inheritance of colorblind chromosome (father)

Obviously, there are more genes than there are chromosomes. In man, the genetic characteristics are numerous and each cell has only 46 chromosomes. It follows that many genes are stored within a single chromosome. Therefore, it is not surprising that groups of genes probably in a single chromosome are transferred simultaneously to the progeny. This is the phenomenon of linkage—a genetic event that in some ways contradicts the original theory of Mendel, who assumed that all genes were single units transferred at random.

The concept of linkage stems mainly from studies on maize carried out in the early 1900's. In his studies on maize, Hutchison [50] found two types of ears, some containing seeds that were full and colored, and others that contained shrunken, colorless seeds. By crossing the first generation of seeds, Hutchison soon discovered that the "full" character was a single dominant over the shrunken character, and that the "color" character was a single dominant over the colorless. According to the original mendelian theory, which postulates that genes are transmitted in a segregated manner, one would expect, after a few crosses, to have four different phenotypes distributed at random: (1) color and shrunken, (2) color and full, (3) colorless and shrunken, and (4) colorless and full. This, however, is not the case. The phenotypes "colored and full" and "colorless and shrunken" are 26 times more frequent than the phenotypes "colored and shrunken" and "colorless and full" (see Table 2-2). This suggested that the genes responsible for color and fullness, on the one hand, or colorless and shrunken, on the other hand, are linked to each other and transmitted together. Such an assumption explained the prevalence of the two phenotypes. However, in these experiments, a small percentage of the seeds, approximately 6%, completely dissociated the characteristics of fullness and color (about 3% of the seeds are full and colorless, and another 3% are shrunken and colored). Thus, even linked genes can be dissociated.

Table 2-2. Linkage of color and fullness in maize (F_2)

Colorful	4,032	96% linkage
Shrunken	149	3% dissociation
Colorless	152	3% dissociation
Shrunken	4,035	96% linkage
	8,368	

The mechanism invoked to explain the dissociation is as follows: the fragments of the chromosomes that contain the two-linked allelic genes occasionally overlap in a crosslike pattern; then, in the course of the meiotic division that follows, they interchange parts between chromosomes. In the example of maize, assume for a moment that the characteristics "colorless" and "shrunken," carried by the same chromosomes, are segregated in such a way that the characteristic "colorless" occupies the upper half of the chromosome, and "shrunken" occupies the lower part of

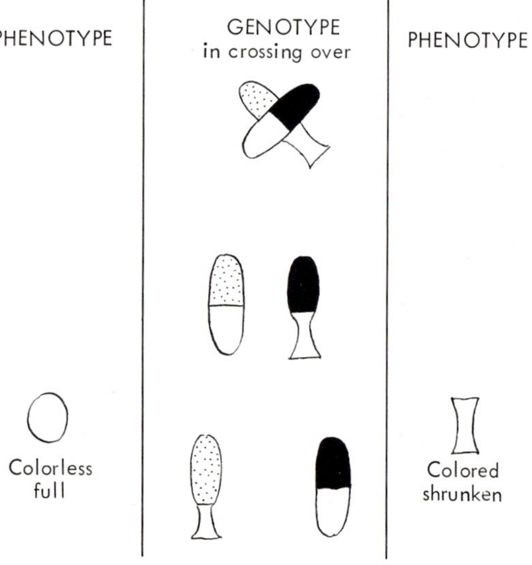

Fig. 2-7. Crossing over

the chromosome. Similarly, another chromosome would carry the color character in its upper part, and the fullness gene in its lower half. Now, if in the course of meiosis the two chromosomes are superimposed, forming a St. Andrew cross, and are then split according to a vertical line passing through the center of the cross, two new types of chromosomes will appear: one whose upper half contains the color gene, and the lower half, the gene for shrunkenness. In contrast, the upper half of the other chromosome will contain the "colorless" gene, and the lower half, the "fullness" gene. If the first of these new chromosomes is paired with the "colorless-shrunken" chromosome, it will lead to a seed of the "color-shrunken" type, since "color" is dominant over the colorless character. In contrast, if a "colorless-shrunken" gene is paired with a colorless-full gene, it will lead to the colorless-full phenotype (see Fig. 2-7).

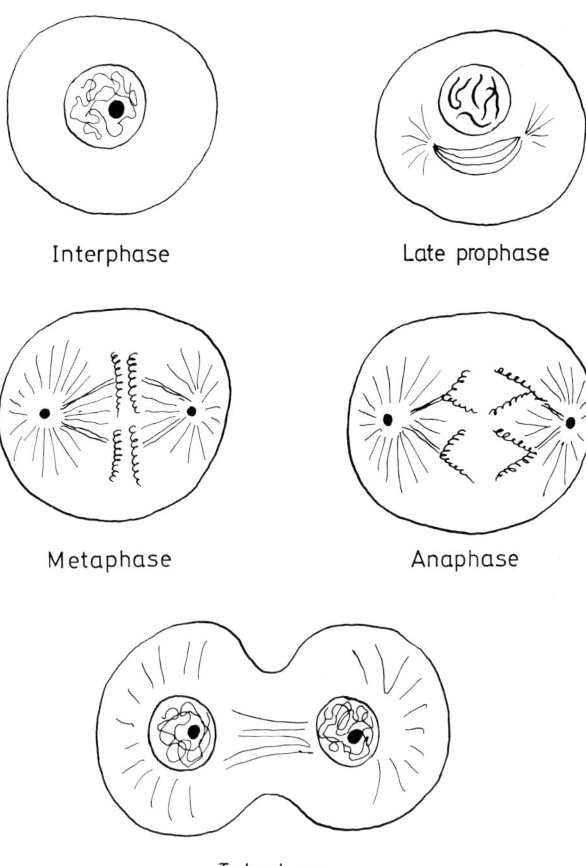

Fig. 2-8. Schematic presentation of mitosis

Mitosis

The chromosome is thus entrusted with storing the hereditary information. This information is transferred from cell to cell during the process of cell division. The reproduction of the chromosomal material and its distribution in equal parts to two daughter cells is a crucial event in the cell cycle [51–53].

Mitosis is a continuous process, the details of which vary considerably from one cell type to another. However, it is convenient to describe four different stages in the mitotic process: prophase, metaphase, anaphase, and telophase (see Fig. 2-8). In the beginning of prophase, the physicochemical properties of the chromosome change. The chromosomes, which in the interphase cell were not identifiable and could hardly be made visible by ordinary methods, become clearly apparent. Changes in the degree of hydration of the chromosome are thought to be responsible for its visibility during prophase, the stage during which the individual chromatids shorten and thicken. It is during prophase that the nuclear membrane disappears and the nucleoli vanish, but in some protozoa, the nucleolus persists during the entire mitotic cycle.

Important cytoplasmic changes also occur during prophase. The interphase cell contains a cytoplasmic structure closely associated with the nuclei, called the centrosome. Although the existence of the centrosome during interphase in cells is sometimes questioned, there seems to be no doubt that the organelle plays an important role during prophase. During that period of the cell cycle, the centrosome divides and the two halves of the centrosome migrate in opposite directions but remain united by a group of fibers that form the spindle. The cytoplasm surrounding each daughter centrosome is modified in a way that leads to the formation of the aster. At the end of the migratory process of the two centrosomes, each of them is located at one of the poles of the spindle. The presence of centrosomes, however, is not a *sine qua non* for the development of the spindle; in some insects and other animals and all higher plants, the spindle may develop in the absence of a pair of centrosomes.

During metaphase, the chromosomes are arranged in the center of the spindle to form the equatorial plate. They are attached to the spindle in a specific area of their structure called the centromere, spindle attachment, fiber insertion, or kinetochore. The centromere can be considered as a permanent organ of the chromosome. It has been recognized at all stages of mitosis or meiosis, provided that the preparations are adequate. The relationship between chromosome and spindle, and the spatial arrangement in the equatorial plate depends largely on the position of the centromere within the chromosome. That position is a constant for a given chromosome.

Centromeres may occupy two main positions with respect to the chromosomal structure: median and submedian. They are never truly terminal. If the centromere is median, the chromosome has a V-shape with two equal branches; and if the centromere is submedian, the chromosome forms a hook. Several experimental observations emphasize the importance of the centromere in the attachment of the chromosome to the spindle. X-Irradiation occasionally breaks chromosomes. If the break is unequal, only that part containing the centromere is able to attach to the spindle; the other remains free in the cytoplasm during mitosis (see Chapter 3, Chromosome Anomalies).

During anaphase, the chromatids separate and are distributed to each daughter nucleus in equal number. Anaphase is characterized by the migration of the sister chromatids in opposite directions along the spindle from the equator to the pole. This migration separates the chromatids of the shorter chromosomes completely, but the chromatids of the longer chromosomes may remain in contact and separate completely only in the later stages of anaphase, a period during which the center of the spindle seems to elongate between the centromeres. Thus there are two stages of anaphase: independent migration of the chromatids and progressive elongation of the spindle.

During telophase, the polar cap of the spindle undergoes dissolution, but the central part remains intact for relatively long periods of time after the onset of telophase. The nuclear membrane reappears, and the steps during the latter half of prophase are reversed; the individual chromosomes vanish and the coiled spiral disappears. (During interphase, the chromatin forms a bundle of coiled Feulgen-positive material. Individual chromosomes cannot be recognized, but this does not necessarily imply that there is a continuous spireme as described by early histologists. In any event, individual chromosomes appear in early prophase.) New nucleoli become apparent in the nucleus, and finally, cytokinesis occurs.

It is obvious from this description that the chromosome is the most important element in mitosis, and that at the end of mitosis, each chromosome has produced an exact replica of itself that is transferred to a daughter cell. What is it in the chromosomal structure that makes such replication possible? An answer to that vital question was sought by studying the ultrastructure and the chemical composition of the chromosome.

The Chemistry of the Chromosome

An adequate understanding of the chromosome's chemical composition would help considerably in expressing heredity in detailed molecular terms. Much of our knowledge of the biochemical composition of the chromosome stems from studies correlated with two different methods: Histochemistry, based on ultraviolet absorption and histophotometry of preparations treated with specific stains, and biochemical analysis of the chromosome after its isolation from cell nuclei.

Miescher was the first to investigate the biochemical composition of cell nuclei. He prepared nuclei derived from pus and discovered a specific nuclear component that he named "nuclein." About 20 years later, Holtman discovered that the compound isolated by Miescher was acidic, and that phosphorus was responsible for the acidic properties; therefore, "nuclein" was renamed "nucleic acid." Later it was found that there were two types of nucleic acids—ribose and deoxyribose. For a long time, ribonucleic acid was thought to be present only in plants, and deoxyribonucleic acid only in animals. By applying fuchsin sulfuric acid reagent (which stains deoxyribose sugars) to nuclei after their hydrolysis with hydrochloric acid, investigators found that deoxyribonucleic acid is associated with the chromonemata of all cells.

At this time, remarkable studies of Brachet [54, 55] proved that ribonucleic acid is present also in animal cells. He stained the preparations with pyronine and methyl green, and by comparing the staining properties of the cells before and after treating the preparation with ribonuclease, Brachet conclusively demonstrated the presence of ribonucleic acid in all cells investigated.

In 1934, Caspersson [56, 57] took advantage of the fact that both ribose and deoxyribose nucleic acid absorb light at 260 mμ and built a microscope with quartz lenses enabling him to take ultraviolet photomicrographs of resting and dividing cells. This method permitted the amount of nucleic acid in the cell to be measured. Of course, ultraviolet microscopy does not distinguish RNA from DNA. However, by combining ultraviolet photometry with histochemical methods, it was possible to demonstrate that some of the nucleic acids found in the cell nucleus are not of the deoxyribose type.

Although they absorb ultraviolet light, the nucleic acids of the nucleoli are not stained with the Feulgen reagent and disappear after treatment with ribonuclease. Small amounts of ribonucleic acid are also associated with the chromosome; and, finally, after chromosomal nucleic acid is digested, the presence of the basic proteins can be revealed by adequate staining methods.

Chromosomal Proteins

At least three biochemical components have been recognized histochemically in the cell chromosome: DNA, RNA, and protein. We will now consider each of these components separately.

There are at least three types of chromosomal proteins: histones, protamines, and residual proteins [58–63]. It has been estimated that the nuclear histones represent 90–92% of the total protein in the chromosome. The remaining portion (8–10%) is composed of the residual protein, or non histone proteins.

Protamines

Two main groups of basic proteins have been found in the cell nucleus: protamines, which are present only in the sperm of fish and lower animals (a protaminelike protein, galline, has been found in rooster sperm), and histones, found in all somatic cells. Protamines are acid-extractable polypeptides with a molecular weight ranging between 4,000 and 10,000. Basic amino acids (particularly arginine) constitute the main amino acids in the protamine polypeptide chain; in fact, one-third of the amino acid residues of protamines are arginine. Proline and alanine are the only amino acids that have been found in the NH_2-terminal position of protamine. One of the differences between protamines and histones is the absence of sulfur amino acids in the former. It was first believed that the main sequence in protamines is of the type AAAXYAAAA, where XY are amino acids other than arginine. However, this hypothesis was disproved when it was found that the sequence of one of the polypeptides (peptide Z), obtained by the tryptic digestion of clupeine, was Ala-Arg_4-Ser-Arg_2-Ala-Ser-Arg-Pro-Val-Arg_4-Pro-Arg_2-Val-Ser-Arg_4-Ala-Arg_4.

To form nucleoproteins, the basic groups of the arginine combine with the phosphorus residues of DNA. When protamines are allowed to recombine with DNA in the presence of sodium salts, a nucleoprotein appears that has properties very similar to nucleoproteins extracted from sperm head. Under appropriate conditions, these nucleoproteins can be precipitated in the form of fibers that can then be examined by X-ray diffraction.

The patterns obtained by X-ray diffraction suggest that the protamine has an extended β-form and is wrapped around the DNA molecule, filling in the small grooves of the DNA. The basic functions of the arginine residues of the polypeptide react with the phosphorus of the DNA. The XY amino acids of the polypeptides remain free and protrude to form loops.

Nucleoproteins are difficult to extract from mammalian spermatozoa; this is not surprising because the nucleus of mammalian spermatozoa is protected against mechanical damage by a special keratinous membrane. The nuclear protein, like keratin, is rich in sulfur, but it contains more arginine than regular keratin. Studies on disrupted bull sperm nuclei revealed that the arginine content of mammalian sperm is intermediate between that of protamine and that of histone. Whether protamines are in fact present in mammalian sperm is not known. Analysis of a protein hydrolysate obtained after extraction of bull sperm does, however, indicate that a basic protein, relatively rich in arginine, is present in mammalian sperm.

At the terminal stage of spermatogenesis of salmon fish, nucleohistones are replaced by nucleoprotamine. Not much is known about the mechanisms by which one kind of basic polypeptide is replaced by another. What seems certain is that protamines are made in the cytoplasm and transferred to the nucleus before the displacement can take place. Prior to the exchange, both types of basic proteins, preformed histones and newly synthesized protamines, are phosphorylated.

The role that phosphorylation plays in the exchange is not clear. Phosphorylation does not seem to modify the affinity of histones for DNA but slightly decreases the affinity of protamines for DNA [64].

Histones

Histones can be extracted from whole cells, whole nuclei, or nucleoproteins. The methods of extraction do not protect histones against degradation by proteolytic enzymes; even the best nuclear preparations are contaminated by proteolytic enzymes. Washing the nuclei with trinitrotoluene and saline solution before extracting the nucleoproteins only partially eliminates the danger of hydrolysis. The whole histone preparation obtained by extraction of nuclei can be further fractionated into a number of subfractions. The first step in the fractionation procedure of histones is either chromatography or selective extraction followed by differential precipitation. The histone fraction can then be further subdivided by the use of starch gel or polyacrylamide gel electrophoresis. When the histone fraction is prepared, the basic protein can be traced in the chromatographic eluents, in extract, or in precipitates by at least three different methods: the ninhydrin test, absorption at 276 mμ, and fluorometry.

Histones can be separated by chromatography using amberlite resins, carboxymethyl-cellulose, and sephadex. The proteins have been eluted with different solutions, depending upon the investigator. The elution may be complete or incomplete. When incomplete, there is of course the possibility that specific histones with important biological properties have been lost. The histone preparations that are put on the column are usually heterogeneous, and several fractions are obtained. Because several laboratory groups have studied the amino acid composition of the various fractions, a comparison may be made between fractions obtained by different methods.

A fraction obtained by a given procedure can also be partially analyzed, and the polypeptide fractions can then be fingerprinted.

Many laboratories have succeeded in fractionating histones by selective extraction and differential precipitation. The selective extraction procedure is based on the difference in the amino acid composition of the various types of histones. For example, a histone-rich lysine can be obtained by extraction from whole thymus with 5% perchloric acid followed by precipitation with 10% trichloroacetic acid. If the histones found in the perchloric acid extract are passed through a carboxymethyl-cellulose column and eluted from the column with borate buffer at pH 9, then different components called 1, 2, and 3 can be recovered.

When whole thymus nuclei are washed with increasing concentrations of sodium chloride, different types of histones are extracted; for example, lysine-rich histone is extracted with the lowest concentration of sodium chloride.

These brief considerations of the various histone fractionation methods indicate that different types of histones are obtained, depending upon the method used. Sometimes a histone isolated by one method has an amino acid composition homologous to a histone isolated by another procedure. Often a procedure will yield a type of histone that cannot be obtained by other procedures. Whether the different fractions are representative of native histones or whether they are the result of degradation remains to be established.

slowly heated, optical density increases at about 65° C and progresses very rapidly to a peak of about 72° C. When DNA is complexed with its native histone, the changes in optical density start at higher temperatures (about 70° C) and progress slowly with increasing temperature to reach a peak at around 90° C. Although a graphic representation of the changes in optical density dramatically illustrates the difference between purified DNA and nucleohistones, immediate comparison is best expressed in terms of TM value—the temperature at which half of the hyperchromicity is manifested. It is 70° C for DNA, but 84° C for the nucleohistone. When native chromatin is heated, two shoulders are observed: one has a TM value of 69.5° C and the other's is 84° C, indicating that some of the DNA in chromatin is present in the form of nucleohistones. From these experiments it seems obvious that the complexing of histones with DNA confers upon the DNA molecule a stability that it does not possess on its own.

The histones added to an incubation mixture containing single-stranded DNA as primer, the four triphosphates (one of which is labeled), and DNA polymerase inhibit the incorporation of the labeled triphosphate into DNA. Whether this is a specific interaction of histones with the primer that operates *in vivo*, or whether it is simply the result of precipitating the primer out of the incubation mixture is not established.

Whether the interaction between histone and DNA is specific and occurs *in vivo* to repress DNA is not certain. However, it cannot be excluded that the histones interfere with the polymerase reaction simply by making the primer unavailable in a fashion common to all basic polymers. Histones and protamines interfere with the hydrolysis of DNA by deoxyribonuclease, and it thus can be established that the interference results from the precipitation of substrate. Injected, histones have been claimed to interfere with DNA biosynthesis in regenerating liver. Such intriguing observations undoubtedly have to be confirmed in several laboratories before their full impact can be appreciated. In conclusion, modern studies suggest that histones may be important in the transfer of genetic information from DNA to messenger RNA, but the molecular details remain to be elucidated.

In the cell, the histones are associated with DNA and nucleic acid, and the proteins are extracted as a nucleoprotein complex. Deoxyribonucleoproteins contain histones and small amounts of nonbasic protein and lipids as well. The cationic groups of the histones most likely neutralize the phosphate groups of the DNA, but there is some evidence that the chromosomes contain DNA molecules not coated by histones.

The nucleoproteins contain relatively large amounts of sodium ions. This is not surprising because it has been known for a long time that polyelectrolytes such as DNA react with the counter ion. The exact amount of sodium associated with DNA is not known. In a nucleoprotein preparation, there is enough sodium to neutralize 20% of the phosphate groups of DNA. Thus, it is possible that although 20% of the phosphate groups of DNA are neutralized by sodium, 70% are neutralized by the lysine and arginine residues of histone.

The configurational relationship between histones and DNA remains ambiguous, mainly because no clear-cut results can be obtained with standard physicochemical analysis. Sharp patterns are obtained with X-ray diffraction of DNA, but X-ray diffraction studies of histones provide only diffuse images in which the ring shape and the orientation of the arc are difficult to define. However, it seems well established that the DNA maintains its structure inside the nucleoprotein. Studies of nucleohistones by infrared spectroscopy or rotatory dispersion have also yielded results that are not easily interpreted.

Clearly, the ionic bonding between DNA acidic-DNA phosphate group and the basic lysine residues of the histones constitutes a major mode of interaction between the two macromolecules, but ionic bonds are not the only type of bonds between DNA and histones. The finding that urea helps to dissociate the DNA-histone complex, especially in the case of the arginine-rich histones, suggests that hydrophobic bonding may be involved as well [74].

A favored model for histone binding to DNA proposes that histone binds only to the major and not the minor groove of DNA. In this arrangement, the basic part of the histone is in an extended configuration, whereas the nonbasic parts are partially or totally condensed into an α-helix [75].

Since the phosphate group plays a principal role in determining the electrolytic properties of DNA, it can be expected to be the primary site of ionic bonding with Na^+. Whether bonding between DNA and the counter cation is site bonding or ion-atmosphere bonding has not been established. The relevance of sodium bonding to DNA and to the role of DNA in genetics is not known. However, it has been established with solutions of DNA that the stability of the DNA molecule as measured by temperature denaturation varies with the ionic strength of the medium [76].

Histone nomenclature

Nomenclature		Molecular Weight	Molar ratio/ F1 of Histone
I	F1	21,000	1
IV	F2Al	11,300	2
IIb1	F2A2	14,500	2
IIb2	F2B	13,700	2
III	F3	15,300	2

All types of histones are found in chromatin in equimolecular amounts except for F1 of which there is only half as much as of the other histones. We have seen that in the classical view, based on X-ray crystallography studies, the DNA histone complexes in chromatin are believed to form a supercoiled helical structure. Although it is certain that chromatin fibers are folded and coiled, the pattern of this coiling and folding has not been discerned. One may, therefore, have the impression that these folds and coils occur at random as they would in an overextended spring whose ends are suddenly set loose, thus permitting the wire to recoil, not with the original helical structure, but in a tangled-up fashion.

Shadow cast samples of chromatin examined under the electron microscope reveal widely spaced nodules alternating within the strand. An accurate molecular interpretation of these pictures is not available and such pictures are not in conflict with the supercoil concept [233].

Roger Kornberg [234] and his associates have presented a different working hypothesis for the interaction of DNA and histones in chromatin. They propose that the chromatin, instead of being made of supercoiled springs, is a flexible chain made of a number of repeating units. The repeated unit is composed of a globular histone tetramer $(F2Al)_2(F_3)2$ which has, in fact, been extracted from chromatin preparations under conditions mild enough that they would seem to preclude artifactual self-aggregation. The core is linked by dimers or oligomeric chains of F_2A_2 and F_2B, which determine therefore the spacing between the tetramer's F_2A_2 and F_2B links. Because chromatin treated with nucleases releases 200 base pair segments of DNA, it is proposed that a set of 200 base pairs is complexed with repeating units, most with the tetramer. The remaining DNA connects these tetramers along a path defined by F_2A_2 and F_2B.

Acidic Proteins. When histones have been extracted from chromosomes, the chromosomal DNA is far from pure; it still contains the so-called residual or acidic proteins. Although there is no doubt that acidic proteins are in the chromosome, the exact amount and composition remain unknown. The amount ascribed to the chromosome varies according to the mode of extraction. Some of the early investigators proposed that as much as 30–70% of the total protein population of the chromosomes are residual, but the more recent data suggest this amount to be between 8% and 10%.

A great variety of methods have been designed to extract acidic proteins from the chromosomes. They are all rather drastic, and it is not certain whether they allow the functional properties of the acidic proteins to survive the procedure. In one method designed in Bonner's laboratory [77] chromatin is prepared and then extracted with acid to exclude histones. After histone extraction, the residue is treated with sodium dodecyl sulfate. The DNA sediments and the proteins are collected in the supernatant. Ammonium sulfate precipitates the proteins, which can then be submitted to polyacrylamide gel electrophoresis. The polyacrylamide gel electrophoresis of such a preparation yields a multiband pattern. The patterns are similar for acid proteins extracted from rat kidney and liver, but different for acid protein extracted from pea bud chromatin and rat liver chromatin.

Bonner and his associates [77] have summarized the minimum requirements for demonstrating gene activation by acidic proteins. One requirement is to exclude the possibility that the effect of the acidic proteins on gene expression results from histone precipitation.

It now appears that the acidic proteins act in concert with histones to activate genes. Thus, DNA covered with acid proteins has priming properties for RNA polymerase similar to those of naked DNA. But, the addition of acidic proteins to transcribing DNA counteracts the inhibitory effect of histones.

The acidic proteins are synthesized in the cytoplasm and located in the major groove of the DNA molecule. Organizing the scanty available knowledge on acidic proteins, Baserga [78] has proposed an interesting mechanism for the gene activation that takes place in a special case of initiation of DNA synthesis. Baserga proposes that stimulants modify the cell membrane by inducing the loss of an important membrane component, which in some ways inhibits translation of acidic proteins from preexisting RNA templates. Once the inhibition is relieved, the acidic proteins are synthesized in the cytoplasm and transported in the resting cell.

Two hypothetical functions—structural and derepressant—have been proposed for the acidic proteins. The acidic nucleoproteins were at first believed to link various DNA strands together in the chromosome. Therefore, it is interesting that the acidic proteins, rather than the histones, become crosslinked to the DNA in cells exposed to bifunctional alkylating agents, and that it has been proposed that the acidic protein is the very one affected by X-irradiation. The damage to the acidic protein caused by X-irradiation is believed to be responsible for breaking and altering the chromosomes.

It has also been suggested that acidic proteins function as derepressors of regulatory genes. This hypothesis is based on the finding that the amount of acidic nucleoproteins is greater in tumors than in normal, rapidly proliferating tissues such as spleen. Russian investigators have claimed that the amount of residual protein in the chromosome increases during liver regeneration. In addition, at least at first approximation, the coincidence of amino acid incorporation into nuclear proteins and of RNA precursors into nuclear RNA appears to provide supportive evidence for the role of acidic proteins in the control of messenger RNA. However, one cannot exclude the possibility that the parallel synthesis of acidic proteins and nuclear RNA indicates that DNA synthesis is paralleled or immediatly followed by the biosynthesis of templates needed for the transcription of new coating proteins. For further details see Chap. 16.

DNA Content per Nucleus

A series of biological observations indicated that DNA is the cellular component most likely to be storing genetic information. Some properties of the substance storing genetic information can be predicted: it must be in the chromosomes; it is probably not altered during interphase; and the exact replica of the compound has to be made during mitosis. In addition, since the stock of genetic information in each cell is derived from replication of the chromosomes of the fertilized ovum, the amount of DNA per chromosome should be constant [79–82].

The evidence supporting DNA's participation in the transfer of genetic information is both indirect and direct. The indirect evidence rests on studies of DNA's cytochemistry and on the effect of mutagens. The direct evidence is based on experimentation with transforming agents and bacteriophages. It was demonstrated histochemically that DNA, with few exceptions, is found in the cell nucleus—specifically, within the chromosome. The DNA content per nucleus is constant in all diploid nuclei of a given species. DNA is the only chromosomal constituent without turnover.

Boivin and his associates [83] were among the first to determine DNA content in nuclei. They used two different methods: chemical analysis of a preparation containing a known number of nuclei, and direct cytophotometric measurements of the DNA content in a single nucleus. The two methods yielded results in reasonable agreement, and led to the conclusion that the DNA content per chromosome is constant in a given species and not influenced by environment. Exceptions to these rules in mammalian chromosomes are observed only during cell division and under some pathological conditions. For example, the DNA per nucleus is increased in some of the nuclei of intestinal polyps or in some cancers. As can be expected, an increase in the DNA content per nucleus is observed during cellular division and, therefore, the DNA content is greater in some nuclei of embryonic tissues and regenerating tissues.

Erythroblasts and megaloblasts differ from ordinary cells with respect to DNA content. The DNA content per erythroblast nucleus decreases during differentiation. In the megaloblast (found in pernicious anemia), nuclear DNA content is greater, probably as a result of polyteny.

DNA and Mutagens

Mutagens can be divided into two main types—chemical and physical—both of which act on the DNA molecule. Chemical mutagens induce mutation in two ways: first, the combination of the mutagen with the DNA molecule modifies its physicochemical properties, and, second, direct incorporation of the mutagen within the primary structure of the molecule alters the base sequence. Physical mutagens, like X-irradiation and ultraviolet rays, are known to block DNA synthesis *in vivo* and *in vitro* and to cause alterations in DNA structure (see Chapter 12).

Such observations constitute only circumstantial evidence for the genetic role of DNA because mutagens affect other cellular constituents as well as DNA. Among all the studies with mutagens, one of the most intriguing is the investigation of the effect of ultraviolet radiation. A source of ultraviolet light can induce mutation in a cell population as a function of the wavelength. The degree of mutation is highest when the cell preparation is irradiated with a source of ultraviolet light with a maximum wavelength of 260 mμ, but is lowest when the maximum wavelength is around 280 mμ (260 mμ is in the maximum absorption range for nucleic acid, and 280 mμ is in the maximum absorption range for protein). Although these experiments with ultraviolet light strongly suggest that nucleic acids are important in mutation, a possible effect of the mutagen on RNA and proteins was not ruled out. Mutations develop after administering ultraviolet light with a wavelength of about 280 mμ; furthermore, proteins will absorb at 260 mμ. Therefore, such experiments do not exclude the possibility of protein participation in hereditary determination.

Bacterial Transformation

Experiments carried out with bacteria and bacterial viruses offer conclusive proof of the nucleic acid nature of the hereditary determinants. A critical experiment was done in 1928 by Griffith [84], who injected mice with two bacterial preparations: a nonencapsulated pneumococcus called rough and referred to as the R type, and an encapsulated pneumococcus referred to as the S type. When mice were injected simultaneously with living R-type pneumococci and heat-killed S-type pneumococci, live S pneumococci appeared in the culture of the peritoneal fluid. This suggested that the R type mutated to an S type under the influence of dead S pneumococci. Obviously, the objection can be raised that in such an experiment some of the S-type pneumococci survived heat and the peritoneal environment favored the growth of the S-type bacteria. Such an objection was overcome by the experiments that demonstrated that water extracts of the S type were capable of transforming the R cell into an S cell, even after passage through a bacterial filter. Years later, it was finally demonstrated that the component responsible for the transformation was DNA.

In 1944, Avery, MacLeod, and McCarty [85] extracted highly polymerized deoxypentose nucleic acid from the active preparation and demonstrated that this purified DNA could transform the R cell into an S cell. Later, transformation was demonstrated with a variety of bacteria, including *E. coli, Shigella paradysenterica, Bacillus proteus, Salmonella*, staphylococci, tubercular bacilli, and other bacteria. Mutations induced by transforming agents cause molecular changes in the elaboration of the bacterial capsule polysaccharide, in the appearance or disappearance of enzymes, and in the development or resistance to antibiotics.

There are at least 70 different types of pneumococci; the most common are types I, II, and III, whereas some less frequent are IV and XIV. All these types differ by the presence of a specific polysaccharide in the capsule. Thus, transforming one type of pneumococcus into another determines the biosynthesis of the specific polysaccharide that is then used to make the capsule. Transformation of one type of *Haemophilus influenzae* into another induces the formation of a specific polysaccharide and also leads to the appearance of a new class of immunologically active compounds—polysugar phosphates. For example, the type B influenza bacteria contains a polyribose phosphate chain, whereas the type A contains a polyglucose phosphate chain, and the type C a polygalactose phosphate chain. Transformation may also involve alterations in the pathway of sugar fermentation, changes in glucose and lactic acid metabolism, and other changes.

Much evidence suggests that DNA is responsible for transformation. The substance extracted from bacteria that is capable of inducing transformation has physicochemical properties identical to those of DNA, and preparations have been obtained that contain less than 0.002% of protein and can transform bacteria. There is also some indication that transforming factors can be synthesized enzymatically by using the Kornberg polymerase system. It is unlikely that a transforming factor is an enzyme. The transforming principles prepared from *H. influenzae* can be heated at temperatures of up to 81° C for 1 hour without losing biological activity. Thus, available evidence suggests that the known transforming agents are nucleic acids rather than proteins.

The various inactivation mechanisms of the transforming principles are compatible with their identity with DNA. Although high ionic strength does not destroy the transforming properties of the principle extracted from *H. influenzae*, low ionic strength interferes with this biological property. Low ionic strength has been postulated to split the hydrogen bonds of the transforming molecule, as it would in DNA. When DNA is dissolved in the absence of salt, the anions in the molecule cause repulsion within the two chains.

Deamination rapidly inactivates the transforming factor; in fact, it has been suggested that all the primary amino groups are essential for biological activity. The transforming properties of *H. influenzae*, for example, can be inhibited by ultraviolet radiation. The dose necessary to inactivate the transforming factor is about that required to inactivate bacterial viruses. Similarly, appropriate concentrations of nitrogen mustards inactivate the transforming factor.

An interesting problem of transformation is the mechanism by which it is induced by DNA. DNA could be responsible in at least two ways. The DNA introduced into the cell could become part of the existing genetic stores of the bacterial cells, and thus directly participate in the transformation; or the introduced DNA could alter the existing DNA molecules without directly participating in the transformation. By using transforming DNA labeled with ^{32}P, workers have shown the transforming principle entering the bacterial cell where it takes one of two forms*: (1) a "reversible" one that is DNase sensitive and readily extractable by washing the cells with saline, and (2) and "irreversible" one that can be extracted only by treating the cells with lysozyme. The number of cells that undergo the transformation is proportional to the amount of "irreversible" DNA in the preparation. The irreversible attachment (DNase-sensitive) stage is followed by extensive degradation of transforming DNA. To explain this event, it has been proposed that the competent cell has more sites for DNA uptake than there are loci for integration of the transforming DNA on the chromosome. Bacteria do not segregate in favor of or against a specific transforming DNA. For example, *H. influenzae* pick up both their own and foreign DNA but transform only in the presence of foreign DNA.

Transforming DNA is quickly incorporated into bacteria, but the transforming properties are not immediately conferred to the native bacterial DNA; before bacterial DNA acquires transforming properties, a latency period of approximately 10–15 minutes must elapse. However, the acquisition of transforming properties precedes DNA replication. Therefore, copying of transforming DNA does not seem to be an absolute requirement for the acquisition of transforming properties, and the incorporation *per se* of transforming DNA into native DNA is sufficient.

The incorporation of transforming DNA in the native bacterial DNA is not arbitrary. When the transforming DNA carries a gene which belongs to the same cistron as another gene in the native DNA, after transformation both the transforming and the native genes are linked. Since no DNA has been synthesized during the acquisition of the transforming properties, it must be assumed that the transforming molecule is incorporated by breakage and recombination of the DNA. These findings suggest that the metabolism of the genetic material involves two separate steps: DNA breakage and recombination, and DNA replication. Although several investigators have recovered single-strand forms of the transforming DNA from the host, it is not clear whether the transforming DNA is integrated in the single- or double-strand form.

Thus, genetic transformation of a bacterial cell results from transfer of new genetic information by free DNA. The transformation process involves: (1) DNA uptake and penetration; (2) association and recombination of the transforming DNA with the genome of the host cell; (3) expression of the genotype in the phenotype; and (4) replication of the genome.

Transformation takes place only when the cell population exposed to the DNA is "competent." Competence is thought to be genetically controlled, and its development is believed to depend on the presence of glucose, divalent cations, and auxotrophic requirements. The development of competence is likely to

* It has been possible to calculate that only 1 out of 120 molecules of DNA in the original transforming principle is responsible for the transformation that induces resistance to streptomycin.

be associated with the elaboration of specific receptor sites for the DNA uptake. Thus, DNA would first attach to a specific membrane protein and then penetrate into a cell by a mechanism that is still not elucidated but must be an active, energy-dependent process [86, 87].

Bacteriophages

Despite the great importance of all the biochemical aspects of the bacteriophage, only the most significant property of that molecular complex is discussed here—the fact that it is capable of self-replication. Structurally, the T-even phage presents two different features, a hexagonal head and a cylindrical tail. The head is composed of a protein shell and a nucleic acid core. The nucleic acid is of the deoxyribose type, and all the cytosine of the deoxyribonucleic acid is replaced by 5-hydroxymethylcytosine. If a culture of *E. coli* is mixed with coliphage, the virus is soon adsorbed on the surface of bacteria. The tail of the bacteriophage contacts the bacterial cell. The first step of the adsorption is reversible and probably results from the formation of salt linkages between the protein of the bacterial membrane and that of the bacteriophage. In a second step, the binding is much tighter. It is assumed to involve the formation of S-S bonds between SH groups of the protein coat of the bacteriophage and the SH groups of the bacterial protein. The second step is probably catalyzed by specific enzymes. The lysozymes and neuraminidases contained in the tail of phage digest the lipopolysaccharide layer of the bacterial wall. Then, by a process that is still not clear but known to involve tail contraction, the nucleic acid core and a small amount of protein from the head of the phage are injected in the bacterial structure, leaving the protein coat attached to the outside of the bacteria. As a matter of fact, these protein ghosts can be harvested and recombined with *E. coli*, thus inducing abortive infection. Under those conditions, the ghosts of the phage kill the bacteria but do not replicate in it.

It has been conclusively demonstrated that it is mainly the nucleic acid that enters the bacteria. Bacteriophage can be prepared with ^{32}P-labeled DNA and ^{35}S-labeled proteins. When such labeled preparations are used, the proteins remain attached to the bacterial wall from which they can be stripped off. In contrast, the ^{32}P-labeled DNA remains so tightly associated with the bacteria that it cannot be washed off. However, small amounts of proteins always accompany the infecting DNA.

The newly introduced phage DNA blocks the specific replication ability of the bacteria without interfering with their bioenergetic pathways. A new pattern of synthesis takes place inside the bacteria geared to the *de novo* synthesis of bacteriophage DNA. These changes can best be illustrated by describing the events following the penetration of T-even bacteriophage in *E. coli* (see Fig. 2-9).

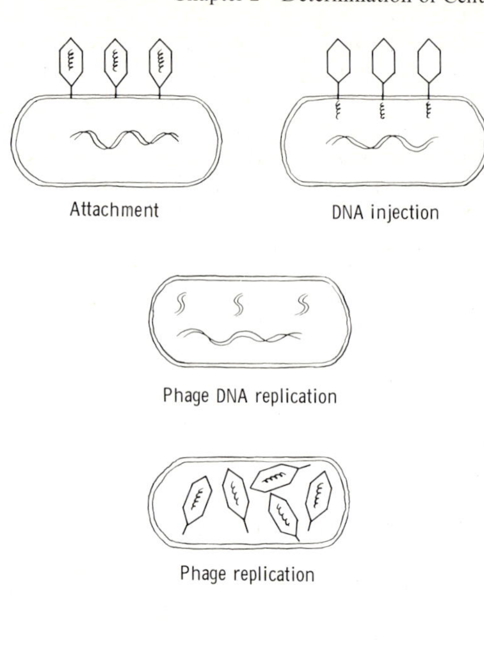

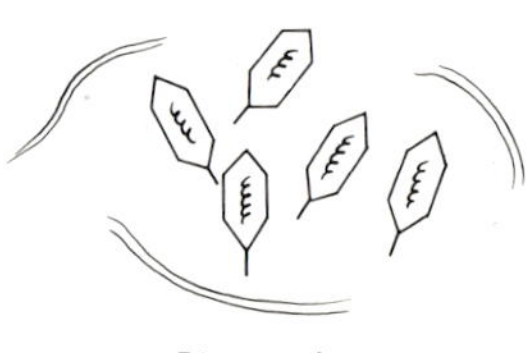

Fig. 2-9. Fate of phage = invaded bacteria

The interaction between T-even phage and bacteria can only be understood if the special properties of T-even phage DNA are taken into account. It was already mentioned that the DNA contained in T-even phage differs from other DNA by the presence of 5-hydroxymethylcytosine instead of cytosine. Furthermore, the 5-hydroxyl groups of the base are substituted by glucose in any of three types of substitution: an α-monoglucoside, a β-monoglucoside, or an α-diglucoside. Consequently, there are four different forms of the 5-hydroxymethylcytosine bases—three glucosylated and one nonglucosylated. The relative proportion of the various forms of 5-hydroxymethylcytosine in a given DNA molecule is determined hereditarily.

Because of the typical characteristics of the T-even phage, its fate can be traced easily once the virus has penetrated the bacteria. First, almost all of the phage DNA and very little of the phage protein penetrate the bacteria. After penetration, the biosynthesis of normal bacterial DNA is blocked, and 5–7 minutes after injection, the biosynthesis of typical phage DNA proceeds rapidly.

Phage DNA synthesis involves the phosphorylation of the four bases (dAMP, dGMP, dHCMP, and dTMP) to yield the respective triphosphates, which are then polymerized in the presence of DNA polymerase and a primer to yield new DNA. In this sequence of reactions, at least two enzymes are required that are not present in the uninfected *E. coli*: one that converts deoxycytidine to hydroxymethyldeoxycytidine, and another that converts the 5-hydroxymethyldeoxycytidine monophosphate to the triphosphate. Because of the presence of hydroxymethylcytosine and glucosidic linkages in the phage DNA, the DNA molecules are more resistant to DNase activity.

As may be anticipated, the destruction of the bacterial genome leads to a block in synthesis of a number of bacterial enzymes. Without these enzymes, infected *E. coli* must synthesize *de novo* the entire enzyme battery necessary for phage DNA synthesis under the direction of the phage DNA. We have already mentioned that a new kinase and polymerase appear, and several enzymes catalyzing the glucosylation of 5-hydroxymethylcytosine of the DNA molecules at the expense of UDP-glucose have been described. A new thymidylic synthetase develops that can be separated on a DEAE-column from the enzyme in the noninfected cell. The activity of the kinase catalyzing the formation of dGTP and dTTP increases more than ten times after infection.

Thus, in addition to being an excellent example of the balance of catabolic and anabolic pathways in directing metabolism, the infection of *E. coli* by the T-even phage constitutes one of the simplest models for determining specificity.

The Chemistry of the DNA Molecule

The elucidation of the molecular structure of the chemical compound containing the genetic information would naturally go far in helping to decode the mechanism of specificity transfer. Since Mieschers' early discoveries, biochemists, chemists, and physicochemists have studied the physicochemical properties of the DNA molecule. If physical and chemical studies of DNA are to have any biological significance at all, the DNA molecule extracted from living material must be identical in every way to that molecule found in the intact cell. At present, the biological properties of a DNA preparation can be evaluated only in terms of its ability to transform. For these reasons, the physicochemical properties of the transforming principles are of prime importance because they can serve as standards for all other DNA preparations. Obviously, many of these studies on the physicochemical properties of the DNA molecule have been carried out on preparations obtained from various sources and for which transforming properties have not yet been demonstrated.

A great number of methods for DNA extraction have been described, but, unfortunately, no universal method exists that can be used to extract DNA from all sources. Therefore, the investigator must develop the technique most suitable to his purpose. The ultimate goal in preparing DNA is to obtain an undenatured product, free of contaminants.

During the preparation of DNA, DNA is denatured by DNase and shearing forces. The deleterious effect of DNases in mammalian tissue is often exaggerated. These enzymes can readily be eliminated by differential centrifugation; adding EDTA to the homogenizing mixture complexes the magnesium ions required for at least some of the DNase activity, and washing the nuclear pellet with sodium dodecyl sulfate and other detergents, such as Triton X-100, eliminates traces of DNase activity. Furthermore, native nuclear DNA is quite resistant to the hydrolytic enzymes for two reasons: first, the presence of the complexing basic protein, and second, the presence of the specific DNase inhibitor. DNA hydrolysis by deoxyribonucleases can also be eliminated by homogenizing the tissues in the presence of phenol, which denatures the proteins.

The molecular weight of DNA prepared from various tissues has been estimated to have values ranging between 6×10^6 and 14×10^6. However, these DNA molecules probably are not representative of those present in the intact cell because shearing forces break the DNA molecules in the center, yielding two equal halves that can themselves be broken in similar ways, until a uniform molecular fragment is obtained that can no longer be separated into equal halves by shearing forces. In fact, this breakage may very well be responsible for the appearance of various types of DNA after chromatographic analysis of the preparation.

Among the contaminants of DNA preparations obtained from mammalian tissue, protein and RNA are the most conspicuous. Proteins can be separated by denaturation with chloroform and amyl alcohol, a procedure that may be followed by repeated washings with sodium dodecyl sulfate. RNA can be eliminated by repeated digestion of the preparation with ribonuclease. (Even crystallized RNase has deoxyribonuclease activity; therefore, the RNase preparation must be boiled for 3 minutes before using it for treating the DNA preparation.)

The requirements for a good DNA preparation emerged from the work of several laboratories. The preparation must be free of proteins, RNA, and inorganic salts. The phosphorus content should be about 9.2%. The maximum ultraviolet absorption for the DNA should fall between 257 and 261 mμ with a molecular absorption for the phosphorus of around 6,000. Any DNA preparation that yields a molecular absorption for the phosphorus with values higher than 7,200 should be considered denatured, according to Chargaff [88]. When precipitated in alcohol, a good DNA preparation yields long white fibers that can then be dissolved in salt solution. The DNA solution should have a high viscosity and streaming birefringence.

The molecules storing genetic information must be organized in such a way that their structure includes enough similarities to account for the constancy of genetic characteristics in a given species, and enough dissimilarities to account for the differences between species. DNA is a polymer composed of at least four different nucleotides. The specific arrangement of the bases in the molecules dictates the hereditary characteristics. This postulate has led chemists to analyze the base ratios of DNA obtained from various sources.

After numerous attempts to establish the base composition of DNA had led to the false theory of the tetranucleotide structure of DNA, Chargaff [88] and his associates and, somewhat later, Mirsky and Wyatt [89] studied extensively the base composition of nucleic acid. After improving the DNA extraction method considerably, Chargaff's group characterized DNA preparations from various biological sources in terms of their base composition. The student of normal or abnormal biology should not fail to read Chargaff's original papers. These austere quantitative analyses are the indispensable cornerstone of the successful construction of the DNA model. The minimum requirements outlined by Chargaff for analyzing the base composition of nucleic acids were as follows: each investigation should involve a complete characterization of at least two preparations isolated separately from two completely independent specimens; duplicate hydrolysis should be performed on these samples; and the product of the hydrolysis should yield a full balance of the recovered compound on a molar basis.

One of the most striking observations in these experiments is the monotonous consistency of the base compositions of DNA obtained from various organs of the same animal. This base composition is practically identical, within, of course, the limitations of the analytical techniques. Neither is there much difference between the base composition of DNA obtained from normal and carcinomatous liver, and of DNA obtained from various species of the same genus. On the other hand, important unifying features were found in all the DNA studied; all DNA molecules were found to contain at least four elementary bases—thymine, adenine, cytosine, and guanine. In all DNA hydrolysates, the sum of the molar concentration of thymine and cytosine is equal to that of guanine and adenine. Furthermore, the hydrolysate contains equal amounts of thymine and adenine and equal amounts of cytosine and guanine. Consequently, in each hydrolysate, the number of amino groups in position 6 of the bases equals the number of keto groups in position 6.

Chargaff did not restrict his studies on DNA base composition to related species. Some important differences in the base compositions of the various biological groups were discovered. They indicated that living beings can be divided into three main groups with respect to the base composition of DNA: the AT type, the CG type, and the intermediate group. The ratio of the sum of the total number of adenine-thymine pairs to the total number of guanosine-cytosine pairs in the hydrolysate is greater than one in the AT type, smaller than one in the CG type, and equals one in the intermediate group.

The chemical or enzymatic hydrolysis of the DNA molecule followed by chromatographic analysis or electrophoresis of the nucleotides is a long, laborious procedure. Our knowledge of the base composition of many DNA preparations would have been delayed for a long time if more convenient methods had not been developed. These newer methods are based on the mathematical relationship that exists between some of the physical properties of the DNA molecules and its base composition. Three such methods are now in use: (1) the determination of the bouyant density of DNA in cesium chloride; (2) the determination of the midpoint of the ultraviolet absorption rise that accompanies increasing temperature; and (3) the ratio of the absorbency of DNA at 260 and 280 mμ at pH 3.

The density of one molecule of DNA rich in guanine-cytosine pairs is greater than that of a DNA rich in adenine-thymine pairs. Thus, if a DNA suspension composed of a mixture of two different DNA's—one rich in AT pairs and another rich in CG pairs—is centrifuged in gradient concentrations of cesium chloride, the two types of DNA band separately at the end of the centrifugation. The position of the band depends on the bouyant density of the DNA molecule. The bouyant density of the DNA molecule was shown to be directly proportional to the amount of CG pairs present. The determination of the bouyant density of DNA has thus provided a convenient method for establishing quickly the base composition of small amounts of DNA.

The analysis of the base composition of DNA in bacteria, algae, protozoa, invertebrates, and vertebrates has progressed due to these new methods. Although the variation is great among bacteria, algae, and some protozoa, it is considerably narrower in invertebrates, and it is even smaller in vertebrates, in which the proportion of CG bases ranges between 40% and 44%. No significant difference in the base composition of DNA was observed between various DNA preparations obtained from the same individual or strain. Thus, the base composition of DNA extracted from tissue obtained from the fetus or from young adults was the same. Compositions of DNA obtained from normal and cancerous tissue did not differ.

Since the genetic code is written in the nucleotide sequence within the DNA molecules, elucidating that sequence is one of the most pressing problems in biochemistry. Attempts to establish the nucleotide assembly have, until now, provided only scanty information. A stepwise degradation of the molecule similar to the degradation procedures used for proteins is not yet available. We can only estimate the frequency at which some short polynucleotide sequences are repeated within the DNA molecule.

For partial-sequence studies, the DNA molecule is hydrolyzed with acid or with specific deoxyribonucleases, and the oligonucleotides are then separated on Dowex columns. By such methods, one-half of the helix has been shown to contain rows of three to five and sometimes more identical deoxypyrimidine nucleotides. For example, three consecutive deoxycytidylic acids and four to five consecutive thymidylic acid residues have been found in DNA obtained from bacteria, plants, and animals.

The base sequence of DNA obtained from different sources can be evaluated by renaturation of hybrid

DNA strands. When DNA solutions are heated, the forces binding the two helixes collapse, and (at least in the case of some viral and bacterial DNA) the two strands separate. When the DNA is cooled slowly, the two complementary strands recombine to form a typical double helix. Thus, DNA from different sources can be denatured, the strands separated, and the complementariness of single strands can be tested by mixing strands obtained from different DNA's. Although the method distinguishes between DNA's that differ markedly in base composition, it is not sensitive enough to recognize the small differences in base composition as they occur, for example, in different strains of *E. coli* that are genetically related. Denaturation of mammalian DNA probably does not completely separate the two DNA strands but only partially unzips the helix, and the renaturation is accompanied by random recoiling of the strands.

Although the base sequence of DNA remains to be elucidated, Watson and Crick [90, 91] proposed a model for its tridimensional structure. Several models had been proposed prior to that enounced by Watson and Crick, but they were all discarded because they were not compatible with the established physical and chemical properties of the DNA molecule. Modern concepts of the tridimensional structure of the DNA molecule are based on four different sources of factual information: (1) the base composition of the DNA molecule, (2) the physicochemical information suggesting the existence of hydrogen bonds; (3) the electron microscopy data, which indicates that the molecule is a long, extended structure about 30,000 A long and 30 A wide; and (4) X-ray diffraction studies of crystalline and paracrystalline DNA molecules from which the molecule is known to have a helicoidal shape. (The term spiral should not be used to describe DNA. A spiral winds around a cone and a helix winds around a cylinder.)

The Watson-Crick Model of DNA

We will not attempt to review the experiments that have led to the description of the model, but will only outline the principal features of the DNA model [90–101].

X-Ray diffraction* studies demonstrated that the DNA molecule has a helical structure and that the distance between two complete turns of the helix is 34 A. This is a distance considerably greater than that separating nucleotides, and on the basis of the size of single nucleotides, it was calculated that the helical structure is repeated every 10 nucleotides. Some of the DNA properties suggested that the molecule contains hydrogen bonds, which are essential to the maintenance of the helical structure.

X-Ray diffraction patterns compatible with the existence of a double helix were obtained for all DNA prepared from a great variety of sources, indicating that the structure proposed by Watson and Crick was common to all DNA molecules examined, with few exceptions. (The DNA of the $\phi\chi$ phage is single stranded.)

The first X-ray diffraction studies were done on purified DNA preparations, and one could argue that the molecular regularity observed under those conditions is an artifact, and that the DNA present in the intact cell may not exhibit such characteristics. This objection was overcome when it was observed that the X-ray diffraction patterns of sperm heads and viruses were compatible with the existence of double-helical DNA in the native state.

The ratio of purines and pyrimidines determined in DNA hydrolysates suggests complementariness between the purine and pyrimidine bases. Because the total number of amino groups and keto groups in position 6 are equal, hydrogen bonds are likely to unite complementary bases. Hydrogen bonds have been demonstrated by physicochemical studies on the DNA molecule. When DNA solutions are exposed to alkali, low ionic strength, or heat, the properties of the solutions change, the ultraviolet absorption spectrum increases, and the optical rotation, the viscosity, and the birefringence drop. The changes are accelerated if compounds facilitating the rupture of hydrogen bonds, such as urea, are added to the medium. It is now established that the forces involved in forming hydrogen bonds are not the only agents holding the two halves together—hydrophobic bonds and the participation of London-van der Waals forces have been demonstrated.

Although there are many ways in which a purine and a pyrimidine base can form hydrogen bonds, most possibilities were ruled out on the basis of X-ray crystallography data.

Because of its dimensions, the DNA helix can be constructed only by pairing adenine with thymine and guanine with cytosine (see Figs. 2-10 and 2-11). In the adenine-thymine pair, only two sites are available for hydrogen bonding: one uniting the amino group in position 6 of the adenine to the keto group in posi-

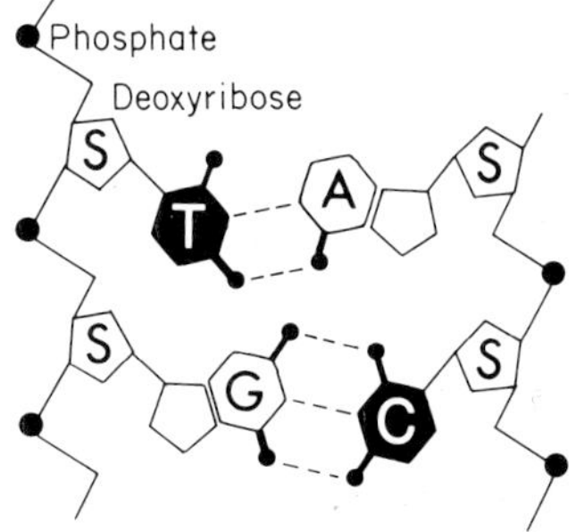

Fig. 2-10. Base pairing in DNA

* X-Ray sources constitute a powerful tool for the resolution of atomic structures. With the proper equipment, X-ray waves 1.2 A long can readily be produced. This is about the average distance separating bound atoms (1.0–0.5 A). The repeating crystalline structure will thus act as a diffraction grid for a narrow monochromatic X-ray beam. Perutz's review is a good introduction to X-ray diffraction studies of biological compounds.

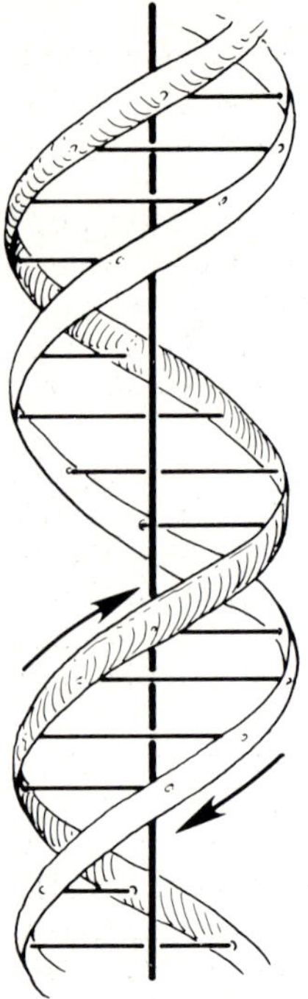

Fig. 2-11. Watson-Crick model of DNA

tion 6 of the thymine; and another between the nitrogens in position 1 of both the purine and the pyrimidine. Six atoms of the cytosine-guanine pair are close enough to form three different types of hydrogen bonds: a first between the amino group in position 6 of cytosine and the keto group in position 6 of the guanine, a second between nitrogen number 1 of both the purine and the pyrimidine, and a third between the amino group in position 2 of the guanine and the keto group in position 2 of the cytosine. Consequently, the ties between guanine and cytosine are greater than those joining thymine and adenine. This is illustrated by denaturation experiments conducted on DNA preparations obtained from bacteria containing various proportions of AT and CG pairs. Such experiments demonstrated that the melting point temperature increases with the proportion of CG pairs. This finding explains why bacterial DNA preparations melt within a narrow temperature range, whereas mammalian DNA melts in a wider range of temperatures. The CG content of bacterial DNA varies considerably depending on the preparation. DNA rich in CG pairs is more resistant to radiation than DNA rich in AT pairs.

DNA Replication

Semiconservative Replication

The genetic information stored in the DNA molecule must be transferred integrally to all daughter cells. Therefore, any molecular model for DNA must provide for the maintenance of the DNA structure during interphase and its exact replication during cell mitosis. The constancy of DNA per nucleus and the absence of DNA turnover have already been discussed. One of the most appealing aspects of the DNA model proposed by Watson and Crick is that it provides the biologist with a simple hypothesis for DNA replication.

During DNA replication, the two strands presumably unwind, and each strand serves as a template for the biosynthesis of a complementary polynucleotide chain. Such a semiconservative mechanism of replication finds support in two types of experimental evidence: studies involving the separation of the two DNA strands during replication, and the mechanism of DNA synthesis *in vitro*.

Meselson and Stahl [102] grew *E. coli* on two different media. First, the cells were cultured in the presence of ^{15}N as the sole source of nitrogen. They were then transferred to a medium containing ^{14}N. Samples of the growing cells were taken at various times during the incubation in the second medium. The DNA was extracted and centrifuged in cesium chloride density gradient. This procedure separates ^{15}N, ^{14}N, and hybrid DNA into three separate bands. The results of these experiments were those anticipated for a semiconservative model for DNA replication. A hybrid DNA was found after the first generation of *E. coli* had been grown on the labeled nitrogen. The DNA of the second generation could be separated into two bands of DNA; one completely unlabeled, the other with half of the DNA labeled and the other half unlabeled. Further studies of Rolfe suggested that the labeled and unlabeled subunits were arranged side by side rather than end to end.

Enzymic Mechanism of Replication

The enzyme mechanism for DNA synthesis was clarified by Kornberg and his associates [103–108] in bacterial systems. A single DNA strand is a polynucleotide chain in which the individual nucleotides are linked by ester bonds between the carbon 3 of one deoxyribose and the carbon 5 of another deoxyribose belonging to an immediately adjacent nucleotide. There are two distinct aspects in the synthesis of a single strand: formation of the ester bond and the alignment of the nucleotide in the proper sequence. On the basis of observations made on the biosynthesis of coenzymes, Kornberg postulated that the precursor used for the formation of the polynucleotide chain was an "activated" 5′-mononucleotide, namely, a triphosphate.

On the basis of Watson and Crick's model for DNA, the sequence of the new DNA synthesized was assumed to be determined by the complementary sequence of a DNA single strand. Kornberg's group synthesized the triphosphates of four deoxyribonucleotides (one of them labeled with ^{32}P), incubated them with *E. coli* extracts, and demonstrated that ^{32}P was incorporated into DNA. The system was functional only if small amounts of DNA were also present as primer. The enzyme responsible for the reaction was then extensively purified and net DNA synthesis could be demonstrated. It was further possible for Kornberg's group to establish that single-stranded DNA was more effective as a primer than double-stranded DNA. The new DNA synthesized had physicochemical properties (birefringence, viscocity, etc.) characteristic of double-stranded DNA, and its base composition was identical to that of the primer DNA even when large amounts of new DNA were synthesized.

Methods for establishing DNA sequence are just becoming available. Kornberg used an indirect method to demonstrate that some similarities exist between the sequence of the primer DNA and that of the DNA newly synthesized. The method, referred to as that of the nearest neighbor, is based on the following principle: when the α-phosphate of deoxyribonucleotide triphosphate is labeled, the label forms an ester bond with the C_3 of the deoxyribose of the adjacent deoxynucleotide during incorporation of the nucleotide triphosphate into DNA. If the new DNA synthesized is then degraded with micrococcus DNase and spleen phosphodiesterase, 3′-deoxynucleotides are formed exclusively. Thus, the ^{32}P of the labeled 5′-nucleotide is recovered in the 3′-nucleotide, the nucleotides can then be separated by column chromatography, and by measuring the radioactivity of each of the four nucleotides, the frequency with which each of these nucleotides is bound to the original 5′-labeled nucleotide used as a substrate can be evaluated.

The detailed results of Kornberg's elegant experiments cannot be outlined here, but because of the compelling logic of these studies and the care of these experiments, a review of Kornberg's work—especially the monograph "Enzymatic Synthesis of DNA" [105]—would be most inspiring. Only the conclusions are presented: (1) all 16 combinations of nucleotides occur; (2) the nucleotide sequence in DNA is not random; (3) the frequency of the adenylic-adenylic sequence equals that of the thymidylic-thymidylic sequence, and, similarly, the frequency of the guanylic-guanylic sequence equals that of the cytidylic-cytidylic sequence; and (4) the two DNA strands exhibit an antiparallel direction.

DNA polymerases have been purified from numerous bacterial and mammalian sources. In few cases have the purification steps yielded a preparation in which the molecular interaction between enzyme and substrate could be studied. Three different DNA polymerases have been found in *Escherichia coli*. They are referred to as DNA polymerase I, II, and III. Only the mode of action of DNA polymerase I—the most extensively investigated polymerase—will be briefly reviewed [109–112].

DNA polymerase takes one nucleotide at a time and orients it in such a way that the 5′-end faces the 3′-OH of the existing polynucleotide chain. The enzyme binds covalently the 3′-OH terminus to the 5′-end of the nucleotide. Thus, the presence on the template of a free 3′-OH is indispensable for activity. The enzyme (mol wt 109,000) is a single polynucleotide chain with a single disulfide bond and a single sulfhydryl group. It contains 0.6 phosphorous atoms per protein molecule.

Obviously, the enzyme can catalyze the elongation of the polynucleotide chain only if it binds to primer and substrate. However, this enzyme is somewhat unusual because it reacts with at least four different substrates—the triphosphates of the four nucleotides that compose the DNA.

The binding of the four triphosphates to the enzymes has been studied in the absence of DNA. Only a single binding site appears to exist for all four triphosphates, but the affinity of the polymerase for the different triphosphate varies greatly (the dissociation constant $K=1.2$ for dGTP, 3.3 for dATP, 8.1 for dTTP, and 14.7 for dCTP).

The mode of binding of DNA polymerase to DNA depends upon whether the DNA is single or double stranded. With an unbroken double-stranded DNA, the polymerase binds only at the ends of the molecules. If strands have broken, the enzyme binds wherever there is a nick, provided that nicking yields a free 3′-OH. In contrast, binding with single-stranded DNA may occur at several different sites along the chain. The existence of another binding site for mono- and di-deoxynucleotides has also been reported.

Kornberg has studied the mode of action of DNA polymerase with dAT and with single-stranded and double-stranded phage DNA. dAT polynucleotides 24 or 40 polynucleotides long were prepared by treating longer chains with deoxyribonuclease and separating the oligonucleotide by gel filtration or polyacrylamide gel electrophoresis.

Two oligomers were used, one 24, the other 40 oligonucleotides long. To mimic a double-stranded DNA, the oligonucleotide was forced into a hairpin configuration, which is conducive to hydrogen bond formation between the bases of each arm of the hairpin. This is achieved by melting the copolymer and cooling it rapidly at low ionic strength.

When such a hairpin-shaped primer is used, the results of the experiments vary with the conditions of the incubation mixture. Two extreme situations are that in which the amount of enzyme molecules exceeds the amount of primer molecules, and *vice versa*—that in which the amount of primer molecules exceeds that of enzyme molecules.

Sedimentation studies performed in sucrose gradient established that all the enzyme is bound in the first case and that one enzyme molecule binds to one molecule of primer in the second case. Thus, it was concluded that the enzyme's affinity for the primer is great and that each molecule of enzyme contains one binding site for the primer.

Kornberg and his associates chose T 174 phage to study the binding of the polymerase to DNA. T 174 phage is a relatively small, circular molecule that may be obtained in a single-stranded or duplex form. Such DNA is relatively simple to label with [^{3}H]thymidine or ^{32}P. Kornberg and his associates also devised methods for preparing of mercury-tagged DNA polymerase (at the SH group).

Studies performed with labeled phage DNA and tagged DNA polymerase revealed that there are 20 binding sites on the single-stranded phage DNA* and no binding sites on the duplex. However, when the duplex is denatured, for example in alkali, 20 binding sites appear for each strand. But the duplex can be made to bind DNA polymerase if nicks are made in the strand. Two types of nicks were introduced in Kornberg's laboratory. In the first, scission between the 3′-OH of one nucleotide and the 5′-phosphate of the other is performed, yielding nicks ending with 3′-OH (DNase); in the second type, the scission occurred between the 5′-OH and the phosphate, yielding 30 PO_4 nicks (micrococcus nuclease). Only DNA in which 3′-OH nicks have been introduced functions as an efficient primer, yet the enzyme appears to bind to the 3′PO_4 nicks.

Like many other enzymes, DNA polymerase exhibits different catalytic properties. As mentioned previously, it catalyzes the polymerization of monodeoxynucleotide triphosphate to yield DNA. In a reaction that is the reverse of the polymerase reaction, DNA polymerase can cleave DNA pyrophosphorolytically to yield the four deoxynucleotide triphosphates. The enzyme can hydrolyze quantitatively native or denatured DNA, synthetic polymers, and oligonucleotides to yield deoxynucleoside-5-monophosphate.

Pyrophosphorylase activity of the DNA has been determined by an indirect and a direct approach. In the indirect one, the PP_i exchange with deoxynucleoside triphosphate is measured. In the direct approach, cleavage of DNA to the respective pyrophosphatase is assayed.

The pyrophosphorylase activity attacks the 3′-OH group and requires primer, template, and triphosphate in the incubation mixture.

In contrast to the polymerization and pyrophosphorolysis that attack 3′-terminal groups, the hydrolytic property (exonuclease II) of *E. coli* DNA polymerase can affect both ends of the molecule (3′-OH or 5′-OH and 5′-PO_4-terminal). However, the exonucleolytic attack does not take place when the polynucleotide chain is terminated by the 3′-PO_4-terminus. DNA polymerase—which is synthesized after *E. coli* is infected by T4 phage—appears to hydrolyze exclusively from the 3′-hydroxyl terminus and is inhibited by conditions conducive to DNA synthesis. The mode of action of DNA polymerase is discussed further in the section on DNA repair.

The experiments done in Kornberg's laboratory illustrate the complexity of the reactions involved in extending the DNA chain, but there are many other aspects to semi-conservative replication of DNA, among them: the unwinding of the intertwined strands, the initiation of DNA synthesis, the formation of deoxyribonucleotide intermediates, and the connection of the intermediates. We can only consider these various aspects of DNA synthesis briefly. For further details, the reader is referred to the book of Wells and Inman [235].

We have mentioned how DNA polymerase I from *Escherichia coli* extends the DNA chain, but how does it initiate the chain? A clue to the mechanism was obtained when it was observed that rifampicin, an inhibitor of RNA polymerase, also inhibits the conversion of a phage single stranded into a double stranded DNA circle. This suggested that DNA synthesis is initiated by RNA priming. It is now established that the initiation of DNA synthesis by DNA polymerase I requires RNA and DNA polymerase, ribose and deoxyribose triphosphates, an unwinding protein and other factors whose roles are still unclear. The initiation step seems to involve a transcriptional operation by RNA polymerase. The newly synthesized RNA serves as primer for DNA polymerase I, yielding an RNA-DNA complex and the primary RNA is excised by a nuclease [236, 237].

Before semi-conservative replication of DNA can take place, the two intertwined parental DNA strands must be untwisted. The mechanism of this unwinding is not clear. Two have been proposed: (1) the formation of a single-strand break which would allow a portion of the strand to unwind; the unwound strand could then be rejoined by a ligase after semi-conservative replication, or (2) the existence of a specific enzyme which would temporarily unwind the DNA has been suggested. A protein capable of unwinding DNA *in vitro* has been isolated from *E. coli* [238].

Two observations led Okazaki to suspect that DNA is replicated in a discontinuous fashion. (1) When newly synthesized DNA is isolated, short fragments are obtained. (2) Inhibition of polynucleotide ligase results in the accumulation of such fragments. Synthesis of the short DNA fragments is initiated on a very short RNA primer which is transcribed on the parental DNA. The RNA segments are removed before the DNA fragments can be put together through the catalytic action of polynucleotide ligase [239].

Because a mutant of *E. coli* referred to as pol A^-, defective in polymerase I, replicates DNA normally, it has been suspected that at least in some cases DNA polymerase I might not be the enzyme responsible for DNA semi-conservative replication. This observation led to a search for other DNA polymerases. In *E. coli*, two other DNA polymerases were discovered; polymerase II and III.

DNA polymerase II has been purified to homogeneity on sodium dodecyl sulfate polyacrylamide gel.

* DNA polymerase purified from *E. coli* is extremely inefficient in initiating DNA synthesis. In the course of studies on the biosynthesis of $\phi\chi$ 174 phage DNA, a factor in a boiled extract prepared from *E. coli* proved to stimulate DNA synthesis. The χ174 is a double-stranded circular phage. In these experiments, single-stranded phage DNA was used as a template for synthesizing the duplex. For completion, the reaction required DNA polymerase and polynucleotide ligase. Moreover, as mentioned above, the replication of single- to yield a double-stranded DNA was stimulated by a boiled extract. Treatment of the boiled extract with DNase and snake venom phosphodiesterase indicated that the activator was a polydeoxynucleotide. The exact mechanism by which the polynucleotide stimulates the polymerase reaction is not clear.

It has a molecular weight of 60,000 to 90,000 daltons and requires NH_4^+, Mg_2, and native DNA for activity. It interacts with the DNA unwinding protein and this reaction accelerates the rate of DNA synthesis. The synthesis proceeds from the 5′ toward the 3′ direction. RNA polymerase generates the primer.

DNA polymerase III is heat labile and has been purified 10,000-fold. Mutants which are temperature sensitive for DNA polymerase III lose their ability to synthesize DNA suggesting that the enzyme is indispensable for DNA replication. The DNA synthesis occurs in the 5′→3′ direction. The enzyme is inhibited by sulfhydryl reagent and increased salt concentration.

Schekman, Weiner and Kornberg [240] have dissected the steps involved in DNA synthesis in a number of bacterial and phage systems and established that in all cases the conversion of a single stranded to a duplex circle involves the coordinated activation of a number of complex enzymes. The details of these elegant experiments cannot be described here. It is sufficient to point out that the authors distinguish essentially five steps in DNA replication: (1) initiation by synthesis of short RNA fragments which is catalyzed by RNA polymerase in *E. coli*, but requires other special proteins in other systems (the unwinding proteins are needed in this step), (2) elongation by DNA polymerase III on enzymes made of several subunits (ATP is required to complex the enzyme with the DNA, but not for elongation), (3) termination by excision of the priming RNA sequence, (4) filling of the gap left by the RNA excision by DNA polymerase I, (5) sealing of the phosphodiesterase bond with polynucleotide ligase.

Because of the importance of DNA synthesis, the properties of DNA polymerases prepared from various sources have now been reported. A number of enzymes, all exhibiting different molecular properties, have been purified from viral, bacterial, and mammalian sources. Certainly in the future the knowledge of the special properties of a battery of polymerases will permit their use as tools in the study of the mechanisms of DNA synthesis in both bacteria and mammals and will also provide the means for studying DNA sequence. It is, however, impossible to catalog all the properties of the polymerases that have been purified and only a few critical points will be made here.

In mammalian cells, although no DNA polymerase which is the equivalent of the bacterial polymerase I has been discovered, polymerases have been partially purified from, among other sources, thymus, liver, and human KB cells. Two points emerge: (1) The enzyme exists in the same tissue in several forms that may be distinguished primarily by their molecular weights. The enzymes do not exhibit exonucleolytic activity although some have endonuclease activity associated with the polymerase activity. Too often the level of purification of the enzyme is too low to decide whether or not the endonucleolytic activity is a contaminant or a genuine part of the molecule. But it has been established that exonucleolytic activity is not indispensable for polymerization. Partial purification of a mitochondrial DNA has also been achieved.

The reasons for the existence of various forms of DNA polymerases in animal tissue are not known. It is of interest that there is an enzyme in liver which sediments at 3.4S and another that sediments at 6 to 8S. After partial hepatectomy, the activity of 6 to 8S DNA polymerase, which is mainly a cytoplasmic enzyme, increases 6- to 7-fold.

Problems of the Replication Model

Watson and Crick's proposed replication mechanism implies that the formation of a new strand is determined by each nucleotide's ability to form hydrogen bonds with a complementary nucleotide in the strand that serves as primer. If new strands are formed in this way, 5-methylcytosine and 5-bromouridine (and the pseudouracil in RNA), when present, must replace cytosine and thymine in newly synthesized DNA molecules at random. Chargaff and his associates have shown that this is not the case. One could argue that the 5-methyl group of 5-methylcytosine is added to cytosine when the DNA molecule has been formed, but this argument cannot be invoked for 5-bromouridine.

A DNA methylase has been purified 400 times from *E. coli* extracts. The enzyme methylates cytosine to yield methylcytosine and adenine to yield 6-methylaminopurine. Whether the same enzyme catalyzes both types of methylation remains to be established.

It has been proposed that forces other than hydrogen bonds are involved in distinguishing analogs from the normal bases. These findings have brought Cavalieri and Rosenberg [93] to conclude that the selection of complementary bases to build the new polynucleotide chain results from van der Waals interactions. These authors proposed a new model for DNA replication in which the double-stranded DNA serves as primer and the biosynthesis of the new DNA is determined by the van der Waals forces attracting a complementary base. During replication, the new bases are intercalated between the base pair in such a way that adenine is superimposed on thymine, and cytosine is superimposed on guanine. Consequently, in this model it is the double helix that acts as template. Although it is not absolutely excluded that such a model may operate *in vivo*, the available evidence renders this alternative to the Crick and Watson model of replication rather unlikely.

Commoner [113] has criticized the replication models proposed by Watson and Crick. On the basis of observations on physical models, Commoner claims that in the case of a linear code as complex as the one storing all the information locked in a single DNA molecule, the individual signals emitted by each complementary base of the primer strand to attract the new base would be lost in the background noise emitted by the molecule as a whole. Furthermore, according to Commoner, the biosynthesis of DNA as proposed by Watson and Crick has no counterpart in polymer synthesis in organic chemistry. Finally, Watson and Crick's theory does not account for all findings made in genetics, for example, epigenetic development, cytoplasmic inheritance, and nonmendelian inheritance. For all these reasons, Commoner has proposed a replication mechanism in which both the base of the primer and some enzymes (possibly several different polymerases) determine the base sequence of the new DNA.

DNA and Chromosome Structure

Although much is known of the structure of the DNA molecule, little of the relationship between DNA and the chromosome has been established. Three theories concerned with the arrangement of the DNA molecule within the chromosomal structure have been proposed: the protein backbone theory, the multistranded or rope theory, and the differential coiling theory [31, 114–118].

In the backbone theory, it is proposed that protein constitutes a sort of central wire to which the DNA molecules are attached to form strands perpendicular to the long axis of the protein, like the teeth of a comb; or the two ends of the DNA molecule may be attached to the protein to form a loop the plane of which is perpendicular to the long axis of the chromosomal protein. It has also been proposed that the DNA and the protein form continuous strands on which the nucleic acid molecules alternate with the protein molecule.

According to the second hypothesis, the chromosome is made of a large number of coiled, tightly packed DNA molecules, forming a multistranded structure. These coiled strands are thought to extend throughout the entire length of the chromosome.

Perhaps the simplest and most plausible hypothesis proposed to describe the relationship between DNA and the chromosomal structure is that of Swift. He has postulated that the chromosome is made of one or two strands that are nothing other than long DNA molecules. In some areas of the chromosome, these strands are excessively coiled, and this coiling is responsible for the morphological appearance of the chromomeres, which are visible at prophase.

Unfortunately, we cannot discuss the pros and cons of each of these theories. The magnitude of the problem can be adequately illustrated by discussing some of the argumentation that pertains to the last theory. If one assumes that the chromosome is made of two long coiled and continuous DNA molecules, the fact that the Feulgen-positive material in the chromosome has a bandlike disposition must be taken into account. It has been claimed that Feulgen-positive material in the chromosome is present only in the form of bands and is completely absent in the interband region. Further studies on polytene chromosomes demonstrated that although some bands are much more deeply stained by the Feulgen reagent than the interband areas, some staining is also present in the interband area.

The continuity of the DNA in the chromosome has been further demonstrated with dyes more sensitive than those classically used in the Feulgen preparation. By carefully determining the amount of DNA in the interband area by cytophotometric methods, Swift [31] ruled out the possibility that each chromosome is made up of a large number of strands, each corresponding to a DNA molecule. He claims that the existence of many strands is physically impossible; many strands could not be stacked within the area occupied by interbands, and a multistranded structure would yield staining characteristics much more marked than those in the interband.

Chromosome Ultrastructure

Two major techniques have been used to study the dimension of chromosome fibers by electron microscopy. The first is the surface spreading method, in which the cells are spread on a water-air interphase, and all the intracellular components are dispersed. But the chromatin and the spindle remain close together and can be picked up on a grid, which can then be prepared for electron microscopic examination. The second technique is the thin-sectioning method. After fixation the tissue is embedded in plastic and cut less than 1000 A thick. Refer to specialized texts for descriptions of these techniques [119]. The important finding is that with the first method, the dimension of the fibers' diameter is estimated to be 200–300 A, whereas it is only 80–180 A with the second.

In 1955, Ris [120], working with sectioned meiotic and mitotic chromosomes of rat, salamander, lily, and onion, was the first to observe that long fibers about 200 A in diameter were fundamental units of structure in both plant and animal chromosomes. During mitosis and meiosis, these 200-A fibers are apparently coiled around each other, forming large bundles. From his studies on honey bee embryonic cells, DuPraw [121] demonstrated that 230-A fundamental fibers can be found *in vivo* and that such fibers are firmly attached to the nuclear envelope at the edges of the annuli.

However, the ultrastructure of these 200–300-A fibers has not yet been clearly established. Ris reported that the substructure of calf thymus chromatin fibers of the 200-A fibers reveals two 100-A fibers coiled around each other. Each 100-A fiber is, in turn, made of two coiled 40-A fibrils. But because of the preparation of the material, the possibility of aggregation could not be excluded and the four-strand concept was abandoned.

Electron micrographs of metaphase chromosomes of honey bee and HeLa cell and human chromosomes reveal that the chromosome is a tortuous bundle of 230-A fibers [122–125], but the substructure of the fiber remains unsettled; although DuPraw has shown that after trypsinization the 230-A fibers contain

DNase-sensitive fiber with dimensions compatible with that of a Watson-Crick double helix.

The sectioning method has revealed a series of light and dark band areas adjacent to the nuclear membrane, and it is believed that these bands are composed of well-ordered, 130-A fibers and that each fiber is itself made of a coil of thinner fibers 20–40 A thick [126].

X-Ray diffraction studies of chromosomes from various sources subjected to environmental alteration—including ionic strength and other factors—have been interpreted to mean that the chromosome fiber is made of a supercoiled DNA duplex with a pitch of 120 A and a diameter of 100 A. Similar reflection patterns were obtained when a complex of DNA and histones was examined by X-ray diffraction.

In conclusion, although the diameter of the chromosome fiber varies with the method used to study it, available data suggest that the chromatin fiber contains DNA condensed in a supercoil pattern.

The pattern of the coiling is, however, unknown. Three different types of supercoiling have been proposed: (1) a compact, but irregular supercoil, (2) a loose supercoil, and, (3) a highly regular three order supercoil. In the first order, the DNA strands are coiled in the shape of a spring. In the second order, the spring is supercoiled and in the third order, the supercoiled spring is regularly and tightly coiled once more [241].

It is likely that histones play a role in the maintenance of the supercoil by interacting with each other.

We have seen that only the basic portion of the histone sequence is directly bound to DNA. When chromatin is subjected to gentle trypsinization, 25–30% of the total lysine and arginine residues are released. Physicochemical studies of the trypsinized chromatin suggest a loss in the supercoiling, a finding compatible with the notion that the nonbasic segments of the histones interact with each other and are responsible for the supercoiling [127].

Moreover, when chromosome fibers are treated with 4 M urea, the chromosome fibers are extended into a thread 30–40 A thick, and no proteins are lost in the process. After urea is removed the diameter of the fibers is increased 2.5 times, and thus supercoiling is restored. The urea experiment suggests that if the nonbasic portion of the histones interacts to generate supercoiling, the type of bonding is certainly not covalent but more likely involves hydrogen and hydrophobic bonding [128].

The Mitotic Apparatus

After telephase each daughter cell inherits a set of chromosomes identical to that of the parent cell. This is possible because each of the parent chromosomes is duplicated and one set is distributed to each daughter cell. The cell manages this orderly distribution by providing, sometime during mitosis, an array of retractable fibers to which chromosomes attach and by which they are pulled from the equatorial plates into the cytoplasmic territory of one of the two daughter cells.

Early microscopists described the spindle long ago. In 1952 it was seen in live cells in the shape of birefringent fibers, thus excluding the possibility of fixation artifacts. Electron microscopists helped to clarify knowledge of the spindle when they showed that the fibers were essentially composed of microtubules.

Several attempts were made to separate the mitotic apparatus, that is, the spindle, the matrix substance found between the fibers, and possibly other chemicals directly associated with the fibers or the matrix, from the remaining nuclear structure. Most of these studies have yielded results that are difficult to interpret. Only recently has it become obvious that the fibers are made of special proteins that constitute the microtubules [242–246].

The microtubule unit is a protein (tubulin). It is a dimer with a molecular weight of 110,000 to 120,000 daltons. The amino acid composition of each monomer is believed to be identical except for a few residues. The monomer has a molecular weight of 60,000 daltons. A number of laboratories have succeeded in isolating a single protein component in the microtubule with a molecular weight of 60,000. After reduction and alkylation this protein migrates as a single band in urea containing acrylamide gels. One mole of guanosine nucleotide is found in association with every mole of monomer.

The tubules have an outer diameter of 240 Å with a dense wall 50 Å thick. Within the wall is a less dense core 140 Å in diameter. The exact number of subunits forming a microtubule is not known; it varies from 11 to 13. It could be that microtubules from different sources contain different numbers of units.

Although it was believed at first that all tubulin proteins are alike, there is evidence accumulating which suggests that the amino acid composition of the tubulins varies depending upon their source.

What is certain is that there are marked differences in turnover of the microtubule arrays. While some, like that of the tail of the spermatozoon, persist during the lifetime of the cell, others make only a transient appearance in the cell cycle. The microtubular system provides the cell with rigid, but elastic units that can be put together as the components of a scaffold whenever the cell needs it, for example, to move chromosomes during mitosis or wave the tail of the spermatozoon.

The function of microtubules is not restricted to mitosis. Microtubules also function in cilia, flagellae, nerve processes, the cortex of meristematic plant cells, elongation during lens formation, cell motility, sensory transduction, and determination of cell shape. Moreover, there is indication that histamine, insulin, and lysosomal enzyme release is controlled by microtubules.

Purified tubulin is phosphorylated on the serine residue in presence of cyclic AMP. The tubulin itself exhibits its phosphoprotein kinase activity. The kinase activity of the tubulin is not inhibited by calcium and colchicine. Colchicine has been known for a long time to block cellular mitosis in metaphase. When it was observed that colchicine interfered with the development of birefringence, it seemed logical to suspect that it might also interfere with the formation of microtubules.

It is now known that the administration of colchicine results in the binding of one mole of colchicine for each 110,000 molecular weight dimer. There is no binding to the monomer. Podophyllotoxin, melatonin, vinblastine, and griseofulvin are believed to act in the same way as colchicine, although it is not excluded that the site of binding might be different.

The assembly of microtubules, especially in the case of the spindle, is of interest. It has long been assumed that the spindle components were present in the cytoplasm and their polymerization resulted from loss of structured water (see chapter on electrolytes). This theory of the assembly of the spindle, which excludes *de novo* synthesis, is referred to as the dynamic equilibrium theory. The theory has been extended to the assemblage of tubules.

It is known that some factors will influence the formation of tubules *in vitro*; for example, the pH, the presence of divalent cations, the equilibrium between polymer and monomer. It is therefore reasonable to assume that similar factors may contribute to the assembly of tubulin *in vivo*. However, other factors are involved. The assembly requires energy; it is believed that GTP might be the energy donor.

Reassembly *in vitro* has been achieved under conditions resembling those that exist *in vivo*.

It is also assumed that the polymerization of microtubules requires an organizer, much like the formation of bacterial flagellae. The addition of fragments of flagellae facilitates the precipitation of flagellar elements *in vitro*. Inasmuch as kinetochores exhibit birefringence and the centriole has a tubular structure reminiscent of the microtubules, they have been suspected of playing a role in organizing the polymerization of microtubules.

Although existing evidence is not irrefutable, there is strong suggestion that the kinetochore may play a role in spindle formation. Irradiation of the kinetochore with a narrow beam of UV light brings about the disappearance of all birefrigence. But, if the beam is aimed at a point on the birefrigent fiber away from the kinetochore, only a small portion of the birefringence disappears. Similarly, flagellae and cilia appear to grow by the opposition of microtubular components to the basal body.

Protein Synthesis

Müller showed great insight when he chose "protein" to be the name of those nitrogenous substances in egg white, milk, and blood serum that precipitate with heat or with strong acids and alkali. The Greek word πϱοτεος means "first," and it is now clear that proteins play a primary role as building blocks or functional units in the cell. Although decades were spent in isolating, purifying, and characterizing proteins, only since the late 1950's has some insight into their mode of biosynthesis been gained [129–134].

As is the case for most biosynthetic pathways, the investigations of protein biosynthesis can be divided into two distinct periods: a period during which protein synthesis was investigated *in vivo*, and a stage during which *in vitro* systems were used and an effort was made to identify the intermediates formed during the biosynthesis of the macromolecule. Experiments *in vivo* were concerned with studying of the incorporation of injected amino acids into specific proteins or with measuring the activity of specific enzymes.

It has long been obvious that protein metabolism is very active. Proteins are continuously synthesized either for growth or for replacement of those lost during cellular renewal. The new proteins are synthesized from a pool of amino acids that include ingested amino acids and those derived from the endogenous breakdown of proteins. This dual amino acid source was clearly demonstrated by remarkable experiments carried out in Schoenheimer's laboratory [135]. These investigators administered labeled amino acids to intact animals and showed that the excreted labeled amino acids were diluted by nonlabeled amino acids, and that 57% of the labeled compound was recovered in the protein fraction of the injected animal. Since the total weight of the animals did not change during these experiments, it was concluded that the new proteins were synthesized to replace destroyed proteins.

Similar *in vivo* labeling studies contributed other important information; they demonstrated that the rate of protein synthesis was different in different tissues. Although the proteins of liver, kidney, intestinal mucosa, spleen, pancreas, bone marrow, and testicles are rapidly labeled, labeling is slow in muscle, brain, erythrocyte, and skin. Within a given tissue, some proteins are labeled faster than others. In the pancreas, the hydrolytic enzymes incorporate amino acids more rapidly than other proteins. The label appears sooner in some cell fractions than in others. After *in vivo* administration of antigens, 50% of the antibodies can be recovered in the microsomal fraction only a few minutes after injection. It was also shown that under conditions of net protein synthesis, labeled amino acids injected in mice are first recovered in the microsomes, then in the supernatant, and later in the zymogen granules of the pancreas.

The determination of the specific activity of a labeled amino acid in various portions of a protein molecule gave some insight into the mechanism of biosynthesis. Assuming (1) that the macromolecule can be split into three fractions—X, Y, and B; (2) that each fraction contains a valine residue; and (3) that after labeled valine is administered the specific activity is the same for all valine residues, then it is highly probable that fractions X, Y, and B were synthesized at the same time, and that the protein is synthesized as a single molecular unit rather than by addition of polypeptide fragments. In contrast, if the valine in X has a specific activity higher than that in Y, the two fragments X and Y might have been synthesized at different times in the presence of amino acid pools different in composition. Thus, if the specific activities of the two fragments are different, it seems logical to assume that at least two polypeptide intermediates are combined in the final step of protein synthesis. Although this reasoning appears logical, the results of such experiments have been somewhat controversial.

Whereas some failed to find differences in the specific activity, others demonstrated unequal labeling of several proteins. Similarly, the specific activities of the glycine residues of rabbit hemoglobin (both *in vitro* and *in vivo*) were found to be unequal. This observation was not confirmed when labeled valine was used as a precursor. Velick and his associates [136, 137] also studied the labeling in aldolase, triose phosphate dehydrogenase, and phosphorylase after isolating and purifying the enzymes and found equal labeling, which suggests a simultaneous incorporation of all the amino acids within the protein.

In conclusion, on the basis of *in vivo* experiments, it was demonstrated that proteins undergo continuous synthesis and breakdown, and that their rate of synthesis is very rapid, although this varies depending upon the specific tissue and protein studied.

In Vitro Protein Synthesis

In the second phase of these investigations on protein synthesis, an attempt was made to study the process *in vitro*. At first, intact cells had to be used. Amino acid incorporation was studied in liver slices; amylase, lipase, and ribonuclease synthesis was investigated in slices of pigeon pancreas and the incorporation of glycine into reticulocyte hemoglobin was measured. The first of the above systems was exploited with exceptional success in Zamecnik's laboratory [138].

Early in the course of those investigations, Zamecnik and his associates demonstrated that protein synthesis is energy dependent. Amino acids are not incorporated in the absence of oxygen or in the presence of dinitrophenol. Furthermore, these experiments confirmed that microsomes are the main site of protein synthesis because after separation by ultracentrifugation, the microsomes of the homogenized slices were found to be more extensively labeled than any other cell fractions. Finally, these studies for the first time provided direct evidence for the role of RNA in protein synthesis in mammalian tissue; the incorporation of amino acid into protein was inhibited by RNase.

Zamecnik and his group [139] extended their studies to cell-free systems. As previously, they selected labeled leucine and valine for incorporation studies because they wanted to avoid using an amino acid extensively used in other metabolic processes. This restriction eliminated a number of amino acids. For instance, aspartic acid, alanine, and glutamic acid enter the citric acid cycle, and glycine is used for purine biosynthesis. Lysine and sulfur amino acids may become attached to proteins by side-chain formation through the ε-amino group or by formation of disulfide bonds.

It was established that the incorporation of labeled leucine reflected true synthesis and did not result from absorption or exchange of amino acids within the protein molecule. Two observations support actual protein synthesis. One is that the incorporation is energy dependent, and the other that the labeled proteins do not lose their label in the presence of large amounts of unlabeled amino acids. The cell-free system required five components: microsomes, a pH 5 enzyme precipitated from the supernatant, ATP, GTP, GDP, and the labeled amino acids.

In conclusion, the early *in vitro* experiments with tissue slices in cell-free systems clearly demonstrated that the incorporation of labeled amino acids into proteins is energy dependent and requires RNA. Polypeptide intermediates are probably not formed.

Although the exact details of the building of the polypeptide chain are not known, we have come a long way in understanding the process. The amino acids are activated, they are bound to tRNA, tRNA binds to ribosomes, codon and anticodon are matched, and peptide bonds are synthesized. Let us consider the steps involved in an attempt to understand their significance in the overall process.

In the living cell, amino acid activation and binding to tRNA probably occurs in a single step, catalyzed by a single macromolecular complex. But, for the sake of clarity, amino acid activation is described as a separate event.

Amino Acid Activation

In the first step of protein synthesis, acyl adenylate is formed. The amino acids are activated in preparations obtained from yeast, liver, pancreas and a number of microorganisms. Activation of an amino acid can be measured by determining the exchange of labeled pyrophosphate with ATP or by measuring the formation of hydroxamate. The reaction is similar to that involved in fatty acid, pantothenic acid, and acetate activation. Acyl adenylate presumably is formed on the enzyme molecule by linking the carboxyl group of the amino acid to the α-phosphate group of the adenylic acid. Isolation of tryptophanyl-AMP and seryl-AMP in substrate amounts has provided direct proof of the acyl adenylate formation. The enzymes catalyzing the activation are precipitated at pH 5 and are found in the soluble fraction of the cell. There is now a long list of highly purified enzymes that specifically catalyze the activation of a single amino acid [141–143].

Within a cell, all amino acids are not necessarily activated at the same rate. This may be explained in various ways: (1) the enzymes specific for a given amino acid may differ in their resistance to hydrolysis;

(2) only some amino acids can be activated by reaction with ATP, others enter the protein by a different pathway; and (3) in protein synthesis, as in other biosynthetic pathways, many enzymes are present in excess of the cell's needs, and only a few of them are rate limiting. One or two of the activating enzymes may be rate limiting for protein synthesis. When an amino acid analog is incorporated in a protein, it also will be activated by this specific enzyme. The amino acids that are not incorporated are not activated, but may act as competitive inhibitors.

The exact specificity of all the amino acid-activating enzymes is still not known. It is important to establish the affinity of any specific activating enzyme for all other natural and unnatural amino acids and amino acid analogs. In general, the amino acid analog is activated by the enzyme that catalyzes the activation of the natural amino acid. But in the case of ethionine, the mechanism of activation is not clear. Although some groups claim that ethionine is activated by the enzyme catalyzing the activation of methionine, others think that ethionine is not activated. When an amino acid-activating enzyme's specificity for other amino acids and their analogs is studied the specificity of the enzyme is found to be rather restricted, but not limited to one substrate. For example, the enzyme catalyzing the activation of isoleucine also activates valine, and the enzyme catalyzing valine activation also activates threonine. In each case, however, the K_m is about twice as great for the "alternate" amino acid as for the "regular" amino acid. Therefore, errors in the activation are unlikely except when the concentration of the normal amino acid is abnormally low. Such a situation occurs in some inborn errors of metabolism, for example, phenylketonuria, in which phenylalanine levels are extremely high and tyrosine levels relatively low. If the activating enzyme for phenylalanine also acts on tyrosine, tyrosine activation could be inhibited in the presence of large amounts of phenylalanine.

The molecular interaction between the activating enzyme and the amino acid is not known, but it probably varies with the type of activating enzyme. Some enzymes (*e.g.*, the alanine-activating enzyme) are unaffected by paramercuric benzoate, but others, like the tryptophan-activating enzyme, appear to be SH enzymes the activity of which depends on the presence of free SH groups in the molecule. Potassium ions are known to activate the tyrosine enzyme.

Amino Acyl RNA Synthetase

The second step in protein synthesis is the binding of the amino acid to RNA. If one places free amino acyl adenylate (labeled in the amino acid) in the presence of the RNA prepared from the supernatant fluid, the RNA will be labeled. The labeling is low because the acyl adenylate degrades rapidly in the absence of its enzyme. The enzymes responsible for transferring the activated amino acid to the tRNA are called amino acyl RNA synthetases.

In each cell synthesizing proteins, there are at least 20 different amino acyl RNA synthetases, one for each amino acid. The overall reaction catalyzed by the synthetase leads to the formation of an ester between amino acid and tRNA. Thus, the reaction takes place in two major steps; a first, already discussed, in which the amino acid is activated and complexed with the enzyme, and a second, in which the activated amino acid is transferred to tRNA. ATP and Mg^{++} are required for the first step. In the first step, the enzyme catalyzes the formation of a mixed anhydride between the carboxyl of the amino acid and the 5'-PO_4 of AMP. Pyrophosphate is released in the course of the reaction. The anhydride never appears as a free intermediate but remains tightly attached to the enzyme.

In the next step, the enzyme acts like a clamp. One arm carries the activated amino acid, the other holds the tRNA. When the activated amino acid contacts tRNA, the activated amino acid is transferred to the tRNA where it forms an ester linkage between the carboxyl of the amino acid and the 2'- or 5'-hydroxyl of the terminal adenine residue of tRNA. This sequence of steps in which the enzyme AA$\sim$AMP is an intermediate has been challenged by Loftfield [144], who believes that under physiological conditions a concerted reaction takes place in which tRNA, amino acid, ATP, and enzymes react in one step to form amino acyl tRNA, AMP, PP_i, and free enzyme.

In any event, in protein metabolism the amino acyl RNA synthetases are located at the gates of specificity. If they attach the wrong amino acid to the wrong tRNA, the mistake will be carried into the protein structure.

In view of the importance of these enzymes in determining specificity, it is not surprising that much effort has been spent investigating their properties. The simplest way to guarantee the specificity of the translation of the nucleic acid's four-letter code into the twenty-letter code of the protein (see below) would be to have one enzyme specific for one amino acid and one tRNA, and a two-armed clamp, with a keyhole in one arm to fit the amino acid and a keyhole in the other arm to fit the tRNA. But nature can afford to take millions of years to evolve a successful biological mechanism in a most wasteful way. The result is often an extremely delicate machinery with elaborate controlling biological mechanisms more complex than all the plans any logician could conceive, but it also has more versatility than one could dream. Yet, as far as is known, in determination of specificity in general and in the specificity of the acyl amino acid synthetase in particular, nature takes few or no chances. There is one synthetase for each tRNA.

A *sine qua non* for studying the specificity of these synthetases was the successful purification of these enzymes. Some of the enzymes, like the tyrosyl RNA synthetase of *E. coli* and *Bacillus subtilis*, have been isolated and partially purified from either bacterial systems or mammalian liver. Yet, in spite of all the efforts in studying these enzymes, it is still too early to appreciate their specificity accurately. Some general conclusions can be reached about their molecular weight, the SH content, the relative kinetics toward amino acid complex formation, the specificity for amino acid and tRNA, and the redundancy of a specific enzyme.

There seems to be one specific acyl amino acid synthetase for each amino acid; thus, there are at least 20 such enzymes, and although the molecular weight of most of the enzymes is around 100,000 (except the phenyl RNA synthetase of *E. coli* and yeast, which has a molecular weight of 180,000), the amino acid composition of enzymes catalyzing identical reactions may vary. Comparisons of the electrophoretic and immunological properties and the amino acid composition of two tryosyl RNA synthetases—one purified from *E. coli* (mol wt 95,000) and another from *B. subtilis* (mol wt 88,000)—reveal significant differences between the two molecules. One of the most remarkable is the relative content of cystine, 15 residues in *E. coli* enzymes and 2 residues in *B. subtilis* enzymes.

Most, if not all, the synthetases appear to be SH enzymes, and at least some of the SH residues seem to be indispensable for activity. For example, each mole of tryptophanyl RNA synthetase purified from beef pancreas contains eight SH groups, and four are involved in the reaction. Synthetases with different specificities from the same sources or synthetases with the same specificity from different sources may vary considerably in sensitivity to SH inhibitors. In fact, *E. coli* tyrosyl RNA synthetase has been reported to have enzyme activity for amino acid activation insensitive to SH inhibitors, even though its molecule contains two SH groups that could be accessible to the substrate (another SH group appears only after treatment with 6 molar urea). Although paramercuric benzoate and 5,5′-dithio bis (2 nitrobenzoic acid) do not affect the $P_1 \rightleftharpoons$ ATP exchange reaction, they interfere with the transfer of lysine to tRNA. ATP cannot be substituted by other nucleotide triphosphates in the synthetase reaction. Studies with a number of purified synthetases have clearly established that the specificity for the amino acid is a rather rigid one.

Some enzymes can also catalyze the activation of an amino acid analog, but usually the reaction is much slower than with the normal L-amino acid. Isoleucyl RNA synthetase is, however, known to activate isoleucine and valine, yet it cannot transfer valine to its tRNA. Apparently, tRNA stimulates the hydrolysis of the valyl enzyme complex. When substrate amounts of isoleucyl synthetase complexed with [^{3}H]ATP and [^{14}C]isoleucine are incubated with purified tRNA, one molecule of enzyme transfers one molecule of amino acid. For reasons that are not clear, the transfer is not complete (60–70% of the amino acid is transferred and 70–80% of the ATP is released).

Purified amino acyl-binding enzyme (mol wt 186,000) has been prepared from rabbit reticulocytes. The preparation was homogeneous on electrophoresis, sedimentation, and immunodiffusion. The purified protein is believed to be made of three identical subunits. The enzyme transfers in the presence of CTP and poly U (phenylalanine transfers RNA to ribosomes), and it exhibits guanosine triphosphatase activity, but only in the presence of amino acyl RNA and ribosomes. Guanosine diphosphate and 5′-guanyl methylene diphosphonate (a CTP analog) are competitive inhibitors. The hydrolysis of ATP appears to be a step essential to amino acyl RNA binding to ribosome, but hydrolysis of ATP can take place without binding.

The code for each amino acid in the protein is stored in a three base sequence of DNA which is transcribed in a complementary triplet in the messenger RNA. The information of the messenger RNA triplet is translated into an amino acid by the apposition on the messenger RNA codon of a complementary triplet anticodon of a specific tRNA. It is therefore important that each tRNA be esterified by the proper amino acid. This means that the binding of tRNA to tRNA synthetase must be rigidly specific, and the specificity is believed to reside in the tertiary structure of a critical portion of the tRNA molecule.

The amino acid-activating protein and the tRNA synthetase enzyme might play key roles not only in the qualitative control, but also in the quantitative regulation of protein synthesis. To understand the molecular mechanisms of the control and regulation of protein synthesis, it is necessary to be able to describe the molecular interaction between the enzyme and its substrate.

In spite of the fact that many tRNA synthetases have been purified and much has been learned about the structural relationship of their substrates, no conclusive generalization about the mode of action of these enzymes has yet been reached. In fact, it has been proposed that the binding site for the amino acid is different in each enzyme so that it could adapt to the requirement of each amino acid. The amino acid binding is thought to introduce a conformational change that reveals a concealed catalytic site.

tRNA General Properties

tRNA is distinct from other RNA populations in molecular size, terminal side chain, and base composition. tRNA, sRNA, soluble RNA, adaptor RNA, or transfer RNA has a molecular weight of 25,000–30,000. It is soluble in sodium chloride, and after differential centrifugation of cell homogenates, it is recovered in the supernatant fluid. The sum of the tRNA in the cell represents approximately 20% of the total RNA. It is not a homogeneous polynucleotide preparation, but is composed of a population of RNA molecules [145–150].

To separate the various tRNAs, numerous methods—including chromatography (on resins, silicic acid, hydroxyapatite, Sephadex, etc.), electrophoresis, and counter current—have been devised. An interesting procedure consists of treating the amino acid-charged RNA with either periodate or converting the terminal amino acid into the benzyl N-carboxyl aspartate (in the case of valine) anhydride. Such procedures facilitate the purification of the RNA charged with the altered amino acid. In fact, the anhydride precipitates together with free β-benzyl-2-aspartate. When the precipitate is extracted with chloroform, the valine tRNA is readily recovered. The yield is relatively low (30%), but considerable purification is achieved in a single step (50–60%).

Evidence suggests that a specific tRNA exists for each of the 20 amino acids. Such specificity is in keeping with the adaptor theory proposed by Crick for protein synthesis.

The base composition of tRNA is different from that of ribosomal and nuclear RNA. One of the most striking differences is the presence of rare bases, such as pseudouridine, thiouridine, dihydrouridine, methyl purines, and inosine. The significance of this is not clear. These bases are assumed to play a determinant role in amino acid selection. Yet Littauer and his associates [151] found no differences between the abilities of methylated and unmethylated transfer RNA to accept leucine, methionine, and isoleucine. Unmethylated RNA, however, is more sensitive to RNase than methylated RNA. Each tRNA almost certainly does not contain all the odd bases. Pseudouridine is the only one that has been found consistently in all the RNA's studied.

tRNA-modifying enzymes capable of catalyzing methylation, thiolation, isopentenylation, pseudouridine formation, and esterification of 5-carboxymethyluridine have been isolated and characterized from various sources [152, 153].

In spite of innumerable attempts to catalog the types of tRNA found in prokaryotic or eukaryotic cells under various physiological or pathological stimuli, generalizations assigning a meaningful role to special bases—except for the possibility that they may modify the conformation of the tRNA—are not available. Neither is there convincing evidence that *in vivo* the types or the rates of the proteins that are synthesized by cells are regulated by adjusting the concentrations of tRNA or by modifying the affinity of the tRNA for ribosome and amino acid [154].

Even the relatively simple matter of classifying the role of the C-C-A terminus of tRNA (see below) has raised more questions than it has answered. The "mature" tRNA found in the cytoplasm which, as we shall see, attaches to ribosomes, is the product of transcription of a large nuclear polynucleotide precursor, which is ultimately cleaved to a small cytoplasmic precursor, longer than the tRNA, and devoid of modified bases except for pseudouridine.

Sequence

Among modern contributions to molecular biology, the complete elucidation of the base sequence of alanine transfer RNA is probably one of the most impressive.

In contrast to DNA and messenger RNA, pure transfer RNA is a relatively small polynucleotide made up of 20–100 ribonucleotide residues. The nucleotide sequence of alanine RNA was first identified by Holley and his associate. As mentioned, transfer RNA contains unusual bases such as inosinic acid, methylinosinic acid, methylguanine, N_2-dimethylguanine, dihydrouracil, and pseudouracil. The presence of unusual bases proved to be extremely helpful in the study of the base sequence of transfer RNA.

Since there are at least 20 different transfer RNA's (one per amino acid), it was necessary to develop reliable methods of purification of a given transfer RNA before any attempt was made to study the base sequence. Transfer RNA was extracted with phenol and submitted to countercurrent distribution in a medium containing formamide, isopropyl alcohol, and a phosphate buffer. By this procedure, different types of transfer RNA (alanine, valine, histidine, and tyrosine) were separated from the bulk of the transfer RNA.

After three years of work, Holley and his associates managed to collect 1 g of alanine transfer RNA, which they used to study the sequence of the polynucleotide. To perform this task, investigators used three different enzymes: pancreatic ribonuclease, taka-diastase (T1), and snake venom phosphodiesterase.

Like all ribonucleases, pancreatic ribonuclease attacks phosphodiester linkages between the 3′ and 5′ positions of the RNA nucleotide moieties to yield 2′,3′-cyclic phosphate derivatives. The cyclic phosphates are then further hydrolyzed to yield the 3′-phosphate derivatives. The specificity of pancreatic ribonuclease is such that it restricts its activity to the phosphodiester bonds that involve 3′,5′-linkages between a pyrimidine and any other base. Therefore, the product of the ribonucleic acid digestion with ribonuclease contains 3′-pyrimidine mononucleotides and oligonucleotides with a 3′-cytidylic or uridylic mononucleotide. Taka-diastase (T1) is a ribonuclease that attacks the polynucleotides to yield 3′ guanosine mononucleotides and oligonucleotides with a terminal 3′-guanylic acid.

The specificity of snake venom phosphodiesterase, unlike that of RNase, is not concerned with the nature of the bases. Therefore, it can release one by one all nucleotides found in the polynucleotide chain. However, in contrast to pancreatic RNase, the snake venom phosphodiesterase acts at the other end of the phosphodiester bond, and 5′-mononucleotides are produced.

To study sequence, a sample of the RNA was divided into two batches, one treated with RNase and the other with taka-diastase. At the end of the incubation, in both cases the RNA is split into mononucleotides and oligonucleotides. Various mono- and oligonucleotides can be separated by chromatography on DEAE-cellulose columns. In the case of dinucleotides, the two nucleotides are split by alkaline treatment, and the mononucleotides are separated by paper chromatography or electrophoresis. The position of the 3′-

Table 2-3. Products of total digestion of yeast RNA

Ribonuclease	Taka-diastase	
A.		
(C) × 13		*Mononucleotides*
(U) × 6		
(Ψ) × 1	P(G) × 1	
(C^{OH}) × 1	(G) × g	
(GC) × 2	(CG) × 4	*Dinucleotides*
(GU) × 4	(CG^m) × 1	
(G^mC) × 1	(UG^m) × 1	
(AC) × 1	(UG) × 1	
($I^m\Psi$) × 1	(AG) × 2	
AGU^h	UAG	*Trinucleotides*
GGU^h		
GGU^h	U^hCG	
G^mGC	U^hAG	
AGC		
GAU		
IGC		
GGT		
GGAC	CI ΨG	*Tetranucleotides*
	TΨCG	
GGGAGAGU*	ACUCG UCUCCG	
	AUUCCG $UCCACC^{OH}$	
	CUCCCUUI	

B. Reconstruction of Sequence

Products: UG^m, UG, CG, G × 3, ^{P}G

Taka diastase

$$^{P}G^{\downarrow}G^{\downarrow}G^{\downarrow}C_{\uparrow}G^{\downarrow}U_{\uparrow}G^{\downarrow}U_{\uparrow}G^{\downarrow}G^{m}\;G^{\downarrow}C$$

↑ *Ribonuclease*

Products: ^{P}G G G C, G U, G U, G^m G C

A	adenylic acid
I	inosonic acid
I^m	i-methylinosonic acid
G	guanylic acid
G^m	i-methylguanylic acid
G^m	N^2 dimethylguanylic acid
C	Cytidylic acid
U	Uridylic acid
U^h	Dihydrouridylic acid
U*	Mixture of uridylic and dihydrouridylic acid
Ψ	Pseudouridylic acid
T	Ribothymidylic acid

In part A. On the right, are the products of the digestion of alanine transfer RNA by pancreatic ribonuclease. All nucleotides end with pyrimidine structures.

On the left, those obtained after digestion with taka-diastase. All nucleotides end with pyrimidine structures.

In part B are given the ribonuclease and taka-diastase sites of action with the first 11 residues of the alanine tRNA (after R.W. Holley).

pyrimidine nucleotide (in the course of RNase treatment) or of the 3′-guanylic acid (when taka-diastase is used) in the dinucleotide sequence is known (see Table 2-3).

Methods were devised long ago for elucidating the sequence of small nucleotides, using a combination of alkaline and enzyme hydrolysis. Holley divised new methods for sequence studies in large oligonucleotides (8 bases). Once the structure of the oligonucleotides had been established, such mono- or oligonucleotides had to be placed in their correct position within RNA. The tRNA molecules were reconstructed by sheer logic. The sequence of the six last nucleotides (UCCAC-C^{OH}) and the four first nucleotides (^{P}GGGG) could easily be predicted.

Some of the rare nucleotides occur only once in the molecule. When the same rare nucleotide is found in an oligonucleotide obtained by digestion with taka-diastase and also in an oligonucleotide obtained by digestion with RNase, the sequence of these two oligonucleotides clearly must overlap within the RNA molecule. For example, RNase yields IGC and T1 yields CUCCCUUI. The following sequence, CGIUUCCCUC, must exist within the tRNA molecule. On the basis of this type of reasoning, 16 partial sequences could be constructed, two of which were known to be located at the ends of the RNA molecules. The positions of the other 14 oligonucleotides in the RNA molecule were then established by controlled digestion of RNA into two halves, or in more frag-

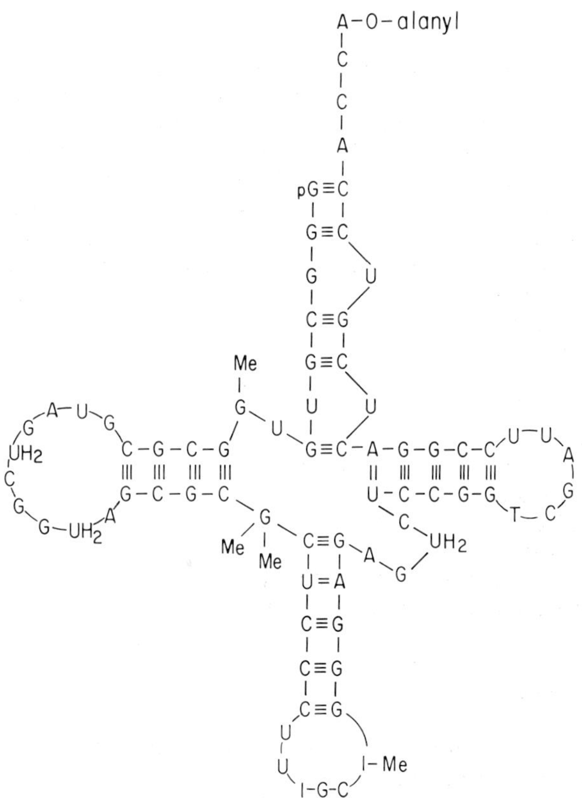

Fig. 2-12. Base sequence in alanyl transfer RNA

ments with T1. Finally, the sequence presented in Fig. 2-12 was established.

Configuration

In the complex macromolecular machinery responsible for elaborating the polypeptide chain, tRNA plays a key role—it carries the anticodon and is charged with the amino acid. Unloading the amino acid from amino acyl tRNA synthetase onto tRNA requires specific binding between that enzyme and tRNA. During polypeptide chain elaboration, tRNA must bind to messengers (therefore, its anticodon must be placed in a suitable position) and to ribosomes. Finally, during peptide bond formation, the amino acid that tRNA carries must be accessible to the enzyme involved in that step. Thus, the tRNA is the "jack of all trades" of protein synthesis. It interacts at the same time with a number of proteins and nucleic acids. This versatility must be reflected in the tridimensional configuration of tRNA; either the molecule has a relatively stable configuration with many specialized arms, or it is extremely flexible and adaptable to various modes of interactions. In any event, a correct description of the three-dimensional structure of tRNA will go far in clarifying our understanding of protein synthesis.

Even before the sequence of L-alanine RNA was known, some of the properties of tRNA had brought Zubay [155] to propose an intriguing model for tRNA structure that fit nicely a more general scheme for protein synthesis. tRNA of rabbit liver is soluble in 0.2 M sodium chloride. Its molecular weight is 23,000, and it is composed of 68 nucleotides per chain. Under the electron microscope, tRNA molecules appear as rods measuring about 100 A long. Physicochemical analysis and X-ray crystallographic studies suggest that the molecule has a helical structure.

In Zubay's model, the polynucleotide sequence is bent in the middle. Under conditions in which all except five bases are paired to form hydrogen bonds of the same type as those found in Watson and Crick's model for DNA, the base pair sequence is twisted to form a regular helix. At the acceptor end of the molecule, the pCpA sequence is free and can therefore accept the amino acid, and the first cytosine of the pCpCpA sequence forms hydrogen bonds with a complementary guanine. The loop of the helix is formed by three free bases and plays an important role in connecting the tRNA molecule with messenger RNA. Zubay's model was soon superceded by new and more sophisticated models that were proposed when the base sequence of transfer RNA became known.

It soon became obvious that the sequence Holley revealed for L-alanine RNA could not fit Zubay's simple model. Furthermore, it was also established that the low molecular weight material whose study had led to the elaboration of the Zubay model was most likely to be fragments of degraded ribosomal RNA (which in fact has a highly ordered tertiary structure). The secondary structure of tRNA is important because it determines activity of the tRNA, but may also determine specificity of the transfer reaction from the synthetase to tRNA. A number of observations suggest that tRNA has an ordered secondary structure; such studies include (1) temperature denaturation; (2) optical rotary dispersion; (3) formylation (which attacks the free amino group of the nucleic acid bases); (4) nitrous acid treatment (which deaminates adenine, guanine, and cytosine); and (5) studies of the rate at which tRNA exchanges tritium with the solvent. Results obtained with these different methods indicate that tRNA contains numerous hydrogen bonds. However, all bases are by no means involved in hydrogen bonding. Depending upon the method and the investigator, 64–82% of the bases have been proposed to be involved in hydrogen bonding.

For example, consider the effects of formylation on the temperature denaturation properties of RNA. Although normal RNA denatures within a narrow temperature range, the temperature denaturation curve of formylated RNA resembles that of poly UA—it is progressive and extends on a wide range of temperatures. To explain the differences observed between formylated and normal RNA, it has been proposed that the natural tRNA contains hydrogen bonds.

Since the sequence of tRNA alanine is known, a specialist can calculate the shape that the optical rotary dispersion curves would assume if the RNA is single stranded and the ordering simply results from stacking of the bases. When such calculations were compared to the actual curves, the values obtained by prediction were much smaller than the measurements. The conclusion is again that hydrogen bonds must contribute to the ordered structure of tRNA.

Whether hydrophobic bonds also affect the ordering of the RNA structure is not known. Most RNA, like DNA, is relatively unstable in the presence of ethylene glycol and methanol, which break hydrophobic bonds. The interpretation of these findings may not be as simple as it might first appear.

We have seen that only a fraction of the bases are involved in hydrogen bonding and that tRNA is only partially helical. Consequently, the molecules harbor the possibility of forming loops of different kinds, which could provide each tRNA with a unique configuration.

As the sequence of more and more tRNA became available and the physicochemical properties of various tRNA's were investigated in some detail, it appeared that all known tRNA's could be represented by a similar two-dimensional pattern resembling a four-leaf clover. The axes of the four arms are made of double helical regions held together by hydrogen bonds. The peripheral ends of three of the arms form loops of nonhelical nucleotide sequences. The north arm carries the anticodon, the east arm the pseudouridine residue, the west arm the dihydrouridine residue, and the south arm extends peripherally with the C-C-A sequence to which the amino acid will attach. In the center, a few bases not involved in hydrogen bonding hold the four arms together. In between the east and north arms, a nonhelical extra arm of variable length emerges. Such a structure provides semirigid arms with ends free to interact and be flexed on the central plates, thus providing a varied, but not unlimited, number of three-dimensional configurations.

Physical chemists, biochemists, and organic chemists each with their own tools (spin label, fluorescent measurements, small angle X-ray scattering, limited enzymic attacks, interaction with suitable chemical reagents) are attempting to delineate the three-dimensional structure of various tRNA's. A number of models have been proposed in which the arms are placed in positions most compatible with the experimental data. For further details refer to the specialized reviews [156–158]. Too little is known of the three-dimensional structure of tRNA to permit a description of the interaction of nucleic acid with enzymes, messenger and ribosome.

That the ordered structure is important for function is indicated by reports from several laboratories. Denaturation of tRNA by formaldehyde nitrous acid, or even denaturation by adding divalent ions, or simply altering the ionic strength prevents tRNA from binding the amino acid. An interesting observation was made with purified isoleucyl RNA synthetase. Remember that isoleucyl synthetase is also active on valine, but the affinity of the enzyme is much less for valine than it is for isoleucine. It was observed that the selective affinity for the two amino acids changed at 50° C. Moreover, at that temperature the enzyme's ability to release the amino acid from the tRNA was also changed.

At 50° C, enzyme-activated valine and isoleucine catalyzed the transfer of isoleucine to isoleucine tRNA, but failed to catalyze the transfer of valine to valine tRNA. As the temperature was increased from 50–80° C, the formation of isoleucyl tRNA decreased whereas that of valyl tRNA began to take place at 65° and increased up to 80° C. To explain these data, it was proposed that the temperature changes induced in tRNA conformational changes, which would either facilitate its reaction with the synthetase (in the case of valine) or interfere with it (in the case of isoleucine).

When the transfer of activated valine from enzyme to tRNA is measured at various temperatures, the transfer is quantitative up to 55° C, but then it drops slowly between 55 and 60° C (80% of transfer at 60° C) and rapidly between 60 and 65° C (no transfer at 65° C). Yet at 65° C the enzyme catalyzes the amino acid acylation at normal rates. Again the findings were interpreted to mean that conformational changes in tRNA interfered with its ability to react with the enzyme. Transfer RNA forms a link between most of the individual components involved in protein synthesis, amino acyl RNA, ribosomes, and messenger RNA.

If tRNA is incubated with pyrophosphate, it loses its capacity to bind amino acids. However, if CTP and ATP are added to the incubation mixture, tRNA regains its activity, and it was demonstrated in Zamecnik's laboratory that active RNA is the original polynucleotide, extended by the addition of CMP and AMP. If the terminal group is removed by short exposure to snake venom phosphodiesterase, the transfer properties of RNA are lost. Herbert has purified the enzyme that catalyzes the addition of CMP and AMP, separated it from the activating enzyme, and demonstrated that the purifed enzyme can catalyze only the addition of ADP and CMP to RNA. By isolating adenosyl amino acid derivatives from the ribonuclease digest of transfer RNA labeled with [^{14}C]leucine and [^{14}C]valine, it was demonstrated that the transferred amino acids are attached by ester bonds to the 2′- and 3′-hydroxyl group of the adenosine residue of tRNA. Since all transfer RNA's have identical terminal groups, it is unlikely that the terminal group selects the amino acids. However, specificity of selection might be determined by the rest of the molecule.

Attempts to determine which segment of the tRNA binds to the enzyme have included chemical modification and partial digestion of the sequence. The results of these experiments are too complex and too preliminary to permit a detailed description of the molecular contacts between enzyme and polynucleotide. Suffice to mention here the conclusion reached by Yarus [157] who reviewed the literature on the subject. There seems to be no unique site of binding between polynucleotide and enzyme; rather, the nucleotides involved in binding are scattered throughout the molecules, except the middle portion of the molecule, which can be digested extensively without altering binding to the enzyme. Of importance is the fact that the specificity of the binding is directed toward the RNA and not toward the amino acid it carries. It does not matter whether tRNA is charged or not, or whether it is charged with an amino acid analog. However, without enzyme, the binding is slow and reversible. There are clearly two binding sites: one nonenzymic, the other enzymic.

A similar but distinct problem is raised when one considers the interaction of tRNA with nucleotidyl transferase, the enzyme responsible for the attachment of the C-C-A terminus to the north arm of the cloverleaf. (It is not certain whether *in vivo* the C-C-A ending is always attached to tRNA after maturation. Inasmuch as maturation results from the splitting of nucleotide segments at both ends of the precursor molecule, the C-C-A ending possibly is coded for by DNA.) The similarity between the interaction of tRNA with the nucleotidyl transferase and the amino acyl tRNA synthetase is that both enzymes must recognize a segment on the tRNA molecule. But although synthetase reacts only with a few types of tRNA, transferase reacts with all tRNA's, whatever their source and whatever their specificity for the amino acid. As Deutscher [158] pointed out: "the amino acyl tRNA synthetases recognize the differences between tRNA molecules, and tRNA nucleotidyl transferases recognize the similarities."

Both bacterial and mammalian tRNA's are synthesized on special DNA genes. The product of gene transcription is a polynucleotide exceeding the size of the tRNA sequence. The precursors are believed to be

4-S or 5-S RNA molecules, which in the presence of a crude *E. coli* enzyme fraction, can be cleaved to yield the tRNA [247].

Because the sequence T-Ψ-C-Cp blocks aminoacyl tRNA binding sites to ribosomes more effectively than any other similar sequence, it is believed that T-Ψ-C-Cp is part of the nucleotide sequence which binds to ribosomes. This conclusion needs, however, to be reconciled with the fact that eukaryotic initiator methionyl-tRNA of yeast and rabbit lacks the sequence G-T-Ψ-C-G(A-) which is replaced by G-A-U-C-G [248–249].

RNA Methylases

During the discussion of tRNA, small amounts of odd bases in the molecule, especially methylated bases, have been emphasized. Our knowledge of the mechanism of methylation and of the biological significance of methylation of purine and pyrimidine bases is still at an early stage. But, in contrast to messenger RNA, both transfer and ribosomal RNA contain methylated bases. The transfer RNA is particularly rich in methylated bases, but, in spite of earlier reports, ribosomal RNA contains one-fourth as many methylated bases as tRNA. The proportion of methylated bases in ribosomal RNA varies with its sedimentation properties. 18 S RNA is 20% richer in methylated bases than 28 S RNA. This is, however, only *in vivo*. When 28 S RNA is used as a substrate for methylation *in vitro*, the degree of methylation that takes place is similar to what is observed with 16 S RNA [159, 160].

Both purine and pyrimidine bases are methylated in RNA. With adenine, the addition of one intact methyl group leads to either the formation of a carbon-to-carbon bond involving the carbon 2 of adenine, or a nitrogen-carbon bond involving the N_6 or N_1 of adenine. Guanine is methylated in N_1, N_2, and N_7; uracil and cytosine form carbon-to-carbon bonds between CH_3 and their carbon 5. Hypoxanthine is methylated on its N_1; dimethyl derivatives of adenine (N_6) and guanine (N_2) have also been observed. Two methylated bases have been found in DNA: 6 methylated aminopurine and 5 methylcytosine. In addition, methyl ribosides have been found in tRNA.

The nucleic acid bases are methylated when the biosynthesis of the polynucleotide chain is complete. This has been demonstrated for tRNA, rRNA, and DNA. In the case of DNA, only double-strand DNA is methylated.

The most elegant demonstration was with tRNA. It was first shown by isotopic techniques that all methyl groups of all methylated bases (including thymine) in RNA were derived from methionine. *In vitro* studies later established that S-adenosyl methionine is an intermediate and that the methyl group is transferred as a whole. Similar observations were made for the methylaminopurine and 5-methylcytosine found in DNA, but, of course, not for DNA thymidine. K_{12} mutants of *E. coli* present the unusual property of having the synthesis of nucleic acid and the methylation of nucleic acid bases dissociated in time. One can thus maintain mutants on methionine-free media and obtain unmethylated nucleic acids, which can be used as substrates for studying methylating enzymes. The substrates of polynucleotide methylases vary in at least 2 ways: by the type of polynucleotide (tRNA, DNA), and by the type of bonding formed in the methylation reaction. As might be expected, the specificity of the enzyme varies with the properties of the substrate, and tRNA, rRNA, and DNA methylases must be distinguished.

There exists one type of uracil, cytosine, and adenine tRNA methylase. Two different guanine tRNA methylases have been discovered—one for N_7 and two for N_1. It is possible but not established that the difference between the two guanine methylases results from the different abilities of the enzyme to recognize neighboring bases.

Two types of rRNA methylases have been purified, one with catalytic properties restricted to rRNA, and another that acts on both rRNA and tRNA. A bacterial preparation capable of methylating either adenine or cytosine of DNA has been purified. But although it is most likely that the enzyme methylating adenine (N-C bond) is different from that methylating cytosine (C-C bond), successful separation has not been accomplished. Almost nothing is known of the enzymes catalyzing 0′ methyl riboside formation.

Study of the biological significance of methylation has barely started. In fact, it has been known for a long time that a small number of DNA and RNA bases were methylated, but since the role of methylation was not obvious and only a small number of the bases were methylated, it was concluded that methylation played only a secondary biological role. It was even proposed that these reactions were only remnants of some more primitive type of function, or that nucleic acid methylation was an unspecific event serving to absorb excess methyl groups. But if every cell contains at least 14 different methylases, it is unlikely that nucleic acid methylation is purposeless.

Moreover, the activity of the crude enzyme is modulated by numerous activators and inhibitors suggesting a delicate regulatory mechanism [161]. The high methylase activity observed in embryonic and tumoral tissues result, at least in part, from the absence of such inhibitors.

Yet, studies in bacteria continue to yield intriguing results; when *E. coli* are grown in a methionine-deficient medium, tRNA is synthesized but not methylated. Such tRNA usually binds to codon enzymes and ribosomes normally. However, there are exceptions and further studies are needed to evaluate the role of methylation in tRNA function.

There is some indication from bacterial studies that methylation of DNA might confer a type of specificity, such as species specificity. Methylation of tRNA has been invoked to play a role in binding to ribosome or to enzyme, acceptance of the appropriate amino acid, or even recognition of the codon in the messenger. But none of these assumptions is supported by studies with nonmethylated tRNA. Whether methylation of rRNA plays a role in aggregation of precursors of ribosome is also in great need of experimental evidence.

Whatever the biological role of methylation may be, the observation is most intriguing, especially when one considers that the number of methylated bases is increased in many tumors and that the activity of methylase (measured *in vitro* with amethylated tRNA) is four times greater in rapidly growing tumors than in normal tissue. Whether this biochemical anomaly of tumors is relevant to carcinogenesis and, if relevant, whether it is closely related to the primary insult remains to be shown.

Amino Acid Code

One of the most fantastic achievements of this scientific era has been deciphering the amino acid code. As Garen, Crick, and Nirenberg pointed out repeatedly, the problem amounted to translating a 4-letter alphabet (adenine, guanine, thymine, and cytosine) into a 20-letter alphabet—the 20 amino acids in the cell proteins—or rather finding how messenger RNA's made of only four different nucleotides dictate the sequence of 21 amino acids in the polypeptide chains [162–170].

At first it appeared that this question would be answered only when the full sequence of a specific messenger could be aligned next to the full sequence of the polypeptide whose sequence is dictated by the messenger. However, ingenious experiments done in Nirenberg and Ochoa's laboratories broke the code much sooner than had been anticipated.

Even before Kornberg demonstrated the mechanism of DNA synthesis, Ochoa and his group studied the steps involved in RNA synthesis. The requirements for that reaction in Ochoa's system were an enzyme (nucleotide phosphorylase), four diphosphonucleotides (ADP, GDP, CDP, UDP), magnesium, and a primer. The proper combination of primer, nucleotides, and enzyme permitted the synthesis of ribose polynucleotides of different base compositions, such as poly U, poly A, etc.

When poly U was added to an *in vitro* system capable of incorporating amino acid into protein (poly U, a source of energy, pH 5 enzyme, the amino acid and ribosomes, and tRNA), a new polypeptide was made, composed entirely of phenylalanine. Similarly, poly A was found to dictate the biosynthesis of polylysine while poly C dictated the biosynthesis of polyproline. Poly G could not act as a "surrogate messenger," because hydrogen bonds bundle up the molecule into a multistranded double-helix configuration that interferes with the priming activities of the nucleotide [171].

Since there are only 4 nucleotides and 21 amino acids, the sequence of the amino acids in the polypeptide chains can obviously not be dictated by a single base. Even a combination of two nucleotides would be insufficient because it would provide only 16 different combinations. Consequently, it was assumed that the "codon" was made up by at least three bases. Today the "codon" can be defined as a sequence of three bases capable of determining the position of one or more amino acids in the polypeptide sequence. The experiments with the polymers established that the codon of phenylalanine is UUU; that for lysine is AAA; and for proline CCC.

Further information was obtained by measuring the incorporation of various amino acids using mixed copolymers as surrogate messenger RNA. For example, poly AC is a random copolymer synthesized by incubating adenine and cytosine with polynucleotide phosphorylase. And it can be prepared with known proportions of A and of C. However, in the earlier experiments the sequence of the bases in the copolymer was unknown. One can predict that the polymer will contain eight different codons: (AAA, CCC, ACC, AAC, CAA, CCA, ACA, CAC). Therefore, it could also be predicted that the AC polymer will stimulate the incorporation of eight amino acids, provided, of course, that each triplet can operate as a codon, and that the code is not degenerate. A degenerated code is one in which the position of an amino acid can be dictated by more than one triplet sequence. CA polymers stimulate the incorporation of proline, lysine, and four other amino acids (asparagine, glutamine, histidine, and threonine).

The rate of amino acid incorporation can be considerably modified by changing the ratio of adenine to cytosine 1:5 in a poly AC 5:1. Proline is generously incorporated. The incorporation of asparagine and glutamine is low, that of lysine is almost nonexistent, but that of histidine is one-fourth the rate of incorporation of proline. In a random polynucleotide such as poly AC, the frequency at which a triplet combination will occur can be predicted by applying the mathematics of probability. For example, in a poly AC 5:1, there will be one AAC triplet for every five AAA triplets. On the basis of these findings, a tentative codon was assigned for each amino acid; for example, AAC for asparagine and glutamine; ACC for histidine, etc.

This procedure tells us what the base composition of the codon is, but it does not provide any information on the sequence of the bases within the codon. The solution of the sequence in a given codon seemed to be a hopeless task until it was solved at least in part by Nirenberg's ingenious approach. Nirenberg found that it is not necessary to use amino acids *per se* in the incubation mixture, nor were long polynucleotide chains necessary to have amino acid RNA bind to microsomes because trinucleotides suffice. Trinucleotides of known sequence can be prepared, and labeled amino acid RNA can be prepared; the trinucleotides and tRNA can be incubated in the presence of ribosomes. As a result, the sequence of the triplet dictating the position of a given amino acid can be established. In fact, this technique is more specific than the technique in which amino acid incorporation is measured.

When phenylalanine is omitted from the incubation mixture, poly U stimulates the incorporation of leucine. In contrast, triuridilate stimulates only phenylalanine tRNA attachment to ribosomes, but is without effect on leucine tRNA.

These procedures have permitted researchers to determine the base composition and the base sequence of the codons for a large number of amino acids and to establish that the code is degenerate, universal, operates *in vivo*, and is not overlapping. The accepted code for each amino acid is presented in Table 2-4. There are 64 sequence combinations of trinucleotides when four different bases are used to build the triplet. A brief look at Table 2-4 shows that 60 of the 64 combinations are involved in amino acid coding. Two sequences have no known coding properties (CUA and CUG) and are therefore called nonsense codons. The intercalation of such a triplet into the sequence of messenger RNA is responsible for nonsense mutation. UAA and UAG are now known to be involved in punctuation.

Table 2-4. Various codons for each amino acid

Alanine	GCA	GCG	GCC	GCU		
Arginine	AGA	AGG	CGA	CGG	CGU	CGC
Asparagine	AAC	AAU				
Aspartic	GAC	GAU				
Cysteine	UGC	UGU	UGA			
Glutamic	GAA	GAG				
Glutamine	CAA	CAG				
Glycine	GGA	GGG	GGU	GGC		
Histidine	CAU	CAC				
Isoleucine	AUA	AUC	AUU			
Leucine	CUC	CUU	UUA	UUG	CUA	CUG
Lysine	AAA	AAG				
Methionine	AUG					
Proline	CCA	CCG	CCC	CCU		
Serine	AGC	AGU	UCA	UCG	UCC	UCU
Threonine	ACA	ACG	ACC	ACU		
Tryptophan	UGG					
Tyrosine	UAC	UAU				
Valine	GUA	GUG	GUC	GUU		
Amber	UAG					
Ochre	UAA					
Umber	UGA					

Lysine (AAA, AAG), glutamine, and cysteine have two codons; isoleucine, leucine and phenylalanine have three; threonine, glycine, alanine, valine, proline, and tyrosine all have four; and serine and arginine have six. Consequently, in many cases more than one codon is able to dictate the position of a given amino acid. The code is therefore referred to as a degenerate code. As a result, many mutations occur which are not accompanied by alterations in the amino acid sequence. It would therefore appear that a degenerate code constitutes an evolutionary advantage, and even if all 64 triplets existed in the RNA template, all triplets need not function in transcription. The choice of a triplet could also be restricted by the number of available transfer RNA's. But if 60 sequences can be used for amino acid coding and all the sequences are used, then it is likely that 60 different transfer RNA's exist, each with a specific anticodon. In the case of leucine, two such RNA's have been found—one with a hypothetical anticodon AAC (for UUG) and another with a hypothetical anticodon GAA (for CUU).

The degeneracy of the code explains results obtained in some point mutations in which a base is deleted. Such mutations can be induced by administering nitrous acid or acridine orange. Nitrous acid attacks the bases with a free amino group in position 6 (adenine and cytosine). By deaminating the bases, it converts adenine into guanine and cytosine into uracil, changing the spelling of at least some of the triplets containing adenine or cytosine.

In vitro experiments done in Khorana's laboratory with synthetic nucleotides of known sequence have clearly established that the code is not overlapping, but that the messenger is translated for a fixed sequence three nucleotides at a time. Thus, when a nucleotide chain of the sequence UCUCUC is used, a polypeptide in which serine alternates with leucine in a regular position is formed:

UCUC UC UCUC UC
Ser Leu Ser Leu

If one could use overlapping triplets,

UCU
UCU
UCU

the code for leucine would never be read, or at least not be read in alternate sequence, CUC or CUU.

Although unlikely, the possibility of ambiguity—a specific codon in the messenger combining with more than one anticodon—has not been irrevocably excluded. The trinucleotide pA pC pA binds to both leucine and threonine tRNA. However, the copolymer ACACACA induces binding only of threonine and histidine RNA. Poly U is also known to stimulate the incorporation of small amounts of leucine in the polynucleotide chain. Whether such ambiguity exists when natural messengers are operating in the intact cell is not known.

Conclusive evidence proving that the code as we have described it works *in vivo* was obtained in Khorana's laboratory, where large polynucleotides were synthesized, the codes of which corresponded to biological polypeptides of known AA sequences. The polynucleotide was shown to serve as template for the synthesis of a protein whose amino acid sequence had the anticipated sequence [172].

The studies in which isolated ribosomes are combined with tRNA and synthetic polynucleotide indicate that both messenger and transfer RNA need to bind to the ribosomes, and also that hydrogen bonding is involved in binding codon and possibly the body of the transfer RNA and ribosome.

Studies of protein synthesis indicate that the codons are the same irrespective of bacterial, plant, or mammalian preparations. Therefore, the code appears to be universal in actual terrestrial life. It cannot, however, be excluded that the code may not have been different at early stages of evolution. Although some codons may, as we know them, occasionally have been misspelled, the code as a whole almost certainly is not an artifact.

These elegant experiments would not be much more significant that a sophisticated game of chess if they did not mirror at least part of reality. The combined techniques of molecular biology and genetics have provided compelling evidence that the code as deciphered *in vitro* is likely to operate also *in vivo*. The basic principle that underlies the *in vivo* experiments is a comparison of changes in the sequence of the RNA molecule with changes in the amino acid sequence in the proteins. Three such systems have been studied: the protein of tobacco mosaic virus (TMV), tryptophan peroxidase, and human hemoglobin.

Tobacco mosaic virus is made of RNA and a protein coat. The amino acid sequence of the protein coat is dictated by the base sequence of the tobacco mosaic RNA (to determine the composition of the coat protein). Obviously, much can be learned about the control of protein synthesis by changing the composition of the template RNA. This is precisely what happens after chemical mutagens are applied. Three different mutagens have been used: nitrous acid, hydroxylamine, and 5-fluorouracil. Among the three, the results obtained with nitrous acid are particularly interesting.

When RNA of tobacco mosaic virus is incubated with nitrous acid, changes occur in the base composition because nitrous acid deaminates cytosine and adenine. (It also deaminates guanine, but not in the intact virus.) Cytosine is converted to uracil and adenine is converted to hypoxanthine (adenylic acid → inosinic acid). Consequently, the six amino groups of adenine and cytosine are replaced by keto groups. Since the complementariness of the newly synthesized RNA chain is dictated by the formation of hydrogen bonds with the bases in the template, the newly synthesized RNA is different from the wild RNA and thus a base mutation has been introduced experimentally.

As may be expected, the changes in the structure of the RNA molecule are reflected in the morphological features of virus. A certain number of TMV mutants also show changes in the secondary structure of the TMV protein. In other words, some amino acids in the sequence of the wild TMV protein have been replaced by others in the mutated TMV proteins.

In the ideal theoretical experiment, the complete base sequence of the stretched out messenger RNA molecule would be specified and superimposed onto the amino acid sequence of the protein whose sequence it dictates. Such an experiment is not possible yet. But the number of base replacements can be counted and compared with the amino acid replacements. In the case of the TMV protein, this procedure was likely to yield quite significant information because the TMV protein had been extensively purified and its complete amino acid sequence was known.

Trypsin splits the protein at 11 different points, yielding 12 different polypeptides that can readily be separated by a combination of chromatography and electrophoresis. The further digestion of each polypeptide according to methods that have now become standard provides the exact amino acid composition and sequence of the polypeptide. By this procedure, Wittman and Wittman and Liebold were able to localize the amino acid changes in a number of mutants.

For example, after treatment with nitrous acid,* the mutated protein contained less threonine residues than the wild TMV protein, but the mutated protein contained new isoleucine residues and new methionine residues. The codons for threonine are ACU and ACA. Under the effect of nitrous acid, one may expect to convert cytosine to uracil and thus the codon would turn into AUU, AUA (see Fig. 2-13). Both codons can dictate the incorporation of isoleucine. Another possibility is that one of the adenines is converted to a guanine leading to GCU, ACG, or GCA (alanine, threonine, alanine).

Amino acid replacements have also been studied in TMV. Considering the large number of base changes that can be introduced, the number of amino acid substitutions is surprisingly low. This can be explained in various ways. For example, the code may be degenerate with the result that some of the base replacements do not yield coding changes. Some of the replacements could also be in some way lethal to the virus. The code cannot be overlapping because if codons for given amino acids overlapped, substitution in neighboring amino acid would be expected, and this was not observed.

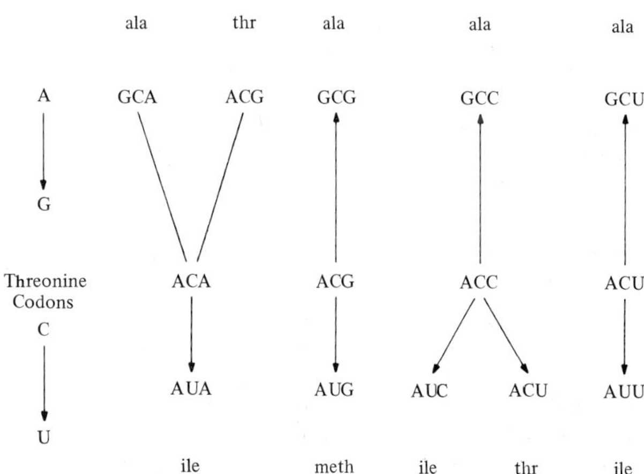

Fig. 2-13. Change caused by nitrous acid in the threonine codon

* When virus RNA is incubated with hydroxylamine at pH 6, cytosine reacts with hydroxylamine to yield a compound that codes like uracil. 5-Fluorouracil is incorporated into the virus RNA and replaces uracil in the sequence. In contrast to uracil, which can form hydrogen bonds with adenine only, 5-fluorouracil seems to be able to form hydrogen bonds with adenine or guanine.

When one computes the amino acid substitution observed in protein of spontaneous or experimental TMV mutants, almost all the tryptic polypeptides clearly are subject to substitution except one—peptide IX. Changes in the tertiary structure of peptide IX are fatal to the protein configuration and prevent reaggregation of the RNA and virus protein. Uncoated RNA is rapidly digested by hydrolases.

Auxotrophs of *E. coli* elaborate the enzyme tryptophan synthetase. The enzyme molecule is composed of two different polypeptides, referred to as polypeptides A and B. Polypeptide A has been extensively purified and its molecular weight is known to be 26,000. Although the full amino acid sequence of polypeptide A is not yet known, it is possible to digest the molecule, separate smaller polypeptidic fragments, and establish the sequence of the small polypeptide fragments. One of these polypeptides, referred to as peptide CP_2, has the following sequence:

Asp-NH_2-Ala-Ala-Pro-Pro-Leu-Glu NH_2-Gly-Phe

The glycine that occurs in the second position from the carboxy terminal amino acid can be replaced in the course of mutation by five different amino acids: arginine, glutamic acid, threonine, serine, and valine. The mutant with a peptide CP_2 containing a substitute for glycine may also, either spontaneously or under specific experimental conditions, revert to a CP_2 containing a glycine residue in the correct position. Consequently, 7 amino acids in the CP_2 polypeptide substitutions can occur. All the amino acid substitutions can be explained in the light of the code as we know it today.

Although the amino acid sequence of hemoglobin is the same in most humans, a few exceptions result from mutation (inborn errors of metabolism). We shall see that these mutations are also explainable in terms of the known amino acid code.

A number of triplets are referred to as nonsense codons because they are not known to code for any specific amino acid. When bacterial strains were developed in which mutations leading to the appearance of such triplets in their messenger were produced, 3 of the so-called nonsense codons, UAG, UGA (amber mutation), and UAA (ocre mutation) were found to act as punctuation codons for termination of the chain.

Indirect evidence with synthetic polynucleotides had already suggested that at least two bases—U and A—might be part of the triplet responsible for termination. With most polypeptides used as "surrogate messenger," only 10% of the polypeptides synthesized are released; however, 60% of the newly made polypeptides are released when the coding polypeptide contains A and U. An important difference between the initiation and termination codon is that initiation requires that an anticodon of tRNA charged with a formyl methionine be lined up, but termination does not require that the codon be coupled with the appropriate anticodon.

Bacterial and Mammalian Messenger RNA; RNA Polymerase

In the preceding paragraphs, we have reviewed the evidence establishing that the genetic information is stored in the DNA molecule, then later described the mechanism of protein synthesis. Now the flow of information from DNA to the protein molecule will be discussed [173–179].

Spiegelman [180] has distinguished three steps in protein formation in mammalian and bacterial systems: duplication of the DNA molecule, transcription of the information included in the DNA sequence into an RNA sequence, and translation of the RNA nucleotide sequence into the protein amino acid sequence.

Morphologists postulated for many years that the transfer of information from DNA to protein involves a form of nuclear RNA that migrates from nucleus to cytoplasm. This hypothesis found some concrete support as molecular biology developed. At first, it was found that a small fraction of bacterial RNA has a base composition identical to DNA in bacteria infected with T^2 phage. The synthesis of DNA-dependent RNA was demonstrated in bacterial preparations and in mammalian nuclei. Triphosphate ribosides (UTP, GTP, CTP, and ATP) are required precursors for the synthesis of the new polynucleotide. Double-stranded DNA serves as a template in the reaction, and an RNA polymerase catalyzing the polymerization of the ribonucleotide was partially purified from both the bacterial and mammalian systems.

Newly synthesized RNA has a base composition complementary to DNA, but different from that of the greater part of cytoplasmic RNA. When this DNA-dependent RNA is synthesized *in vitro* using single-stranded DNA as primer, hybridization with the mother DNA readily occurs. In contrast, there is no hybridization if DNA strands different from the mother DNA are used for the combination.

This small amount of RNA synthesized using DNA as template is now referred to as "messenger" RNA because it is involved in transferring information from DNA to protein and can stimulate amino acid incorporation. The RNA synthesized using single-stranded DNA as primer cannot stimulate amino acid incorporation [181].

As we shall see later, viral, bacterial, and mammalian messenger RNA's have been isolated. But before we can discuss their preparation, we must review some of the properties of the enzyme catalyzing DNA transcription—RNA polymerase.

RNA polymerases* not only catalyze the biosynthesis of messenger RNA, but the two types of ribosomal RNA and transfer RNA also seem to be synthesized on a double-stranded DNA template. This was

* Although it is generally accepted that *in vivo* various types of polynucleotide polymerases exist (some using DNA as primer for RNA synthesis, others using RNA as primer), *in vitro* the activity of the enzymes is not necessarily restricted by the nature of the primer. *In vitro* both DNA and RNA may act as primers for certain RNA polymerases.

suggested by hybridization experiments in which the various RNA's were annealed with denatured DNA. These experiments indicated that the base composition of ribosomal and transfer RNA are complementary to that of DNA. Thus, part of the cells' DNA stores serves as template for ribosomal and transfer RNA, and part serves as template for messenger RNA. It seems that all the polynucleotide chains used to dictate the base composition of transfer and ribosomal RNA are located in a single area of the DNA molecule, whereas the template sequences for messenger RNA are found throughout the deoxypolynucleotide chains.

DNA-dependent RNA synthesis is intriguing—especially messenger RNA synthesis in which it is almost certain that a single-stranded product is formed using a double-stranded primer.

At least four steps in the enzymic transcription of DNA into messenger RNA can be distinguished: (1) the binding to a promoter with specific sequence by special factors referred to as T or auxiliary factors (M, Ψ, and CAP), (2) formation of the first phosphodiester bond, (3) chain elongation, and (4) chain termination. Obviously, the molecular properties of the RNA polymerase must be known before the molecular interactions that occur at each of these steps can be clarified.

Knowledge of the molecular properties of the RNA polymerase has only begun to be accumulated. Several bacterial polymerases have been partially purified. *E. coli* RNA polymerase was found to be composed of several subunits ($\sim$5) referred to as α (mol wt 39,000), β (mol wt 155,000), and β' (mol wt 165,000), σ or γ (mol wt $\sim$90,000), and ω (mol wt 10,000). A molecule of RNA polymerase (core enzyme) contains two α-subunits, one each of β- and β'-subunits, and one or less γ- or ω-subunit. In addition, molecules of RNA polymerase made of the subunits described above appear to agglomerate to yield dimers (holoenzyme) or higher polymers. Because of aggregation, molecular weight measurements have proven difficult, but it has been estimated that the active monomeric size is 330,000–390,000.

The role played by the various subunits in the polymerase reaction has only begun to be clarified. A σ factor has been isolated from *E. coli* which specifically stimulates the initiation of ribosomal cistron transcription. β'-Factor is believed to constitute the binding site of the core enzyme to DNA. Whether the β'-unit also contains the active site remains to be demonstrated. The σ-unit recognizes the initiation codon and facilitates specific binding.

The importance of the σ-factor in determining transcription specificity is illustrated by the events in T4 phage. One minute after infection of *E. coli* by T4 phage the σ units of the *E. coli* polymerase are inactivated. A few minutes later, the σ factor disappears altogether and the subunit is electrophoretically altered. New initiation factors are made. Thus, virus infection is elicited simply by subverting the normal process of gene expression—in some cases by modifying the subunit structure of existing polymerases, in others by making a new polymerase altogether.

The molecular steps of the initiation reaction are believed to include the binding of two nucleoside triphosphates to the enzyme and the release of one molecule of pyrophosphate, leaving a dinucleoside tetraphosphate bound to the enzyme.

Elongation of the messenger RNA chain proceeds by the stepwise addition of nucleoside monophosphates to the 3′-terminus of the new RNA chain. It is believed that at some stage of elongation the σ factor is released, possibly freeing it to engage in a new initiation step. Several mechanisms for chain termination have been described. Among them are: (1) messenger synthesis using fd RF or Φ 80 phage DNA is stopped when the RNA polymerase encounters a unique DNA sequence that signals termination; (2) *in vitro* transcription of T4, T7, and φ DNA is terminated by the addition of a protein factor [182, 183] sometimes referred to as the T factor.

It is not known whether the preparation of the T factor is composed of a number of similar molecules, each capable of binding to a specific promoter site or whether other T factors not yet discovered also exist.

Since the discovery of the T factor, at least three other factors have been found that stimulate the activity of RNA polymerase—the CAP, M, and Ψ factors. Little is known about the M and Ψ factors except that they work only in the presence of the T factor and the remaining portion of the polymerase to stimulate the activity of the holoenzyme.

The discovery of the cyclic AMP receptor protein (CAP) in *E. coli* is most intriguing because it further illustrates the complexities of gene expression even in a relatively uncomplicated system. In *E. coli*, cyclic AMP stimulates the elaboration of catabolic enzymes involved in the breakdown of arabinose and maltose [184, 185]. A mutant of *E. coli* has been found that fails to grow on maltose and arabinose. The mutants lack a protein receptor for the cyclic AMP which is believed to bind probably to DNA, but possibly to RNA, which must be present for the induction of the catabolic enzymes. It has also been shown that in order to secure optimal transcription of the lac operon (see below), both the inducer—for example, isopropyl thiogalactoside—and cyclic AMP must be added.

Rutter and his associate [186, 187] have solubilized mammalian RNA polymerase by sonication of nuclear preparations in high salt followed by dilution of the suspension. After dilution, most free proteins aggregate with the DNA, but the polymerase is removed in solution. The enzyme is then purified more extensively, and as a result three chromatographic peaks are separated. The peaks can be distinguished by their response to various salt concentrations, activity on various primers, requirements for manganese and magnesium, and the effects of inhibitors.

For example, peak III requires higher salt concentration for activity than peak II, and peak II requires higher salt concentration for activity than peak I. Peak I is active in the presence of manganese and magnesium. Peak II has low activity with magnesium and high activity with manganese. α-Amanitin, a cyclic peptide, inhibits the enzyme found in peak II, but not that found in peak I. Differences in the readouts for the templates have also been observed, especially with respect to SV 40. Peak II reads the simian virus DNA well, but peak I reads it poorly. Of course, these differences in templates are difficult to interpret unless one works with a very pure protein; otherwise, differences may simply result from an increased incidence of nicks.

There is indication that peak I is primarily a nucleolar enzyme that reads ribosomal genes, whereas peaks II and III are found in the nucleoplasm and are involved in the biosynthesis of most other types of RNA.

The three peaks have been prepared from five different sources: amphibian eggs, sea urchin eggs, rat liver, calf thymus, and yeast.

Attempts to study the molecular structure of the protein included in peak II suggest that it is formed of a complex of symmetrical subunits. However, factors resembling sigma factors have not yet been discovered in mammalian or eukaryotic systems. These studies are of considerable interest because they suggest that the specificity for the templates resides with the enzyme rather than with the DNA molecules.

RNA polymerase has now been purified in various laboratories from various mammalian sources [188], and RNA polymerase of eukaryotic nuclei always exists in multiple forms that can be distinguished by the cation requirement, their sensitivity to α-amanitin, and their ability to react with specific templates. One enzyme (polymerase I) is found in the nucleolus, the others in the nucleoplasm. All mammalian ribonucleases are complex proteins formed of several subunits. Three types of RNA polymerases have been purified from ascites tumor cells by chromatography on carboxymethyl-cellulose. Two are nucleolar and one is nucleoplasmic. Protein factors of unknown nature that stimulate all three enzymes have been found in calf thymus, rat liver, and ascites cells. These factors can be separated into two classes: heat stable and heat labile. Both types stimulate the activity of the Novikoff RNA polymerase several-fold, but only with native DNA as templates. The factors have no effect on *E. coli* RNA polymerase [266–267]. For further information, refer to the review of Jacobs [189].

Although DNA molecules store all genetic information available to a cell, only the DNA that is transcribed into messenger RNA determines special functions, structure, metabolic cycles, and reproduction rhythm in a given cell type. Consequently, the fate of messenger RNA must be clarified for an understanding of cellular adaptation to the environment, growth, and differentiation.

In extreme situations in which all templates are continuously transcribed, the selection of the template that will be translated must result from either selective destruction of messenger or selection of messenger by the translator. However, a mechanism regulating the specificity of the cell's protein mosaic at the level of transcription seems to better safeguard specificity.

Although regulation at other levels cannot be excluded, regulation of transcription is known to occur in bacteria. Various types of messengers are produced at various ratios under different circumstances. In fact, on the basis of somewhat preliminary kinetic data, molecular biologists have assumed that the expression of the genome is roughly equal to the rate of initiation of transcription. Thus, once the transcription of the message has been initiated, elongation and termination follow inevitably, and the production rate of either type of messenger does not need to be regulated, at least as long as translation is pursued. In fact, transcription and translation are concurrent in bacteria as shown by measurements of transcription with the tryptophan or the β-galactosidase operon. Kinetic studies have shown that ribosomes are bound to the messenger before transcription is completed, and that ribosomes travel on the messenger behind the RNA polymerase.

Although fundamentally similar in bacteria and eukaryotic cells, transcription and translation must differ in details because of the difference in structural organization and of the surrounding environment. The environment of bacterial cells may change suddenly and considerably, but that of mammalian cells is stable because of homeostatic control. Therefore, metabolic patterns are much more stable in mammalian than in bacterial cells, and ordinarily there is no need for the sudden appearance of enzyme clusters like those that appear in bacteria challenged by inducers. Thus, it may not be necessary to produce polycistronic messenger in mammalian cells. Moreover, the mammalian cell may find it advantageous to maintain stable templates.

Hayashi and Hayashi have prepared a polycistronic messenger of the Φ X 174 phage—a DNA phage containing approximately 10 cistrons, the functions of 9 of which have been identified [190]. Several laboratories have prepared messenger RNA from viral, bacterial, and even mammalian sources.

In 1967, DeVries and Zubay synthesized *in vitro* a fragment of β-galactosidase which, when added to complementary fragments synthesized *in vivo,* restored enzymic activity. Similarly active molecules of lysozyme have been synthesized by Salser and his associates [191]. In this system, a lysozyme messenger is used. Inasmuch as lysozyme is composed of 165 amino acids, one may expect the messenger to be 500 nucleotides long. On the basis of the length of the molecule, one may predict that its sedimentation properties would be around 9.3 S, although the exact size of the lysozyme messenger remains to be established. The messenger activity sediments in sucrose gradients in a peak just about that of the 16-S RNA.

In reticulocytes, 80% of protein synthesis is directed toward hemoglobin. Consequently, one might expect that hemoglobin messengers would be abundant and relatively easy to purify from rabbit reticulocytes.

Hunt and Laycock [192] have separated two different messenger RNA's from hemoglobin—one sedimenting at 10 S, the other sedimenting at 8 S. The reason for the existence of two messengers in reticulocytes is still not clear.

Avian interferon messenger RNA preparation was found to stimulate the appearance of an avian-type interferon when incubated with mouse cells [193]. An RNA fraction obtained from the hen oviduct supports the biosynthesis of ovalbumin when incubated with an *in vitro* translation system obtained from rabbit reticulocytes [194].

The messenger RNA coding for mouse heavy chain immunoglobulin has been isolated from myeloma cells [250].

When the globin messenger RNA is released from reticulocyte polysomes it is associated with two proteins: one with a molecular weight of 78,000 daltons, the other with a molecular weight of 52,000 daltons. Similarly, when the heterogeneous messenger RNAs of rat hepatocytes or L cells are released from polysomes a 78,000 dalton protein is bound to the adenylate rich region of the messenger RNA. The significance of this protein-messenger RNA association is not known [265].

Nuclear Heterogeneous RNA—A Precursor of Mammalian Messenger RNA

Mammalian nuclear RNA can be divided into four classes: nucleolar RNA, heterogeneous nuclear RNA (hnRNA), chromosomal RNA, and a small molecular weight RNA. The hnRNA constitutes only 3% of the total RNA population in the nucleus. Although the precursors of nucleolar RNA form a homogeneous class of RNA, the heterogeneous RNA contains a population of RNA molecules with molecular weights ranging from 10^5–2×10^7 daltons [195]. Two other characteristic features of hnRNAs, are that their bases are not methylated and the molecules have an extremely rapid turnover. Four types of nucleotide sequences in the hnRNA can be distinguished: (1) a sequence related to messenger RNA, (2) poly A sequences, (3) mixed-base RNase-resistant regions, and (4) oligo U sequences.

Several laboratories have measured the ability of hnRNA and mRNA to hybridize to frequently repeated DNA base sequences. The results clearly demonstrate that hnRNA contains more repetitive sequences than mRNA. When heterogeneous RNA is treated with ribonuclease, a ribonuclease-resistant core is obtained, much of which is made of polyadenylic acid, the remaining double-stranded mixed sequences, and oligo U sequences.

Heterogeneous RNA contains various regions resistant to RNase treatment; one is the poly A sequence attached to the tail (see below), the others are various base sequences. The amount of RNase-resistant sequence varies considerably from species to species, and such portions of the hnRNA are referred to as the dsRNA. The resistance to RNase is believed to result from the fact that the resistant portions of the RNA are bent into a hairpin configuration forming hydrogen bonds between complementary bases that have been brought face to face in the bending, thus imparting a double-stranded configuration to these portions of the hnRNA. When these RNase-resistant portions of the hnRNA are unwound and hybridized with DNA, they are found to hybridize with the repetitive sequences.

The poly A sequence hybridizes with poly T, and the portions have a sedimentation constant around 10 S. They are located at the 3′ end of the hnRNA. The size of the poly A chain varies from one hnRNA molecule to another and usually ranges from 150–200 residues; but some heterogeneous RNA contains no poly A chain.

Poly A does not exist in an isolated form in the nucleus and must, therefore, be added to preexisting polynucleotide chains. Thus, when hnRNA molecules have been transcribed, they serve as primer for the biosynthesis of their poly A tail at the 3′-end. Inasmuch as they are no poly (dT) or poly (dA) regions in the DNA, it seems clear that poly A is not synthesized by the traditional transcription mechanisms. Moreover, the enzymes that synthesize poly A are sensitive to cordycepin, which has no or little effect on mRNA synthesis. The enzymes responsible for poly A synthesis have been purified to gel electrophoretic homogeneity from calf thymus in Mary Edmonds' laboratory [196].

The oligo U sequence contains about 30 UMP sequences with one GMP residue at the 3′-end and 2 or 3 AMP and CMP in undetermined positions in the sequence. Finally, the hnRNA contains triphosphate groups at its 5′-end.

Although the exact role of the hnRNA in cellular metabolism is far from completely elucidated, the evidence available strongly suggests that it is a precursor of messenger RNA. Perhaps the most compelling evidence in favor of such a notion is the fact that hybridization studies have established that the hnRNA contains stretches of base sequences also found in the messenger RNA. In the conversion of hnRNA to messenger RNA, the poly A sequences are maintained, but the dsRNA, large amounts of repeated sequences found in the hnRNA, and the terminal triphosphates are lost.

The role of the low molecular weight stable nuclear RNA (snRNA) remains a mystery. In addition to the nucleolar and hnRNA, mammalian nuclei contain substantial amounts of discrete species (A, B, C, ..., L) of low molecular weight RNA (96–220 nucleotides) with relatively long half-lives (1 day compared to 10–30 for the hnRNA). Like tRNA and rRNA, the snRNA is methylated. snRNA is believed to be synthesized by DNA transcription. The role of the snRNA in cellular metabolism eludes us completely.

Busch and his associates [197] have partially sequenced one of the hnRNAs, the H type, and have already established that (1) in contrast to other species of snRNA, the H type is not methylated; (2) the 5′-half is rich in purine and includes a G tetraphosphate; (3) the 3′-half is rich in purine and terminates with 4–5 uracil residues in sequence.

The function of snRNA must be elucidated before the significance of the nucleotide sequence can be completely evaluated.

Visualization of Transcription

Miller and his associates have developed electron microscopic techniques that have for the first time permitted visualization of transcription with amphibian, bacterial, and mammalian systems. We will briefly summarize here only those observations made with eukaryotes.

The oocytes of the newt *Triturus veridens* mature for months and grow to 2 mm in diameter with nuclei 1 mm in diameter. By micromanipulation, the chromosomes and nucleoli can be extracted. After appropriate treatment of the nucleoli, it can be shown under the electron microscope that their core is made of an axial fiber of deoxyribonucleoprotein from which ribonucleoprotein fibrils emerge in clusters of 80–100. Between the clusters are portions of nontranscribed DNA referred to as spacer DNA. A distinct granule 125 A in diameter links the DNP to the RNP fibers. The granule is believed to be RNA polymerase. Somewhat similar pictures were obtained when portions of the lampbrush chromosomes were examined.

When HeLa cell chromatin was dispersed by treating the cells with detergent and preparing their content for electron microscopy, gene activity was similar to that observed with amphibian genes. Although it is too early to expect an accurate interpretation of these remarkable pictures, what they have revealed to date is not in conflict with available biochemical data.

Morphological Features of RNA Transfer from Nucleus to Cytoplasm

Although we know a great deal about the sequence of macromolecular events that take place in the nucleus and the nucleolus and we have detailed descriptions of the changes that take place in the structure of these organelles during development and after injuries, there remains a wide gap of ignorance between molecular and ultrastructural observations.

We also know little about the mode of transfer of information from nucleus to cytoplasm. What is certain is that the RNA's (messenger, ribosomal, and transfer) that are made in the nucleus are transferred to the cytoplasm. Moreover, it seems logical to assume that the pores in the nuclear membrane and the annulated figures may have something to do with such transfer, and that, therefore, changes in the appearance of these structures may have consequences for the passage of macromolecules from nucleus to cytoplasm. Electron microscopists have described a number of changes that occur at the periphery of the nucleus. However, so little is known of either their molecular origin or their significance that we shall content ourselves to list their occurrence. Refer to specialized articles for further information [198, 199].

Pathology of the nuclear membrane
Thickening:
Aged Leydig cell
Viral infection
Burkitt's lymphoma
Human leukemia
Invagination and tortuosity:
Neoplasia
Viral infection
Muscular dystrophy
Thermal injury
Hepatic cells in cholestasis
Evagination and bleb formation:
Cells with increased metabolic activity
Starvation
Amino acid deprivation
Thermal injury
Lymphoproliferative disease
X-Irradiation injury

Pathology of nuclear pores
Decreased nuclear pore concentration:
Aged Leydig cell
Cells with low metabolic activity
Increased nuclear concentration:
Many neoplasms
Cells with increased protein synthesis
Cells treated with thioacetamide
Cells treated with 2-acetylaminofluorene
Cells regenerating after ethionine intoxication
Annulate lamellae
Stack of parallel cisternae filled with porelike structures probably of membranous origin observed in:
Oocytes
Endometrial cells
Sertoli cells
Bronchial epithelial cells

The release of previously labeled messenger and ribosomal RNA from nuclei has been achieved *in vitro* without nuclear lysis in presence of dialyzed cytosol and spermidine. The release induced by the cytosol is energy dependent. The release from regenerating liver nuclei in presence of regenerating liver cytosol is twice what it is in presence of normal liver cytosol. It is postulated that the cytosol contains nondialysable factors that modulate the processing and transport of RNA to cytoplasm after transcription of the RNA has taken place [251].

Elaboration of Polypeptide Chains

Protein synthesis is a complex biosynthetic process in which the information stored in the DNA bases is translated into a specific amino acid sequence. We have already described both ends of this process: DNA and the formation of amino acyl tRNA. We will now consider the intricacies of the reaction that brings the individual tRNA charged with this amino acid into proper alignment to form a specific polypeptide chain.

The actual elaboration of the chain occurs at some distance from the nucleus, somewhere in the cytoplasm. Therefore, the DNA message must be carried to the cytoplasm; messenger RNA performs this function.

The complexity of the reaction involved in protein synthesis is so great that these reactions could not be efficiently carried out with substrate (charged tRNA) and template (messenger RNA) circulating free in solution. Successful protein synthesis thus requires that substrate and template be brought together in close proximity. The ribosomes are the pillars that support this intricate structural organization (see Fig. 2-14).

The schematic presentation of the mechanism of protein synthesis that we have just presented immediately raises a number of problems.

If the reaction occurs on ribosomes, transfer RNA must first attach to the ribosome. The amino acids that are attached to adjacent tRNA's become involved in peptide bond formation only after attachment. Thus, we need to discuss the mode of attachment of the tRNA to the ribosome and the mechanism of peptide bond formation.

Even if we assume that secondary and tertiary structures are automatic consequences of the amino acid sequence, the synthesis of a new protein involves more than the simple alignment of the amino acid. The new protein also contains a specific N- and C-terminal amino acid. Thus, in addition to discussing the translation of the base message into the amino acid sequence, we must consider the direction of the reading of the message and the molecular mechanisms of punctuation that signal the beginning and end of the message.

Ribosomes and Polyribosomes

A good deal of *in vivo* and *in vitro* evidence indicates that ribosomes constitute the principal if not the exclusive site of protein synthesis. *In vivo* experiments demonstrate that labeled amino acids are incorporated first into the microsomal fraction, which of course contains most of the cell's ribosomes. Newly synthesized proteins, such as RNase, β-glucuronidase, acid phosphatase, and serum albumin, appear in the microsomal fraction before they appear in other cell fractions [200–206].

Although the role of the membranes of the endoplasmic reticulum has not yet been extensively studied, it is now clear from *in vitro* studies on bacteria, plants, and animals that ribosomes are essential for protein synthesis. Therefore, some of the properties of these cellular structures should be reviewed before their participation in protein synthesis is described further.

The ribosomes are small ribonucleoproteic particles that stud the membranes of the endoplasmic reticulum. These particles are found free in the cytoplasm of bacteria. In mammalian microsomal preparations, ribosomes can readily be separated from the membranes of the endoplasmic reticulum by treatment with deoxycholate. Electron microscopic examination suggests that inside, the cell ribosomes tend to agglomerate in packages of five or more individual particles. A helical arrangement of ribosomes has been described in the cytoplasm of some mammalian tissues. The complexes of particles are called polyribosomes. (The significance of the arrangement of ribosomes into polyribosomes is discussed later.)

Each particle has a spheroid shape and is composed of two subunits held together by magnesium. If two-thirds of the magnesium content of the ribosomes is eliminated, a larger, cup-shaped fragment separates from a smaller, cap-shaped segment. Each subunit of the ribonuclear protein particle is usually referred to by its sedimentation characteristic.

Bacterial and mammalian ribosomes have many similarities. However, the mammalian ribosomes (80 S, mol wt $3.5–5 \times 10^6$) are a little larger than the bacterial ribosomes, and the latter dissociate more easily than the mammalian ribosomes. (After dissociation of liver ribosome, 4 particles differing in size can be obtained: 32 S, 47 S, 50 S, and 56 S. It is believed that 50 S is a dimer of the 32 S subparticle, whereas 56 S is a 47 S particle that includes the nascent protein.) The ribosome is composed of protein and a considerable amount of ribonucleic acid. In bacteria, about 80% of the cell's total ribonucleic acid is in the ribosomes. The base compositions of ribosomal RNA obtained from various sources are, in general, remarkably similar. But this base composition differs from the base composition of soluble RNA inasmuch as it does contain a smaller amount of the rare bases found in soluble RNA. The base composition of ribosomal RNA also differs from that of messenger RNA_3.

After extraction, the RNA obtained from mammalian ribosomes separates into two groups—a 28 S and an 18 S RNA. The 28 S is derived from the two-thirds portion, and the 18 S from the one-third portion of the ribosome. Whether or not each of these two RNA's constitutes a single polynucleotide chain is not certain, but if they do, the molecular weight of the 28 S would be about twice that of the 18 S (1.2×10^6 and 0.6×10^6, respectively). Similarly, extraction of nucleic acid from ribosomal segments of *E. coli* yields 23 S and 16 S RNA; base composition studies show significant differences between the two types of RNA.

The secondary structure of the ribosomal RNA is not known. The present data suggest that half of the molecules or molecule of ribosomal RNA has a certain

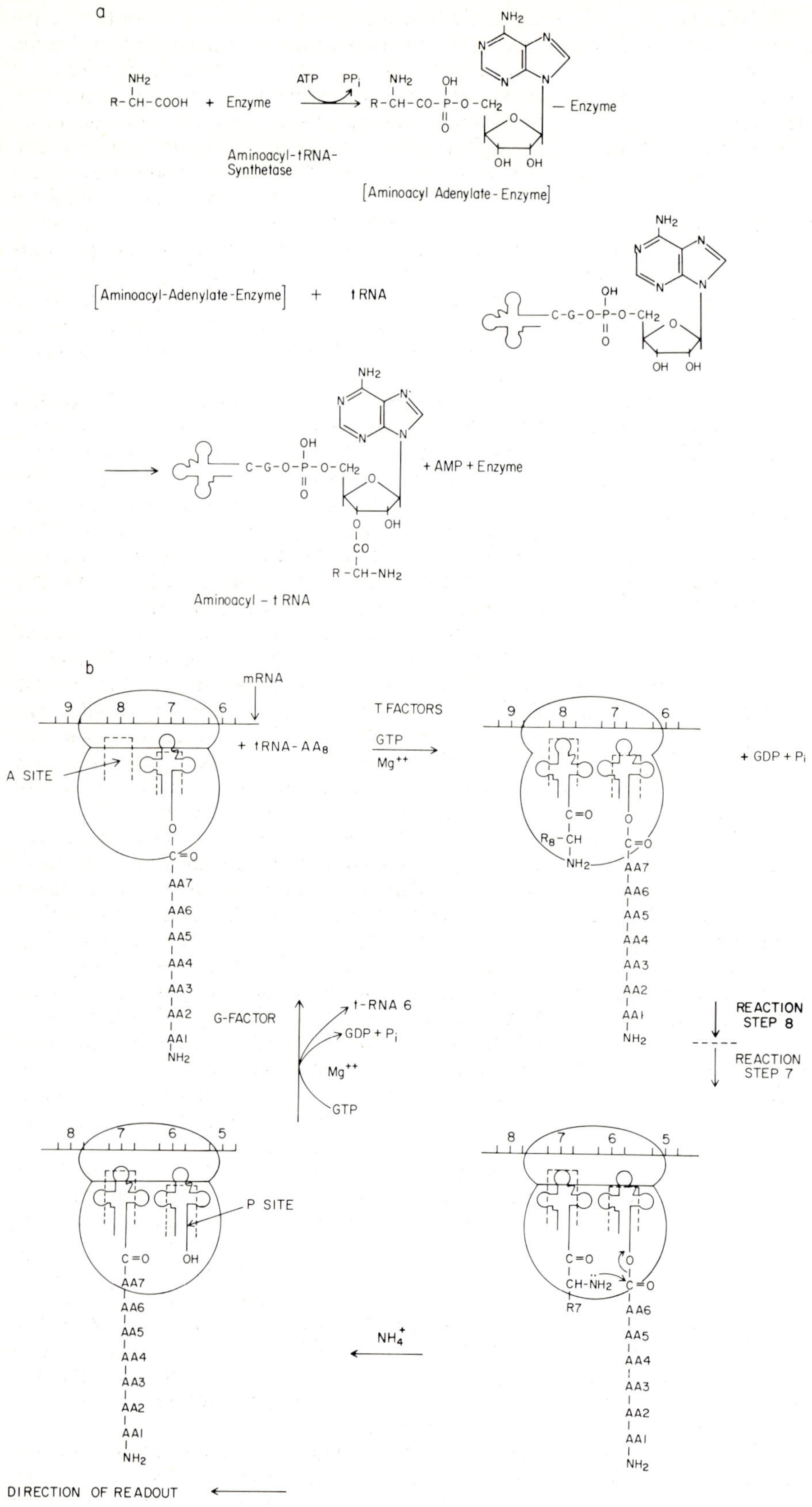

Fig. 2-14. Protein synthesis

degree of secondary structure—probably in the form of the double helix—resulting from hydrogen bond formation between complementary base pairs.

Sixty-five per cent of the ribosome is ribonucleic acid, and the remainder is protein. The proteins associated with the ribosomal preparations are of two main types: the genuine ribosomal and the newly synthesized polypeptide chains.

The identification of the protein of *E. coli* ribosomes has barely begun. The approach to the study of such proteins includes molecular dissection of the ribosome with either chemical or enzymic methods and preparation of mutants that lack some proteins or possess altered ribosomal proteins.

The ribosome proteins were at first believed to be mainly represented by one or a few structural (possibly basic) proteins whose main role was to maintain the ribosomal RNA in the proper configuration.

The more recent investigation in which *E. coli* ribosomes were separated by column chromatography on carboxymethyl cellulose by acrylamide gel electrophoresis in the absence and presence of sodium dodecyl sulfate have revealed that the ribosomes contain as many as 54 proteins, 20 in the 30 S and 34 in the 50 S component.

The role of these proteins is not known except that some are essential for the maintenance of the ribosome's structural integrity and others play, as we shall see, catalytic roles in polypeptide chain formation.

To locate the ribosomal proteins within the ribosomal structure, Hultin and his associate [207, 208] extracted the ribosomal proteins before and after treating the particle with proteolytic enzymes. The extraction is done with either 0.2 normal hydrochloric acid or 3 molar lithium chloride and urea. Either method yields a protein extract that can be resolved by gel electrophoresis into a group of 10 proteins that migrate toward the cathode, and into 5 proteins that migrate toward the anode. Hultin refers to each protein of the first group with arabic numbers, and that of the second with Roman figures.

Trypsin and chymotrypsin were used to predigest the ribosomes. Hultin observed that some of the protein fractions disappear with both enzymes, but others disappear with only one of the enzymes, and some are not affected at all by the proteolytic enzymes. To interpret these results, Hultin argues that those proteins that are digested by both enzymes are readily accessible and must be located at the surface of the ribosome. In contrast, those proteins unaffected by either enzyme are, according to Hultin, buried within the ribosomal structure, and those affected by only one of the enzymes have a susceptible portion located at the surface, whereas the remaining is protected because it is within the ribosomal structure. Although the approach seems promising, the results are still too preliminary to permit a correlation between the distribution of the proteins on the ribosome and their function.

A group of Dutch investigators [209] has devised an ingenious procedure to study the mode of attachment of ribosomal RNA to ribosomal protein. They treated unfolded ribosomes with nuclease and thereby released 85% of the RNA. The remaining 15% of ribosomal RNA sediments into a pellet that contains most of the ribosomal protein. These ribose polynucleotides have an average chain length of 30 nucleotides. Base analysis of the residual protein-bound RNA indicates that that class of RNA is richer in C-A sequence than the remainder of the ribosomal RNA. This finding was in agreement with these authors' previous observations, which showed that polylysine selectively protects RNA rich in C-G sequence against nuclease attacks. In contrast to the remainder of the rRNA, the portion of the rRNA that binds to protein appears to be nonhelical.

Ribosomal proteins can be acetylated or phosphorylated both *in vivo* and *in vitro*. The administration of [^{3}H] acetate results in acetylation of ribosomal protein of rat liver, heart, kidney, and thymus within 5 minutes to form two major radioactive compounds, N-α-acetyllysine and N-ζ-acetyllysine [252–257]. The acetyl transferase is found in the cytosol and acetylation of ribosomal proteins takes place *in vitro* in presence of cytosol or purified acetyl transferase.

Two protein kinases can phosphorylate ribosomal proteins *in vitro;* one of the enzymes is attached to the ribosome and the other is found in the cytosol. Both enzymes require cyclic AMP for activity. The γ-phosphoryl group of ATP or GTP is transferred to primarily serine and, to a lesser degree, threonine residues. Four proteins of the 40 S unit and 10 proteins of the 60 S units are phosphorylated *in vitro*.

In mammary glands, prolactin and insulin act synergetically, to activate two protein kinases; one cyclic AMP dependent, the other functioning independently of the presence of cyclic AMP. Although these findings suggest a link between hormonal stimulus and phosphorylation, it is not clear how the protein phosphorylation helps to implement mammary gland proliferation and milk secretion. Although phosphorylation of ribosomal protein may be critical, the function of the phosphorylated proteins is unknown.

Most of the proteins synthesized on the surface of the ribosomal structure are used in other parts of the cell. The ribosomes' relationship with nascent protein is discussed later. The relationship between "messenger" and ribosomal RNA cannot be understood without first considering the size of the unit involved in protein synthesis. For example, Rich calculated that in hemoglobin the size of the messenger RNA responsible for synthesizing hemoglobin polypeptide chains is 1,500 A. This is contrasted with a ribosome's diameter of 230 A. It seemed likely, therefore, that the messenger RNA chain that serves as template for the biosynthesis of a hemoglobin polypeptide is attached to more than one ribosome.

Proof of this assumption was provided by centrifuging in sucrose gradient a ribosome population prepared from reticulocytes. A number of peaks could be detected by ultraviolet spectrophotographic analysis at 260 mμ. The preparations separated by gradient centrifugation were examined with the electron microscope, and clusters of ribosomes (polysomes) were found. The first peak contained mainly single ribosomes; the second contained clusters of two; the third, clusters of three; the fourth, clusters of four; and the fifth, clusters of five ribosomes (pentosomes). The pentosome preparation was the most efficient in stimulating amino acid incorporation in an appropriate *in vitro* system. These observations were confirmed with other mammalian, and with bacterial and yeast, ribosomes.

What holds the ribosomes together to form "polysomes"? From Rich's original preparation, one would suspect it to be messenger RNA. Ribonuclease separates polysomes that are effective in protein synthesis into ineffective smaller fragments. Furthermore, actinomycin, which blocks the synthesis of messenger RNA, also breaks the polyribosomes into smaller units in the organs of animals injected with the antimetabolite.

The addition of synthetic polynucleotides to isolated ribosomes further suggests that messenger RNA holds ribosomes together to form polyribosomes. Poly U is a synthetic polynucleotide exclusively made up of uracil residues. A triplet made of 3-uracil residues codes for the incorporation of phenylalanine in polypeptide chains. When poly U is added to free *E. coli* ribosomes, the incorporation of phenylalanine into TCA-precipitable material is markedly increased.

It has been possible to define the site of attachment of the synthetic messenger on the ribosomes. The treatment with ribonuclease of ribosome clusters held together by poly U yields individual ribosomes to which small oligonucleotides remain attached. When the ribosome is separated into its one-third and two-thirds components, the oligonucleotides are attached to the 30 S segment.

Takanami and his colleagues [210] have studied the binding of the F2 phage RNA to *E. coli* ribosome and have established that one molecule of RNA binds per ribosome and that the binding involves a specific site on the RNA molecule. The binding site is believed to involve a 25–30 nucleotide sequence. The positions of the various bases in the binding site are not known except for the first three, which have the sequence of an initiating triplet (AUG). The base composition of the binding site is: A, 25%; C, 25%; G, 35%; U, 20%.

Binding of Messenger and Transfer RNA to Ribosomes and Peptide Bond Synthesis

Now that we are more familiar with the polyribosome, the framework that supports the translation of the code stored in the four bases in the protein's amino acid sequence, we are ready to consider the mechanism of the reactions involving amino acyl tRNA, ribosome, and messenger RNA that lead to the formation of the peptide bond. The reactions include attachment of the tRNA to ribosomes, the formation of the peptide bonds, and the mechanism of initiation and termination of the chain. The molecular intricacies of all these steps of protein synthesis are far from elucidated, and any description of these events is necessarily incomplete and possibly inaccurate.

Up to now, we have simply described noncovalent, nonenzymic binding of messenger RNA or synthetic polynucleotide and have said nothing about its relation to binding and peptide bond formation. Such a relationship can best be understood by describing the results of experiments in which poly U was incubated with reticulocyte ribosomes. When poly U is added to washed ribosomes in the presence of acyl phenylalanyl tRNA, one molecule of phenylalanyl tRNA is attached per ribosome, but no peptide bonds are formed. If an enzymic preparation partially purified from the cell sap and GTP are added, GDP is formed, and a second phenylalanyl tRNA molecule is bound to the ribosome. Another enzymic preparation must be added to the incubation mixture for peptide bond formation. (We shall see later that the catalytic molecules involved in these reactions have been identified.)

When washed *E. coli* ribosomes are incubated with tRNA in the presence of Mg^{++} (10^{-2} M), tRNA binds to ribosome. One ribosome can bind only to one tRNA under those conditions, and the binding takes place in the 50 S portion of the ribosome. This type of binding is believed to be noncovalent. There are no energy or enzymic requirements; Mg^{++} and K^+ alone are needed. The attachment of the tRNA to the ribosome is believed to involve the formation of magnesium bridges between the phosphate of the tRNA and that of the ribosome. The specificity of the requirements is not stringent; ribosomes bind to tRNA whether charged or not, provided that the 3′-C-C-A sequence is present in that RNA molecule.

Thus, it appears that the specificity of the ribosome binding site is directed toward the tRNA rather than toward the acyl amino acid. This observation has found further support from studies made with tRNA charged with either modified amino acids or amino acid analogs. Consequently, tRNA seems to be bound to the mRNA by hydrogen bonds and to ribosome by Mg^+ bridges. Experiments in which binding of tRNA to ribosome was determined in the presence of varying concentrations of K^+ and NH_4 suggest that two sites on the ribosome might be capable of binding tRNA, because only part of the binding properties of the ribosome is altered by K^+ and NH_4, as if K^+ and NH_4 inhibit binding at one site but not at another.

Studies with puromycin have provided further evidence for the existence of two binding sites on reticulocyte ribosomes. Apparently, puromycin can act as an acceptor and thus can form a peptide bond with an amino acyl tRNA in the donor position. Puromycin cannot, however, be bound enzymatically to ribosomes and as a result cannot act as a donor. Consequently, the elaboration of the polypeptide chain ends with puromycin. When the polypeptide is released from the ribosome, the latter becomes the C-terminus of the unfinished polypeptide chain.

If an additional degree of complication is introduced into the system, and an artificial polynucleotide—for example, poly U—is added to the ribosomes and the tRNA, the tRNA carrying the proper anticodon binds to the ribosome. The binding is reversible, independent of energy sources, and nonenzymic. The messenger-directed binding is specific for the tRNA carrying the proper anticodon (phenyl tRNA in the case of poly U), and binding takes place whether the tRNA is charged or not. The messenger-specific binding involves the 30 S units and is inhibited by tetracycline. Whether tetracycline binds to the naked 30 S units or to the 30 S units complexed with mRNA is not known.

Some of the proteins involved in the formation of the polypeptide bond have now been identified.

The first step in the elongation of a polypeptide chain consists in the binding of the amino acyl tRNA to the ribosome on which the messenger is already attached and in the process of being transcribed. At low magnesium concentration, such binding requires that 100,000 × *g* *E. coli* cytosol and GTP be added to the incubation mixture. A factor referred to as the EF-T factor has been purified from the *E. coli* cytosol, which carries out the binding of amino acyl tRNA. Passage of EF-T through DEAE-cellulose columns has yielded two components—EF-T_U and EF-T_S with molecular weights of 47,000 and 34,000, respectively. EF-T_U has a great affinity for EF-T_S, and both compounds form a complex that is dissociated by the addition of GDP. EF-T_U also readily binds to GTP. The EF-T_U-GTP complex binds to the amino acyl tRNA, one amino acyl tRNA per GTP molecule, but the amino acyl tRNA is unable to bind to the EF-T_U complex without GTP. Also, the ternary complex GTP-EF-T_U-amino acyl tRNA is believed to be an intermediate between unbound amino acyl-tRNA and amino acyl tRNA bound to ribosomes on the A site. One molecule of GTP is hydrolyzed for every amino acyl tRNA bound, and after binding and hydrolysis an EF-T_U-GDP complex is released in the cytosol, and the release of EF-T_U is followed by the peptidyl transfer. Evidence that cannot be reviewed here indicates that GTP hydrolysis is not indispensable for the binding of amino acyl tRNA to the ribosome. The role of EF-T_S in the process has not been clarified, but it is believed to activate the EF-T_U and accelerate the binding process by stimulating the formation of EF-T_U GTP from EF-T_U GDP. In fact, phosphoenolpyruvate, which will phosphorylate the GDP complex, can replace EF-T_S in the reaction.

Thus, EF-T_U is a protein interacting with four different components—EF-T_S, amino acyl tRNA, GTP, and ribosomes. Its amino acid sequence and tertiary structure should be most interesting. Nothing is known of the binding sites except that 2 SH groups are necessary for activity. Binding factors resembling those found in bacteria have been isolated from mammalian systems, but little else is known about the mode of binding of amino acyl tRNA to eukaryotic ribosomes.

Once the amino acyl tRNA (aa-tRNA) is bound to ribosome, the α-amino group of the newly bound aa-tRNA reacts with the carboxyl terminal residue of a previously bound aa-tRNA to form a peptide bond. This reaction is catalyzed by a peptidyl transferase, which is an integral part of the ribosome. The transferase has not as yet been purified to the extent that its molecular properties and mode of action can be described.

At this point in the elongation process, on the ribosome are the initiating aa-tRNA (A) and the first aa-tRNA (B) of the polypeptide sequence united by a polypeptide bond. Since there are only two sites for binding of tRNA on the ribosome (the acceptor site for A and the donor site for B), to further elongate the chain, the aa-tRNA in the donor site must first be translocated to the acceptor site, and the aa-tRNA in the acceptor site must be released from the ribosome. In addition, elongation cannot proceed until the messenger moves on the ribosome in the direction of the detached amino acid, a distance permitting a new codon to be placed in correspondence with the donor site. Finally, a new aa-tRNA must be placed in the donor site of the ribosome with its anticodon facing the codon of the messenger.

The central event of this complicated molecular ballet is the translocation of the codon and of the aa-tRNA occupying the donor to the acceptor site.

Experiments done with AUG-UUU-UUU have shown that:

(1) the first codon is translated in the presence of purified initiator factors (see below);
(2) the second codon does, in addition, require the EF-T factor and GTP for translation;
(3) the third codon also requires the EF-G factor and GTP for translation.

EF-G has been highly purified from *E. coli*. The purified factor binds to ribosome, but only in the presence of GTP, GDP, or GDPCD. The binding of EF-G in the presence of GDPCP (which is not hydrolyzed) suggests that the binding of EF-G to ribosomes does not require GTP hydrolysis. Yet, EF-G acts as a GTPase, but only after it is bound to ribosomes.

Thus, it seems clear that GTP is not required for binding of any of the elongation factors (EF-T_U, EF-T_S, or EF-G). Yet, GTP is necessary for translocation, and, as a matter of fact, two molecules of GTP are hydrolyzed for each peptide bond formed during elongation through two different mechanisms: one involving EF-T, the other EF-G (see Table 2-5).

The fate of EF-G after translocation is not clear, but the protein is believed to be released and recycled. The role of GTP hydrolysis in protein synthesis remains unknown. To date, no functional phosphorylation or guanylation of any of the factors involved has been described.

These findings and others have led to the conclusion that the G factor is necessary for translocation. In addition to being a translocase, the G factor is also

Table 2-5. Some properties of elongation factors

	Degree of purification	Molecular weight	Special properties	Catalytic properties
T_u	Crystallized	39,000–42,000	3 SH	Binds GDP very strongly; CTP weakly; forms complex with T_s
T_s	Purified extensively	19,000		Forms complex with T_u
G	Crystallized	72,000–80,000	free thial group	GTPase activity; translocase activity
T_1	Homogeneity, in acrylamide gel and ultracentrifuge	186,000	3 subunits (each 62,000 mol wt)	
T_2	Homogeneity, in acrylamide gel and ultracentrifuge	60,000		GTPase activity; translocase activity

a potent GTPase. The role of the GTPase activity in the formation of the peptide bonds is not clear.

A plausible role for GTP hydrolysis is that the energy derived is used for introducing a conformational change in the ribosomal structure which would facilitate translocation of the polypeptide tRNA from acceptor to donor site.

Two elongation factors have also been purified from eukaryotes; an EF-T factor and a protein similar to the G factor. The EF-T factor has properties similar to that of the EF-T_U-EF-T_S complex in bacteria. In fact, EF-T_U-EF-T_S can replace the eukaryote EF-T with mammalian ribosome although the eukaryotic factor does not function with bacterial ribosome. A protein factor with properties similar to the G factor has also been found in eukaryotes, but it does not replace the EF-G factor when added to bacterial ribosome. Mammalian ribosome also contains a peptidyl transferase inserted in its 60 S unit.

In this movement, the polypeptide tRNA is believed to remain attached to the codon in the messenger. Thus, when a nascent polypeptide chain moves from donor to acceptor site on the ribosome, it carries with it the portion of the RNA messenger that is already coded for. As a result, the next uncoded triplet is aligned with the empty acceptor site. The uncoded codon attracts the aa-tRNA with the proper anticodon. Codon and anticodon form hydrogen bonds, and the newly coded aa-tRNA falls into the acceptor site and extends its carboxyl to form a peptide bond with the peptidyl tRNA, which is now in the donor site. When the polypeptidyl tRNA has been lengthened by one or more aa-tRNA units, a new molecule of GTP is hydrolyzed. The ribosome contracts again, and the last member of the polypeptide chains is again transferred from acceptor to donor site.

Direction of the Reading of the Message

The transfer of specificity from messenger to polypeptide can take place only if the two chains move in phase. In-phase displacement of the message (polynucleotide chain) and of its translation (polypeptide chain) requires that: (1) the attachment of the ribosome to the messenger is specific; (2) the direction for reading the message is always the same, and (3) included in the message are signals indicating the beginning and end of the polypeptide chain elaboration. The evidence available indicates that the messenger is read in the 5′-nucleotide toward the 3′-nucleotide, and that the polypeptide chain synthesis starts with the amino acid terminal and ends with the carboxy terminal. The mechanism for the unidirectional extension of the polypeptide chain is based on *in vivo* experiments and on the studies with synthetic polynucleotides *in vitro*. When lysozyme of growing bacteria was labeled with radioactive leucine for short periods of time and then purified, the specific activity of the various leucine residues was not the same, but the relative specific activity increased from the amino terminal to the carboxy terminal. The incorporation of labeled asparagine and lysine into acid-insoluble material was studied in the presence of an amino acid-incorporating system and poly AAA—AAC, in which 16 and 20 residues are made of A.

Such a polynucleotide chain theoretically can stimulate the incorporation of five lysines and one asparagine in the acid-precipitable fractions. In fact, it was observed that only lysine and asparagine incorporation take place, and the incorporated asparagine is always found in the N-terminal position. These initial experiments were extensively confirmed with more sophisticated methods also involving synthetic polynucleotides. The unidirectional reading of the N-terminal amino acid implies that the 5′-phosphate either coincides with or is very close to the codon for the N-terminal amino acid.

The binding of natural messenger RNA to ribosomes differs considerably from the mode of binding of synthetic polynucleotides to ribosomes. Whereas the natural messenger RNA binds to the ribosomes at a point probably corresponding to the 5′-terminal phosphate of a cistron, the binding of polynucleotides to ribosomes occurs at random. Therefore, it is unlikely that the synthetic polynucleotides can dictate the biosynthesis of specific polypeptides. It is interesting that the direction of the translation from the 5′-nucleotide toward the 3′-nucleotide is the same as the direction of transcription. This allows translation to start before transcription is completed.

Initiation of the Message

As was mentioned previously, in order for the polypeptide and the polynucleotide to move in phase, the polynucleotide must contain "traffic signals" indicating when the biosynthesis of the polypeptide chain must be started. Interesting investigations done in *E. coli* have begun to reveal information on the mechanism of punctuation in the reading of the code. Marcker and Sanger [211] established that there were two types of methionyl tRNA in *E. coli* extract. Both bind methionine, but only one methionine is readily formylated. N-10-formyl-tetrahydrofolic acid is the formyl donor, and formylation occurs only after methionine has been attached to tRNA. When methionine composition of the protein was determined after *in vitro* biosynthesis of phage F2, the N-terminal amino acid of the phage code was found to be N-formyl methionine. In contrast, when the protein coat is extracted from virus grown *in vivo*, the N-terminal is found to be alanine.

The discrepancy between the *in vivo* and *in vitro* results with respect to the biosynthesis of F2 phage protein led to the discovery of the role of N-formyl methionine RNA in initiation of the polypeptide chain. When synthetic N-formyl methionine RNA is used in a cell-free system, it is found in the N-terminal position of the coat protein, and fingerprinting of the protein reveals that the N-terminal fragment includes an N-formyl methionine-alanine sequence. These findings suggest that *in vivo* methionine is split by a specific enzyme. Thus, the combination of *in vivo* and *in vitro* results suggests that the following sequence of events takes place in initiation. Some methionine molecules react with a special type of tRNA. The complex methionyl tRNA is accessible to an enzyme for formylation of the methionine. This methionyl tRNA has a special anticodon. The cell contains an enzyme that, at least in some cases, would split the formyl methionine from the polypeptide chain.

Further studies of the N-terminal amino acid of ribosomal and soluble proteins of *E. coli* indicated that 90% of these proteins end with methionine. Alanine, serine, and threonine were the N-terminal amino acids in the other proteins. In spite of the great uniformity in N-terminal amino acids, the amino acid composition of these proteins was found to be quite heterogeneous. Studies of the codon of the methionyl tRNA that can be formylated and of the methionyl tRNA that cannot be formylated indicated that AUG functions as a codon for both. A small amount of methionyl tRNA that could be formylated was also coded by GUG and UUG. The fact that AUG codes for both methionyl tRNA's is not irreconcilable with an eventual role of N-formyl methionyl tRNA as initiator of the chain. Methionyl tRNA could preferentially bind to the acceptor site. Under these conditions, the formyl methionyl tRNA would occupy the donor site first and could therefore function as an initiator of the chain. The preferential binding of formyl methionyl tRNA to the donor site of the ribosomes using AUG as codon could result from a special configuration of the tRNA.

In *E. coli* cell-free systems deficient in transformylase activity, protein synthesis directed by viral RNA is stimulated if formyl tRNA or N-10-formyl tetrahydrofolic acid is added to the medium.

Thus, the initiation process in bacteria involves the recognition of a special codon by the anticodon of an fMet-tRNA. Yet a number of other factors are needed for initiation, including at least three protein initiator factors—F_1 (mol wt 9,000), F_2 (mol wt 65,000), F_3 (mol wt 24,000)—and GTP.

IF_3 exists in two forms referred to as $IF_{3\alpha}$ and $IF_{3\beta}$. Either form attaches to the 30S subunit of the ribosome and promotes the attachment of the messenger RNA to that subunit. It is not certain whether the IF_3 attaches to the ribosome or to the messenger first or *vice versa*. Neither is it known what the molecular recognition sites are in the ribosomes or the messenger. IF_2 has also been found to contain two classes of polypeptide chains, each of which has the properties of the heterologous IF_2 preparation. IF_2 directs the binding of fMet-tRNA to ribosomes in the following sequence. After the 30S ribosome has been charged with IF_3 and messenger RNA, IF_2 binds to the ribosome, and this is followed by the binding of GTP and fMet-tRNA. GTP is bound at the rate of one mole per mole of fMet-tRNA and is not hydrolyzed during the process.

IF_1 incites IF_2 to become catalytically active. When 30S and 50S units of the ribosome join, IF_1 and IF_2 are dropped from the 30S unit. The release of IF_2 requires hydrolysis of GTP to GDP. Whether or not ribosome proteins are also involved in the process of initiation remains to be established.

Progress has also been made in our understanding of initiation of translation in eukaryotes. Two species of methionine RNA have been found in guinea pig liver. Whether formylation of one of the species of Met tRNA is necessary for initiation has not been shown. Methionine (instead of the traditional threonyl) occupies the N-terminus of a number of cytochrome c variants obtained by mutation of yeasts. Elegant experiments using rabbit reticulocytes support the assumption that Met tRNA is involved in polypeptide chain initiation. By manipulating the *in vitro* system one can selectively inhibit initiation with sodium fluoride, thus preventing new polypeptide chain formation while that of the old chains continues. It is also possible to inhibit elongation with sparsomycin, which inhibits peptide bond formation, and thus prevent further aggrandizement of existing polypeptide chains without interfering with the initiation of new ones.

The old (long chains) can then be separated from the new short chains. In such experiments, it was found that the new short chain contains methionine as an N-terminus. The results are interpreted to mean that a methionine residue is needed to start the polypeptide chain, but this initial residue is cut off the original chain early in the course of elongation. Although preliminary evidence suggests their existence, much remains to be learned about initiating factors in eukaryotic cells.

Since no formylating mechanism is known, it is not surprising that in eukaryotic cells Met tRNA rather than fMet-tRNA seems to be the initiator tRNA.

However, when mammalian Met tRNA is formylated with the appropriate bacterial system, the loaded transfer RNA can serve as initiator. A protein factor necessary for initiation, found in reticulocytes and referred to as M_3, seems to have properties similar to those of the IF_3 factor in bacteria. M_2A and M_2B protein factors have also been isolated from reticulocytes and are said to be indispensable for the AUG-dependent binding of Met tRNA. A third binding factor M has also been discovered in reticulocytes; its exact role is not clear.

In *E. coli,* mutation of a given triplet to a UAG or UAA sequence leads to the premature termination of the polypeptide chains. These data and many other genetic studies led to the conclusion that UAG and UAA constitute termination codons in *E. coli.* Once such codons have been reached in the course of translation of a messenger RNA ribbon, a molecular mechanism is put in gear which releases the messenger from ribosomes, detaches the polypeptide chain from ribosomes and the tRNA that carries the carboxyl terminus, and releases the remaining tRNA from ribosomes.

Much remains to be learned about the complex termination process, even in *E. coli,* in which most of the data has been compiled. Nevertheless, a number of protein factors have been identified. The R factor (releasing factor) is a protein with a molecular weight of 40,000–50,000 daltons which is likely to be composed of two different proteins—R_1 and R_2—and which is indispensable for release of the polypeptide chain.

The S factor has no releasing properties of its own but accelerates release by the R factor. Peptidyl transferase is believed to be converted into a hydrolase under the influence of the R factor. The hydrolase activity transfers the peptidyl moiety of the carboxyl terminal tRNA to water, instead of transferring it to another aa-tRNA, as in the classical peptide bond synthesis. A TR factor is believed to release tRNA from the acceptor site into the soluble portion. The TR factor is different from the G factor and does not require GTP for activity.

A release factor with properties reminiscent of the R factor has been found in a number of mammalian systems. No S factor has been found, and, in contrast to termination in bacteria, mammalian termination requires GTP hydrolysis.

Operon Theory

While molecular biologists were investigating the structure of the molecules that store and serve to transfer specificity from DNA to protein, microbiologists investigated the spontaneous and induced mutations that occur in bacteria. Jacob and Monod used all the modern discoveries in molecular biology and bacterial genetics in a theory on the organization of genes and on the mechanism of protein synthesis. The theory is now called the operon theory [212].

Much of the experimental evidence on which the operon theory is based was obtained by studying the induction and repression of bacterial enzymes. "Induction" is indicated by an increase in enzyme activity, and "repression" by a decrease in that activity. The inducer is usually a substrate of the enzyme; the repressor is usually a product of the enzyme reaction. The changes in enzyme activities result from alterations in the rate of protein synthesis.

Studies on bacterial mutation have shown that the induction of a specific enzyme is usually accompanied by an increase in activity of several other enzymes involved in the same catabolic or anabolic pathways. When *E. coli* are grown in the presence of galactosides, the formation of β-galactosidase is induced, and it has been established that the new enzyme formed (1000 times the activity present in the wild type of *E. coli*) is not derived from the activation of preexisting proenzyme, but from the net synthesis of new enzyme. In 1953, Monod [213] discovered that the activity of tryptophan synthetase is inhibited by tryptophan and certain of its analogs. It was established that the inhibition or repression resulted from a block of the enzyme biosynthesis.

To explain the phenomena of induction and repression, the operon theory distinguishes between three types of genes: structural genes, operator genes, and regulator genes. Structural genes are deoxypolyribonucleotide sequences, which dictate the specific sequence of the amino acid in the polypeptide chain. Operator genes control the expression of the structural gene. Regulator genes control the formation of substances that prevent the structural gene from expressing itself. According to the theory, these three different types are organized in a single unit called the "operon" (see Table 2-6).

Table 2-6. Structural genes in the lactose operon in *E. coli*

Thiogalactonide transacetylase	Galactoside permease	Galactosidase
x	y	z

Studies on bacterial mutations have shown that the induction or repression of a specific enzyme is usually accompanied by change in activity of several other enzymes, which catalyze reactions of the same catabolic or anabolic pathway.

Studies on the biosynthesis of histidine in Salmonella by Ames and Hartman [214] established the simultaneous repression of several enzymes. In the series of reactions (at least 10 different steps) involved in the biosynthesis of the histidine molecule, the 5-carbon chain of phosphoribosyl pyrophosphate is converted to the 5-carbon chain of histidine. In Salmonella, the biosynthetic pathway for histidine is a straight pathway in which all the intermediates are known and none of the intermediates are used in branching pathways. As a result, if a mutation produces a block in one or more steps of the biosynthesis of histidine, a new strain of bacteria that require histidine for growth is obtained. More than 1000 histidine-requiring mutants have been discovered, and by carefully studying the enzyme mosaic of the mutant strain, it was possible to map, within the chromosome, the position of the gene responsible for each enzyme's amino acid sequence. By these procedures it was established that all the structural genes determining the biosynthesis of the enzymes required for the elaboration of the histidine molecule are found within a cluster in the chromosome. Table 2-7 shows the sequence of the structural genes controlling the amino acid sequence of each enzyme within the histidine operon. Each enzyme is referred to with an Arabic number, and a capital letter designates each section of the chromosome determining the amino acid sequence.

If on the basis of the studies of mutants, one places all the segments of the operon in sequence as they appear in the chromosome, the sequence of Table 2-7 is obtained. Although the genes controlling histidine synthesis are arranged in a cluster, they are clearly not inserted in the operon in a sequence that corresponds to the sequence of steps in the metabolic pathway.

Operator

Table 2-7. Schematic representation of the operon of the histidine system in *Salmonella typhimurium*

		Cyclase	Isomerase	Amido-transferase	Phos-phatase	Trans-aminase	Dehydro-genase	Pyrophos-phorylase	Operator
E	I	F	A	H	B	E	D	G	O
2	3	6	4	5	7–1	8	10	1	

By studying the activity of all the enzymes involved in histidine biosynthesis in a mutant of *Salmonella typhimurium* in which the gene for the amino transferase (H) was deleted, it was observed that although the activity of all the enzymes to the right of the deleted gene (phosphatase, transaminase, dehydrogenase, pyrophosphorylase) were unaffected by the mutation, the activities of the enzymes with structural genes located at the left of the mutant gene were markedly reduced. The simultaneous decrease in activity of all the enzymes with structural genes located at the left of the mutated gene is referred to as "polarity mutation." This type of mutation is explained by the theory that information stored in the operon is transcribed from the operator end and procedes linearly to the left.

Studies of mutation in the histidine pathway of *S. typhimurium* have revealed another interesting relationship between genes and operon—coordinate repression. When the wild type of *S. typhimurium* is grown in a medium containing histidine, the activities of the enzymes involved in histidine biosynthesis are depressed. The activities of all enzymes are depressed equally so that in the mutant, the ratio of the activity of one enzyme to that of another enzyme is the same as in the wild type. To explain coordinated repression it is proposed that the entire base sequence in the operon is translated into a single molecule of messenger RNA.

Before the histidine system was used, much of our knowledge of bacterial genetics was derived from studies on the lactose system in *E. coli*—a system much simpler than the histidine system. Only three enzymes are involved in lactose utilization–β-galactosidase, galactoside permease, and thiogalactoside transacetylase. The organization of the operon for the galactose system is presented in Table 2-6. Studies of mutation in the lactose system led to the discovery of regulator, operator, and coordinator genes.

Jacob and his collaborators have described mutants of *E. coli* that are incapable of synthesizing permease and transacetylase. Genetic studies with these mutants show that their X and Y genes are intact. Such mutants are referred to as operator O mutants, because in this type of mutation the production of all of the enzymes of the operon is abolished. This type of mutation could result from two molecular alterations of the operon: an alteration of the operator gene, and a polarity mutation of the structural gene closest to the operator gene. The distinction between these two possibilities is not simple. In the case of the operator O mutation of the lactose system in *E. coli,* Jacob and Monod have proposed that the mutation involves the Z gene (β-galactosidase) because genetic recombination of the mutant caused operator O mutation to revert; and the reverted mutant contained a β-galactosidase with physicochemical properties somewhat different from those of wild type.

The wild type of *E. coli* has low levels of all the enzymes involved in the lactose system, but when lactose or β-galactosides are added to the culture medium, the activity of all these enzyme increases to 1000 times the normal values. Since all the enzymes of the system are induced at the same time, this type of change is called coordinated induction. It has been established after purification of the enzymes of the mutant that 25 molecules of β-galactosidase are synthesized for every molecule of transacetylase.

Mutants have been observed for many bacterial systems (lactose, histidine, etc.) in which the levels of activity of all the enzymes of the system are increased in absence of induction. This type of mutation is likely to result from a constitutive alteration of the operator gene.

There is another type of constitutive mutation in which the activity levels of all the enzymes in the system are increased in the absence of an inducer; this mutation results from the deletion of a gene (I-mutant).

When mapping of the deleted gene is accomplished on the basis of genetic recombination, the deleted gene falls outside of the lactose operon. These findings have been interpreted by postulating the existence of a regulator gene (present in the I+wild type) that controls the biosynthesis of repressors that combine with the operator gene and prevent the expression of the operon unless the repressor is inactivated by an inducer.

Gilbert and Müller-Hill [215] have isolated a tetrameric, slightly acidic protein with a molecular weight of 150,000. *In vitro* studies indicated that the protein has properties compatible with those of a repressor. It combines readily with isopropyl-β-D thiogalactoside (a most powerful inducer of β-galactosidases), as well as with double-stranded DNA.

The complete amino acid sequence of the lac repressor has been determined. The protein is a single peptide chain composed of 347 amino acids [258]. Available evidence suggests that the NH_2 terminal region of the repressor is required for binding to the lac operator gene, but is not indispensable for binding of the residues [259].

The protein is believed to be a product of the lac gene; nonsense mutation of that gene leads to the constitutive elaboration of these enzymes. Elaboration under the control of a gene of a protein that in fact prevented expression of the operon obviously constitutes a negative form of operon control.

Thus, the lac repressor is a protein that binds specifically to one site of the genome, probably the operator gene. The affinity of the DNA for the repressor is so great that only a few molecules of repressor are needed to keep the operon in the repressed state. In *E. coli* it has been estimated that there are only ten repressor molecules per lac operon. The repressor is believed to act by inhibiting promoter-bound RNA polymerase and thus preventing the enzyme from transcribing the promoter. The promoter is the site at which the RNA polymerase binds to initiate transcription in the lac operon.

The operator gene (the site of attachment of the repressor) is located between the promoter and the first structural gene. Once transcription has taken place, adding repressor molecules has no effect.

As we already mentioned, the identified components necessary for the transcription of the lac operon include (1) a lac promoter (a sequence susceptible to bind to the T factor), (2) an operator gene sequence, (3) a holo RNA polymerase, including the T factor, (4) a lac repressor, (5) lac inducers, (6) a cAMP receptor, and (7) cAMP itself. This is complex molecular machinery, indeed, for transcribing one of the simplest operons known. There is no way to predict what intriguing scheme nature has evolved for gene expression in the complex mammalian cell. For further information on the lac operon the reader is referred to the articles of Beek and those of Gilbert and Müller-Hill [215]. Mutations that alter the activity of enzymes involved in many other metabolic pathways (leucine, tryptophan, pyrimidine, arabinose, isoleucine, valine, arginine, cysteine, and other enzymes) have largely confirmed the findings made with the galactose and histidine systems.

The operon theory can be expressed in molecular terms as follows. The operator, structural, and regulator genes of a given pathway are organized in a continuous unit called the operon, which is a continuous polydeoxynucleotide sequence. The operator gene, which forms a small region at one end of the sequence, is the combination that permits the operon's full sequence to be unlocked. Without the operator gene, the operon cannot be read. The operon's base sequence is translated into a complementary sequence of ribonucleotides to form the "messenger" RNA, which was originally postulated to be a short-lived RNA. The sequence of the messenger RNA is read, starting with that of the operator gene. Regulator genes elaborate substances that can inhibit the expression of the information stored in the DNA.

Implicit to the theory are the following assumptions: (1) the operon is polycistronic; (2) all proteins with structural genes in a single operon are synthesized in equimolecular amounts after induction; (3) the information stored in the operon is transcribed into short-lived, single, high molecular weight messenger RNA molecules; (4) repressor substances are formed under the direct control of the chromosomes. Gene mapping and studies of the changes in enzyme activity after induction or repression have provided strong evidence for the existence of a polycistronic operon. However, polarity mutations have indicated that all genes in the operon are not equal, and that equimolecular production of all the enzymes dictated by a given operon does not occur. In fact, the elaboration of equimolecular amounts of all the enzymes involved in a single pathway would be surprising, because it would deprive the organism of an important mechanism controlling the rate at which the pathway functions.

The observations on the histidine operons described above clearly show that all the enzymes with amino acid sequences determined by the histidine operon are not synthesized in equimolecular amounts. But the rate at which each of the proteins is synthesized is determined by the position of the cistron within the operon. The amount of protein synthesized decreases the further the cistron is from the operator gene. Furthermore, mutations that interfere with the biosynthesis of one enzyme of the histidine pathway lead to a reduction of the rate of synthesis of all the enzymes with structures dictated by cistrons distal to the mutated cistron.

This polarity suggests that the chances of transcribing the DNA sequence of the polycistronic operon into a complementary sequence of messenger RNA decrease as the DNA sequence is more distal to the operator gene.

A molecular explanation for polarity must exist. The existence of modulating triplets has been invoked to explain polarity. The polycistronic operon dictates the biosynthesis of a single ribbon of messenger RNA. The ribosome attaches to the messenger RNA by means of the anticodons of the tRNA. Messenger and transfer RNA-laden ribosomes form the polyribosomes. The ribosomes move along the messenger like the ball of a typewriter and decipher the code starting from the operator end of the messenger. When ribosomes have read the full sequence of a given enzyme, the protein detaches from the ribosomes; at this point, the relationship between ribosome and messenger depends on the anticodon sequence of the transfer RNA that is next in line for reading the messenger.

The anticodon may either correspond to a triplet sequence normally involved in dictating the position of a given amino acid, or it is a "modulating triplet," which facilitates the disconnection of the ribosome from the messenger. Thus, the chances for a ribosome to fall off its mRNA increase as the distance between the reading point and the operator increases. The position of the cistron within the polycistronic operon and the existence of modulating triplets could well explain the "modulation of protein synthesis," or rate at which different proteins are synthesized. When the "polarity" is modified as a result of a mutation, it is assumed that a coding triplet is replaced by a modulation triplet.

A number of findings suggest that the operon sequence is translated into a single polycistronic messenger RNA. After Spiegelman had provided some indications that this might be the case, Martin claimed to have identified, by double-labeling techniques, the messenger RNA of *Salmonella typhimurium*. In these experiments, Martin labeled the RNA of one strain essential for histidine synthesis with [^{14}C]uridine and another with a long deletion in the histidine operon with [^{3}H]uridine.

After pulse labeling, Spiegelman stopped the growth of the cultures, mixed the cells, and isolated the RNA by chromatographic analysis on a methyl-cellulose column. The radioactivity profiles of the two RNA's corresponded closely, except that there was approximately 4% more ^{14}C-than ^{3}H-counts. The higher radioactivity in the constitutive gene corresponded to an RNA increase of 1%. The constitutive strain also synthesized 1% more protein than the depressed strain. Furthermore, the RNA fraction, in which a difference in ^{14}C- and ^{3}H-labeling was observed, has a sedimentation coefficient of 34 S, precisely that expected for an RNA polycistronic for the histidine biosynthetic pathway. Similar observations were made for the lactose and galactose operons.

Apparently, only the protein configuration's versatility explains complexing between inducer and repressor. For that reason, Monod and his collaborators had proposed that the "repressors were proteins" the specific amino acid sequence of which was dictated by the regulator gene.

The regulation of alkaline phosphatase synthesis in *E. coli* involves the elaboration of a proteic substance referred to as the R2A protein. The presence of the R2A protein is determined genetically and is repressed by inorganic phosphate. Galactoside elaboration is in part regulated by a protein substance.

The molecular structures of only a few repressors are known. On the basis of the operon theory, apparently the simplest and most efficient way to make a repressor is to make it directly on the chromosome. An RNA the sequence of which is dictated by a specific DNA sequence satisfies such a role.

The Endoplasmic Reticulum

The endoplasmic reticulum is an intracytoplasmic structure involving the entire cytoplasmic volume. If the mitochondria still conceal many of their secrets from the morphologist and biochemist, the intracytoplasmic system—composed of membranes, vesicles, and fine particles and called either endoplasmic reticulum, ergastoplasm, or cytomembranes—remains an intricate web of mysteries [216–218].

The complexity of the nomenclature strikingly illustrates how little is known about this cytoplasmic system. The term ergastoplasm originates from the Greek work εϱγαϱομαι, meaning to elaborate. It is frequently used by French writers, possibly because it was originally coined by Garnier. Because of its functional connotation, the term is rejected by some Anglo-Saxon and Swedish authors. The latter went so far as to deprive the nomenclature of any functional or descriptive implication by calling the system alpha, beta, and gamma cytomembranes. The term generally accepted by American workers—endoplasmic reticulum—is used here, although some writers think the term inadequate because the system is neither endoplasmic nor reticular.

Morphological Features

To understand the controversy that exists about the interpretation of the tridimensional structure of the endoplasmic reticulum, it is necessary to describe the observations made in both cell cultures and thin sections. When a living cell is examined with the light microscope, the cytoplasm usually appears as a homogeneous mass, resembling translucent glass and containing fine granules in suspension. The particles are mainly of two types: refringent grains or rods called mitochondria, and secretory droplets called paraplasm or ectoplasm by the earlier investigators. The translucent background maintaining these particles in suspension is called hyaloplasm. Early histologists considered the translucency of the hyaloplasm quite characteristic of life, probably because dead cells are not translucent.

The hyaloplasm has been arbitrarily divided into a peripheral zone called the ectoplasm and a more central zone called the endoplasm. But before we describe the appearance of the endoplasmic reticulum in fixed preparations, remember that the cytoplasm is not inert. The cytoplasm of living cells has the consistency of a gelatinous liquid, and its viscosity changes depending upon the site or the physiological state. These changes in viscosity are accompanied by, or lead to, the formation of currents in the cytoplasmic gel that keep it in constant motion. This dynamic aspect of the cytoplasm should warn us against hasty interpretations drawn from the static pictures obtained after fixation, however they may further knowledge in biology. Although pictures obtained after fixation reflect some events in living cells, there is no way of measuring the gap between the living cell and the fixed material.

In electron micrographs of spread cultured cells or macrophages, the hyaloplasm forms a background of low electron density with a fine lacework appearance. Mitochondria appear against the ground substance as small spheres or rods with marked electron density. The less dense strands of the hyaloplasm delimit small vesicular or tubular areas because in tissue culture cells, the lacework structure is more abundant in the endoplasm than in the ectoplasm. The vesiculotubular structure was called endoplasmic reticulum by Porter and his associates. Later, when ultrathin sections of a variety of tissues became available, this fundamental cytoplasmic structure was observed in all tissues studied, except the cytoplasm of mature erythrocytes. The structure observed in sections is not exactly like that seen in thin-spread cells; instead of a continuous reticular structure, a discontinuous system of membranes delimits a tubular or vesicular structure the outer aspect of which may or may not be lined with fine granules.

Thus, electron microscopic examination of the ground substance in thin sections reveals three major constituents: the membranes, the space delimited by the membranes, and fine granules. The membranes, also called filaments, lamellae, or fibrillae, are on the average 40 A thick and are frequently disposed in parallel arrays, delimiting canaliculi or tubular or vesicular spaces. The inner aspect of the cavities may be rough or smooth depending upon the presence of fine, dense granules (130–150 A in diameter). Inasmuch as the reticulum described in the thin-spread cell and the membranous system described in the ultrathin sections were both found in the ground substance of the cytoplasm, the two systems were, by analogy, called endoplasmic reticulum. This was despite the fact that in thin-spread cells there is no such membranous structure, and the system is not restricted to the endoplasm of the cell, but is found throughout the section from the cell membrane to the nuclear membrane.

Although there is no disagreement concerning the appearance of the endoplasmic reticulum on thin section, the tridimensional reconstruction of this planar structure is debated. The controversy is easier to understand if the proposed roles of the endoplasmic reticulum in biology are kept in mind. Among these hypothetical roles are: (1) the endoplasmic reticulum provides compartments within the cytoplasm; (2) it separates multiple-enzyme systems or separates the enzyme from its substrate; (3) it provides a support for enzymes and substrates, thereby facilitating reaction; and (4) it furnishes an intracellular circulatory system in connection with the exterior, which promotes rapid diffusion of cellular products not only within the cell but also from one cell to another.

When the role of the endoplasmic reticulum is taken into account, the importance of its tridimensional structure becomes obvious. To fulfill the first three of these physiological roles, a system of individual tubules or vesicles, coated or not with widely interconnected granules may be sufficient. But if the endoplasmic reticulum is also to function as an intracellular circulatory system, it should probably constitute a tridimensional network of canaliculi extending throughout the entire volume of the cytoplasm from the cell membrane to the nuclear membrane.

The arguments favoring the second system are not completely convincing. They take two forms: (1) the fact that in the thin-spread culture cells the endoplasmic reticulum appears as a lacework or continuous reticular system, and (2) the occasional observation in ultrathin sections of connections between the membranes of the endoplasmic reticulum and the nuclear or cell membrane. Although many accept the connections between the endoplasmic reticulum and the nuclear membrane, the connection between the endoplasmic reticulum and the cell membranes remains doubtful; probably because it is difficult to obtain sections in which the endoplasmic reticulum and the cell membrane are in the proper planar relationship to provide a picture that can be interpreted irrefutably.

Evagination of the cell membranes, or "caveola intracellulares," have been observed, and some consider them the origin of canalicular structures connecting the endoplasmic reticulum with the cell membrane. Not everyone accepts this view; some consider these caveola intracellulares as stages of micropinocytosis. These structures have been observed in the membrane of the vascular endothelium. They are assumed to be filled with extracellular fluid and then, by narrowing their necks, become vesicles that travel across the endothelial cell to reach the vascular pole where they empty their content into the capillary lumen.

The tridimensional structure of the endoplasmic reticulum remains one of the most intriguing problems of biology; extensive serial section studies must be done for it to be solved. In discussing the tridimensional structure of the endoplasmic reticulum, we have pointed out its possible connection with the nuclear membrane and the cell membrane; however, the problem of its relationship with other cell structures is not restricted to the membrane. Connections between the endoplasmic reticulum and the Golgi apparatus and microbodies have also been described.

Role in Protein Synthesis

One of the major functions of the cell is to elaborate structural and catalytic proteins for replacement as a result of turnover for replication and for export. If most of the cell's energy is generated in autonomous granules, the mitochondria, protein elaboration is the function of a complex nuclear and cytoplasmic interaction. The endoplasmic reticulum is the cytoplasmic framework on which intracellular and extracellular proteins are produced.

Intriguing as they may be, available results do not even partially answer one of the major questions about the endoplasmic reticulum's function. The membranes of the endoplasmic reticulum provide a sort of scaffold on which protein biosynthesis takes place. The template and a prefabricated assembly line (the ribosome) come from the nucleus. The energy is derived from ATP generated in mitochondria and the cytosol. Can any part of the scaffold be used for the biosynthesis of any protein, or are specific templates directed toward a specific portion of the cytoplasm? Is the specificity of the interaction between assembly line and membrane rigidly restrictive, such as the hole that restricts introduction and turning of the key, or are the conditions of the interaction much more permissive and simply conditioned—for example, by availability of empty membrane surface, precursor gradients, and sources of ATP? What directs the movement of template and ribosome away from the nucleus?

If recognition of the nuclear message is restrictive, the membrane must include a code that reorganizes the message. On the other hand, if the recognition of the nuclear message is permissive, a combination of rotating movements of either the nucleus or the cytoplasm and the existence of chemical gradients could explain the templates' orderly distribution in the cytoplasm. A rotating movement of the nucleus would be simplest to conceive, but one could not exclude that the cytoplasm itself is rotating.

The study of the molecular steps involved in elaborating the membrane of the endoplasmic reticulum (ER) has barely begun. Two major approaches have been used to study the formation of ER—proliferation of the ER after administration of phenobarbital, and developing ER in liver cells post partum. Both types of studies indicate that the phospholipids and the enzymes appear first in the rough membrane and later in the smooth membranes of the ER, suggesting that the rough is a precursor of the smooth ER. The discovery of a DNA associated with the endoplasmic reticulum and unlikely to result from contamination by nuclear and mitochondrial DNA may require reexamining the mechanism of producing the ER membrane [219].

Electron microscopy has revealed the existence of two types of polyribosomes—some bound to the membrane and others free. It is believed that the bound polyribosomes are involved in elaboration of export proteins, while nonbound polyribosomes are involved in the biosynthesis of proteins to be used for local consumption. In the pancreas, it has been shown that chymotrypsin is formed on bound ribosomes. Redman studied the incorporation of labeled amino acid *in vivo* and *in vitro* in purified serum proteins and ferritin of rat liver and showed that the radioactivity incorporated in albumin is associated with the bound ribosome, whereas that of ferritin is associated with free ribosomes [220].

Almost nothing is known of the mode of elaboration of the endoplasmic reticulum membrane, but two fundamental mechanisms can be contemplated. The first postulates the existence of growing points, like tree buds, that lead to the formation of new membranes on the old support; the other proposes that once a basic framework is deposited, the entire structure is simultaneously adorned with new enzymes or other proteins [221] (see Chap. 16).

Studies of the distribution of glucose-6-phosphatase in hepatocytes after birth support the second view. During hepatocyte development in the fetus and the newborn rat, the rough endoplasmic reticulum appears first. The smooth endoplasmic reticulum expands a few days before birth, and glucose-6-phosphatase appears after birth. Electron microscopic localization of the glucose-6-phosphatase shows that the enzyme appears simultaneously in most hepatocytes in all parts of the endoplasmic reticulum. Therefore, there are no growing points for glucose-6-phosphatase, and the new enzyme molecules appear to be continuously inserted within the old framework.

When the cell is homogenized and nuclei, mitochondria and lysosomes are separated from the homogenate which is centrifuged at high speeds, a microsomal pellet is obtained (see Chapter 1). The pellet contains membranes of the endoplasmic reticulum, plasma membranes, and the Golgi apparatus. Isopycnic equilibration and centrifugation of the microsomal pellet have yielded several groups of membranes referred to as a 1 containing primarily monamine oxidase and believed to be fragments of the outer mitochondrial membrane; a2, containing 5′nucleotidase alkaline phosphodiesterase I, alkaline phosphatase, and cholesterol, and believed to originate primarily from the plasma membrane; a3 containing galactosyltransferase, N-acetylglucosamine transferase, and sialyltransferase, believed to represent membranes of the Golgi complex; and b containing NADH cytochrome reductase, NADPH cytochrome reductase, aminopyrine demethylase, cytochrome b_5, cytochrome P_{450}, and phospholipids which probably constitute the smooth and rough part of the endoplasmic reticulum [260].

Since there is considerable overlap of most enzymes measured in the various fractions of the isopycnic gradient, it can not be excluded that at least some of the enzymes are found in several of the morphological components of the cell; for example, Golgi, plasma membrane, and endoplasmic reticulum.

The Golgi Apparatus

The Golgi apparatus was described by early microscopists as an intracellular canalicular system connected with vacuoles. The peripheral vacuoles were believed to be capable of producing smaller vacuoles. The Golgi apparatus has both osmophilic and osmophobic components, and in the living cell it is near the nucleus at the secretory pole. Some early investigators claimed that the Golgi apparatus took supravital stains, but this was later contested. To the electron microscopist, the Golgi apparatus appears as a U-shaped or platelike structure, often located near the nucleus, consisting of densely packed membranes surrounded by smaller vacuolar or granular structures 50 mμ in diameter.

Some electron microscopists still consider the Golgi apparatus a separate entity within the cytoplasmic structure; many believe it to be an area of differentiation of the endoplasmic reticulum. Continuity between the structures of the Golgi apparatus and those of the endoplasmic reticulum is thus assumed. It is true that the essential structures of the Golgi apparatus (membranes, cavities, and vacuoles) resemble those of the endoplasmic reticulum. Furthermore, intimate connections between the membranes of the endoplasmic reticulum and the Golgi apparatus—particularly in cells of the hypophysis and in neuron cells—have been observed by various groups of investigators.

The Golgi apparatus is a most elusive cellular organelle. Cajal may well have been the first to recognize this structure in silver-stained cells, but his results were so inconsistent that he preferred not to report them. It was left to Golgi to demonstrate the existence of an intracellular system composed of filaments and plates stainable with the silver stain. For decades, morphologists argued whether the Golgi apparatus was an artifact or a real intracellular entity. Only when the electron microscope revealed the presence of well-developed Golgi bodies in secretory cells were the skeptics convinced [222–224].

The exact role of the Golgi apparatus is not clear, but it is believed to be involved in secretion. Radioautographic studies in Leblond's laboratory have shown that after they are synthesized on the ribosomes, the new proteins elaborated for cellular export are rapidly concentrated in the saccules of the Golgi apparatus. The formation of mucoprotein in the goblet cells remarkably illustrates this event. In that cell, the Golgi apparatus is found at the secretory pole close to the nucleus. It has a horseshoe shape with its convexity oriented toward the nucleus. The horseshoe is made of stacked lamellae and saccules. The saccules of the lower layers contain only small amounts of mucoprotein, but more and more mucoprotein is found toward the upper layers. In fact, the saccules of the upper layers are so filled with mucoprotein that they become swollen and spherical and detach from the Golgi apparatus to move freely in the hyaloplasm toward the cell membrane where they are excreted. At least in rats, the Golgi apparatus seems to be similarly involved in the formation of zymogen granules.

Leblond and his associates also proposed that the Golgi apparatus is the major center for the last step of glycoprotein biosynthesis—namely, the addition of the sugar to the polypeptide moiety. Biochemists have shown that the sugar molecules are inserted in glycoprotein only when the biosynthesis of the polypeptide chain has been completed. Histochemical and radioactive studies have suggested that the proportion of sugar increases within the protein molecules that concentrate in the Golgi apparatus. The exact participation of the Golgi apparatus in glycoprotein biosynthesis is not clear, but biochemical evidence suggests that it is not the exclusive center for the addition of sugar molecules to protein moiety. Moreover, it is unlikely that only glycoproteins are concentrated and packaged in the Golgi for export.

However, data on the biosynthesis of a number of glycoproteins clearly suggest that the Golgi apparatus serves as an assembly line along which sugar moieties are sequentially added to such proteins as thyroglobulin and immunoglobulin [261].

Electron microscope radioautography preparation of a number of tissues obtained from rats and mice injected with $^{35}H_2SO_4$ established that the sulfate was in every case detected in the smooth membranes of the Golgi vesicles. Investigators therefore believe that the Golgi complex is involved in sulfation of mucopolysaccharides, glycoproteins, steroids, lipids, phenols, thyroxine derivatives and arylamines. In these reactions the sulfate is known to be activated by ATP in presence of ATP sulfurylase to yield adenosine 5′-phosphosulfate which is, in turn, in presence of a specific phosphokinase, phosphorylated by ATP to 3′-phosphoadenosine 5′-phosphosulfate. Sulfatransferases catalyze the transfer of the activated sulfate to the acceptor. Although the sulfate does appear in the Golgi complex, it cannot be excluded that most of the enzymes involved in the sulfation are also located in other cell fractions [262].

The Golgi apparatus is also believed to concentrate lysosomal enzymes and to be at the source of small vacuoles filled with acid hydrolases, which constitute the primary lysosomes. The evidence invoked in support of this hypothesis is based primarily on histochemical and electron microscopic observations that have demonstrated the presence of such lysosomal enzymes as acid phosphatase, β-glucuronidase, sulfatase, and esterases in the Golgi apparatus.

Biochemists have concentrated Golgi apparatus by equilibration in sucrose gradient and found acid phosphatase in the crude preparation but no β-glucuronidase or esterase. Histochemists have used demonstrations of nucleoside diphosphatase on thiamine pyrophosphatase and, to a lesser degree, guanosine triphosphatase to identify the apparatus. Several electron microscopists have shown a direct connection between Golgi membranes and ER, suggesting that the endoplasmic reticulum is a tridimensional system of vesicles and canaliculi encompassing the Golgi and the nuclear membranes [225–229].

Cleaner preparations of Golgi apparatus were obtained from rat liver and epididymis by zonal centrifugation of a crude Golgi prepared by the method of Kuff and Dalton. Electron microscopic examination of the fraction shows osmophilic membranes arranged like Golgi membranes. The fraction is rich in galactosyl transferase, an enzyme found only in negligible amounts in other cell fractions except the smooth endoplasmic reticulum, which is believed to be contaminated with Golgi membrane. The Golgi membrane preparation is devoid of ATPase, 5′-nucleotidase glycosidase, glucose-6-phosphatase, and acid phosphatase.

Microbodies—Peroxisomes

Rhodin described in 1954 a newcellular organelle that he discovered in the proximal tubular epithelium of the kidney and called a microbody [263]. At first microbodies were believed to exist only in a few tissues such as liver kidney, and tetrahymina. It is now apparent that they exist almost in every vertebrate tissue and in many plants.

One characteristic of many, but not all peroxisomes, is the presence of a nucleoid and, therefore, three different types of microbodies have been distinguished; those with one crystalloid nucleoid, those with several smaller nucleoids (subcrystalloids), and those without nucleoid. The physiological meaning of the presence or absence of nucleoid is not clear except for the fact that the nucleoid contains urate oxidase. Apparently those microbodies which do not contain urate oxidase are also devoid of nucleoid.

Microbodies cannot be seen with the light microscope; on electron micrographs they appear like small round bodies (5 μ in diameter in liver) with a single membrane surrounding a homogeneous electron-dense mass.

It has been shown that these organelles contain peroxidase, catalase, uricase, and a number of other oxidases, such as D amino oxidase, L-hydroxy acid oxidase, and even isocitrate dehydrogenase [264]. Because of their content in peroxidase, microbodies have been renamed "peroxisomes." Nothing is known of their role in cellular physiology. Usually on the basis of circumstantial evidence they have been suspected to play a role in purine (because of the presence of xanthine oxidase and allantoinase in the kidney and liver of chicken microbodies), cholesterol, lipid, and steroid metabolisms, gluconeogenesis, photosynthesis, and respiration.

Among organelles, microbodies have an unusually short half life; 1 to 5 days or about $^1/_{10}$ of that of mitochondria. The reasons for the rapid turnover of microbodies are not known. The enzymes found in microbodies are synthesized in the endoplasmic reticulum. The mechanism of transfer of the enzymes from endoplasmic reticulum to microbodies has not been discovered. In view of their rapid turnover, microbodies must constantly be eliminated from the cell. Although entrapment in areas of focal cytoplasmic degradation is certainly involved, it appears that other still unknown mechanisms of degradation exist.

In contrast to mitochondria which are believed to be autonomic organelles, microbodies are highly interdependent and closely associated with the endoplasmic reticulum, thus in the living they form pools of enzymes that are in continuity with each other and with at least some of the cisterns of the endoplasmic reticulum. Again the meaning of these interactions is not understood.

Sometimes the role of a metabolic pathway or of a cellular organelle in metabolism is better understood by the observation of pathological conditions in which there is either an increase or a decrease of the activity of the object under investigation. Unfortunately, this has not been the case for microbodies. A number of factors, including cellular growth, administration of salicylate and of some lipogenic agents lead to an increase in the number of microbodies, in some animals (mouse, rat, hamster, and dog), but not in others (guinea pig, chicken, rabbit, squirrel, monkey). The studies have told us nothing about the metabolic function of microbodies. Usually whenever microbodies increase in normal animals, the increase is associated with an increase in catalase activity. The new catalase is, in most, if not all cases, synthesized *de novo*.

The understanding of the relationship between the organelle and cellular metabolism or even the enzyme associated with it is further complicated by the observation that the induction of microbodies with a lipogenic agent occurs in acatalastic mice or in mice in which inhibitors of catalase synthesis (like aminotriazole or allylisopropylacetamide) have been injected, in spite of the lack of a concomitant increase of catalase activity.

References

1. Hughes, A.: The interphase nucleus. In: The mitotic cycle. The cytoplasm and nucleus during interphase and mitosis, 1st ed., p. 30–69. New York: Academic Press 1952
2. Dounce, A.L.: The isolation and composition of cell nuclei and nucleoli. In: The nucleic acids: chemistry and biology (Chargaff, E., and Davidson, J.N., eds.), vol. II, p. 93–153. New York: Academic Press 1955
3. Monty, K.J., Litt, M., Kay, E.R.M., Dounce, A.L.: Isolation and properties of liver cell nucleoli. J. biophys. biochem. Cytol. **2**, 127–145 (1956)
4. Briggs, R., King, T.J.: Nucleocytoplasmic interactions in eggs and embryos. In: The cell (Brachet, J., and Mirsky, A.E., eds.), vol. I, p. 537–617. New York: Academic Press 1959
5. Mitchell, J.S.: The cell nucleus. London: Butterworth and Co., Ltd. 1960
6. Picken, L.E.R.: The organization of cells and other organisms, p. 97. Oxford: Clarendon Press 1960
7. Maggio, R., Siekevitz, P., Palade, G.E.: Studies on isolated nuclei. I. Isolation and chemical characterization of a nuclear fraction from guinea pig liver. J. Cell Biol. **18**, 267–291 (1963)
8. Maggio, R., Siekevitz, P., Palade, G.E.: Studies on isolated nuclei. II. Isolation and chemical characterization of nucleolae and nucleoplasmic subfractions. J. Cell Biol. **18**, 293–312 (1963)
9. Brachet, J.: The role of the cell nucleus in morphogenesis. In: The cell nucleus—metabolism and radiosensitivity. Proc. Int. Symp., p. 3–14. London: Taylor and Francis 1966
10. Swift, H.: Studies on nucleolar function. In: A symposium on molecular biology (Zirkle, R., ed.), p. 266–303. Chicago: Chicago University Press 1959
11. Davis, J.M.G.: The ultrastructure of the mammalian nucleolus. In: The cell nucleus (Mitchell, J.S., ed.), p. 3–14. London: Butterworth and Co., Ltd. 1960
12. Stowell, R.E.: Nucleic acids and cytologic changes in regenerating rat liver. Arch. Path. **46**, 164–178 (1948)
13. Errera, M., Hell, A., Perry, R.P.: The role of the nucleolus in ribonucleic acid and protein synthesis. II. Amino acid incorporation into normal and nucleolar inactivated HeLa cells. Biochim. biophys. Acta (Amst.) **49**, 58–63 (1961)
14. Perry, R.P., Errera, M.: The influence of nucleolar ribonucleic acid metabolism on that of the nucleus and cytoplasm. In: The cell nucleus (Mitchell, J.S., ed.), p. 24–29. London: Butterworth and Co., Ltd. 1960
15. Vincent, W.S., Baltus, E.: The ribonucleic acids of nucleoli. In: The cell nucleus (Mitchell, J.S., ed.), p. 18–23. London: Butterworth and Co., Ltd. 1960

16. Vincent, W.S.: Heterogeneity of nuclear ribonucleic acid. Science **126**, 306–307 (1957)
17. Litt, M., Monty, K.J., Dounce, A.L.: Isolation and properties of rat liver cell nucleoli. Cancer Res. **12**, Sci. Proc., p. 279 (1952)
18. Busch, H., Desjardins, R., Higashi, K., Jacob, S.T., Muramatsu, M., Ro, T.S., Smith, S.J., Schwartz, S.M., Steele, W.J.: The cell nucleus—metabolism and radiosensitivity. In: Proc. Int. Symp., p. 181–202. London: Taylor and Francis 1966
19. Busch, H., Byvoet, P., Smetana, K.: The nucleolus of the cancer cell: A review. Cancer Res. **23**, 313–339 (1963)
20. Busch, H., Lane, M., Adams, H.R., DeBakey, M.E., Muramatsu, M.: Isolation of nucleoli from human tumors. Cancer Res. **25**, 225–233 (1965)
21. Smetana, K., Unuma, T., Busch, H.: Ultrastructural studies on nucleic acids of nucleolar granular components in Novikoff hepatoma cells. Exp. Cell Res. **51**, 105–122 (1968)
22. Brown, D.D., Gurdon, J.B.: Absence of ribosomal RNA synthesis in the anucleolate mutant of *Xenopus laevis*. Proc. nat. Acad. Sci. (Wash.) **51**, 139–146 (1964)
23. Hadjiolov, A.A.: Ribonucleic acids and information transfer in animal cells. Prog. Nucleic Acid Res. Mol. Biol. **7**, 195–242 (1967)
24. Edström, J.E.: Composition of ribonucleic acid from various parts of spider oocytes. J. biophys. biochem. Cytol. **8**, 47–51 (1960)
25. Edström, J.E.: Chromosomal RNA and other nuclear RNA fractions. In: The role of chromosomes in development (Locke, M., ed.), p. 137–152. New York: Academic Press 1964
26. Burdon, R.H.: Ribonucleic acid maturation in animal cells. Prog. Nucleic Acid Res. Mol. Biol. **11**, 33–79 (1971)
27. Siebert, G., Villalobos, J., Jr., Ro, T.S., Steele, W.J., Lindenmeyer, G.A., Adams, H.R., Busch, H.: Enzymatic studies on isolated nucleoli of rat liver. J. biol. Chem. **241**, 71–78 (1966)
28. White, M.J.D.: The chromosomes, 2nd ed. New York: John Wiley & Sons 1952
29. Heitz, E.: Das Heterochromatin der Moose, I. Jb. wiss. Bot. **69**, 762–818 (1928)
30. McClintock, B.: The relation of a particular chromosomal element to the development of the nucleoli in *Zea Mays*. Z. Zellforsch. **21**, 294–328 (1934)
31. Swift, H.: Nucleic acid and cell morphology in dipteran salivary glands. In: The molecular control of cellular activity (Allen, J.M., ed.), p. 73–125. New York: Mc Graw-Hill Book Co. 1962
32. Siebert, G.: Enzymes of cancer nuclei. Exp. Cell Res., Suppl. **9**, 389–417 (1963)
33. Siebert, G.: Energy yielding reactions in nuclei. In: The cell nucleus—metabolism and radiosensitivity. Proc. Int. Symp., p. 265–272. London: Taylor and Francis 1966
34. Roodyn, P.B.: A survey of metabolic studies on isolated mammalian nuclei. Int. Rev. Cytol. **8**, 279–344 (1959)
35. Allfrey, V.G., Mirsky, A.E.: Some effects of substituting the deoxyribonucleic acid of isolated nuclei with other polyelectrolytes. Proc. nat. Acad. Sci. (Wash.) **44**, 981–991 (1958)
36. Osawa, S., Allfrey, V.G., Mirsky, A.E.: Mononucleotides of the cell nucleus. J. gen. Physiol. **40**, 491–513 (1957)
37. Stern, H., Timonen, S.: Position of the cell nucleus in pathways of hydrogen transfer: Cytochrome c, flavoproteins, glutathione and ascorbic acid. J. gen. Physiol. **38**, 41–52 (1954)
38. Betel, I., Klouwen, H.M.: Oxidative phosphorylation in nuclei isolated from rat thymus. In: The cell nucleus-metabolism and radiosensitivity. Proc. Int. Symp., p. 281–293. London: Taylor and Francis 1966
39. Klouwen, H.M., Appelman, A.W.M., Betel, I.: The biochemical basis of cell death in interphase. In: The cell nucleus—metabolism and radiosensitivity. Proc. Int. Symp., p. 295–306. London: Taylor and Francis 1966
40. Errera, M., Ficq, A., Logan, R., Skreb, Y., Vanderhaeghe, F.: Nucleocytoplasmic relationships in irradiated cells. Exp. Cell Res., Suppl. **6**, 268–276 (1959)
41. Leblond, C.P., Everett, N.B., Simmons, B.: Sites of protein synthesis as shown by radioautography after administration of S^{35}- labelled methionine. Am. J. Anat. **101**, 225–271 (1957)
42. Leblond, C.P., Amano, M.: Symposium in synthetic processes in the cell nucleus. IV. Synthetic activity in the nucleolus as compared to that in the rest of the cell. J. Histochem. Cytochem. **10**, 162–174 (1962)
43. Allfrey, V., Mirsky, A.E., Osawa, S.: The nucleus and protein synthesis. In: The chemical basis of heredity (McElroy, W.D., and Glass, B., eds.), p. 200–231. Baltimore: The Johns Hopkins University Press 1957
44. Allfrey, V.G., Mirsky, A.E.: Amino acid transport into the cell nucleus and reactions governing nuclear protein synthesis. In: Protein biosynthesis (Harris, R.J.C., ed.), p. 49–81. London: Academic Press 1961
45. Allfrey, V.G.: Nuclear ribosomes, messener-RNA, and protein synthesis. Exp. Cell Res., Suppl. **9**, 183–212 (1963)
46. Smellie, R.M., Gray, E.D., Keir, H.M., Richards, J., Bell, D., Davidson, J.N.: Studies on the biosynthesis of deoxyribonucleic acid by extracts of mammalian cells. III. Net synthesis of polynucleotides. Biochim. biophys. Acta (Amst.) **37**, 243–250 (1960)
47. Wilson, E.B.: The cell in development and heredity. New York: Macmillan 1925
48. Morgan, T.H.: An attempt to analyze the constitution of the chromosomes on the basis of sex-limited inheritance in Drosophila. J. exp. Zool. **11**, 365–413 (1911)
49. Morgan, T.H.: The physical basis of heredity. Philadelphia: J.B. Lippincott Company 1919
50. Hutchison, C.B.: Heritable characters of maize. VII. Shrunken endosperm. J. Hered. **12**, 76–83 (1921)
51. Mazia, D.: Mitosis and the physiology of cell division. In: Thc cell (Brachet, S., and Mirsky, A., eds.), vol. III, p. 77–412. New York: Academic Press 1961
52. Stern, H., Hotta, Y.: Facets of intracellular regulation of meiosis and mitosis. In: Cell growth and cell division (Harris, R.J.C., ed.). Symp. Int. Soc. Cell Biol., vol. II, p. 57–76. New York: Academic Press 1963
53. Scherbaum, O.H.: Chemical prerequisites for cell division. In: The cell in mitosis (Levine, L., ed.), p. 125–157. New York: Academic Press 1963
54. Brachet, J.: The biological role of the pentose nucleic acids. In: The nucleic acids (Chargaff, E., and Davidson, J.N., eds.), vol. II, p. 476–519. New York: Academic Press 1955
55. Brachet, J.: Le role des acids nucleiques dans la vie de la cellule et de l'embryon. Paris: Masson 1952
56. Caspersson, T.: Cell growth and cell function. New York: W.W. Norton & Company 1950
57. Caspersson, T.: Ultraviolettmikroskopien och dess anvandningsomraden. Nord. Med. **7**, 337–340 (1934)
58. Dixon, G.H., Smith, M.: Nucleic acids and protamine in salmon testes. Prog. Nucleic Acid Res. Mol. Biol. **8**, 9–34 (1968)
59. DeLange, R.J., Smith, E.L.: Histones: Structure and function. Annu. Rev. Biochem. **40**, 279–314 (1971)
60. Hnilica, L.S.: Proteins of the cell nucleus. Prog. Nucleic Acid Res. Mol. Biol. **7**, 25–106 (1967)
61. Neidle, A., Waelsch, H.: Histones: Species and tissue specificity. Science **145**, 1059–1061 (1964)
62. Harbers, E., Vogt, M.: Studies on the properties of nucleohistones. In: The cell nucleus—metabolism and radiosensitivity. In: Proc. Int. Symp., p. 165–177. London: Taylor and Francis 1966
63. Stellwagen, R.H., Cole, R.D.: Chromosomal proteins. Annu. Rev. Biochem. **38**, 951–990 (1969)
64. Marushige, K., Ling, V., Dixon, G.H.: Phosphorylation of chromosomal basic proteins in maturing trout testis. J. biol. Chem. **244**, 5953–5958 (1969)
65. DeLange, R.J., Fambrough, D.M., Smith, E.L., Bonner, J.: Calf and pea histone IV. II. Complete amino acid sequence of calf thymus histone IV. Presence of E-N-acetyllysine. J. biol. Chem. **244**, 319–334 (1969)
66. DeLange, R.J., Fambrough, D.M., Smith, E.L., Bonner, J.: Calf and pea histone IV. III. Complete amino acid sequence of pea seedling histone IV; comparison with the homologous calf thymus histone. J. biol. Chem. **244**, 5669–5679 (1969)
67. Ogawa, Y., Quagliarotti, G., Jordan, J., Taylor, C.W., Starbuck, W.C., Busch, H.: Structural analysis of the glycine-rich arginine-rich histone. III. Sequence of the amino-terminal half of the molecule containing the modified lysine residues and the total sequence. J. biol. Chem. **244**, 4387–4392 (1969)
68. Kleinsmith, L.J., Allfrey, V.G., Mirsky, A.E.: Phosphoprotein metabolism in isolated lymphocyte nuclei. Proc. nat. Acad. Sci. (Wash.) **55**, 1182–1189 (1966)
69. Stevely, W.S., Stocken, L.A.: Phosphorylation of rat-thymus histone. Biochem. J. **100**, 20c–21c (1966)
70. Allfrey, V.G.: The role of chromosomal proteins in gene activation. In: Biochemistry of cell division (Baserga, R., ed.), p. 179–205. Springfield: Charles C. Thomas Publisher 1969
71. Pogo, B.G.T., Allfrey, V.G., Mirsky, A.E.: RNA synthesis and histone acetylation during the course of gene activation in lymphocytes. Proc. nat. Acad. Sci. (Wash.) **55**, 805–812 (1966)
72. Pogo, A.O., Allfrey, V.G., Mirsky, A.E.: Evidence for increased DNA template activity in regenerating liver nuclei. Proc. nat. Acad. Sci. (Wash.) **56**, 550–557 (1966)
73. Allfrey, V.G., Pogo, B.G.T., Littau, V.C., Gershey, E.L., Mirsky, A.E.: Histone acetylation in insect chromosomes. Science **159**, 314–316 (1968)
74. Bartley, J.A., Chalkley, R.: The binding of deoxyribonucleic acid and histone in native nucleohistone. J. biol. Chem. **247**, 3647–3655 (1972)
75. Richards, B.M., Pardon, J.F.: The molecular structure of nucleohistone (DNH). Exp. Cell Res. **62**, 184–196 (1970)
76. Felsenfeld, G., Miles, H.T.: The physical and chemical properties of nucleic acids. Annu. Rev. Biochem. **36**, 407–448 (1967)
77. Elgin, S.C.R., Froehner, S.C., Smart, J.E., Bonner, J.: The biology and chemistry of chromosomal proteins. Advanc. Cell. Mol. Biol. **1**, 1–57 (1971)
78. Baserga, R.: Induction of DNA synthesis by a purified chemical compound. Fed. Proc. **29**, 1443–1446 (1970)
79. Vendrely, R.: The deoxyribonucleic acid content of the nucleus. In: The nucleic acids (Chargaff, E., and Davidson, J.N., eds.), vol. II, p. 155–180. New York: Academic Press 1955
80. Leslie, I.: The nucleic acid content of tissues and cells. In: The nucleic acids (Chargaff, E., and Davidson, J.N., eds.), vol. II, p. 1–50. New York: Academic Press 1955
81. Swift, H.: Cytochemical techniques for nucleic acids. In: The nucleic acids (Chargaff, E., and Davidson, J.N., eds.), vol. II, p. 51–92. New York: Academic Press 1955

82. Leuchtenberger, C.: Cytochemistry of nucleic acids. In: New approaches in cell biology (Walker, P.M.B., ed.), p. 155–172. New York: Academic Press 1960
83. Boivin, A., Vendrely, R., Vendrely, C.: L'acide deoxyribonucléique du noyau cellulaire, dépositaire des caractères héréditaires: arguments d'ordre analytique. C.R. Acad. Sci. [D] (Paris) **226**, 1061–1063 (1948)
84. Griffith, F.: The significance of penumococcal types. J. Hyg. (Lond.) **27**, 113–159 (1928)
85. Avery, O.T., MacLeod, C.M., McCarty, M.: Studies on the chemical nature of the substance inducing transformation of pneumococcal types. J. exp. Med. **79**, 137–158 (1944)
86. Hotchkiss, R.D.: The biological role of the deoxypentose nucleic acids. In: The nucleic acids (Chargaff, E., and Davidson, J.N., eds.), vol. II, p. 435–473. New York: Academic Press 1955
87. Tomasz, A.: Cellular factors in genetic transformation. Sci. Amer. **220**, 38–44 (1969)
88. Chargaff, E.: Isolation and composition of the deoxypentose nucleic acids and of the corresponding nucleoproteins. In: The nucleic acids (Chargaff, E., and Davidson, J.N., eds.), vol. I, p. 307–371. New York: Academic Press 1955
89. Wyatt, G.R.: The purine and pyrimidine composition of deoxypentose nucleic acids. Biochem. J. **48**, 584–590 (1951)
90. Watson, J.D., Crick, F.H.C.: Molecular structure of nucleic acids. Nature (Lond.) **171**, 737–738 (1953)
91. Watson, J.D., Crick, F.H.C.: Genetical implications of the structure of deoxyribonucleic acid. Nature (Lond.) **171**, 964–967 (1953)
92. Crick, F.H.C.: The structure of DNA. In: The chemical basis of heredity (McElroy, W.D., and Glass, B., eds.), Symp., p. 532–539. Baltimore: The Johns Hopkins University Press 1957
93. Cavalieri, L.F., Rosenberg, B.H.: Nucleic acids and information transfer. Prog. Nucleic Acid Res. Mol. Biol. **2**, 1–18 (1963)
94. Cavalieri, L.F., Rosenberg, B.H.: Nucleic acids: Molecular biology of DNA. Annu. Rev. Biochem. **31**, 247–270 (1962)
95. Wilkins, M.H.F.: X-ray diffraction studies of the molecular configuration of nucleic acids. In: Aspects of protein structure (Ramachandran, G.N., ed.), p. 23–37. London: Academic Press 1963
96. Wilkins, M.H.F.: Molecular structure of deoxyribonucleoproteins. In: Nucleoproteins (Stoopes, R., ed.), p. 45. New York: Interscience Publishers 1959
97. Wilkins, M.H.F., Arnott, S., Marvin, D.A., Hamilton, L.D.: Some misconceptions on Fourier analysis and Watson-Crick base pairing. Science **167**, 1693–1694 (1970)
98. Hershey, A.D.: Idiosyncrasies of DNA structure. Science **168**, 1425–1427 (1970)
99. Perutz, M.F., Thompson, L., Randall, J.T., Wilkins, M.H.F., Franklin, R.E., Gosling, R.G., Watson, J.D.: DNA helix. Science **164**, 1537–1539 (1969)
100. Sinsheimer, R.I.: The structure of DNA and RNA. In: The molecular control of cellular activity (Allen, J.M., ed.), p. 221–243. New York: McGraw-Hill Book Company, 1962
101. Doty, P.: Inside nucleic acids. Harvey Lect. Series **55**, 103–139 (1961)
102. Meselson, M., Stahl, F.W.: The replication of DNA in *Escherichia coli*. Proc. nat. Acad. Sci. (Wash.) **44**, 671–682 (1958)
103. Kornberg, A.: Pathways of enzymatic synthesis of nucleotides and polynucleotides. In: The chemical basis of heredity (McElroy, W.D., and Glass, B., eds.), Symp., p. 579–608. Baltimore: The Johns Hopkins University Press 1957
104. Kornberg, A.: Biologic synthesis of deoxyribonucleic acid. In: The molecular control of cellular activity (Allen, J.M., ed.), p. 245–257. New York: McGraw Hill Book Company, 1962
105. Kornberg, A.: Enzymatic synthesis of DNA. New York: John Wiley & Sons 1961
106. Kornberg, A.: The synthesis of DNA. Sci. Amer. **219**, 64–78 (1968)
107. Singer, M.F.: In vitro synthesis of DNA: A perspective on research. Science **158**, 1550–1551 (1967)
108. Mitra, S., Kornberg, A.: Enzymatic mechanisms of DNA replication. In: Macromolecular metabolism (New York Heart Association Symp., ed.). Proc. Symp. N.Y. Heart Assoc., p. 59–79. Boston: Little, Brown and Company, 1966
109. Englund, P.T., Deutscher, M.P., Jovin, T.M., Keely, R.B., Cozzarelli, N.R., Kornberg, A.: Structural and functional properties of *Escherichia coli* DNA polymerase. Cold Spr. Harb. Symp. quant. Biol. **33**, 1–9 (1968)
110. Goulian, M.: Initiation of the replication of single-stranded DNA by *Escherichia coli* DNA polymerase. Cold Spr. Harb. Symp. quant. Biol. **33**, 11–20 (1968)
111. Bollum, F.J.: "Primer" in DNA polymerase reaction. Prog. Nucleic Acid Res. Mol. Biol. **1**, 1–26 (1963)
112. Keir, H.M.: DNA polymerases from mammalian cells. Prog. Nucleic Acid Res. Mol. Biol. **4**, 81–128 (1965)
113. Commoner, B.: Is DNA a self-duplicating molecule? In: Horizons in biochemistry (Kasha, M., and Pullman, B., eds.), p. 319–334. New York: Academic Press 1962
114. Ris, H.: Chromosome structure. In: The chemical basis of heredity (McElroy, W.D., and Glass, B., eds.), Symp., p. 23–69. Baltimore: The Johns Hopkins University Press 1957
115. Swift, H.: Nucleoprotein localization in electron micrographs: Metal binding and radioautography. In: The interpretation of ultrastructure (Harris, R.J.C., ed.). Symp. Intl. Soc. Cell Biol., vol. 1, p. 213–232. New York: Academic Press 1962
116. Swift, H.: Cytochemical studies on nuclear fine structure. Exp. Cell Res., Suppl. **9**, 54–67 (1963)
117. Kihlman, B.A.: Molecular mechanisms of chromosome breakage and rejoining. Adv. Cell Mol. Biol. **1**, 59–107 (1971)
118. Demerec, M.: The fine structure of the gene. In: The molecular control of cellular activity (Allen, J.M., ed.), p. 167–177. New York: McGraw-Hill Book Company 1962
119. Huberman, J.A.: Structure of chromosome fibers and chromosomes. Annu. Rev. Biochem. **42**, 355–378 (1973)
120. Ris, H.: The submicroscopic structure of chromosomes, p. 121. In: Fine structure cells. Groningen: Nordhoff 1955
121. DuPraw, E.J.: The organization of nuclei and chromosomes in Honeybee embryonic cells. Proc. nat. Acad. Sci. (Wash.) **53**, 161–168 (1965)
122. DuPraw, E.J.: The cell and molecular biology. New York: Academic Press 1968
123. Abuelo, J.G., Moore, D.E.: The human chromosome; electron microscopic observations on chromatin fiber organization. J. Cell Biol. **41**, 73–90 (1969)
124. Benzer, S.: The fine structure of the gene. Sci. Amer. **206**, 70–84 (1962)
125. Hearst, J.E., Botchan, M.: The eukaryotic chromosome. Annu. Rev. Biochem. **39**, 151–182 (1970)
126. Hamkalo, B.A., Miller, O.L., Jr.: Electronmicroscopy of genetic activity. Annu. Rev. Biochem. **42**, 379–396 (1973)
127. Simpson, R.T.: Modification of chromatin by trypsin. The role of proteins in maintenance of deoxyribonucleic acid conformation. Biochemistry **11**, 2003–2008 (1972)
128. Varshavsky, A.Y., Ilyin, Y.V., Kadyckov, V.I., Senchenkov, E.P.: Collapse of extended deoxyribonucleoprotein molecules upon increase of the ionic strength of solution. Biochim. biophys. Acta (Amst.) **246**, 583–588 (1971)
129. Attardi, G.: The mechanism of protein synthesis. Annu. Rev. Microbiol. **21**, 383–416 (1967)
130. Schweet, R., Heintz, R.: Protein synthesis. Annu. Rev. Biochem. **35**, 723–758 (1966)
131. Moldave, K.: Nucleic acids and protein biosynthesis. Annu. Rev. Biochem. **34**, 419–448 (1965)
132. Lengyel, P.: Problems in protein biosynthesis. In: Macromolecular metabolism. Proc. Symp. New York Heart Assoc., p. 305–330. Boston: Little, Brown and Company 1966
133. Vogel, H.J., Vogel, R.H.: Regulation of protein synthesis. Annu. Rev. Biochem. **36**, 519–538 (1967)
134. Nisman, B., Pelmont, J.: *De novo* protein synthesis *in vitro*, Prog. Nucleic Acid Res. Mol. Biol. **3**, 236–297 (1964)
135. Barnes, F.W., Jr., Schoenheimer, R.: On the biological synthesis of purines and pyrimidines. J. biol. Chem. **151**, 123–139 (1943)
136. Simpson, M.V., Velick, S.F.: The synthesis of aldolase and glyceraldehyde-3-phosphate dehydrogenase in the rabbit. J. biol. Chem. **208**, 61–71 (1954)
137. Heimberg, M., Velick, S.F.: The synthesis of aldolase and phosphorylase in rabbits. J. biol. Chem. **208**, 725–730 (1954)
138. Zamecnik, P.C.: An historical account of protein synthesis, with current overtones—a personalized view. Cold Spr. Harb. Symp. quant. Biol. **34**, 1–16 (1969)
139. Zamecnik, P.C., Keller, E.B., Hoagland, M.B., Littlefield, J.W., Loftfield, R.B.: Studies on the mechanism of protein synthesis. In: Ionizing radiations and cell metabolism (Wolfstenholme, G.E.W., and O'Connor, C., eds.). Ciba Found. Symp., p. 161–173. Boston: Little, Brown and Company 1956
140. Haselkorn, R., Rothman-Denes, L.B.: Protein synthesis. Annu. Rev. Biochem. **42**, 397–438 (1973)
141. Mehler, A.H.: Induced activation of amino acid activating enzymes by amino acids and tRNA. Prog. Nucleic Acid Res. Mol. Biol. **10**, 1–22 (1970)
142. Novelli, G.D.: Amino acid activation for protein synthesis. Annu. Rev. Biochem. **36**, 449–484 (1967)
143. Neidhardt, F.C.: Roles of amino acid activating enzymes in cellular physiology. Bact. Rev. **30**, 701–719 (1966)
144. Loftfield, R.B.: The mechanism of aminoacylation of transfer RNA. Prog. Nucleic Acid Res. Mol. Biol. **12**, 87–128 (1972)
145. Loftfield, R.B.: The mechanism of aminoacylation of transfer RNA. In: Progress in nucleic acid research and molecular biology (Cohn, W.E., Davidson, J.N., eds.) Vol. 12, p. 87–128. New York: Academic Press 1972
146. Brown, G.K.: Preparation, fractionation, and properties of sRNA. Prog. Nucleic Acid Res. Mol. Biol. **2**, 259–310 (1963)
147. Holley, R.W.: The nucleotide sequence of a nucleic acid. Sci. Amer. **214**, 30–39 (1966)
148. Arnott, S., Fuller, W., Hodgson, A., Prutton, I.: Molecular conformations and structure transitions of RNA complementary helices and their possible biological significance. Nature (Lond.) **220**, 561–564 (1968)
149. Sueoka, N., Kano-Sueoka, T.: Transfer RNA and cell differentiation. Prog. Nucleic Acid Res. Mol. Biol. **10**, 23–55 (1970)
150. Cramer, F.: Three-dimensional structure of tRNA. Prog. Nucleic Acid Res. Mol. Biol. **11**, 391–421 (1971)
151. Rodeh, R., Feldman, M., Littauer, U.Z.: Properties of soluble ribonucleic acid methylases from rat liver. Biochemistry **6**, 451–460 (1967)

152. Hall, R.H.: $N^6(\Delta^2$-isopentenyl) adenosine: Chemical reactions, biosynthesis, metabolism and significance to the structure and function of tRNA. Prog. Nucleic Acid Res. Mol. Biol. **10**, 57–86 (1970)
153. Littauer, U.Z., Inouye, H.: Regulation of tRNA. Annu. Rev. Biochem. **42**, 439–470 (1973)
154. Gefter, M.L., Bikoff, E.: Studies on synthesis and modification of transfer RNA. Cancer Res. **31**, 667–670 (1971)
155. Zubay, G.: Molecular model for protein synthesis. Science **140**, 1092–1095 (1963)
156. Gauss, D.H., Haar, F. von der, Maelicke, A., Cramer, F.: Recent results of tRNA research. Annu. Rev. Biochem. **40**, 1045–1078 (1971)
157. Yarus, M.: Recognition of nucleotide sequences. Annu. Rev. Biochem. **38**, 841–880 (1969)
158. Deutscher, M.P.: Synthesis and functions of the -C-C-A terminus of transfer RNA. Prog. Nucleic Acid Res. Mol. Biol. **13**, 51–92 (1973)
159. Borek, E., Srinivasan, P.R.: The methylation of nucleic acids. Annu. Rev. Biochem. **35**, 275–298 (1966)
160. Starr, J.L., Sells, B.H.: Methylated ribonucleic acids. Physiol. Rev. **49**, 623–669 (1969)
161. Kerr, S.J.: Absence of a natural inhibitor of the tRNA methylases from fetal and tumor tissues. Proc. nat. Acad. Sci. (Wash.) **68**, 406–410 (1971)
162. Crick, F.H.C.: On the genetic code. Science **139**, 461–464 (1963)
163. Crick, F.H.C.: The recent excitement in the coding problem. Prog. Nucleic Acid Res. Mol. Biol. **1**, 163–217 (1963)
164. Goldberg, A.L., Wittes, R.E.: Genetic code: Aspects of organization. Science **153**, 420–424 (1966)
165. Garen, A.: Sense and nonsense in the genetic code, 3 exceptional triplets can serve as both chain-terminating signals and amino acid coding. Science **160**, 149–159 (1968)
166. Jukes, T.H., Gatlin, L.: Recent studies concerning the coding mechanism. Prog. Nucleic Acid Res. Mol. Biol. **11**, 303–350 (1971)
167. Crick, F.H.C.: The genetic code—Yesterday, today, and tomorrow. Cold Spr. Harb. Symp. quant. Biol. **31**, 3–9 (1966)
168. Bernfield, M.R., Nirenberg, M.W.: RNA codewords and protein synthesis, the nucleotide sequences of multiple codewords for phenylalanine, serine, leucine, and proline. Science **147**, 479–484 (1965)
169. Nirenberg, M., Caskey, T., Marshall, R., Brimacombe, R., Kellogg, D., Doctor, B., Hatfield, D., Levin, J., Rottman, F., Pestka, S., Wilcox, M., Anderson, F.: The RNA code and protein synthesis. Cold Spr. Harb. Symp. quant. Biol. **31**, 11–24 (1966)
170. Salas, M., Smith, M.A., Stanley, W.M., Jr., Wahba, A.J., Ochoa, S.: Direction of reading of the genetic message. J. biol. Chem. **240**, 3988–3995 (1965)
171. Bennett, J.C.: Genetic coding for protein structure. Annu. Rev. Biochem. **33**, 205–234 (1964)
172. Khorana, H.G., Büchi, H., Ghosh, H., Gupta, N., Jacob, T.M., Kössel, H., Morgan, R., Narang, S.A., Ohtsuka, E., Wells, R.D.: Polynucleotide synthesis and the genetic code. Cold Spr. Harb. Symp. quant. Biol. **31**, 39–49 (1966)
173. Barondes, S.N., Nirenberg, M.W.: Fate of a synthetic polynucleotide directing cell-free protein synthesis. I. Characteristics of degradation. Science **138**, 810–813 (1962)
174. Lipmann, F.: Messenger ribonucleic acid. Prog. Nucleic Acid Res. Mol. Biol. **1**, 135–161 (1963)
175. Harris, H.: Nuclear ribonucleic acid. Prog. Nucleic Acid Res. Mol. Biol. **2**, 19–59 (1963)
176. Hadjiolov, A.A.: Ribonucleic acids and information transfer in animal cells. Prog. Nucleic Acid Res. Mol. Biol. **7**, 195–242 (1967)
177. Henshaw, E.C.: Messenger RNA in mammalian cells. In: The cell nucleus—metabolism and radiosensitivity. In: Proc. Int. Symp., p. 53–60. London: Taylor and Francis 1966
178. Chantrenne, H., Burny, A., Marbaix, G.: The search for the messenger RNA of hemoglobin. Prog. Nucleic Acid Res. Mol. Biol. **7**, 173–194 (1967)
179. Penman, S.: Ribonucleic acid metabolism in mammalian cells. N. Engl. J. Med. **276**, 502–511 (1967)
180. Spiegelman, S.: Information transfer from the genome. Fed. Proc. **22**, 36–54 (1963)
181. Wood, W.B., Berg, P.: Studies on the "messenger" activity of RNA synthesized with RNA polymerase. Cold Spr. Harb. Symp. quant. Biol. **28**, 237–246 (1963)
182. Richardson, J.P.: RNA polymerase and the control of RNA synthesis. Prog. Nucleic Acid Res. Mol. Biol. **9**, 75–116 (1969)
183. Gumport, R., Weiss, S.B.: Transcriptional and sedimentation properties of ribonucleic acid polymerase from *Micrococcus lysodeikticus*. Biochemistry **8**, 3618–3628 (1969)
184. Magasanik, B.: Catabolite repression. Cold Spr. Harb. Symp. quant. Biol. **26**, 249–256 (1961)
185. Loomis, W.F., Jr., Magasanik, B.: The relation of catabolite repression to the induction system for β galactosidase in *E. coli*. J. molec. Biol. **8**, 417–426 (1964)
186. Roeder, R.G., Rutter, W.J.: Multiple forms of DNA-dependent RNA polymerase in eukaryotic organisms. Nature (Lond.) **224**, 234–237 (1969)
187. Roeder, R.G., Rutter, W.J.: Multiple ribonucleic acid polymerases and ribonucleic acid synthesis during sea urchin development. Biochemistry **9**, 2543–2553 (1970)
188. Sugden, B., Keller, W.: Mammalian deoxyribonucleic acid-dependent ribonucleic acid polymerases. I. Purification and properties of an α-amanitin-sensitive ribonucleic acid polymerase and stimulatory factors from HeLa and KB cells. J. biol. Chem. **248**, 3777–3788 (1973)
189. Jacobs, S.T.: Mammalian RNA polymerases. Prog. Nucleic Acid Res. Mol. Biol. **13**, 93–126 (1973)
190. Hayashi, Y., Hayashi, M.: Fractionation of ΦX174 specific messenger RNA. Cold Spr. Harb. Symp. quant. Biol. **35**, 171–177 (1970)
191. Salser, W., Gesteland, R.F., Ricard, B.: Characterization of lysozyme messenger and lysozyme synthesized *in vitro*. Cold Spr. Harb. Symp. quant. Biol. **34**, 771–780 (1969)
192. Hunt, J.A., Laycock, D.G.: Characterization of messenger RNA for hemoglobin. Cold Spr. Harb. Symp. quant. Biol. **34**, 579–584 (1969)
193. De Maeyer-Guignard, J., De Maeyer, E., Montagnier, L.: Interferon messenger RNA: Translation in heterologous cells. Proc. nat. Acad. Sci. (Wash.) **69**, 1203–1207 (1972)
194. Means, A.R., Comstock, J.P., Rosenfeld, G.C., O'Malley, B.W.: Ovalbumin messenger RNA of chick oviduct: Partial characterization, estrogen dependence and translation *in vitro*. Proc. nat. Acad. Sci. (Wash.) **69**, 1146–1150 (1972)
195. Weinberg, R.A.: Nuclear RNA metabolism. Annu. Rev. Biochem. **42**, 329–354 (1973)
196. Winters, M.A.S.C., Edmonds, M.: A poly (A) polymerase from calf thymus. J. biol. Chem. **248**, 4756–4762 (1973)
197. Busch, H., Tae Suk Ro-Choi, Prestayko, A.W., Shibata, H., Hooke, S.T., El-Khatib, S.M., Choi, Y.C., Mauritzen, C.M.: Low molecular weight nuclear RNAs. Perspect. Biol. Med. **15**, 117–139 (1971)
198. Blackburn, W.R.: Pathobiology of nucleocytoplasmic exchange. Pathobiol. Annu. **1**, 1–31 (1971)
199. Harris, H.: Nucleus and cytoplasm, 2nd ed. Oxford: Clarendon Press 1970
200. Nomura, M.: Ribosomes. Sci. Amer. **221**, 28–35 (1969)
201. Kurland, C.G.: Ribosome structure and function emergent. Science **169**, 1171–1177 (1970)
202. Attardi, G., Amaldi, F.: Structure and synthesis of ribosomal RNA. Annu. Rev. Biochem. **39**, 183–226 (1970)
203. Schlessinger, D., Apirion, D.: *Escherichia coli* ribosomes: Recent developments. Annu. Rev. Microbiol. **23**, 387–426 (1969)
204. Hultin, T.: Ribosomal functions related to protein synthesis. Int. Rev. Cytol. **16**, 1–36 (1964)
205. Birnstiel, M.L., Chipchase, M., Spiers, J.: The ribosomal RNA cistrons. Prog. Nucleic Acid Res. Mol. Biol. **11**, 351–389 (1971)
206. Burdon, R.H.: Ribonucleic acid maturation in animal cells. Prog. Nucleic Acid Res. Mol. Biol. **11**, 33–79 (1971)
207. Hultin, T., Ostner, U.: Specific unmasking of ribosomal proteins under the influence of chelating agents and increased ionic strength. Biochim. biophys. Acta (Amst.) **160**, 229–238 (1968)
208. Ostner, U., Hultin, T.: The use of proteolytic enzymes in the study of ribosomal structure. Biochim. biophys. Acta (Amst.) **154**, 376–387 (1968)
209. Borst, P., Ruttenberg, G.J.: Mitochondrial DNA. IV. Interaction of ribopolynucleotides with the complementary strands of chick-liver mitochondrial DNA. Biochim. biophys. Acta (Amst.) **190**, 391–405 (1969)
210. Takanami, M., Yan, Y., Jukes, T.H.: Studies on the site of ribosomal binding of f2 bacteriophage RNA. J. molec. Biol. **12**, 761–773 (1965)
211. Marcker, K., Sanger, F.: N-Formyl-methionyl-S-RNA. J. molec. Biol. **8**, 835–840 (1964)
212. Jacob, F., Monod, J.: Genetic regulatory mechanisms in the synthesis of proteins. J. molec. Biol. **3**, 318–356 (1961)
213. Monod, J.: Inducteurs et inhibiteurs spécifiques dans la biosynthèse d'un enzyme. R.C. Ist. sup. Sanità **16**, 74–81 (1953)
214. Ames, B.N., Hartman, P.E.: The histidine operon. Cold Spr. Harb. Symp. quant. Biol. **28**, 349–356 (1963)
215. Gilbert, W., Müller-Hill, B.: The lac operator is DNA. Proc. nat. Acad. Sci. (Wash.) **58**, 2415–2421 (1967)
216. Swift, H.: Cytoplasmic particulates and basophilia. In: The chemical basis of development (McElroy, W.D., and Glass, B., eds.), Symp., p. 174–213. Baltimore: The Johns Hopkins University Press 1958
217. Haguenau, F.: The ergastoplasm: Its history, ultrastructure and biochemistry. Int. Rev. Cytol. **7**, 425–483 (1958)
218. Palade, G.E.: Electron microscopy of mitochondria and other cytoplasmic structures. In: Enzymes: Units of biological structure and function (Gaebler, O.H., ed.). Henry Ford Hosp. Int. Symp., p. 185–215. New York: Academic Press 1956
219. Bond, H.E., Cooper, J.A., 2nd, Courington, D.P., Wood, J.S.: Microsome-associated DNA. Science **165**, 705–706 (1969)
220. Redman, C.M.: Biosynthesis of serum proteins and ferritin by free and attached ribosomes of rat liver. J. biol. Chem. **244**, 4308–4315 (1969)
221. Leskes, A., Siekevitz, P., Palade, G.E.: Differentiation of endoplasmic reticulum in hepatocytes. I. Glucose-6-phosphatase distribution *in situ*. J. Cell Biol. **49**, 264–287 (1971)
222. Sjöstrand, F.S., Hanzon, V.: Ultrastructure of Golgi apparatus of exocrine cells of mouse pancreas. Exp. Cell Res. **7**, 415–429 (1954)
223. Lacy, D., Challice, C.E.: The structure of the Golgi apparatus in vertebrate cells examined by light and electron microscopy. Symp. Soc. exp. Biol. **10**, 62–90 (1957)
224. Rambourg, A., Hernandez, W., Leblond, C.P.: Detection of complex carbohydrates in the Golgi apparatus of rat cells. J. Cell Biol. **40**, 395–414 (1969)

225. Kuff, E.L., Dalton, A.J.: Biochemical studies of isolated Golgi membranes. In: Subcellular particles (Hayashi, T., ed.), p. 114–127. New York: The Ronald Press Company 1959
226. Fleischer, B., Fleischer, S., Ozawa, H.: Isolation and characterization of Golgi membranes from bovine liver. J. Cell Biol. **43**, 59–79 (1969)
227. Maul, G.G., Brinkley, B.R.: The Golgi apparatus during mitosis in human melanoma cells *in vitro*. Cancer Res. **30**, 2326–2335 (1970)
228. Leelavathi, D.E., Estes, L.W., Feingold, D.S., Lombardi, B.: Isolation of a Golgi-rich fraction from rat liver. Biochim. biophys. Acta (Amst.) **211**, 124–138 (1970)
229. Trelstad, R.L.: The Golgi apparatus in chick corneal epithelium: Changes in intracellular position during development. J. Cell Biol. **45**, 34–42 (1970)
230. Franke, W.W., Scheer, U.: Structures and functions of the nuclear envelope. In: The cell nucleus (Busch, H., ed.), vol. 1, p. 219–347. New York: Academic Press 1974
231. Kasper, C.B.: Chemical and biochemical properties of the nuclear envelope. In: The cell nucleus (Busch, H., ed.), vol. 1, p. 349–384. New York: Academic Press 1974
232. Phillips, D.M., Phillips, S.G.: Repopulation of postmitotic nucleoli by preformed RNA; II. Ultrastructure. J. Cell Biol. **58**, 54–63 (1973)
233. Kornberg, R.D., Thomas, J.O.: Chromatin structure: Oligomers of the histones. Science **184**, 865–868 (1974)
234. Kornberg, R.D.: Chromatin structure: A repeating unit of histones and DNA. Science **184**, 868–871 (1974)
235. Wells, R.D., Inman, R.B. (eds.): DNA synthesis *in vitro*. Baltimore: University Park Press 1973
236. Schekman, R., Wickner, W., Westergaard, O., Brutlag, D.: Initiation of DNA synthesis. In: DNA synthesis *in vitro* (Wells, R.D., and Inman, R.B., eds.), p. 175–183. Baltimore: University Park Press 1973
237. Gefter, M.L., Kornberg, T., Molineux, I.J., Khorana, H.G., Mendich, L., Hirota, Y.: Studies on DNA polymerases II and III of Escherichia coli. In: DNA synthesis *in vitro* (Wells, R.D., and Inman, R.B., eds.), p. 71–82. Baltimore: University Park Press 1973
238. Wang, J.C.: Protein ω: A DNA swivelase from Escherichia coli? In: DNA synthesis *in vitro* (Wells, R.D., and Inman, R.B., eds.), p. 163–174. Baltimore: University Park Press 1973
239. Okazaki, R., Sugino, A., Hirose, S., Okazaki, T., Imae, Y.: The discontinuous replication of DNA. In: DNA synthesis *in vitro* (Wells, R.D., and Inman, R.B., eds.), p. 83–106. Baltimore: University Park Press 1973
240. Schekman, R., Weiner, A., Kornberg, A.: Multienzyme systems of DNA replication. Science **186**, 987–993 (1974)
241. Solari, A.J.: The molecular organization of the chromatin fiber. In: The cell nucleus (Busch, H., ed.), vol. 1, p. 493–535. New York: Academic Press 1974
242. Olmsted, J.B., Borisy, G.G.: Microtubules. Annu. Rev. Biochem. **42**, 507–540 (1973)
243. Bucher, N.L.R.: Microtubules. New Engl. J. Med. **287**, 195–196 (1972)
244. Hartmann, J.F., Zimmerman, A.M.: The mitotic apparatus. In: The cell nucleus (Busch, H., ed.), vol. II, p. 459–486. New York: Academic Press 1974
245. Lee, J.C., Frigon, R.P., Timasheff, S.N.: The chemical characterization of calf brain microtubule protein subunits. J. biol. Chem. **248**, 7253–7262 (1973)
246. Dustin, P.: Some recent advances in the study of microtubules and microtubule poisons. Arch. Biol. (Brux.) **85**, 263–288 (1974)
247. Chen, G.S., Siddiqui, M.A.: Biosynthesis of transfer RNA: in vitro conversion of transfer RNA precursors from Bombyx mori to 4S RNA by Escherichia coli enzymes. Proc. nat. Acad. Sci. (Wash.) **70**, 2610–2613 (1973)
248. Ofengand, J., Henes, C.: The function of pseudouridylic acid in transfer ribonucleic acid; II. Inhibition of amino acyl transfer ribonucleic acid-ribosome complex formation by ribothymidylylpseudouridylyl-cytidylylguanosine 3′-phosphate. J. biol. Chem. **244**, 6241–6253 (1969)
249. Simsek, M., Petrissant, G., Rajbhandary, U.L.: Replacement of the sequence G-T-Ψ-C-G(A)- by G-A-U-C-G- in initiator transfer RNA of rabbit-liver cytoplasm. Proc. nat. Acad. Sci. (Wash.) **70**, 2600–2604 (1973)
250. Stevens, R.H., Williamson, A.R.: Isolation of messenger RNA coding for mouse heavy-chain immunoglobulin. Proc. nat. Acad. Sci. (Wash.) **70**, 1127–1131 (1973)
251. Schumm, D.E., Morris, H.P., Webb, T.E.: Cytosol-modulated transport of messenger RNA from isolated nuclei. Cancer Res. **33**, 1821–1828 (1973)
252. Sherton, C.C., Wool, I.G.: Determination of the number of proteins in liver ribosomes and ribosomal subunits by two-dimensional polyacrylamide gel electrophoresis. J. biol. Chem. **247**, 4460–4467 (1972)
253. Liew, C.-C., Gornall, A.G.: Acetylation of ribosomal proteins. J. biol. Chem. **248**, 977–983 (1973)
254. Traugh, J.A., Mumby, M., Traut, R.R.: Phosphorylation of ribosomal protein by substrate-specific protein kinases from rabbit reticulocytes. Proc. nat. Acad. Sci. (Wash.) **70**, 373–376 (1973)
255. Majumder, G.C., Turkington, R.W.: Hormone-dependent phosphorylation of ribosomal and plasma membrane proteins in mouse mammary gland in vitro. J. biol. Chem. **247**, 7207–7217 (1972)
256. Eil, C., Wool, I.G.: Function of phosphorylated ribosomes. J. biol. Chem. **248**, 5130–5136 (1973)
257. Eil, C., Wool, I.G.: Phosphorylation of liver ribosomal proteins. J. biol. Chem. **248**, 5122–5129 (1973)
258. Beyreuther, K., Adler, K., Geisler, N., Klemm, A.: The amino-acid sequence of lac repressor. Proc. nat. Acad. Sci (Wash.) **70**, 3576–3580 (1973)
259. Platt, T., Files, J.G., Weber, K.: Lac repressor; specific proteolytic destruction of the NH_2-terminal region and loss of the DNA-binding activity. J. biol. Chem. **248**, 110–121 (1973)
260. Amar-Costesec, A., Wibo, M., Thinès-Sempoux, D., Beaufay, H., Berthet, J.: Analytical study of microsomes and isolated subcellular membranes from rat liver; IV. Biochemical, physical, and morphological modifications of microsomal components induced by digitonin, EDTA, and pyrophosphate. J. Cell Biol. **62**, 717–745 (1974)
261. Whaley, W.G., Dauwalder, M., Kephart, J.E.: Golgi apparatus: Influence on cell surfaces. Science **175**, 596–599 (1972)
262. Young, R.W.: The role of the golgi complex in sulfate metabolism. J. Cell Biol. **57**, 175–189 (1973)
263. Svoboda, D.J., Reddy, J.K.: Some biologic properties of microbodies (peroxisomes). In: Pathobiology annual (Ioachim, H.L., ed.), vol. 4, p. 1–32. New York: Appleton-Century-Crofts 1974
264. De Duve, C., Baudhuin, P.: Peroxisomes (Microbodies and related particles). Physiol. Rev. **46**, 323–357 (1966)
265. Blobel, G.: A protein of molecular weight 78,000 bound to the polyadenylate region of eukaryotic messenger RNAs. Proc. nat. Acad. Sci. (Wash.) **70**, 924–928 (1973)
266. Froehner, S.C., Bonner, J.: Ascites tumor ribonucleic acid polymerases. Isolation, purification, and factor stimulation. Biochemistry **12**, 3064–3071 (1973)
267. Lee, S-C., Dahmus, M.E.: Stimulation of eukaryotic DNA-dependent RNA polymerase by protein factors. Proc. nat. Acad. Sci. (Wash.) **70**, 1383–1387 (1973)
268. Jarasch, E-D., Franke, W.W.: Is cytochrome oxidase a constituent of nuclear membranes? J. biol. Chem. **249**, 7245–7254 (1974)

Chapter 3

Inborn Errors of Metabolism

Hereditary Disease

Sickle Cell Anemia and Other Hemoglobinopathies

Sickle-shaped red cells were first observed in a Negro patient by Herrick, a Chicago physician [1]. Instead of classifying this finding as a curiosity, he pursued his investigation and established a correlation between sickling and hemolytic anemia. When the family histories of individuals presenting such anomalies were studied, it became evident that sickle cell anemia is a hereditary disease transmitted by an autosomal dominant gene. In 1949, Pauling, Itano, Singer, and Wells, in a classic paper entitled "Sickle Cell Anemia, A Molecular Disease" [2], described different electrophoretic patterns for hemoglobin of normal individuals and hemoglobin of patients with sickle cell anemia. This was the first demonstration of a genetically controlled abnormal protein in the phenotype. This alteration in a single molecule is responsible for the complex symptomatology of sickle cell anemia.

Normal and Abnormal Hemoglobins

Different hemoglobins have been recognized in normal human blood for a long time. In 1866, it was demonstrated that placental blood was more difficult to denature by alkali or acid than adult blood. The existence of fetal hemoglobin was confirmed with a variety of techniques.

Since the original observation of Pauling and his associates, a multitude of normal and abnormal hemoglobins have been discovered (providing a most fertile field of investigation for students of molecular genetics). A complete review of the different hemoglobins and the physicochemical properties of their primary, secondary, and tertiary structures is not possible, and only those facts relevant to the pathogenesis of hereditary anemia will be discussed here.

At first normal adult hemoglobins were referred to by the letter A, and abnormal hemoglobins by other letters of the alphabet: S for sickle cell anemia, and C, D, E, G, etc., for other abnormal hemoglobins. Combined chemical and genetic studies on individuals with abnormal hemoglobin indicate that whereas both adult and fetal hemoglobins are controlled by different loci on the gene, all adult hemoglobins, normal or abnormal, are alleles.

As more and more abnormal hemoglobins were discovered, the newly found ones were named after the discoverer, the patient, or most often after the place of origin. The chemical nomenclature is, of course, the most logical because it directly refers to the amino acid substitution in the peptide chain; thus, hemoglobin S is represented by

$$\begin{array}{c} 6\ \text{Glu}\rightarrow\text{Val} \\ \alpha_2\beta_2 \end{array}$$

hemoglobin Stanleyville II

$$\begin{array}{ccc} & 78\ \text{Asn}\rightarrow\text{Lys} & \\ \alpha_2 & & \beta_2 \end{array}$$

Now there are close to 50 known hemoglobins with an amino acid substitution in the α chain and almost 100 with amino acid substitutions in the β chain.

Only a few examples of such substitutions are given in Table 3-1. Other types of abnormal hemoglobins include substitutions in the γ chain (e.g., F Texas $\alpha_2\gamma_2$ 5 Glu→Lys) and those including additional residues at the C terminal (e.g., hemoglobin Tak).

Normal and Abnormal Hemoglobin

Hemoglobin A is a huge ferroprotein containing a prosthetic group (heme) and a large protein (globin). It has a molecular weight of approximately 68,000. X-Ray studies have shown that the hemoglobin mole-

Table 3-1. Differences between some abnormal hemoglobins and hemoglobin A

Amino acid position	Substitution α-chain variants	Name	Amino acid position	Substitution β-chain variants	Name
57	Gly→Asp	Norfolk			
58	His→Tyr	M_{Boston}	63	His→Tyr	$M_{Saskatoon}$
60	Lys→Asn	Zambia			
			67	Val→Glu	$M_{Milwaukee}$
			67	Val→Asp	Bristol
68	Asn→Lys	$G_{Philadelphia}$			
			76	Ala→Glu	Seattle
78	Asn→Lys	Stanleyville II			
85	Asp→Asn	$G_{Norfolk}$			
			94	Asp→Asn	Oak Ridge
			113	Val→Glu	New York
114	Pro→Arg	Chiapas			
			146	His→Asp	Hiroshima
141	Arg→Pro	Singapore			

cule has an ellipsoid shape and is made of two halves. The smallest diameter of the ellipse measures 55 A and the longest measures 70 A. Thus, the size of the molecule is about 0.01 the size of the red cell (see Fig. 3-1, *upper left corner*).

The molecule contains four heme groups, which suggests that it is made of four subunits. If all the subunits were identical, the number of amino acid residues obtained by digestion of the molecule would be a multiple of four. But this is not the case for at least some of the amino acids. The ferroprotein contains two identical halves, and each half is subdivided into two subunits called chains. The subunits differ in amino acid composition and sequence. One is called the α- and the other the β-chain (see Fig. 3-1). The β-chain ends with valine and leucine, but the α-chain ends with valine. The α-chain can be separated from the β-chain by chromatography or by countercurrent techniques.

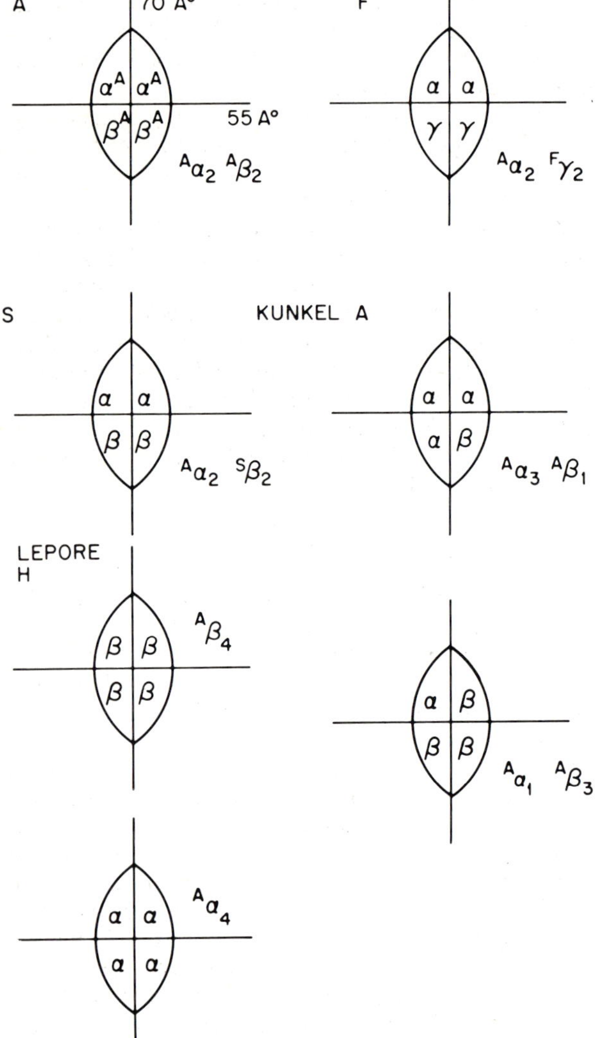

Fig. 3-1. Chain combinations in some abnormal hemoglobins

The amino acid sequences of the α- and β-chains of human hemoglobin A have now been elucidated. Anomalies in the molecular structure of hemoglobin may involve alterations in the amino acid sequence of the α- or the β-chains, or abnormal recombinations of α- and β-chains in the entire hemoglobin molecule. Some known β-chain alterations involve the peptide containing amino acids 1–8, or the peptide composed of amino acids 18–30, of the β-chain. In the first group are included hemoglobins S, C, and G. Although these first abnormal hemoglobins are characterized by an alteration of amino acid 6 in the β-chain, in hemoglobin G the alteration involves amino acid 7. Thus, in S a glutamic acid is replaced by valine, in C by lysine, and in G by glycine. Hemoglobin D and hemoglobin E present changes in the polypeptide composed of amino acid residues 18–30 of the β-chain. The 26th residue, glutamic acid in hemoglobin A, is replaced by lysine in hemoglobin E. The nature of the molecular alteration is not clear in hemoglobin D. Changes in the amino acid sequence have also been observed in the α-chain. Such changes have been described in the polypeptides forming the residue ranging from 13–31. Thus, in hemoglobin I, the 17th residue—a lysine in hemoglobin A—is replaced by a glutamic residue. Hemoglobin D (see Fig. 3-2) has alterations of both α- and β-chains [3, 4].

In sickle cell anemia, the electrophoretic pattern of hemoglobin differs from the pattern of normal adult hemoglobin in that a peak called the S-peak moves more rapidly than the normal A-peak (see Fig. 3-3). Insofar as available information suggests that all varieties of human hemoglobin have a similar molecular weight and shape, and since differences in mobility are constant despite changes in pH, at a given pH the relative mobility indicates a difference in the content of ionizable molecules. Thus, on the basis of the electrophoretic pattern, it was calculated that hemoglobin S contains two more positive charges than hemoglobin A. The differences between the amino acid sequences of hemoglobin S and hemoglobin A were known many years before the amino acid sequence of hemoglobin A was deciphered by comparing "fingerprints" of the polypeptides obtained from trypsin digestion of hemoglobin A and hemoglobin S [5–7].

The specificity of trypsin is such that it splits all peptide bonds, including either lysine or arginine, but no others. There are 56 such bonds in each hemoglobin molecule. Since the hemoglobin molecule is made of two identical halves, the tryptic digest yields 28 different polypeptides. Whereas the charged polypeptides can be separated by electrophoresis, the uncharged polypeptides move by chromatography. Therefore, when the mixture is successively submitted to paper electrophoresis and paper chromatography, a sort of fingerprinting of the hemoglobin molecule is provided.

The distribution of the polypeptides on the paper is the same for hemoglobin A and hemoglobin S except for a single peptide. One polypeptide (4) in the hemoglobin S digest does not migrate to the same spot as its homolog in the hemoglobin A digest. The abnor-

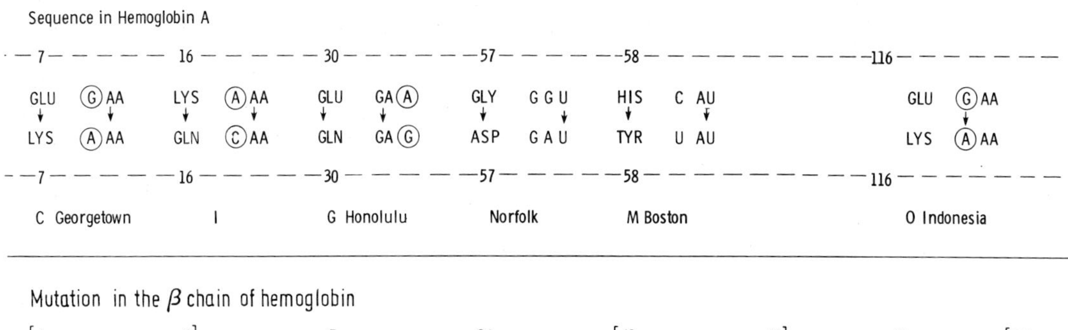

Fig. 3-2. Mutation in the α- and β-chains of hemoglobin

mal polypeptide of hemoglobin S and peptide 4 of hemoglobin A were purified and their amino acid sequence was investigated. One glutamic residue found in A appears as a valine residue in S; this minor alteration in the amino acid sequence in the huge protein molecule explains the insolubility of nonoxygenated hemoglobin S.

Proteins are made of amino acids linked together by peptide bonds. As shall become more obvious as we learn more about protein structure, the amino acid sequence not only determines the molecule's configuration, but it also modulates the interaction between subunits or reacting molecules, such as oxygen in the case of hemoglobin. Therefore, it is not unexpected that changes in the amino acid sequence will modify the interaction between hemoglobin chains. During the sickling process, the S hemoglobin molecule forms columnar aggregates called tactoids. The stacking results from the modified interaction between molecules. The two valine residues that replace two glutamic residues at 6 result in the introduction of two apolar residues, in contrast to polar residues in the β-chain.

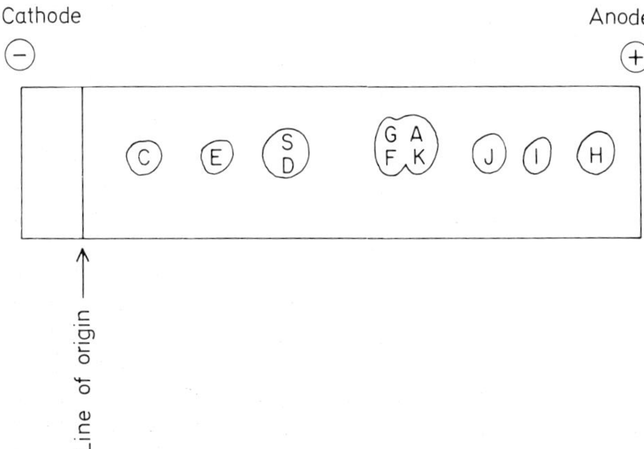

Fig. 3-3. Relative electrophoretic mobility of some hemoglobins (paper electrophoresis, pH 8.6 veronal buffer)

Such a change in molecular structure in the reduced molecule of S hemoglobin makes a polar contrast that causes the molecules to stack together into aggregates. Oxygenated molecules do not adhere to adjacent molecules and the cells unsickle.

It has been proposed that the sickling crisis is triggered by prostaglandin E_2 which impairs the red cell's oxygen and stimulates its calcium uptake—events which facilitate sickling [197].

An understanding of the molecular mechanism of sickling extends beyond its relevance to the interpretation of the pathogenesis of sickling. It may also be at the source of a rational therapy of sickle cell anemia. The formation of tactoid aggregates requires that these two contacts take place between adjacent molecules. The presence of non-S hemoglobin interferes with such contacts and decreases the incidence of sickling. Unfortunately, transfusion has had little effect on the sickling process [7].

Pathology of Sickle Cell Anemia

The presence of abnormal hemoglobin in the red cell leads to several pathological manifestations (Table 3-2): (1) anemia due to the reduced life span of the sickle cell; (2) occlusion of the small vessels by hemoglobin precipitation; and (3) altered susceptibility to infection [8–11]. The anemia is of the hemolytic type, and its severity varies from case to case. The red blood cell count occasionally drops to 1,000,000 per mm^3 during a crisis, and a blood smear reveals the presence of normoblasts, reticulocytes, polychromatophils, and basophilic stippling. The normal shape of the red cell is preserved or lost depending upon the fixative. For instance, in Zenker-fixed material, red cells have a normal shape, whereas in formaldehyde-fixed material, sickle cells are present. Obviously, since only the reduced form of sickle cell hemoglobin is insoluble, merely admitting oxygen to a fresh preparation or a test tube containing sickle cell anemia blood will

Table 3-2. Pathogenesis of sickle cell anemia

Dominant autosomal gene
↓
Replacement of glutamic residue by valine β-chain
↓
S (sickle cell) Hb insoluble when reduced

↙	↘
Sickle cells Reduced life span? ↓ Hemolysis ↓ { anemia, hematuria, jaundice } Hyperplasia of bone marrow	Obstruction of small capillaries ↓ { Thrombosis (spleen), Ulcers (legs), Pains (general) }

1. Increased susceptibility to Salmonella
2. Resistance to malaria

restore the shape of the red cell. Associated with the anemia are hemolytic jaundice and hemosiderosis; the latter is often present in the spleen, liver, bone marrow, lymph nodes, kidneys, and reticuloendothelial cells. There is no satisfactory explanation for the reduced life span of the erythrocyte in sickle cell anemia. The change in the solubility of hemoglobin is not sufficient to reduce the life span, since the existence of many other abnormal hemoglobins is compatible with normal hemoglobin levels. Furthermore, nonanemic individuals whose red cells contain 70% or more S hemoglobin have been observed. Concomitant alteration of the red cell stroma has been suggested to explain the reduced life span.

Sickle cell rigidity leads to obstruction of small capillaries, which is probably responsible for most of the other symptoms of sickle cell anemia—changes in bone structure, transitory muscle pains, ulceration of the leg, and thrombosis, especially of the spleen. The transitory pains probably can be explained by local ischemia due to temporary or permanent capillary occlusion. Pains, which can occur anywhere in the body, often simulate acute visceral conditions, such as appendicitis, cholecystitis, nephritis, poliomyelitis, and myocardial infarcts. The leg ulcerations are probably due to poor circulation leading to deficient oxygenation, thus yielding more rigid sickle cells, which in turn occlude the small capillaries. The spleen, enlarged in the young patient, becomes smaller and fibrotic with increasing age due to repeated thrombosis leading to small infarcts (see Fig. 3-4). These thromboses occur more frequently at high altitudes where oxygen tension is reduced. Infarcts are not responsible for all the scarring of the spleen in sickle cell anemia. Hemorrhage about the follicles is frequently observed in young patients, and these small hematomas are progressively replaced by connective tissue containing macrophages filled with hemosiderin pigment. Kidney lesions include severe congestion, focal hemorrhage (see Fig. 3-5), necrosis with fibrosis, and regeneration.

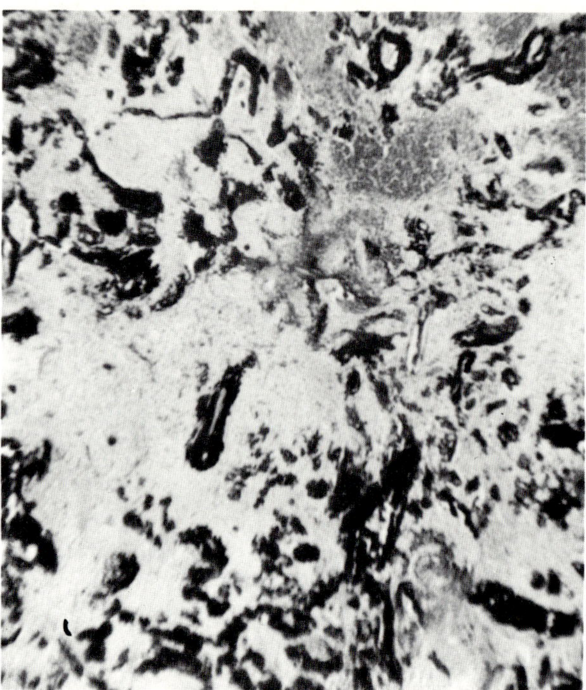

Fig. 3-4. Iron accumulation in spleen of patient with sickle cell anemia

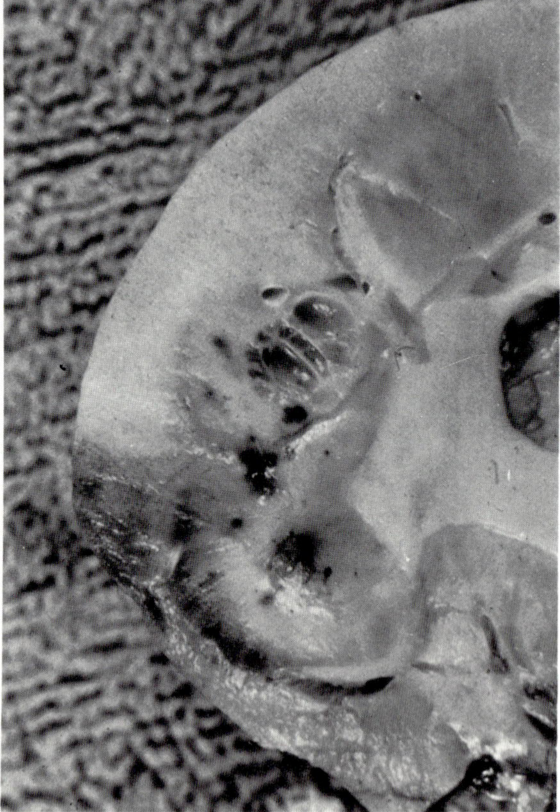

Fig. 3-5. Hemorrhagic necrosis in kidney of patient with sickle cell anemia.(From M. Jackson)

These lesions are associated with hematuria and defective concentration by the renal tubule.

The changes in bone structure are not satisfactorily explained. They are probably the result of both the compensatory hyperplasia of the marrow and capillary occlusion. The bone marrow is often hyperplastic, and fatty marrow is replaced by erythroblastic and granulocytic elements. The bone alterations lead to a nonspecific radiological picture, usually more prominent in the bones of the skull and characterized by both osteoporosis and bone condensation. The skull bones are usually thickened, especially in the parietal region. The general thickening of the diploë is associated with thinning of the outer table and active proliferation of the diploë at the outer surface of the skull to form spicules. These changes lead to radiological pictures referred to as "tower skull" and "hair-on-end." Varied lesions of the long bone are less frequent; trabecular markings, irregularities in thickness of the cortex, abnormal bone formation within the medullary cavity, cortical thickening with narrowing of the medullary cavity, and complete obliteration of the marrow cavity have been described.

There is a definite increase in the incidence of salmonellosis among patients with sickle cell anemia, suggesting that this type of anemia interferes with the normal development of resistance to Salmonella. This increased susceptibility to salmonellosis cannot be ascribed to a general lowering of resistance to infection; the susceptibility of patients with sickle cell anemia to other infections, for example, those frequent in leukemia or aplastic anemia, is not increased. Since necrosis and ischemia are frequent in the bones of sickle cell anemic patients, Salmonella osteomyelitis can be expected in these patients. Indeed, such infections are frequent among sickle cell anemia patients who have contracted salmonellosis. No less interesting is the increased resistance against malaria observed among patients affected with sickle cell anemia.

Sickle cell anemia is almost exclusively found in Negroes. When it affects whites, sickle cell anemia usually suggests a Negro ancestry. It can be detected along the equator from Africa to India. In Africa south of the Sahara desert, the incidence of sickle cell anemia may be as high as 40% of the local population. The disease has also been described in localized areas of Greece and Sicily. Sickle cell anemia results from a mutation that would normally be deleterious to the victims. Yet, the disease has succeeded in implanting itself in many areas of the world. This paradox has been explained by geneticists in terms of balanced polymorphism. Allison [11] observed that the recessive gene is particularly common in areas where malaria is prominent, suggesting that heterozygotes for sickle cell anemia are more resistant to malaria than are homozygotes for normal hemoglobin. This hypothesis was confirmed by administering the agent of malaria to volunteers homozygous for sickle cell anemia and homozygous for normal hemoglobin.

Although a patient with classical sickle cell anemia is homozygous for the sickle cell gene, the "sickle cell trait" is a condition that develops in the heterozygous person. The hemoglobin of these individuals is made of 50% S and 50% A. The condition is usually benign, the red cells have a normal life span, and only minor anomalies are observed—such as spleen lesions that occur when the patient has been exposed to reduced oxygen tension.

Protein Structure and Clinicomolecular Correlation in Hemoglobin Disease

A considerable number of amino acid substitutions have now been described in both the α- and β-chains of human hemoglobin. Because of the development of knowledge of the amino acid sequence of human hemoglobin, it is possible to correlate symptoms and molecular alterations and in many cases to predict dominant symptomatology on the basis of the molecular distortion, and *vice versa*. To appreciate this deployment of molecular biology into medicine, some basic concepts of protein structure must be briefly reviewed [12–17].

The Structure of Proteins. Around 1830, the Dutch chemist Geraldus Mulder described a class of nitrogen-containing substances found in plant and animals alike. Because Mulder suspected that these substances played a role in regulating cellular metabolism, he stated that these were the most important substances in cells, and at the suggestion of Berzelius, he named them proteins (from the Greek πρωτοσ, of first rank).

The names of many great chemists are associated with the study of proteins. The famous organic chemist Baron Justus von Liebig maintained for a long time that all proteins stemmed from only one type of molecule, and all variations had the same radical. But the composition of these large molecules was to remain mysterious until Emil Fisher established that proteins were made of up to 20 different amino acid residues linked by polypeptide bonds.

Although methods for hydrolyzing proteins into their unit compound, the amino acids, were available, the exact composition of the protein was not simple to determine because in most cases the amino acids could not be separated. In spite of the methods developed by organic chemists to separate amino acids and measure their amounts in the hydrolysate—for example, differential precipitation using aromatic sulfonic acid—it was only when Martin and Singer applied the paper chromatographic technique to the study of protein hydrolysates that the components of the hydrolysate could be qualitatively identified. The development of column chromatography by Stein and Moore first using starch (1949), later exchange resins (1951), permitted the separation of amounts of amino acids sufficient for quantitative analysis of the hydrolysate of an equivalent of 3 mg of protein.

Amino Acid Sequence. The protein chain can be compared to a long string of interlocked plastic beads. Each bead has a short peg on one side and a small cavity on the other. To form the chain, the peg of one bead is introduced into the cavity of the other. The beads come in 20 different colors, and thus a large number of different strings of beads can be made. The problem that the chemist faces in attempting to establish a protein's amino acid composition is similar to that of identifying the proportion of each type of bead in, for instance, a string of 200 beads. A large number of molecules are involved, and there is no way to release amino acids one by one and leave the rest of the molecules intact. However, it is possible to identify the first and the last amino acid of the polypeptide chain—the C- and N-terminal amino acid.

In 1945, Sanger demonstrated that the amino acid, which in the intact polypeptide chain has a free amino group, reacts with dinitrofluorobenzene to yield a yellow dinitrophenyl derivative. By this means it is possible to identify the N-terminal amino acid in a polypeptide chain. Edman introduced the phenylthiocarbamyl method, in which the polypeptide is treated with phenylisothiocyanate to yield a phenylthiocarbamyl peptide derivative. Hydrolysis by hydrochloric acid in organic solvent (nitromethane) releases the N-terminal amino acid to yield an organosoluble phenylthiohydantoine of the N-terminal amino acid. N-terminal groups have also been identified by treating the polypeptide with thiocarbamate (pH 3–4) to yield the 2 thio-5-thiozolidone derivatives, and bromoacetate to yield the N-carboxymethyl derivative.

One method for identifing the C-terminal amino acid is to treat the protein with hydrazine (NH_2NH_2) at 100° C. Such a procedure splits all peptide bonds and converts all amino acid except the C-terminal end of the chain into hydrazides ($CONHNH_2$). The terminal carboxyl group of the polypeptide chain can be reduced to the carbinol (CH_2OH) by treatment with lithium bromide.

Enzymes, carboxypeptidases, or amino peptidases have been found that can release C- and N-terminal amino acids selectively. Chromatography is used to separate the product of enzymic or organic hydrolysis.

By determining the N- and C-terminal amino acids, an investigator identifies the first and last component of the polypeptide chain, but he learns nothing about the sequence of the amino acids in between. Yet his problem is to identify the exact position of each amino acid in the polypeptide chain. If strings of beeds are shaken gently in a drum, the strings will break and yield numerous fragments that can be classified according to their size. Moreover, the bead sequences might overlap enough to make it possible to deduce the entire sequence of the bead string. Of course, the longer the original string, the more difficult it is to determine the exact bead sequence. In this description of the approach used to determine protein sequence—which was inspired by papers by Pauling, Perutz, and Sanger—only the reasoning and general approach, rather than specific details, used in establishing amino acid sequences in proteins are discussed.

How, then, were these principles presented and exploited by Sanger to establish the amino acid sequence of insulin? He first separated the alpha from the beta chain, then submitted each chain to controlled chemical or enzymic hydrolysis, separated the polypeptide fragments by chromatography and electrophoresis, analyzed their amino acid composition, and established their amino acid sequence after end-group analysis.

Let's consider, for example, the elucidation of the sequence of the first 8 amino acids of the A chain of the insulin molecule

Gly-Ile-Val-Glu-Glu-Cys-Cys-Ala.

The hydrolysate contained the following polypeptides:

1. Ile-Val-Glu-Glu
2. Glu-Cys-Cys-Ala
3. Gly-Ile-Val-Glu
4. Ile-Val-Glu
5. Cys-Cys-Ala
6. Glu-Glu
7. Val-Glu
8. Glu-Cys
9. Ile-Val

Since the end-group analysis established that glycine is the first amino acid in the sequence, clearly the sequence of the first four amino acids must be

Gly-Ile-Val-Glu

The sequence Ile-Val-Glu is repeated in polypeptide 1, which also includes an additional glutamic residue; therefore, the 5 first amino acids of the sequence must be

Gly-Ile-Val-Glu-Glu

The sequences of peptides 4 and 6 confirm this assumption, because glutamic acid is found only twice in the sequence. Since the hydrolysate yields a dipeptide Glu-Cys, the 6th amino acid in the sequence must be cysteine:

1	2	3	4	5	6	7	8
Gly	Ile	Val	Glu				
	Ile	Val	Glu	Glu			
		Val	Glu				
			Glu	Glu			
				Glu	Cys		
				Glu	Cys	Cys	Ala
					Cys	Cys	Ala

The sequence of two other amino acids in the octopeptide chain must be Cys-Ala because of peptide 2 and peptide 5.

Once a protein's amino acid sequence is established, it is possible to write a formula of the protein, similar to that of any organic chemical and which accounts for each atom and each covalent bond. Such formulae are symbolic representations of the primary structure of the protein. However, proteins are not stretched out molecules. They have a very specific stereoconfiguration that determines their secondary and tertiary structures.

Protein Configuration. The polymerization of a number of amino acids to yield a polypeptide chain generates a molecule with properties of the sum of the parts and new properties that transcend those of the individual amino acids and confer a new dimension to the molecule. Many secrets of the protein configuration were revealed with the aid of physicochemical tools, among which X-ray crystallography proved to be one of the most useful. This method combined with model building permitted reconstruction of such complex molecules as myoglobin and hemoglobin.

A combination of forces is generated within the molecules as a result of the amino acid assemblage and outside the molecule, when the protein is in solution. However, the effect of these forces is restricted by the bond distance within the polypeptide chain. A number of rather simple principles, taken together, contribute to an understanding of protein configuration.

The Amino Acids. A quick look at the formula of each amino acid (as pointed out by Perutz and others) reveals that all amino acids are substitutes of glycine. Consequently, the backbone of proteins is formed by the repetition of $H_2N—CH_2$ COOH units linked by peptide bonds. The 20 amino acids that compose the polypeptide chains are made of a unit core

```
H
  \   /
    C
  /   \
HOOC   NH2
```

and a side chain R. In glycine, the side chain is an atom of hydrogen, but in other amino acids the size, configuration, and reactivity of the side chain vary, and, as one may anticipate, the properties of the polypeptide chain depend largely upon its amino acid composition and sequence. The reactivity of amino acid side chains is classified as one of four groups: (1) nonpolar (alanine, valine, leucine, and isoleucine); (2) aliphatic hydroxyl (serine and threonine); (3) those ionized at a physiological pH (anionic—aspartic and glutamic acid; cationic—lysine and arginine); (4) aromatic.

These fundamental groups influence the overall structure of the polypeptide chain in at least the following ways: (1) by acting as spacers in the interior of the polypeptide chain (*e.g.*, nonpolar side chains); (2) by contributing to hydrogen bond formation (*e.g.*, aliphatic hydroxyl groups, amide of aspartic and glutamic acid); (3) by affecting electric charge.

Because of their special properties, some amino acids—proline, cysteine, histidine, and serine—confer unique properties to the polypeptide chain. In proline, four carbons are on the side chain; each carbon except the first (α-carbon) is saturated with two hydrogen atoms. The last carbon (δ-carbon) forms a bond with the iminonitrogen, thus generating the formation of an imino group (—NH—), rather than the amino group (—NH_2—) formed by all other amino acids and thereby preventing the participation of the imino group in hydrogen bonding. Consequently, proline is a misfit in the amino acid sequence and is found where the polypeptide chain abandons its helical form to change direction.

Cysteine possesses a highly reactive sulfhydryl group that forms a disulfide bridge when it is close to another cysteine residue of the same or a different polypeptide chain. Histidine forms coordination complexes and carries an extra proton below pH 6. Serine forms ester bonds with phosphate. Both histidine and serine are part of the catalytic site of many enzymes.

The Bond in the Protein Backbone. Amino acids are amphoteric molecules made of a glycine backbone that contains an NH^+ and COO^- reactive group (the two groups that make the amino acid amphoteric) and various side chains with different types of reactive groups. The amino and the carboxyl groups of the backbone contribute the peptide bond. In the condensation process, a molecule of water is formed, the hydrogen atom of which is derived from the amino group, whereas the carboxyl unit contributes the hydroxyl groups. Consequently, as soon as the peptide bond is formed, the individual amino acids lose their amphoteric properties, and the backbone no longer contributes to the protein's total charge except, of course, for those amino acids at the beginning and end of the polypeptide chain.

The peptide bonds are the target of proteolytic enzymes. Provided that the amino or carboxyl end groups are not masked by acetylation or involved in an interchain-type linkage (as is the case for chymotrypsin), the end groups are susceptible to attack by amino or carboxy peptidases.

Because it binds the repeating unit together, the peptide bond plays a key role in the structure of proteins. Studies of the X-ray crystallographic pictures of polyglycine or similar peptide chains have yielded fundamental parameters of the polypeptide chain: the distance between the atoms, the angle of the bonds, the coplanarity of the carbon atoms involved in bonding, and the transrelationship of the alpha carbon of the polypeptide backbone (glycine groups).

The C—C=O and the C—N H groups must be in the same plane because of resonance stabilization of the double bond. Thus, on the basis of the measurements of bond distances resonance stabilization was suggested to occur. Indeed, while the normal C—N bond distance is 1.47 A, the actual distance determined by X-ray crystallography in the peptide bond is much shorter than expected (1.32 A). In the peptide bond, a π electron is associated with the C=O bond and an isolated electron on the N atom.

To understand the resonance of the peptide bond, consider first the two extreme forms in which this double bond may exist: the covalent and the ionic. In active polypeptide bonds, neither the covalent nor the ionic form exists, but the two types form a resonating unit, as is expected from the resonance theory.

Thus, the actual bond is a compromise between the two extremes.

$$-\overset{\overset{\displaystyle O}{\|}}{C}-\underset{|}{N}- \longleftrightarrow -\overset{\overset{\displaystyle O^-}{|}}{C}=\underset{|}{N^+}-$$

The amide bond is said to have a 40% double-bond character. The molecular orbital theory requires that the atoms involved in resonance all be present in a single plane, because resonance requires p-orbital overlap and sp^2 hybridization of nitrogen and carbonyl carbon atoms.

A consequence of the resonance of the peptide bond is that it may exist in either a *cis* or *trans* isomeric form. Each carbon atom adjacent to the nitrogen next to the peptide bond's carbon atom may be located either in the plane of the bond and on one side of it *(trans)*, or in the same plane but on opposite sites of the bond *(cis)*. An example of a molecule in which the *cis* form exists is ketodipiperazine.

Spectroscopic and other evidence suggests that natural peptides contain only the *trans* variety of the bond, which is the most stable. In passing from one mesomeric form to another *(trans* or *cis)*, the carbons involved in determining mesomerism must pass through a point at which both carbons are in planes perpendicular to each other. At that point, the potential energy is so great that it offers an effective energy barrier to the conversion of one form to the other. In such a position, the resonance energy is lost because overlap between carbonyl and nitrogen p-orbital is impossible. The potential energy barrier is at least 10 Cal/mole in passing from the *cis* to the *trans* form.

The other bonds involved in forming the protein backbone are regular bonds with normal length that are capable of rotating freely (1.53 A for C—C and 1.47 A for C—N bond). Therefore, one might expect the peptide chain to form a random coil in which the distance between individual atoms would be that calculated from the sum of the van der Waals radii. But polypeptide chains are not freely flexible, and the distance between atoms as revealed by X-ray crystallography is smaller than expected from calculations. Thus, calculations of the van der Waals radii of O and N yield a value of 3.05 A, whereas the values found by X-ray crystallography are 2.95 A. The difference between calculated and observed values indicates that other forces must be involved in attracting these atoms. These are intramolecular hydrogen bonds.

Hydrogen Bonds and Secondary Structure. When a covalent bond is formed between a hydrogen atom and an electronegative atom (*e.g.*, N), the hydrogen atom acquires a partial positive charge. If another electronegative atom is near, an attractive force develops between the partially positive hydrogen atom and the electronegative atom. Thus, hydrogen bonds can be defined as the attraction that develops between two nearly electronegative atoms separated by a proton. In other words, the hydrogen bond is a noncovalent bond in which a hydrogen atom is shared by two electronegative groups (in the case of the backbone of the polypeptide chain, a CO and an NH group). The bonding energy (5 Cal/mole) is much lower than that of covalent hydrogen (40–100 Cal/mole).

In addition to X-ray crystallography, a number of other techniques have demonstrated the existence of intramolecular hydrogen bonds in the polypeptide chain. Hydrogen bonds form between the carbonyl and amide groups of nearly all the polypeptide chains. They are responsible for the secondary structure of proteins. Thus, intramolecular interactions (hydrogen bonds and others—see below) impose a new superstructure upon the primary structure of the polypeptide, and this first level of suprastructure is referred to as the protein's secondary structure.

Using X-ray crystallographic analysis, Corey and Pauling [18, 19] concluded that the secondary structure can best be explained if one assumes that the polypeptide chain does not exist in nature in an extended form, but is twisted to form a spiral. After attempting to fit a variety of calculated spirals within the limitation of the data obtained by X-ray crystallography, Corey and Pauling proposed that the secondary structure of at least some proteins assumes the shape of an α-helix.

At least four parameters define the α-helix—the number of amino acids per turn, and the pitch, diameter, and length of the chain. The helix was calculated to contain 18 residues in 5 turns, or 3.600–3.625 residues per turn. The distance between two consecutive turns, the pitch, is 5.4 ± 0.1 A. The diameter of the molecule varies somewhat with the type of side chain; the length depends upon the length of the polypeptide chain.

The secondary structure of the polypeptide chain is stabilized by hydrogen bonding between imino and carboxyl groups, hydrogen bonds between approaching amino acids (e.g., between hydroxyl of tyrosine and neighboring carboxyl), van der Waals forces, and electrostatic attraction and mutual repulsion of solvent by hydrophobic groups.

In conclusion, the extended backbone is regularly coiled by rotation around the axis and translation parallel to the axis of the elementary structure of the helix. The elements are composed of a planar amide group and a tetrahedral carbon. The shape of the coil is that of the thread of a screw; each turn comprises 3.6 amino acids. The residues are separated from each other by a distance of approximately 1.5 A. The helix is held together by hydrogen bonds, van der Waals forces, and hydrophobic forces, and each amide group is bonded to the third amide group from it on the chain.

Hydrogen bonds vary somewhat in length, depending upon whether the hydrogen is shared by 2 oxygens or nitrogen and oxygen

$$\text{O-H} \ldots \text{O} = 2.6\ \text{A}$$
$$\text{N-H} \ldots \text{O} = 2.8\ \text{A}$$

Tertiary Structure and the Structure of Myohemoglobin. If all that is known of the structure of proteins were included above, all polypeptide chains must be thought to form coiled threads. Such a structure might be adequate for nonfunctioning proteins. However, it would not satisfy the special requirements of enzymes or other functional proteins such as hemoglobin. Enzymes must bind substrates to specific sites and release products. Therefore, enzyme structure requires a unique combination of specificity and flexibility, which would not be provided by the coiled thread. Protein specificity is provided by a second form of superstructure, a form of supercoiling of the original structure referred to as tertiary structure. We know little about the tertiary structure of most proteins, but in myoglobin and hemoglobin it has been studied extensively.

The knowledge of the structure of myoglobin derives primarily from X-ray crystallographic studies combined with chemical analysis of the amino acid sequence. Myoglobin is made of a single polypeptide chain composed of 53 amino acid residues. The chain is folded to form a unique but somewhat irregular three-dimensional figure. Only portions of the molecule have the classical secondary structure of the right-handed helix; those form straight stretches (see Fig. 3-6).

There are seven such straight stretches united by corners and irregular regions of the chain. As can be anticipated, proline residues occupy at least some of the corners, but all bends in the molecule do not result from the presence of a proline residue. In fact, in the entire myoglobin molecule only four of the corners are occupied by such residues. Other forces are involved in bending the molecules. For example, one corner is stabilized by a residue that is hydrogen bonded to the free peptide imino group of the amino end of the helix.

In the myoglobin molecule, the heme is important in stabilizing the molecule. The heme forms two kinds of bonds with the myoglobin polypeptide chain. It is inserted into a hole of the polypeptide chain with the vinyl groups facing the crevice. The propionic groups form hydrogen bonds with the amino acids of the polypeptide chain. The iron of the heme can

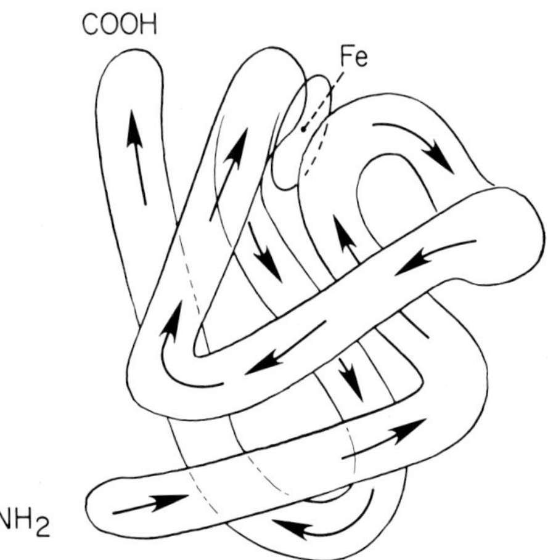

Fig. 3-6. Schematic representation of myoglobin

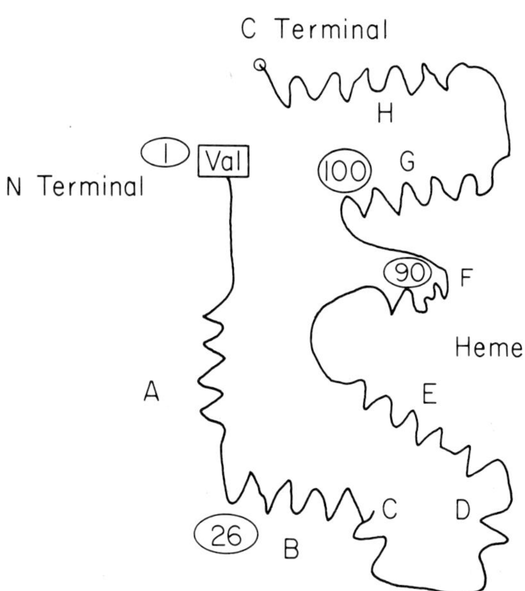

Fig. 3-7. Tertiary structure of β-chain hemoglobin

form a covalent bond with the nitrogen 1 of the vinyl residue.

Thus, among the forces that impose the tertiary structure are hydrogen, hydrophobic, ionic, and covalent bonds (including S-S bonds). Hydrophobic bonding plays a major role in stabilizing protein molecules. Since in proteins most of the nonpolar groups are tucked inside of the molecule and most of the polar groups are at the surface of the molecule, it seems that protein molecules are folded so that all hydrophobic bonds are turned inside, but all hydrophilic bonds are directed toward the medium. Now, hydrophobic side chains tend to organize the surrounding water molecule in a pentagonal or hexagonal crystallike arrangement. As a result, when two hydrophobic side chains are close to each other, the total volume of the organized water becomes smaller than that of the sum of the volumes of the organized water surrounding each molecule. Free energy is released as the total volume of organized water is decreased and greater molecular stability is achieved.

The Structure of Hemoglobin and Quaternary Structure. The tertiary structure of hemoglobin's β-chain (see Fig. 3-7) resembles that of myoglobin. It is also composed of alternating helical and nonhelical segments. It has been divided into 8 segments referred to by alphabetic letters from A to H. In spite of their similarity in tertiary structure, myoglobin and hemoglobin show little resemblance between the amino acid sequence of the polypeptide chains. The iron, which is in a crevice between E and F helices, forms a bond with histidine 92 of the β-chain. The N-terminal 6 amino acids are not helical, and the residue in position 5, starting with the 7 N-terminal amino acid, is a proline residue. The three-dimensional structure of the α-chain resembles that of the β-chain. The iron of the heme is linked covalently to the histidine of amino acid 8 in the F helix (amino acid 87 in the α-chain).

The configuration of the hemoglobin molecule presents a new dimension over that of myoglobin. Hemoglobin is a tetramere composed of two complimentary α- and β-chains, thus there are some close contacts between portions of the chains. In addition, there are contacts between the two dimers. Inasmuch as there is little interaction between the two α-chains or the two β-chains, the contacts between the two dimers involves α_1, and β_1 or α_1, and β_2. The α_1-β_1 contact involves 34 amino acids, in contrast to the α_1-β_2 contact, which involves only 19 amino acids. As may be expected, there is much more flexibility in the α_1-β_2 contact than in the α_1-β_1 contact; therefore, it is not surprising that the molecular displacements caused by oxygenation take place at the α_1-β_2 contacts.

Clinicomolecular Correlation in Hemoglobin Diseases. The hemoglobin molecule is a large, globular molecule made of 4 closely interlaced subunits, each of which contains a heme. Several types of amino acid substitutions have evolved. With respect to the overall hemoglobin configuration, one can distinguish at least 6 types of substitutions: those that occur (1) at the surface of the molecule; (2) inside one of the subunits; (3) in the heme pockets; (4) at the α_1-β_1 contacts; (5) at the α_1-β_2 contacts; and (6) inside the cavity. In most cases, how such substitutions will modify molecular function can be predicted. Substitution at the outside may lead (depending upon the type of substitution) to the formation of aggregates, to decreased solubility, or to instability. Substitutions inside the subunits lead to instability and hemolytic anemia. Substitution at the heme pocket interferes with hemoglobin oxygenation. Substitution at the α_1-β_1 contact results either in no disease or in disassociation of the hemoglobin into subunits. Substitution at the α_1-β_2 contact modifies the oxygen disassociation properties and abolishes heme-to-heme interactions. Substitution in the internal cavities between like chains leads to polycythemia, high oxygen affinity, reduced heme-heme interaction, and reduction of the Bohr effect.

Sickling Mechanisms. A critical problem is the molecular mechanism of sickling. The answer was predicted by Pauling who assumed that deoxygenated hemoglobin S molecules exhibit regions of complementariness which permit the molecules to fit together much in the way antigens and antibodies react (see Chapter 14).

Further studies established that sickling is inhibited at low temperature, a finding which indicated that the forces holding the hemoglobin molecules together were weak (electrostatic, hydrogen, hydrophobic, or van der Waals bonds).

It had been demonstrated that deoxygenation causes a rearrangement of the β-chain configuration such that the distance between the two β-chains' heme groups increases by 6.5 A. This overall change in the configuration of the S hemoglobin molecule should be sufficient to bring complementary surfaces of individual molecules close enough together that they can be linked through the weak forces described above and form fibers.

The basic unit of the fiber is a circular arrangement of six molecules of S hemoglobin with a diameter of approximately 170 A and a central core of approximately 40 A. These disks made of six molecules of hemoglobin are stacked on top of each other to form the tubular fiber. The tubular fiber is then twisted and thereby acquires a helical configuration. Viewed lengthwise, six strands of stacked hemoglobin S molecules are twisted and form a hollow tube with a total diameter of 170 A and a central cavity with a diameter of 40 A.

It is not known which portions of the surface of hemoglobin S molecules contact each other, but it would seem logical to assume that the β_6 region is involved.

If it became possible to draw an exact map of all interacting amino acids, it might be possible to explain the cogellification of hemoglobin A and hemoglobin C with hemoglobin S and the inability of hemoglobin F to affect sickling [198–199].

Therapy for Sickle Cell Anemia

Knowledge of the molecular alteration in sickle cell anemia has contributed to an understanding of the pathogenesis of the disease and has also generated some hope for therapy. Two compounds have been used in the treatment of sickle cell anemia: urea and cyanate.

Urea is believed to disrupt hydrophobic bonds supposedly formed between S hemoglobin molecules during the gellification of the abnormal hemoglobin. Although concentrations of 1–8 molar urea are usually needed to unfold a protein, unexpected success was observed in the treatment of a sickle cell anemia crisis by the administration of relatively small amounts of urea in a solution of inverted sugar. Therefore, the beneficial effects of urea on sickling did not result from urea itself, but from the cyanate normally found in equilibrium with urea in solution. The suspicion of the lack of effect of urea was confirmed by investigations carried out on the sickling of red cells *in vitro* even in presence of concentrations of urea as high as 0.2 M (concentrations which would yield, if secured *in vivo,* a BUN of 560 mg/100 ml).

Not only is there no observable effect on sickling morphologically, but there are also no significant changes in the release of potassium from the red cells. When sickling occurs, cellular permeability increases, which can be correlated with a loss of cellular potassium.

When [^{14}C] cyanate is administered to animals, most of the radioactivity is accounted for by carbamylation of NH_2-terminal residues of hemoglobin. No detectable carbamylation of lysine or cystine residues is observed. The exact mechanism by which cyanate protects against sickling is not known. Because cyanate carbamylates reduced hemoglobin, the beneficial effects of carbamylation may not result as much from blocking sickling as they do from increasing the cells' oxygen affinity. Increased oxygen affinity would then decrease deoxyhemoglobin levels, and consequently decrease sickling.

Aspirin has been known to acetylate albumin. Since the chemical blocking of the amino terminal valine chains appears to reduce the degree of sickling, aspirin was incubated *in vitro* with hemoglobin S and was found not only to acetylate the protein, but also to increase the affinity of the hemoglobin for oxygen. Whether or not acetylation occurs *in vivo* is not known.

Methemoglobinemia

There are both acquired and congenital forms of methemoglobinemia. Although the acquired is of considerable clinical significance, especially in the young child, hereditary methemoglobinemia seldom constitutes a clinical problem. Hereditary methemoglobinemia is most interesting because of its pathogenesis and therapy. There are at least three different classes of congenital methemoglobinemia [20–24]: one is characterized by the absence of methemoglobin reductase; another results from structural alteration of the globin portion of the hemoglobin; and the third group of methemoglobinemias is of unknown origin.

The enzyme defect is handed down as a mendelian recessive gene. The abnormal globin is transferred as a mendelian dominant character. The different types of methemoglobinemias illustrate, at the molecular level, a phenomenon familiar to the pathologist. In morbid anatomy and histology, identical morphological alterations in response to different injurious agents are frequently seen. Similarly, in methemoglobinemia, mutations in two different genes—the one responsible for synthesizing reductase and the one directing the synthesis of globin—ultimately result in almost identical symptoms.

Physiological Properties of Methemoglobinemia. Because hemoglobin combines with oxygen reversibly, it transports oxygen from air to tissues. Oxygen uptake is assured only if the iron of hemoglobin exists in the ferrous state (Fe^{2+}). But just as Fe^{2+} (divalent, ferrous) is oxidized to Fe^{3+} (trivalent, ferric), oxidizing agents convert ferrous hemoglobin into ferric hemoglobin or methemoglobin. The ferric iron is bound by two of its valences to the nitrogen of the pyrrole ring.

Methemoglobin has a pK value of 8.5. It acts as an indicator—it is brown in acid and red in alkaline solutions.

The replacement of the ferrous iron by a ferric iron harms the function of the respiratory pigment. Unlike hemoglobin, methemoglobin does not readily associate with oxygen. A familiar sigmoid curve is obtained when one plots the oxygen saturation in the blood versus the oxygen tension in the medium (in mm O_2). With normal hemoglobin, the saturation of the blood drops rapidly as the oxygen tension decreases. This drop indicates a rapid dissociation of oxygen from hemoglobin. With methemoglobin, the dissociation is slow, and as a result of oxygen transfer to the tissues is slowed down and hypoxia results.

The symptoms of methemoglobinemia can easily be predicted from this brief description of the properties of methemoglobin. A moderate methemoglobinemia induces dyspnea, cyanosis, and compensatory polycythemia.

Even in the normal individual, hemoglobin is spontaneously converted to methemoglobin. As a result, steady levels of methemoglobin (1%) are normally present (see Fig. 3-8).

$$\text{Hb} \xrightarrow[1\%]{99\%} \text{Hb O}_2$$
$$\downarrow$$
$$\text{MetHb}$$

An enzymic mechanism catalyzes the reduction of methemoglobin. The reduction of methemoglobin in red cells *in vitro* is accelerated when glucose or lactate is added to the incubation mixture, and pyruvate is formed in the process.

This finding indicates that the reduction of methemoglobin is linked to glucose oxidation through the glycolytic pathway. The oxidative step of glucose occurs in the triose dehydrogenase reaction and is accompanied by the reduction of NAD to yield NADH. Normally, NADH is reconverted to NAD in the lactic dehydrogenase reaction, in which pyruvic

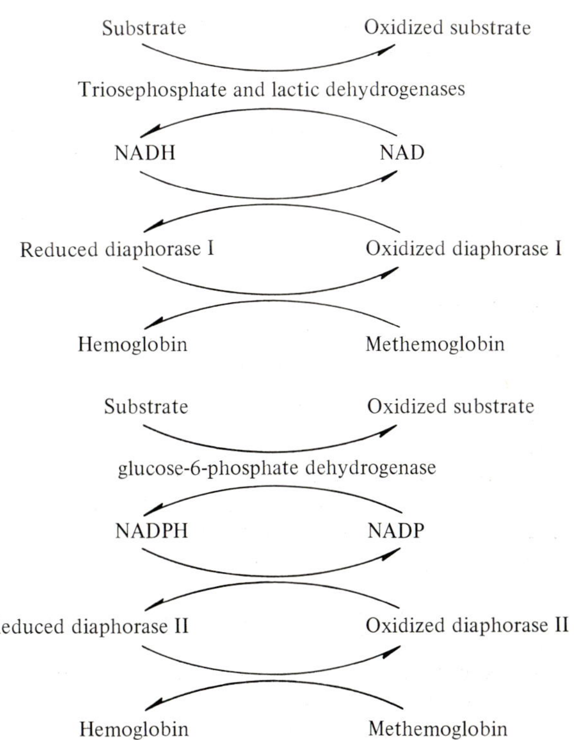

Fig. 3-8. Conversion of hemoglobin to methemoglobin

Polypeptide chains with abnormal amino acid sequences are not formed in thalassemia. The disease results from an inability to synthesize one or both α- or β-chains of hemoglobin. The abnormal hemoglobin reported in the thalassemia results from unusual chain combinations rather than from abnormal sequences.

There are two major molecular variants of thalassemia, the α- and β-type. In each type, the genetically determined block interferes with biosynthesis of either the α- or β-chain. β-Thalassemia is the most common and is associated with a reduction in A ($\alpha\beta$) hemoglobin and an increase in A_2 hemoglobin. An α_4 tetramer does not usually appear in the blood; it is too unstable. In contrast, no A_2 is formed in α-thalassemia, and tetramers of either the γ-hemoglobin Bart (γ_4) or β-hemoglobin H (β_4) may appear in the blood.

The β-chain tetramer that forms in thalassemia is useless as a respiratory pigment. It is unstable and readily oxidized within the red cell. The Heinz bodies observed in patients with α-thalassemia are thought to result from the precipitation of hemoglobin H. Heinz bodies are conspicuous in the blood of splenectomized patients, suggesting that the spleen may selectively degrade those red cells that contain Heinz bodies.

Thalassemia provides the student of disease mechanism with an excellent model for studying control of protein synthesis and eventual feedback between the synthesis of one polypeptide chain and that of others, and between hemoglobin synthesis and red cell maturation or lysis. Such studies of control and regulation of protein synthesis and erythrokinetics are still at a preliminary stage, and only fragmentary information is available. The cause of reduced hemoglobin chain synthesis in thalassemia is not clear. It could result from the deletion of one of several structural genes, or from a mutation of the structural gene that would lead to the formation of a messenger RNA that is ineffective, rapidly destroyed, or incapable of attaching to ribosomes [30–33].

Available evidence strongly suggests that at least in some cases of thalassemia the messenger RNA for either the α or the β chain is quantitatively deficient; thus, Housman *et al.* have reported ratios of α/β mRNA of $^{1}/_{6}$ in α-thalassemia and of $^{10}/_{15}$ in β-thalassemia [200, 201]. Whether this quantitative deficiency results from a structural mutation that produces an unstable mRNA [202] or gene deletion is not known. It is likely that both forms of molecular injuries occur.

A mechanism involving inadequacies of the ribosome itself seems unlikely because one would then also expect overall protein synthesis to be altered in the thalassemia cell. Yet ribosomal alteration has not been excluded.

Hemoglobin synthesis involves both polypeptide chain formation and heme biosynthesis. In normal individuals, the biosynthesis of the polypeptide chain and heme is delicately balanced, but in thalassemia that balancing mechanism is distorted [34, 35].

An illustration of such distortion is provided by studies of the simultaneous incorporation of radioactive iron and radioactive amino acids into heme and globin, respectively, in thalassemia patients. The specific activity of the globin of the hemoglobin A is lower than that of hemoglobin F, since reticulocyte ribosomes incorporate less amino acids in hemoglobin A in patients with β-thalassemia. But, surprisingly, the specific activity of the heme is also greater in F hemoglobin than in A hemoglobin.

As may be anticipated, the erythrokinetics in thalassemia are somewhat different from what they are in other types of hemolytic anemia, in that much of the red cell destruction seems to involve red cell precursors rather than mature erythrocytes. In this, thalassemia resembles pernicious anemia. In either case, it is not known what factor is involved in determining death of the immature cells, or whether the same factor is responsible for death of mature and immature cells [36].

Inborn Errors of Plasma Proteins

Anomalies or deficiencies of plasma proteins are numerous; only those that cause clinical symptoms are discussed here. Many congenital coagulation defects should be included in this study, but because the molecular nature of the defect remains unknown in many cases, they are discussed in a separate chapter.

Properties of Plasma Proteins

The importance of plasma proteins needs no emphasis. They are polyvalent electrolytes responsible for many physicochemical and biological properties of blood, such as its buffering capacity [37]. The plasma protein fraction maintains red cells in suspension and lubricates their passage through capillaries. The viscosity of the plasma proteins is responsible for the resistance to blood flow, and the osmotic effect is an important factor in regulating water distribution between blood and interstitial tissue. Antibodies and carriers of normally poorly soluble compounds (lipids, lipid-soluble vitamins, cholesterol, metal, etc., and some nutrients) are plasma proteins.

Most plasma proteins are synthesized by the liver. Liver synthesis accounts for 100% of the fibrinogen, 95% of the albumin, and 85% of the globulin found in blood. Plasma protein biosynthesis in the liver is under the influence of a variety of physiological or pathological conditions. Reduction of osmotic pressure of the blood accelerates the biosynthesis of albumin, massive loss of blood leads to acceleration of globulin synthesis, and systemic inflammation stimulates fibrinogen biosynthesis. Both liver and reticuloendothelial systems participate in the biosynthesis of α- and β-globulins. Thus, hepatectomized animals synthesize serum globulins, but not albumin. Of course, some of the plasma components found in small amounts, like insulin or thyroglobulin, are synthesized in highly specialized tissues.

As can be expected, plasma proteins have a rapid turnover. Isotopic studies have shown that the half-life of a plasma protein molecule is about 15 days, and that plasma proteins are an important source of proteins in tissues. Animals put on a protein-free diet maintain a perfect nitrogen balance for 3 months when plasma is administered to them. During starvation, plasma proteins remain intact for a long time, in spite of rapid use of tissue proteins. These facts explain the salutary effects of plasma proteins administered for chronic malnutrition, or to surgical patients who cannot be fed. Since there is no correlation between the rate of plasma and tissue protein use in malnutrition, serum protein determination is of little value in estimating a patient's nutritional status.

Human plasma contains a large population of proteins, with molecular weights ranging from 4,500,000–15,000,000. Plasma proteins are composed either exclusively of amino acids or of amino acids and lipids or carbohydrates. As already mentioned, plasma proteins may circulate independently or be conjugated to other plasma constituents, such as cholesterol or uric acid. Human plasma contains about 100 proteins, 40 of which are present in relatively large quantities and form 90% of the total plasma. The other 60 proteins constitute only 10%, and their function is frequently an enigma. Each of the second group of proteins is present in microquantities.

The level of plasma proteins in the blood changes with age. Lowest at birth, it gradually rises to reach a peak at the age of 5. The slow decrease between the ages of 5 and 25 brings the plasma protein level to normal adult values. Groups of plasma proteins have been separated by a variety of techniques, including differential precipitation at various pH's, ionic strength, ultracentrifugation, and electrophoresis. Plasma proteins can also be separated by precipitation with alcohol and by the Cohn dialysis technique. Electrophoresis, the method of separation most frequently used, is based on the anionic properties of the plasma proteins. Thus, if plasma proteins are submitted to the action of an electric current in the presence of a buffer solution (pH 7.1–8.6), the proteins migrate to the anodes at different rates, separating six principal groups of proteins. The most rapidly moving plasma proteins are the albumins, and gamma globulin is the slowest. The peak of the albumin is followed by two peaks, one named α- and the other named β-globulin. Then follows a small peak of fibrinogen, and finally the γ-globulin peak. Each peak obtained by electrophoresis is made of a population of proteins different in their amino acid sequence, but identical in mobility, and, consequently, charge. The various plasma proteins are present in the following percentages: albumin 55%; α_1 and α_2 13%; β-globulins 14%; fibrinogen 7%; and γ-globulins 11%.

Gammaglobulin, as mentioned before, forms a 98% electrophoretically homogeneous peak of slow-moving plasma proteins. If immunological methods are applied to such a preparation, at least 20 different antibodies can be detected: those for diphtheria, measles, mumps, influenza, streptococcic antitoxin, typhoid (H) agglutinin, and others. Obviously, γ-globulin preparations constitute a population of different proteins. Gammaglobulin preparations are stable in solution for years, which makes them very useful for therapy, provided, of course, that no proteolytic enzymes are present. Several inborn errors of metabolism lead to absence of plasma proteins, among them are many coagulation defects, several enzyme deficiencies, agammaglobulinemia, and Wilson's disease.

Agammaglobulinemia

Various types of hereditary agammaglobulinemia [38, 39] have been described. They will be reviewed more appropriately in the chapter on immunopathology. Only Wilson's disease and acatalasia will be discussed here.

Wilson's Disease

Wilson's disease was described for the first time in 1912. Since then, physicians have been faced with a puzzle of apparently unrelated symptoms transmitted hereditarily. Among these symptoms, cirrhosis and hepatolenticular degeneration are the most prominent. Two discoveries have provided some clues with respect to the pathogenesis of the disease. In 1930, Haurowitz observed that liver and brain of patients with Wilson's disease contained increased amounts of copper. Later, Homberg and his collaborators discovered ceruloplasmin, and subsequently demonstrated that this protein is decreased in the blood of most patients with Wilson's disease. The symptoms and pathology of the disease will be briefly reviewed, then the clinical and pathological findings will be correlated with the biochemical alterations.

The neurological symptoms are the most striking manifestations of Wilson's disease. A young child or sometimes an adult develops a tremor that resembles the tremor observed in Parkinson's disease, except that it is not an intention tremor. Later, the tremor develops into rigidity with contracture in flexion. Choreoathetoid movements involving the muscles of the limbs, articulations, and organs of speech and deglutition develop, leading to rigidity and spasticity with the open-mouth smile, distorted attitudes of the limbs, and convulsive laughter. Speech is slurred and swallowing is difficult. The neurological symptoms are always associated with mental deterioration. The patient, at first irritable or somewhat euphoric, progressively develops dementia. In cases of suspected Wilson's disease, ophthalmoscopic examination of the eyes reveals a golden brown ring behind the limbic border of the cornea called the Kayser-Fleischer ring.

Life expectancy is considerably reduced in patients with Wilson's disease. In acute cases, a child lives only a year. In the chronic, progressive, hepatolenticular degeneration, the child may survive as long as 10 years.

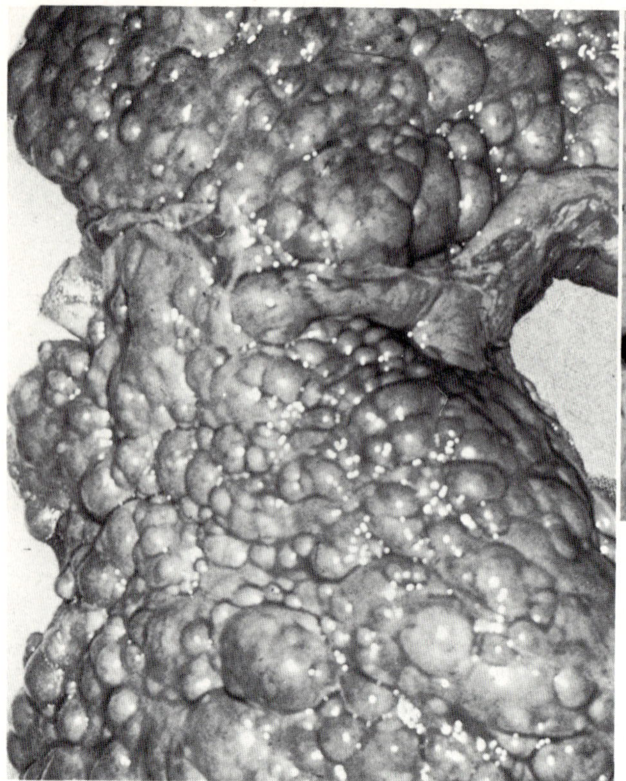

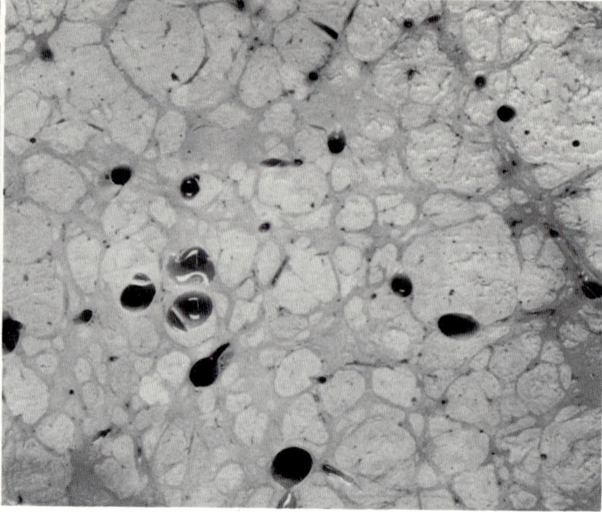

Fig. 3–10. Gross appearance of liver in Wilson's disease. Surface of liver (left) sagittal section (right)

A severe amino aciduria is usually present. All amino acids normally found in the urine, as well as 1-methylhistidine and 3-methylhistidine, are excreted in excess in advanced cases. The amino acid level in the plasma, however, remains normal [40–45].

The plasma can also be tested for ceruloplasmin by determining either direct or indirect reacting copper (see below) or by measuring the plasma's oxidative activity. Indeed, a decrease in serum copper oxidase activity is a frequent and striking abnormality in some forms of hepatolenticular degeneration.

A simple assay for the serum paraphenylenediamine oxidase present in serum was developed and proved to be useful in the early detection of hepatolenticular degeneration. It is usually advisable to examine all the children of a family in which Wilson's disease is discovered because the disease is apparently transmitted by an autosomal recessive gene and might therefore be manifest with various degrees of severity in several members of the same family.

At autopsy, increased amounts of copper are mainly in the liver and in the basal ganglia. This abnormal accumulation of metal has been demonstrated chemically and histochemically. In the liver (see Fig. 3-10), there is usually a lobular cirrhosis with pseudolobules that can hardly be distinguished from those seen in nutritional cirrhosis. In the brain although lesions of the gray matter are the most conspicuous, the subcortical white matter may also be involved.

In the basal ganglia (see Fig. 3-11), the putamen usually has a peculiar brown cast, the nerve cells are shrunken, and the nerve fibers are demyelinated. Damaged fibers in the internal capsule are responsible for part of the motor symptoms described. In longstanding cases, the substance of the putamen and the caudate nuclei is progressively lost, leading to the formation of small cavities. Often large astrocytes, called Alzheimer cells, with bizarre nuclei and a narrow rim of either pigmented or nonpigmented cytoplasm are in the affected areas. The lesions of the putamen extend beyond that structure and involve the thalamus, the body of Luys, and occasionally the nuclei of the cerebellum.

Copper is detectable histochemically using rubianic acid, which reacts with the metal to yield a black precipitate. Such techniques have shown that the copper accumulates in the hepatic cell first leaving the Kupffer cell free of copper. In the hepatic cell, at first the metal is diffused throughout the cytoplasm, but later it accumulates to form a metallic halo around the nucleus. Copper accumulation in the liver cell precedes the development of cirrhosis. The copper is widely distributed among the brain cells, but the largest concentration of metal is usually found in the basal ganglia. The type of nerve cell involved in copper accumulation in Wilson's disease has not been unequivocally identified.

Copper Metabolism. To describe the pathogenesis of Wilson's disease as currently understood, it is necessary to review briefly copper metabolism in mammals, its absorption in the intestine, its transport in the

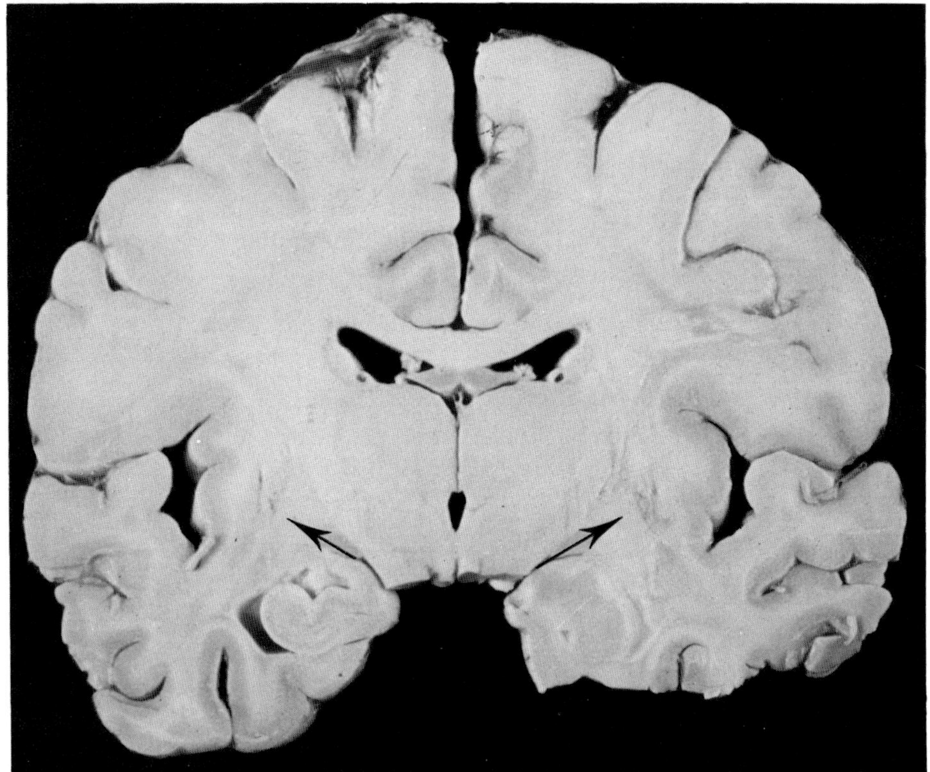

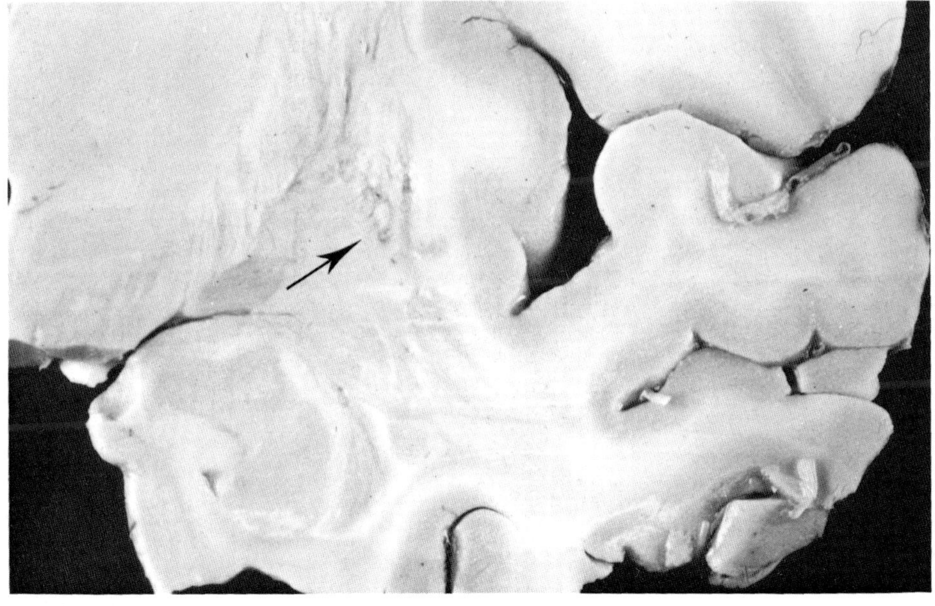

Fig. 3-11. Gross appearance of brain in Wilson's disease

blood, and its role in tissues [46]. Two factors seem to influence copper absorption in the gastrointestinal tract: the presence of molybdenum in the diet, and the chemical form of the copper present in the food. Molybdenum's role in copper absorption became clear when it was established that a cattle disease resulting from excessive molybdenum intake could be prevented by copper sulfate administration. Copper seems to be antagonistic to the intestinal absorption of molybdenum. A reciprocal effect of molybdenum on copper absorption has also been demonstrated.

High copper storage with intoxication is observed in animals fed a diet rich in copper sulfate with low molybdenum. In contrast, copper deficiency is observed in animals fed a normal-copper, high-molybdenum diet. The mechanism of interaction between the two metals is not clear because although copper administration to animals with high-molybdenum diets relieves the principal symptoms (diarrhea, scouring, and weight loss), the molybdenum concentration in the blood and tissues is not necessarily altered. Furthermore, in animals with high molybdenum intake, copper metabolism is not altered. However, the interaction seems to involve molybdenum and copper sulfate.

The importance of the chemical nature of the copper contained in food in determining the intestinal absorption of copper was demonstrated by the interesting experiments on copper contents of pasture herbage. Copper is present in herbage in at least three different forms—organic complexes not extractible by water, organic complexes readily extracted by water, and ionic copper. The stable, water-soluble copper complexes are more efficient in promoting copper storage in liver of copper-deficient rats than either the ionic copper (copper sulfate) or the water-soluble copper complexes.

These findings suggest that the copper transported through the intestinal membrane is bound to polypeptides. The exact structure of these polypeptides is not clear, but, as is mentioned later, copper polypeptides have been found in the urine of patients with Wilson's disease. The relation between intestinal absorption and urinary excretion remains to be investigated.

In blood, copper is found in both red cells and plasma. Copper of the red cells is tightly bound to a colorless protein (mol wt approximately 33,000) containing 2 atoms of copper per molecule (erythrocuprein). Erythrocuprein has an isoelectric point of 5.3 and presents no oxidative properties. Its function in the physiology of the cell is not clear, although there is a rapid exchange between the red cell copper and the plasma copper [47].

Plasma copper exists in two forms: 5% is loosely bound to albumin, and the remainder exists in the form of copper complexes. Copper forms ionic bonds with either an imidazole or a carboxyl group of the amino acid of albumin. Loosely bound copper reacts readily with dithiocarbamate, and therefore has been called the directly reacting copper. It is generally assumed but not established that the albumin that binds this copper plays an important role in transporting copper in the blood.

The remaining plasma copper is strongly bound to protein—ceruloplasmin [48–50]. Human ceruloplasmin, an α_2-globulin, was purified and crystallized. It is a blue protein with a molecular weight of about 160,000, containing 8 atoms of copper per molecule of protein, and 40% of the copper is in the cupric state. The purified protein is labile and is readily altered by copper removal or by the action of reducing agents. The exact mode of attachment of the copper to the protein is unknown. Histidyl, tyrosyl, and lysyl residues have been implicated. Although crystalline ceruloplasmin yields a homogenous peak in the ultracentrifuge and moves as a single band on starch gel electrophoresis, it is immunologically heterogeneous, and the injection of human ceruloplasmin into rabbits leads to the development of at least three antibodies. The physiological role of ceruloplasmin is not clear, but it is known to catalyze the oxidation of paraphenylenediamine. Although the exact significance of this property is not understood, it has led to a convenient assay for ceruloplasmin in the blood of patients suspected of having Wilson's disease.

Studies of the turnover of labeled porcine ceruloplasmin show that the half-life of ceruloplasmin is 3.2 days. The labeling studies also revealed that a large extravascular pool of ceruloplasmin must exist.

The copper present in the plasma is used to synthesize copper proteins in practically all tissues. Kidney, brain, and liver seem to be particularly rich in copper on a weight basis. However, because of their mass, skeleton and muscle contain about 50% of the total copper in the body.

In tissues, copper is associated either with noncatalytic proteins or with specific enzymes—oxidase, phenoloxidase, ascorbic acid oxidase, uricase, and also cytochrome oxidase (see section on electron transport). These copper enzymes usually have an absorption maximum between 320 and 330 mμ, and when exposed to oxygen, they develop a sharp absorption band at approximately 600 mμ. This intense absorption of the oxygen complex is explained by charge transfer between metal and oxygen.

Pathogenesis of Wilson's Disease. In Wilson's disease, known distortions in copper metabolism are as follows: (1) accumulation of copper in all tissues, particularly in liver, brain, kidneys, and Descemet's membrane of the cornea; (2) increased urinary excretion of copper*; (3) increased direct copper reaction (copper bound to albumin) in the plasma with a decreased indirect copper reaction (ceruloplasmin) and low serum paraphenylenediamine oxidase activity; and (4) increased intestinal absorption of copper. A number of theories have been advanced to explain the metabolic alterations in copper metabolism in Wilson's disease [51–54].

Increased absorption through the intestinal mucosa explains the high level of free copper, the accumulation of the metal in tissues, and the increased urinary excretion of copper, but it doesn't explain the low levels of indirect copper in the plasma and the low levels of ceruloplasmin. The fact that intestinal absorption of copper is increased in patients with Wilson's disease was established by oral administration of radioactive copper.

Wilson's disease is often attributed to a hereditary deficiency of ceruloplasmin. Ceruloplasmin deficiency can be demonstrated by direct colorimetry of the blue protein, by determination of phenylenediamine oxidase activity, and by immunological techniques. The role of ceruloplasmin in copper metabolism is not known, but it is assumed to serve as an acceptor of the free copper in the plasma. The changes in cerulo-

* Since rapid-screening tests for detecting copper in the urine are now available, the urinary excretion of copper should be checked in all cases of suspected Wilson's disease. Amino acid excretion will be increased when the disease exists.

plasmin concentration, however, are not necessarily responsible for Wilson's disease. The transfusion of ceruloplasmin does not affect the high copper levels of the plasma. There is no correlation between the levels of ceruloplasmin in the blood of patients with Wilson's disease and the duration and severity of the disease. Wilson's disease has been described in individuals with normal ceruloplasmin levels, but in such cases the ceruloplasmin molecule may be abnormal and may not accept "free" copper at the normal rates. However, there is no difference between the immunoelectrophoretic properties of ceruloplasmin purified from normal individuals and that obtained from patients with Wilson's disease. In any case, low ceruloplasmin levels explain the increased intestinal absorption of copper only if the increase is assumed to result from a compensatory mechanism.

Uzman and his associate [55] proposed that the disease results from a general disturbance in protein metabolism (possibly increased proteolytic activity) leading to the appearance in tissues of substances of great binding affinity for copper. The evidence in favor of this concept is twofold: first, copper peptides were found in the urine of patients with Wilson's disease; and, second, electrophoretic studies of liver proteins revealed the presence of abnormal copper proteins. This interesting hypothesis merits further confirmation. It will also be necessary to explain why abnormal copper polypeptides lead to lower ceruloplasmin levels.

Amino aciduria without cupriuria has been described in some normal relatives of patients with Wilson's disease. It is assumed that such individuals are heterozygotes, and that amino aciduria is the primary defect in Wilson's disease rather than a secondary effect due to the disturbance in copper metabolism.

Interpretation of Wilson's disease on the basis of disrupted copper metabolism is far from complete. The accumulation of copper in tissues is generally considered responsible for the lesions of the liver (cirrhosis), renal tubules (amino aciduria), and basal ganglia (neurological disorder), but the special affinity of these various tissues for copper remains to be explained, as does the mechanism of copper toxicity in these organs. Penicillamine, a copper-chelating substance, has been administered for treatment of Wilson's disease with some degree of success. In fact, it has even been claimed that asymptomatic victims of Wilson's disease can be treated preventively [56].

Some patients with Wilson's disease have episodes of acute hemolytic anemia associated with massive excretion of copper in urine. Jaundice ensues, and some patients die because of renal impairment.

Sheep and cattle ingesting large amounts of copper have a similar disease. It is believed that hemolysis results from the sudden release of copper from the tissues into the blood. Ionic copper forms a copper-protein complex with red cells, which ultimately hemolyzes.

How copper induces hemolysis is not known, but at least two mechanisms have been proposed. Copper could interfere with the transport mechanism of the erythrocyte membrane (this has been shown to occur with glycerol), or it could interfere with the activity of the glutathione reductase.

It is unlikely that the mere accumulation of copper explains all the symptoms because the experimental administration of large amounts of copper does not lead to Wilson's disease. Furthermore, in some patients with Wilson's disease and liver cirrhosis, the amount of copper on a unit basis is close to that found in fetal liver. Even if copper were responsible for the injuries described, it still would be necessary to pinpoint the metabolic steps in copper metabolism that are responsible for accumulation of the metal in the body and to describe the chain of events that leads from copper accumulation to cell necrosis.

Acatalasia

Acatalasia is a hereditary disease probably transmitted through an autosomal recessive mendelian gene characterized by the absence of catalase in the mucosa, erythrocytes, liver, muscle, and bone marrow. The disease is very rare and has been described in man only in Japan, in other countries in dogs and guinea pigs. The victim is usually completely asymptomatic. However, when the mucosa, particularly the oral mucosa, is infected by bacteria producing hydrogen peroxide, ulcerating gangrenous necrosis develops, which can usually be corrected rapidly by extracting the teeth.

Inborn Errors of Carbohydrate Metabolism

Carbohydrate metabolism involves several main pathways: (1) the interconversion of sugars, (2) the anaerobic oxidation of glucose through glycolysis, (3) the hexose monophosphate shunt and the glucuronic pathway, and (4) the aerobic oxidation of the products of glycolysis through the Krebs cycle (5), glycogen synthesis, and (6) gluconeogenesis.

Except in the red cell, few inborn metabolic errors resulting from the absence of an enzyme concerned with glycolysis or the Krebs cycle are known. These pathways are so essential to cell life that one can expect any mutations leading to metabolic distortions to be lethal early in embryonic life. The known inborn errors of carbohydrate metabolism involve glycogen breakdown or synthesis, sugar interconversion, and the hexose monophosphate shunt. But although the glycogen storage diseases, galactosemia, and glucose 6-phosphate dehydrogenase deficiency constitute major pathological problems, pentosuria and fructosuria, for example, are mainly clinical curiosities.

Glycogen Storage Disease

Glycogen storage disease refers to a group of hereditary diseases characterized by the abnormal accumulation of glycogen in a variety of tissues. The nature of the metabolic block leading to glycogen storage is known in most cases. And the chemical properties of the stored glycogen vary, depending on the chemical block. For example, the structure of glycogen is normal in von Gierke's disease, but it is abnormal in Forbes' disease.

Glycogen Metabolism

Although glycogen metabolism has been discussed previously, it might be worthwhile to review the most important metabolic steps involved in glycogen anabolism and catabolism. Glycogen is a branched polymer of glucose. Along the straight-chain portions, the glucose molecules are all linked by an oxygen atom between the carbon 1 of one molecule of glucose and the carbon 4 of another. At the branch points, the carbon 1 of the first glucose molecule of the branch is linked by an oxygen bridge to the carbon 6 of a molecule of the main stem. Inasmuch as there are two different linkages, it is not surprising that at least two enzymes are necessary for the biosynthesis of the glycogen molecule; one catalyzes the formation of the 1-4 linkages and another, the branching enzyme, is responsible for the biosynthesis of the 1-6 linkages. Similarly, glycogen catabolism requires two enzymes. Although the 1-4 linkages are cleaved by phosphorylase, the 1-6 linkages are hydrolyzed by amylo-1,6-glucosidase. The product of the phosphorylase reaction is glucose-1-phosphate, which in the presence of phosphoglucomutase is transformed into glucose-6-phosphate, a compound that may be used for glycolysis or may be further hydrolyzed by glucose-6-phosphatase to yield glucose.

Several metabolic blocks [57] cause glycogen to accumulate in parenchyma or muscle; some blocks affect glycogen breakdown and others affect glycogen synthesis. The absence of glucose-6-phosphatase, or von Gierke's disease, will be used as a model for all forms of glycogen storage disease. The other types will be mentioned only briefly.

Von Gierke's Disease

Von Gierke's disease is a rare condition transmitted by an autosomal recessive gene. (There is, of course, no evidence that the original case described by von Gierke resulted from glucose-6-phosphate deficiency, but this seems very likely on the basis of the symptoms described.) It is difficult to evaluate its frequency because not even all observed cases are reported. The disease results from a lack of glucose-6-phosphatase, which leads to increased deposition of normal glycogen in the liver and kidneys. No symptoms are observed until several months after birth, at which time the child, already presenting flabby muscles and a tendency toward obesity, fails to grow normally, develops abdominal enlargement, and vomits. The most striking symptom is a greatly enlarged liver, smooth to palpation, and extending at the inferior edge as low as the umbilicus or the right iliac crest. The spleen is not enlarged, an important point in differential diagnosis. The kidneys are considerably enlarged, but they are not palpable because of the enlarged liver. Fasting hypoglycemia and failure of the blood sugar to rise after parenteral administration of epinephrine are other important diagnostic findings in addition to hepatomegaly (in the absence of splenomegaly). A biopsy may then confirm the diagnosis.

The child usually shows a degree of dwarfism. At autopsy, the significant alterations are primarily in the liver and in the kidneys. The liver is soft, pale, and enlarged to as much as five times the normal size. On section, the normal liver lobulation is blurred, and the organ seems to contain a large amount of fat. Histological sections stained with Best's carmine or treated by the periodic acid-Schiff reaction demonstrate the presence of large amounts of glycogen. Before diagnosing glycogen storage disease, one should keep in mind that large amounts of glycogen are also found in cases of islet adenoma of the pancreas or after intravenous administration of large amounts of glucose. In addition to glycogen accumulation there is also much fatty infiltration of the hepatic cells, and occasionally the beginning of cirrhosis. Glycogen is also found in the renal convoluted tubules. None is present in the glomeruli. Small amounts of glycogen can also be demonstrated in the cardiac and skeletal muscles, in the circulating red cell, and occasionally in the myelogenic cells of the marrow.

Glucose-6-phosphatase is a microsomal enzyme that catalyzes the hydrolysis of glucose-6-phosphate to yield glucose. Found only in a few organs—liver, kidneys, and intestines—the enzyme plays an important role in maintaining the normal blood glucose level by gluconeogenesis.

A look at the metabolic map of glycogen synthesis and breakdown indicates the possibility of at least two alterations in the metabolic pathway occurring as a result of the absence of glucose-6-phosphatase: (1) prevention of normal glycogen breakdown in those organs where glucose-6-phosphatase is present, leading to accumulation of glycogen in liver and kidneys, and (2) impairment of the supply of glucose to the blood, causing starvation, hypoglycemia, and lack of response to epinephrine.

The convulsions often observed in patients with von Gierke's disease probably result from this susceptibility to develop hypoglycemia. Because glucose-6-phosphatase deficiency involves no block in glycogen synthesis or breakdown, anomalies in the glycogen structure are not expected, and the glycogen that accumulates has a molecular structure identical to that of tissues of normal individuals.

The mechanism of the ketosis and hyperlipidemia observed in von Gierke's disease is not known, and only tentative explanations can be provided. Several pathways are available for the utilization of glucose-6-phosphate: glycogen synthesis, glycolysis, oxidation through the hexose monophosphate shunt, and glucuronic acid synthesis. Normally, part of the glucose-6-phosphate synthesized in liver is returned to the circulating blood in the form of glucose after hydrolysis by glucose-6-phosphatase. The hydrolysis of glucose-6-phosphate is the main source of the glucose available to muscle. In von Gierke's disease, glycogen synthesis becomes the main pathway for glucose-6-phosphatase use, and, consequently, the muscles are deprived of their main source of chemical energy. This could lead to fatty acid mobilization from adipose tissue, acceleration of fatty acid oxidation in liver, and ketosis. These metabolic distortions may account for the fat accumulations in von Gierke's disease. Without glucose-6-phosphatase, more glucose-6-phosphate may be available to the hexose monophosphate shunt, an important source of NADPH. Fatty acid synthesis requires NADH and NADPH, and, therefore, it is possible that the rate of fatty acid synthesis and consequently of triglyceride synthesis is accelerated in liver of patients with von Gierke's disease. Such a sequence of events could thus explain the hyperlipidemia and the fatty livers of von Gierke's disease; however, experimental evidence supporting these views is not available at the present.

The retarded growth observed in von Gierke's disease could well result from the reduced levels of glucose available to peripheral tissues.

Because it was observed that portocaval transposition reduces glycogen levels in liver, such therapy has been used in treating patients with von Gierke's disease, and successful results have been reported. After portocaval transposition, the glucose absorbed in the intestinal tract is directed toward the general circulation before it reaches the liver, and as a result the amount of glucose available for glycogen synthesis is reduced in liver.

Other Forms of Glycogen Storage Disease

Pathologists and clinicians occasionally encounter forms of glycogen storage disease that cannot be classified as typical von Gierke's disease because the symptomatology is atypical, glycogen is deposited in organs other than liver, or the glycogen structure is abnormal. Enzyme activity determinations and structural analyses of glycogen in these diseases have shed some light on their pathogenesis.

Forbes [58] described a case in which the glycogen of both liver and muscles contained an abnormally short outer chain. In Forbes' disease, glycogen degradation cannot proceed beyond the action of the phosphorylase in the outer branches; thus, the defect was suspected to involve the amylo-1,6-glucosidase (see Fig. 3-12). Cori and coworkers [59] confirmed this in 1956 by directly assaying a liver fragment for amylo-1,6-glucosidase. Anomalies in the structure of glycogen similar to those of Forbes' disease were observed in certain cases of glycogen storage in muscle. Although

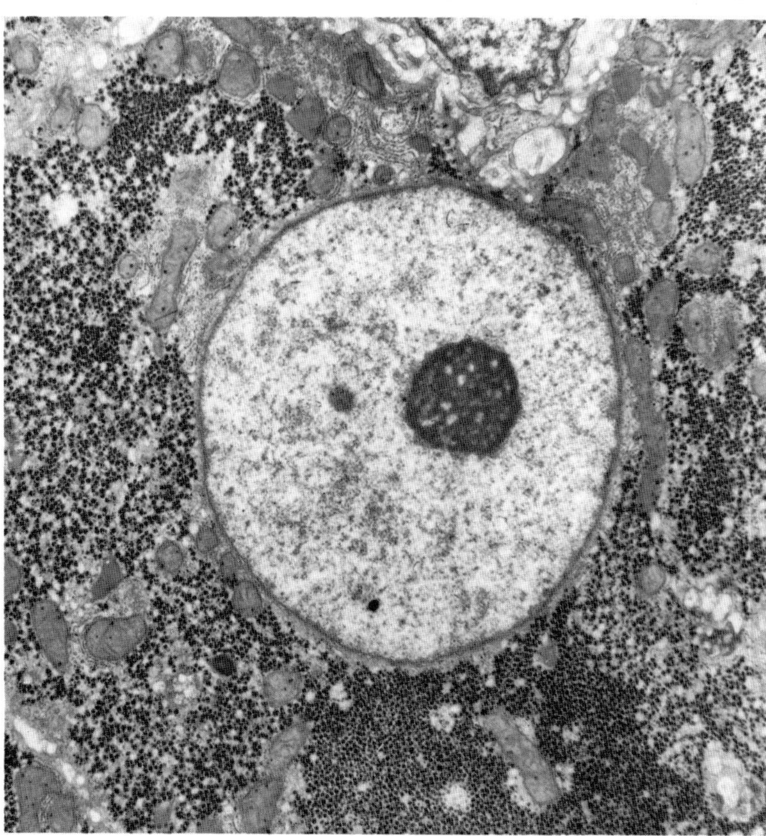

Fig. 3-12. Pools of cytoplasmic granules in liver cell ($\times 8{,}000$) of patient with Type III glycogenosis (Zamboni)

the same enzyme block was postulated, there is no direct evidence for such an explanation. In glycogen storage disease in muscle, the patient is weak and has an enlarged heart and tongue due to glycogen accumulation in the muscle fibers. The skeletal muscles are pale and flabby but have a normal volume. Such forms of glycogen disease are also hereditary; however, little is known of the mode of transmission of these diseases since only few cases have been reported.

In Andersen's disease there is a generalized glycogen infiltration with liver cirrhosis. The organs most affected by the glycogen deposition are the liver, spleen, lymph nodes, and intestinal mucosa. Small amounts of glycogen are also found in the diaphragm and kidneys. The structure of the glycogen is abnormal; 50% of the molecule is degraded by phosphorylase, indicating that the external chains are abnormally long, and that the branch points are relatively few. It is therefore postulated and in some cases it has been shown that the disease results from an absence of the branching enzyme.

Phosphorylase defects have been observed in muscle and liver. Multiple defects have been observed in glycogen storage disease: for example, deficiency in glucose-6-phosphatase combined with either phosphorylase or amylo-1,6-glucosidase deficiency. McArdle's disease is a mild myopathy characterized by a minimal accumulation of glycogen in muscle and resulting from the absence of phosphorylase. In McArdle's disease, liver phosphorylase activity is present in normal amounts, which indicates that the genes controlling myophosphorylase must be different from those controlling hepatophosphorylase.

In Hers' disease, glycogen accumulates in liver, and only small amounts of hepatic phosphorylase (25% of normal) activity are present. This suggests that the concentration of phosphorylase is rate limiting for glycogen breakdown. The pathogenesis of the disease in the cases described by Hers is not settled for at least two reasons: (1) the method used to determine phosphorylase did not distinguish between active and inactive enzymes, and (2) the levels of phosphorylase activity vary considerably among normal individuals. In some of the cases described by Hers [57], the phosphorylase activities fall within the normal range. However, since Hers' original report, several papers in the literature have described patients with increased hepatic glycogen of normal structure and with normal levels of glucose-6-phosphate activity. Liver phosphorylase activity was one-sixth of normal. Furthermore, low phosphorylase activities were found in circulating leukocytes [56].

Another form of glycogen storage disease with low phosphorylase activity was described by Hers and his associates. In this patient, the simple addition of phosphorylase B kinase to the liver homogenates restored phosphorylase activity to normal. Therefore, it was established in that case that the low phosphorylase activity resulted not from a decrease in the molecular concentration of active phosphorylase or a molecular alteration of phosphorylase, but from a lack of phosphorylase kinase. The conversion of active to inactive phosphorylase is illustrated in Fig. 3-13.

Pompe's disease is a generalized form of glycogen storage disease with predominant accumulation in the heart or in the neuromuscular system. The symptoms become manifest soon after birth. In the heart form,

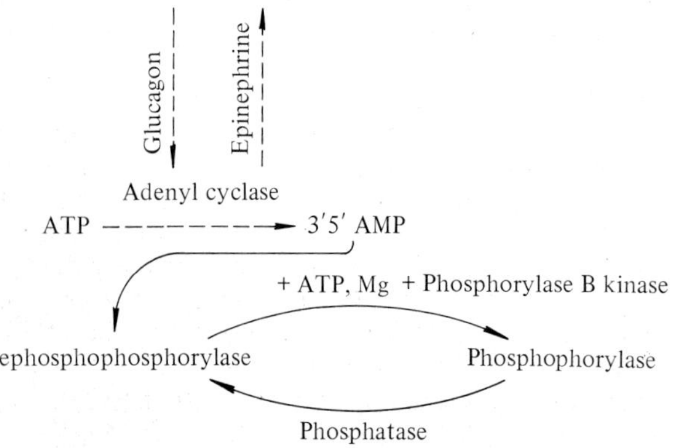

Fig. 3-13. Conversion of inactive to active phosphorylase

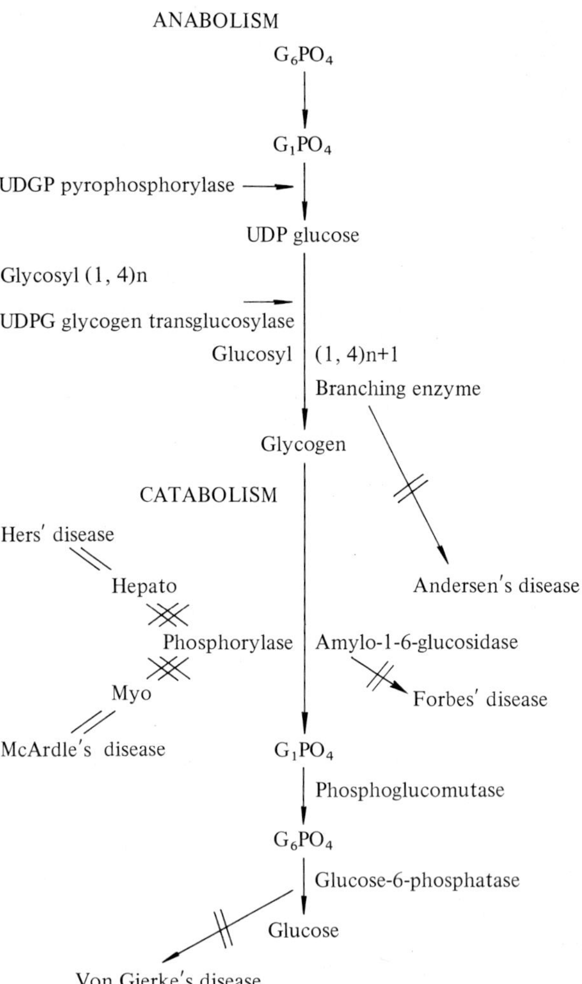

Fig. 3-14. Pathogenesis of glycogen storage diseases

the infant eats poorly, does not gain weight, and soon has crises of cyanosis and polypnea. He dies during the first two years of life from heart failure. Clinically, the main finding is cardiomegaly and electrocardiographic alterations. At autopsy, the heart is found to be loaded with glycogen, which is deposited in a lacework pattern throughout the myocardium. In the neuromuscular form, muscular hypotonia and mental retardation develop. Many of the viscera—especially liver and kidneys—contain glycogen. The glucose-6-phosphatase of liver and kidney is normal, and the glycogen deposited is structurally normal.

Because in the liver glycogen accumulations are sometimes surrounded by a single membrane, it has been suggested [60] that Pompe's disease is a lysosomal disease, and that glycogen accumulation results from the absence of a specific lysosomal enzyme involved in glycogen breakdown—namely, acid α-glucosidase (see Chapter 10, Necrosis). The enzyme deficiencies in the various types of glycogen storage disease are summarized in Fig. 3-14.

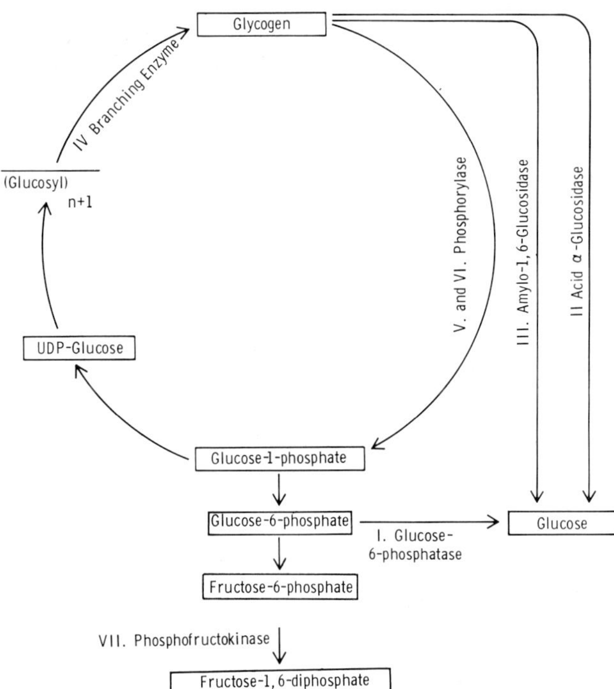

Fig. 3-15. Types of glycogen storage diseases

Table 3-3. Classification of glycogenosis

Type	Enzymic defect	Principal site of glycogen storage
I	Glucose-6-phosphatase	Liver, kidney
II	Acid α-glucosidase	Generalized disease
III	Amylo-1,6-glucosidase	Liver, muscle (erythrocytes)
IV	Branching enzyme	Liver
V	Phosphorylase	Muscle
VI	Either no detectable lesion or low hepatic phosphorylase	Liver
VII	Phosphofructokinase	Muscle

Recently, attempts have been made to simplify the classification of glycogen storage diseases by abolishing the use of eponyms and replacing them with Roman numerals, as is indicated in Table 3-3 and Fig. 3-15. As the table shows, all cases in which no detectable lesion or low hepatic phosphorylase activity was found were pooled in a single group. As a result, this becomes the group in which most cases are found, but it is very unlikely that the pathogenesis of glycogen storage is the same in all these forms of the group.

Galactosemia

History and Clinical Pathology

Galactosemia, accompanied by galactosuria, results from an inherited inability to use galactose. Not until the intermediary metabolism of galactose was elucidated was the metabolic block involved in galactosemia understood. The reader is already familiar with the kinase, transferase, and epimerase reactions of galactose metabolism, and with the role of UDP-glucose as a necessary coenzyme. The reactions involved are summarized in Fig. 3-16. Schwarz and coworkers [61] observed in 1956 that galactose-1-phosphate is a normal metabolite that accumulates in the blood of galactosemic patients. This suggested a block beyond the kinase reaction. Since it was later discovered that both epimerase and the pyrophosphorylase were present in normal amounts in galactosemic patients, a block of the transferase reaction was suspected. This was elegantly demonstrated by Kalckar and his associate [62]. These investigators showed a reduced transferase activity in liver homogenates and umbilical cord blood in galactosemia, and their work also suggested that this reduction is not due to an inhibitor, since the mixing of normal blood with galactosemic blood does not affect transferase activity in normal blood. Furthermore, they found that the rate of transferase activity cannot be correlated with the severity of the disease.

The untreated patient, clinically normal at birth, starts to vomit and lose weight as soon as he is given human or cows' milk, and then usually refuses to eat. At that time, a test of the urine for galactose reveals galactosuria. If the child is fed on a galactose-rich diet, he will develop hepatomegaly, polyuria, aminoaciduria, and acetonuria. Subsequently, the child either dies or recovers and many of the symptoms disappear. However, he is by no means cured because the galactosuria and galactosemia persist; the victim is retarded both physically and mentally, and eventually develops cirrhosis of the liver and cataracts. By contrast, complete regression of symptoms occurs if the child is given a lactose-free diet early enough in the course of the disease. Of course, the clinicopathological syndrome recurs as soon as the young child is again exposed to sources of galactose. However, a small

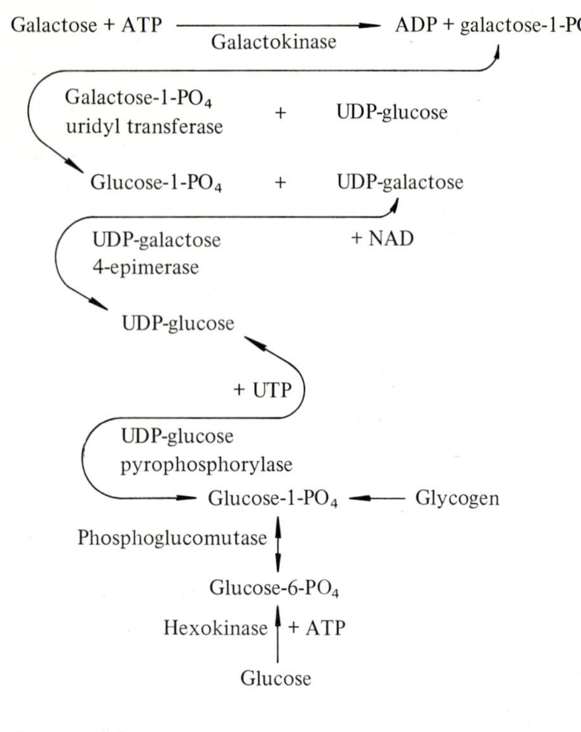

Alternative Pathway

UTP + Galactose-1-PO_4 ⟶ UDP-galactose + PP

UDP-galactose
Pyrophosphorylase

Fig. 3-16. Pathways involved in galactose synthesis

amount of galactose can be adequately metabolized by the child at a later age.

Pathogenesis

The pathogenesis of injury to the liver (fatty liver or even cirrhosis), nervous system (mental deficiency), kidney (albuminuria, aminoaciduria), and lens (cataracts) occurring in galactosemia must be explained. It was formerly assumed, without much evidence, that galactose itself is responsible for the liver damage and cataracts, but more sophisticated explanations have been proposed by biochemists [63, 64]. It is now accepted that galactose-1-phosphate is the metabolite exerting a toxic effect in galactosemia. A mechanism explaining most of the symptoms observed in galactosemia has been proposed: competitive inhibition of phosphoglucomutase.

Various theories have been proposed to explain the development of hypoglycemia in galactosemic patients: (1) an insulinic effect of galactose accumulated in the blood (however, there is no correlation between the degree of hypoglycemia and the severity of the galactosemia); (2) a block in glucose-6-phosphatase activity; and (3) a block in phosphoglucomutase activity leading to a reduction in glucose-1,6-diphosphate.

A competitive inhibition of crystalline phosphoglucomutase obtained from rabbit muscle can be demonstrated when galactose-1-phosphate is added to the system. Some of the intricacies of the phosphoglucomutase reaction are known. The enzyme is a phosphorylated molecule that reacts with the hexose monophosphate. During this reaction, the phosphate attached to the enzyme is transferred from the protein to hexose monophosphate to yield a hexose diphosphate, glucose-1,6-diphosphate. In the second step of the reaction, the protein is rephosphorylated and a monophosphate is formed, this time the glucose-6-phosphate. Although either galactose-1-phosphate or galactose-6-phosphate may react with the enzyme to yield the galactose-1,6-diphosphate and the dephosphoenzyme, the 1,6-diphosphate does not react readily with the dephosphoenzyme. Thus, even in presence of 1,6-diphosphogalactose, the dephosphoenzyme can no longer act unless large amounts of glucose-1,6-diphosphate are present in the system to rephosphorylate the dephosphoenzyme. The levels of glucose-1,6-diphosphate are low and will rapidly be exhausted in the presence of large amounts of galactose-1-phosphate.

Two reactions in mammalian tissues are capable of generating glucose-1,6-diphosphate. One is a kinase reaction activating glucose-1-phosphate in the presence of ATP to yield glucose-1,6-diphosphate and ADP. The other is a transphosphorylase reaction, which in the presence of two molecules of glucose-1-phosphate yields one molecule of glucose and a molecule of glucose-1,6-diphosphate. Thus, in this reaction, the phosphate is transferred from the position 1 of a glucose molecule to the position 6 of another. Neither of these reactions is known to yield large pools of glucose-1,6-diphosphate.

An inhibition of phosphoglucomutase could explain many of the symptoms observed in galactosemia, including jaundice (indirect bilirubinemia), mental deficiency, cataracts, and hypoglycemia. In the absence of phosphoglucomutase, the glucose-6-phosphate produced in the hexokinase reaction at the expense of glucose is no longer converted to glucose-1-phosphate, and the amount of UDP-glucose is therefore reduced. UDP-glucose is an important precursor or coenzyme in the biosynthesis of cerebrosides from UDP-galactose, and in the biosynthesis of lens proteins at the expense of UDP-glucosamine, UDP-galactose, and UDP-mannose.

UDP-glucose is necessary for the conjugation of bilirubin to form bilirubin glucuronide. Consequently, reduction in the levels of UDP-glucose could explain the increased indirect van den Bergh reaction and the jaundice observed in galactosemia.

Cerebrosides constitute 4–6% of the total brain and 16% of the white matter. They contain one molecule of 1-galactose per molecule of cerebroside. Cerebrosides are present only in low concentrations in the brain of the newborn, but they are rapidly synthesized during brain development, and their synthesis parallels that of myelin. In absence of UDP-glucose, the precursors of cerebrosides are not made, and, therefore, mental retardation observed in galactosemia could result from the inability to synthesize basic brain components.

A similar reasoning might apply to the development of cataracts. It has been proposed that the opacification of the lens observed in galactosemia results from an interference with the biosynthesis of lens glycoproteins. It was shown that the development of cataracts is associated with a reduction in the soluble fraction of the lens proteins, but there were no changes in the proportion of lens-insoluble proteins. In interpreting these results, investigators suggested an interference with the biosynthesis of the precursors of lens proteins.

A "precursor-product" relationship between soluble and insoluble lens proteins remains, however, to be established. Significantly, lens opacification in galactosemic patients starts in the anterior suture line where the glycoprotein concentration is highest. Other mechanisms have been invoked to explain cataractogenesis in galactosemia. The uptake of galactose and glucose by the lens is competitive. Thus, in the presence of excess galactose, the amount of glucose entering the lens is reduced. Since the lens does not possess the required enzymic machinery to use galactose, glucose remains the main source of chemical energy, and, therefore, the rate of the bioenergetic pathways is bound to be reduced. The synthesis of ATP is indeed

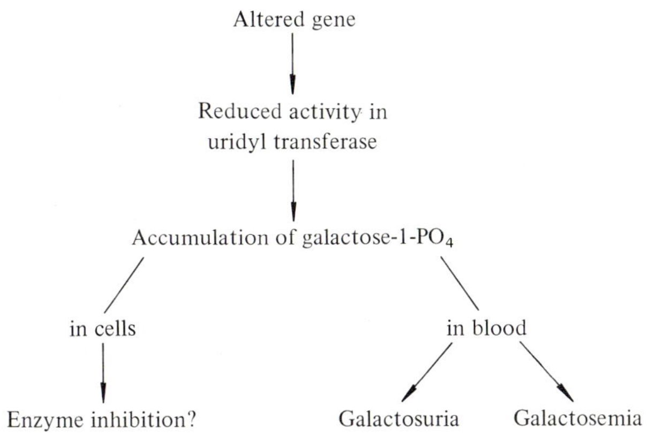

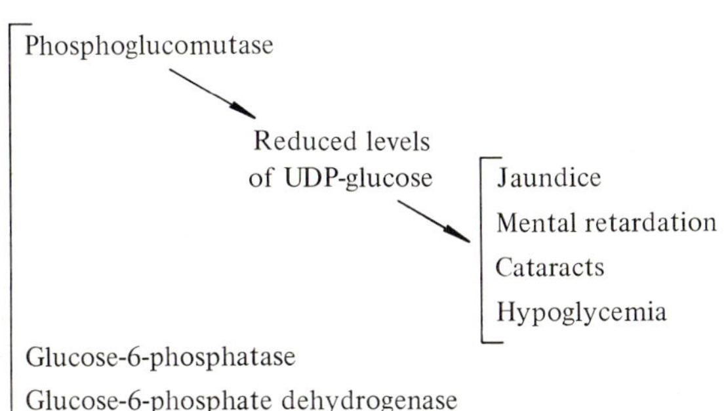

Fig. 3-17. Pathogenesis of galactosemia

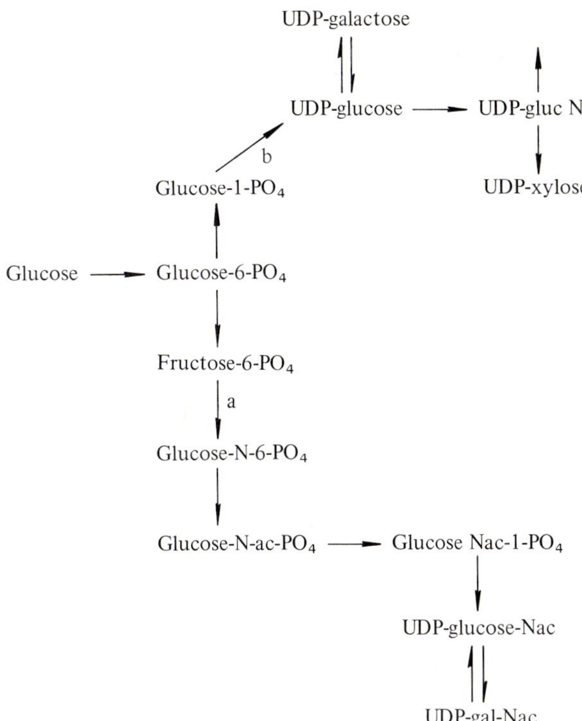

Fig. 3-18. Regulation of galactose metabolism. *a* Feedback inhibited by UDPN acetylglucosamine; *b* Feedback inhibited by UDP-xylose

impaired in the lenses of galactosemics. It has also been proposed that galactose-1-phosphate inhibits the glucose-6-phosphate dehydrogenase activity of the lens, thereby reducing ATP and NADPH production. This hypothesis is difficult to reconcile with observations made on leukocytes, in which galactose-1-phosphate is without effect on the hexose monophosphate shunt. Fig. 3-17 summarizes the pathogenesis of the most common type of galactosemia.

In addition to the classical galactosemia, there is a form of transferase deficiency that seems to result from alteration of the regulation (see Fig. 3-18) of the enzyme biosynthesis. Thus, in this form of galactosemia (referred to as the Duarte form), the homozygous have 50% and the heterozygous 75% of the normal levels of transferase activity [65]. Another form of galactosuria has been described in a few cases with normal transferase and epimerase activity but low galactokinase [66, 67].

Fructosuria

Fructosuria is a disease characterized by the excretion of fructose in the urine. There are two types of fructosurias: a benign form with no symptoms; and a form associated with fructose intolerance. Both forms are transmitted as mendelian recessive traits.

In the benign form, fructose administration results in the accumulation of fructose in the patient's blood. Inasmuch as most fructose is not metabolized, there

one and glucose-6-phosphate dehydrogenase have been compared in patients affected with favism and those with primaquine sensitivity, the results have been the same. In addition, patients with favism are also sensitive to primaquine.

The gene of glucose-6-phosphate dehydrogenase is carried on the X chromosome. The female has two X chromosomes, but the gene stored in only one of them is expressed. Therefore, there are two populations of red cells in victims of glucose-6-phosphate deficiency: one that carries the deficiency and another that is normal. This rare situation has been taken advantage of for the study of tumor origin.

Inborn Errors of Aromatic Amino Acid Metabolism

Despite the relatively few steps in intermediary metabolism of the aromatic amino acids, a large number of metabolic errors have been reported. Some, like phenylketonuria, lead to dramatic clinical situations, whereas others, like alkaptonuria, are less clinically important, and still others, like tyrosinosis, have no clinical significance. The reasons for many errors in a given pathway are not understood. Metabolic errors may be as common in other metabolic pathways, but they are not observed either because they do not lead to clinical alterations or because they are lethal. However, the genes concerned with the appearance of each enzyme involved in a specific metabolic pathway may be located on adjacent loci in the chromosome, and these loci are perhaps more prone to spontaneous mutation.

Phenylketonuria

Phenylketonuria or phenylpyruvic oligophrenia is a disease transmitted as a typical mendelian recessive character and manifested by a specific biochemical and neurological syndrome. Biochemically, phenylketonuria is characterized by increased levels of phenylalanine in the blood, cerebrospinal fluid, and urine. In addition to phenylalanine derivatives, a few indole derivatives are found in the urine of phenylketonuric patients.

Most patients continuously excreting phenylpyruvic acid and other related metabolites in the urine are feeble minded; usually they fall into the class of idiots or imbeciles. However, they show no paralysis or increase in muscular tone, despite markedly accentuated superficial and deep reflexes.

Phenylalanine Metabolism. As the metabolic map in Fig. 3-20 shows, phenylalanine is an essential amino acid located at the crossroads of various metabolic pathways. In the normal individual, it is used mainly for hydroxylation to tyrosine, a reaction catalyzed by phenylalanine hydroxylase. The conversion of phenylalanine to tyrosine was demonstrated by the isotopic experiments of Moss and Schoenheimer [70], who added deuterium-labeled DL-phenylalanine to the diet and recovered the label in L-tyrosine. Phenylalanine hydroxylase has been found in the soluble fraction of mammalian liver. The mechanism of the reaction is not completely known. It was first established that the enzyme is composed of two parts: a labile Fraction I found only in liver; and a stable Fraction II also found in kidney and heart. Pyrimidine nucleotides were required for the hydroxylation. When crude preparations were used, NADH was the most effective hydrogen donor, but in purified preparations, NADPH was more effective than NADH. In a crude system supplemented with NADP, both Fraction I and Fraction II were required for the hydroxylation of phenylalanine [71, 72].

Kaufman and his associates [73] reinvestigated the role of the two enzyme fractions in phenylalanine hydroxylation. The reaction was carried out *in vitro* in the presence of extensively purified rat and sheep liver enzymes. Two cofactors were necessary for the overall reaction: a pteridine and NADPH. Dihydropteridine [74, 75] has been established as the cofactor of phenylalanine hydroxylase (see Fig. 3-21).

The authors demonstrated further that tyrosine was readily formed from phenylalanine in the presence of the liver enzyme only, provided that adequate amounts of pteridine were available. Thus, the sheep enzyme was not directly involved in the hydroxylation reaction, but simply restored the pteridine in a reduced form after it had been oxidized in the hydroxylation reaction.

The phenylalanine oxidase reaction illustrates oxidation of cyclic compounds in higher animals. Although most oxidation occurs by removal of hydrogen atoms (or hydride ions), the oxidation of cyclic compounds often results from the direct addition of molecular oxygen. Thus the reaction can be written as follows:

$$\text{phenylalanine} + {}^1/_2 O_2 \rightarrow \text{tyrosine}$$

In this reaction, a tetrahydropteridine acts as a specific electron donor, and oxygen acts as the ultimate electron acceptor.

The enzyme has been purified from rat and human livers. The properties of the human enzyme are essentially similar to those of the rat enzyme [204, 205]. Phenylalanine hydroxylase has also been purified from rat kidney. Available methods of investigation suggest that the kidney enzyme is similar to that of liver [206]. The human liver enzyme exists in a soluble and particulate form. The particulate and soluble enzymes have identical pH optima and K_m's apparent for phenylalanine and 2-amino-4-hydroxy-6,7-dimethyl-5,6,7,8-tetrahydropteridine [207].

The particulate enzyme can be released in the supernatant by treatment with bile salts or by sonication. This suggests that the enzyme is bound to a membrane by a lipoprotein. A variety of substances are capable of stimulating phenylalanine hydroxylase activity: α-chymotrypsin, lysolecithin, and a protein factor which

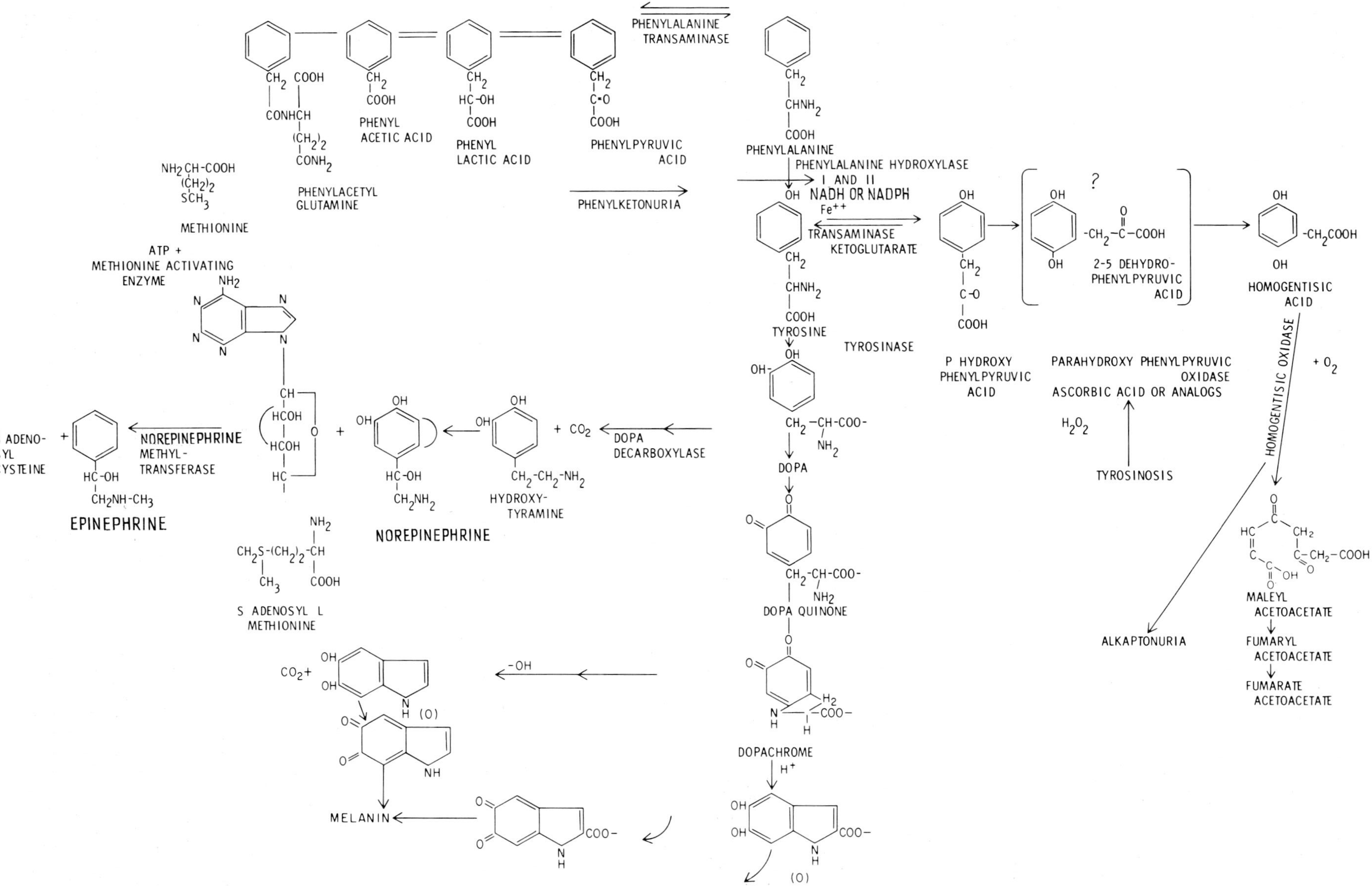

Fig. 3-20. Pathology of the intermediary metabolism of the aromatic amino acids

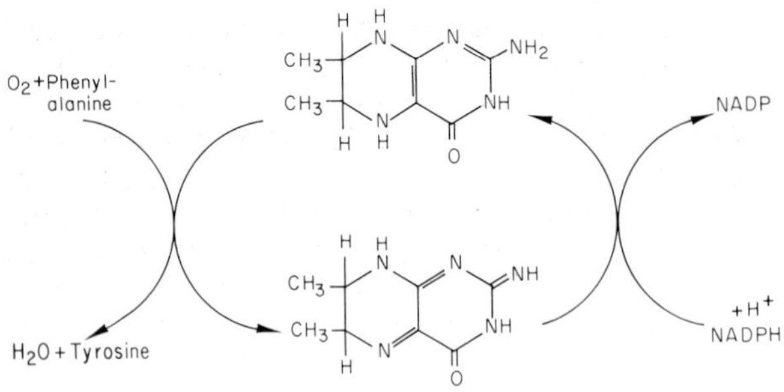

Fig. 3-21. Oxidation of tetrahydropteridine to dihydropteridine in the phenylalanine hydroxylase reaction

is believed to be an enzyme that breaks down an intermediate released during the hydroxylase catalyzed reaction [208]. In no case has the exact mechanism of stimulation been clarified.

It has been demonstrated that phenylalanine hydroxylase, found in liver, is absent in phenylketonuria.

Phenylalanine hydroxylase is also capable of hydroxylating tryptophan. Insofar as decarboxylation of 5-hydroxytryptophan is well known, all steps of serotonin biosynthesis would seem to be known. The role of phenylalanine hydroxylase in tryptophan metabolism is not clear, however, because available evidence is confusing. For example, in patients with carcinoids, the serotonin level in the blood is a hundred times the level in normal blood, but this is not accompanied by an increase in phenylalanine hydroxylase activity in the intestine. (This discrepancy between the enzyme activity and the amounts of serotonin produced in the tumor should not be distressing; quite possibly, the concentration of phenylalanine hydroxylase is not rate limiting for tryptophan hydroxylation.) In contrast, serotonin levels are decreased in phenylketonuria, which, at first approximation, leads one to suspect that phenylalanine hydroxylase plays a physiological role in the hydroxylation of tryptophan. However, a more detailed examination of phenylketonuric patients argues against this interpretation. First, the reduction in serotonin levels is only about 40%, which means that large amounts of tryptophan are still hydroxylated. Yet, phenylalanine, a compound with a much greater affinity for the enzyme than tryptophan, is present in the blood in large excess and could be expected to totally inhibit the formation of serotonin.

Tyrosine is either used for the biosynthesis of proteins, thyroxine, epinephrine, or melanin, or catabolized to yield fumaryl acetoacetate. The biosynthesis of proteins and thyroxine is discussed elsewhere; this discussion is restricted to epinephrine and melanin synthesis and tyrosine catabolism. Dopa 3,4-dihydroxyphenylalanine is an intermediate common to epinephrine and melanin. To yield epinephrine, dopa is first decarboxylated by an enzyme called dopa decarboxylase. This enzyme is present in several mammalian tissues, including the adrenal medulla, where the reaction yields hydroxytryptamine chloride. From this point the pathway is unknown, but norepinephrine has been postulated as an intermediate.

Dopamine β-oxidase is an enzyme catalyzing the hydroxylation of dopamine to norepinephrine. However, its specificity is not restricted to dopamine. Several compounds (phenylethylamine derivatives) can act as substrate: epinine, M-tyramine, and a branched α-methyl derivative, methoxytyramine. Many sympathomimetic drugs are good substrates for the dopamine oxidase reaction. The products of the reaction are pharmacological compounds with sympathomimetic activities higher than that of the original substrates. Therefore, dopamine oxidase could be important in determining the pharmacological activities of the sympathomimetic drugs.

Norepinephrine is methylated in the presence of a specific enzyme (S-adenosylmethionine transferase) and a cofactor (S-adenosylmethionine) to form epinephrine. This reaction is analogous to the methylation of guanidinoacetic acid, which occurs in liver. Two distinct enzyme reactions are involved: methionine is converted to S-adenosylmethionine in the presence of a methionine-activating enzyme and ATP, in which reaction all the phosphates of ATP are lost, the terminal phosphorus of ATP is liberated as inorganic phosphate and the two internal phosphoryl groups yield pyrophosphate; and (2) norepinephrine is then methylated in the presence of a specific enzyme S-adenosylmethionine transferase to form epinephrine.

Our limited knowledge of the biosynthesis of melanin leaves us much in the dark. However, it is fairly well established that the reaction starts with the hydroxylation of tyrosine in the presence of tyrosinase to yield 3,4-dihydroxyphenylalanine or dopa. Mammalian tyrosinase has not been purified, but a similar enzyme has been prepared from potatoes. The protein preparation contains copper, and after extensive dialysis against cyanide, the copper is lost and with it the catalytic properties of the enzymes. The mechanism by which the copper participates in the catalytic reaction is not entirely elucidated, but it is generally accepted that during the reaction, a cuprous form of the enzyme is transformed into a cupric form, which is in turn reoxidized by oxygen. The reaction of tyrosinase (a polyphenoloxidase) with tyrosine (a monophenol) is rather sluggish, but the addition of ascorbic acid or dopa to the reaction mixture reduces the lag period. Melanin is thought to be a polymer of indole 5,6-quinone, a compound derived from dopa through a complex series of reactions not yet elucidated. Except for the first step in the chain of reactions, in which a specific dopa oxidase is probably involved, the enzymes responsible for the reaction are unknown, and it is believed that in the potato some of the steps may even be nonenzymic.

In the presence of α-ketoglutarate and pyridoxal phosphate, tyrosine loses its amino group to yield hydroxyphenylpyruvate. The enzyme involved in that reaction—tyrosine transaminase—has been extensively purified from dog liver, and some aspects of its specificity have been established. Only α-ketoglutaric acid will act as acceptor of the amino group. The donors require a phenolic hydroxyl, an α-amino group and an α-carboxyl group. Furthermore, only amino acids of the L-type are attacked. Copper is closely associated with transaminase but is not required for enzyme activity. The product of the reaction, parahydroxyphenylpyruvate, strongly inhibits the transaminase reaction. In addition to its interesting role in aromatic amino acid metabolism, tyrosine transaminase is an inducible enzyme that can be elicited in rodents by administering tyrosine, hydrocortisone or a mixture of these two biochemicals. The enzyme activity is also stimulated by a variety of other components, such as RNA, ATP, DL-methionine, DL-aspartate, chondroitin sulfate, heparin, and a mixture of nucleotides. Stimulation by nonspecific agents requires hydrocortisone. Stimulation by tyrosine, hydrocortisone, or by some other nonspecific compound is inhibited by ethionine. The ethionine inhibition can be reversed by methionine. The inductions both by hydrocortisone and tyrosine are sensitive to protein depletion. It is not clear whether the tyrosine and hydrocortisone inductions of tyrosine transaminase activity are the same. Indeed, the former is more sensitive to protein depletion than the latter.

The next metabolic step involves the transformation of P-hydroxyphenylpyruvate into homogentisic acid. This reaction is accomplished in two steps: (1) side chain migration and the addition of the hydroxyl group to the ring; and (2) oxidative decarboxylation of the side chain. Ascorbic acid or ascorbic acid derivatives activate the system in an unknown manner. A study of the enzymes involved in parahydroxyphenylpyruvic oxidation demonstrated that two protein fractions, neither of which is active alone, are needed. One of the protein fractions was identified as catalase; however, the formation of hydrogen peroxide during the reaction was not established.

Parahydroxyphenylpyruvate oxidase is one of the enzymes studied in the course of the development of rats. The enzyme activity is low in guinea pigs and rats *in utero*, but increases soon after birth to reach adult levels slowly in the rat and rapidly in the guinea pig. The absence of enzyme activity in the fetal liver is not due to a lack of the enzyme molecule, but to the absence of reducing agents, particularly glutathione. Thus, by preincubating the enzyme with 2,6-parachlorphenol, indophenol, or glutathione, one can increase the activity in the fetal liver several times. No such increase is observed if similar preincubation techniques are applied to preparations from adult liver. Similarly, the increase in tyrosine α-ketoglutarate transaminase activity in rats soon after birth does not result from a *de novo* synthesis, but from activation of preformed inactive fetal enzyme.

Homogentisic acid is further oxidized in another reaction which consumes one molecule of oxygen and requires the presence of both ferrous ions and sulfhydryl compounds to yield maleylacetoacetate. The latter compound can be transformed into its isomer in the presence of a transisomerase to yield fumarylacetoacetate. This is an irreversible reaction catalyzed by a sulfhydryl enzyme in the presence of glutathione.

In addition to this main degradation pathway of phenylalanine via tyrosine, there is an alternative reversible pathway via phenylpyruvic acid, of which only the intermediates are known. In normal as well as in phenylketonuric individuals, phenylalanine and phenylpyruvic and phenyllactic acids are interconvertible.

Metabolic Alterations. Many of the biochemical anomalies of phenylketonuria can be explained by the inhibition or absence of phenylalanine hydroxylase. The evidence in favor of this mode of pathogenesis is both direct and indirect. The indirect evidence is based on the metabolic experiments of Jervis, who administered a variety of amino acids to phenylketonuric patients and demonstrated that only phenylalanine increases the urinary excretion of phenylalanine and all its derivatives: phenylpyruvic, phenyllactic, and phenylacetic acids. These experiments suggest that the block in the phenylalanine-tyrosine pathway is located beyond phenylalanine. Since tyrosine administration to phenylketonuric patients did not produce such effects, the block was assumed to be located between phenylalanine and tyrosine. This was confirmed by isotope experiments. When labeled phenylalanine is administered to normal patients, radioactivity is soon recovered in tyrosine and in proteins. In phenylketonuric patients, most of the radioactivity is recovered in the urine, and practically none is found in the protein. The administration of phenylpyruvic acid and phenylacetic acid demonstrated that in the phenylketonuric patient there is no block in the conversion of these compounds into phenylalanine. However, the most conclusive evidence was obtained when Jervis studied autopsy specimens from normal and phenylketonuric individuals and demonstrated that phenylalanine hydroxylase activity is practically nonexistent in phenylketonuric patients [76].

Pathogenesis of Symptoms. This primary metabolic alteration explains most of the clinicopathological manifestations in phenylketonuric patients [77, 78]. A block in the hydroxylation of phenylalanine to tyrosine leads to an accumulation of phenylalanine in the blood and cerebrospinal fluid. When the concentration in the blood rises above the renal threshold for phenylalanine, phenylalaninuria occurs. In addition, since the normal oxidative pathway of phenylalanine through tyrosine is blocked, the patient apparently compensates for the inability to hydroxylate phenylalanine by increasing the transamination process and transforming phenylalanine to phenyllactic, phenylacetic, and phenylpyruvic acids. Since most of the phenylacetic acid is conjugated to glutamine to form phenylacetoglutamine, this glutamine derivative should be found in the urine of phenylketonuric patients.

In view of the role played by tyrosine in melanin and epinephrine biosynthesis, the light skin pigmentation and low levels of epinephrine in the blood could be explained by a lack of tyrosine. However, enzymes involved in epinephrine and melanin biosynthesis could be inhibited by abnormal metabolites.

Phenyllactic and phenylpyruvic acids have an inhibitory effect on dopa decarboxylase and glutamic decarboxylase activity, but phenylalanine does not. Phenylacetic acid in equimolar concentrations had only half the inhibitory power of the two other metabolites. These findings might explain why low-phenylalanine diets restore pigmentation to normal in phenylketonuric individuals.

Parahydroxyphenylpyruvic, phenylacetic, and phenyllactic acids inhibit tyrosinase, but whereas the first of these compounds is a potent inhibitor, the others are only weak inhibitors. It seems that if this inhibitory effect were important in phenylketonuria, pigment metabolism would be more apparently altered in tyrosinosis than in phenylketonuria, which seems not to be the case. However, the enzyme block might explain why small doses or dietary amounts of tyrosine have no effect on the pigmentation of patients with phenylketonuria. Only when large doses of the amino acid are administered are pigmentation and epinephrine biosynthesis restored to normal, probably because tyrosine competes with phenylalanine metabolites for melanocyte tyrosinase and dopa decarboxylase.

The presence of indol derivatives in the urine of phenylketonuric patients is more difficult to understand. The experiments of Tyler and Armstrong [79] suggest that there are side effects of the main metabolic block. By keeping the patient on a diet containing only sufficient amounts of phenylalanine to maintain normal growth and normal protein synthesis, these authors demonstrated that all these patients' biochemical symptoms disappeared, including the excretion of indole derivatives. Furthermore, it was demonstrated that hydroxytryptophan decarboxylase (an enzyme identical to dopa decarboxylase) is inhibited by the abnormal metabolites in a way analogous to that for dopa decarboxylase. This may explain both the low levels of phenyltryptamine in patients with phenylketonuria and the accumulation of unidentified indole compounds [80, 81].

Both the absence of phenylalanine hydroxylase and the block of the 5-hydroxytryptophan decarboxylase could account for the low plasma and brain serotonin observed in patients with phenylketonuria.

Pathogenesis of Neurological Lesions. It is not understood how the single enzyme defect can be responsible for the drastic neuropathological syndrome in phenylketonuria. The experiments of Wolff and Griffiths and Armstrong and Tyler suggest that an abnormal accumulation of metabolites is responsible for mental deficiency in phenylketonuric patients. These investigators have demonstrated that the metabolism of phenylketonuric patients maintained on a diet low in phenylalanine is eventually restored to normal. In most cases, the epileptic syndrome often associated with phenylketonuria also recedes. Low-phenylalanine diets are now routinely used as therapy for phenylketonuria [79, 82].

It is often assumed that the accumulation of the amino acid or its derivatives interferes with important metabolic acitivites in the nervous system. At least four different theories have been proposed. The first postulates that phenylalanine metabolism directly affects the molecular system responsible for the maintenance of normal mental performance. However, one should not eliminate the possibility of an indirect effect by which the phenylalanine metabolites interfere with biochemical conversions in other pathways, thereby causing mental deficiency. An example of such an indirect mechanism is the chain reaction that links the distortion of phenylalanine metabolism to that of indole metabolism. Low serotonin levels in phenylketonurics could also cause mental retardation, but there is no correlation between serotonin levels and severity of mental retardation. Moreover, the role of serotonin in mental functions is not known.

Phenylpyruvic and phenyllactic acids inhibit glutamic acid decarboxylase *in vitro*. If the inhibition exists also *in vivo*, no γ-isobutyric acid is produced. This compound seems to play an important role in the brain-blood barrier regulation.

The accumulation of phenylalanine and its metabolites may interfere with the metabolism of other amino acids. Stein and Moore, and later Knox, showed that in phenylketonuric patients the amount of other amino acids in the plasma is decreased while phenylalanine accumulates. This interaction between the amino acids' metabolism acquires particular significance in view of the mode of amino acid uptake in the brain. The investigators demonstrated that phenylalanine inhibits tyrosine uptake in the brain. Thus, in the presence of large amounts of phenylalanine, protein synthesis in the brain might be inhibited. Furthermore, because of the absence of tyrosine, the biosynthesis of well-known neuroregulators derived from tyrosine, such as norepinephrine and 3,4-dihydroxyphenylethylamine, could also be reduced in the brain.

Phenylalanine and its metabolites have also been shown to inhibit competitively hexokinase and pyruvic kinase *in vitro*; therefore, it has been proposed that the brain damage in phenylketonuria may develop because of interference with glycolysis. However, direct evidence that glycolysis is inhibited *in vivo* in phenylketonuria is not available.

The levels of monoenoic lipids in brain are reduced in phenylketonuric patients, and it has been postulated that the absence of phenylalanine hydroxylase is associated with reduced levels of fatty acid desaturase. The desaturase is truly a mono-oxygenase, but there is no indication that mono-oxygenases are generally depressed in phenylketonuria.

Whatever the mechanisms, the neurological findings are quite striking, as demonstrated by macroscopic and microscopic examinations of the central nervous system of patients affected with phenylketonuria. The disease is characterized by deficient myelination and gliosis. Several parts of the nervous system are in-

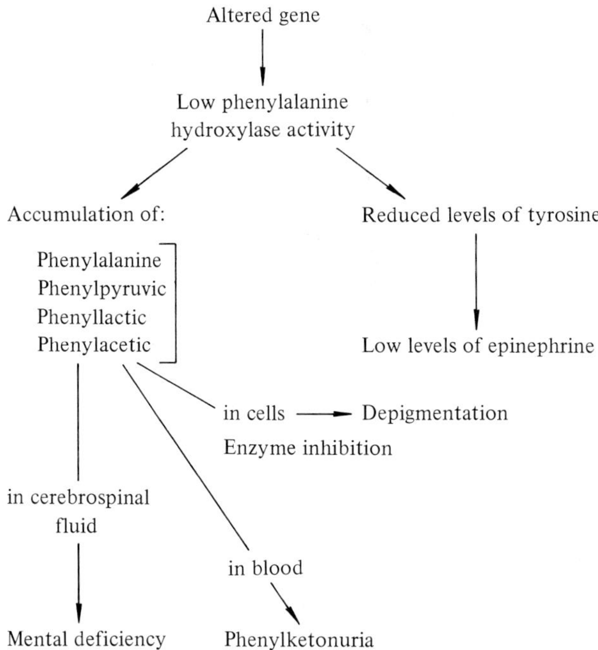

Fig. 3-22. Pathogenesis of phenylketonuria

volved, but lesions are most conspicuous in the optic, corticospinal, and corticopontocerebellar tracts. Increase in fat around blood vessels and the presence of rod-shaped microglial cells are also observed. There is occasionally a decrease in the number of Purkinje cells in the cerebellum.

Myelination is considerably more complete in adults than in young phenylketonuric patients, suggesting that the process is retarded but not completely absent in phenylketonuria. An addition to containing lipids, myelin also contains proteins—specifically neurokeratin—and the metabolic block responsible for demyelination might inhibit protein synthesis in general, and that of neurokeratin in particular. The change in myelin that occurs in phenylketonuria will be most easily understood after the mechanisms of demyelinization are discussed. The pathogenic mechanism of phenylketonuria is summarized in Fig. 3-22.

Tyrosinosis

Tyrosinosis is a rare metabolic disease described in 1932 by Medes in a patient under observation for myasthenia gravis. On examination of the urine for phosphorus with Fiske and SubbaRow's method, Medes observed spontaneous reduction of the phosphomolybdic acid. Medes [83] first thought that the reducing power of the urine was characteristic for patients with myasthenia gravis, but after more extensive investigation of other affected patients, it soon became clear that these exceptional properties of the urine were unique for this patient. Instead of rejecting this finding as a biochemical curiosity, Medes investigated the compound excreted in the urine and the metabolic block responsible for this anomaly. By so doing, she identified a new disease characterized by the excretion of large quantities of parahydroxyphenylpyruvic acid.

Medes observed further that the excretion of this compound paralleled the patient's protein or tyrosine intake and that administration of homogentisic acid did not affect urine composition. These findings suggested that the disease is due to a block of the transformation of tyrosine to homogentisic acid. Although this is the most probable interpretation, it has not yet received direct confirmation by analysis of enzyme activity in the liver. Furthermore, determination of the abnormal metabolite in the blood has not been included in the studies of tyrosinosis. The possibility of an abnormal excretion of the compound at the level of the renal tubules cannot be excluded entirely.

The disease has been described only once, but it is generally accepted that tyrosinosis is a hereditary disease. Cases of acquired tyrosinosis have been described among patients with liver disease. Although patients with liver disease do not usually excrete parahydroxyphenylpyruvic acid, cases of cirrhosis and hepatitis with associated tyrosinosis have been described. In those cases, the spleen was enlarged. Tyrosinosis was also observed in a patient with hemolytic anemia, but only after splenectomy had been performed. These findings suggest that the liver is essential in tyrosine oxidation, but when the liver is diseased, the spleen takes over this function. Consequently, after splenectomy or injury to the spleen in patients with liver disease, tyrosine is no longer oxidized and parahydroxyphenylpyruvic acid is excreted in the urine. Whatever the mechanism of acquired tyrosinosis may be, these observations point out that one cannot consider metabolic disease, or metabolic pathways for that matter, as restricted to a single organ.

Alkaptonuria

Alkaptonuria is another disease of intermediate metabolism of aromatic amino acids. It was the first of a group of diseases discovered by Sir Archibald Garrod [84] called "inborn errors of metabolism." The transformation of homogentisic acid into maleylacetoacetate is blocked in alkaptonuria. Direct evidence for the location of the metabolic block was obtained by researchers who demonstrated a specific enzymatic block in the liver of alkaptonurics. These studies also indicated that the metabolic block was due to an absence of the enzyme rather than to the presence of an inhibitor or the lack of a cofactor. Consequently, whereas the urine of normal individuals contains no homogentisic acid (the renal threshold for that compound is low in both normal and diseased individuals), large amounts are recovered from the urine of alkaptonuric patients, and the amount excreted parallels the patients' protein and tyrosine intake.

Homogentisic acid is the only compound that is increased in the urine of alkaptonuric patients. It is normally a white compound, but on oxidation in an alkaline medium, it becomes a brown-black polymer. Thus, the disease is often apparent in infants by dark staining of the diapers. Alkaptonurics are usually asymptomatic for the first two decades of life, but later a typical pigmentation of cartilage and connective tissue occurs. Virchow first described this change, which he called ochronosis. It is more than a pathological curiosity, for severe articular mutilation often results from it. The skin is also discolored through accumulation of pigment in the endothelial cells of the vessels and in the connective tissue. Thus, these patients acquire a bluish-black complexion. The pigmentation is more pronounced in the nose and ears where the cartilage pigmentation accentuates the discoloration. In the sclera, triangular brown patches appear with their bases directed toward the cornea [85, 86].

At autopsy, the most striking finding is the ubiquitous pigment and predominant discoloration of the fibrous cartilages and tendons. Cooper and Moran [87] estimated that 11% of pigment was found in the cartilages (on a wet basis), but only 2% was found in the liver and 3% in the kidney. The most severe discoloration is usually observed in tracheal rings, intervertebral discs, and articular cartilages. However, closer examination often reveals pigmentation of the endocardium, the intima of the aorta, the subendothelial connective tissue of the endothelial walls, and the walls of the vessels of the kidney and liver.

With advancing age, in addition to the discoloration of the skin, the most severe clinical consequences are due to the deposition of alkaptones in the articular cartilages. The presence of the pigment renders them friable, and fragments of cartilage are soon broken off during the articular movements; they then erode the synovia and may even become embedded in the synovial layer. This process, in the chronic state, leads to the loss of cartilage; both synovia and cartilage are replaced by connective tissue, which soon becomes pigmented itself. In the vertebrae, for instance, the intervertebral discs, the nucleus pulposus, and the ligaments are all loaded with pigment. The progressive destruction of the articular structures leads to rigidity and distortion, with fusion of the vertebral bodies and articular abnormality in the knee and hip (Charcot's hip). As can be expected, the X-ray pictures are not specific and reveal only abnormal calcification of the intravertebral discs, the tendon sheaths, the synovial membranes, and the costal cartilages. The affinity of cartilages and connective tissue for the pigment is not understood, nor is the mechanism by which pigment deposition leads to tissue destruction.

Individuals with alkaptonuria are usually homozygous for a rare abnormal gene transmitted as a mendelian recessive. However, a rare second form of alkaptonuria is transmitted by a dominant gene. Both diseases are characterized by excretion of homogentisic acid, but it is not known whether the biochemical alteration is the same in both cases. It has not been established if these two genes are alleles or are located on a different loci of the chromosome.

Sometimes alkaptonuria is associated with abnormal excretion of uric acid in the urine, and alkaptonuria might be associated with defective uric acid metabolism. However, it seems more likely that the high incidence of uricemia reported is attributable to errors in the colorimetric analysis—probably due to the high levels of homogentisic acid—because cases of hyperuricemia have been reported with the colorimetric method, but not when uricase was used [88].

Albinism

Albinism is a condition characterized by total or partial absence of melanin at its normal sites—skin, sclera, etc.* Melanin is elaborated by the melanocyte, a special type of cell that originates from the neural crest in the embryo. The cells are found in the skin, in the uveal tract, and in the retina [89, 90].

Albinism may involve all melanocytes in the body or may be restricted to a few groups of cells. Clinicians thus describe oculocutaneous and ocular albinism.

The condition of oculocutaneous albinos is so striking, particularly among highly pigmented races such as Indians and Negroes, that it was recognized thousands of years ago. An 1699, Wafer vividly described moon-eyed Indians. Albinos were revered by their fellow tribesmen in some African countries. At first glance, it seems highly improbable that such a striking morphological alteration should result from a single molecular disturbance. Nevertheless, decades ago Sir Archibald Garrod included albinism among the inborn errors of metabolism.

The clinical symptoms of oculocutaneous albinism are of two types: hypersensitivity of the skin to sunlight and photophobia. The skin is white and does not tan when exposed to ultraviolet radiation; instead, the skin rapidly burns with erythema and desquamation. The incidence of skin cancer, normally high among blond individuals, is even higher among albinos. The iris has a blue-grey translucency.

The disease does not result from the absence of melanocytes. In fact, the following evidence suggests that these cells are present: (1) demonstration of melanocytes in skin of albinos by histochemical methods (gold impregnation techniques); (2) electron microscopic recognition of melanocytes in albinos' skin; and (3) the occurrence of amelanotic nevi, or benign tumors. The inability of the albino's melanocyte to produce melanin is not understood.

A deficiency in tyrosinase is often suspected to be responsible for the disease, but the site of the biochemi-

* The composition of pigment found in *locus ceruleus* and the *substantia nigra* is not known, but this pigment remains unchanged in albinism. Furthermore, melanomas never originate at these sites. Aromatic amino acids, however, affect the composition of the *substantia nigra* and the *locus ceruleus*. The pigment of the nuclei is decreased in phenylketonuria.

cal block has proved difficult to establish for at least two reasons: (1) the mechanism of the tyrosinase reaction is not clear; and (2) tyrosinase assays are difficult—they are based on the manometric determination of oxygen uptake in the presence of tyrosine as a substrate.

In albinism, histochemical methods proved more useful than biochemical techniques in determing the molecular nature of the disease. Tyrosinase activity can be detected histochemically by using dopa as a precursor (this method is complicated by the instability of the dopa compounds), and by radioautography after administration of [^{14}C]tyrosine. In the presence of tyrosine, the labeled amino acids are incorporated into melanin, which is then bound in the melanin granules of melanocytes. Since tyrosinase is assumed to be the only enzyme involved in the conversion of tyrosine to melanin, the absence of melanin even when labeled tyrosine is present in large excess suggests that tyrosinase is absent. Insofar as there is no change in the concentration of the pressor amines in albinism, and tyrosine is converted to melanin and epinephrine through a common intermediate, another pathway for epinephrine synthesis must exist [91].

There are two forms of oculocutaneous albinism; a tyrosinase negative due to a tyrosinase defect and a tyrosinase positive of unknown origin (probably caused by inavailability of tyrosine to the melanocyte). Both forms are inherited as autosomal recessive traits, but the genes for each type are found in different loci. Ocular albinism is inherited as an X-linked recessive trait. It is associated with vision defects, nystagmus, and head nodding. The metabolic defect is unknown.

HISTIDINEMIA

Histidinemia is a rare hereditary disease characterized by high levels of histidine in blood and urine [92–94]. Affected patients often have speech deficiencies, and excessive amounts of imidazole pyruvic acid are found in the urine.

In mammals, histidine is an essential amino acid. Consequently, any inborn error leading to histidine accumulation in the blood is likely to result from an alteration in the breakdown pathway. The breakdown pathway of histidine starts with an enzyme called histidase, a histidine deaminase. The enzyme in liver catalyzes the conversion of the histidine to urocanic acid. The requirements of the purified liver enzymes are not clear. Although the rabbit enzyme requires folic acid, a divalent cation, and glutathione for activity, the guinea pig enzyme requires only the metal and the SH group.

Urocanic acid (imidazole acrylic acid) is converted to the imidazole propionic acid in the presence of urocanase by oxidation of the ring carbon and reduction of the side chain (see Fig. 3-23). Urocanase has been found in the liver of mammals; its acivity varies considerably, depending upon the species. Urocanase has been purified extensively from beef liver [95]. The product of the urocanase reaction was clearly established to be the 4(5)-imidazolone-(5)4-propionic acid. The imidazolone propionic acid, a key metabolite in histidine degradation, lies at the crossroad of several metabolic conversions.

Further oxidation of the imidazole ring is catalyzed by imidazolone propionate oxidase, an enzyme partially purified from guinea pig liver. (The purified enzyme was free of xanthine oxidase, crystalline xanthine oxidase also catalyzes the oxidation of urocanate.)

Imidazolone propionate oxidase is thought to be a flavoprotein. [^{14}C]hydantoinpropionate is excreted without further conversion in monkeys and man. It is thus a terminal metabolite.

Imidazolone propionate hydrolase catalyzes the enzymatic cleavage of the imidazole ring to yield formiminoglutamate. The rat liver enzyme has been partially purified. In addition to the enzymic conversion, two nonenzymic spontaneous reactions yield N-formylisoglutamine and 4-oxoglutamic acid. In addition to the oxidative pathways for histidine, there exist three other pathways for its use: protein synthesis, decarboxylation to yield histamine (see Inflammation), and transaminase. The activity of histidine pyruvic transamination in rat liver is three times that of histidase. The product of the transaminase reaction is imidazole pyruvic acid, which in turn is converted to imidazole acetic acid.

In histidinemia, the missing enzyme is histidase. This was established in two ways: administration of urocanic acids, which is followed by histidinemia and histinuria; and direct assay for the enzyme. The metabolic block leads to histidine accumulation in the blood and excessive excretion in the urine.

The accumulating metabolite is metabolized through alternative pathways, mainly the transaminase reaction. Consequently, imidazole pyruvic acid accumulates in the urine. The imidazole pyruvic, like phenylpyruvic, acid reacts with ferric chloride to yield a blue compound. As a result, the diaper test does not distinguish between phenylketonuria and histidinemia. Yet the diagnosis is of considerable importance because histidinemia is a much more benign disease. Furthermore, histidinemia is not alleviated by withdrawal of phenylalanine from the diet.

Undoubtedly, the most intriguing aspect of the disease is the selected impairment of mental development that results from the metabolic block. Some children with histidinemia trail behind other children in learning to talk; and when they talk, they exhibit speech difficulties. One patient communicated with only single words. The elucidation of the role of histidine metabolism in the development of speech could be an important landmark in our understanding of brain chemistry (see Fig. 3-24).

Maple Syrup Urine Disease

Maple syrup urine disease was first described in England by Mackenzie and Woolf [96]. Affected patients

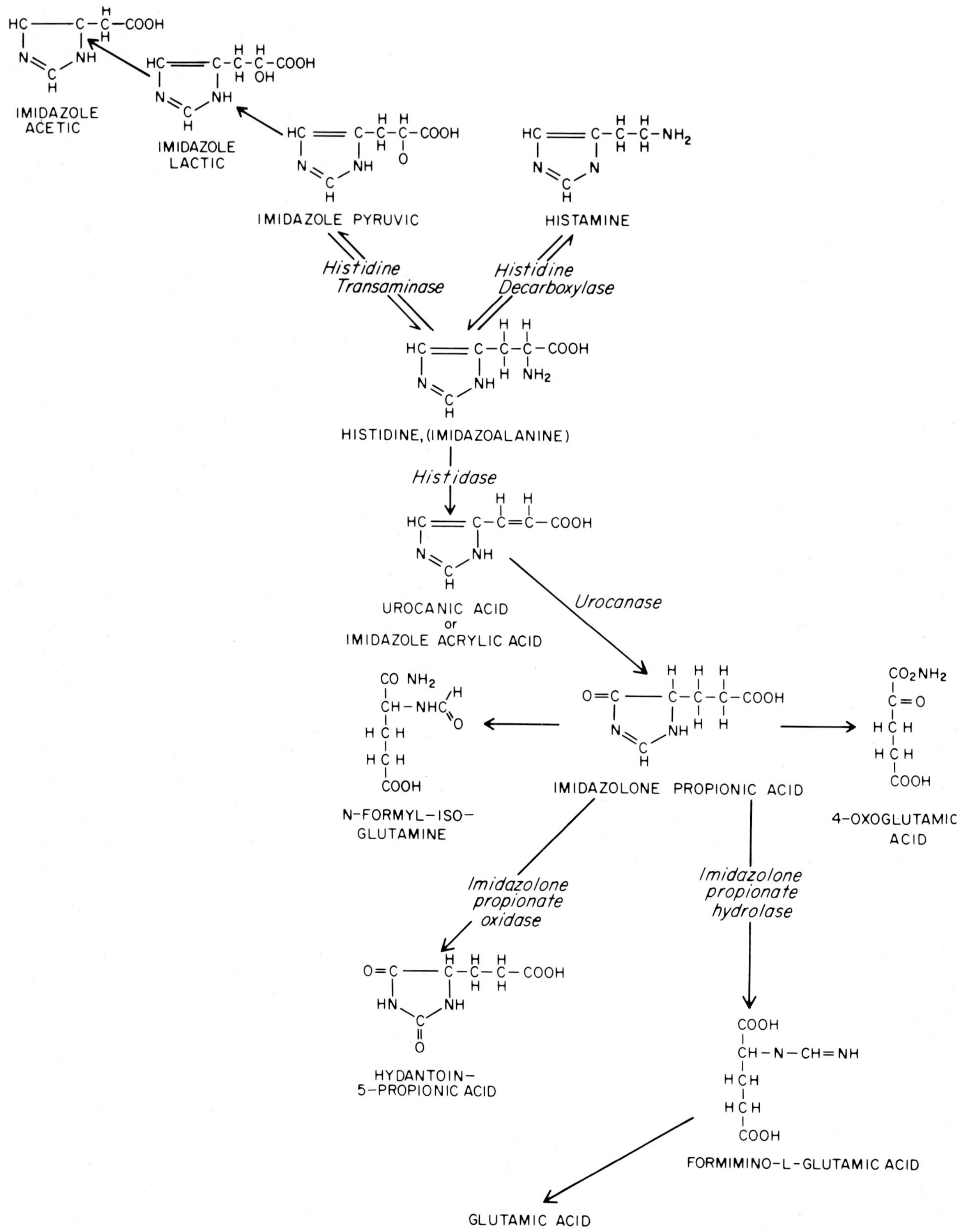

Fig. 3-23. Histidine degradation

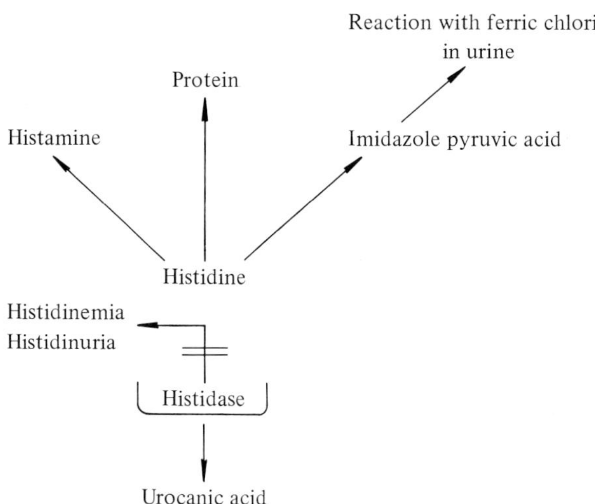

Fig. 3-24. Pathogenesis of histidinemia

have a special type of amino aciduria involving the branched amino acids leucine, valine, isoleucine, and alloisoleucine. Corresponding keto acids—indolacetic and indollactic acids—are excreted, and the branched amino acids and the keto acids accumulate in most body fluids including blood, saliva, and cerebrospinal fluid. An increase in the blood level of methionine and a decreased cystine concentration have also been observed. Unidentified compounds also probably accumulate in maple syrup urine disease. The origin of the maple syrup odor is not known.

Patients with maple syrup urine disease are mentally retarded, have seizures, and show abnormal electroencephalographic patterns. The disease is probably hereditary, but the exact mechanism of transmission is not established. Although the pathogenesis of maple syrup urine disease is still unknown, the metabolism of the branched amino acids provides the key to understanding the proposed mechanism [97].

Again, the branched amino acids are valine, leucine, isoleucine, and alloisoleucine. All four amino acids are made up of a main chain that carries a terminal carboxyl group, an α-amino group, and a side chain made of one or two methyl groups. The main chain of the valine is made of three carbons, and two methyl groups are attached to the terminal β-carbon. Leucine has four carbons in its main chain, and two methyl groups are attached to the α-carbon. Isoleucine has five carbons in the main chain, and the methyl group is attached to the β-carbon. Alloisoleucine is a steric isomer of isoleucine.

As are most amino acids, branched amino acids are used through two main pathways: the biosynthesis of proteins and polypeptides and their breakdown after oxidative deamination or transamination. The oxidative deamination is catalyzed by an amino acid oxidase and yields α-keto acids and ammonia. Monamine oxidases are widely distributed in biological material.

There are at least two types of amino acid oxidases: one is specific for the D-, the other for the L-amino acids. The D-amino acid oxidase, a flavoprotein free of iron with a molecular weight of 182,000, has been crystallized from pig kidney. The L-amino acid oxidase has been crystallized from *Crotalus adamanteus*, and it is a flavoprotein with a molecular weight of 130,000 (2 moles of FAD per mole of protein).

The mechanism of the reaction is not clear, but the enzyme reacts in presence of O_2 and yields the α-keto acid, ammonia, and hydrogen peroxide. In spite of the restricted stereochemical specificity, the specificity for the amino acids is rather broad. The affinity of the monoamine oxidase is low for short-chain amino acids, but the enzyme is very effective on intermediate and long-chain amino acids. The enzyme attacks straight- and branched-chain amino acids, as well as amino acids containing aromatic ringlike phenylalanine, epinephrine, etc.

The second reaction involved in the catabolism of the branched amino acid is the transaminase reaction, during which the amino group of the amino acid is transferred to a keto acid without liberating ammonia. A great number of transaminases have been discovered. Glutamic pyruvic transaminase, glutamic keto glutaric transaminase, and tyrosine oxaloacetic transaminase have been purified. Each cell probably has its own unique transaminase pattern.

In the transaminase reaction, the L-amino acid is the most common donor, but D-amino acids have also been described in that role. Studies on the specificity of transaminase suggest that the enzyme is specific for both the donor of the amino group and the acceptor keto acid. Transaminase requires pyridoxal phosphate as coenzyme for activity (the role of the pyridoxal phosphate in the reaction is discussed further in the section on vitamins).

Several metabolic blocks could account for the biochemical distortion observed in maple syrup disease. A deficiency in amino oxidase could lead to accumulation of amino acids. Because the enzyme has such a broad specificity, whenever it is completely deleted a more complex aminoaciduria can be expected to develop. The deletion of a specific transaminase could hardly explain the keto acid accumulation. Therefore, it seems more likely that the metabolic block involves a step between the keto acid and the simple acids, possibly the oxidative decarboxylation of the keto acid. This reaction requires coenzyme A, NAD, lipoic acid, and thiamine pyrophosphate, and it was described in some detail in the chapter devoted to the bioenergetic pathways. Leukocytes of at least some patients with maple syrup disease have been shown to contain normal transaminase activity but are defective in the oxidative decarboxylase.

Multiple deletion cannot be excluded as a possible cause of maple syrup urine disease because a deficiency in a specific decarboxylase does not adequately explain the accumulation of indole derivatives and methionine. Of course, the branched amino acids or the corresponding keto acids could inhibit other metabolic pathways.

OXALURIA

Oxaluria is a rare disease occuring mainly between one and four years of age and most often in white males. The patient is usually afflicted with a chronic renal disease associated with normal blood pressure. Renal failure results from the deposition of pure calcium oxalate calculi into the pelvis and calices. The lithiasis causes chronic inflammation of the kidney with parenchymal involution, fibrosis, and formation of thick capsules that strip off with difficulty. Microscopic examination of the kidney demonstrates the presence of oxalate calculi in the tubules. Calcium oxalate is also frequently found in the arterial walls, but usually the glomeruli remain free of salt. Histologically, the calculi appear as globular or rhomboidal crystals; they are slightly yellowish and doubly refractile under polarized light [98–104].

Deposition of calcium oxalate is not restricted to the kidney. The salt has been found in many other visceral and endocrine organs: parathyroid, pancreas, testes, adrenals, lymph nodes, and the wall of the veins, arteries and arterioles, cerebrospinal fluid, etc. In oxaluria the kidney contains 50–70 mg and the heart contains over 1 mg of oxalic hydrate per gram of dry weight of tissue. This is between 40–50 times as much as the amount found in normal tissue.

The exact mode of transmission of the disease is not clear, but there is good indication that it is a hereditary disease, transferred as an autosomal recessive trait.

Nine to sixty milligrams of oxalic dihydrate is excreted daily by normal individuals maintained on an average diet. Urinary oxalate originates from both exogenous and endogenous sources. Some foods (spinach, rhubarb, chocolate, cocoa, peas, meats) are rich in oxalate. However, little vegetable oxalate is absorbed by the intestines because the dietary oxalate precipitates the calcium contained in food, forming insoluble calcium oxalate and therefore depriving the individual of his normal calcium requirements. Consequently, with a diet containing oxalate in amounts 5–20 times greater than normal, only a small fraction (2.5–5%) is absorbed and excreted, and oxalate intake contributes little to the oxalate excretion under normal conditions. Oxalate poisoning has been observed only in a few exceptional cases in which excessive amounts of rhubarb were ingested.

Oxalate is a terminal product of normal metabolism. When [^{14}C]oxalate is injected, most of the labeled compound is excreted, and only a small fraction of the radioactivity is recovered in bone and muscle. There are only two immediate precursors of oxalic acid in normal metabolic pathways: glyoxylic acid and 2,3-diketo-L-gulonic acid. The first of these compounds is the product of amino acid oxidation (serine and glycine), and the second is derived from the oxidation of ascorbic acid (see Fig. 3-25).

Glycine is converted to glyoxylic acid by two different pathways: oxidation and transamination. Glycine oxidase, an enzyme found in liver and kidney, deaminates and oxidizes glycine in the presence of molecular oxygen to yield ammonia, hydrogen peroxide, and glyoxylic acid. In the presence of specific transaminases, the amino group of glycine is transferred from the monocarboxylic acid to a dicarboxylic acid, yielding glutamic and glyoxylic acids. Similarly, serine decarboxylation in the presence of serine decarboxylase leads to the formation of CO_2 and ethanolamine. Further oxidation of ethanolamine leads in successive steps to the appearance of glycoaldehyde and glycolic acid. Glycolic acid oxidase catalyzes the oxidation of glycolic acid to glyoxylic acid using FAD as coenzyme. Glyoxylic acid has a multiple fate—glyoxylic acid is converted to oxalate by the catalytic action of aldehyde oxidase, an oxidative enzyme which in the presence of an aldehyde and molecular oxygen yields the acid and hydrogen peroxide oxalate. The catabolism of glyoxylic acid to form formic acid and CO_2 is complex. In the presence of glyoxylic acid dehydrogenase, two coenzymes (thiamine pyrophosphate and NAD), and manganese, a molecule of glyoxylic acid condenses stoichiometrically with a molecule of glutamic acid to yield a 7-carbon compound, N-glyoxylglutamic acid. In the next step, the carboxyl—originally present in the glyoxylic acid molecule—is oxidized to yield CO_2, and the N-carbon is converted to an aldehyde group, yielding N-formylglutamic acid. The 6-carbon compound is then hydrated and split to yield glutamic acid and formic acid.

Glyoxalate can be reduced to glycolate either by specific NADH or NADPH glyoxalate reductases or by lactic dehydrogenase.

Transamination of glyoxalate takes place with several amino acid donors; glycine, glutamate, ornithine, arginine, and methionine.

Glyoxalate can synergistically be decarboxylated in the presence of α-ketoglutarate to yield α-hydroxy-β-ketoadipate. The reaction requires thiamine pyrophosphate, Mn^{++}, L-glutamate, and NAD. The carboligase catalyzing the reaction has been partially purified from pig liver mitochondria and is known to exist in humans.

When [^{14}C]ascorbic acid labeled in the carboxyl group is administered, ^{14}C-labeled oxalate is formed, and as much as 20% of the label is recovered in the oxalate.

The intricacies of the metabolic pathway that leads from ascorbic acid to oxalate are not clear. Ascorbic acid is oxidized to yield dehydro-L-ascorbic acid; this compound is further hydrated to yield 2,3-diketo-L-gulonic acid. The intermediates between 2,3-diketo-L-gulonic acid and oxalate are not known. L-Xylulose, L-xylose, and L-xylosone have been eliminated. An enzyme was partially purified from rat kidney; it catalyzes the decarboxylation of 2,3-diketo-L-gulonic acid to yield L-xylonic and L-lyxonic acids [105].

It is further postulated that L-xylonate and L-lyxonate are then dehydrated to yield L-xylonolactone, which is further oxidized to yield L-erythroascorbate. It is not known where this metabolic pathway branches off to yield oxalic acid.

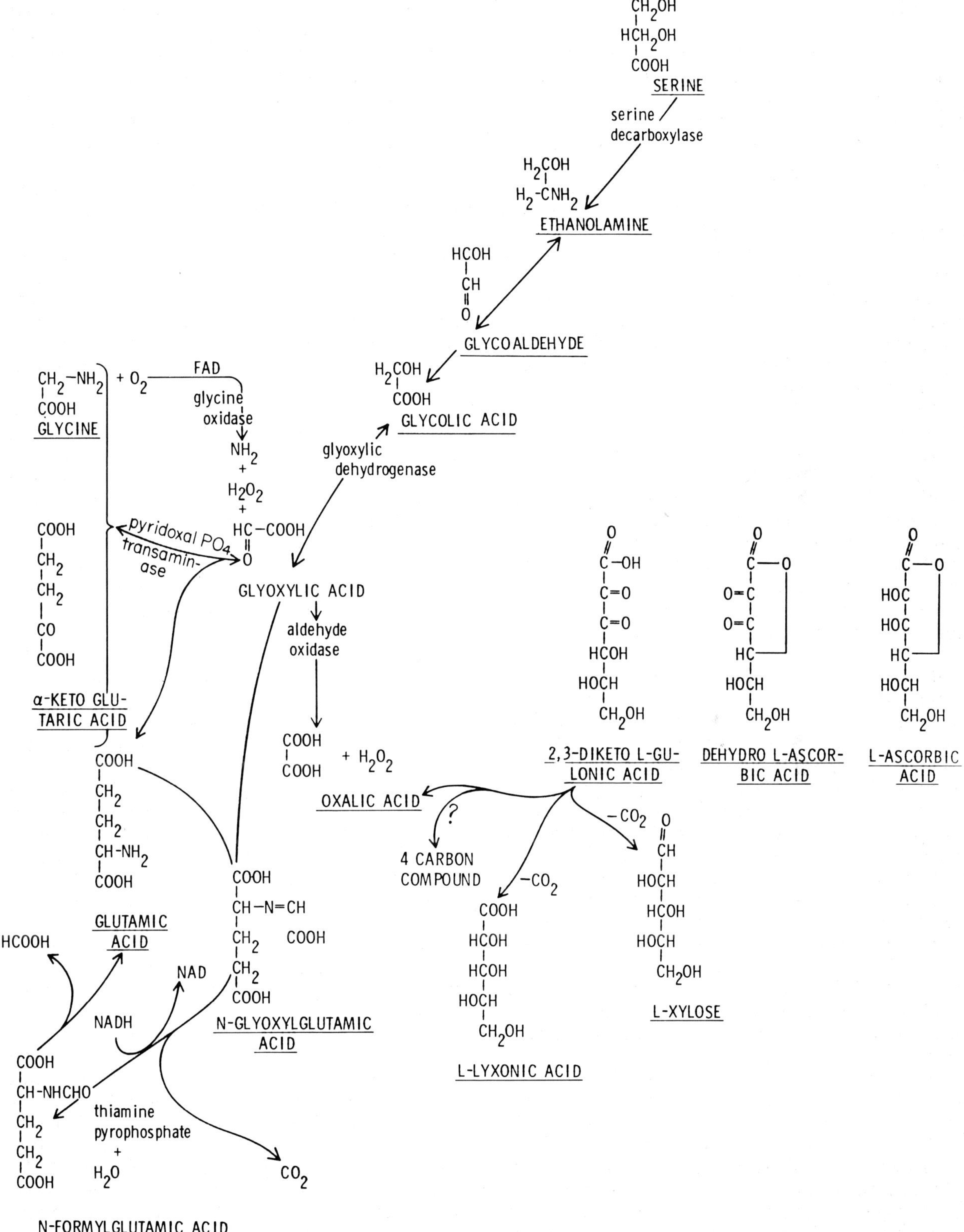

Fig. 3-25. Pathways involved in oxaluria

If an accurate knowledge of the individual steps of the metabolic pathways leading to oxalic acid formation is necessary to understand oxaluria, an evaluation of the relative importance of each of these pathways in metabolism is of no less interest. In that respect, it is significant that under normal conditions the addition of 25–100 g of glycine to the diet does not affect of oxalic acid excretion. In contrast, the inclusion of 40 g of gelatin (corresponding to 10 g of glycine) doubles the amount of oxalic acid excreted.

Two major forms of oxaluria have been detected; one results from a deficiency of the glyoxalate carboligase. The disease is associated with increased excretion of glycolic acid, thus the inability to synergistically decarboxylate glyoxalic acid coupled to α-ketoglutarate leads to an increase in the levels of glyoxalate (the immediate precursor of oxalate and glycolic acid) found in the urine.

The second type of oxaluria is associated with the urinary excretion of L-glyceric acid. Hydroxypyruvate is a substrate for both D-glyceric and lactic dehydrogenases. In presence of the first it will yield D-glycerate, in presence of the second L-glycerate.

Hydroxypyruvate is primarily derived from serine through a transamination catalyzed by an alanine hydroxypyruvate transaminase. It is believed that excessive amounts of hydroxypyruvate are converted to L-glycerate (the reaction catalyzed by lactic dehydrogenase) because of a lack of D-glyceric dehydrogenase which catalyzes the conversion of hydroxypyruvate to D-glyceric acid.

It is impossible to distinguish these two types of oxaluria without careful biochemical investigation of the blood and the urine, and without direct attempts to identify the enzyme defect. The crucial difference between glycolic and glyceric aciduria is precisely that in addition to oxalate, glycolic acid accumulates in primary oxaluria, whereas glyceric acid accumulates in the second type [106, 107].

Hyperoxaluria and occasionally calcium oxalate stones develop in patients with diseases of the ileum or after resection of that portion of the intestine. A number of sophisticated biochemical pathways, all involving overproduction of glycine and glyoxalate, have been invoked to explain the hyperoxaluria. Recently, however, a study from the Hamersmith Hospital in England has suggested that the hyperoxaluria results from increased absorption of oxalate and consequently that it can be readily corrected by reducing oxalate intake in the diet [209].

Lipoidosis

Several diseases—Gaucher's, Tay-Sachs, Niemann-Pick, and Fabry's diseases, and the leukodystrophies—are characterized by accumulation of lipoid material in cells. These diseases differ in the composition and site of the lipoid accumulation. Whereas gangliosides accumulate in Tay-Sachs disease, sphingosides are found in Niemann-Pick disease, and cerebrosides in Gaucher's disease. Gangliosides accumulate primarely in the cells of the nervous system, but sphingosides and cerebrosides pile up in the reticuloendothelial cells [108, 109].

Infantile Amaurotic Familial Idiocy (Tay-Sachs Disease)

Infantile amaurotic familial idiocy is a hereditary disease transmitted as an autosomal recessive trait resulting in the accumulation of gangliosides in the brain cells. The main clinical manifestations are dementia, paralysis, and blindness. The disease has been most frequently described among Jews, and it is generally restricted to children 6 months to 3 years old. However, late infantile and even juvenile forms have been described.

In a typical case, a child, healthy until almost 6 months of age, presents progressive weakness, mental retardation, and slow motor deterioration with feeding difficulties. Signs of blindness, such as fixed gaze and inattentiveness, can be detected. As the disease progresses, epileptic seizures, convulsions, and in some cases decerebrate rigidity develop, and deep reflexes disappear. An ophthalmoscopic examination reveals atrophy of the optic nerve, and a cherry-red spot in place of the macula lutea in the middle of a yellowish-grey zone of deterioration. The brain shows greater alteration than any other organ affected by the disease. It is atrophic, with ventricular dilatation and leathery consistency. On histological examination, the most striking changes are found in the ganglion cells, which are blown up by a foaming material that pushes the nucleus to the periphery of the cytoplasm. The nucleus is often pyknotic or karyolytic. Ultimately, the Nissl substance disappears and the ganglion cell dies. As a result, myelination of the fibers does not occur, and the dead nervous cells are progressively replaced by proliferating glial cells which may be loaded with lipoid material (see Figs. 3-27, 3-29, 3-30). Changes similar to those observed in the brain cells are observed in the ganglion cells of the retina, the site of the cherry-red spot. Other tissues are seldom affected by the disease. However, vacuolization of the ganglion cells of the spinal cord and lipid accumulation in liver cells and lymphocytes have been reported.

The lipoid compounds that accumulate in the ganglion cells are gangliosides. In infantile amaurotic familial idiocy, the grey matter contains 10–20 times the amount of gangliosides found in normal individuals. In contrast, the levels of cerebrosides and sphingomyelins are normal. (A case has been described in which the levels of both cerebrosides and gangliosides of the brain were increased.)

The biochemical buildup of the brain is far from complete at birth, and the biosynthesis of brain lipids is essential to the proper development of the child. Among the lipids found in brain are gangliosides, cerebrosides, and sphingomyelins. The synthesis of these lipids assumes particular importance when the nerve fibers are myelinated during the first and second year of age.

Ganglioside Structure and Metabolism. Although cerebrosides are principally found in the myelin sheet, gangliosides are preponderant in the ganglion cell. The structure of gangliosides is far from elucidated. Gangliosides obtained from brain tissue contain long-chain fatty acids, the base sphingosine, hexoses (glucose or galactose), hexosamines (N-acetyl galactosamine), and N-acetyl neuraminic acid (see Figs. 3-26 and 3-33).

The basic pathways for fatty acid biosynthesis and palmitic acid biosynthesis have already been discussed. But the biosynthesis of fatty acid in brain raises special problems, especially with respect to the biosynthesis of long-chain, even- and odd-numbered saturated or unsaturated fatty acids.

Ceramide (1←1) GLU (4←1) GAL (4←1)-galnac (3←1) GAL

Galactosidase (blocked) — (3←2) NANA — Generalized gangliosidosis

Tay-Sachs ganglioside
CER GLU GAL – GAL NAC + Galactose
NANA

Hexosaminidase (blocked) — Tay Sachs

CER GLU GAL + Galactosamine
NANA

Neuraminidase

CER GLU GAL + Sialic acid

Galactosidase

CER GLU + Galactose

Glucosidase

CER + Glucose

Ceramidase

Sphingosine + Fatty acid

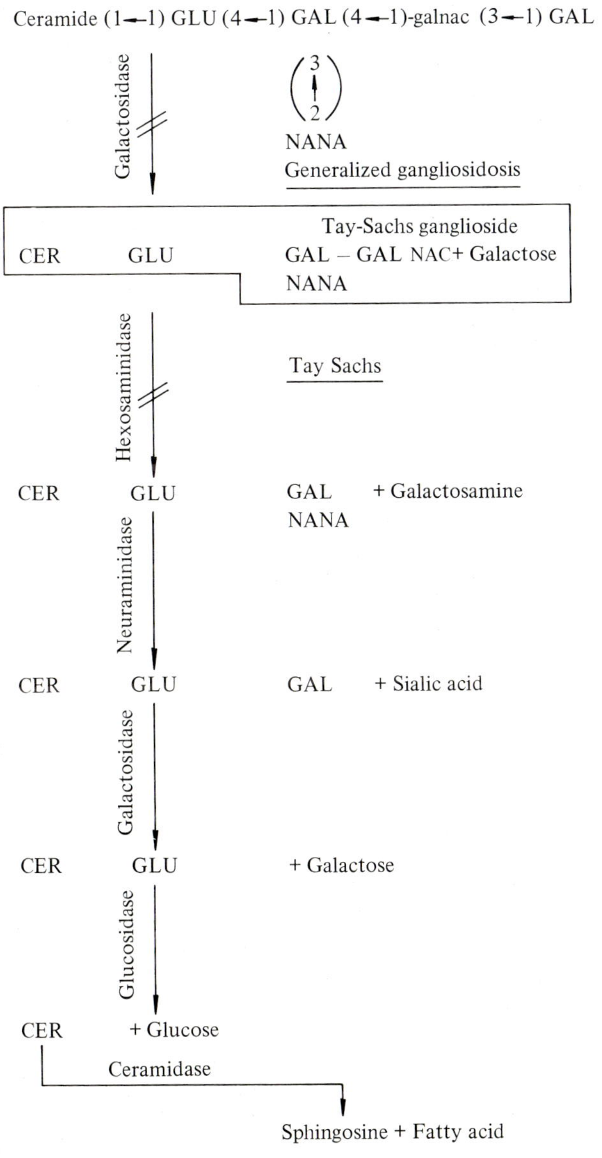

Fig. 3-26. Breakdown of gangliosides

The detailed molecular mechanism of the conversion of saturated into unsaturated fatty acid is not known. But already in 1936, Schoenheimer and Rittenberg demonstrated that deuterium-labeled stearic acid was converted to oleic acid in mammalian tissues. The aerobic conversion of stearate to oleate or palmitate to palmitoleate was studied in yeast. The conversion requires O_2 and NADPH and the active substrate is stearyl CoA or palmityl CoA.

$$\underset{\text{stearic}}{CH_3(CH_2)_{16}\overset{O}{\overset{\|}{C}}-SCoA} \xrightarrow{\text{NADPH} \to \text{NADP}} \underset{\text{oleic}}{CH_3(CH_2)_7CH=CH(CH_2)_7\overset{O}{\overset{\|}{C}}-SCoA}$$

The enzymes involved in this reaction and the structure of the intermediate are not known. Because 9-hydroxystearic acid or 10-hydroxystearic acid can replace oleate for the growth of yeast, it was proposed that hydroxystearic acid is an intermediate in the formation of oleate. It is now clear that this is true.

The brain sphingolipids contain even-numbered, long-chain saturated and unsaturated fatty acids (C_{18}, C_{20}, C_{22}, C_{24}). The long-chain fatty acids are probably synthesized by elongation of the saturated or unsaturated 16 (palmitic or palomitoleyl) or 18 (stearic or oleic) fatty acids.

In addition to even-numbered, long-chain fatty acids, the brain also contains odd-numbered fatty acids, which may contain one or more unsaturated bonds. The ω^7, ω^9 isomers are related to palmitoleate and are therefore assumed to be derived from palmitoleate by elongation and α-oxidation. The precursors of the ω^{10} and ω^8 isomers are unknown.

Since the double bond of the 2-hydroxy fatty acids is in the same position as the double bond of the unsaturated fatty acids in brain, it is assumed that these fatty acids are also derived from the unsatured fatty acids.

Sphingosine is a base with a technical name of D-erythro-1,3-dihydroxy-2-amino-4-*trans*-octadecene. In gangliosides, the base forms an acid amide linkage between the amino group in position 2 of the base and the carboxyl group of the fatty acid. This complex of base and fatty acid is named ceramide.

Our knowlege of the biosynthesis and breakdown of gangliosides is still fragmentary. Sphingosine is synthesized by the condensation of palmitic aldehyde with serine. In the presence of NADPH and a brain cell-free preparation, palmityl CoA is reduced to the aldol derivative yielding CoA and NADP. Manganese and pyridoxal phosphate are required for the next reaction in which serine reacts with fatty acid. The serine is decarboxylated in the process and dihydrosphingosine is formed. The steric configuration of the dihydrosphingosine formed has been studied. Erythrodihydrosphingosine is the major product, but little or no threodihydrosphingosine is present. The oxidation of dihydrosphingosine to sphingosine is catalyzed by a flavoprotein (see Figs. 3-28 and 3-31).

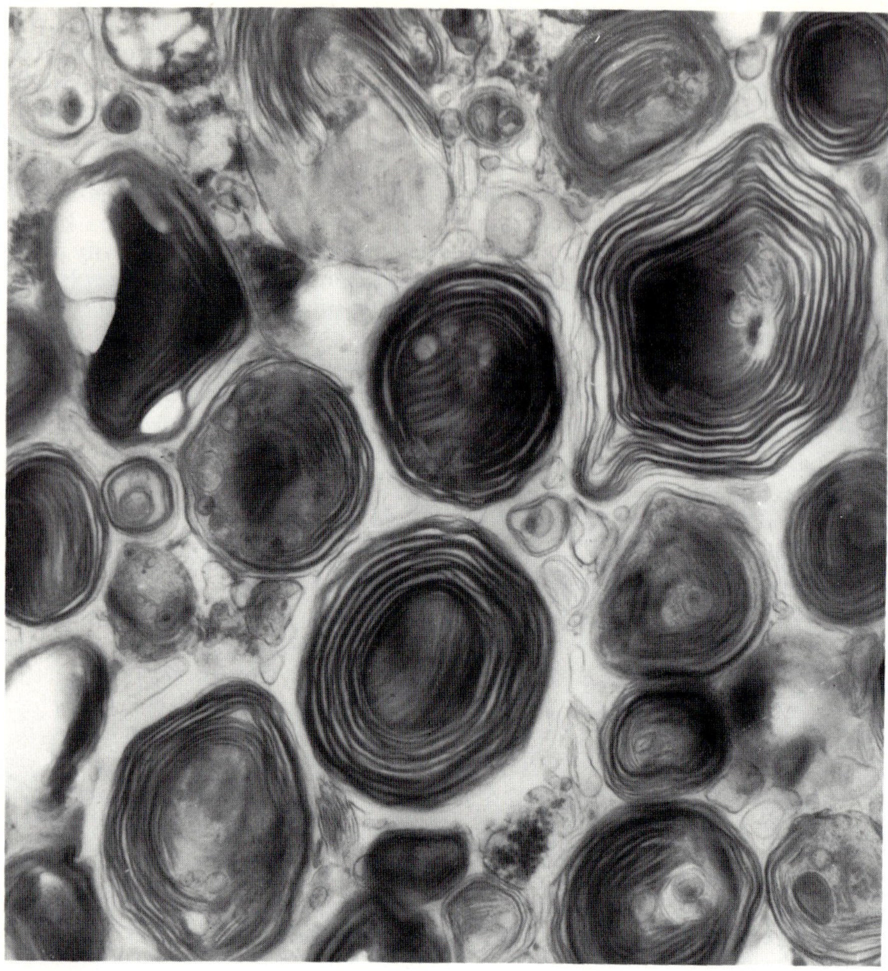

Fig. 3-30. Ganglion cell from the anterior horn of spinal cord of a three-year-old girl affected by Tay-Sachs disease. The cell cytoplasm displays considerable numbers of abnormal inclusion bodies made up of concentrically arranged membranes. These bodies are known to consist prevalently of lipid (GM_2 ganglioside) and also to contain acid phosphatase activity (+35,000) (Zamboni)

NANA are attached to each other, and the position 2 of the sialic acid forms a bond with the middle galactose. The third sialic acid moiety binds to the position 3 of the terminal galactose. And in addition to these mono-, di-, and trisialogangliosides, there is a second disialoganglioside with a complex of two NANA molecules attached to the position 3 of the middle galactose.

In addition to these four gangliosides, which constitute 40% of the ganglioside population of human brain, at least four other gangliosides are present only

Sphingosine

Acylase + R–C(=O)–S–CoA
Fatty acyl CoA

$CH_3–(CH_2)_{12}–CH{=}CH–CH(OH)–CH(NH–C(=O)–R)–CH_2OH$ + CoASH

Ceramide

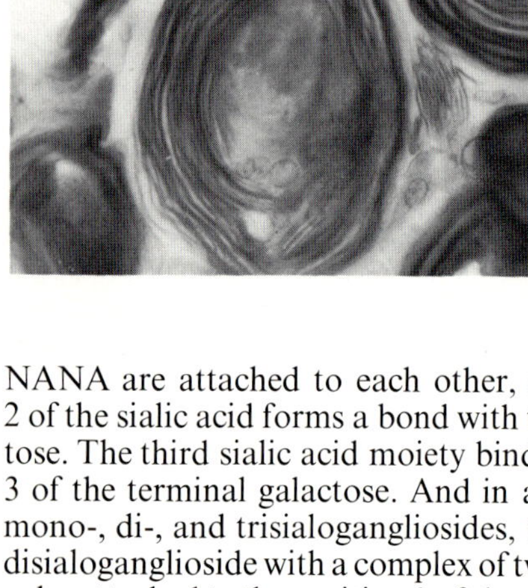

Fig. 3-31. Biosynthesis of sphingosine and ceramide

in small amounts. They vary from the parent ganglioside sequence by alteration of the skeletal structure: one disialoganglioside has lost the terminal galactose; another disialoganglioside and a monosialoganglioside are short a galactose and an N-acetylgalactosamine but contain a lactose moiety. The fourth minor component is the following hypothetical structure: galactosyl (1 → 3) galactosyl (1 → 3) galactosyl (1 → 1) ceramide.

Gangliosides have been found in many tissues besides brain—particularly in spleen and erythrocytes. In brain, most of the gangliosides are found in larger amounts in grey than in white matter.

Information on the intracellular distribution of gangliosides is somewhat controversial. On the basis of the reports available, it seems difficult to decide whether gangliosides are located in the microsomal, the "lysosomal," or the "nerve ending" vesicles.

After differential centrifugation of brain homogenates, gangliosides are associated with the microsomal fraction. Treatment of microsomes with deoxycholate and EDTA established that the gangliosides are present in the membranes of the endoplasmic reticulum rather than in the ribonucleoprotein particles. These findings were confirmed by density gradient cen-

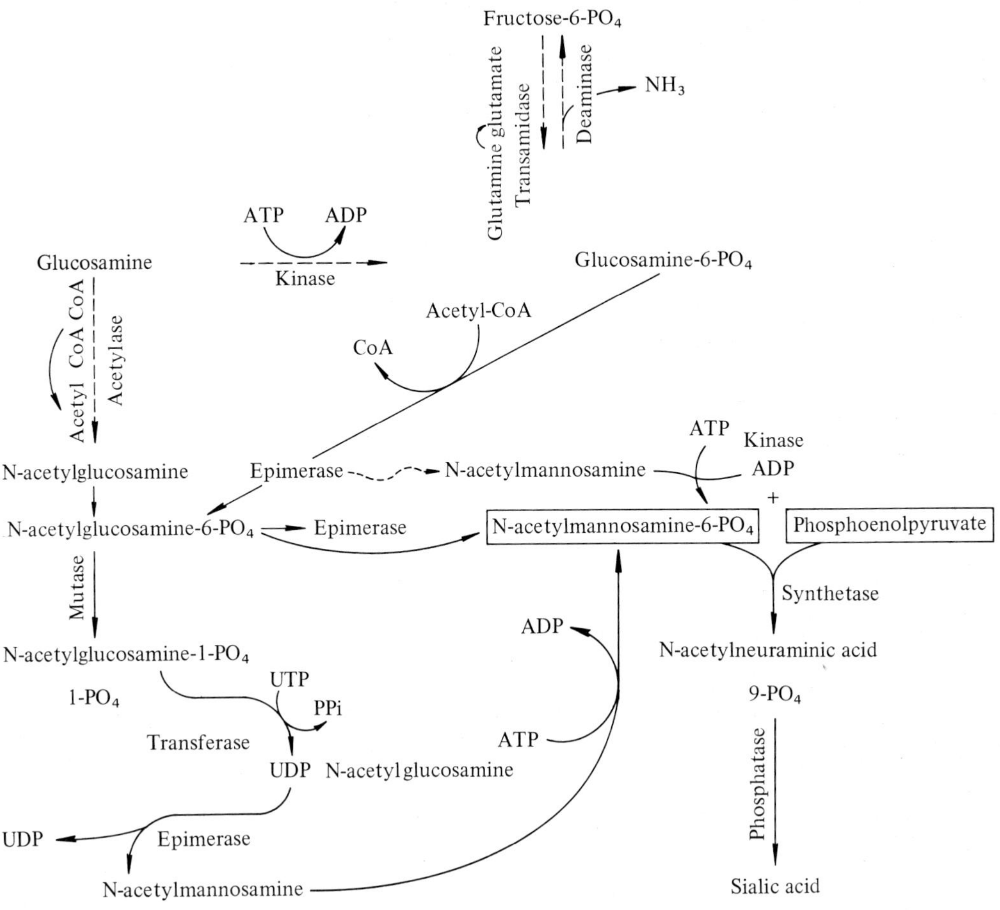

Fig. 3-32. Biosynthesis of sialic acid

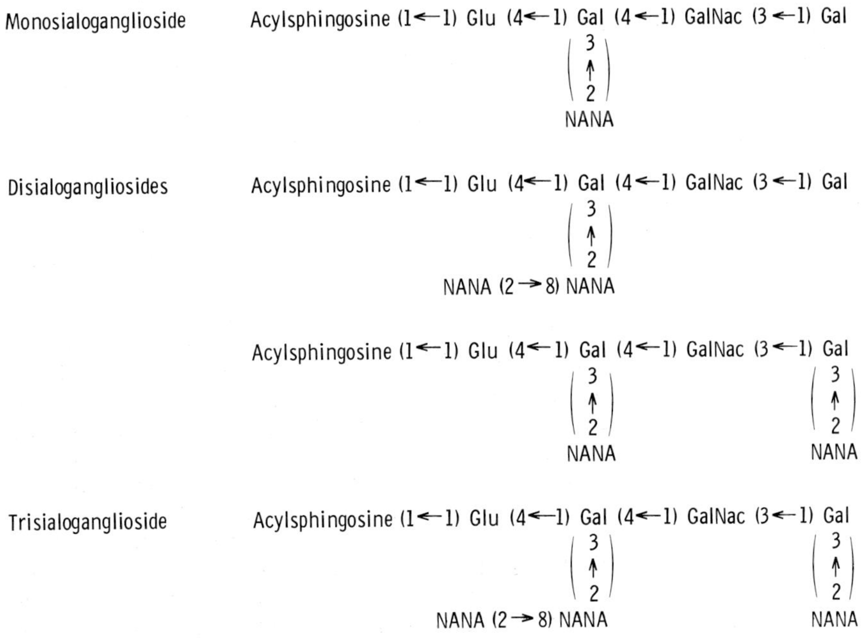

Monosialoganglioside

Acylsphingosine (1←1) Glu (4←1) Gal (4←1) GalNac (3←1) Gal
with NANA (2→3) on the first Gal:
(3 ↑ 2) NANA

Disialogangliosides

Acylsphingosine (1←1) Glu (4←1) Gal (4←1) GalNac (3←1) Gal
(3 ↑ 2) on the first Gal:
NANA (2→8) NANA

Acylsphingosine (1←1) Glu (4←1) Gal (4←1) GalNac (3←1) Gal
(3 ↑ 2) NANA on the first Gal; (3 ↑ 2) NANA on the terminal Gal

Trisialoganglioside

Acylsphingosine (1←1) Glu (4←1) Gal (4←1) GalNac (3←1) Gal
(3 ↑ 2) on the first Gal: NANA (2→8) NANA; (3 ↑ 2) NANA on the terminal Gal

Fig. 3-33. Schematic representation of the four most common gangliosides

trifugation and electrophoresis of the microsomal preparation. The association of gangliosides with the membranes of the endoplasmic reticulum is so convincing that ganglioside determination was proposed as a biochemical marker for the identification of membrane structures [111].

The findings made with tissue fractionation techniques conflict with histochemical studies of Koenig, who claims that gangliosides are mainly in lysosomes. These discrepancies may result from a variety of causes: (1) the staining method used to identify the glycolipid may be nonspecific and stain sialic acid-containing proteins; (2) brain lysosomes may not be separated from microsomes; and (3) possibly most of the glycoprotein is in the microsomes, but only that in the lysosomes is detectable by histochemical techniques.

Tissue fractionation studies of guinea pig brain by differential centrifugation followed by gradient density separation have led to the claim that gangliosides are present in the "nerve ending" vesicles with acetylcholine. It is claimed that gangliosides form 10–12% of the dry weight of the synaptic vesicles [112].

Pathogenesis. In view of our fragmentary knowledge of ganglioside metabolism, it is not at all surprising that the pathogenesis of Tay-Sachs disease is not completely clarified. The absence of an enzyme involved either in using gangliosides in other biosynthetic pathways (such as those involved in synthesizing sphingomyelins and cerebrosides), or in the catabolism of the glycoprotein could explain the storage of lipoid material. A deficiency in ganglioside breakdown could also result from an anomaly in the biosynthetic pathway leading to the formation of abnormal gangliosides.

On the basis of the analysis of complex gangliosides that have been obtained from normal brain or from brains of patients with Tay-Sachs disease, Gatzkewitz, Pitz, and Sanhoff have proposed a tentative scheme for ganglioside breakdown. Neuraminidase, an enzyme found in most animal tissues, removes sialic acid from the complex ganglioside to yield a new complex A, which contains the ceramide, four sugars, and one molecule of N-acetylneuraminic acid (sialic acid) (ceramide, glucose, galactose NANA, N-acetylgalactosamine, galactose). An enzyme capable of splitting one galactose molecule from A yields a compound B, which contains the ceramide, glucose, galactose NANA, and N-acetylgalactosamine. The ceramide tetrasaccharide (complex A) or the ceramide trisaccharide (complex B) may lose the second molecule of N-acetylneuraminic acid and yield compound C (ceramide, glucose, galactose, N-acetylgalactosamine, and galactose), or complex D (ceramide, glucose, galactose, N-acetylgalactosamine) under the catalytic influence of a special enzyme. It is believed that in Tay-Sachs disease, the B and D gangliosides accumulate because of a block in their further degradation.

On the basis of enzymic studies carried out in several laboratories, a tentative scheme for the complete hydrolysis of ganglioside has been proposed (see Fig. 3-26). In this scheme, neuraminidase converts di- and trisialogangliosides to the monosialogangliosides. In the second step, a β-galactosidase splits one galactose residue from the monosialoganglioside to yield ceramide glucose, galactose-NANA-galactosamine. Third, the latter monotrihexose monosialoceramide loses the acetylgalactosamine under the specific catalytic effect of an N-acetylhexosaminidase. Fourth, a special neuraminidase, which is apparently incapable of acting on the trihexose monosialoceramide is, however, able to attack the dihexose monosialoceramide to yield the dihexose ceramide. Finally, the successive attack of the dihexose ceramide by a galactosidase, a β-glucosidase, and a ceramidase splits the trihexose ceramide into glucose, galactose, sphingosine, and fatty acid. If this scheme described in rats also functions in man, it would appear that the Tay-Sachs ganglioside accumulates in that disease because of deficiency in hexosaminidase [113, 114]. Such a conclusion would seem to be in conflict with what we described concerning the accumulation of compounds D and B, which suggests that the missing enzyme is a β-galactosidase.

Whether such a block constitutes the only mechanism by which ganglioside accumulates in brain is by no means certain. In fact, reports have appeared claiming an accumulation of the ceramide tetrasaccharide complexes (A and C) in some cases of Tay-Sachs disease. Furthermore, the coexistence of Tay-Sachs disease and abnormal gangliosides cannot be excluded (see above). Brain lipoidosis, like glycogenesis, could result from several different metabolic alterations.

Studies in two patients with generalized gangliosidosis clearly established that β-galactosidase was low in liver, spleen, kidney, and brain. In this hereditary disease (autosomal recessive), gangliosides accumulate in large excess in the brain, histiocytes of liver, spleen, and bone marrow, and in the cells of the glomerular epithelium. The disease has complex symptoms, in which skeletal alterations reminiscent of Hurler's disease are associated with progressive mental degradation, hepatomegaly and splenomegaly, and renal failure.

Okada and O'Brien showed that the ganglioside that accumulates is the {galactosyl-(1→3)-N-acetylgalactosaminyl(1→4)-[(2→3)-N-acetylneuraminyl]-galactosyl-(1→4)-glucosyl-(1→1)-[2-N-acyl]-sphingosine}. However, the same authors were unable to demonstrate a decrease in galactosidase activity in patients with Tay-Sachs, Niemann-Pick, Fabry's, or Gaucher's diseases [115].

The substances stored in liver are not likely to be identical to those stored in brain. Wolfe *et al.* [210] have reported the accumulation of a glycoprotein probably of erythrocyte origin in liver of patients with generalized gangliosidosis.

A neurological disorder with clinical, morphological and biochemical symptoms similar to those seen in humans has been observed in Siamese cats [211]. This animal model may help to clarify the sequence of steps that link the metabolic defect to morphological alterations and clinical symptomatology.

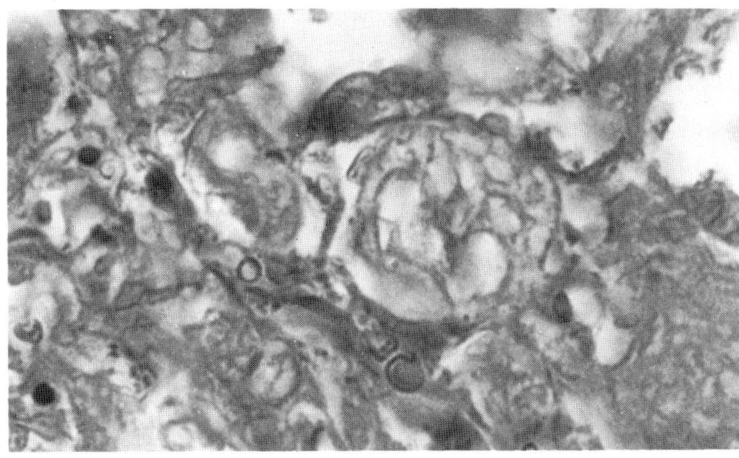

Fig. 3-34. Photomicrograph of liver cells in Gaucher's disease

Gaucher's Disease

Gaucher's disease is a hereditary disease transmitted by an autosomal recessive or dominant gene and characterized by the accumulation of cerebrosides in reticuloendothelial cells.

In contrast to what is observed in Tay-Sachs and in Neimann-Pick diseases, Gaucher's disease is not restricted to young children, but develops also in adults and even in old people. The first symptoms of the disease have been reported to occur in patients of 70 years; however, at least one-third of Gaucher's disease cases are infants or young children.

Clinical Pathology. When Gaucher described the disease for the first time, he emphasized three symptoms: splenomegaly, hepatomegaly, and skin pigmentation. Later, lipoid material was found in the reticuloendothelial cells, and 1924 it was shown that the compounds that accumulate in those cells are cerebrosides [116–119].

The pathognomonic finding in Gaucher's disease is the Gaucher cell—a large reticuloendothelial cell 20–100 microns in diameter with a typical microscopic appearance. Irregular, streaklike waves can be recognized in the cytoplasm of these cells, giving them the appearance of wrinkled tissue paper (see Fig. 3-34). Unlike Neimann-Pick and Tay-Sachs diseases, Gaucher's disease is characterized by accumulations of lipoid material primarily in the reticuloendothelial cell, and the nervous system is only indirectly affected.

Ultrastructural studies on spleen biopsies of the Gaucher cell have shown the presence of ovoid, irregular cytoplasmic bodies, which correspond to the striation described by the light microscopist (see Fig. 3-35). Degenerate mitochondria were found in close association with these bodies, and it was postulated that the accumulating material was in some way derived from mitochondria. Vesicles of the endoplasmic reticulum were also present in these areas, and the possibility that the lipoid material was incorporated in areas of focal cytoplasmic degradation was not excluded [120].

Cerebrosides accumulate in spleen, liver, bone marrow, and lymph nodes. The spleen is so full of cerebroside-loaded cells that its is enlarged, making it palpable and of soft consistency. Similarly, the liver is diffusely infiltrated with Gaucher cells; on gross examination, one finds it enlarged with translucent spots. The involvement of the reticuloendothelial cell causes compression of liver capillaries and resultant necrosis of the hepatic cells. However, neither loss of liver function nor massive hepatomegaly is often observed. The normal structure of the bone marrow is replaced by lipid-laden cells, which in some areas invade the bone and lead to fracture with all its consequences. The long bones, the phalanges, the ribs, the vertebrae, and the pelvis are preferred sites for the conglomeration of Gaucher cells, and are indeed more frequently affected by the disease than are skull bones. The lymph nodes are often enlarged because of cerebroside accumulation in the reticular cells. The oversized thoracic nodes can sometimes be detected radiographically. In infancy, the ganglion cells of the brain are also affected by the disease, and central nervous dysfunction is often dominant.

Twenty-five per cent of patients with Gaucher's disease have a typical eye injury—a wedge-shaped, yellowish streak with its base directed inward close to the cornea and its apex directed outward toward the canthus. Although one would suspect that this anomaly results from the conglomeration of Gaucher cells in the sclera, no such cells have been detected on histological examination.

Irregularly distributed patches of skin pigmentation in the exposed area of the body accompanies Gaucher's disease. The cause of this pigmentation is unknown.

Cerebroside Structure and Metabolism. The structure of cerebrosides is particularly important because it has been claimed that those that accumulate in Gaucher's disease are abnormal. The general formula of cerebrosides includes equimolecular amounts of hexose, sphingosine, and fatty acids.

In cerebrosides, the primary hydroxyl group of ceramide is linked by a glycosidic bond to either a monosaccharide or oligosaccharide chain.

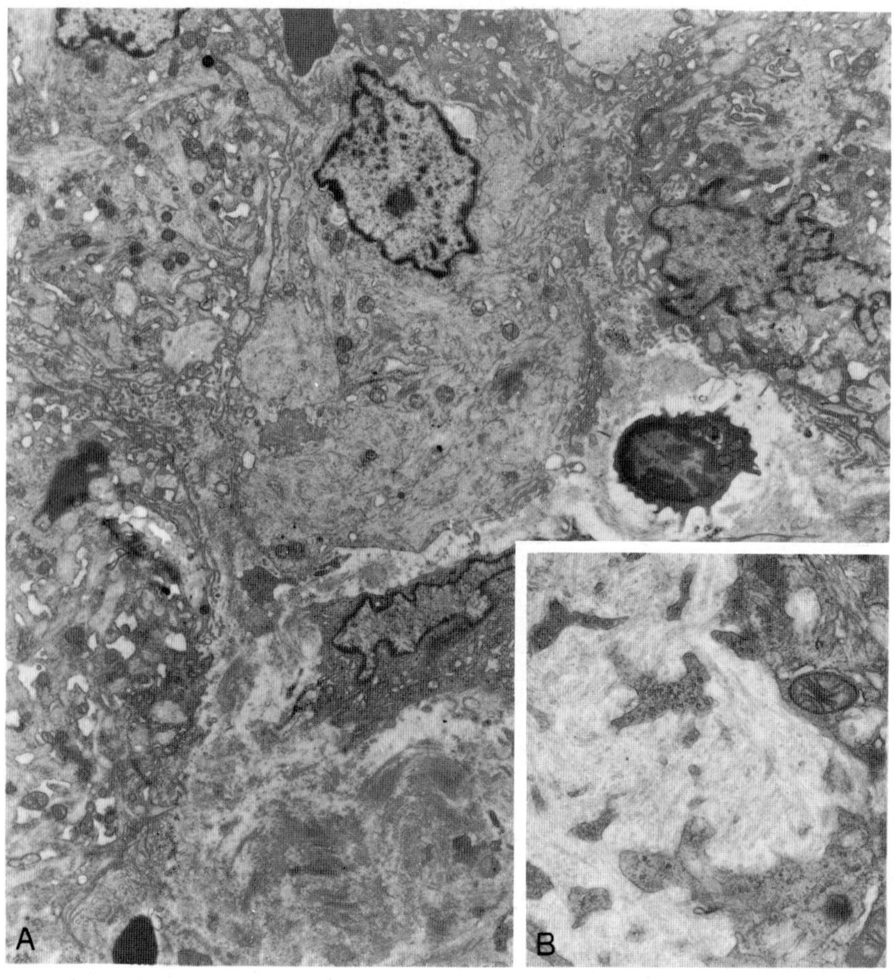

Fig. 3-35. Electron micrograph of liver cell in patient with Gaucher's disease. *A* (×3,400) *B* (×10,600) (Zamboni)

The hexose most abundantly found in brain cerebrosides is galactose, therefore, cerebrosides are sometimes referred to as galactolipids. It is not clear what determines the presence of a galactose moiety in a cerebroside molecule. The cerebrosides extracted from brains of growing pigeons fed with glucose contain glucose only, so one could assume that the type of hexose in cerebrosides is determined only by the amount of that hexose in the diet. However, these findings made in pigeons were not confirmed in growing kittens, in which the composition of the brain cerebrosides was not affected by replacing galactose by glucose in the diet.

Cerebrosides differ also in the composition of the constituent fatty acids. Alpha-hydroxy fatty acids with chain lengths ranging between 18 and 25 carbon atoms have been obtained from cerebroside hydrolysates. The saturated fatty acids most frequently found are stearic and lignoceric acids. The unsaturated fatty acids found mainly in cerebroside hydrolysates are 24 carbons long. Sphingosine is present in 95% of the cerebrosides, but cerebrosides containing dihydrosphingosine have also been reported.

The biosynthesis of the ceramide has already been discussed. The exact molecular mechanism for the incorporation of the hexose in the cerebroside molecule is not completely known. Radioisotopic studies have, however, established that all six carbons of the hexose are incorporated at once and that glucose or galactose *per se* is not a good precursor for cerebrosides. Only after their conversion to UDP derivatives can they be used effectively for cerebroside biosynthesis.

Now it seems established that the biosynthesis of cerebrosides occurs in two distinct steps. A brain or spleen microsomal enzyme catalyzes the first: the reaction of sphingosine and UDP-galactose to yield UDP and galactopsychosine. A second enzyme, about which little is known, catalyzes the reaction between the galactopsychosine and a fatty acyl-CoA to yield Co ASH and galactocerebroside (see Fig. 3-36).

It is important to know when in the course of brain development cerebrosides are synthesized because such information might give some clues on the pathogenesis of Gaucher's disease and also some forms of mental deficiency. In rodents, cerebrosides are not actively synthesized during embryonic life or at birth. In mice, the incorporation of labeled hexoses in cerebrosides occurs approximately four days after birth and in rats, seven days after birth. In either case, the synthesis of brain cerebrosides seems to precede that

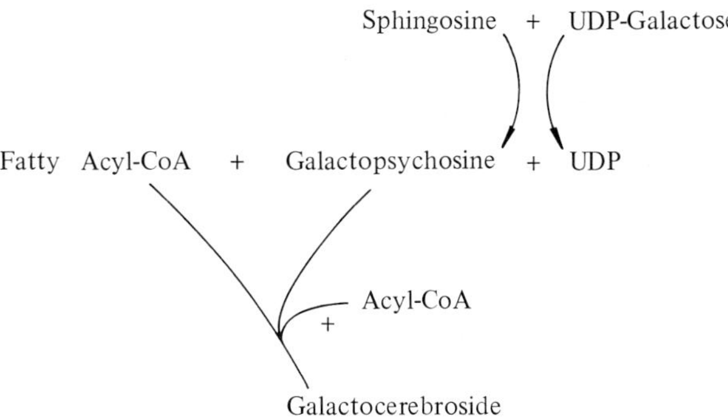

Fig. 3-36. Biosynthesis of cerebrosides

of gangliosides, thus it has been proposed that cerebrosides are precursors of gangliosides.

Breakdown and Pathogenesis. Not much is known of the breakdown pathway of cerebrosides. Results obtained in *in vivo* studies are controversial. Although isotopic studies of Radin and his associates indicated that the half-life of brain cerebrosides is approximately 13 days, Davison claimed that there was no turnover of cerebrosides.

Two enzymes have been found that catalyze cerebroside breakdown; one is a brain enzyme, the other a spleen enzyme. The brain enzyme has a broad specificity and is capable of splitting the hexose moiety of most cerebrosides to yield galactose, or glucose as the case may be, and ceramide. Ceramide lactoside can also serve as substrate for the brain enzyme. In contrast, the spleen enzyme is specific for glucocerebrosides and is believed to be missing in patients with Gaucher's diesease. In fact, work from the National Institutes of Health has demonstrated the existence of low enzyme activities in spleens of patients with Gaucher's disease. The enzyme activity was 30–800 times lower than normal in spleen of patients with Gaucher's disease. Inasmuch as isotope incorporation studies suggest that the rate of cerebroside synthesis is normal in patients with Gaucher's disease, it seems compelling to conclude that the disease results from the lack of catabolic enzyme.

Cerebrosides accumulate in the reticuloendothelial cells in Gaucher's disease. The cerebroside content of spleen and liver is increased tenfold; however, an increase in the cerebroside content of brain has not been convincingly demonstrated.

Dead red cells or senescent polymorphonuclears are believed to be the source of the cerebrosides that accumulate in the reticuloendothelial cells of patients with Gaucher's disease. One of the products of red cell degradation is "globoside," a galactosamine-containing glycolipid. When such globosides are injected, the hexosamine is split off under the influence of a hexosaminidase. The hexosamine-free globosides are rapidly degraded further by the action of glucosidase. In the absence of glucosidase, one may expect the accumulation of glucocerebroside in the reticuloendothelial cells, which is precisely what happens in Gaucher's disease.

The activity of at least some of the enzymes involved in cerebroside breakdown is increased when the life span of red cells is reduced. Senescent and injured cells are selectively destroyed in the spleen. Part of the degradative process consists of hydrolyzing the membrane's neutral lipids: lecithin, phosphatidyl ethanolamine, sphingomyelin, phospholipids, and the cerebroside globoside. The increased red cell destruction by phenylhydrazine is accompanied by a significant increase in activity of spleen glucocerebrosidase and sphingomyelinase [121].

Kattlove and his associates believe that the glucocerebroside which accumulates in Gaucher's disease is principally derived from polymorphonuclear cells which have a rapid turnover and are rich in ceramide lactoside [122]. (This explains the occasional presence of Gaucher cells in marrow of patients with chronic myelogenous leukemia.)

An interesting alteration in Gaucher's disease is the increase in serum acid phosphatase. The increased enzyme activity can be detected only if phenylphosphate is used as a substrate. The enzyme activity is not altered by tartrate, formaldehyde, or copper, and it was therefore suggested that the acid phosphatase found in the serum of patients with Gaucher's disease does not originate in erythrocytes, liver, prostate, spleen, or bone marrow. However, histochemical studies of the Gaucher cell led investigators to conclude that the acid phosphatase increase results from its release from proliferating Gaucher cells.

In addition to an increase in cerebroside, ganglioside content of the spleen cell also increases in Gaucher's disease. The gangliosides are probably a normal constituent of the spleen. The molecular composition (not the structure) of one of the accumulating gangliosides has been established. And the ceramide moiety of the accumulating gangliosides resembles the ceramide of the accumulating cerebroside, which suggests that the metabolism of gangliosides and of cerebrosides may be related.

Niemann-Pick Disease

Niemann-Pick disease is a lipoidosis resulting from the accumulation of sphingomyelin in reticuloendothelial cells, parenchymal cells, and in ganglion cells. The lipid-laden cell typically is large, stuffed with droplets all of the same size, and its nucleus, often pyknotic, is shoved to the periphery close to the membrane. These changes give the cell a foamy or mulberry appearance by which it can be distinguished from a Gaucher cell [123, 124].

The reticuloendothelial cells of the spleen and liver are principally affected by the disease (see Fig. 3-37). The spleen is so filled with reticuloendothelial cells loaded with sphingomyelin that it is markedly enlarged and becomes palpable. In the liver (see Fig. 3-38), the lipoid material accumulates mainly in the Kupffer

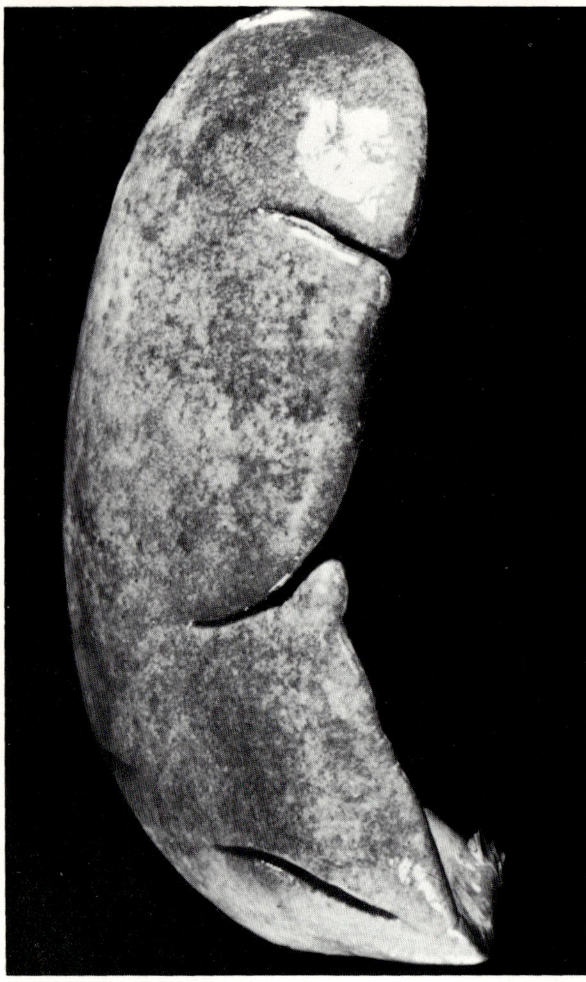

Fig. 3-37. Enlarged spleen in patient with Niemann-Pick disease

cells, causing enlargement and yellowish discoloration. Sometimes vacuoles are also found in hepatic cells, probably the result of the accumulation of neutral fats or glycogen rather than sphingomyelin. The abnormal accumulation of lipoid material in the reticuloendothelial cells lining the sinusoids of the liver may obstruct the smaller bile ducts and lead to jaundice. Only in rare cases does jaundice result from the occlusion of a major bile duct by an enlarged lymph node. Although the cells of the lymphoid tissues of the thymus and lymph nodes may store abnormal amounts of lipoid material, the accumulation is seldom excessive. In some patients, the liver and spleen are so enlarged that compression of the inferior vena cava with all its consequences ensues.

In contrast to what happens in Gaucher's disease, the accretion of lipoid material in the liver does not induce fibrous proliferation. Although spleen and liver are the primary targets of the disease, the lipoid material sometimes accumulates in alveolar cells of the lung and in the endothelial cells of the small capillaries of the lung. These lesions appear on a radiological picture as mottling or miliary nodules. Careful examination of tissues obtained at autopsy may demonstrate foam cells in bone marrow, thyroid, pituitary, pancreas, salivary glands, kidney, and other organs.

In addition to the spleen and the liver, the nervous system and the eye are also preferred sites for the accumulation of sphingomyelin in Niemann-Pick disease. Lipoid droplets are found principally in the glial cells, which acquire a foamy appearance. On histological examination of brain material, foam cells are found in the brain substance, in the leptomeninges, the choroid plexus, and even in the connective tissue that surrounds the vessels. The ganglion cells of the brain also become loaded with lipoid material, a process that leads to cellular degeneration followed by demyelination. These cellular alterations are responsible for the macroscopic changes observed in the brain. Because of gliosis, the brain is harder and may even have a leathery consistency. The degeneration of the grey matter is probably responsible for the widening of the sylvian and interparietal fissures.

The clinical picture is varied; all types of motor, sensory, and mental deterioration may occur, with eventual complete collapse of the central nervous system. Cherry-red spots develop in the eyes of patients with Niemann-Pick disease just as in Tay-Sachs disease.

The striking biochemical alteration in Niemann-Pick disease is the accumulation of lipoid material in the involved tissues, where the total lipid content may be two, three, or even four times the normal. After the accumulating lipid was fractionated, it was established that sphingomyelin loads the cells.

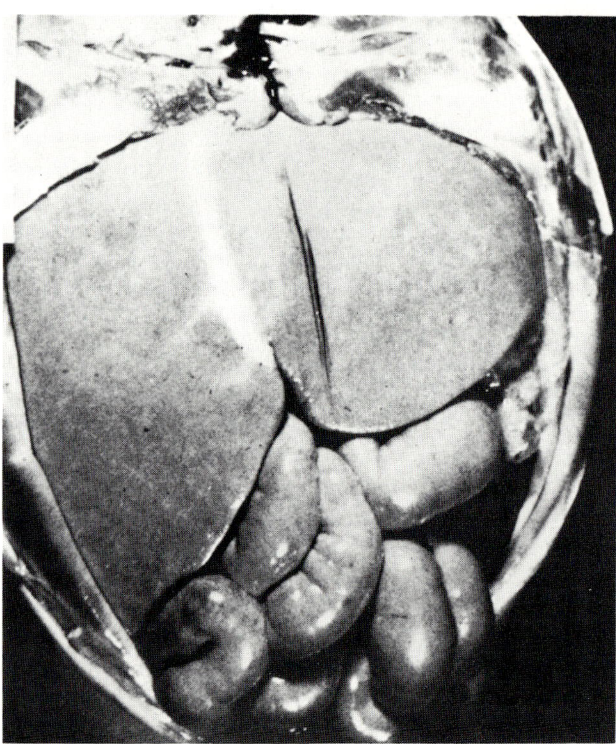

Fig. 3-38. Enlarged liver in patient with Niemann-Pick disease

Although our present knowledge of the chemistry and metabolism of sphingomyelin has not provided an answer to the pathogenesis of Niemann-Pick disease, the known facts will briefly be summarized with the hope that they will prepare the reader for further developments in the field of lipoidoses.

The sphingomyelins constitute a group of substances containing equimolecular amounts of sphingosine or dihydrosphingosine, choline, and phosphoric acid. The fatty acid residue is attached to the base by the formation of an amide linkage between the carboxyl group of the fatty acid and the amino group of the sphingosine. Sphingomyelins are present in most tissues and in plasma. In liver, they constitute 5–20% of the total phospholipids. In the red cell they are a structural part of the stroma; in brain and spinal cord they are found in cells, but they are also an important constituent of the myelin sheets. Fifteen to twenty per cent of the plasma phospholipids are sphingomyelins. Sphingomyelins have been extracted from various sources; those prepared from liver and spleen contain palmitic and lignoceric acids. Traces of nervonic acid and an unidentified C_2O fatty acid are present in sphingomyelins obtained from brain. Behenic acid was in association with the sphingomyelin of horse spinal cord. It is not known how much sphingomyelins vary by the type of fatty acid contained in these molecules, nor is the significance of these variations understood, but a more accurate knowledge of the composition of sphingomyelins and the quantitative determination of the various types of sphingomyelins will probably uncover new clues as to the pathogenesis of Niemann-Pick disease.

The biosynthesis of sphingosine and the formation of ceramides have already been discussed. The last step in the biosynthesis of sphingomyelins (see Fig. 3-39) involves cytidine diphosphate choline as a phosphate donor, in a reaction similar to that involved in the biosynthesis of lecithin. Scribney and Kennedy have found a chicken liver enzyme, now called the PC ceramide transferase, which catalyzes the following reaction:

CDP-choline + N-acyl sphingosine
⇆ CMP + sphingomyelin

Little is known of sphingomyelin catabolism. The existence of an enzyme capable of splitting the phos-

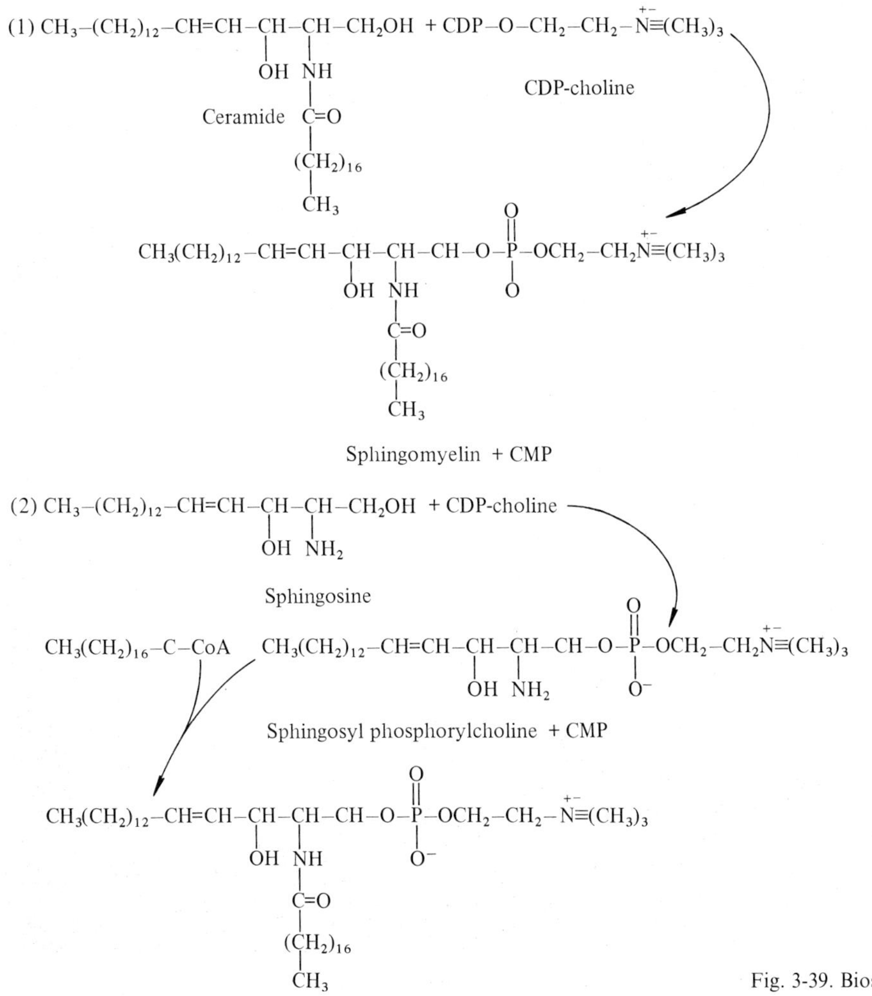

(1) $CH_3-(CH_2)_{12}-CH{=}CH-CH(OH)-CH(NH-C{=}O-(CH_2)_{16}-CH_3)-CH_2OH$ + $CDP-O-CH_2-CH_2-\overset{+-}{N}{\equiv}(CH_3)_3$

Ceramide CDP-choline

$CH_3(CH_2)_{12}-CH{=}CH-CH(OH)-CH(NH-C{=}O-(CH_2)_{16}-CH_3)-CH-O-P(=O)(O)-OCH_2-CH_2\overset{+-}{N}{\equiv}(CH_3)_3$

Sphingomyelin + CMP

(2) $CH_3-(CH_2)_{12}-CH{=}CH-CH(OH)-CH(NH_2)-CH_2OH$ + CDP-choline

Sphingosine

$CH_3(CH_2)_{16}-C-CoA$ $CH_3(CH_2)_{12}-CH{=}CH-CH(OH)-CH(NH_2)-CH-O-P(=O)(O^-)-OCH_2-CH_2\overset{+-}{N}{\equiv}(CH_3)_3$

Sphingosyl phosphorylcholine + CMP

$CH_3(CH_2)_{12}-CH{=}CH-CH(OH)-CH(NH-C{=}O-(CH_2)_{16}-CH_3)-CH-O-P(=O)(O^-)-OCH_2-CH_2-\overset{+-}{N}{\equiv}(CH_3)_3$

Fig. 3-39. Biosynthesis of sphingomyelin

phorylcholine contained in sphingomyelin molecules to yield ceramides has been reported. Evidence now indicates that the fatty acid moiety of sphingomyelin can be split by one of the known or as yet unknown phosphorylases. Studies of Davison and his associates demonstrate that the lipid fraction of the myelin sheet incorporates a radioactive precursor during brain development and that the precursor persists in this lipid fraction after development. Such findings indicate that once the myelin structures have incorporated the lipids during brain development, these structures remain stable for life.

Our knowledge of the fate of sphingomyelin in cells is still so fragmentary that all explanations for the pathogenesis of Niemann-Pick disease can only be hypothetical.

The accumulation of sphingomyelin is likely to result from an alteration in the metabolic pathways using sphingomyelin. A block in the use of sphingomyelin for the biosynthesis of larger macromolecules or for breakdown explains the storage disease, but does not answer the questions raised by the metabolic alterations. Why does sphingomyelin accumulate in tissues without leading to an increase of the sphingomyelin levels in blood? What are the reasons for the accumulation of cholesterol and neutral fats often associated with sphingomyelin in Niemann-Pick disease?

Partial answers to the pathogenesis of Niemann-Pick disease have been obtained recently. An enzyme, sphingomyelinase, which hydrolyzes sphingomyelin to yield phosphorylcholine and ceramide has been found and partially purified from liver. Drastic reduction in the activity of this enzyme has been found in at least some patients with Niemann-Pick disease (see Fig. 3-40). In other cases the biochemical defect remains unknown [125], especially in those forms of the disease in which cholesterol accumulates (Nova Scotia variant).

Sphingomyelinase is believed to be a lysosomal enzyme. It has been partially purified from human liver and spleen. Isoelectric focussing of a partially purified preparation of the enzyme has permitted identification of two isozymic forms; one with a pH optimum of 3.8 and another with a pH optimum of 5 [212].

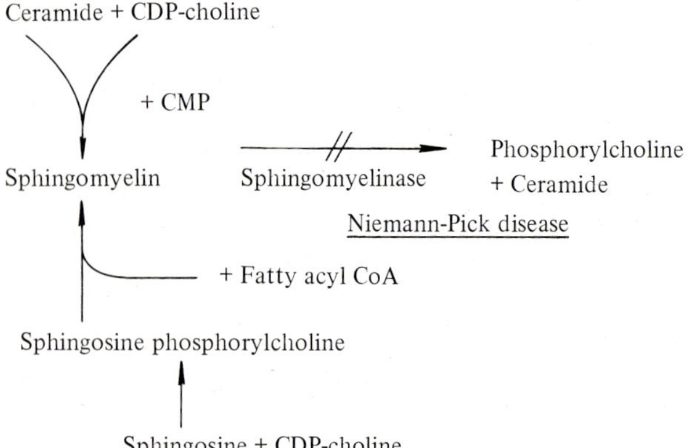

Fig. 3-40. Pathogenesis of Niemann-Pick disease

Leukodystrophies and Fabry's Disease

In addition to the ordinary cerebrosides already described, the brain contains cerebroside sulfates, or sulfatides—lipid esters composed of a ceramide galactose and sulfate. The exact structure of the sulfatides is not known, but it has been proposed that the carbon 3 of galactose is involved in the formation of an ester bond with sulfuric acid. Sulfatides have been found in a number of tissues (liver, kidney, lung, heart, muscle, spleen), but brain seems to contain the highest concentration of sulfatides. Sulfatides are likely to accumulate in two forms of hereditary lipoidosis: metachromatic leukodystrophy and Fabry's disease. Both these diseases are rare and will be considered only briefly.

Little is known of the biosynthesis, breakdown, or role of sulfatides. Two observations seem established: the biosynthesis of sulfatides coincides with the appearance of myelin; and studies with the aid of a radioactive precursor have established a slow but definite turnover of sulfatides in the adult brain. Consequently, the brain constantly synthesizes and degrades sulfatides. Similar anabolic and catabolic processes are likely to occur in other tissues.

Two enzymes have activities related to the metabolism of sulfatides: a sulfatase, which is capable of hydrolyzing sulfatides to yield cerebroside and sulfate, and a "sulfatide synthetase," which catalyzes the formation of cerebroside and a sulfate donor (phosphoadenosine phosphosulfate). The exact role of these enzymes in the overall sulfatide metabolism is not clear. A galactosylglucosyl ceramide esterfied with sulfuric acid in position 3 of the galactose has been found in kidney. The dihexose ceramide sulfatide is believed to be synthesized from the neutral ceramide in presence of phosphoadenosine phosphosulfate. The synthetase is a microsomal enzyme. The sulfatase is believed to be a lysosomal enzyme.

In 1928, Bielschowsky and Henneberg coined the term "leucodystrophia" to designate a group of hereditary diseases (mendelian recessive) characterized by the progressive degeneration of the cerebral white matter. The degeneration, which is diffuse and symmetrical, involves the centrum ovale, the gyral cores, and the cerebellar white matter. Three pathological alterations are observed: demyelinization, gliosis, and lipoid accumulation. Depending upon the staining properties of lipoid material, leukodystrophies have been divided into sudanophilic and nonsudanophilic lipoidoses.

Krabbe's Disease, or Globoid Lipoidosis. The brain of these patients usually appears normal, except for a reduction in size. On section, leukodystrophy of the centrum ovale, the brain stem, and the cerebellum is detectable. Histologically, demyelinization, astrocytic proliferation, and accumulation of histiocytes can be seen. Some histiocytes have an epithelioid appearance, and others are large, globoid cells. The histiocytes contain a substance not readily soluble in ordinary lipid solvent; it does not stain with Sudan black, but does stain strongly with para-aminosalicylic acid (PAS). Andrews and Cancilla [126] have studied the ultrastructural manifestations of the lipoidosis (see Figs. 3-41 and 3-42).

Cerebrosides and sphingomyelin are believed to accumulate in the globoid bodies. In fact, the injection of cerebrosides into rats has led to the appearance in the white matter of cells that resemble globoid cells. The biochemical defect in Krabbe's disease is still unknown, but two clues are available. There are no defects in sphingomyelin breakdown, there is a shift in the ratio of cerebrosides to sulfatides (from 3 to 1 in the normal individuals to 12 to 1 in those with leukodystrophy. These observations have led to the suggestion [127–130] that the lipidosis results from a deficiency of a sulfate-transferring enzyme (see Fig. 3-43).

A report from Wenger and his associates indicates that brain, liver, tissue, and fibroblast cultures from children with Krabbe's disease are deficient in the galactosidase that normally degrades lactosyl ceramide [213]. In these patients the breakdown of galactosyl ceramide and monogalactosyl diglyceride is also low. The decrease in β-galactosidases is in conflict with previous findings suggesting that β-galactosidase activities were normal or increased in patients with Krabbe's disease [214]. Clearly more data are needed before a definite description of the metabolic error in Krabbe's disease will be available.

In *metachromatic leukodystrophy*, the brain is often heavier than normal. The leukodystrophy is diffuse, as in the globoid type. Demyelinization, astrocytosis, and histiocytosis are found in the areas of leukodystrophy. Metachromatic globules are dispersed in the white matter and within histiocytes. Groups of neurons containing metachromatic lipids are also often found (see

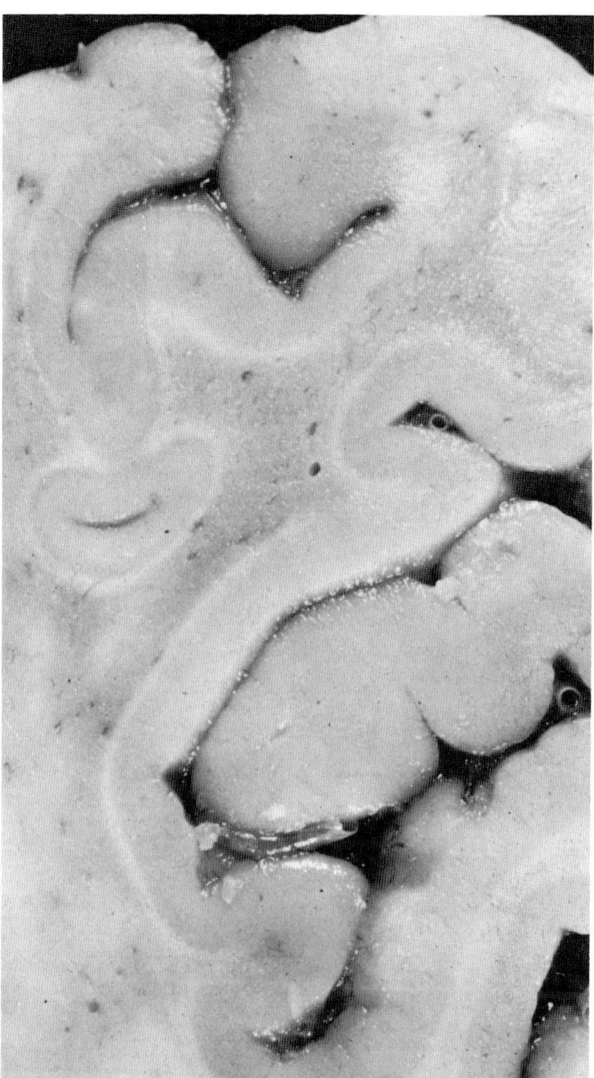

Fig. 3-41. Gross appearance of brain of patient with Krabbe's disease (Cancilla)

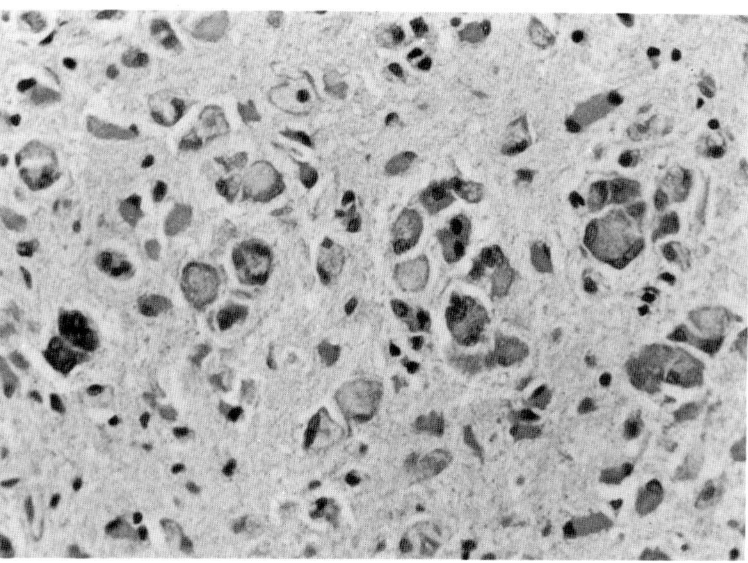

Fig. 3-42. Histological appearance of globoid cell leukodystrophy (Cancilla)

Galactocerebroside + Phosphoadenosine phosphosulfate
(protein bound?)

Sulfate transferase

Krabbe's disease

Phosphoadenosine phosphate + Cerebroside sulfate (sulfatide)

Sulfatase + H_2O ← Metachromatic leukodystrophy

Galactocerebroside + H_2SO_4

Sulfatide metabolism

Fig. 3-43. Pathogenesis of metachromatic leukodystrophy and Krabbe's disease

Fig. 3-44). Careful examination of all organs might reveal the presence of metachromatic granules in the renal tubules, the gall bladder, and the intrahepatic bile ducts. In metachromatic leukodystrophy, the levels of sulfatides increase more (2 to 4 times) in the white than in the grey matter. The increase in sulfatides is associated with an increase in cerebrosides. The concentration of the other lipids is nearly normal. The proportion of sphingolipids containing long-chain fatty acids (18 or more carbons) is decreased in metachromatic leukodystrophy, and the ceramides found in the accumulating lipids contain mainly short-chain fatty acids. The relationship between the lipid accumulation and the demyelinization and the symptoms of leukodystrophy remains to be elucidated.

But it is significant that in metachromatic leukodystrophy the myelin fibers have an abnormal composition—they are rich in sulfatides (mono- and dihexose

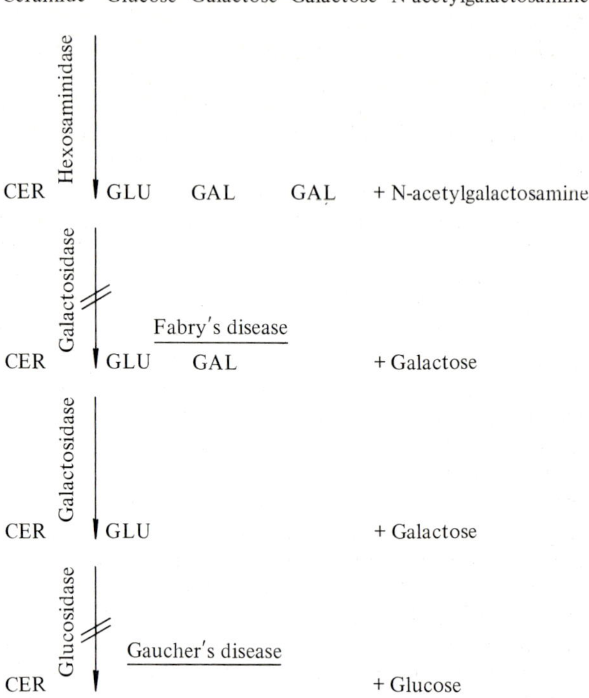

Fig. 3-45. Pathogenesis of Fabry's and Gaucher's diseases

sulfatides) and low in cerebrosides. Consequently, the myelin sheet in leukodystrophy must contain an abnormal amount of charged lipids. This alteration is bound to lead to changes in the surface properties of the myelin sheet. It is believed that the sulfatase which splits the sulfate moiety of either mono- or dihexose cerebroside or a cofactor necessary for enzyme activity is missing in metachromatic leukodystrophy (see Fig. 3-43).

Fabry's syndrome, or angiokeratoma corporis diffusion, is a disease affecting several organs, and although

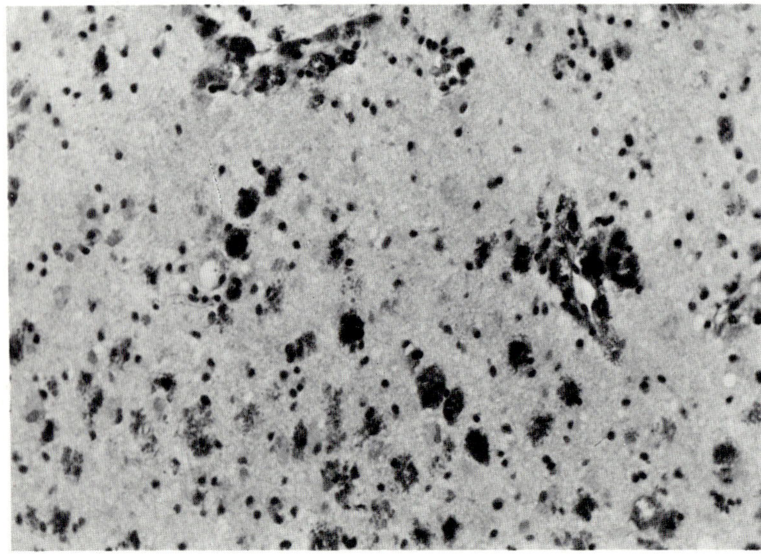

Fig. 3-44. Microscopic appearance of metachromatic leukodystrophy (Cancilla)

the most obvious lesions are those of the skin and mucosa, kidney lesions often lead to death.

The skin lesions are often mistaken for petechiae or telangiectasia and are deep red or black. They may be barely visible or reach diameters of 2–4 mm. Lesions usually appear in clusters on the lower abdomen (below the umbilicus), the hips, and thighs. Palms, soles, ears, and scalp are usually not involved. The mucosa of mouth and conjuctiva also present typical Fabry's lesions. The kidney involvement is responsible for the development of uremia, which often is the cause of death. During the fourth or fifth decade of his life, the patient with Fabry's syndrome is also prone to cardiovascular and cerebrovascular accidents.

Studies by workers of the National Institute of Neurological Diseases and Blindness demonstrated that the major component that accumulates in Fabry's disease is galactosyl-galactosyl-glucosyl ceramide. The ceramide trihexoside is a breakdown product of globoside (ceramide glucose galatose galactose—N-acetylgalactosamine). A hexosaminidase splits the hexosamine from the globoside to yield the trihexoside (see Fig. 3-45). The trihexoside is converted to the lactosyl ceramide by a galactosidase. The elimination of another galactose moiety yields the glucosyl ceramide. Glucosidase hydrolyzes the glucosyl ceramide to form ceramide and glucose. Finally, ceramidase separates the sphingosine and the fatty acid.

We have mentioned already that glucosidase is missing in Gaucher's disease. In Fabry's disease, evidence stemming from the study of the structure of the accumulating product and direct measurement of enzyme activity in kidney indicates that the ceramide trihexosidase is missing. The interrelationship between the metabolism of the various sphingolipids is summarized in Fig. 3-46, and the pathogenesis of the various lipoidoses is outlined in Fig. 3-47.

Inasmuch as some of the enzymes missing in inborn errors of metabolism are acid hydrolases, Hers has developed the concept of lysosomal disease. Whenever a lysosomal enzyme is missing it leads to a storage disease inside the vacuoles where the stored material is usually heterogeneous. The disease, which can affect all tissues, is progressive because at birth the lysosomes have not yet had time to accumulate large amounts of material. It is possible that this interesting generalization will, over the years, prove to be correct. At present it leaves a lot to be desired. Even in the most classical of all lysosomal diseases, Pompe's disease caused by a deficiency in acid maltase, all the glycogen is not found in lysosomes. In von Gierke's disease in which the glycogen deficiency results from the absence of glucose-6-phosphatase, an enzyme located in the endoplasmic reticulum, portions of the glycogen are often found in areas of focal cytoplasmic degradation, or secondary lysosomes. In many cases variants of an enzyme defect have been described, some of which do not involve a lysosomal enzyme. Yet, the electron microscopic picture is not essentially different whether the defect is caused by a lysosomal enzyme

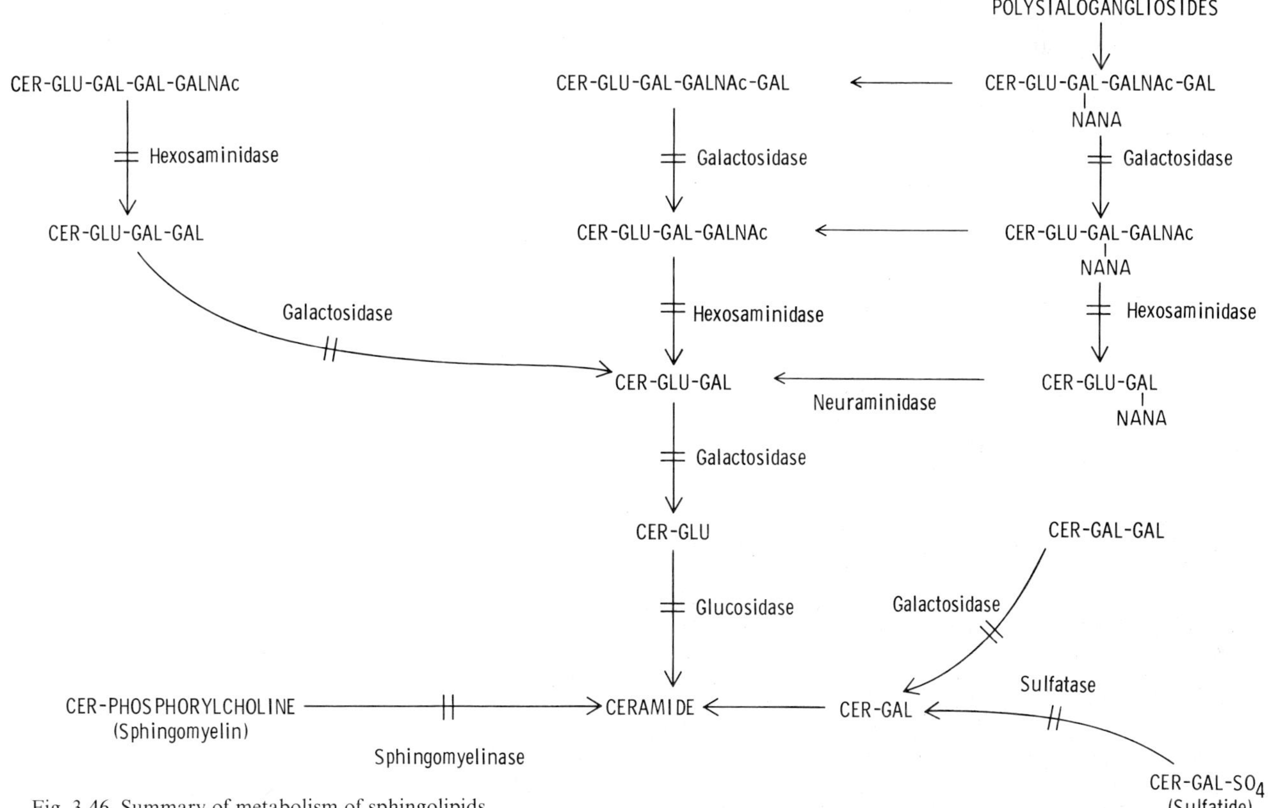

Fig. 3-46. Summary of metabolism of sphingolipids

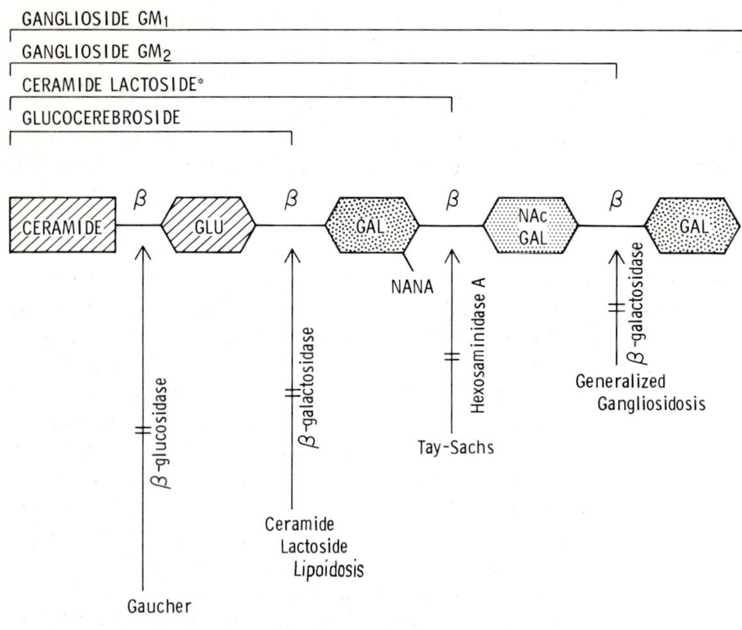

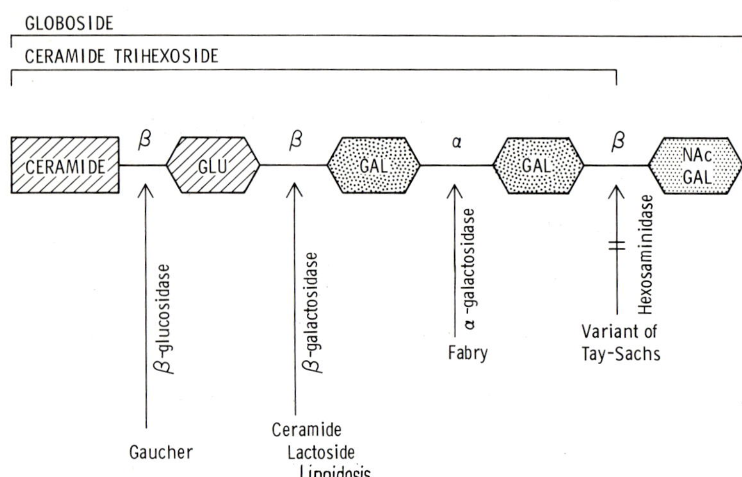

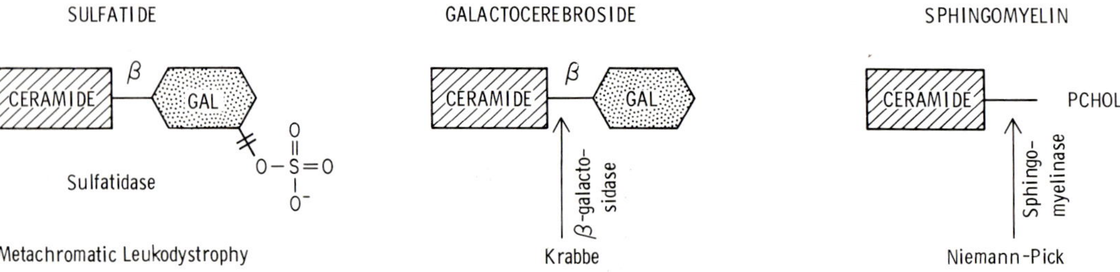

Fig. 3-47. Accumulating compounds and missing enzymes in sphingolipoidosis

or not. Studies on the intracellular distribution of the missing enzymes are very often lacking or incomplete. All lysosomal enzymes do not yield insoluble substrates. Therefore one can predict that some of the defects in lysosomal enzymes will not lead to storage diseases. In many cases it remains to be proven that the substrate is capable of entering the lysosomal membrane to be degraded.

In conclusion, it cannot be excluded that the surrounding of the storage material by membranes results from their segregation in a process akin to focal cytoplasmic degradation (see chapter on necrosis) and is a secondary, rather than a primary, event of the disease [224].

Porphyrin Metabolism and Porphyria

Chemistry of Porphyrins

The structure of the porphyrin molecule is so complex that a discussion of porphyrin biosynthesis should be preceded by a brief review of the chemistry of the biological porphyrins. Porphyrins form a large group of components derived from porphin, a cyclic compound made of four pyrrole rings (referred to as ABCD) held together by methylene bridges. Because the porphyrin nucleus has eight carbons in which substitution may occur, the number of porphyrins and the number of possible isomers of porphyrins are enormous.

The formula indicates that porphin is a highly conjugated system. Stability in the molecule is secured by distribution of the energy in the molecule. Maximum stability is achieved when the energy is distributed equally among all components of the molecule. Equal distribution of the energy in the case of porphin is achieved only if the molecular components are disposed on a single plane. Then the charges of one pair of electrons associated with each double bond are removed from that bond and redistributed in the conjugated system. Thus, the porphin molecule, like the benzene molecule, is stabilized by resonance and exists in the form of resonating hybrids. The graphic representation of the molecule as it appears in the formula

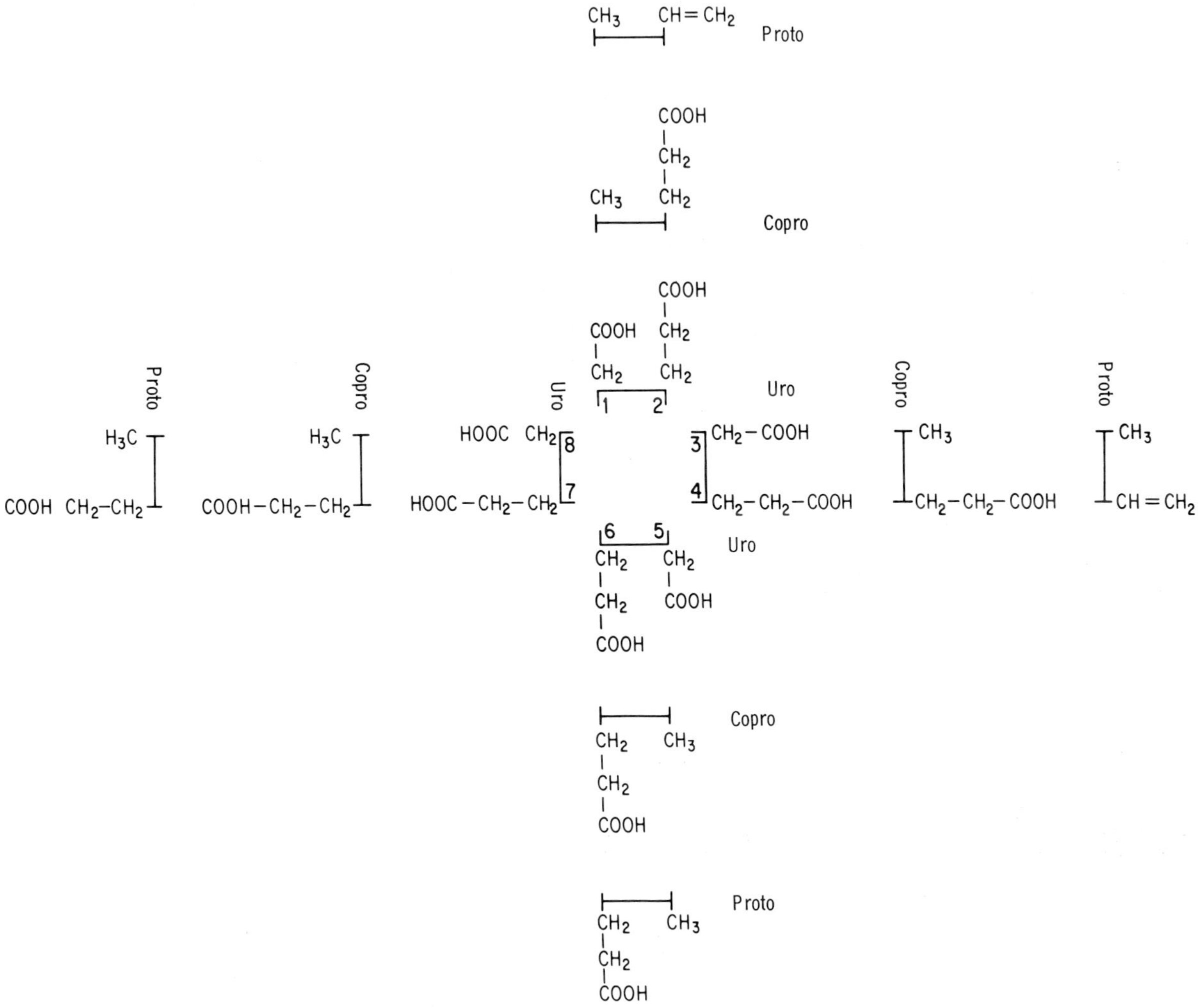

Fig. 3-48. Side chains of natural porphyrin

necessarily includes fixed double bonds, but the picture does not truly represent the porphin molecule in which a continuous circuit of moving electrons exists within the flat molecule.

The energy developed by these moving electrons, the resonance energy, undoubtedly determines the physicochemical and biological properties of the porphin. A detailed interpretation of these properties in terms of electronic activities of the molecules is not always available, but refer to specialized texts for such information.

Porphin and its substituted derivatives exhibit characteristic absorption spectra by which they can be identified. Porphyrin in neutral solution has an absorption peak near 400 mμ, and a Soret and four other bands between 550 and 600 mμ. Porphyrins are stable and resist the action of strong acids and bases. They are not altered by sulfonation or bromidation. However, porphyrins react readily in two ways: they are oxidized by peroxides and form chelates with transition metals.

In natural porphyrins, the 8 carbons at the periphery of the carbon ring are substituted by various side chains. Protoporphyrin has four methyl, two vinyl, and two propionic acid side chains. These could theoretically be arranged into 15 different isomers of protoporphyrin. Protoporphyrin IX formula is the isomeric form found in heme of hemoglobin, catalase, peroxidase, cytochrome c, etc. The side chain stabilizes the heme molecule and attaches the porphyrin to the protein. Thus, in hemoglobin, the propionic acid groups are bound by coulombic forces to the amino groups of the protein. In cytochrome c, the vinyl groups form covalent thioester bonds with cysteine. The formulas of a few other porphyrins of biological importance, uroporphyrin I and II, coproporphyrin I and III, are given in Fig. 3-48. Porphyrins form chelate complexes with iron to yield heme. In heme, the central iron atom forms a coordination complex with the nitrogen of the pyrrole ring. The properties of the heme molecule are discussed in another chapter.

Biosynthesis of Porphyrins

The brilliant work of Shemin and his associates opened the field of porphyrin synthesis [131]. They studied the incorporation of the labeled atoms of single compounds, such as glycine and acetate, into each carbon and nitrogen atom of the porphyrin ring. Labeled precursors were administered, and the porphyrin nucleus was disrupted by various organic degradation methods. The contribution of each precursor in labeling a specific atom of the porphyrin ring was then determined. It was observed that the methylene carbon of glycine forms the bridge between two pyrrole rings and that the amino group of glycine provides the heterocyclic nitrogen of each pyrrole ring. The remaining 26 carbons of the porphyrins were all derived from acetate. Therefore, it seemed logical to assume that another member of the Krebs cycle acted as a precur-

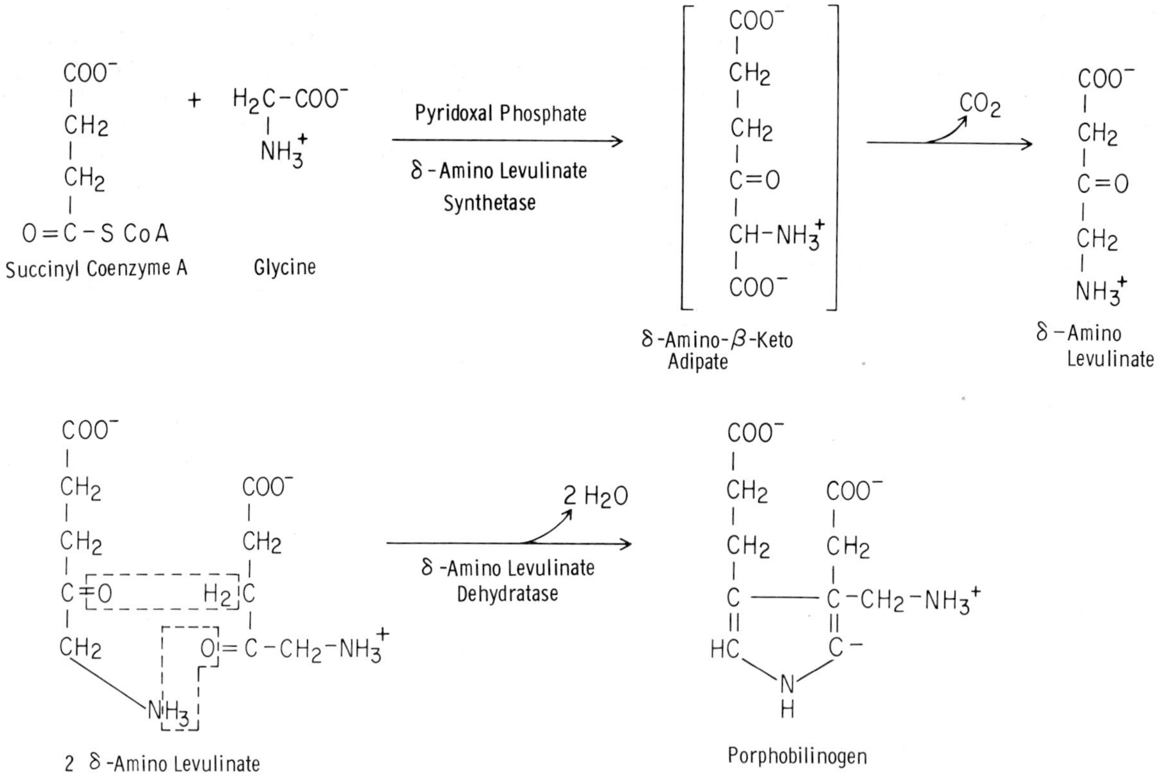

Fig. 3-49. Biosynthesis of porphobilinogen

sor, and indeed, when ^{14}C-labeled succinate, citrate,or α-ketoglutarate was used as precursor, the labeled carbons were incorporated into the porphyrin ring as predicted. Therefore, it was postulated that the condensation of two molecules of succinate with one of glycine led to the formation of a common pyrrolic precursor. It is now known that the condensation involves succinic CoA rather than succinate. Porphobilinogen was discovered at that time in the urine of porphyrinuric patients. Porphobilinogen satisfied all the assumptions made by Shemin with respect to the structure of the compound that could act as a precursor of the tetrapyrrole ring. It was soon proved that porphobilinogen is an active precursor in porphyrin synthesis of red cells. Later, another intermediate of the porphyrin pathway was discovered—δ-amino levulinic acid, which was found to be a precursor of porphobilinogen. Thus, the first stage of experimentation on the biosynthesis of porphyrins established that the molecule is synthesized *de novo*, starting with simple compounds such as glycine and acetate. The acetate is then transformed to succinyl CoA by way of the Krebs cycle, and succinyl CoA condenses with glycine to form an amino keto adipate that yields δ-amino levulinate by decarboxylation (see Fig. 3-49).

Experiments made on pyridoxal-deficient animals suggested that pyridoxal phosphate and iron are required for the biosynthesis of porphyrin. It was later established that pyridoxal phosphate is required for the formation of δ-amino levulinic acid, probably by participating in the formation of "active glycine." The condensation of succinate and glycine leads to the formation of a very labile α-amino-β-keto adipic acid. The participation of α-amino-β-keto adipic acid as an intermediate in the reaction was established in experiments proving the acid to be an efficient precursor of porphyrin biosynthesis *in vitro*. An enzyme system capable of catalyzing the succinyl CoA-glycine condensation and the decarboxylation of the intermediate to yield amino levulinic acid has also been obtained from a particular fraction of chicken erythrocyte. In liver, an enzyme has been found in the mitochondria [132].

The exact mechanism of action of the δ-amino levulinic acid synthetase is not clear. It involves the activation of glycine by pyridoxal phosphate in a Schiff base formation on the enzyme. A proton is lost in the process, and a stable carbanion is formed. In the presence of succinyl CoA, the activated glycine is transferred to succinate, but it is not clear whether the decarboxylation occurs while the succinate and glycine are still attached to the enzyme or later. The activity of the δ-amino levulinic acid synthetase is inhibited by L-penicillamine, thiosolidine, and L-cysteine. The compounds presumably react with pyridoxal phosphate forming a thiazolidine ring. The enzyme molecule is itself inhibited by the addition of para-chloro-mercuribenzoic acid. Insofar as succinate is a precursor of δ-amino levulinic acid, the pathway for succinyl CoA synthesis acquires particular significance (refer to the first part of this book). In the reticulocyte, the tricarboxylic acid cycle is probably the main pathway of succinyl CoA formation. Therefore, the disappearance of the aerobic oxidative pathway in the erythrocyte is associated with a block in the formation of δ-amino levulinic acid, and by an interruption of the biosynthesis of heme.

The fate of δ-amino levulinic acid is dual. It may be converted to porphobilinogen by a pathway to be described below, or under the influence of a transaminase it may yield α-ketoglutaraldehyde, which in turn produces α-ketoglutarate or succinate (see Fig. 3-50). Thus, δ-amino levulinic acid occupies a key position between the citric acid cycle and the porphyrins' biosynthetic pathway. The significance of δ-amino levulinic acid in metabolism is illustrated in Fig. 3-50 showing the metabolic conversions involved in the so-called Shemin succinate glycine cycle.

The condensation of two molecules of levulinic acid leads to the formation of a pyrrole ring with two substituted carbons, one containing a propionic and the other an acetic acid side chain. This condensation of two molecules of δ-amino levulinic acid occurs in two steps—condensation and dehydration. Each step involves a different point of the molecule, but both reactions are apparently brought about by a single enzyme called δ-amino levulinic acid dehydrase. This enzyme is abundantly distributed in nature, and the active form of the enzyme contains SH groups. Gibson studied the distribution of δ-amino levulinic acid dehydrase in various tissues. The enzyme is very active in liver and is present in relatively high concentrations in both kidney and bone marrow. However, little enzyme is found in the blood, brain, heart, or small intestine. The concentration in tissues and blood increases slightly in anemia and markedly in liver and kidneys of patients with porphyrinuria. The liver enzyme has been purified, and it appears to contain copper and magnesium, although the activity of the bone marrow

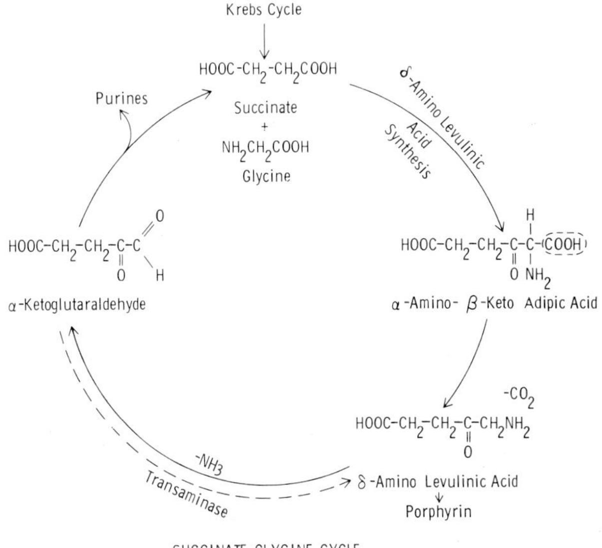

Fig. 3-50. Succinate-glycine cycle

enzyme is not affected in copper deficiency. Chelating agents, such as ethylenediaminetetraacetic acid, inhibit the enzyme [133].

Before porphyrin biosynthesis is completely understood, several processes must be elucidated: (1) the transformation of the side chain of porphobilinogen into the side chain of proto-, uro-, and coproporphyrins; (2) the transformation of the monopyrrole into the tetrapyrrole ring; (3) the mechanism of incorporation of iron in the tetrapyrrole ring to yield heme; and (4) the nature of the porphyrin intermediates that serve as precursors for the heme of cytochrome or hemoglobin.

The exact mechanism of formation of the various porphyrins is not known. It seems that the series of reactions that alter the side chain of porphobilinogen may lead to the formation of the pyrrole derivatives of the natural tetrapyrrole. In the case of protoporphyrin, for example, there are two methyl and two vinyl side chains. A decarboxylation of the acetic side chain of porphobilinogen yields the methyl group, and the decarboxylation and oxidation of the propionic groups of porphobilinogen yield the vinyl groups. If one were to assume that the tetrapyrrole ring is synthesized by the condensation of four molecules of porphobilinogen, the intermediate would be a continuous side chain in which each pyrrolic ring has 1 acetic group on C_4 and 1 propionic group on C_3. Yet, in the natural protoporphyrin, one monopyrrole has its side chains in opposite positions (acetic on C_3 and propionic on C_4). Several mechanisms leading to the formation of natural porphyrins have been proposed. It is impossible to review here the advantages and disadvantages of the various possible mechanisms of condensation; the reader should refer to specialized reviews [134].

The most likely pathway for porphyrin biosynthesis is assumed to involve two enzymes: (1) a deaminase that splits the amino group of porphobilinogen and condenses the four pyrrole rings, yielding a tetrapyrrole ring with four reduced methane chains holding the individual pyrrole rings together (uroporphyrinogen); and (2) an isomerase acting on the polypyrrole

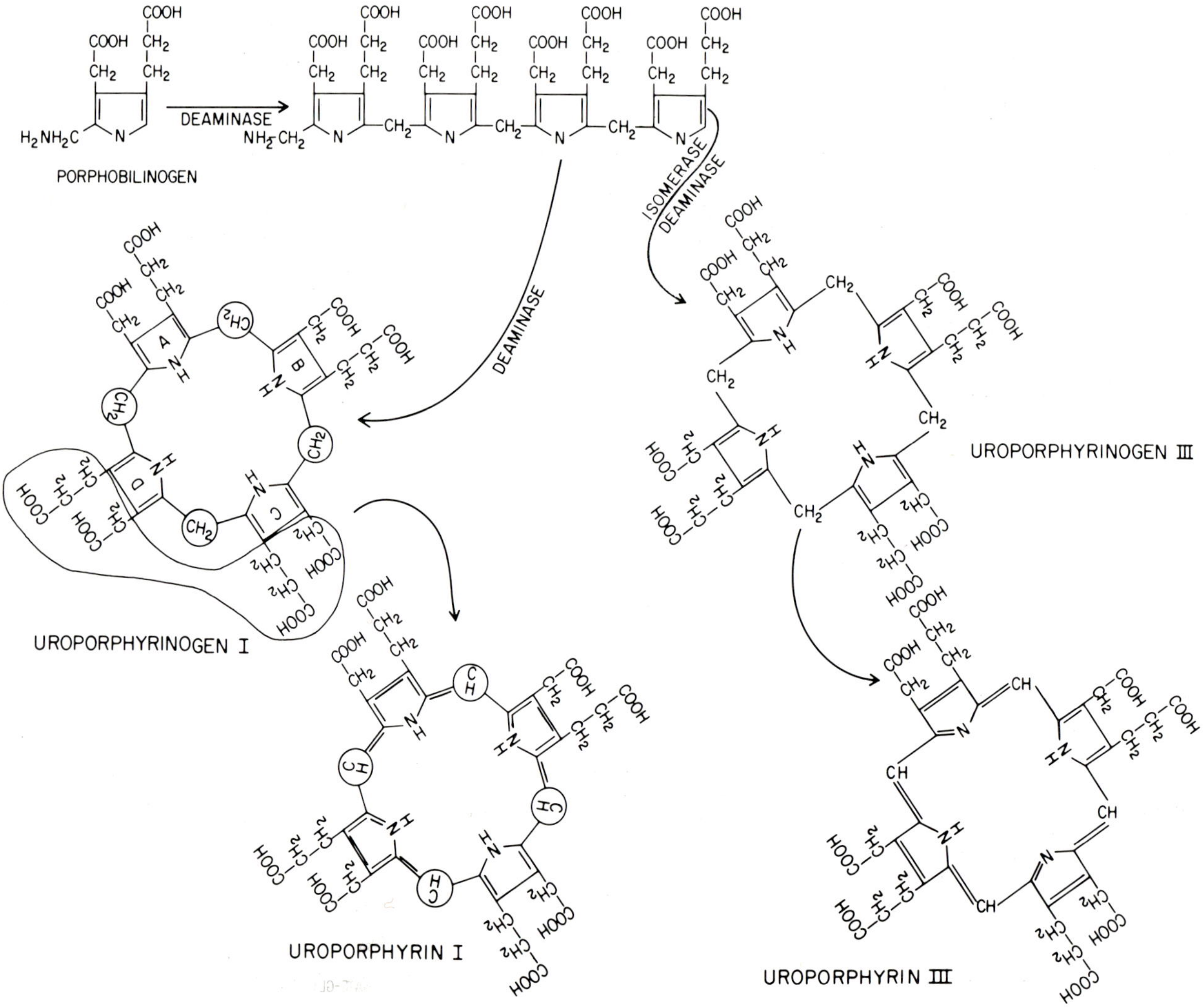

Fig. 3-51. Formation of uroporphyrin

methane chain to catalyze the transfer of acetic acid and the propionic acid side chains on the pyrrole ring that form the ring C of the tetrapyrrole nucleus. This yields a precursor which by condensation gives rise to uroporphyrinogen III (see Fig. 3-51).

The uroporphyrinogens can undergo decarboxylation or oxidation. The decarboxylation of the acetic side chains of the uroporphyrinogens I or III yields coproporphyrinogen I or III in which all the acetic side chains are replaced by methyl groups. A decarboxylase catalyzing the conversion of uroporphyrinogens to coproporphyrinogens has been partially purified from reticulocytes. Only reduced substrates are decarboxylated, and the enzyme is thus without effect on uroporphyrin I and III. It is not known whether all the acetic side chains are decarboxylated at once or one by one. Intermediates with 5, 6, or 7 carboxyl groups have not been isolated, yet it is believed that decarboxylation proceeds stepwise. The conversion of uroporphyrinogens to uroporphyrins results from the oxidation of the 4-methane(-CH_2) into a methine (-CH) bridge. The enzyme catalyzing the oxidation has not been identified, but the reaction proceeds stepwise since various cyclic tetrapyrroles with 1–3 methane bridges have been isolated. Coproporphyrinogen III is converted to protoporphyrinogen III by the oxidation-decarboxylation of two propionic acid side chains and formation of two vinyl side chains. The enzyme catalyzing that reaction is probably associated with mitochondria and is quite specific for coproporphyrinogen III. The reaction requires oxygen but is not inhibited by cyanide. These conversions are all illustrated in Fig. 3-52.

Heme Synthesis

Protoporphyrin is the precursor of heme, the prosthetic group of hemoglobin, cytochrome, and other metalloproteins. The iron occupies the center of the porphyrin ring and forms bonds with the four nitrogen atoms of the pyrrole ring. In the process, two protons are displaced from the nitrogen atoms leading to the appearance of two negative charges, which are then redistributed among the nitrogens and neutralized by

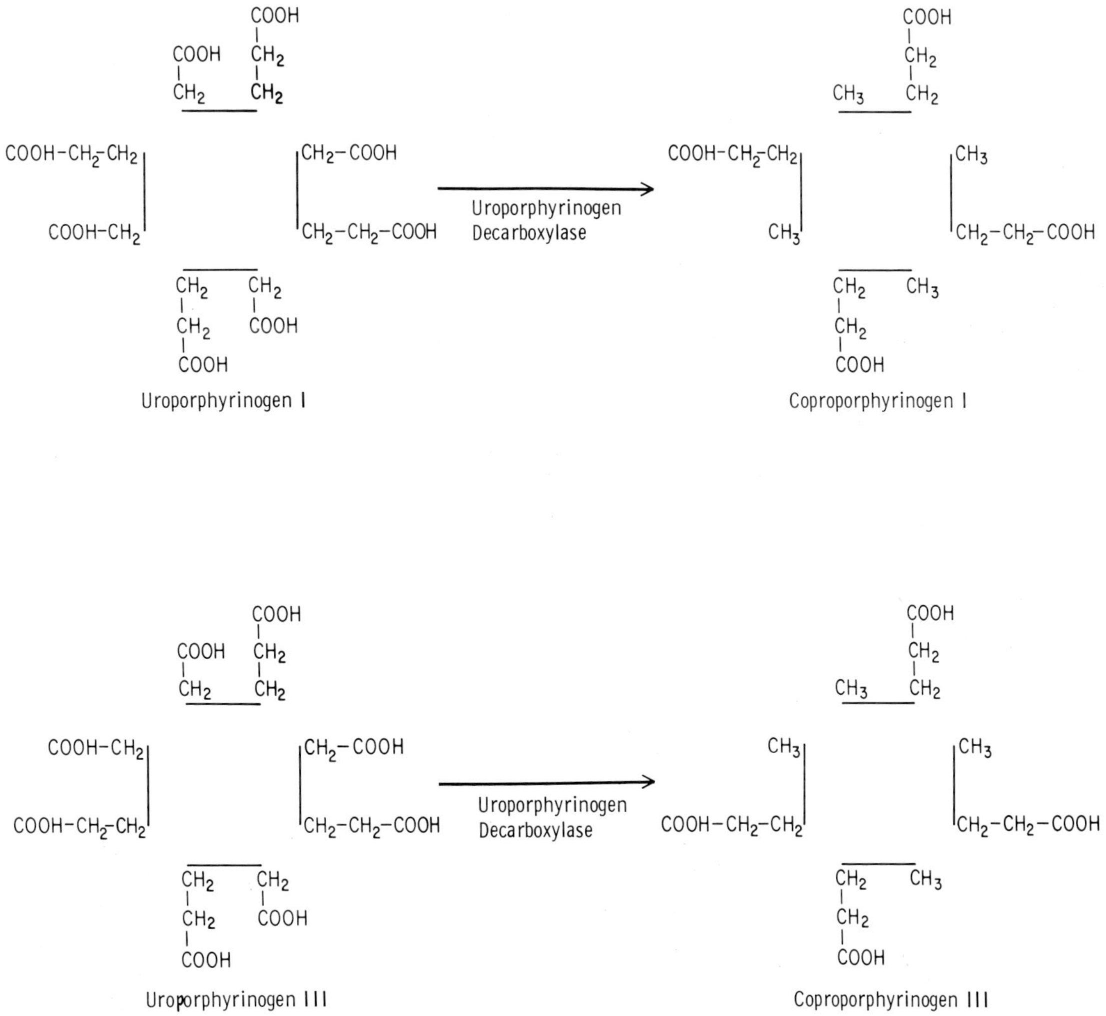

Uroporphyrinogen I → (Uroporphyrinogen Decarboxylase) → Coproporphyrinogen I

Uroporphyrinogen III → (Uroporphyrinogen Decarboxylase) → Coproporphyrinogen III

Fig. 3-52. Conversion of porphyrinogens

two positive charges of the iron molecule. Because of this rearrangement of charges within the resonating ring, the bonds between the iron and the nitrogen are equivalent. Since ferrous iron is bipositive, all charges are neutralized in the heme molecule.

The mechanism by which iron is incorporated into the porphyrin ring is not completely elucidated [135]. It is proposed that the reduction of ferric iron to ferrous iron is coupled to the oxidation of ligand. The enzyme catalyzing that reaction—sometimes called ferrochetalase, or iron-chelating enzyme—has been partially purified from pig liver and requires ascorbic acid, cysteine, or glutathione for activity. It is assumed that impaired iron uptake in red cell hemoglobin during development may explain some refractory anemias which develop in spite of large iron intake [136–138].

Clinical Pathology of Porphyria

The two major types of porphyria, classified according to pathogenesis, are porphyria erythropoietica and porphyria hepatica.

Erythropoietic porphyria is a disease resulting from an inability of the red cells' stem cells to convert porphobilinogen into uroporphyrinogen III, the precursor of protoporphyrin IX. The metabolic block leads to the accumulation of uroporphyrinogen I and its decarboxylated and oxidized derivative in bone marrow, spleen, red cells, urine, and feces. The accumulation of the porphyrin derivatives is associated with photosensitivity and hemolysis. The disease is hereditary and is transmitted as a mendelian recessive trait. Fluorescent microscopic examination of the unstained bone marrow of patients with congenital erythropoietic porphyria demonstrates fluorescence in granules within or close to the nucleus. The cells containing such granules are sometimes called fluorocytes. All types of immature red cells except the reticulocytes contain such granules. However, fluorescence does not appear in all bone marrow cells, as if in porphyria the bone marrow produced two lines of cells: a normal cellular line and one unable to metabolize porphyrin. The bone marrow of patients with porphyria often shows erythroid hyperplasia as a result of moderate hemolytic anemia. Chemical analysis of bone marrow biopsy material has demonstrated uroporphyrin I in the bone marrow. The uro- and coproporphyrin content of the plasma is not markedly increased, but the erythrocytes contain uroporphyrin I in large amounts. Uroporphyrin III and coproporphyrin III are found in smaller amounts. Protoporphyrin III may also be elevated, but the levels are not greater than in many other hemolytic anemias. The patient usually has normochromic anemia with a high reticulocyte count. The anemia results from a reduced life span of the red cells. The pathogenesis of the hemolytic anemia is not clear, but red cell sensitization to the hematoporphyrin is postulated to cause hemolysis.

The spleen is constantly enlarged, possibly by vicarious erythropoiesis. The spleen contains large amounts of uroporphyrin I. The porphyrin content of the urine is elevated. The urine is discolored and in severe cases may look like rich red wine. Uroporphyrin I is the porphyrin found in largest amounts in the urine, which also contains coproporphyrin I and small amounts of uroporphyrin III and coproporphyrin III. When hemolytic anemia exists, urobilinogens are also found in the urine, but more urobilinogen seems to be excreted than can be accounted for by hemolysis; it is likely that at least part of the urinary urobilinogen is not derived from the degradation of hemoglobin contained in the circulating red cells. Degradation of immature marrow red cells of the erythropoietic lineage has been invoked to explain the high levels of urobilinogen found in the urine.

The most striking clinical symptom in patients with erythropoietic porphyria is photosensitivity, which the patient develops at an early age. Occasionally, photosensitivity exists at birth, but often it becomes apparent during the first years of life, in contrast to the photosensitivity of porphyria hepatica which develops later in life.

People protect their skin from the sun because normal skin is sensitized when it absorbs electromagnetic energy. This sensitization will be best understood if we first consider some classical experiments done with dyes. In 1905, Lumiere and his associates demonstrated that gelatin containing absorbed erythrosin hardened when exposed to light. The discovery later inspired many procedures used in photochemistry and photocopy. Half a century later, it was shown that the intercalation of eosin between DNA bases leads to the reduction in the viscocity of the eosin DNA solution when exposed to light. In this case, the hydrogen bonds holding the two DNA strands together are probably ruptured. If erythrosin is adsorbed on the actinomycin and the complex is then exposed to light, the actinomycin contracts. From these and similar experiments, it could be concluded that dyes complex with macromolecules and that the exposure of the complex to light transfers the excitation of the photosensitive dyes to the macromolecule. In fact, it was shown that the dyes are bound to the protein by ionic and ion dipole bonds, and that the excitation of the dye by light releases electrons that are transferred to the protein and can be captured by an electroscope [139].

The photosensitization of biological compounds is assumed to lead to physiological or pathological reactions by a mechanism similar to that described *in vitro*. Thus, in the presence of a quantum of electromagnetic energy, a compound X in the occurrence of light enters an excited state and transfers the excitation to a biological macromolecule playing a vital role in metabolism. Blum assumed that in the presence of oxygen, the excited macromolecule is more readily oxidized than the unexcited macromolecule. However, photosensitization does not always require oxygen.

These simplified basic concepts can now be applied to the skin exposed to light. Under ordinary conditions, the radiant energy that hits the exposed parts of the skin is in the form of electromagnetic waves with wavelengths in the ultraviolet or visible range (approximately 200–700 mμ). Although sunlight emits a large spectrum of radiant energy, all wavelengths shorter than 290 mμ are filtered by the atmosphere. Therefore, wavelengths between 200 and 260 mμ are readily absorbed by the keratinized epithelial cells of the skin (stratum corneum) and could cause erythema, but these wavelengths need not be considered in the pathogenesis of porphyria. Furthermore, the production of erythema by wavelengths greater than 320 mμ requires energy levels considerably greater than those needed to produce erythema with wavelengths of around 300 mμ (500–1,000 times greater). Consequently, radiant energy with wavelengths ranging between 290 and 320 mμ are critical in inducing erythema in normal skin.

Radiant energy is thought to activate photosensitive substances when the quantum of energy is absorbed within the electronic shell and raises the energy level of some valence electrons. Molecules rich in π electrons tend to absorb radiant energy and, therefore, are probably important in photosensitization. The molecular changes that follow the sensitization are not known, although it is assumed that excitation is transferred to a macromolecule and it alters some part of the macromolecule essential to its role in metabolism. The postulated molecular changes induced by photosensitization include the oxidation of SH groups or the opening of the imidazole ring of histidines. Both reactions are fatal if they involve amino acids that form the active center of an enzyme or lead to disruption of the hydrogen or sulfhydryl bonds responsible for protein configuration. In tissue, photosensitive substances may modify the absorption spectrum and alter the efficiency of a given wavelength in producing skin reaction. Porphyrins have a highly conjugated system of double bonds with several absorption peaks, some of which fall in the range of visible light. Therefore, porphyrins can be expected to act as photosensitizing substances. The photosensitizing properties increase as follows: protoporphyrin, amino levulinic acid, coproporphyrin, uroporphyrin. In porphyria erythropoietica, sunlight excites the uro- and coproporphyrins in the capillaries of young children's thin skin, thereby eliciting a chain of still unknown molecular changes which in turn are responsible for the development of erythema, edema, and necrosis that appear in the skin of these patients.

Parts of the skin exposed to light develop a marked erythema followed by numerous bullae, which may rupture and become secondarily infected. If the process is repeated often enough, scarring of the skin and deformities of the ears, lips, and nose may develop. Hypertrichosis and pigmentation of the exposed part of the face and extremities are sometimes observed.

The metabolic block in erythropoietic porphyria is not known, but the clinical situation can be summarized as follows: uro- and coproporphyrin type I accumulate in bone marrow, spleen, blood, and urine. Yet, there is no marked depression of the ability to synthesize protoporphyrin IX. The metabolic distortion is likely to result from an alteration of the pathway leading from porphobilinogen to uroporphyrinogen I and uroporphyrinogen III. The conversion of porphobilinogen to uroporphyrinogen requires a deaminase, and that of porphobilinogen to uroporphyrinogen III requires a deaminase and an isomerase. The absence of the isomerase could explain the accumulation of uroporphyrin I, but then a decrease in the rate of biosynthesis of porphyrin IX should also be expected, yet there is no indication that the biosynthetic rate of hemoglobin or other heme proteins is reduced. An acceleration of the deaminase activity would explain both the accumulation of uroporphyrin I and the formation of normal heme, but type III porphyrin would then accumulate (except if the isomerase reaction is not rate limiting in the formation of protoporphyrin IX). Studies on red blood cells of porphyric patients have shown that they contain normal amounts of a stable deaminase, but the isomerase found in the patients' red cells is unusually labile to heat. The role of the spleen in this disease also requires further study because the favorable effect of splenectomy is difficult to explain if the metabolic block primarily affects bone marrow cells. Finally, the fact that the metabolic block is restricted to only a few bone marrow cells even though porphyrins are synthesized in practically all tissues is puzzling.

Porphyria Hepatica

Porphyria hepatica differs from erythropoietic porphyria in that the metabolic defect is found only in the hepatic cell, and therefore metabolites accumulate only in liver tissue.

Porphyria hepatica can be subdivided into two extreme types: acute intermittent porphyria characterized by the development of abdominal and nervous symptoms without photosensitivity; and cutanea tarda characterized by photosensitivity without abdominal and nervous symptoms. Both diseases are hereditary and are probably transmitted by mendelian dominant genes (see Table 3-4).

Acute intermittent porphyria was first described in Sweden by Waldenström. The disease manifests itself in the form of a crisis during which the patient has porphyria, acute abdominal pains, and a complex neurological syndrome. The attacks are precipitated

Table 3-4. Clinical manifestations in various forms of porphyria

	Erythropoietic porphyria (infants)	Hepatic porphyria (adults)	
		Acute intermittent	Cutanea tarda
Photosensitivity	Yes	No	Yes
Abdominal pain	No	Yes	No
Neurologic signs	No	Yes	No
Mental disturbances	No	Yes	No

by the administration of drugs (barbiturates, sulfonal, ergot preparation, alcohol) or by emotional disturbances. The urine mainly contains precursors of the porphyrins (δ-amino levulinic acid and porphobilinogen). In contrast to the uroporphyrin I excreted in the urine of victims of porphyria erythropoietin, the chromogens excreted in hepatic porphyria are colorless. (Both porphobilinogen and urobilinogen react with Ehrlich's aldehyde reagent; porphobilinogen aldehyde is insoluble, but urobilinogen aldehyde is soluble in chloroform. This difference in solubility is at the basis of the clinical test used to separate them. Furthermore, porphobilinogen has two absorption bands, a faint one at 520 mμ and a strong one at 560 mμ, but urobilinogen absorbs only at 560 mμ.) Other chromogens—porphyrinogens with 1, 2, or 3 methylene chains—are also present in small amounts in the urine of patients with porphyria hepatica. Since all these precursors can be oxidized spontaneously to uroporphyrin, it is not surprising that uroporphyrins are also found in those patients, yet the striking alteration in the urine is the presence of δ-amino levulinic acid and porphobilinogen.

The porphyrin content of stools is not increased, which suggests that porphobilinogen and α-amino levulinic acid, in contrast to uro- and coproporphyrins, are not excreted in the bile.

The patient with acute intermittent porphyria suffers a severe acute abdominal pain not definitely localized and without rigidity or tenderness of the abdominal wall. Moderate fever and leukocytosis develop. If the physician is not aware of the porphyria, he is likely to be confused and suspect appendicitis, renal or biliary colics, pancreatitis, perforated ulcer, acute bowel obstruction or another common cause of abdominal pain. The differential diagnosis of porphyria and bowel obstruction is further complicated because the attacks of porphyria hepatica are often associated with severe constipation. Abdominal X-rays of porphyric patients show colonic distension. The pathogenesis of the abdominal symptoms is not known. They could result either from a direct effect of porphobilinogen or porphyrin on the intestinal mucosa or be the consequence of an increased excitability of the autonomic system.

The neuropsychiatric symptoms are complex and include manifestations attributable to disturbances of the peripheral, central, and autonomic nervous system. Nothing is known of the biochemical and ultrastructural alterations responsible for the neuropsychiatric manifestations. The electronic circuits of porphyrin may disturb the electronic impulses transferred from one nerve fiber to another.

As pointed out previously, the abdominal pains are likely to result from disturbances of the autonomous nervous system since ganglion-blocking agents and abdominal splanchnicectomy may reliev the abdominal distress. Pain in the extremities—especially in the legs—quadriplegia, paresthesia, and paralysis have all been observed. Paralysis of bulbar origin affects the vocal chords and the muscles involved in deglutition and regurgitation. Tachycardia and even blockage of respiration may occur. The patient may show signs of hysteria, schizophrenia, or manic depression.

At autopsy, few findings are striking. The body is often emaciated, and the cause of death may have been uremia, cachexia, or respiratory paralysis. The most conspicuous and consistent findings are in the liver and nervous system. The liver is congested, with central lobular necrosis and occasionally fatty degeneration. An iron-free lipochrome pigment is found in the hepatic cells of some patients. Chemical analysis of the liver demonstrates chromogen, but no porphyrins are found in spleen or bone marrow.

A careful microscopic examination of the spinal chord and brain demonstrates patches of demyelinization. Degeneration of the axis cylinders of the peripheral nerves, the dorsal, and the autonomic system have been described. Neuronal degeneration has also been observed in the dorsal nuclei of the vagus nerve and the celiac plexus, the cerebellum, the dorsal root ganglia, and the anterior horn cells of the spinal chord.

The metabolic alterations are restricted to liver porphyrin metabolism. There are no changes in porphyrin biosynthesis in bone marrow and no hematological changes. The biochemical lesions in the intermittent type of porphyria are probably different from that responsible for erythropoietic porphyria, since in the intermittent type mainly α-amino levulinic acid and porphobilinogen accumulate, whereas mainly uroporphyrin I accumulates in erythropoietic porphyria. The lesions that develop in acute intermittent porphyria could result either from a lack of porphobilinogen use for further metabolic conversion or from an overproduction of α-amino levulinic acid or porphobilinogen.

Under normal conditions, the main pathway for porphobilinogen is its conversion to protoporphyrin IX and heme. Heme is used for heme protein synthesis in the liver (cytochromes, and such enzymes as catalase). An alternative pathway for porphobilinogen is its transformation to uro- and coproporphyrin I, which are not further used by cellular metabolism. A block in porphobilinogen use for heme synthesis is likely to divert the porphobilinogen into the alternative pathway, and then uroporphyrin I accumulates. This does not occur in acute intermittent porphyria.

A deaminase deficiency would block heme and uroporphyrin I production. It is unlikely that such a deficiency exists in acute intermittent porphyria because there is no reduction in the amount of heme protein synthesized in the liver. Uroporphyrin I production is slightly increased. Thus, an increase in amino levulinic acid production is the only explanation for the development of acute intermittent porphyria.

After carefully establishing optimal conditions for determining the activities of δ-amino levulinic acid synthetase in livers of normal and porphyric humans, a group of investigators demonstrated that the activity of the enzyme was increased in the livers of porphyric patients [140]. Most of the known inborn errors of metabolism result from a decrease rather than an increase in enzyme activity. The mechanism that causes an increase in synthetase activity is not clear, primarily because the regulation of porphyrin biosynthesis has not been elucidated. Many macromolecular alterations could lead to increased synthetase activity: shifts in the intracellular distribution of the enzymes, abolition of feedback inhibition, elimination of inhibitors, activation of natural activators, interference with repression at the level of the genome, and derepression of redundant genes for amino levulinic acid synthetase. There are many potential regulators of the porphyrin biosynthetic pathway. For example, although most enzymes of the pathway are found in the cytosol, mitochondrial enzymes function at the origin and at the end of the assembly line. Consequently, the regulation of the penetration of amino levulinic acid into mitochondria could play a considerable role in the regulation of porphyrin synthesis. Furthermore, since many of the enzymes involved in porphyrin synthesis require reduced SH groups in their molecule, the intracellular levels of reduced or oxidized glutathione could markedly influence the rate of porphyrin biosynthesis [141].

It is believed that at least in some cases of porphyria a partial block of the biosynthetic pathway is associated with the increased activity of ALA synthetase. Since heme seems to be a natural inhibitor of the synthetase, it is conceivable that in those cases the primary defect results in a deficiency in heme synthesizing enzyme [215].

The second major type of porphyria hepatica—porphyria cutanea tarda—was described among the descendants of a Dutch immigrant couple who were married in South Africa in 1688. The disease is associated with photosensitivity of the skin and usually no abdominal or neurological symptoms. Large amounts of uro- and coproporphyrins are continuously excreted in the feces, even between crises of photosensitivity. The urine contains large amounts of δ-amino levulinic acid and porphobilinogens when the patient shows clinical symptoms, but these precursors are not found in urine during remission. Precursors are also found in the liver cell. The photosensitive reaction, which consists mainly of erythema and increased fragility of the skin, usually develops in the summer. Apparently, ultraviolet light with wavelengths ranging from 400–450 mμ produce these injuries most effectively. It has been claimed that the exposure to light, in addition to photosensitizing the skin, also produces alterations in porphyrin metabolism, leading to a rise in the uroporphyrin content of urine and stools. In both porphyria cutanea tarda and acute intermittent porphyria, amino levulinic acid and porphyrinogen are excreted in the urine. The acute type is not associated with photosensitivity, while the cutanea tarda is. The reasons for these differences are not understood; however, irradiation of the skin of patients with the acute type with large amounts of a monochromatic light (440 mμ) evokes the skin lesions artificially [142].

In addition to the two hepatic types of porphyria already reviewed, clinicians have described a number of porphyrias in which the symptoms are intermediate between those described in the acute type and in those described in porphyria cutanea tarda. Refer to specialized clinical textbooks for a classification of porphyria [143, 144].

Acquired Porphyria

A form of porphyria has been described in Turkey and among the Bantus of South Africa and Southern Rhodesia. Large amounts of uro- and coproporphyrins are in the urine, the porphyrin level is elevated in blood, but no porphyrins are found in the stools. There is no elevation of δ-amino levulinic acid or porphobilinogen contents of tissues or body fluids. In the type of porphyria reported in Turkey the patient develops severe photosensitivity, bullae formation and superinfection, hyperpigmentation, and hypertrichosis. The factor triggering the disease is assumed to be the consumption of wheat seeds treated with hexachlorobenzene-containing fungicides. The Bantus report symptoms ranging from photosensitivity to darkening of the skin. The development of porphyria is often associated with some degree of liver failure. The cause of the disease is not clear. It has been reputed to occur in association with siderosis, but the role that siderosis may play in the porphyria is not known [145].

A group of drugs not related structurally, but which include some sedatives, cause porphyria in patients with no obvious hereditary propensity for the disease [146]. The mechanism of this form of acquired porphyria has in large part been resolved. Porphyria results from the increased activity of δ-amino levulinic acid synthetase. The increased activity of the enzyme results from *de novo* synthesis. The formation of light-sensitive precursors is *in vitro* prevented by heme [147]. Granick proposed that the drug interferes with the mechanism of repression and that heme is part of the repressor of a gene (operator, structural, or regulator) involved in coding for the biosynthesis of δ-amino levulinic acid synthetase. The changes in coproporphyrin levels in the urine in various diseases are summarized in Table 3-5.

Table 3-5. Coproporphyrin levels in normal and diseased individuals

	μg/day	Isomer type
Normal	75–300	I > III
Lead poisoning	500–3,000	III > I
Poliomyelitis	300–900	III > I
Liver disease	300–800	III > I
Anemia	200–500	I > III
Hodgkin's disease	200–1,000	I > III
Acute alcoholism	250–500	III > I

Inborn Errors of Purine and Pyrimidine Metabolism

One can hardly exaggerate the importance of purines and pyrimidines in cellular metabolism. These bases are the essential building blocks of DNA and RNA.

Purines are an integral part of ATP and of many coenzymes—such as NAD, FAD, and UDP—involved in important bioenergetic or biosynthetic reactions. The stepwise elucidation of the biosynthesis of purine and pyrimidine nucleotides from 1940 to 1960 might well be one of the most remarkable contributions to the development of modern biochemistry.

Chemical Properties of the Bases

Purine nucleotide biosynthesis will be more easily understood if some of the chemical properties of the bases are first briefly reviewed (see Fig. 3-53). Only two purines are consistently found in all living beings: 6-aminopurine or adenine and 2-amino-6-hydroxypurine or guanine. In 1885, Kossel isolated adenine from acid hydrolysates prepared from nuclein extracted from beef pancreas, and its structure was established by synthesis from trichloropurine by Traube in 1904. Guanine was first extracted from bird excreta (guano) found in the mountains of Peru. The chemical structure of the purine was identified well before its relation with nucleic acid metabolism was established; as a matter of fact, the French with their talents for glorifying the most trivial objects, used guanine during the reign of Louis XIV to cover glass beads and make them shine like pearls. Only small amounts of guanine are present in mammalian tissue; an exception is pigs affected by gout, in which large amounts accumulate in bone and other tissues.

Uric acid (2,6,8-hydroxypurine) is formed in small amounts in man, and traces are found in blood, urine, and feces. In birds and reptiles uric acid constitutes the normal metabolic end product of nitrogen and is found in large amounts in guano.

Xanthine, or 2,6-dihydroxypurine, was first associated with caffeine in certain plants. It was later found in urine, blood, and liver. In 1817, Marcet detected xanthine in renal calculi. Xanthine was synthesized from chloropurine or from 4-aminouracil. An addition to its biological significance as the product of nucleic acid degradation (deamination of guanine or oxidation of hypoxanthine), xanthine is also important as an organic compound because when its nitrogen atoms are methylated, natural compounds of the greatest significance in pharmacology are formed: 1,3-dimethylxanthine, or theophylline; 1,3,7-trimethylxanthine, or caffeine; and 4,7-dimethylxanthine, or paraxanthine.

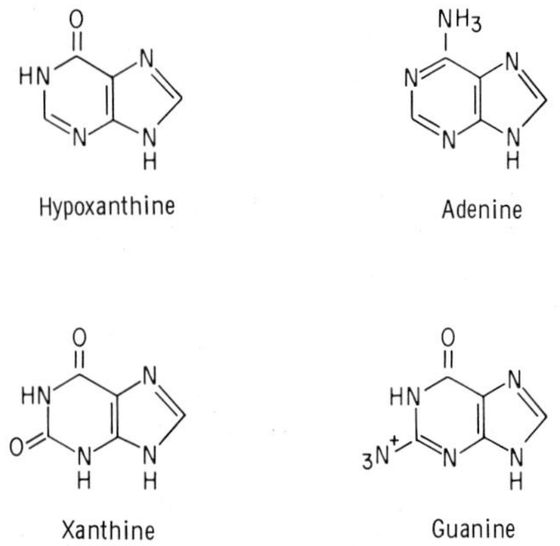

Hypoxanthine, or 6-hydroxypurine, is also a product of nucleic acid dissimilation. It is produced by the deamination of adenine. In addition to the classical purine bases that we have just described, a group of methylated or hydroxylated purines have been discovered: 6-methylaminopurine, 2-methyladenine, 2,6-dimethylaminopurine, 2,2-dimethylaminopurine, 1-methylguanine, and 2-methylamino-6-hydroxypurine. Most of these compounds have been found in association with RNA in the form of a nucleotide linked by a 5′-phosphodiester linkage. It may be significant that these bases, present only in small amounts in mammalian tissues, are found almost exclusively in the tRNA. A small number of the bases are associated with ribosomal RNA.

Hypoxanthine

Adenine

Xanthine

Guanine

Fig. 3-53. Formulaes of purine bases

Purine Biosynthesis

Nutritional experiments in which the effects of postulated precursors on uric acid excretion were investigated yielded controversial results until isotopes were used.

Experiments with labeled amino acids, purines, etc., compounds that were postulated to be precursors of purine, demonstrated that the alleged precursors were not readily used for synthesis. In contrast, a more classical experiment of Barnes and Shoenheimer [148] established that the purine ring could be synthesized from simple compounds such as ^{15}N-labeled ammonium salts. These pioneering studies were followed by a systematic investigation of the origin of each nitrogen and carbon atom in the purine ring. In these experiments, a simple labeled precursor (^{14}C glycine, etc.) is administered, usually to pigeons. The excreted uric acid is extracted and purified, and the ring is degraded by the methods of organic chemistry to yield a compound containing a specific atom participating in the formation of the purine ring. The atom that contains the label originally present in the precursor is the one derived from that precursor [149].

Uric acid can be obtained in good yields from bird excreta, and the compound is easily degraded. German chemists have known degradation methods since the beginning of the nineteenth century. Mainly two procedures were used for identifying the carbon units of the purine ring. The first degradative method involved two steps. Step I involves oxidation with alkaline manganese dioxide (see Fig. 3-54), a process that ultimately releases carbon 6 in the form of CO_2, carbons 5 and 4 in the form of glyoxylic acid, and carbons 2 and 8 in the form of urea. In step II the carboxyl carbon and the aldehyde carbon of glyoxylic acid can be separated by converting the glyoxylic acid to the semicarbazone, which is then oxidized with acid permanganate (see Fig. 3-55).

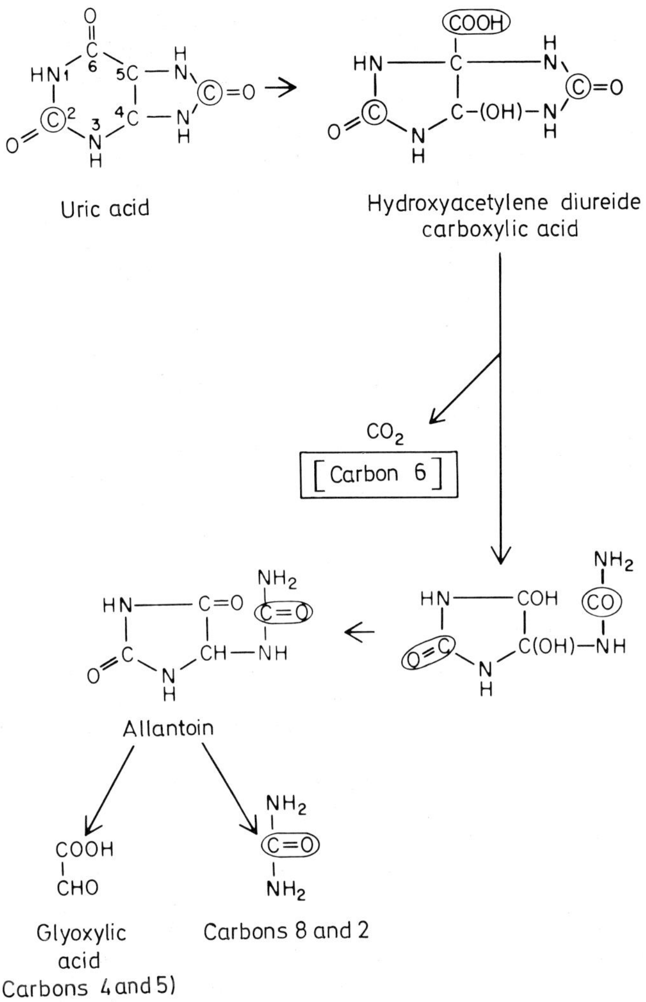

Fig. 3-54. Degradation of uric acid with alkaline manganese dioxide

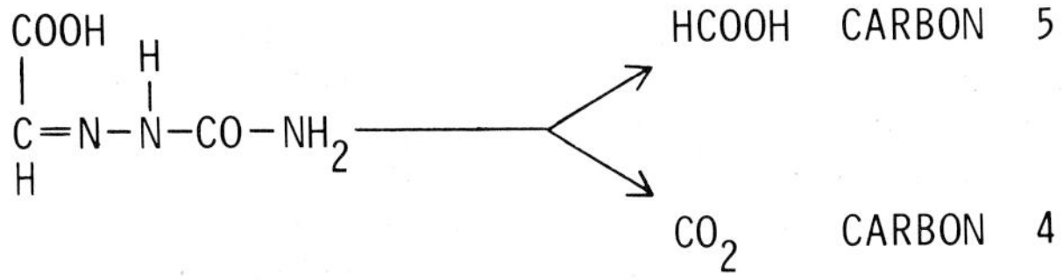

Fig. 3-55. Degradation of the semicarbazone of glyoxylic acid with acid permanganate

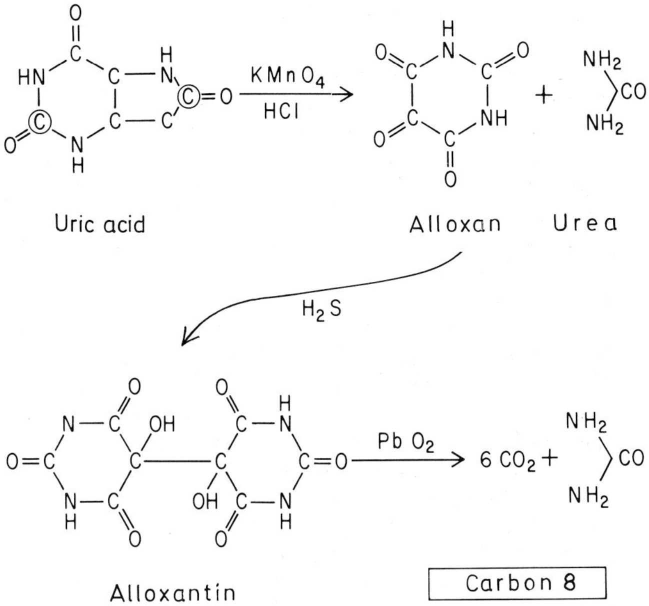

Fig. 3-56. Degradation of uric acid to alloxan, alloxantin and urea

The purpose of using a second method of uric acid degradation is to separate the carbon 8 from the carbon 2. This is achieved by treating uric acid with potassium chlorate in strong hydrochloric acid. This treatment yields alloxan and urea (see Fig. 3-55). The urea contains the carbon 8. Alloxan can be further degraded after its reduction to alloxantin with hydrogen sulfide, which after oxidation with lead dioxide yields two urea molecules, each containing a carbon 2 of the original purine rings (see Fig. 3-56).

Similar degradative procedures permitted the separation of the various nitrogen atoms that form the purine ring. We have already seen that the acid oxidation of uric acid yields the nitrogens 1 and 3 in the form of alloxan and nitrogens 7 and 9 in the form of urea, and alkaline oxidation of uric acid yields hydroxyacetylene diureide carboxylic acid. The treatment of that compound with peroxide yields two types of oxonic acid molecules: one containing nitrogens 7 and 3, and another containing nitrogens 3 and 9. In addition, ammonia is formed at the expense of nitrogens 1 and 7. Hydrolysis of uric acid with hydrochloric acid yields glycine containing nitrogen 7 and 3 molecules of ammonia (N^1, N^3, and N^9). By submitting uric acid molecules with one of their nitrogen molecules labeled to several of these degradation procedures, it is possible to determine which nitrogen atom in the ring is labeled. The administration of simple labeled precursors—such as $^{15}NH_4$, $^{14}CO_2$, [^{15}N]glycine, or [^{14}C]glycine—followed by the chemical degradation of the excreted uric acid has indicated the origin of each single atom of the purine ring. These preliminary studies were later followed by the identification of the precursor and the products of partial purification of the enzyme and determination of the requirements in each of the enzymic reactions involved in the biosynthesis of the purine ring [150–152].

In the first steps of purine biosynthesis, PRPP (phosphoribosyl pyrophosphate) reacts with glutamine, a classic donor of amino groups, to yield ribosylamine and pyrophosphate. (According to Wyngaarden, ribosylamine formation is the rate-limiting step in the sequence of reactions leading to the formation of finished purine.) The enzyme catalyzing this reaction is called the ribosylamine pyrophosphorylase (see Fig. 3-57).

PRPP synthetase has been purified from red cells and Ehrlich's ascites cells. The enzymes from both sources require inorganic phosphate and magnesium

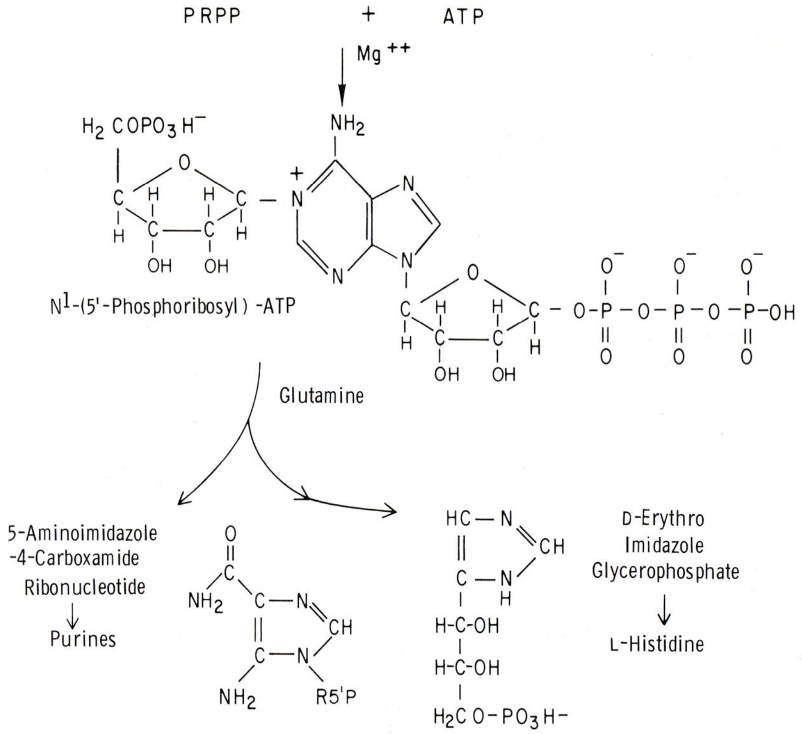

PRPP + ATP → N^1-(5'-Phosphoribosyl)-ATP → Glutamine → 5-Aminoimidazole-4-Carboxamide Ribonucleotide → Purines; D-Erythro Imidazole Glycerophosphate → L-Histidine

Fig. 3-57. Purine biosynthesis

for activity. ADP, GDP, and 2,3-diphosphoglycerate certainly inhibit the red cell enzyme and are believed to inhibit the ascites cell enzyme [153].

At least two factors regulate PRPP formation: the availability of the precursor, ribose-5-phosphate; and the activity of the enzyme, PRPP synthetase. Present evidence suggests that depending upon the tissue, the amount of PRPP formed is determined by either the amount of ribose-5-phosphate produced or the feedback regulation of the enzyme.

For example, in red cells and ascites cells, PRPP formation cannot depend entirely upon the levels of ribose-5-phosphate because in both types of cells these levels are high enough to saturate PRPP synthetase, yet the amount of PRPP made is not greater than what is observed with much lower levels of ribose-5-phosphate. Tight regulation of PRPP synthesis at the level of synthetase has been demonstrated.

The biosynthesis of both purine and pyrimidine nucleotides—and consequently the biosynthesis of ATP, coenzymes, and nucleic acids—depends upon the availability of PRPP. Therefore, it is important to establish whether in mammalian tissues the PRPP supplies outweigh the needs, or whether they are produced as needed. If the latter is true, PRPP biosynthesis will play a key role in regulating nucleotide biosynthesis. The question that we have just raised is not completely answered. No direct evidence indicates that PRPP is truly limiting, but indirect observations suggest that such limitations may exist.

We shall give only two examples here. First, the fact that the administration of a diet containing 1% orotic acid to rats depresses the biosynthesis of purine nucleotides possibly by inhibiting PRPP synthesis. Second, children have been observed with concomitant hyperuricemia and glucose-6-phosphate deficiency. It is believed that in those cases, the hyperuricemia would at least in part result from overproduction of PRPP, presumably as a result of increased production of ribose-5-phosphate. The role that increased PRPP plays in gout and in the Lesch-Nyhan syndrome is discussed in another section of this chapter. The sequence of steps involved in the biosynthesis of inosine 5′-monophosphate is presented in Fig. 3-58.

The 5′-ribosylamine phosphate is converted in the presence of glycine to the glycinamide ribotide in a reaction catalyzed by glycinamide ribonucleotide synthetase. This reaction provides four of the atoms forming the imidazole ring of the purine base—nitrogens 7 and 6 and carbons 4 and 5. The imidazole ring is completed by the addition of carbon 8, which is provided by the formylation of the amino group of the glycinamide ribosylphosphate to yield the formylglycinamide ribotide. The reaction is catalyzed by a pigeon liver enzyme (glycinamide ribotide transformylase) and folic acid is required. Of course, it is not folic acid *per se* that acts in this formylation reaction, but an active coenzyme named N^5, N^{10}-anhydroformyltetrahydrofolic acid. The biosynthesis and role of this coenzyme is discussed in greater detail in the section on vitamins. This formylation reaction is particularly important because it is a target for the antifolic acid compounds.

In the next reaction, formylglycinamide ribotide acquires the nitrogen 3 of the purine ring. The molecular origin of the nitrogen donor was detected by the use of an antimetabolite. When L-azaserine was added to a pigeon liver enzyme system capable of synthesizing

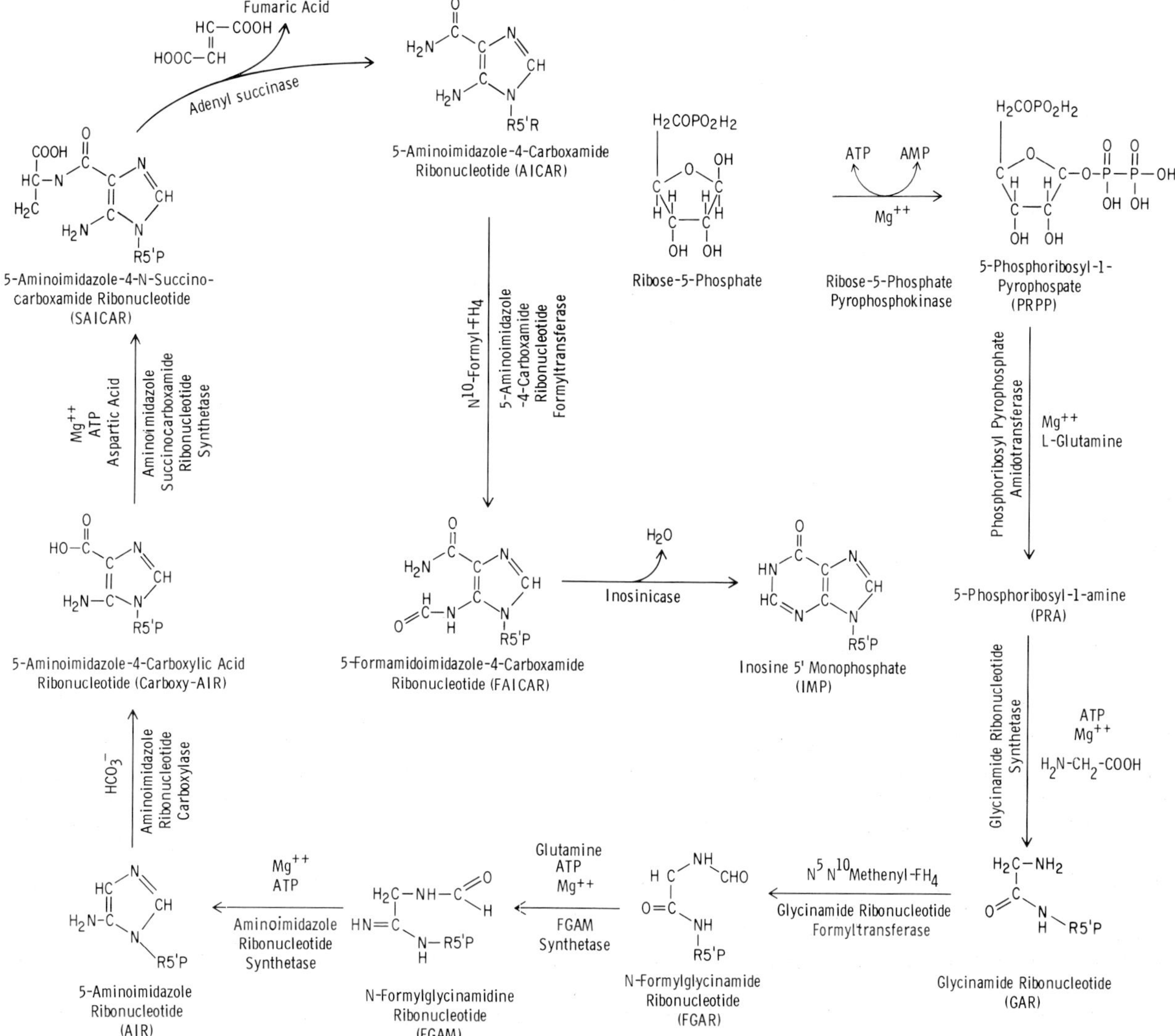

Fig. 3-58. Purine biosynthesis

inosinic acid, formylglycinamide ribotide accumulated. This suggested that glutamine acts as nitrogen donor, and it was later established that formylglycinamide ribotide is converted to the formylglycinamidine ribotide in the presence of glutamine, ATP, H_2O, and Mg^+. The imidazole ring is then closed to yield an arylamine aminoimidazole ribotide. A pigeon liver enzyme preparation can convert the aminoimidazole ribotide into the 5-amino-4-imidazole carboxamide ribotide in the presence of CO_2, aspartic acid, and ATP. The 5-amino-4-imidazole carboxylic acid ribotide and 5-amino-4-imidazole N-succinocarboxamide ribotide are intermediates in this reaction. In addition to 5-amino-4-imidazole carboxamide ribotide, fumarate and ADP are also produced. Up to the present, the enzyme aminoimidazole ribotide carboxylase catalyzing the conversion of formylglycinamidine ribotide to the 5-amino-4-imidazole carboxylic acid ribotide (5-aminoimidazole) has not been separated from the enzyme that catalyzes the formation of 5-amino-4-imidazole N-succinocarboxamide ribotide from the 5-amino-4-imidazole carboxylic acid ribotide. The enzyme complex catalyzing the above reaction has, however, been separated from a purified enzyme that acts in the cleavage of a fumarate moiety from the 5-amino-4-imidazole N-succinocarboxamide ribotide. There is some evidence, at least in yeast, that this enzyme is identical to that cleaving adenosylsuccinic acid (adenylosuccinase).

5-Amino-4-imidazole carboxamide ribotide is converted to 5-formamido-4-imidazole carboxamide ribotide in a formylation reaction using N^{10}-formyltetrahydrofolic acid. The formylation is catalyzed by 5-aminoimidazole-4-carboxamide ribotide transformylase. Isonicase then closes the purine ring, forming inosinic acid in the absence of ATP. Fig. 3-59 illustrates the origin of each atom of the purine ring.

Inosinic acid is an intermediate for the biosynthesis of both adenylic and guanylic acids. These nucleotide interconversions are complex reactions, and their

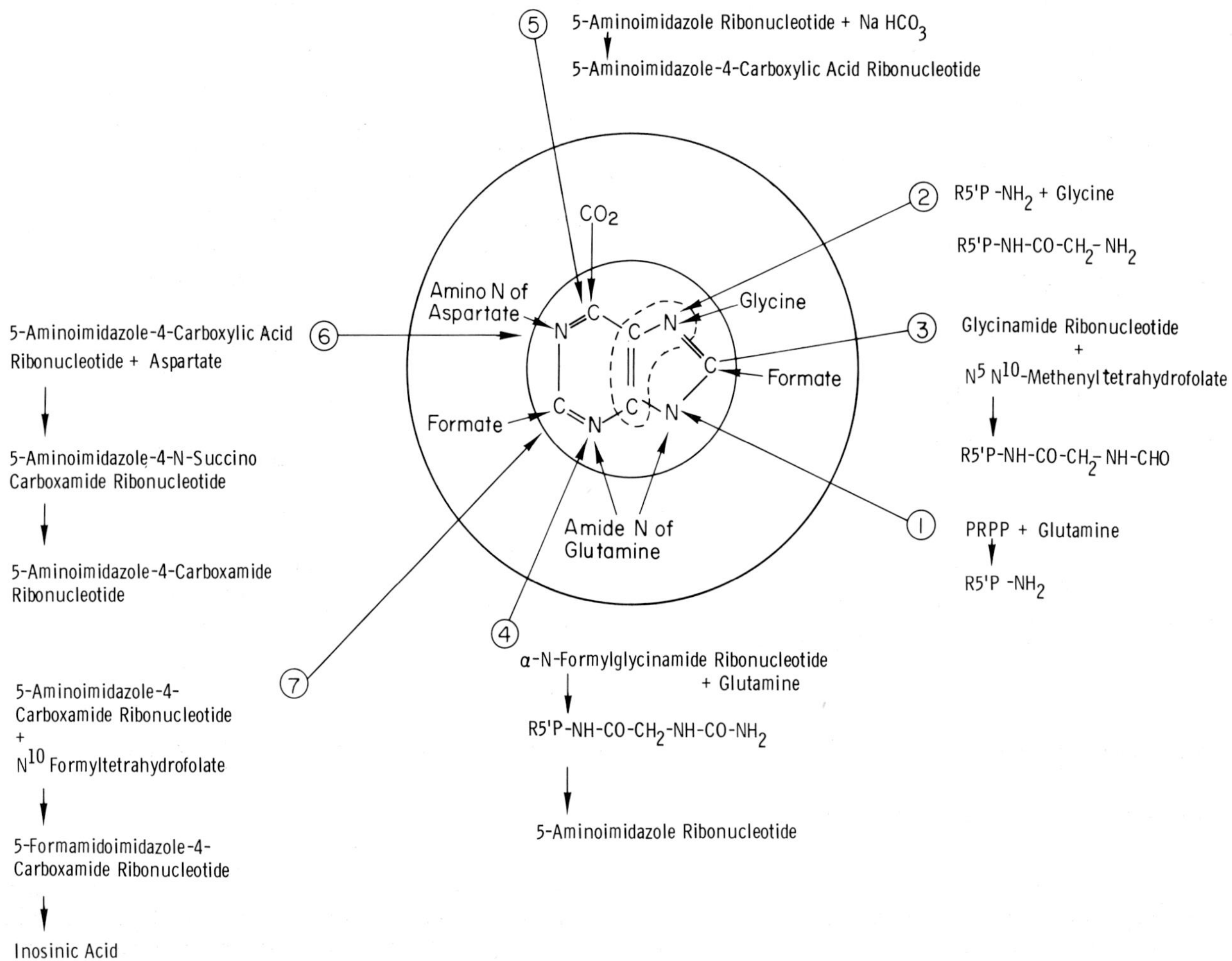

Fig. 3-59. Origin of the various atoms of the purine ring

detailed mechanism is still not completely understood. In the presence of adenylosuccinic synthetase, inosinic acid reacts with aspartic acid to yield adenosylsuccinic acid. GTP provides the required energy, and GDP and inorganic phosphate are formed. This specific requirement for GTP in the biosynthesis of AMP—a precursor of ATP—is of special interest, and one may well wonder how the levels of GTP in the cell could regulate the amounts of AMP and possibly ATP synthesized. Unfortunately, the evidence for the requirement for GTP is somewhat controversial.

Adenosylsuccinic acid is further cleaved to yield fumaric acid and adenylic acid. The reaction is catalyzed by adenylosuccinase. It is not clear what intermediates are formed in the course of the conversion of inosinic acid to adenylosuccinic acid, but at least two mechanisms have been proposed: the formation of 6-phosphorylinosine 5′-phosphate as an intermediate; and a simultaneous reaction between GTP, the amino group of aspartic acid, and the carbon 6 of inosinic acid without the formation of phosphorylinosinic acid.

The conversion of inosinic acid to guanylic acid has been demonstrated in bone marrow preparations and with acetone powder of pigeon livers. The conversion occurs in two steps: oxidation of inosinic acid to xanthosine-5′-monophosphate (XMP) and amination of XMP to yield GMP. The formation of adenylxanthosine-5-phosphate as an intermediate has been postulated (see Fig. 3-60). NAD, ATP, and glutamic acid are required for these two reactions. Thus, NAD acts as the hydrogen acceptor in the conversion of IMP to XMP by inosinic dehydrogenase, and glutamine provides the amino group needed in the conversion of XMP to GMP. As may be expected, the formation of GMP from XMP is blocked by azaserine. Although both AMP and GMP have an amino group on their purine ring, the pathway for the biosynthesis of the amino group of AMP is different from that responsible for the biosynthesis of GMP.

In summary, the biosynthesis of the purine ring occurs by the piecemeal addition of small molecular fragments (NH_2, CO_2, glycine, HCO) to a skeleton of ribose-5-phosphate. Nature has preferred to synthesize these important building blocks in a stepwise fashion rather than by borrowing large molecular moieties, such as the purine base, amply provided by the ordi-

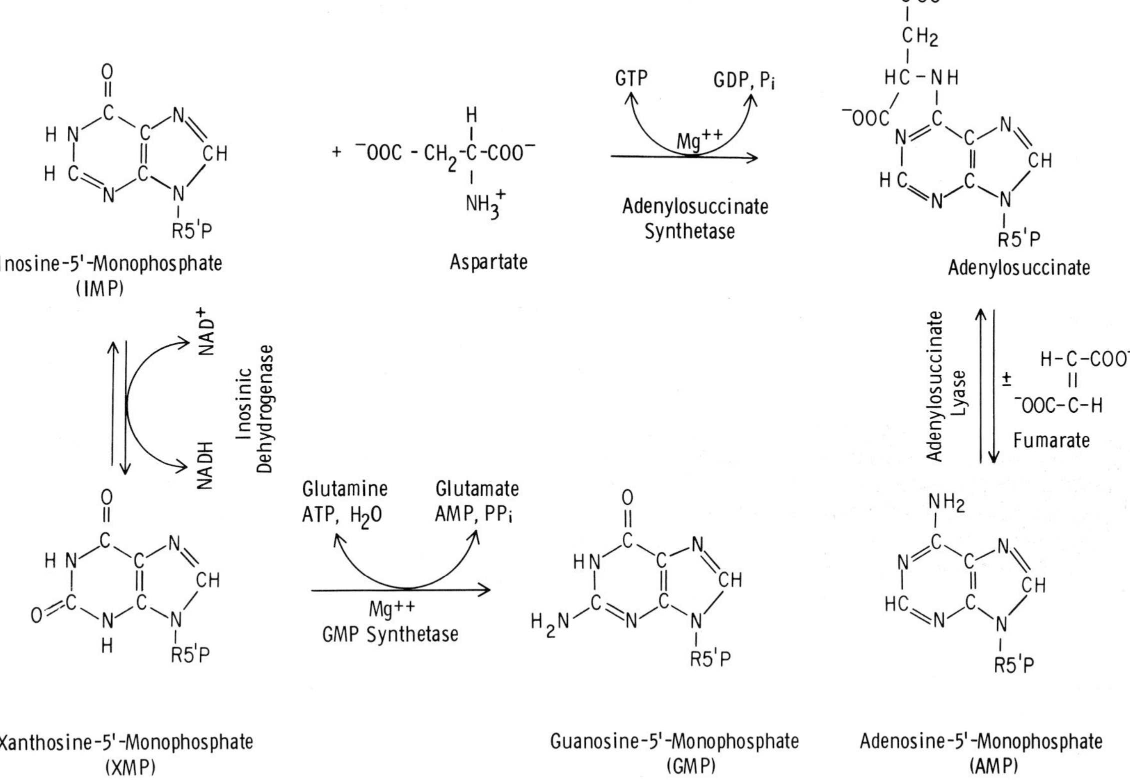

Fig. 3-60. Purine nucleotide interconversions

nary diet. Because of the *de novo* pathway, purine biosynthesis is virtually independent of dietary purines. However, it should be kept in mind that some vitamins, like folic acid, are required for biosynthesis of purine bases, and, therefore, some vitamin deficiencies can lead to severe distortion in purine biosynthesis.

Although there is little doubt that *de novo* synthesis provides most of the purine nucleotides used in metabolism, alternative pathways have been described. They assume particular importance because they can be used to bypass metabolic blocks caused by antimetabolites.

5-Amino-4-imidazole carboxamide ribotide, a precursor only two steps removed (formylation and cyclization) from inosinic acid, can be synthesized by the direct condensation of the imidazole with 5-phosphoribosyl pyrophosphate. The enzyme catalyzing this reaction was purified from an acetone powder of beef liver. The same enzyme (AMP pyrophosphorylase) catalyzes the condensation of adenine, guanine, and hypoxanthine. Nucleoside phosphorylase is an enzyme that catalyzes the formation of a ribose nucleoside from a purine base and ribose-1-phosphate. Guanine, adenine, xanthine, hypoxanthine, 2,6-diaminopurine, and aminoimidazole carboxamide are known to be converted to their respective nucleosides by such a mechanism. In the presence of a specific kinase and ATP, the nucleoside is then phosphorylated to the corresponding nucleotide.

Purines are not only the essential building blocks for nucleic acid and coenzyme biosynthesis, but purine metabolism is also closely interrelated with metabolism of flavins, pteridines, and histidine. In bacteria, this interrelation between various metabolic pathways may be important in regulating the cellular metabolism. Plaut [154, 155] presented a comprehensive discussion of the role of purine in the biosynthesis of flavin and pteridine. Since riboflavin and folic acid are vitamins and histidine is an essential amino acid, the cell in mammalian tissue is dependent on the diet for these compounds, rather than on *de novo* biosynthesis. Therefore, the interaction between purine metabolism and that of riboflavin, pteridine, and histidine assumes less importance in mammals than in bacteria. The pathway of histidine biosynthesis illustrates such interactions.

The role of purines in the biosynthesis of the imidazole ring of histidine has been investigated in bacteria. The use of labeled purine bases (guanine and adenine) followed by the chemical breakdown of newly formed histidine established that the N_1 and the C_2 of the purine base were incorporated in the N_3 and C_2 of the imidazole ring of histidine. The N_3 of the imidazole ring was shown to be derived from glutamine. On the basis of further studies on cell-free preparations, the biosynthetic pathway of the imidazole ring of histidine can now be outlined. Ribose-1-phosphate reacts with ATP to yield phosphoribosyl pyrophospate and AMP: PRPP reacts with ATP to form a new complex assumed to be phosphoribosyl ATP. Phosphoribosyl (PR) ATP is then converted to PR AMP by a specific pyrophosphohydrolase. A cyclohydrolase converts PR MP to the N-(5′-phosphoribosyl formimino)-5-amino-1 (5′phosphoribosyl)-4-imidazole carboxamide), which is referred to as BBMII, by opening the pyrimidine ring between the nitrogen and the carbon 6. BBMII is converted to N-(5-phospho-D-1′-ribulosyl formimino)-5-amino-1-(5′phosphoribosyl)-4-imidazole carboxamide (BBMIII) by the action of an isomerase. BBMIII inherits the amino group of glutamine, and the product of the reaction is then split in the presence of glutamine to yield imidazole glycerol phosphate and 5-amino-4-imidazole carboxamine ribotide, a precursor of purines.

Importance of the Salvage Pathway

The quantitative importance of the salvage pathway in purine metabolism is difficult to estimate [156]. Comparison between uric acid production in normal children and in children affected with a deficiency in hypoxanthine guanine phosphoribosyl transferase has, however, permitted researchers to approximate how much the salvage pathway contributes to purine metabolism. Thus, whereas in normal children uric acid excretion in the urine (per 24 hours and per kilogram of body weight) is of the order of 10 mg, in children affected with the Lesch-Nyhan syndrome, uric acid excretion is around 47 mg. The difference between enzyme-deficient and normal children is believed to reflect the amount of uric acid normally used in the salvage pathway.

The biochemical mechanism regulating the balance between *de novo* and salvage pathways remains to be discovered. But observation of patients afflicted with gout or the Lesch-Nyhan syndrome suggests that hypoxanthine guanine phosphoribosyl transferase may play an important role in maintaining the balance between *de novo* and salvage pathways.

The contribution of the salvage pathway to purine metabolism is of considerable practical significance. Some tissues contain both the *de novo* and the salvage pathway, others contain only the salvage pathway (*e.g.*, bone marrow and blood leukocytes) and therefore depend on exogenous purine for nucleotide and nucleic acid production. Therefore, it is interesting that [^{14}C]formate incorporation in bone marrow DNA is depressed in partially hepatectomized mice. Similarly, thymus regeneration following irradiation is impaired by partial hepatectomy. Neurons have been claimed to depend upon purine synthesis by microglia [157].

The liver is believed to be the main source of noningested purine bases. (Muscle is suspected to contribute to the endogenous purine pool, but this is not proven.) Thus, purine nucleotides are dephosphorylated (by 5′-nucleotidase), nucleosides are broken down to bases, and the bases are transported by red cells to other sites. Hypoxanthine and xanthine are then released from the red cell pool and transferred to tissue. The mechanism of release of purines from red cells to other tissues is unknown. Of course, red cells use some of the transported purine for their metabolic welfare. Hypoxanthine guanine phosphoribosyl transferase converts hypoxanthine into IMP, which at least in rabbit red cells can be converted to AMP and GMP. In contrast, human red cells depend upon either the salvage of adenine through the PRPP reaction or on the phosphorylation of adenosine through the adenosinic kinase reaction for their supply of AMP. Adenine is not believed to be salvaged extensively because adenine levels are low in normal tissues.

Purine Catabolism

Depending upon the nature of the enzyme involved in the breakdown of polynucleotides (DNA and RNA), two types of mononucleotides will appear: 5′- and 3′-mononucleotides. The purine mononucleotides are degraded by various enzyme mechanisms that catalyze dephosphorylation of the nucleotides; deamination of nucleotides, nucleosides, and bases; and cleavage of N-glycosidic bonds of nucleosides. The purine ring is then oxidized to yield uric acid and opened for further degradation. Obviously, the enzymes involved in purine degradation are legion, and it is not surprising that their properties are not all known.

The phosphate group of a purine nucleotide can be lost in at least three different types of reactions catalyzed by phosphatases with broad specificity, nucleotidases, and transferases. Intestinal and liver phosphatase, prostatic acid phosphatase, and bone phosphomonoesterase hydrolyze purine 5′- and 3′-mononucleotides. A 5′-nucleotidase acting on 5′-AMP, 5′-UMP nicotinamide 5-nucleotide and R_5P prepared from bull has been studied more extensively. It acts optimally at pH 8.5 and is activated by magnesium and inhibited by fluorides and borates.

5′-Nucleotidases active on the four deoxynucleotides (dAMP, dCMP, dTMP, and dGMP) have been described in normal and regenerating rat liver. Studies on the intracellular distribution of enzymes have established that dTMP nucleotidase activity is in the lysosomal fraction of the cell. The enzyme activity was optimal at pH 5, and therefore dTMP nucleotidase activity could result from the presence of an acid phosphatase with broad specificity. Segal and Brenner have studied the intracellular distribution of 5′-nucleotidase and found it to be associated with the microsomes of the cell. The enzyme has a broad pH optimum, ranging from 6.5–8.0, and acts on 5-AMP, 5-CMP, 5′-UMP, 5-IMP, 5′-dCMP, and dAMP. The enzyme is inhibited by inorganic phosphorus and by a natural inhibitor found in mitochondria and nuclei. A 3′-nucleotidase with a narrow pH optimum around pH 7.5 acting on 3′-AMP, 3′-IMP, 3′-GMP, 3′-UMP, and 3′-CMP has been prepared from various germinating plants. The affinity of the enzyme is greatest for 3′-AMP and lowest for 3′-CMP. Phosphate, arsenate, cysteine, glutathione, and potassium cyanide inhibit the enzyme.

Various phosphatases can transfer phosphate groups from a nucleotide donor to a nucleoside acceptor in a reaction:

$$YMP + ZDR \rightarrow YMR + ZMP,$$

where Y and Z are two purine bases.

Mammalian tissues do not contain adenine deaminase, but specific enzymes able to deaminate adenosine and 5′-adenylic acid have been found in a variety of mammals. Adenosine deaminase activity has been detected in intestine, liver, spleen, brain, kidney, heart, and skeletal muscle. Adenosine deaminase has been partially purified from the intestinal mucosa. The enzyme has a great affinity for adenosine and deoxyadenosine and only low affinity for 2′-AMP and 3′,3′-cyclic AMP. Its activity is lost on dialysis. The enzyme acts optimally at pH 7.

A 5′-adenylic deaminase found in muscle acts on 5′-AMP and 5′-dAMP. The affinity of the enzyme is much greater for AMP than for dAMP. The pH for optimum activity is 5.9.

A guanase that deaminates guanine, guanosine, 5-GMP, and azaguanine has been prepared from liver and intestine. This enzyme is rigidly specific for guanine and its derivatives, and its pH optimum is around 8.

The N-glycosidic bonds of nucleosides and nucleotides are split by enzymes called nucleoside phosphorylases or nucleoside hydrolases. These enzymes have been prepared from various sources, and only the properties of some of the mammalian enzymes will be considered here. A nucleoside phosphorylase purified from rat and calf liver acts on various nucleosides and nucleotides (guanosine, deoxyguanosine, and hypoxanthine) but is without effect on adenosine and xanthosine. The mechanism of the reaction is as follows:

$$\text{Nucleoside} + \text{Phosphate} \rightleftharpoons \text{Base} + R_1PO_4$$

Nucleoside phosphorylase

Nucleoside hydrolase has been partially purified and has been obtained free of contaminating nucleoside phosphorylase. In the presence of water, the enzyme separates the base from the ribose moiety.

$$\text{Nucleoside} + H_2O \rightleftharpoons \text{D-Ribose} + \text{Base}$$

Nucleoside hydrolase

Xanthine oxidase is the enzyme that catalyzes the oxidation of purine, adenine, hypoxanthine, and xanthine to yield uric acid and formic acid, and the oxidation of other aldehydes to yield the corresponding carboxylic acid.

Xanthine oxidase, found in liver and in milk, has been purified from chicken and calf liver and has been crystallized from milk. The crystalline enzyme (mol wt 290,000) is homogeneous in the ultracentrifuge and on electrophoresis. Each protein molecule contains two molecules of FAD. The enzyme also contains iron and molybdenum. The role of xanthine oxidase is generally assumed to be to oxidize xanthine. However, an almost complete deficiency of xanthine oxidase in xanthinuria has been described, and no symptoms were observed except for the formation of xanthine stones.

Because purified xanthine oxidase constitutes a simple model for the study of biological oxidation, many laboratories have tried to elucidate its mechanism of action. A detailed description of the various mechanisms that have been proposed is beyond the scope of this text. However, the use of modern physicochemical techniques will probably be helpful in the study of xanthine oxidase's mechanism of action. Examination of xanthine oxidase frozen in liquid air by the electron spin resonance technique and the use of chemoluminescent substance to detect free radicals have demonstrated that the reduced enzyme is oxidized in the presence of oxygen in a sequence of steps during which free radicals are formed.

Uric Acid Metabolism

The uric acid formed from nucleic acid degradation of endogenous or exogenous nucleic acids is excreted as such without chemical alteration in higher apes and man. (In the dalmatian uric acid is partly oxidized and partly excreted.) In all other mammals, uric acid is further oxidized to yield allantoin, and uricase is the enzyme that catalyzes that reaction. In amphibians, fish, and many invertebrates, two more enzymes exist: one involved in the oxidation of allantoin to allantoic acid (allantoinase), the other catalyzing the splitting of allantoic acid to yield urea and glyoxylic acid.

In the blood, uric acid is distributed between red cells and serum. The proportion found in each fraction of the blood is determined by the Gibbs-Donnan equilibrium. Two types of methods are available for determining uric acid in blood—colorimetric and enzymic analyses. The colorimetric methods require preliminary precipitation of blood proteins, which entails coprecipitation of urates. In addition, the color tests are never rigidly specific for urates. For these reasons, accurate uric acid determinations are obtained only by treating the serum with uricase. The disappearance of uric acid from the sample can be determined by a spectrophotometric reading at 292 mμ before and after enzymic treatment.

The reported uric acid levels in human blood necessarily depend on the method used for the determination. But whatever the method, individual variations are found; the upper limits of normal uric acid concentrations are around 8 mg/100 ml in men and 6 mg/100 ml in women. This sex difference is a constant finding in adults but is not seen in children. The higher levels of blood urates in men remain unexplained, but it has been proposed that the renal clearance for urate is greater in women than in men. It has also been suggested that concentration of uric acid in blood is controlled hormonally, and that the higher rates of 17-ketosteroid secretion in men could be correlated with the higher uric acid levels in blood. A shift in hormonal balance has also been invoked to explain the rise in blood uric acid observed after menopause. The uric acid levels of the blood are affected by factors other than hormones [158].

Serum uric acid is elevated in renal disease and whenever tissues are rapidly broken down—for exam-

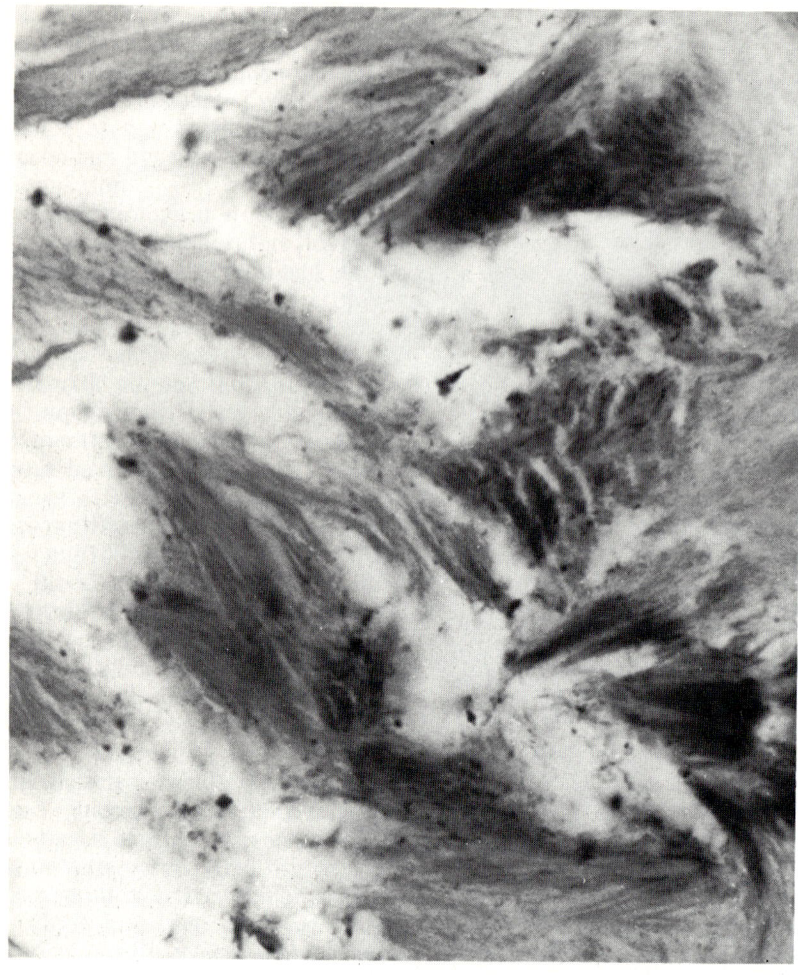

Fig. 3-62. Urate crystals in gouty tophi

result with ulcerations and discharge of the urate crystals in the form of chalky material. (Since urates are water soluble, material fixed in alcohol must be used to demonstrate them histologically.) Tophi are usually found in the soft tissue surrounding the joints, particularly those of the toes and fingers. The tophi are responsible for the knobby deformities about the joints (see Fig. 3-61) of the toes, fingers, knees, elbows, and along the tendons of the ankles and wrists. The helix of the ear is also a preferred site for tophic deposition.

Tophi should not be confused with the subcutaneous nodules of rheumatic fever. The differential diagnosis is easily achieved by needling the tophi, collecting the urate crystals, and demonstrating a positive murexide test in the case of gout. In the joints, the urate crystals are likely to be brought in the cavity through the synovial fluid.* Synovitis ensues, with progressive erosion of the synovia and deposition of the urates into the articular cartilage. There the crystal deposits are surrounded by an area of necrosis. The progressive disintegration of the articular cartilage ultimately leads to severe osteoarthritis, which may be followed by fibroankylosis.

In the kidney, the urate crystals (see Fig. 3-62) are usually in the pyramids as diffuse, pale yellow, chalky deposits. They induce some interstitial inflammation, which may develop into a foreign body granuloma similar to that observed in the soft tissue tophi. The urate accumulation may lead to obstruction of the renal papillae with functional impairment, but usually the development of chronic pyelonephritis is responsible for renal failure.

The morbid anatomy of acute gout is not well known, and only the clinical appearance can be described. The joints exhibit all the characteristics of an acute inflammatory process; they are swollen, painful, warm, and reddish. Rapid accumulation of urates in the synovial fluid is thought to cause these changes. The medical literature abounds in dramatic descriptions of the acute attack of gout. In 50% of the cases, the initial attack involves the metatarsophalangeal joint, but soon other joints of the toes and the fingers are involved. The attacks, widely spaced in time in the beginning, become more frequent when the patient grows older.

* The Greeks described gout under the name *ποδαγϱα* from *ποδος*, meaning foot, and *αγϱα*, meaning attack. Roman physicians described it as *gutta*, which became *goute* in French and gout in Englisch. This name was selected because drops of highly concentrated solution of urates were found at the surface in the synovia of gouty patients.

Pathogenesis of Gout

Many aspects of the pathogenesis of gout remain unsettled—the causes of the acute attack, the prevalence of the urate deposition in some joints, and the mechanism of the hyperuricemia. Our ignorance of the nature of the mechanism responsible for triggering the acute attack of gout is illustrated by the variety and number of explanations (emotional upsets, severe purgation, administration of blood, surgical operation) that have been invoked. The factors most frequently implicated are food and drink. It is often claimed that general excesses or the ingestion of purine-rich foods contribute to the acute attack. But rigid clinical studies of affected patients have not supported these assertions. Similarly, excessive alcoholic intake as a causal factor in gouty attacks has not been satisfactorily demonstrated. It has been proposed that the role of foods in gouty attacks does not depend upon their chemical composition but on their ability to induce allergic reactions, which in turn lead to urate deposition in a hyperuricemic patient.

The acute attack is usually associated with the precipitation of urates. The solubility of the urates decreases with the pH. In the intact animal, lactic acid accumulation is the most likely responsible factor for acidification of the tissue. Lactic acid accumulation results from increased anaerobic glycolysis.

During muscular activity, lactic acid probably accumulates in muscle and some body fluids (possibly synovial fluid) before it reaches the blood. The lowering of the tissue and tissue fluid pH generates a pH gradient between blood and tissue. The urate salts, which are soluble at the higher pH of the blood, precipitate at the lower pH of the synovial fluid. Then they trigger a inflammatory reaction with lymphocyte accumulation. Since lymphocytes are notorious for their high rate of anaerobic glycolysis, their presence is likely to further encourage the precipitation of urates.

Such a mechanism goes far in explaining the role of some of the factors that trigger the crisis, such as exertion, infection, and allergy. It also provides an explanation for the involvement of the feet (especially the big toe) in gout attacks on the basis of the poor circulation in that area. One might also expect urates to precipitate in leukemia and cancer patients, since anaerobic glycolysis is low in the cells of such patients. This possible explanation has not been adequately documented. In fact, very little is known about the mechanism of urate deposition in specific articulations. Trauma has often been invoked, and the susceptibility of the big toe to trauma is said to be responsible for its vulnerability to gout.

Obviously, the most crucial problem yet to be solved in the pathogenesis of gout is the source and cause of hyperuricemia. Uric acid, the terminal metabolite of the breakdown of endogenous and exogenous purines, is excreted through the kidney. It passes freely through the glomeruli, but is actively reabsorbed in the tubules. Hyperuricemia results either from an overproduction of uric acid or from defective renal excretion.

Uric acid accumulation could result from underexcretion of purines, either as a result of excessive reabsorption in the tubules (see above), or because of the presence of a noncompetitive inhibitor of the "carrier" of urates through the epithelial membrane in the process of secretion. There is no direct evidence for such a hypothesis, but the plotted curves representing the changes in uric acid content in the urine for a given level of uric acid in the blood are compatible with it. Relevant to this assumption is an observation made on uric acid excretion in salivary glands: the secretion of urate in saliva is lower in patients with gout than in normal individuals.

An overproduction of uric acid could be brought about in at least three ways: increased breakdown of nucleic acids, acceleration of an unknown shunt pathway for uric acid, and acceleration of the regular pathway for purine biosynthesis.

Accelerated breakdown of nucleic acid is the mechanism likely to operate in secondary hyperuricemia, *e.g.*, in leukemia, as demonstrated by isotopic studies with labeled precursors. In gout, precursors are incorporated in uric acid so rapidly that the existence of a more direct pathway between purine precursors and uric acid has been postulated. Such hypothetical pathways are often referred to as shunt pathways. On the basis of his studies with [^{15}N]glycine, Stetten has pointed out that the turnover of nucleic acids is too slow to account for the rapid incorporation of [^{15}N]glycine into uric acid, and therefore the dietary glycine must enter uric acid by bypassing nucleic acid purines. The notion of a shunt pathway is still speculative and based on circumstantial data obtained from studies with precursor incorporation (using as precursor aminoimidazole carboxamide ribotide) and from studies on birds indicating the existence of a direct pathway between IMP and uric acid. 4-Amino-5-imidazole carboxamide (AIC) ribotide labeled with ^{13}C in the carbon 4 was fed to human subjects. By following the transfer of the aminoimidazole carboxamide ribotide label in urine uric acid and in the AIC of the urine, researchers found that AIC could reach the urine by two major routes: a rapid one responsible for converting about 23% of the AIC into urinary uric acid and 20% of direct transfer of AIC into the urine; and a slow one responsible for slowly converting the AIC ribotide to urine uric acid. The slow pathway is likely to involve the conversion of aminoimidazole carboxamide to purine nucleotides and nucleic acids. The rapid pathway was suspected to involve an unknown shunt pathway. Attempts were made to establish the existence of such a shunt pathway by using [^{15}N]glycine and [^{13}C]aminoimidazole carboxamide concomitantly in patients with gout in whom the [^{15}N] glycine incorporation was known to be normal. Although the results were inconclusive, they suggested a more rapid incorporation of the [^{13}C]aminoimidazole carboxamide into the urine uric acid of gouty patients.

The concept of a shunt pathway also stems from studies in which [^{15}N]glycine was used. Although a shunt pathway has not been demonstrated in mammals, in birds the waste nitrogen is converted by some mechanism the details of which are unknown, to inosine monophosphate, which is further degraded to yield hypoxanthine and uric acid. In birds the waste nitrogen can be eliminated in the urine in the form of hypoxanthine and uric acid after a rapid conversion into inosine monophosphate. Further experimentation with [^{14}C]glucose has also suggested an increase in phosphoribosyl pyrophosphate turnover in patients with gout.

In some cases of gout a twofold increased activity of the phosphoribosyl pyrophosphate synthetase has been demonstrated in fibroblasts with increased production of PRPP. The possibility of an inhibitor causing the increased activity was excluded. This is another example of an enzyme increase in hereditary disease [217].

Adequate concentration of ATP seems to be needed for the maintenance of a proper balance between anabolism and catabolism of purines. The trapping of ATP by 2-deoxyglucose in ascites tumor cells causes an increase in inosine and a decrease in IMP with accelerated breakdown of ribonucleotides to uric acid.

In man, intravenous injections of fructose deplete ATP, ADP, and AMP by trapping the ATP in the form of fructose-1-PO_4; this is associated with hyperuricemia.

Thus, it would appear that whenever a decrease in intracellular levels of ATP occurs it results in at least transient hyperuricemia. Whether this is the mechanism by which anoxemia or muscular exercise helps to trigger the crisis in gout remains to be shown.

A biochemical pathway can be compared to an assembly line in which a skeletal structure is progressively adorned with new conveniences. In passing from one step of a metabolic pathway to the next, the product of the first reaction becomes the substrate of the second. Normally, the "skeleton" (first substrate in the chain of reactions) is converted to the "finished product" (last product in the sequence of reactions) at a steady pace. When the conditions of the environment change in such a way that the requirements for the "finished product" are either greater or smaller, then the pace of the entire pathway changes. A control mechanism could hardly accelerate or decelerate simultaneously all the steps of the pathway and maintain them in synchrony; therefore, the control of a metabolic pathway often acts at one step. The rate of a single reaction of the sequence is expressed by the amount of product that is formed at the expense of the substrate in a unit of time. The conversion of the substrate to the product is a function of the concentrations of substrate, cofactor, activator, inhibitor, etc.

Assays of the individual activities of each enzyme involved in the sequence of steps of a nonbranched pathway reveal that the activity (the amount of product produced per unit time under defined conditions) varies considerably. Some enzymes work fast, some are slow. The rate of production of the "terminal product" will obviously be limited by the activity of the slowest enzyme of the chain. The transformation catalyzed by the slowest enzyme is called "the rate-limiting step." Experiments, done mainly in bacteria, have shown that in an unbranched chain of reactions, the rate-limiting step is usually the first metabolic transformation of the reaction.

The activity of the enzyme that catalyzes the transformation of the rate-limiting step can be modified in at least two ways: derepression and negative feedback inhibition. Derepression affects one or more genes involved in directing the biosynthesis of one or more enzymes of the pathway. For example, in bacteria, which depend on purines in the medium for growth, minimal amounts of purine derepress the genes dictating purine biosynthesis. In contrast, large amounts of purines derepress these genes, and *de novo* purine synthesis is minimized. Many examples of "repression" and induction have already been presented.

The best illustration of the negative feedback mechanism is related to pyrimidine metabolism; only the basic concept is presented in this chapter. The first step of the pathway is usually inhibited. The inhibitors are the terminal product of the pathways or one or more of its derivatives. The enzyme inhibition results from the direct attachment of the inhibiting product to a special site of the inhibited enzyme. The site of attachment of the inhibitor is different from the site of attachment of the substrate.

The reaction that catalyzes the conversion of ribosyl pyrophosphate to 5′-phosphoribosylamine is likely to be the rate-limiting step in purine biosynthesis. Of course, it is difficult to pinpoint a rate-limiting step in an intact mammal, but *in vitro* experiments have established a feedback inhibition of glutamine phosphoribosyl pyrophosphate amino transferase by adenylic and guanylic nucleotides (ATP, ADP, GMP, GDP, and IMP).

At least two findings indicate that such an inhibition is the result of the allosteric binding of the inhibitors with the enzyme. The enzymes, sensitivity to the inhibitors is modified by changes in the environment (pH, temperature, dialysis, passage through Sephadex G25), which introduce conformational changes in the molecule. There are indications that the molecule must contain at least two different binding sites—one for the aminopurine derivatives and one for the hydroxypurine derivatives. Inhibition of enzyme activity by mixtures of GMP and AMP is greater than expected for binding of both nucleotides at a single site. The observation that the site of attachment of the 6-aminopurine is different from that of the 6-hydroxypurine is interesting because it is an example of a specific binding between nucleotide and protein, which is precisely the problem posed by the nature of the repressor.

To understand how the activity of the rate-limiting step could be disturbed in gout, consider the various components involved in the reaction.

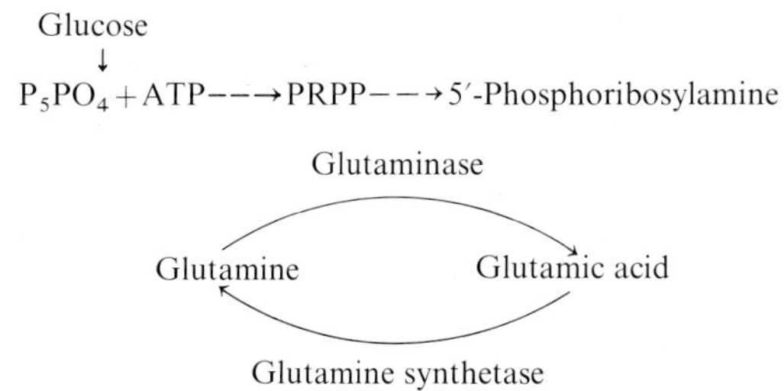

A metabolic alteration leading to overproduction of either PRPP or glutamine would cause an overproduction of 5′-phosphoribosylamine. Excess PRPP could result from overproduction of its precursor glucose, or from reduced use in other pathways, such as pyrimidine biosynthesis. Increased incorporation of [^{14}C]glucose and accelerated turnover of PRPP have been reported. A case of gout with low orotidylic acid pyrophosphorylase and decarboxylase activity has also been described. Whether these metabolic changes constitute the primary injury or are several steps removed from it remains to be established.

When [^{15}N]glycine is administered to patients with gout, the incorporation into uric acid is above normal in patients with severe hyperuricemia. If the ^{15}N-labeled uric acid of these patients is extracted and the molecule is carefully dissected, ^{15}N is located mainly in the N_3 and N_9 positions the purine ring, although glycine contributes the N_7 of the purine ring. These findings made with [^{15}N]glycine corroborate the results obtained with glycine labeled in both carbon 1 and nitrogen. The incorporation of nitrogen is two or three times greater than that of carbon (ratio $^{15}N/^{14}C=2:3$).

These experiments suggest that the nitrogen of glycine is not directly transferred to purine, but instead, is transferred to an NH_3 pool, where it is used for glutamine synthesis. The glutamine then donates its ammonia to form carbons 3 and 9 of the purine ring. Glutamine seems to be preferentially used for purine biosynthesis.

An increase in the amount of glutamine available for purine biosynthesis could be due to accelerated synthesis, decreased use in other pathways, or decreased breakdown. The third possibility is unlikely, since the activity of glutaminase was found to be normal in a patient with severe hyperuricemia. The possibility of an increase in glutamine synthetase activity cannot be excluded. Inborn errors resulting from increased enzyme activities are rare; yet an increase in the activity of δ-aminolevulinic acid occurs in patients with porphyria. All of the metabolic alterations described in gout would satisy a situation in which glutamine use in pathways other than purine synthesis is reduced, but no such alterations have been conclusively demonstrated.

Studies of the regulation of glutamine metabolism in birds (urecotelic) and mammals (ureotelic) may be relevant to the pathogenesis of gout. The major function of glutamine in mammals is to serve as an intermediate for the detoxification of ammonia generated in metabolism, while in birds it is to serve as a precursor for purine synthesis. The difference in the utilization of the metabolite must in some way be expressed by variation in the activities of the enzymes involved in glutamine metabolism.

As may be expected, glutaminase activity is low in birds compared to mammals. Moreover, in birds the glutaminase is markedly product-inhibited by glutamate and is not inducible by high protein diets as is the case in mammals.

In contrast, glutamine synthetase is inducible by a high protein diet in birds, but not in mammals.

We have already seen that in mammals glutamine PRPP amidotransferase, an enzyme which is at the gate of the purine biosynthetic pathway, is feedback inhibited by the terminal products of the reaction, AMP, IMP, and GMP. This regulatory mechanism prevents the formation of excessive amounts of purines. Such feedback inhibition is much less marked with the bird enzyme.

In conclusion, the enzymic machinery of the bird is modulated in such a fashion that glutamate is converted to glutamine which will carry the NH_2 with it in a straight-forward excretory pathway for uric acid. While in the mammal, the glutamine plays a dual role: the major portion is split and the NH_2 is used to form urea, a smaller portion enters the purine biosynthetic pathway and the activity of the key enzyme involved is delicately regulated by allosteric mechanisms.

Clearly, a metabolic error, which would in a human favor the "bird" pathway, would make such humans more urecotelic [217].

At the start of this discussion on the pathogenesis of gout, it was pointed out that the increased incorporation of purine precursors into uric acid suggested that hyperuricemia results from the overproduction of uric acid; however, some patients with gout have been described whose incorporation of labeled glycine into uric acid is normal. In such cases, it has been proposed that the primary injury is in the kidney. A role of the kidney in the pathogenesis of at least certain forms of gout is also suggested by the observation of Nugent and Tyler, who claimed to have demonstrated that the tubular reabsorption of uric acid is greater in patients with gout than in normal individuals (93% in gout versus 91% in normal individuals).

A form of hyperuricemia has been described which results from the partial deficiency of phosphoribosyl transferase. In this type of gout, the deficiency appears to result from the elaboration of an abnormal molecular form of the enzyme. Thus, the enzyme has been shown to alter electrophoretic properties and heat stabilities.

In conclusion, gout—like glycogen storage diseases—is the ultimate manifestation of several forms of metabolic alterations among which must be included increased nucleotide catabolism; increased sensitivity of PRPP aminotransferase to feedback inhibition; molecular alteration of hypoxanthine guanine phosphoribosyl transferase; and decreased uric acid clearance in the kidney.

Therapy for Gout

Although it is not possible to describe here the intricacies of the management of gout, a brief outline of the therapeutic approach in gout is worthwhile because it illustrates how knowledge of disease metabolism can be helpful in planning therapy.

The therapeutic approach differs depending upon whether the patient is in the acute or chronic phase. In acute gout, the drug of choice has for a long time been colchicine. The exact mode of action of colchicine is not known, but it is believed to inhibit leukocyte phagocytosis and thereby interfere with the inflammatory cycle. Phenylbutazone and indomethacin are also used in the treatment of the acute phase of gout. The administration of anti-inflammatory corticosteroids enhances the effect of colchicine.

The treatment of the chronic gouty patient aims at maintaining the lowest possible blood uric acid level. This can be achieved either by preventing reabsorption of uric acid in the proximal tubules or by blocking uric acid biosynthesis through normal metabolic pathways.

Benamid is the model of the uricosuric drugs. It is nontoxic and has an immediate and sustained effect on uric acid reabsorption, thus leading to an increase of urine and a decrease of blood uric acid levels. Benamid was discovered through a search for an agent capable of blocking the secretion of antibiotics, such as penicillin. Benamid is believed to act by blocking an enzyme involved in uric acid reabsorption. The fact that Benamid does not interfere with the reabsorption of glucose and amino acids strongly suggests that it does not act as an unspecific toxin for the tubular cells. Salicylate and aspirin also are excellent uricosuric agents. Their site of action in metabolism is unknown, but their effect differs from that of Benamid. The uricosuric effect of salicylates is delayed and gradual and, in fact, antagonizes that of Benamid.

The inhibitor of xanthine oxidase, allopurinol, is rather effective in decreasing the levels of uric acid in blood and urine. The decrease of uric acid in the urine exceeds what is expected from the increased levels of hypoxanthine and xanthine in the blood, thus suggesting that these metabolites must be in part used through alternative metabolic pathways. Allopurinol is best indicated in those patients who have calculi and crystalluria after the administration of Benamid. In general, however, allopurinol offers no advantage over Benamid in the treatment of gout except, of course, in patients with renal insufficiency.

Lesch-Nyhan Syndrome

In 1964, Lesch and Nyhan described a new form of hyperuricemia associated with uricosuria. The disease was observed in a 1-year-old child with spastic cerebral paralysis, choreo-athetosis, and an unusual inclination to aggressiveness and self-mutilation. This form of hyperuricemia associated with mental retardation and neurological symptoms has now been described in several boys. The disease is familial and apparently is transmitted through a sex-linked recessive gene, probably located on the X chromosome. Affected children are normal at birth and during early infancy. Later they develop unusual irritability without convulsion, followed by an inability to roll over or sleep, spasticity of extremities, and choreo-athetotic movements. Between 6 and 12 months of age, all the early symptoms intensify, and mental retardation becomes obvious. The tendency to self-destruction develops between the second and third years and is shown in destructive arm swinging, finger chewing, teeth grinding, and lip biting. The spinal fluid and blood uric acid levels are elevated. The patients may improve after the administration of drugs, such as probenecid (a uricosuric agent), that reduce the uric acid levels in blood. Some of the patients have died as a result of gouty nephropathy [161].

The biochemical alteration includes overexretion of uric acid in urine (two or three times above normal). The hyperuricuria results from a massive overproduction of uric acid (two or three times above what is observed in gout patients). Studies of the incorporation of [^{14}C]glycine into urinary uric acid clearly illustrate the uric acid overproduction, which is 100 and sometimes 200 times as great as in normal individuals.

At least three biochemical changes are now established to be part of this disease: (1) decreased activity of hypoxanthine guanine phosphoribosyl transferase activity*; (2) increased levels of PRPP; and (3) increased activities of adenine phosphoribosyl transferase. The enzyme defect and increased activity of phosphoribosyl transferase was demonstrated in erythrocytes, skin fibroblasts, liver, and brain tissues of patients. Similarly, high levels of PRPP have been demonstrated in red cells and fibroblasts (4 to 20 times normal).

From these observations, one may reason that the absence of the salvage pathway resulting from the deficiency in hypoxanthine guanine phosphoribosyl transferase makes available more PRPP, which is used for purine biosynthesis through the *de novo* pathway. Consequently, the total pool of degradation products of purines from both endogenous and exogenous origin (hypoxanthine, xanthine, and uric acid) is increased because exogenous precursors are not used for biosynthesis, but at the same time more endogenous nucleotides are made.

The mechanism by which a decrease in activity of the hypoxanthine adenine phosphoribosyl transferase is responsible for the neurological symptoms observed in the Lesch-Nyhan syndrome is not known. At least two possibilities have been proposed. Some portions

* In some cases the disease has been diagnosed before birth by collecting embryonic cells by amniocentesis, culturing them, and assaying for hypoxanthine incorporation [162, 163].

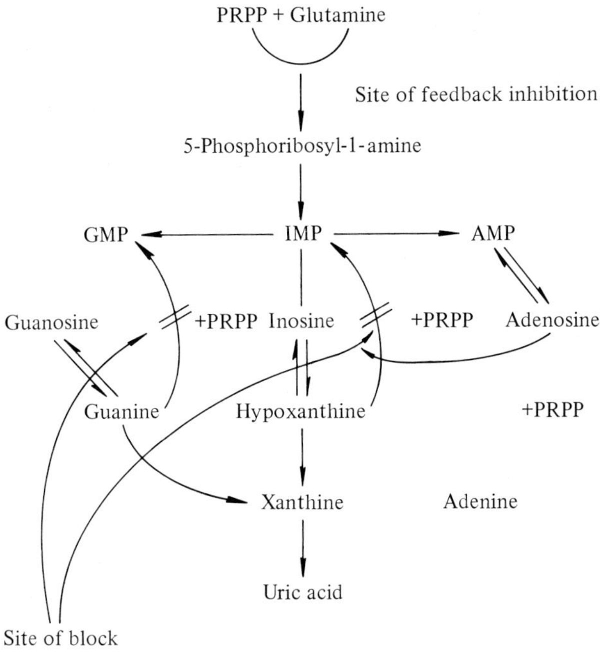

PRPP + Glutamine
Site of feedback inhibition
5-Phosphoribosyl-1-amine
GMP IMP AMP
Guanosine +PRPP Inosine +PRPP Adenosine
Guanine Hypoxanthine +PRPP
Xanthine Adenine
Uric acid
Site of block

Fig. 3-63. Guanine hypoxanthine phosphoribosyl transferase deficiency

or special cell arrangements of the brain may be lacking the *de novo* pathway, and thus without hypoxanthine guanine phosphoribosyl transferase, which is at the origin of the salvage pathway, these brain areas could not make purine nucleotides. Hypoxanthine, or some derivative of it, could also be toxic to the brain (see Fig. 3-63).

Caffeine or theophylline injection in the spinal canal of rats reportedly leads to self-mutilative behavior. Why this happens is not known, but it has been proposed that the effect of both methylated purines results from their inhibition of cyclic nucleotide phosphodiesterase.

An elevated plasma dopamine-β-hydroxylase is associated in some cases of Lesch-Nyhan syndrome with the hypoxanthine guanine phosphoribosyl transferase activity. The hydroxylase is found, among other sites, in the synaptic vesicles where it converts dopamine to norepinephrine. Elevation of the hydroxylase occurs when the patients are exposed to stress that generates acute sympathetic stimulation. Blood pressure does not increase as it does in normal individuals. Those individuals with low transferase, but normal hydroxylase activity do not mutilate themselves.

The mechanism by which the association of elevated plasma dopa hydroxylase and defective response to the sympathetic stimulus is achieved is unknown. The hydroxylase could be readily released either from synaptic vesicles and adrenals prematurely so that its action on the substrate is prevented, catecholamines themselves could be overproduced and unable to bind to the proper receptor, or the receptor could be defective [219].

Pyrimidine Metabolism and Orotic Aciduria

Although orotic aciduria has been described in only a few patients, pyrimidine metabolism is so important in the biochemistry of growth that the biosynthetic and breakdown pathways of the various pyrimidines will be described in some detail [150–152].

Uracil, cytosine, thymine, 5-methylcytosine, 5-hydroxymethylcytosine, and pseudouracil are some of the most important pyrimidine bases. Pyrimidines are heterocyclic compounds called diazines that contain two nitrogens in the ring. In pyrimidines, the two nitrogens are in position meta.

Uracil, or 2,6-dihydroxypyrimidine, was discovered in 1901 in yeast nucleic acids, and its structure was established by Fisher's organic methods. Uracil is a universal constituent of ribonucleic acids. In general, it is absent from DNA, although some bacterial DNA's have been reported to contain small amounts of uracil deoxyribosides. Cytosine, or 2-hydroxy-6-aminopyrimidine, was discovered in 1894 by Kassel and Newman. Except for the T-even bacteriophages, all DNA and RNA specimens in which the pyrimidine has been characterized contain cytosine.

Thymine (2,6-hydroxy-5-methylpyrimidine or 5-methyluracil) was first isolated in 1893 by Kassel and Newman from nucleic acid hydrolysates obtained from calf thymus. Thymine is found in all DNA and is rarely present in RNA. It has never been found free in nature except in sponges in the form of xyloside.

5-Methylcytosine was first found in tuberculinic acid obtained from tubercule bacilli. Later it was isolated from thymus DNA and found to be a minor pyrimidine constituent of mammalian DNA and a major pyrimidine compound in wheat germ DNA. However, contrary to the original reports, 5-methylcytosine could not be found in bacterial DNA, including tubercule bacilli.

5-Hydroxymethylcytosine, which is found in T-even phage, was discussed in the chapters on molecular genetics.

Pseudouracil is found in pseudouridylic acid, a nucleotide present in small amounts in the soluble RNA. On a molar basis, pseudouridylic acid contains amounts of carbon, nitrogen, hydrogen, and oxygen identical to those found in uridylic acid. The main difference between pseudouridylic acid and uridylic acid is in the position of the glycosidic bond between the base and the ribofuranose. When the base was first discovered, the exact position of that linkage was not clear, but ultraviolet-absorption studies of pseudouridylic acid yielded a spectrum similar to that of 5-hydroxymethyluracil. This led to the assumption that a carbon-to-carbon bond was formed between the carbon 5 of the uridine base and the carbon 1 of the sugar. Methylation of the positions 1, 2, 3, and 6 of the pseudouracil molecule eliminated these positions as possible sites for the glycosidic linkage. More direct evidence for the existence of the glycosidic bond involving the carbon 5 of the uracil was provided by Kohn, who used both classical chemical techniques and nuclear magnetic resonance studies. The periodate oxidation of pseudouridylic acid followed by reduction with bromides led to the formation of a compound chromatographically and spectrophotometrically identical to 5-hydroxymethyluracil. Nuclear magnetic resonance revealed the presence of a proton on the base's carbon 4 and the absence of a proton on carbon 5. Such findings suggested that the carbon 4 was not involved in the glycosidic linkage. The configuration of the proton of the carbon 1′ of the sugar was precisely as is expected when the carbon 1′ is involved in a carbon-to-carbon rather than in a carbon-to-nitrogen bond.

Phosphoribosyl pyrophosphate is of capital importance in the biosynthesis of both purine and pyrimidine nucleotides. But since it was discovered in the course of investigations on pyrimidine biosynthesis, its properties and biosynthesis will be reviewed here. Kornberg and his associates [164] were the first to isolate 5-phosphoribosyl pyrophosphate. They further demonstrated that this phosphorylated compound is essential in converting purine and pyrimidine bases to their respective nucleotides. PRPP is labile to heat and decomposes at low pH. The exact configuration of sugar in the PRPP molecule is not known, but some comparative studies of synthetic and natural ribose suggest that the ribose-1-phosphate derived from guanosine is an α-D-ribofuranose-1-phosphate, and it has therefore been assumed that the same configuration exists in the PRPP molecule. PRPP is formed by the reaction of ATP with ribose-5-phosphate in the presence of an enzyme, the ribose pyrophosphorylase. In the course of this reaction, the carbon 1 of ribose is phosphorylated, and AMP is produced. Ribose pyrophosphorylase was purified, and the enzyme was demonstrated to require magnesium. The steps involved in the biosynthesis of orotidylic acid are presented in Fig. 3-64.

Carbamyl phosphate is the first molecular complex formed in the course of pyrimidine biosynthesis. In bacteria, carbamyl phosphate can be formed from CO_2, ammonia, and ATP. This overall reaction has also been described in mammalian tissue, where carbamyl phosphate acts as a carbamyl donor in at least two metabolic pathways—urea and pyrimidine biosynthesis. Thus, the carbamyl group may be transferred to the delta amino group of arginine and form citrulline, or it may be transferred to the alpha amino group of aspartic acid and form carbamyl aspartic acid. In each reaction, the carbamyl group is transferred to an amino group, and a peptide linkage is formed. The formation of carbamyl aspartic acid from carbamyl phosphate and aspartic acid is catalyzed by an enzyme, or carbamyl aspartic acid synthetase, which has been found in pigeon liver and in several rat tissues. Another pathway for the formation of carbamyl aspartic acid has been proposed in mammalian tissue. It is said to start with citrulline, a metabolite normally used in the urea cycle, which then yields acetyl glutamate in the presence of ATP. The carbamyl acetyl glutamate would then act as carbamyl donor and lead to the formation of carbamyl aspartic acid. The formation of carbamyl aspartic acid is discussed in greater detail in the chapter devoted to the urea cycle.

The role of carbamyl aspartic acid in the biosynthesis of pyrimidine was suggested by indirect and direct evidence. The addition of carbamyl aspartic acid to bacterial culture maintains and stimulates growth efficiently, suggesting that the compound is a precursor of either proteins or nucleic acids. Isotopic studies with labeled carbamyl aspartic acid demonstrated that the label was incorporated in the pyrimidines found in nucleic acid. An enzyme system capable of converting carbamyl aspartic acid to orotic acid was also purified. The enzyme preparation is involved in at least two different chemical reactions: ring closure and dehydration. The ring closure involves the formation of a carbon-nitrogen bond between the carbon of the carboxyl group and the nitrogen of the amino group of carbamyl aspartic acid. The closing of the ring is accompanied by the loss of a molecule of water. The water is derived from the hydroxyl group that is removed from the carboxyl, and a hydrogen is removed from the amino group of the carbamyl aspartic acid. The enzyme catalyzing the oxidation of dihydroorotic acid is called dihydroorotic acid dehydrogenase.

As early as 1949, it was demonstrated that injected ^{15}N- or ^{14}C-labeled orotic acid was readily incorporated into DNA and RNA of mammalian tissue, indicating that orotic acid is a precursor of nucleic acid pyrimidine. The next step in pyrimidine biosynthesis is the formation of the first nucleotide in the sequence. It involves the reaction between ribosyl pyrophosphate and orotic acid to yield 5′-orotidylic acid; the reaction is catalyzed by orotidylic pyrophosphorylase. Thus, the first steps of pyrimidine biosynthesis differ from the early steps of purine biosynthesis in at least two ways. Orotic acid, instead of being synthesized atom by atom as is the case for the purine ring, is made from the condensation of rather large molecules, namely, carbamyl phosphate and aspartic acid. Furthermore, all the steps of purine biosynthesis occur at the level of the nucleotide, but the the pyrimidine ring is closed at the level of the base.

Orotidine-5′-monophosphate was isolated by adding to the appropriate incubation mixtures antimetabolites which cause it to accumulate by blocking further steps of pyrimidine biosynthesis.

Uridylic acid, the nucleotide found in RNA, is formed by decarboxylation of orotidylic acid in the presence of orotidine-5′-phosphate decarboxylase, an enzyme purified from yeast. This is an irreversible reaction that has been observed in bacteria, birds, and several mammalian tissues. The antimetabolite 6-azauracil blocks orotidylic acid decarboxylase.

The formation of cytidylic acid from uridylic acid was studied in a Novikoff hepatoma and *E. coli.* In *E. coli,* glutamine appears to be the primary amino donor for cytidylic acid synthesis, and the synthesis occurs after uridylic acid has been converted to the triphosphate in the presence of ATP. The detailed mechanism of the amination of UTP to yield CTP

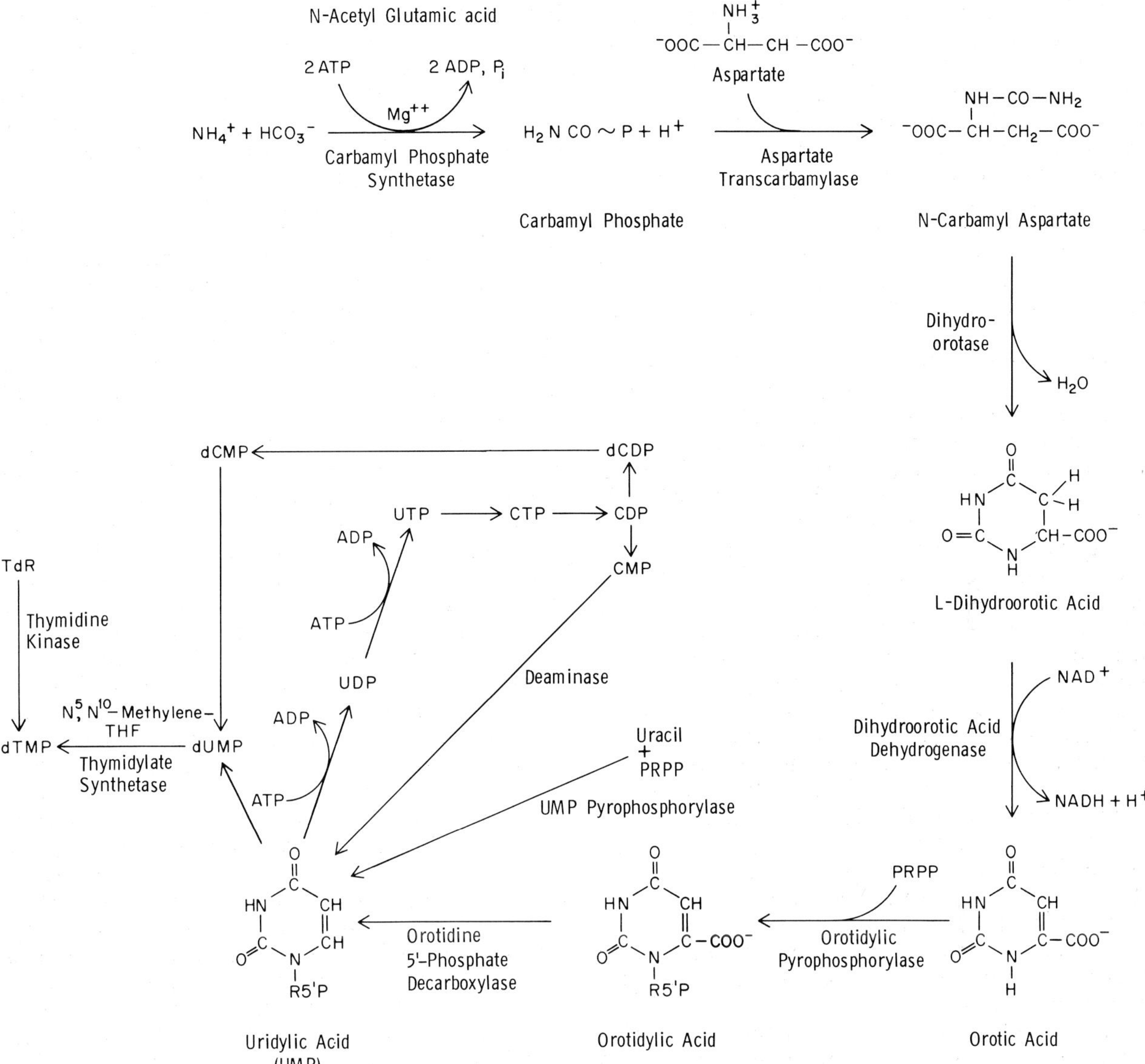

Fig. 3-64. Pyrimidine biosynthesis

is not clear. In the presence of soluble enzyme of the Novikoff hepatoma, uridine nucleotide conversion to cytidine nucleotide depends on the presence of glutamine and is inhibited by the addition of 6-diazo-5-oxo-L-norleucine. Small concentrations of guanosine nucleotides stimulate the conversion of uridine nucleotides to cytidine nucleotides in mammalian tissues but do not affect the bacterial system.

Thymidine is formed by methylation of the carbon 5 of uracil. *In vivo* experiments demonstrated that the methyl donor is either formate or the beta-carbon of serine. The conversion of uracil to thymidine *in vitro* was first demonstrated using bone marrow or embryo cell suspensions. Three main points were established: (1) the deoxyriboside of the 2[^{14}C]uracil is in their system incorporated into DNA; (2) the incorporation of the deoxyriboside is inhibited by the addition of aminopterin; (3) the incorporation is diluted or interfered with when cold thymidine is added to the system, which suggests that thymidine is an intermediate in this reaction. It was later shown that the addition of deoxycytidine to the medium did not interfere with the incorporation of deoxyuridine into DNA. This finding indicates that there is no biosynthetic pathway for thymine utilizing deoxycytidine as a precursor. The formation of thymidine from deoxyuridine is catalyzed by an enzyme called thymidylate synthetase, which is found in bacteria, bacteriophage-infected *E. coli*, and thymus. The calf thymus enzyme, which has been partially purified, is a heterogenous preparation containing at least four different active protein components. It is not known whether this heterogeneity is due to the existence of four different thymidylic synthetases or to the combination of one or more thymidylic synthetases with contaminating proteins. As may be anticipated, folic acid derivatives are required in the reaction. The role of thymidylate synthetase in regenerating liver and the mechanism of the reactions are discussed elsewhere.

Uridylic acid is converted to thymidylic acid after uridine monophosphate is converted to the deoxyuridine monophosphate. The conversion of ribose to deoxyribose is discussed in detail in the chapter devoted to vitamin B_{12}.

Phosphodiesterases, phosphatases, 5′-nucleotidases, 3′-nucleotidases and phosphotransferases have been discussed in the section on purine breakdown. Some of these enzymes also act on pyrimidine nucleotides, yielding either the nucleoside or the free base. Cytosine deaminases have been found in yeast and *E. coli*. Cytidine and cytidylic acid deaminases are present in extract of most mammalian tissue. The properties of these enzymes are still poorly understood.

From unfertilized eggs and embryos of sea urchins, Scarano has extracted an enzyme catalyzing the deamination of 5′-deoxycytidylic acid to yield 5′-deoxyuridylic acid. The exact role of that enzyme in biology is not clear, but it is assumed to be involved in the formation of dUMP, which can be used for TMP biosynthesis. This assumption is based mainly on the fact that deoxycytidylic deaminase was found to be increased in growing tissues, including regenerating liver. (Some of the biological properties of deoxycytidylic deaminases are discussed in the chapter on regeneration.)

Since 1901, diets rich in thymine or nucleic acid have been known to increase the amount of urea excreted in the urine. A diet rich in DNA or thymine was later found to lead to the appearance of α-aminoisobutyric acid in the urine. (5-Methylcytosine administration does not increase aminoisobutyric acid in the urine.) Dihydrothymine was much more efficient than thymine in stimulating the appearance of 5-aminoisobutyric acid. Thus, a degradative pathway for thymine was postulated in which dihydrothymine and aminoisobutyric acid were intermediates (see Fig. 3-65). It is now known that in the presence of a hydrogen donor, thymine is converted to dihydrothymine. The pyrimidine ring of dihydrothymine is split in the presence of water to yield ureidoisobutyric acid. Uracil's degradation pathway is analogous to that of thymine. Uracil is converted to dihydrouracil, then to ureidopropionic acid, which is in turn split to yield β-alanine, CO_2 and NH_3. The conversion of thymine or uracil to the dehydropyrimidine is catalyzed by a dihydropyrimidine dehydrogenase that has been purified about 20-fold from beef liver acetone powder. The dehydrogenase is unstable and requires NADP as a hydrogen acceptor. In addition to uracil and thymine, 5-bromouracil and 5-iodouracil act as substrates for the dehydropyrimidine dehydrogenase. The enzyme involved in the conversion of dihydrothymine to β-aminoisobutyric acid was purified 200-fold from beef liver acetone powder, dihydrothymine hydrolase. The reaction is reversible and pH dependent. The optimum pH for opening the ring is 8.5, and that for its closure is 5. The enzyme involved in that reaction is dihydrothymine hydrolase. A dihydrouracil hydrolase capable of catalyzing the conversion of ureidopropionic acid to the dihydrouracil is present in the soluble fraction of calf, rat, and pigeon livers.

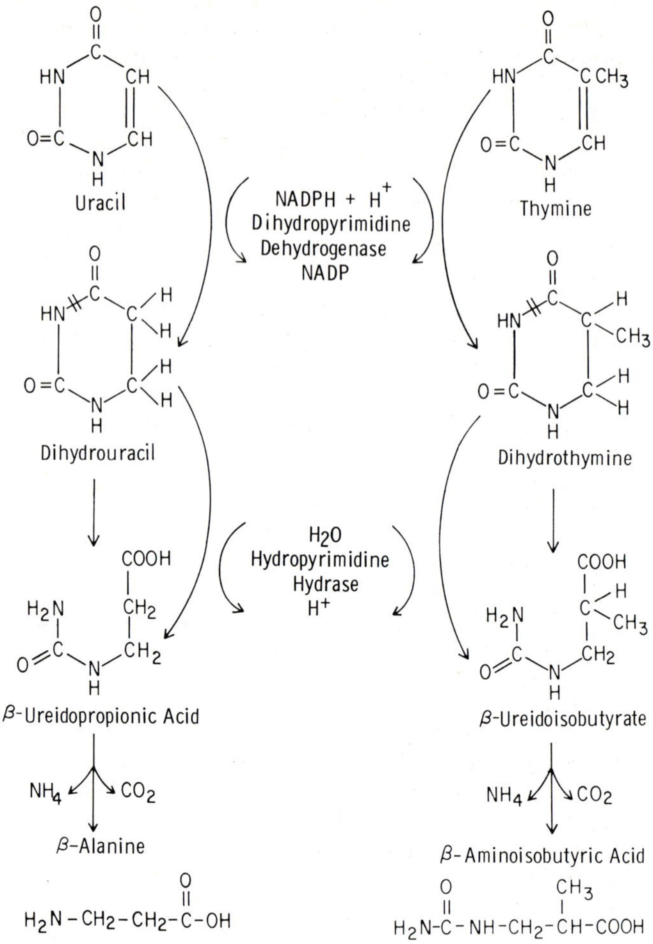

Uracil | Thymine

NADPH + H^+ Dihydropyrimidine Dehydrogenase NADP

Dihydrouracil | Dihydrothymine

H_2O Hydropyrimidine Hydrase H^+

β-Ureidopropionic Acid | β-Ureidoisobutyrate

NH_4 CO_2

β-Alanine $H_2N-CH_2-CH_2-C(=O)-OH$ | β-Aminoisobutyric Acid $H_2N-C(=O)-NH-CH_2-CH(CH_3)-COOH$

Fig. 3-65. Pyrimidine catabolism

Orotic Aciduria. Although orotic aciduria is a rare disease, the study of the few patients afflicted with it has provided a great deal of information on the role of pyrimidines in disease and the *in vivo* metabolism of pyrimidine nucleotides. The first patient was studied when he was 9 months old. At that time, he had two main groups of symptoms: a rather unique hematological picture and massive excretion of orotic acid in the urine. The blood smear contained hypochromic erythrocytes and revealed anisocytosis and poikilocytosis. A leukocyte count demonstrated moderate leukopenia and monocytosis. Bone marrow examination established hyperplasia of the marrow with a marked increase in the number of megaloblastic cells.

Apparently, both the granulopoietic and the erythropoietic lineage participated in the megaloblastic proliferation. Thus, the most salient hematological feature was a hyperchromic megaloblastic anemia. The only other abnormality was found in the urine, where large amounts of pyrimidine bases, later demonstrated to be orotic acid, were excreted. Such observations suggest either a block in orotic acid use for further biosynthesis of pyrimidine and nucleic acid, or interference with renal orotic acid excretion.

The accumulation in blood and the excessive urinary excretion of orotic acid are postulated to result from the absence of either decarboxylase or the pyrophosphorylase. These enzyme activities could not be directly assayed in the patient's tissues, but assays in the parents and two of the siblings indicated that the activities of both decarboxylase and phosphorylase were defective. In the patients described, the orotic aciduria and the megaloblastic anemia that it causes could be relieved by injecting nucleotides (cytidylic and uridylic acid). And in addition to its therapeutic significance, this observation also provides some invaluable information on the functioning of the pyrimidine metabolic pathway *in vivo*.

These findings suggest that under normal dietary conditions, the bone marrow derives its nucleotides from the *"de novo"* pathway, rather than from the salvage pathway, even if the nucleotides are in such shortage that a megaloblastic anemia develops. It is therefore unlikely that mechanisms controlling the rate of DNA synthesis operate at the level of thymidine kinase. However, some negative feedback must function when nucleotides prevent the accumulation of orotic acid. It must operate at a biosynthetic step preceding the formation of orotic acid. Obviously, there is no direct evidence of the site of action of the feedback control of orotic acid synthesis in patients with orotic aciduria, but a remarkable example of such a control mechanism is seen in bacteria. This negative feedback mechanism was discovered by Pardee and his associate [165, 166].

The first step in pyrimidine biosynthesis is the reaction catalyzed by aspartic transcarbamylase, a reaction in which the carbamyl group of carbamyl phosphate is transferred to aspartic acid to yield ureidosuccinic acid. In *E. coli*, the end products of the pyrimidine pathways, UTP and especially CTP, inhibit the transcarbamylase. Pardee and his associate, who discovered this important event, established that the site of action in the transcarbamylase molecule of the inhibitor is different from the site of action of the substrate.

The native enzyme is believed to be a large protein molecule (mol wt 310,000) containing 24 sulfhydryl groups, made of regulatory and catalytic subunits. There are four regulatory subunits, each with a molecular weight of 30,000, and two catalytic subunits (mol wt 96,000 each). All the sulfhydryl groups are found in the regulatory subunit. Treatment of the native enzyme with parachloromercuribenzoate (PCMB) not only desensitizes the enzyme to inhibition by CTP, but it also physically separates the native molecule into the two kinds of functional subunits. Elimination of PCMB is followed by reaggregation of all subunits and restoration of native enzyme.

Each regulatory subunit can bind eight molecules of CTP, whereas each catalytic subunit is capable of binding four molecules of substrate. However, the four catalytic binding sites of the catalytic subunit react strongly with each other, and as a result the affinity for the substrate is low. But as a soon as the first molecule of substrate is bound to the catalytic subunit, the interaction between the four binding sites weakens. The affinity of the catalytic subunit for the second molecule is greater than its affinity for the first molecule. The cooperative effect of substrate binding is further enhanced as the third molecule of substrate is bound to the catalytic subunit. In conclusion, the catalytic subunit contains four binding sites for the substrate and there is a cooperative effect on the successive binding of the molecules of substrate to the enzyme.

Consequently, substrate saturation curves are sigmoid instead of hyperbolic. CTP exerts its inhibitory effect by increasing the interaction between the four catalytic binding sites, which decreases the affinity of the enzyme for the substrate.

ATP also binds to the regulatory site but does not induce conformational changes in the molecule. Thus, ATP does not directly affect the affinity for the substrate. Yet, by acting as a competitor for CTP, ATP desensitizes the enzyme to CTP inhibition.

The existence of four catalytic sites has also been invoked to explain the unusual effects of metabolic analogs of aspartate, succinate, and malate on the affinity of the enzyme for the substrate. Small doses of analog activate the enzyme, but large doses inhibit it. Small amounts of the analog are believed to attach to only a small number of the binding sites, leaving enough sites for binding with aspartate. But in binding with the catalytic subunit, the analog exerts the same cooperative effect as the substrate; thus small doses of analog enhance the affinity for the substrate. In contrast, large amounts of the analog would simply occupy all binding sites and abolish all the enzyme's affinity for the substrate [167].

CYSTINURIA

Cystinuria is a hereditary disease characterized by the excessive excretion of cystine, lysine, ornithine, and arginine in the urine, probably resulting from a deficiency in the renal tubular transport mechanism. Sir Archibald E. Garrod postulated that cystinuria resulted from a metabolic block involving the oxidation of cystine to sulfate. Later investigations of the pathogenesis of cystinuria demonstrated that the hereditary deficiency does not involve a metabolic block. If a metabolic impairment existed, cystine would be expected to accumulate in the plasma of cystinurics, but plasma levels of cystine are normal or low in cys-

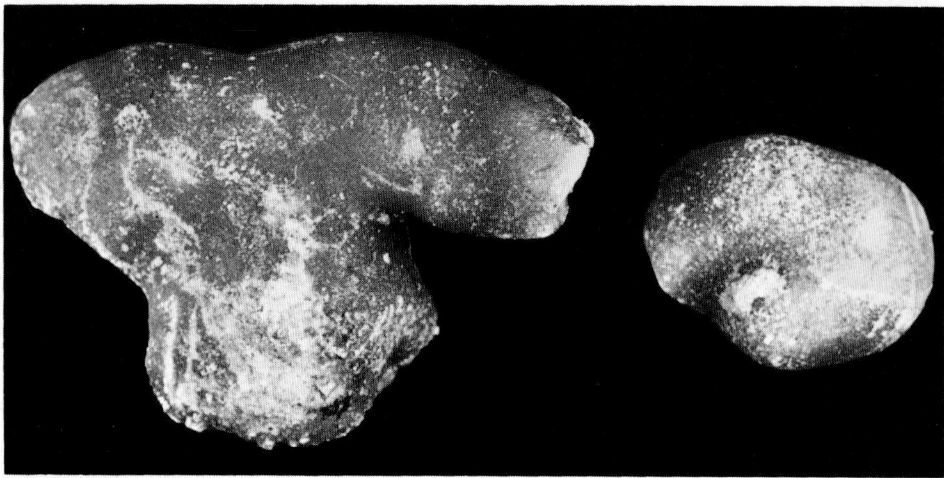

Fig. 3-66. Cystinuria. Cystine stone assuming shape of calices

tinuric patients, and cystinuria is not increased by the administration of cystine [168, 169].

In contrast, the renal clearance of cystine is 30 times greater in cystinurics than in normal individuals. Ornithine, arginine, and lysine are spilled in the urine along with cystine, and the renal clearance of these four amino acids is increased.

The administration of cystine to cystinurics has no further effect on the renal clearance of cystine. In contrast, when similar amounts of cystine are administered to normal individuals, the renal clearance of cystine is considerably increased. These findings suggest that although normal individuals can adapt their renal reabsorption of cystine according to the blood concentration of cystine, cystinurics cannot.

For these reasons, it is accepted that a deficiency in renal absorption is responsible for the increased urinary excretion of amino acids in cystinuria. A defect in the reabsorption mechanism of four amino acids can be explained by the deletion of a single protein. Examination of the formula of the four amino acids spilled in cystinuria reveals a common denominator: all four are diamino acids, therefore, it seems logical that their reabsorption involves a single carrier protein rich in dicarboxylic acids (aspartic and glutamic).

A number of experimental procedures have established that the dibasic monocarboxylic acids lysine, arginine, and ornithine are transported in the intestines and kidneys by a common system. Moreover, the system that transports the dibasic monocarboxylic acid also transports cystine [170–172]. The transport mechanism just described is the one that is deficient in patients with cystinuria.

However, although cystine clearance in some cystinurics equals that of inulin, clearance of lysine, ornithine, and arginine is markedly below that of inulin. These discrepancies between the reabsorption of cystine and that of the dibasic monocarboxylic acid are explained by the fact that the kidney possesses only one transport system for cystine, but it possesses two for lysine [171].

Because cystine does not accumulate in the blood of cystinurics, no severe metabolic distortions are observed. But cystine is an amino acid that is poorly soluble, especially at the pH of urine, and therefore it crystallizes in the renal tubules, calices, and pelvis, forming calculi that obstruct the flow of urine and cause hydronephrosis and secondary pyelonephritis (see Fig. 3-66). Cystine precipitation can be prevented by reducing the amount of cystine precursors in the diet, administering alkalinizing substances, and diluting the urine.

Various combinations of phenotypes exist among cystinurics. In a first group, the homozygote has severe amino-aciduria while the heterozygotes are all symptomless and excrete no cystine in their urine. In this group, the disease is transmitted as an autosomal recessive gene. In another group, the homozygote has cystinuria and forms stones. The heterozygous fall in various categories; some excrete only high amounts of cystine, others excrete cystine and arginine. One interpretation of these observations is that the disease is transmitted by an incomplete recessive gene responsible for elaborating a compound involved in the resorption of cystine, lysine, arginine, and ornithine. It is postulated that the substance controlling amino acid absorption has a greater affinity for arginine and ornithine than for lysine.

Cystinuria is the prototype of a number of inborn errors of metabolism believed to result from the deficiency of carrier proteins involved in transporting molecules through the cell membrane. Transport defects in kidney, intestine, or both have also been described for other amino acids—glycine, cystine, tryptophan, methionine—for glucose and galactose, and even for electrolytes such as calcium, chloride, and sodium [173]. Some of these inborn errors are described in other chapters.

A partial list of other inborn errors of metabolism appears in Table 3-6.

Table 3-6. Some other inborn errors of metabolism

Hereditary hemolytic anemia	Phosphoglycerate kinase deficiency in erythrocytes and leukocytes	X chromosome linked (Chinese)	Valentine, W.N., Hsieh, H., Paglia, D.E., Anderson, H.M., Baughan, M.A., Jaffe, E.R., and Garson, O.M.: Hereditary hemolytic anemia associated with phosphoglycerate kinase deficiency in erythrocytes and leukocytes: A probably X-chromosome-linked syndrome. N. Engl. J. Med. **280**, 528–534 (1969)
Hydroxyprolinemia: hydroxyprolinuria, mental retardation, no obvious skeletal or collagen abnormality	Hydroxyproline oxidase deficiency	?	Efron, M.L., Bixby, E.M., Palattao, L.G., and Pryles, C.V.: Medical intelligence – hydroxyprolinemia associated with mental deficiency. N. Engl. J. Med. **267**, 1193–1194 (1962)
Hyperlysemia: convulsion, mental retardation, delayed development, retarded growth, muscle and ligamentous weakness, anemia	Inability to degrade lysine?		Woody, N.C., Hutzler, J., and Dancis, J.: Further studies of hyperlysinemia. Am. J. Dis. Child. **112**, 577–580 (1966)
Hypersarcosinemia: sarcosinuria, difficulty in swallowing	Decrease in sarcosine dehydrogenase?		Gerritsen, T., and Waisman, H.A.: Hypersarcosinemia. An inborn error of metabolism. N. Engl. J. Med. **275**, 66–69 (1966)
Calcium oxalate kidney stones L-glyceric aciduria	Deficiency in D-glyceric acid dehydrogenase		Resnick, M., Pridgen, D.B., and Goodman, H.O.: "Calcium oxalate kidney stones"; genetic predisposition to formation of calcium oxalate renal calculi. N. Engl. J. Med. **278**, 1313–1318 (1968)
Saccharase intolerance: diarrhea	Low activity in intestinal maltase-III, IV, V (but not I or II)		Clément, R.: Intolérance au saccharose. La Presse Médicale **75**, 969–970 (1967)
Cystinosis	Accumulation of cystine crystals in RE cells, amorphous cystinuria, cystine accumulation in fibroblasts, leukocytes, and mucosa cells		Baar, H.S., and Bickel, H.: Part 8: Morbid anatomy, histology and pathogenesis of Lignac-Fanconi disease (Cystine storage disease with aminoaciduria). Acta Paediat. 42, Suppl. **90**, 171–237 (1952)
Homocystinuria: excretion of homocystine in urine, mental retardation, dislocation of ocular lenses, sparse blond hair, genu valgum, convulsions, fatty liver	Deficiency in cystathione synthetase activity		Schneider, J.A., Bradley, K., and Seegmiller, J.E.: Increased cystine in leukocytes from individuals homozygous and heterozygous for cystinosis. Science **157**, 1321–1322 (1967) Hummeler, K., Zajac, B.A., Genel, M., Holtzapple, P.G., and Segal, S.: Human cystinosis: Intracellular deposition of cystine. Science **168**, 859–860 (1970) Carson, N.A.J., Cusworth, D.C., Dent, C.E., Field, C.M.B., Neill, D.W., and Westall, R.G.: Homocystinuria: A new inborn error of metabolism associated with mental deficiency. Arch. Dis. Childhood **38**, 425–436 (1963) Gerritsen, T., and Waisman, H.A.: Homocystinuria, an error in the metabolism of methionine. Pediatrics **33**, 413–420 (1964) Finkelstein, J.D., Mudd, S.H., Irreverre, F., and Laster, L.: Homocystinuria due to cystathionine synthetase deficiency: The mode of inheritance. Science **146**, 785–787 (1964)
Hyper β-alaninemia: excess β-alanine, β-aminoisobutyric acid, β-taurinic in urine, excess β-alanine in blood, somnolence and convulsion	Interaction between β-alanine and a specific transport system for β-amino acid		Scriver, C.R., Pueschel, S., and Davies, E.: Hyper-β-alaninemia associated with β-aminoaciduria and γ-aminobutyricaciduria, somnolence and seizures. N. Engl. J. Med. **274**, 635–643 (1966)
Lactase intolerance: irritable colon (increased intestinal peristalsis), abdominal cramp, distension, flatulence, diarrhea	Lactase deficiency		Weser, E., Rubin, W., Ross, L., and Sleisenger, M.H.: Lactase deficiency in patients with the "irritable-colon syndrome." N. Engl. J. Med. **273**, 1070–1075 (1965)

Table 3-6 (continued)

Cystathioninuria: increased levels of cystathione in tissues, additional abnormalities are usually associated with the cystatonuria	Deficiency in cystathionase and homoserine		Harris, H., Penrose, L.S., and Thomas, D.H.H.: Cysthathioninuria. Ann. Human Genet. **23**, 442–453 (1959) Perry, T.L., Robinson, G.C., Teasdale, J.M., and Hansen, S.: Concurrence of cystathioninuria, nephrogenic diabetes insipidus and severe anemia. N. Engl. J. Med. **276**, 721–725 (1967)
Xanthinuria: low uric acid excretion in urine	Xanthine oxidase deficiency		Ayvazian, J.H., and Skupp, S.: The study of purine utilization and excretion in a Xanthinuric man. J. Clin. Invest. **44**, 1248–1260 (1965)
Pentosuria: symptomless	NADP Xylitol dehydrogenase deficiency	Mendelian recessive trait	Touster, O.: Essential pentosuria and the glucuronatexylulose pathway. Fed. Proc. **19**, 977–983 (1960)
Iminoglycinuria: associated with congenital nerve deafness	Renal tubular defect glycine, proline, hydroxyproline	Autosomal	Rosenberg, L.E., Durant, J.L., and Elsas, L.J.: II. Familial iminoglycinuria; an inborn error of renal tubular transport. N. Engl. J. Med. **278**, 1407–1413 (1968)
Methylmalonic aciduria: metabolic acidosis, developmental retardation, long chain ketonuria intermittent hyperglycinemia	Defect in conversion of methyl malonyl CoA to succinyl CoA		Rosenberg, L.E., Lilljeqvist, A.C., and Hsia, Y.E.: Methylmalonic aciduria; an inborn error leading to metabolic acidosis, long-chain ketonuria and intermittent hyperglycinemia. N. Engl. J. Med. **278**, 1319–1322 (1968)
L-glyceric aciduria: urinary excretion of oxalic acid, urinary L-glyceric acid oxalate stones	D-glyceric dehydrogenase deficiency	Autosomal recessive	Williams, H.E., and Smith, L.H., Jr.: L-glyceric aciduria; a new genetic variant of primary hyperoxaluria. N. Engl. J. Med. **278**, 233–239 (1968)
Familial iminoglycinuria: iminoglycinuria	Renal tubular defect for reabsorption of glycine, proline, and hydroxyproline		Rosenberg, L.E., Durant, J.L., and Elsas, L.J.: II. Familial iminoglycinuria; an inborn error of renal tubular transport. N. Engl. J. Med. **278**, 1407–1413

Therapy for Inborn Errors of Metabolism

Inborn errors of metabolism are caused by a distortion of gene expression which ultimately leads to the inadequate synthesis of proteins. Although in some cases—*e.g.*, sickle cell anemia—the primary insult can be traced to DNA, in others—such as thalassemia and most enzyme deficiencies—it is not known whether the primary injury results from damage to the DNA molecules, instabilities of the messenger, inadequacies in translation, or interference with other mechanisms still not discovered.

An early form of therapy involves eliminating the substrate either by excluding the substrate from the diet, as in phenylketonuria, or by administering drugs—such as penicillamine in Wilson's disease or allopurinol in gout. Orotic aciduria can be corrected by the administration of uridine, which provides the substrate for the biosynthesis of the nucleosides used in RNA and DNA synthesis and is also a substrate for the biosynthesis of inhibitors of the carbamyl aspartate synthetase, the first enzyme in the formation of orotic acid. By this feedback inhibition, the levels of orotic acid in the urine are reduced by the administration of uridine.

Another mechanism by which the substrate can be eliminated from the medium is by depriving the organ that uses the substrate. This has been achieved in some cases of glycogen storage diseases. Ingested glucose is absorbed by the intestine and carried through the portal vein directly to the liver where it is converted to glycogen. If glucose is given intravenously, a great deal of it is used by the muscles. Ingested glucose can be prevented from reaching the liver by deviating the path of the portal vein. This is done by performing a portocaval shunt—an operation in which the portal vein is anastomosed side-to-side with the inferior vena cava—or by interchanging the positions of the portal vein and the inferior vena cava. The inferior vena cava, which drains the blood of the lower limbs, is attached to a stump of the portal vein, thus supplying glucose-free blood to the liver. The distal portion of the portal vein is connected to the stump of the inferior vena cava.

Just as orotic aciduria results from an increase in orotic acid accumulation and a uridine defect, homocystinuria is caused by accumulation of homocystine and low levels of cystine. The most effective therapy for that disease is a diet low in thiamine and supplemented with cystine.

A number of inborn errors of metabolism—vitamin dependent genetic diseases—result from the inability of the enzyme to bind to the coenzyme, which is often a metabolite of a vitamin. Since the binding defect is not absolute, the administration of large doses of vitamins may relieve the symptoms. For example, the administration of 100 mg of pyridoxine per day corrects cystathioninuria.

Protein replacement has long been used in the therapy of hemophilia and hemorrhage. A direct enzyme replacement has been successful in animals where it has been possible to correct UDP-glucuronyl transferase deficiencies in mice or rats; but in humans, enzyme administration has met with little success. Enzymes can also be administered by transfusion of leukocytes or by bone marrow grafts, which have been particularly successful in some cases of agammaglobulinemia.

A promising form of therapy could involve genetic engineering or direct replacement of the defective gene by introducing in the host viruses that contain the gene coding for the missing enzymes. For example, attempts have been made to correct argininuria using Shope papilloma virus, which increases arginase activity in infected cells. Human galactosemic fibroblasts have been corrected by the administration of a λ-phage carrying the bacterial DNA coding for the missing enzyme; namely, galactose-1-phosphate uridyl transferase.

Replacing the missing enzyme has many drawbacks; the enzyme is hard to prepare and purify; the enzyme's half-life inside the body is sometimes short, and there is no way to assure that the enzyme reaches the site where it is needed. In Fabry's disease, renal transplantation has been performed for the dual purpose of (1) correcting the renal failure that may result from the disease and (2) providing a source of new enzymes. Liver transplantation has been proposed in cases of Wilson's disease.

Because of the clinicians' hopes that enzyme replacement therapy might either reduce the symptomatology or cure lipoidosis, there have been several attempts to purify the enzymes involved in the breakdown of gangliosides. This is usually achieved by using affinity columns adsorbance. Arylsulfatase A, sphingomyelinase, glucosyl ceramide hydrolase, and β-galactosidase have been partially purified. In the cases of sphingomyelinase and α-galactosidase [220, 221] several isozymic forms of the enzymes have been found. This observation raises a number of new questions: Which form of the enzyme is missing in each individual case? Can any form of the enzyme be administered to cure the hereditary disease?

Chromosomal Anomalies

Hereditary Anomalies

Observation of the repeated occurrence of disease among relatives of the same family—especially in cases of inbreeding, incest, and polygamy—followed by a careful pedigree of the descendants permitted researchers to conclude that some diseases were transmitted hereditarily and also that various modes of hereditary transmission exist. Thus, the gene carrying the disease trait can be associated either with sex chromosomes (for example, color blindness and hemophilia) or with autosomal genes.

The two major modes of autosomal transmission of hereditary character are dominant and recessive. When dominant, the gene is fully expressed in the phenotype even when the genotype is heterozygous. Consequently, one of the parents has the disease, and the disease appears in half of the children in the first generation. In contrast, when transmitted as an autosomal recessive character, the gene is fully expressed in the phenotype only if the genotype is homozygous for that character. An autosomal recessive character can be transmitted to the children of the first generation only in two ways. In the first mode, one parent carries the gene for the disease and is homozygous. Half the children of the first generation can be expected to be heterozygous and become carriers themselves. In the second mode, parents are heterozygous, and one-quarter of the siblings can be expected to be homozygous and exhibit the disease.

In contrast to what happens with diseases linked to the sex chromosomes and therefore exhibited only in one of the sexes, diseases associated with the autosomal chromosomes are equally distributed in both sexes.

The terms dominant and recessive do not refer to actual molecular situations, but to the mode of expression of the genotypic characteristics into the phenotype. Therefore, the mode of transmission may vary with the point of view of the observer. A case in point is sickle cell anemia. To a clinician, sickle cell anemia is recessive because symptoms develop only in the homozygous individual. For the hematologist, sickle cell anemia is dominant because in the absence of oxygen sickling occurs both in the homozygous and the heterozygous. For the biochemist, genes dictating the biosynthesis of the normal β-chain of normal or s hemoglobins are neither recessive nor dominant because each allele can dictate in the heterozygous the synthesis of the polypeptide chain it codes for.

With these concepts in mind we can now attempt to describe two major groups of hereditary diseases on the basis of pathogenesis: those associated with detectable chromosome alterations, and those exhibiting no obvious chromosome alterations.

Appreciation of the role of chromosomes in disease may well have started in 1903 with the work of Sutton. While he was still a student, Sutton demonstrated parallelism between chromosome behavior and inheritance of genetic traits. This approach had been used extensively in insects, which have large and few chromosomes. Yet human chromosomal behavior and the distribution of genetic traits in the progeny could not be convincingly correlated until 1956 simply because no way had been discovered to observe individual chromosomes.

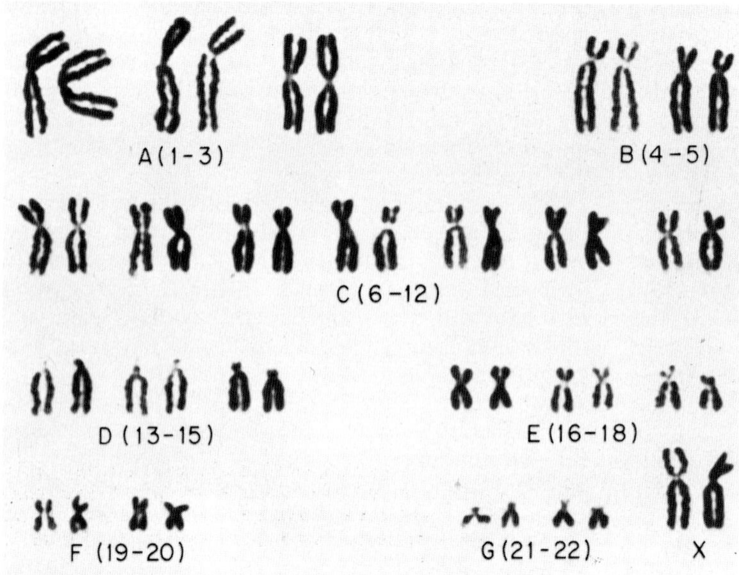

Fig. 3-67. Normal female karyotype using standard stain. Note the difficulty in identifying specific chromosomal pairs in groups B, C, D, F, and G. (From R. Sparkes)

The individualization of the chromosome was achieved in 1952 by T.C. Hsu [174] by placing human cells locked in metaphase into a hypotonic solution. Once chromosomes could be visualized and accurately counted, knowledge of chromosomes and their relationship to disease in humans exploded. Lymphocytes are usually cultured to study the chromosomes in human cells. Culture of lymphocytes had been reported by Kruchev in 1932, but a discovery by Peter Nowell [175] made the procedure practical. Nowell demonstrated that the addition of phytohemagglutinin to the culture medium stimulates mitosis in lymphocytes. When colchicine is added to a culture medium containing dividing cells, it blocks cell division in metaphase. If cells in metaphase are then treated with hypotonic saline, the chromosomes disperse, become individualized, and are easily recognizable. As they appear in the cell under the microscope, the chromosomes are dispersed in a disorderly fashion and are hard to identify. To simplify identification, a picture of all the chromosomes is taken, then each individual chromosome is cut out of the picture and reordered according to size and shape.

Chromosomes are made of two chromatids held together by a centromere, which may be in the middle (epicentric), at one end (acrocentric), or between the midline and the end (metacentric) of the chromosome. The chromatids, which adhere to each other before treatment with saline, separate except for contact in the area occupied by the centromere. Consequently, after hypotonic treatment, epicentric chromosomes look like X's, acrocentric chromosomes like V's, and all other chromosome like Y's. The length of the stem of the Y depends upon the position of the centromere between the midline and the end of the chromosome.

Human cells contain 44 autosomal chromosomes and 2 sex chromosomes. In women, these two sex chromosomes are of the X type, and in men one is of the X and the other the Y type. Half of the 44 human chromosomes come from the father and half from the mother. Father and mother chromosomes look alike; consequently, the chromosomes can be grouped into 22 pairs. Such an arrangement according to shape and size yields a karyotype (see Figs. 3-67 and 3-68). As we shall see, the study of human karyotypes has often established a correlation between chromosome anomalies and hereditary diseases. Of course, karyotyping would be much more useful if the chromosomes of each pair could always be identified. Unfortunately, this could at first not be done. Therefore, chromosomes were assembled in groups (see Figs. 3-67 and 3-68), but some pairs—for example, pairs 4 and 5 of group B and pairs 13, 14, and 15 of group D—cannot be distinguished. Attempts have been made to identify chromosomes on the basis of other morphological changes, such as constrictions. Such a procedure is unfortunately seldom useful because constrictions are inconsistent in chromosomes 1 and 16, for example. Chromosome 9 is the only one that consistently has a constriction at the source.

The study of chromosome satellites, which are small, inherited bodies on the short arms of the acrocentric chromosomes (except the male Y chromosome), has helped little in identifying chromosomes.

A more effective technique has been the study of the incorporation of tritium-labeled thymidine. Tritium-labeled thymidine is not incorporated regularly in all chromosomes at the same time, and it has been possible to identify chromosomes by determining kinetics of thymidine incorporation. For example, in group D, replication occurs late throughout the length

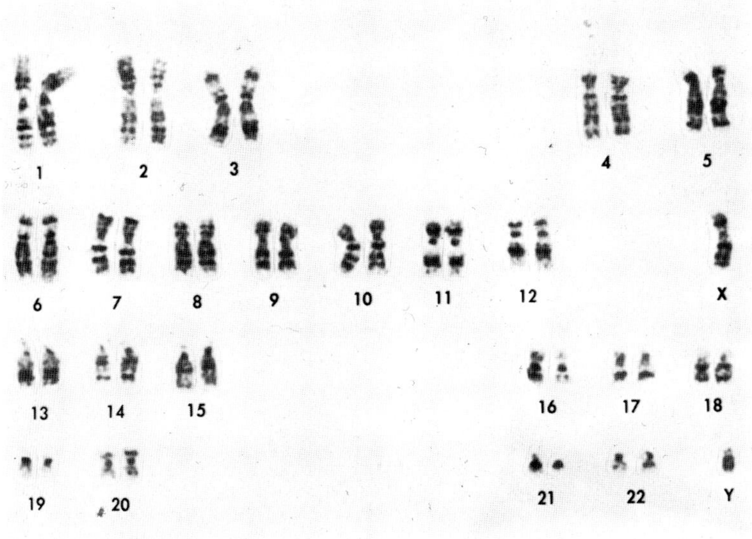

Fig. 3-68. Normal male karyotype using a differential Giemsa banding stain. Each of the chromosome pairs has a distinctive pattern permitting specific chromosome identification. (From R. Sparkes)

of chromosome 13, in contrast to chromosome 14, in which late replication takes place only in the region surrounding the centromere. Chromosome 15 differs from chromosomes 13 and 14 in that it replicates early throughout its full length.

Fortunately, in the early 1970's, two relatively simple staining methods were developed which permitted unequivocal identification of each chromosome; a method using a quinacrine fluorescent stain and another using the familial Giemsa stains [222].

These techniques yield chromosome preparations in which each chromosome exhibits a characteristic and reproducible banding pattern. The molecular basis for the banding remains unknown [223].

All mammalian cells do not contain 46 chromosomes (diploid). Sperm cells and ova are haploid and contain only 23 chromosomes—one of each pair of autosomes plus an X chromosome in the ovum (since all other cells of the female contain two X chromosomes) and an X or a Y in the sperm cell. The haploid reproductive cells, called gametes, develop as we have seen previously through a special form of cytokinesis called meiosis (see Chap. 8).

Trisomy

Various types of chromosomal defects have been associated with hereditary disease. Trisomies are among the most spectacular. In cases of trisomies, instead of having 22 pairs of identical chromosomes, the cells contain 22 pairs and three identical chromosomes. As a result, the total number of chromosomes in the cell is increased by one [176–180]. Autosomal trisomies are usually associated with severe damage to the fetus and are therefore often incompatible with life. Thus, although trisomies affecting every identifiable chromosome have been observed, most occur in fetal material as a result of spontaneous abortion. Postnatal trisomies involve chromosomes 13, 18, and 21. The multiple defects associated with trisomies 13 and 18 are so severe that most of the victims die within the first year. Trisomy 21 results in Down's syndrome (see Figs. 3-69, 3-70). A hereditary disease leading to the development of facial anomalies including downturned mouth and microphthalmia and several bone anomalies has been ascribed to chromosome 22.

Down's syndrome, or mongolism, is the most common of the trisomies. Trisomies can be produced in at least three different ways: nondisjunction, mosaicism, and translocation. The majority of trisomies are believed to result from mitotic nondisjunction. We have already seen that the gametocyte (diploid) is converted to the gamete (haploid) during meiosis. Diploidy is reconstructed during fertilization. If during meiosis autosomal chromosomes are unevenly distributed to the two daughter cells and one receives one too few and the other one too many, one of the daughter cells will contain 23 autosomals, but the other will contain only 21 autosomal chromosomes. When a gamete, which contains 2 identical chromosomes, is fertilized by a normal gamete, the resulting cell will contain 3 identical chromosomes (trisomy). In contrast, when a gamete that is one chromosome short combines with a normal gamete during fertilization, the resulting cell is short one chromosome and its karyotype contains only 21 pairs plus one isolated autosomal (monosomy). Theoretically, the incidence of monosomy and trisomy would be expected to be equal since they both result from the same distortion of meiosis. However, trisomies are the most commonly observed because monosomies are almost inevitably lethal.

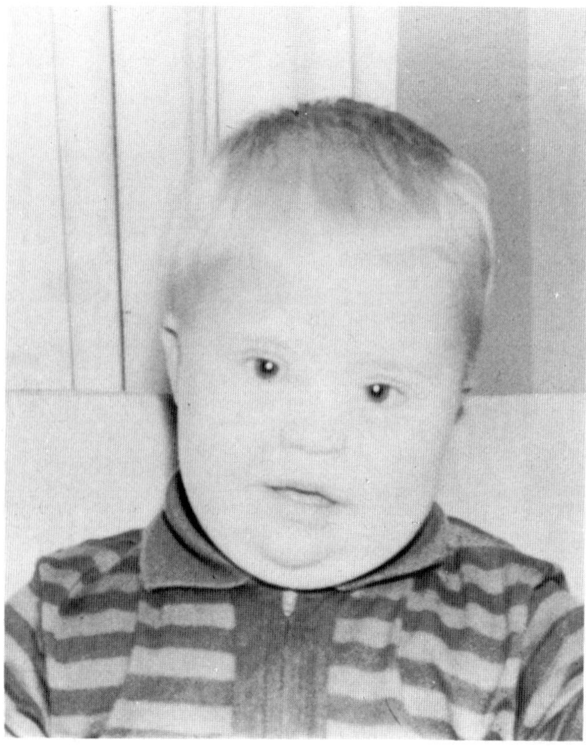

Fig. 3-69. Young boy with facial features typical of Down's syndrome (mongolism). (From R. Sparkes)

Studies of patients in which Down's syndrome occurred in association with color blindness have established that the trisomy is usually derived from nondisjunction in the mother. The incidence of mongolism increases considerably with the age of the mother—from 1 in 3,000 for mothers 20 years old to 1 in 40 for mothers 45 years old. The reason for the increased incidence of trisomy with increasing mother's age is unknown. Environmental factors have been suggested to influence the incidence of nonmeiotic disjunction; thus older women would be more prone to bear trisomic children because they have been exposed longer to environmental hazards. If this were true, the incidence of mongolism would increase linearly with the mother's age. But the increase is exponential. Delayed fertilization has also been invoked to explain the increased incidence of trisomy. In humans, the interval between ovulation and fertilization is believed to increase with age because intercourse is less frequent in older women. Animal experiments have established that delayed fertilization of the mature ovum can be associated with chromosomal anomalies in the zygote. Nevertheless, this appealing hypothesis has hardly survived the storm of mathematical equations thrown at it.

A legitimate question is whether all chromosomal defects develop during meiosis when sperm cells and ova are produced. Chromosomal aberrations may appear at any time during embryonic development, although meiotic trisomies account for 70% of the trisomies observed in chromosomes 13, 18, and 21.

If meiotic nondisjunction occurs during the first meiotic division of the zygote, the embryo develops into a typical trisomic since the monosomic cells die. If meiotic nondisjunction occurs later, it results in mosaicism.

When chromosomal aberrations develop after the first meiosis, two or more cell populations appear in the body. The presence of several cell populations containing different chromosomal contingents is called mosaicism. Some cases of mosaicism are associated with trisomy. The trisomic cell appears as a result of meiotic nondisjunction after several normal mitoses have taken place. In such cases of mosaicism the body's cell population includes cells with the normal karyotype and others with 47 chromosomes. When trisomy 21 is associated with mosaicism, the patient often manifests a minor form of Down's syndrome in which the physical symptoms are less apparent and mental retardation is less severe. In fact, some mosaic patients with trisomy 21 are normal. In such a case it may be necessary to culture tissues other than blood to detect the abnormal karyotype. Moreover, the parents' karyotypes should be tested if the mother has borne children with Down's syndrome.

Patients with mosaicism can marry and have children. If they are fertile, as is the case in some patients whose mosaic patterns include trisomy 21, half of the offspring will be trisomic because nondisjunction inevitably occurs in half of the gametocytes. This phenomenom is referred to as obligatory nondisjunction.

Chromosomal anomalies may develop by mechanisms other than nondisjunction. An important such mechanism is translocation. During their life span, chromosomes of gametocytes or somatic cells may break spontaneously or as a result of attacks by environmental agents, such as X-irradiation or chemicals. (The consequences of such breaks will become clear when we discuss the chromosomal alterations caused by X-irradiation.) In translocation, a portion of the chromatin that has been detached from the chromosomal structure as a result of breaks is transferred to another chromosome. Such translocations of genetic material are usually of little consequence to the cell in which they occur because the total complement of genetic material remains unchanged. However, when such cells divide, the chromatin is distributed irregularly between the daughter cells.

A form of Down's syndrome results from translocation between chromosomes 21 and 15 in the oocyte. In such oocytes the two chromosomes—each belonging to a different pair—form a single chromosome, and the total karyotype contains only 45 chromosomes; pairs 21 and 15 are disrupted, and only one normal 21 and one normal 15 chromosome are present. When such oocytes undergo meiosis, chromosomal segregation may result in 6 different combinations in the ova: a single 21, a single 15, 15–15, 21 trisomy 21, 21 and 15 trisomy, a single 15–21 transloca-

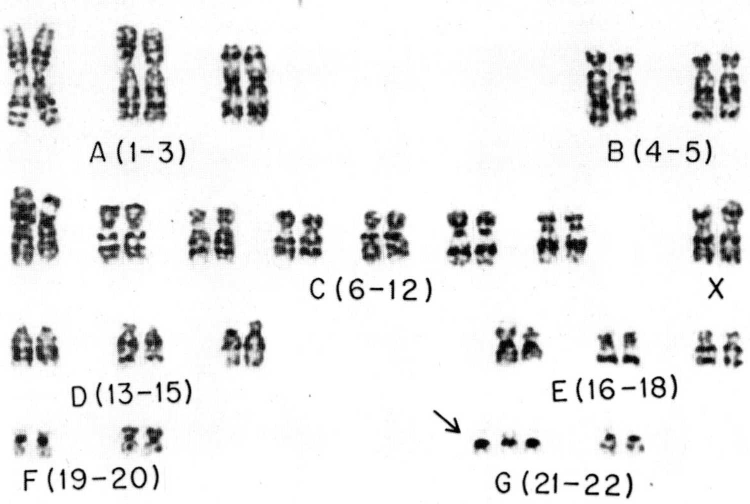

Fig. 3-70. Chromosome anomalies in Down's syndrome. (From R. Sparkes)

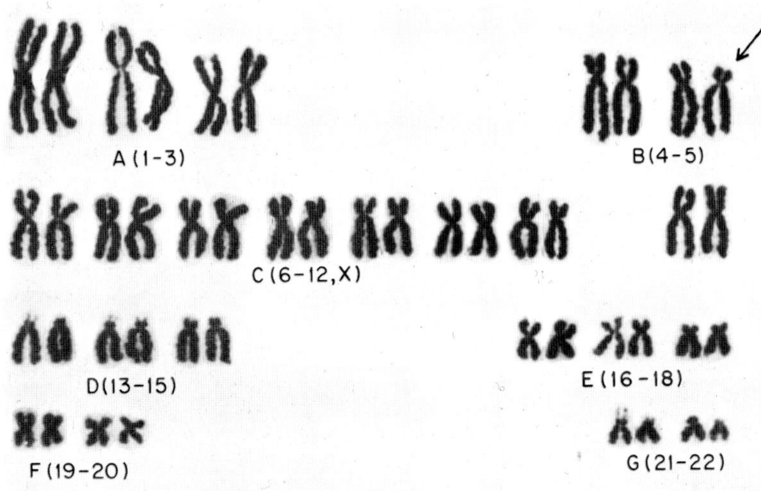

Fig. 3-71. Karyotype of female with partial deletion of the short arm of chromosome 5, the usual change in the cri-du-chat (cat's cry) syndrome. (From R. Sparkes)

tion, and a normal karyotype. After fertilization, the following karyotypes can be expected: monosomy 15, monosomy 21, trisomy 15, trisomy 21, a balanced translocation carrier, and a normal individual. The first three of these possibilities do not produce viable progeny.

Trisomy 21 leads to the classical Down's syndrome; balanced translocation results in the development of an apparently normal individual who can produce offspring with trisomy 21. Theoretically, the chances for a woman with a 15–21 translocation to produce a baby with Down's syndrome are 1 in 3; in reality the chances are only 1 in 5 because of the reduced viability of trisomic embryos. Translocation seldom causes trisomy, which most frequently results from nondisjunction. However, translocation is associated with a number of other congenital anomalies generally less severe than those observed in Down's syndrome.

Very rarely translocation may occur between two identical chromosomes, such as 21–21. This is referred to as isochromosome translocation. When it involves chromosome 21, a trisomic offspring is formed because there are only two possibilities—trisomy 21 and monosomy, and monosomic embryos are not viable.

It is generally accepted that chromosome translocation occurs as a result of a double chromosomal break: one at the side of the donor chromosome, and one at the side of the receptive chromosome. Such breaks are believed to yield sticky ends that facilitate the agglomeration of the translocated portion of the chromosome. Occasionally, during translocation a chromosome or a portion of the chromatin is lost. For example, cri-du-chat syndrome is believed to result from the loss of the short arm of chromosome 5. Loss of the short arm of chromosome 4 and of the short arm of chromosome 18 have been described and are associated with symptoms somewhat different from what is observed in the typical cri-du-chat syndrome (see Fig. 3-71).

Acquired Chromosomal Anomalies

In the study of the mechanism of hereditary disease production, it is not possible to separate gene alteration from environmental effects on the genetic material. Indeed, most if not all chromosomal (or DNA) alterations that are transmitted from parent to progeny may have resulted from environmental injuries. A number of environmental factors known to injure chromosomes are ionizing and ultraviolet radiation, viruses, chemicals—including drugs, and vitamin deficiencies.

Anomalies Caused by Viruses

In 1961, Hampar and Ellison reported chromosomal alterations in Chinese hamster culture cells [181] infected with herpesvirus, and Nichols [182, 183] demonstrated an abnormal incidence of chromosome breaks in cultured leukocytes obtained from children with measles. Since then the list of publications on the effect of viruses on chromosomes has grown into an innumerable bibliography, and chromosomal anomalies have been described after infection with rubella, mumps, chicken pox, and even after vaccination for smallpox or yellow fever.

Studies of the role of viruses in chromosomal anomalies are of two types: the studies in which the infection takes place *in vivo* and the infected cells, usually leukocytes, are cultured *in vitro* to study chromosomal anomalies [184–187]; and studies in which the effect of viruses on the mammalian karyotype has been investigated *in vitro* using either primary cell cultures or cell lines (cells capable of indefinite division—in humans, at least 70 consecutive divisions with 3-day intervals between cell subculture). When primary cell cultures are used (cells freshly derived from a living organism), usually lymphocytes stimulated with phytohemagglutinin are infected with the virus. Since only one cell division can be observed in primary cell cultures, established cell lines, such as HeLa cells, would seem to be preferable for studying the effect of viruses on the karyotype. But these cultures are heteroploid, and therefore most cell lines are inadequate for chromosomal studies. Fortunately, diploid cell lines are now available.

Hayflick and associates [179] were the first to provide researchers with human diploid cell lines—cells with a fibroblastic appearance and a karyotype identical to that of the original donor. One interesting feature of these diploid human cultures is that they are not indefinite, they cannot be expanded beyond 55–60 subcultures and still live. In contrast, animal diploid cell lines may grow indefinitely, but morphological and karyotypical alterations take place in subcultures.

Conclusive studies on the effect of viruses on the karyotype must of course be conducted in a system in which spontaneous alterations of the karyotype do not occur and no other agents capable of inducing chromosomal alterations are present.

Since the discovery of antibiotics, it has been possible to avoid contamination with bacteria. However, mycoplasms and viruses are not always so simple to exclude from the culture cells. Although obvious virus infection is seldom encountered in human cultured cells, virus infection (Sv 40) is the norm in monkey cells. Moreover, although electron microscopy may help to exclude infection by visible virus, it does not permit the identification of latent viruses that might be nestled in the genome.

We have already seen that Nichols [182] detected chromosome breaks in primary cell lines of lymphocytes obtained from children infected with measles. When children were vaccinated with attenuated life virus, the incidence of chromosome breaks was reduced. Similarly, chromosome breaks occur in primary cell lines cultured from aborted fetuses of mothers infected with German measles (rubella) virus (14% incidence of break in rubella cases compared with 7.3% of break in normal fetuses). Chromosomal breaks have been found in 1-year-old children who were infected with rubella *in utero*. A significant observation made by Boué [187] suggests that in cells cultured from fetuses infected *in utero*, the breaks affect the small, acrocentric chromosomes 21 and 22, in contrast to random breaks observed in human cell lines injected *in vivo*.

In human cell lines infected with virus—for example, measle virus—the virus kills the cells, and giant cells are often formed in the process. After infection, several of the culture cells conglomerate to form a giant cell. The incidence of cell division is generally reduced after infection, but when one of these giant cells divides, its chromosomal pattern becomes markedly distorted in that it exhibits polyploidy and numerous breaks. In some cases chromosomal identification is possible despite the breaks, but in many cases the chromosomes have been broken into such small fragments that the chromatin becomes dispersed in a powderlike fashion, and chromosomal structure is unrecognizable. Of course, such changes are inevitably associated with cellular death.

All viruses that reach the fetus do not kill it. Some cause abnormal development and are therefore called teratogenic. The best-known teratogenic viruses are rubella, cytomegalic virus, and herpesvirus. Fetal infection by rubella virus may be either generalized or localized. In the generalized form, the virus infects every organ, and it is estimated that 1 cell in 1,000 is infected. Even in generalized rubella, the virus may accumulate in some tissue preferentially. For example, chondrocytes usually harbor a relatively larger number of viral particles than other cells. Generalized viral infection usually results in abortion within a few months after infection or death early after delivery. In localized infection, the virus, for reasons still unexplained, concentrates in some tissues such as brain, eye, and heart.

The viral infection elicits an immunological reaction, but in spite of measurable circulating antibodies, infection persists and virus particles can be detected by the fluorescent antibody technique.

Cytologically, the major alteration caused by viruses in localized infection is chromosome breaks. As a result of infection, the virus cell ceases to proliferate and dies. Whether this is a direct effect of chromosome break or an indirect event resulting from ischemia caused by swelling and proliferation of capillary endothelial cells remains to be seen. But if chromosome breaks are responsible for cellular death, then the sequence of molecular events that leads from such damage to interference with cellular proliferation and death remains to be clarified.

Localized infection is often compatible with survival of the fetus. But the infant has a number of congenital anomalies including microcephaly, cataracts, glaucoma, patent ductus arteriosus, and other defects.

The consequences of rubella infection of the fetus vary greatly depending upon the stage of pregancy at the time of infection. If infection occurs early after conception, the fetus usually dies, but as the pregancy is more advanced, infection is compatible with survival of the fetus, and chances of developing congenital anomalies increase. After the first trimester, complications of infection—fetal death or congenital anomalies—decrease markedly.

Radiation and Chromosomal Breaks

Physical agents that cause chromosomal breaks include ionizing radiation [188, 189], ultraviolet light, magnetic fields, sound waves, and cold shocks. The number of breaks produced by ionizing radiation increases linearly with the dose administered and is independent of the rate at which the dose is administered. Unless they are produced in large numbers, breaks do not seriously threaten cellular survival since they do not lead to a loss of genetic material. But the breaks disrupt the structural organization of the genome, which often leads to unequal distribution of the genetic material in the daughter cell.

The types of chromosomal alterations that appear as a consequence of the break depend upon the site of the break in the chromosome. Thus, inasmuch as the distribution of the breaks is believed to be random, irradiated cells can be expected to exhibit a broad panoply of chromosomal alterations. Therefore, ionizing radiation has long been a most useful tool for investigating chromosomal damage. At first, the studies were done on cells with large chromosomes such as those of Drosophila or *Vicia faba,* but the development of new techniques in chromosomal studies has permitted direct observation of chromosomal damage in mammalian cells including human cells.

For clarity, we will distinguish between the consequences of single breaks in one chromosome, multiple breaks in a single chromosome, and breaks in more than one chromosome.

When a break occurs in one of the arms of a single chromosome, the break may be repaired (rejunction) or may not be repaired. If the break is not repaired, the two fragments differ fundamentally in that one contains the centromere (centric fragment), and the other is devoid of a centromere (acentric fragment). Thus, during mitosis both chromosomes divide but only one moves in the equatorial plate during anaphase. After telophase, the genetic material is distributed unequally, and one daughter cell contains the small fragment and has an excess of chromosomal material, whereas the other misses the small fragment and has too little of it.

In some cases, the fragment containing the centromere may be even more complex. After replication of the centric fragment, the newly made and old fragments may join to form a chromosome with two centromeres (dicentric). The dicentric fragment exhibits a form of chromosomal schizophrenia. In anaphase, the chromosomal material must join one cell or the other. If the dicentric bridge does not break, one cell contains excessive amounts of genetic material, and the other, less than normal amounts. Such cells either die or divide again, leading to a population of abnormal cells, usually with aberrant gene expression and thus a distorted phenotype. Sometimes the dicentric bridge breaks, and then fragments of the dicentric chromosomes are distributed in one of the two daughter cells. At the next mitosis, each fragment divides again, the new and the old fragment join, forming another population of dicentric cells that form dicentric bridges in the anaphase of the next division.

When two breaks occur on a single chromosome, three fragments are formed—two end fragments and one middle fragment—and only one of the fragments contains the centromere. When double breaks take place on a single chromosome, the potential consequences are increased; first, all parts may rejoin, reconstituting a chromosome that is normal with respect to its gene complement and alignment. Second, the three fragments may remain separate, and only the centric portion may be transferred to one of the daughter cells with deletion of the small fragments. Depending upon the importance of the genetic contingent associated with the deleted fragments, the daughter cell may or may not be viable, or may or may not show distorted gene expression. One acentric fragment may rejoin with a centric fragment, leading to recombination with terminal or interstitial deletion. When the breaks occur at both ends of the chromosomes, leaving a central portion containing the centromere, the cut ends of the central portion may join to form a ring chromosome. Such circular chromosomes may divide, and if the circle is not twisted, each daughter cell inherits a ring chromosome. In contrast, if the ring chromosome is twisted, the original and the newly formed ring chromosome form two interlocked rings. The interlocked ring chromosome constitutes a special form of dicentric chromosome.

Genes are arranged in the chromosome in a specific sequence, which in Drosophila, for example, has been mapped out by studying experimental and spontaneous mutation. Such mapping has revealed that although a chromosome reconstituted after multiple breaks may contain the normal genetic information, the alignment of the information on the chromosome is inverted because in the process of reconstitution one of the fragments has rotated 180 degrees before it rejoins the other fragments. Such chromosomal alterations are called inversions.

When a single chromosomal break occurs in two different chromosomes (A and B), a new and important type of redistribution of genetic material—a translocation—takes place. In a translocation, a fragment of A may rejoin with a fragment of B. Thus, after breaking, each chromosome yields acentric fragments A_a and B_a and centric fragments A_c and B_c. Translocation may yield two types of combinations: A_a and B_c and A_c and B_a; two monocentric chromosomes, or A_a and B_a and A_c and B_c; and an acentric and a dicentric chromosome.

Chemicals and Chromosomal Breaks

In his attempts to eradicate pest-borne disease, protect himself from bacterial, viral, and parasitic infections, preserve his food, fight pain and anxiety, and secure comfort, man has contaminated land, water, air, and almost all living things that the earth supports—including his body—with a myriad of chemicals. This was or should have been a calculated risk. Yet widespread environmental catastrophes have occurred or are predictable unless the risk is minimized. In some cases, such as the use of DDT and acetyl aminofluorene as insecticides, cyclamates as sweeteners, thalidomide (in the United States) as a tranquilizer, and hexachlorophene as an antiseptic, government has acted most effectively. But many dangerous contaminants from the environment remain ignored. Man's concern for protecting his environment goes far beyond such ecological considerations as the survival of fish and pelicans. Mankind's genetic stock itself is threatened. The slow extinction of animal species may be only a prodrome of what can be expected. The 17th century poet John Donne could not have imagined how universal his statement was to be when he said "No man is an island, entire of itself. ... And therefore never send to know for whom the bell tolls. It tolls for thee."

Because they are aware of these dangers, geneticists have attempted to design simple tests to identify chromosomal damage [190–192]. The basic chromosomal lesion caused by chemicals is again chromosomal breaks, the consequences of which have been described. These consequences are easily observable when acentric, dicentric, or ring chromosomal fragments appear, but those structures are unstable changes that disappear in the course of repeated mitosis or meiosis. Geneticists are concerned with stable chromosomal alterations, such as discrete inversion and translocation, which unfortunately are difficult to recognize. But one may assume that every drug causes chromosomal breaks that are not repaired. The breaks indicate that repairable, stable alteration may occur as well, and therefore the assessment of the incidence of chromosomal breaks is an indirect, qualitative way to estimate the incidence of stable chromosomal alterations. However, the significance of chromosomal breaks that occur *in vitro* [193] is harder to assess.

Many chemicals—caffeine, theophylline, aspirin, and others—cause chromosomal breaks *in vitro*. Yet, there is no evidence that these drugs cause chromosomal alterations *in vivo* in adults or in embryos. Nevertheless, some may lead to undetectable damage. For example, in spite of the severe and varied congenital anomalies thalidomide causes in newborns of mothers who took the drug during pregnancy, no chromosomal anomalies are detectable in rat or human embryos. However, thalidomide induces chromosomal breaks in human lymphocytes. Although there is no compelling evidence that LSD causes chromosomal breaks *in vivo* it is unquestionably teratogenic when given to pregnant rats or hamsters and causes chromosome breaks *in vitro*. To list all drugs (antibiotics, nucleic and folic acids, antimetabolites, analgesics, hallucinogens, alkylating agents, acridine, food ingredients, additives and contaminants [caffeine, theophylline, theobromine, cyclamates, sodium nitrate, glutamate, nitric acid, sodium nitrite, alcohol and its derivatives, aflatoxins], water and air pollutants (benzene-pyrene, nitrous oxide, ozone, sodium hypochloride) that cause chromosomal breaks *in vitro* would be exhaustive and out of place in this book; the reader is referred to specialized reviews. Only a few examples will be presented here.

Many analogs of nucleic and folic acids, which interfere with DNA synthesis, can be expected to cause chromosomal breaks. Similarly, breaks are caused by substances that alkylate DNA, such as bifunctional mitomycin C—which causes the formation of interstrand breaks—or the monofunctional streptronigrin, actinomycin, and neomycin—which form covalent bonds with DNA strands and intercalate between the strands (see Necrosis). Photodynamic dyes, like acridine or its derivative quinacrine, intercalate between DNA strands and in the presence of light and oxygen cause the strands to break. The fungicides Captan and maleic hydrazide (used to retard growth of ivy and grass) cause chromosomal breaks *in vitro*.

Molecular Mechanism of Chromosomal Breaks

Clearly one or more of the molecules that constitute the chromosome must be the target of the agent that causes chromosomal breaks. To understand the mechanism of chromosomal breaks, two sets of information are necessary; a detailed description of the molecular organization of the chromosome, and a description of the damage the various breaking agents do to each molecular component of the chromosome. Obviously, too little is known about either process to expect a detailed description of chromosomal anomalies; yet, on the basis of what we already know of chromosomal molecular arrangement and chromosomal molecular alterations caused by viral, physical, and chemical agents, a plausible working hypothesis can be presented. A modern view of chromosomal macromolecular arrangement professes that each chromatid is composed of a double strand of DNA molecules surrounded by protein and RNA (see Chapter 2, Determination of Cellular Specificity). Physical, chemical, and viral agents are known to cause breaks—single and sometimes double-strand breaks—in the DNA molecule, either directly or as a result of secondary endonucleolytic attacks on, for example, ultraviolet-irradiated DNA or DNA containing acetyl aminofluorene. Therefore, Kihlman [194], among others, proposed that DNA constitutes the substratum for the chromosomal break. This notion would imply that reassociation of chromosomal fragments may involve enzymic mechanisms similar to those seen in dark repair of DNA (see section on irradiation).

Amniocentesis

In 1967, colorimetric determination of bilirubin content of amniotic fluid was found to provide a reliable and sensitive measurement of the degree of erythroblastosis in the fetus. The discovery that the uterine content could be collected without endangering survival of the fetus led to using similar techniques in the study of hereditary disease, in particular detectable chromosomal anomalies and identifiable inborn errors of metabolism [195, 196].

The amniotic fluid, which is elaborated by the amniotic membrane, contains only cells derived from fetus and amniotic membrane and no maternal cells. Thus, the sex of a male fetus has always been determined correctly by studying cells collected by amniocentesis. If amniotic fluid were significantly contaminated with maternal cells, this obviously would not have been possible.

In amniocentesis, a syringe mounted with a long special needle is used to puncture the uterine wall and collect approximately 8 ml of amniotic fluid. The fluid is centrifuged and the cells are collected for either direct study or culture. Direct study has included detection of sex chromatin (Barr bodies), staining with toluidine blue for detection of storage granule (*e.g.*, in mucopolysaccharidosis or cystic fibrosis), and staining with nile blue sulfate to test for maturity. Development of orange staining properties is thought to correlate with maturity. Cultured cells have been used for detecting chromosomal anomalies, karyotyping, and biochemical analysis. Karyotyping is particularly valuable in the case of a parent carrying chromosomal anomalies such as a 13-12 translocation. This chromosomal alteration is of no consequence for the carrier but harms the offspring if the defect is transferred to the fertilized egg. Biochemical studies have proved helpful in some enzyme defects, such as galactosemia and Lesch-Nyhan syndrome.

Although amniocentesis is promising, it is not free of problems. Amniocentesis can be performed without complications even during the later part of pregnancy, but then interruption of pregnancy becomes dangerous. In contrast, amniocentesis is not without risks during the early stage of pregnancy. The earlier amniocentesis is performed, the greater the risk to the fetus. The procedure carries little risk 15 weeks after the onset of pregnancy, but it usually takes at least 6 weeks to culture amniotic cells and study anomalies. Consequently, if pathological findings are made, interruption of pregnancy cannot usually take place earlier than the 21st week after onset of pregnancy. At that time abortion is still safe, but physicians would undoubtedly feel more comfortable if the time between collection of cells and scheduling of therapeutic abortions could be reduced.

References

1. Herrick, J.B.: Peculiar elongated and sickle-shaped red blood corpuscles in a case of severe anemia. Arch. intern. Med. **6**, 517–521 (1910)
2. Pauling, L., Itano, H.A., Singer, S.J., Wells, I.C.: Sickle cell anemia: A molecular disease. Science **110**, 543–548 (1949)
3. Jonxis, J.H.P.: Hemoglobinopathies. Annu. Rev. Med. **14**, 297–322 (1963)
4. Perutz, M.F.: Structure and function of hemoglobin. Harvey Lect. Series **63**, 213–261 (1967–68)
5. Ingram, V.M.: The genetic control of protein specificity. In: The molecular control of cellular activity (Allen, J.M., ed.), p. 179–188. New York: McGraw-Hill Book Company 1962
6. Pauling, L., Itano, H.A. (eds.): Molecular structure and biological specificity. Washington, D.C.: American Institute of Biological Sciences 1957
7. Heller, P.: The molecular basis of the pathogenicity of abnormal hemoglobins—Some recent developments. Blood **25**, 110–125 (1965)
8. Margolies, M.P.: Sickle cell anemia: A composite study and survey. Medicine (Baltimore) **30**, 357–443 (1951)
9. Edington, G.M.: The pathology of sickle-cell disease in West Africa. Trans. roy. Soc. trop. Med. Hyg. **49**, 253–267 (1955)
10. Smith, E.W., Conley, C.L.: Clinical features of the genetic variants of sickle cell disease. Johns Hopkins Hosp. Bull **94**, 289–318 (1954)
11. Allison, A.C.: Malaria in carriere of the sickle-cell trait and in newborn children. Exp. Parasit. **6**, 418–447 (1957)
12. Kopple, K.D.: Peptides and amino acids. New York: W.A. Benjamin 1966
13. Klotz, I.M., Langerman, N.R., Darnall, D.W.: Quaternary structure of proteins. Annu. Rev. Biochem. **39**, 25–62 (1970)
14. Antonini, E., Brunori, M.: Hemoglobin. Annu. Rev. Biochem. **39**, 977–1042 (1970)
15. Perutz, M.: Proteins and nucleic acids; structure and function. Amsterdam: Elsevier 1962
16. Huehns, E.R.: Diseases due to abnormalities of hemoglobin structure. Annu. Rev. Med. **21**, 157–178 (1970)
17. Stamatoyannopoulos, G., Bellingham, A.J., Lenfant, C., Finch, C.: Abnormal hemoglobin with high and low oxygen affinity. Annu. Rev. Med. **22**, 221–234 (1971)
18. Pauling, L., Corey, R.B.: Configuration of polypeptide chains with favored orientations around single bonds: two new pleated sheets. Proc. nat. Acad. Sci. (Wash.) **37**, 729–740 (1951)
19. Pauling, L., Corey, R.B.: Compound helical configuration of polypeptide chains: Structure of proteins of the α-keratin type. Nature (Lond.) **171** (4341), 59–61 (1953)
20. Keitt, A.S., Smith, T.W., Jandl, J.H.: Red-cell "Pseudomosaicism" in congenital methemoglobinemia. New Engl. J. Med. **275**, 399–405 (1966)
21. Gerald, P.S., Efron, M.L.: Chemical studies of several varieties of hemoglobin Proc. nat. Acad. Sci. (Wash.) **47**, 1758–1767 (1961)
22. Jones, R.T., Coleman, R.D., Heller, P.: The chemical structure of hemoglobin M_{Iwate} ($M_{Kankakee}$). Fed. Proc. **23**, 173 (1964)
23. Morrison, D., Williams, E.F., Jr.: Methemoglobin reduction by glutathione or cysteine. Science **87**, 15–16 (1938)
24. Scott, E.M., Duncan, I.W., Elkstrand, V.: Purification and properties of glutathione reductase of human erythrocytes. J. biol. Chem. **238**, 3928–3933 (1963)
25. Kiese, M.: Die reducktion des hämoglobins. Biochem. Z. **316**, 264 (1944)
26. Huennekens, F.M., Caffrey, R.W., Basford, R.E., Gabrio, B.W.: Erythrocyte metabolism. IV. Isolation and properties of methemoglobin reductase. J. biol. Chem. **227**, 261–272 (1957)
27. Fishberg, E.H.: Excretion of benzoquinoneacetic acid in hypovitaminosis C. J. biol. Chem. **172**, 155–163 (1948)
28. Cooley, T., Lee, P.: Series of cases of splenomegaly in children with anemia and particular bone changes. Trans. Amer. pediat. Soc. **37**, 29 (1925)
29. Valentine, W.N., Neel, J.V.: Hematologic and genetic study of the

transmission of thalassemia (Cooley's anemia; Mediterranean anemia). Arch. intern. Med. **74**, 185–196 (1944)
30. Delivoria-Papadopoulos, M., Oski, F.A., Gottlieb, A.J.: Oxygen-hemoglobin dissociation curves: Effects of inherited enzyme defects of the red cell. Science **165**, 601–602 (1969)
31. Kreimer-Birnbaum, M., Bannerman, R.M.: Heme and globin synthesis control: Observations *in vivo* in beta thalassemia. Science **155**, 1116–1118 (1967)
32. Kan, Y.W., Nathan, D.G.: Beta thalassemia trait: Detection at birth. Science **161**, 589–590 (1968)
33. Marks, P.A., Burka, E.R.: Hemoglobins A and F: Formation in thalassemia and other hemolytic anemias. Science **144**, 552–553 (1964)
34. Wilt, F.H.: Heme and regulation of embryonic hemoglobin synthesis. Biochem. biophys. Res. Commun. **33**, 113–118 (1968)
35. Weatherall, D.J.: Control mechanisms in human haemoglobin synthesis. Annu. Rev. Med. **19**, 217–232 (1968)
36. Malamos, B., Belcher, E.H., Gyftaki, E., Binopoulos, D.: Simultaneous radioactive tracer studies of erythropoiesis and red-cell destruction in thalassaemia. Brit. J. Haemat. **7**, 411–429 (1961)
37. Kugelmass, I.N.: Blood as a carrier of regulatory compounds—The enzymes. In: Biochemistry of blood in health and disease, p. 245–262. Springfield, Illinois: Charles C. Thomas 1959
38. Bruton, O.C.: Agammaglobulinemia. Pediatrics **9**, 722–727 (1952)
39. Good, R., Kelly, W.D., Rötstein, J., Varco, R.L.: Immunological deficiency diseases; agammaglobulinemia, hypogammaglobulinemia, Hodgkin's disease, and sarcoidosis. Progr. Allergy **6**, 187–319 (1962)
40. Bickel, H., Neale, F.C., Hall, G.: A clinical and biochemical study of hepatolenticular degeneration (Wilson's disease). J. Med. **26**, 527–558 (1957)
41. Reilly, R.W.: The pathophysiology of Wilson's disease. Med. Clin. N. Amer. **47**, 207–217 (1963)
42. Scheinberg, I.H., Sternlieb, I.: The dual role of the liver in Wilson's disease. Med. Clin. N. Amer. **47**, 815–824 (1963)
43. Ogle, R.G., Gibilisco, J.A., Goldstein, N.P., Randall, R.: Oral roentgenographic changes in Wilson's disease. Lancet **87**, 464–466 (1967)
44. McIntyre, N., Clink, H.M., Levi, A.J., Cumings, J.N., Sherlock, S.: Hemolytic anemia in Wilson's disease. New Engl. J. Med. **376**, 439–444 (1967)
45. Lygren, T., Sorensen, E.W., Bernhardsen, A.: Hepatolenticular degeneration (Wilson's disease). A case diagnosed biochemically before clinical manifestations. Lancet **1959I**, 276–277
46. Frieden, E.: The complex copper of nature. In: Horizons in biochemistry (Kasha, M., and Pullman, B., eds.), p. 461–496. New York: Academic Press 1962
47. Markowitz, H., Cartwright, G.E., Wintrobe, M.M.: Studies on copper metabolism. XXVII. The isolation and properties of an erythrocyte cuproprotein (erythrocuprein). J. biol. Chem. **234**, 40–45 (1959)
48. Morell, A.G., Van Den Hamer, C.J.A., Scheinberg, I.H., Ashwell, G.: Physical and chemical studies on ceruloplasmin. IV. Preparation of radioactive, sialic acid-free ceruloplasmin labeled with tritium on terminal D-galactose residues. J. biol. Chem. **241**, 3745–3749 (1966)
49. Gibbs, F.A. (ed.): Molecules and mental health. Philadelphia: J.B. Lippincott Company 1959
50. Brown, F.C., Mitchell, R.J.: The distribution and elimination of tritiated porcine ceruloplasmin in the rat. Canad. J. Biochem. **43**, 1455–1464 (1965)
51. Maytum, W.J., Goldstein, N.P., McGuckin, W.F., Owen, C.A., Jr.: Copper metabolism in Wilson's disease, Laennec's cirrhosis and hemachromatosis: Studies with radiocopper. Mayo Clin. Proc. **36**, 641–660 (1961)
52. Wintrobe, M.M., Cartwright, G.E., Hodges, R.E., Gubler, C.J., Mahoney, J.P., Daum, K., Bean, W.B.: Copper metabolism in Wilson's disease. Trans. Ass. Amer. Phycns **47**, 232–241 (1954)
53. Scheinberg, I., Sternlieb, I.: Wilson's disease. Annu. Rev. Med. **16**, 119–134 (1965)
54. Bush, J.A., Mahoney, J.P., Markowitz, H., Gubler, C.J., Cartwright, G.E., Wintrobe, M.M.: Studies on copper metabolism. XVI. Radioactive copper studies in normal subjects and in patients with hepatolenticular degeneration. J. clin. Invest. **34**, 1766–1778 (1955)
55. Uzman, L., Denny-Brown, D.: Amino-aciduria in hepatolenticular degeneration (Wilson's disease). Amer. J. med. Sci. **215**, 599–611 (1948)
56. Sternlieb, I., Scheinberg, I.H.: Prevention of Wilson's disease in asymptomatic patients. New Engl. J. Med. **278**, 352–359 (1968)
57. Hers, H.G.: Glycogen storage disease. Advanc. Metab. Disord. **1**, 1–44 (1964)
58. Forbes, G.B.: Glycogen storage disease. Report of a case with abnormal glycogen structure in liver and skeletal muscle. J. Pediat. **42**, 645–653 (1953)
59. Cori, G.T.: Biochemical aspects of glycogen deposition diseases. Mod. Probl. Paediat. **3**, 344–359 (1958)
60. Van Lancker, J.: Hydrolases and cellular death. In: Metabolic conjugation and metabolic hydrolysis (Fishman, W., ed.), vol. I, p. 355–430. New York: Academic Press 1970
61. Schwarz, V., Goldberg, L., Komrower, G.M., Holzel, A.: Some disturbances of erythrocyte metabolism in galactosaemia. Biochem. J. **62**, 34–40 (1956)
62. Kalckar, H.M., Maxwell, E.S.: Biosynthesis and metabolic function of uridine diphosphoglucose in mammalian organisms and its relevance to certain inborn errors. Physiol. Rev. **38**, 77–90 (1958)
63. Sidbury, J.: The role of galactose-1-phosphate in the pathogenesis of galactosemia. In: Molecular genetics and human disease (Gardner, L.I., ed.), p. 61–82. Springfield, Illinois: Charles C. Thomas Publisher 1961
64. Donnell, G.N., Bergren, W.R., Ng, W.G.: Galactosemia. Biochem. Med. **1**, 29–53 (1967)
65. Beutler, E., Baluda, M.C., Sturgeon, P., Day, R.: A new genetic abnormality resulting in galactose-1-phosphate uridyltransferase deficiency. Lancet **1965I**, 353–354
66. Gitzelmann, R., Auricchio, S.: The handling of soya alpha-galactosides by a normal and a galactosemic child. Pediatrics **36**, 231–235 (1965)
67. Gitzelmann, R., Curtius, H.C., Müller, M.: Galactitol excretion in the urine of a galactokinase-deficient man. Biochem. biophys. Res. Commun. **22**, 437–441 (1966)
68. Kirkman, H.: Primaquine sensitivity. In: Molecular genetics and human disease (Gardner, L., ed.), p. 106–133. Springfield, Illinois: Charles C. Thomas 1961
69. Anstall, H.B., Trujillo, J.M.: Glucose-6-phosphate dehydrogenase. Its purification and properties. Amer. J. clin. Path. **47**, 296–302 (1967)
70. Moss, A.R., Schoenheimer, R.: The conversion of phenylalanine to tyrosine in normal rats. J. biol. Chem. **135**, 415–429 (1940)
71. Udenfriend, S., Bessman, S.P.: The hydroxylation of phenylalanine and antipyrine in phenylpyruvic oligophrenia. J. biol. Chem. **203**, 961–966 (1953)
72. Kaufman, S.: Phenylalanine hydroxylation cofactor in phenylketonuria. Science **128**, 1506–1508 (1958)
73. Kaufman, S.: The role of pteridines in the enzymatic conversion of phenylalanine to tyrosine. Trans. N.Y. Acad. Sci. **26**, 977–983 (1964)
74. Kaufman, S.: The structure of the phenylalanine-hydroxylation cofactor. Proc. nat. Acad. Sci. (Wash.) **50**, 1085–1093 (1963)
75. Fisher, D.B., Kirkwood, R., Kaufman, S.: Rat liver phenylalanine hydroxylase, an iron enzyme. J. biol. Chem. **247**, 5161–5167 (1972)
76. Jervis, G.A.: Studies on phenylpyruvic oligophrenia: The position of the metabolic error. J. biol. Chem. **169**, 651–656 (1947)
77. Menkes, J.H.: The pathogenesis of mental retardation in phenylketonuria and other inborn errors of amino acid metabolism. Pediatrics **39**, 297–308 (1967)
78. Klavins, J.V.: Pathology of amino acid excess. VII. Phenylalanine and tyrosine. Arch. Path. **84**, 238–250 (1967)
79. Armstrong, M.D., Tyler, F.H.: Studies on phenylketonuria. I. Restricted phenylalanine intake in phenylketonuria. J. clin. Invest. **34**, 565–580 (1955)
80. McKean, C.M., Schanberg, S.M., Giarman, N.J.: A mechanism of the indole defect in experimental phenylketonuria. Science **137**, 604–605 (1962)
81. Godin, C., Dolan, G.: Tryptophan metabolism in normal and phenylketonuric rats. Biochim. biophys. Acta (Amst.) **130**, 535–537 (1966)
82. Armstrong, M.D., Low, N.L., Bosma, J.F.: Studies on phenylketonuria. IX. Further observations on the effect of phenylalanine-restricted diet on patients with phenylketonuria. Amer. J. clin. Nutr. **5**, 543–554 (1957)
83. Medes, G.: New error of tyrosine metabolism: Tyrosinosis, intermediary metabolism of tyrosine and phenylalanine. Biochem. J. **26**, 917–940 (1932)
84. Garrod, A.: Inborn errors of metabolism. London: Oxford University Press 1909
85. Brookler, M.I., Martin, W.J., Underdahl, L.O., Worthington, J., Mathieson, D.: Alkaptonuria and ochronosis; further experiences. Mayo Clin. Proc. **39**, 107–117 (1964)
86. Moran, T.J., Yunis, E.J.: Studies on ochronosis. II. Effects of injection of homogentisic acid and ochronotic pigment in experimental animals. Amer. J. Path. **40**, 359–369 (1962)
87. Cooper, J.A., Moran, T.J.: Studies on ochronosis. I. Report of case with death from ochronotic nephrosis. Arch. Path. **64**, 46–53 (1957)
88. Indenbaum, S., Ward, L.E., Tauxe, W.N.: Spuriously increased urate values in alkaptonuria. Mayo Clin. Proc. **40**, 127–131 (1965)
89. Quevedo, W.C., Jr., Youle, M.C., Rovee, D.T., Bienieki, T.C.: The developmental fate of melanocytes in murine skin. In: Structure and control of the melanocyte (Della Porta, G., and Mühlbock, O., eds.), p. 228–241. Berlin-Heidelberg-New York: Springer 1966
90. Fitzpatrick, T.B., Quevedo, W.C., Jr.: Albinism. In: The metabolic basis of inherited disease (Starbury, J.B., Wyngaarden, J.B., and Fredrickson, D.S., eds.), 2nd ed., p. 324–340. New York: McGraw-Hill Book Company 1966
91. Chian, L.T., Wilgram, G.F.: Tyrosinase inhibition: Its role in suntanning and in albinism. Science **155**, 198–200 (1967)
92. Ghadimi, H., Partington, M.W., Hunter, A.: A familial disturbance of histidine metabolism. New Engl. J. Med. **265**, 221–224 (1961)
93. Auerbach, V.H., DiGeorge, A.M., Baldridge, R.C., Tourtellotte, C.D., Brigham, M.P.: Histidinemia: A deficiency in histidase resulting in the urinary excretion of histidine and of imidazolepyruvic acid. J. Pediat. **60**, 487–497 (1962)
94. Shaw, K.N.F., Boder, E., Gutenstein, M., Jacobs, E.E.: Histidinemia. J. Pediat. **63**, 720–721 (1963)
95. Rao, D.R., Greenberg, D.M.: Studies on the enzymic decomposition of urocanic acid. II. Properties of products of urocanase reaction. Biochim. biophys. Acta (Amst.) **43**, 404–418 (1960)
96. Mackenzie, D.Y., Woolf, L.I.: "Maple syrup urine disease"; an inborn error of the metabolism of valine, leucine, and isoleucine associated with gross mental deficiency. Brit. med. J. **1**, 90–91 (1959)
97. Dancis, J., Levitz, M., Miller, S., Westfall, R.G.: Maple syrup urine disease. Brit. med. J. **1**, 91–93 (1959)

98. Jeghers, H., Murphy, R.: Practical aspects of oxalate metabolism. New Engl. J. Med. **233**, 208–215 (1945)
99. Davis, J.S., Klingberg, W.G., Stowell, R.E.: Nephrolithiasis and nephrocalcinosis with calcium oxalate crystals in kidneys and bones. J. Pediat. **36**, 323–334 (1950)
100. Chou, L.Y., Donohue, W.L.: Oxalosis; possible "inborn error of metabolism" with nephrolithiasis and nephrocalcinosis due to calcium oxalate as predominating features. Pediatrics **10**, 660–666 (1952)
101. Burke, E.C., Baggenstoss, A.H., Owen, C.A., Jr., Power, M.H., Lohr, O.W.: Oxalosis. Pediatrics **15**, 383–391 (1955)
102. Koten, J.W., Van Gastel, C., Dorhoutmess, E.J., Holleman, L.W.J., Schuiling, R.D.: Two cases of primary oxalosis. J. clin. Path. **18**, 223–229 (1965)
103. Weber, A.L.: Primary hyperoxaluria: Roentgenographic, clinical and pathologic findings. Amer. J. Roentgenol. **100**, 155–161 (1967)
104. Frederick, E., Rabkin, M.T., Smith, L.H., Jr.: Primary hyperoxaluria: A defect in glyoxylate metabolism. J. clin. Invest. **41**, 1358 (1962)
105. Kanfer, J., Ashwell, G., Burns, J.J.: Formation of L-lyxonic and L-xylonic acids from L-ascorbic acid in rat kidney. J. biol. Chem. **235**, 2518–2521 (1960)
106. Koch, J., Stokstad, E.L., Williams, H.E., Smith, L., Jr.: Deficiency of 2-oxo-glutarate: Glyoxylate carboligase activity in primary hyperoxaluria. Proc. nat. Acad. Sci. (Wash.) **57**, 1123–1129 (1967)
107. Williams, H.E., Smith, L.H., Jr.: L-glyceric aciduria. A new genetic variant of primary hyperoxaluria. New Engl. J. Med. **278**, 233–238 (1968)
108. Thannhauser, S.: Lipidoses; diseases of the intracellular lipid metabolism, 3rd ed. New York: Grune & Stratton 1958
109. Bartsch, G.: The nature of the lipids in the brain in Niemann-Pick's disease. In: Cerebral lipidoses. A Symposium (Cumings, J.N., and Lowenthal, A., eds.), p. 159–163. Springfield, Illinois: Charles C. Thomas Publisher 1957
110. Sribney, M.: Enzymatic synthesis of ceramide. Biochim. biophys. Acta (Amst.) **125**, 542–547 (1966)
111. Hanahan, D.J., Thompson, G.A., Jr.: Complex lipids. Annu. Rev. Biochem. **32**, 215–240 (1963)
112. Carter, H.E., Johnson, P., Weber, E.J.: Glycolipids. Annu. Rev. Biochem. **34**, 109–142 (1965)
113. Leibovitz, Z., Gatt, S.: Enzymatic hydrolysis of sphingolipids. VII. Hydrolysis of gangliosides by a neuraminidase from calf brain. Biochim. biophys. Acta (Amst.) **152**, 136–143 (1968)
114. Brady, R.O.: Cerebral lipidoses. Annu. Rev. Med. **21**, 317–334 (1970)
115. Okada, S., O'Brien, J.S.: Generalized gangliosidosis: Beta-galactosidase deficiency. Science **160**, 1002–1004 (1968)
116. Fredrickson, D.S.: Cerebroside lipidosis: Gaucher's disease. In: The metabolic basis of inherited disease (Stanbury, J.B., Wyngaarden, J.B., and Fredrickson, D.S., eds.), 2nd ed., p. 565–585. New York: McGraw-Hill Book Company 1966
117. Aballi, A.J., Kato, K.: Gaucher's disease in early infancy: A review of literature and report of case with neurological symptoms. J. Pediat. **13**, 364–380 (1938)
118. Frisell, E.: Gaucher's disease in infants. Acta paediat. scand. **30**, 470–486 (1943)
119. Banker, B.Q., Miller, J.Q., Crocker, A.C.: The cerebral pathology of infantile Gaucher's disease, in Cerebral Sphingolipidoses (Aronson, S.M., and Volk, B.W., eds.), p. 73–99. New York: Academic Press 1962
120. Fisher, E.R., Reidbord, H.: Gaucher's disease: Pathogenetic considerations based on electron microscopic and histochemical observations. Amer. J. Path. **41**, 679–692 (1962)
121. Kampine, J.P., Kanfer, J.N., Gal, A.E., Bradley, R.M., Brady, R.O.: Response of sphingolipid hydrolases in spleen and liver to increased erythrocytorhexis. Biochim. biophys. Acta (Amst.) **137**, 135–139 (1967)
122. Kattlove, H.E., Williams, J.C., Gaynor, E., Spivack, M., Bradley, R.M., Brady, R.O.: Gaucher cells in chronic myelocytic leukemia: An acquired abnormality. Blood **33**, 379–390 (1969)
123. Crocker, A.C., Farber, S.: Niemann-Pick disease: A review of 18 patients. Medicine (Baltimore) **37**, 1–95 (1958)
124. Pick, L.: II. Niemann-Pick's disease and other forms of so-called xanthomatosis (Dunham Lecture). Amer. J. med. Sci. **185**, 601–616 (1933)
125. Brady, R.O., Kanfer, J.N., Mock, M.B., Fredrickson, D.S.: The metabolism of sphingomyelin. II. Evidence of an enzymatic deficiency in Niemann-Pick diesease. Proc. nat. Acad. Sci. (Wash.) **55**, 366–369 (1966)
126. Andrews, J.M., Cancilla, P.A.: Ultrastructure of human globoid cell leukodystrophy. In: Proc. VIth International Congress of Neuropathology, p. 1051–1052. Paris: Masson & Cie. 1970
127. Austin, J., Suzuki, K., Armstrong, D.A., Brady, R., Bachhawat, B.K., Schlenker, J., Stumpf, D.: Studies in globoid (Krabbe) leukodystrophy (GLD). V. Controlled enzymic studies in ten human cases. Arch. Neurol. **23**, 502–512 (1970)
128. Suzuki, Y., Suzuki, K.: Krabbe's globoid cell leukodystrophy: Deficiency of galactocerebrosidase in serum, leukocytes, and fibroblasts. Science **171**, 73–75 (1971)
129. Brady, R.O., Gal, A.E., Bradley, R.M., Martensson, E., Warshaw, A.L., Laster, L.: Enzymatic defect in Fabry's disease, ceramidetrihexosidase deficiency. New. Engl. J. Med. **276**, 1163–1167 (1967)
130. Bachhawat, B.K., Austin, J., Armstrong, D.: A cerebroside sulphotransferase deficiency in a human disorder of myelin. Biochem. J. **104**, 15C–17C (1967)
131. Shemin, D.: The succinate-glycine cycle: The role of δ-aminolevulinic acid in porphyrin synthesis. In: Porphyrin biosynthesis and metabolism (Wolstenholme, G.E.W., and Millar, E.C.P., eds.). Proc. Ciba Symp., p. 4–22. Poston: Little, Brown and Company 1955
132. Granick, S., Urata, G.: Increase in activity of α-aminolevulinic acid synthetase in liver mitochondria induced by feeding of 3,5-dicarbethoxy-1,4-dihydrocollidine. J. biol. Chem. **238**, 821–827 (1963)
133. Gibson, K.D.: Some properties of δ-aminolaevulic acid dehydrase. In: Porphyrin biosynthesis and metabolism (Wolstenholme, G.E.W., and Millar, E.C.P., eds.). Proc. Ciba Symp., p. 27–39. Boston: Little, Brown and Company 1955
134. Margoliash, E.: Porphyrins and hemoproteins. Annu. Rev. Biochem. **30**, 549–578 (1961)
135. Coons, C.M.: Iron metabolism. Annu. Rev. Biochem. **33**, 459–480 (1964)
136. Cline, M.J., Berlin, N.I.: Studies of the anemia of multiple myeloma. Amer. J. Med. **33**, 510–525 (1962)
137. de Gruchy, G.C., Loder, P.B., Hennessy, I.V.: Haemolysis and glycolytic metabolism in hereditary elliptocytosis. Brit. J. Haemat. **8**, 168–179 (1962)
138. de Leeuw, N.K.M., Shapiro, L., Lowenstein, L.: Drug-induced hemolytic anemia. Ann. intern. Med. **58**, 592–607 (1963)
139. Ling, G. Ning: A physical theory of the living state, p. 385. New York: Blaisdell 1962
140. Ingelfinger, F.J.: The digestive system: The gut. Year Book Med. 337–412 (1967–68)
141. Budinger, J.M.: Hematology: Hemoglobin, erythrocytes, and anemia, Year Book Pathol. and Clin. Path., 320–342 (1961–62)
142. Fitzpatrick, T.B., Pathak, M.A., Magnus, I.A., Curwen, W.L.: Abnormal reactions of man to light. Annu. Rev. Med. **14**, 195–214 (1963)
143. Watson, C.J.: The problem of porphyria; some facts and questions. New Engl. J. Med. **263**, 1205–1215 (1960)
144. Eames, L.: The porphyrins and the porphyrias. Annu. Rev. Med. **12**, 251–270 (1961)
145. De Matteis, F.: Disturbances of liver porphyrin metabolism caused by drugs. Pharmacol. Rev. **19**, 523–557 (1967)
146. Petrozzi, C., Nixon, R.K.: Hemochromatosis and porphyria cutanea tarda. Henry Ford Hosp. med. Bull. **13**, 285–288 (1965)
147. Granick, S.: Porphyrin biosynthesis, porphyria disease, and induced enzyme synthesis in chemical porphyria. Trans. N.Y. Acad. Sci. **25** (series 2), 53–65 (1962)
148. Barnes, F.W., Jr., Schoenheimer, R.: On the biological synthesis of purines and pyrimidines. J. biol. Chem. **151**, 123–139 (1943)
149. Buchanan, J.M.: Biosynthesis of the purines. J. cell. comp. Physiol. **38**, (Suppl. 1), 143–171 (1951)
150. Carter, C.E.: Metabolism of purines and pyrimidines. Annu. Rev. Biochem. **25**, 123–146 (1956)
151. Heppel, L.A., Rabinowitz, J.C.: Enzymology of nucleic acids, purines, and pyrimidines. Annu. Rev. Biochem. **27**, 613–642 (1958)
152. Hartman, S.C., Buchanan, J.M.: Nucleic acids, purines, pyrimidines (nucleotide synthesis). Annu. Rev. Biochem. **28**, 365–410 (1959)
153. Murray, A.W.: The biological significance of purine salvage. Annu. Rev. Biochem. **40**, 811–826 (1971)
154. Plaut, G.W.E.: Water soluble vitamins, part II (Folic acid, riboflavin, thiamine, vitamin B_{12}). Annu. Rev. Biochem. **30**, 409–446 (1961)
155. Plaut, G.W.: The biosynthesis of flavin derivatives. In: Metabolic pathways (Greenberg, D., ed.), 2nd ed., vol. 2, p. 673–712. New York: Academic Press 1961
156. Murray, A.W., Elliot, D.C., Atkinson, M.R.: Nucleotide biosynthesis from preformed purines in mammalian cells. Regulatory mechanisms and biological significance. Progr. Nucleic Acid Res. Mol. Biol. **10**, 87–119 (1970)
157. Smoot, A.O., Van Lancker, J.L.: Effect of partial hepatectomy on nucleic acid metabolism of regenerating thymus. Radiat. Res. **19**, 659–669 (1963)
158. Gutman, A.B.: The biological significance of uric acid. Harvey Lect. Series **60**, 35–55 (1964–65)
159. Hall, A.P., Barry, P.E., Dawber, T.R., McNamara, P.M.: Epidemiology of gout and hyperuricemia. A long-term population study. Amer. J. Med. **42**, 27–37 (1967)
160. Wyngaarden, J.B.: Gout. Advanc. Metab. Disord. **2**, 1–78 (1965)
161. Michener, W.M.: Hyperuricemia and mental retardation with athetosis and self-mutilation. Amer. J. Dis. Child. **113**, 195–206 (1967)
162. DeMars, R., Sarto, G., Felix, J.S., Benke, P.: Lesch-Nyhan mutation: Prenatal detection with amniotic fluid cells. Science **164**, 1303–1305 (1969)
163. Seegmiller, J.E., Rosenbloom, F.M., Kelley, W.N.: Enzyme defect associated with a sex-linked human neurological disorder and excessive purine synthesis. Science **155**, 1682–1684 (1967)
164. Kornberg, A., Lieberman, I., Simms, E.S.: Enzymatic synthesis and properties of 5-phosphoribosylpyrophosphate. J. biol. Chem. **215**, 389–402 (1955)
165. Gerhart, J.C., Pardee, A.B.: The enzymology of control by feedback inhibition. J. biol. Chem. **237**, 891–896 (1962)
166. Gerhart, J.C., and Pardee, A.B.: Aspartate transcarbamylase, an enzyme designed for feedback inhibition. Fed. Proc. **23**, 727–735 (1964)
167. Stadtman, E.R.: Allosteric regulation of enzyme activity. Advanc. Enzymol. **28**, 41–154 (1966)
168. Dent, C.E., Rose, G.A.: Aminoacid metabolism in cystinuria. Quart. J. Med. **20**, 205–219 (1951)
169. Dent, C.E., Heathcote, J.G., Joron, G.E.: The pathogenesis of cystinuria.

I. Chromatographic and microbiological studies of the metabolism of sulfur-containing amino acids. J. clin. Invest. **33**, 1210–1226 (1954)
170. Thier, S.O., Segal, S., Fox, M., Blair, A., Rosenberg, L.E.: Cystinuria: Defective intestinal transport of dibasic amino acids and cystine. J. clin. Invest. **44**, 442–448 (1965)
171. Rosenberg, L.E., Albrecht, I., Segal, S.: Lysine transport in human kidney: Evidence for two systems. Science **155**, 1426–1428 (1967)
172. Segal, S., Crawhall, J.C.: Characteristics of cystine and cysteine transport in rat kidney cortex slices. Proc. nat. Acad. Sci. (Wash.) **59**, 231–237 (1968)
173. Scriver, C.R.: Genetic defects in membrane transport mechanisms. Hosp. Pract. **5**, 92–100 (1970)
174. Hsu, T.C.: Mammalian chromosomes *in vitro*. 1. The karyotype of man. J. Hered. **43**, 167–172 (1952)
175. Hungerford, D.A., Donnelly, A.J., Nowell, P.C., Beck, S.: The chromosome constitution of a human phenotypic intersex. Amer. J. hum. Genet. **11** (3), 215–236 (1959)
176. Grant, V.: The development of a theory of heredity. Amer. Sci. **44**, 158–179 (1956)
177. Redding, A., Hirschhorn, K.: Guide to human chromosome defects. Birth Defects **4**, 1–16 (1968)
178. Bearn, A.G., German, J.L.: III. Chromosomes and disease. Sci. Amer. **205**, 66–76 (1961)
179. Hayflick, L.: Chromosomes and human disease. Hosp. Pract. **2**, 54–63 (1967)
180. Hirschhorn, K.: Chromosomal abnormalities. I. Autosomal defects. Hosp. Pract. **5**, 39–49 (1970)
181. Hampar, B., Ellison, S.A.: Chromosomal aberrations induced by an animal virus. Nature (Lond.) **192**, 145–147 (1961)
182. Nichols, W.W.: The role of viruses in the etiology of chromosomal abnormalities. Amer. J. hum. Genet. **18**, 81–92 (1966)
183. Nichols, W.W., Levan, A., Hall, B., Ostergren, G.: Measles-associated chromosome breakage, Preliminary communication. Hereditas **48**, 367–370 (1962)
184. Cohen, M.M., Shaw, M.W.: Specific effects of viruses and antimetabolites on mammalian chromosomes. In Vitro **1**, 50–66 (1965)
185. Sever, J.L.: Viral teratogens: A status report. Hosp. Pract. **5**, 75–83 (1970)
186. Kornones, S.B.: Congenital rubella syndrome: advances and new concepts. Gen. Pract. Clin. **35**, 78 (1967)
187. Boué, A., Boué, J.G.: Virus et chromosomes humains. Path. Biol. (Paris) **16**, 677–690 (1968)
188. Purdom, C.E.: Genetic effects of radiations. London: Newnes 1963
189. Pizzarello, D.J., Witcofski, R.L.: Basic radiation biology. Philadelphia: Lea & Febiger 1967
190. Shaw, M.W.: Human chromosome damage by chemical agents. Annu. Rev. Med. **21**, 409–432 (1970)
191. Carr, D.H.: Chromosome anomalies as a cause of spontaneous absorption. Amer. J. Obstet. Gynec. **97**, 283–293 (1967)
192. Epstein, S.S., Legator, M.S.: The mutagenicity of pesticides; concepts and evaluation. Cambridge, Massachusetts: The MIT Press 1971
193. Nowell, P.C.: Biological significance of induced human chromosome aberrations. Fed. Proc. **28**, 1797–1803 (1969)
194. Kihlman, B.A.: Molecular mechanisms of chromosome breakage and rejoining. In: Advances in cell and molecular biology (DuPraw, E.J., ed.), vol. 1, p. 59–107. New York: Academic Press 1971
195. Dancis, J.: The prenatal detection of hereditary defects. Obstet. gynec. Surv. **24**, 1351–1353 (1969)
196. Milunsky, A., Littlefield, J.W., Kanfer, J.N., Kolodny, E.H., Shih, V.E., Atkins, L.: Prenatal genetic diagnosis. III. New Engl. J. Med. **283**, 1498–1504 (1970)
197. Wolf, P.L., Rabinowitz, I., Berman, S., Shikuma, N.: Prostaglandin E_2 may be trigger to sickle crisis. Lab. Management (May), 14 (1974)
198. Bertles, J.F.: Hemoglobin interaction and molecular basis of sickling. Arch. intern. Med. **133**, 538–543 (1974)
199. Moffat, K.: Gelation of sickle cell hemoglobin: Effects of hybrid tetramer formation in hemoglobin mixtures. Science **185**, 274–277 (1974)
200. Housman, D., Forget, B.G., Skoultchi, A., Benz, E.J., Jr.: Quantitative deficiency of chain-specific globin messenger ribonucleic acids in the Thalassemia syndromes. Proc. nat. Acad. Sci. (Wash.) **70**, 1809–1813 (1973)
201. Forget, B.G., Benz, E.J., Skoultchi, A., Baglioni, C., Housman, D.: Absence of messenger RNA for beta globin chain in β°-thalassaemia. Nature (Lond.) **247**, 379–381 (1974)
202. Dow, L.W., Terada, M., Natta, C., Metafora, S.: Globin synthesis of intact cells and activity of isolated mRNA in β-thalassaemia. Nature [New Biol.] **243**, 114–115 (1973)
203. Yoshida, A.: Hemolytic anemia and G6PD deficiency. Science **179**, 532–537 (1973)
204. Fisher, D.B., Kaufman, S.: The stimulation of rat liver phenylalanine hydroxylase by lysolecithin and α-chymotrypsin. J. biol. Chem. **248**, 4345–4353 (1973)
205. Friedman, P.A., Kaufman, S.: Some characteristics of partially purified human liver phenylalanine hydroxylase. Biochim. biophys. Acta (Amst.)**293**, 56–61 (1973)
206. Ayling, J.E., Pirson, W.D., Al-Janabi, J.M., Helfand, G.D.: Kidney phenylalanine hydroxylase from man and rat. Comparison with the liver enzyme. Biochemistry **13**, 78–85 (1974)
207. Connellan, J.M., Danks, D.M.: Demonstration of two forms of phenylalanine hydroxylase in human liver obtained at autopsy. Biochim. biophys. Acta (Amst.) **293**, 48–55 (1973)
208. Huang, C.Y., Max, E.E., Kaufman, S.: Purification and characterization of phenylalanine hydroxylase-stimulating protein from rat liver. J. biol. Chem. **248**, 4235–4241 (1973)
209. Chadwick, V.S., Modha, K., Dowling, R.H.: Mechanism for hyperoxaluria in patients with ileal dysfunction. New Engl. J. Med. **289**, 172–176 (1973)
210. Wolfe, L.S., Senior, R.G., Ying Kin, N.M.K. Ng: The structures of oligosaccharides accumulating in the liver of G_{M1}-gangliosidosis, Type I. J. biol. Chem. **249**, 1828–1838 (1974)
211. Baker, J.J., Jr., Lindsey, J.R., McKhann, G.M., Farrell, D.F.: Neronal GM_1 gangliosidosis in a siamese cat with β-galactosidase deficiency. Science **174**, 838–839 (1971)
212. Callahan, J.W., Khalil, M., Gerrie, J.: Isoenzymes of sphingomyelinase and the genetic defect in Niemann-Pick disease, Type C. Biochem. Biophys. Res. Commun. **58**, 384–390 (1974)
213. Wenger, D.A., Sattler, M., Hiatt, W.: Globoid cell leukodystrophy: Deficiency of lactosyl ceramide beta-galactosidase. Proc. nat. Acad. Sci. (Wash.) **71**, 854–857 (1974)
214. Austin, J., Suzuki, K., Armstrong, D., Brady, R., Bachhawat, B.K., Schlenker, J., Stumpf, D.: Studies in globoid (Krabbe) leukodystrophy (GLD). V. Controlled enzymic studies in ten human cases. Arch. Neurol. **23**, 502–512 (1970)
215. Tschudy, D.P.: Enzyme aspects of acute intermittent porphyria. Molec. Cell. Biochem. **2**, 63–70 (1973)
216. Fox, I.H.: Purine ribonucleotide catabolism: Clinical and biochemical significance. Nutr. Metabol. **16**, 65–78 (1974)
217. Becker, M.A., Meyer, L.J., Wood, A.W., Seegmiller, J.E.: Purine overproduction in man associated with increased phosphoribosylpyrophosphate synthetase activity. Science **179**, 1123–1126 (1973)
218. Katunuma, N., Matsuda, Y., Kuroda, Y.: Phylogenic aspects of different regulatory mechanisms of glutamine metabolism. In: Advances in enzyme regulation (Weber, G., ed.), vol. 8, p. 73–81. New York: Pergamon Press 1970
219. Rockson, S., Stone, R., Van der Weyden, M., Kelley, W.N.: Lesch-Nyhan syndrome: Evidence for abnormal adrenergic function. Science **186**, 934–935 (1974)
220. Mapes, C.A., Sweeley, C.C.: Preparation and properties of an affinity column adsorbent for differentiation of multiple forms of α-galactosidase activity. J. biol. Chem. **248**, 2461–2470 (1973)
221. Mapes, C.A., Suelter, C.H., Sweeley, C.C.: Isolation and characterization of ceramide trihexosidases (form A) from human plasma. J. biol. Chem. **248**, 2471–2479 (1973)
222. Hirschhorn, K.: Chromosome identification. In: Annual review of medicine (Creger, W.P., Coggins, C.H., and Hancock, E.W., eds), vol. 24, p. 67–74. Palo Alto, California: Annual Reviews Inc. 1973
223. Crossen, P.E.: Factors influencing Giemsa band formation of human chromosomes. Histochemie **35**, 51–62 (1973)
224. Hers, H.G.: The role of lysosomes in the pathogeny of storage diseases. Biochimie **54**, 753–756 (1972)

Chapter 4

Malnutrition

General Malnutrition

Considering the variety of foods flooding the supermarkets of the western world, a chapter on malnutrition should be antiquated in a modern textbook of pathology. To be sure, most of the obvious nutritional deficiencies have successfully been eliminated in Europe and North America. However, flagrant protein and vitamin deficiencies continue to plague large segments of the populations of the world, and even in some western countries subtle forms of mineral and vitamin deficiencies continue to exist. Furthermore, because of the progress in food technology, some of the oldest diseases resulting from nutritional imbalance (obesity and arteriosclerosis) have become more frequent, and new problems of nutrition have developed.

Irradiation of food has led [1, 2], at least in animals, to discrete forms of vitamin deficiency, and the addition of "protective" agents to foods has sometimes had dramatic results because of the additives unsuspected toxicity. When the toxic effect is immediate and dramatic, the additive is rapidly eliminated from food; but when the toxic effect is manifest only after long latent periods, the classical methods for evaluating toxicity often fail, and the real toxicity of the drug is known only after large segments of the population have been harmed. The press has justifiably informed the general population of the danger of spraying cranberry plants with a potent carcinogen, acetylaminofluorene.

Nutritional deficiencies are of two main types: highly specific deficiencies (vitamin, fatty acid, and amino acid deficiencies) and general lack (malnutrition).

Malnutrition During Pregnancy and Lactation

Life is maintained by the continuous intake of food. A properly balanced diet provides the substrates and the coenzymes needed for the elaboration of structural and catalytic molecules. With an unbalanced diet, the delicate cellular machinery is damaged; its performance is inadequate and disease develops [3–7].

The cells are most susceptible to malnutrition during periods of maximal activity (growth, pregnancy, lactation) or when several stresses act synergetically (fever, infection, diabetes, aging). Consequently, the requirements for all food in general, and for some food in particular, vary with age and sex. Therefore, the overall changes in the fetus, in the mother during pregnancy, and in the growing child must be considered.

Soon after fertilization of the ovum, a woman's body undergoes considerable changes geared at fetal development. These changes are triggered by dramatic hormonal shifts, which stimulate proliferation of the uterus and the placenta and the growth of the fetus. The gonadotropins, secreted during pregnancy, appear in the serum and urine a few days after the first missing menstruation. Gonadotropin levels increase steadily, reach a plateau, and then drop at approximately the 11th, the 100th, and 110th days, respectively, after the onset of pregnancy. Gonadotropins stimulate fetal and placental growth. Progesterone is secreted by the corpus luteum, which secretes progesterone up to the 3rd or 4th month of pregnancy, and the placenta, which starts to secrete progesterone approximately 3 weeks after the onset of pregnancy. During the 3rd month of pregnancy, the placenta secretes almost all the progesterone of the pregnant woman, and from then on the progesterone secretion levels are high until a maximum is reached sometime between the 220th and the 240th day of pregnancy. From the 241st day until the end of pregnancy, progesterone levels decrease steadily, and they drop to zero almost immediately after delivery. Progesterone prepares and maintains the uterus for the implantation of the ovum. Estrogen secretion is low in the beginning of pregnancy, begins to rise appreciably on the 100th day of pregnancy, and reaches maximum levels during the second half of pregnancy.

The Placenta

The placenta is a rapidly proliferating organ that communicates between mother and fetus and keeps them separate at the same time—the placenta is a barrier and a filter. It is an endocrine gland and a source of food for the fetus [4–7]. Among the problems raised by the placenta are its permeability to gases and solutes, and its enzymatic structure responsible for nutritional and endocrine roles. Only gases and possibly some foreign substances are transported by passive diffusion. Solutes are carried by active transport mechanisms, which require that the placenta possess active bioenergetic and biosynthetic pathways. Consequently, it is not surprising that the placenta contains all the enzymes of glycolysis, the hexose monophosphate shunt, and the Krebs cycle, in addition to the enzymes necessary to glycogen, lipid, protein, and nucleic acid synthesis and breakdown.

Winick [8] has distinguished two stages in the development of the placenta: hyperplasia and hypertrophy. Interference with cell division causes permanent damage; interference with hypertrophy is reversible. It is not known whether malnutrition in the human alters placental cell division. Diabetes influences both stages of placental growth. Placentas from diabetic or prediabetic pregnant women contain more DNA, RNA, and protein than do normal placentas. Diabetic factors stimulate placental growth most during the earliest stages of gestation.

The enzyme pattern of the placenta varies with the stage of development. For example, although early in development the placenta (in contrast to the fetal liver) has high glucose-6-phosphatase activity and glycogen content, placental glycogen and glucose-phosphatase activities are low at term.

The overall pattern of glucose metabolism changes as gestation progresses. During the early stages of gestation, placental glucose metabolism resembles that of the liver; at the late stages, it resembles that of muscle. Similarly, placental mitochondria seem to be less able to couple oxidative phosphorylation as term approaches.

One intriguing aspect of the enzymology of the placenta is the molecular properties of the enzymes synthesized in the placenta. The placenta contains all the enzymes that make cholesterol and progesterone. But the placental enzyme arsenal needed to convert progesterone to other steroids is incomplete, as is that of the fetus. However, if fetus and placenta are considered as a single unit, such a unit can synthesize all needed steroid hormones. The enzyme mosaic of the fetalplacental unit is discussed in greater detail in the chapter on steroid hormones [9].

The placenta and fetus have the same genetic composition. Therefore, much can be learned of the fetal phenotype by studying the placenta. This aspect of placental enzymology is in its infancy, and placental and adult enzymes have been compared only crudely. One major difficulty in such studies is that the placenta contains large amounts of the mother's blood; therefore, unless the placenta is repeatedly washed—a procedure that is bound to be deleterious to at least some of the placental enzyme [10, 11]—excessive contamination will occur. Several enzymes (17-β-hydroxysteroid dehydrogenase, glutamic dehydrogenase) have been partially purified, and information on their physical, chemical, and catalytic properties suggests that the molecular structure of the placental enzyme is similar or identical to that of the adult enzyme. In contrast, homogeneous preparations of placental alkaline phosphatase differ from a similar preparation of adult intestinal phosphatase. The affinity for β-glycerophosphate and the response to inhibitors distinguish the two enzymes. Although during the early months of pregnancy the placenta performs for the fetus most of the functions that the liver performs for the adult organism, the placenta is devoid of urea cycle enzymes, and the fetal liver depends upon the mother's liver for urea formation. When the placenta is small, fetal growth is disturbed. Effects of this range from mild microcephaly to dwarfism.

Malnutrition in the Fetus

From fertilization until delivery of the mature fetus, cells proliferate and differentiate. The morphological changes in the development of the embryo into a mature infant have been described in detail by embryologists. In contrast, little is known of the detailed molecular alterations associated with each stage of embryonic development. As a result, the biochemistry of growth and development can be described only superficially.

Obviously, the intimate chemical composition and morphological appearance of a cell depend on the cell's enzyme mosaic. Consequently, proteins are central to the biochemical changes that accompany growth and differentiation. Thus, the development of the fertilized egg into the mature infant is largely due to the sequential synthesis of specific proteins. Protein synthesis requires energy, is dictated by the base sequence in nucleic acids, and results in quantitative and qualitative changes in the protein population of the embryo [12].

The egg and the embryonic cell are well endowed with bioenergetic pathways. The multiple-enzyme systems involved in glycolysis, the hexose monophosphate shunt, the Krebs cycle, the electron transport chain, and oxidative phosphorylation have all been found in the vertebrate embryo. In the embryonic and in the mature cell, oxidation through the Krebs cycle, electron transport, and coupling of oxidation and phosphorylation occur in mitochondria. The chemical energy provided by these pathways is needed for normal development because if either glycolysis, Krebs cycle, or electron transport chain inhibitors are administered *in vivo* or added to explanted chick or sea urchin embryos, embryonic development is arrested.

Bioenergetic pathways shift at certain stages of development. These changes are reflected in the use of substrates, the formation of product, and enzyme activities. Three examples of shifts in substrate use during development include changes in oxygen consumption, increased oxidation of glucose through the hexose monophosphate shunt, and a shift from carbohydrates to fats, and later to proteins as a source of energy.

During development, oxygen uptake does not increase gradually but evolves through successive peaks. In the sea urchin egg, oxygen uptake rises immediately after fertilization. The oxygen uptake gradually increases until hatching and is maintained at hatching level until gastrulation, when it rises again. The ratio of ATP to ADP has been postulated to regulate oxygen uptake. Just before fertilization, the respiration rate is low because the ATP/ADP ratio is high. By a feedback mechanism, the high ATP level inhibits the further phosphorylation of ADP and slows down electron transport and oxygen consumption.

Interesting observations were made during the development of tryptophan pyrrolase. In the adult, the activity of that enzyme can be increased considerably by administering either the substrate or cortisone. Cortisone administration doesn't affect the 1-week-old rat, but increases the trytophan pyrrolase activity of the 2-week-old rat to adult levels. The 3-week-old rat responded to cortisone exactly like an adult.

The postnatal development of alkaline phosphatase in the intestinal mucosa of rats and mice is similar to that of tryptophan pyrrolase. Much work in mammalian developmental biochemistry was done on developing liver. Studies on prenatal liver include investigations on the bioenergetic pathways and studies of the formation of enzyme found exclusively in liver. Burch [13] measured the activity of several enzymes involved in glycolysis, the hexose monophosphate shunt, glycogenolysis, and gluconeogenesis. The results of these studies show that the biochemical development of the liver can be divided into three periods: prenatal, from 0 to 21 days, and after 21 days.

Hexokinase activity, which controls the entry of glucose into the glycolytic pathway and the hexose monophosphate shunt, is very high during gestation. Five days before delivery, its activity is five times that of adult liver. The activity of liver hexokinase decreases rapidly after birth and reaches adult levels 21 days after delivery. Such a finding suggests that glucose use through glycolysis and the hexose monophosphate shunt decreases with maturation. The varying ratios of traditional hexokinase and glucokinase during the development of the fetal liver are discussed in other sections of this book.

Phosphofructokinase, the enzyme that phosphorylates fructose-1-phosphate to yield the diphosphate, the precursor of the triose phosphates, has a fate similar to that of hexokinase, except that its prenatal activity is only three times greater than that of adult liver, and the prenatal activity drops to adult values within 9 days after birth. Fructose-1,6-diphosphate, triose-P-isomerase, and glyceraldehyde phosphate dehydrogenase all have high fetal activities that slightly increase at the adult levels in the newborn. Thus, in the fetal liver the activity of these enzymes seems to favor the formation rather than the use of lactic acid.

The fate of lactic dehydrogenase differs from that of the glycolytic enzymes. The activity of that enzyme is the same in fetus and adult, but lactic dehydrogenase activity increases immediately after birth. During the first five days after birth, the enzyme's activity is about 250 times that of the adult. The activity slowly drops to the adult value between the 5th and 21st day after birth.

In the early chick embryo, lactic dehydrogenase activity goes through three major stages. The activity of the enzyme decreases during the first two days of development; from the second to the tenth day, the activity slowly increases; but after the tenth day it rises suddenly. Lactic dehydrogenase activity increases unequally in all organs. For example, on the 12th day of the development of the chick embryo, activity rises sharply in liver and gradually in muscle and brain. Chick embryo tissues contain two different types of lactic dehydrogenase: one found principally in muscles (muscle lactic dehydrogenase), the other in heart (heart lactic dehydrogenase). The former is inhibited by pyruvate and is therefore relatively inefficient in an anaerobic environment. When pyruvate accumulates during muscle development in the chick embryo, there is a progressive shift from pure heart lactic dehydrogenase to pure muscle lactic dehydrogenase.

In contrast to the enzymes involved in glycolysis, some enzymes involved in glycogenolysis (phosphorylase) and gluconeogenesis (glucose-6-phosphatase) are low or nonexistent in fetal liver, and these two functions develop with maturation.

The fate of glucose-6-phosphatase in newborn liver has been investigated in some detail in several laboratories. The enzyme is absent from fetal liver of rodents and monkeys, but enzyme activity appears soon after birth by *de novo* synthesis of the enzyme molecules. Ethionine and puromycin, but not X-irradiation, prevent the formation of the enzyme. The factors that determine the formation of glucose-6-phosphatase are clearly related to the birth process because the enzyme activity appears soon after birth in premature liver.

The appearance of glucose-6-phosphatase in the newborn can also be prevented by the administration of glucose or insulin. On the basis of all these findings, the following mechanism for the control of glucose-6-phosphatase activity was proposed. When placental circulation ceases immediately after birth, the blood glucose levels of the newborn fall because of peripheral use. The drop in blood glucose is reflected in the liver, which reacts by increasing glycogen breakdown and by making glucose available for peripheral utilization. This is achieved by converting glucose-6-phosphate to glucose. The mobilization of glycogen in newborn liver is independent of the changes in glucose-6-phosphatase activity, and, consequently, the mechanisms controlling the development of gluconeogenesis and glycogenolysis must be different.

A number of enzymes found in adult liver (*e.g.* tryptophan pyrrolase, glucuronide synthetase, phenylalanine hydroxylase) are absent in fetal liver and appear only after birth. Many of these enzymes present in the adult but absent in the fetus have been found to be lacking in various hereditary diseases. As was the case for glucose-6-phosphatase, the appearance of these enzymes is related to the process of birth. Nemeth has proposed that some factor *"in utero"* represses the expression of the genes geared to the production of greater amounts of RNA. Although many of the enzymes involved in gluconeogenesis are present in fetal liver, gluconeogenesis appears only at birth. The fetal liver's inability to use pyruvate to make glucose results from the absence of the enzyme phosphopyruvate carboxylase. The activity of this enzyme develops only after birth as a result of *de novo* synthesis, and the capacity of the liver to make glucose from pyruvate rises with the increase in enzyme activity [14, 15].

As soon as the egg engages in *de novo* synthesis ATP is consumed, and the breakdown of ATP to ADP drives the phosphorylation of ADP, the electron transport chain, and oxygen consumption. The increased energy consumption probably precipitates a cascade of events; new proteins are synthesized, among them proteins that catalyze the production and use of oxidizable substrate.

From gastrulation to neuroectodermal development, the amphibian embryo appears to oxidize mainly carbohydrates. Later, during entodermal development, fats are also used as substrates. It is not known whether all the cells of the embryo participate in changing substrate, or if the change results from the proliferation of a new type of cell.

Studies of the incorporation of amino acids and ^{35}S into proteins of sea urchin eggs have established that fertilization is immediately followed by increased protein synthesis. In contrast, the unfertilized egg synthesizes little or no protein. The inhibition of protein synthesis blocks cellular proliferation. Detailed study of the machinery involved in protein synthesis has established that the increase in protein synthesis is accompanied by changes in the activity of amino acid-activating enzyme in RNA and in messenger RNA. In embryonic tissues of Synopsus larvae, the activation level decreases from gastrulation to hatching. In contrast, the amino acid activation of metamorphosing frog liver does not change. Amino acid activation is increased four times during the pubal life of the fly *Lucilia cuprina* and during the formation of the silk gland. Ribosomes and tRNA of unfertilized sea urchin eggs can stimulate amino acid incorporation. Increased protein synthesis observed after fertilization is thought to depend upon the formation of a specific messenger RNA. If labeled phenylalanine and leucine are provided in addition to polyuridylic and polycytidylic acids and added to a system containing the tRNA's and ribosomes of unfertilized eggs, the labeled amino acids are incorporated into protein.

As a result of this active protein synthesis, the total protein mass is increased, and new proteins appear. The development of new proteins has been established by immunological methods and by determination of enzyme activity. During the development of the chick embryo lens, seven antigenic proteins appear. The antigen reactive groups of myosin appear in the heart-forming area of the chick embryo. Fluorescent antibody techniques have demonstrated that myosin is diffusely distributed in the early embryo; it is later restricted to the heart and muscle-developing areas [16].

These are only a few examples of the molecular alterations that occur during development; the mechanisms of gene expression are discussed elsewhere. Even a comprehensive list of these changes would not permit us to give a synthetic view of the cellular and molecular alterations that take place during the development of the embryo.

Yet, these few examples demonstrate how the developing cell weaves the delicate patterns of these multiple-enzyme systems by elaborating specific proteins, which requires the dictation of specific messenger RNA's by DNA and the formation of ATP by the bioenergetic pathway. Just as the morphological changes in the embryo are rigidly scheduled, the biochemical changes occur in sequence. Therefore, the withdrawal of a specific cofactor (for example, a vitamin), the addition of antimetabolites, or the lack of total proteins or essential amino acids understandably affects the molecular development of the embryo inducing congenital alterations, death of the fetus, or a host of other misfortunes. Harmonious growth depends upon adequate nutrition before, during, and after pregnancy.

Competent pediatricians feel that the decrease of neonatal death is still too slow, and that many fetal and neonatal deaths and incurable damage to infants result from inadequate nutrition. Experiments in animals have clearly established that malnutrition of the mother results in congenital malformation, death of the fetus, or premature delivery. Obviously, no deliberate experiments have been made in humans that would permit us to decide what degree of malnutrition causes fetal malformation.

The importance of adequate nutrition during pregnancy is well illustrated by comparing the deliveries of pregnant women in Holland and in Leningrad during World War II. Although the children of the Dutch war mothers were somewhat smaller than the children delivered in peacetime, the incidence of prematurity, stillbirth, and congenital malformations was not increased in the Dutch children born between October 1944 and May 1945. In contrast, during the siege of Leningrad, which lasted from August 1941 until January 1943, the stillbrith rate doubled and the incidence of prematurity increased almost 50%. In the latter part of the siege, the diet was so unsatisfactory that amenorrhea and sterility became the rule. It has been proposed that the differences in the effect of the dietary deficiencies in the Dutch and the Russians resulted from the fact that in spite of severe rationing, the Dutch mothers continued to receive an adequate and balanced diet, but the diet of the Russian mothers was inadequate. A study of indigent pregnant women in Iowa established that the incidence of prematurity was three times higher among poorly fed mothers than among mothers who received a standard American diet.

During pregnancy, mother and fetus compete for nutrients—the mother may take some mainly for herself, while others are primarily used by the fetus. The levels of vitamins A and E are lower in the fetus than in maternal blood. In contrast, the fetus stores twice as much iron and vitamin C as the pregnant mother, whose stores drop to half the levels of the nongravid woman.

Nutrition in Infancy

The importance of proper nutrition during infancy is illustrated by the infant's nutritional requirement. On a body weight basis, the infant requires far more essential amino acids than the adult. The infant must receive histidine, which is not essential in the adult diet. The 2-month-old premature infant and the 3-month-old term infant may require as much as 8 mg of iron a day. Such an amount is usually not found in the diet, and supplementation is necessary. An adequate protein intake is necessary for the maintenance of health in the infant and growing child, but the exact protein requirement is not known.

In evaluating food requirements in infancy, one must consider that the infant's enzymic arsenal is not equivalent to that of the adult. We have already mentioned that in guinea pigs and rats, glucose-6-phosphatase is practically nonexistent until shortly before term, but after term it rises rapidly. In humans, phenylalanine hydroxylase is absent from liver until several weeks after term. If the formation of the enzyme molecules is completely inhibited because of a genetic defect, an inborn error of metabolism develops. In rats, tyrosine transaminase increases immediately after birth and reaches a maximum 12 hours after birth.

In the same animal the activity of tryptophan pyrrolase is low at birth and starts to increase only two weeks after delivery, reaching a maximum the third week after delivery.

In chick embryos, sugar glucose and galactose are actively transported through the interstitial epithelium near the time of hatching and are accompanied by increases in alkaline phosphatase and succinic dehydrogenase in the villi. Similarly, some fundamental physiological functions do not develop until several months after birth. Infants' renal function does not mature until four or even six months after birth; before that, the kidney has below normal concentration ability.

Infants' biochemistry and physiology gain particular significance when mothers' milk must be substituted by cows' milk. Pediatricians often claim that infants fed entirely or partially with mothers' milk have lower mortality and morbidity rates. The incidence of gastrointestinal, respiratory, and ear infections is reduced in infants fed mothers' milk. In infants fed cow milk, the growth rate and nitrogen and salt retention may be increased.

Although cow and human milk resemble each other superficially, there are many quantitative and qualitative differences between the two types of milk (see Table 4-1). The water and fat content of cow milk is the same as that of human milk, but cow milk contains three times as much minerals and proteins as human milk. In contrast, the lactose content of cow milk is only two-thirds that of human milk. At first approximation, it seems that a product similar to human milk can be obtained simply by diluting cow milk with two volumes of water and by supplementing it with carbohydrates. But a more careful comparison of the composition of cow milk and human milk indicates that cow milk is low in iron, copper, vitamins A and E, essential fatty acids, and essential amino acids. Therefore, the dietary qualities of human milk are not completely matched by the administration of cow milk.

Table 4-1. Composition of cow milk relative to human milk (in % of human milk = 100%)

	Human	Cow
Minerals		
Calcium	400	
Phosphorus	300	
Sulfur	200	
Iron		50
Copper		33
Magnesium	300	
ASH	346	
Fat		
Total	100	
Fatty acid		
Saturated		
C_4–C_{10}	300	
C_{12}–C_{20}	110	
Unsaturated		
Total		92
Lauric acid	300	
Essential		
Total		27
Linoleic		19
Arachidonic		12
Linolenic		
Vitamins		
Liposoluble		
A		64
D	540	
E		1
K	400	
Water soluble		
Ascorbic acid		37
Biotin	1000	
Folic acid	127	
Niacin		50
Pantothenic	180	
Pyridoxine	430	
Riboflavin	670	
Thiamine	670	
B_{12}	500	
Nitrogen		
Nonprotein		
Total		82
Urea		47
Uric acid		41
Creatine		95 (9.5)?
Creatinine		41
Amino acid		
Essential		
Arginine	110	
Histindine		69
Isoleucine	150	
Leucine	130	
Lysine	146	
Methionine	150	
Phenylalanine	140	
Threonine	125	
Valine	180	
Protein		
Total	270	
Casein	700	
Lactalbumin	135	
Lactoglobulin	100	
Whey protein	100	
Water	100	
Lactose		68

In addition to the differences in mineral, vitamin, protein, carbohydrate, and fat composition, more subtle differences exist between cow and human milk. Workers have claimed that human milk contains antibodies that are absent from cow milk, and that human milk contains a factor that stimulates the growth of the *Lactobacillus bifidus*. Human milk is 50 times more active than cow milk in that respect, and, therefore, the intestinal bacterial flora depends largely upon the type of milk that is ingested. The importance of that factor is illustrated by the therapeutic effect of human milk on acrodermatitis enteropathica, a disease characterized by a vesicular dermatitis (at the extremities and around the body orifices) and diarrhea. The disease occurs after weaning or in infants that are fed cow milk and disappears soon after the administration of human milk.

Malnutrition in the Growing Child

After birth, growth proceeds by cell multiplication and enlargement. Cell multiplication predominates in the fetus, but cell enlargement is more important after birth. It has been estimated that 44 cell divisions occur successively during fetal life, bringing the total number of cells to 15×10^{12}. Postnatally, each cell is thought to divide only 4 times. This low average results from the fact that some cells (muscle, brain, liver) almost never enter mitosis after birth and grow only by increasing in size. The development of polyploidy often accompanies the increase in the size of liver cell. In contrast to what occurs in muscle, parenchyma, and brain, mitosis continues in epithelia and in hematopoietic tissue. After delivery, a period of rapid growth of the infant body is followed by a slower period of growth regulated by hereditary and environmental factors.

The influence of heredity and environment (nutrition, emotional stress, geophysical factors, climate) may have determined differences in body size in time. The human body size has often been claimed to have increased during the last two centuries. A Norwegian study of 1941 indicates that the size of man has increased 1 inch per generation, and that the maximum stature is reached in men at 18 or 19 years of age rather than at 20 and in 16 and 17 years of age in women. These differences could result from the reduced inbreeding (the invention of the bicycle has been claimed to be responsible for the marriages of people from different villages) and better nutrition. The small size of the armor of the knights of the Middle Ages has often been invoked as an argument in favor of the increase in size of the last centuries. Skeptics have claimed that only the smaller armor could be conveniently preserved in museums.

Interferences with growth may result from an interference with the genotype; inadequate oxygenation depriving the cell of chemical energy; hormonal imbalances; and malnutrition resulting from deficient diet or disease. Growth retardation (see Table 4-2) is not the only form of malnutrition that is encountered in the growing child. Obesity resulting from overeating, inactivity, or hormonal imbalance is also frequently encountered. (The pathogenesis of this condition is discussed in the section on obesity.)

Table 4-2. Growth retardation

Hereditary	
Mongolism	
Morquio's disease	
Hurler's disease	
Hypophosphatasia	
Galactosemia	
Fructose intolerance	
Phenylketonuria	
Maple syrup disease	
Cretinism	
Hormone imbalances	
Lack of growth hormone	
Excess cortisone	
Diabetes	
Pheochromocytoma	
Hypothyroidism	
Lack of sex hormones	
Nutritional	
Dietary diseases	Rickets; Kwashiorkor
Gastrointestinal diseases	Atresia; Celiac syndrome; Regional enteritis
Renal tubular acidosis	(Leakage of glucose, amino acids, and minerals)
Inadequate oxygenation	
Congenital cardiovascular disease	
Placental insufficiency	

Starvation

Starvation results from an insufficient caloric intake, which may range from insufficient nutrition to complete deprivation of food. Complete deprivation of food is observed in professional "hunger artists," prisoners or zealots who submit themselves to hunger strikes, among castaways, and occasionally in patients suffering from anorexia nervosa. Partial food deprivation is observed in children and adults during famines and among the prisoners of concentration camps [17–23].

The sequence of events following complete food deprivation will be discussed first; then we will consider low-calorie diets administered for a prolonged period of time. The most drastic change that follows total food deprivation is a rapid loss of body weight. Humans can safely lose 30% of their body weight, but losses on the order of 40% are considered fatal. The starvation victim's skin becomes dry and pale, the cutaneous tissue hangs over the subcutaneous tissue and loses its normal elasticity, hyperkeratosis usually develops at the upper surface of the thighs and on the upper arms, a brownish pigmentation may appear anywhere on the body surface, and the hair becomes dull, dry, and friable. As starvation progresses, the victim may develop cardiovascular symptoms, including decreased cardiac output, bradycardia, weak pulse, and a drop in blood pressure. If starvation is maintained for prolonged periods of time, the victims lose their libido and become apathetic and depressed.

When food intake is suppressed, the stores of the body are progressively depleted. The victim loses first glycogen then fat reserves, and at the same time breaks down his muscle protein to generate gluconeogenic substrates. The combustion of fat in the absence of carbohydrates results in the development of ketosis (see Diabetes). While the fat and carbohydrate reserves are consumed, nitrogen excretion is reduced. Once all other reserves are exhausted, protein reserves must be tapped for maintenance of caloric requirements, and the excretion of nitrogen in the urine is then increased.

Fasting leads to a decrease in the activity of lypogenic enzymes. Reductions in the activity of glucokinase, acetyl-CoA carboxylase, glucose-6-phosphate dehydrogenase, long-chain acyl-CoA desaturase, the citric-cleaving enzyme, and fatty acid synthetase have been described. In fact, the activity of many of these enzymes is also decreased in alloxan diabetes. Studies with fatty acid synthetase have clearly demonstrated that depletion results from active decrease in the absolute amount of enzyme molecules, and that activity was restored by *de novo* synthesis of the enzyme.

Glucose is the major source of aerobic and anaerobic energy. In starvation, the major source of glucose is stored glycogen. Once the glycogen is exhausted, proteins provide the only source of gluconeogenic substrates. Lactate and pyruvate which are substrates derived from glucose can be reconverted into glucose by the reversal of the glycolytic pathway, but such products are scarce and quickly exhausted.

The enzymes capable of converting even-chain fatty acids are missing in mammals, and odd-chain fatty acids which could be used in gluconeogenesis constitute only 5% of the fatty acid contingent. Thus, proteins from muscle are the major source of glucose in prolonged starvation.

The proteins are broken down in a discrete fashion that remains unknown, but certainly does not involve lysosomal enzyme release or even lysosome formation in the muscle. The amino acids are freed, most are reutilized for protein synthesis, but alanine which constitutes only 5% of the muscle amino acid component and to a smaller extent glutamine are transported from muscle to liver where the amino acids are used in the process of gluconeogenesis.

Therefore one speaks of a "glucose alanine cycle" involving the conversion of alanine to pyruvate through the catalytic action of the proper transaminase. If starvation is prolonged, the alanine production in muscle is reduced and gluconeogenesis decreases in spite of the fact that the activities of the liver enzyme remain adequate.

At autopsy, the most typical findings are the loss of subcutaneous and internal fat and atrophy of muscle and parenchyma. In severe starvation, 33% of the muscle and 50–60% of the liver mass may be lost. The weight of the heart may be decreased, and the mass of the spleen and kidney may be considerably reduced. The bones show osteoporosis because of a lack of osteoblasts, and the mucosa of the gastrointestinal tract is atrophic. Usually, brain and nerve sizes are not decreased by starvation, and even under severe starvation, brain weight is reduced by only 3%. Histological examination of the muscle reveals signs of atrophy. In the liver, the endoplasmic reticulum becomes dispersed and vesiculate, the mitochondria acquire bizarre shapes and may be surrounded by the membranes of the endoplasmic reticulum, and in terminal stages foci of cytoplasmic degradation may appear.

Starvation ultimately results in hypoproteinemia. When an individual is fed a diet of 1,600 calories for a prolonged period, he depletes his glycogen and fat stores and loses weight. At first, the weight loss is rapid and is accompanied by loss of nitrogen in the urine. Later, the starvation victim becomes stabilized at a low weight and adapts to the low caloric intake by reducing his basal metabolism and muscular activity.

Protein Deficiency

Dietary Proteins

Among the various forms of malnutrition that continue to afflict the human race, protein deficiency is one of the most common. Typical of protein deficiency is kwashiorkor. Although the condition is frequently seen, its pathogenesis remains poorly understood and requires appropriate knowledge of the fate of dietary proteins.

Dietary proteins are a source of nitrogen, essential amino acids, and calories. The nitrogen of the ingested protein is retained in the body in the form of newly synthesized protein, but as new proteins are synthesized, some of the endogenous proteins are catabolized. Consequently, after the protein is digested, the amino acids enter a dynamic pool to which both endogenous and exogenous proteins contribute.

Nitrogen Balance

Nitrogen retention by the body is expressed by the nitrogen balance, the difference between the ingested nitrogen (I) and the nitrogen excreted in the feces (F) and in the urine (U). $B=I-(U+F)$. The balance is positive when the intake is greater than the loss, and negative when the loss is greater than the intake. When the balance is positive, the overall rate of protein synthesis is greater than the overall rate of protein and amino acid breakdown. The fact that the term "balance" refers to the total organism should be emphasized. Sometimes nitrogen retention in an animal excedes nitrogen excretion, but only because nitrogen is retained in a single tissue at the expense of the nitrogen stores in the rest of the body. This is the case in rats with sarcoma. Retention in the sarcoma is high, and part of the endogenous nitrogen pool is used to form the sarcoma reserve.

When the nitrogen balance is negative, the labile protein stores are depleted. In a person fed a nitrogen-deficient diet, the labile protein stores drop rapidly at first, but as the deficiency persists, the rate of protein catabolism decreases and the rate of reduction of the labile protein stores decreases. Consequently, the labile protein stores provide the mechanism for adaptation to various levels of protein intake.

All proteins are not equally digestible; therefore, the nitrogen that they contain is not equally absorbed. Little of the nitrogen of soya bean is absorbed, but the nutritive value of the bean can be increased considerably by proper cooking. The digestibility of a protein is expressed by the ratio of the absorbed nitrogen to the ingested nitrogen ($D = A/I$). The amount of absorbed nitrogen can be estimated by measuring the difference between the ingested nitrogen and the fecal nitrogen of dietary origin. The fecal nitrogen of dietary origin is equal to the difference between the total fecal nitrogen and the fecal nitrogen of endogenous origin. The total fecal nitrogen can be determined directly by analyzing the feces after administration of the protein under investigation. The nitrogen of endogenous origin is usually determined by administering a protein-free diet: $A = I - (F + M)$.

The nitrogen balance index is the rate (K) of change in the nitrogen balance (B) versus the absorbed nitrogen (A). When catabolism of tissues is constant and independent of the diet, $B/A = K$. The nitrogen balance index reflects the biological value of a protein, which corresponds to the fraction of ingested nitrogen absorbed by the body. In man, $\delta B/\delta A$ of egg protein is close to unity (0.94), indicating that most of the absorbed nitrogen is retained. The K value for gluten is 0.42, indicating that gluten proteins are not used for anabolic processes. The low nutritive value of gluten results from its low lysine content. Zein and gelatin are also proteins of low nutritive value. They are completely devoid of essential amino acids.

The nutritive value of a protein resides in its amino acid composition. Thus, if a protein provides all the amino acids in exactly the required amount, the protein will be used effectively. Egg protein contains 60% of the indispensable and semi-indispensable amino acids. In contrast, whole wheat proteins contain only 40% of the indispensable amino acids. Consequently, even if whole wheat protein is supplemented with lysine, which is almost completely lacking, it still provides less indispensable amino acids per gram of nitrogen than all egg protein. Thus, the efficiency of a protein depends not only on the actual indispensable amino acids it contains, but also on the proportion in which these amino acids occur [24–26].

It has been calculated that if humans were fed a protein that contained ideal amounts and proportions of the various amino acids, the daily requirement of that protein would be 0.5 g/kg of body weight. On the basis of this value and the biological value determined for various proteins, the daily requirement for egg powder is 0.53 g/kg of body weight, and the requirement for gluten, 1.12 g/kg of body weight. With an ordinary diet, an average of 1 g of protein per kilogram of body weight per day is estimated as adequate.

From the analysis of the protein biological values, protein requirements would appear to be satisfied equally well by administering daily 0.53 g of egg powder or 1.12 g of gluten per kilogram of body weight. However, when the requirements are satisfied by large amounts of protein with poor nutritive values, the organism does not function optimally.

Ingesting large amounts of protein with a low nutritive value increases the amount of some nonessential amino acids, and amino acid imbalance results. The effect of this imbalance is demonstrated by the inefficiency of the dietary protein in restoring depleted protein stores or in maintaining normal growth rates. Amino acid imbalances express themselves for the amino acid that is most limiting for growth. For example, when a diet low in valine is supplemented with threonine, growth is retarded. These amino acid imbalances must be distinguished from the toxic effect resulting from the excessive intake of individual amino acids (methionine or tyrosine).

As may be anticipated, protein requirements are accelerated when the anabolic processes increase, as during growth, pregnancy, lactation, and menstruation. Trauma, fever, infection, and other forms of disease are often associated with negative nitrogen balances. It is not known whether it is suitable under those conditions to restore the nitrogen balance to normal. During menstruation, the nitrogen losses are of the order of 1.5–3.3 g per day, and naturally these losses must be replaced by a dietary protein.

In view of the low protein content of the diet in some countries, it is of paramount importance to be able to evaluate the biological efficiency of a given protein. To express this property mathematically for each protein, one can determine the essential amino acid index, which is the geometric mean of the ratio of the amino acid in the test protein versus the amino acid in a standard protein most efficiently utilized (egg protein). On the basis of the amino acid index, an equation was developed which expresses the biological value of the protein.

Whenever the calculated biological value was compared to experimental determinations of the biological value, the two values were found to be extremely close. Such mathematical formulas are limited by the fact that the proteins are not all digested at equal rates. This difficulty has been overcome by estimating the amino acid composition not directly after total hydrolysis of the protein, but on the product of peptic and tryptic digestion of the protein. Such experiments have revealed that whereas the amino acids of the proteins contained in roast beef or wheat or peanut flour are readily available, the amino acids of cotton seed flour are not [27].

Protein Digestion

Saliva

The digestion of food requires the participation of several different segments of the gastrointestinal tract: salivary gland, stomach, intestine, liver, and pancreas. Much before "enzyme presoakers" had flooded the market, little boys knew of the cleaning power of saliva and used it profusely to remove blood stains from their shirts or trousers. Even poets have been intrigued by the powers of saliva. Baudelaire expressed his own concern in an exquisite verse in *Les Fleurs du Mal:* "Je voudrais connaître les secrets de la salive qui mord."

True salivary glands are found only in mammals. Humans have three pairs of salivary glands—the parotid (20–30 g each), the submaxillary (8–10 g each), and the sublingual (2–3 g each)—that vary in anatomical location, size, and relative proportion of serous versus mucous cells. All three types of glands have essentially similar macroscopic, submicroscopic, and ultrastructural architecture. All salivary glands have a racemose structure, which means that the overall appearance of the decapsulated gland resembles that of a bunch of grapes. The unit of the bunch is the acinus (Latin for grape).

The acini are built by secretory cells, serous cells in the parotid, and serous and mucous cells in the submaxillary (ratio 4:1) and sublingual (ratio 1:4) glands. The secretions of the acini are collected in small ducts that converge into larger ones. Smaller ducts converge into larger ducts several times, and this gives the gland its racemose appearance. Finally, in the parotid and the submaxillary glands, all ducts converge to form a single canal: Stensen's duct in the former (35–40 mm long), and Wharton's duct in the latter (40–50 mm long). In contrast to the parotid and submaxillary glands, which are drained by a single canal, the sublingual glands are drained by 5 to 15 small canals (ducts of Rivinus).

The serous cells are large, clear cells that contain refringent granules on fresh preparations similar to those described long ago by Claude Bernard in the pancreas. On stained preparations, the cell presents a large central nucleus, a well-developed Golgi apparatus, and a fine basophilic background described in some detail in Garnier's thesis on the ergastoplasm.

When examined with the electron microscope, the serous cells appear identical to pancreatic cells (see below). The unique feature of these cells is of course the zymogen granule, a structure that we will discuss in more detail later.

Saliva is a hypotonic, slightly acid, watery excretion of the salivary gland. Man excretes 1 ml of saliva per minute, about 1,000–1,500 ml daily. The submaxillary gland is responsible for 10% of salivary secretion, the parotid only for 25%, and the sublingual for approximately 5%. Saliva secretion is blocked by metabolic inhibitors that interfere with the bioenergetic pathways, and is therefore believed to be an active process. The only solid components of saliva that are relevant to this discussion on the role of saliva in digestion are the enzymes. The electrolyte composition of saliva is discussed in another chapter.

The principal enzyme found in saliva is a starch-digesting hydrolase, α-amylase. The catalytic properties and mechanisms of production of the enzyme are discussed elsewhere*. Other important protein components of saliva are lysozyme and kallikrein. Saliva has three functions: it keeps the mouth humid; it lubricates food during mastication and swallowing; and it partially digests food, thereby increasing the stimulus of the taste buds.

Gastric Juice

The first major food breakdown takes place in the stomach. Protein digestion is the result of the action of secretions from the stomach—an expanded pouch of the gastrointestinal tract located between the esophagus and the duodenum—the intestine, and the pancreas.

Examination of the stomach wall from inside out with the microscope at low power reveals a thick mucosa, a thin muscularis mucosa, and several well-developed muscular layers covered with peritoneal serosa. The mucosa is lined by tall, columnar mucous cells, which secrete mucus constantly, have a relatively short life span, and are continually replaced. When the surface epithelium is washed and the thick, viscous mucus that covers it is removed, small depressions of the mucosa—the gastric foveolae—can be seen with the magnifying lens. In each foveola, three or four gastric glands open.

Gastric glands are tubular structures that extend from foveola to submucosa. The most typical gastric glands are found in the fundus of the stomach, that large portion (60–80%) of the organ located between cardia and pylorus. The three parts of the gastric gland are the neck (closest to the foveola), the body, and the fundus (closest to the muscularis mucosa). The tubular structure, which is perpendicular to the surface of the mucosa, is lined by a glandular epithelium resting on a basal membrane. The fundic gland contains four kinds of cells: mucous, chief, parietal, and argentaffin.

The chief cells, believed to secrete pepsinogens (renin in calves and gelatinase in pigs), line the inner surface of the basal membrane. In the fasting animal, chief cells are cuboidal or columnar with a clear central nucleus, a stainable Golgi apparatus, and large granules. The Golgi and low-density granules have been confirmed by electron microscopy, which also

* A few patients have been reported with severe malabsorption, emaciation, and diarrhea whose serum contained high levels of amylase. Ultracentrifugation of the purified enzyme showed that the amylase, which sedimented at 11 S, has a molecular weight much greater than ordinary serum amylase, which sediments at 4 or 5 S. The high molecular amylase results from the formation of a molecular complex between the regular enzyme and IgA globulin [28, 29].

revealed a dense intracellular population of mitochondria and rich arborization of the secreting membrane to form microvilli. After a meal, microvilli are less conspicuous, and granules tend to disappear.

Between the lining of chief cells and the basal membrane, clusters of large polygonal cells are interdispersed. Their cytoplasm stains readily with acidophilic stains. Electron microscopy has revealed that the plasma membranes of these cells form invaginations and evaginations. The invaginations penetrate deep inside the cell, giving the appearance of intracellular canaliculi.

The evaginations of the membrane extend between chief cells toward the lumen of the gastric gland, thus forming canaliculi that communicate between lumen and the deep-seated parietal cells. The parietal cells are believed to secrete the hydrochloric acid and most of the water found in the gastric juice.

Mucous cells, found in the neck of the gland, secrete a mucus with staining properties different from those of the surface epithelium. Argentaffin cells are not seen with ordinary staining or fixation techniques. After fixation with potassium bichromate and silver or chromium impregnation, they stand out as cells scattered singly between the lining of chief cells and the basal membrane.

The pyloric glands contain only one type of cell. The function of these cells is not clear, but they are believed to be mucous cells of the type found in the neck of the fundic glands. The cardial glands also are made primarily of mucous cells, although a few pepsinogen cells may be found.

Each day, 1000–1500 ml of gastric juice is secreted in humans by the 35,000,000 glands of the stomach. Pure gastric juice is a clear, colorless fluid, primarily a dilute solution (0.4–0.5%) of hydrochloric acid in water. The most important organic constituent of the juice is pepsin.

The molecular mechanism involved in the secretion of hydrochloric acid secretion is discussed in the section devoted to body fluids; the molecular nature of pepsin and the mechanisms regulating gastric secretion are covered here.

The pancreas contains a number of inactive zymogens, which, once activated, have specific proteolytic activity. Such zymogens include chymotrypsin, trypsin, carboxypeptidase, and elastase. Zymogens are activated by limited but highly specific proteolysis.

Proteolytic Enzymes

Proteins are digested in the stomach and intestine under the influence of proteolytic enzymes. The stomach contains at least two enzymes—pepsin and renin—which are secreted by the mucosal cells in the form of zymogens. The gastric cell secretes pepsinogen, a long polypeptide molecule with no catalytic property. Pepsin is activated to pepsinogen within the gastric lumen under the influence of at least two factors: the low pH of the medium, which is a result of hydrochloric acid secretion, and the enzymic breakdown of the pepsinogen molecule.

Pepsinogen and pepsin were first crystallized by Northrop [30]. The molecular weight of pepsinogen is around 42,500; that of pepsin is between 39,930 and 27,000, with an average of 34,000. A number of pepsinogens differing in electrophoretic mobility have been discovered in extracts of stomach of pig, beef, rabbit, and dog. Whether these represent various degradative forms of the same mother molecule, or whether this finding illustrates genetic variation in the amino acid sequence and composition of various pepsinogens remains to be established. Fortunately, crystalline bovine pepsinogen yields a single N-terminal and is homogeneous electrophoretically. In contrast, it has not been possible to prepare homogeneous preparations of pepsin. Pepsinogen is a single amino acid chain containing 363 amino acid residues stabilized by disulfide bridges. The conversion of pepsinogen to pepsin involves the liberation of 40–42 N-terminal amino acids. The split fragments contain glycine, two histidines, and 2 arginines. Consequently, the change in molecular weight is associated with a change in the electrophoretic mobility of the protein. Pepsin is much more acidic than pepsinogen, and the changes in molecular weight, charge, and pH result from the loss of a polypeptide chain rich in basic amino acids. In the conversion of pepsinogen to pepsin, nine peptide bonds per molecule of pepsinogen are split, yielding a number of polypeptides.

One of the smaller polypeptides is an inhibitor of pepsin, and it remains attached to the pepsin molecule until pH values below 5.4 are reached. In addition to this inhibitor, five neutral peptides are produced, each containing an average of 10 amino acids; thus, the overall conversion of pepsinogen to pepsin can be written:

$$\text{Pepsinogen} \rightarrow \text{Pepsin} \sim \text{Inhibitor} + 5\ \text{Peptides}$$

$$\text{Pepsin} \sim \text{Inhibitor} \xrightarrow[\text{pH} < 5.4]{} \text{Pepsin} + \text{Inhibitor}$$

$$\text{Inhibitor} \xrightarrow[\text{Pepsin}]{} 4\ \text{Peptides}$$

The pepsin inhibitor itself is split in the presence of pepsin into four different peptides. As crystalline pepsinogen is made of a single peptide chain with a valyl, leucyl, alanine C-terminal polypeptide and leucine N-terminal, the pepsin molecule contains the same C-terminal polypeptide, but its N-terminal is isoleucine instead of leucine. Such a finding suggests that pepsin is the C-terminal moiety of pepsinogen. The determination of the N- and C-terminal amino acids of pepsin is a remarkable achievement in view of the difficulty encountered in obtaining pepsin free of polypeptide. This difficulty stems from the rapid breakdown of the original enzyme.

The purity of crystalline pepsin, however, remains questionable. Although the crystalline enzyme behaves as a single protein in the ultracentrifuge, at electrophoresis or on chromatography on a DEAE-column, the presence of a family of protein in this preparation has not been excluded. This is of capital importance for interpreting the multiple catalytic properties attributed to pepsin. In addition to splitting peptide bonds, pepsin is claimed to catalyze transpeptidation. Therefore, it is significant that contaminants are known to be associated with crystalline pepsin. Indeed, hydroxyproline has been found in purified pepsin, and it is highly probable that the presence of this amino acid in the preparation results from its contamination by smaller polypeptides derived from the breakdown of collagen.

The pepsin molecule contains 44 aspartic acids, 27 glutamic acids, 15 prolines, a large number of hydroxy-amino acids and nonpolar residues, and only 4 basic residues—1 lysine, 1 histidine, and 2 arginines. Apparently, three disulfide bonds form crosslinks within the single amino acid chain. The pepsin molecule supposedly has little helicoidal content and is a random coil. It contains hydrogen bonds of the type CO=NH and OH—OOC, but they seem to have little or no relationship with the catalytic properties of the molecule. Partial hydrolysis of pepsin reveals the presence of phosphoserine and of a phosphopolypeptide composed of serine, threonine, glutamic acid, and phosphoric acid. The phosphorus is not necessary to enzymic activity because if pepsin is treated with intestinal phosphatases and the released phosphorus is dialyzed, the protein does not lose its activity. However, the dephosphorylation of pepsin is responsible for a certain degree of loss of activity. It has been proposed that the phosphorus molecules crosslink the single peptide chain into a loop. Treatment of pepsin by urea induces a loss of activity associated with changes in ultraviolet absorption, electrophoretic pattern, and solubility in TCA. The changes involve the tertiary structure, probably by affecting the carboxyl of tyrosine.

The molecular weight of the pepsin inhibitor is assumed to range from 3,100 to 3,242, depending upon the method used for determination. The pepsin inhibitor has been isolated on ion exchange resin, and its N-terminal (leucine) amino acid composition is known; it contains a high proportion of lysine.

The specificity of pepsin has been investigated using synthetic substrates and proteins of known sequence as substrate. From these studies, the following conclusions have been reached. Pepsin attacks peptide linkages involving L-amino acids. The enzyme is specific for both components of the peptide bonds. The attack of the enzyme on the polypeptide chain is facilitated by the presence of aromatic residues in the chain. For example, in the insulin molecule, highly susceptible bonds are those between Leu 13 and Tyr 14 in the A chain, or Phe 24-Phe 25, and Phe 25-Tyr 26 in the B chain. However, the enzyme specificity is complex, and in insulin the Leu 11-Val 12 peptide bond in the B chain proves to be susceptible to the enzyme, and numerous other points of the molecule are the site of a slower attack by the protease [31].

Walsh [32] and his associates have published the almost complete sequence of bovine trypsinogen (see Fig. 4-1). The conversion of trypsinogen to trypsin is catalyzed by trypsin and consists of the splitting of a hexapeptide that occupies the amino terminal group of the chain. The sequence of the peptide is valine-asparagine-4-lysine. The remaining polypeptide chain has isoleucine-valine-glycine-protein in the N-terminal part of the molecule. During the conversion from trypsinogen to trypsin, the molecular weight of trypsin does not change considerably but remains around 23,000.

The role of calcium in the conversion of trypsinogen to trypsin has not been clarified. In the presence of calcium, trypsinogen is quantitatively converted to trypsin. Without calcium, an inactive product is made in addition to trypsin. Calcium appears to accelerate the splitting of the lysine-isoleucyl bond and protect other sites of the molecule susceptible to attack by the enzyme.

The specificity of the trypsin is very high. Trypsin attacks peptide bonds formed by the carboxyl group of a basic amino acid argine and lysine with the amino group of another amino acid. Consequently, when the number of basic amino acids in a substrate is known, the exact number of polypeptides that will appear in trypsin can be predicted.

Two chymotrypsinogens have been purified from beef pancreas: chymotrypsinogen A was crystallized by Kunitz and associates [33], and chymotrypsinogen B by Laskowski and coworkers [34]. Chymotrypsinogin C has been prepared from pig pancreas. The complete amino acid sequence of chymotrypsinogen A has been established separately in a British and a Czechoslovakian laboratory. Only fragments of the sequence of chymotrypsinogen B are known, but the amino acid composition of chymotrypsinogen B was found to resemble that of chymotrypsinogen A except for 54 out of the 248 amino acids that form the polypeptide sequence. Little is known about chymotrypsinogen C, except for its amino acid composition, which reveals that the zymogen is unusually rich in tryptophan. Trypsin and chymotrypsin are zymogens or inactive proteins which are converted into active proteolytic enzymes under the influence of trypsin first and chymotrypsin itself in subsequent steps. The active products obtained after activation of chymotrypsinogen have been crystallized. A large number of these compounds are referred to as α-, β-, γ-, δ-ENP chymotrypsin. An ε-ENP chymotrypsinogen also exists, but it has not been crystallized. Although the α-, β-, and γ-chymotrypsinogens have been crystallized in the polypeptide form, the δ-chymotrypsinogen has been crystallized in the form of the di-isopropylphosphoryl derivative.

Two sources of information have permitted workers to establish the sequence of transformation of one type of chymotrypsin into another and to outline the general structure of the protein molecules; namely, the study of the terminal amino acids in each type

—Val—Asp—Asp—Asp—Asp—Lys—Ile—Val—Gly—Gly—Tyr—Thr—Cys—Gly—Ala—
1 5 10 15

—Asn—Thr—Val—Pro—Tyr—Gln—Val—Ser—Leu—Asn—Ser—Gly—Tyr—His—Phe—
16 20 25 30

—Cys—Gly—Gly—Ser—Leu—Ile—Asn—Ser—Gln—Trp—Val—Val—Ser—Ala—Ala—
31 35 40 45

—His—Cys—Tyr—Lys—Ser—Gly—Ile—Gln—Val—Arg—Leu—Gly—Gln—Asp—Asn—
46 50 55 60

—Ile—Asn—Val—Val—Glu—Gly—Asn—Gln—Gln—Phe—Ile—Ser—Ala—Ser—Lys—
61 65 70 75

—Ser—Ile—Val—His—Pro—Ser—Tyr—Asn—Ser—Asn—Thr—Leu—Asn—Asn—Asp—
76 80 85 90

—Ile—Met—Leu—Ile—Lys—Leu—Lys—Ser—Ala—Ala—Ser—Leu—Asn—Ser—Arg—
91 95 100 105

—Val—Ala—Ser—Ile—Ser—Leu—Pro—Thr—Ser—Cys—Ala—Ser—Ala—Gly—Thr—
106 110 115 120

—Gln—Cys—Leu—Ile—Ser—Gly—Trp—Gly—Asn—Thr—Lys—Ser—Ser—Gly—Thr—
121 125 130 135

—Ser—Tyr—Pro—Asp—Val—Leu—Lys—Cys—Leu—Lys—Ala—Pro—Ile—Leu—Ser—
136 140 145 150

—Asn—Ser—Ser—Cys—Lys—Ser—Ala—Tyr—Pro—Gly—Gln—Ile—Thr—Ser—Asn—
151 155 160 165

—Met—Phe—Cys—Ala—Gly—Tyr—Leu—Glu—Gly—Gly—Lys—Asp—Ser—Cys—Gln—
166 170 175 180

—Gly—Asp—Ser—Gly—Gly—Pro—Val—Val—Cys—Ser—Gly—Lys—Leu—Gln—Gly—
181 185 190 195

—Ile—Val—Ser—Trp—Gly—Ser—Gly—Cys—Ala—Gln—Lys—Asn—Lys—Pro—Gly—
196 200 205 210

—Val—Tyr—Thr—Lys—Val—Cys—Asn—Tyr—Val—Ser—Trp—Ile—Lys—Gln—Thr—
211 215 220 225

—Ile—Ala—Ser—Asn—
226 229

Fig. 4-1. Amino acid sequence of bovine trypsinogen

of chymotrypsin and the investigation of the structure of the α-chymotrypsin.

In the conversion of chymotrypsinogen A to the active enzyme, rapid and slow activations must be distinguished. During rapid activation (see Fig. 4-2) chymotrypsinogen A, under the influence of trypsin, is converted to a very active polypeptide chymotrypsinogen II. This change involves the splitting of an isoleucyl-16-arginine-15 bond, yielding additional N- and C-terminals isoleucine and arginine. This first hydrolysis is then followed by autodigestion of the molecule by chymotrypsin itself. The second bond of one of the C-terminal sequences (leucyl-13-serine-14) is split and δ-chymotrypsin and a seryl-arginine dipeptide are formed.

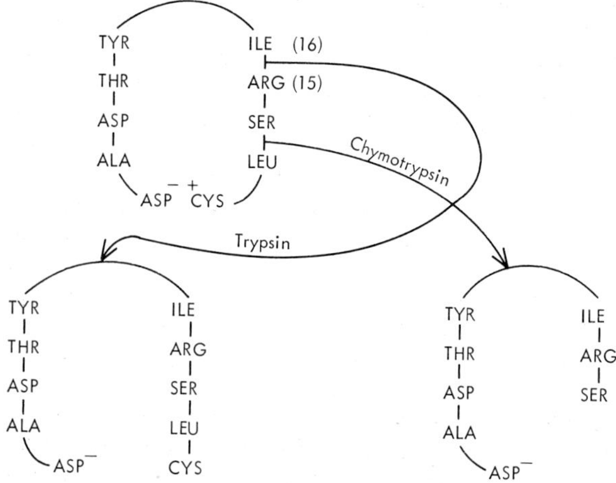

Fig. 4-2. Activation of chymotrypsinogen

When chymotrypsin is digested slowly, four peptide bonds instead of two are split. The slow activation yields δ-chymotrypsin, which acts on chymotrypsin A, splitting first a tyrosine-45-threonine-46 bond, and then an asparagine-alanine peptide bond (Asp-147-Ala-148). The products of these hydrolyses are inactive compounds (neochymotrypsinogen) and a dipeptide.

The activation of neochymotrypsinogen to chymotrypsin [35] involves the catalytic action of trypsin on an arginine-isoleucine peptide bond (Arg-15-Ile-16).

The activation of chymotrypsinogen to chymotrypsin involves splitting a polypeptide bond somewhere between the first and the fifteenth amino acid in the N-terminal sequence. As was the case for trypsinogen, the activation involves the rupture of an isoleucyl bond. In fact, in the activation of both trypsinogen and chymotrypsinogen, the first accessible bond in the sequence is split, yet both chymotrypsinogen and trypsinogen contain more than one polypeptide bond that is susceptible to proteolytic attack, and only one (isoleucyl-serine for chymotrypsinogen and isoleucyl-lysine for trypsinogen) is attacked. The resistance of the other susceptible bonds to proteolytic activity can be explained only on the basis of the internal configuration which probably makes these other susceptible bonds inaccessible to the proteolytic enzyme.

The degree of specificity of chymotrypsin is intermediate between that of trypsin (highly specific for basic amino acids) and pepsin (relatively broad specificity). Chymotrypsin preferentially splits peptide bonds in which the carboxyl group of an aromatic amino acid forms a covalent bond with the amino group of another amino acid. Yet the enzyme also splits peptide bonds in which leucine, glutamine, asparagine, and methionine are involved.

Two procarboxypeptidases are found in bovine pancreas. Both of these hexopeptidases are zinc enzymes, and the presence of zinc in the molecule is essential. Indeed, if zinc is eliminated by dialysis, the enzyme loses its activity, but activity may be restored by reintroducing zinc into the molecule. Both enzymes exist in the form of propeptidases which can be activated by splitting a polypeptide chain. The molecular weight of carboxypeptidase A ranges between 94,000 and 96,000, and its isoelectric point is below pH 4.5.

The A form is unusually large, and its activation by trypsin, chymotrypsin, or other proteases yields an active enzyme about one-third the size of the original proenzyme.

Breakdown of proenzyme yields three subunits (I, II, and III), of which only I is the active enzyme. Subunit II is a protease that is capable of activating the proenzyme. The role of subunit III is unknown.

One major property of living beings is the ability to speed up chemical reaction by the catalytic effect of specific proteins. An accurate understanding of the molecular interaction that takes place in these catalytic events has been the goal of many investigators. The mode of action of chymotrypsin has been understood more than that of most other enzymes. The information was obtained by determining enzyme activity at various pH's in the presence of substrates or their analogs, by mapping the electron density of the molecule, and by analyzing images obtained by X-ray crystallography.

As mentioned above, the hydrolysis by trypsin of a single peptide bond between arginine 15 and isoleucine 16 converts bovine chymotrypsin (245 amino acids) to active α-chymotrypsinogen. Of course, this is followed by the autocatalytic cleavage of other bonds leading to other active chymotrypsins.

The details of this fascinating excursion in molecular biology cannot be described here and only the conclusions that were reached by the workers in the field will be presented. The rupture of the peptide bond relieves the chymotrypsinogen of some restraints imposed on its tertiary structure. Thus, in the zymogen, the clusters of amino acids—asparagine 102, histidine 51, serine 195, asparagine 194, histidine 40, methionine 192—are tucked inside the molecule and covered by two surface amino acids, arginine 15 and isoleucine 16. This cluster includes the amino acids involved in the enzyme activation—isoleucine 196, histidine 40, asparagine 194, methionine 192. Thus, in the zymogen the aspartic residue 194 is tucked inside the molecule and hydrogen bonded with the Nε of histidine 40. In the active enzyme, a portion of the polypeptide chain (from residue 181 to 193) rotates as a result of the rupture of the polypeptide bond between amino acids 15 and 16. The aspartic acid 194, which was bonded to histidine 40 in the zymogen, forms hydrogen bonds with the amino group of isoleucine 16. This movement brings methonine 192 to the surface of the molecule, and, as a result, at the surface of the molecule a pocket forms which contains a cluster of amino acids involved in catalysis. This system of hydrogen bonds is believed to facilitate the withdrawal of a proton from the oxygen of serine 195. The serine oxygen is then believed to attack the carboxyl carbon of the substrate. The removed proton is transferred first to the polarizable imidazole ring of histidine 57, and from there to the amide nitrogen of the substrate.

Control of Gastric Secretion

Students of the regulation of gastric secretion have included such famous names as Claude Bernard, Pavlov, and Lindstrom-Lang. Many ingenious surgical techniques were performed on animals to provide the experimentalists with gastric juice uncontaminated by food. Three classic surgical approaches used in these studies are the Pavlov, the Heidenhain, and the Dragstedt pouches.

Heidenhain completely separated a portion of the fundus and cannulated it through the skin. Pavlov made a similar pouch, but made sure that its vagal innervation was intact. In the Dragstedt technique, the entire stomach becomes the pouch. The stomach is cut off from the esophagus and the pylorus, an esophagojejunostomy is performed, and a cannula is placed in the stomach through the skin.

Occasional accidents or therapeutic procedures have provided experimentalists with patients with fistulas, permitting the study of mechanisms controlling the secretion of gastric juice in humans. The most impressive collection of data in humans was made by Beaumont, a United States Army surgeon, on Alexis St. Martin. Few books inspire as much respect for medical research as that of Beaumont [36]. His warm interest in his patient's welfare, his detailed observations, his objectivity in interpreting the results, and his courage in challenging accepted but not proven theories are qualities, if not unique, then surely unusual.

Pavlov said "Appetite spells gastric juice." The two stages in the stimulation of gastric secretion before a meal are the cephalic and the gastric stages. French cooks, television advertising, and gourmet magazines are well aware that the sight of food stimulates the appetite. Sales at popcorn stands in the United States and fried chips stands in England depend largely on the diffusion of captivating smells into the surrounding air. The slow-cooking Thanksgiving turkey may bring large families close to neurotic feelings of hunger before dinner is served. The good soup and appealing appetizers so artistically presented at parties are designed to stimulate gastric secretion.

Smell, sight, and taste elicit a nonconditioned reflex the impulses of which converge from cortex to the vagus nucleus, where efferent stimuli are sent to the gastric mucosa. Many a good meal has been spoiled by inappropriate conversation, vulgar eating habits, or the finding of a hair or worm in what was a most delicious soup or salad. This psychological interference with one's appetite results from inhibition of the secretion of gastric juice. Because gastric secretion is stimulated without direct contact of food with the gastric mucosa, this type of stimulation is called cephalic or psychic stimulation.

Beaumont showed that the introduction of a thermometer into the gastric fistula of Alexis St. Martin induced the secretion of gastric juice. Similarly, gastric juice was secreted even when food surrounded by gauze was introduced into the stomach of St. Martin, who went on with his daily tasks not knowing what had been put in his stomach. Pavlov later established that food directly introduced into the pouch of a dog stimulates gastric secretion even if the dog does not see, smell, or taste the food.

These experiments convincingly established that direct contact of food with the mucosa also stimulates gastric secretion. Direct stimulation of secretion by contact with the mucosa is referred to as the gastric phase. Pavlov was the first to show that when food is introduced directly into the intestine, avoiding contact with the stomach, gastric juice is secreted.

Later a new type of gastric pouch was devised which connected the esophagus with the duodenum, and transplanted the stomach pouch under the skin, thus completely depriving the pouch of nervous connection with the vagus. The introduction of food into the intestine elicited the secretion of gastric juice in the stomach pouch. In fact, simple distention of the small intestine with a rubber balloon elicited secretion in the denervated pouch. It was later established that the same effect could be brought about by injecting a histamine-free intestinal extract. This positive feedback between intestinal stimulation and gastric secretion is referred to as the intestinal phase of gastric secretion.

What are the mechanisms that stimulate the secretion of gastric juice as a result of either the gastric or intestinal stimulation? In 1906, Edkins proposed that the stimulus of the stomach mucosa was chemical in nature. A substance that he called gastrin was postulated to be secreted by the gastric mucosa, transferred through the blood to the gland that it excites to secrete. Because of the postulated role of a chemical mediator, the gastric phase was also referred to as the chemical phase of digestion.

For many years, gastrin was believed to be an unspecific stimulant of gastric secretion, but perseverance coupled with clairvoyance finally culminated in Gregory and Tracy's discovery of new polypeptide hormones gastrin I and gastrin II. Both hormones are heptadecapeptides. The amino acid sequence of gastrin I is shown in Fig. 4-3. The difference between gastrin I and II is that a sulfate group is attached to the tyrosine residue in II [37–40].

Once the sequence of the gastric hormone was known, the polypeptide and polypeptide analogs were synthesized. Such studies permitted identification of the active fragments of the hormone. All the physiological properties of gastrin are invested in the C-terminal tetrapeptide sequence: tryptophan, methionine, aspartic acid, and phenylalanine NH_2. Of the four amino acids, three—tryptophan, methionine, and phenylalanine—are involved in binding. Aspartic acid is believed to be involved in a protein transfer reaction, which in some way triggers the physiological effects. The role of the remaining tridecapeptide is not clear; it has been proposed to promote effective transport, to protect the tetrapeptide against degradation, or both.

The human gastrins are also heptadecapeptides with a sequence similar to that in pig gastrin. A leucine residue replaces methionine in position 5. A number of smaller polypeptides with gastrin effects have been described.

Gastrin strongly stimulates hydrochloric acid secretion in the stomach and zymogen secretion in the pancreas, whereas it weakly stimulates pepsin secretion in the stomach and bicarbonate secretion in the pancreas.

Mechanisms of Gastrin Release

Although earlier physiological experiments suggested that the mechanism of stimulation of gastric secretion was different in the cephalic and gastric phases of secretion, it now appears that known physiological mechanisms for the cephalic and gastric phases involve only acetylcholine as a mediator.

Acetylcholine is released in the mucosa by postganglionic nerve endings found either in the fundus in the area of the fundic gland, which secretes hydrochloric acid, or in the pylorus in the area of the endocrine gland, which secretes gastrin.

Acetylcholine is known to stimulate directly the fundic gland to secrete hydrochloric acid; to sensitize parietal cells to the action of gastrin; and to stimulate directly gastrin secretion in the pyloric region.

Various types of reflexes can lead to the release of acetylcholine. Those stimuli resulting from tasting, seeing, smelling, or even thinking of food originate in the cortex. The impulses are transferred from the cortex to the vagus nucleus, and from there to postganglionic endings in the mucosa.

It is believed that insulin hypoglycemia brings about gastrin secretion by stimulation of the long vagal reflexes. Distension of the stomach is likely to release

Human

Pyro* — Gly — Pro — Trp — Leu — Glu_5 — Ala — Tyr(R**) — Gly — Trp — Met — Asp — Phe — C(=O) — NH_2

Hog

Pyro — Gly — Pro — Trp — Met — Glu_5 — Ala — Tyr(R) — Gly — Trp — Met — Asp — Phe — C(=O) — NH_2

Dog

Pyro — Gly — Pro — Trp — Met — Glu_2 — Ala — Tyr(R) — Gly — Trp — Met — Asp — Phe — C(=O) — NH_2

Cow and sheep

Pyro — Gly — Pro — Trp — Val — Glu_4 — Ala — Ala — Tyr(R) — Gly — Trp — Met — Asp — Phe — C(=O) — NH_2

Cat

Pyro — Gly — Pro — Trp — Leu — Glu_4 — Ala — Ala — Tyr(R) — Gly — Trp — Met — Asp — Phe — C(=O) — NH_2

Fig. 4-3. Amino acid sequences of the gastrin peptides. * Pyro = pyroglutamyl; ** R = H or SO_3H

acetylcholine in the mucosa by long vagovagal reflexes, whereas short intramural reflexes are responsible for the release of acetylcholine brought about by chemical stimulation of the stomach wall. Consequently, both cephalic and gastric phases result in acetylcholine release, which directly stimulates the oxyonic gland and elicits the release of gastrin by the pyloric gland.

Regulation of Gastrin Secretion

Gastrin secretion is regulated by a short feedback loop. The simple presence of acid in the pyloric region inhibits gastrin secretion, but the mechanism of the inhibition is not known. When the gastric mucosa is exposed to 0.1 M hydrochloric acid, it reportedly secretes one or more inhibitory hormones called gastrones.

Thus, modern knowledge of the stimulation of gastric juice secretion in response to a meal can be summarized as follows. Sight, smell, and taste generate cortical stimuli that are transferred to the vagus. Distention and chemical stimulation of the stomach generate long vagovagal reflexes. Both cortical and vagovagal stimulation result in release of acetylcholine in the fundic and pyloric regions. Acetylcholine directly stimulates the oxyonic gland and the endocrine gland that produces gastrin. This dual effect of acetylcholine is responsible principally for hydrochloric acid secretion. Acidification activates pepsinogen, prevents bacterial growth in the stomach, and inhibits gastrin secretion.

Many other substances besides acetylcholine and gastrin stimulate the secretion of gastric juice. One of the most potent of these substances is histamine, which vigorously stimulates the secretion of pepsin; whether or not it plays a role in physiological digestion is disputed.

Chymotrypsin, trypsinogen, and procarboxypeptidase are excretory secretion products of two organs combined into one anatomical structure: the endocrine pancreas, which elaborates insulin, and the exocrine pancreas, which secretes the pancreatic juice. Histologically, the exocrine pancreas resembles the salivary glands. The unique feature of the exocrine pancreatic cell is the zymogen granules, the formation and secretion of which are discussed in another section of this book.

For the present, some aspects of the composition of the pancreatic juices and regulation of their secretion will be considered. Pancreatic juice is alkaline and contains zymogens. The alkalinity (pH 7–8) results from the excretion of blood bicarbonate (HCO_3) against a concentration gradient. Thus, the intravenous administration of radioactive bicarbonate is rapidly followed by concentration of the radioactivity in the pancreatic juice.

Pancreatic secretion is controlled neurologically and hormonally. When food is given to a dog whose stomach has been disconnected from the duodenum and directly connected to the skin of the abdomen, and a pancreatic fistula is produced, the act of eating stimulates the secretion of pancreatic juice. Vagotomy suppresses this effect. Direct stimulation of the vagus also stimulates pancreatic secretion. In addition stimulation of the sympathetic system influences pancreatic secretion, but this is more likely an indirect effect of vasodilation of the gland and the musculature of the duct, rather than a direct effect on the secretory elements.

Two hormones are involved in pancreatic secretion: secretin and pancreozymin. In the early 1900's, researchers observed that dilute hydrochloric acid placed in the duodenum stimulated pancreatic secretion. Bayliss and Starling prepared an acid extract of the intestinal mucosa, injected it, and observed pancreatic stimulation. This was the first instance of a nonnervous transmission of a stimulus. Bayliss and Starling boldly coined the word "hormone" to refer to this type of new agent and called the first hormone "secretin."

Since then, secretin has been purified and found to be a strongly basic polypeptide hormone (mol wt 5,000) composed of 27 amino acids. Its amino acid sequence resembles that of gastrin and includes 4 argine units. A specific enzyme (secretinase) in blood seems to be capable of hydrolyzing secretin selectively.

In contrast to secretin, which stimulates only the secretion of bicarbonate and the volume of pancreatic juice, pancreozymin stimulates the secretion of the pancreatic enzyme. Although a hormone distinct from secretin that stimulated pancreatic enzyme secretion was suspected, it was not convincingly established until Wang and Grossman [41] transplanted pancreatic tissue in the mammary glands of dogs. Pancreozymin has been purified, and its amino acid sequence is known. Its molecular weight is approximately 3,884, it is composed of 33 amino acids, and the 5 CO_2-terminal amino acids are the same as in gastrin. Pancreozymin has many of the properties of gastrin. Until now, pancreozymin's effects have been inseparable from those of cholecystokinin, and, therefore, pancreozymin is often referred to as PZCCK [42].

The intestinal mucosa secretes an alkaline, watery juice containing several enzymes. This secretion mixes with pancreatic juice and bile to form the intestinal juice. The succus entericus, a pure intestinal secretion, is a thin liquid consisting of water, bicarbonate, and a few enzymes. Among the enzymes found in the intestinal secretion are exopeptidase, aminopeptidase, amylase, maltase, invertase, lactase, lipase, and enterokinase, which activates trypsinogen.

Kwashiorkor

Definition and Symptoms

"Kwashiorkor" is a word coined by the natives of the Gold Coast, and it means red boy. Kwashiorkor develops in the weaned child as a result of dietary protein deficiency, and it is characterized by digestive disturbance, edema, dermatosis, and hepatomegaly. Kwashiorkor must have been known for a long time, but it was not described in detail until 1933. Soon after the original description of kwashiorkor, reports came from various parts of Africa, Asia, and South America which proved that kwashiorkor is found in many countries where the diet is inadequate. The syndrome [43–46] originally described by Miss Williams received new and sometimes inappropriate names (infantile pellagra, pellagroid syndrome, M'waki, etc.).

The disease develops in children 4–48 months old. These children are usually in very good health while they are breast fed, but they become severely ill as soon as they have been weaned. Gastrointestinal symptoms (vomiting, diarrhea*, anorrhexia) appear first. Vomiting occurs early in the course of the development of kwashiorkor, and although no pattern can be traced, the victim often vomits immediately after a meal. The child also develops severe diarrhea with four to eight emissions a day. The feces are very liquid, yellowish or greenish, and may contain undigested food particles. Microscopic examination of the feces demonstrates the presence of starch granules, desquamated epithelial cells, and lymphocytes. If the dietary deficiency is not corrected, the digestive symptoms are followed by the appearance of edema, dermatosis, and hepatomegaly.

The total body water content increases in patients with severe protein malnutrition. The increase involves both the extracellular fluid, in which the water content may rise up to 400 ml/kg compared to 250 ml/kg in normal individuals, and in the intracellular fluid, in which the water content may reach values above 80%, compared to the normal 67% values in normally fed individuals. Fluid accumulation in the extracellular tissues leads to edema. The severity of the edema may be masked somewhat by the loss of body solids and fats. However, in the later stages of the disease, edema becomes obvious. The edema fluid is not distributed uniformly throughout the body of the victim—swelling usually starts in the inferior limbs, probably as a result of gravity and deficient circulation. The back of the hand and the face are frequently swollen. A patient with kwashiorkor—particularly an older patient—may have edema in the lower part of the body and be dehydrated in the upper part. In younger children, this does not occur because the upright position is not maintained constantly and does not play such an important role in the accumulation of fluid in the inferior limbs. The pathogenesis of the edema in protein deficiency is discussed in the section on body fluids.

Although the sodium content in the body fluid and the tissues varies considerably in children with protein malnutrition, potassium depletion of various degrees of severity is a constant finding in severely malnourished children. Alterations in mineral metabolism observed in malnutrition may result from increased metabolic breakdown with a decrease in available energy sources.

The dermatosis of kwashiorkor is quite characteristic. It mimics the dermatosis observed in pellagra, yet it does not disappear with the administration of vitamin PP. The dermatosis is sometimes referred to as

* The diarrhea is not likely to result from parasites (ascariasis and helminthiasis), which usually contaminate the intestines of African children. These worms do not invade the intestine until a later age.

the "crazy pavement" dermatosis. Black hyperpigmented patches appear on the buttocks, back, and perineum. The skin of the hyperpigmented area dries out and scales off, and the white underlying subepithelial layer becomes visible. The subepithelial layer soon sluffs off and gives way to the reddish denuded derma.

The skin alterations are responsible for the typical reddish appearance of the sick child. The areas of desquamation often become infected, and purulent ulcers develop. The alteration of pigment metabolism affects the skin and also the hair, which acquires a reddish hue. In addition, the hair becomes friable, and patches of baldness develop.

Hepatomegaly is a constant finding in severe cases of kwashiorkor, but it is less common in patients suffering from a moderate form of the disease. It is a uniform type of hepatomegaly. The inferior border of the liver may project three or four fingers below the costal margin, and the liver is hard on palpation. Splenomegaly has been observed in association with the hepatomegaly, but splenomegaly is not a regular finding in kwashiorkor. When it occurs, however, splenomegaly cannot be ascribed to malaria, which develops only in children 18 to 25 months old. Other symptoms frequently found in kwashiorkor are anemia, lower basal metabolism, and low plasma albumin levels. The drop in plasma albumins may be considerable (40–60 g/liter and even 10 g/liter). Persistent malnutrition does not necessarily cause early death, but patients with kwashiorkor fail to grow and are mentally retarded; although in areas where malnutrition is endemic, children have been reported to have mental ages more advanced than those of European children during the first few years after birth. This precocity is not maintained after the first year. During the second year, the mental age of these malnourished children drops to three-fourths of that of well-fed westerners.

Pathology

In kwashiorkor [47], the pancreas has a reduced number of acini, cells are atrophic and often vacuolated, and there is fibrosis of the stroma. These changes in the acinar structure of the pancreas are associated with alterations in enzyme secretion. In liver, severe fatty infiltration can be seen in the hepatic cell, starting at the periphery of the lobule and progressively involving all hepatic cells. Depending upon the duration of the disease, the injury to the hepatic cells may or may not be followed by fibrosis. The fibrosis is usually discrete, but it insidiously proliferates between every cell of the liver parenchyma. As might be expected, the incidence of cirrhosis is high in countries where kwashiorkor prevails, yet the consumption of alcohol is low.

Pathogenesis

When children in underprivileged countries are withdrawn from breast feeding, they are put on a poorly digestible, low-protein diet. Their diet consists mainly of boiled or steam-cooked rice, millet, and maize. Peanut oil is the only supplement of such an unbalanced diet, and the child never eats eggs, milk, or meat. If kwashiorkor is the consequence of a deficiency in essential amino acids, it seems likely that the deficiency concerns most essential amino acids and is the consequence of protein deficiency. The symptoms of kwashiorkor are very much the same throughout the world, yet the composition of the diet varies considerably from one country to another. A study of the amino acid composition of the diet in areas where kwashiorkor is prevalent would undoubtedly help to elucidate the pathogenesis of the disease.

Nitrogen absorption is difficult to determine in patients with kwashiorkor. Nitrogen intake can easily be overestimated because of vomiting and spilling during feeding; on the contrary, the amount of nitrogen lost in feces and urine is often underestimated because of loss during collection. Consequently, systematic errors are likely to be introduced in these balance studies. Therefore, safe conclusions regarding nitrogen absorption in patients with kwashiorkor cannot be reached. But such results would be of interest for two reasons: they would provide information on the rate of overall protein synthesis in these patients, and they would also indicate which proteins are preferentially used. Failure to digest specific proteins may be significant in determining symptomatology [48–59].

Such balance studies have suggested that although the ability to absorb nitrogen from highly digestible proteins (milk) is not altered in kwashiorkor, the use of plant proteins (corn and beans) is very poor. When a protein diet is administered to children with kwashiorkor, the nitrogen retention rises steadily, indicating that such infants have a normal protein absorption. Nevertheless, the possibility that the children who die malnourished go through a stage, before death, of irreversible nitrogen absorption cannot be excluded.

The distribution of amino acids in the blood and urine of patients with kwashiorkor is modified. The total amino acid content of the serum is decreased. The reduction affects some amino acids more than others. For example, the serum is low in arginine, leucine, and threonine, but has a normal content of phenylalanine and tyrosine. The decrease in the amino acid content of the plasma varies considerably with the individual. Mexican authors have claimed that the total amino acid content of the plasma may drop to half of the normal values.

In contrast to what happens in plasma, the amino acid content of urine is increased. Increased amounts of β-aminoisobutyric acid, ethylalanine, isoglutaric acid, taurine, and histidine have been reported. The increase results from reabsorption failure of the kidney, either because of renal damage or because the amount of amino acid in the glomerular filtrate exceeds tubular thresholds. Neither of these interpretations is quite satisfactory since there is usually no renal damage in kwashiorkor, and the amino acid level is reduced in plasma.

Ethylalanine excretion can be reversed by administering methionine; therefore, increased excretion of ethylalanine appears to result from the lack of methyl groups. The total urinary nitrogen content is increased in patients with kwashiorkor. This results from increased levels of urinary amino acids. But the increase in urinary amino acid cannot explain the overall increase in urinary nitrogen, and it has been suggested that part of this increase results from increased purine excretion. The structure of the excreted purine is not known.

The protein deficiency in kwashiorkor is likely to be responsible for reduced synthesis of key functional structural proteins, such as immunoglobin, hemoglobin, myoglobin, and others. Although protein deficiency plays an important role in the development of kwashiorkor, the interpretation of the pathogenesis of the various symptoms of kwashiorkor is often complicated by concomitant vitamin deficiencies and infections. Reports from India indicate that the incidence of kwashiorkor increases during the monsoon season, a period during which the incidence of diarrhea is also known to increase considerably. Diarrhea is usually treated by depriving the patient of food in general, and by the administration of starchy foods. Such a treatment accentuates the symptoms of existing (and precipitates the development of latent) kwashiorkor. Therefore, it is sometimes difficult to decide whether diarrhea precipitates the onset or is the result of the development of kwashiorkor. The deficient diet modifies the intestinal flora and predisposes the patient to gastrointestinal infection (bacterial and helminthic). Furthermore, because of the victim's debilitated state, a gastrointestinal infection that would be benign in a well-nourished child is fatal to a patient with kwashiorkor. The lack of resistance to infection results in part from inadequate antigenic response. This has been established among malnourished individuals in Germany between 1939 and 1945. A lack of antigenic response toward Salmonella typhus has been observed in malnourished children in Brazil. In Mexico, debilitated children were found to have diminished production of diphtheria antitoxin.

In kwashiorkor, the total level of circulating hemoglobin is reduced, probably as a result of protein deficiency. Occasionally, the low-protein diet is also associated with folic acid deficiency. The anemia is usually of the normocytic-normochromic type, except when folic acid deficiency is also present, in which case the anemia is megalocytic, but it does not disappear with the administration of vitamin B_{12}.

The changes in hair texture and color in patients with kwashiorkor remain unexplained. A deficiency in sulfur amino acid was postulated as the cause of changes in the hair. However, when cysteine content of hair of affected patients was compared with that of normal hair, contradictory results were obtained. Some investigators find that the cysteine content is decreased in kwashiorkor; others find no changes. Another possibility is that the deficiency responsible for the hair changes in kwashiorkor is similar to that observed in phenylketonuria, in which lack of tyrosine causes the light pigmentation of the hair.

There is no correlation between the severity of the protein deficiency and the incidence and severity of fatty livers, which vary from place to place. For reasons that are unknown, fatty livers occur frequently in Jamaica. The fatty livers disappear only slowly after the administration of a protein diet. Changes in the pancreas correlate with the presence of fatty livers, and pancreatic changes are absent when fat deposition in the liver is minimal.

Mental retardation may well result from protein deficiency and be the consequence of apathy affecting behavior, particularly language. However, there is no rigid correlation between mental deficiency and the severity of kwashiorkor, and the possibility cannot be excluded that mental retardation results from causes other than low protein intake.

In evaluating the changes in protein content in patients with kwashiorkor, one must be careful not to confuse an absolute with a relative loss of protein. A relative loss of protein occurs in the presence of edema. Although in such cases the total amount of protein doesn't change, an apparent decrease in protein content is found if the protein concentration is expressed per gram of tissue. Similarly, an apparent increase in protein concentration is observed in tissues in which large amounts of fat are lost. Both edema and loss of body fat occur in kwashiorkor. Therefore, to evaluate the protein content of tissues accurately, it is necessary to express the protein concentration in function of some constant components. The DNA of undividing cells (liver, kidney, muscle, brain) is more satisfactory for that purpose. In liver and kidney, the protein content decreases in both the parenchymal and nonparenchymal cells. In general, the decrease is greater in the parenchymal cells than in fixed tissues.

Changes in activity of some enzymes have also been measured in the liver of patients with kwashiorkor. The results are somewhat discouraging and have shed little light on the pathogenesis of the disease. The activities of a number of enzymes (cholesterase, cytochrome oxidase, several dehydrogenases, transaminases) are unchanged. Catalase and alkaline phosphatase activities have been reported to be increased, and succinoxidase activity of liver is decreased. The loss in enzyme activity is greater in liver than in kidney. There is no loss of succinoxidase activity in heart, muscle, or brain.

Early malnutrition in rats and humans, however, interferes with cellular proliferation in the developing brain, delays cell migration and retards myelin synthesis [155].

More important are the changes in the enzyme content of the pancreatic juice. During the seige of Budapest, lipase, trypsin, and amylase activities of the pancreatic juices were measured in several individuals. No milk was available during the seige. A marked reduction of the activity of these three enzymes was observed. This finding, which was later confirmed under different conditions of protein malnutrition by other investigators, correlates with the atrophy of the exocrine pancreas observed in patients with kwashiorkor. Yet protein malnutrition is the only form of malnutrition in which this drop in pancreatic enzymes occurs. Attempts to find such enzymic changes in patients with general malnutrition have failed.

The protein content of muscle is also reduced in kwashiorkor, and the same proportion is lost in the sarcoplasm and the fibrillae. In contrast, there is a relative increase in muscle collagen. Inasmuch as creatinine is formed mainly from muscle proteins, measurement of the creatinine content in the urine is an indirect measurement of the loss of muscle mass. The increase of creatinine output correlates with the muscle thickness (calculated from measurements of limb circumference and skinfold thickness). In fact, the increase in creatinuria is $1^1/_2$ times greater than the loss in body weight. These results indicate that the loss in muscle protein can be greatly underestimated if this loss is evaluated on the basis of changes in the body weight.

The exact causes of the water retention in protein malnutrition are not clear. It has been suggested that the wasting of cellular tissues provides more places for water, and therefore contributes to the edema. In malnutrition, the hypoproteinemia is about 45 g/100 ml (albumin 1.7 g/100 ml); this suggests that the Starling law of reduced colloid osmotic pressure in the plasma may apply in protein malnutrition. However, it is believed that hypoproteinemia is a modifying rather than a determinant factor in water retention due to malnutrition. Potassium depletion may lead to the water accumulation; however, the potassium loss in gastroenteritis (a disease usually not associated with edema) is more severe than that observed in malnourished children.

Anomalies in the secretion of the antidiuretic hormone and aldosterone have been implicated in the pathogenesis of this protein-deficient edema. However, little convincing information is clinically available, and the experimental data are still controversial. An increase in the available amounts of antidiuretic hormone has been suggested to occur in malnutrition as a result of the inability of the liver to break down the antidiuretic hormone.

Blood carbohydrate metabolism alterations occurring in protein deficiency have never been studied extensively. Three findings in animals are pertinent: (1) blood glucose levels drop from values of 60 mg/100 ml under normal conditions to 13–34 mg/100 ml under conditions in which a sort of hypoglycemic coma precedes death; (2) the glucose tolerance curve to carbohydrates is increased (this is suggested by the glucose tolerance test); and (3) liver glycogen increases. Hypoglycemia has been suggested as the cause of death in these animals, but this is improbable because the administration of glucose does not increase survival.

The changes in lipids include a reduced level of plasma lipids, greater for the α-lipoproteins than for the β-lipoproteins, and a rapid increase in plasma lipids during the period of restoration. It has been suggested that this rapid increase results from mobilization of the lipids from the liver. The question of lipid absorption is unresolved; some believe it is markedly decreased, but others see no significant changes.

The exact concentration and distribution of the amino acids within the cell are not known. These data would be valuable because it has been claimed that in potassium loss, there is a loss in dicarboxylic acids and an increase in free basic amino acids. As a matter of fact, lysine has been claimed to substitute for potassium in potassium losses. In protein malnutrition, a disproportion in the amino acid distribution in the cell would result, thus interfering with protein synthesis. The study of overall protein synthesis using labeled amino acid is full of pitfalls. Indeed, the rate of incorporation of an amino acid into a protein may reflect the rate of circulation in that given tissue, the rate of absorption of the amino acid in the cell, the rate of synthesis of a given protein. If the total amino acid pool is not known, it is difficult to evaluate how a specific amino acid has been diluted by preexisting unlabeled amino acid. Moreover, the rate of the protein breakdown also interferes with the determination of protein synthesis by isotope studies if kinetic studies are not included. The body pool of a given compound is the total amount of that compound in the body which is available for exchange with additional or newly synthesized molecules of that specific compound. The turnover half time of a compound is the amount of time required for half of the body pool of that protein to be lost. The results on protein turnover in kwashiorkor are still controversial. However, a working hypothesis has emerged, enouncing (1) that in protein malnutrition, there is a greater tendency to use the amino acid liberated by tissue catabolism than there is in the normal individual; (2) the overall rates of protein synthesis have changed, and protein synthesis is now restricted to more essential organs and more essential components in a given organ. Some animal experiments are relevant to the biochemical changes that occur in kwashiorkor.

The effect of a diet free of one of the essential amino acids has been studied in rats. When the animals are simply presented with a diet devoid of one of the amino acids, the rats regulate their food intake accordingly by reducing consumption. If the diet devoid of one of the essential amino acids (except arginine) is force-fed, the rats develop a malnutrition syndrome that has sometimes been compared to that observed in human kwashiorkor.

The response to the dietary challenge varies in these animals depending upon whether it is investigated in muscle or liver. Protein synthesis is decreased in muscle but increased in liver. However, the increased rate of protein synthesis in liver is selective. For example, the level of serum albumin synthesized by liver is unchanged, whereas that of glucoproteins is markedly increased. This increase in protein synthesis in associated with an elongation of the size of polysome chains, acceleration of RNA synthesis, and an increase in RNA polymerase activity. The meaning of these observations remains to be understood.

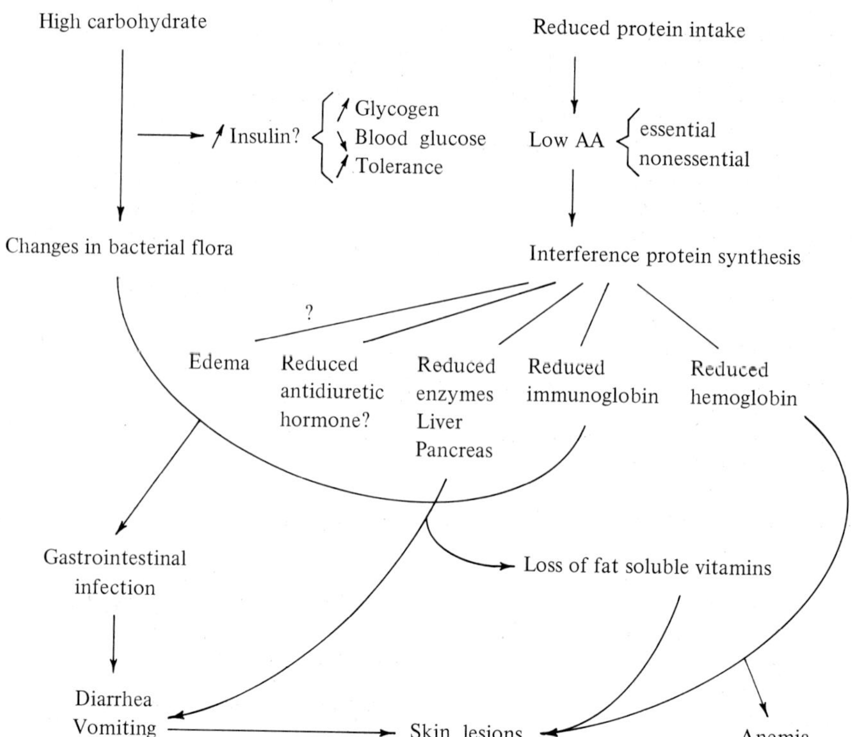

Fig. 4-4. Pathogenesis of kwashiorkor. ↗ = increase; ↘ = decrease

Allen and collaborators measured the effect of changes in the amino acid composition of the diet. The "charging levels" were found to be 100% in fed as well as in fasted animals. When the animals were force-fed a diet deficient in one amino acid (x), the amount of acyl xtRNA was markedly reduced. Tryptophan deficiency is one of the essential amino acid deficiencies that causes the most serious decrease in the levels of tryptophanyl tRNA. Whether this observation is related to the deleterious effect of tryptophan deficiency and in particular to its effect on liver polysome aggregation remains to be seen [60].

The intricacies of the biochemical changes that take place in kwashiorkor are not known, and the pathogenesis can only be hypothesized. The proposed steps in the pathogenesis of kwashiorkor are summarized in Fig. 4-4.

Vitamin Deficiency

Beriberi and Thiamine Deficiency

A syndrome called beriberi, characterized by edema and polyneuritis, has been known for a long time. Although beriberi is a multifactor deficiency, thiamine deficiency is mainly responsible for the disease.

Thiamine deficiency exists in an endemic and sometimes an epidemic form in countries of the Far East where diet is based on polished rice. Remarkable experiments performed by a Dutch physician in 1889 demonstrated that chickens fed polished rice present a neuroparalytic syndrome similar to that described in cases of beriberi. Already in 1883, Takaki, a Japanese, protected his countrymen against the disease by adding small amounts of meat to the pure rice diet. Twelve years later, Funk extracted the curative principle—thiamine—from the husk of rice. Jansen and Donath crystallized the chloride form of thiamine in 1926 [61]. Williams and his group elucidated the structure of the vitamin and synthesized it completely.

Thiamine is found in yeast, rice husk, and wheat germ (which contains more than 1 mg/100 g), but many fruits and vegetables also contain thiamine (vitamin B_1). Freezing does not affect the vitamin, but prolonged cooking destroys a large portion of a food's vitamin content. Small amounts of vitamin B_1 are also found in meats, but liver and kidney (called *rognon* in French culinary jargon) are particularly rich in thiamine. In contrast, only small quantities of the vitamin are contained in milk and eggs. Because the diet in the Western world is rich in thiamine, only a few forms of the vitamin deficiency are observed. These are usually associated with a general nutritional deficiency, as is seen in alcoholism or in newborn children fed an exclusive milk diet. Thiamine content of breast milk depends on the individual diet. Indeed, in poorly fed Indian women, breast milk averages 12.2 mg of thiamine in 100 ml of blood, whereas it averages 25 mg/ml in American or English women.

Of course, diet is not the only source of vitamine B_1. Some normal intestinal saprophytes, like the *Bacillus bifidus*, synthesize large amounts of vitamin B_1 which are then reabsorbed in the intestine. Therefore, the elimination of bacteria from the gastrointestinal tract by severe diarrhea or by prolonged administration of antibiotics may induce thiamine deficiency. The mechanism of thiamine use is not quite clear. Indeed, Wernick has observed that humans excrete less thiamine when a diet made up exclusively of potatoes provides one-third of the vitamin than when bacon is included in the potato diet.

Thiamine deficiency can also be produced bacterially. Numerous bacteria, such as the *Bacillus thiaminolyticus,* contain a thiaminase that breaks thiamine either by catalyzing a base transfer or by splitting the pyrimidine portion from the thiazole moiety of the thiamine. The disease, first observed in cats, chickens, and foxes, has also been recognized in humans with beriberi in Japanese clinics. In such cases thiaminase is in the distal portion of the intestinal tract, and the disease can be cured by the administration of sulfanilamides or antibiotics followed by the administration of lactic acid bacteria. The agent responsible for the human disease is either *Bacillus thiaminolyticus* or *Clostridium thiaminolyticus*. Antibiotics may stimulate growth in some animals by interfering with the growth of bacteria that secrete thiaminase, thus it is interesting that penicillin given orally is more efficient than penicillin given parenterally.

A case of thiaminase disease due to a fungus has been reported. The agent was *Trichosporon aneurolyticum* which grew in the patient's mouth. Experimentally, the administration of thiamine analogs, such as pyrithiamine and oxythiamine, may also produce thiamine deficiency. The mechanism by which they act is discussed later (see formulas in Fig. 4-5).

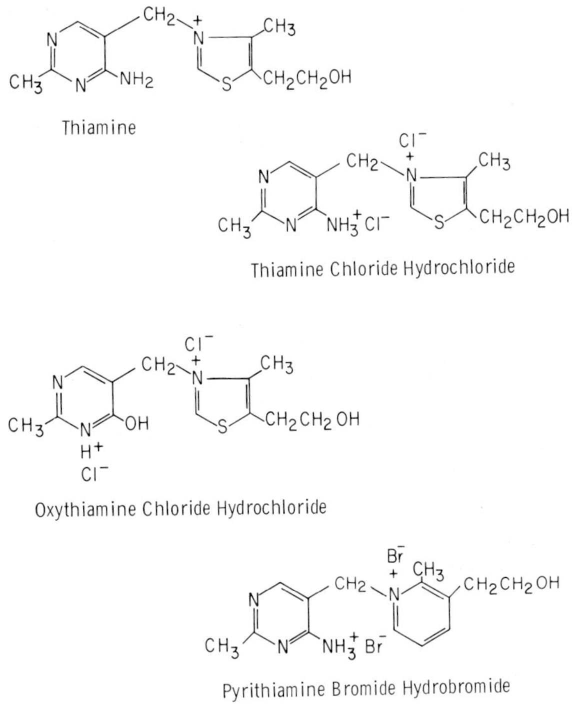

Thiamine

Thiamine Chloride Hydrochloride

Oxythiamine Chloride Hydrochloride

Pyrithiamine Bromide Hydrobromide

Fig. 4-5. Thiamine forms and analogs

Clinicopathological Aspects of Beriberi

A variety of clinical forms of beriberi have been described, but the disease is usually classified as wet or dry. The wet form is usually a combination of edema (probably a consequence of congestive heart failure) and polyneuritis [62–64].

The first manifestation of the deficiency is a mild form of neurosis with irritability, neurasthenia, loss of memory, uncoordination, and inability to concentrate. These ill-defined neurological symptoms soon develop into a severe polyneuritis, which usually starts in the lower limbs. The sensory nerves are affected first by hyperesthesia, paresthesia, loss of vibration, and loss of the superficial sensations. Later, motor anomalies appear—such as weakness of tendon reflexes, cramps (especially in the calf muscles)—and paralysis occurs, followed by muscular atrophy. If, as is often the case among younger victims, especially in the Far East, the disease is associated with heart enlargement, the enlargement is usually more marked on the right than on the left side of the heart. Enlargement is due mainly to dilatation. Hypertrophy is rare but is seen occasionally. Histological examination of the myocardial fibers shows loss of muscular striation and hydropic and fatty degeneration. The dilatation of the heart leads to lowered blood pressure and cardiac failure resistant to cardiotonics.

In classical cases, the peripheral nerves, cranial nerves, and nerves of the autonomous system are all affected. There are no changes at gross examination, but microscopically fragmentation of the axis cylinder with demyelination and a typical wallerian degeneration are observed. Thus, the nerve degeneration begun at the periphery progressively reaches the neuron. Obviously, if therapy is started before the neuron has degenerated, the injuries caused by thiamine deficiency can probably be reversed. The changes in the central nervous system are generally inconspicuous and are characterized by demyelination, particularly at the level of the fibers of the posterior columns, and chromatolysis of the ganglion cells.

At least two factors determine which part of the motor system is affected in thiamine deficiency: the length of the nerves and the degree of activity of the muscle involved. For instance, the left vocal cord, which is supplied by the longest of the two laryngeal nerves, is more frequently paralyzed in thiamine deficiency than the right. The professional activities of the victim play an important role in determining which limb is affected by polyneuritis in thiamine deficiency.

The cause of the peripheral nerve degeneration is not clear. No alteration in the peripheral nerves was found after prolonged thiamine deficiency associated or not with pantothenic deficiency. Furthermore, Follis [64] claims that in experimental deficiencies in animals, all lesions of the peripheral nerves can be explained on the basis of inanition.

Muscular degeneration and mild hyalinization of the muscles have also been described. In addition to the alteration described in the nervous system and muscles, changes in the endocrine glands frequently observed in experimental thiamine deficiency include hypertrophy of the thyroid, anterior hypophysis, adrenal, and islet of the pancreas.

Of course, the thiamine content of the blood is reduced in thiamine deficiency, but the most conspicuous alteration is an increase in blood pyruvic and lactic acids. Thiamine deficiency may be diagnosed by measuring blood pyruvic and lactic acid levels. However, the acids accumulate only in severe beriberi and usually cannot be measured in mild deficiencies observed in the Western countries. Therefore, a pyruvate tolerance test was devised which measures the increase in blood pyruvate after administration of a large amount of glucose. The glucose is oxidized to pyruvate, which then accumulates in the blood of the thiamine-deficient patient. The test loses some of its diagnostic value because of its unspecificity. Several forms of polyneuritis also lead to abnormally high blood pyruvic acid content. The reasons for this are not understood, but it is often assumed that in those cases, polyneuritis is due to the action of toxic agents that affect the pyruvic acid decarboxylase.

Pathogenesis of Beriberi

The thiamine molecule contains two cyclic rings—a pyrimidine and a thiazolium—linked by a methylene group and carrying side chains to make the formula: 3,4-amino-2-methylpyrimidyl-5-methyl-4-methyl-5-hydroxyethyl thiazolium hydrochloride.

Thiamine is the precursor of the coenzyme cocarboxylase, which is the pyrophosphate derivative of the vitamin (see Fig. 4-6). Although thiamine biosynthesis has been extensively investigated in yeast and bacteria, little is known of the anabolism and catabolism of the coenzyme in mammals.

The biosynthesis of the vitamin was studied with purified baker's yeast enzymes. The pyrimidine and thiazole rings are synthesized separately and are then complexed to yield thiamine monophosphate as the final product. Enzymes capable of catalyzing the condensation of the thiazole and the pyrimidine moieties have been found in rabbit liver and kidney.

In humans, thiamine is absorbed in the duodenum and the small intestine by a mechanism still unknown. However, it has been suggested that the absorption of thiamine involves an active process.

Rat liver extracts use thiamine rather than thiamine monophosphate for the formation of thiamine pyrophosphate. Therefore, the formation of coenzyme appears to be a one-step reaction between thiamine and ATP. This assumption is supported by experiments using ATP labeled in the terminal phosphate. When such a precursor is added to the incubation mixture, thiamine pyrophosphate labeled in the terminal phosphate only is formed. A thiamine pyrophosphokinase has been purified 120-fold from rat liver. It is activated by Mg^{++}, Mn^{++}, Co^{++}, Ni^{++}; UTP and ATP (but not GTP, ITP, or CTP) act as pyrophosphate donors.

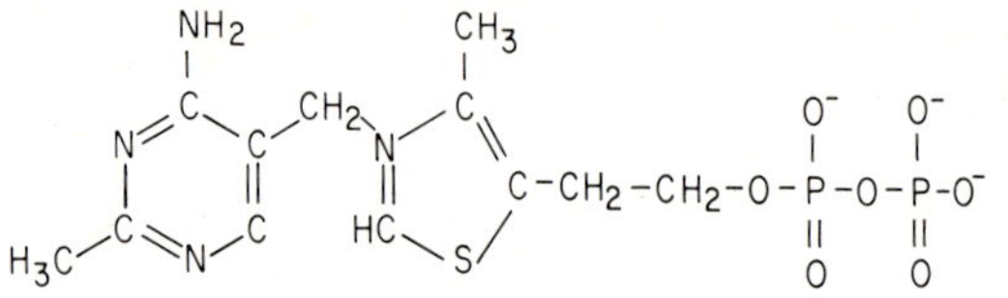

Fig. 4-6. Thiamine pyrophosphate

The content of phosphorylated derivatives of thiamine has been investigated in a variety of tissues using Dowex-1 ion-exchange columns. The following proportions of thiamine and thiamine mono-, di-, and triphosphates were found in rat brain, heart, liver, and kidney: thiamine 5%, monophosphate 10%, diphosphate 80%, and triphosphate 5%. The high levels of diphosphate are not surprising—diphosphate is the form in which the vitamin is active. But the presence of triphosphate remains unexplained because the role of triphosphate is not clear, and it has been suggested that triphosphate appears only when excessive amounts of thiamine are administered.

Thiamine pyrophosphate serves as a coenzyme in several biochemical reactions. Among them are the decarboxylation of pyruvic and α-ketoglutaric acids, and the transketolase reaction.

Before the molecular mechanism of the pyruvic decarboxylase reaction is discussed, it might be appropriate to summarize briefly modern concepts on the mechanism of action of coenzymes. Modern research in that field suggests that the coenzyme functions in biochemical reactions by virtue of its unique molecular properties, and that the apoenzyme acts mainly as a basic or acidic catalyst and provides a structural skeleton that brings substrate and coenzyme in close contact.

To elucidate the molecular mechanism of action of coenzymes, biochemists often use chemical models. The first such model in the case of thiamine was based on the observation that heat decarboxylates pyruvic acid, and that amines catalyze this reaction. However, thiamine is not the most effective catalyst in this type of reaction.

Mizuhara [65] developed a more adequate model by demonstrating that thiamine catalyzes the decarboxylation of pyruvic acid in basic aqueous solution (pH 8.8). Acetoin is the final product of the reaction. Breslow [66] later showed that the hydrogen in position 2 of the thiazole ring of the coenzyme is exchanged with deuterium when deuterium oxide (D_2O) is added to the incubation system. Thus, carbon 2 of the thiamine appears to react in this chemical process. The hydrogen in position 2 is acidic and is thus readily ionized to an anion in basic media (see Fig. 4-7). On the basis of these findings, researchers have proposed the following sequence of reactions to explain the catalytic action of carboxylase. The carbon 2 of the thiazo-

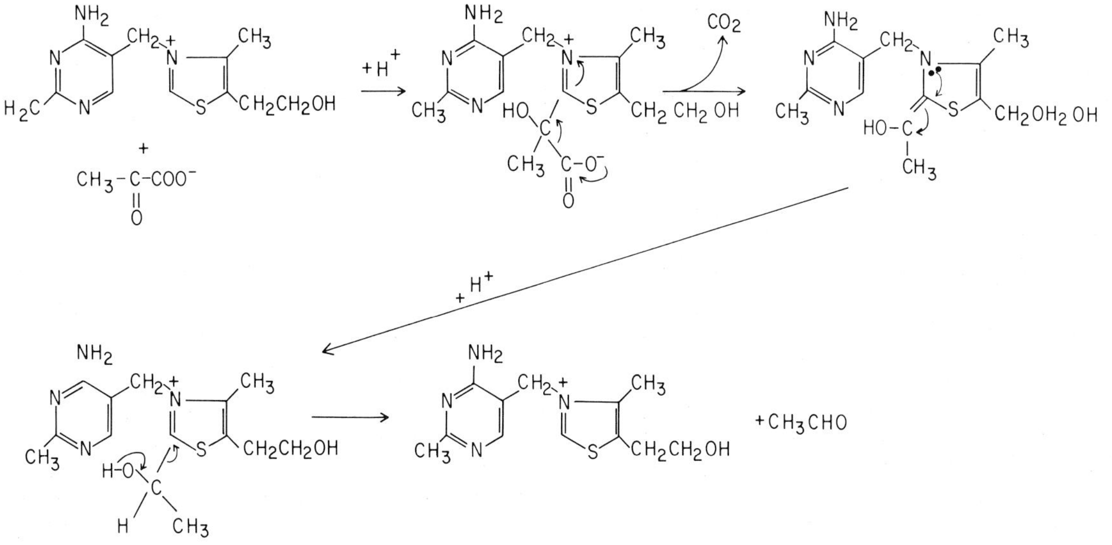

Fig. 4-7. Hydroxyethyl thiamine (HET): active acetaldehyde

line ring is involved in a ketol condensation with pyruvate to form a pyruvic-TPP complex, which is then decarboxylated to yield a hydroxyethyl intermediate that is split to yield acetaldehyde and thiamine pyrophosphate (see Fig. 4-8). Hydroxyethyl thiamine pyrophosphate is a key compound in this reaction. The existence of this intermediate has been established in at least two ways: (1) by synthesizing the intermediate and using the intermediate instead of thiamine for bacterial growth; and (2) by isolating the active acetaldehyde from incubation mixtures and characterizing it by chemical methods [67].

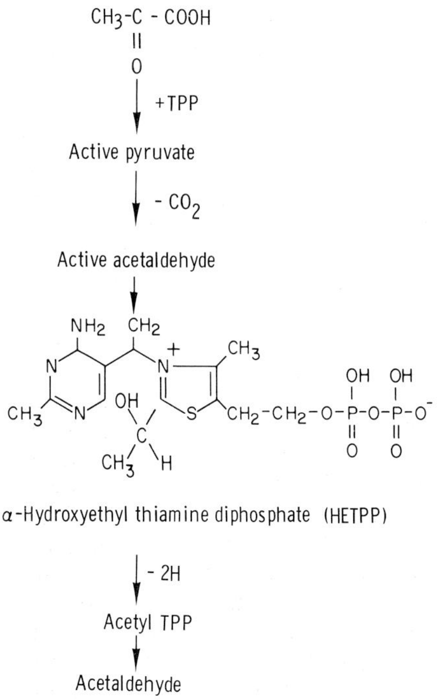

Pyruvic decarboxylase controls the entry of the end products of glycolysis into the Krebs cycle. Therefore, thiamine deficiency must have dramatic consequences if no alternative pathway is available for pyruvic acid oxidation. Understandably, in the absence of an alternative pathway, thiamine deficiency leads to a block of pyruvic decarboxylation, which is the first of the two reactions of the Krebs cycle requiring thiamine. In addition, half of the thiamine content of the brain is used in that reaction. The maintenance of the integrity of the Krebs cycle is probably more important to the cell than that of the hexose monophosphate shunt.

The most logical explanation for the increase in blood pyruvic acid is that the increase results from the blocking of pyruvic decarboxylase. However, in experiments in which blood pyruvic acid levels in thiamine-deficient animals were correlated with the changes in the adrenals, it was concluded that the increase in blood pyruvate can be correlated more closely to adrenal hypertrophy than to thiamine deficiency. Furthermore, in thiamine-deficient adrenalectomized and hypophysectomized animals, pyruvemia was much less than in the controls. Finally, cortisone administration increases the level of blood pyruvate. Consequently, pyruvemia in thiamine-deficient animals may result from immediate stress caused by the deficiency and from progressive block of pyruvic decarboxylase due to the slow depletion of coenzyme.

The fact that rats maintained on a carbohydrate-free diet can survive for a long time, even if their diet is completely devoid of thiamine, demonstrates that the main effect of thiamine deficiency is on carbohydrate metabolism. As soon as 5–10% carbohydrates are added to the diet, polyneuritis and death occur. On the contrary, if sorbitol is added as a source of carbohydrate, the animals survive, and this is

CH_3-C-COOH (C=O)

↓ +TPP

Active pyruvate

↓ $-CO_2$

Active acetaldehyde

↓

α-Hydroxyethyl thiamine diphosphate (HETPP)

↓ -2H

Acetyl TPP

↓

Acetaldehyde

Fig. 4-8. Nonoxidative decarboxylation of pyruvate

explained by the stimulation of growth of an intestinal bacteria flora capable of thiamine synthesis.

α-Ketoglutarate decarboxylase inhibition was demonstrated in heart and breast muscle in pigeon, and it is thus not surprising that keto acids other than pyruvic acid accumulate in the blood of thiamine-deficient animals.

The changes in α-ketoglutarate decarboxylase and pyruvic acid decarboxylase activities are not the only biochemical alterations observed in thiamine deficiency. The activities of some other enzymes were found to be altered, for example, erythrocyte transketolase and adenosine-5′-phosphatase activities are decreased in chicken brain.

Beriberi is associated with low transketolase activity in the blood, and determination of transketolase activity may be a helpful diagnostic tool in detecting mild forms of thiamine deficiencies. The fact that adding thiamine pyrophosphate to the system restores transketolase activity confers specificity to this procedure. Although the decrease in transketolase activity is easy to explain in view of the role that thiamine plays in that reaction, the mechanism of the alteration in adenosine-5′-phosphatase is not so obvious. The total activity is decreased despite a slight increase of adenosine phosphatase in the nuclei, and the decrease in total activity results from a marked decrease of adenosine phosphatase in the fibers. Increased thiamine pyrophosphate and alkaline phosphatase have also been observed in the optic tectum of the deficient chicken, but these changes remain unexplained.

An explanation for the pathogenesis of the lesions observed in thiamine deficiency would seem to follow logically from these biochemical observations, for in the thiamine-deficient animal, at least two enzymes involved in the Krebs cycle are blocked. The block of pyruvic decarboxylase prevents the entry of the products of glycolysis into the Krebs cycle. The block of α-ketoglutarate decarboxylase restricts the oxidation of both carbohydrates and fatty acids. A severe metabolic distortion follows, and one of the main manifestations of the distortion is a reduction of the amount of chemical energy available in the form of ATP. Clearly, those organs that suffer the most from such alterations are those that are metabolically most active, and the heart and the peripheral nervous system surely qualify as such.

Unfortunately, such a reasoning constitutes a gross oversimplification. First of all, if the lesions result from reduced chemical energy, why don't pantothenic acid (a precursor of coenzyme A), nicotinic acid (a precursor of NAD), and riboflavin (a precursor of the flavin nucleotide) lead to similar lesions?

Also, how the block in pyruvic acid decarboxylation and α-ketoglutaric decarboxylation leads to the morphological changes observed in muscle and in the nervous system and why the injuries are restricted to those organs remain to be explained. The detailed mechanism of the development of these morphological lesions, which appear to be distant consequences of the block in the decarboxylation reaction, has not been investigated. Do the degenerative lesions of muscle and nervous system result from a metabolic adaptation in which the smaller components of the constituent proteins and lipids, such as amino and fatty acids, are thrown into the metabolic pool as a vicarious source of energy? Is energy so necessary for the maintenance of cellular structure that a deficient supply of it leads to cellular disruption? Does the metabolic block lead to the accumulation of toxic compounds, pyruvate, α-ketoglutarate, and methylglyoxal, or does it deprive the tissues of some metabolites which they greatly need?

We have seen that the morphological lesions are restricted to the muscle and peripheral nervous system. Other organs, metabolically very active, such as liver, pancreas, and brain, remain unaffected. Again, there is no definite answer to such selectivity. It could be explained in many ways: in a given organ, this selectivity could depend upon the amount of vitamin present at the time of onset; the activity of the enzymes involved in the breakdown of the vitamin; or the mechanism of elimination of the vitamin from the organ. But it is also possible that in some of these organs, alternative pathways for pyruvic acid and α-ketoglutaric decarboxylation are available. Pyruvic acid may enter the Krebs cycle by bypassing the decarboxylation. These reactions involve CO_2 fixation; they include the Wood-Werkman reaction in which pyruvic phosphate combines with CO_2 in the presence of NADP to yield malic acid. Although the equilibrium of the latter reaction is slightly in favor of the breakdown of malic acid to yield pyruvic acid and CO_2, the free energy of the reaction is so low (–80 calories) that it obviously is readily reversible.

The inherited deficiencies in pyruvate decarboxylase described by Blass and his collaborators [156] might shed some light on the pathogenesis of thiamine deficiency. The most striking abnormality in the patients affected by the enzyme defect is cerebellar ataxia. The authors found that the activity of the enzyme varied from one portion of the cerebellum to another. The enzyme activity was much lower in the rostral cerebellar vermis and rostral cerebellar hemisphere than in any of seven areas of brain gray matter examined. Significantly, in patients with the Wernicke-Kozsokoff syndrome (see alcoholism) in which alcoholism is associated with thiamine deficiency, the anatomic lesions affect the rostral cerebellar vermis with occasional lesions in the rostral cerebellar hemisphere. These findings suggest that those areas of the brain containing the lowest levels of activity of pyruvate decarboxylase are those most affected in thiamine deficiency.

An example of the difficulties encountered in evaluating the role of a vitamin deficiency is provided by those cases in which thiamine deficiency and type I hyperoxaluria are combined. Type I hyperoxaluria is caused by a deficiency in α-ketoglutarate: glyoxylate carboxyligase; thiamine acts as a cofactor in the reaction and one would expect, therefore, that thiamine deficiency would enhance the severity of the disease. Yet, in patients with Wernicke encephalopathy and type I oxaluria there is no increase in the incidence of oxaluria [157].

Fortunately beriberi victims did not have to wait to be saved from their ill fate until the coordination of morphological and biochemical research unraveled the pathogenesis of the disease. But despite the fact that an effective therapy for beriberi now exists, the pathologist must solve the puzzle of its pathogenesis because such information would undoubtedly improve diagnostic methods and may help us understand the pathogenesis of other diseases that have similar morphological alterations, such as demyelination.

Pellagra and Nicotinic Acid Metabolism

Even though the physicians of the Middle Ages did not recognize pellagra, the disease was later known in Spain, and in 1735 Casal described it under the name "mal de rosa." He suspected that the disease was due to a nutritional deficiency. The Italians were responsible for calling the disease "pelle agra," which means rough skin.

Nicotinic acid was known to play an important role in electron transport much before its role in the therapy of pellagra was suspected. When Funk isolated the group of B vitamins, he suspected that nicotinic acid deficiency was an important factor in the genesis of pellagra. In his remarkable experiments on the black tongue of dogs, Goldberger demonstrated that the disease was due in part to the absence of vitamin PP in the diet. In 1938 Elvehjem and his associates [68] isolated nicotinic acid from a liver extract that cured the black tongue observed in experimental pellagra. During the same year, the therapeutic properties of nicotinamide, or nicotinic acid, were demonstrated in man.

Clinicopathology of Pellagra

Pellagra is a multivitamin deficiency, called the 3-D disease because it is characterized in its culminating form by the appearance of dermatitis, dementia, and diarrhea. In other words, the vitamin deficiency is marked mainly by alteration of the superficial teguments and nervous tissue.

Initially, an erythema usually appears in localized patches in areas of the skin predominantly exposed to the sun. The erythema is often followed by erosion or ulceration of the skin. When niacin is reintroduced in the diet either spontaneously or therapeutically, the erythema disappears, and previously erythematous areas often become pigmented. In countries where pellagra exists in a chronic form, erythema appears every spring, then disappears. After a certain number of years, it leaves the skin thickened, roughened, and pigmented in areas affected by the disease. Microscopic examination of the skin shows alteration of the epithelium and derma. The epithelium reveals acanthosis, parakeratosis, hyperkeratosis, and increased amounts of melanin at the base of the epithelium. The connective tissue is edematous, the collagen fibers are swollen, and the capillaries are dilated. There is little lymphocytic infiltration of the corium.

The mucosa most frequently affected in pellagra are those of the digestive and genitourinary tracts. Stomatitis, gastritis, and enteritis affect the digestive tract. The mouth, tongue, and pharynx are red. Cracks appear, especially at the junctures of the lips, and small, greyish ulcerations develop in several areas of the mouth. Frequently, tonsilitis occurs as a superinfection by the bacillus of Vincent.

The gastric mucosa is hyperemic and edematous. Achlorhydra is associated with the alteration of the mucosa in 50% of pellagrins. Similar lesions occur in the intestines (mainly in the colon) where ulcerations are present; 75% of patients develop bloody diarrhea. Microscopically, the fundamental skin injuries are also present in the gastric mucosa. The mucosa is atrophied and connective tissue is congested and edematous.

The lesions of the nervous system start in the cortex and progressively invade the midbrain and the medulla. The first symptoms are paresthesias, pain in the back, vertigo, headaches, general fatigue, and sensorial perversion associated with depression, melancholy, and suicidal tendencies. These symptoms may develop into a more characteristic psychiatric syndrome, with hallucinations, schizophrenia, and maniac dementia. At autopsy, the brain reveals edema, congestion of the cortex, and loss of ganglion cells. The neurons show chromatolysis.

In long-standing cases of vitamin deficiency, pellagra may also affect the spinal cord; the lateral and posterior columns are demyelinated leading to ataxia, altered tendon reflexes, and loss of the sense of position.

Nicotinic Acid Deficiency

Nicotinic acid controls the blood lipid level in chicken and several mammals, including man. Nicotinic acid administration prevents the development of cholesteremia that normally occurs after 48 hours of starvation. The mechanism by which the administration of nicotinic acid reduces lipemia is not clear. Existing experimental evidence suggests that nicotinic acid stimulates the oxidation of sterols and lipids and deviates potential precursors of sterols and lipids via the oxidative pathway. Nicotinic acid blocks the conversion of acetate to cholesterol *in vitro* in a step preceeding the biosynthesis of mevalonic acid.

The main dietary sources of nicotinamide are meat, fish, and egg yolks. During World War II, when the restrictions mainly involved these substances, the primary sources of vitamin PP were bread, tomatoes, spinach, and fruit. Some cereals—rice, wheat, bran, and corn—have a low biological activity with respect to niacin activity because these cereals contain niacin in the form of niacinogen. The vitamin is not freed by the gastric or intestinal juices and is therefore not absorbed. Mild alkali treatment of the cereal before consumption releases niacin from the complex and permits its use. The exact structure of niacinogen is not known, but the complex contains a chromophoric moiety and a peptide chain of 17 different amino acids.

Before the role of nicotinic acid was discovered, pellagra existed all over the world, especially in the southern countries of Europe, Rumania, and Egypt. In the United States it was prevalent particularly in the southern states, where it was recognized only in the early years of this century. Usually, the incidence of pellagra is high in countries where the diet contains nothing but corn or maize. In India, where the diet is low in vitamin PP, pellagra is not observed because corn or maize is not a major part of the diet. The reasons for this will become clear later.

The nicotinic acid content in human liver is 15 mg/100 g. Human blood contains 8 mg/liter; a drop to 6 mg/liter of blood is a sign of deficiency.

Nicotinic Acid Metabolism

The mode of action of vitamin PP can be understood only if its site of action in metabolism is clarified. It is therefore necessary to review the biosynthesis of nicotinic acid, its role in the synthesis of NAD, and the breakdown mechanism of nicotinic acid and NAD.

A diet low in nicotinic acid does not necessarily lead to the development of pellagra. If sufficient amounts of proteins are absorbed, the disease does not develop, suggesting that proteins contain a precursor of nicotinic acid. Soon after the discovery of vitamin PP, tryptophan was shown to replace nicotinic acid in the diet.

The conversion of tryptophan into nicotinic acid involves a series of molecular transformations. The original tryptophan molecule not only loses its alanine side chain and its imidazole ring, but it also includes nitrogen in its benzoic ring to form a pyridine ring [69–71].

Research on tryptophan metabolism can be divided into three stages: a first in which the end products of tryptophan metabolism were studied by investigating the composition of urine; a second in which the intermediates of the various metabolic pathways were studied by following the urinary excretion of injected radioisotopes; and a third in which an attempt was made to identify the enzymes involved in tryptophan metabolism. In the middle of the 19th century, it was demonstrated that kynurenine is excreted in dogs' urine. Fifty years later, it was shown that kynurenic acid excretion is stimulated by the administration of tryptophan. A variety of other compounds were also found in the urine, including xanthurenic acid and quinolinic-2-carboxylic acid.

Laborious studies in different laboratories elucidated the pathway used by tryptophan to yield nicotinic acid. The intermediates are outlined in the metabolic map (see Fig. 4-9). The first of these reactions is catalyzed by tryptophan pyrrolase, a heme-containing enzyme acting as a peroxidase when the iron is in the ferric form and as an oxidase when it is in the ferrous form. During the oxidation, hydrogen peroxide is formed. Thus, in the transformation of tryptophan into formylkynurenine, two atoms of oxygen are added directly to the substrate. Although no intermediate has been detected, it is assumed that hydrogen peroxide is formed as an intermediate because the reaction is inhibited by catalase. The reaction is assumed to proceed in two distinct steps. In the first, tryptophan reacts with H_2O_2 to yield an unknown intermediate which is then oxidized in the presence of molecular O_2, and formylkynurenine and H_2O_2 are formed. The H_2O_2 does not directly participate in the reaction, but it converts an inactive ferric form of the enzyme into the active ferrous pyrrolase. As can be anticipated from the mechanism involved, cyanide and catalase inhibit the conversion of ferric to ferrous protein.

Besides its role in nicotinamide synthesis, tryptophan pyrrolase is interesting in another respect. The enzyme is absent in embryonic liver, and in adults hepatic activity is increased by the administration of tryptophan. The developmental biochemistry and the induction of tryptophan are discussed in the section on the biochemistry of growth.

Kynurenine formamidase catalyzes the hydrolysis of formylkynurenine. A variety of aromatic formamides will react, but the enzyme is more active with its natural than with the synthetic substrates. The product of the reaction, kynurenine, may then either be hydroxylated or decarboxylated.

Kynurenine is hydroxylated to hydroxykynurenine by an enzyme (kynurenine-3-hydroxylase) found in rat liver mitochondria. The reaction requires NADPH and molecular oxygen. In the presence of pyridoxal phosphate, hydroxykynurenine is hydrolyzed by an enzyme (kynurenase) found in liver and kidney. The product of this reaction is 3-hydroxyanthranilic acid. The same enzyme catalyzes the cleavage of the side chain of kynurenine to yield alanine and anthranilic acid. Studies made with labeled 3-hydroxyanthranilic acid demonstrated its role as an intermediate of the biosynthesis of nicotinic acid. These studies established that the label of the carbon 3 of 3-hydroxyanthranilic acid is transferred to the α-carbon of quinolinic acid and is lost as $C_{14}O_2$ during the conversion of quinolinic to nicotinic acid. The details of the metabolic conversion of 3-hydroxyanthranilic acid to nicotinic acid are known.

An enzyme catalyzing the oxidation of 3-hydroxyanthranilic acid was found in liver and kidney. The 3-hydroxyanthranilic oxidase has been partially purified from beef liver, and requirements for ferrous ions and sulfhydryl groups have been demonstrated.

In the presence of O_2, the oxidase opens the ring of 3-hydroxyanthranilate and oxidizes both the carbon bound to the hydroxyl group to yield a carboxyl group and the adjacent carbon to yield an aldehyde group. 2-Acroleyl-3-aminofumarate is the final product of the reaction. The nitrogen is then incorporated into a heterocyclic ring of quinolinic acid (or pyridine-2,3-

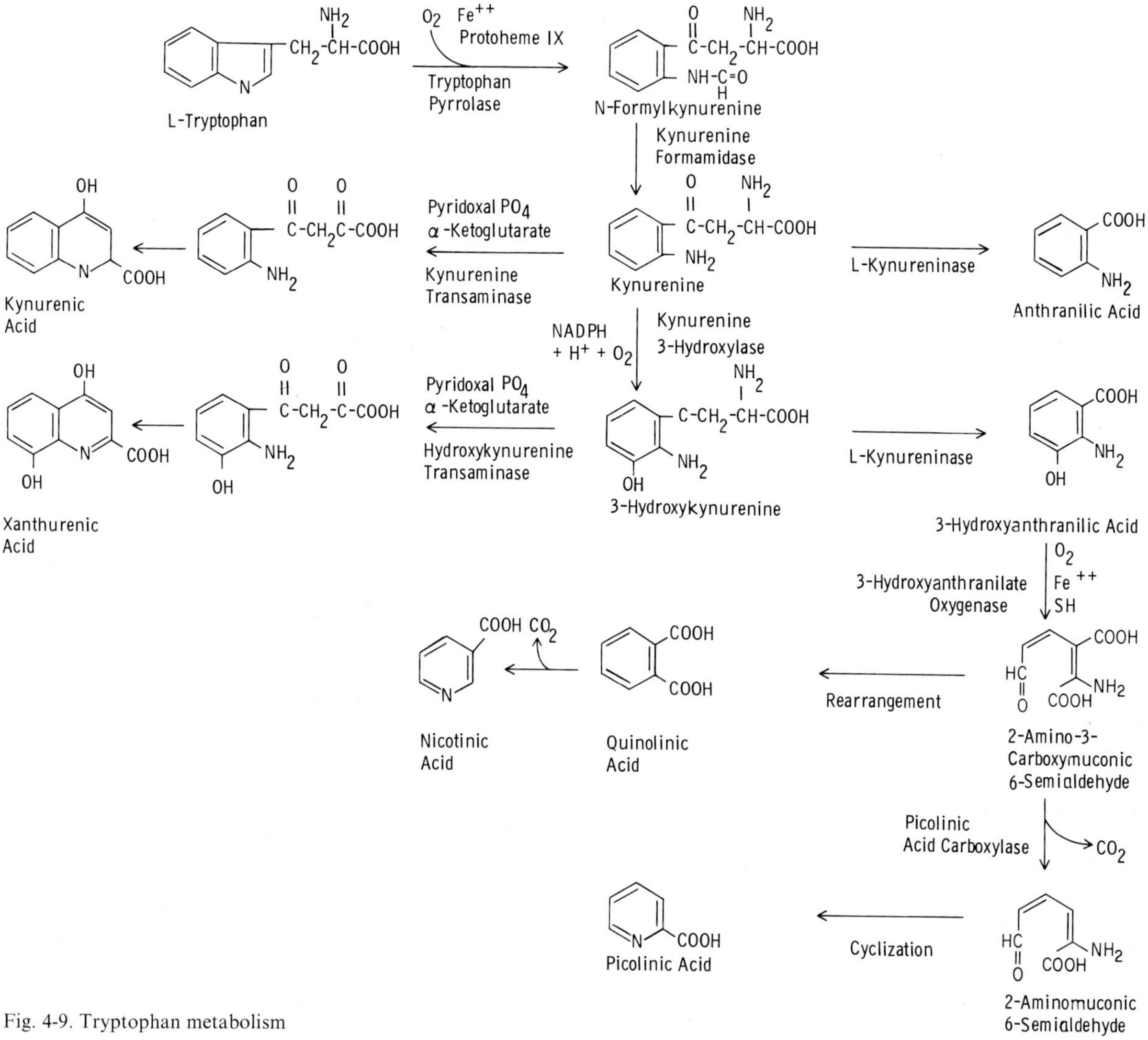

L-Tryptophan
O_2 Fe^{++} Protoheme IX
Tryptophan Pyrrolase
N-Formylkynurenine
Kynurenine Formamidase
Kynurenine
Pyridoxal PO_4 α-Ketoglutarate
Kynurenine Transaminase
Kynurenic Acid
L-Kynureninase
Anthranilic Acid
NADPH + H^+ + O_2
Kynurenine 3-Hydroxylase
3-Hydroxykynurenine
Pyridoxal PO_4 α-Ketoglutarate
Hydroxykynurenine Transaminase
Xanthurenic Acid
L-Kynureninase
3-Hydroxyanthranilic Acid
3-Hydroxyanthranilate Oxygenase
O_2 Fe^{++} SH
2-Amino-3-Carboxymuconic 6-Semialdehyde
Rearrangement
Quinolinic Acid
CO_2
Nicotinic Acid
Picolinic Acid Carboxylase
CO_2
2-Aminomuconic 6-Semialdehyde
Cyclization
Picolinic Acid

Fig. 4-9. Tryptophan metabolism

dicarboxylic acid) by spontaneous condensation of 2-acroleyl-3-aminofumarate.

An enzyme preparation obtained from mammalian liver requiring Mg^{++} or Mn^{++} catalyzes both the conversion of quinolinic acid to the corresponding 5-ribonucleotide (5′-QMP) and the decarboxylation of 5′-QMP to nicotinic ribonucleotide. (In diabetes, the rate of synthesis of nicotinic acid from tryptophan is reduced. It is assumed that this is due to development of an alternative pathway: the decarboxylation of the 3-hydroxyanthranilic group to yield picolinic acid.)

The breakdown pathways for NAD and NADP were discussed in the section on electron transport. Studies in Pardee's laboratory have shed some new light on the control of NAD synthesis in *E. coli* (see Fig. 4-10). Pardee was able to demonstrate that in the sequence of reactions, the NAD mononucleotide pyrophosphorylase reaction is the rate-limiting step. The levels of pyrophosphorylase are low in the wild type of *E. coli,* but they are even lower in mutants requiring nicotinic acid for growth when the bacteria are maintained in a medium saturated with nicotinic acid. In contrast, the pyrophosphorylase activity increases 200 times when the nicotinic acid in the medium is brought to low levels. From these observations, it seems logical to conclude that the rates of NAD synthesis are regulated by the levels of nicotinic acid mononucleotide pyrophosphorylase activity, and that the levels of the enzyme activity are regulated by a genetically controlled repressor mechanism.

Kynurenic acid and xanthurenic acid, side products of the reaction, are the products of the transamination of the α-amino group of kynurenine and 3-hydroxykynurenine to α-ketoglutaric acid in the presence of pyridoxal phosphate and an enzyme found in mammalian liver and kidney, kynurenine transaminase. The keto acid resulting from the transamination reaction condenses spontaneously. Liver homogenate also decarboxylates 3-hydroxykynurenine to yield 4,8-dehydroxyquinoline. Kynurenase may catalyze the cleavage of the side chain of kynurenine or 8-hydroxykynurenine and lead to the formation of alanine and

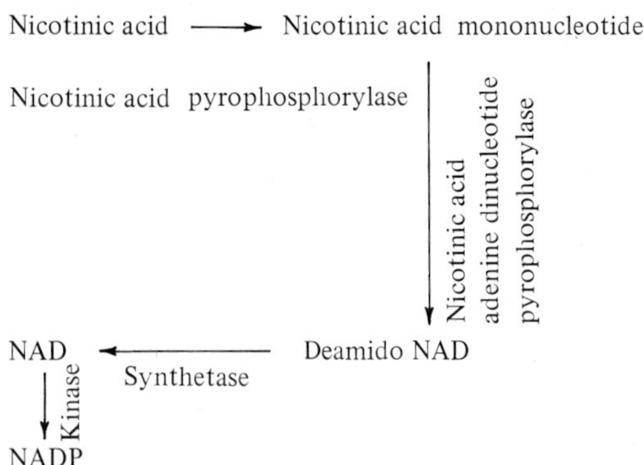

Fig. 4-10. Pathway of NAD synthesis in *E. coli*

anthranilic and 3-hydroxyanthranilic acids to yield kynurenic and xanthurenic acids.

Decarboxylation of 3-hydroxyanthranilic acid in the presence of picolinic carboxylase leads to the formation of picolinic acid. The steps involved in this transformation are not clear, nor are the enzymes involved known.

As already pointed out, quinolinic acid is not directly converted to nicotinic acid. Instead, it reacts with PRPP to yield the 5′-nucleotide. This is followed by decarboxylation of the ribose nucleotide of quinolinic acid. The same protein, quinolinate transphosphorylase, may catalyze both the formation of the glycosidic bond and the decarboxylation.

Nicotinic acid mononucleotide reacts with ATP in the presence of nicotinic acid adenine dinucleotide pyrophosphorylase, a magnesium-dependent enzyme, to yield the deamido derivative of NAD (see Fig. 4-11). Deamido NAD, in the presence of ATP, glutamine, Mg^{++}, K^{+}, and an NAD synthetase, is converted to NAD. In this reaction, the amino group of glutamine is transferred to the carboxyl group of the nicotinic acid moiety of deamido NAD. Yet the nicotinamide moiety of NAD synthetase is found in liver supernatant and, as may be expected, it is inhibited by azaserine.

A scheme in which nicotinamide was used as a precursor of NAD was described before the presently accepted biosynthetic pathway for NAD was elucidated. It is not certain whether nicotinamide can serve as a precursor of NAD. The administration of small amounts of nicotinic acid leads to a marked increase in NAD coenzyme levels in the blood. In contrast, the intake of high doses of nicotinamide has little effect on the blood concentration of NAD. Such results may be interpreted two ways: either nicotinic acid is a better precursor for NAD coenzymes than nicotinamide, or the conversion of nicotinamide to the coenzyme is restrained.

The conversion of nicotinamide to nicotinic acid requires the activities of a deaminase and a ribonucleotide pyrophosphorylase. Although a detailed description of mechanisms controlling NAD coenzyme synthesis in mammalian tissues is not available, it has been suggested that the regulatory restraints are exerted at the level of these two enzymes. Little direct evidence is available showing a direct role of the phosphorylase in regulating the biosynthesis rates of NAD coenzymes. But one inhibitor of the pyrophosphorylase has been found, and the activity of that enzyme appears to be sensitive to variations in ATP concentrations. Moreover, phosphorylase may be subject to feedback control by NAD. The deamidase is only found in liver, and it has been proposed that it could play a role in the interorgan distribution of coenzyme precursor.

From the role of tryptophan in the biosynthesis of nicotinic acid, it is obvious that the nutritional studies on nicotinic acid deficiency must take tryptophan intake into account. Indeed, 60 mg of tryptophan in the diet is as effective as 1 mg of nicotinic acid. Since 70 g of protein yields 720 mg of tryptophan, the intake of such an amount of protein corresponds to 12 mg of nicotinic acid in preventing niacin deficiency. Since the requirements for niacin, like those of thiamine, depend essentially on the caloric intake, it is useful to express the requirements in niacin equivalents per 1000 calories. The optimum requirement is 4.4 mg niacin per 1000 calories.

The reasons for niacin deficiency in individuals nourished on a corn or maize diet are still not clear. Corn is low in niacin and tryptophan, but it is also possible that the niacin in corn forms a chemical complex that cannot be used; therefore, the lime treatment of the corn used in Central American diets may be important in preventing pellagra. Niacin is not the active molecule in the body. The importance of niacin, as we have previously seen, stems from its role in NAD and NADP synthesis, and the deficiency can be expected to lead to reduced NAD and NADP concentrations in liver and blood serum. However, the exact significance of this reduction is not clear. Up to now, it has been impossible to establish a direct correlation between the growth rate and the amount of NAD in the liver.

Pathogenesis of Pellagra

NAD and NADP are involved in glycolysis, aerobic oxidation, fatty acid oxidation, glucuronide, pyrimidine, amino acid, and steroid biosynthesis (see Table 4-3). Therefore, it is not surprising that niacin deficiency leads to a complex pathological picture. Any attempt to explain the lesions observed in pellagra by a single metabolic defect would constitute an oversimplification. However, interference with pathways concerned with energy supply are probably responsible for some of the lesions observed. Therefore, those organs depending mainly on aerobic oxidation are more severely affected; thus, brain lesions are understandably most conspicuous in niacin deficiency.

The interpretation of the pathogenesis of injuries developing in pellagra is further complicated by the fact that victims of niacin deficiency are also deficient in other vitamins, and pellagra is usually assumed to

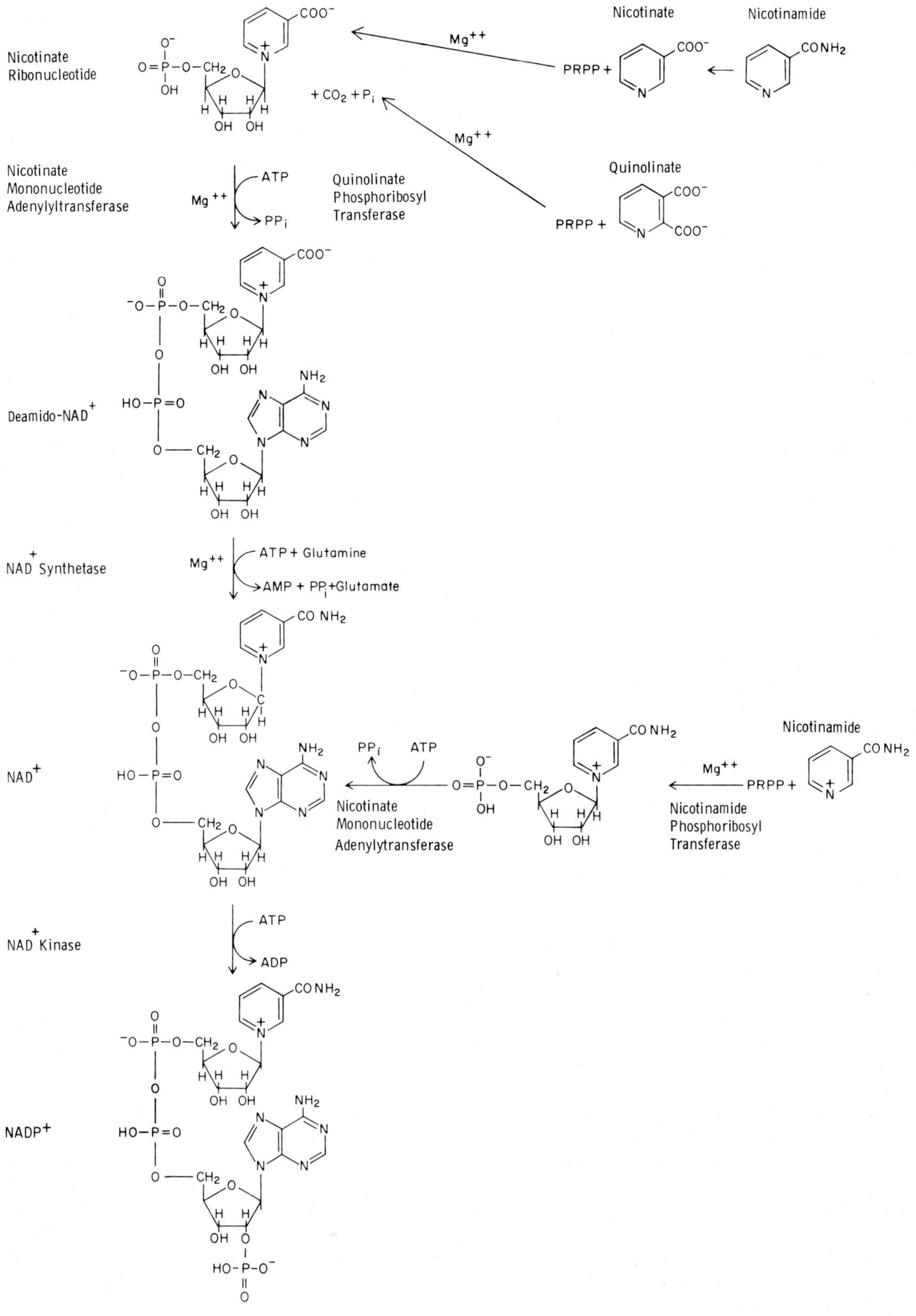

Fig. 4-11. NAD metabolism

Table 4-3. Some reactions requiring NAD

Reaction
Glyceraldehyde-3-P + H_3PO_4 = + NAD^+ $\rightleftharpoons$1,3-diphosphoglycerate + NADH + H^+
Lactate + NAD^+ $\rightleftharpoons$pyruvate + NADH + H^+
Malate + $NAD(P)^+$ $\rightleftharpoons$oxaloacetate + NAD(P)H + H^+
Glutamate + NAD^+ + H_2O $\rightleftharpoons$α-oxoglutarate + NH_4 + NADH + H^+
Isocitrate + NAD^+ + H^+ $\rightleftharpoons$α-oxoglutarate + CO_2 + NADH + H^+
Dihydrolipoate + NAD^+ $\rightleftharpoons$lipoate + NADH + H^+
Glucose-6-P + $NAD(P)^+$ $\leftrightarrows$Gluconolactone-6-P + NAD(P)H + H^+
UDP-glucose + 2 NAD^+ + H_2O →UDP-glucuronate + 2 NADH + 3H^+
Sorbitol + NAD^+ $\rightleftharpoons$Fructose + NADH + H^+
$RCH_2CHOH\ CH_2\ COCoA$ + NAD^+ $\rightleftharpoons$$RCH_2$–CO CH_2 COCoA + NADH H^+
Dihydroorotate + NAD^+ $\rightleftharpoons$orotate + NADH + H^+
Homoserine + NAD^+ $\rightleftharpoons$Aspartate-β-semialdehyde + NADH + H^+
Alcohol + NAD^+ $\rightleftharpoons$acetaldehyde + NADH + H^+
$Retinol_1$ + $NAD^+(P)$ $\rightleftharpoons$$Retinal_1$ + NAD(P)H + H^+
Cytochrome b_5 reduced + NAD^+ $\rightleftharpoons$cytochrome b_5 oxidized + NADH + H^+

result from a multiple-enzyme deficiency in which riboflavin is responsible for skin injuries and thiamine deficiency is responsible for at least some of the symptoms.

Pantothenic Acid Deficiency

Pantothenic acid 3-(2′,4′-dihydroxy-3′,3′-dimethylbutyramido) propionic acid is a precursor of coenzyme A. It is a vitamin, and if pantothenic acid is synthesized at all by mammalian organisms, the amount synthesized in most animals is too small for adequate metabolic performance. Pantothenic acid is abundant in the diet, especially in egg yolk, fresh vegetables, and meats. The vitamin survives most methods of cooking and food storage. For these reasons, pantothenic acid deficiency is unknown in man.

Pantothenic acid is not catabolized by human tissues because they lack the enzyme that splits the peptide bond; therefore, the excess vitamin is excreted unaltered.

Pathology of Pantothenic Acid Deficiency

Although symptoms clearly attributable to pantothenic acid deficiency have not been demonstrated in humans, studies on a variety of animals—including rats, calves, hens, and turkeys—have established that the vitamin is required in the diet. In pantothenic acid-deficient animals, a variety of pathological manifestations occur, among which are growth retardation, congenital anomalies, and duodenal ulcers (rats maintained on a pantothenic acid-deficient diet develop duodenal ulcers and increased pediculosis). For example, severe pantothenic acid deficiency may be produced in the calf. Most teguments are affected: dermatitis develops, the nasal mucosa is congested with increased secretion, the gastric mucosa is atrophic and the animal loses its appetite. There is also impaired growth, muscle edema, and alteration of the nervous fibers. These nerve injuries are responsible for changes in conditional reflexes. In dogs, nervous symptoms are predominant and are characterized by changes in conditional reflexes and behavioral disturbances.

Although the growth rate of rats is not affected by the administration of pantothenic acid, the addition of pantothenic acid to the culture media stimulates the growth of myoblasts, fibroblasts, and embryonic skin. Pantothenic acid also reduces some of the effects of 6-mercaptopurine on tissue culture, such as blocking of mitosis, interference with lipid synthesis, and fragmentation of the mitochondria.

When a pantothenic acid-free diet is administered to pregnant rats, the progeny develop various types of congenital anomalies, such as cerebral and ocular defects, cardiovascular anomalies, cleft palate, hydronephrosis, and hydroureter.

Biochemical Role of Pantothenic Acid

The biochemical role of pantothenic acid is intimately associated with that of coenzyme A (see Fig. 4-12). The role and mode of action of the coenzyme have been considered in other chapters. Thus, it is logical to assume that the morphological alterations in the different tissues result from interference with reactions requiring coenzyme A.

The corollary of the above postulate is that all biochemical reactions that are altered in pantothenic acid deficiency require coenzyme A. For example, pantothenic acid-deficient rats secrete much less nicotinuric acid than normal rats. This suggests that coenzyme A is necessary for the conjugation of nicotinic acid with glycine in the same way as it is necessary for the conjugation of benzoic acid and glycine to yield hypuric acid. Urinary excretion studies also seem to indicate that pantothenic acid is involved in the conversion of glucose to ascorbic acid. The incorporation of [^{14}C]glycine 2 and [^{14}C]succinate 2 into heme of duckling whole blood is reduced in pantothenic acid-deficient animals. Adding pantothenic acid restores glycine incorporation to normal.

Although pantothenic acid is assumed to be a vitamin, it is not completely nontoxic. When administered intraperitoneally in mice at doses of 5–7.5 g/kg of body weight, it kills 50% of the animals.

Several pantothenic acid analogs inhibit those reactions that require coenzyme A (acetylation of sulfanilamide and acetylcholine synthesis). Among these anti-

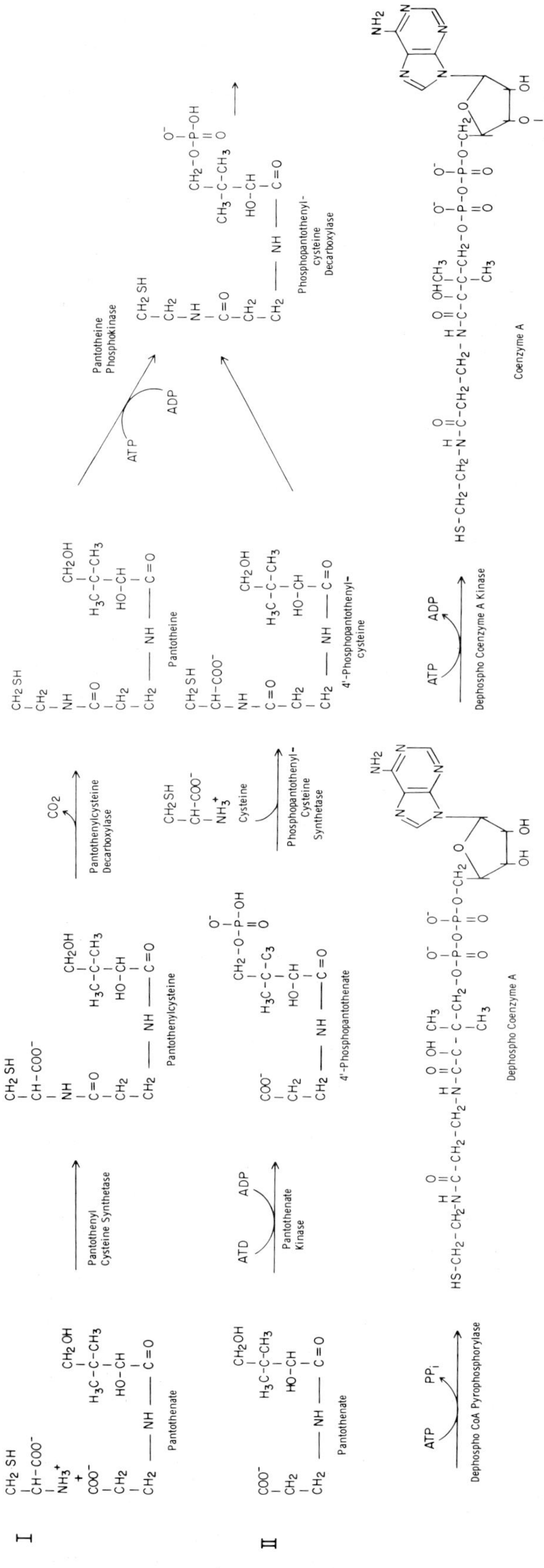

Fig. 4-12. Biosynthesis of coenzyme A

vitamins, one of the most effective is α-p-biphenylbutyric acid.

The importance of pantothenic acid as a precursor of coenzyme A has motivated investigators to determine the amounts of the vitamin in critical organs under various physiological conditions. The amount of pantothenic acid in the ovaries of codfish has been studied at various stages of the reproductive cycle. In the first stages of regeneration, the pantothenic acid content is high, but during the reproductive cycle, the pantothenic acid content of the ovaries steadily decreases, reaching a minimum before spawning. The pantothenic acid content of colostrum and breast has also been investigated. It is low the first day of lactation, reaches a maximum around the tenth day, and then remains rather high for about eleven months. Those observations are significant when compared with studies on the fate of coenzyme A in mammary tissue.

Coenzyme A is rapidly degraded in homogenate of lactating mammary tissue. However, prolactin injection to guinea pigs considerably increases coenzyme A levels. Consequently, the coenzyme A accumulation in the mammary gland during pregnancy results from the inhibitory effect of prolactin on coenzyme A catabolism. The levels of coenzyme A and the succinoxidase activity of liver are reduced by 20% in pantothenic acid-deficient animals. The changes in coenzyme A concentration probably result from interference with its biosynthesis, since the incorporation of acetate and mevalonate in coenzyme A is only 50% of normal in pantothenic acid-deficient animals.

Although pantothenic acid deficiency is unknown in humans, the effect of the vitamin has been studied in various pathological conditions, such as hypercholesterolemia, diabetes, and liver disease. Hypercholesterolemia or hyperlipemia does not occur in pantothenic acid deficiency when blood lipids are increased by the injection of Triton-X-100. The administration of β-biphenylbutyric acid alleviates both hypercholesterolemia and hyperlipemia.

When excess pantothenic acid is administered to diabetic rats, the excretion of sugar and ketone bodies is markedly reduced, and cholesterol accumulates in the liver, adrenals, and blood. Therefore, pantothenic acid administration to diabetic rats appears to stimulate the use of acetate and acetoacetate for cholesterol and fatty acid synthesis. The role of coenzyme A in diabetes is discussed in greater detail in the chapter devoted to that disease.

Pantothenic acid deficiency induced either by a deficient diet or by the administration of pantothenic acid antimetabolites causes adrenal deficiency in rats. The ACTH content of the adrenal venous blood is reduced, and the concentration of coenzyme A is decreased in the gland. Pantothenic acid administration normalizes the coenzyme A concentration. The administration of growth hormone to pantothenic acid-deficient rats that have also been hypophysectomized increases the death rate and the incidence of adrenal necrosis. How the vitamin deficiency affects the humoral interrelationship remains to be investigated. The mechanism of these altered hormonal interactions is not clear.

Pantothenic acid excretion and concentration are altered in liver disease. Patients with liver injury excrete less pantothenic acid than normal individuals, and the amount of pantothenic acid excreted correlates with a given liver function test in which pantothenic acid or coenzyme A is known to play a role. Pantothenic acid concentration is reduced in regenerating liver.

Seborrheic Dermatitis and Biotin Deficiency

In the 1930's, a compound named bios was isolated from egg yolk and found to stimulate the growth of yeast. At about the same time, an extract of *Rhizobium trifolii,* which stimulated growth and respiration, was discovered. This new factor was called vitamin R. It has also been known for some time that raw egg whites inhibit the growth of young rats. This effect was shown to result from a compound called avidin in raw egg.

In 1939, Giorgy isolated from liver a vitamin called vitamin H, which prevented the development of the deleterious effects observed after the administration of avidin. After several years, the biochemists succeeded in relating these isolated findings and demonstrating that bios, vitamin R, and vitamin H are all the same compound, the official name of which became biotin. Du Vigneaud established that biotin is the d-*cis*-hexahydro-2-keto-1 H-thieno (3,4) imidazole-4-valeric acid, an optically active monocarboxylic acid. Biotin is found in a bound form in many animal and plant tissues [72].

Biotin Deficiency

Biotin deficiency is seldom observed in humans, because under normal conditions the vitamin is synthesized in sufficient amounts by the bacterial flora of the intestine. Only when his flora is destroyed by the administration of antibiotics or when the antivitamin avidin is administered does a vitamin deficiency develop.

The ingested avidin combines with biotin in stoichiometric amounts to yield a biotin derivative resistant to the action of the enzyme secreted in the lumen of the gastrointestinal tract. The vitamin can be liberated from the complex it forms with avidin only by heat treatment or by acid hydrolysis. Biotin deficiency is observed mainly in children under four years of age. In biotin deficiency, a scaling seborrheic dermatitis affects mainly the chest, neck, umbilicus, and groin. This infantile seborrheic dermatitis has reportedly been cured after biotin administration. Yet biotin doesn't affect seborrheic dermatitis in older children and adults. An exceptional form of biotin deficiency was reported in an Italian laborer who ate nothing but 10–20 raw eggs a day. The victim developed exfoliative dermatitis and conjunctivitis, which disappeared as soon as a more balanced diet was followed [73].

The Role of Biotin in Metabolism

All aspects of the biochemical role of biotin have not yet been clarified. The vitamin has been implicated in the metabolism of carbohydrates, lipids, proteins, and nucleic acids. Available evidence indicates that biotin acts as a CO_2 carrier in a number of carboxylation and decarboxylation reactions connected with carbohydrate and fatty acid metabolism. A number of experimental procedures are used to establish the participation of biotin in a given biochemical reaction: (1) the study of enzyme activity in biotin-deficient animals; (2) the effect of avidin administered *in vivo* or added to the incubation mixture on the activity of the enzyme under study; and (3) purification of the enzyme and demonstration of the existence of enzyme-bound biotin. Studies of this kind have established that biotin is required for the carboxylases of β-methylcrotonyl CoA, acetyl-CoA, propionyl CoA, and oxaloacetic transcarboxylase. Only some of the results are presented here [74–76].

A coenzyme role of biotin in the reaction catalyzed by oxaloacetate carboxylase has been proposed on the basis of studies made with the purified enzyme, which suggest a close association between the enzyme's biotin content and the catalytic properties of the protein. The purified enzyme preparation contains 3 mμg of biotin per mg of protein. Biotin as a CO_2 carrier in the CO_2 fixation reaction was further supported in studies made on enzyme preparations obtained from birds and mammalian livers. These preparations catalyze the reversible conversion of pyruvate into oxaloacetate in the presence of ATP.

$$\mathrm{CH_3COCOOH + CO_2} \underset{}{\overset{\mathrm{ATP \leftrightarrow ADP + P_i}}{\rightleftharpoons}} \mathrm{HOOC\ CH_2COCOOH}$$

Adding avidin to the incubation mixture blocks this reaction, but it can be restored to normal by adding biotin.

The metabolic pathway for propionic acid use in mammals involves the carboxylation of propionyl CoA in the presence of a specific carboxylase and ATP. The products of the reaction are methylmalonyl CoA and ADP. The ultimate product of propionate metabolism is succinate, which is oxidized via the Krebs cycle. These reactions all occur in the mitochondria. The ability of the mitochondria of biotin-deficient animals to use propionic acid is reduced. This metabolic defect of the deficient mitochondria cannot be corrected by adding biotin *in vitro,* but the rate of propionic acid use by mitochondria is restored to normal if the deficient animals are fed a normal diet containing biotin.

Propionyl CoA carboxylase has been crystallized, and biotin has been demonstrated in the enzyme molecule. Biotin was also found to be covalently bound to acetyl CoA carboxylase, an enzyme that catalyzes the carboxylation of acetyl-CoA to yield malonyl CoA in the presence of ATP. Methylmalonyl oxaloacetic transcarboxylase is also a biotin-containing enzyme. The enzyme catalyzes transfer of CO_2 from methylmalonyl CoA to pyruvic acid.

Some of the pioneering studies of biotin's mechanism of action in carboxylation reactions were done in Lynen's laboratory on β-methylcrotonyl CoA carboxylase [74]. Lynen's early findings on that coenzyme have since, either in his own or in other laboratories, been extended to many other carboxylases.

β-Methylcrotonyl CoA carboxylase has been prepared in a pure form yielding a single boundary in the ultracentrifuge ($S_{20,\omega}=19.4$; 1 mole of biotin for 184,000 g of protein). Lynen and his collaborators found that, contrary to other biotin-requiring carboxylases, β-methylcrytonyl CoA carboxylase carboxylates free biotin. The product of the reaction can be esterified to yield a compound identifiable by the methods of organic chemistry. The esterified compound was identified as 1′-N-methoxycarbonylbiotin. Therefore, it was deduced that enzyme-bound 1′-N-carbonylbiotin is the intermediate in the carboxylation reaction (see Fig. 4-13).

A CO_2 biotin-enzyme complex was demonstrated by reacting the purified apoenzyme with biotin, $^{14}CO_2$, and ATP. The protein complex formed during the reaction is freed from the smaller molecular component in the incubation mixture by passing the mixture through a Sephadex column. The biotin CO_2 moiety of the complex is then stabilized by esterification with diazomethane. The methylated complex is then digested with a proteolytic enzyme, such as trypsin, papain, or pronase. Proteolysis of the methylated CO_2 biotin-protein complex yields 1′-N-methoxy-C_{14}-carbonyl biocytin. Biocytin, known for some time, is found in extract of active yeast. It is formed by one molecule of lysine and one molecule of biotin linked through an amide linkage between the carboxyl of biotin's side chain and the ε-amino group of lysine. The binding is similar to that uniting lipoic acid to the ε-amino group of lysine of pyruvic dehydrogenase.

Treatment of the biocytin derivative with biotinidase, an enzyme that splits the bond between biotin and lysine, yields 1′-N-methoxy-C_{14}-carbonylbiotin. This mode of attachment of CO_2 to biotin (in the 1′-N position) and of biotin to the apoenzyme (6-N-(+) biotinyl lysyl) proved to be identical for β-methylcrotonyl CoA carboxylase, propionyl CoA carboxyylase, and oxaloacetic transcarboxylase. The binding of CO_2 with biotin is a high-energy binding (−4.7 kilocalories per mole) and the CO_2 bond to the 1′-N-methoxy-C_{14}-carbonylbiotin can be considered as a form of activated CO_2.

Biotin apotranscarboxylase synthetase catalyzes the formation of the 6-N-(+)-biotinyl-L-lysyl carbon-nitrogen bond in methylmalonyl CoA oxaloacetic transcarboxylase. The conversion of the apoenzyme into the holoenzyme requires biotin, ATP, and Mg^{++}. A (+)-biotinyl-5′-AMP enzyme complex seems to be formed in the course of that reaction (see Fig. 4-14).

The complex is a mobile carrier of carboxyl groups according to the following general scheme:

$$\text{E-biotin} + HCO_3^- + \text{ATP} \xrightleftharpoons{Mg^{++}} \text{ADP} + \text{Pk} + \text{E-biotin-}CO_2$$
$$\text{E-biotin-}CO_2 + \text{acceptor} \rightleftharpoons \text{E-biotin} + \text{acceptor-}CO_2$$

Thus, the biotin is carboxylated in the first step and transcarboxylation takes place in a second step from coenzyme to acceptor.

Biotin, once a mysterious compound needed to breed healthy rats, is now known to participate in a vital biochemical reaction. In spite of this striking progress in the molecular biology of biotin, the pathogenesis of the seborrheic dermatitis of biotin-deficient children remains unexplained. The molecular pathologist must link the dermatitis with biotin's role in the carboxylation reaction.

Scurvy and Ascorbic Acid Metabolism

Scurvy has plagued seamen and explorers since antiquity. It was responsible for some of the ghost ships that wandered on the sea with their dead crew. Already in 1593, the use of oranges had been proposed as a cure for scurvy. When Jaques Cartier explored Canada, he and his party developed scurvy. The Indians told him that eating pine buds would cure the disease, and so it did. In 1795, Lind, a Scottish physician, revived the use of lime juice as a preventative of scurvy, and introduced an admirality rule prescribing the use of orange and lemon juice in the Royal Navy. From then on, scurvy was wiped out from the Royal Navy, and British sailors became so committed to the use of lemons that they were called limeys. Funk suspected that scurvy resulted from a vitamin deficiency. Then, in the early 1930's, Szent-Györgyi [77] crystallized hexuronic acid from an adrenal preparation without suspecting that the new compound was precisely the vitamin that is deficient in the diet of scurvy victims. Several years later, the compound isolated from the adrenal gland was shown to be identical to the vitamin contained in the fruits (see Fig. 4-15).

Clinicopathology of Scurvy

The typical injury of scurvy consists of an incapacity of the mesenchymal cells—such as fibroblasts, osteoblasts, or odontoblasts—to produce their normal fundamental substance. In other words, the cell can proliferate, but it cannot form collagen, osteoid, or dentin. This basic alteration is responsible for increased fragility of the capillaries, retarded wound healing, abnormal bone growth, and eventually abnormal tooth growth [78].

Scurvy is most frequently seen in children today; it is rare in adults. In children, the first symptom is limb tenderness, which is manifested as soon as the child is handled. Tenderness may be do to hemorrhage in muscle or periosteum or to fractures. Bone lesions are quite characteristic. Normally, when both diaphyseal and epiphyseal endochondral ossification is completed, a thin cartilaginous band remains and prolifer-

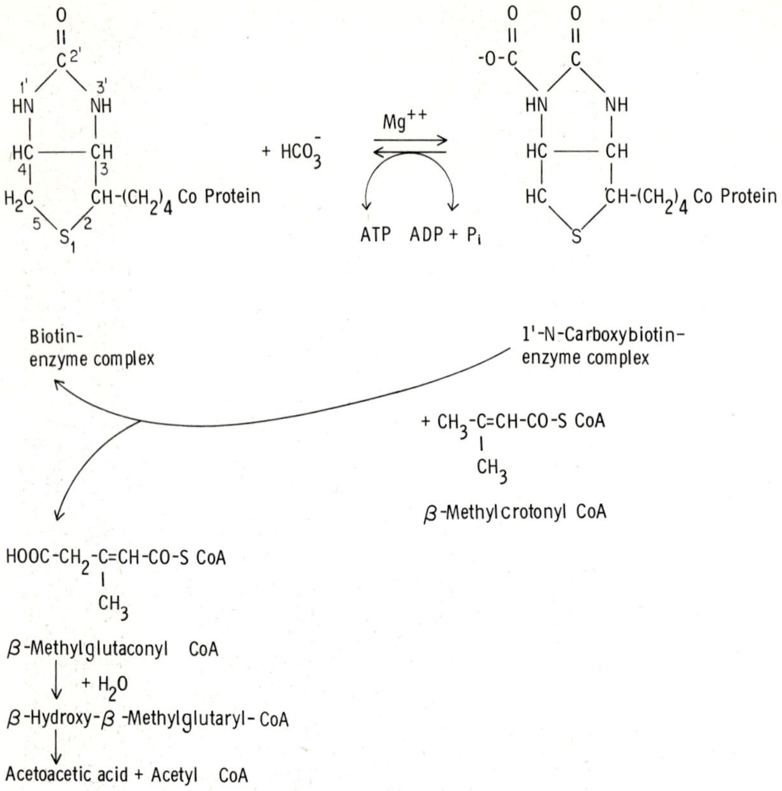

Biotin-enzyme complex + HCO_3^- $\underset{ATP \quad ADP + P_i}{\overset{Mg^{++}}{\rightleftharpoons}}$ 1'-N-Carboxybiotin-enzyme complex

(Biotin ring: O=C2', 1' HN, 3' NH, 4 HC—CH 3, H_2C 5, S 1, 2 CH-$(CH_2)_4$ Co Protein → -O-C(=O)-N ring … CH-$(CH_2)_4$ Co Protein)

+ CH_3-C(-CH_3)=CH-CO-S CoA

β-Methylcrotonyl CoA

HOOC-CH_2-C(-CH_3)=CH-CO-S CoA

β-Methylglutaconyl CoA

↓ + H_2O

β-Hydroxy-β-Methylglutaryl-CoA

↓

Acetoacetic acid + Acetyl CoA

Fig. 4-13. Mode of action of β-methylcrotonyl-CoA carboxylase

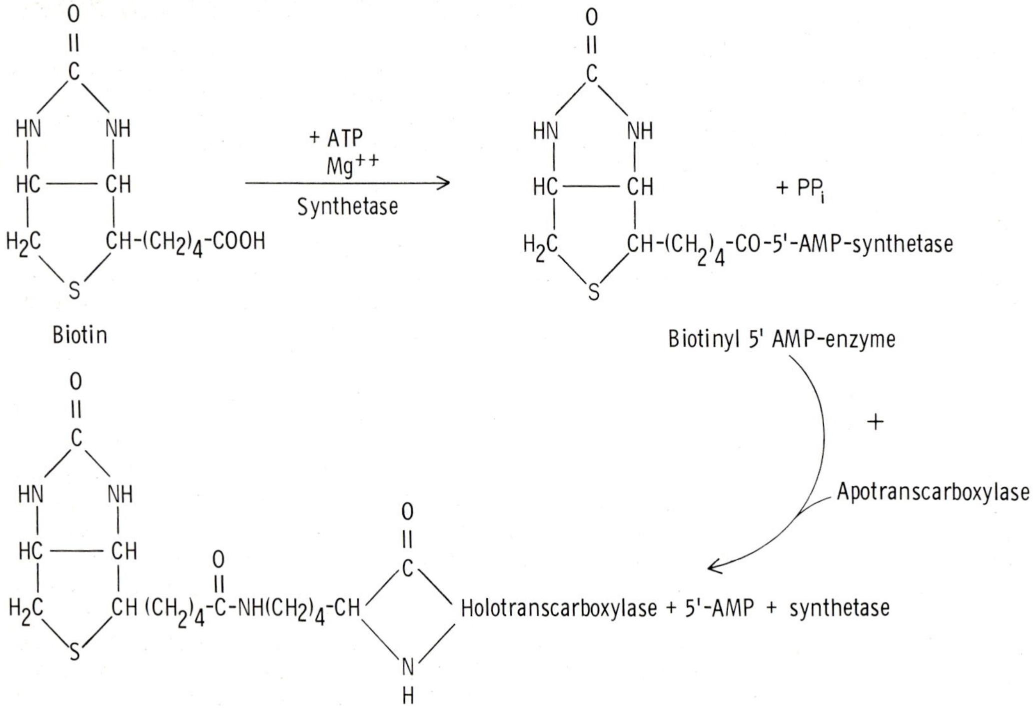

Biotin (CH-$(CH_2)_4$-COOH) $\xrightarrow[\text{Synthetase}]{+\text{ ATP, } Mg^{++}}$ CH-$(CH_2)_4$-CO-5'-AMP-synthetase + PP_i

Biotinyl 5' AMP-enzyme

+

Apotranscarboxylase

→ CH $(CH_2)_4$-C(=O)-NH$(CH_2)_4$-CH Holotranscarboxylase + 5'-AMP + synthetase

Fig. 4-14. Mechanism of conversion of apotranscarboxylase into holotranscarboxylase

```
O = C ——┐
        │
HO — C  │
     ‖  O
HO — C  │
     |  │
 H — C ——┘
     |
HO — C — H
     |
     CH2OH
```

2, 3 Enediol-L-Gulono-1, 4-lactone

Fig. 4-15. L-Ascorbic acid

ates toward the diaphysis, making the bone grow longitudinally. This maturation proceeds in parallel rows separated by capillaries and connective tissues. The chondroblasts of the hyalin cartilage mature into the large cells of the serial cartilage (the last cells of the row), which are finally impregnated with calcium salt to form the calcified cartilaginous spicules. The osteoblasts accompany the vasculoconjunctive bands and deposit osteoid material at the surface of the calcified cartilage.

In vitamin C deficiency, there is no interference with the proliferation or maturation of the cartilaginous cells and the osteoblasts, but those cells fail to form osteoid material. Thus, instead of proceeding to true ossification, the scurvy bone produces only a rather friable cartilaginous matrix. Fractures ensue, and soon the entire histological structure is distorted at the site of the osteochondral junction. This distorted area, called the "Trummerfeld zone," contains irregularly distributed calcified cartilaginous spicules, foci of hemorrhage, colonies of osteoblasts not surrounded by osteoid, and macrophages containing hemosiderin.

These long bone lesions (see Fig. 4-16) can be easily identified radiologically. The epiphyseal plate is thickened and slightly irregular. Instead of measuring 0.1–0.3 mm as in normal bone, it is only 1 mm. Within the epiphysis, the center of ossification is surrounded by a dense rim forming a so-called halo, which results from the accumulation of calcified cartilage. The Trummerfeld zone appears as a zone of rarefaction near the white line of the diaphysis, in which the trabecular markings are thin and less numerous than normal. In the rib, lesions similar to those observed in the long bones lead to fracture or dislocation with enlargement of the osteocartilaginous junction. Animals deficient in vitamin C also fail to produce dentin and enamel in the odontoblasts of the tooth germs. In contrast, no such alterations have been observed in scorbutic children, but in experimental scurvy in human adults, the alveolar bone atrophies and is replaced by fibroblasts incapable of producing collagen. Any symptoms observed in adults probably result from secondary effects, such as local stress or poor hygiene. In experimental scurvy, the only objective observations are a decrease in the level of vitamin C in the blood and retarded wound healing; the latter does not appear until 2 months after the deficiency begins.

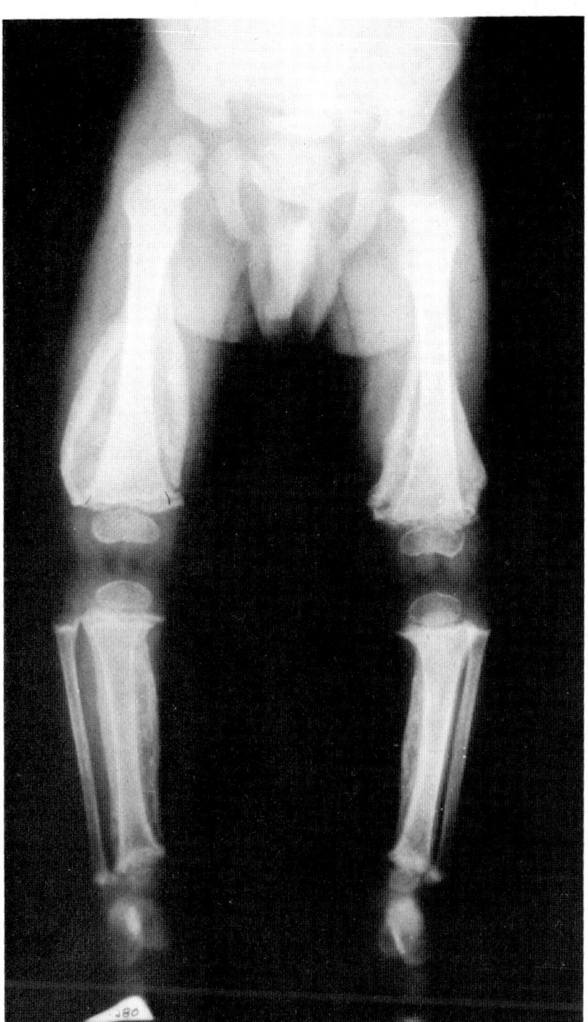

Fig. 4-16. Scurvy (nutrition), showing subperiosteal hemorrhage. The cortex of the diaphysis is thickened, and the lower extremities of the femoral diaphysis and the metaphysis are cuffed by radiolucent masses lifting the periosteum. Due to subperiosteal hematoma, the periosteum is thickened. (From S. Collins)

The incidence of scurvy in humans is difficult to evaluate because a good clinical test for early diagnosis is not available. Furthermore, the disease, at least in children, is often confused with rickets. Only a few years ago, 11% of all the children autopsied during a 10-year period at the Johns Hopkins Hospital presented histological changes in bone characteristic of scurvy. The frequency in adults is unknown.

The role of vitamin C in the genesis of collagen has been studied by various means. One of the most interesting approaches is to study wound healing capacity in normal and scorbutic individuals. When the wound is artificially produced in a scorbutic animal or in an experimentally scorbutic human, healing is retarded. Histologically, fibroblasts proliferate, but no collagen is formed. Microscopic slides prepared from such wounds show a pinkish homogenous material instead of the typical collagenous fiber. The resistance of the regenerated tissue to stress has been measured by air injections in the peritoneal cavity. The pressure necessary to rupture the repairing tissue in a scorbutic animal is half that necessary to rupture the repair in normal individuals. A similar situation occurs in the formation of abscesses. After local injection of pyogenic bacteria in scorbutic animals, the inflammation process becomes localized and there is no retardation in polymorphonuclear and fibroblast proliferation, but a solid fibrous capsule like the one that appears in normal individuals fails to develop.

Vitamin C—Sources and Requirements

Vitamin C (ascorbic acid) is a powerful reducing agent, and when oxidized it forms dehydroascorbic acid. The vitamin is abundant in nature; green vegetables and fruit juices, citrus fruits, oranges, and tangerines are rich in the vitamin. Of course, an adequate supply of vitamin C can only be provided if the vitamin is stable in fruit. Therefore, its stability has been extensively investigated in a variety of fruit juices produced both in this country and abroad. Addition of calcium carbonate to cabbage to increase the calcium content of vegetables does not reduce the vitamin content.

In animals, practically all organs contain vitamin C, but the adrenal gland contains an exceptionally large amount. Vitamin C deficiency occurs only in guinea pigs and primates. Other mammals and invertebrates synthesize their own ascorbic acid from glucose through the glucuronic pathway. The vitamin, a water-soluble substance, is rapidly absorbed by the intestine and transported to the various viscera. It is not stored in large amounts, therefore the body's vitamin C supply is rapidly depleted when the diet is deficient. Furthermore, if the vitamin is administered biweekly or bimonthly instead of daily, the requirements increase considerably, indicating that no capacity for prolonged storage is available.

The optimum requirement for vitamin C is about 75 mg daily in adults. Values below 10 mg result in vitamin deficiency, and when more than 100 mg is administered, the vitamin is excreted in the urine. However, the requirements change with age. This may be related to the changes in vitamin concentration that occur with age in various organs. The content of vitamin C decreases consistently with age in such tissues as pituitary gland, cerebral cortex, myocardium, and pectoral muscles. Whereas in some tissue, such as the cerebral cortex, the decrease is continuous until death, it stops in others after 20 years (myocardium) or 40 years (skeletal muscles).

Mental patients often manifest signs of ascorbic acid deficiency despite an adequate intake of ascorbic acid. They present tissue unsaturation in vitamin C, reduced levels of ascorbic acid in the plasma, and increased capillary fragility. Although the administration of ascorbic acid to such patients restores the blood level to normal, it has no effect on the capillary fragility.

Vitamin C and Metabolism

A large amount of work has been devoted to determining the role of vitamin C in metabolism. In view of the symptoms of scorbutic animals, a specific role of vitamin C has been envisaged in both bioenergetic and biosynthetic pathways. Thus, vitamin C has often been proposed to play an important role in carbohydrate metabolism or electron transport. The high concentration of the vitamin in the adrenals and the drastic changes in the bodily distribution of vitamin C in stress suggest that the vitamin may be involved in adrenal hormone metabolism. The incapacity of the scorbutic animal to synthesize normal collagen indicates that it is directly or indirectly involved in protein metabolism.

Among mammalian tissues, the adrenals are the richest in ascorbic acid. Szent-Györgyi first crystallized the vitamin from the adrenal gland. This high concentration of ascorbic acid in the adrenals has intrigued investigators for decades. The vitamin is apparently present in the adrenals in both bound and free forms. Cortisone injection releases the bound ascorbic acid, which is then excreted from the adrenal vein in a free form. The concentration of ascorbic acid in the vein is not affected by the injection of ACTH, and therefore continuous diffusion seems to exist between the adrenal and the vein. Furthermore, ascorbic acid in venous blood is in the oxidized rather than the reduced form, suggesting that the vitamin is excreted from the adrenals only after it has participated in some metabolic step. The reason for vitamin C in the adrenals is not clear, but it has been suggested to protect epinephrine against oxidation. The rates of oxygen consumption and epinephrine oxidation are increased in scorbutic animals as compared to normal guinea pigs. Another possibility is that ascorbic acid participates in a specialized electron transport pathway. In the pig adrenal microsome, a respiratory chain is insensitive to cyanide. It involves NAD and cytochrome b_5, and it is stimulated by the addition of small amounts of cytochrome c. It is therefore assumed that the L-ascorbic acid and monodehydroascorbic acid systems act as electron carriers between NADH and cytochrome b_5.

It is well established that stress affects the ascorbic acid concentration in various components of the body. What mechanism causes stress is unimportant because similar effects are obtained with prolonged exposure to cold, heat, or X-irradiation, excessive oxygen tension, or even a simple injection of saline. Stress increases the excretion of ascorbic acid in the urine and the plasma levels of ascorbic acid in the blood (in the early stages of stress), and reduces the ascorbic acid level of the adrenal. Increased urinary excretion of dehydroascorbic and diketonic acids usually is associated with ascorbic acid in the urine.

The rise of ascorbic acid in the plasma in temporary stress is considered by some to constitute the most reliable index of stress, although it is somewhat less sensitive than eosinophilia. Ascorbic acid concentration in tissue changes after X-irradiation, but the changes depend on the nature of the organ—although such organs as the thymus, the spleen, and the lymph nodes have a reduced amount of ascorbic acid, that compound increases in the intestines. There are usually no changes in other tissues. It has been concluded from such observations, without convincing evidence, that ascorbic acid is involved in mitosis.

In the scorbutic guinea pig, the excretion of 17-hydroxycorticosteroid increases and the concentration of ascorbic acid in the adrenals decreases. If vitamin C is added to the diet, both phenomena rapidly return to normal. The injection of cortisone reduces the amount of ascorbic acid in the adrenals. This reduction is absolute, and the amount of vitamin C per gram of tissue decreases. In contrast, the injection of ACTH reduces the total amount, but there is no reduction in the concentration in the adrenal, which indicates that the ACTH from the reaction reduces the size of the organ and doesn't directly affect vitamin release. (A more detailed description of the effects of vitamin C on the corticosteroid hormones appears in chapters in which the metabolism of these hormones is discussed.)

In scorbutic animals, blood glucose levels are increased, glucose tolerance is lowered, and liver and glycogen content drop. Insulin restores everything to normal. Therefore, it was concluded that vitamin C deficiency also leads to insulin deficiency, and that the latter is responsible for the deviation in carbohydrate metabolism.

The accumulation of pyruvic acid and α-ketoglutarate in vitamin C deficiency may be explained by the reduced concentration of CoA in ascorbic acid-deficient animals. The coenzyme is involved in the decarboxylation of both metabolites, but again, it has been suggested that the effect of ascorbic acid is secondary. The increased levels of pyruvic and lactic acids in blood could be caused by the insulin deficiency that develops in scorbutic animals. This conclusion finds support in the observation that fructose administration normalizes pyruvic and lactic acid levels. Whether the increased levels of α-ketoglutarate in scorbutic animals could result from decreased use in protein hydroxylation remains to be seen.

Ascorbic acid deficiency also reduces the activity of several dehydrogenases involved in the Krebs cycle. The mechanism by which the vitamin leads to such alteration is not clear, but the effect is reversed by insulin administration. Thus, miscellaneous observations on the effect of ascorbic acid on carbohydrate metabolism have been made, but they are difficult to interpret because no specific coenzyme effect of ascorbic acid has been demonstrated. Again, ascorbic acid is assumed to be directly involved in an electron transport chain that involves cytochrome b_5 and NAD. The vitamin may also affect the electron transport chain indirectly because a decrease in NADH concentration has been observed in vitamin C deficiency. But this decrease may also result from an interference with the insulin production because it is relieved by fructose administration. Lipoic acid also can alleviate some of the symptoms induced by vitamin C (as well as by vitamin E) deficiency. The mechanism by which this occurs is not clear, but lipoic acid is thought to protect the dehydro derivatives of both vitamins.

Several laboratories have been concerned with the effect of ascorbic acid deficiency on amino acid metabolism. Two main aspects have been investigated: (1) a rather unspecific effect on the amino acid content in muscle and in blood; and (2) a more specific action of vitamin C on the hydroxylation of some amino acids or intermediates in amino acid biosynthesis. Vitamin C deficiency alters the ratio of the various amino acid concentrations in muscle and blood. In muscle, while glutamic acid, leucine, valine, and methionine levels increase, glutamine and aspartic acid concentrations decrease. In blood of scorbutic guinea pigs, the concentrations of most of the amino acids decrease, while phenylalanine, leucine, and histidine levels rise. There is no definite explanation for these changes; they indicate only that ascorbic acid is directly or indirectly involved with amino acid metabolism.

It was observed some years ago that in vitamin C-deficient guinea pigs the hydroxylation of tyrosine is inhibited and hydroxypyruvic acid, the deamination product, is excreted. These observations are of little practical interest because the enzymatic block is not manifested unless large amounts of tyrosine are administered to the individual. Since this original observation was made, ascorbic acid has been thought to be involved in several other dehydroxylation reactions, but the mechanism by which it acts is not clear. It probably involves the oxidation of ascorbic acid in the presence of a suitable metal. However, several reducing substances (glucoascorbic acid, isoascorbic acid, dichlorophenylindophenol, and hydroquinone) are active.

When the purified and dialyzed enzyme that catalyzes the conversion of kynurenine into trihydroxykynurenine is put in the presence of its substrate, no product is formed until ascorbic acid is added to the system. But the addition of dichlorophenylindophenol is much more effective, which means either that the effect of ascorbic acid is unspecific or that ascorbic acid acts through dichlorophenylindophenol. The latter hypothesis is supported by the fact that after more extensive purification of the hydroxyphenylpyruvic hydroxylase, the addition of vitamin C alone does not restore the activity of the enzyme. Only when 2,6-dichlorophenylindophenol is added is activity restored. In conclusion, the hydroxylated product appears during the reversible oxidation of ascorbic acid in the presence of a suitable metal, usually copper or iron. The nature of the hydroxylating agent is not known, but it is often assumed to be a quinone, and in many of these hydroxylation reactions, ascorbic acid may act merely as a reductant of the hydroxylating agent rather than as a specific coenzyme. (The intraperitoneal administration of L-tyrosine to scorbutic guinea pigs in conjunction with sodium butyrate in-

Proline

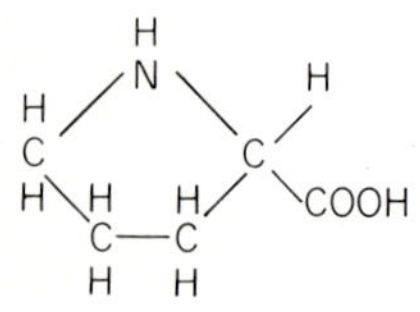

```
       H
    /N\     H
H /     \ C/
C
H\H   H/  \COOH
  C—C
  H  H
```

Hydroxyproline

```
       H
      N     H
H /     \ C/
C
H\H   H/ \
  C—C    COOH
  OH  H
```

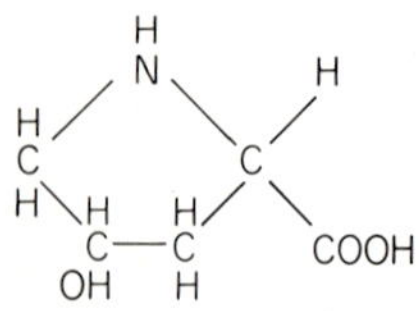

Fig. 4-17. Proline and hydroxyproline

duces cataracts. This is apparently related to the formation of benzoquinone; the administration of benzoquinone acetic acid also leads to the formation of cataracts.)

One of the most interesting aspects of ascorbic acid is its role in hydroxyproline synthesis (see Fig. 4-17). Connective tissue grows subcutaneously in polyvinyl sponge implants. Ascorbic acid causes rapid growth of the connective tissue under those conditions, and connective tissue proliferation is associated with a rapid incorporation of hydroxyproline.

The conversion of proline to 4-hydroxyproline is still under investigation. It is known to occur after the incorporation of proline in a polypeptide chain, and ascorbic acid is required for the reaction.

The reaction has been studied in a cell-free system. Microsomes obtained from chick embryos containing proline-rich polypeptides were incubated with ascorbic acid and a soluble protein fraction. If the incubation took place anaerobically, no hydroxyproline was formed. In contrast, when it was done aerobically, 4-hydroxyproline appeared. Hydroxyproline can be released from the polypeptide into the acid-soluble fraction with the aid of collagenase.

Proline hydroxylase is an oxygenase similar to γ-butyrobetaine hydroxylase. It uses α-ketoglutarate as a hydrogen donor and transfers an oxygen to the α-carbon of α-ketoglutarate, eliminating the adjacent carboxyl group and converting the α-keto group into a carboxyl group.

The mechanism of hydroxylation of proline to hydroxyproline has been elucidated at least in part. In the presence of molecular oxygen, a hydroxyperoxide is formed at the 4 (*trans*) position of the appropriate proline residue of the polypeptidyl proline. The anionic form of the hydroxyperoxide is then believed to react with α-ketoglutarate to yield a complex. In the new reaction, a proton is transferred to the oxygen in position 4 (*trans*) of proline, and the α-ketoglutarate moiety of the complex is decarboxylated, thus yielding hydroxyproline, succinate, and CO_2 in stoichiometric amounts (see Fig. 4-18).

Although there seems to be no doubt that ascorbic acid facilitates collagen synthesis *in vitro* and *in vivo*, the exact role of vitamin C in the proline hydroxylase is still debated. When purified proline hydroxylase is used, reducing substances other than ascorbic acid can participate in the reaction (e.g., tetrahydrofolate or dithiothreitol). Inasmuch as an inactive prolyl hydroxylase has been found, it has been suggested, but not convincingly established, that ascorbic acid participates in the conversion of an inactive to an active form of the enzyme.

Although it is well accepted that hydroxyproline is synthesized after proline is incorporated into proteins, a tRNA specific for hydroxyproline has been discovered [79–82].

The mucopolysaccharide content of the costochondral cartilage and the skin of scorbutic guinea pigs is reduced 15–20%. In addition, the rate of sulfate incorporation is greatly reduced in the costochondral cartilage after cortisone administration, but is restored to normal by the addition of vitamin C. Finally, ^{35}S-incorporation is also reduced in the healing wounds of scorbutic animals.

In conclusion, in spite of the simplicity of the structure of the molecules, the exact role that ascorbic acid plays as a coenzyme remains unknown. The vitamin is suspected to influence reactions in numerous metabolic pathways (see Fig. 4-19), and it also interacts in some way with hormones. Whether the role of ascorbic acid is a specific one as a coenzyme or results from its potent reductive properties is not clear. In any case, the metabolic alteration which seems to be most closely related to the clinical manifestation of scurvy is the conversion of proline to hydroxyproline.

Pernicious Anemia

Chemistry

Adison described the clinical manifestation of a chronic anemia the course of which was so infallibly fatal that it was called pernicious. The disease remained incurable until 1926 when Minot and Murphy introduced the oral administration of fresh liver extract. The success of this therapy encouraged attempts to isolate the active factor. These investigations resulted in the crystallization of vitamin B_{12}, or cyanocobalamin. The elucidation of the complex structure of vitamin B_{12} constitutes one of the most remarkable feats of modern biochemistry. The empirical formula is $C_{63}H_{88}N_{14}O_{14}PCo$. The isolation and identification of the products of vitamin B_{12} hydrolysis led to the identification of a new basic component, the ribofuranosyl-5,6-dimethylbenzimidazole (see Fig. 4-20). The ring of this compound differs from the purine ring in that the pyrimidine moiety is replaced by a benzoic ring. The nucleotide is attached

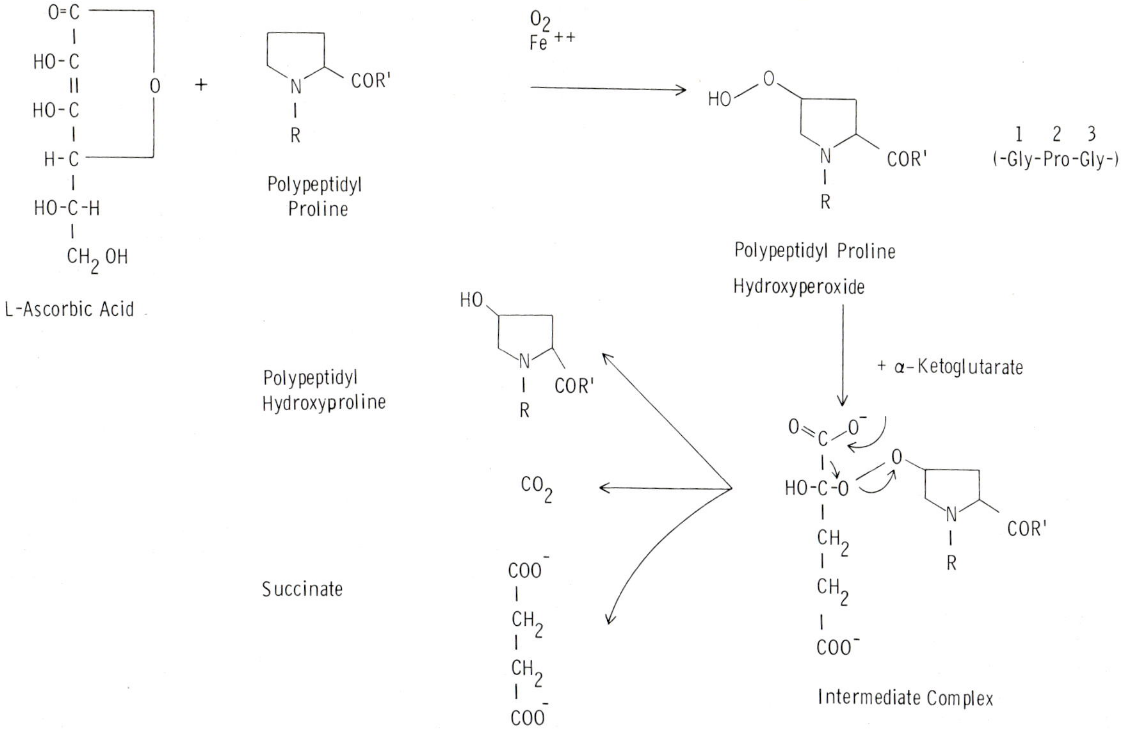

L-Ascorbic Acid + Polypeptidyl Proline → (O_2, Fe^{++}) Polypeptidyl Proline Hydroxyperoxide

1 2 3
(-Gly-Pro-Gly-)

+ α-Ketoglutarate → Intermediate Complex → Polypeptidyl Hydroxyproline + CO_2 + Succinate

Fig. 4-18. Hydroxylation of proline by collagen proline hydroxylase

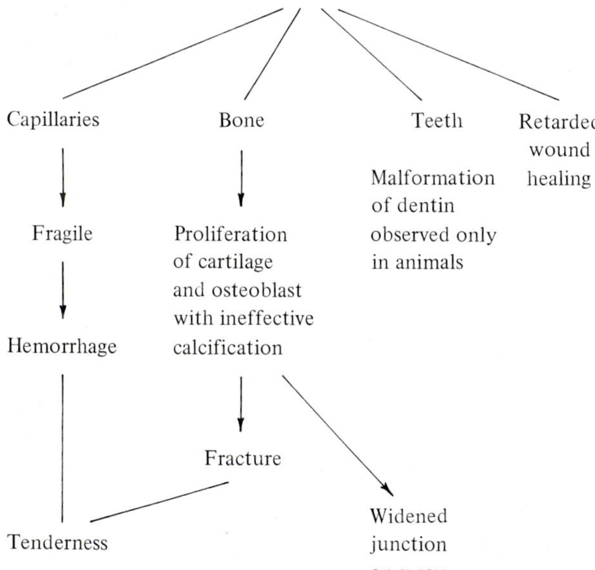

Fig. 4-19. Pathogenesis of vitamin C deficiency

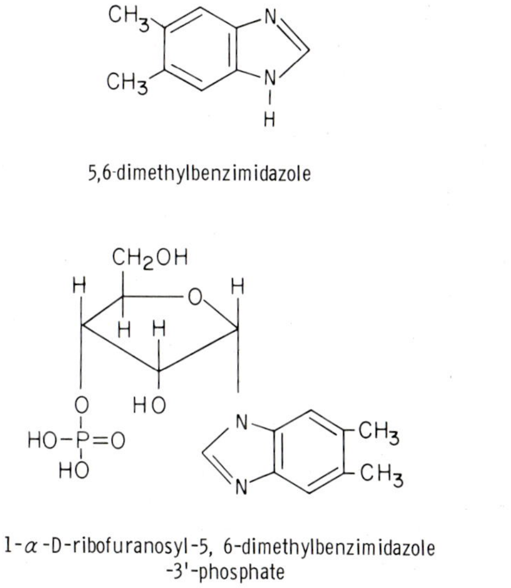

5,6-dimethylbenzimidazole

1-α-D-ribofuranosyl-5, 6-dimethylbenzimidazole -3'-phosphate

Fig. 4-20. Ribofuranosyl-5,6-dimethylbenzimidazole

to a highly substituted tetrapyrrole ring that contains four propionamide, four acetamide, and eight methyl side chains.

A propionic moiety links the amino group of one of the acetamide side chains to the phosphate of the nucleotide. In addition, the free nitrogen of the imidazole moiety of the nucleotide complexes with a cobalt atom, which is also complexed with the four amino groups of the tetrapyrrole ring and a CN moiety. Once the basic molecular constituents of vitamin B_{12} had been identified, it remained for X-ray crystallography to establish the tridimensional structure of the entire molecule as it appears in Fig. 4-21.

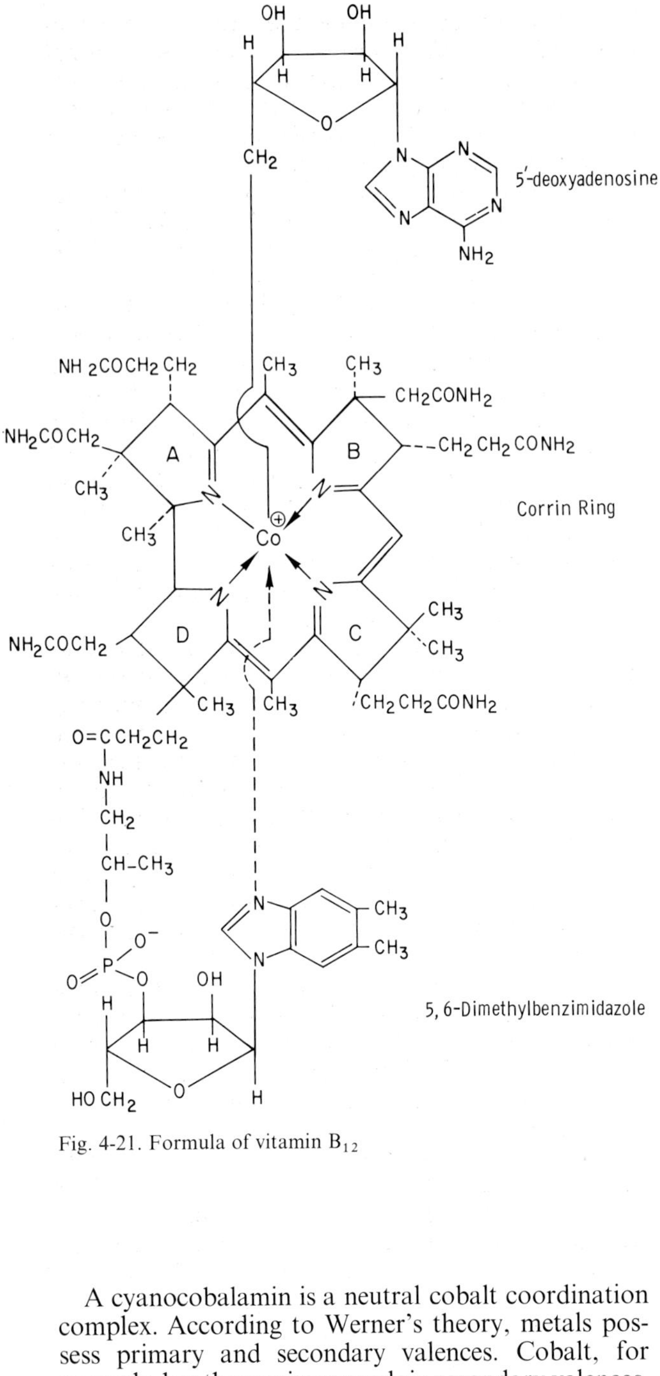

Fig. 4-21. Formula of vitamin B_{12}

A cyanocobalamin is a neutral cobalt coordination complex. According to Werner's theory, metals possess primary and secondary valences. Cobalt, for example, has three primary and six secondary valences, so its coordination number is six. Although primary valences can be satisfied only by negative ions, secondary valences can be satisfied by negative and neutral groups alike. Often a given negative group satisfies both a primary and secondary valence. In cyanocobalamin, cobalt is trivalent (three positive charges are neutralized by three negative charges), and its coordination number is six. Two of the valences are satisfied by the negative charge of a cyano group (hence the name cyanocobalamin) and the negative charge of one of the nitrogens of the porphyrin ring. The four other valences are satisfied by the three other nitrogens of the porphyrin ring and the free nitrogens of the benzimidazole ring. Since cyanocobalamin is neutral, one positive charge remains to be neutralized, which is accomplished by the phosphate anion. One of the characteristics of these coordination valences is that they are all directed in space about the central metal ion. In a metal with a coordination number of six, the space delimited by the valence forms a regular octahedron, which explains the tridimensional structure of vitamin B_{12} in which the porphyrin ring forms the plane basis that surrounds the cobalt atom. The cyanide anion is located above the cobalt atom, perpendicular to the plane of the porphyrin, and the neutral nitrogen group of the benzimidazole ring is below (see Fig. 4-21).

Vitamin B_{12} Deficiency and the Intrinsic Factor

In most cases, pernicious anemia does not result from the absence of vitamin B_{12} in the diet. The vitamin is abundant in diets that include meat, milk, and eggs. However, plant leaves are low in vitamin B_{12}, and therefore symptoms suggestive of vitamin B_{12} deficiency have been observed among confirmed vegetarians. Generally, however, the main cause of vitamin B_{12} deficiency is interference with its use [83–88].

In dog and man, vitamin B_{12} is absorbed mainly in the ileum. This explains the interference with the absorption of vitamin B_{12} observed in surgical patients whose ileum has been resected or short-circuited. But parts of the gastrointestinal tract other than the ileum may participate in vitamin B_{12} absorption. Before the vitamin is transferred to the blood of the portal vein, it may be temporarily stored in the mucosa.

The absorption of vitamin B_{12} requires the presence of the intrinsic factor. When Minot and Murphy discovered the effect of liver extract on pernicious anemia in 1926, they observed it to be effective only when administered parenterally. As in many other diseases, the first clues in the pathogenesis of pernicious anemia came from careful observations of patients by physicians. In 1929, Castle summarized his observations in a remarkable paper, "Ethiological Relationship of *Achylia Gastrica* and Pernicious Anemia." The association of achlorhydria and macrocytic anemia prompted Castle to investigate the effects of gastric juice on the disease. He found that the concomitant administration of gastric juice and beef muscle cured the anemia. In contrast, neither of these components was active alone. It has since been well established that a factor in the stomach, probably in some cellular granule, binds vitamin B_{12}. Consequently, gastrectomy in men and rats leads to pernicious anemia. Furthermore, rat stomach extract increases the uptake of the vitamin by the small intestine, and rat gastric juice increases the uptake of vitamin B_{12} in gastrectomized rats. Thus, the intrinsic factor undoubtedly exists and is involved in vitamin B_{12} absorption, despite some statements to the contrary. The factor has been prepared in a crude form from dried powder of hog stomach, and is nondialyzable. It is resistant to alkali but is destroyed by acid and precipitates with ammonium sulfate. Therefore, this factor is undoubtedly a protein.

Cell fractionation studies of homogenates of the gastric mucosa have demonstrated that the intrinsic factor is found both in mitochondria and the supernatant fluid. The factor remains difficult to assay; a variety of methods have been used, but none is entirely satisfactory. These methods include: (1) administration of intrinsic factor preparation to gastrectomized animals (rats or hogs); (2) measurement of vitamin B_{12} binding capacity of the preparation; (3) measurement of vitamin B_{12} uptake in tissue; and (4) evaluation of the inhibition of cobamide coenzyme activity in the glutamate isomerase reaction.

In humans, the two principal methods used for assaying the intrinsic factor are the reticulocyte response after administration of the intrinsic factor to pernicious anemia patients, and studies of the effect of intrinsic factor on urinary or fecal excretion, on hepatic uptake, on blood plasma radioactivity, and total body radioactivity after the administration of labeled vitamin B_{12}.

The factor was first purified from hog stomach and human gastric juice. Hog stomach intrinsic factor was prepared by treating the pyloric mucosa with pancreatin. The extract was then submitted to ion-exchange chromatography, passed through a calcium phosphate gel, and placed in the ultracentrifuge in the presence of vitamin B_{12}. This procedure yields a homogeneous preparation that combines with vitamin B_{12}. Only tentative attempts to purify the intrinsic factor from human gastric secretions have been made. In these experiments, the human gastric secretions are submitted to electrophoresis, and peaks obtained are assayed for their ability to stimulate the binding of vitamin B_{12} to tissue slices. Usually several peaks are obtained, but not all of them stimulate the binding of vitamin B_{12}. Richmond and his associates have obtained a peak free of blood-group contaminants and peptic enzymes. It is important to establish the specificity of the various types of intrinsic factors. Furthermore, the elucidation of the molecular structure of human intrinsic factor may shed new light on the pathogenesis of pernicious anemia. At present, the possibilities cannot be excluded that the disease results either from an inborn error of metabolism responsible, for example, for elaborating an intrinsic factor of poor activity, or from an autoimmune disease in which the body builds antibodies antagonizing its own protein.

The purified human factor has a molecular weight of 60,000 and may exist in the form of a dimer (mol wt 120,000). The factor is a mucoprotein with great affinity for vitamin B_{12}. The factor and the vitamin form a complex in the presence of calcium at pH values larger than 6. At least three forms of vitamin B_{12} malabsorption resulting from the distorted elaboration and use of the intrinsic factor have been described [83–86]. These include congenital absence of intrinsic factors; secretion of adequate quantities of immunologically active but physiologically inert intrinsic factors; and secretion of adequate quantities of normal intrinsic factors with a defect in intestinal reabsorption. The complex is then attached to specific receptor sites on the microvillous membranes of the ileum to be absorbed by pinocytosis [89].

The binding of corrinoids has been studied under various circumstances. It appears that the 5,6-dimethylbenzimidazole is involved in the formation of the cobalamin intrinsic factor complex [158].

Undoubtedly, the intrinsic factor is required for vitamin absorption. The total amount of vitamin B_{12} absorbed by the gastrointestinal tract is not determined by the amount of vitamin present there, but by some intestinal factor. Thus, 1 mg of vitamin B_{12} is absorbed when 5 mg of the vitamin is ingested, but only 1.5 mg is absorbed when 50 mg of vitamin B_{12} is ingested. The intestinal control of vitamin B_{12} absorption reminds one of the mechanism controlling the absorption of iron by the intermediate of apoferritin. Evidence suggests that at some stage of absorption, the vitamin is bound to the intrinsic factor.

The injection of purified intrinsic factor into rabbits induces the formation of specific antibodies. Although those antibodies have been of little help in purifying the factor, the immunological properties of the intrinsic factor are important for at least three reasons: (1) the antibodies can be used for testing the homogeneity of the preparation; (2) the formation of antibodies probably explains resistance to the intrinsic factor which develops in some patients treated with purified intrinsic factor; and (3) pernicious anemia may result from an autoimmune reaction involving the intrinsic factor.

In patients under prolonged therapy with the intrinsic factor, antibodylike substances can be demonstrated in the sera. The antibodies are specific for a given species of intrinsic factor, and their level decreases when the administration of intrinsic factor is interrupted.

Numerous laboratories have attempted to demonstrate a relation between the clinical efficiency of the intrinsic factor and its capacity to combine with the vitamin. If the purified intrinsic factor separates into several peaks, either by electrophoresis or by chromatography on a cellulose column, the fractions with the greatest capacity for binding the vitamin have the most active intrinsic capacities when measured clinically.

All these observations suggest that the binding capacity of the intrinsic factor and its therapeutic efficiency are closely correlated. Furthermore, anything that interferes with binding also interferes with the therapeutic efficiency. Acetylation of the intrinsic factor inhibits its activity without interfering with the binding of vitamin B_{12}. Therefore, it was concluded that in addition to those parts of the molecule involved in the binding, other components of the molecular structure are needed to confer biological activity. The intrinsic factor is not the only protein with which vitamin B_{12} may combine. Indeed, the vitamin is also bound by serum, cerebrospinal fluid, and saliva. In serum, under normal conditions the binding involved is β_2-globulin almost exclusively, and only when large excesses of vitamin B_{12} are administered are gammaglobulins also involved in the binding.

Cooper and Castle have proposed a three-step sequence to explain the vitamin B_{12} absorption in the gastrointestinal tract. In the first step, vitamin B_{12} binds with the intrinsic factor in gastric juice. The affinity of the vitamin for the intrinsic factor is thought to be greater than its affinity for proteins in the intestinal content; consequently, the intrinsic factor (IF) successfully displaces the vitamin from its weaker bonds with other proteins. Calcium facilitates and EDTA inhibits the absorption of vitamin B_{12} by the everted intestine. On the basis of these and related findings, workers proposed the second step in vitamin B_{12} absorption. At that stage it is assumed that the intrinsic factor-vitamin B_{12} complex is trapped in the intestinal wall by the intermediate of calcium bonds and absorbed by pinocytosis. This stage of the absorption process probably is interfered with in sprue and steatorrhea where calcicum soaps are formed in the intestinal lumen.

The mechanism of interference with the absorption of vitamin B_{12} in the presence of normal levels of IF and in the absence of IF antibodies is still unclear. It was first proposed that the defect results from the absence of ileal receptors for the vitamin B_{12}-IF complex. Yet, observations made on homogenates of intestines obtained from a child with vitamin B_{12} malabsorption clearly suggested that the IF-vitamin B_{12} complex was perfectly capable of binding to the mucosa. Consequently, it was concluded that the absorption of the vitamin must be interfered with at a step located beyond that of the binding of the complex to the receptor.

In the last step of vitamin B_{12} absorption, the vitamin is assumed to be released from the intrinsic factor by releasing factors specific to each species. The "enzymic" breakdown of the intrinsic factor-vitamin B_{12} complex has been demonstrated using intestinal juice and intestinal homogenates. The releasing factor has not yet been isolated, and other tissues (kidney, liver, muscle, heart, spleen, and lung) also contain factors capable of releasing vitamin B_{12}. As a matter of fact, the megaloblastic anemia that develops when the gastrointestinal tract is infected by the fish tapeworm *Diphyllobothrium latum* results from the elaboration of a "releasing factor" by the parasite.

The possibility that the intrinsic factor-vitamin B_{12} complex is absorbed *per se* cannot be excluded simply on the basis that the intrinsic factor is a protein, since it is known that proteins such as ribonuclease enter the cell.

In addition to the intrinsic factor, human tissues contain a number of proteins capable of binding vitamin B_{12} derivatives. These proteins include a plasma transcobalamin that facilitates B_{12} uptake in various tissues and a so-called R type group with unknown function whose binding capacities for vitamin B_{12} seem to vary with their carbohydrate content [159]. Many of the proteins binding to vitamin B_{12} derivatives have been partially purified by affinity chromatography [160–163].

Clinicopathologic Correlations

Pernicious anemia is a chronic macrocytic anemia, usually associated with atrophy of the gastric mucosa, with achlorhydria and degeneration of the nervous system and most often affecting people over 30 years of age. Its incidence is higher in temperate climates. Scandinavians and Anglo-Saxons seem more prone than members of other ethnic groups. To what extent heredity is involved is not known.

A hereditary transmission of pernicious anemia was suspected long ago on the basis of clinical observation of relatives of patients with pernicious anemia. The assumption that pernicious anemia is inherited has received more scientific support from the careful observations of 106 relatives of 34 patients affected with pernicious anemia. The individuals under observation were submitted to the Schilling test for vitamin B_{12} absorption. A significantly large proportion of individuals had low test values among the relatives of pernicious anemia patients. Workers concluded that a predisposing factor in pernicious anemia is transmitted as a dominant autosomal gene. Before such a concept is definitely accepted, more careful studies of the pedigrees of pernicious anemia patients must be made. A test more easily controlled than the Schilling test is also desirable; the Schilling test is susceptible to alteration by a variety of factors, such as certain intestinal conditions. Patients who have gastrointestinal or renal diseases score low on the test.

The study of the incidence of achlorhydria and vitamin B_{12} absorption in relatives of patients affected with pernicious anemia has shown that, although there is no increase in the incidence of achlorhydria, the absorption of vitamin B_{12} is impaired in 40% of the relatives tested. On the basis of these studies, it has been suggested that an autosomal dominant gene is responsible for vitamin B_{12} absorption. These relatives with reduced ability to absorb vitamin B_{12} must then be heterozygous for the gene controlling the absorption. Individuals with below normal ability to absorb vitamin B_{12} are more prone to pernicious anemia, although the precipitating factor is not known (age atrophy of the mucosa has been suggested).

Usually, long before the hematological and neurological symptoms develop, a typical atrophic gastritis exists. Gastritis is diffuse, but it is most manifest in the area of the fundus, the very area where the intrinsic factor is formed. Grossly, the mucosa of the stomach is reddish and thin. The rugae are poorly developed and sometimes may be completely absent. Microscopically, the most striking finding is the reduced number of chief, parietal, and argyrophil cells in the glandular mucosa. Other signs of chronic gastritis are fibrosis of the mucosal stroma and infiltration with lymphocytes. A frequent observation is the thickening of the muscularis mucosa. Of course, without chief cells, histamine-resistant achlorhydria develops, and all the symptoms associated with the defect occur—indigestion, constipation, looseness of the bowels, gas. The tongue is also frequently affected in pernicious anemia. Its mucosa is thin and reddish, and it is sometimes ulcerated. Gastric smears obtained from pernicious anemia patients have occasionally been diagnosed incorrectly as malignant because of the cytological abnormalities. Squamous epithelial cells from the mouth are swallowed and can be collected with the stomach content. In addition, the histiocytes and the columnar cells of the gastrointestinal tract are larger than normal. Similar giant cells with fine reticular chromatin structure may be found in the mouthwashings of patients with pernicious anemia, and examination of mouthwash is considered by some as a good diagnostic procedure.

However, the hematological changes are the most typical findings in pernicious anemia. They consist mainly of a severe macrocytic and hyperchromic anemia. For reasons that are not clear, the disease is not progressive; it evolves in a succession of intermissions and relapses during which the number of red cells drops to onefifth of its normal count. But some cells are bigger than normal, therefore, the amount of hemoglobin per cell, not per unit volume, is greater than in iron deficiency anemia. Thus, the anemia has been called macrocytic and hyperchromic.

Not all the cells are bigger than normal; some may even be smaller. The size of the mature erythrocyte varies widely, a phenomenon called anisocytosis. The best way to demonstrate anisocytosis is to measure on smear preparations the size of a large number of red cells and plot a frequency curve that presents several peaks in pernicious anemia instead of a sharp peak as in the normal range. If appropriate stainings are made, immature red cells are found in the blood smear. In other words, red cells containing basophilic remnants are commonly found in the peripheral blood of pernicious anemia patients. Also, nucleated red cells of the normoblastic or megaloblastic type are present. Although the changes in the red blood cells are the most conspicuous, the absolute number of polymorphonuclears decreases, leading to a decrease in the white blood cell count and a relative increase in lymphocytes. Furthermore, polymorphonuclear maturation is altered; many of them develop nuclei with an increased number of lobes (5–8). Insofar as the life span of pernicious anemia red cells is somewhat reduced, as demonstrated by studies with radioactive iron, a certain degree of hemolytic jaundice is always associated with the anemia. Thus, the blood bilirubin is increased and the combination of marked pallor and slight jaundice leads to the particular complexion of the patient affected with pernicious anemia which the French have called "la couleur jaune paille de l'anemie pernicieuse."

Pernicious anemia is not of the anaplastic type; the stem cells of the red cell can still proliferate. But some step of the maturation process is blocked or slowed down, and the bone marrow tends to compensate for this deficiency by hyperplasia. Consequently, the fatty spaces decrease, the cellular density increases, and the proportion of cells of the erythropoietic lineage is greater than that of the myelopoietic lineage. But the characteristic findings in the bone marrow of pernicious anemia patients are megaloblasts, polychromatophils, and hemosiderin loaded histyocytes (see Fig. 4-22). The megakaryocytes are usually reduced in number. This leads to a reduction of the number of thrombocytes in blood, which explains the occurrence of small hemorrhages in the kidney, spleen, and liver.

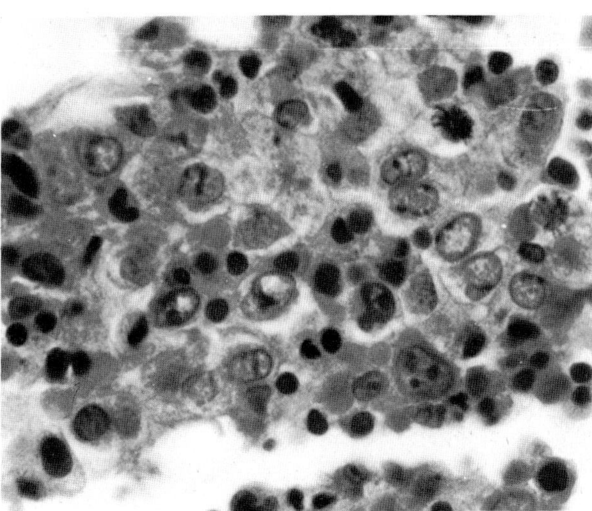

Fig. 4-22. Bone marrow smear from a patient with pernicious anemia with numerous megaloblastic erythroid precursors (Wright's stain)

In advanced pernicious anemia, diffuse neurological degeneration primarily affects the fibers of the posterior and lateral tracts of the spinal cord and of the direct and indirect pyramidal tracts, with progressive demylination of all those structures. Sometimes the ganglia of the dorsal root and peripheral nerves also degenerate. When the posterior columns are the first to degenerate, sense of position and deep reflexes are lost, as in tabes dorsalis. But later, when the pyramidal tracts are also affected, the superficial reflexes increase, the deep reflexes decrease, and pathological reflexes typical of disease of the pyramidal tract appear. The neurological symptoms may sometimes appear before anemia or may even be completely independent of the anemia, as was the case in a group of vegetarians observed recently in England.

Role in Metabolism

Like thiamine, vitamin B_{12} is not the active coenzyme. The active coenzyme was identified a decade after the vitamin was discovered [90–93].

Barker discovered the need for cobamide coenzymes during his studies on the conversion of glutamate to

β-methylaspartate in *Clostridium tetanomorphum*. It was later shown that the coenzyme differed from cyanocobalamin in two ways: the absence of cyanide and the presence of 5-deoxyadenosine in the molecule. The 5′-carbon of the deoxyribose of 5-deoxyodenosine is linked to the central cobalt atom by a covalent bond.

X-Ray diffraction studies have established the position of the adenosine with respect to the plane of the corrin ring. The plane of the deoxyribose moiety is perpendicular to that of the corrin ring; that of the adenine moiety is parallel to that of the corrin ring.

These considerations on the structure of the molecule of coenzyme B_{12} may seem rather trivial. Although the physician can practice medicine adequately without being thoroughly familiar with the structure of the molecule of vitamin B_{12}, he should be aware that the elucidation of that structure is vital to an adequate understanding of the vitamin's mechanism of action. Misconception on the structure of vitamin has delayed the organic synthesis of the coenzyme. At first the adenine was assumed to be attached to the cobalt by the intermediate of one of its nitrogens, and it was also believed that the cobalt in the coenzyme was divalent rather than trivalent. Only when X-ray diffraction and other physicochemical investigations suggested that the cobalt was trivalent rather than divalent was it possible for organic chemists to seek a new approach to the biosynthesis of the coenzyme, and this time their endeavors were successful.

The coenzyme can be synthesized by reacting the cobamide with adenosine esterified at C-5′ with a strong acid. But it is difficult for the organic chemist to make such esters because cyclization readily occurs between the carbon 5′ of the deoxyribose and the nitrogen of the base, thus making this nitrogen quaternary. Furthermore, in these attempts to esterify the carbon 5′ of deoxyadenosine with strong acid, investigators find that carbons 2′ and 3′ are esterified with the carbon 5′.

These difficulties were overcome in various ways, two of which were (1) isopropylidene tosyladenosine is condensed with vitamin B_{12}; the acetone is removed from this compound by mild acid hydrolysis, and (2) the 5′-trityl adenosine is first synthesized and then converted to the 2′,3′-diacetyl trityl derivative, which is detritylated by dilute acid. The 2′,3′-diacetyl coenzyme is then prepared by tosylation and condensation. The compound is then deacetylated to give coenzyme B_{12} in good yields.

Such methods have permitted the synthesis of coenzyme B_{12} and coenzyme B_{12} analogs that can now be used to study the mode of action of coenzyme B_{12} in biochemical pathways. It is not known by what mechanism the 5′-deoxyadenosine group is transferred to cobamide in the cell. It has been suggested that an enzymic cobamide reaction requiring ATP as an adenosine donor is involved.

In tissues, vitamin B_{12} is converted into the 5′-deoxyadenosyl derivative in a reaction involving cyanocobalamin and ATP. Mg^{++}, K^{+}, FMN, and sulfhydryl groups are also required.

The ATP acts as adenosyl donor. Ribose is reduced to the deoxyribose in the process.

Comparative studies of the physiological properties of vitamin and coenzyme B_{12} conducted in humans have suggested a more rapid uptake by erythrocytes and greater efficiency of the coenzyme in stimulating the hematopoietic response. Coenzyme B_{12} has now been established to play a role in a number of enzymic transformations, most of which occur in bacteria, but a few have been described in mammalian cells [94].

In a number of reactions catalyzed by coenzyme B_{12} and the appropriate apoenzyme, the carbon skeleton of the molecule is drastically rearranged. Such reactions include the conversion of glutamate to β-methylaspartate, succinyl CoA to methylmalonyl CoA, and α-methylene glutarate to methylitaconate (see below).

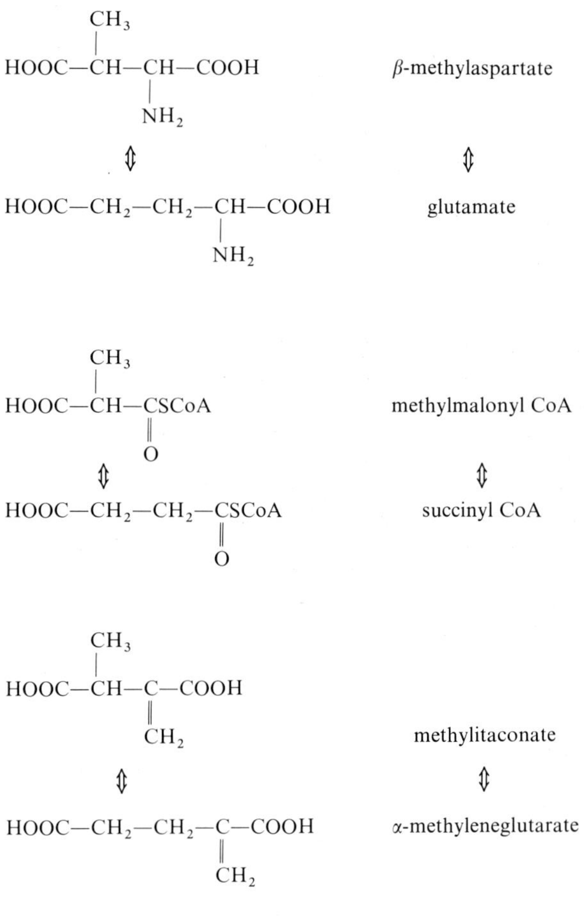

$$\mathrm{HOOC{-}CH(CH_3){-}CH(NH_2){-}COOH} \quad \beta\text{-methylaspartate}$$
$$\Updownarrow$$
$$\mathrm{HOOC{-}CH_2{-}CH_2{-}CH(NH_2){-}COOH} \quad \text{glutamate}$$

$$\mathrm{HOOC{-}CH(CH_3){-}C({=}O)SCoA} \quad \text{methylmalonyl CoA}$$
$$\Updownarrow$$
$$\mathrm{HOOC{-}CH_2{-}CH_2{-}C({=}O)SCoA} \quad \text{succinyl CoA}$$

$$\mathrm{HOOC{-}CH(CH_3){-}C({=}CH_2){-}COOH} \quad \text{methylitaconate}$$
$$\Updownarrow$$
$$\mathrm{HOOC{-}CH_2{-}CH_2{-}C({=}CH_2){-}COOH} \quad \alpha\text{-methyleneglutarate}$$

In others, amino groups are transferred from one position of the carbon chain to another. Such reactions are catalyzed by aminomutases. At least three B_{12} coenzyme-dependent aminomutases are known: L_1-β-lysine, D_1-lysine, and ornithine aminomutase. Coenzyme B_{12} also catalyzed in the presence of the appropriate apoenzyme the conversion of ethylene glycol and 1,2-propanediol to acetaldehyde and propionaldehyde, respectively. The apoenzyme involved in these reactions is a diodehydrase (see below).

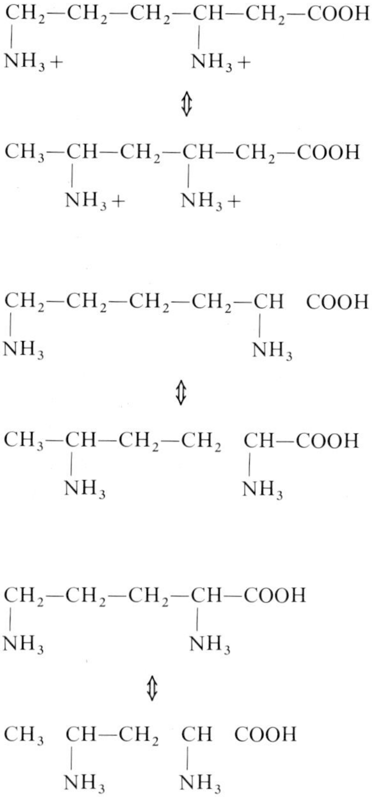

$$\underset{NH_3+}{CH_2}-CH_2-CH_2-\underset{NH_3+}{CH}-CH_2-COOH$$ lysine (3,6-diaminohexanoate)

⇕

$$CH_3-\underset{NH_3+}{CH}-CH_2-\underset{NH_3+}{CH}-CH_2-COOH$$ 3,5-diaminohexanoate

$$\underset{NH_3}{CH_2}-CH_2-CH_2-CH_2-\underset{NH_3}{CH}\quad COOH$$ D-lysine (2,6-diaminohexanoate)

⇕

$$CH_3-\underset{NH_3}{CH}-CH_2-CH_2\quad \underset{NH_3}{CH}-COOH$$ 2,5-diaminohexanoate

$$\underset{NH_3}{CH_2}-CH_2-CH_2-\underset{NH_3}{CH}-COOH$$ ornithine

⇕

$$CH_3\quad \underset{NH_3}{CH}-CH_2\quad \underset{NH_3}{CH}\quad COOH$$ 2,4-diaminovalerate

Another B_{12} coenzyme-dependent dehydrase converts glycerol to 3-hydroxypropionaldehyde. A B_{12}-dependent ethanolamine deaminase catalyzes the irreversible conversion of ethanolamine to acetaldehyde and ammonia. A corrinoid protein participates in methyl transfer reactions and methane formation by bacteria producing methane. Corrinoid proteins are also involved in acetate synthesis from carbon dioxide.

In some bacteria, vitamin B_{12} can be replaced in the culture medium by deoxyribosides, thus suggesting that vitamin B_{12} plays a role in the conversion of ribose to deoxyribose. Diphosphate ribose nucleotide usually serves as the substrate for the enzyme. The 2-hydroxyl group of the ribose nucleotide is directly replaced by a hydrogen. The exact mechanism of the hydrogen transfer is not known, but it is believed to occur in the form of a hybrid ion, which is derived from a reducing agent.

In *Lactobacillus leischmanii,* reduced thioredoxin is the natural reductant, but dithiols (dihydrolipoate, dithiothreitol) and monothiols (2-mercaptoethanol and glutathione) have been found to be active reductants. In this reaction, coenzyme B_{12} is believed to act as a hydrogen carrier.

In vitamin B_{12}-deficient rats, the urinary excretion of methylmalonate is markedly increased, and the levels of methylmalonyl CoA mutase activity in liver and kidney homogenates, as well as the concentrations of deoxyadenosylcobalamin, are decreased in various tissue. The levels of deoxyadenosylcobalamin may be inadequate for full saturation of the methylmalonyl CoA mutase even in normal tissue. Thus, adding the coenzyme to homogenate markedly increases the mutase activity.

Whether the increase in methylmalonate levels in patients with vitamin B_{12} deficiency and pernicious anemia is significant in the pathogenesis of the disease is not certain, but because of the structural similarity between methylmalonic CoA and malonyl CoA, it has been suspected that methylmalonyl CoA might replace malonyl CoA in the biosynthesis of fatty acids, thus yielding methyl-branched rather than straight-chain fatty acids. This was found to occur *in vitro,* but no branched-chain fatty acid could be detected in the liver of vitamin B_{12}-deficient pigs. A study of the effects of vitamin B_{12} deprivation made with rat glial cells grown in culture indicated that the cells made increasing amounts of unbranched 17 and 15 carbon long fatty acids. As the ability to metabolize propionic acid through the methylmalonyl (CoA) mutase reaction increased, increased activities of fatty acid synthetase and acetyl coenzyme A carboxylase were also found in the liver and brain of rats deprived of vitamin B_{12} [164–165]. However, in at least one case of hereditary methylmalonic aciduria, methyl-substituted hexodecanoic acids were found in the nervous tissue [95].

As was already pointed out, methylmalonic aciduria occurs in children as a hereditary disease. Some patients are responsive to large doses of vitamin B_{12}. The biochemical pathogenesis of the disease has been clarified at least in part. Cultures of fibroblasts obtained from such patients oxidize propionate and methylmalonic acid to CO_2 much more slowly than do fibroblasts obtained from normal cells. But if large amounts of hydroxycobalamin are added to the culture medium, methylmalonate and propionic acid oxidation is restored to normal, thus excluding a defect in the methylmalonate mutase levels. Assays for deoxyadenosylcobalamin in fibroblasts obtained from methylmalonic aciduria patients and normal individuals revealed that the concentration of coenzyme is in the mutant only 10% of that in the normal fibroblast. Inasmuch as the mechanism of vitamin B_{12} conversion to the deoxyadenosylcobalamin coenzyme is not known, except for the fact that several enzymic steps are involved, the exact nature of the defect in methylmalonic aciduria cannot be ascertained.

In the reaction, the cobalamin coenzyme acts as the methyl carrier. The reaction has been described in two steps. In the first, the holoenzyme (a salmon pink protein) is, in the presence of THF, adenosylmethionine, and the reducing system is converted to comethyl holoenzyme (red protein). In the second step, the methylated holoenzyme attacks the homocysteine molecule, and methionine and demethylated holoenzymes are formed.

Some patients with methylmalonic aciduria also have increased levels of homocysteine and cystathione in the urine and decreased levels of methionine in tissues. Again, the defect can be overcome in fibroblasts by adding large amounts of hydroxycobalamin to the incubation mixture, suggesting that the defect results either from inadequate uptake of the vitamin or from inadequate conversion of the vitamin to the active coenzyme.

Since interference with the metabolism of methylmalonic acid can be corrected by adding large doses of hydroxycobalamin to the medium, absence of the enzyme as a cause of the disease is excluded. Three other possibilities are a reduced permeability to the vitamin, an inability to convert the vitamin to coenzyme, and an inability of the apoenzyme to bind the coenzyme.

Inasmuch as the levels of deoxyadenosine B_{12} are low in fibroblast, the defect in methylmalonic aciduria appears to result from an inability to convert the vitamin to the coenzyme.

Buchanan and his collaborators [96] have studied the role of coenzyme B_{12} in methionine synthesis. They have demonstrated that methionine is synthesized by transfer of the methyl group of 5,10-methyltetrahydrofolate to homocysteine (see Folic Acid).

5,10-methylene tetrahydrofolate + NADH + H^+

→ (5,10-methylene tetrahydrofolate reductase)

5-methylfolate H_4 + homocysteine

→ ($FADH_2$ S-Adenosylmethionine transferase)

methionine + folate H_4

In bacteria, two different types of transferases are found: (1) one prepared from *Escherichia coli* contains vitamin B_{12} and requires $FADH_2$ and S-adenosylmethionine for activity; and (2) another enzyme present in *Aerobacter aerogenes* does not contain vitamin and has no requirements other than magnesium and P_i for activity.

5,10-Methyltetrahydrofolate is an intermediate for methionine biosynthesis in mammalian tissues. ATP is required when crude mammalian transferase preparations are used, but ATP can be replaced by S-adenosylcysteine. Furthermore, a requirement for $FADH_2$ was established. Purification of the transferase of pig liver strongly suggests that the enzyme contains vitamin B_{12}.

Inasmuch as TTP, one of the four triphosphate nucleotides needed for DNA synthesis, is formed through the methylation of UMP to TMP which is then further phosphorylated, one may expect that a reduction in methyl transfer would reduce the levels of TTP and thereby cause interference with DNA synthesis and maturation of the red cell. Yet, the pools of TTP in lymphocytes were normal in untreated patients deficient in vitamin B_{12}. In contrast, in patients treated with methotrexate a marked drop in the TTP pool is found [166]. A drop in thymidylate synthetase activity in phytohemagglutinin stimulated lymphocytes of patients with pernicious anemia has been described. In summary, the relationship between the block of methylation, TMP synthesis, and cell maturation requires much more work before it can be understood [167].

Studies of the effect of vitamin B_{12} on the overall rate of DNA and RNA synthesis have yielded contradictory results. Although there have been reports suggesting, for example, that vitamin B_{12} deficiency reduces the rate of DNA and RNA synthesis in regenerating liver, in better controlled experiments in which ^{32}P-incorporation was measured in the nucleic acid of a variety of tissues, an effect of vitamin B_{12} seems to be excluded. In the bone marrow of pernicious anemia patients, the DNA per cell is not altered, but the capacity of the bone marrow stem cells to mature into normoblasts is affected. When vitamin B_{12} is added to the culture medium, cells mature normally.

The megaloblast differs from the normoblast in its increased amount of cytoplasmic RNA. The DNA and the nuclear RNA are apparently unchanged, and therefore the thymine-uracil ratio is decreased. From such observations, it was concluded that the megaloblasts are red blood cell precursors whose interphase is abnormally prolonged because of interference with DNA synthesis. On the basis of the alteration of thymine-uracil ratios, it was assumed that the metabolic block would be located on the passage from deoxyuridylic acid to thymidylic acid. Such conclusions are unwarranted at present because the role of vitamin B_{12} in the methylation of thymidine is not established. Furthermore, it is not certain that cytoplasmic RNA acts as a reservoir for DNA as the above hypothesis implies.

The effect of B_{12} on several steps of protein synthesis—including synthesis of the methyl group of choline and amino acid activation—has also been investigated. Vitamin B_{12} stimulates the incorporation of the α-carbon of glycine into choline, but it doesn't affect the incorporation of glycine into serine or ethanolamine or the transmethylation of betaine to homocysteine. Furthermore, glycine serves as a precursor for choline only when vitamin B_{12} is present. The conclusion of the experiment is that although vitamin B_{12} is not needed for transmethylation, it is required for the synthesis of the methyl group of choline. This is interesting in view of choline's role in preventing of fatty livers. Indeed, in rats submitted to a soybean diet, the addition of 20% lard oil increases 20-fold the requirement for vitamin B_{12}, despite the fact that the vitamin is normally absorbed and stored. In relation to the properties of vitamin B_{12} in stimulating methyl group synthesis, it is remarkable that the vitamin stimulates the production of cancer induced with paradimethylaminoazobenzene. The demethylated compound is noncarcinogenic, and therefore an excess of methyl donors preserves the carcinogen.

In vitamin B_{12} deficiency, the incorporation of serine into protein is reduced, and the amount of albumin, α-globulin, and γ-globulin in the blood is also reduced. Of course, such experiments do not conclusively prove that vitamin B_{12} directly affects protein synthesis, yet

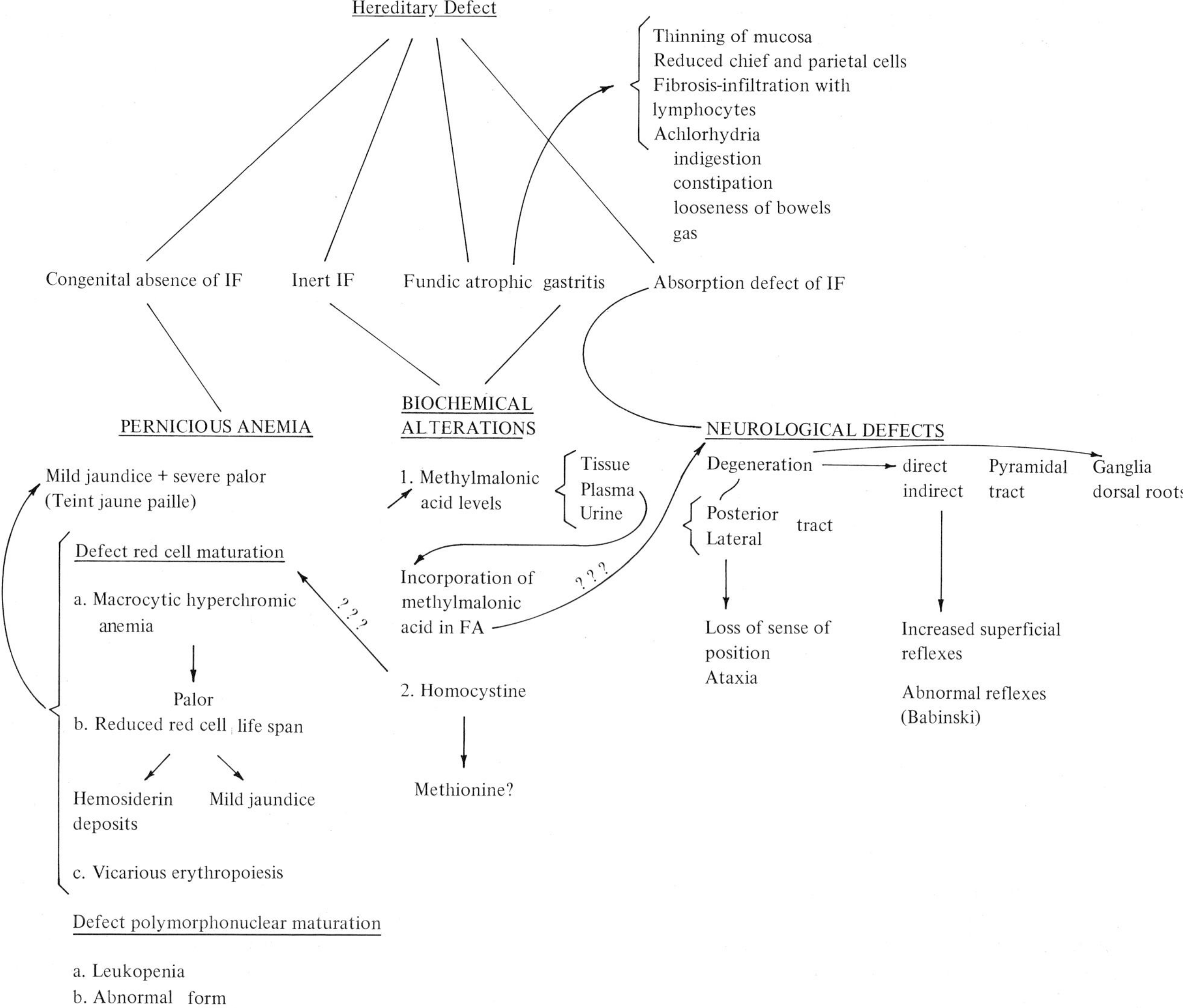

Fig. 4-23. Pathogenesis of pernicious anemia

this concept has been extensively supported. Amino acid incorporation in the microsomes *in vitro* was shown to be decreased in B_{12}-deficient animals. When ^{60}Co-labeled vitamin B_{12} was administered, it was found to be associated with the liver pH 5 enzyme, which is responsible for activating amino acids, and is found in the liver supernatant fluid in a fraction that precipitates with 40–60% ammonium sulfate. Furthermore, when this enzyme fraction was prepared from deficient rats and placed in the presence of normal microsomes, no decrease in the incorporation was observed. On the contrary, when the enzyme was prepared from normal rats and incubated with microsomes from deficient animals, amino acid activation definitely decreased. Therefore, investigators postulated that vitamin B_{12} stimulates amino acid activation.

Pernicious anemia is one of the many diseases for which an adequate form of therapy has been developed, thanks to the competent observations of physicians and the assiduous efforts of chemists. But the pathogenesis of the disease remains unsolved. Vitamin B_{12} was found to interfere with two key reactions: donation of methyl groups, and the conversion of methylmalonyl CoA to malonyl CoA. Logically, interference with the first reaction should block the biosynthesis of TMP, the precursor of TTP, a precursor of DNA and thereby interfere with DNA synthesis and cell maturation. A defect in the second reaction could distort the biosynthesis of fatty acids and be responsible for the nervous symptoms in the brain. Yet, the sequence of steps that lead from the biochemical injury to the clinical symptoms is far from elucidated. If it were, not only would we have learned much about the mechanism of red cell maturation, but we might also have a clue to the mechanism of some forms of demyelinization. The pathogenesis of pernicious anemia is summarized in Fig. 4-23.

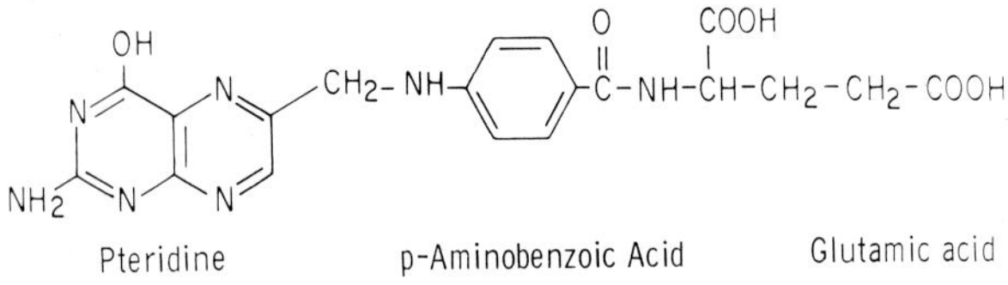

Pteridine p-Aminobenzoic Acid Glutamic acid

Fig. 4-24. Pteroylglutamic acid (folic acid)

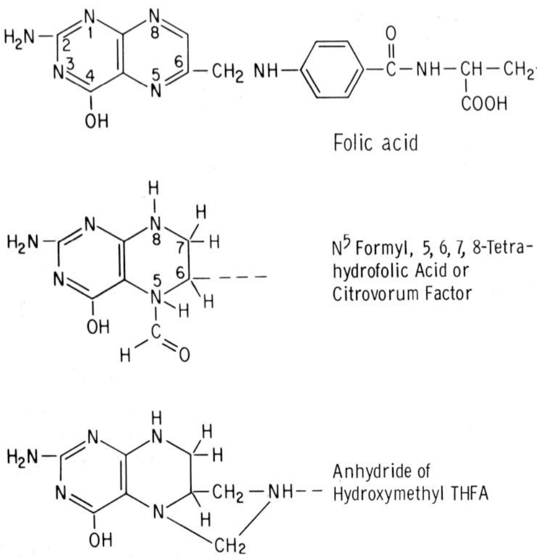

Folic acid

N^5 Formyl, 5, 6, 7, 8-Tetrahydrofolic Acid or Citrovorum Factor

Anhydride of Hydroxymethyl THFA

Fig. 4-25. Folic acid and its derivatives

Folic Acid

When efforts were made to isolate the maturation factor present in liver extracts, folic acid rather than vitamin B_{12} was discovered. The discovery of folic acid generated great hopes. Unfortunately, they soon faded away when it became evident that the initial cure of pernicious anemia induced by folic acid could not be maintained by prolonged therapy, and that folic acid was without effect on the nervous lesions associated with the anemia. But the discovery of folic acid was an important landmark in biochemistry because the vitamin is involved in numerous steps of amino acid, purine, pyrimidine, and nucleic acid metabolism [97, 98]. Folic acid is N-(4-[(2 amino-4-hydroxy-6-pteridyl) methyl] amino benzoyl) glutamic acid. The molecule is thus made of a pteridyl ring, a benzoic ring, and glutamic acid (see Fig. 4-24). The pteridyl ring has an amino group in position 2 and a hydroxyl group in position 4. The benzoyl ring is attached to the carbon 6 of the pteridyl ring by the intermediate of the methyl amino group. Folic acid is not the molecular form that is active in metabolic reactions. It was suspected that folic acid was converted to more effective coenzymes when liver extracts were found to be 10–100 times as active as folic acid in overcoming the toxic effects of folic acid antimetabolites. The presence of folic acid derivatives in addition to folic acid explains this beneficial effect of liver extracts. These folic acid derivatives are all reduced compounds derived from 5,6,7,8-tetrahydrofolic acid. At least four of these compounds have been identified in mammals: N^{10}-formyltetrahydrofolic acid, N^5, N^{10}-methenyltetrahydrofolic acid, N^5, N^{10}-methylenetetrahydrofolic acid, and N^5-methyltetrahydrofolic acid (see Fig. 4-25). N^5-Formyltetrahydrofolic acid (citrovorum factor) is not found in mammals, but it can be converted to the N^5, N^{10}-methenyltetrahydrofolic acid. In addition, a variety of animal and bacterial preparations contain pteroyltriglutamic acid. A pteroylheptaglutamic acid has been found in yeast.

In bacteria, two enzyme systems catalyze the reduction of folic acid: folic acid reductase and dihydrofolic acid reductase (see Fig. 4-26). In rat liver these two enzymic steps are catalyzed by a single reductase. The mammalian reductase has been purified from sheep liver and cow thymus. With the sheep liver enzyme, both NADH and NADPH may serve as hydrogen donors. The pH of the incubation medium determines which of the two hydrogen donors acts optimally. At pH 5 it is NADH, and at pH 6.5 it is NADPH. Curiously enough, the effect of pH on the efficacy of the hydrogen donor is reversed when thymus reductase is used for catalyzing the reduction of folic acid; the thymus enzyme requires NADPH at pH 5.4 and NADH at pH 7.5. The specificity of the reductase is not restricted to folic acid or dihydrofolate. The reductase also catalyzes the reduction of triglutamate derivatives, pteridine derivatives lacking the glutamate residues, pteroate, N^{10}-methylfolate, and pteridine-6-aldehyde.

With aminopterin, the reductase forms a complex with a low dissociation constant. Thus, in the presence of aminopterin many of the reductase molecules are trapped in an inactive form. The coenzymes necessary for purine biosynthesis are not formed, and that pathway is blocked. In this manner aminopterin interferes with the progress of leukemia and with the proliferation of normal bone marrow. Unfortunately, the cells of individuals treated for leukemia overcome the metabolic block by building up a resistance to the antimetabolites by increasing the level of the reductase. We will now consider the mechanism of action of each of these coenzymes separately.

It is not known how N^5, N^{10}-methylenetetrahydrofolate is formed in tissues. In the laboratory, formyladehyde mixed with tetrahydrofolate results in spontaneous formation of the N^5, N^{10}-methylenetetrahydrofolate. In tissues, a formaldehyde-activating enzyme probably catalyzes the condensing of formaldehyde with tetrahydrofolate.

N^{10}-Formyltetrahydrofolate is formed by the condensation of THFA with formate in the presence of ATP. Formyltetrahydrofolate synthetase has been found in bacteria and in mammalian tissues. The erythrocyte enzyme has been partially purified.

N^{10}-Formyltetrahydrofolate is rapidly converted to the N^5,N^{10}-methylenetetrahydrofolate by simple acidification of the incubation mixture. The enzyme N^5,N^{10}-methenyltetrahydrofolate cyclohydrolase catalyzes the reversible conversion of N^5,N^{10}-methenyltetrahydrofolate to N^{10}-formyltetrahydrofolate in the presence of water.

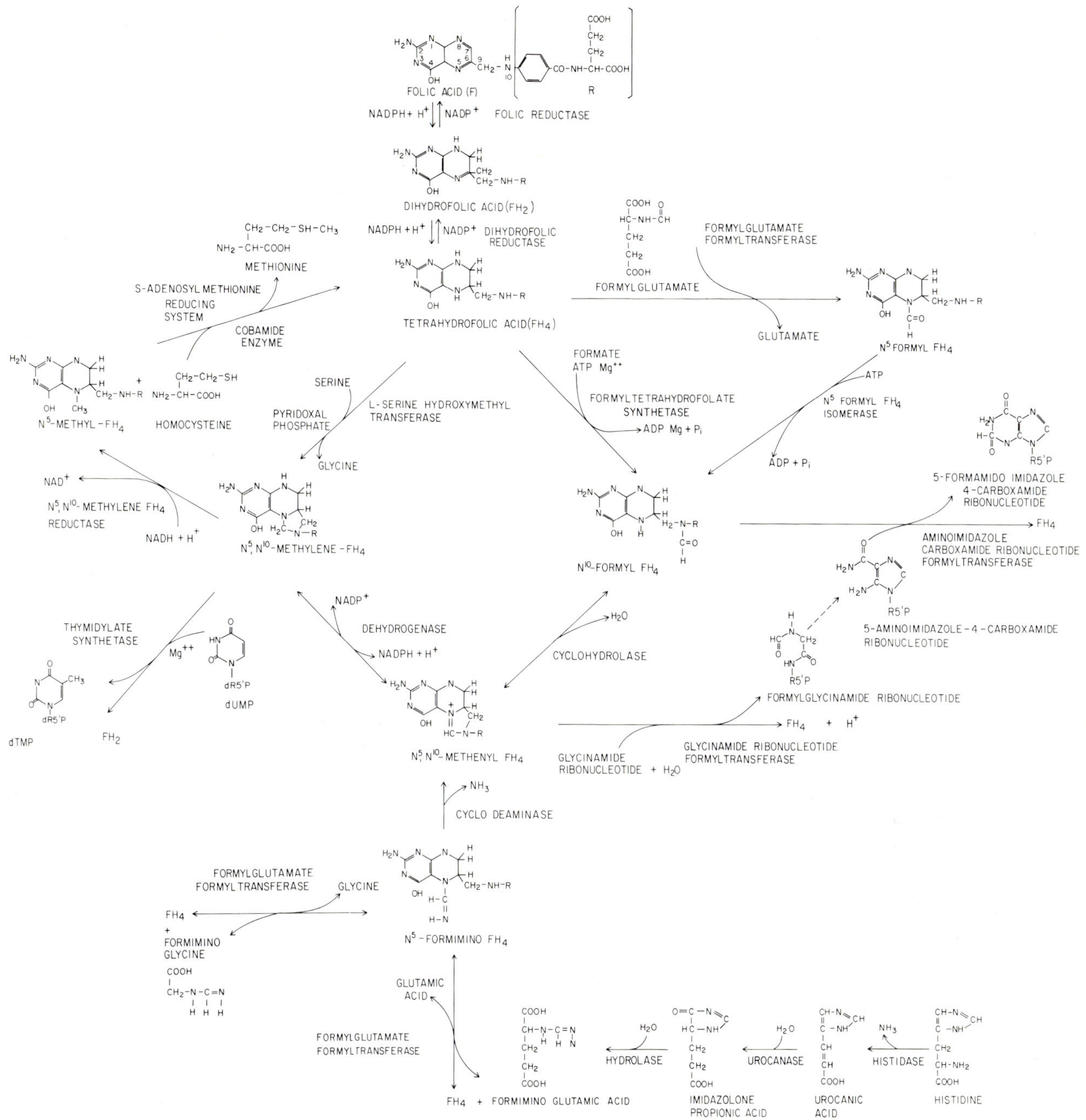

Fig. 4-26. The role of folic acid in cellular metabolism

A formiminotetrahydrofolate cyclodeaminase catalyzes the deamination of N^5-formiminotetrahydrofolate to yield the N^5,N^{10}-methenyltetrahydrofolate and ammonia. N^5,N^{10}-Methenyltetrahydrofolate can also be formed at the expense of N^5-formyltetrahydrofolate and ATP in the presence of N^5-formyltetrahydrofolate isomerase (cyclodehydrolase), an enzyme found in sheep liver, chicken liver, and bacteria.

Formiminoglutamate or formiminoglycine reacts with tetrahydrofolic acid in the presence of formiminoglutamate, formiminotransferase, or formiminoglycine formiminotransferase to yield N^5-formiminotetrahydrofolate. Both enzyme reactions are reversible. The first of these enzymes occurs in a variety of mammalian tissues and has been partially purified from hog liver, and the second has been found in *Clostridium cylindrosporum* and in *Clostridium acid urici*.

Bacteria and liver contain an N^5,N^{10}-methylenetetrahydrofolate reductase that converts N^5,N^{10}-methylenetetrahydrofolate to yield the N^5-methyltetrahydrofolate. The reductase is a flavoprotein that requires NADH for activity.

From this outline, the different forms of folate coenzyme appear to be readily interconvertible (Fig. 4-26).

Pyridoxine phosphate may either be used unchanged in enzymic reactions, or it may be split by nonspecific phosphatases to yield pyridoxine and inorganic phosphate. Pyridoxine is then further oxidized to yield 4-pyridoxic acid in the presence of a purified preparation of liver pyridoxic oxidase and aldehyde oxidase.

Clinicopathological Correlations

The deleterious effects of pyridoxine deficiency have been observed by experimentation on animals, by observations of people with deficient diets, and by experimentation with antivitamins in humans. Depending upon the animals used in the experiments, at least four major groups of lesions have been attributed to pyridoxine deficiency: dermatitis involving the extremities, microcytic hypochromic anemia, nervous injuries associated with convulsions, and vascular lesions resembling those of arteriosclerosis.

In rats, the skin injuries are the most prominent. They start with a scaly dermatitis involving first the dorsal and later the plantar aspects of the hind legs. The dermatitis later spreads to the forelimb, the nose, the ear, and other areas. The skin lesions are characterized histologically by the development of hyperkeratosis and acanthosis of the epithelial layers, with vasodilatation and edema of the corium.

Microcytic hypochromic anemia develops in pyridoxine-deficient rats, dogs, and pigs. The anemia is most developed in deficient pigs. It is likely to result from a block in hemoglobin synthesis rather than from acceleration of hemoglobin breakdown, since none of the symptoms of hemolytic jaundice are observed. The hyperferremia and hemosiderosis probably result from defective absorbed iron use. A microcytic hypochromic normoblastic anemia responsive to the administration of pyridoxine has been described in humans. It is associated with bone marrow hyperplasia, elevated iron levels in serum, and hemosiderosis. In these cases, the administration of large doses of pyridoxine intravenously induces a marked reticulocyte response, but a clear-cut pyridoxine deficiency cannot be demonstrated.

In pyridoxine-deficient monkeys, vascular lesions have been described which resemble those seen in human atherosclerosis in gross appearance and anatomical distributation. Histological examinations reveal areas of severe fibrosis separating the intima from the internal elastic lamellae. Similar lesions have been described in pyridoxine-deficient dogs. No vascular lesions have been reported in rats, but hypertension has been reported.

Dental caries and nonalignment of the teeth have also been observed in deficient monkeys. Caries probably results from alterations in tooth development; for example, cases have been described in which the central incisors of the mandible inclined backward and cut into the gums and upper jaw.

Rats, chicks, pigs, cows, puppies, and human newborns affected with pyridoxine deficiency have developed convulsions. In 1954, an epidemic of a convulsive disease developed among children fed an artificial diet composed of vegetable and animal fats and defatted cow's milk. The diet was supplemented with all the required vitamins, but sterilization destroyed the natural content of pyridoxine. The convulsive children could rapidly be relieved either by being placed on a more natural diet or by receiving a 100-mg pyridoxine injection. The neuroanatomical changes associated with these convulsions have not been identified. Although the physiological disturbances seem to precede the development of anatomical changes, the brachial and sciatic nerves may be demyelinated at first. Later, demyelination involves the fibers of the dorsal roots and even those of the dorsal column. The changes in the myelin sheets may or may not be associated with axonal degeneration and atrophy of the cell body.

In cats, pyridoxine deficiency is associated with the formation of calcium oxalate calculi in the kidneys. The magnesium and pyridoxine levels in the diet are related. Diets low in magnesium are responsible for increased incidence and severity of the oxalate lithiasis in the kidney, and the effect of the low-magnesium diet is counteracted by the administration of pyridoxine. These observations made in animals may have some relevance to the development of lithiasis in humans. Patients with recurring calcium oxalate stones in the kidney secrete more xanthurenic and pyridoxic acid than normal individuals, suggesting that lithiasis may result from deficient pyridoxine metabolism, possibly due to accelerated breakdown of the coenzyme. If other signs of pyridoxine deficiency develop, one must assume that the accelerated breakdown occurs only in a few organs, probably only in the kidneys.

Mode of Action

Pyridoxal phosphate is a necessary coenzyme for a number of different biochemical reactions: transamination, amino acid oxidation, amino acid decarboxylation, glycogen breakdown, and racemization of D- and L-amino acids.

The molecular participation of the coenzyme in transamination has been partially elucidated. Transaminases are enzymes that catalyze the reversible transfer of an amino group from an amino acid to a keto acid. The most common transaminases are involved in the transfer of the α-amino group of an α-amino acid to the α-keto acid, but a number of unusual transaminases have been described.

The mechanism of reaction of pyridoxine in transamination was elucidated by three groups of studies: studies on nonenzymic transamination in the absence of pyridoxine, studies on nonenzymic transamination in the presence of pyridoxine, and studies on enzymic transamination. The studies are briefly reviewed here because they illustrate the continuity of the mechanism regulating organic and biological reactions. Long ago, chemists described reactions that resemble biological transamination. Approximately equimolecular amounts of benzaldehyde CO_2 and alanine are formed

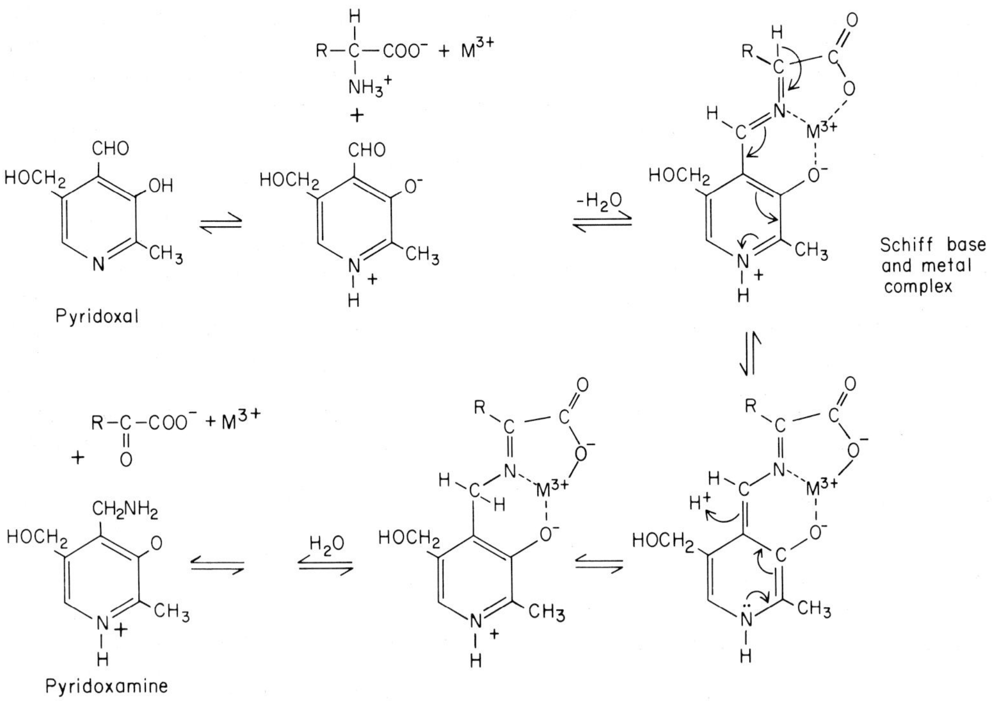

Fig. 4-28. Nonenzymic transamination

when pyruvic acid and α-aminophenylacetic acids are brought to react with each other by boiling them in water. When these studies are repeated using amino acids labeled with deuterium on the α-carbon, deuterium is not exchanged with the hydrogen of the solvent; when the reaction is performed using ^2H_2O as a solvent, part of the deuterium is recovered in the α- and the β-positions of the newly formed alanine. These observations help to outline a model for the molecular mechanism of transamination. It was proposed that the aromatic aldehyde reacts with the amino acid to yield a Schiff base and undergoes intramolecular hydrogen transfers yielding an isomeric Schiff base, which is in turn decarboxylated and hydrolyzed. The reaction just described differs from a classical transamination in that the amino acid donor is decarboxylated as well as deaminated. Esterifying the carboxyl group and using hydrosolvent induces transamination between the α-amino and the keto acid without decarboxylation of the amino acid donor. In such reactions, a mechanism similar to the one postulated above can be invoked.

The role of pyridoxal in the transamination reaction has been extensively investigated by comparing the mechanism of action of nonenzymic transamination (see Fig. 4-28) in the presence of metal (iron) acting as a catalyst to the transamination catalyzed by the specific proteins. The mechanism of the nonenzymic transamination was deduced from a series of rather simple observations: (1) when an amino acid is added to a dilute solution of pyridoxal or pyridoxine phosphate, the absorption maximum of the solution shifts from 345 mμ to 430 mμ. This change in absorption indicates that an imine bond is formed between pyridoxal and the amino acid. (2) The incubation of the amino and the keto acids in the presence of pyridoxal or pyridoxal phosphate and the metal causes the transfer of the α-amino group to the keto acid. (3) The addition of metal ions to the incubation mixture increases the rate of transamination and induces the formation of a chelate complex, which can be detected spectrophotometrically and electrophoretically. (4) A study of the structural features required for transamination with the aid of synthetic pyridoxal derivatives permitted the identification of the atoms of the pyridoxal molecule that are directly involved in the transamination reaction.

A comprehensive discussion of the activities of all the compounds used in these studies is not possible, and only the conclusions pertinent to an understanding of transamination are presented. The studies demonstrated that in order to be active, pyridoxal must have its formyl group in an ortho or para position of the heterocyclic nitrogen atom, and in a position ortho with respect to a free phenolic group. From these observations, researchers concluded that in the nonenzymic transamination catalyzed by metal ions, an imine bond is formed between the α-amino group of the donor acid and the carbon of the pyridoxal that carries the formyl group. The Schiff base thus produced is stabilized by chelate formation with the metal ions. Chelate formation gives the molecule planarity, which facilitates electron shifts. Thus, a molecular complex is produced that favors intramolecular electron transfer because a conjugated system of double bonds extends from the α-carbon on the amino acid

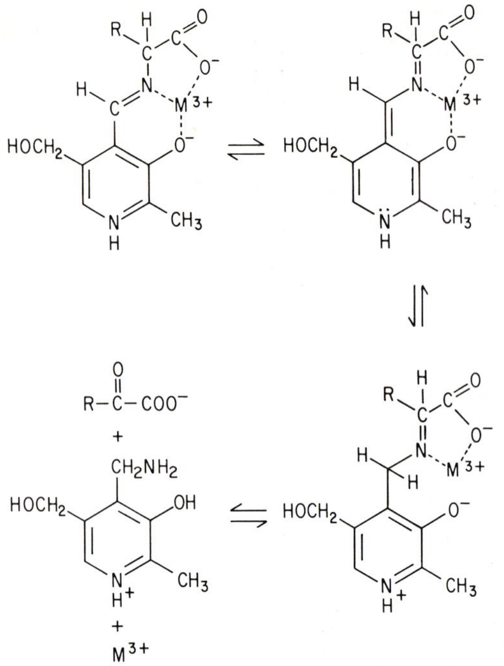

Fig. 4-29. The role of Schiff base in nonenzymic transamination. *1* pyridoxal+amino acid in presence of catalyst (metal or enzyme) yields a Schiff base. *2* Hydrogen and double-bond shifts lead to the formation of an isomeric Schiff base. *3* The base splits and leads to the formation of the keto acid and pyridoxamine. *4* The reaction is reversible. In the reversible step, the amino group of pyridoxamine is transferred to the keto acceptor

to the heterocyclic nitrogen of the pyridoxal (which acts as an electron-negative group); thus, in nonenzymic transamination, amino acid, pyridoxal phosphate, and the catalyst form a complex that has the properties of a Schiff base. The Schiff base then undergoes a series of intramolecular changes with shifts in hydrogen and double bonds leading to the formation of an isomeric Schiff base in which the bond between the amino group and the donor keto acid is more labile. The donor keto acid (see Fig. 4-29) is released from pyridoxamine, and the amino group of the pyridoxamine is in turn transferred to the acceptor keto acid [99].

Obviously, the elucidation of the enzymic mechanism required the preliminary purification of at least one of the transaminases. An 85–90% pure glutamic aspartic transaminase was obtained and found to contain 2 moles of pyridoxal phosphate per mole of enzyme. But pyridoxal is not the active coenzyme. Gunsalus, Bellamy, and Umbreit discovered that the addition of pyridoxal to a culture medium of a strain of *Streptococcus faecalis* grown on a pyridoxal-deficient medium has little effect on the ability of the bacteria to decarboxylate tyrosine. When the culture was supplemented with pyridoxal and adenosine triphosphate, or with phosphorylated derivatives of pyridoxal, the tyrosine decarboxylation activity was greatly enhanced. It was later established that the active form of the vitamin was the 4-formyl-3-hydroxy-2-methyl-5-pyridylmethyl phosphoric acid.

Kinases catalyzing the phosphorylation of pyridoxine, pyridoxamine, and pyridoxal have been found in microorganisms and in mammalian tissues. The kinases require zinc and magnesium, and ATP acts as phosphate donor. The affinity for the substrate varies depending on the source; pyridoxine and pyridoxamine are the preferred substrates with yeast enzyme, but the mammalian enzyme has a greater affinity for pyridoxal.

The pioneering studies on the role of pyridoxal phosphate in the glutamic aspartic transamination reaction were carried out in Snell's laboratory, and a series of observations led these investigators to conclude that the mechanism of the enzymic transamination is similar to the transamination catalyzed by metal (iron). The enzyme solution is yellow at pH 6.2 and becomes colorless at basic pH. This shift in absorption is comparable to that observed with pyridoxamine subjected to varying pH. This observation suggested that the formyl group of pyridoxal forms an imine bond with a free amino group of the enzyme. When the pyridoxal enzyme is incubated with glutamate, the absorption spectrum of the pyridoxal enzyme complex changes. The peak of maximum absorbancy is shifted toward lower wavelengths at both acidic and basic pH. This shift results from the formation of a pyridoxamine-coenzyme complex. Direct evidence for the existence of the complex was provided by isolating the pyridoxamine enzyme by elution chromatography on Dowex-l of a mixture of glutamate and pyridoxal enzyme. Attempts to determine the role of the metal in the glutamic aspartic transaminase reaction have failed. The role of the metal in the reaction is still controversial. Some glutamic aspartic transaminases are inactivated by chelation and extensive dialysis. After dialysis, activity is restored by adding iron and pyridoxal phosphate.

On the basis of these observations, pyridoxal phosphate is thought to be bound to the protein by two types of bonds: an imine bond that links the formyl group of the coenzyme and an amino group of the apoenzyme, and ionic bonds that link the phosphate and the heterocyclic nitrogen of the coenzyme to complementary groups of the apoenzyme. Instead of being stabilized by chelate formation as in the nonenzymic transamination, the Schiff base so formed is stabilized by hydrogen bonding. Isomerization of the imine bond yields pyridoxamine, which in turn reacts with the keto acid (see Fig. 4-30).

Pyridoxal phosphate is a necessary coenzyme for many amino acid decarboxylations. In these reactions, the carboxyl group adjacent to the α-amino group is split from the amino acid molecule. The detailed molecular mechanism of the reaction is not known, but studies with glutamic decarboxylase obtained from *Escherichia coli* have shed light on that mechanism. Ultracentrifugation and electrophoretic examination of the purified enzyme indicated that it has been obtained 90% pure and that the molecular weight was

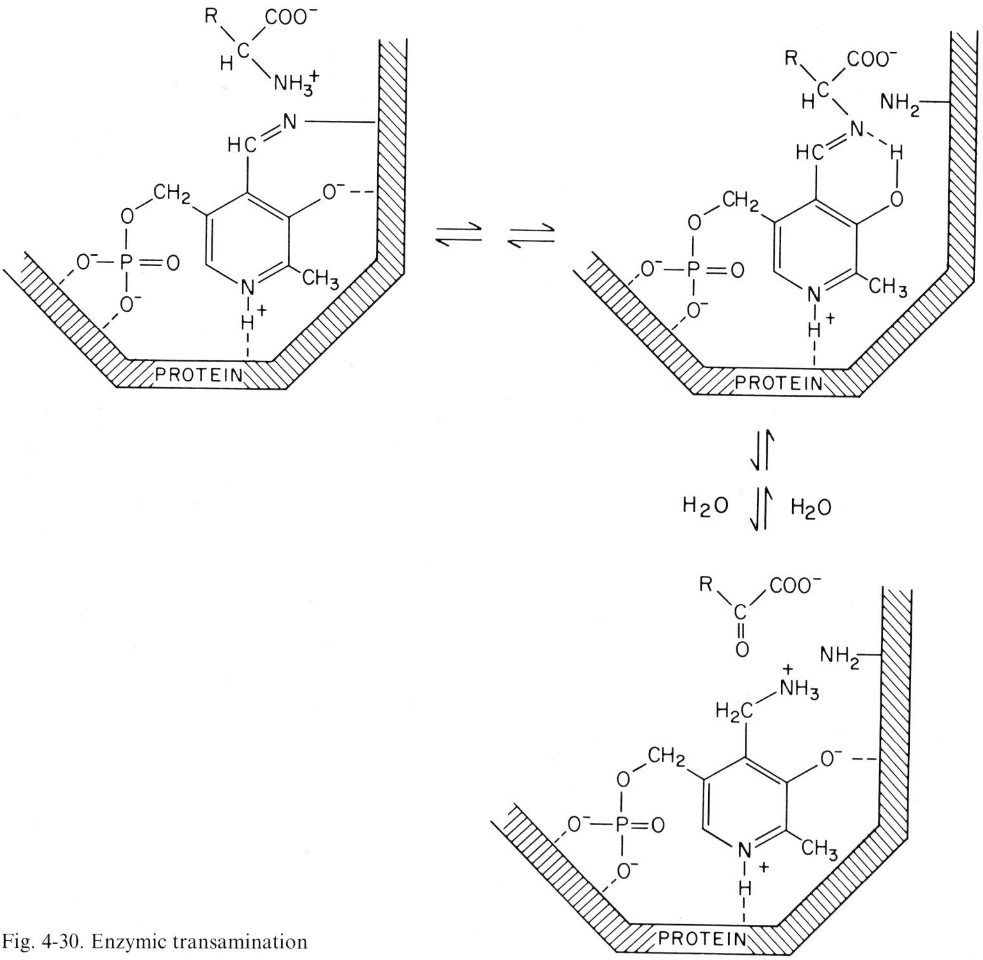

Fig. 4-30. Enzymic transamination

300,000. Analysis of pyridoxal phosphate demonstrated that the enzyme contained 2 moles of the coenzyme per mole of apoenzyme. Spectroanalysis of the holoenzyme suggested that the formyl group of the pyridoxal phosphate forms imine bonds with the amino groups of the apoenzyme, and that when glutamate is added, a Schiff base is formed. In mammalian tissues, metabolites of considerable biological significance are formed by amino acid decarboxylation. Only a few of these reactions are presented here; more detailed descriptions appear in other chapters [100].

Mammalian tissues contain enzymes that catalyze the nonoxidative deamination of serine, threonine, and homoserine. Since the postulated reaction mechanism involves a dehydration before the deamination, these enzymes are called dehydrases. L-Serine, L-threonine, and L-homoserine dehydrases have been partially purified and all are specific for the L-amino acid. Serine and threonine dehydrases require pyridoxal phosphate, ATP, and glutathione for activity. Pyridoxal phosphate requires the homoserine enzyme, but the need for ATP and glutathione has not been demonstrated. The reaction is likely to involve the formation of a Schiff base. The homoserine dehydrase has been crystallized. It is a yellow enzyme that can be resolved into pyridoxal phosphate and inactive apoenzyme. In addition to homoserine, the enzyme also deaminates cystathione. Pyridoxal phosphate is also required for the metabolism of urocanic 4(5)-imidazolone-5(4)-propionic acid. The product of the reaction has been isolated and identified by spectrophotometric and chemical methods. Urocanase has been extensively purified. It is likely to be an SH enzyme requiring pyridoxal phosphate for activity. Pyridoxal phosphate is also required in the biosynthesis of amino levulinic acid from succinyl CoA and glycine (see chapter on porphyria).

Again, muscle phosphorylase contains pyridoxal phosphate, and the pyridoxal is bound to the protein molecule by an imine group, although the formyl radical is not indispensable to phosphorylase activity. Reduction of the pyridoxal form of the phosphorylase with sodium borohydride does not inactivate the enzyme.

Pyridoxal phosphate is required for deamination, transamination, and decarboxylation of amino acids and also is involved in D-amino oxidation and in racemization of the D- to the L-amino acid. Pyridoxal phosphate seems also to be required for diamine oxidase activity, an enzyme widespread in nature identical with histaminase.

Racemases are enzymes capable of interconverting D- to L-amino acids. Pyridoxal phosphate has been claimed to play a role as a cofactor in bacterial racemases for alanine, glutamic acid, and methionine, but not in others. It has also been claimed that in mammals the administration of pyridoxine facilitates the use of D-amino acids.

In discussing the L-serine hydroxymethyl transferase reaction in the section on folic acid, it was pointed out that the purified enzyme contains pyridoxal phosphate, which is bound to the protein amino group forming a Schiff base. The maximum absorption of the purified enzyme (450 mμ) shifts after the addition of cysteine (330 mμ), a change that suggests the formation of a thiazolidine of pyridoxal phosphate. The thiazolidine pyridoxal phosphate is readily removed by dialysis, and an inactive apoenzyme is obtained which can be reactivated by the addition of pyridoxal phosphate.

Pyridoxal phosphate plays an important role in the intermediary metabolism of amino acids and also seems to influence the intestinal absorption of certain amino acids. The intestinal absorption of L-methionine, L-histidine, and L-lysine is markedly decreased in pyridoxine-deficient rats, but absorption can be normalized by adding pyridoxine to the diet. The administration of deoxypyridoxine to rats interferes with the intestinal absorption of L-methionine.

Although it is possible to describe the clinicopathological manifestations of pyridoxine deficiency and the metabolic role of pyridoxal phosphate, each pathological alteration cannot be explained by a specific metabolic alteration. Deficiency of a vitamin involved in several steps of the intermediary metabolism of amino acids is bound to be associated with severe clinicopathological changes, but the specific metabolic alterations responsible for the anemia and convulsions in pyridoxine deficiency have not been identified. γ-Amino butyric acid, cystathione, sphingosine, and 5-hydroxytryptamine are compounds abundant in the brain. Pyridoxal phosphate is involved in their metabolic formation. Is there any correlation between the role of pyridoxal phosphate in the metabolism of these compounds and the development of convulsions and ataxia in pyridoxine deficiency? Is the role of pyridoxine phosphate in the intermediary metabolism of sulfur amino acid related to the development of seborrheic dermatitis?

The observations made in oxaluric patients suggest that the lithiasis that pyridoxine-deficient cats develop might be related to the role that pyridoxine plays as a cofactor of the transaminase. Anemia developing in pyridoxine-deficient animals probably can be explained at least in part by an interference with heme synthesis. Indeed, heme synthesis is reduced to one-third of normal in reticulocyte-rich blood obtained from pyridoxine-deficient animals. The rate of heme synthesis reaches normal values when pyridoxal phosphate is added to the incubation medium. This effect of pyridoxal phosphate deficiency on heme synthesis in reticulocytes probably results from its role in amino levulinic acid formation.

The fact that pyridoxal phosphate is involved in so many different metabolic pathways presents a major difficulty in correlating the metabolic effects of the vitamin with the clinical manifestations of the deficiency. The use of inhibitors has been of little help in that respect because their effect is seldom restricted to a single reaction.

Riboflavin

In 1879, Blythe, an English chemist, extracted a yellow pigment from milk. The significance of this discovery was not appreciated until new observations made in several laboratories focused the attention of nutritionists and biochemists on the yellow pigment. In 1932, Warburg and Christian discovered the now famous yellow enzyme and demonstrated that the coenzyme part of the holoenzyme was a freely dialyzable, water-soluble yellow pigment. Kuhn, György, and Wagner-Jauregg isolated from milk a new flavin that resembled the coenzyme discovered by Warburg and also manifested properties attributed to the vitamin B complex. The new compound was named riboflavin, and two years after the discovery by Kuhn and his associates, the compound had been identified and synthesized. At about the same time, a yellow vitamin was isolated from egg and another from liver, called ovoflavin and hepatoflavin, respectively. Kuhn, Karrer, and associates then established the structure of the vitamin as the 6,7-dimethyl-9-(1′-D-ribityl)-isoalloxazin (see Fig. 4-31).

It is now accepted that the requirement of riboflavin for the average adult is about 1.2 mg per day. The requirement changes during rapid growth, pregnancy, lactation, and in certain pathological conditions. People working in hot climates excrete excessive amounts of riboflavin. Riboflavin requirements are also influenced by diet and the rate of oxygen consumption; thus, there is a close correlation between riboflavin requirements and protein use and oxygen consumption.

Patient with second- and third-degree burns excrete twice the normal intake of riboflavin. It is assumed that the increased excretion results from decreased use. Although the cause of such decreased use is not known, it has also been suggested that the massive nitrogen loss that occurs in extensive burns is responsible for the decreased vitamin consumption. When the burns heal, amounts of riboflavin greater than normal are needed to promote adequate cellular restoration.

The administration of riboflavin to anoxemic rats stimulates glycogen synthesis, and the glycogen content may rise to values up to 10 times above normal. A number of tissues have been analyzed for their riboflavin content. For example, the liver contains 25–33 μg riboflavin per gram, and the aorta contains 1.9 μg riboflavin per gram. The riboflavin content of

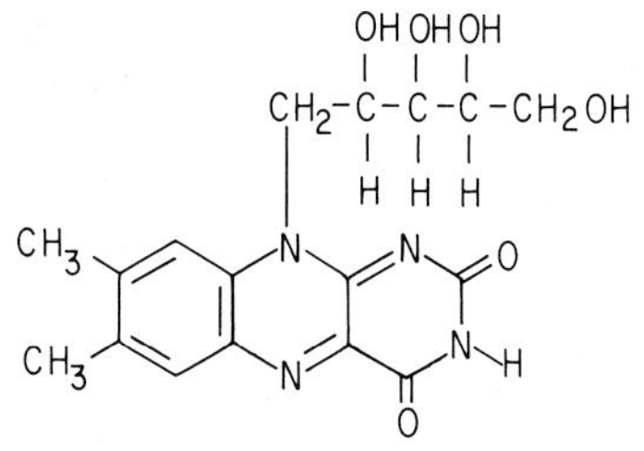

6,7-Dimethyl-9-(D-1'ribityl)-isoalloxazin

Fig. 4-31. Riboflavin

the aorta decreases with age, but it is not known if this change is related to the pathogenesis of atherosclerosis or if it is secondary to the histological alterations that occur in the aged aorta.

Riboflavin is readily absorbed in the intestinal tract. Some workers have proposed that the absorption involves the conversion of riboflavin to the 5′-monophosphate. The data supporting this concept are still controversial. At least two different mechanisms of phosphorylation have been proposed. One is a reaction in which riboflavin is phosphorylated to the 5′-monophosphate in the presence of ADP and a flavokinase; in fact, the conversion of riboflavin to 5′-riboflavin monophosphate in the presence of intestinal extract was demonstrated. The second mechanism involves a transphosphorylation at the expense of several phosphate esters, such as β-glycerophosphate, thiophosphate, or adenosine monophosphate. In this type of reaction, β-glycerophosphate reacts with riboflavin in the presence of an unspecific phosphatase to yield glycerol and riboflavin-5′-monophosphate. These findings are significant because they yield new information on a vitamin's mechanism of action and also shed some new light on the possible physiological role of such hydrolases as phosphatase. Phosphatases are included in some secretory granules of the intestine and are usually assumed to affect cellular digestion, but these findings suggest that these phosphatases may be involved in much more active processes, such as the absorption of the vitamin and its conversion to an active coenzyme.

The concentrations of riboflavin in the blood range between 0.3 and 0.6 γ per cc of whole blood. There is no correlation between the levels of riboflavin in the blood and the severity of riboflavin deficiency. Placental blood has a concentration of riboflavin four times that of maternal fetal blood.

Little is known of the catabolism of riboflavin in humans, but riboflavin is a normal constituent of the urine. Therefore, all the excess vitamin could be excreted without breakdown of the molecule. The vitamin is not stored; consequently, administration of excessive doses of riboflavin is accompanied by increased excretion in the urine. Riboflavin is also excreted in the milk, and the riboflavin concentration in milk reflects the vitamin intake.

Pathogenesis of Riboflavin Deficiency

The symptoms attributable to riboflavin deficiency were difficult to identify clinically in humans because pure riboflavin deficiency is seldom observed. Endemic carency disease usually results from a combination of deficiencies—thiamine, riboflavin, vitamin PP, etc. Symptoms specific to riboflavin have been distinguished from those resulting from other vitamin deficiencies by experimentation with animals and humans; alterations of the epithelia of the mouth, eye, and skin develop in riboflavin deficiency. In about the fourth month of a deficient diet, the lips become pale and scaly, and later the epithelium atrophies and the lips turn red. Fissures appear, especially at the angle of the mouth. The lips and the skin at the angle of the mouth ulcerate, and secondary infection by streptococci and staphylococci may develop. These injuries of the lips are often referred to as angular stomatitis or cheilitis. Glossitis has been observed in animals. The tongue is painful and becomes magenta colored and pebbly because of the alterations of areas of epithelial renovation with bulging, sometimes keratinized, papillae. Histologically, the epithelia of the lips and tongue appear hyperkeratotic at first, but as the deficiency continues, the epithelia atrophy. The epithelial lesions are not restricted to the mouth; seborrheic dermatitis and scaly scrotal skin lesions develop in 12 out of 15 humans maintained for 9 to 17 months on a riboflavin-deficient diet. Histological studies of the skin in animals suggested that the epithelium and all its annexes atrophy as the deficiency develops. Thus, the sebaceous and sweat glands are swollen at first, soon atrophy, and may even completely disappear. The hair pedicles degenerate and epilation occurs.

Corneal vascularization by outgrowth of new capillaries originating from the vessels of the limbic plexus has been described in rats, mice, and dogs. Leukocytes also appear in the cornea while the capillaries are being invaded. This cellular exudation is later followed by vascular degeneration of the superficial cell layer of the cornea, which becomes cloudy and ulcerates. All these changes recede rapidly after adequate doses of riboflavin are administered. Cataracts have been described in riboflavin-deficient rats and swine, but experimental riboflavin deficiency in man did not induce eye lesions.

In the first part of this book devoted to cellular metabolism, it was pointed out that riboflavin is converted to monophosphate and diphosphate, yielding the active coenzymes of a large number of enzymes involved in the cellular oxidation-reduction reaction. As can be expected, the enzyme activity for which these flavoproteins are responsible is decreased in riboflavin deficiency. This has been clearly demonstrated for amino acid oxidase. Extensive studies on the effect of riboflavin deficiency on a large number of proteins have not been carried out; but it seems logical to assume that the clinicopathological changes that develop in riboflavin deficiency result from the reduced activity of the flavoprotein enzyme. Yet, many molecular changes that link the primary biochemical injury to the pathological manifestation remain unknown. As a matter of fact, the mere observation of the symptoms of riboflavin deficiency in animals and man raises the question: Why are the epithelia of the skin more sensitive to the deficiency than other cellular structures. Why are some parts of the skin more sensitive than others? Indeed, one would expect a deficiency in a vitamin involved in a reaction as important as electron transport to be rapidly fatal. Yet, in humans, even after months of deficiency, only epithelial involvement is observed. Does this sensitivity of the epithelium result from its greater anabolic activity?

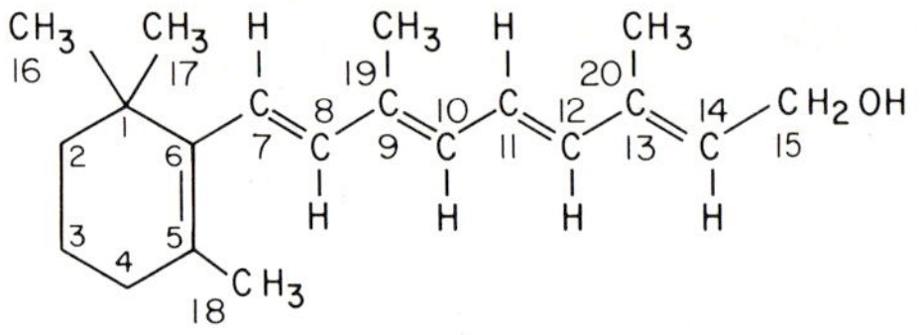

Vitamin A Deficiency

Whereas most deficiency diseases disappear as soon as the deficient factor has been identified and reinstituted, vitamin A deficiency is somewhat different. Although the vitamin has been chemically identified, crystallized, and synthesized, the ocular defects resulting from vitamin A deficiency are still prevalent in many parts of the world [101, 102].

The ancients knew about night blindness, but they apparently never suspected that it resulted from a dietary deficiency. In contrast, the Sepoys of India were aware that an absence of fruit in the diet is responsible for night blindness, a symptom still frequently observed in some parts of India.

At the end of the last century, a new child's disease was observed for the first time in Japan during a period of depression. It was called Hikan and was characterized by hemeralopia, xerophthalmia, and keratomalacia. Untreated, the victims soon died, but they could be cured almost instantly by the administration of cod liver oil.

In 1913, Osborne and Mendel in one laboratory and McCowan and Davis in another demonstrated that young rats deprived of some lipid fraction of the diet stopped growing. In 1920, Steenbock discovered axerophthol and determined its empirical formula: $C_{20}H_{21}OH$. The compound was then synthesized by the Dutchmen Ahrens and VanDorp in 1947. But it was Karrer who elucidated the structural formula of carotene, a precursor of vitamin A, and of vitamin A itself [103, 104].

Chemistry

Several chemical compounds possess the biological properties of vitamin A; vitamin A_1 can be considered as the prototype (see Fig. 4-32). Karrer elucidated its structure and found the empirical formula to be $C_{20}H_{30}O$. It is made of a benzoic ring, of which three carbons are substituted. The carbon 6 carries two methyl groups; another methyl group is attached to carbon 2, and carbon 1 carries a long aliphatic side chain of nine carbons. The side chain is highly unsaturated, and vitamin A_1 is thus a polyene containing four

Retinol$_1$ (Vitamin A_1)

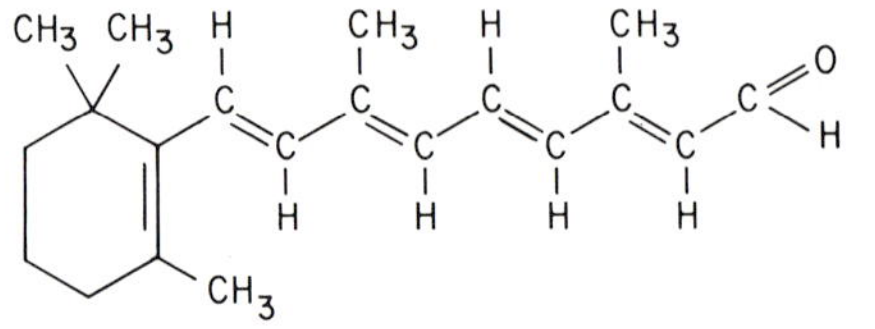

Retinal (Vitamin A, Aldehyde)

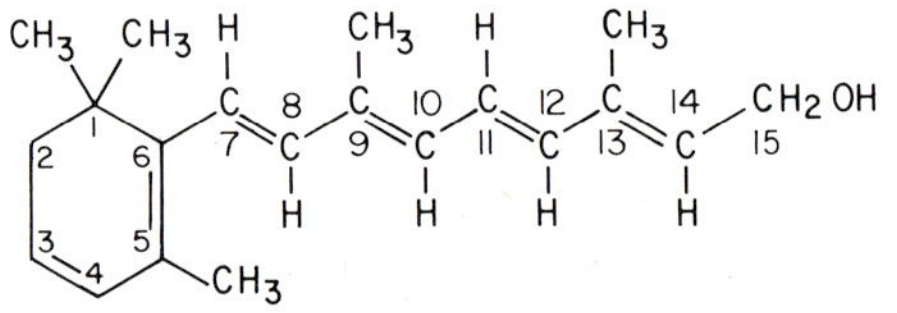

3-Dehydroretinol (Retinol$_2$, Vitamin A_2)

Fig. 4-32. Compounds with the biological properties of vitamin A

double bonds. Carbons 7 and 3 of the side chain carry a methyl group. The terminal carbon, or carbon 1 of the side chain, may participate in various different types of endings. It may form an aldehyde, an alcohol, or an anhydride group. The polyenes with vitamin A activity differ in the structure of their molecules in at least three different ways. In vitamin A_2, for example, the ring is altered—it is dehydrogenated. Vitamin A_1 aldehyde, anhydrous vitamin A, differs from the ordinary vitamin A_1 by the nature of the chemical group on the terminal carbon. Also, certain compounds possess the biological activities of vitamin A_1, but differ from it because they are geometrical isomers.

In contrast to what happens when only a single covalent bond connects two carbons, a double bond restricts the rotation around the carbon-to-carbon bond. This gives rise to geometric isomers. Consider a compound much more simple than the polyene chain of vitamin A, namely, dichlorethylene. The two carbons, the two hydrogens, and the two chlorides are located in the same plane, so there are two possible relationships for the two chlorides: they can both be in the upper part of the plane (*cis* form); or one chloride can be above and the other below (the median plane, or *trans*, form).

Vitamin A has four such double bonds; therefore, 32 different geometrical isomers are possible. Only those four isomers that exist in nature will be considered here. The *trans* form (see Fig. 4-33) is the most stable form, so most natural compounds contain mainly *trans* vitamin A. However, it is possible to con-

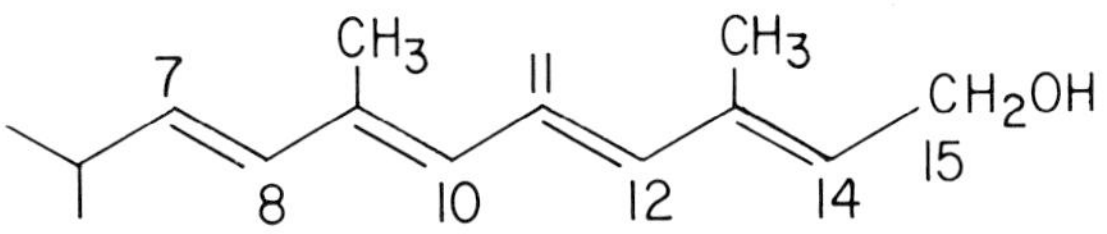

Vitamin A (All *trans*-Retinol)

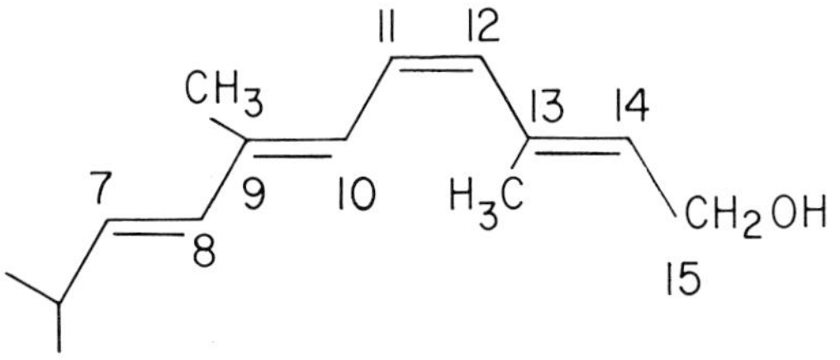

Neovitamin A_b (11-*cis*-Retinol)

Fig. 4-33. Isomers of vitamin A

vert the *trans* into the *cis* form by supplying energy: by irradiating the preparation with ultraviolet or visible light in the presence of iodine. Although these isomers have vitamin A activity, the biological activity differs considerably from one compound to another. The *trans* compound is usually more active than the *cis* compound.

In rats, enzymes catalyze the conversion of the various naturally occurring *cis* forms in vitamin A. When the *cis* form of vitamin A is administered to rats, a mixture of all *trans* and *cis* vitamin A is stored in the liver, including the 6-*cis* isomer, a compound with very low biopotency. The 4-monocis and the 2,4-dicis vitamin A aldehydes are about twice as active biologically as the corresponding acetates. This supports the conclusions that the aldehyde group labilizes the 2,3 and 4,5 double bonds, enabling the rat to isomerize the aldehydes before absorption.

Vitamin A is essentially a liposoluble vitamin, and therefore a diet that contains little animal fat—such as a pure rice diet—often leads to vitamin A deficiency. In contrast, the deficiency is rare in countries where animal fats are abundant in the diet, as in the United States and Western European countries. Butter, milk, cream, fat cheeses, cod liver oils, egg yolks, liver, and brain are among the animal sources richest in vitamin A. An interesting relationship between vitamin A deficiency and its sources in the diet was observed during the First World War in Denmark. The Danish farmers, who produced much of the western world's milk and butter during that time, were so anxious to sell their products to foreign countries that little of it was consumed within the country. This resulted in the development of an epidemic of night blindness and xerophthalmia.

Apparently, only animal tissues contain vitamin A. Bacteria and plants contain carotene (see Fig. 4-34), but no vitamin. Plants, therefore, often provide a good source of carotene or provitamin A. Carrots, tomatoes, maize, peas, and other vegetables are rather rich in carotenes. In fact, there seems to be a correlation between the intensity of a fruit's yellow color and its biopotency as a vitamin A provider. The stability of provitamin A and vitamin A in food is of great importance, especially for the preparation of cattle feed. Usually cattle feed is stored for long periods of time during the winter, and therefore, extensive studies have been carried out on the stability of vitamin A in grass of good quality. For example, the addition of 5% animal fats or animal oils to dehydrated alfalfa meal decreases the loss in carotene considerably during storage. In chicken food, iron, copper, and cobalt salt adversely affect vitamin A conservation. It was also found necessary to study the loss of vitamin A during cooking in countries where the diet is relatively poor in vitamin A, as in India.

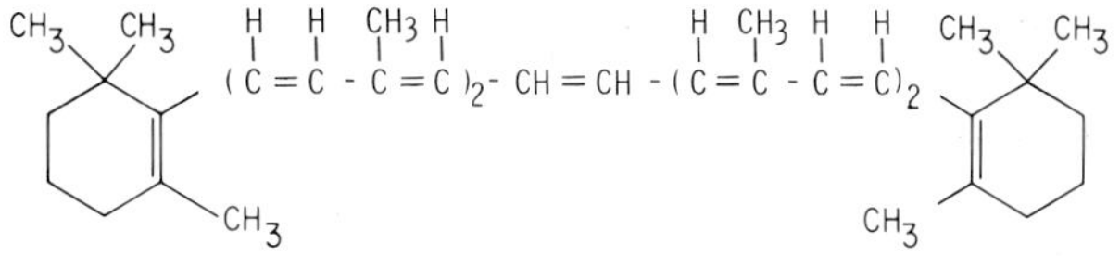

Fig. 4-34. β-carotene

Since carotenes and vitamin A are liposoluble, any condition that interferes with lipid absorption—for example, steatorrhea—will cause vitamin A deficiency. The vitamin is also easily oxidized, and the oxidized form is inactive. Therefore, antioxidants or tocopherol in the intestinal content improves vitamin A absorption.

Deficiency

The diet contains vitamin A esters that are hydrolyzed in the lumen of the intestine or in the brush border. Before absorption, however, free vitamin A is reesterified within the intestinal cell, and then the esterified molecule is secreted in the lymph and transported to tissues in the esterified form [105, 106].

Enzymes that catalyze the synthesis and splitting of the ester must be in a variety of tissues, because both the hydrolysis of the ester and the esterification of the alcohol occur rapidly after intravenous injection.

The use of carotenes for vitamin A biosynthesis raises two problems: (1) the site of transformation of carotenes into vitamin A; and (2) the biochemical pathway involved in the transformation. The breakdown is generally thought to occur in the intestine, but there is some debate as to whether it occurs within the lumen of the intestine or within the cells. Using techniques capable of detecting fluorescence, investigators have shown that vitamin A in the intestinal lumen is in close contact with the surface of intestinal villi before it can be detected inside of the cells of the intestinal epithelium. It is also known that intravenously administered carotene is broken down to vitamin A and stored as such in the liver and in the kidney; consequently, tissues other than the intestine can also split carotene. Carotene breakdown to vitamin A occurs even in hepatectomized and eviscerated rats. It has been suggested that carotene conversion to vitamin A is then taken over by the lungs.

Three different pathways for the breakdown of carotene to vitamin A have been proposed: (1) splitting the molecule in two equal halves; (2) nibbling half of the molecule by repeated oxidation; and (3) splitting half of the molecule at various sites, leading to intermediates of various lengths. In β-carotenes, the two parts of the molecule are symmetrical. It would seem logical, therefore, that the molecule is split in the middle, yielding two halves that can each be used. This was believed to be the case because the β-carotenes were found to be twice as effective as the α-carotenes in correcting vitamin A deficiency (the two halves are asymmetrical in α-carotene). However, when the vitamin properties of equal weights of β-carotene and vitamin A were compared, β-carotene was always much less effective than vitamin A. This suggests that the use of β-carotene is less efficient than the use of vitamin A, indicating either that different pathways of utilization exist, or that at least half of the molecule is broken down. The enzyme that breaks down β-carotene after its absorption in the intestinal tract is β-carotene-15, 15′-deoxygenase. The enzyme requires oxygen for activity [107].

To our knowledge, no direct evidence supports the nibbling theory. It is based mainly on the exclusion of the first of these theories, and on the appearance of $^{14}CO_2$ after the administration of carotene.

Fifty to seventy per cent of the ingested ^{14}C-labeled carotene is absorbed in the intestine. Only 2% of the absorbed isotope is eliminated in the form of CO_2, and although the highest concentration of the radioactive material is recovered in the liver in the form of stored vitamin, this represents only 30% of the total absorbed radioactivity. The remaining ^{14}C was recovered in sterols and fatty acids, which indicates that vitamin A occurs by splitting bonds at various points of the halved molecule yielding intermediate chains of different lengths.

The presence of these labeled compounds after the administration of labeled carotene does not constitute a serious argument against the nibbling theory. The carbon 2 compound that results from repeated oxidation could very well be reused for sterol or fatty acid synthesis.

Many factors are suspected to lead to an increased rate of conversion of carotene to vitamin A. For example, aureomycin increases the biological effectiveness of vitamin A, not by stimulating the absorption of the vitamin through the intestinal epithelium, but by altering the bacterial flora in such a way that the conversion of carotenes to vitamin A is favored. The administration of thyroxine increases the conversion of carotenes into vitamin A in chickens. Numerous other experiments suggest that various dietary and hormonal factors can facilitate the conversion of carotenes into vitamin A in the intestine. However, many of these experiments do not exclude other mechanisms that could also facilitate the vitamin A uptake, such as alteration of the rate of absorption or decreased breakdown of the carotenes.

Little is known of the mechanism of breakdown of vitamin A in humans or in other animals. Some insight into vitamin A catabolism was achieved by injecting ^{14}C-labeled vitamin in rats. Some water-soluble ^{14}C-compounds were excreted, demonstrating that part of the vitamin must be catabolized. Five per cent of the vitamin-labeled carbon was expired in the form of CO_2.

In contrast to the water-soluble vitamins, vitamin A can be stored within the organisms. Liver is the main site of storage, but other organs—such as the adrenal, retina, and kidney—are also rich in vitamin A. Some investigations of Dowling and Wald have established a close relationship between vitamin A storage and night blindness. They observed the following sequence of events in vitamin A-deficient animals: the first measurable manifestation, occurring after two weeks, was a drop of the vitamin A content in the retina. This finding was associated with a drop in the blood levels of vitamin A and the occurrence of night blindness. The amount of opsin in the retina was still normal. Although after two weeks of deprivation, the vitamin level of the retina had already dropped to low values, the vitamin content of the liver remained unchanged, and only after two more weeks did storage in the liver fall to abnormal values. Four weeks after the deficiency began a general drop in protein also occurred, and a reduced concentration of opsin in the retina was observed.

Vitamin A is assumed to be contained mainly in the Kupffer cells of the liver. The intracellular distribution of the vitamin has also been studied in chicken liver. Twenty-one per cent of the vitamin is found in the nucleus, 7% in the mitochondria, and 72% in the supernatant. Therefore, the vitamin seems to be concentrated essentially in the supernatant fluid, and the amount of vitamin associated with the nucleus and the mitochondria is probably due to contamination of the fraction by the supernatant vitamin A. This intracellular distribution might be peculiar to the liver—the main site of vitamin A storage; therefore, the cellular distribution of the vitamin gives no clue as to its mode of action. The liver vitamin is bound to protein. The protein that is bound to the alcohol derivative of vitamin A is probably different from that bound to the ester derivative of vitamin A.

Vitamin A is also present in the blood. As already mentioned, the level of vitamin A in the blood depends on the degree of deficiency. However, it will decrease only when the liver storage is exhausted. The vitamin is transported in the blood in the form of a protein complex.

Vitamin A is found in the blood in the form of retinol bound to a specific protein called the retinol binding protein (RBP). RBP has been purified from human plasma; it has a molecular weight of 21,000 and binds one molecule of retinol per molecule of protein and moves as an α_1 protein on electrophoresis.

In plasma it circulates as a protein complex with prealbumin [167].

Clinicopathological Correlations

In the privileged countries of the western world, the physician is no longer accustomed to vitamin deficiencies. The medical student is tempted to skip over these chapters in his textbooks, but many of these diseases are still endemic or epidemic in more than three-quarters of the world. And most doctors in their lifetime, even in the most prosperous countries, encounter more or less severe cases of vitamin deficiency [108].

Among the changes attributed to vitamin A deficiency, only those affecting the eye can be related to vitamin A deficiency without hesitation. In the eye, the vitamin A deficiency leads to a decrease in the amount of pigment in the rods and cones of the retina, causing nyctalopia, or night blindness, and alterations of the epithelia, cornea, and conjunctiva.

Nyctalopia is often the first symptom of the deficiency. The symptom, common in underdeveloped countries, seems to have disappeared from the United States. It manifests itself with greater intensity when the patient has been exposed to brilliant light. For example, the fishermen of New Foundland often develop nyctalopia after spending sunny days at sea. In some parts of Africa, older women are credited with the supernatural gift of predicting summer storms because, in contrast to the men of the tribe who spend the day in brilliantly lighted fields, the old women remain in the shade of their huts, thus protecting their retinal pigment, and this allows them to recognize the summer lightning at night in the distant skies.

Xerophthalmia, the second most common symptom of vitamin A deficiency, is often seen among younger children. Dr. Oman, who has had great experience with such patients, describes a typical case as a school child who has bumped his head because he did not see a pole in the dim light of the evening, or a young child who is brought in by his mother because he has not opened his eyes for ten days. In vitamin A deficiency, the superficial epithelium of the cornea and the conjunctiva differentiates abnormally. This leads to a condition called xerophthalmia, which means "dry eye." The epithelium of the cornea and conjunctiva, which do not normally form keratin, becomes keratinized without vitamin A. Keratinization and scaling of the conjunctiva usually preceed corneal lesions. The white of the eye loses its shiny porcelain appearance to take on a milky quality that obscures the underlying vascular pattern. In the cornea this leads to scaling, loss of transparency, and grey, opaque spots. The scaling leads to infection by bacteria, which in turn perforate the cornea, a condition called keratomalacia. The substantia propria of the cornea becomes vascularized. The corneal and conjunctival conditions are usually aggravated by scaling of the lacrimal duct, which prevents the normal flow of tears. The entire condition may lead to bilateral panophthalmia.

Vitamin A deficiency is often thought to lead to a general epithelial condition characterized by keratinization. The skin, the epithelial lining of the bronchi, the pelvis, the bladder, the pancreatic duct, and vagina are all assumed to be affected. Although there is no doubt that such alterations occur in patients affected with vitamin deficiency, some clinicians are reluctant to attribute these changes specifically to a vitamin A deficiency, but prefer to think them due to general malnutrition. Even though many epithelial injuries can be induced experimentally in vitamin A-deficient animals, it is difficult to arrive at any definite conclusions as to the pathogenesis of those epithelial lesions observed, since pure vitamin A deficiency is practically never experimentally induced in humans.

The effect of vitamin A on the epithelia is, however, of great interest, because in animal deficiences, specialized epithelia—such as the ciliate epithelium of the lung or the transitional epithelium of the pelvis and kidney—lose their capacity but keep all their potentialities to differentiate normally. As soon as vitamin A is restored, the tissue recovers its normal structure.

Whatever their pathogenesis may be, the epithelial lesions must be taken into account for an adequate understanding of the symptomatology associated with vitamin A deficiency. In the skin, hyperkeratosis of the hair follicles occurs, leading to obstruction of the sebum glands. Rough, dry skin results, characterized by the formation of small, elevated cornified papules, which measure from 1–2 mm in diameter. The etiology of the skin condition attributed to vitamin A deficiency is not clear. Some feel strongly that it results from the vitamin A deficiency, but others believe that various concomitant but unrelated factors might be involved, such as a fat or vitamin B deficiency or inadequate skin hygiene.

In the bronchi, the deficiency results in the disappearance of the normal ciliated epithelium, which normally prevents the penetration of bacteria in the bronchial wall. Thus, in vitamin A deficiency the infectious agents find their way deep within the ramification of the bronchial tree, and severe bronchitis, bronchiolitis, and lobular pneumonia result.

In the urinary tract there is localized, sometimes marked, keratinization of the epithelium of the base of the bladder. The epithelial changes are followed by obstruction, bacterial infection, and stone formation.

In vitamin A-deficient male rats, the germinal epithelium also atrophies, but in contrast to what happens in vitamin E deficiency, as soon as vitamin A is administered the animal rapidly recovers to normal. The vaginal smear from rats shows keratinization.

Among the alterations caused by vitamin A deficiency in animals, bone growth retardation is one of the most characteristic. It affects long, epiphyseal, and flat bones. The retarded growth of the cranial bone is followed by degeneration of the cranial nerves, and sometimes lesions in the white and grey matter of the brain also develop. Vitamin A undoubtedly affects bone growth directly or indirectly because, as we shall later see in cases of hypervitaminosis, one can make the bone grow in 15 days as much as in an entire year. But, of course, under those conditions, the new bone is very friable and susceptible to fractures.

Hydrocephalus is a frequent complication of vitamin A deficiency. It would be tempting to attribute these changes to epithelial alteration of the choroid plexus (the structure that secretes cerebrospinal fluid), but there are no histological changes of the choroid plexus. Conversely, hydrocephalus due to vitamin A deficiency only occurs in children, and is therefore always associated with a retardation in bone growth. Hence, it is often assumed that the hydrocephalus results from anatomical interference with the development of the central nervous system, due to retarded bone growth. In any case, hydrocephalus due to vitamin A deficiency responds rapidly to the administration of the vitamin.

Vitamin A-deficient pigs have severe congenital anomalies, such as hair lip, cleft palate, accessory ears, subcutaneous cysts, and various malformations of the eyes.

Vitamin A is often claimed to possess anti-infection and detoxifying properties. The evidence on which such conclusions are based is often controversial, and alleged beneficial properties are also likely to result from secondary effects of the vitamin.

Role in Metabolism

As is the case for other vitamin deficiencies, it is difficult, in the face of the complex symptomatology that results from the absence of a single molecule, to conceive of a single or a few biochemical injuries that could be responsible for the broad clinicopathological manifestations of vitamin A deficiency. Vitamin A, like some other vitamins, may act as a coenzyme in one or more biochemical reactions vital to the maintenance of cellular integrity. One of the major difficulties in detecting the site of action of the vitamin in metabolism stems from our inability to distinguish primary from secondary effects of the vitamin. Many of the injuries described above may not be directly related to the primary metabolic lesions caused by the vitamin deficiency, but may be simply manifestations of a chain of deleterious metabolic alterations triggered by the primary injury. Because of these complex interactions, *in vivo* metabolic alterations associated with the vitamin deficiency which have been reproduced *in vitro* are of interest in interpreting the vitamin's mode of action.

Fell and his collaborators studied the effect of excess vitamin A on the development of explants on undifferentiated chick embryo bones and on cultures of undifferentiated embryonic chick ectoderm. Histological examination of the embryonic bones exposed to the excess vitamin suggested that chondroitin sulfate disappears under the influence of the vitamin. As for the ectodermic cells, instead of developing into a squamous epithelium topped by a keratinizing layer, they became a mucus-secreting epithelium.

These effects of excess vitamin A are opposite from those observed in vitamin deficiency, and the changes induced *in vitro* by the addition of the vitamin mimic hypervitaminosis *in vivo*. On the basis of these observations made in cultures, it seems safe to conclude that the effects of vitamin A on the metabolism of chondroitin sulfate and on the formation of mucus are direct and do not result from the stimulation of hormonal secretion by the vitamin. The effects of vitamin A on bone and epithelium suggest that the vitamin in some way affects the biosynthesis of mucopolysaccharides, yet the effect on the bone is opposite to that on the epithelium; whereas excess vitamin A accelerates the dissolution of chondroitin sulfate, it enhances mucus synthesis.

Wolf and his associates [109, 110] studied the effect of vitamin A on mucus synthesis by investigating the effect of vitamin A on mucopolysaccharide synthesis by colon segments and colon homogenates of vitamin deficient rats. The incorporation of ^{35}S in the mucus polysaccharides of colon segments of vitamin deficient rats is about half of that observed in colon segments obtained from rats fed a diet containing vitamin A. The rate of ^{35}S incorporation into the deficient mucosa was restored to normal by adding vitamin A to the incubation medium.

These studies were repeated using colon homogenate of vitamin A-deficient rats, and it was further established that the metabolic block resulting from vitamin A deficiency was located beyond those steps involved in uridine nucleotide biosynthesis, and that it did not involve the conversion of glucose to hexosamine. The homogenates prepared from colon of vitamin A-deficient rats partially lost their ability to form phosphoadenosine phosphosulfate; therefore, it was suggested that vitamin A is involved in mucopolysaccharide synthesis, possibly by acting as a coenzyme in the reaction activating sulfate.

Although Wolf's experiments have opened a new avenue, a molecular explanation of the mechanism of epithelial keratinization observed in vitamin A deficiency is still not available. Russian investigators have proposed that the effect of the vitamin on keratinization is secondary to its effect on oxidative processes. They claim that the oxidation of cysteine to cystine is a rate-limiting step in keratin formation, that copper catalyzes this oxidation, and that carotinoids inhibit the oxidative role of copper, thereby preventing the oxidation of SH groups into a disulfide bond.

Altered glycolysis has been demonstrated in vitamin A deficiency. The synthesis of glycogen from acetate, lactate, and glycerol is depressed. However, it is unlikely that vitamin A deficiency is responsible for the block in the glycogen synthetic pathway, because the rate of glycogen synthesis in the deficient rat can be restored to normal by cortisone administration. In contrast, the deoxycorticosterone administration doesn't affect the rate of glycogen synthesis in the deficient rat. Such a finding suggests that vitamin A deficiency might be associated with a block in the biosynthesis of steroid hormones. This assumption finds support in the following observations: (1) the total steroid content of the adrenals is reduced in the vitamin-deficient rat, but is restored to normal within 4 hours after the administration of the vitamin; (2) although progesterone administration can restore the estrus cycle and increase the number of living newborns in vitamin A-deficient rats, pregnenolone has no such effect. The possibility that vitamin A acts as a coenzyme for Δ^5-3β-hydroxysteroid dehydrogenase has been raised [111].

Because other liposoluble vitamins (K and E) have often been implicated in the electron transport chain, the effects of vitamin A on mitochondria and mitochondrial enzymes have also been investigated. Although the mitochondrial respiration rate may be decreased in vitamin A-deficient animals, there is no change in the activity of succinoxidase or the enzymes of the Krebs cycle. Some investigators claim that cytochrome c reductase activity is decreased in tissues of vitamin A-deficient animals. A decrease in transhydrogenase activity of liver mitochondria and an increase in NADPH cytochrome c reductase has also been described in vitamin A deficiency. Ubiquinone is consistently in excess in deficient rats but not in chickens.

Interesting observations have been made on the corneas of vitamin A-deficient mice. The corneas contain mitochondria that in vitamin A carency become progressively loaded with electron-dense material.

The effect of vitamin A deficiency on the performance of several biosynthetic pathways has also been investigated. Vitamin A deficiency does not affect cholesterol, fatty acid, or protein synthesis, but it has been claimed to interfere with DNA synthesis. This is probably an indirect effect resulting from the general retardation of growth that accompanies vitamin A deficiency. Plasma and liver cholesterol levels are not changed in vitamin A deficiency; however, cholesterol intake reduces the vitamin A stored in rat liver.

On the basis of studies of the effect of excessive doses of vitamin A on culture bones and on the tail of *Xenopus laevis*, a theory has been put forward suggesting that vitamin A acts by rupturing lysosomes and by releasing hydrolytic enzymes in the media. *In vitro* experiments done in Dingle's laboratory [112, 113] and in De Duve's laboratory [114], where the hydrolases were released under the influence of vitamin A, are often called upon to support this theory. This theory meets with a number of serious objections: (1) the release of a large number of hydrolases can hardly be reconciled with the effect of the vitamin on mucus synthesis and steroid hormone biosynthesis; and (2) the experiments done *in vivo* on the *Xenopus laevis* were carried out with doses of vitamin A that killed 30% of the larvae, so that much of the effect may be due to necrosis rather than to a specific effect of the vitamin. Similar objections can be made with respect to experiments done *in vitro* with organ culture. The validity of the interpretation of an effect of vitamin A on the isolated particles is discussed in more detail in the section on cellular necrosis.

Vitamin A could control differentiation by decreasing enzyme activities, by stimulating enzyme synthesis, or both. The activity of a number of enzymes (gulonolactone oxidase, codeine demethylase, squalene cyclodehydrase, ATP-sulfurylase, and sulfate transferase) decreases in vitamin A-deficient animals. The significance of these observations remains obscure, and no specific coenzymic role for vitamin A has as yet been discovered.

De Luca and his associates have studied the effects of vitamin A on protein synthesis in the intestinal mucosa and have observed that vitamin A interferes with the biosynthesis of protein by membrane-bound polyribosomes, but not with the biosynthesis of proteins by free polyribosomes [115].

During the last three decades, many investigators have attempted to express in molecular terms the role of vitamin A in vision, and these studies have now developed into one of the most fascinating chapters of modern biology. The retina is a highly differentiated nerve structure which develops early in embryonic life in the proencephalon out of bilateral invaginations called primary optic vesicles. After full development, the retina, like the brain, is composed of an epithelium (pigment epithelium) and a chain of four different types of neurons: the photoreceptor neuron, the hori-

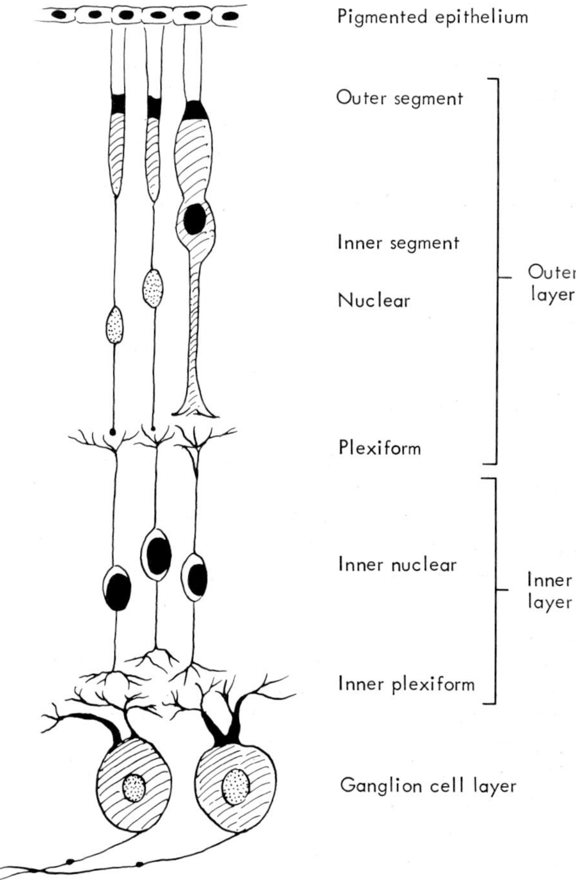

Fig. 4-35. Schematic representation of the retina

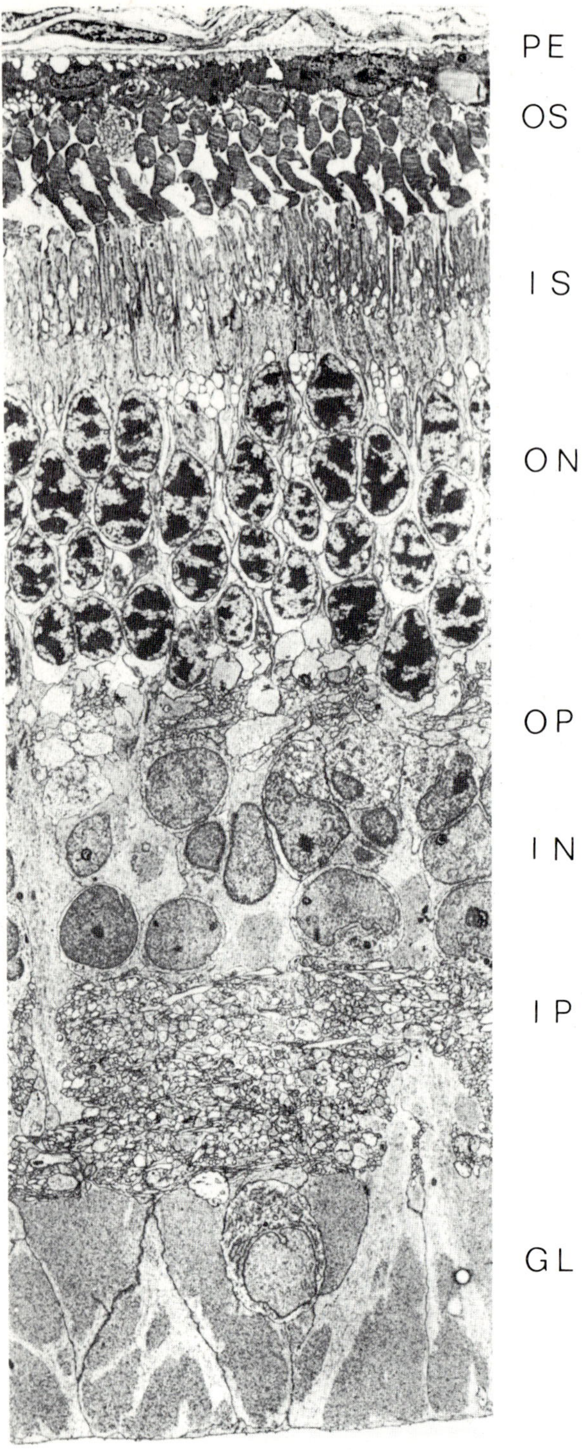

Fig. 4-36. Survey picture of cross section through the rabbit retina. From top to bottom: choroid layer and pigment epithelium (PE); layer of the outer segments (OS) of the receptor cells; the inner segments (IS); outer nuclear layer (ON); outer plexiform layer (OP); inner nuclear layer (IN); inner plexiform layer (IP); and ganglion cell layer (GL), represented by a single ganglion cell embedded in the foot processes of the Müller cells, the equivalent of the glial cells in the central nervous system. Note the extensive convergence in the retinal circuity in this part of rabbit retina, reflected in the successive reduction of conducting elements in the direction of receptor cells to ganglion cells. × 1,700. (From Sjöstrand)

zontal cells, the bipolar cells, and the ganglion cells. In most of the retina (except the central fovea and the point of emergence of the optic nerve), the epithelium and the cellular and fibrous components of the neuron are concentric, forming the layers of the retina described by histologists (see Figs. 4-35 and 4-36).

Although the pigmented cells are not directly related to the role of vitamin A in vision, these cells are of interest because they contain pigment granules that move under the influence of light from the body of the cell into the thin protoplasmic expansions. This pigmented protoplasmic film prevents diffusion of light from one photoreceptor to another.

The cells most relevant to the present discussion are those forming the photoreceptive neurons. The dendrites of the neurons assume a special shape well suited to photoreception. The synapses transfer the impulse through their connections with dendrites originating from the bipolar and the horizontal cells. The photoreceptor cell is made up of cone or rod, body, and efferent fibers with the synapse.

The dendrites of the photoreceptor cell form the rods and cones, which are morphological variations of a basic structure common to both types of photoreceptors. To describe them more conveniently, morphologists have subdivided the rods and the cones into inner and outer segments. The two segments are connected by a cilium, usually composed of nine pairs of filaments arranged in two concentric circles. The outer segment of both the rods and the cones is closely associated with the pigment epithelium. Under the light microscope, the outer segment of the rod appears as a smooth, elongated cylinder with a rounded end, and that of the cone assumes the shape of an elongated slender cone. The substance of the rod is homogeneous and brilliant and contains the visual purple, which is responsible for the red color of the retina during life.

The electron microscopist has devoted much effort to observing and describing the ultrastructure of the outer segment of the rod. Although the ultrastructural appearance varies somewhat with the fixative used to prepare the specimen, some dependable conclusions are now available. Sjöstrand can be credited for having made the pioneering observations in the field. After fixation with osmium tetroxide two basic components can be distinguished in the outer segment of the rod. At the periphery, a typical cell membrane encloses a stack of small discs in the center of the rod. These discs in the center of the rod have an electron-transparent center and an electron-dense double membrane. They are piled up with their largest diameter perpendicular to the long axis of the rod and are usually not in intimate contact with the cell membrane.

In some animals, like the toad, the guinea pig, the rat, and the rabbit, but not the monkey, two additional features have been described in connection with the discs found in the outer segment of the rod: a deep incision from the periphery to the center of the disc and tubules connecting individual discs. The outer segment of the cone has a structure similar to that in the rod, except that no incision or tubules have been described, and the disc and cell membranes have often been connected. Such connections have brought the morphologist to suggest that the discs in the outer membrane are infolds of the cell membrane rather than individual, flattened bags piled on top of each other. The exact location of visual pigment in the rod is not known. Some have assumed that it is enclosed between the membranes of the discs, but others believe that the disc membranes contain the lipoprotein moiety of the visual pigment (opsin), whereas the retinene moiety is found between the discs [116–119].

The inner segment of the rod contains two specialized regions in its outer part called the elipsoid. Elongated mitochondria are stacked parallel to the long axis of the cell; the number of mitochondria in the rods varies with the species (6 in mouse and 20–30 in man). A well-developed endoplasmic reticulum and a Golgi apparatus are in the inner part of the rod's inner segment. The inner segment of the cone differs from that of the rod only in that it contains many more mitochondria. The inner segments of the rod and cone are connected to the cell body by the intermediate of a nonmyelinated fiber. The body contains a nucleus surrounded by scanty cytoplasm with endoplasmic reticulum and a Golgi complex. From the body of the rod emerges a nonmyelinated fiber which enlarges in the shape of a sphere at its proximal end forming a synapse, which is connected to a dendrite of a bipolar cell. The cone synapse is more complex and receives dendrites from a number of cells.

During visual excitation, the light beam that passes the refractile portion of the eye reaches the cones and rods of the photoreceptive neurons. Then the radiant energy is converted into chemical energy, which in turn generates an electrical impulse carried through the demyelinated fibers of the photoreceptor cell to its synapse. The horizontal neuron computes the impulses generated in several cones and rods, and the impulses generated in the photoreceptive system are finally transferred to the brain through the bipolar and ganglion cells of the retina. Physiologists have clearly demonstrated the development of action potentials in the efferent fibers of the retina, both in the isolated eye and *in vivo* by stimulating the organ with light. These studies have established that the eye, like other sensory organs, obeys the Weber-Fechner law. Thus, the size of the voltage changes recorded in the efferent fibers is proportionate to the logarithm of the intensity of the light source used to excite the eye.

This brief review of the physiology and morphology of the retina raises several problems: the site of conversion of radiant to chemical energy in the retina; the molecular change associated with the conversion; and the site of conversion of chemical energy into electrical impulses.

We have seen that the rods and cones contain the visual pigment and that the pigment bleaches when exposed to light. The retina of the intact animal recovers its bright color when placed in the dark. Such observations indicate that the photochemical reaction of vision takes place in the rods and cones, and that rhodopsin is the photosensitive molecule. This fact was firmly established by the comparison of the spectral sensitivity curve of human rod vision and the absorption curve of rhodopsin in the visible region of the spectrum. The two curves coincide and reach a peak at 510 mμ. The curves failed to agree, however, in the extreme short wave end of the spectrum. This has been ascribed to difficulties in correcting for preretinal absorption.

Most of our knowledge on the molecular aspect of the photochemical reaction stems from studies in Wald's laboratory [120, 121], and the results of many years of work in that field can be summarized as follows: under the influence of light, the visual pigment undergoes a series of molecular transformations, and the ultimate products of the photochemical reaction are retinene and opsin. The exact number and the molecular structure of the intermediates formed during the photochemical reaction are not known, but the reaction is probably complex. Intermediates referred to as lumirhodopsin and metarhodopsin have been reported. The rods and the cones contain two retinenes, retinene 1 and retinene 2. Morton demonstrated that retinene 1 is the aldehyde of vitamin A_1. The retina contains an enzyme catalyzing the reversible conversion of retinene 1 to vitamin A_1. This enzyme requires NADH as a coenzyme, and the enzyme responsible for the reaction is identical to the alcohol dehydrogenase of liver. The equilibrium of the reaction is largely in favor of the formation of vitamin A_1, and the equilibrium can be shifted in favor of the formation of retinene only when retinene is trapped

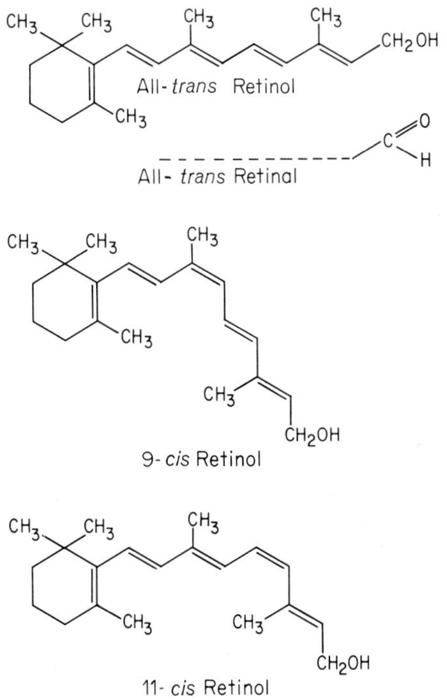

All-*trans* Retinol

All- *trans* Retinal

9- *cis* Retinol

11- *cis* Retinol

Fig. 4-37. *Trans*- and *cis*-isomers of retinol and retinal

as soon as it is formed. In the retina, retinene is withdrawn from the medium by its rapid recombination with opsin.

Bleaching of the visual pigment always yields all *trans*-retinene 1 (see Fig. 4-37). When the combination of all *trans*-retinene 1 with opsin is attempted *in vitro,* no visual pigment is formed, but if a mixture of stereoisomers of vitamin A is incubated with opsin, a certain amount of visual pigment is formed. This suggested that the retinene that combines with opsin is a *cis* isomer of vitamin A. Wald and his associates had suspected that a stable *cis* isomer of vitamin A was involved in the reaction. Molecules with a *cis* configuration at the double bond 9 or 13 were likely to be stable, while those containing a *cis* configuration at double bonds 7 and 11 would be unstable because the conversion of the all-*trans* molecule to the *cis* molecule would meet with steric hindrance. For example, the shift from the *trans* to the *cis* configuration at the level of the double bond 11 would bring a methyl group to collide with a hydrogen. As a result, the molecule would be twisted, its planarity lost, and stabilization through resonance impossible. Since resonance provides for stability of the molecule, sterically hindered molecules will be unstable. Yet, the 11-*cis* isomer combines with opsin. The peculiar stereospecificity of the retinene that forms the visual pigment raises many questions, few of which have been answered.

The 11-*cis* retinene is in a higher energy state than the all-*trans* retinene, and therefore it needs energy for its formation. Light can provide the energy necessary for the shift from the *trans* to the *cis* configuration *in vitro*. When retinene is dissolved in a polar organic solvant such as ethanol, in addition to forming other isomers, up to 25% of the all-*trans* can be converted to the 11-*cis* isomer in the presence of light. In the retina, rapid isomerization is needed because vision depends on the continuous resynthesis of the bleached pigment. Hubbard has found an isomerase that catalyzes the conversion of all-*trans* retinene to 11-*cis* retinene (see Fig. 4-38). But whereas 32% of the retinene is convered to the 11-*cis* in the presence of light, at equilibrium in the dark 90% of the retinene is in the all-*trans* form. Although the mechanism by which the 11-*cis* isomer is formed in the dark is still unknown, it is well established that the isomer appears in the retina in the dark.

The exact molecular relationship between retinene and opsin has not yet been clarified, but it is generally agreed that the structural geometry of the molecule is essential in determining the relationship between the retinene and the protein molecule and also in the photochemical reaction. It is assumed that the prosthetic group, retinene 11-*cis,* fits in a groove of the opsin molecule like a key in a keyhole (see Fig. 4-39). It has also been established that the retinene is bound to an ε-NH_2 group of lysine of opsin by the intermediate of the aldehyde group. The binding of retinene to opsin thus forms a Schiff baselike linkage. The amino acid sequence of the decapeptide adjacent to the ε-NH_2 amino acid has been elucidated.

Light has a triple effect on the rhodopsin molecule. It changes the geometrical configuration of retinene by shifting it from the 11-*cis* to the all-*trans,* thereby changing the charge distribution and releasing part of the retinene molecule from the protein molecule, which in turn, takes a new configuration. Thus, the protein molecule could shift from a configuration best fitted to accept the 11-*cis* retinene to a configuration in which the all-*trans* is loosely attached to the protein molecule. The combination of the all-*trans* retinene with these various configurations of the protein in the lumirhodopsin can be stabilized at low temperature. In lumirhodopsin, the all-*trans* retinene is associated with an opsin molecule that has kept the configuration typical for that which accepts the 11-*cis* retinene, while the metarhodopsin is an intermediate in which all-*trans* retinene is associated with a protein that has modified its configuration. However, these relationships between the retinene molecule and the

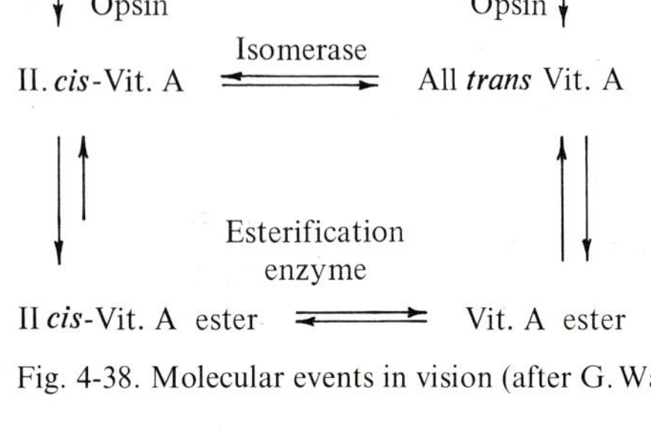

Rhodopsin
Light
II. *cis*-Retinal ⇌ (Isomerase) All *trans* Retinal
+ Opsin | + Opsin, Alcohol Dehydrogenase
II. *cis*-Vit. A ⇌ (Isomerase) All *trans* Vit. A
II *cis*-Vit. A ester ⇌ (Esterification enzyme) Vit. A ester

Fig. 4-38. Molecular events in vision (after G. Wald)

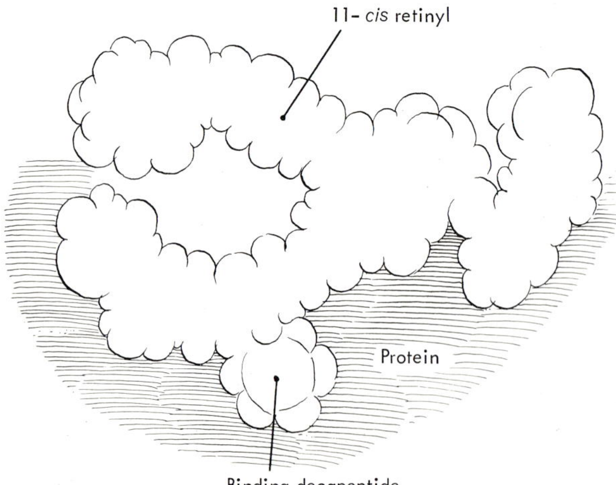

Fig. 4-39. Schematic model of the chromophoric site of rhodopsin

protein molecule remain speculative, and further investigation is awaited with the greatest interest.

Clearly, except in the case of night blindness, there is no way to correlate the symptoms observed in vitamin A deficiency with molecular alterations or, for that matter, with a specific molecular role of vitamin A in metabolism. Coenzyme roles in hydroxysteroid dehydrogenase, squalene cyclodehydrase, sulfate transferase, codeine demethylase, and gulonolactone oxidase have been proposed but not established conclusively. Effects on membrane, protein synthesis, and electron transport have been described, but none satisfactorily explains the pathogenesis of vitamin A deficiency [122].

Fortunately, the deficiency is readily cured by administration of the vitamin, and one would therefore expect that vitamin A eficiency would have been eliminated from the earth. Yet, it continues to plague mankind in a large segment of the world, while the most fortunate countries indulge in sophisticated research on the vitamin's molecular mechanism of action. For the medical student whose primary goal is to cure the sick with the means at his disposal, this is a shockling paradox. Yet the cure of vitamin deficiencies in underdeveloped countries is often out of reach of the physician because of sociological and political considerations. The student's duty as a scholar is also to keep abreast of the developments in modern medicine, because adequate understanding of the mechanism of action of vitamin A may help to elucidate the pathogenesis of many diseases of the skin, the connective tissue, and the eye.

Vitamin E Deficiency

A dietary factor needed for normal reproduction in rats was discovered in 1922 by Evans and Bishop [123]. Fourteen years later, Evans succeeded in isolating from wheat germ pure alcohols with two marked biological activities. They were named α- and β-tocopherol. The molecular structure of α-tocopherol was identified by degradation and synthesis as 2,5,7,8-tetramethyl-2-(4,′8′,12′-trimethyl-decyl)-6-chromanol; it was later discovered that several isomers of this compound are biologically active. The term vitamin E is used in reference to a number of biologically active isomers the nomenclature of which has necessitated using the first seven letters of the Greek alphabet. The formulas of these various isomers are given in Fig. 4-40.

A similar pattern has been followed in the discovery of most vitamins. Clinicopathological observations are made in humans or animals receiving a deficient diet; a factor capable of correcting the symptoms caused by the deficiency is partially purified; the factor is purified and crystallized; and the molecular structure of the factor is identified. The last of these endeavors is often unappreciated by the physician, yet it involves painstaking collaborative efforts of biologists and organic chemists and always constitutes the indispensable cornerstone for all further investigation on the vitamin's molecular mode of action [124].

Tocopherol was identified by degradation and synthesis. Several methods of degradation were used, each

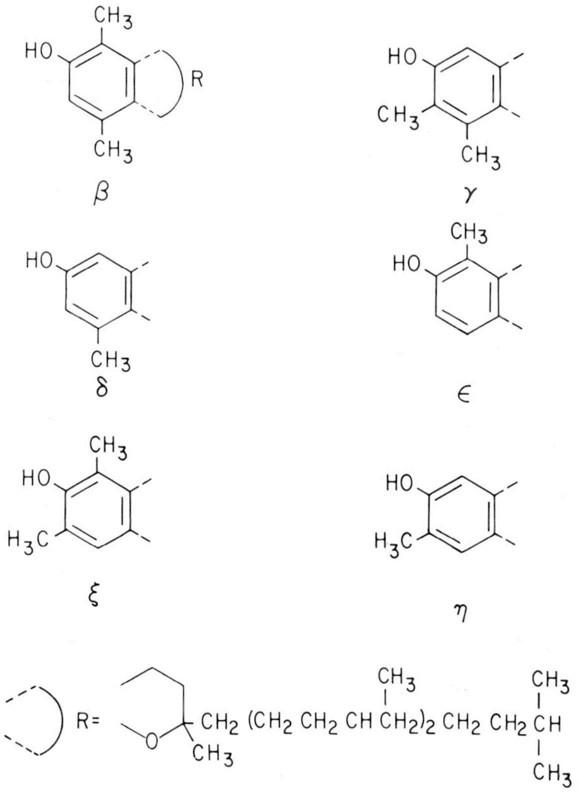

Fig. 4-40. Formulas of various tocopherols

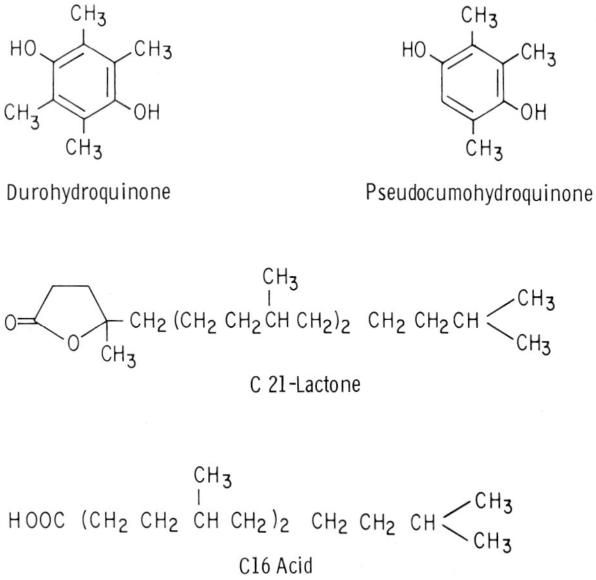

Fig. 4-41. Vitamin E: organic degradation

yielding different organic compounds. Pyrolysis of α-tocopherol at 350° yielded durohydroquinone, an aromatic compound with a six-carbon ring, all the carbons of which are substituted. Four of the carbons carry methyl groups and two carry hydroxyl groups. When β-tocopherol was pyrolyzed in the presence of hydriodic acid, a different hydroquinone was obtained, pseudocumohydroquinone, in which only 5 carbons in the aromatic ring are substituted (3 methyl, 2 hydroxyl).

Different degradation methods established the structure of the remaining part of the molecule. Thus, a C_{21} γ-lactone was obtained by mild chromic acid oxidation of the α-tocopherol. More drastic degradation of α-tocopherol yields a C_{16} acid (see Fig. 4-41).

After reconstructing the products of degradation like the pieces of a puzzle, the organic chemist proposes the most likely structure for the biological compound. The exact structure is usually confirmed by synthesis. In the case of α-tocopherol, this was achieved by Karrer and his associates who first synthesized tocol (tocopherol without alkyl substitution on the phenyl ring) by condensing hydroquinone and phytol, and later α-tocopherol was synthesized by condensing trimethylhydroquinone with phytylbromide (see Fig. 4-42).

Under the proper conditions, a chromane is obtained, as indicated in the figure, and the synthetic compound has biological properties identical to that of the natural vitamin. Once the structure of α-tocopherol was established, the structures of β-, γ-, and δ-tocopherol were soon elucidated. The degradation products were identical in all cases, with the exception of the substituted hydroquinone. The γ- and δ-isomers were shown to have the substitution illustrated in Fig. 4-40. Later, the ε- and β-tocopherols were purified by two-dimensional chromatography.

Thus, the different active tocopherols differ by the number and position of the methyl substitution on the aromatic ring. α-, β-, and γ-tocopherols have only two methyl groups on their aromatic ring, while the ε- and η-isomers carry only one methyl group on their aromatic ring. All these isomers of tocopherol have an identical side chain.

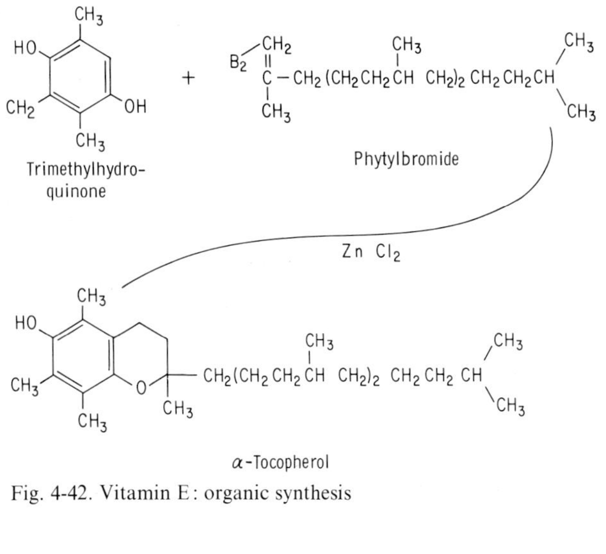

Fig. 4-42. Vitamin E: organic synthesis

Although the phytyl side chain is indispensable for activity, isomerization within the side chains doesn't affect vitamin activity. Furthermore, esterification of the alcohol or replacement of the two methyl groups of the aromatic ring with ethyl or propyl groups alters the biological activity very little. This greatly oversimplified description of the methods of identifying active tocopherol illustrates the approach used in identifying many biologically active compounds.

Distribution and Deficiency

Vitamin E is widely distributed among plants and animals. Vegetables are generally a good source of the vitamin. The concentration of a given isomer depends on the origin of the source. For example, American wheat germ is rich in α-tocopherol, but European germ is richer in β-tocopherol. The daily requirements of vitamin E for humans are not known since the deficiency is not endemic and is not readily produced experimentally in man; yet some clinical cases with symptoms likely to be attributable to vitamin E deficiency have been described. An estimated 15 mg of α-tocopherol is present in the average daily American diet.

The vitamin is absorbed through the intestine by a mechanism similar to that involved in the absorption of other liposoluble vitamins; consequently, intestinal disorders leading to lipid malabsorption are associated with malabsorption of vitamin E. Humans with such malabsorption syndromes have reduced levels of vitamin E in their tissues. They do not show classical symptoms of muscular dystrophy, although a certain degree of muscular fatigue may be observed.

Horwitt [125] has investigated vitamin E requirements in humans by administering a vitamin E-deficient diet supplemented with unsaturated fats for the purpose of removing vitamin E. He found that the amount of tocopherol in tissue is inversely related to the amount of unsaturated fat that has been administered. A significant observation was that 30% of the individuals receiving the unsaturated fats also developed duodenal ulcers. Hemorrhagic endarteritis similar to that observed in vitamin E-deficient chicks has been described in a child who, for therapeutic purposes, was fed a diet rich in milk and unsaturated fat, but devoid of vitamin E.

Although vitamin E deficiency is rare in humans, it is worthwhile to describe the lesions that develop in deficient animals since they resemble diseases observed in humans. Elucidation of the pathogenesis of injuries observed in animals might help to understand the pathogenesis of similar human diseases observed in the absence of vitamin E deficiencies.

The lesions produced in avitaminosis E vary greatly, depending upon the animal used for the experimentation. The pathological observation can be classified into four main groups: interference with reproduction; muscular degeneration; pigment formation in adipose tissue; and liver necrosis.

Interference with reproduction has been observed in mice, rats, guinea pigs, and swine. Only those lesions observed in rats will be described here. The failure of vitamin E-deficient rats to reproduce normally stems from injuries that develop in the male and in the fetus.

Although the avitaminosis has no detectable effect on the female genital tract, it interferes with the normal course of pregnancy. The fetus appears to develop normally during the first week of pregnancy, but it is clearly retarded on the eighth day. The embryonic blood vessels and the blood islets of Pander-Wolf are not properly developed, and the ectodermic cavity fails to appear. These are only some of the most striking changes among other alterations in the embryonic development. The embryo is dead on the 12th day and is completely resorbed within approximately 2 weeks after conception.

The details of the pathogenesis of the fetal death are not known, but at least two different concepts prevail. It has been suggested that the primary structural injury involves failure of the placental vessels to develop normally, so that the embryo is killed by anoxemia. Others have proposed that the first structural injury is observed in the embryo itself where blood vessels develop inadequately.

By the prolonged administration to rats of a vitamin E-deficient diet, the testicles are ravaged. In the seminal tubules, the successive layers representing the various stages of spermatogenesis progressively disappear. At first, sperm mobility is impaired; later sperm production becomes abnormal with formation of irregular spermatids and coalescence of spermatocytes to form giant cells. Later, spermatogenesis ceases completely, and the empty seminal tubules, surrounded by a single layer of sertolian cells, shrink and collapse. The testicular lesions are not reversed by the administration of vitamin E.

Although muscular injuries are observed in most animals maintained on the vitamin E-deficient diet, the severity of the lesions varies considerably depending upon the animal. Whereas the rabbit is prone to muscular injury, the rat and monkey are rather resistant to muscular atrophy produced by vitamin E deficiency. The lesion affects the skeletal and the heart muscle. In the susceptible animal, the entire muscle cell swells, the sarcolemma becomes edematous, and the dissociated fibers fragment and become hyalinized leading to a morphological appearance resembling the Zenker degeneration described in humans. Fiber degeneration is accompanied by accumulation in the sarcolemma of fat droplets and golden pigment inclusions. In later stages, the sarcolemma nuclei proliferate, and a leukocytic exudate invades the degenerative area. In some animals, like the calf, the degenerated muscle become studded with small foci of calcification. A feedback degeneration of the motor end plate has been described in the rat.

The gold pigment inclusions resembling ceroid observed in the dystrophic muscle are also found in many other tissues of the vitamin E-deficient animal: testes, ovaries, lymph nodes, and others. The pigment resembles ceroid and reacts positively with the acid-fast staining. The significance of this pigment will become clearer when the biochemical role of vitamin E is discussed.

Vitamin E Metabolism

Studies of the metabolism of vitamin E were triggered by the observations of Alaupovic and coworkers [126, 127]. When ^{14}[C]-D-α-tocopherol-5-methyl was administered to rats or pigs and attempts were made to detect metabolic derivatives, two compounds were separated by chromatography. One of the compounds is ^{14}[C]-D-α-tocopherol quinone; the other is either a dimer or a trimer of α-tocopherol. The dimer and trimer are terminal oxidation products of α-tocopherol and are excreted in the bile. α-Tocopherol quinone can be converted to α-hydroquinone. α-Tocopherol hydroquinone may be esterified in liver and eliminated in the feces after concentration in the bile and excretion in the intestine, or it may be oxidized in the kidney to α-tocopheronic acid, which may be converted into an α-tocopheronolactone conjugate, which is excreted in the urine. In conclusion, vitamin E is excreted as such in the urine or the bile after conversion to a dimer or a trimer, in the form of a conjugated hydroxyquinone or tocopheronic acid (see Fig. 4-43).

Among the metabolites identified, α-tocopherol quinone seems to be the active compound. That compound is known to participate in photosynthetic electron transport in chloroplasts, but there is no evidence that it functions on an "active vitamin E." In fact, no available evidence indicates that a metabolite of the vitamin constitutes the active form.

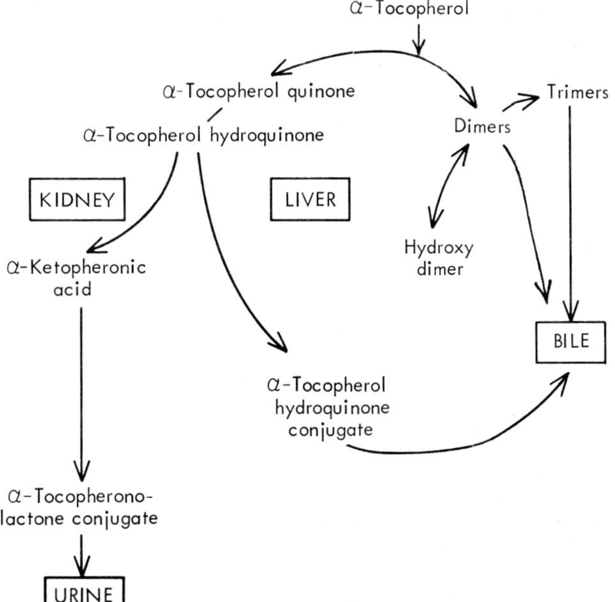

Fig. 4-43. Vitamin E metabolism

The role of vitamin E in metabolism is still unknown. Two different mechanisms of action have been proposed. One implicates its antioxidant properties, and the other assumes that the vitamin affects a specific metabolic pathway.

Each of the active tocopherols possesses antioxidant properties. However, the antioxidant properties and the biological activity are inversely related. Thus, if the active forms of vitamin E are classified with respect to their biological activity, the sequence appears as follows: $\delta<\gamma<\beta<\alpha$. The order of the sequence is reversed if the active forms are classified according to their antioxidant properties.

In spite of this lack of correlation between biological and antioxidant activities, a number of arguments favor the hypothesis that the primary action of vitamin E is to prevent peroxidation of lipids. (1) The lipid peroxide content of vitamin E-deficient animals is claimed to be higher than in normal animals. (2) Tissue extracted from vitamin E-deficient animals is more rapidly peroxidized when exposed to air than tissue obtained from normal animals. (3) Tocopherol prevents the peroxidation of linoleic acid catalyzed by hemoglobin *in vitro*. (4) Synthetic antioxidants, such as N-N′-diphenyl-phenylenediamine and 1,2-dihydroxy-6-etoxy-2,2,4-trimethylquinoline, exhibit biological properties similar to those of vitamin E. (5) The feeding of readily oxidizable compounds, such as polyunsaturated fats, raises the requirements for tocopherol in rats, chicks, calves, and sheep. Many of the arguments supporting the theory that vitamin E acts as an antioxidant are not accepted by all investigators without serious reservations.

The difficulties encountered in determining the peroxide content in tissues should be kept in mind. Although polyunsaturated fatty acids accelerate the production of some lesions attributed to vitamin E deficiency, such as exudative diathesis and encephalomalacia in chicks, they do not affect the reproductive functions in rats. Antioxidants cannot correct all of the pathological manifestations resulting from vitamin E deficiency. The alteration of the fetus and the testicles of vitamin E-deficient rats is not prevented by the administration of methylene blue to an animal maintained on a vitamin E-deficient diet. On the basis of such observations, it has sometimes been proposed that antioxidants do not replace vitamin E but spare the existing vitamin. Even if it is accepted that vitamin E acts by virtue of its antioxidant properties, the nature of the critical compounds that are oxidized in the deficiency remains to be elucidated.

Oxidation could inhibit specific enzymes or alter the permeability properties of structural membranes. The activities of several oxidases have been reported to be markedly reduced in liver homogenates of vitamin E-deficient rats. Among them are alcohol dehydrogenase, gulonolactone oxidase, and the oxidation of such Krebs cycle substrates as α-ketoglutarate and succinate. It has been proposed that the alteration in the activity of these enzymes does not result from changes in the molecule *per se,* but rather from the formation of inhibitors in the microsomes. It is tempting to speculate that these inhibitors are products of peroxidation.

Tappel and his associates [128] have claimed that such peroxidation products are present in mitochondria and microsomes, and that the amount in microsomes is greater than that in mitochondria. However, the results of such experiments are far from conclusive because individual results vary considerably. Furthermore, if the difficulties in determining the peroxidation product in cell fractions and homogenates are taken into account, it seems unwise to incriminate peroxidation products in inhibiting the different oxidases, especially since the dystrophic muscles in vitamin E-deficient animals contain no more peroxidation products than do muscles of control animals. Even if one accepts the quantitative analysis for peroxidation products, the values obtained for the microsomes are controversial, because although it is claimed that the total amount of peroxidation products is increased in vitamin E-deficient animals when the amount is expressed on a nitrogen basis, no differences between the microsomes of the deficient animals and the microsomes of the control animals can be detected. Finally, in brain and heart of vitamin E-deficient animals, the peroxidation products are mainly associated with the cell supernatant.

This critical analysis of ingenious working hypotheses and experiments designed to solve difficult problems is not intended to berate the quality of the work or to discourage new investigation in the field, but to focus the reader's attention on the limitations of available methods and to warn against unconditional adherence to the prevailing theory.

A lack of correlation between the intracellular content of peroxidation products and the cellular susceptibility to vitamin E deficiency does not exclude peroxidation products as the cause for the lesions observed in the deficiency. A few altered molecules of great biological significance may be lost among a great proportion of peroxidation products without biological effects. Yet, although the formation of peroxidation products could explain, for example, the blocking of gulonolactone oxidase, it could not satisfactorily account for the interference with the α-ketoglutarate oxidation because this inhibition cannot be overcome by adding vitamin E *in vitro* at doses that stop all peroxidation. The peroxidation that occurs in vitamin E deficiency could attack some molecules vital to the maintenance of cellular structure, and it has been proposed that alterations of the mitochondrial membrane might cause some of the modifications in cellular respiration observed in vitamin E deficiency. These manifestations of the deficiency are discussed below.

Tappel and his associates [129] have proposed that vitamin E deficiency is associated with alteration of the lysosomal membrane, which induces the release of a number of acid hydrolases. Such lysosomal rupture would then cause the muscular dystrophy observed in vitamin E deficiency. It now appears that the increase in acid hydrolase observed by Tappel and his associates in muscles of vitamin E-deficient chicks results from invasion by macrophages.

The observation that erythrocytes of vitamin E-deficient newborn infants and rats are more susceptible to hemolysis by hydrogen peroxide than normal erythrocytes is of particular significance with regard to the role vitamin E plays in maintaining the integrity of the cellular membrane. But the molecular mechanisms responsible for the alteration remain unknown.

The second working hypothesis on tocopherol's mechanism of action on metabolism is that the vitamin acts as cofactor in some enzyme reactions. Much effort has been devoted to demonstrating a role of vitamin E in the electron transport chain. The first suggestion for such a role came from comparing enzyme assays made on tissue obtained from normal and vitamin E-deficient rats. The results of such studies were often controversial due to variation in the individual data and overlapping of the values obtained on controls and experimental animals. Furthermore, observations made in one laboratory were often not confirmed in another. Yet the general impression was gained that the oxygen consumption in muscle of vitamin E-deficient animals was up to 60% greater than that in muscles of normal animals. It has not been established whether this alteration constitutes the primary injury induced by vitamin E deficiency or is a step common to all forms of muscular degeneration removed from the primary injury.

The increase in oxygen consumption found in muscle strips was not demonstrable in muscle homogenates or in rat liver, but an increase in oxygen consumption has been demonstrated in hamster kidney. The belief that vitamin E affects the cell's oxygen use found little consistent support in assays of enzyme activities in tissues obtained from vitamin E-deficient animals. The results obtained with cytochrome oxidase illustrate the difficulties encountered in interpreting such data. Depending upon the investigator, cytochrome oxidase activity of dystrophic muscle in vitamin E-deficient animals was reported unchanged, decreased, or increased. Close reexamination of the experimental conditions would probably clarify these discrepancies, yet conclusive interpretations are not possible for the moment. Although studies *in vitro* have not elucidated the role of vitamin E, they have yielded more promising results than the *in vivo* investigations. Slater and his associates showed vitamin E in the Keillen-Hartree preparation, and although the analytical techniques have been criticized, it seems established that some vitamin E is in the electron transport chain.

Vitamin E is in the NADH cytochrome c reductase system. When the system was extracted with isoectane, NADH oxidation was blocked; normal oxidation rates can be restored by adding α-tocopherol, but many other lipid substances are also active in that respect: *e.g.*, vitamin K, ubiquinone, phytol. Of course, such observations do not exclude the possibility that *in vivo* vitamin E is in fact the only substance capable of maintaining the integrity of the NADH cytochrome c reductase system.

Isoectane extracts only a small fraction of the vitamin E associated with the NADH cytochrome c reductase system. Added lipids could restore the activity of the system because they fulfill nonspecific functions proper to many lipids, including vitamin E. Therefore, added lipids could spare the vitamin E remaining in the isoectane-extracted system for participation in the electron transport chain. This possibility is in keeping with observations made in Mason's laboratory, in which the vitamin E content of the NADH cytochrome c reductase system was completely extracted; in that case the activity of the system is restored by α-tocopherol, and apparently by α-tocopherol only. Unfortunately, activity is not restored in all the preparations tested, probably because many uncontrollable factors can influence the preparations. In these experiments, antimycin A acts as a competitive inhibitor of α-tocopherol. Therefore, it appears that vitamin E is in some way involved in electron transport between cytochrome b and cytochrome c. Furthermore, among a number of lipids tested, only tocopherol can prevent antimycin A inhibition of NADH cytochrome c reductase.

In vitamin E-deficient rabbits, the activity of creatine phosphokinase increases, apparently in muscle first and in blood later. The administration of α-tocopherol normalizes the serum CPK levels. Because ethionine mimicks this effect of vitamin E, it has been proposed that vitamin E acts as a repressor of the gene responsible for the biosynthesis of creatine phosphokinase. It was also proposed that a similar mechanism could explain the increase in acid hydrolases observed in muscles of vitamin E-deficient animals. In the latter case, such an interpretation is unlikely because the appearance of the hydrolase is associated with the appearance of macrophages.

In experimental animals, many injuries caused by vitamin E deficiency can be corrected completely or in part by selenium, coenzyme Q, or some sulfur amino acids.

If weaning rats are maintained on a diet free of vitamin E and selenium, liver necrosis develops, but it can be prevented by the administration of either selenium or α-tocopherol.

In vitamin E-deficient rabbits, muscular dystrophy can be corrected by the administration of α-tocopherol but not selenium. Tocopherol-deficient monkeys develop an anemia that is reversed by the administration of vitamin E or coenzyme Q chromanols, but not by coenzyme chromanols or selenium.

Little is to be gained by listing further instances in which these various compounds can be interchanged because the interactions between vitamin E, coenzyme Q, and selenium are still unknown [130–135]. There are two major theories: the first proposes that all these compounds function in metabolism as antioxidants, protecting, for example, the integrity of the lipid membrane; the second holds that these compounds react with certain apoenzymes in specific but still unknown metabolic reactions.

Malnutrition Due to Disease

Malnutrition may also result from disease of the gastrointestinal tract. Among the malabsorption syndromes, carbohydrate malabsorption due to inborn errors of metabolism, the dumping syndrome resulting from gastrectomy, and steatorrhea will be discussed [136].

Carbohydrate Malabsorption

Food contains one polysaccharide (starch) and three disaccharides (maltose, sucrose, and lactose). Salivary and pancreatic amylase digests starch to yield maltose and sucrose, and lactose to yield maltose and sucrose. Sucrose, maltose, and lactose are split by invertase, maltase, and lactase, respectively. The products of the disaccharidase reactions are fructose, glucose, and galactose. Whenever amylase or one of the disaccharidases is absent from the intestinal content, the undigested sugars pass in the lower part of the intestinal tract and are fermented by the bacterial flora. As a result, lactic acid and volatile acids are formed and stimulate peristalsis and fluid secretion by the intestinal mucosa. Liquid foaming acid and foul-smelling feces are emitted. Amylase may be absent in pancreatic disease. Inborn errors characterized by the absence of intestinal lactase, maltase, and invertase have been described.

The Dumping Syndrome

The gastric mucosa is a source of compounds (hydrochloric acid, pepsin, mucin) involved in food digestion. The epithelium of the antrum secretes a hormone, gastrin, which stimulates the parietal cells of the body of the stomach to secrete acid. The elaboration of a gastric hormone that depresses hydrochloric acid secretion is controversial. The mucin secreted by the glandular cells found throughout the stomach coats the gastric mucosa and protects it against autodigestion.

Gastrectomy is followed by nutritional impairment in at least 50% of the patients, and 2–10% of the gastrectomized patients are severely disabled as a result of the operation. The most common symptoms are those associated with the dumping syndrome, in which gastrointestinal and cardiovascular symptoms develop 15–20 minutes after ingestion of a meal. The patient experiences a sensation of fullness in the epigastric region followed by increase of gastrointestinal motility and diarrhea. At the same time, he becomes hypotensive, pale, weak, and may sweat profusely and develop tachycardia. As a result, the patient is not inclined to eat, and his daily caloric intake is reduced from approximately 2,200 to 1,800. The hypermotility of the intestine and the diarrhea are responsible for malabsorption of food, and both nitrogen and fat intake may be defective in gastrectomized patients. The pathogenesis of the steatorrhea (see below) associated with the dumping syndrome is not clear, but it has been proposed that after gastrectomy the reservoir function of the stomach is lost, and the food, instead of being released in the intestine intermittently in smaller amounts, is abruptly dumped in the intestinal tract, where it is not properly mixed with digestive enzyme and bile. Since after gastrectomy part of the vagus innervation has been eliminated, and since the removal of the acid-secreting glands of the stomach eliminates the secretin response, it is also possible that decreased stimulation of pancreatic and biliary secretion accounts for the postgastrectomy steatorrhea [137].

The pathogenesis of the dumping syndrome is not known either. The accepted theory is that of Robert and Randall, which proposes that a massive and rapid hydrolysis of ingested nutrients in the jejunum makes the jejunal content hypertonic. The hypertonic hydrolysate is diluted by fluid drawn from the plasma and extracellular fluid. The loss of fluid leads to decreased blood volume and reduced cardiac output, which in turn trigger a vasomotor response: sweating, tachycardia, electrocardiographic change, etc. The following findings support this concept: (1) The symptoms of the dumping syndrome can be produced by ingestion of hypertonic glucose, hypertonic protein hydrolysate, or hypertonic sodium sulfate; and (2) in the dumping syndrome, plasma volume drops 300–1 000 cc and hematocrit levels increase.

Experiments in dogs submitted to a three-stage operation have, at least in part, explained the pathophysiology of the vascular events in the dumping syndrome. In the first stage, a portion of the proximal jejunum is exteriorized together with the vascular and nerve attachment and covered by skin. Once the animals are eating well and are again in good health, they are ready for the second stage—an electromagnetic probe is introduced into the aorta above the heart and the aorta is catheterized. Finally, in the third stage a flow probe is installed in the superior mesenteric artery near the aorta, and a catheter is also placed in the portal vein. By these means it is possible to directly inject solutions of known composition in the intestinal pocket, to monitor cardiac output, and to measure mesenteric pressure, input, and outflow.

The injection of 50% glucose in the pocket of these animals suggests that the cardiovascular symptoms occurr in two stages, described as primary and secondary. The primary event starts with a decrease in mesenteric resistance, followed by an increase in superior mesenteric artery flow, splanchnic pooling of blood, and increased portal venous pressure. The secondary events probably result from a compensatory reaction to the changes in the blood flow of the mesenteric bed. The reaction involves increased heart rate, peripheral resistance, and blood flow. Consequently, splanchnic pooling further increases, and cardiac output and blood flow to the brain decrease.

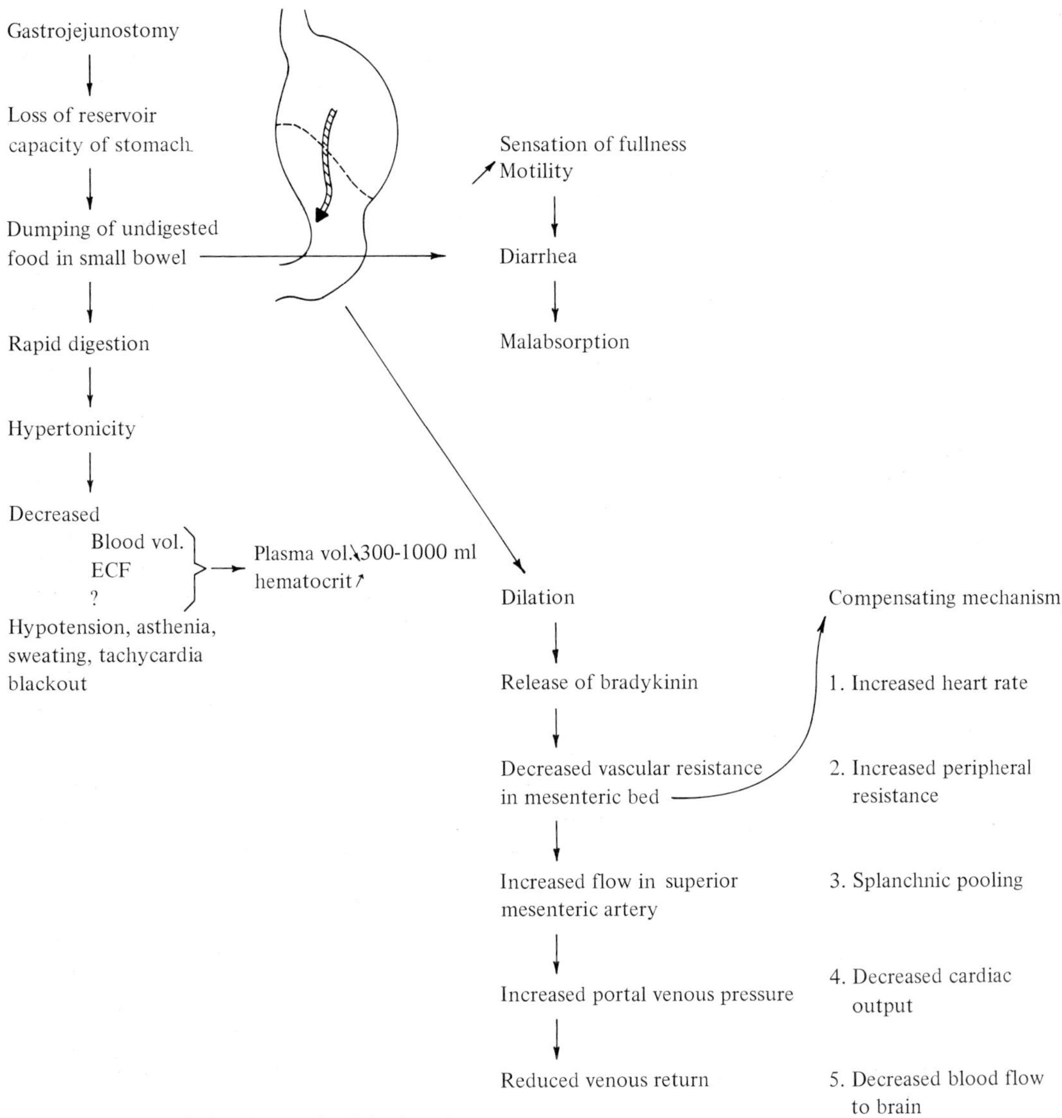

Fig. 4-44. Hypothetical pathogenesis of the dumping syndrome

Whether an increase in bradykinin secretion is responsible for the decreased resistance to flow in the mesenteric artery remains to be established.

These experimentations further suggest that it is not likely to be the osmotic pressure of the solution injected into the jejunum that induces the dumping syndrome, but rather the distension of the intestinal loop by the sudden dumping of large volumes of fluid in it.

Indeed, the injection of 5% glucose and even of polyethylene glycol leads to symptoms similar to those caused by higher concentrations [138]. The pathogenic mechanism of the dumping syndrome is summarized in Fig. 4-44.

Steatorrhea

Steatorrhea is characterized by the emission of fat in the stool. The feces typically are pale, greasy, and foul smelling. In occult steatorrhea, fat is lost in the stool but there is no diarrhea. In frank steatorrhea the presence of fat is easily detected by direct examination of the stool, but in more subtle cases, it may be necessary to resort to microscopic examination of the stool, to fat balance studies, or to the tracing of radioactive tagged lipids. The presence of excessive amounts of fat in the stool may result from a failure to digest the fats properly or from malabsorption (see Table 4-4). These two causes of steatorrhea can be distinguished by the use of ^{131}I-labeled triolein and iodine labeled oleic acid. In normal subjects, the ^{131}I-labeled triolein is rapidly recovered in the blood, and less than 5% is found in the stool. In frank steatorrhea the blood levels of ^{131}I are depressed, and the radioactivity of the stools is increased. The ^{131}I-labeled oleic acid is depressed in blood only in patients with malabsorption. In digestive defects, the appearance of radioactivity in the blood is normal. Steatorrhea is asso-

Table 4-4. Causes of steatorrhea

Impairment of digestion	
Pancreatic insufficiency	Lithiasis Chronic relapsing pancreatitis Carcinoma of the pancreas Cystic fibrosis Kwashiorkor
Liver insufficiency	Cirrhosis Hepatitis Carcinoma
Obstruction of common duct	
Surgical	Gastrectomy Gastroileostomy Gastroenterostomy Pyloroplasty
Absorption defects	
Surface defects	Surgical secretions of intestine Intestinal fistula Tuberculosis Regional enteritis Amyloidosis Scleroderma X-Irradiation
Cellular defects	Celiac disease Sprue
Block of lymphatic drainage	Carcinoma in lymphoma Whipple's disease
Defect of digestion	
Lack of pancreatic lipase	Cystic fibrosis of pancreas Chronic relapsing pancreatitis Carcinoma of the pancreas Pancreatic lithiasis Pancreatic fistula Pancreatectomy
Lack of bile salts	Obstructive jaundice Hepatic insufficiency Cirrhosis Hepatitis

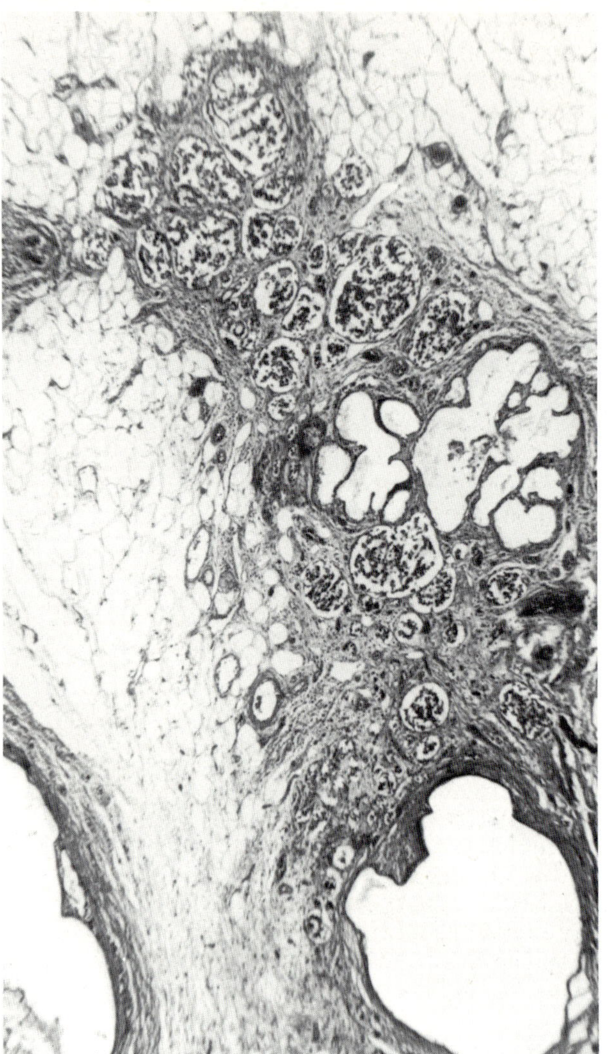

Fig. 4-45. Histological changes in pancreas of patients with cystic fibrosis

ciated with reduced caloric intake and weight loss, calcium loss, osteomalacia, loss of liposoluble vitamins A, D, K, and E, and avitaminosis.

Two diseases causing steatorrhea deserve further attention: cystic fibrosis and sprue. Cystic fibrosis is a hereditary dysfunction of the mucous and serous exocrine glands (lung, pancreas, salivary gland, sweat glands, etc.) and is sometimes associated with cirrhosis of the liver. The disease has been described mainly in whites, rarely in Negroes. Cystic fibrosis is probably transmitted by an autosomal recessive mammalian gene.

Three types of exocrine gland anomalies have been described in cystic fibrosis: (1) secretion of abnormal mucus that occludes the exocrine channels (pancreas, intestinal glands, intrahepatic bile ducts, gall bladder, prostate, sublingual salivary glands); (2) hypersecretion of mucus with distension of the glandular lumen (glands of the tracheobronchial tree); and (3) abnormal electrolyte secretion (serous, salivary, and lacrimal glands).

The most typical defect is observed in the pancreas (see Fig. 4-45). At first, each acinus is filled with some thick, coagulated eosinophilic material, and no changes in the acinar lining or in the islet structure occur. Later, the mucus material accumulates and part of it is extruded in the ramification of the pancreatic duct, and the lining of the acini shows atrophy and proliferation. Later both acini and ducts are filled with mucus material. The acini show cystic distension, and fibrosis develops between the acini. If the patient lives long enough, the islet may atrophy and diabetes may develop. Similar lesions are found in the mucous salivary gland. In patients with cystic fibrosis, the intestinal glands secrete a mucus material that accumulates in the lumen of the intestine, obliterates it, and causes symptoms of obstruction. The block is usually in the terminal ileum, but it can occur in any part of the small or large intestine.

Intestinal occlusion has been observed in the fetus causing meconium plugs before birth, at birth, or sometimes in older infants, but the typical injury occurs at birth. A plug of thick greyish mucus with the consistency of putty blocks the terminal ileus. The part of the intestine proximal to the block is distended and may contain fecal material.

If the child has been fed, the part of the intestine distal to the plug is atrophic. Often the ileal occlusion is associated with volvulus.

The patient may have two liver defects—fatty liver and multilobular cirrhosis. Fatty liver results from malnutrition; multilobular cirrhosis results from the occlusion of the intrahepatic bile ducts, followed by atrophy and necrosis of the hepatic cells with regeneration and fibrosis. When cirrhosis is severe, hepatic insufficiency and portal hypertension (hypersplenism, esophageal varices, ascites) develop.

The primary alteration of the respiratory tract of patients with cystic fibrosis is the hypersecretion of mucus. It is not known whether the mucus is normal or contains excessive amounts of electrolytes. In any case, the mucus secretion is not properly evacuated. It first occludes the smaller bronchi. If the occlusion is incomplete, the alveoli become distended, the inspired air cannot be expired, and emphysematous blebs develop. If the obstruction is complete, atelectasis occurs. The changes in the bronchi and the pulmonary parenchyma favor infection, especially by staphylococci. Pulmonary abscesses, bronchial pneumonia, bronchitis, and pneumonia develop; in fact, children with cystic fibrosis often die from respiratory infection. Pneumothorax, hemoptysis, and emphysema are rare complications of the pulmonary disturbance observed in cystic fibrosis (see Fig. 4-46).

Although no histological changes are detectable in the serous glands of patients with cystic fibrosis, the sweat, lacrimal, salivary, and stomach glands excrete excessive amounts of electrolytes (potassium and sodium chloride). The volume of sweat, however, is within normal limits.

The administration of a diet restricted in sodium chloride induces a normal response of the kidney and reduced electrolytes in urine, but the sweat glands continue to excrete excessive amounts of electrolytes. As a result, patients with cystic fibrosis are oversensitive to heat in that they lose large amounts of sodium chloride in the sweat. The loss of sodium leads to a reduction of the extracellular fluid with vascular collapse. The situation is readily reversed by the intravenous administration of saline [139].

The pathogenesis of cystic fibrosis is unknown. One working hypothesis proposes that the genetic defect is responsible for the elaboration of an abnormal glycoprotein as a result of an enzymic defect. Another postulates that mucus accumulates because of an interference with the regulatory mechanism of mucus secretion. Two laboratories have reported that mucus obtained from the duodenums of patients with cystic fibrosis presents molecular anomalies. Duodenal mucus contains two glycoproteins that differ by their solubility in 67% alcohol and their composition, especially in fucose and sialic acid. The more sialic acid and the less fucose a glycoprotein contains, the more soluble is the glycoprotein. The effect of these two compounds on glycoprotein solubility is explained in the following way. Sialic acid and fucose are N-group constituents. The sialic acid is linked to the carbohydrate chain by its carbonyl group. Such a molecular arrangement leaves the hydrophilic carboxyl group of

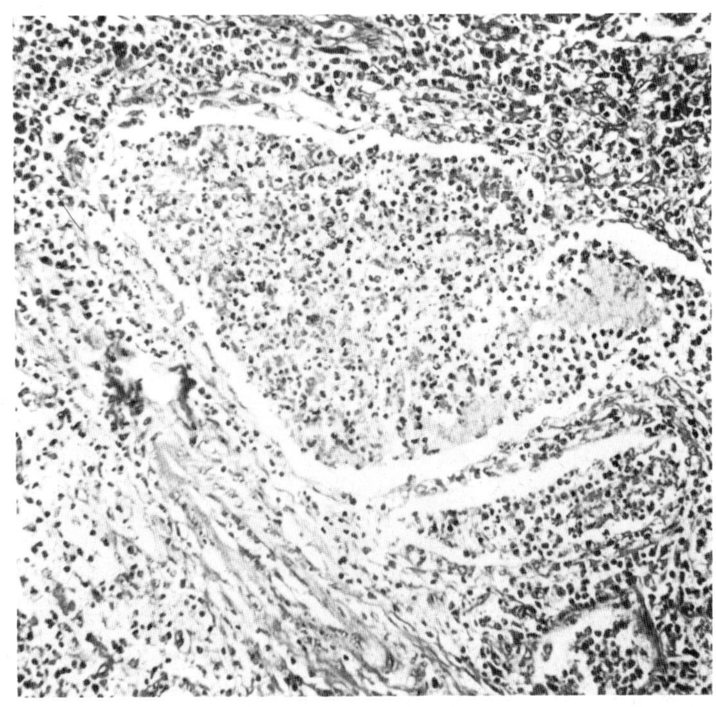

Fig. 4-46. A case of cystic fibrosis showing bronchodilation and acute bronchitis

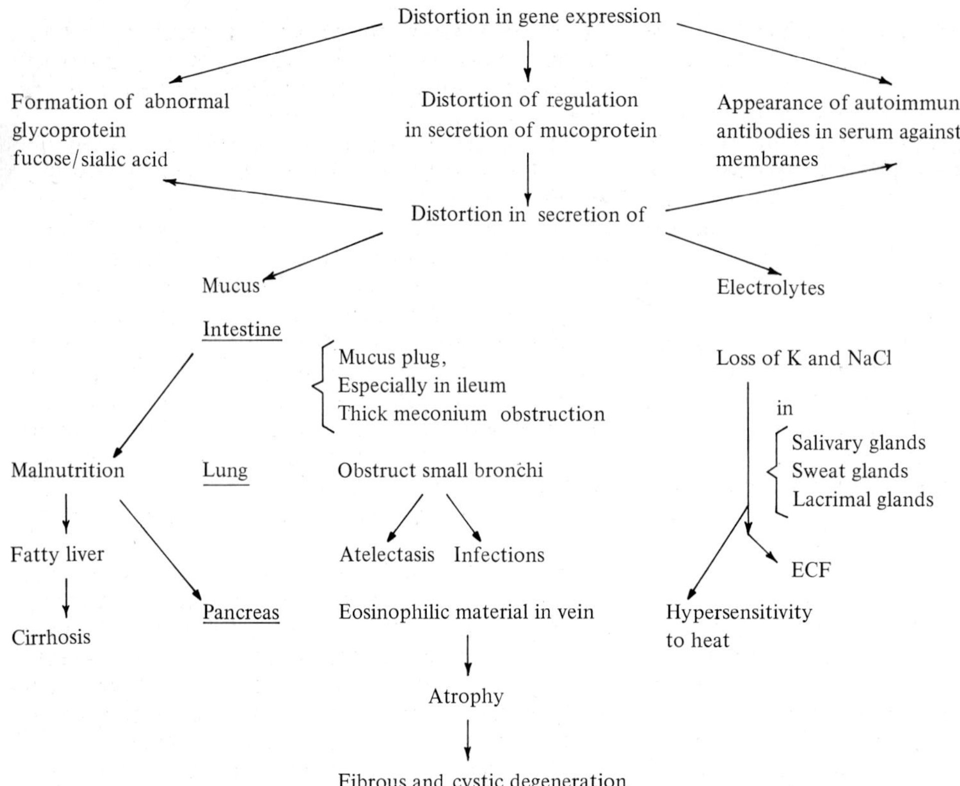

Fig. 4-47. Hypothetical pathogenesis of cystic fibrosis

sialic acid and the hydrophobic methyl group of fucose free. Thus, the free group of sialic acid attracts while the free group of fucose repels water. In cystic fibrosis, a water-soluble protein was found to be rich in fucose and relatively poor in sialic acid [140, 141].

Serum of patients with cystic fibrosis contains a protein factor that inhibits ciliary movements in oyster gills, fresh water mussels, and rabbit trachea.

The protein factor elutes with the immunoglobulin G on diethylaminoethyl-cellulose with a molecular weight of 125,000–200,000. The protein's mode of action is not known, but it has been postulated to act as an autoantibody or a polycation that destroys membranes. Although this hypothesis is suggestive, there is no evidence of a link between the presence of the protein and the pathogenesis of cystic fibrosis [142].

Because of the availability of antibiotics and more adequate general care of patients with cystic fibrosis, their survival has been markedly enhanced. The incidence of glucose intolerance appears to increase as the patients' longevity increases. Whether this is due to a form of hereditary diabetes or to damage to the islets as a result of histological distortion remains to be seen. At first glance, histological distortion seems to be sufficient to explain the metabolic alteration. The steps in the pathogenesis of cystic fibrosis are summarized in Fig. 4-47.

Idiopathic Steatorrhea

To understand idiopathic steatorrhea, it is necessary to review some of the basic properties of the gastrointestinal tract—the rate of proliferation of the epithelium, the biochemistry of the intestinal epithelium, and the bacterial flora contained in the intestinal lumen [143–146].

The origin and fate of the intestinal epithelial cells have been traced using radioautography. The absorptive cell proliferates in the crypts of Lieberkühn. The new cells migrate toward the apex of the villi where they are extruded. Cellular proliferation and cellular extrusion are not rigidly linked; for example, irradiation blocks cellular proliferation but doesn't affect cellular extrusion, and the villi atrophy. In sprue the reverse occurs; cellular proliferation persists, but cellular extrusion seems to be accelerated.

The chemical composition of the cell progressively changes as it passes from the proliferative to the absorptive stage. Whereas the proliferative cells are highly basophilic and rich in RNA particles, the absorptive cells contain little RNA. In contrast, lipid droplets are practically absent in the proliferative cells but are large and abundant in the absorptive cells.

An interesting aspect of the differentiation of the cells of the intestinal tract is the appearance of alkaline phosphatase. The enzyme appears abruptly in the cells at the junction of the crypts and the villus. The role of alkaline phosphatase in absorption remains to be determined.

In humans, proliferating epithelial cells of the stomach, colon, ileum, and rectum are produced at the rate of 1–2 cells per 100 cells per hour. At this rate, most of the epithelial lining of the gastrointestinal tract is replaced within 3–6 days. The S phase (time for DNA synthesis) equals 10–15 hours; g-2 (premitotic time) equals 1–2 hours; m (mitotic time) equals 1 hour; g-1 (postmitotic time) equals 7–12 hours.

The cells of the intestinal epithelium are metabolically active—their rate of proliferation is among the most rapid in the human organism. They secrete large amounts of mucus, proteins, and electrolytes, and they are critical in the absorption of digested food and reabsorption of fluid. Many of the absorptive processes are active and require energy. The intestinal mucosa contains the classical biosynthetic and bioenergetic pathways necessary for RNA, DNA, protein, glycogen, and lipid synthesis, and for glucose oxidation through the glycolytic pathway and the hexose monophosphate shunt. They also possess a well-developed aerobic pathway for oxidation of Krebs cycle substrates with an electron transport chain coupled to the oxidative phosphorylation of ADP to ATP. The amount of glucose oxidized through the hexose monophosphate shunt is greater than that in liver. This observation is in keeping with the active proliferation of the intestinal mucosa. Except for some transamination and decarboxylation reactions, the intestinal mucosa does not actively degrade amino acid. Glutamic acid, however, is actively metabolized in the intestinal mucosa. Two specific transaminases, which convert glutamine to alanine and aspartic acid, respectively, and an intestinal glutamic dehydrogenase convert the amino acid to α-ketoglutaric acid and make it available to the Krebs cycle. The intestine contains no glycerol kinase. Consequently, the α-glycerophosphate necessary for triglyceride synthesis must be derived from glycolysis.

Glutamic decarboxylase generates γ-aminobutyric acid, and glutamine synthetase converts glutamic acid to glutamine in the presence of ATP and NH_3. Glutaminase 1 and glutaminase 2 convert glutamine to glutamic acid. The reasons for the various metabolic pathways using glutamic acid are not known, but glutamic acid may be important in intestinal function. The content of glutamine in gliadin, the toxic component of gluten, will be discussed below. Several steps of the urea cycle have been identified in the intestine, and the intestinal epithelium contains arginase.

The intestinal lumen of animals and man always contains a rich bacterial flora. The exact composition of the bacterial flora is known only qualitatively and includes lactobacilli, bacteroides, enterobacilli, Proteus, and Pseudomonas. The lactobacilli and the bacteroides constitute the most important fraction of the cellular population, and the size of that bacterial population remains constant. In contrast, *E. coli,* Proteus, and Pseudomonas are much less abundant, and the size of their population varies considerably. Furthermore, it appears that the latter group of intestinal bacteria is responsible for the deleterious effect of the intestinal flora.

Some intestinal bacteria are harmful to the individual; others are beneficial, such as those with morphogenetic effects, those that compete with pathogenic bacteria, and those that elaborate vitamins. Among the harmful effects are destruction of nutrients, elaboration of toxins, and interference with absorption.

The morphogenetic effect of the intestinal flora becomes evident when the intestinal mucosa of a conventionally reared animal is compared with that of a germ-free animal. In a germ-free animal, the lymphoid tissue of the Peyer's patches is poorly developed; the intestinal epithelium remains undifferentiated and resembles the epithelium of the prenatal stages. The intestinal flora synthesizes vitamin K and B group vitamins.

Severe nutritional deficiencies, X-irradiation, the administration of antifolic (see Fig. 4-48) or bactericidal agents allow some bacteria of the intestinal lumen to pass through the mucosa and enter the bloodstream to cause septicemia. Furthermore, the endotoxins normally present in the intestinal lumen enter the bloodstream under certain pathological conditions and transform reversible shock into irreversible shock. All gram-negative bacteria produce endotoxins—large molecules with a molecular weight of 1,000,000 formed of a complex of proteins, lipids, and polysaccharides.

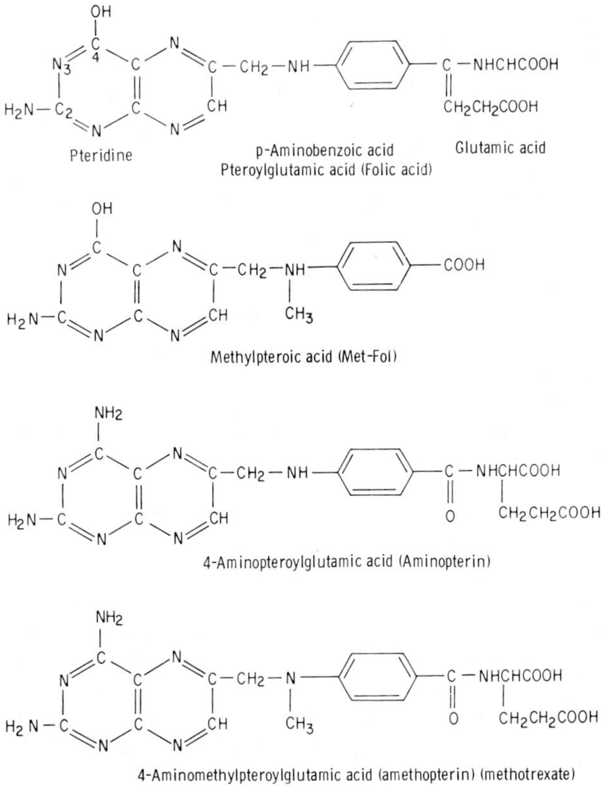

Pteridine — p-Aminobenzoic acid — Glutamic acid

Pteroylglutamic acid (Folic acid)

Methylpteroic acid (Met-Fol)

4-Aminopteroylglutamic acid (Aminopterin)

4-Aminomethylpteroylglutamic acid (amethopterin) (methotrexate)

Fig. 4-48. Some folic acid antimetabolites

These molecules are continuously absorbed by the intestine. The amount of endotoxin absorbed is proportional to the size of the endotoxin pool in the lumen. The endotoxins are known to block the activity of the reticuloendothelial system and to precipitate the transition from reversible to irreversible shock. In shock, the effect of the endotoxin gives the final blow to the reticuloendothelial system, which has been damaged by hypoxia and reduced blood flow.

The intestinal flora may manifest lipolytic activity. A property of the bacterial flora, however, is of little importance to fat absorption because bacterial lipolysis occurs in that part of the intestine where lipid absorption does not take place.

The accumulation of undigested food in the intestinal tract is responsible for changes in the distribution of the bacterial flora. The bacterial flora is quantitatively and qualitatively altered, and its geographical distribution in the gastrointestinal tract is modified. Such changes are responsible for some discomfort, such as gas production, accumulation of short-chain fatty acids, lack of conjugation of bile salt, and vitamin deprivation. In normal subjects the bacteria grow mainly in the large intestine. When food accumulates in the intestinal tract as a result of obstruction or malabsorption, bacteria invade the jejunum.

The fermentation products of the bacteria include some lactic acid and some of the short-chain fatty acids such as butyric, acetobutyric, and propionic acids. Bacteria contain enzymes that split the bile salt, and as a result the detergent properties of bile are lost adding to malabsorption.

In addition, the bacterial flora consumes some of the needed vitamins, in particular vitamin B_{12}. This depletion and a deficiency of the intrinsic factor are responsible for pernicious anemia.

A groups of investigators from Tufts University tried to study the distribution of bacteria in the human jejunum mucosa. They investigated both the quantitative and the qualitative aspects of the bacterial flora. The large intestine contains 2 billion organisms per gram, and the bacterial population in the normal content (both the fluid and the mucus covering the mucosa) of the small intestine is of the order of 1 million per milliliter. All specimens of upper intestinal juice these investigators examined contained bacteria, and the ratio of anaerobic to aerobic was 3 to 1. The microflora of the ileoileostomy eluent contain greatly increased amounts of bacteria, although the relative distribution of the various types of bacteria is not changed after ileostomy.

Celiac Disease or Nontropical Sprue

Primary steatorrhea is referred to as celiac disease in children and nontropical sprue in adults. In severe nontropical sprue, the delicate pattern designed by the slender villi is replaced by a flattened knobby epithelium [147–149]. Thus, the distended crypts open directly in the flat luminal surface. These changes result from alteration in the rate of replacement and from abnormal differentiation of the epithelial cells. In nontropical sprue, mitosis, instead of being limited to the crypts, occurs throughout the epithelium. In sprue the mean mitotic index of the epithelium is twice that of normal epithelium, which suggests that cells proliferate more rapidly in epithelium of tropical sprue than in normal epithelium. It has therefore been proposed that in nontropical sprue the rate of cellular elimination is greater than normal, and epithelial atrophy results (see Figs. 4-49 and 4-50).

The pathogenesis of nontropical sprue is not known, yet the disease improves when a gluten-free diet is administered. At first nontropical sprue was thought to be caused by hypersensitivity of the intestinal

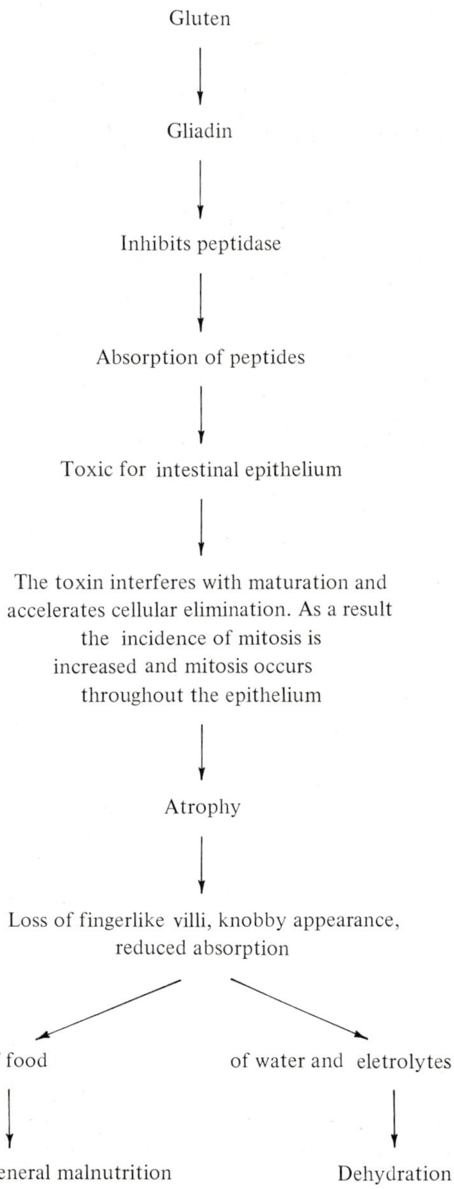

Fig. 4-49. Pathogenesis of nontropical sprue

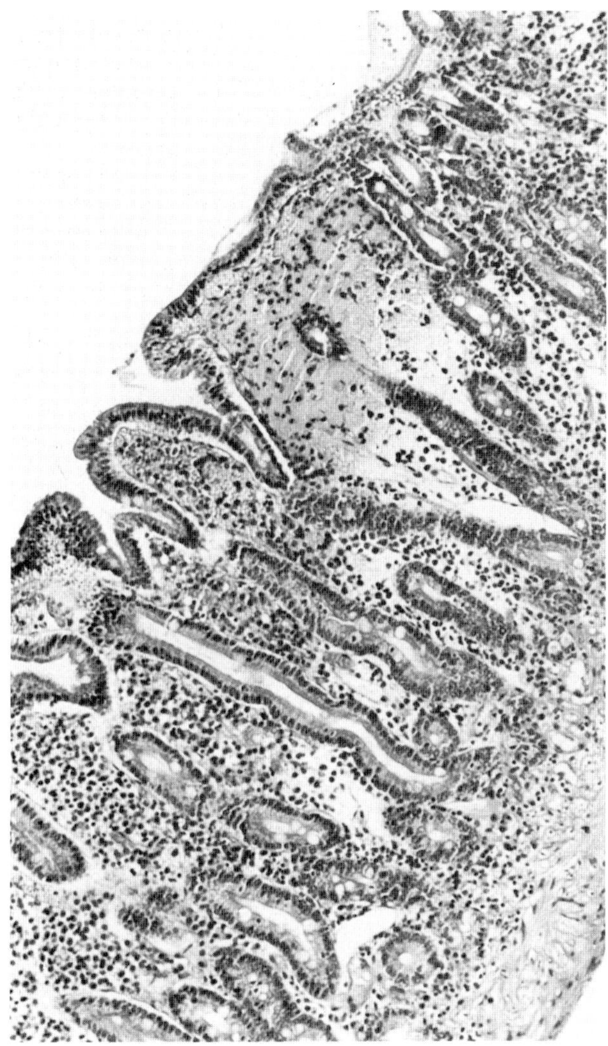

Fig. 4-50. Histology of nontropical sprue. (From Ament)

mucosa to wheat, but it was later established that it results from the toxic effect of a product of protein digestion. It has been proposed that patients with celiac disease and sprue are deficient in a peptidase which normally breaks down peptides into free amino acids, and that the undigested peptides are then absorbed and exert a toxic effect on the intestinal mucosa [150].

The nefarious fraction contained in gluten is gliadin a protein that contains as much as 43% of glutamine and a high proportion of proline. In support of the toxic theory is the observation that the histology of the intestinal mucosa returns to normal in some, but not all, patients fed a gluten-free diet.

There are a few fields of pathology in which the biopsy technique has been more fruitful to clarify pathogenesis than in sprue. When a gross specimen of intestine obtained at surgery or by biopsy is examined with a magnifying lens, the surface of the mucosa reveals a floral pattern which is due to the extension above the surface of delicate fingerlike villi. Ingelfinger has compared the appearance of the fresh intestinal mucosa to that of the myriad of tentacles on the underside of starfish. This extensive folding and refolding of the surface mucosa increases the absorption surface many times above what it would be if the submucosa were simply lined by a flat mucosa. In a frank case of sprue, the fingerlike villi are reduced in number and are blunted, thus the absorptive surface is reduced.

In the normal intestine, the slender cylindrical cell of the intestine shows morphological and functional polarity. It absorbs at one end and releases its content at the other. At its absorptive surface, each mucosal cell has numerous slender folds of the cell membrane called microvilli, which entrap the nutrients. In sprue, the cell is cuboidal. The morphological polarity has disappeared, and the microvilli are flattened. In the normal intestinal epithelium, granules resembling zymogen granules containing digestive enzymes accumulate under the so-called microvillar brush border. The zymogen granules are reduced in number in the epithelial cells of patients affected by sprue.

The consequences of such injury are, of course, not only malabsorption because of the reduction in the surface epithelium and in microvilli, but ineffective digestion because of reduction in enzyme secretion and breakdown of conjugated bile acids. The result is total malabsorption of lipid, protein, sugar, electrolytes, calcium, and both lipid-soluble and water-soluble vitamins.

Thus, nontropical sprue expresses itself by a complex blend of general malnutrition with vitamin deficiencies and dehydration (see Fig. 4-49). The dehydration results from lack of water absorption because of the reduced mucosal surface and interference with sodium absorption (see Electrolyte Metabolism).

Obesity

Obesity results from excess triglyceride deposition in the subcutaneous and internal adipose tissue. The exact incidence of obesity in the population is difficult to estimate because there are no established standards to determine when a person is obese, and most methods available for determining the body's fat content are complicated or unreliable. Olson states that a person may be considered obese when his body weight exceeds by 15% that of a healthy individual of 25 of the same height.

The fat content of the human body cannot be measured directly, and all heavy persons are not necessarily fat—in football players, for example, the excessive weight is made up by the skeleton and muscle. One method of measuring fat is based on the fact that fat has a lower density than the rest of the tissues. By weighing a person under water and then again in air it is possible to determine the density of the fat-free body, provided that the density of fat is known. Consequently, it is possible to determine the proportion of fat in the body. Using this method, Benkhe demonstrated that athletes are heavier because of muscle development.

Weighing under water is a complicated procedure, and clinicians usually resort to much simpler techniques, such as the radiological evaluation of the thickness of fat layers and measurement of the thickness of the skin folds. Since there is a constant relationship between bone, lean body mass, and extracellular water content, the fat content of the body can also be estimated indirectly by measuring the extracellular water.

On the basis of the standard presented above, the incidence of obesity is estimated as low before 20, rising sharply after 30, and reaching a maximum in the sixth decade. The incidence of obesity drops after the sixth decade because obese people die younger, and older people spontaneously reduce their food intake. Approximately 30% of the middle-aged American population is obese. In addition to the cosmetic disadvantages, the hazards of obesity are numerous. The incidence of diabetes, atherosclerosis, and hypertension is greater in the obese. The relationship between these diseases and obesity is discussed in other sections of this book.

To equate obesity with gluttony is in most cases a crude oversimplification. Obesity is more likely to result from some functional defect, yet the pathogenesis of obesity remains unknown, and today the mechanism of production of obesity can be discussed only in theoretical terms. To understand such a discussion it is necessary to review briefly the mechanisms controlling food intake.

Hunger is a vague craving for food. The sensation of hunger probably results from vigorous contractions in the stomach. Food intake usually follows the sensation of hunger. Satiety is a sensation of fullness that originates in the stomach and is related to the volume of food taken in rather than the caloric intake. Yet, if an animal is given a diet diluted with cellulose, it slowly increases its threshold for satiety until an adequate caloric intake is achieved.

The control of food intake raises two questions: What nervous centers dictate the sensations of hunger and satiety, and what signals are given to these centers? Two centers in the hypothalamus are credited for controlling food intake: a centromedial center referred to as the satiety center, and a lateral one referred to as the hunger center. Food intake is increased by electrical stimulation of the lateral hypothalamus in cats. Stimulation of the ventromedial nucleus abolishes the sensation of hunger. In contrast, lesions of the ventromedial nucleus in rats, mice, guinea pigs, cats, dogs, and monkeys induce obesity from overeating.

A number of mechanisms have been invoked to explain the activation of the hypothalamic centers. These mechanisms can be categorized as changes in body temperature, changes in the concentration of blood metabolites (glucose, fats, amino acids, and even water), and efferent nervous impulses originating in the viscera. The most popular theory is that proposed by Mayer [151–153] and referred to as the "hypothalamic glucostatic control of gastric hunger contraction." According to the theory, falling blood glucose stimulates the hunger center, and rising blood glucose stimulates the satiety center in the hypothalamus. The theory stems from a number of observations, including the fact that an intravenous injection of glucagon inhibits gastric hunger contractions, and hyperphagia and obesity occur in rats after a single injection of gold thioglucose—a compound known to damage severely several central nervous structures, including the ventromedial nucleus. In monkeys, low blood glucose levels stimulate the oxygen uptake in the hunger center, whereas high blood glucose levels stimulate the oxygen uptake of the satiety center. Perhaps the most elegant experiments designed to demonstrate the glucostatic response of the hypothalamic center are those in which electrodes were implanted bilaterally in various areas of the hypothalamus and cortex of monkeys. Naturally, electrodes were implanted in the hunger and satiety centers. Then the investigators studied the electroencephalographic effect of various nutrients on blood levels.

Hyperglycemia increased (by 300%) the frequency of the waves generated in the satiety center and drastically decreased the frequency of the waves generated in the hunger center. Hypoglycemia caused by insulin injection had the reverse effect. No changes were observed in the electroencephalographic patterns of other parts of the brain. Changes in the blood concentration of amino acids or lipids were without effect on the hypothalamic centers. These experiments clearly established that glucoreceptors are in the hypothalamus, but the changes in blood glucose concentration needed to stimulate these centers are so great that it is doubtful that these glucoreceptors control food intake at all. Although the hypothalamus undoubtedly controls food intake, researchers are not sure that the glucoreceptors participate in the control.

On the basis of these theoretical concepts, experimental obesity has been classified as regulatory and metabolic obesity. In the regulatory type, the primary lesion involves the hyporegulator centers (*e.g.*, destruction of the ventromedial nucleus by surgery or gold thioglucose). Metabolic obesity involves an acquired or inborn metabolic distortion, such as obesity caused by grafting ACTH-secreting pituitary tumors or that observed in mice with hereditary hyperglycemia. Whether this classification can be extrapolated to the etiology of obesity in man remains to be established. At least some forms of obesity observed in humans can be explained in terms of the mechanism of control of food intake described above; namely, those forms of obesity associated with hypothalamic lesions.

The pathogenesis of most cases of human obesity remains unexplained. Efforts to demonstrate differences in the metabolism of the obese and the normal individual have been unsuccessful. Oxygen consumption and other metabolic activities, as far as it is known, are the same in the obese and in the normal individual. There is no evidence in man that lipogenesis is more efficient in obese than in normal individuals, but reduced ketogenesis and improved nitrogen retention have been reported.

When one compares the energy spent by the obese and the normal individual to accomplish a specific physical effort, it appears that the obese person does not spend less energy in performing that exercise than the normal individual. In fact, the obese person exerts more energy than normal in sitting, standing, and walking because he must displace a larger mass. It seems unlikely that the basal metabolism of the obese is lower than that of the normal individual. For all these reasons, it appears that most cases of obesity are not of metabolic origin, but are of a regulatory nature. The mechanism by which the control of food intake is lost in the obese is not clear. Psychological factors seem to be important. This is not surprising since numerous nerves connect the cortex and the hypothalamic centers that control food intake. Often new experience, marriage, pregnancy, or loss of love precipitates obesity, possibly by abolishing the normal regulatory role of the satiety center.

To say that a person is obese because he has a low level of physical activity does not shed any new light on the pathogenesis of obesity. Indeed, a normal individual is expected to regulate his caloric intake according to his needs, and if his physical activity is low his food intake should be reduced.

Jean Mayer reviews observations made in families, ethnic groups, and twins which suggested that hereditary factors may predispose an individual to obesity. The weight of natural children correlates well with that of parents, but there is no such correlation between that of adopted children and their legal parents. As in most studies of this kind, it is difficult to distinguish environmental from hereditary factors.

Physiological and behavioral studies of obese and normal individuals have provided some insight on the part played by psychological factors in the pathogenesis of obesity. Food deprivation brings about changes in the physiological chemistry of the body that are signaled to the hunger center. Stimulation of the hunger center brings the "hungry" to search for food. In addition, the urge to eat (appetite) is also stimulated by exterior factors, such as smell and taste. Schachter has proposed that the response of the obese to internal and external clues to hunger differs from that of the normal [154].

To test the hypothesis, workers initiated experiments in which a balloon was introduced in the stomach of obese and normal volunteers to measure gastric motility. The subject was then asked every 15 minutes whether or not he was hungry. In normal individuals a positive answer coincided with gastric contractions in 71% of the cases, but in the obese the coincidence was only 38%. The results of that experiment were interpreted to mean that normal and obese individuals respond differently to hunger signals. Hunger signals can be manipulated in various ways. For example, food deprivation stimulates contractility of the stomach and induces hypoglycemia, whereas fear inhibits contractility and induces glucose release. Schachter measured the number of crackers eaten by two groups of experimental subjects, some fed roast beef sandwiches, others deprived of food. Each group included obese and normal subjects. He found that although a previous meal interfered with cracker eating by normal individuals, it had no effect on the obese, who seemed to eat more crackers even after they had a big meal, often bigger than the meal eaten by the normal subjects.

In another experiment, some of the subjects were frightened by telling them that the cracker eating would be followed by a painful stimulation of their taste buds with the aid of an enormous electrode hooked to a gruesome and mysterious machine. This promise discouraged the so-called normal group from eating the crackers, but had little effect on the appetite of the obese subjects.

Since the obese do not respond to internal hunger signals, it is not surprising that manipulating the time elapsed between meals does not effect their food intake. When the observer deceives the normal eater about the time elapsed since his last meal simply by lying about the hour of the day, the normal individual will not respond by eating a big meal at an earlier time. In contrast, the obese is too glad to eat all he can at any time. Similar studies yielded results in concordance with the conclusion that obese individuals are relatively insensitive to variation in physiological hunger signals but are highly sensitive to environmental, food-related clues. These studies tested: the influence of taste by feeding ice cream of various qualities; the effect of prolonged fasting in the absence of exterior stimuli (by observing the behavior of obese and nonobese orthodox Jews during Yom Kippur fasting in the synagogues); and the participation and cooperation of obese and nonobese in feeding programs in colleges and universities.

At a first glance these studies appear to provide merely a quantitative framework for common notions. However, because these experiments were planned to test a specific hypothesis they became part of a systematic approach to the elucidation of a problem. The purpose of briefly describing these experiments was not to substitute a working hypothesis whose merit must obviously be weighed against less subjective and better controlled experimentation, but to illustrate the complexity of the variables influencing the incidence of obesity.

References

1. Mellette, S., Leone, L.: Influence of age, sex, strain of rat and fat soluble vitamins on hemorrhagic syndromes in rats fed irradiated beef. Fed. Proc. **19**, 1045–1049 (1960)
2. Richardson, L., Ritchey, S., Rigdon, R.: A long-term feeding study of irradiated foods using rats as experimental animals. Fed. Proc. **19**, 1023–1027 (1960)
3. Proc. Symp. IV.: Nutrition in pregnancy, council on foods and nutrition. Chicago: American Medical Association 1958
4. Proc. Symp. VII.: Infant nutrition, council on foods and nutrition. Chicago: American Medical Association 1961
5. Platt, B.S.: Digestion in infancy. Fed. Proc. **20**, 188–195 (1961)
6. Beaton, G.H.: Nutritional and physiological adaptations in pregnancy. Fed. Proc. **20**, 196–201 (1961)

7. Chase, H., Martin, H.: Undernutrition and child development. New Engl. J. Med. **282**, 933–939 (1970)
8. Winick, M.: Cellular growth of the fetus and placenta. In: Fetal growth and development (Waisman, H., and Kerr, G., eds.), p. 19–27. New York: McGraw-Hill Book Company 1970
9. Villee, C.A.: Enzymic development of the placenta in relation to fetal growth. In: Fetal growth and development (Waisman, H., and Kerr, G., eds.), p. 11–17. New York: McGraw-Hill Book Company 1970
10. Hagerman, D.: Enzymatic capabilities of the placenta. Fed. Proc. **23**, 785–790 (1964)
11. Diczfalusy, E.: Endocrine functions of the human feto-placental unit. Fed. Proc. **23**, 791–798 (1964)
12. Herrmann, H., Tootle, M.: Specific and general aspects of the development of enzymes and metabolic pathways. Physiol. Rev. **44**, 289–371 (1964)
13. Burch, H.B.: Biochemical changes in the perinatal rat liver. Ann. N.Y. Acad. Sci. **111**, 176–182 (1963)
14. Yeung, D., Oliver, I.: Development of gluconeogenesis in neonatal rat liver. Biochem. J. **105**, 1229–1233 (1967)
15. Moog, F.: Enzyme development and functional differentiation in the fetus. In: Fetal growth and development (Waisman, H., and Kerr, G., eds.), p. 29–48. New York: McGraw-Hill Book Company 1970
16. Levi-Montalcini, R., Angeletti, P.U.: Growth and differentiation. Annu. Rev. Physiol. **24**, 11–56 (1962)
17. Passmore, R., Meiklejohn, A.P.: Disorders of nutrition. In: Biochemical disorders in human disease (Thompson, R.H.S., and King, E.J., eds.), p. 544–577. New York: Academic Press 1957
18. Keys, A., Brozek, J., Henschel, A., Mickelsen, O., Taylor, H.: Human starvation. Minneapolis: University of Minnesota Press 1951
19. Consolazio, C., Matoush, L., Johnson, H., Nelson, R., Krzywicki, H.: Metabolic aspects of acute starvation in normal humans (10 days). Amer. J. clin. Nutr. **20**, 672–683 (1967)
20. Hanson, P., Johnson, R., Zaharko, D.: Correlation between ketone body and free fatty acid concentrations in the plasma during early starvation in man. Metabolism **14**, 1037–1040 (1965)
21. Spencer, H., Lewin, I., Samachson, J., Laszlo, J.: Changes in metabolism in obese persons during starvation. Amer. J. Med. **40**, 27–37 (1966)
22. Knox, W., Auerbach, V., Lin, E.: Enzymatic and metabolic adaptations in animals. Physiol. Rev. **36**, 164–254 (1956)
23. Van Riet, H., Schwartz, F., Kinderen, P. de: Metabolic observations during the treatment of obese patients by periods of total starvation. Metabolism **13**, 291–302 (1964)
24. Harper, A.E., Kumta, U.S.: Amino acid balance and protein requirement. Fed. Proc. **18**, 1136–1142 (1959)
25. Proc. Symp. II.: Relation of proteins and amino acids to nutritional health, council on foods and nutrition. Chicago: American Medical Association 1956
26. Fomon, S.J. (ed.): Amino acid and protein metabolism. Report of the 30th Ross Conference on Pediatric Research, Ross Laboratories, Columbus, Ohio (1958)
27. Oser, B.L.: An integrated essential amino acid index for predicting the biological value of proteins. In: Protein and amino acid nutrition (Albanese, A.A., ed.). New York: Academic Press 1959
28. Wilding, P., Cooke, W.T., Nicholson, G.I.: Globulin-bound amylase. A cause of persistently elevated levels in serum. Ann. intern. Med. **60**, 1053–1059 (1964)
29. Levitt, M.D., Cooperband, S.R.: Hyperamylasemia from the binding of serum amylase by an 11S IgA globulin. New Engl. J. Med. **278**, 474–479 (1968)
30. Northrop, J.H.: Crystalline pepsin: Isolation and tests of purity. J. gen. Physiol. **13**, 739–780 (1930)
31. Bovey, F., Yanari, S.: Pepsin. In: The enzymes (Boyer, P., Lardy, H., and Myrbäck, K., eds.), 2nd ed., vol. 4, p. 63–92. New York: Academic Press 1960
32. Walsh, K., Neurath, H., Kauffman, D.: Peptides isolated from tryptic and chymotryptic hydrolysates of S-sulfotrypsinogen. Biochim. biophys. Acta (Amst.) **65**, 540–543 (1962)
33. Northrop, J.H., Kunitz, M., Herria, R.: Crystallinic Enzymes, 2nd ed. New York: Columbia University Press 1948
34. Keith, C.K., Kazenko, A., Laskowski, M.: Studies on proteolytic activity of crystalline protein B prepared from beef pancreas. J. biol. Chem. **170**, 227–238 (1947)
35. Desnuelle, P.: Chymotrypsin. In: The enzymes (Boyer, P., Lardy, H., and Myrbäck, K., eds.), 2nd ed., vol. 4, p. 93–118. New York: Academic Press 1960
36. Beaumont, W.: Experiments and observations on the gastric juice and the physiology of digestion, facsimile ed. New York: Dover 1939
37. Morley, J.: Structure-activity relationships. Fed. Proc. **27**, 1314–1317 (1968)
38. Hirschowitz, B.I.: Gastrin I, pentagastrin and histamine in the fistula dog. Fed. Proc. **27**, 1318–1321 (1968)
39. Grossman, M.: Physiological role of gastrin. Fed. Proc. **27**, 1312–1313 (1968)
40. Farrar, G., Jr., Bower, R.: Gastric juice and secretion: Physiology and variations in disease. Annu. Rev. Physiol. **29**, 141–168 (1967)
41. Wang, C.C., Grossman, M.I.: Physiological determination of release of secretin and pancreozymin from the intestine of dogs with transplanted pancreas. Amer. J. Physiol. **164**, 527–545 (1951)
42. Dupre, J.: Regulation of the secretions of the pancreas. Annu. Rev. Med. **21**, 299–316 (1970)
43. Brock, J., Autret, M.: Kwashiorkor in Africa, WHO Monograph Series No. 8. Geneva: World Health Organization 1952
44. Bjørnesjø, K., Belew, M., Zaar, B.: Biochemical study of advanced protein malnutrition in Ethiopia. Scand. J. clin. Lab. Invest. **18**, 591–602 (1966)
45. Scrimshaw, N.: Malnutrition in underdeveloped countries. New Engl. J. Med. **272**, 137–144 (1965)
46. Wolf, C.B.: Kwashiorkor on the Navajo Indian reservation. A discussion of the disease as seen throughout the world and a report of three cases in Navajo children. Henry Ford Hosp. med. Bull. **9**, 566–571 (1961)
47. Follis, R., Jr.: Deficiency disease, p. 315. Springfield, Illinois: Charles C. Thomas Publisher 1958
48. Hegsted, D.: Protein requirement in man. Fed. Proc. **18**, 1130–1135 (1959)
49. Pearson, W., Darby, W.: Protein nutrition. Annu. Rev. Biochem. **30**, 325–346 (1961)
50. Waterlow, J.: Protein nutrition and enzyme changes in man. Fed. Proc. **18**, 1143–1155 (1959)
51. Brock, J.F.: Dietary proteins in relation to man's health. Fed. Proc. **20**, 61–65 (1961)
52. Stare, F.J. (ed.): Nutrition and learning behavior. Nutr. Rev. **25**, 20–22 (1967)
53. Wharton, B.A., Howells, G.R., McCance, R.A.: Cardiac failure in kwashiorkor. Lancet **1967 II**, 384–387
54. Brenton, D.P., Brown, R.E., Wharton, B.A.: Hypothermia in kwashiorkor. Lancet **1967 I**, 410–413
55. Umaña, R.: Changes produced by protein malnutrition in the proteolytic activity of human liver. Canad. J. Biochem. **45**, 1353–1356 (1967)
56. Chase, H., Dorsey, J., McKhann, G.: The effect of malnutrition on the synthesis of a myelin lipid. Pediatrics **40**, 551–559 (1967)
57. Metcoff, J.: Biochemical effects of protein-calorie malnutrition in man. Annu. Rev. Med. **18**, 377–422 (1967)
58. Allison, J.B.: Protein malnutrition. Trans. N.Y. Acad. Sci. **25**, 293–306 (1963)
59. Shinozuka, H., Verney, E., Sidransky, H.: Alterations in the liver due to acute amino acid deficiency. An electron microscopic study of young rats force-fed a threonine-devoid diet. Lab. Invest. **18**, 72–85 (1968)
60. Allen, R., Raines, P., Regen, D.: Regulatory significance of transfer RNA charging levels. I. Measurements of charging levels in livers of chow-fed rats, fasting rats, and rats fed balanced or imbalanced mixtures of amino acids. Biochim. biophys. Acta (Amst.) **190**, 323–336 (1969)
61. Jansen, B., Donath, W.: Isolation of anti-beriberi vitamin. Med. dienst d. volksgenzondh. in Nederl. Indie **1**, 190 (1927)
62. Kinney, T., Follis, R.: Beriberi. Fed. Proc. **17**, 3–56 (1958)
63. Bhuvaneswaran, C., Sreenivasan, A.: Problems of thiamine deficiency states and their amelioration. Ann. N.Y. Acad. Sci. **98**, 576–601 (1962)
64. Follis, R., Jr.: Deficiency disease, p. 405. Springfield, Illinois: Charles C. Thomas Publisher 1958
65. Mizuhara, S., Handler, P.: Mechanism of thiamine-catalyzed reactions. J. Amer. chem. Soc. **76**, 571–573 (1954)
66. Breslow, R.: On the mechanism of thiamine action. IV. Evidence from studies on model systems. J. Amer. chem. Soc. **80**, 3719–3726 (1958)
67. Krampitz, L., Suzuki, I., Greull, G.: Mechanism of action of thiamine diphosphate in enzymic reactions. Ann. N.Y. Acad. Sci. **98**, 466–478 (1962)
68. Woolley, D.W., Elvehjem, C.A., Madden, R.J., Strong, F.M.: Anti-black tongue activity of various pyridine derivatives. J. biol. Chem. **124**, 715–723 (1938)
69. Moline, S.W., Walker, H.C., Schweigert, B.S.: 3-Hydroxyanthranilic acid metabolism. VII. Mechanism of formation of quinolinic acid. J. biol. Chem. **234**, 880–883 (1959)
70. Chaykin, S.: Nicotinamide coenzymes. Annu. Rev. Biochem. **36**, 149–170 (1967)
71. Price, J., Brown, R., Yess, N.: Testing the functional capacity of the tryptophan-niacin pathway in man by analysis of urinary metabolites. Advanc. Metab. Disord. **2**, 159–225 (1965)
72. Wagner, A.F., Folkers, K.: The vitamins. In: Medicinal chemistry (Burger, A., ed.), 2nd ed., p. 133–286, p. 166. New York: Interscience 1960
73. Follis, R., Jr.: Deficiency disease, p. 387. Springfield, Illinois: Charles C. Thomas Publisher 1958
74. Lynen, F.: Participation of acyl-CoA in carbon chain biosynthesis. J. cell. Physiol. **54**, 33–49 (1959)
75. Calvin, M., Pon, N.: Carboxylations and decarboxylations. J. cell. Physiol. **54**, 51–74 (1959)
76. Knappe, J.: Mechanism of biotin action. Annu. Rev. Biochem. **39**, 757–776 (1970)
77. Szent-Györgyi, A.: Observations on the function of peroxidase systems and the chemistry of the adrenal cortex; description of a new carbohydrate derivative. Biochem. J. **22**, 1387–1409 (1928)
78. Follis, R., Jr.: Deficiency disease, p. 387. Springfield, Illinois: Charles C. Thomas Publisher 1958
79. Gottlieb, A.A., Peterkofsky, B., Udenfriend, S.: Use of collagenase to identify enzymatically formed collagen and a hydroxyproline-deficient intermediate. J. biol. Chem. **240**, 3099–3103 (1965)
80. Peterkofsky, B., Udenfriend, S.: Conversion of proline to collagen hydroxyproline in a cell-free system from chick embryo. J. biol. Chem. **238**, 3966–3977 (1963)

81. Peterkofsky, B., Udenfriend, S.: Enzymatic hydroxylation of proline in microsomal polypeptide leading to formation of collagen. Proc. nat. Acad. Sci. (Wash.) **53**, 335–342 (1965)
82. Broquist, H., Trupin, J.S.: Amino acid metabolism. Annu. Rev. Biochem. **35**, 231–294 (1966)
83. Ellenbogen, L., Highley, D.R.: Intrinsic factor. Vitam. and Horm. **21**, 1–49 (1963)
84. Jerzy Glass, G.: Gastric intrinsic factor and its function in the metabolism of vitamin B_{12}. Physiol. Rev. **43**, 529–849 (1963)
85. Mollin, D.L., Baker, S.J., Doniach, I.: Addisionian pernicious anaemia without gastric atrophy in a young man. Brit. J. Haemat. **1**, 278–290 (1955)
86. Spurling, C.L., Sacks, M.S., Jiji, R.M.: Juvenile pernicious anemia. New Engl. J. Med. **271**, 995–1003 (1964)
87. Grasbeck, R., Gordin, R., Kantero, I., Kuhlbäck, B.: Selective vitamin B_{12} malabsorption and proteinuria in young people: a syndrome. Acta med. scand. **167**, 289–296 (1960)
88. Katz, M., Lee, S.K., Cooper, B.A.: Vitamin B_{12} malabsorption due to a biologically inert intrinsic factor. New Engl. J. Med. **287**, 425–429 (1972)
89. Donaldson, R.M., Jr., Mackenzie, I.L., Trier, J.S.: Intrinsic factor-mediated attachment of vitamin B_{12} to brush borders and microvillous membranes of hamster intestine. J. clin. Invest. **46**, 1215–1228 (1967)
90. Perlman, D. (ed.): Vitamin B_{12} coenzymes. Ann. N.Y. Acad. Sci. **112**, 547–921 (1964)
91. Wagner, F.: Vitamin B_{12} and related compounds. Annu. Rev. Biochem. **35**, 405–434 (1966)
92. Jaenicke, L.: Vitamin and coenzyme function: Vitamin B_{12} and folic acid. Annu. Rev. Biochem. **33**, 287–312 (1964)
93. Barker, H.A.: Structure and function of cobamide coenzymes. Fed. Proc. **20**, 956–961 (1961)
94. Barker, H.A.: Corrinoid-dependent enzymic reactions. Annu. Rev. Biochem. **41**, 55–90 (1972)
95. Kishimoto, Y., Williams, M., Moser, H., Hignite, C., Biemann, K.: Branched-chained and odd-numbered fatty acids and aldehydes in the nervous system of a patient with deranged vitamin B_{12} metabolism. J. Lipid Res. **14**, 69–77 (1973)
96. Buchanan, J.M., Elford, H.L., Loughlin, R.E., McDougall, B.M., Rosenthal, S.: The role of vitamin B_{12} in methyl transfer to homocystein. Ann. N.Y. Acad. Sci. **112**, 756–773 (1964)
97. Herbert, V.: Folic acid. Annu. Rev. Med. **16**, 359–370 (1965)
98. Luhby, A., Cooperman, J.: Folic acid deficiency in man and its interrelationship with vitamin B_{12} metabolism. Advanc. Metab. Disord. **1**, 263–334 (1964)
99. Snell, E., Jenkins, W.: The mechanism of the transamination reaction. J. cell. comp. Physiol. **54**, 161–177 (1959)
100. Fasella, P.: Pyridoxal phosphate. Annu. Rev. Biochem. **36**, 185–210 (1967)
101. Aykroyd, W.R., Pathwardhan, V.N., Oomen, H.A.P.C., Mason, K.E., Ramajingaswami, V., Handler, P., McLaren, D.S.: Hypovitaminosis A. Fed. Proc. **17**, 103–143 (1958)
102. György, P.: Protein-calorie and vitamin A malnutrition in Southeast Asia. Fed. Proc. **27**, 949–953 (1968)
103. Karrer, P., Morf, R., Schöpp, K.: The synthesis of perhydrovitamin A. I and II. Helv. chim. Acta **16**, 557–625 (1933)
104. Karrer, P., Morf, R., Schöpp, K.: Vitamin A from fish oils. I and II. Helv. chim. Acta **14**, 1036–1040, 1431–1436 (1931)
105. Pitt, G., Morton, R.: Fat-soluble vitamins. Annu. Rev. Biochem. **31**, 491–514 (1962)
106. Fidge, N.H., Shiratori, T., Ganguly, J., Goodman, D.S.: Pathways of absorption of retinal and retinoic acid in the rat. J. Lipid Res. **9**, 103–109 (1968)
107. Goodman, D.S., Huang, H.S., Kanai, M., Shiratori, T.: The enzymatic conversion of all-trans β-carotene into retinal. J. biol. Chem. **242**, 3543–3554 (1967)
108. Follis, R., Jr.: Deficiency disease, p. 125. Springfield, Illinois: Charles C. Thomas Publisher 1958
109. Wolf, G., Varandi, T., Johnson, B.C.: Vitamin A and mucopolysaccharide synthesizing enzymes. Biochim. biophys. Acta (Amst.) **46**, 59–67 (1961)
110. Wolf, G.: Vitamin A in adrenal hormone and mucopolysaccharide biosynthesis. Amer. J. clin. Nutr. **9**, 36–42 (1961)
111. Juneja, H.S., Moudgal, N.R., Ganguly, J.: Studies on metabolism of vitamin A. The effect of hormones on gestation in retinoate-fed female rats. Biochem. J. **111**, 97–105 (1969)
112. Dingle, J.T.: Studies on the mode of action of excess of vitamin A. 3. Release of a bound protease by the action of vitamin A. Biochem. J. **79**, 509–512 (1961)
113. Dingle, J.T., Lucy, J.A., Fell, H.B.: Studies on the mode of action of excess of vitamin A. I. Effect of excess of vitamin A on the metabolism and composition of embryonic chick-limb cartilage grown in organ culture. Biochem. J. **79**, 497–500 (1961)
114. De Duve, C., Wattiaux, R., Wibo, M.: Effects of fat-soluble compounds on lysosomes *in vitro*. Biochem. Pharmacol. **9**, 97–116 (1962)
115. De Luca, L., Little, E.P., Wolf, G.: Vitamin A and protein synthesis by rat intestinal mucosa. J. biol. Chem. **244**, 701–708 (1969)
116. Creutzfeldt, O., Sakmann, B.: Neurophysiology of vision. Annu. Rev. Physiol. **31**, 499–544 (1969)
117. Dartnall, H., Tansley, K.: Physiology of vision: Retinal structure and visual pigments. Annu. Rev. Physiol. **25**, 433–458 (1963)
118. Crescitelli, F.: Physiology of vision. Annu. Rev. Physiol. **32**, 525–578 (1960)
119. Glickstein, M.: Organization of the visual pathways. Science **164**, 917–926 (1969)
120. Wald, G.: Molecular basis of visual excitation. Science **162**, 230–239 (1968)
121. Wald, G.: The receptors of human color vision. Science **145**, 1007–1016 (1964)
122. Wasserman, R., Corradino, R.: Metabolic role of vitamins A and D. Annu. Rev. Biochem. **40**, 501–532 (1971)
123. Evans, H., Bishop, K.: On the relations between fertility and nutrition. II. The ovulation rhythm in the rat on inadequate nutritional regimes. J. Metab. Res. **1**, 335–356 (1922)
124. Wagner, A., Folkers, K.: The vitamins. In: Medicinal chemistry (Burger, A., ed.), 2nd ed., p. 133–286, p. 160. New York: Interscience 1960
125. Horwitt, M.K.: Interrelations between vitamin E and polyunsaturated fatty acids in adult men. Vitam. and Horm. **20**, 541–558 (1962)
126. Alaupovic, P., Johnson, B., Crider, Q., Bhagawan, H., Johnson B.: Metabolism of alpha-tocopherol and the isolation of a nontocopherol-reducing substance from animal tissues. Amer. J. clin. Nutr. **9**, 76–88 (1961)
127. Draper, H., Csallany, A.: Metabolism and function of vitamin E. Fed. Proc. **28**, 1690–1695 (1969)
128. Tappel, A.: Vitamin E as the biological lipid antioxidant. Vitam. and Horm. **20**, 493–510 (1962)
129. Tappel, A.L., Sawant, P.L., Shibko, S.: Lysosomes: Distribution in animals, hydrolytic capacity and other properties. In: Lysosomes (de Reuck, A.V.S., and Cameron, M., eds.). Ciba Found. Symp., p. 78–125. Boston: Little, Brown and Company 1963
130. Olson, R.E.: Introductory remarks; Interrelationships among vitamin E, coenzyme Q, and selenium, Nutrition Symp. Fed. Proc. **24**, 55–57 (1965)
131. Schwarz, K.: Role of vitamin E, selenium, and related factors in experimental nutritional liver disease. Fed. Proc. **24**, 58–67 (1965)
132. Horwitt, M.K.: Role of vitamin E, selenium, and polyunsaturated fatty acids in clinical and experimental muscle disease. Fed. Proc. **24**, 68–72 (1965)
133. Tappel, A.L.: Free-radical lipid peroxidation damage and its inhibition by vitamin E and selenium. Fed. Proc. **24**, 73–78 (1965)
134. Folkers, K., Smith, J.L., Moore, H.W.: On the significance of biological activities in the coenzyme Q and vitamin E groups. Fed. Proc. **24**, 79–84 (1965)
135. Olson, R.E.: Anabolism of the coenzyme Q family and their biological activities. Fed. Proc. **24**, 85–92 (1965)
136. Van De Kamer, J.H., Weijers, H.A.: Malabsorption syndrome. Fed. Proc. **20**, 335–344 (1961)
137. Welch, C.E., Ellis, D.S.: Physiology of the surgically altered stomach. Annu. Rev. Med. **12**, 19–38 (1961)
138. Stahlgreen, L.H.: The dumping syndrome: a study of its hemodynamics. Hosp. Prac. **5**, 59–64 (1970)
139. Johnstone, M.C. (ed.): Problems in cystic fibrosis. Ann. N.Y. Acad. Sci. **93**, 485–623 (1962)
140. Dische, Z., Pallavicini, C., Cizek, L.J., Chien, S.: Changes in the control of the secretion of mucous glycoproteins as possible pathogenic factor in cystic fibrosis of the pancreas. Ann. N.Y. Acad. Sci. **93**, 526–540 (1962)
141. Johansen, P.G.: Some observations on mucous secretions in cystic fibrosis of the pancreas. Ann. N.Y. Acad. Sci. **106**, 755–756 (1963)
142. di Sant'Agnese, P., Talamo, R.: Pathogenesis and physiopathology of cystic fibrosis of the pancreas. New Engl. J. Med. **277**, 1399–1408 (1967)
143. Lipkin, M.: Cell proliferation in the gastrointestinal tract of man. Fed. Proc. **24**, 10–15 (1965)
144. Dubos, R., Schaedler, R.W., Costello, R.: Composition, alteration, and effects of the intestinal flora. Fed. Proc. **22**, 1322–1329 (1963)
145. Kalser, M., Cohen, R., Arteaga, I., Yawn, E., Mayoral, L., Hoffert, W., Frazier, D.: Normal viral and bacterial flora of the human small and large intestine. I and II. New Engl. J. Med. **274**, 500–505, 558–563 (1966)
146. Rosenberg, I., Hardison, W., Bull, D.: Abnormal bile-salt patterns and intestinal bacterial overgrowth associated with malabsorption. New Engl. J. Med. **276**, 1391–1397 (1967)
147. Crane, R.: Digestive-absorptive surface of the small bowel mucosa. Annu. Rev. Med. **19**, 57–68 (1968)
148. Rubin, C.: Malabsorption: Celiac sprue. Annu. Rev. Med. **12**, 39–54 (1961)
149. Rubin, W., Ross, L., Sleisinger, M., Weser, E.: An electron microscopic study of adult celiac disease. Lab. Invest. **15**, 1720–1747 (1966)
150. Reid, A., Brunser, O.: Pathogenesis of small intestine changes in celiac disease. Arch. Path. **77**, 525–528 (1964)
151. Mayer, J.: Obesity. Annu. Rev. Med. **14**, 111–132 (1963)
152. Mayer, J.: Some aspects of the problem of regulation of food intake and obesity. I, II, and III. New Engl. J. Med. **274**, 610–616, 662–673, 722–731 (1966)
153. Mayer, J., Thomas, D.: Regulation of food intake and obesity. Science **156**, 328–337 (1967)

154. Schachter, S.: Obesity and eating. Science **161**, 751–756 (1968)
155. Winick, M.: Nutrition and nerve cell growth. Fed. Proc. **29**, 1510–1515 (1970)
156. Blass, J.P.: Metabolic hypothesis for selective cerebellar damage in deficiencies of pyruvate decarboxylase. (Unpublished manuscript) (1974)
157. Salyer, W.R., Salyer, D.C.: Thiamine deficiency and oxalosis. J. clin. Path. **27**, 558–559 (1974)
158. Lien, E.L., Ellenbogen, L., Law, P.Y., Wood, J.M.: Studies on the mechanism of cobalamin binding to hog intrinsic factor. J. biol. Chem. **249**, 890–894 (1974)
159. Burger, R.L., Allen, R.H.: Characterization of vitamin B_{12}-binding proteins isolated from human milk and saliva by affinity chromatography. J. biol. Chem. **249**, 7220–7227 (1974)
160. Allen, R.H., Majerus, P.W.: Isolation of vitamin B_{12}-binding proteins using affinity chromatography; II. Purification and properties of a human granulocyte vitamin B_{12}-binding protein. J. biol. Chem. **247**, 7702–7708 (1972)
161. Allen, R.H., Majerus, P.W.: Isolation of vitamin B_{12}-binding proteins using affinity chromatography; I. Preparation and properties of vitamin B_{12}-sepharose. J. biol. Chem. **247**, 7695–7701 (1972)
162. Allen, R.H., Mehlman, C.S.: Isolation of gastric vitamin B_{12}-binding proteins using affinity chromatography; I. Purification and properties of human intrinsic factor. J. biol. Chem. **248**, 3660–3669 (1973)
163. Allen, R.H., Mehlman, C.S.: Isolation of gastric vitamin B_{12}-binding proteins using affinity chromatography; II. Purification and properties of hog intrinsic factor and hog nonintrinsic factor. J. biol. Chem. **248**, 3670–3680 (1973)
164. Barley, F.W., Sato, G.H., Abeles, R.H.: An effect of vitamin B_{12} deficiency in tissue culture. J. biol. Chem. **247**, 4270–4276 (1972)
165. Frenkel, E.P., Kitchens, R.L., Johnston, J.M.: The effect of vitamin B_{12} deprivation on the enzymes of fatty acid synthesis. J. biol. Chem. **248**, 7540–7546 (1973)
166. Hoffbrand, A.V., Ganeshaguru, K., Lavoie, A., Tattersall, M.H.N., Tripp, E.: Thymidylate concentration in megaloblastic anaemia. Nature (Lond.) **248**, 602–604 (1974)
167. Muto, Y., Goodman, D-W.S.: Vitamin A transport in rat plasma; isolation and characterization of retinol-binding protein. J. biol. Chem. **247**, 2533–2541 (1972)
168. Haurani, F.: Vitamin B_{12} and the megaloblastic development. Science **182**, 78–79 (1973)

Chapter 5

Calcium and Phosphorus Metabolism

Importance of Calcium

By its participation in the structural buildup of the skeleton, calcium forms the metallic frame of the body. More than a kilogram of calcium is found in the human body; 99% of it is in bone, and the remainder is in the body fluids and soft tissues.

The circulating and intracellular calcium participate in many vital metabolic reactions: muscle contraction, neuromuscular excitability, blood coagulation, adenosine triphosphatase activations, and other reactions.

Sources of Calcium

Milk and milk products are the best sources of calcium because they are rich in calcium and the calcium that they contain is in a readily absorbable form.

Other foods contain little calcium. In addition to the calcium that is complexed with protein, food contains several calcium salts, phosphates, carbonates, tartrates, oxalates, and the calcium salts of the hexaphosphoric acid ester of inositol acid. In most vegetables, calcium is in the form of carbonate, oxalate, citrate, malate, and tartrate. If it is not lost in the cooking water, the calcium contained in vegetables penetrates the intestinal mucosa only with difficulty because vegetables contain large amounts of oxalates that precipitate calcium.

Calcium Absorption

In the stomach, all the calcium salts formed from calcium and weak acids are readily solubilized, and the calcium is converted to calcium chloride. Because the pH of the contents of the small intestine is more alkaline, most of the ingested calcium salts are insoluble, except for calcium phosphate—the only readily absorbed calcium salt. Thus, in the duodenum the ingested calcium is in the form of two salts: $CaHPO_4$ and $Ca(H_2PO_4)_2$. Normally, the content of the upper part of the small intestine is maintained slightly acid by the passage of gastric juice from the stomach into the duodenum. The low pH of the contents of duodenum and ileum favors calcium absorption. Most of the calcium phosphate is absorbed in the small intestine by diffusion in the duodenum and by active transport against a concentration gradient in the ileum. This absorption process is influenced by vitamin D and parathormone (see below).

Local conditions in the intestinal lumen also influence calcium absorption. The concentration of the metal in the intestinal content increases calcium absorption. Saccharides and disaccharides accelerate calcium absorption in the ileum by a mechanism that is not yet understood. This effect is of physiological significance only in the case of milk, which is rich in lactose—a sugar that, in contrast to other sugars, is not digested and absorbed before it reaches the ileum. Small amounts of organic acid, such as lactic acid or fatty acids, in the gastric content facilitate calcium absorption in the duodenum by maintaining a low pH. In contrast, if the intestinal content includes large amounts of acids forming insoluble salts with calcium (acetic, citric, or oxalic acid), the metal is excreted in the feces rather than absorbed. When lipid absorption is impaired, much of the dietary calcium is lost in the feces because of interference with vitamin D absorption and formation of insoluble calcium soaps. Calcium absorption is stimulated by bile salts in the intestinal lumen. The bile salts probably prevent the formation of insoluble soap.

Little is known of the chain of molecular events that brings the calcium from the luminal to the serosal side of the intestinal cells. Certainly calcium must enter the luminal and be excreted at the serosal side. Studies in which metabolic inhibitors were used have suggested that calcium uptake may involve an energy dependent accumulation of calcium in mitochondria and that calcium efflux at the serosal site against the chemical and electrical potential gradients is an active transport process.

A number of factors have been suspected in calcium transport; ionic exchange, a calcium sensitive ATPase, a calcium binding protein, vitamin D, parathormone, cortisone. Calcium uptake is depressed by sodium and enhanced if one-third of the sodium is replaced by mannitol. It appears that sodium is not required for calcium uptake or transepithelial transfer. The intestinal brush border contains a calcium-sensitive magnesium dependent ATPase. In contrast to the K^+, Na^+ ATPase, the enzyme is not inhibited by ouabain. Vitamin D induces its appearance. The exact role of the ATPase remains unresolved.

A good correlation between calcium absorption and the concentration of a calcium binding protein found in various animals including mammals has been demonstrated. The protein is found at the surface of the microvilli of the goblet cells. The mammalian protein has a molecular weight of 10,000 to 12,000 daltons. The carrier protein can be induced to appear by the administration of vitamin D to vitamin D deficient rats. The role of the carrier protein in calcium transport remains dubious. Cortisone or antiepileptic drugs inhibit calcium absorption without affecting the concentration of the carrier protein. The role of vitamin D and parathormone in calcium absorption is discussed in other sections of this chapter. Cortisone may depress calcium uptake by altering mitochondria which are known to be markedly enlarged after the administration of the hormone [35].

Serum Calcium

In normal individuals, the calcium content of serum is rigidly maintained between 8.5 and 11.5 mg/ml. The average values tend to shift toward the lower limit after early childhood. The serum calcium exists in a diffusible and a nondiffusible form. These two forms can be separated by dialysis and ultrafiltration through collodion membranes. Sixty-five per cent of the serum calcium is ultrafiltrable. The diffusible calcium exists in two form—the largest fraction is ionized, and a small fraction forms complex ions with bicarbonate, phosphate, and citrate. The concentration of the calcium complexes cannot be determined directly. It is assumed that in the serum, all the phosphate, bicarbonate, and citrate anions are complexed with calcium. Insofar as it is possible to measure the serum concentration of these anions, the amount of complexed calcium can be estimated indirectly. By subtracting the values obtained for the complex calcium from those obtained for the diffusible calcium, one can approximate the proportion of ionized calcium.

Calcium is known to form chelate complexes with a number of compounds. Polycarboxylic hydroxylated compounds—such as malate, isocitrate, citrate, and phosphorus-containing compounds—are potent calcium chelators (polymeric phosphate complexes calcium as efficiently as citrate). In contrast, amino groups (glycine) or monocarboxylic acids (acetate) do not form complexes with calcium. In view of these observations made on simple molecular compounds, it is not surprising that the ability of proteins to chelate calcium varies considerably depending upon the amino acid composition; although the ordinary carboxyl group plays no role in chelating calcium, those carboxyl groups located near hydroxyl groups effectively chelate the metal. Similarly, the inclusion of phosphate groups into the protein molecule enhances its chelating properties.

Approximately 35% of the serum calcium is bound to protein. Serum albumins are twice as effective (10.5 moles of calcium bound to 10^5 grams) as globulins (5.6 moles of calcium bound to 10^5 grams) in the chelating calcium.

The homeostasis of calcium in the blood is partly maintained by calcium excretion. Calcium is excreted in bile, urine, feces, and milk. Except during lactation, the kidney and intestine provide the main excretory paths for calcium. The exact amount of calcium that is excreted by way of the intestine is difficult to evaluate because of the large amounts of unabsorbed calcium normally present in the feces (6.5–8 mg/100 ml). Under normal conditions, it seems that little calcium is excreted through the intestine. The renal threshold for calcium is below the minimal levels of calcium in the blood, so the glomeruli constantly remove calcium from the circulating blood. Most of the excreted calcium is reabsorbed in the tubules. In chronic renal failure, excessive amounts of calcium are lost in the urine, leading to decalcification of the bones and a form of secondary hyperparathyroidism.

The normal calcium requirements for the adult human are 0.5 g a day. However, in order to guarantee an adequate calcium balance, it is safer to include 1 to 1.2 g of calcium in the daily diet. Obviously, the requirements for calcium change during growth, lactation, and pregnancy. The diet of breast fed infants should contain 45 mg of calcium per kilogram of body weight. Three times as much calcium is required if the infant is fed cow's milk because the calcium contained in cow's milk is not as efficiently absorbed as that of breast milk. The calcium requirements of pregnant or nursing women are considerable. If the diet does not provide enough calcium to maintain the calcium balance, she will deplete her own calcium reserve and ultimately decalcify her bones. The calcium losses are particularly acute during lactation because, in addition to the loss of the calcium excreted in milk, a large proportion of calcium is also eliminated in the feces. This calcium loss continues even several months after lactation.

Ossification and Bone Metabolism

Calcium metabolism in bone cannot be properly appreciated without reviewing the histology of growing and adult bones. The morphological unit of the bone, the various types of bone, bone development, and bone involution will be described.

Bone is composed of two major structural elements: the fundamental substance and the cell. The fundamental substance includes collagen fibrils, an amorphous mucoid substance—the ossein, and calcium salts. At least three types of cells are found in bone tissues: osteoblasts, osteocytes, and the osteoclasts.

The osteoblast is found in young bones and is associated with bone development. The origin of the osteoblast is still debated. It is assumed to be derived from the differentiation of fibroblasts. The quasitotal absence of mitosis in osteoblasts is in agreement with such a hypothesis. The osteoblast looks like a metabolically active cell. It contains a large nucleus with a well-developed nucleolus. Its endoplasmic reticulum is made of densely packed lamellae studded with ribosomes. Occasionally, the endoplasmic lamellae are dilated to form cisternae. The Golgi apparatus can usually be readily recognized, and the osteoblast usually contains only a few mitochondria.

Filamentous processes emerge from the cell surface to extend through fine canaliculi within the newly formed bone substance where they anastomose with similar processes emerging from the surrounding osteoblasts. The plasma membrane of the osteoblast is separated from the calcium-impregnated matrix by a wide intercellular space.

Two histochemical findings are significant with respect to the alleged role of osteoblasts in bone formation: the presence of glycogen vacuoles and alkaline phosphatase. (The properties of the osteoblasts will be further discussed when the mechanism of calcification is considered.) When the osteoblast is completely surrounded by the calcified matrix, it becomes an osteocyte, which is a smaller cell resembling a fibroblast. The plasma membrane of the osteocyte is almost in contact with the calcified matrix. Although the osteocyte does not appear to elaborate new fundamental substance, its endoplasmic reticulum is readily recognizable.

The osteoclast is a multinucleated plasmoidal giant cell involved in bone tissue digestion. The origin and physiological properties of the osteoclast have intrigued several generations of histologists, yet the role of the osteoclast in histophysiology remains unknown. Depending upon the osteoclast's size, the number of nuclei varies from 2 or 3 up to 100. The cells at some distance from the bone lamellae usually are round or oval, but when they are close to the bone, they are flat and elongated and either wrap themselves around the bone spicules or are plastered at the surface of the bone lamellae. Electron microscopy and histochemistry of the osteoclast cytoplasm have not revealed highly specific structures which could be responsible for the specialized role of the osteoclast. Vacuoles have been described, especially in degenerating osteoclasts whose nuclei have become pyknotic or karyorrhexic. It is not known whether these vacuoles contain secretion products or result from cytoplasmic degradation. There is no evidence that bone is phagocytized by the osteoclast. It has been proposed that the plasma membrane of the osteoclast can send out pseudopods that engulf surrounding osteocytes within a cytoplasmic vacuole.

Inclusions, possibly of a polysaccharide nature, have been found in osteoclasts, but they cannot be differentiated from inclusions found in osteoblasts and osteocytes [1]. One of the most typical structures associated with the osteoclast is the so-called brush border. When a specialized staining procedure is used, a fibrillar pattern can be recognized between the plasma membrane of the osteoclast and the underlying bone. These fibrillae are perpendicular to the bone lamellae, the origin of the fibrillae is still debated, and at least three different theories have been proposed: the fibrillae are formed by cytoplasmic processes originating in the osteoclast; the fibrillae are collagen fibers of decalcified bone; and the fibrillae have a dual origin—osteoclastic and collagenic. The role of the fibrillae in the histophysiology of the osteoclast remains to be clarified. It is not certain whether fibroblasts, histiocytes, or osteoblasts coalesce to form the osteoclast.

There are two types of bone tissue, cancellous and compact. Cancellous bone is made of a network of interlaced calcified trabeculae that surround large medullary spaces. The calcified matrix contains a number of small lacunae in which the osteocytes are lodged. The osteocytes generate thin fibrous processes that find their way into the delicate network of canaliculi connecting the lacunae. Cancellous bone does not have high resistance to trauma. On the contrary, it is easily crushed because it is formed of relatively thin calcified layers separated by richly vascularized cavities. Cancellous bone is more vulnerable than compact bone to pathological and physiological decalcification, and therefore constitutes a labile calcium reserve.

The histological structure of compact bone is more complex than that of cancellous bone, and it is therefore not surprising that compact bone appears only later in the ontogenic order. Compact bone is made of a system of tubes, disposed parallel to the long axis of bone. Each tube (Havers canal) is made of a series of tightly packed concentric lamellae. The lamellae contain lacunae with osteocytes and fine canaliculi containing the filamentous processes derived from the osteocyte. The center of the Havers canal forms a small cavity containing connective tissue and one or two small blood vessels [2].

Ossification is a process of differentiation involving the calcification of the ossein that is elaborated in a framework of connective or cartilaginous tissue. Ossification on a connective tissue model is observed during the development of flat bones and the bones of the face. In this form of calcification, the apatite salts are deposited on the collagen fibers, and the structural orientation of the collagen fibers determines the orientation of the apatite crystals.

In the second type of ossification, a cartilaginous model develops before the calcium salts are deposited. In the ossification of a human phalanx, the first model is made of hyalin cartilage consisting of cells (endochondrocytes) with small nuclei surrounded by abundant fundamental substance. During ossification, the cartilaginous cells in the center of the model hypertrophy, and the large cells are soon dissociated by the invasion of vasculoconjunctive pedicles that originate at the periphery of the cartilage (perichondron). The vessels probably bring in osteoblasts that elaborate a fundamental substance deposited between the hypertrophied cartilaginous cell. The fundamental substance is later impregnated with calcium salts. Ossification progresses from the center of the diaphysis toward the periphery of the bone leading to the typical appearance of the serial cartilage to be described in the section on the histological alterations resulting from vitamin D deficiency.

While the bone lengthens by this ossification on a cartilaginous model, it thickens by the apposition of bone lamellae at the periphery of the diaphysis. This form of ossification results from the differentiation of the connective tissue elaborated by the periosteum, and it constitutes another form of ossification on a connective tissue model.

The bone framework formed in the center of the diaphysis by the differentiation of hyalin cartilage, and at the periphery by the differentiation of the periosteum, constitutes only a temporary pattern which during development is completely reabsorbed and replaced by new bone with the typical haversian pattern. Vasculoconjunctive pedicles originating from the medullary cavity penetrate into the primary bone, bringing with them osteoclasts that dig deep galleries the long axis of which is usually parallel to the long axis of the bone. The lumina of the galleries are lined with osteoblasts, which participate in the elaboration of the typical concentric lamellae of the haversian canal. Even this new framework made of packed tubules is not permanent. During development these first haversian models are eroded, and deep galleries are formed that constitute the guideline for the elaboration of new concentric lamellae.

These considerations on bone development indicate that ossification is a dynamic phenomenon involving the participation of cellular and humoral factors, and therefore, it is unlikely that calcification and decalcification can be explained only in terms of simple physicochemical laws. Yet reviewing the physicochemical properties of the calcium salts forming bone facilitates the understanding of some of the most elementary factors involved in calcification. But the complexity of the composition of the bone salts raises many problems. Analytical determination of the mineral composition of bone has yielded the following results: calcium, 36%; carbonate, 7%; phosphorus, 50%; chloride, 0.04%; sodium, 0.81%; potassium, 0.15%; magnesium, 0.63%. Thus, calcium phosphate constitutes by far the main mineral component of bone. The questions that arise are: In which molecular form are the calcium and phosphate ions deposited in the bone matrix? Are all these chemical elements arranged in a single mineral phase, or is bone made of a mixture of insoluble salts?

The Apatite Crystal

Since specialists cannot agree on the mineral structure of bone, it would be unwise for the uninitiated to make dogmatic statements in that respect; yet it now seems safe to state that the mineral part of bone is composed of impure apatite molecules [3–7].

Apatite minerals are calcium phosphate salts of the type $Ca_{10}(PO_4)$ 6 X_2 where X is fluorine, chlorine, or hydroxyl. The apatites form a group of closely related minerals containing phosphorus. They are widely distributed in nature, and one or another mineral of the apatite family ($Ca_{10}[PO_4,CO_3]$ 6 [F, Cl, $OH]_2$) forms the bulk of the major types of natural phosphate deposits, among the most important of which are included the igneous apatite deposit ($Ca[PO_4]_6$ S_2) and also ($Ca_{10}[PO_4]$ $6[OH]_2$) (CO_4) 10 (PO_4) 6 (CO_2), which results from the differentiation of the constituents of a cooling magna. Marine phosphorites (Ca_{10} [PO_4, $CO_3]6F_{2\text{-}3}$) in the sea floor result from the precipitation of carbonate fluoroapatite. The chemical properties of apatites have puzzled mineralogists and chemists for centuries. Because of this frustration in his attempts to elucidate the chemical properties of apatite, Wenner in 1790 gave it the symbolic name "apatite," a word derived from the Greek verb (απαταω) meaning to deceive. Apatites usually form such minute crystals that they are readily mixed with impurities, and therefore a detailed study of their crystalline structure is difficult. Thus, the exact atomic arrangement is not known for all apatites; fluoroapatite occurs naturally in the form of macrocrystals with a hexagonal cross section. The fluoroapatite crystals (see Fig. 5-1) are thus suitable for accurate chemical analysis and detailed X-ray diffraction studies. Therefore, the arrangement of the different atoms within the fluoroapatite molecule is known. Since most other apatite molecules have X-ray diffraction patterns similar to that of fluoroapatite, the intramolecular arrangement is assumed to be similar to the one described for fluoroapatites.

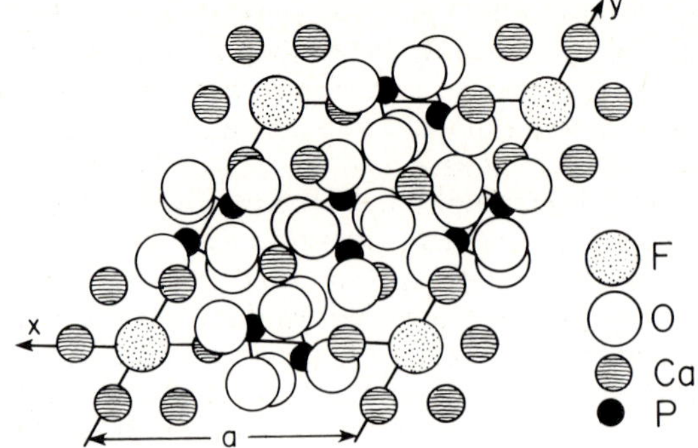

Fig. 5-1. Apatite crystal

Fluoroapatite

The arrangement of the atoms that constitute the crystalline molecule is constantly repeated in space. If an identical point is selected in each of the repeating units, and parallel lines joining these points are drawn in three-dimension, the space inside the crystal is divided into identical parallel epipeds. The three-dimensional pattern outlined by these lines forms the lattice. A single parallel epiped forms the unit cell. Each unit cell can be defined by the length of the edges (a), (b), and (c) of the unit cell and the angles α, β, and γ between the sides. For example, in a simple cubic crystal, a=b=c, and the angles between the sides α, β, and γ all equal 90°. In a hexagonal crystal, two of the angles are equal, but one differs from the other two. α equals 90°, β equals 90°, and γ equals 120°, and thus whereas the sides a and b are equal, c differs from a and b. In an orthorhombic crystal, the three angles equal 90°, but all three sides differ from each other.

X-Rays have been used to study the interior of crystals since 1912. Exactly as a diffraction grating produces interference effects with light, crystals diffract the incident X-ray beam. Indeed, as already stated, the lattice of the crystal can be divided into an indefinite number of parallel and equidistant planes called the lattice planes, and each of these lattice planes is to the X-ray beam what a line of the diffraction grating is to the light beam. When an X-ray source passes through the crystal, the beam is reflected in different directions, and interference effects are obtained. The direction of the outgoing beams is determined by the direction of the incident rays, their wavelength, and the distance between two lattice planes. In the crystal, the simple relationship between the interplanar spacing and the changes incurred by the X-ray beam that passes through the crystals is expressed by Bragg's equation: $n\lambda = 2d \sin \theta$ where n is the integral number of wavelengths, d the distance between two lattice planes, and θ the angle between the lattice planes and the incident beam. When an X-ray beam reaches the crystal's surface, some of the rays (a) are reflected at the surface of the crystal, while others (b) penetrate inside the crystal and are reflected by the underlying lattice plane. Thus, the path traveled by the photons of ray (a) equals OPQ; that traveled by ray (b) equals ORQ. If the difference between OPQ and ORQ equals an integral number of wavelengths, then in view of the laws of interference, the intensity at Q will be increased. But if the diffracted rays are not in phase, they cancel each other out. Thus, the principle of X-ray diffraction is simple: if a crystal is rotated in the center of the path of a monochromatic beam of X-rays, the incoming rays are diffracted. It the outgoing rays are intercepted by a photographic plate, a series of spots appears. The densities and positions of the spots depend on the intensity and the angles of the outgoing rays. The study of the intensities and positions of these spots allows the reconstruction of the atomic structure of simple crystals. In practice, however, the technique becomes intricate when the interior of the complex crystals is investigated.

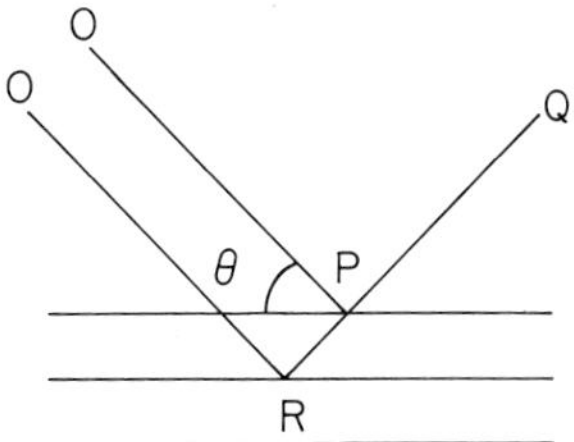

The description of these techniques would be laborious, and only the results will be briefly presented. The fluoroapatite unit cell contains one molecule of $Ca_{10}(PO_4)_6F2$. It has two equal edges: a and b=9.37 (±).001 A. The two equal edges are inclined toward each other 120°, and c (=6.88 (±).001 A) is perpendicular to the two other edges.

Bone Apatite

X-ray diffraction studies of bones and enamel reveal that the salts contained in the tissues have a crystalline structure identical to that of apatite. Although the ultimate shape assumed by the crystal is not known, there is, according to Posner, no doubt that the unit cell of bone salts is identical to that of calcium apatite.

Although the apatite crystals undoubtedly are fundamental to bone structure, a number of problems remain unsolved. Among them are the lack of stoichiometry in the composition of the mineral phase of bone, and the exact morphological appearance of the crystal. It is important that these problems be understood, because the physical properties of the apatite crystals influence the rate of calcium deposition and removal and the composition of the bone salts in bone.

In a perfect crystalline preparation, the atomic arrangement and composition of each molecule are identical and, therefore, the stoichiometry will be the same in any sample of the crystalline preparation. This is not the case for the apatite samples taken from bones. The expected stoichiometry for hydroxyapatite should yield upon analysis a calcium/phosphorus ratio of 1.67, but when various samples of apatites with identical X-ray diffraction patterns are analyzed for their calcium and phosphorus contents, the ratio ranges between 1.3 and 2.0. Two types of irregularities have been invoked to explain the lack of stoichiometry: distortion in the crystalline structure and surface exchanges. Three types of modifications of the crystalline structure have been described, each of witch can alter the preparation's stoichiometry: (1) the calcium ions forming the lattice of the crystals are substituted in some crystals by a hydronium (H_3O^+); (2) some crystals have lattice defects; and (3) ions are missing, leading to the formation of holes in the lattice structure. Since the apatites form only minute crystals, a large part of the mineral phase, approximately two-thirds, exists in the form of unshared surfaces.

The surface ions of the apatite crystals are therefore readily exchanged with ions of the medium capable of replacing calcium (*e.g.*, magnesium), phosphate (*e.g.*, carbonate or citrate), or the hydroxyl (*e.g.*, fluorine). As we shall see, the surface reaction assumes particular importance in the process of mineralization, and some investigators claim that the lack of stoichiometry in apatite preparations can be entirely explained by such surface reactions.

Because of the lack of stoichiometry, a question that is often raised by those who study the mineral composition of bone is whether the mineral phase is single or multiple. If the crystalline defects of the apatites are considered, it seems quite justified to describe the mineral composition of the bone in terms of a single phase. Thus, the bone salts would be made essentially of an impure hydroxyapatite, in which the impurities found their way either within the lattice or at the surface of the lattice. Such a concept is in keeping with the mineral composition of the body fluids, which are undersaturated in calcium carbonate, calcium citrate, magnesium carbonate, and magnesium bicarbonate.

It would seem that X-ray diffraction or electron microscopic studies of the bone minerals could help in describing the mineral phase of the bone. Unfortunately, specialists do not agree on a uniform description of the crystal morphology of bone. At least three different interpretations of the observations made by X-ray diffraction or by electron microscopy prevail. A brief outline of the three theories is presented. On the basis of electron microscopic examination, the mineral phase of bone is described as a honeycombed mineral framework. The organic substance is enclosed within the cavities that are delimited by the apatite wall. The second and third theories differ from the first one in that it is assumed that the mineral phase is composed of individual crystals instead of a continuous framework. One group claims on the basis of studies with low-angle X-ray scatter diagrams that the crystals appear as rods or hexagonal prisms. Another group, on the basis of electron microscopic examination, claims that the crystals form minute tablets.

The apatite crystals of bone are not inert. They are in contact with the body fluids, and the ions of the crystal exhange with those in the circulating medium. Obviously, all of the ions of the crystals do not participate equally in this exchange. The ions forming the surface of the crystals are much more susceptible to be exchanged than those found in the lattice. Because the effective surface of the crystal is determined by its size and shape, it is important to describe the properties of the crystal exactly. Only if the exact size and shape of the crystal are known can the interaction between crystal and medium be expressed mathematically. Such knowledge is not available, and the crystal can be described only qualitatively. Two basic properties of the crystal are generally agreed on: apatite crystals are extremely small (of the order of a few angstroms), and they are elongated (tables or rods). Thus, the surface of the apatite crystal is considerable, and it has been estimated that two-thirds of the ions composing the crystal are at the surface. Consequently, extensive exchange between the apatite ions and the medium ions can be anticipated.

Part of the forces determining the formation of an apatite crystal are ionic in nature. In the interior of the crystals, the charges carried by each ion are neutralized by those carried by ions of opposite charge. In contrast, the charges of the surface ions are not neutralized, at least not by ions constitutive of the apatite crystal. When the crystal is suspended in water, a double-layer potential (Helmholtz double layer) develops at the solid-liquid boundary. The field surrounding each crystal can be divided into two different compartments: one immediately adjacent to the crystal surface where the drop in potential is sharp, and another removed from the surface where the potential decreases gradually.

The first of these compartments is made of a single layer of ions, the second is several ion layers thick. The ions of the first compartment are fixed, whereas the ions of the second compartment can move, but their movement is restricted by the electric field generated in the first compartment, therefore the charges of the mobile ions are unequally distributed. The description of the electric field at the liquid-solid boundary is expressed mathematically by assuming that the double-layer potential is equivalent to an electrical condenser with two parallel plates. The difference of potential between the two plates equals:

$$\frac{\zeta = 4\pi e d}{D}$$

where e equals the charge per square centimeter at the surface of the plate, d equals the distance separating the plates, and D refers to the dielectric constant of the surrounding medium. Thus, it can be deduced from this formula that it is much easier to dilute the charges at the solid-liquid boundary when the dielectric constant of the medium is great, as it is for water.

The significance of the surface properties of the apatite crystals stems from the fact that these properties help regulate the movement of ions from the crystal to the medium and *vice versa*. Such movements occur in the steady state during calcification and during decalcification in bone. In the steady state, exchange between the ions of the crystal and those of the medium is balanced. During calcification, the equilibrium of the ion movement is shifted toward accretion of new apatite crystals. In demineralization, the equilibrium is shifted toward the release of the crystal ions into the medium.

Studies on model systems in which a suspension of apatite crystals is put in the presence of solutions

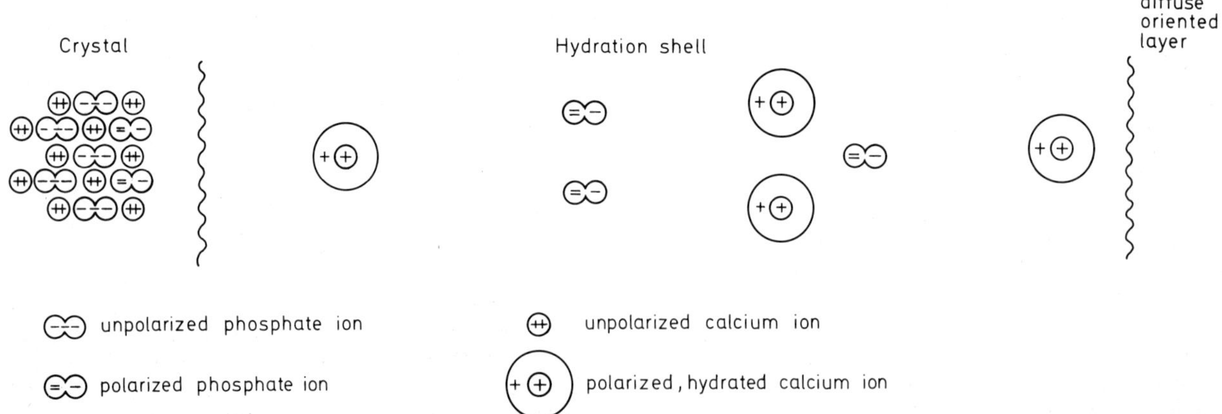

Fig. 5-2. A diagrammatic representation (ignoring OH^-) of the hydroxyapatite crystal-solution interface. The quantitative relations between crystal and bound ions as represented here have been established experimentally (from Neuman, W.F. and Neuman, M.W. [3])

containing carrier-free labeled ions (*e.g.*, [^{32}P]phosphate) have revealed four types of interrelation between ions of the medium and those of the crystal. Before these interrelations are considered, remember that the normal medium surrounding the crystal contains: sodium and magnesium ions capable of replacing the calcium ions of the crystal, and citrate and carbonate ions capable of replacing the phosphate ions. The medium may also contain fluoride ions capable of replacing the hydroxyl ions of the crystals. (1) Some ions penetrate the hydration shell without ever reaching the crystal itself, they may then either simply diffuse in and out of the hydration shell (monovalent cations Na^+ and K^+, and anions Cl^- and S^-); (2) other ions concentrate in the hydration shell and neutralize the surface charges of the crystal (divalent anions CO_3, $COOH^-$, CH_2—CH_2COO^-, and cations Mg^{++}, Sr^{++}, Ra^{++}, and $UO_2{}^{++}$); and (3) other ions penetrate the hydration shell and reach the crystals, where they may either exchange with surface ions (Sr^{++}, R^{++}, $CO_3{}^{--}$, Na^+, S^{+-}) or find their way inside the crystal lattice and fill the defective parts of the crystal. Except when the ions of the medium become part of the lattice, all ion exchanges between medium and crystal reach a steady state rapidly and are reversible (see Fig. 5-2).

Mineralization

In describing the process of ossification, we emphasized two facts: (1) that calcification occurs on a cartilaginous and connective tissue model; and (2) that the ossified tissue is continuously remodeled. Thus, one should clearly distinguish the processes of ossification and calcification. Ossification is a form of differentiation involving the elaboration of a matrix, calcium deposition, calcium resorption, and remodeling of the matrix followed by new calcium deposition. Thus, calcium deposition (calcification) and calcium removal (decalcification) are two processes essential to the differentiation of bone tissue.

Calcification involves the seeding of the calcium apatite crystal and the growth of the crystals. It would be more convenient to review the second of these mechanisms first. This can best be done by assuming for the moment that the bone matrix, connective or cartilaginous, is bathed in a medium supersaturated in calcium phosphate. Intrinsic to the matrix, there is a mechanism that precipitates calcium phosphate in the form of tiny apatite crystals. If there is no interference with the process of calcification, either by deficient absorption of calcium (vitamin D deficiency) or by active dissolution of the calcium crystals (parathormone), most of the process of mineralization can probably be explained by the physicochemical properties of the apatite crystals. Thus, the divalent cations and anions, PO_4 and calcium, penetrate the hydration shell reaching the surface of the crystals where they are crystallized, thereby increasing the size of the crystals. As mineralization proceeds, the amount of bound water contained in the bone tissue necessarily decreases. Most of the water found in dentin and bone tissue (the water content of the lumen of the haversian canal excepted) is bound, and the proportion of bound water decreases as mineralization proceeds. While the content in the fundamental substance remains unchanged, in dentin, when mineralization is completed, almost all of the bound water is replaced by a mineral. Similarly, the proportion of bound water in fully mineralized bone is minimized, according to Robinson.

Mineralization probably occurs both by increasing the size of the seed and by the formation of new crystals. It is not known how much each one of these processes contributes to mineralization. Some indirect evidence for the increase in crystal size has been obtained. Animals whose bones contain radium in the mineral phase exhale less radon—a gas derived from the disintegration of radium—as they increase in age. The radium in the bone is found within the lattice of the crystal. If the crystal size is considerably increased, then instead of blasting its way through the crystal, the radon remains trapped inside the crystal.

The fact that in fully mineralized bone the hydration shell is minimized considerably influences the mineral exchanges between bone and plasma and bone composition.

On the basis of these observations, some of the dynamics of bone mineral can be predicted. In young bone, mineralization is incomplete, the hydration shell is large, and the surface exchanges (penetration in the hydration shell, neutralization of the surface charges, filling of the defect, crystallization and dissolution of the crystal) occur at a maximum rate. When ingested, bone-seeking elements (magnesium, fluorine, or strontium and radium) reach the plasma because they have been included in the diet. They tend to concentrate in the growing parts of the bone—the epiphyseal plate, the inner surface of the endostum, and the inner surface of the periostum. As the bone ages, mineralization reaches a maximum; and because the plasma is supersaturated with bone minerals, the crystals increase in number and size. Therefore, the bone-seeking elements that were easily exchanged with the same elements in the plasma in the young animal are now trapped within the lattice of the crystal of the fully mineralized bone. Even chelating agents are not able to remove the bone-seeking isotopes. Furthermore, bone-seeking isotopes are no longer rapidly incorporated after ingestion. However, if the bone is exposed for weeks or months to plasma containing radioactive bone-seeking elements, the deposition of the radioactive material in the bone is no longer dictated only by the surface exchanges, but it is dictated mainly by the growth rate and bone remodeling. Because of these complex interactions between surface reaction, growth, and remodeling, the studies of bone dynamics by the isotope dilution method, the constant specific activity method, or kinetic analyses are so difficult to interpret. These studies involve complex mathematics, and the reader is referred for that discussion to specialized reviews. At this point the fate of some minor components of bone, such as magnesium, fluorine, strontium, and radium [3], will be discussed.

Other Minerals of Bone

From the surface reaction described above, one would anticipate that most of the sodium is probably present in the hydration shell or at the surface of the apatite crystals; yet, 50% of bone sodium is not available for interchange with the metabolic pool, and therefore it appears that part of the sodium is either closely associated with the matrix of the bone or is precipitated in the bone with citrate.

The fluorine solubility of the apatite crystal, and therefore its hardness, depends upon the composition of the bone. Whereas a high citrate content increases bone solubility and facilitates demineralization, fluorine hardens the bone and prevents demineralization. This is the basis for using fluorinated water in preventing dental caries [8, 9]. When fluoride is ingested, the levels of fluoride associated with the bone are considerable, and it is likely that fluoride both participates in the exchange reaction at the apatite crystal's surface and penetrates within the lattice of the crystal replacing the hydroxyl ion. Whereas the blood contains only a few micrograms of fluoride per 100 ml, the skeleton contains as much as 100 mg/100 g. Of course, the fluoride incorporated within the crystal lattice is stable and not readily mobilizable. Long-term exposure to high concentrations of fluoride leads to mottling of the enamel (fluorosis).

Strontium participates in the formation of new strontium apatite crystals. It may either fill the lattice defects or neutralize the surface charges of the old crystals. Thus, in animals given strontium-containing foods for prolonged periods, the distribution of the radioisotopes in the bone and the exchange of the bone isotope with minerals contained in the body fluid will be determined by the rate of bone growth and by the surface reaction of the crystal. Thus, in the young child, strontium is readily incorporated into the growing bone, and as mineralization proceeds to completion, strontium is trapped within the crystal.

Consequently, the amount of strontium available for exchange decreases as the animal ages. In old bone, the strontium finds its way either within the lattice defects or it is deposited in the form of new crystals in those areas where bone remodeling takes place. The mechanism of incorporation of radium in bone resembles that of strontium. The content in bone citrate will be discussed in the section on decalcification.

The seeding of the apatite crystals triggers the entire mineralization process. All tissues are irrigated with a plasma supersaturated by calcium and phosphate. Yet crystallization occurs only in five different structures. Does the cartilage or the connective tissue of the bone contain specific molecules that precipitate calcium phosphate, or does the osteoblast secrete an enzyme that triggers crystallization? In 1923, Robinson incubated slices of rachitic rat tibial epiphysis. Using various incubation media, he followed calcium deposition in the matrix by staining the preparation with the von Kossa method. At the pH of the blood plasma, the phosphate and carbonate are present mainly in the form of HPO_4, H_2PO_4, and HCO_3 with low levels of PO_4 and CO_3. To induce calcification, the local condition must in some way raise the level of PO_4 and CO_3. Robinson thought that this was accomplished by the action of alkaline phosphatase (pH optimum: 9), which is found in large quantities in the osteoblasts but is absent in the matrix and the osteoclast. But alkaline phosphatase is rather unspecific except for the fact that it attacks only monophosphates.

Glycogen breakdown has been suggested as a source of the organic monophosphates, which act as substrates for alkaline phosphatase. The cartilage cell contains large amounts of glycogen that disappear just before or during mineralization. Furthermore, glucose stimulates calcification, whereas iodoacetate, inhibitor of 1,3-diphosphoglyceric dehydrogenase, and fluoride, inhibitor of enolase, inhibit calcification. The presence of alkaline phosphatase in nonossifying tissue and in rachitic bone is often used to deny a role of that enzyme in calcification. Such an objection is of little value because any theory that involves alkaline phosphatase in calcification does not necessarily imply that the enzyme is the only factor involved in the calcification process, and it is irrelevant whether or not the alkaline phosphatase content of bone increases in Paget's disease or in hyperthyroidism. Follis claims that alkaline phosphatase is not present in the cartilaginous model for ossification in the area of proliferating cells where new matrix is produced, but it has been proposed that alkaline phosphatase is released under those conditions into matrix, while the cartilaginous cells die.

Much before a role for alkaline phosphatase in calcification was proposed, it was suggested that the calcification process is triggered by some property inherent in the ground substance. After a long period of discredit, this theory has been revived. Sobel found that the cartilage obtained from rats with experimental rickets will not calcify when placed in a calcifying medium unless the cartilage is first vigorously shaken in the presence of calcium. These and other experiments brought the investigators of calcification to conclude that the bone matrix contains specific sites that readily combine with calcium. Such a concept illustrates the significance of the bone matrix. In all calcified tissue, the formation of an organic matrix precedes the deposition of calcium salt. The physicochemical properties of the matrix are discussed in more detail in the chapter on connective tissue. The problem pertinent to this discussion concerns the molecular differences between noncalcifying cartilage and connective tissue and calcifying matrixes. This problem still remains unsolved.

In contrast to a crystal where a repeating unit appears in all directions, in a fibrous structure repeating units appear only in one direction. The collagen fibers are made of fibers 2,900 A long and 14 A wide.

The collagen molecule is made of three polypeptide chains coiled to form a triple-coiled helix. The sections of the helix are held together by hydrogen bonds. Collagen is thus a highly organized structure with rigidly repeating units. Not only is the integrity of the structure necessary for calcification, but calcium salts are not deposited at random within the collagen fiber. Some very specific part of the fiber seems to be involved in precipitating calcium salt.

Collagen from various sources (fish bladder, calf skin, and rat tail tendon) can be extracted, solubilized, and purified. The purified preparation can be made to reaggregate in the form of the native fibrils (640–700 A) with the typical axial repeat and intraperiod fine structure. By appropriate physicochemical treatment, these longer fibers can be broken down to smaller fibrils (220 A). The fibers can then be dispersed into a calcifying medium, and the deposition of the calcium salt, or nucleation, can be followed either by X-ray crystal diffraction methods or electron microscopy. These studies demonstrated that only the long (640–700 A) axial repeat fibers could induce nucleation.

Electron microscopy permitted to a certain extent the correlation between the nucleation and the fine ultrastructural periodicity of the collagen fibers. Originally, the collagen fibers were divided by the electron microscopist into band and interband areas, a band corresponding to a complete axial repeat, but as the techniques improved a large number of bands within both the original band and interband areas could be distinguished. These new bands are sometimes extremely narrow, measuring no more than a few angstroms in thickness. The exact site of nucleation within a collagen fiber is not known, but Glimcher has suggested that the area of the fiber previously called the Nth band is the first to precipitate calcium salt. Nucleation presumably depends on a specific but unknown amino acid sequence within the polypeptide chain. Experiments unfortunately somewhat difficult to interpret have suggested that the ε-amino groups of lysine interact with phosphate ions to form either electrostatic or covalent bonds.

An intriguing observation made with this *in vitro* system is that all collagen seems to be able to induce nucleation, yet in the living tissue only the collagen in bone tissue induces mineralization. This paradox can be explained by assuming: (1) that the preparation procedures modify the collagen fiber of noncalcifying tissue to make them calcify collagen, or (2) that the noncalcifying tissue *in vivo* contains inhibitors that are removed while the fibers are prepared.

Some new insight in the mechanism of calcification has been derived from electron microscopic observations which revealed the presence of extracellular vesicles in the matrix of cartilage. Although the vesicles are dispersed in the matrix, they are believed to be buds of chondrocytes. The vesicles that, after proper fixation, appear surrounded by a trilamellar membrane, contain apatite crystals. Anderson and his associates have succeeded in isolating the vesicles by differential centrifugation of collagenase treated cartilage and have shown that the vesicles not only contain apatite crystals, but also a few enzymes; among them alkaline phosphatase, pyrophosphatase, ATPase and 5′ nucleotidase. The vesicles are also rich in lipids including phospholipids whose affinity for calcium is known. The vesicles are capable of concentrating hydroxyapatite *in vitro*.

On the basis of the findings made in his and other laboratories, Anderson has proposed a plausible hypothesis for the priming of mineralization. The enzymes found in the vesicle hydrolyze inorganic pyrophosphate, yielding phosphate that reacts with calcium to form hydroxyapatite. The vesicles serve to concentrate the minerals and once the membrane is dissolved by mechanisms which remain to be discovered, the crystals serve as seed for further mineralization [39].

Rickets

Chemistry of Vitamin D

Nations are often identified with disease. For centuries, the French and the exuberant people of Naples were thought to have spread the dreadful diseases that accompany lovemaking. As for the austere British, they have to be content with being identified with rickets. In 1650, Glisson Jonville wrote a thesis for his doctorate in medicine entitled, *"De morbo puerili anglorum quem patrio idiomate indigenae vocant de rickets."* Rickets was also known in Central Europe as the English Krankheit.

Since 1782 it has been known that rickets can be cured by the administration of cod liver oil. Yet it was only in 1912 that Fung proposed the vitamin hypothesis, which postulated that separate dietary compounds were required to cure beriberi, scurvy, rickets, and pellagra. At first, it was thought that cod liver oil cured rickets because of the presence of vitamin A in the oil. Collum succeeded in separating an active antiricket factor from vitamin A. In 1890, Palm observed the importance of sunshine in preventing the development of rickets. The active vitamin is only sparsely distributed in nature, and fish oils usually constitute one of the best sources of the vitamin. In 1924, it was discovered that rickets could be cured either by irradiating the skin with ultraviolet light or by the administration of irradiated food. It was concluded that irradiation converts a vitamin precursor into the active vitamin.

Formula of Calciferol

Spectroscopic analysis of the lipids and sterols yielding vitamin D activity under the influence of ultraviolet irradiation provided useful clues for detecting the active sterols. A preparation acquiring vitamin D activity manifests characteristic absorption spectra not observed in sterols (such as pure cholesterol) that cannot be activated by ultraviolet irradiation. For example, upon irradiation with ultraviolet light, ergosterol yields a compound with spectral characteristics that permitted the compound to be successfully crystallized in 1931. The new vitamin was called calciferol. Ergosterol is converted to a calciferol in two molecular alterations: (1) the inversion of an asymmetric carbon atom carrying one of the angular methyl groups, and (2) the rupture of the ring B.

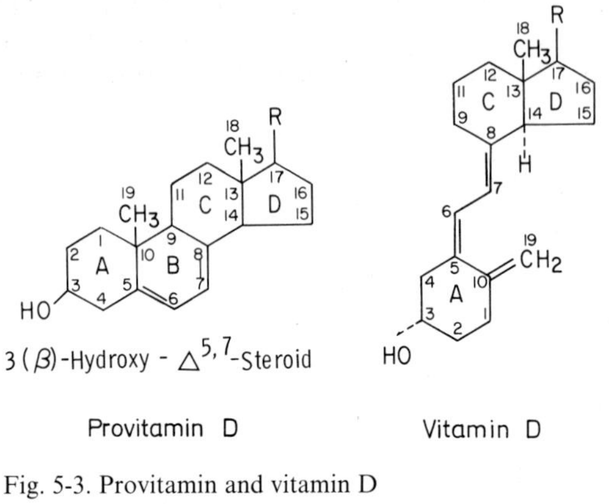

3 (β)-Hydroxy - $\Delta^{5,7}$-Steroid

Provitamin D   Vitamin D

Fig. 5-3. Provitamin and vitamin D

Animal and plant tissues contain several provitamin D's (see Fig. 5-3). They all are composed of a steroid ring substituted by a hydroxyl on carbon 3, a methyl on carbons 18 and 19, and a side chain on carbon 17. All provitamin D's have the same steroid derivative: 3(β)-hydroxy-Δ-5,7-steroid. Ergosterol was the first provitamin the structure of which was elucidated. It carries an extended side chain on carbon 17 as indicated by the formula, and the configuration of the side chain of ergosterol has been established by studies with the aid of X-ray crystallography. All other provitamins have a different side chain. There is no vitamin D_1 because the compound originally referred to as vitamin D_1 was found to be vitamin D_2.

Irradiation of the provitamin with moderate doses of ultraviolet light yields a mixture of compounds from which pure active vitamins can be isolated. In the active product, the β-ring has been opened between carbons 9 and 10, and a double bond is formed between the 10 and 19 positions. The other compounds formed during irradiation—luminosterol, tachysterol, precalciferol—were at first thought to be intermediates between the provitamin with its intact steroid ring and the active vitamin with the open ring (see Fig. 5-4).

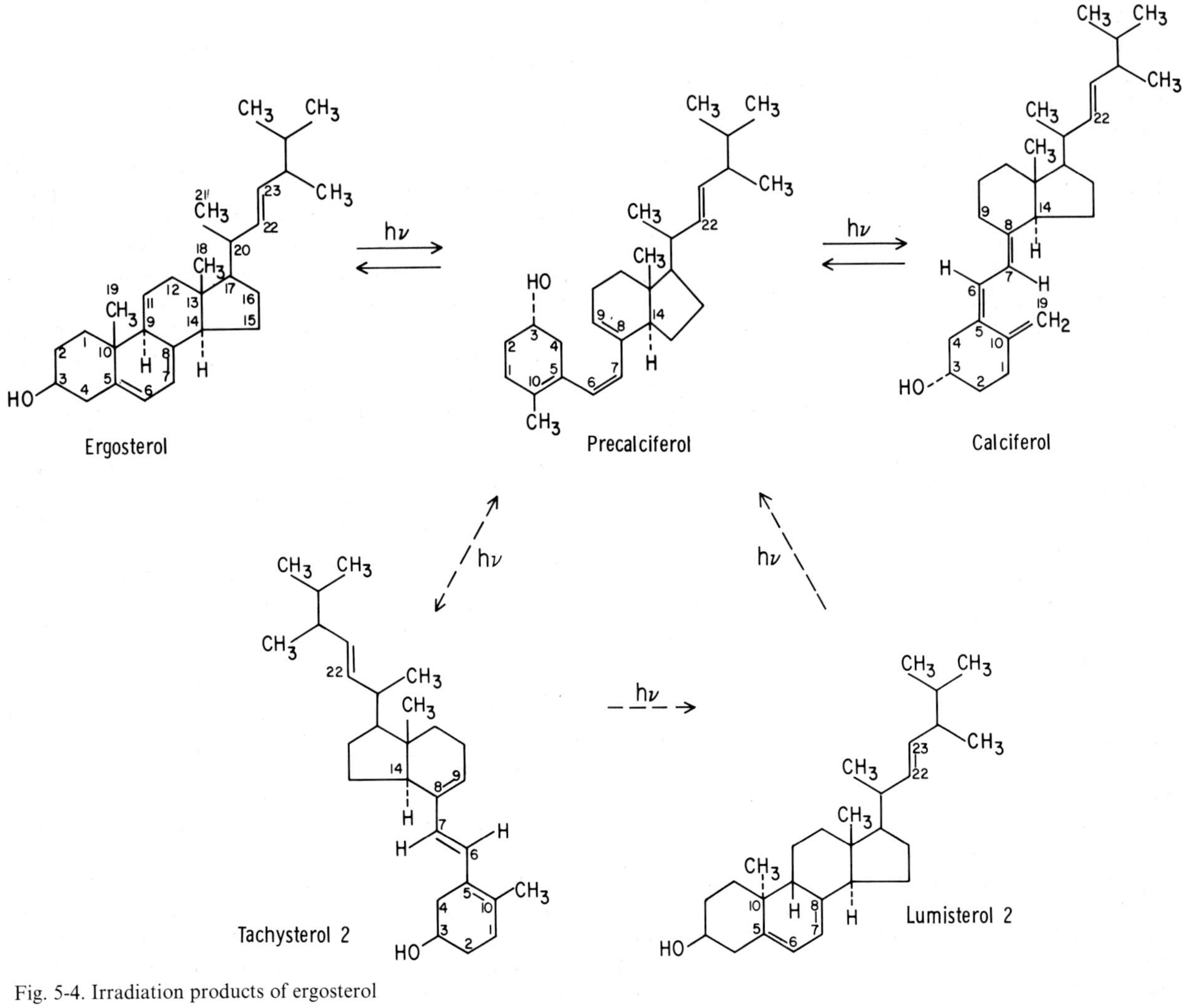

Ergosterol   Precalciferol   Calciferol

Tachysterol 2   Lumisterol 2

Fig. 5-4. Irradiation products of ergosterol

However, it is now accepted that luminosterol and tachysterol arise from a side reaction. Excessive irradiation of the provitamin leads to the formation of toxic compounds, which in the case of ergosterol are toxisterol 2, supersterol 2-1, and suprasterol 2-2. The exact structure of these compounds is not known.

Provitamin and vitamin D are liposoluble compounds, and their absorption in the intestine depends on lipid absorption. Intestinal diseases accompanied by pronounced steatorrhea result in vitamin D deficiency with osteomalacia and rickets. Although vitamin D was one of the first vitamins to be identified, it is only recently that its role in metabolism has been clarified. The mechanism of action of vitamin D raises problems at the histological and the molecular level: Which tissues are the target of vitamin D? What are the intricacies of the molecular reaction involving vitamin D in calcification?

Metabolism and Mode of Action

Vitamin D is absorbed in the lower portion of the intestine after uptake of the sterol by the mucosal tissue, and it is transferred to the lymph. Like vitamin A, vitamin D is esterified in the intestinal mucosa. Esterification occurs with a number of fatty acids, predominantly palmitate, but also stearate, linoleate, oleate, and myronate. Again like vitamin A, vitamin D is stored in the liver in the form of the ester. Two other groups of vitamin D metabolites are also known: excretory products (glucuronide and bile acid derivatives) and active forms (hydroxylated derivatives). A portion of vitamin D is excreted in the bile where it appears conjugated either to glucuronides or to taurocholic acid [10].

The administration of vitamin D restores normal calcification in animals and humans affected by rickets by stimulating the intestinal absorption of calcium or by directly acting on bone mineralization. Although it is well accepted that vitamin D increases the intestinal absorption of calcium, a direct role of vitamin D in bone mineralization and kidney excretion is still in doubt.

In vivo isotope studies using labeled calcium suggested that vitamin D accelerates calcium absorption through the intestinal mucosa. Yet these experiments were susceptible to the criticism that the increased absorption observed *in vivo* could result from a direct effect of vitamin D on the bone itself. Because studies *in vivo* did not reveal whether the vitamin acts on the bone matrix or on the intestinal mucosa, attempts were made to study directly the effect of vitamin D on *in vitro* preparations of intestine or bone.

In vitro studies on the effect of vitamin D on the intestinal mucosa are performed with the aid of intestinal segments. Intestinal mucosal sacs are capable of transferring calcium against a concentration gradient from the mucosal to the serosal site. The active transport through the intestinal walls is energy dependent and is inhibited by agents blocking oxidative phosphorylation. The rate of the absorption through the intestinal sacs depends upon the calcium requirements, and the vitamin D status of the animal from which the sac is obtained. This active transport of calcium was found to be decreased if the sac is obtained from vitamin D-deficient rats. The administration of vitamin D to the deficient animals restores to normal the ability of the intestinal mucosa to concentrate calcium. The findings made with intestinal sacs were later confirmed using slices of rat intestine. It was demonstrated that slices of vitamin D-deficient rats accumulate much less [^{45}C]calcium than slices obtained from normal rats.

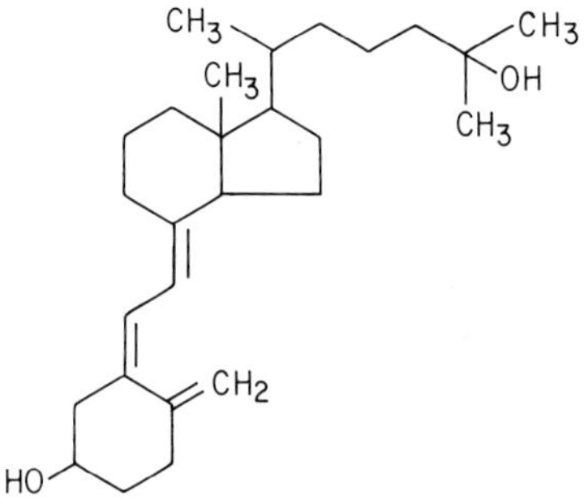

Fig. 5-5. 25-Hydroxycholecalciferol

As is the case with most vitamins, it is not the vitamin but one of its metabolites that is responsible for the physiological effect. DeLuca [11] discovered a vitamin metabolite that is highly active on the intestine. The metabolite has been identified as 25-hydroxycholecalciferol (see Fig. 5-5); it is produced in liver exclusively. The hydroxylation of vitamin D appears to occur in liver by the catalytic effect of a hydroxylase that requires NADPH but is insensitive to traditional inhibitors of lipid peroxidases (N,N′-diphenyl, p-phenylenediamine) [12].

Although the 25-hydroxy derivative is the major metabolite circulating in the blood in the form of a protein complex with an α_2-globulin, it is not the effective compound.

A metabolite formed in the kidney, 1,25-dihydroxycholecalciferol, is now believed to be the active compound. If vitamin D administration to vitamin D-deficient animals is preceded by actinomycin administration, the effect of the vitamin on calcium absorption in the intestine is abolished. *In vitro* experiments using intestinal sacs have excluded a direct effect of actinomycin on intestinal absorption. Actinomycin's effect results from interference with transcription of enzymes which in the kidney convert 25-hydroxycholecalciferol to 1,25-dihydroxycholecalciferol [13]. The kidney is the only organ that can convert the monohydroxy to the dihydroxy derivative, and therefore the monohydroxy derivative has no effect on calcium absorption in anephric animals. Chronic renal damage [14, 15] impairs the biosynthesis of the active vitamin, and this explains the vitamin D-resistant uremia observed in some patients with advanced renal failure.

The cellular target of the 1,25-hydroxylated derivative is unknown. It is still debated whether the hormone acts at the mucosal or the serosal site of the intestinal cell. A number of observations have suggested that the metabolite might stimulate calcium absorption by modifying gene expression: *e.g.*, the 7 hour time lag between metabolite administration and the increase in calcium absorption, and the intranuclear localization of the metabolite. Vitamin D is known to induce the formation of the calcium binding protein and the calcium-sensitive ATPase. However, this view must be reconciled with the fact that actinomycin D does not interfere with calcium absorption while it is known to block the conversion of the 25-hydroxylate to the 1,25-hydroxylate derivative.

Moreover, the 1,25-hydroxylated derivative found in the nucleus is not associated with chromatin, but with the nuclear membrane.

Several other types of investigations suggest that 25-hydroxycholecalciferol or one of its derivatives is the active agent. When injected in vitamin-D deficient rats, the hydroxylated compound is four times as active as vitamin D in curing rickets, and it exhibits its effect more rapidly than when vitamin D is injected. Trummel and DeLuca demonstrated an effect of the metabolite on calcium transport *in vitro* [16].

The concept that the hydroxylated compounds are more active than the nonhydroxylated compounds has, however, been challenged on the basis of experiments in which the direct additions of vitamin D_3 to intestinal sacs elicited a calcium absorption response more rapid and more intense than when the vitamin was injected intravenously [17].

The factors regulating the activities of the hydroxylase are important [35, 36]. Contrary to early reports, it seems clear that neither hydroxylase is regulated by enzymic induction.

Administration of vitamin D_3 depresses the activity of the liver 25-hydroxylase and the toxic effects of hypervitaminosis D have been explained on that basis. The activity of the 1,25-hydroxylase is believed to be regulated by the levels of extracellular calcium. The enzyme is activated by hypocalcemia and inhibited by hypercalcemia. In contrast, activity of the 24,25-hydroxylase which yields an inactive metabolite is increased in hypocalcemia and decreased in hypercalcemia.

Low levels of plasma phosphate stimulate the conversion of the 25-hydroxyl derivative to the 1,25-hydroxyl compound, while high levels of plasma phosphate enhance the formation of the 24,25-hydroxylated derivative. The mobilization of calcium from bone seems to require both parathormone and the 1,25-hydroxylated derivative. An effect of calcitonin on

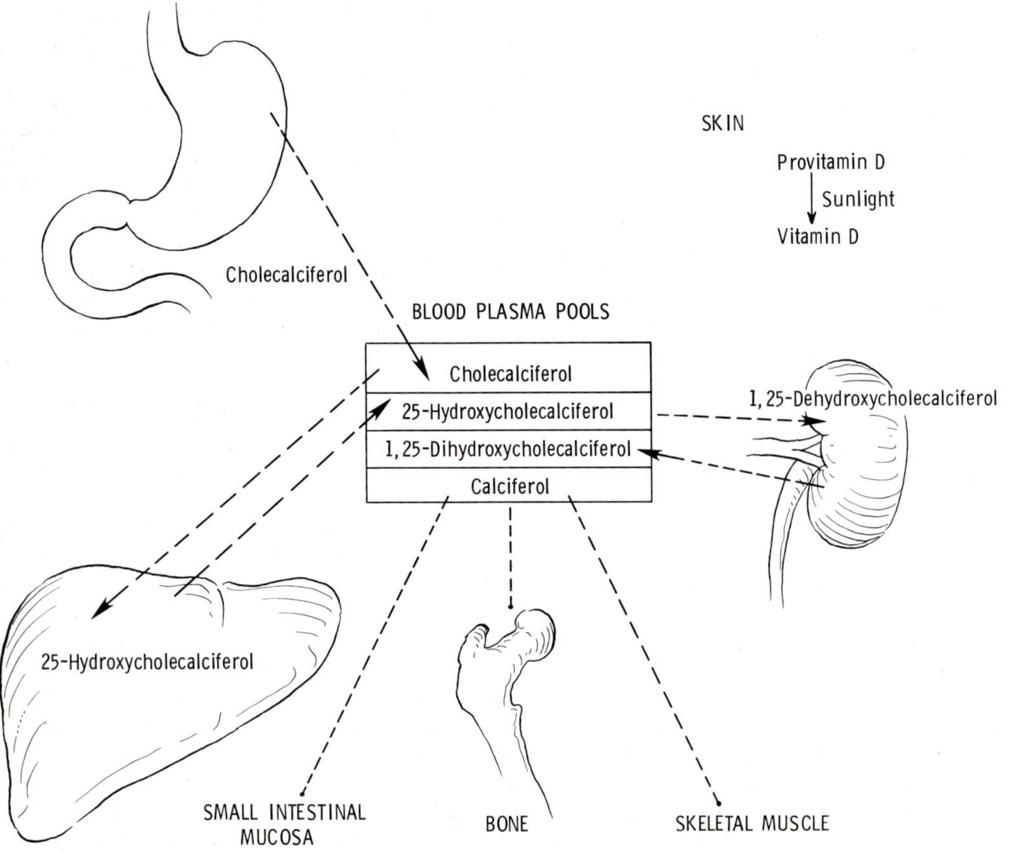

Fig. 5-6. Vitamin D metabolism

either the hepatic or the kidney enzyme has not been conclusively established.

In conclusion, vitamin D yields a number of hydroxylated derivatives; some are active, some are inactive. The active pathway involves elaboration of a 25-hydroxyl metabolite in liver which is converted to an active compound in kidney. The 1,25-hydroxylated derivative, the latter compound, can be looked upon as a hormone whose production is delicately regulated by the serum levels of calcium and phosphorus. It is conceivable that the use of the derivatives might help to cure some form of vitamin D resistant bone diseases (see Fig. 5-6).

It is often stated that in addition to its effect on calcium absorption through the intestinal tract, vitamin D directly affects calcification in bone. But there is no compelling evidence for this assertion. The increased mineralization observed in bone of rachitic rats that received vitamin D could very well be explained by increased intestinal absorption of calcium.

Vitamin D is postulated to stimulate bone calcification by accelerating the conversion of organic to inorganic phosphate and by stimulating cartilage biosynthesis. It has been claimed that vitamin D stimulates the incorporation of ^{35}S in the chondroitin sulfate of bone cartilage, and that the concentrations of serine, glycine, glutamic acid, and threonine increase in the matrix of rachitic rats treated with vitamin D.

If vitamin D has no effect on calcium absorption, it does affect calcium mobilization and thereby restores plasma concentrations of calcium. This finding explains why vitamin D deficiency was associated with hypocalcemia. The effect on the bone seems to require a synergetic action of vitamin D and parathormone. The 1,25-hydroxylated derivative seems to be the major active compound causing calcium release from the bone. The 25-hydroxyl derivative has, however, been shown to be active as well. Finally, vitamin D increases renal proximal tubular reabsorption of phosphate in normal and vitamin D deficient animals. Consequently phosphate excretion is decreased. Inasmuch as this effect occurs in parathyroidectomized animals, the effect of vitamin D or its metabolites must be direct. Again, the active metabolites are the 25 and 1,25-hydroxy derivatives. A calcium binding protein has been isolated from the kidney cortex, but its role in renal reabsorption is not known.

Vitamin D active metabolites have a triple action: (1) to increase the absorption of calcium in the intestine; (2) to facilitate resorption of calcium in bone; (3) to increase reabsorption of phosphate in the kidney. This triple action increases the levels of plasma calcium and phosphorus and thereby facilitates mineralization.

Effects of vitamin D on specific enzymes have been frequently described. One such effect of particular interest is the inhibitory effect of large doses of vitamin D on the purified pig heart aconitase. If such an effect were physiological, citrate—a metabolite important in calcium metabolism—would accumulate. However, citrate accumulation cannot be reconciled with the mineralization effect of vitamin D. The effect of vitamin D on specific enzymes is usually investigated with the aid of large doses of the vitamin, and it is therefore difficult to decide whether the effect is specific for vitamin D, or simply could be reproduced with high concentrations of any sterol.

Rickets

Now that the mechanism of action of vitamin D has been discussed, it will be easier to understand the morbid anatomy and the clinical aspect of rickets [18, 19]. Vitamin D deficiency leads to low calcium absorption through the intestinal tract, and thereby to deficient mineralization of bone matrix. When rickets develops in the young child, all bones soften with typical malformation. The long bones are curved, especially the tibia, which often acquires a saber-blade shape. The thorax is distorted as a result of scoliosis and kyphosis and enlargement of the lower part of the thoracic cage. The pelvic bones are often underdeveloped and lead to narrowing of the pelvic lumen, which may be responsible for dystocia of the female patient. The bones of the skull are soft and flattened and present an appearance described as crania tabes.

Remember that the long bones grow on a cartilaginous model. The proliferating cartilage usually forms a thin line a few millimeters thick, separating the epiphysis from the diaphysis of the long bone. This is referred to as the epiphyseal line. Histologically, it is composed of at least six different layers from the extremity toward the center: (1) The normal hyalin cartilage, (2) the serial cartilage with proliferating chondroblast, (3) the hypertrophied cartilage, (4) the calcified cartilage, an area in which some fundamental substance is elaborated between the cartilaginous cells and where the fundamental substance is itself impregnated with calcium salts, (5) the erosion line, which coincides with the vascular loops that penetrate between the cartilaginous layers, and (6) the osteoid zone made of irregular rows of calcified cartilage surrounded by numerous osteoblasts.

In rickets the epiphyseal line, instead of being a thin and sharp dividing line, is thickened and irregular, and nodular structures may develop at the edges of the proliferating cartilage (see Fig. 5-7). This nodular formation is seen most frequently at the end of the long bone and at the conjugation cartilage of the ribs where they form the structures referred to by the French authors as the *chapelet rachitique*. Histologically, rickets is also obvious; because of the inability of the osteoid material to become mineralized, a regular differentiation of the cartilaginous model into bone tissues becomes impossible. The narrow epiphyseal line, which forms this cartilaginous model, may be ten times increased in thickness. The proliferating cartilaginous cells and their accompanying vascular connective pedicles penetrate deep in the metaphysis of the bone.

from the effect of another hormone, calcitonin, which depresses calcium levels in the blood when they exceed normal values.

A peritoneal lavage technique was developed which permitted study of the effect of calcium and phosphate on the parathyroid. Parathyroid activity is measured by counting the development of osteoblasts in bone. Using calcium-free and phosphate-free lavage, investigators established that the calcium levels of serum control parathyroid secretion directly, and that any effect of the phosphate is indirect on parathormone secretion.

If the level of parathormone is to be maintained constant in blood, as suggested by the feedback mechanism, then a pathway for parathormone breakdown must be available. It is not known where and how parathormone is destroyed. The liver is assumed to participate in parathormone breakdown because transplantation of parathyroid into the spleen prevents parathormone secretion. But such an experiment does not exclude the participation of the spleen in parathormone breakdown.

Nothing is known of the degradation process or the mechanism of transport of the hormone from its site of secretion toward the target organ.

Effect of Parathormone

The major effect of parathormone is to maintain the calcium levels of serum constant by acting on at least three target organs: bone, kidney, and the barrier between the plasma and extracellular fluid. It is now generally accepted that parathormone primary affects bone.

A number of experiments conclusively established that parathormone accelerates bone resorption. Bone resorption is observed after transplantation of parathyroids in contact with parietal bone or in tissue cultures of bone supplemented with parathormone.

The role of bone as a calcium reservoir has already been discussed. Parathormone acts by mobilizing bone calcium, but contrary to the original belief, parathormone does not act on the small exchangable fraction of bone calcium, but on the relatively stable calcium reserve of bone. This conclusion is based on experiments performed by Talmage and his group using the peritoneal lavage technique. Animals were injected with radioactive calcium either 24 hours or 2 or 3 weeks before parathyroidectomy. As expected, parathyroidectomy reduces the total calcium levels of the lavage in either case, but does not affect the radioactivity content of the lavage in animals operated on 24 hours after the injection of ^{45}Ca. In contrast, if the labeled calcium is injected 2 or 3 weeks before the operation, then the levels of the total calcium and of the radioactivity in the lavage drop simultaneously.

Although an effect of parathormone on the calcium reserves of bone can no longer be contested, the metabolic mechanism that triggers the demineralization or the site of action of the hormone in the bone remains to be established. Whether or not the hormone acts primarily on the matrix or on the mineral component of the bone is still debated. A number of observations suggest that parathormones stimulate the release of matrix constituents in the bone. Evidence suggests that parathormone releases components primarily found in the matrix, such as SO_4, hydroxyproline, and hexosamine. After labeled SO_4 is administered to animals injected with parathyroid extracts, the levels of labeled sulfur increase in plasma and urine. Although these experiments suggest that the matrix is affected by demineralization, it does not establish (1) whether the effect on the matrix is specific to parathormone, or (2) whether the effect on the matrix is primary or secondary to demineralization.

Citric acid is a metabolic product of all respiring tissue. The skeleton constitutes an important source of circulating citric acid. The injections of labeled citric acid demonstrate that the turnover of citric acid in the blood is rapid, yet the levels of citric acid in the blood are constant; however, the factors that determine citric acid homeostasis are not known. The amount of calcium in the circulating blood is likely to play an important role in regulating blood levels of citric acid. The plasma concentrations of citrate and calcium correlate consistently in the nephrectomized rat. Nephrectomy markedly elevates plasma citrate and causes a corresponding rise in the diffusible nonionized fraction of plasma calcium. Parathyroid extract produces a decrease in the plasma citric response to nephrectomy, and parathyroidectomy is much more effective in male than in female animals.

The theory proposing that the parathyroid mobilizes calcium by interfering with citrate metabolism is based on an observation made by Sgöström in the late 1930's that increases in parathyroid activity were accompanied by an increase in the blood citric acid level. This original observation was largely confirmed by a variety of experimental techniques, and it was later established that large doses of parathyroid extracts increased the citric release in bone in intact dogs. Studies on resorbing bone incubated in tissue culture confirmed the impression that parathormone stimulates citrate formation during demineralization.

The metabolic source of the citric acid that accumulates during bone resorption seemed to have been elucidated when it was observed that isocitric dehydrogenase activity is low in bone compared to that of the condensing enzyme and aconitase. Unfortunately, further investigation with more delicate methods did not confirm these findings, and, in fact, it has been claimed that the isocitric dehydrogenase activity in resorbing bone is higher than that of aconitase and the condensing enzyme. In the face of these contradictions, the mechanism of citric acid accumulation in bone remains enigmatic.

During bone resorption, lactate accumulates in addition to citrate. Newman has proposed a theory of bone resorption based on organic acid accumulation. Newman's "acid theory of bone resorption" postulates that the solution of hydroxyapatite results from a combination of reduced pH and the chelating effect of citrate. However, it is disturbing that the amount of calcium mobilized does not correlate with the amount of acid (especially lactate) produced. When sections of mouse calvaria are incubated *in vitro*, and the system is supplemented with fluoroacetate, citrate accumulates. Still there is no effect on calcium mobilization.

Parathormone also increases DNA, RNA, and protein synthesis in osteoclasts, but decreases such synthetic processes in osteoblasts. These findings are in keeping with the role of the hormone in bone resorption. The intramitochondrial calcium granules in bone cells are also increased after the administration of parathormone. Whether this effect is related to calcium release remains to be shown. The exact molecular sequence that leads to calcium resorption in bone is not known. But it has been established in experiments using bone calvaria and isolated bone cells that parathormone elicits the formation of a second messenger; namely, cyclic AMP (see chapter on hormones) through the stimulation of adenylcyclase.

In addition to stimulating bone demineralization, parathormone has a collagenolytic effect on bone *in vivo* and in tissue culture. The collagenolytic effect is associated with increased succinic dehydrogenase activity and glucose utilization, and with reduced incorporation of labeled proline into collagen. But the specificity of the effect of parathormone *in vitro* has been questioned because heparin was found to mimic the effect of parathormone on bone in tissue culture.

The experiments described in the previous paragraphs suggest that parathormone releases calcium from the bone by modifying bone metabolism. What type of metabolic changes are responsible for calcium release is not known, but whatever they are they must be mediated through *de novo* protein synthesis. Actinomycin administration prevents bone demineralization under the influence of parathormone *in vivo* as well as *in vitro*. Whether *de novo* protein synthesis is related to new enzyme synthesis in the existing bone cells or relates to the proliferation of the bone cells remains to be established. Actinomycin does not alter the kidney's response to parathormone administration, a finding that suggests that the hormone acts on kidney and bone by different mechanisms.

Although the major effect of parathormone is to mobilize calcium from bone, effects on other organs have also been described. Whereas the effect of parathormone on calcium mobilization is independent of any action of the hormone on the kidney, the phosphaturic action of parathormone is likely to result from a direct effect of the hormone on tubular excretion of phosphate. Parathyroidectomy leads to a decrease in urinary excretion of phosphate, and parathormone administration increases phosphate excretion in rats. An effect on glomerular filtration of phosphate was excluded, but it was established that parathormone acts on the tubular excretion of phosphate.

Again the parathormone acts by binding to a renal receptor and stimulating adenyl cyclase with the subsequent increase in intracellular concentrations of cyclic AMP. The increased phosphaturia is associated with a decreased reabsorption of sodium and phosphorus, and a decreased calcium clearance. As a result, phosphate and bicarbonate excretion are increased. Sodium excretion is, however, not increased appreciably. That is because the sodium rejected in the proximal tubule is reabsorbed in the distal tubules.

Inasmuch as thiazides, which decrease reabsorption in the distal tubules, depress calcium excretion only in presence of parathormone, it would appear that the decreased calcium clearance caused by the hormone is a direct effect on the distal tubule.

Electrolyte movements in general, but calcium movements in particular, are important in mitochondrial physiology. (The movement of electrolytes in and out of mitochondria will be discussed in another chapter.) However, parathormone's effect on calcium movements in kidney and liver mitochondria is relevant to the possible mode of action of the hormone on the kidney. Parathormone has a complex effect on isolated liver and kidney mitochondria. Not only does the hormone stimulate calcium release from mitochondria, but this effect on the divalent cation is associated with all sorts of changes in the electrolyte composition of the mitochondria, and with altering mitochondrial metabolism. When incubated with the hormone, the mitochondria swell and take up potassium, ATP is hydrolyzed, and the ATP-P_i exchange is blocked.

Parathormone stimulates respiration when glutamate is used as a substrate, and the oxidation of NADH and succinate is accelerated by the hormone. Such results confirm a general impression gained from previous studies of electrolyte transport in mitochondria, namely, that ionic transport is in some way related to energy metabolism (see chapter on electrolyte metabolism). Yet these experiments do not indicate that parathormone manifests its effect on the kidney by modifying ionic transport in mitochondria. Two groups of objections can be raised against such conclusions: (1) the effect of parathormone is not specific; and (2) the physiological effects of parathormone do not correlate with the effects *in vitro*.

Although most polypeptide hormones are unable to elicit energy-linked calcium movements in isolated mitochondrial preparations, some basic proteins share this property with parathormone. The doses of parathormone needed to affect mitochondria *in vitro* largely exceed the doses needed to modify kidney calcium transport *in vivo*. One could, however, argue that the *in vitro* conditions are not favorable for proper attachment of the hormone to the target, and therefore much more hormone is needed *in vitro* than *in vivo*.

Studies of the effect of parathormone on the gastrointestinal tract, the mammary glands, the lens, and the distribution of electrolytes have all yielded rather conflicting results.

Hyperparathyroidism

Hyperparathyroidism is an endocrine syndrome. It is sometimes suggested that hyperparathyroidism is a familial disease, and it has been proposed that chief cell hyperplasia is transmitted as an autosomal dominant trait. Hyperparathyroidism results from hypersecretion of parathormone, but increased levels of parathormones in the blood have never been demonstrated.

The disease may be primary—resulting from a tumor, a hyperplasia of the parathyroid—or secondary—the consequence of other disease.

The tumors of the parathyroid that can cause hyperparathyroidism are of four different histological varieties: adenoma, wasserhelle cell hyperplasia, chief cell hyperplasia, and carcinoma (see Fig. 5-8).

The most common cause of hyperparathyroidism is a single benign adenoma (80–90% of the cases). Adenomas develop in a single area of one of the parathyroids. After dissection, they appear as brown nodular masses, often browner than normal parathyroids. The weight of these tumors varies considerably—from not more than a normal parathyroid to as much as 100 g. Adenomas may be under the capsule of the thyroid gland or deeply embedded in the thyroid tissue. Histologically, the most typical characteristic of these tumors is the pleomorphism of the cell type. Oxyphil, chief, and clear cells all participate in the proliferative process.

In rare cases, hyperparathyroidism results from multiple adenomas involving two, three, and even four parathyroid glands. The size of the adenomas and the severity of the clinical symptoms seem to correlate crudely. Adenomas accompanied by renal complications are usually smaller than those associated with bone injury. Occasionally, an adenoma is discovered surgically at autopsy in the absence of renal or osseous disease. It is not known in those cases whether the patient would have developed symptoms if he had lived, but in a few cases in which calcium determinations were made, hypercalcemia was observed.

A syndrome referred to as polyendocrine adenomas has also been described. In this disease, adenomas of the parathyroid, the islet cells in the pancreas, and the pituitary are found.

Some cases of hyperparathyroidism are associated with diffuse hyperplasia of all four parathyroid glands. The size of an affected gland varies considerably, and on dissection it appears lobulated and presents a deep brownish coloration. The weight of hyperplastic parathyroids may be 100 times that of the normal gland. Often the two upper parathyroids are more hyperplastic than the lower gland. Histologically, the hyperplasia usually results from a proliferation of large clear cells referred to as the wasserhelle cells. Rarely the hyperplasia is due to the proliferation of the chief cell type. Thus, the major differences between the hyperplastic parathyroid gland and the adenomas of the parathyroid are: (1) grossly, the hyperplastic parathyroid is diffusely enlarged, whereas the enlargement in the adenomas is localized, and (2) histologically, pleomorphism is exhibited in the adenoma, in contrast to cellular uniformity in the hyperplastic parathyroid.

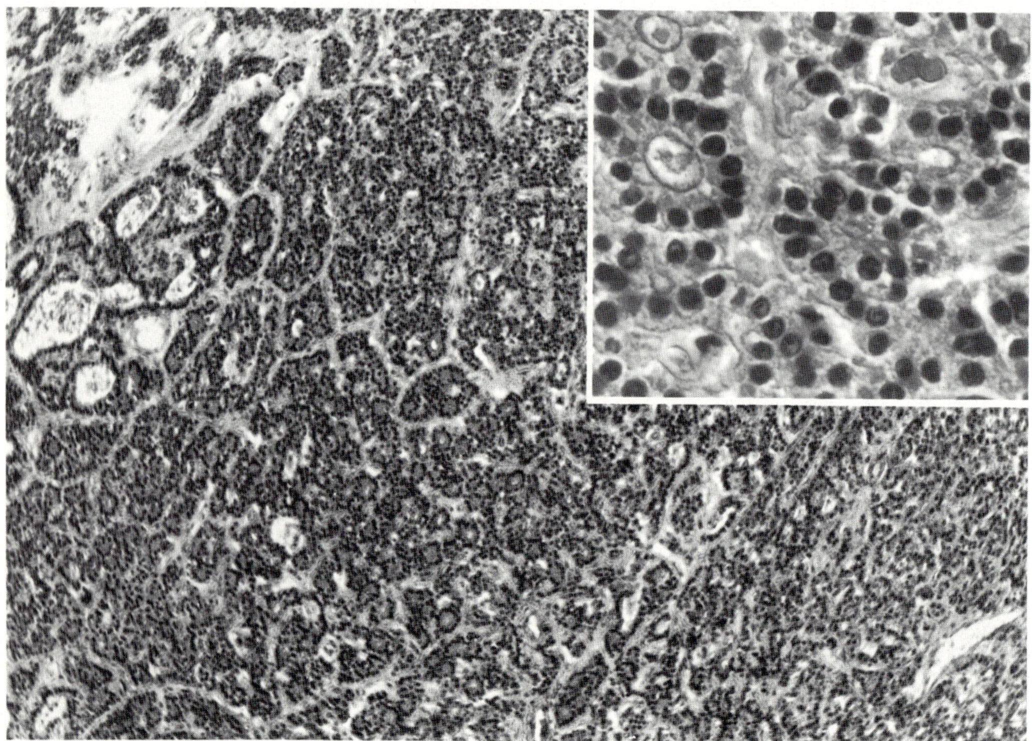

Fig. 5-8. Low- and high-power views of parathyroid hyperplasia

Carcinomas of the parathyroid are rare. Only 1–4% of all cases of hyperparathyroidism are associated with the development of a carcinoma. The criteria for malignancy in cancer of the parathyroid are invasion, recurrence, and metastasis. Metastases are most common in the local lymph nodes, and the tumor, instead of being easy to decapsulate, strongly adheres to the surrounding tissue. Without clear-cut invasion of the capsule of the parathyroid or the development of metastasis, carcinomas may be difficult to diagnose.

Hyperparathyroidism disturbs calcium metabolism, and leads to hypercalcemia, bone atrophy, and nephrolithiasis. These three symptoms are usually present together, yet one of them may dominate the clinical picture. In recent decades, the incidence of renal disease seems to have increased, and that of bone disease decreased among patients with reported hyperparathyroidism.

Hypercalcemia is observed in 95% of the cases of hyperparathyroidism, and therefore measuring the

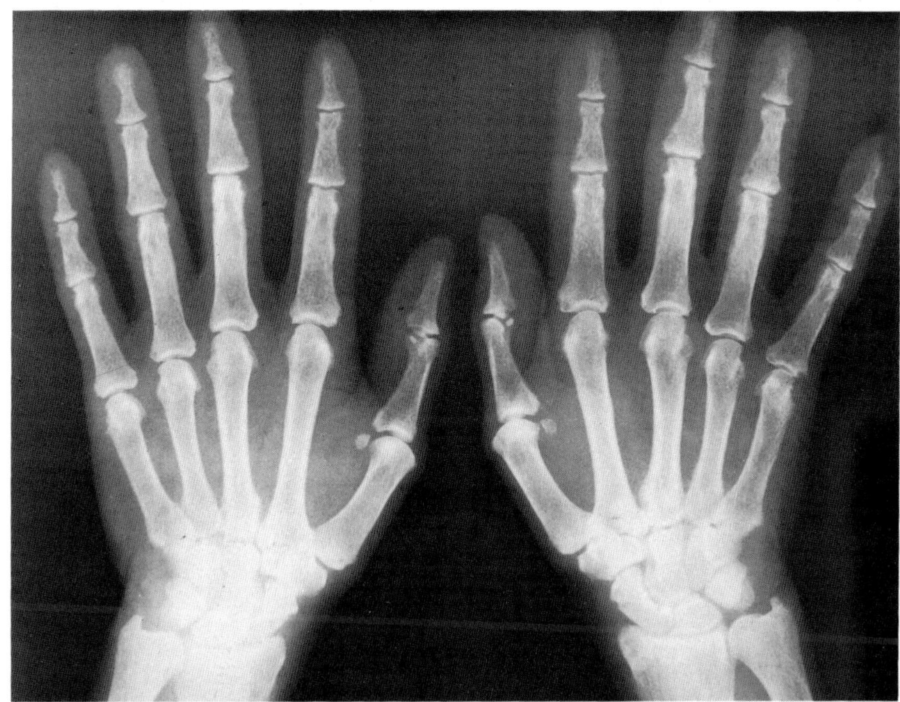

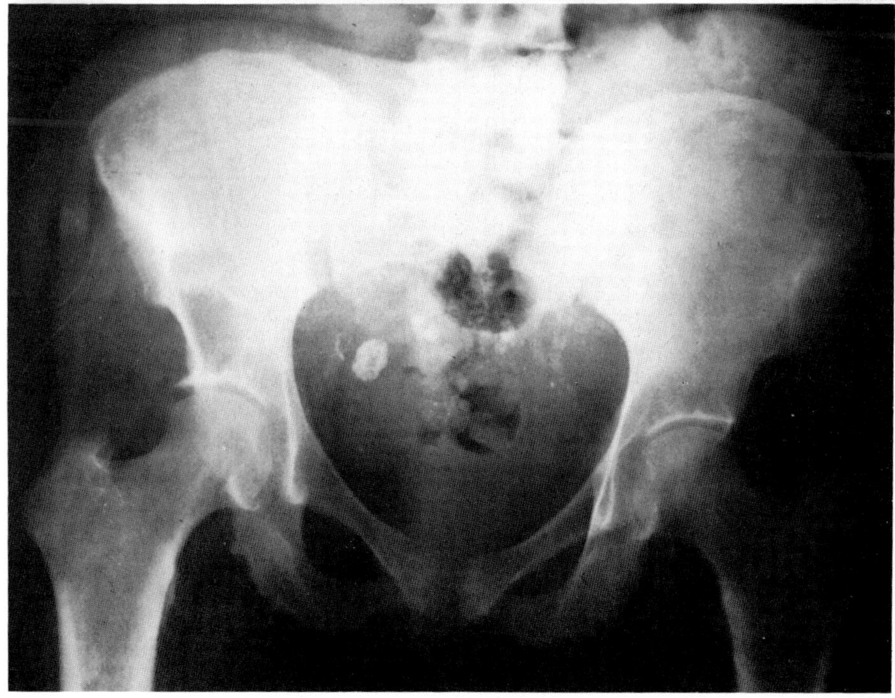

Fig. 5-9. Hands and pelvis of patient with primary hyperparathyroidism. At surgery, 3 parathyroids were removed which proved to be hyperplastic on histological examination. *Hands*. Marked subperiosteal resorption of the proximal and middle phalanges bilaterally, causing a concave narrowing of the diaphysis. Marked motheaten osteoporosis of the phalanges. *Pelvis*. Marked subperiosteal resorption of the proximal and middle phalanges, ischia pubi ramus, and greater trocanter. Large radiolucent defect in left ileum. (From J. Collins)

blood levels of calcium is a useful diagnostic tool, especially among patients with recurrent nephrolithiasis. Hypercalcemia resulting from hyperparathyroidism is sometimes difficult to distinguish from hypercalcemia due to extensive bone metastasis. Occasionally, hyperparathyroid patients have normal levels of blood calcium.

Hypercalcemia is associated with decreased excitability of the nerve muscle apparatus. This disturbance is manifested clinically by generalized fatigue and weakness, vague gastrointestinal disturbances, cardiac irregularities, and the late onset of labor in pregnant women. Occasionally, hypercalcemia may reach very high values (above 20 mg/100 ml), and severe neurological symptoms develop leading to coma and death.

The typical bone disease caused by hyperparathyroidism is von Recklinghausen's disease of bone. The histopathogenesis of the bone disease can be classified as decalcification, loss of organic matrix, and fibrous proliferation. In those areas of the bone affected by the disease, the bone salts are lost first. Demineralization is followed by a loss of matrix substance and invasion of the area by osteoclasts. The proliferation of the osteoclast may be so considerable that it may lead to the formation of brown giant cell tumors. Fibrous tissue proliferates in the digested areas, and the new connective tissue may or may not become calcified. These areas of demineralization and fibrosis become susceptible to fracture, creating a typical radiological picture.

The loss of the calcium content of the bone is responsible for a decreased density of all the bones of the skeleton, particularly long bones. There is marked subperiosteal resorption of the cortical bone and disappearance of the lamina dura about the teeth. Small cysts can be found near the ends of the diaphysis and in the bones of the skull, which may acquire a groundglass or moth-eaten appearance. The vertebrae may have a striated fibrocystic or a mottled appearance. Spinal decalcification may cause kyphosis or scoliosis, and vertebral deformities with biconcave vertebral bodies may result from pressure of the expanded nucleus pulposus on the softened vertebral bodies (see Fig. 5-9).

The bone resorption is responsible for a high alkaline phosphatase level in the serum. The activity of the alkaline phosphatase in the serum can be correlated with the degree of bone demineralization. Of course, alkaline phosphatase is of diagnostic value only if diseases of the hepatic system have been ruled out. The serum inorganic phosphorus levels are reduced in hyperparathyroidism.

Two-thirds of the patients with hyperparathyroidism present nephrolithiasis. In the early stages, hypercalcemia leads to hypercalciuria and hyperphosphaturia with polyuria and occasionally polydypsia. Later, calcium phosphate stones form in the kidney, and nephrolithiasis with typical renal colics, hematuria, and a complicating pyelonephritis develop. The stones may reach sizes that make them visible on X-rays and make them responsible for hydronephrosis. Renal function may be seriously impaired, with uremia, edema, and hypertension. In 10–15% of cases of hyperparathyroidism, peptic ulcers and pancreatitis develop. The reason for these gastrointestinal ailments in hyperparathyroidism is not clear. Other gastrointestinal symptoms that may be predominant among patients with hyperparathyroidism are nausea, vomiting, and abdominal pain.

Excessive amounts of parathyroid hormones can cause focal pancreatic necrosis. It is not known whether pancreatitis precedes hyperparathyroidism or *vice versa*, but it has been suggested that the pancreatitis leads to steatorrhea with hypocalcemia and secondary hyperparathyroidism.

The incidence of bilateral renal vein thrombosis is increased in acute hyperparathyroidism, possibly as a result of generalized dehydration and hypercalcemia. In fact, renal vein thrombosis observed in acute parathyroidism exemplifies the increased tendency to blood coagulation and thrombosis in hypercalcemic patients. An increased tendency to thrombosis is observed in other cases of hypercalcemia associated with plasmacytoma or metastatic myelomas and metastasizing tumors.

Secondary Hyperparathyroidism or Uremic Osteodystrophy

A number of renal diseases induce hypertrophy of all four parathyroid glands with increased bone resorption and increased, lowered, or normal calcium levels in the blood. In secondary hyperparathyroidism, the four parathyroids are enlarged, firmer than normal, and grey rather than the normal brown. The size of the parathyroid correlates with the extent of bone damage. The examination of histological preparations of the hyperplastic gland reveals a relative decrease in stroma and fat cells and a relative increase in glandular cells. All the cells normally found in the parathyroid are hyperplastic, but the proportion of oxyphil cells is, according to some, increased. Chronic nephritis and congenital malformation of the kidney are among the most frequent causes of secondary hyperparathyroidism.

Although all cases of renal insufficiency may lead to moderate hyperplasia of the parathyroid associated with mild signs of demineralization of the skeleton, only in those cases in which slow and progressive destruction of the renal parenchyma occurs does renal osteitis fibrosa cystica generalisata (a disease resembling von Recklinghausen's disease) develop. In these advanced cases, metastatic calcification in soft tissue may also be found. Renal biopsy may prove useful in the differential diagnosis of primary and secondary hyperparathyroidism.

Histological examination of the bone of patients with secondary hyperparathyroidism demonstrates that the edges of the bone trabeculae are sometimes eroded. Rows of osteoclasts are lined along the edges. Some areas of the bone which have been completely

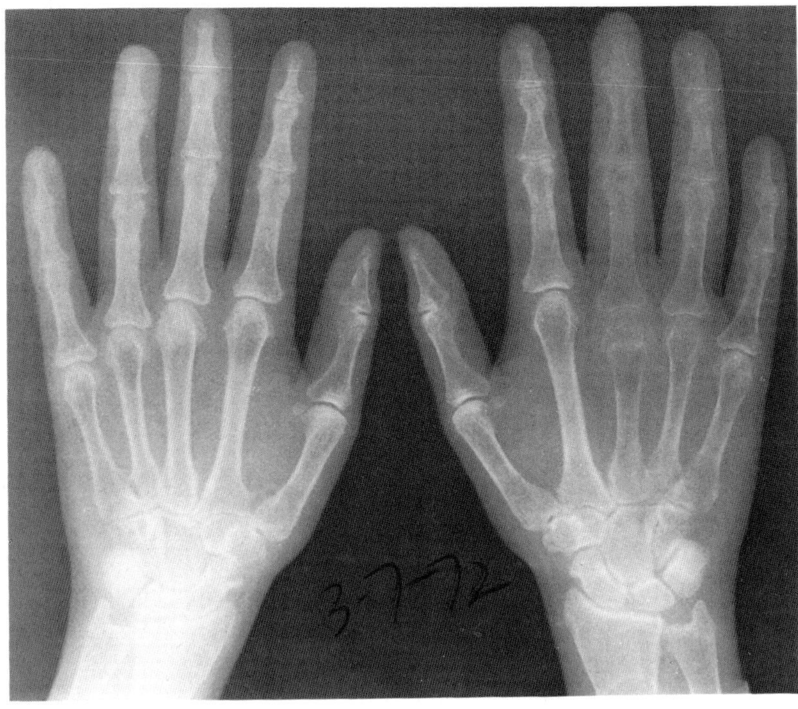

Fig. 5-10a

Fig. 5-10. Secondary hyperparathyroidism in 26-year-old woman. a) The hand shows cystic lesions in the right third and fourth metacarpal. b) In the view of the right knee, notice the well-delineated radiolucency of the patella, an unusual location for manifestation of hyperparahyroidism. (From J. Collins)

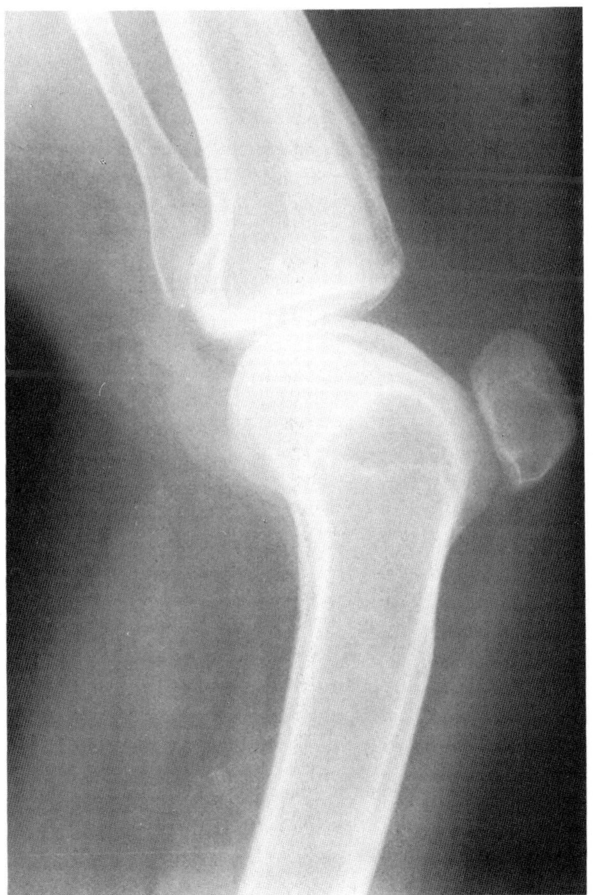

Fig. 5-10b

demineralized are replaced by fibrous tissues. Hypocalcemia, phosphate reduction, and acidosis have been invoked to explain the parathyroid hyperplasia in renal insufficiency. Hypocalcemia could result from a loss of calcium in the urine or feces or from a shift of the blood calcium toward the extracellular fluid. Hypocalcemia could conveniently explain the parathyroid response, but calcium levels are not necessarily low in secondary hyperparathyroidism (see Fig. 5-10).

The phosphorus levels of the blood are high in secondary hyperparathyroidism. This is likely to result from phosphorus retention due to tubular damage. Reduced phosphorus filtration through the glomeruli has been eliminated. In renal insufficiency, the renal tubules either resorb more phosphorus than normal, or the phosphorus secretion is blocked. Since tubular secretion of phosphorus has not been established conclusively, one must assume that the phosphorus retention results from accelerated resorption. Calcium and phosphate ions are maintained in equilibrium in the blood. The calcium levels in the blood will be reduced in the presence of higher levels of phosphate ions, but again, the theory postulates that calcium levels are low in patients with renal insufficiency.

Renal disease is often accompanied by acidosis, and the role of organic acid in bone demineralization has already been discussed. Although the acidosis might explain the demineralization, it does not conveniently explain the hyperplasia of the parathyroid.

The changes in the parathyroid histology in renal insufficiency have been conveniently investigated experimentally by following the parathyroid after bilateral nephrectomy. Hypertrophy and hyperplasia of the parathyroid gland is marked at 48 hours after the operation, and this is accompanied by increase in cytoplasmic basophilia and hypertrophy of the Golgi apparatus. The mitochondria are larger and more numerous, the membranes of the endoplasmic reticulum are more abundant and more tightly packed, and the number of ribonucleoprotein particles is increased. Although these studies seem to have provided interesting observations on the process of cellular hyperplasia, they have given no new clues as to the nature of the cellular granules that contain the parathormones.

Hypoparathyroidism

Hypoparathyroidism is an endocrine disease resulting from an inadequate secretion of parathyroid hormone. Hypoparathyroidism may be idiopathic and result from the absence or hypoplasia of the parathyroid, or be surgical and result from the accidental destruction of the parathyroid during thyroid surgery.

Hypoparathyroidism, whatever its cause, is associated with decreased bone resorption and diminished renal clearance of phosphate; as a result, the serum calcium levels are low and the serum phosphorus levels are high.

The hypocalcemia is accompanied by tetany, electrocardiographic changes, and convulsions. The tetany may exist in a latent or in an overt form. In latent tetany, there is muscular weakness, palpitation and numbness in feeling of the extremities. The latent tetany can be elicited by special procedures that bring into action the hyperexcitable motor nerves; a tap over the facial nerves in front of the ear or just below the zygomatic process leads to the contraction of the facial muscles, especially those of the upper lip (Chvostek's sign).

A tourniquet applied to the upper arm induces carpopedal spasm due to contraction of the flexors of the arm and wrist, the extensors of the phalanges, and the abductor of the thumb. These combinations of muscle contractions lead to a position of the hand referred to as Trousseau's sign. The application of a mild galvanic current (below 6 milliamps) to the motor nerves leads to contraction of the muscles controlled by these nerves (Erb's sign).

In tetany, generalized contraction of the muscle may occur, but the contraction is preponderant in some areas of the body. Some of the most typical spasms are the carpopedal and laryngeal spasms. The carpopedal spasm has already been described. The laryngeal spasm may lead to cyanosis and respiratory failure. Sudden death may occur in tetany due to cardiac spasm, but palpitation tachycardia is more common. The electrocardiographic pattern in cases of tetany reveals a prolongation of the Q-T interval and occasionally T-wave changes.

The convulsions of tetany resemble those observed in the grand mal seizures and are usually accompanied by electroencephalographic abnormalities. Convulsions may be generalized or restricted to one-half of the body (hemitetany) and may result in death if they are prolonged and repeated frequently.

In addition to these more typical forms of tetany, contractions of the bronchial muscles, the muscles of the iris, the ciliary bodies, the esophagus, the urinary bladder, and the entire gastrointestinal tract have also been described. The hyperexcitability of the muscle of the small arteries may result in localized palor (of the fingertip, for example), dermographia, and muscular pain. Spasms of the vessels irrigating the skin and the nail bed are believed to be responsible for the changes observed in the irrigated areas. In some areas, the hair becomes coarse, the skin thickens and roughens, and the nails become brittle.

The X-ray pictures of the skeletons of patients with hypoparathyroidism are not necessarily typical. In the typical cases, the bone appears dense, but cases of hypoparathyroidism with focal decalcification of the skeleton have also been described.

The teeth are often hypoplastic and show defects in the enamel in the form of transverse furrows, horizontal grooves, or punctated holes.

The ectopic calcification in the basal ganglia and in the subcutaneous tissues that is occasionally found in patients with hypo- or pseudohypoparathyroidism is difficult to explain. And it is not known why cataracts develop in both hypo- and pseudohypoparathyroidism. Mental retardation is sometimes described in patients with hypoparathyroidism, but its etiology remains unknown.

Pseudohypoparathyroidism

Pseudohypoparathyroidism is a hereditary disease characterized by hypocalcemia and hypophosphatemia. In pseudohypoparathyroidism, the electrolyte disturbances are not brought to normal by the administration of parathormones. The number of cases studied is too small to permit the exact mode of transmission to be established, but it has been suggested that the disease is transmitted from man to man by a sex-linked dominant gene.

In addition to the classical symptoms of hypocalcemia, the patients affected with pseudohypoparathyroidism also have skeletal anomalies, including shortening of metacarpals and metatarsals. The pathogenesis of the disease is unknown, but it is assumed that the target organ of the parathormone is unable to respond to the hormonal stimulus.

Patients with pseudohypoparathyroidism are stocky, short, and have broad faces. Their metatarsal and metacarpal bones are shorter than normal (short hands, short fingers, and short toes). The parathyroid may be of normal or enlarged size. Ectopic bones form in the skin and in the fascial planes more frequently than in idiopathic hypoparathyroidism.

In pseudohypoparathyroidism, the abnormal metacarpal phalangeal line is pathognomonic, provided that other congenital abnormalities have been ruled out. The line is normally slightly convex, but in pseudohypoparathyroidism the line becomes concave, straight, or irregular as a result of shortening of the metacarpal or metatarsal bones. Mental retardation and abnormal electroencephalographic patterns have been reported in patients with pseudohypoparathyroidism.

Osteoporosis

Systemic osteoporosis is a disease characterized by a lowered bone mass per unit volume of skeleton that results from faulty calcification. In fact, as much as 50% of the bone mass may be lost before the diagnosis of osteoporosis becomes obvious. It is therefore not surprising that attempts are made to quantitate the bone mass on biopsy material for the purpose of securing the early diagnosis of osteoporosis. In osteoporosis there are no known alterations in the mineral composition of the bone, but a slight reduction in the Ca/PO_4 ratio proportional to the demineralization has been reported. A patient is arbitrarily classified as osteoporotic when the specific gravity of the bone drops from 1.075 (normal bone) to 1.050.

In areas of osteoporosis, the calcified bone is progressively replaced by the softer components of the skeleton (fibrous tissue, fat, and marrow). The pathogenesis of osteoporosis has not been established and therefore an adequate classification is not available. Cases of osteoporosis are grouped in two main categories: those patients in whom osteoporosis results from endocrine or vitamin imbalance, and the idiopathic cases of osteoporosis, which cannot be traced to a well-established pathogenesis [26–28].

If radiological pictures of the bone of an osteoporotic individual are compared with those of normal bone, it becomes obvious that the density of the diseased bone is considerably decreased, but the structure of the bone is not changed, and in fact the trabecular appearance may even be exaggerated, as if the outline and the fine structure of the bone had been drawn out with a thick dark pencil. As osteoporosis pro-

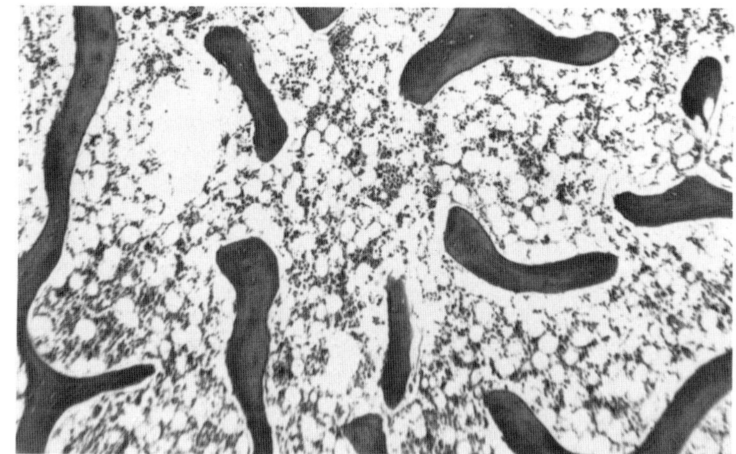

Fig. 5-11. Histological view of osteoporotic bone showing thinning of the bone lamellae

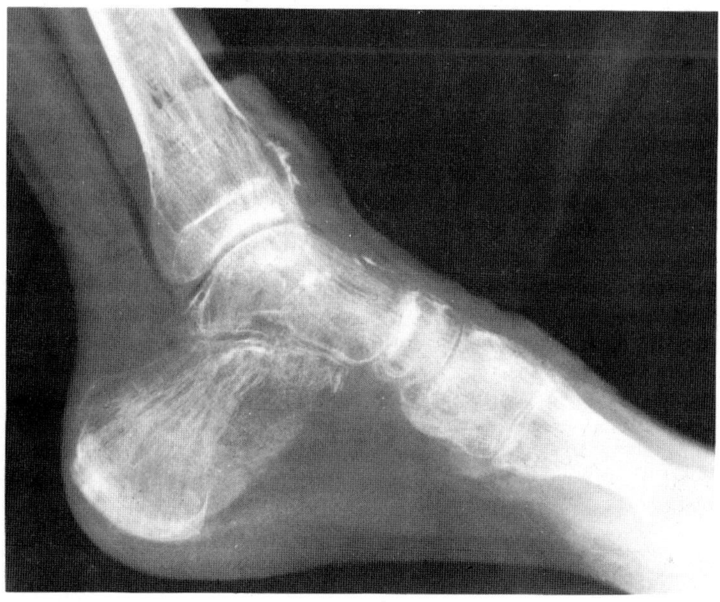

Fig. 5-12. Osteoporosis; notice the radiolucency of the bones of the feet

gresses, the trabeculae of the substantia spongiosa of the bone become thinner, and may even disappear completely. The medullary spaces are enlarged and are filled with fibrous tissue. In the compact bone, the lumina of the haversian canals are enlarged because of progressive disappearance of the bone structure (see Figs. 5-11 and 5-12).

Bone demineralization leads to increased fragility of the bone. A typical deformation occurs in the thinned vertebrae where the nucleus pulposus depresses the softened bone giving a biconcave appearance to the vertebral body. Small herniations in bone of adjacent cartilage occur and are responsible for a deformation known as Schmorl's nodules.

Fractures due to osteoporosis frequently occur in the vertebra, hip, or wrist. In fact, a frequent symptom in osteoporosis is a backache, which results from a fracture of the vertebra. In such cases, the fracture is not always visible radiologically, but changes typical for demineralization of the vertebra are detectable (thinning of the cortex, widening of the intervertebral space, and increase in the vertical trabecular pattern).

It has been claimed that 76% of patients with fracture of the hip presented osteoporosis. The incidence of fracture of the hip increases markedly with age, and fracture of the wrist is even more common in older persons. Mechanical osteosynthesis often fails in older people because the nails and screws are not held in place in the demineralized bone.

Osteoporosis of the Aged

The exact incidence of osteoporosis in the general population is not known, but women are twice as susceptible as men. Although age undoubtedly is often associated with osteoporosis, the cause of osteoporosis of the aged is not known.

Data obtained by using radioactive precursors of bone minerals suggest that net bone formation is not altered by age in patients with osteoporosis. In contrast, bone resorption seems to increase with age, and the osteoclastic activity is about four times greater in osteoporotic patients than in normal individuals. It is not known whether increased bone resorption results from a primary degenerative injury in the bone itself or from prolonged negative calcium balance. A negative calcium balance because of inadequate calcium intake is rare, however, and bone metabolism is not accelerated in older persons. Histological, ultrastructural, and biochemical studies of the bone have provided a detailed description of the disease without revealing the pathogenic mechanism.

Histologic features of bone change in older persons include a decrease of the mass and density of the bone. Although of diagnostic value, such changes may be the cause or the result of osteoporosis. Some of the bone change typical of aging occurs at a younger age in osteoporotic patients. Bones of osteoporotic patients have also been examined by microradiographic techniques. These studies have indicated that in osteoporotics, the time required to form an osteon is markedly increased. Although osteon formation takes only 3 months in normal bone, this process is seldom completed in bones of osteoporotic patients.

X-Ray diffraction studies have established that the hydroxyapatite crystals of osteoporotics were identical to those found in normal individuals. In contrast, alteration of the collagen has been reported in osteoporosis. Irregularities in the fiber pattern have been observed with the electron microscope.

The organic matrix of normal bone and osteoporotic bone was examined in unfixed decalcified bone. Whereas bundles of fibrillae form the organic matrix of normal bone, the matrix of osteoporotic bone is structureless, and it was concluded that in osteoporotic patients, the osteocytes lose their ability to elaborate normal collagen.

Osteoporosis after Immobilization

The mechanism responsible for osteoporosis resulting from immobilization is not clear. One theory suggests that muscular activity stimulates osteoblastic activity in the bone. Another theory proposes that the rate of bone destruction is accelerated in osteoporosis. In immobilized innervated rat limb, there is an increased destruction in the matrix, reflected by a more rapid turnover of [^{14}C]glycine into the soft tissue. The increased turnover of amino acids into the protein of soft tissue is paralelled by an increased turnover by the calcium in the bone and the ^{35}S in the chondroitin sulfate.

Calcitonin—Thyrocalcitonin

Hormones have recently been discovered that are likely to act on bone by either preventing resorption or facilitating deposition of calcium [29–31]. One of these hormones was found in the parathyroid gland (calcitonin), the other, in the thyroid gland (thyrocalcitonin). Thyrocalcitonin, which has been purified from pig thyroids, is a polypetide hormone with a molecular weight of 3,000 and composed of 32 amino acids (see Fig. 5-13). The intact sequence is indispensable for activity. The amino terminal segment of the molecule contains an intrachain disulfide bridge.

The amino acid sequences of porcine, bovine, human, and salmon calcitonin are known. That of porcine calcitonin is shown in Fig. 5-13. Studies relating structure and activity have begun, but it is too early for any definite conclusion. Of the 32 amino acids, 18 in the human calcitonin sequence differ from those in the porcine hormone, and salmon calcitonin is more active in mammals than all known mammalian calcitonins. Moreover, the effect of salmon calcitonin is much longer than that of mammalian calcitonin.

The homologies between all forms of calcitonins involve the amino terminal segment of the molecule (1–9).*

A proline terminal occupies the carboxy terminal position in all cases. Thus the variations affect the amino acids occupying positions 10–30 in the sequence. Such findings suggest that the amino terminal segment is directly linked to activity, and that the sequence between amino acid positions 10 and 30 regulates the level and duration of activity. In all cases (except one), all shifts in the sequence involve a single-base mutation in the triplet codon.

The infusion of calcitonin to rats induces increased calcium deposition in the bone and hypocalcemia. Thus, the hormone has properties antagonistic to those of parathormone. Fluorescent antibody techniques have demonstrated that the hormone is secreted by the C cells of the thyroid gland. A light cell hyperplasia outside the follicular walls of the thyroid has been observed in mice with hereditary hypocalcemia.

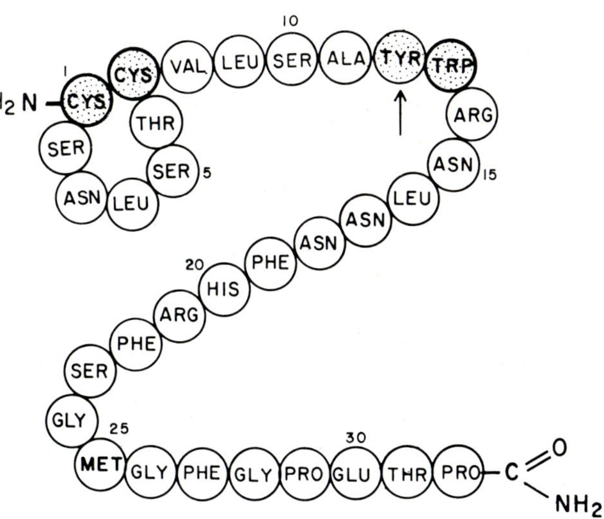

Fig. 5-13. Covalent structure of porcine calcitonin. (dotted circle) —residues important for function

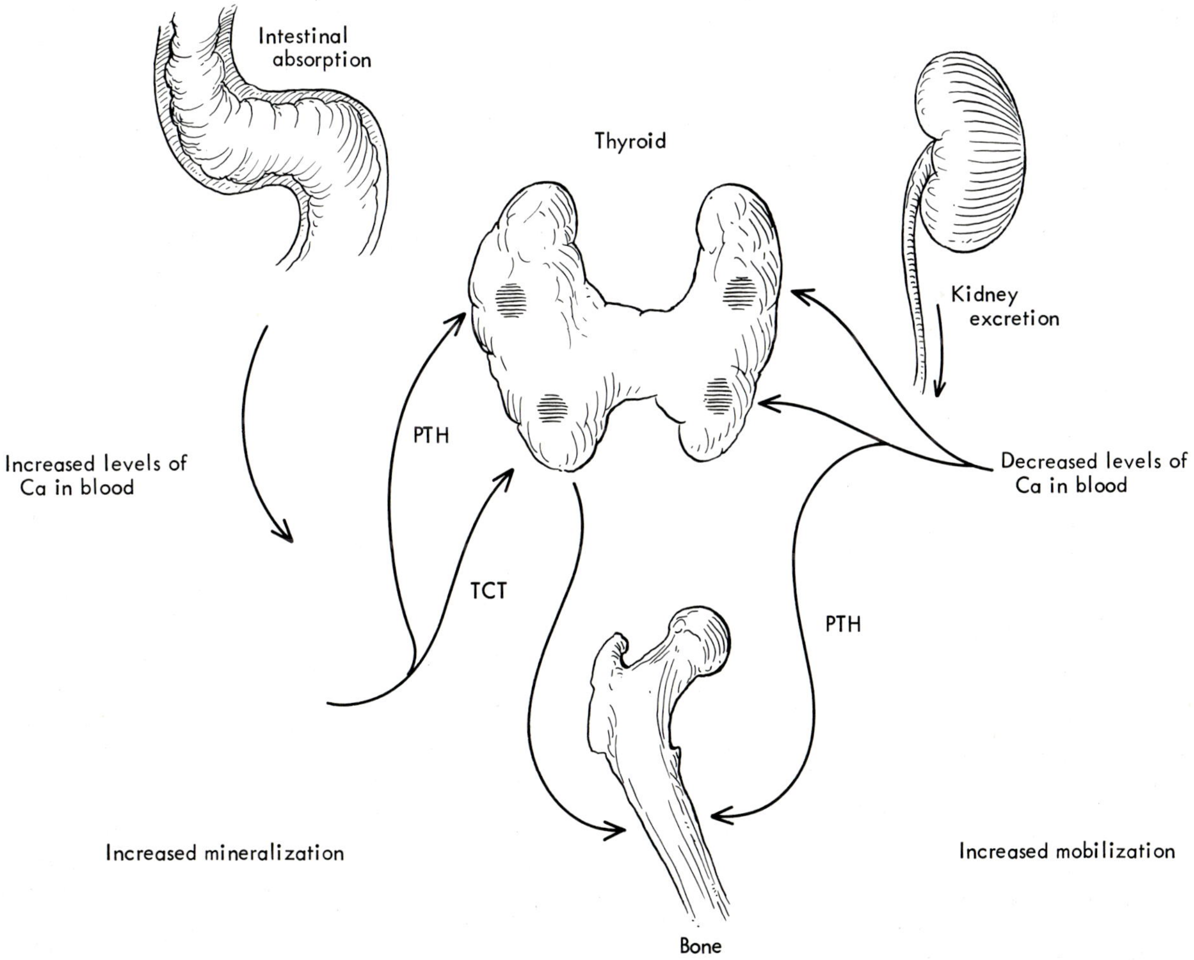

Fig. 5-14. Regulation of calcium levels in blood. A decrease in the calcium levels in blood stimulates parathormone (PTH) secretion and inhibits calcitonin (TCT) secretion and *vice versa* when the blood calcium increases

* H_2N Cys-Ser-Arg-Leu-Ser-Thr-Cys-Val-Leu in pig, beef, and salmon. The sequence is identical in human calcitonin except that the serine in position 3 is replaced by a glycine residue.

Calcitonin is active in nephrectomized and parathyroidectomized animals, indicating that the hormone does not act through the kidney or the parathyroid. Thyrocalcitonin causes hypocalcemia even in rats fed a low-calcium diet. Consequently, the hormone does not seem to influence intestinal absorption of calcium. Calcitonin facilitates calcium retention in bone cultures. It is likely that the hormone acts by interfering with calcium resorption in bone, but the exact mechanism by which the interference takes place is not known.

The exact cellular target of calcitonin in bone has not been identified, but the hormone seems to affect both osteoblasts, osteoclasts, and osteocytes. Mitochondrial granules are increased in osteocytes and osteoblasts. The membranes of the osteoclasts are modified in at least two ways: the borders are ruffled where bone resorption takes place and the membrane potentials are reduced.

In kidney, calcitonin increases the secretion of calcium, phosphate, and sodium. The natriuretic effect of calcitonin is associated with water loss, increased urine volume, and weight loss. The effect of calcitonin on sodium excretion is different from that of parathormone. While parathormone appears to favor a sodium for hydrogen exchange, which takes place in the distal tubules, calcitonin acts on the proximal tubule by blocking sodium and chloride reabsorption [37].

Kidney and bone receptors for calcitonin have been discovered. Competitive binding studies have shown that the receptors are the same in both organs, but different from those that bind parathormone or vasopressin [32]. The high biological potency of salmon calcitonin is explained by its high affinity for receptors.

Elucidation of thyrocalcitonin's mechanism of action hopefully may provide new clues on the pathogenesis of osteoporosis (lack of hormone) and such diseases as Albers-Schönberg disease (osteopetrosis; marble disease).

The role of calcitonin in the pathogenesis of these diseases remains to be proven. It has been suggested that osteopetrosis is caused by an increased secretion of calcitonin; in contrast, osteoporosis could be caused by a defect in calcitonin. Certainly some medullary carcinomas of the thyroid are associated with elevated levels of calcitonin and these tumors are not usually associated with osteopetrosis. The regulation of calcium levels in blood by parathormone and calcitonin is summarized in Fig. 5-14. For further information on the relationship between calcium metabolism and bone disease, read the book of Rasmussen and Bordier [38].

Pathological Calcification

Pathologists frequently find foci of calcification in tisues that normally do not concentrate calcium. Calcium salts may accumulate in abnormal sites either from focal degeneration—dystrophic calcification—or as a consequence of a general disturbance [33, 34] of calcium metabolism—metastatic calcification (see Fig. 5-15).

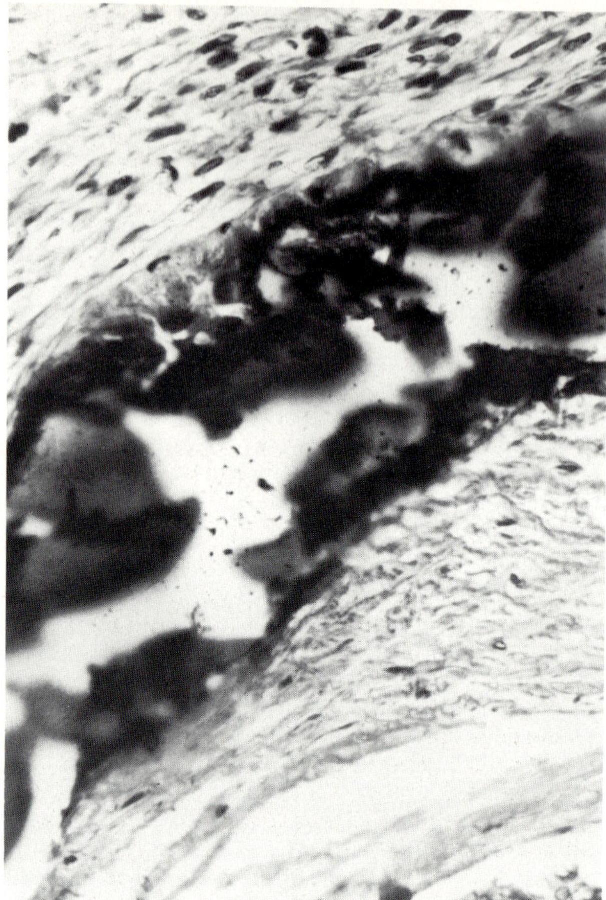

Fig. 5-15. Pathological calcification in arterial wall

The site of calcium accumulation in tissues is readily identifiable on hematoxylin-eosin slides because the protein in the calcified area develops a strong affinity for hematoxylin. But sometimes special stains must be used to identify calcium. Among those techniques, the von Kossa staining method is one of the most popular. The method is based on the precipitation of silver salt by the anionic groups present in the calcium deposit. Calcium can be demonstrated directly either by microincineration or by visualization of the hydroxyapatite crystals with the aid of the electron microscope.

Since calcium is deposited only in selected organs, the nature of the mechanism that triggers the abnormal precipitation of calcium salts is most intriguing. In metastatic calcification, one may well assume that the change in blood calcium, protein, and polysaccharide levels that accompany bone resorption may play a significant role in inducing the foci of calcification. Yet, although all tissues are bathed in the serum, only some selected areas become calcified. Consequently, calcium salt that precipitates undoubtedly comes from the serum, but calcification can be explained only by assuming that local factors trigger the precipitation of the salt. The factors invoked to explain the calcification of normal bone matrix are also invoked to explain the formation of metastatic calcifications. pH changes, citric acid accumulation, increase in alkaline phosphatase activity, and changes in the matrix ultrastructure have each been made responsible for the abnormal precipitation of calcium salts. At the present, the *primum movens* of calcium precipitation is unknown.

The sequence of events that lead to abnormal calcium deposition clearly indicates that local factors are involved in calcium precipitation. For example, if one produces hypercalcification by administering excessive amounts of vitamin D, all organs are not calcified uniformly, and in a given organ all components are not calcified simultaneously. In the heart, the internal elastic membrane of the myocardium is readily calcified, while the fibroelastic tissue of the endocardium is rather resistant to calcification. In the gastrointestinal tract, the stomach is a more common site of metastatic calcification than the esophagus. The kidney is almost always involved in experimental hypervitaminosis D. The calcification starts in the juxtaglomerular zone and spreads toward the center of the organ.

Dystrophic calcification occurs in two stages: calcium deposition and the reaction of the tissue to deposited salts. Again, it is not known what precipitates calcium. It is sometimes assumed that changes in the matrix will lead to crystallization of the calcium salts because the altered matrix acts as a seed or template for hydroxyapatite crystallization. We have seen that in bone the hydroxyapatite crystal is bathed in a supersaturated solution of calcium and phosphorus and serves as a seed for further crystallization of the calcium phosphate salt.

In dystrophic calcification, metals other than calcium (iron, magnesium, and manganese) are often found in the deposits. The reaction of the surrounding tissue toward the calcification varies considerably depending upon the tissue in which the calcium is deposited. The deposit may be sequestrated and surrounded by a capsule of fibrous tissue that isolates it from the rest of the organ. An epithelial layer may grow and surround the calcium deposits completely. Or vasculoconjunctive buds may invade the calcium deposit. According to Eisenstein, the intensity of that vascular proliferation may determine the fate of the deposit. If the vascular proliferation is intense, it will lead to resorption of the deposit if it is not marked, the calcium deposit remains an amorphous mass. A critical degree of vasculoconjunctive proliferation remodels the calcium deposit and converts fibroblasts to osteoblasts, and bone is formed. The differentiation of this ectopic bone may even produce bone marrow.

References

1. Scott, B.L.: The occurrence of specific cytoplasmic granules in the osteoclast. J. Ultrastruct. Res. **19**, 417–431 (1967)
2. Vincent, J.: Recherches sur la constitution de l'os adulte. Bruxelles: Editions Arscia 1955
3. Neuman, W.F., Neuman, M.W.: The chemical dynamics of bone mineral. Chicago: University of Chicago Press 1958
4. Young, R.A.: Dependence of apatite properties on crystal structural details. Trans. N.Y. Acad. Sci. **29** (series 2), 949–959 (1967)
5. Bourne, G. (ed.): The biochemistry and physiology of bone. New York: Academic Press 1956
6. Glimcher, M., Andrikides, A., Kossiva, D.: Studies of the mechanism of calcification. I. The alteration of amino acid side chain groups of collagen and its effect on *in vitro* calcification. In: Structure and function of connective and skeletal tissue. Proc. Symp. NATO Advanced Study Institute, p. 342. London: Butterworth and Co., Ltd. (1964)
7. Glimcher, M.: The macromolecular aggregation state of collgen and biological specificity. In: Calcification, connective tissue, thrombosis and atherosclerosis (Page, I., ed.), p. 97–141. New York: Academic Press 1959
8. Bell, N.H.: Dynamics of bone metabolism. Annu. Rev. Med. **18**, 299–312 (1967)
9. Saville, P.D.: Water fluoridation: effect on bone fragility and skeletal calcium content in the rat. J. Nutr. **91**, 353–357 (1967)
10. Wasserman, R.H., Corradino, R.A.: Metabolic role of vitamins A and D. Ann. Rev. Biochem. **40**, 501–532 (1971)
11. Olson, E.B., DeLuca, H.F.: 25-Hydroxycholecalciferol: direct effect on calcium transport. Science **165**, 405–407 (1969)
12. Horsting, M., DeLuca, H.F.: *In vitro* production of 25-hydroxycholecalciferol. Biochem. biophys. Res. Commun. **36**, 251–256 (1969)
13. Tanaka, Y., DeLuca, H.F., Omdahl, J., Holick, M.: Mechanism of action of 1,25-dihydroxycholecalciferol on intestinal calcium transport. Proc. nat. Acad. Sci. (Wash.) **68**, 1286–1288 (1971)
14. Brickman, A.S., Coburn, J.W., Norman, A.W.: Action of 1,25-dihydroxycholecalciferol, a potent kidney-produced metabolite of vitamin D in uremic man. New Engl. J. Med. **287**, 891–895 (1972)
15. Mawer, E.B., Taylor, C.M., Backhouse, J., Lumb, G., Stanbury, S.: Failure of formation of 1,25-dihydroxycholecalciferol in chronic renal insufficiency. Lancet **1**, 626–628 (1973)
16. Trummel, C.L., Raisz, L.G., Blunt, J.W., DeLuca, H.F.: 25-Hydroxycholecalciferol: stimulation of bone resorption in tissue culture. Science **163**, 1450–1451 (1969)
17. Schachter, D., Kowarski, S., Finkelstein, J.D.: Vitamin D_3: direct action on the small intestine of the rat. Science **143**, 143–144 (1964)
18. Dent, C.: Rickets and osteomalacia. In: Bone metabolism in relation to clinical medicine (Sissons, H., ed.). Proc. Symp., p. 78–87. London: Pitman Medical Publishing 1963
19. Arnstein, A.R., Frame, B., Frost, H.M.: Recent progress in osteomalacia and rickets. Ann. intern. Med. **67**, 1296–1330 (1967)
20. Greep, R.O., Talmage, R., eds.: The parathyroids. Proc. Symp. on Advances in Parathyroid Research. Springfield, Illinois: Charles C. thomas Publisher 1961
21. Park, E.A.: The imprinting of nutritional disturbances on the growing bone. Pediatrics **33** (Suppl.) 815–862 (1964)
22. Aurbach, G.D., Potts, J.T., Jr.: The parathyroids. Advanc. Metab. Disord. **1**, 45–93 (1964)
23. Aurbach, G.D., Potts, J.T., Jr.: Parathyroid hormones, editorial. Amer. J. Med. **42**, 1–8 (1967)
24. Stoerk, H.C., Carnes, W.H.: Relation of dietary calcium-phosphorus ratio to serum calcium and to parathyroid volume. J. Nutr. **29**, 43–50 (1945)
25. Aurbach, G.D., Chase, L.R.: Cyclic 3′, 5′-adenylic acid in bone and the mechanism of action of parathyroid hormone. Fed. Proc. **29**, 1179–1182 (1970)
26. Urist, M.R.: Osteoporosis. Ann. Rev. Med. **13**, 273–286 (1962)
27. Editorial: Nutrition and metabolic bone disease in the elderly. Nutr. Rev. **25**, 71–72 (1967)
28. Nordin, B.: Osteoporosis. Advanc. Metab. Disord. **1**, 125–151 (1964)
29. Potts, J.T., Jr.: Recent advances in thyrocalcitonin research. Fed. Proc. **29**, 1200–1205 (1970)
30. Munson, P.L., Gray, T.K.: Function of thyrocalcitonin in normal physiology. Fed. Proc. **29**, 1206–1208 (1970)
31. Munson, P.L., Hirsch, P.F., Brewer, H.B., Reisfeld, R.A., Cooper, C.W., Wasthed, A.B., Orimo, H., Potts, J.T., Jr.: Thyrocalcitonin. Recent Progr. Hormone Res. **24**, 589–650 (1968)
32. Marx, S.J., Woodard, C.J., Aurbach, G.D.: Calcitonin receptors of kidney and bone. Science **178**, 999–1001 (1972)
33. Gatter, R.A., McCarty, D.J.: Pathological tissue calcifications in man. Arch. Path. **84**, 346–353 (1967)
34. Giacomelli, F., Spiro, D., Wiener, J.: A study of metastatic renal calcification at the cellular level. J. Cell Biol. **22**, 189–206 (1964)
35. Borle, A.B.: Calcium and phosphate metabolism. In: Annual review of physiology (Comroe, J.H., Jr., Sonnenschein, R.R., and Zierler, K.L., eds.), vol. 36, p. 361–390. Palo Alto, Calif.: Annual Reviews Inc. 1974
36. Holick, M.F., DeLuca, H.F.: Vitamin D metabolism. In: Annual review of medicine (Creger, W.P., Coggins, C.H., and Hancock, E.W., eds.), vol. 25, p. 349–367. Palo Alto, Calif.: Annual Reviews Inc. 1974
37. Haymovits, A., Rosen, J.F.: Calcitonin in metabolic disorders. In: Advances in metabolic disorders (Levine, R., and Luft, R., eds.), vol. 6, p. 177–212. New York: Academic Press 1972
38. Rasmussen, H., Bordier, P. (eds.): The physiological and cellular basis of metabolic bone disease. Baltimore: The Williams & Wilkins Company 1974
39. Anderson, H.C.: Calcium-accumulating vesicles in the intercellular matrix of bone. In: Hard tissue growth, repair and remineralization. Ciba Symp. 11, p. 213–226. Amsterdam: Associated Scientific Publishers 1973

Chapter 6

Iron and Bile Pigment Metabolism

Iron Metabolism

The Greeks, it is said, made their weak children drink water in which a sword had been dipped. Whether this was done because the Greeks had some notion of the role of iron in hematopoiesis or because iron was a symbol of strength is not certain. My grandmother, who was less martial in therapeutic approach, made her grandchildren drink water in which nails had been left to rust.

Iron is a metal essential to metabolism. Many complex proteins are unable to function unless iron atoms in the proper state of oxygenation are inserted in the molecule. Without iron, globin and porphyrins are useless and may accumulate and cause disease. Yet only a half-century ago, the pathogenesis of a now well-known iron deficiency disease (chlorosis) was attributed to the tightness of the corsets young ladies wore.

Requirements

Iron requirements in men are very low [1, 2], ranging between 1.5 and 2 mg in the adult. The requirements are higher in the infant (3 mg daily) or the growing child (2–3 mg daily). The adult requirements vary considerably depending upon physiological conditions. For example, much more iron is needed during lactation, menstruation, and pregnancy. The additional daily requirements for iron are 0.72–1.6 mg during lactation, 1 mg during menstruation, and may reach as much as 200–400 mg during pregnancy. Obviously, if these physiological conditions are coupled with dietary deficiencies, malabsorption, or exaggerated iron loss, severe iron deficiency may result. The incidence of iron deficiency in the United States is surprising in view of the adequacy of the average American diet. Yet it has been claimed that 20% of pre-school children in New York City have low levels of hemoglobin due to iron deficiency. The deficiency results from poor dietary habits prevailing because of inadequate economical conditions. After kwashiorkor, iron deficiency is the most common nutritional disease.

Absorption

None of the ingested iron is absorbed in the stomach. Absorption starts in the duodenum and continues in the jejunum. Several factors control intestinal absorption of iron—hydrochloric acid in the stomach, reducing agents in the intestine, and a delicate control mechanism that originates in the intestinal mucosa or in the bone marrow [3–6].

The iron in food is either in the form of salt, mainly ferric hydroxide, or combined with organic compounds—for example, porphyrins. Most of the ingested iron is insoluble, and hydrochloric acid in the gastric juice facilitates the solubilization of iron. Furthermore, ferrous iron is more readily absorbed than ferric iron. Because the reduced iron is absorbed at a rate 50 times greater than is the oxidized form, the presence of reduced glutathione, cysteine, and ascorbic acid in the intestinal lumen is of the greatest importance for iron absorption.

The more complex the iron salt, the more difficult its absorption. Thus, iron salts like ferric verseneate and ferrous pyrophosphate are only slightly absorbed. In contrast, ferrous sulfate and ferrous succinate are readily absorbed in the intestine.

The passage of iron through the cell membrane was investigated *in vitro* with everted intestinal pouches and liver slices. The results obtained for the intestine are conflicting. One group of investigators (using the middle and lower portion of rat intestine) reports that iron diffuses passively from the lumen into the intestinal cell, another group (using the upper intestine) claims that iron transport through the cell membrane is active. The second group found that iron transport was inhibited by the usual antimetabolites—fluoride, azide, malonate, cyanide, and 2,4-dinitrophenol.

Of even more interest, especially with respect to the control of iron absorption, was the observation that iron transport is also inhibited by anoxemia. According to a recent theory, the intestinal mucosa secretes a sugar of unknown nature, which complexes with the intraluminal iron and facilitates its penetration through the cell membrane [7, 8].

Studies by Dowdle and his collaborators describe two steps in iron transport into the intestine: mucosal uptake and serosal transfer. Both the uptake and the transfer seem to depend on oxidative metabolism and take place mainly in the duodenum. Uptake is rapid and involves both ferrous and ferric ions; transfer is slow and involves ferrous ions only [9].

Reduced iron in the intestinal cell has at least a dual fate—it either induces the appearance of a specific protein, called apoferritin, to which it then combines to yield ferritin, or it is oxidized to yield colloid ferric hydroxyphosphate.

Ferritin Structure and Biosynthesis

Ferritin* is a protein that is 17–20% iron, all in the ferric state. It has been characterized by crystallography, ultracentrifugation, and isoelectric and electrophoretic methods, and the molecule has been seen with the aid of the electron microscope. Ferritin is an ellipsoid molecule. The inorganic core is coated by the protein shell. The inorganic component of ferritin can be dissociated from the protein moiety by reduction of the metalloprotein with cysteine, ascorbic acid, or sodium dithionite.

* One should not exclude the possibility that various types of ferritin exist. Human bone marrow contains two types of labeled ferritin after injection of radioiron. One is a fast ferritin with greater electrophoretic mobility, and the other is a slow ferritin that has electrophoretic mobility identical to that of the ferritin found in the spleen.

Studies on the amino acid composition of ferritin could have permitted the prediction that the helical conformation of the molecule could be very high. Indeed, ferritin is low in valine, isoleucine, threonine, serine, and especially proline residues (1.2% of all amino acid), which would seriously disrupt helical regions. Studies of the optical rotary dispersion properties of ferritin established that the molecule is likely to contain 50% helical conformation. Once iron is removed from ferritin, the optical changes indicate further folding of the molecules, and although one can use ammonium sulfate and chromotographic analysis to separate ferritins with various iron content the changes in iron content do not affect the rotation properties of ferritin.

Apoferritin (mol wt 465,000–480,000) is a polymer composed of 20–24 subunits. The subunits can be separated by several means, including the use of detergents. Each subunit contains 19 different amino acids, and the amino acid composition of apoferritin has been elucidated. But, the N- or C-terminal amino acids have not been identified because of technical difficulties.

To form ferritin, micelles of ferric hydroxide phosphate are included in the apoferritin molecule. If ferritin is reacted with Fe^{++} in an HCO_3 buffer, a phosphorus-free ferritin is obtained. The reaction between iron and apoferritin requires that the iron be in the ferrous form, but after its inclusion into the apoferritin molecule, the ferrous iron is oxidized to ferric iron by air.

In the ferritin molecule, iron has only three unpaired electrons, and consequently its coordination number is 4 (instead of 6, as is the case for most other biological compounds). Each coordination number is attached to four points of the protein ligand in a square shape.

Already in 1955, when studies on the induction of protein synthesis in mammalian tissue were in their infancy, Fineberg and Greenberg demonstrated that iron administration induced apoferritin formation [10]. Later, Loftfield and Harris [11] demonstrated that the amino acids valine, isoleucine, and leucine were incorporated in the newly synthesized protein. These experiments are in fact of considerable historical significance to modern knowledge of protein synthesis.

In the late 1940's and early 1950's, considerable efforts were made to elucidate the pathway of protein synthesis. Most experiments done in that realm involved the use of labeled amino acids. This raised a magistral problem. Were the amino acids used for the synthesis of new proteins, or were they used for exchange with amino acids already present in the molecule? In the first case, one expects the specific activity of the amino acid released from the synthesized protein to be identical to the specific activity of the amino acid used in the experiment. In the second case (exchange), it is not likely that all amino acids with molecular structures identical to the one injected will be replaced. Consequently, after the protein molecules are digested and the amino acids have been separated, the injected amino acid molecule (labeled) will be mixed with those unexchanged amino acid molecules (unlabeled), and the specific activity will be reduced. Loftfield and his associates established that incorporation in the case of ferritin was indicative of *de novo* synthesis.

A delicate control mechanism for the biosynthesis of apoferritin has been proposed. The cell does not contain free apoferritin—iron induces the synthesis of apoferritin. Apoferritin exerts a feedback inhibition on its own biosynthesis, which is relieved when apoferritin is converted to ferritin. Of course, the mechanism of induction at the nucleic acid level remains to be studied.

The conversion of apoferritin to ferritin probably involves the chelation of iron molecules, a reaction in which the apoferritin acts as a biological ligand. The molecular arrangement of the inorganic component of ferritin is not known with certainty. It is postulated to be a complex of FeOOH and small amounts of phosphate $(FeOOH)_8$ and $(FeOO\text{-}PO_3H_2)$. Ferritin can be formed *in vitro* simply by mixing apoferritin with a ferrous salt [12].

Ferritin formation in mammalian tissues has been investigated in liver slices and homogenates. In liver slices, the incorporation of labeled plasma iron into ferritin is stimulated by adding a Krebs cycle substrate to the medium and is inhibited by iodoacetamide, cyanide, and anoxemia. These findings suggest that the conversion of apoferritin to ferritin is energy dependent.

Release of Iron from Ferritin

Crystallized ferritin was incubated with labeled plasma-bound iron and liver homogenate. The exchange between the ferritin and plasma-bound iron was found to require ATP and ascorbic acid. The *in vivo* mechanism by which iron is released from ferritin to the plasma is not known. Two different *in vitro* mechanisms have been described: (1) ferritin iron is reduced and then released by biological reducing agents (ascorbic acid, glutathione, and NADH); (2) two observations have led to the suggestion that the xanthine content and xanthine oxidase activity of blood, liver, and the intestinal tract may play an important role in controlling the release of iron from ferritin.

Xanthine oxidase reduces the ferric iron of ferritin to yield a ferrous ion. Hypoxanthine injection leads to eleveted levels of plasma iron, probably by inducing the appearance of xanthine oxidase.

Some puzzling observations need to be explained before xanthine oxidase can be said to have a role in the *in vivo* release of iron. Among these objections is the fact that even *in vitro*, only a small amount of iron is released from ferritin. *In vitro*, oxygen stimulates the release of iron from ferritin. This is intriguing in view of the fact that *in vivo* anoxemia seems to facilitate the release of iron from ferritin and to accelerate iron absorption.

Transferrin

Ingested radioiron appears in plasma within 30 minutes. In blood, iron does not circulate free but is chelated by specific proteins, the transferrins. The ferrous iron transferred from the intestinal cell to the plasma (1–2 mg/day) is oxidized to yield ferric iron, which is chelated by an "apotransferrin" to yield a pink ferroprotein complex.

Transferrins or siderophilins are a group of iron binding glycoproteins found in plasma. They can be concentrated by electrophoresis or alcohol precipitation. The electrophoretic mobility of transferrins is that of β_1-globulin; after alcohol fractionation of plasma, transferrins are found in fraction IV. Transferrins contain 0.13% iron, and it is estimated that each protein molecule (mol wt 90,000) forms strong ionic bonds with two atoms of iron to yield a salmon pink complex with an absorption maximum at 460–470 mμ.

Thus, the transferrin molecule contains two chelating sites. In chelating iron, the protein releases three protons. Titrimetry and ESR measurement have provided some clues as to the amino acid residues involved in iron binding. ESR studies show that an imidazole ring is involved; titrimetry has shown that the three protons are released from tyrosyl residues. Starch-gel electrophoresis of transferrin preparations has revealed the existence of at least eight different protein molecules with transferrin properties.Correlation of the dissociation of transferrin on starch-gel and genetic studies has demonstrated that the appearance of a transferrin with a given mobility is determined by a specific gene, and each of the eight types of transferrin is determined by a single allele [13].

Reducing the pH releases the two iron atoms that are chelated by one molecule of transferrin. Ferritransferrin can readily be regenerated, but if transferrin is placed in the presence of copper instead of iron, a cupritransferrin is formed, which also contains two metal atoms. The iron found in transferrin can also be released by ascorbic acid and ATP. Whether these compounds facilitate iron release *in vivo* is not known.

For each Fe atom that is bound to transferrin a bicarbonate anion is also bound. The exact role of the bicarbonate in the binding of iron to transferrin is unknown. It is, however, established that the binding and the release of iron and bicarbonate from transferrin are tightly coupled.

Iron Secretion

Only small amounts of iron are lost in bile, feces, desquamating cells, sweat, and urine. Consequently, the amount of iron lost daily is no greater than 10–15 mg. Since iron intake is largely in excess of these losses, iron deficiency occurs only under conditions of pathological or physiological stress. This one-way metabolic pathway is rather unique for iron—a metal that enters the body and practically never leaves it. Thus, in contrast to most other metabolites the homeostasis of which is regulated at least in part by the level of excretion, iron's homeostasis is regulated by the level of its absorption.

Iron excretion through the kidney can be stimulated by the administration of chelating agents, such as BAL and EDTA. The mechanism by which the iron stores are mobilized by the chelating agents is not clear, but the effect is useful for clearing the iron accumulated in tissues (for example, after repeated transfusions). The mechanism of iron excretion by the cells of the mammary gland into the milk is not known. The iron concentration in milk is higher than in plasma, and the milk concentrations decrease with the maternal iron stores [4].

Tissue Iron

From blood, iron is transferred to tissue, where it may be: (1) used for hemoglobin biosynthesis (hematopoietic tissue); (2) stored in a variety of tissues (about 100–300 mg of the iron); or (3) used for the biosynthesis of such enzymes as cytochromes, peroxidases, and catalases.

The iron content of many tissues has been determined. These data will be considered in the section on hemosiderosis and hemochromatosis. The iron content of the fetus and of the bone marrow is particularly significant. Determination of the iron content of the bone marrow has become valuable in detecting iron deficiency anemia. Compared to iron stores in other tissues, the bone marrow iron is the last to be depleted in deficiency, and the last to be restored after iron administration [4].

The iron concentration in fetal blood is $1^1/_2$ times greater than that in maternal blood. This can be explained only if an active mechanism in the placenta transfers iron from maternal to fetal blood against a concentration gradient. The placental barrier is also known to actively absorb vitamins B_6 and B_{12}. It is unnecessary to emphasize the pressing need for a better

Table 6-1. Data on body iron

Dietary iron	10–15 mg/day
Iron absorbed (about 10%)	1.0–1.5 mg/day
Iron excreted or lost	
Normal adult male	0.5–1.5 mg/day
Normal adult female	1.0–2.0 mg/day
During growth	0.8–2.1 mg/day
During pregnancy	3.8 mg/day
Body iron (Adults)	
Hemoglobin	2000 (1500–3000) mg
Storage iron (Ferritin and hemosiderin)	1250 (1000–1500) mg
Tissue iron complexes (Catalase, cytochromes peroxidase, myoglobin, etc.)	200 (100–300) mg
Transferrin	1 (0.5–2) mg
Total	3451 (2600–7500) mg

understanding of the "placental pump." It is vital to the adequate interpretation of the physiology and pathology of the fetus. The distribution of iron in a man weighing 70 kg is given in Table 6-1.

The transfer of iron from blood to hemoglobin synthesizing cells has not been investigated extensively. The work of Fielding and Speyer suggests that the iron transferrin attaches with a surface receptor to form (B2) a complex which has been isolated from the reticulocyte membrane and has a molecular weight of 230,000. Before it reaches hemoglobin, the final acceptor, it appears that the iron is transferred to two intermediate iron acceptors (B1 and C); one located in the membrane, the other in the cytosol. Thus p-hydroxymercuribenzoate does not interfere with the formation of the iron transferrin B2 complex, but blocks the transfer to the intermediates with accumulations of the iron transferrin B2 complex [45].

Iron Metabolism and the Red Cell

Hematopoietic System

Alterations of iron metabolism are intimately related to diseases of the hematopoietic system. Therefore, it is necessary to review some aspects of hematopoiesis before we discuss iron deficiency in detail.

The hematopoietic system consists of several organs distributed throughout the body and mainly comprises bone marrow, lymph nodes and lymphoid centers, spleen, and thymus. Its most obvious role is to provide the organism with the cellular constituents of the blood. In comparison with liver parenchymal cells, the life span of the circulating blood cell is relatively short. To maintain the steady state of the blood, hematopoiesis needs to be active even throughout adult life [14, 15].

During embryonic life, hematopoiesis can be divided into three main stages: prehepatic period, hepatosplenic, and bone marrow periods. During the prehepatic period, small islets of cells (the islets of Pander-Wolf) develop rapidly. The most superficial cells flatten, become the angioblasts of Hiss, and form tubules that are at the origin of the vascular endothelium. In contrast, cells occupying a central position in the islets acquire a round shape, become isolated, and form the primitive erythroblast. The erythroblast is rich in cytoplasm and has a large nucleus with coarse chromatin and prominent nucleoli. The cells formed in the embryo are larger than those found in the adult and are therefore often called megaloblasts and megalocytes, depending upon the degree of maturation. The megalocytes disappear later and are replaced by the secondary erythrocytes or normocytes.

The hepatosplenic period starts 3 months after the beginning of pregnancy and is characterized by an intensive hematopoiesis in spleen, thymus, and liver, leading to the formation of white cells and red cells identical to those found in adult blood. Before this hepatosplenic period is over, active hematopoiesis starts in the lymphatic nodes and in the bone marrow.

Five months after the onset of pregnancy, only the bone marrow retains hematopoietic properties under normal conditions. This stage is called the period of myeloid hematopoiesis. Thus, the bone marrow, which at first provided the cells of the white lineage primarily, soon becomes the main source of all the cellular elements in the blood. However, the lymph nodes and the spleen may reactivate their myelopoietic properties under specific physiological or pathological conditions.

The hematopoietic organs contain a network of richly interconnected fibers. The chemical nature of the fiber is not established, but the fiber has the staining properties of reticulin. Cells resembling fibroblasts are found along the network, and a great number of different cells are in the meshes of these webs. All these cells are hematopoietic cells at one or another stage of maturation: mature cells (lymphocytes, polymorphonuclears) and transitional forms between the mother and the mature cells. In adult humans, only the bone marrow presents all hematopoietic potentialities. Under normal conditions, the lymph nodes, the spleen, and the thymus are responsible for only lymphopoiesis.

The importance of the erythron is illustrated by its size and its capacity to regenerate. In fact, the erythron is the largest organ of the body. It is composed of 2 liters of circulating blood and 1.5 liters of bone marrow. The cell population of blood is completely renewed every 4 months at the rate of approximately 1% a day. Three phases can be distinguished in the life of a red cell: (1) production and maturation in the bone marrow; (2) function in the circulating blood; and (3) disappearance into the reticuloendothelial system.

Properties of the Red Cell

The red cell is a circular, biconcave disc measuring 8.6 μ in diameter and 2 μ in thickness. This shape presents considerable advantages. It provides a surface-volume ratio that is 9 times that of a sphere of equal volume. This is particularly suitable for rapid transfer of oxygen. The erythrocyte is so soft and flexible that it readily passes through the most narrow capillaries, even though the lumen of some capillaries is narrower than the diameter of the red cell itself. The red cell is not truly a cell as commonly defined. It does not contain a nucleus except in fish, amphibians, reptiles, birds, and some mammals, such as the camel and the llama. Red cells are intensely stained by acidophilic stain due to the presence of hemoglobin. However, basic stains are often useful because they permit detection of nuclear remnants. The size of the red cell varies from one animal to another; it generally decreases in animals that occupy a higher position in the scale of evolution. Thus, whereas the erythrocytes of man measure 7–8 μ in diameter, those of frogs measure

23 μ; of Tritons, 30 μ; and of salamanders, 40 μ. In man, the size of the red cells varies with age; for instance, a certain degree of anisocytosis is found in newborn children whose blood may contain erythrocytes of 5–6 μ in diameter. Of course, anisocytosis is a common symptom in many forms of anemia.

The number of red cells also depends on a variety of conditions. Animals higher on the evolutionary scale generally have more red cells. For instance, while the frog has 250,000 red cells per mm^3, the leopard has 1,375,000 red cells per mm^3, and man has 5,000,000 red cells per mm^3.

The red cell can be defined as a highly specialized structure containing hemoglobin in solution. Hemoglobin represents 32% of the wet weight of the red cell. In fact, the concentration of hemoglobin in the red cell is so great that further increase would hinder the rate of the reaction between hemoglobin and oxygen. The hemoglobin concentration apparently is not uniform throughout the red cells—the concentration increases from the periphery to the center.

In addition to hemoglobin, the red cell contains 65% water (see Tables 6-2 and 6-3). Thus, except for water, hemoglobin is by far the most important biochemical. In spite of the large amounts of hemoglobin found in the red cell, one should not underestimate the importance of the remaining 3–7%, which is made up mainly of proteins and lipids. The proteins provide several multiple-enzyme systems whose functions are vital to the erythroblast. For these reasons, the histochemical and biochemical properties of the red cell at various stages of maturation will briefly be reviewed.

Table 6-2. Type of proteins found in red cell

1. Stromal	2. Catalytic Glycolytic Catalase Carbonic anhydrase Cholinesterase, etc.	3. Antigenic

Table 6-3. Chemical composition of the red cell

1. H_2O	65%	3. Proteins	3–7%
2. HbO_2	32%	4. Lipids	3–7%

During the mother cell's transformation into a mature erythrocyte, various different morphological types appear. The morphological differences are of capital importance for the diagnosis of hematological disorders. Therefore, the characteristic of each stage has been summarized in the Table 6-4 and presented schematically in Fig. 6-1. But the recognition of these characteristics sheds no new light on the mechanism of such differentiation. Cytochemical studies have given more precise information on the chemical nature of these changes without clarifying the mechanism of differentiation. Thorell has examined the cells of the erythropoietic lineage with the ultraviolet microscope and has studied light absorption of various wavelengths after submitting the preparation to specific staining procedures, such as the Feulgen reagent, which reacts with deoxyribonucleic acid. As could be anticipated, he found DNA in the nucleus and RNA in the nucleolus and the cytoplasm of the immature red cells. Both nucleic acids progressively disappear during maturation. Nucleic acid elimination is accompanied by an increase in hemoglobin concentration. Despite their importance, these histochemical studies only crudely represent the biochemical events that accompany red cell maturation.

The immature red cell has a complete arsenal of cellular organelles—nucleus, mitochondria, and endoplasmic reticulum. These structures can perform their expected biochemical functions, and thus the immature red cell synthesizes DNA, RNA, and proteins. In most cells oxidation of metabolites through the Krebs cycle and transfer of energy through the electron transport chain provide most of the energy necessary for these biosynthetic pathways. Most of the hemoglobin found in the mature red cell is synthesized at that stage. (The biosynthesis of globin and of heme has been discussed in the chapter on inborn errors of metabolism.)

During transformation of the stem cell into mature erythrocytes, many of these cellular structures are lost and their basic biochemical function is lost with them. The mechanism by which nuclei, endoplasmic reticulum, and mitochondria are eliminated is not known. The elucidation of that process would, however, be vital to an understanding of differentiation and cellular death.

The biochemistry of reticulocytes is of particular interest [16]. The reticulocyte is intermediate between the stem cell and the mature erythrocyte. Remnants of the basic structure found in the stem cell can be detected histochemically. Our knowledge of the biochemistry of the reticulocyte is still incomplete.

While the reticulocyte matures, important changes in its lipid composition take place. These changes are great enough to lower specific gravity as the reticulocyte is converted into an erythrocyte. Several investigators have taken advantage of the differences in the specific gravity of the reticulocyte and the mature erythrocyte to separate the two types of cells by differential centrifugation. This is not the only available method of separation; indeed, the mature erythrocyte is much more susceptible to disruption by osmotic shock than the reticulocyte, which is more resistant to hypotonic solutions. The appearance of reticulocytes in the blood can readily be induced by rendering the animal anemic. This is achieved experimentally either by administering toxins, such as phenylhydrazine, or by repeated bleeding.

The difference between the enzyme compositons of the reticulocyte and the erythrocyte will become more apparent if we briefly consider the enzyme composition of the erythrocyte. The erythrocyte seems to have lost most of its biosynthetic pathway. Indeed, it no longer makes DNA, RNA, and proteins. Even its bioenergetic pathways seem to have been reduced to the bare mini-

Table 6-4. Maturation of red cell

Cell type	Nucleus	Nucleolus	Cytoplasm	Percent in marrow
Hemocytoblast	Large, delicate chromatin	1+	Light hyaline blue	Rare
Proerythroblast (Pronormoblast; rubriblast)	Large, fine reticular chromatin	1–2	Scanty, deep basophilia	0–2
Early erythroblast (Basophilic normoblast; prorubricyte)	Large, coarser chromatin	0–1	More abundant, moderately basophilic	1–5
Late erythroblast (Polychromatophilic normoblast; rubricyte)	Smaller, coarse chromatin condensation	0	More abundant, patchy, irregular, acidophilic-basophilic	3–10
Normoblast (Orthochromatic normoblast; metarubricyte)	Smaller, condensed chromatin; pyknotic	0	Relatively abundant, largely hemoglobinized, acidophilic	7–20
Polychromatophilic erythrocyte (Diffusely basophilic erythrocyte)	Absent	0	Mixed acidophilic-basophilic, supravital stains show granulo-filamentous reticulum	1–5

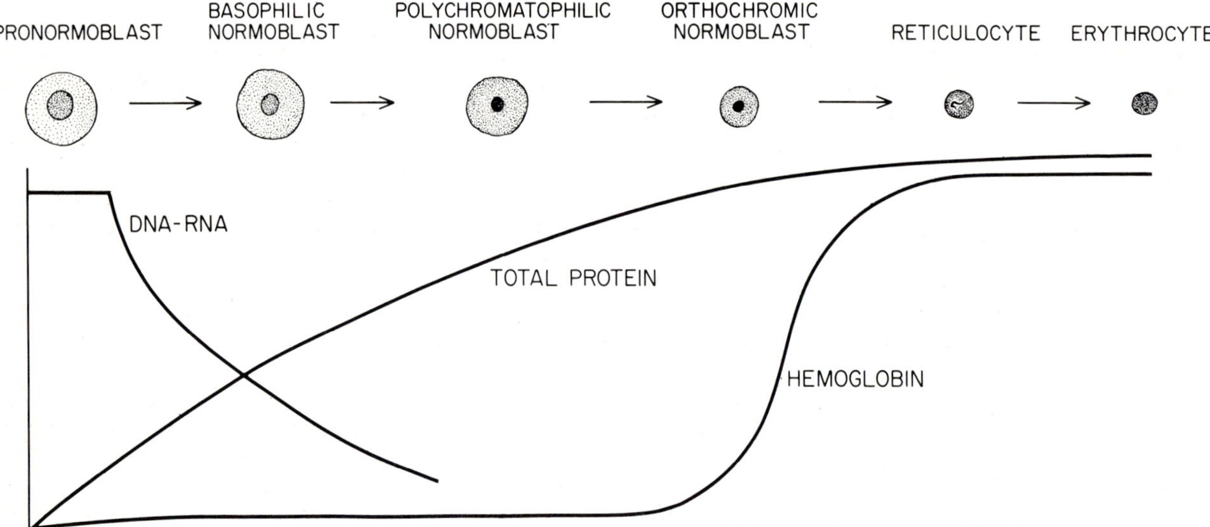

Fig. 6-1. Maturation of a red cell

mum. There is no Krebs cycle, no classical electron transport chain, and no oxidative phosphorylation. However, all the enzymes required for glycolysis are present in the red cell. Thus, anaerobic glucose oxidation is the main source of energy in the mature red cell. The penetration of glucose in the red cell is not controlled by insulin, and probably occurs by free diffusion. The membrane of the red cell is thus freely permeable to glucose, the concentration of which in the erythrocyte may reach up to 600 mg/100 ml. After phosphorylation, through the catalytic action of hexokinase, the glucose is further catabolized through all the classical steps of the Embden-Meyerhof pathway, yielding lactic acid as the final product. Of course, ATP is formed as glucose is oxidized.

In the normal glycolytic pathway, 3-phosphoglycerate may either be further phosphorylated to yield 1,3-diphosphoglycerate, which is then converted to pyruvate, or it may be converted to 2-glycerophosphate in a reaction catalyzed by phosphoglyceromutase, which, like the mutase reaction that catalyzes the conversion of glucose-6-phosphate to glucose-1-phosphate, requires catalytic amounts of a diphosphate. Normally, small amounts of 2,3-diphosphoglycerides are formed.

In the red cell, up to 50% of the organic phosphorus exists in the form of 2,3-diphosphoglycerate. In fact, most tissues synthesize more 2,3-diphosphoglycerate than is necessary for adequate phosphoglyceromutase activity; and therefore the 2,3-diphosphoglycerate is assumed to play a specific biological role. Ropaport and his associates have proposed that in the red cell, 2,3-diphosphoglycerate stores high-energy bonds.

Organic phosphates decrease the oxygen affinity for hemoglobin. Since 2,3-diphosphoglycerate is the most abundant of the organic phosphates, variations in its concentration will modulate oxygen affinity. Increasing levels of 2,3-diphosphoglycerate therefore facilitate the unloading of oxygen to tissues. This effect is achieved by binding of 2,3-diphosphoglycerate to deoxyhemoglobin, stabilizing the quaternary structure of deoxyhemoglobin which in contrast to oxyhemoglobin quarternary structure has a low affinity for oxygen. One molecule of organic phosphate binds to one tetramer of hemoglobin at the entrance of the central cavity between the N terminal amino group of the β chain. It involves the formation of ionic bounds between 8 cationic groups of hemoglobin (4 on each β chain) and the anionic groups of 2,3-diphosphoglycerate. The cationic groups of the β chain are donated by the amino group of valine 1, the basic amino acid, histidine 2, lysine 82 and histidine 143. Whether variations of the concentration of 2,3-diphosphoglycerate affect health is not certain. A relationship between the risk of developing arteriosclerosis and high oxygen affinity has, however, been suggested [46, 47].

The ATP requirements of the red cells are low because most biosynthetic pathways are nonexistent, and ATPase is not very active. All the enzymes of the hexose monophosphate shunt are found in the red cell, and it has been estimated what while 90% of glucose that enters the red cell is oxidized through the Embden-Meyerhof pathway, 10% is oxidized through the hexose monophosphate shunt. In the red cell, the shunt pathway is the only known source of NADPH.

If 90% of the glucose that enters the red cell is metabolized by the glycolytic pathway, and 10% is oxidized through the hexose monophosphate shunt, the end product of glucose metabolism in the red cell is primarily lactic acid. Some compounds divert the use of glucose from the glycolytic pathway to the pentose phosphate shunt. They include, for example, methylene blue, the degradation product of primaquine, ascorbic acid, and cysteine. For each molecule of glucose consumed, two molecules of lactic acid and four molecules of ATP are produced, with a net production of two molecules of ATP. Part of the ATP produced is probably used to transport sodium and potassium across the cell membrane. The NADPH formed by the conversion of lactic acid to pyruvic acid is needed for the reduction of methemoglobin. The main product of the pentose phosphate shunt in the red cell is NADPH, which compound is necessary for the reduction of glutathione through the catalytic action of glutathione reductase, and for the reduction of methemoglobin to hemoglobin.

Although the glycolytic pathway and the hexose monophosphate shunt constitute the core of the enzyme mosaic of the red cell, other enzymic systems, the roles of which sometimes are less obvious than those involved in the anaerobic oxidation of glucose, have also been found.

In spite of the absence of integral mitochondria, the red cell contains some of the enzymes (fumerase, isocitric dehydrogenase, malic dehydrogenase, and cytochrome oxidase) functioning in the Krebs cycle and electron transport. These enzymes probably represent mitochondrial remnants, and their presence in the mature erythrocyte may be a consequence of their greater stability. Similarly, enzymes concerned with amino acid metabolism—such as glutamic, oxalacetic, and pyruvic transaminases, and glutaminase—are also found in the red cell.

Hydrolases, such as acid and alkaline phosphatases, esterase, and cholinesterase, have also been found in the red cell. As in most tissues their role is unknown, nor has it been established whether they are like some other tissue hydrolases associated with cellular organelles.

Glutathione metabolism might be of particular importance to the metabolic integrity of the red cell [17]. Glutathione is a rather simple compound; it is a tripeptide containing glycine, cysteine, and glutamic acid (GSH). During oxidation, two molecules of glutathione are linked together through the formation of an S–S bridge (GS–S–G). The oxidized glutathione can be reduced by the catalytic action of an enzyme, glutathione reductase, in the presence of NADPH as a hydrogen donor. Because in the red cell NADPH is produced by the shunt pathway, glutathione and glucose metabolism are closely linked in the red cell. We shall see later that glutathione has often been implicated in red cell diseases.

Reductase has not been extensively purified from erythrocytes, and the best preparations have been obtained from liver. In that organ, the enzyme is assumed to contain SH groups and is postulated to act as follows:

$$\text{Enzyme SH} + \text{GSSG} \rightarrow \text{Enzyme} - \text{SSG} + \text{GSH}$$
$$\text{Enzyme SSG} + \text{NADPH} \rightarrow \text{Enzyme} - \text{SH} + \text{GSH} + \text{NADP}$$

Glutathione reductase has been purified. The protein contains flavin and sulfhydryl groups. The enzyme retains 10% of its activity after hydrolysis of 60% of the peptide bonds by pronase.

The exact number of sulfhydryl groups in the enzyme molecule is not known. One school holds it to be two, another claims that it is three (one active in the oxidized enzyme, one revealed by the addition of NADPH, and one that becomes accessible to N-ethylmaleimide after exposure of the protein to high concentrations of urea.

The sharp flavin absorption band at 462 mμ is replaced by a broader absorption (between 520 and 700 mμ) when the enzyme is reacted with NADPH.

All the proteins found in the red cell are probably not catalytic; some play a structural role, particularly in building up the stroma and membrane. The stroma of the erythrocyte, which constitutes 2–5% of the wet weight of the cell, is made of proteins and lipids. The protein is called stromatin, and it contains at least nine different amino acids. Stromatin apparently is related to keratin and carries the blood group antigenic character. Ten per cent of the stroma seems to be composed of lipids; among them are cephalin, lecithin, and cholesterol.

The red cell is undoubtedly surrounded by a lipoprotein membrane that has been separated from the soluble components of the erythrocytes. The membrane material is assumed to represent 1% of the cell weight, and its thickness was calculated as 52 A, a value in agreement with electron microscopic observations. (The structure of the red cell membrane is discussed in the chapter on membranes.)

In addition to protein and lipids, the red cell also contains electrolytes: sodium, potassium, chlorides, carbonate, and others. Their exchanges are important in maintaining the integrity of the red cell structure and physiology.

The reticulocyte differs from the erythrocyte in that: (1) it contains small amounts of DNA; (2) it contains amounts of RNA nearly equal to those in the stem cell; (3) it contains intact tricarboxylic acid cycle and electron transport chain; and (4) it can synthesize protein, but at a rate much slower than that observed in the stem cell. Some other enzymes that are also present in erythrocytes have been claimed to have higher activities in reticulocytes. Such is the case for glucose-6-phosphate dehydrogenase, 6-phosphogluconic dehydrogenase, phosphohexose isomerase, transketolase, aldolase, and lactic dehydrogenase.

The Fate of the Molecular Component of the Red Cell

The red cell is relatively short-lived in humans; its normal life span ranges between 110 and 130 days. Its graveyard is the reticuloendothelial system, mainly that of the spleen. The terminal fate of the red cell raises many problems, among them are: (1) the nature of the biochemical modification that determines cellular death; (2) the nature of the mechanism by which the reticuloendothelial cells separate the dying cells from all other cells that come in contact with the reticuloendothelial cell surface; and (3) the mechanisms of breakdown of the intracellular components. Of these questions, only the last two have been partially answered, but the fate of the heme of the hemoglobin molecule will be more appropriately discussed when we review jaundice. The importance of understanding the nature of the molecular mechanism that leads to the death of the red cell should not be underestimated. If it could be elucidated, better methods for blood storage could possibly be devised.

If blood is stored without adding protective agents, the red cells deteriorate rapidly, and there is then a reduction in the rate of glucose oxidation, associated with a decline in organic phosphates and an increase in inorganic phosphates. This inability of the stored red cell to maintain the integrity of its phosphorylated derivatives leads to an incapacity to provide the chemical energy necessary to maintain its structural integrity—particularly that of the membrane, which soon loses its ability to concentrate potassium. Only the glycolytic enzymes are affected by storage; the enzymes of the pentose phosphate shunt remain active even after prolonged storage periods. The addition of inosine rapidly restores the activities of the glycolytic pathway; therefore, it increases the proportion of organic phosphate and reduces the accumulation of inorganic phosphate.

Little is known of the biochemical changes that occur in the red cell during aging *in vivo*. These studies on the aging red cells are usually done on transfused red cells that are separated from the red cells of the recipient by specific agglutination.

The first detected change, which occurs after approximately 30 days of age, is a 50% drop in the activity of glyceraldehyde-3-phosphate dehydrogenase. During the period, the activity of glucose-6-phosphate dehydrogenase also decreases, but the half-life of the latter enzyme is around 50 days.

Hexokinase activity decreases progressively after the 30th day and reaches 60% of the original activity approximately when the erythrocyte is 90 days old. It is not known whether the loss of enzyme activity results from changes in the rate of enzyme breakdown or from the appearance of inhibitors. A dual mechanism seems to be responsible for hexokinase's loss of activity; namely, a net reduction in the number of enzyme molecules and the accumulation of intracellular ADP leading to inhibition of the enzyme.

Indeed, concomitantly to the loss of enzyme activity, there is, as might be expected, a drop in ATP and other phosphorylated compounds. In addition to a loss of dehydrogenase and hexokinase, the concentration of NAD drops rapidly. It is again not known if this results from accelerated breakdown or from a reduced rate of biosynthesis. Those that postulate that the reduced NAD in red cells results from interference with its synthesis have proposed several mechanisms to explain the reduced rate of biosynthesis of NAD, including: (1) a reduction in ribose-5-phosphate resulting from the impaired activity of the hexose monophosphate shunt; and (2) a decrease in the activity of the enzymes involved in the biosynthesis of the coenzymes.

Whatever the mechanism, the NAD concentration of the aging red cell drops to 20% of its original value between the 50th and 90th day—a highly deleterious situation. NADH is necessary to reduce methemoglobin; consequently, a red cell deprived of NAD is soon an inefficient oxygen carrier.

In conclusion, in spite of the fact that compared with other cells the erythrocytes have a low metabolic activity, they do generate considerable amounts of energy. Information obtained from pathological and physiological observations of the red cell indicate that the generation of energy has a triple role: (1) to maintain membrane integrity and thus prevent the development of steep changes in gradients or distortion of osmolarity; (2) to preserve the discoid shape of the cell against forces that try to make it spheroidal; and (3) to maintain hemoglobin in a reduced state. Because oxidized hemoglobin cannot carry oxygen, methemoglobin formation must be prevented.

Iron Metabolism in the Erythron

A description of the erythron in terms of a sequence of morphological stages represents only a small part of the whole picture. The erythron is a dynamic organ that produces erythrocytes with a limited life span; therefore, the lost erythrocytes must constantly be replaced by active division and differentiation of the bone marrow cells. A comprehensive description of these events requires detailed knowledge of the kinetics of cellular replacement. This is a problem that cannot be resolved by biochemical methods because of the heterogeneity of the cellular population. The use of radioautographic techniques has contributed considerably to our understanding of the kinetics of the erythron.

A stem cell is converted to an erythrocyte in a series of cellular modifications including proliferation, cytoplasmic maturation, function, and death. During differentiation, the red cell mobilizes all its resources mainly to synthesize as much hemoglobin as can be concentrated in the volume occupied by the erythrocyte. To achieve this goal, the red cell eliminates most of its components (nucleus, endoplasmic reticulum) and retains only hemoglobin and small amounts of enzymes.

During the functional stage, hemoglobin carries oxygen to the tissues. Proliferation, specialization, and function occur in sequence, but the cellular stages overlap; proliferating cells or reticulocytes, for example, synthesize hemoglobin.

In studying the kinetics of the erythron, one is concerned mainly with the proliferation and the specialization stages. During proliferation, the leading metabolic events are synthesis of DNA and RNA, which can be traced by radioautography after injection of tritiated precursors. During specialization, hemoglobin synthesis is the major event, a metabolic process that can be followed by tracing ^{59}F incorporation in the cytoplasm.

By correlating morphological observations, histochemical analysis of the cell's DNA, RNA, and HbO_2 contents, mitotic counts, grain counts for ^{3}H and ^{59}F incorporations, it is possible to describe the kinetic events in the erythron in considerable detail. From smears of bone marrow, it has been established that many immature cells of the erythropoietic lineage divide actively.

Thus, the proerythroblast, the basophilic erythroblast, and the polychromatophilic erythroblast all contribute to the proliferation of the erythron. The DNA content of the erythroblast varies in dividing cells; as might be expected, it doubles before mitosis. Thus each proliferating cell can be divided into three stages: a postmitotic (G_1), a synthetic (S), and a premitotic (G_2). By measuring the incorporation of tritium-labeled thymidine into bone marrow cells grown in tissue culture *in vitro* and by determining mitotic indices *in vivo*, researchers measured the time that lapses between two mitoses. The intermitotic cycle time varies with the degree of maturation of the "erythroblast." Intermitotic time for the proerythroblast and the basophilic erythroblast is 20 hours. Intermitotic time for the polychromatophilic erythroblast increases with the concentration of hemoglobin in the cell, from 30 hours for the early to 50 hours for the middle and late polychromatophil. On the basis of these findings, it is obvious that in the marrow, the tritium labeled thymidine will be found in the proerythroblast (70% labeled), the basophilic erythroblast (50% labeled), and the polychromatophil (33% labeled). Furthermore, the grains will be denser in the nuclei of the proerythroblast since the overall rate of DNA synthesis is greatest in that cellular compartment.

When a labeled precursor of protein or RNA is injected, all the cells of the marrow are labeled because RNA and protein are synthesized during interphase. But the rate of RNA synthesis decreases as the cell matures. Furthermore, staining with basophilic dyes reveals a progressive decrease in basophilia, which indicates that the RNA content of the cell decreases as the cell matures. This finding is supported by observations made with the ultraviolet microscope.

The rate of iron uptake by the various precursors of the red cells is an excellent measure of the rate of hemoglobin synthesis. Iron uptake occurs in almost every cell of the erythron, except for a small percentage of proerythroblasts (5% of that cell compartment) and a few dying cells. The uptake of radioactive iron per cell remains constant in the proerythroblast and in the basophilic erythroblast; this fact points to a constant rate of hemoglobin synthesis. Yet direct measurements of the hemoglobin content per cell indicate that the proerythroblast and basophilic erythroblast contain various levels of hemoglobin.

By correlating the radioautographic and ultraviolet absorption results, it is possible to conclude that hemoglobin does not turn over but is concentrated in the proerythroblast by steady synthesis.

The rate of hemoglobin synthesis decreases in the polychromatophil and the reticulocyte as the concentration of hemoglobin in the cell increases. Injected radioiron is not directly exchanged with the iron in the circulating red cell. But the iron reaches the bone marrow, is used for HbO_2 synthesis, and is slowly released into the circulation inside the newly formed erythrocytes. The release is slow in the first 20 hours after injection, but it becomes linear between 20 and 120 hours after injection.

Radioautographic examination of blood smears after injection of ^{59}F reveals the number of labeled red cells and the degree of labeling in each cell (number of grains). Although few cells are labeled during the first 20 hours after injection, a large number of highly labeled cells appear between the 20th and 40th hours. After that, the number of the labeled cells and the grain count per cell decrease slowly between 50 and 60 hours after injection, and rapidly after 60 hours. By following the appearance of ^{59}F in the circulating erythrocyte and estimating the number of labeled cells in the circulating blood, one can calculate the rate at which newly formed erythrocytes are transferred from marrow to blood.

By correlating histochemical and radioautographic information, researchers have constructed kinetic models for erythropoiesis. Several such models have been described, but they differ in quantitative details. Only the one described by Lajtha will be presented here [18, 19].

A bone marrow stem cell is converted into a proerythroblast, a large immature cell that soon divides and yields two smaller immature cells, the basophilic erythroblasts.

One of the basophilic erythroblasts leads a full life and proliferates to maximum capacity. The other enters a semiabortive pathway. First, consider the fate of the basophilic erythroblast that leads a normal life. For 20 hours the basophilic erythroblast synthesizes porphyrins and globin and picks up iron very actively. In the process, the cell enlarges and can be described as a proerythroblast. Thus, during the first 60 hours after a stem cell is converted to a proerythroblast, three divisions occur, and the original stem cell has yielded 14 new cells. Only the fate of one half of them will be considered now. As more hemoglobin is synthesized, each basophilic erythroblast is slowly converted into a polychromatophil, and after 30 hours the polychromatophil divides, progressively differentiating into the orthochromic normoblasts of the bone marrow, which finally become erythrocytes.

Thus, in 160 hours, one daughter of the stem cell has yielded 16 erythrocytes. The other daughter also divides after 20 hours. Thus two "descendents" differentiate and divide. But after the second generation of cells, there are alternative pathways different from the regular erythropoiesis. Two of the four basophilic erythroblasts divide and yield four polychromatophils that hasten to differentiate into normoblasts without further division. Two other basophilic erythroblasts enter a physiological cul-de-sac referred to as ineffective erythropoieses. On the basis of this model, one stem cell yields 20 erythrocytes in a period of 130 hours.

Lajtha [18] claims that when the model is checked against the quantitative information already available, the experimental data can readily be explained by the theoretical scheme. For example, on the basis of the model, one can calculate the theoretical number of normoblasts in human marrow. The calculated and the experimental data were found to be in good agreement. Similarly, mitotic indices, measured by other hematologists, fully satisfy the requirements of the model for the erythron.

In conclusion, laborious radioautographic and histochemical studies add a new dimension to our knowledge of the erythron. They tell us how many cells of specific morphology appear and disappear in a given period. This information is indispensable for adequately describing the mechanism controlling the rate of erythropoiesis.

Red Cell Function

The function of the red cell is to transport oxygen from the lungs to the tissues. This vital function cannot be impaired for long without catastrophic results. Under adequate oxygenation, the red cell mass and consequently the concentration of hemoglobin (the molecule responsible for the oxygen exchanges) remain constant. Because the life span of a red cell is relatively short compared to that of the entire organism, the bone marrow must continuously replace the dead red cells by active proliferation. Since the red cell mass does not change, cellular proliferation in the marrow must be rigidly controlled. When the oxygen tension of tissues is reduced because of rarified atmospheric conditions (high altitude), massive loss of blood (hemorrhage), or inadequate oxygen exchange in the lungs (lung or heart disease), the erythron responds by increasing the red cell mass and consequently the total amount of hemoglobin. On the contrary, transfusion of blood causes transitory polycythemia, which is followed by a reduction of bone marrow proliferation. Thus, erythropoiesis is expandable. The rate of proliferation may vary from a standstill to a sevenfold increase above the normal rate.

All these observations long ago led to the conclusion that the oxygen tension in tissues controls the rate of red cell production in the bone marrow. A feedback control circuit exists between the function of the red cell (oxygenation of tissue) and the proliferation of the red cell. Changes in oxygen tension could directly affect bone marrow proliferation, or the effect of oxygen could manifest itself indirectly by regulating the distribution of humoral factors. Brecker and Stohlman claimed that red cell production is also controlled by the "red cell mass" *per se*. This is accomplished by the elaboration of a bone marrow depressant in the aging red cell. The young red cells in circulation do not contain the inhibitor. The evidence for the existence of such an inhibitor is only indirect. The assumption is based on observations made in hemolytic anemia, in which erythropoietic activity is increased in spite of the normal hemoglobin concentration. This was explained by postulating that in hemolytic anemia, inhibitors containing red cells are removed from the circulation.

In vitro studies of bone marrow proliferation (heme and hemoglobin synthesis) have established that those processes are not accelerated by increasing the oxygen tension. In fact, normoblastic mitosis is reduced when the oxygen tension is reduced in the entire medium.

The existence of a humoral factor was already postulated in 1906 by Carnot and Deflandre, probably on the basis of erroneous results [20].

Direct evidence that decreased oxygen tension acts on erythropoiesis through the release of a humoral factor was obtained in experiments with parabiotic rats. In these experiments, one rat received low oxygen levels; the other was maintained at normal levels of oxygenation. Erythropoietic activity was increased in the anoxemic parabiont and the well-oxygenated partner as well.

Progress in our knowledge of erythropoietin has been hindered by the fact that it is present in plasma only in small quantities and cannot easily be concentrated and purified from plasma. Furthermore, available assays are only semiquantitative and are often unreliable. The finding that erythropoietin is present in the urine of anemic humans or dogs has provided new hopes for the elucidation of the molecular structure of erythropoietin. The hormone has been concentrated from those sources by chromatographic analysis on kaolin, ultrafiltration, and absorption of collodion.

In spite of these difficulties, several major observations have already been made. In humans, erythropoietin is found only in the urine of anemic patients. The partially purified hormone is likely to be mucoprotein; strangely enough, erythropoietin does not exhibit any species specificity, and the human hormone stimulates erythropoiesis in the monkey and the guinea pig [21–25].

Erythropoietin has no direct effect on the intermitotic time or on the duration of maturation of the normoblast *in vitro*. However, bone marrow culture has demonstrated that erythropoietin stimulates the proliferation of the erythropoietic cell.

The target cell for erythropoietin has not been identified with certainty. When anemia is produced experimentally, pluripotent cells migrate from the bone marrow to the spleen where they become the parent cell for an erythropoietic colony. These pluripotent cells are therefore referred to as colony forming units or CFU. It seems well-established that erythropoietin does not affect migration or the proliferation of CFU. Erythropoietin is more likely to act on an erythropoietic "stem cell" derived from the pluripotent cell and which is committed to differentiation into an erythrocyte. When the demand for erythropoiesis decreases there seems to be a rise in the CFU cells in the marrow with subsequent increase in myeloid cells [47]. The role of the pluripotent stem cell will be considered in the chapter on inflammation.

It is not known for certain in which organ erythropoietin is elaborated; the kidney has often been suspected of playing this role [26]. The juxtaglomerular apparatus has been claimed as the site of erythropoietin secretion. However, there is no correlation between the disappearance of the vacuole in the cells of the juxtaglomerular apparatus and the appearance of erythropoietin.

The kidney is certainly not the only source of erythropoietin. Nephrectomized rats subjected to hypoxia have a 10% of normal erythropoietin response indicating that an extrarenal source of erythropoietin exists (most likely the liver). The association of erythrocytosis and hepatoma could be explained on the basis of elaboration of erythropoietin by the hepatoma cells. The synthesis of erythropoietin is believed to occur in two steps: the formation of an inactive precursor, erythrogenin, which is converted to active erythropoietin by an unidentified plasma factor [48–49].

Plasma of anemic rabbits was found to stimulate reticulocytosis in normal rabbits. Stohlman and Brecker have claimed that the levels of erythropoietin in blood increase with the severity of the anemia.

Although anoxemia seems to exert its effect on bone marrow proliferation through the intermediate effect of erythropoietin, it has not been possible to demonstrate a decrease in erythropoietin activity in cases of adaptive polycythemia, yet transfusion even of normal blood blocks bone marrow proliferation. Erythropoietin levels were thought to be increased in polycythemia vera. The injection of human polycythemic plasma to rats stimulates their bone marrow cell proliferation. An increase of erythropoietin levels in polycythemia may result either from increased production or from elimination of natural inhibitors. Of course, the possibility of a basic disturbance in the bone marrow stem cell precluding an adequate response to erythropoietin has not been excluded.

The dominating biochemical event in erythrocyte formation is hemoglobin biosynthesis, which has three stages: globin biosynthesis, porphyrin biosynthesis, and the insertion of iron into the porphyrin to yield heme. Globin synthesis follows the regular pattern for protein synthesis and has been studied extensively because reticulocytes provide such a convenient material for the study of a specific protein. Porphyrin biosynthesis was discussed in the chapter on porphyrias. The conversion of porphyrin to heme involves the insertion of an iron atom and is catalyzed by an iron-chelating enzyme that has been found in liver and bone marrow. The molecular mechanism of the reaction is still unknown, but the enzyme requires the presence of reducing agents (ascorbic acid, cysteine, or glutathione).

A more adequate knowledge of the molecular changes that take place when iron is incorporated into the porphyrin molecule to convert it to heme might be of considerable help in interpreting some aspects of the pathogenesis of anemias. For example, some anemias that resist any form of therapy might result from an inability to convert porphyrin into heme.

Iron, of course, is not used only for hemoglobin synthesis; it is used for the biosynthesis of other proteins, cytochromes, catalases, peroxidases, and myohemoglobins.

Red Cell Breakdown

Hemoglobin breakdown is covered in the section on hemolytic anemias. The released iron is not excreted, but is stored in the form of ferritin or hemosiderin.

Control of Iron Metabolism

Inasmuch as almost no iron is excreted, the mechanism controlling the levels of the iron reserves in the body must operate at the level of absorption [27, 28]. For many years the most popular theory invoked to explain the control of iron uptake was the theory of the mucosal block mechanism. The theory postulated the existence of a single, short-loop feedback mechanism within the cell of the intestinal mucosa. The absorbed iron would induce the *de novo* synthesis of apoferritin, and the apoferritin would then chelate iron to yield ferritin, which in turn would inhibit apoferritin synthesis. This conclusion was based essentially on an experiment in which large doses of nonradioactive iron were administered prior to a single dose of radioactive iron. The experiment demonstrated that when absorption is prevented, the body is loaded with iron. The workers concluded that iron absorption was not unlimited but was in some way carefully regulated by the body. However, the experiment did not give any precise clues as to the mode of operation or the location of the control mechanism.

However, the postulated mechanism provides a simple explanation for the effect of anoxemia on iron uptake. The conversion of apoferritin to ferrritin is coupled with the reduction of ferric to ferrous iron. The rate of ferritin formation would be regulated by the redox levels of the cell. Whether this theory for the control of iron uptake will survive a concentrated tide of quantitatively well-controlled *in vivo* experiments remains to be seen.

Whatever the short feedback loop may be, a longer feedback loop also seems to control iron uptake by the intestine. In spite of the fact that iron is easily traced because of the availability of radioactive isotopes, the methods for measuring iron absorption were somewhat unreliable before absorption could be measured by total body counting. The basic error underlying the earlier experiments was the assumption that virtually all absorbed iron is incorporated into hemoglobin. In reality, absorbed iron is rapidly distributed to other compartments of the body. Therefore, determination of the appearance of orally administered iron in blood provides inaccurate information on iron absorption in the intestine. Theoretically, it should be possible to measure the amount of absorbed iron by subtracting the amount of radioactive iron in the stools. This method presents many difficulties because the amount of iron absorbed is so small compared to the amount ingested. Somewhat more reliable results have been obtained in studies in which small amounts of iron were administered to reduce the ratio of ingested to absorbed iron.

This inconvenience is obviated by administering two different iron isotopes to the same individual (^{55}F and ^{59}F)—one by injection, the other administered orally. The amount of iron used to form hemoglobin can be accurately determined by following the appearance and disappearance of the injected isotope. When rates of incorporation into the blood of the injected and the ingested isotope are compared, the proportion of absorbed iron can be estimated accurately. This method is valid only if the distribution of injected and ingested iron is the same throughout the body. Whether this assumption is correct, especially under pathological conditions, remains to be established.

In view of all the complications inherent in these methods, it is not surprising that attempts were made to investigate the mechanism controlling iron absorption using the more reliable method of total body counting. Studies with humans revealed that in addition to iron loading or iron deficiency, the factors that modify the rate of the erythropoietic activity also affect iron absorption. Anemia and severe anoxemia (resulting either from massive hemorrhage or from massive hemolysis) stimulate iron absorption. In contrast, transfusion, polycythemia, and bone marrow hyperplasia reduce iron uptake. Of course, iron absorption is increased in iron deficiency and decreased when the iron stores of the body are overloaded. The correlations between bone marrow activity and iron absorption lead to the assumption that intestinal absorption of iron is directly or indirectly regulated by the rate of erythropoiesis. The proliferation of the cells of the erythropoietic lineage in the bone marrow would bring more mature erythrocytes in the circulating blood and thus provide for better oxygenation of all tissues, including the intestinal mucosa. An increase in oxygen tension in the mucosal cell interferes with the conversion of ferric to ferrous iron.

A multiple-step mechanism could also control iron absorption through the intestine. For example, it has been proposed that the bone marrow controls the secretion of a specific humoral factor responsible for regulating the intestinal absorption of iron. If such a hormone exists, it is unlikely to be erythropoietin. Erythropoietin administration to animals has no effect on iron absorption. Conversely, intestinal absorption and the iron stores influence the rate of erythropoiesis. Iron loading of anemic dogs considerably stimulates hemoglobin synthesis.

Studies of iron stores and iron uptake in the intestinal mucosa have provided direct evidence that absorption is commensurate with iron needs. Radioautographic examination of the intestinal villi after the administration of radioactive iron indicates that the iron absorbed in the intestinal villi is temporarily stored in the epithelial cells, and that is removed from these stores as the metabolic need increases. The epithelial iron is rapidly transferred to the blood after repeated bloodletting, but it is eliminated into the feces through a normal epithelial sluffing process when the metabolic requirements are satisfied. From these experiments it appears that the serosal transfer of iron is the step most influenced by the controlling mechanism.

The role of transferrin has been investigated by measuring iron absorption by isolated intestinal loops. Although it appears that transferrin accelerates iron absorption, the rate of intestinal absorption of iron is not limited by transferrin levels in blood. A number of hormones influence iron absorption by the intestine. The molecular mechanism of these hormonal effects is not known, and only the ultimate response to the hormonal stimulus can be described. In many if not all cases, the hormonal effect on iron absorption is probably secondary to the action of the hormone on other physiological or biochemical functions. Pituitary and thyroid hormones stimulate iron absorption; thyroid hormones probably act by increasing erythropoiesis and oxygen consumption.

Pancreatin restores iron absorption to normal when administered to patients suffering from chronic pancreatis, whose iron absorption is increased. Estrogen increases iron absorption, and iron absorption is reduced in hysterectomized women.

Dietary factors other than the type of the iron salt used in the diet also influence iron absorption. Rats on corn diets, high-methionine diets, or ethionine-supplemented diets absorb more iron than can be effectively used.

Possibly one decisive factor in iron absorption is the use of the metal by the erythropoietic tissue. Conditions that stimulate erythropoietic activity (hypoxemia and some hemolytic anemias) stimulate iron absorption through the intestinal mucosa.

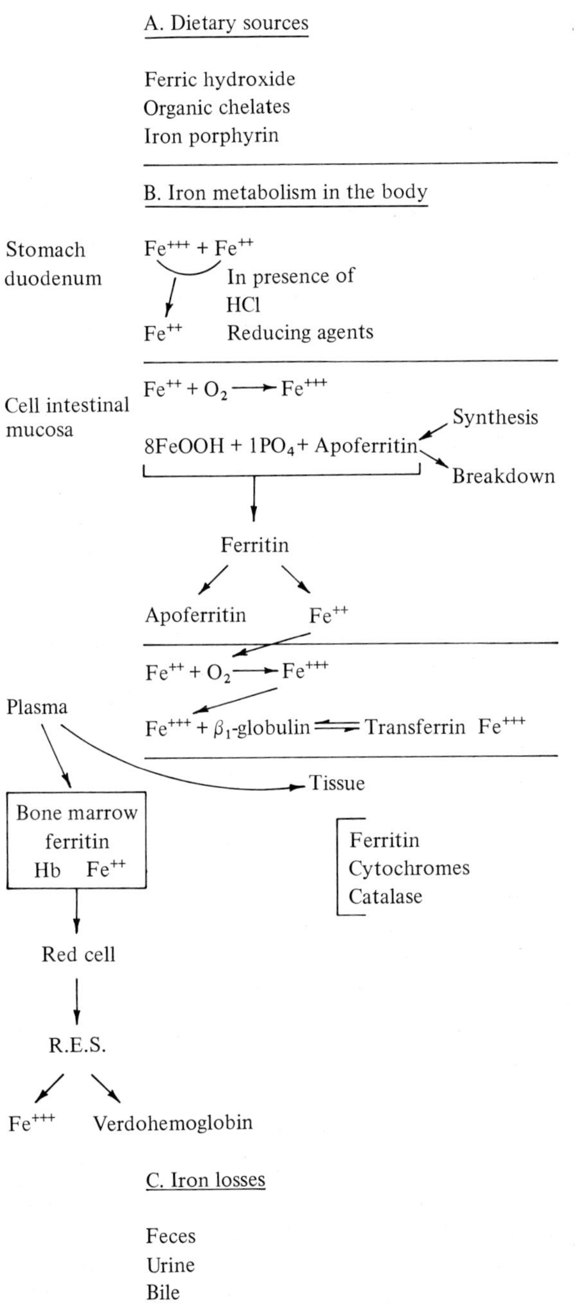

Fig. 6-2. Iron metabolism. Binding is regulated by the partial pressure of O_2 and the concentrations of Fe^{+++} in blood

A new hypothetical mechanism for the control of iron absorption from the intestinal mucosa has been proposed. It is based on two observations: (1) the inverse relationship between iron content in the intestinal mucosa and the absorption of iron; (2) radioautographic studies following injections of labeled iron, which indicate that the iron concentrates in the newly formed cell of the intestinal crypt. In the presence of large iron stores, the iron content of the crypt cell remains unchanged during the entire life of the cell. But if iron is depleted, the iron content of the crypt cell decreases. Thus, as the newly formed crypt cell matures and moves toward the villi, it can absorb new iron. Iron metabolism is summarized in Fig. 6-2.

The Biological Role of Iron and Other Metals

The role of metals, particularly iron, in biology has been intensively investigated for almost a century. But in spite of the considerable advances made in our knowledge of molecular biology of the ferroproteins, the details of the mode of action of iron atoms in oxygen transport, electron transport, and other enzymic activities are not known. A discussion of the biological role of iron must start with a brief review of the chelates in living organisms.

In the periodic table, the elements are arranged in ascending order according to their atomic number. In this listing, the properties of the elements tend to repeat themselves in the following sequence: inert gases, active metals, active nonmetals. The arrangement of the periodic table, as invented by Mendeljeff, places the elements with similar arrangement of the outer electrons (and, therefore, similar chemical properties) in vertical groups. The nonmetallic elements, which tend to be reduced (donate electrons), are found at the right side of the chart, while the strongly metallic elements, which tend to undergo oxidation (accept electrons), are in the left-hand portion of the chart. In the center of the chart is a group of compounds—the transitional elements—with properties intermediate between strong metallic and nonmetallic elements.

In aqueous solutions, metal ions can hold a specific number of water molecules (usually 4–6). The tendency of a cation to become hydrated depends upon its size and charge. The tendency for hydration is least when the radius is greatest and the charge smallest. If other molecules or ions are also present in the solution containing the cation, they may displace the molecules of water involved in the hydration of the cation. For example, a cation Fe^{++++} may combine with 4 chloride ions to yield Fe Cl_4; or Fe^{++++++} may combine with 6 cyanide molecules to yield $FeCn_6$. The combination of cation with several other ions or molecules produces a complex ion. The cation moiety of the complex ion is the central ion.

The molecules, or ions, that displace the water molecule are called ligands. The binding of the cation to the ligand is of particular importance.

Consider the F^{+++} ion. The electronic composition of the superficial shells (3d, 4s, 4p) is as follows:

		3d	4s	4p
Fe	atom	(↑↓) (↑) (↑) (↑) (↑)	(↑↓)	○ ○ ○
Fe^{+++}	ion	(↑) (↑) (↑) (↑) (↑)	○	○ ○ ○

The valence shells of the ion have unfilled orbitals, and the ion is therefore apt to accept electrons.

The bonding between ligand and central ion is primarily covalent (sharing of electrons). Furthermore, since the two electrons involved in bond formation are contributed by the ligand, the bonds that appear in complex ions are of the coordinated type. A coordinated covalent bond is one in which both electrons of the shared pair are contributed by one atom (the ligand in complex ions, the sulfur atom in SO_2).

But the bonding between ligand and central ion is not purely covalent. Complex ions often manifest a high degree of polarity. We must digress to describe what is meant by polarity. Each hydrogen atom contains only one electron in its orbital: H: $\overset{1s}{\oplus}$. The stable electronic structure is achieved in the helium molecule (He ⊕), which contains two electrons in its 1 s orbital with antiparallel spin (Pauly principle). When two atoms of hydrogen are united to form a molecule of hydrogen, the two 1s orbitals of each hydrogen atom combine to form a new orbital, the molecular orbital. In hydrogen molecules the electron density is greatest between the two nuclei. Indeed, the atomic nuclei tend to repel each other, and the high negative charge between the nuclei holds the two atoms together.

In a diatomic molecule composed of two identical atoms, the charge distribution is the same at both ends of the molecule; consequently, the molecule exhibits no polarity. When different atoms are involved in the formation of molecular orbitals, the electron density is not always distributed equally around the two atomic nuclei; the electron density may be greater around one nucleus than around the other. As a result, the new molecule is more electronegative than the other.

A chelate is a special type of complex ion in which the cation, usually a transitional element, reacts with at least two sites of a single organic molecule. The imprisonment of the cation in the chelating ligand has been compared to that of a crab's prey between its claws. (In fact, chelate derives from the Greek word χηλη, meaning claw.)

Like most metals, iron forms three types of biological chelates, those involved in: metal transport (ferritin and transferrin), metal storage (hemosiderin), and biochemical reactions (metalloenzymes, myoglobin, and hemoglobin).

Distribution of iron in the ferroproteins:

Hemoglobin	73%
Ferritin	16%
Myoglobin	3%
Heme enzymes	2%

In reality, very little is known about the mode of attachment of the metal to the ligand in metalloproteins that store or transport the metal. Most likely, the metal binds with a polar side chain of the protein (imidazole, carboxyl, aminohydroxyl, sulfhydryl, etc.).

Among the catalytic metalloproteins, those in which the metal is bound directly to the protein must be distinguished from those in which the metal is attached to the protein by the intermediate of a prosthetic group [29].

All metal binding does not involve chelation. Sometimes the metal atom is bound only to one site of the protein molecule. In such a case, the metal can be removed by electrodialysis, and the bound metal readily exchanges with the radioactive metal molecules that are introduced into the solution. Furthermore, when one-site binding per atom is involved, the metal removed from the protein readily reattaches to the molecule even if the protein molecule is denatured. Because chelation (when two or more binding sites per metal atom are involved) requires specific configurational arrangements, it is difficult to remove the metal atoms from the protein molecule. Once removed, the metal is slow to find its niche in the protein configuration, and binding of the metal to the protein is impossible when the protein is denatured.

These remarks can best be illustrated by describing the binding of zinc to enzymes [30]. Zinc is an integral part of a number of enzymes: 3-phosphoglycerokinase, enolase, carboxypeptidase, and carbonic anhydrase. The binding of zinc to 3-phosphoglycerokinase is similar to its binding to albumin. In serum albumin, zinc binds loosely, mainly to form complex ions with the imidazole group of the histidine residues. Treatment of albumin or 3-phosphoglycerokinase with urea does not interfere with metal binding. Thus in 3-phosphoglycerokinase, as in albumin, the zinc binding does not involve chelation.

Zinc associated with purified enolase molecules appears to involve complex formations with or without chelation. Most of the zinc associated with the enolase molecules can be removed by electrodialysis and restored to the molecule, even after denaturation. In contrast, one of the zinc atoms is tightly bound and cannot be removed by electrodialysis, nor can it be replaced in the enolase molecule after its denaturation with 4 M urea. The last of these observations illustrates one important property of the protein ligand in biological chelation.

The two branches of the chelating claw are not located on two adjacent amino acids of the sequence. Instead, they are brought close together as a result of the special configuration of the protein molecule. To erase this configuration with 4 M urea is to eliminate the chelating properties of the protein.

The zinc atom in carbonic anhydrases is as strongly attached to the enzyme molecule as that of enolases. It cannot be removed by electrodialysis or exchanged with the radioactive zinc of the medium. But if the enzyme is incubated at pH 5 in the presence of a strong chelating agent, the zinc is detached from the protein and the enzyme is inactivated. The enzyme is reactivated by incubating the metal-free protein with Zn^{++} or Co^{++}. However, the activity of the "cobalt enzyme" is only half that of the "zinc enzyme."

However, enzyme activity is restored slowly; equilibration between metal and enzyme is not instantaneous but takes several weeks in some cases. The protein seems to oppose a barrier that the metal ion must overcome. The barrier is likely to involve two components: pure steric hindrance resulting from the specific protein configuration, and charge interaction between the protein and the metal ion.

Studies on the removal of the metal by chelating agents have provided evidence for steric hindrance. For example, ethylenediaminetetraacetic acid has a binding affinity for metal greater than that of 1–10 phenanthroline; yet the former often cannot remove the metal from the protein, whereas the latter can.

The properties of carbonic anhydrase suggest that the slow equilibration between metal and protein cannot result from steric hindrance only. The turnover number of carbonic anhydrase is very high, which means that the substrate is rapidly converted to its products, and the enzyme molecule is quickly reused in a new conversion of the substrate into product. Consequently, the product of the reaction must rapidly diffuse from the active site out of the protein molecule.

Since the zinc molecule in carbonic anhydrase is found at the active site, it must also penetrate into the molecule through the same path used by the substrate and the product. But in spite of the fact that the Zn^{++} ion is smaller than the HCO_3 or CO_2, zinc diffuses more slowly than HCO_3 and CO_2.

Consequently, steric hindrance alone cannot explain the slow penetration of zinc. It has been postulated that the charge distributation around the active site of the enzyme is such that it interferes with the penetration of cations but has no effect on the penetration of anions. Unfortunately, this is purely hypothetical. The charge configuration of the active site will not be known until the amino acid sequence of the ligand is elucidated. Such information might lead to the description of the molecular orbitals involved in the binding between metal and ligand.

More information is available when the ligand is a prosthetic group of the enzyme, as is the case in the hemoproteins.

In the hemoproteins, the central ion is iron. In protoporphyrins, central ion and ligand form the heme. The same heme is inserted into different protein molecules, yielding different molecular complexes with different catalytic properties. For example, cytochromes facilitate electron transfer, catalase converts H_2O_2 into H_2O, and hemoglobin stores and transports oxygen. The same chelates (central ion and ligand) can perform such varied tasks only if the protein framework in some way modifies the electron configuration of the active site, which, in all of the cases cited above, is known to include the metal. This effect of the ligand (porphyrin and protein in hemoprotein) on the central ion (iron) electronic configuration is explained in light of the crystal field theory. In complex ions, the crystal field is an electric field generated by the ligand and acting on the electrons of the central ions [31].

As the central ion approaches the ligand, the ligand electrons tend to repel the electrons of the valance shell of the central ion: as a result, these electrons become polarized.

The degree of polarization conferred upon the central ion is reflected in the ionic character of the bond that unites the central ion with the ligand. The poles that repel all electrons in s and p orbitals are equal, but among the electrons in the d orbitals, some are more repelled by the ligand field than others.

We shall now see how these complex interactions apply to the iron atom. In hemoprotein, the electronic configurations of Fe III and Fe II are:

	shell I K	shell II L	shell III M	shell IV N
Fe III	$(1s)^2$	$(2s)^2$ $(2p)^6$	$(3s)^2$ $(3p)^6$ $(3d)^5$	$(4s)^2$
Fe II	$(1s)^2$	$(2s)^2$ $(2p)^6$	$(3s)^2$ $(3p)^6$ $(3d)^6$	$(4s)^2$

	3d	4s
Fe II	(↑↓) (↑) (↑) (↑) (↑) (↑)	(↓↑)
Fe III	(↑) (↑) (↑) (↑) (↑) (↑)	(↓↑)

Thus Fe III has 6 and Fe II has 5 valence electrons in the 3d orbitals (the two 4s electrons do not need to be included in the discussion that follows). Before we discuss the properties of the iron atom, it may be appropriate to review briefly the structure of the electron shells of atoms [29].

Although it is not possible to determine the exact position of an electron relative to the nucleus, it is possible, with the aid of Schroedinger's equation, to calculate the boundaries of the area in which the electron will travel. Orbitals are those areas in which the probability of finding the electron is high.

The s orbital is unique and spherical. p orbitals are described in a three-dimensional figure 8. Two juxtaposed spheres are separated by a node. The sublevel p can accommodate three different orbitals disposed around three axes (x, y, z) perpendicular to each other. Both s and p orbitals are symmetrical.

The d subshell can accomodate five orbitals that differ from each other in shape and direction. Four of the d orbitals have the shape of four-leaf clover, and one is shaped like a propeller. The position of the orbitals can be described relative to the cartesian coordinates z, x, and y. The long axis of the propeller is parallel to z (dz^2 orbital). The cloverleaf may occupy two different positions: all four leaves may be in the plane containing x and y (dx^2 y^2) coordinates; or all four leaves may be between x and y (see Fig. 6-3).

The number of electrons increases in the electronic shells as the element considered occupies a higher posi-

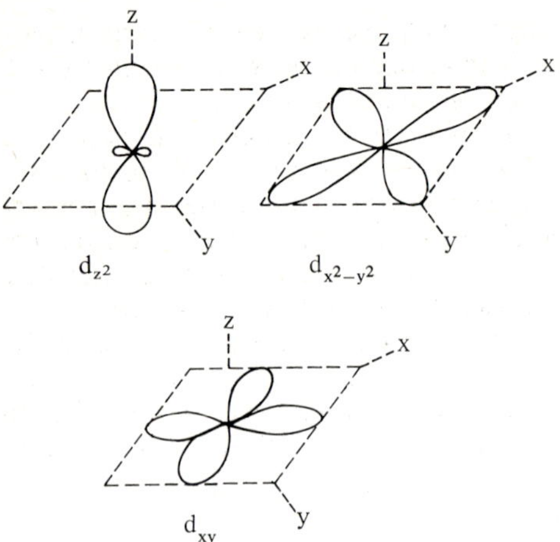

Fig. 6-3. Spatial orientation of iron orbitals

tion in the periodic table. New electrons are added according to strict rules which are expressed mathematically by quantum wave mechanic equations.

Each orbital can accommodate two electrons, but each of these electrons must spin in opposite directions. Consequently, a d subshell with its five orbitals filled will accommodate 10 electrons. When two electrons occupy the same orbital, they tend to repel each other. Because of this repulsion, each electron of the pair spends more energy in traveling the orbital than it would spend if it traveled by itself in its own orbital. The less energy required to maintain an electronic configuration, the more stable that electronic configuration will be. The more stable configurations always prevail over the less stable configurations. Thus, as one progresses in the periodic table of elements, each of the five (3d) orbitals is occupied by one electron before a pair of electrons appears on a (3d) orbital.

With this information in mind, let us see what happens when the iron atom, with its six coordination number, is trapped by the valences of the ligand (porphyrin and protein). In all hemoproteins, four of the coordination positions of the iron atom interact with the nitrogens of the tetrapyrrole ring. Chelation places some stress on the electronic configuration of the valence shell (3d) because of the existence of an octahedral field. Among the five orbitals, two are directed (dz^2 and dx^2-y^2) toward the ligand; and the electrons traveling these orbitals are vigorously repelled by the ligand field. Therefore, stability is achieved, only at the expense of great amounts of energy. In contrast, the dxy, dxz, and dyz orbitals do not interfere with the ligand electronic configuration, and stability can be achieved with relatively low levels of energy.

Thus, the ligand field splits the d orbitals into two groups, some of relatively high and some of lower energy. The high-energy group of d orbitals is referred to as cg; the low energy groups as t_2g.

The increase of energy (Δ) required to pass from t_2g to cg can be measured by analyzing the visible spectra of various central ion-ligand combinations. The "orbital separation" has been found to be of about 40–70 Cal/mole.

Obviously this disturbance of the energy levels of the orbitals may alter the distribution of the electrons among the orbitals. If the energy separating two orbitals is greater than the energy of the repulsion between the two electrons traveling the same orbital, then two electrons will be found on a single d orbital, rather than on two separate d orbitals, and of course the t_2g orbitals will be filled before the cg orbitals are filled.

Thus Fe II and Fe III may exist under two different electronic configurations. Depending upon the strength of the ligand field or the strength of the binding between ligand and central ions, the electronic structure of the central ions correlates well with its biological function.

Spin		t_2g	cg	Field
High	Fe III (3d)	↑ ↑ ↑	↑ ↑	weak
Low		⇅ ⇅ ↑		strong
High	Fe II $(3d)^6$	↑ ↑ ↑	↑ ↑ ↑	weak
Low		⇅ ⇅ ⇅		strong

The hemoproteins are involved in electron transfer, reversible peroxidation, and oxygen and CO_2 transport and exchange. The electronic configuration of the central ion is expected to change with the function of the hemoprotein.

Cytochromes are involved in electron transfer but, with the exception of cytochrome c, little is known about the molecular structure of these hemoproteins. Thanks mainly to the efforts of Margoliash, the complete amino acid sequence of cytochrome c of various species can now be written. The heme moiety of cytochrome is solidly inserted into the apoprotein. The attachment involves four different points of the heme molecule. Two of the bonds are coordination functions of the iron molecule; the two other bonds between heme and protein involve the vinyl side chains of the tetrapyrrole ring. In this bond the vinyl carbon, the one that unites the methyl group to the pyrrolic ring, forms a covalent bond with the sulfur atom of cysteine residues of the protein moiety. A coordination bond links the central iron atom of the heme to a nitrogen atom of the imidazole ring of two histidine residues of the protein moiety. Because of this type of attachment, the heme moiety forms a platform that is solidly gripped in the claws of the protein. The tight binding of the central ion to the ligand facilitates the pairing of electrons in the t_2g orbitals, an electronic configuration conducive to electron transfer. Electron transfer requires that electrons of the cg orbital be transferred to the t_2g orbital, and that all the coordination numbers of the metal be stable at all times.

If a similar tight bond between ligand and metal existed in catalase and hemoglobin, it would interfere with the function of these ferroproteins. One of the

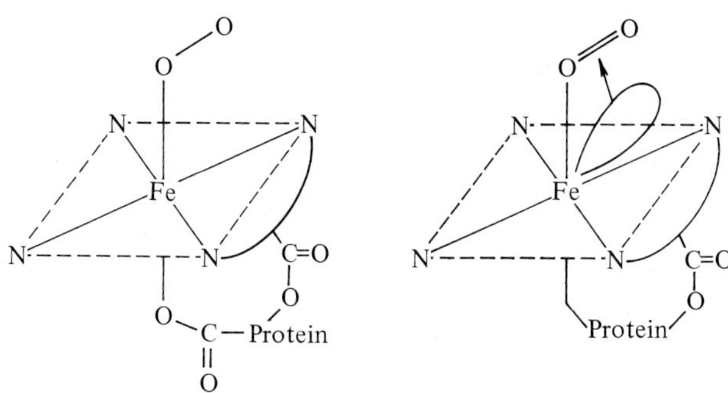

Fig. 6-4. Chelate compounds in biological systems. Ligand attachment in peroxidase (left) and hemoglobin (right), from G.L. Eichhorn, Federation Proceedings, **20**, 48 (1961)

coordination functions (z coordinate) of catalase should remain available for binding with peroxide. Consequently, in catalase, the ligand field must be weaker and the spin higher than in cytochrome. These theoretical predictions are corroborated by experimental observations.

Similarly, hemoglobin iron must be free to combine with oxygen; therefore, reduced hemoglobin should be in a high-spin state with partially impaired electrons. Once bound, the oxygen is specifically, transported in the blood, and the complex (oxygen, electron, and ligand) must be stable. Significantly, the attachment of oxygen to the iron molecule modifies the ligand field and induces the paring of electrons in the t_2g orbitals (see Fig. 6-4).

In conclusion, when metals become part of catalytic proteins, they are usually part of the active site. As such, in the biological reaction, they play a role similar to their role in organic chemistry. However, chelation with the protein not only gives the reaction specificity, but it also accelerates the rate of the reaction considerably. Thus, the transamination reaction can be carried out with iron alone, but the rate of the reaction is much more rapid when iron becomes part of the coenzyme-apoenzyme complex.

The introduction of a metal in the protein molecule often gives it catalytic properties that the apoenzyme does not possess, and, in addition, the metal often modifies the configuration of the protein molecule. For example, removing the metallic components of yeast alcohol dehydrogenase under the influence of a chelate modifies the rotary dispersion properties of the protein molecule.

This brief discussion of the role of metal in biology introduces a vast problem. Many examples of these interactions between protein and metals are found in other chapters.

Siderosis, Hemosiderosis, and Hemochromatosis

Before iron deposition in tissue can be discussed, the terms pathologists use in referring to these deposits must be defined: siderosis, hemosiderosis, and hemochromatosis. Siderosis is a general term usually referring to all iron deposition not associated with the development of clinical symptoms.

Hemosiderosis is a form of siderosis in which the iron deposits result from hemoglobin breakdown. Hemosiderosis may be focal (a consequence of hemorrhage) or generalized, resulting from a reduced life span of the red cell (repeated blood transfusion, hemolytic anemia, etc.). Hemochromatosis is characterized by cirrhosis, diabetes, and skin pigmentation and probably results from some molecular alteration of iron metabolism. Hemochromatosis will be described in more detail after a discussion of the iron pigments in tissues.

Siderosis

Siderosis may result either from excessive intake or from local accumulation of endogenous sources, mainly hemoglobin. Depending on the mechanism that is at its origin, siderosis may be generalized or localized.

At least two forms of generalized siderosis have been described: cytosiderosis and Kaschin-Beck disease. Cytosiderosis occurs among undernourished Bantus who ingest large amounts of iron [32]. Kaschin-Beck disease has been observed in Manchuria among people drinking iron-rich water. Both these diseases are associated with high iron contents in tissues (spleen and liver) and mild hepatic fibrosis. In addition, because patients with Kaschin-Beck disease usually start to drink the iron-rich water in early childhood, skeletal and articular deformities accompany the iron deposition. Local siderosis is sometimes observed in the lungs of those who work in the iron mines.

A special form of neural siderosis is referred to as marginal siderosis of the central nervous system. The disease is characterized by iron deposition either in a granular or a diffused form in the meninges at the origin of the cranial nerves and at the periphery of the central nervous system. Iron accumulation is associated with gliosis, demyelinization, and frequently necrosis of the cortex of the cerebellum. In addition to hemosiderin, one also finds small amounts of hemofuscin in the affected parts of the brain. Clinically, the disease is marked by progressive deafness and disequilibrium. The pathogenesis is not clear, but repeated meningial hemorrhage has been invoked as one mechanism. It has also been proposed that the accumulation may result from trapping of the iron by a siderophilic compound in the brain with an affinity for iron, which would normally be lower than that of the serum transferrin. However, when iron protein binding is for some reason low in the serum, iron accumulates in the brain.

Hemosiderosis

Hemosiderin, the compound that accumulates in hemosiderosis, is a product of hemoglobin degrada-

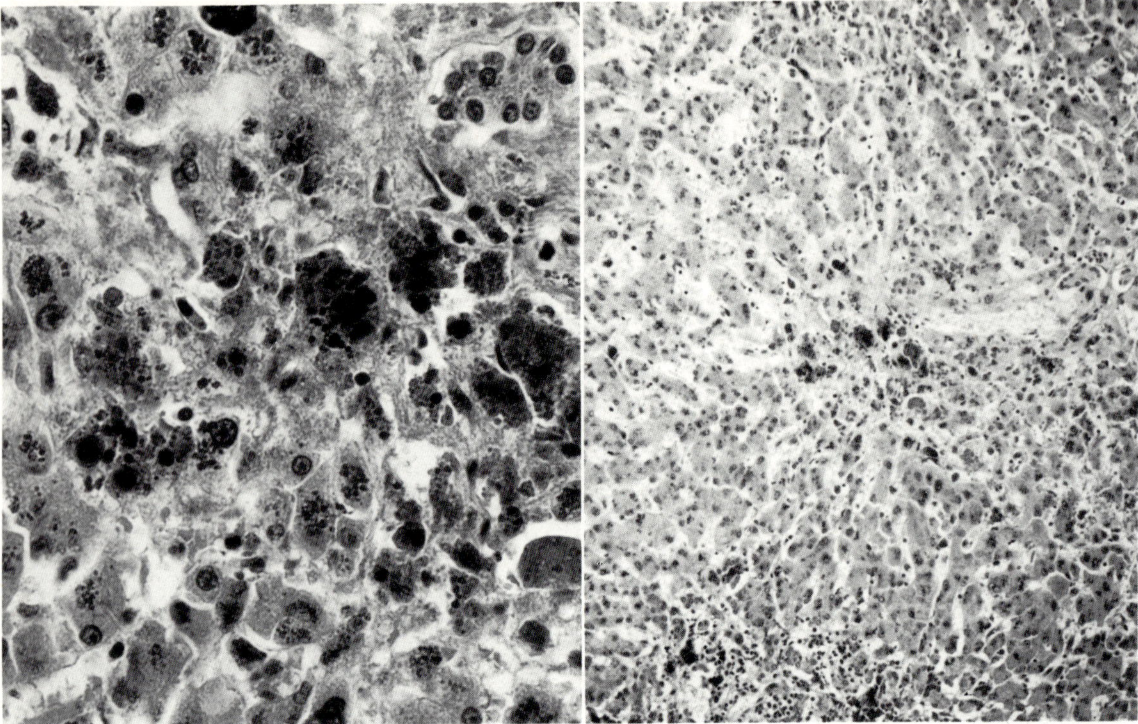

Fig. 6-5. Hemosiderosis of liver showing iron pigment and hepatic cells

tion. It is an iron protein of unknown structure. Hemosiderin is resistant to alkali and fat solvents, but is acid soluble. The mean iron content of hemosiderin is 36.9%; the nitrogen content is 8.9%. Pure hemosiderin crystals have not been obtained, and therefore reports on the amino acid or carbohydrate composition of hemosiderin are subject to criticism. Hemosiderin granules are probably insoluble aggregates of ferritin crystals. Usually more iron is in the granules than in the ferritin, possibly as a result of partial digestion of the protein component. The iron of hemosiderin, like the iron of ferritin, is mobilizable to satisfy metabolic requirements. The iron of hemosiderin is free to react with acid potassium ferrocyanide to yield the typical Prussian blue precipitate by which iron is histochemically detected.

Hemosiderin appears as a result of hemoglobin breakdown. (Hemoglobin catabolism is described in the section on hemolytic anemias.) But in view of its origin, hemosiderin obviously is formed inside cells capable of degrading hemoglobin, and only when the digesting cell degenerates does hemosiderin appear in the intercellular spaces. The presence of hemosiderin in some cells is physiological. In the infant, hemosiderin is found in cells with reticuloendothelial properties. Hemosiderin is found in the reticuloendothelial system less frequently in the young adult than in the infant.

Italian investigators compared the physicochemical, immunological, and ultrastructural characteristics of the product of ferritin with hemosiderin. These investigators concluded that hemosiderin is a degradating product of ferritin. They proposed that ferritin, which at the origin is a homogeneous protein, is freely distributed in the cell cytoplasm until it contacts the oxidizing agent. It is then sequestered into vacuoles converting the ferritin into hemosiderin. One of the events—an attack by protolytic enzymes—results in an increase in the iron-nitrogen ratio in hemosiderin as compared to ferritin.

Studies in weaning rats, in which the spleen was stained with the Prussian blue technique, indicate that the uptake of hemosiderin by splenic macrophage does not occur until adult-age histological development is achieved (58–60 days postpartum). Moreover, the ability of the spleen macrophage to store iron increases with the age of the animal.

Examination of liver slices obtained from unselected autopsy cases demonstrated that stainable hemosiderin is in 58% of normal livers in both the Kupffer and hepatic cells. No correlation could be established with age, sex, hemoglobin levels, iron therapy, or transfusion.

Hemosiderin accumulates in various tissues under pathological conditions, including hemorrhage, congestion, and hemolysis. Figs. 6-5, 6-6, and 6-7 demonstrate hemosiderosis in liver, kidney, and spleen, respectively. When blood leaves the vessels and contacts living cells, hemoglobin is degraded. The protein and the porphyrin are ultimately digested, but the iron accumulates at the site of the hemorrhage. Thus, if a hemorrhagic focus is examined histologically a few days after the onset of the hemorrhage, some brown granular pigment is found, not only in the macrophages, but also in the fixed cells of the connective

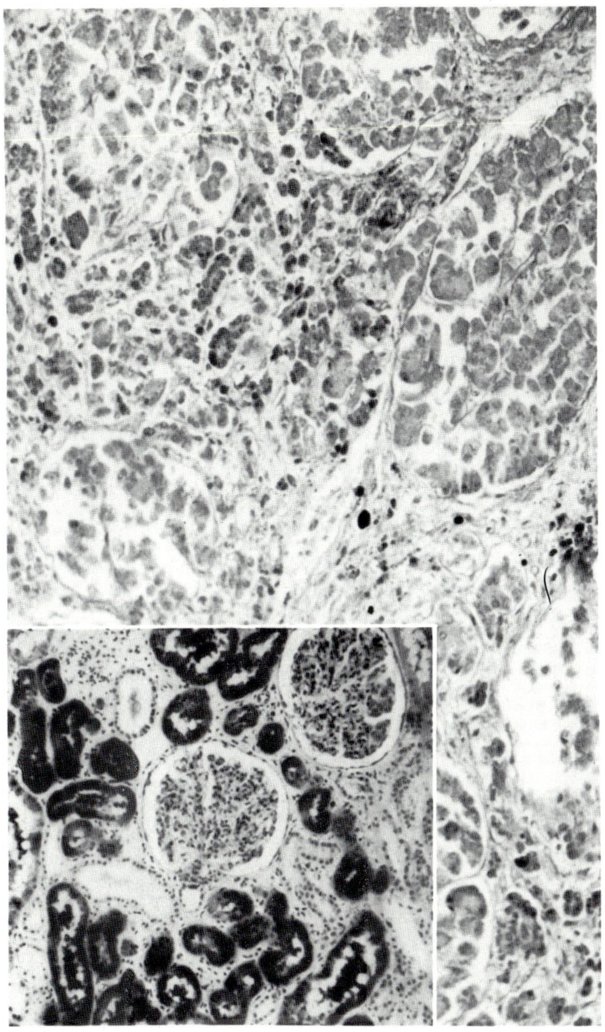

Fig. 6-6. Hemosiderosis in kidney after repeated transfusions, showing iron pigmentation in tubules (low and high power)

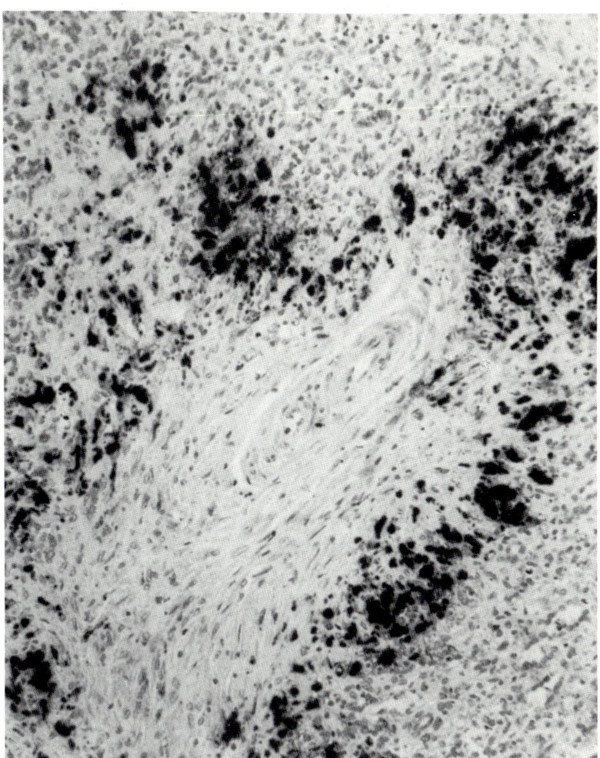

Fig. 6-7. Iron accumulation in reticuloendothelial cells of spleen of a patient with hemosiderosis

tissue and parenchyma. The mineral composition of the pigment can be brought out by staining with potassium ferrocyanide.

If hemorrhage is extensive, some surrounding tissue may be destroyed and later replaced by dense connective tissue. Such a process leads to pigmented fibrotic scars, which can be found in spleen, bone marrow, brain, and a variety of other tissues.

In cardiac failure, chronic passive congestion develops. The red cells that accumulate in the interalveolar sinusoids are phagocytized by the epithelial cells from the lining of the alveolar wall or in histiocytes. Again, after hemoglobin is degraded, stainable iron pigment accumulates in the alveolar cells. The iron-loaded alveolar cells are called heart failure cells. When expectorated with the mucus secreted by the bronchi, these cells cause the brownish discoloration observed in sputum of old cardiac patients.

As the passive lung congestion develops into a chronic condition, more pigment accumulates and the lung becomes fibrotic. At autopsy, the organ is smaller and firmer than normal. On sectioning, the tissue grates when the knife contacts it, and the surface of the sectioned lung presents a brownish discoloration instead of the typical salmon pink. As the condition worsens, the macrophage carries the pigment to the mediastinal nodes, which themselves become discolored. This sequence of events then leads to the development of small, brown, fibrous lungs so often found, for example, in patients with rheumatic fever.

A type of "idiopathic" pulmonary hemosiderosis of unknown etiology and pathogenesis is characterized by iron accumulation in the lung, in macrophage, and along collagen fibers. The source of the iron is not known.

Marchiafava-Micheli syndrome is a rare hemosiderosis of the kidney. The disease results from the appearance in the blood of a hemolysin that acts on the red cells at the low pH values reached only in the kidney. As a result, hemosiderin is deposited in the epithelial cells of the tubules, calices, and pelvis. Hemosiderin appears in the urine as a result of the sloughing of the epithelial lining.

Conceivably, hemosiderosis could result from the blood's inability to transfer iron from the site of absorption or storage to sites where it is used. Low blood levels of transferrin may result in iron accumulation. Three cases of transferrinemia have been described; one was acquired and two seem to have been inherited. The most typical case is that reported by

Heylmeyer and associates. After 7 years, patients with hypochromic anemia, low levels of iron and transferrin in blood, had increased iron stores in liver, spleen, and bone marrow. The total binding capacity of the plasma for iron was markedly reduced, and the ability of the plasma to transport iron was far below normal since the half time for clearance of radioactive iron from the plasma had fallen to less than 0.1 of the normal value.

Hemosiderin accumulates in the macrophages and in the cells lining the sinusoids of the spleen after repeated transfusion. In hemolytic anemias and in chronic passive congestion, brownish fibrotic nodules develop. In hemoglobinuria, iron pigment accumulates inside the cells lining renal tubules. Brown iron pigment granules are also found in the giant cell tumor (myeloplax) of the bone. These are usually hemorrhagic tumors, and the hemosiderin is probably derived from phagocytized red cells.

Alcoholism may, at its mature stage, be associated with the development of cirrhosis and pancreatic fibrosis. Under those conditions, approximately 50% of the patients have increased hemosiderin levels. Hemosiderosis is usually associated with increased stores of lead and calcium and reduced zinc and aluminum. When iron is stored, the symptoms and histological picture are similar to what is observed in familial hemochromatosis. Of course, a hereditary factor involving iron absorption or transport may be responsible for iron storage in the liver in a number of chronic alcoholics.

Hemochromatosis

In hemochromatosis, hemosiderin accumulates in many tissues. The origin of the hemosiderin found in hemachromatosis is not established, but it does not come from hemoglobin breakdown, and hemochromatosis probably is a disease of iron metabolism [33–37].

Hemochromatosis is a familial disease possibly transferred by an autosomal dominant gene. First it was believed that the mode of transfer was sex linked because of the high incidence of hemochromatosis in men. However, when the female sibling and relatives of patients with hemochromatosis were examined for iron content in their liver, it was clear that they developed the disease, but to a lesser degree. This is believed to result from the regular blood loss during menstruation and childbirth. On the basis of this interpretation, it is understandable that women do not develop symptoms until they are between 60 and 80 years old. Hemochromatosis is not such a rare disease as one might assume—records in the United States suggest that it occurs in 1 among 20,000.

In hemochromatosis, another hemoglobin pigment, hemofuscin, is also found, and hemosiderosis is extremely marked and extensive. The appearance of the organs at autopsy is most typical.

The organs are markedly discolored; some sections of the liver, spleen, pancreas, and lymph nodes, and the entire kidney, thyroid, and salivary glands have a rusty, copperlike color. The iron loading involves the liver and the pancreas especially. In older patients, the liver, in addition to its typical discoloration, exhibits the nobnailed appearance of Laennec's cirrhosis. The liver contains so much iron that iron can sometimes be demonstrated by the increased density of the liver on radiographic plates.

Histologically, parenchymal necrosis and regeneration and fibrosis are found in the liver, which in addition often contains enormous amounts of stainable iron and also resembles a typical cirrhotic liver. The pancreas is often smaller than normal and fibrotic; both the exocrine and the endocrine glands are necrotic, and the destroyed parenchyma is replaced by connective tissue fibrils.

Three symptoms dominate the clinicopathological picture of hemochromatosis: cirrhosis, diabetes, and skin pigmentation. Cirrhosis is severe and is followed by its classical complications, portal hypertension and hepatic insufficiency. The portal hypertension leads to ascites, esophageal varices, and gastrointestinal hemorrhage. The destruction of the islet is responsible for diabetes, which is sometimes severe enough to lead to death in coma if insulin therapy is not instituted. The skin acquires a marked tan, sometimes reminiscent of the color of bister. The skin discoloration is due to melanin formation rather than iron accumulation.

Testicular atrophy and hypogenitalism are often observed in the later stages of hemochromatosis. They result partly from accumulation of iron in the endocrine glands, and partly from hepatic insufficiency. Patients with hemochromatosis often die from progressive cardiac failure as a consequence of the metabolic overload and iron deposition in the myocardium.

French and German clinicians (Trousseau, Fraisier, and von Reklinghausen) described the now wellknown aspects of the pathological features and symptoms of hemochromatosis in the late 1800's. In fact, the French gave the disease most suggestive names, such as "diabète bronzé" or "cirrhose pigmentaire du diabète sucré." But the pathogenesis of the disease is still baffling. The pathogenesis of hemochromatosis raises two problems: the source of the iron, and the mechanism by which the accumulated iron leads to necrosis; nothing is known of the latter.

Hemochromatosis does not result from accelerated hemoglobin breakdown because the life span of the red cell is normal, and there is no anemia—hemolytic or otherwise. Since iron excretion is normally very low, it seems unlikely that hemochromatosis results from impaired iron excretion. Increased iron absorption seems to be the most probable cause of hemochromatosis. Whereas iron absorption is about 0.5–1.5 mg/day in normal individuals, it is 2.5–4 mg in patients with hemochromatosis. Even then, the daily iron overload is not great, and only after 40–60 years of accumulation are iron stores typical of the disease.

The mechanism controlling iron absorption in the intestinal mucosa is complex. Among the factors regulating iron absorption are interconversion between ferrous and ferric iron; interconversion between apoferritin and ferritin; iron binding to and release from transferrin molecules; iron storage in the hemosiderin molecule; feedback regulation between iron absorption in the intestine; and hemoglobin synthesis in the bone marrow. It is not known which step of the machinery controlling iron absorption is disrupted in hemochromatosis, but it seems safe to predict that the elucidation of the pathogenesis of hemochromatosis will closely follow upon knowledge of the molecular details of iron absorption.

In rats, the hepatic xanthine oxidase is absent in the fetus and the newborn and appears only after birth. Normally, ferritin iron content in liver is high before birth and decreases after birth when the xanthine oxidase activity appears in the liver. The ferritin iron stores remain low during weaning but increase when an iron-containing diet is taken.

When the plasma iron-binding capacity of rats is saturated by an intravenous infusion of iron, the further administration of iron causes the metal to be deposited in the liver. The liver iron is tightly bound and cannot be released into the circulation. Similar observations were made in patients with spontaneous hemochromatosis. In these individuals a large amount of the intestinally absorbed iron goes directly to the liver without appearing in the plasma, suggesting that the iron that is not bound to transferrin finds its way to the liver and remains trapped there. Whether low liver xanthine oxidase activity or low plasma levels of transferrin could play a role in the pathogenesis of hemochromatosis remains to be seen [38].

Some recent observations on the role of the pancreas in iron absorption may be relevant to the pathogenesis of hemochromatosis. Iron absorption is increased four times in patients with repeated attacks of pancreatitis. A heat- and acid-stable ultrafiltrable inhibitor of iron absorption has been found in the pancreas; its chemical composition is not known.

There is no apparent reason for iron to exert a toxic effect on the cell; yet, as indicated by the histological appearance of the liver, the liver cells degenerate, and the degenerative process is followed by regeneration and fibrosis. Injury and reaction to injury are responsible for the complex histological picture of cirrhosis.

The pathogenesis of cirrhosis in hemochromatosis is complex and probably involves a combination of iron overload and dietary deficiency. The combination of alcohol toxicity and iron overload has been incriminated in the hemochromatosis observed in heavy wine drinkers in areas where the wine (white or red) seemed to be rich in iron.

Two forms of therapy for hemochromatosis have been used: repeated bleeding to facilitate the use of blood stores for erythropoiesis, and administration of chelating agents.

Energetic phlebotomy has been claimed to prevent iron accumulation in a patient with hemochromatosis and even cure histological hepatic cirrhosis and reduce insulin requirements [39].

Iron Deficiency

One of the main consequences of iron deficiency is the reduced synthesis of hemoglobin, leading to the appearance of small (microcytic) erythrocytes poor in hemoglobin (hypochromic) in the blood. Since there is no interference with division and maturation of the mother cell of the erythrocyte, it is not surprising that the bone marrow cells attempt to compensate for the reduced concentration of hemoglobin per cell by increasing the number of cells. Thus, the bone marrow becomes hyperplastic. In fact, vicarious hematopoiesis often appears in spleen and liver of patients with iron deficiency anemia.

Reduction in the total amount of hemoglobin leads to pallor, which may occasionally be very severe, as in chlorosis. Pallor is most obvious on the gums and tongue. The patients manifest general weakness, fatigue, and adynamia. The reduction in blood viscosity often leads to the development of a mitral murmur.

If anemia is sustained for extensive periods, prolonged anoxemia develops, which may result in progressive heart failure with eventual fatty degeneration of the muscle fibers.

Iron deficiency is also associated with epithelial atrophy. The epithelial cells of the pharynx are first affected; the mucosa becomes reddish and infection often sets in, leading to dysphagia. This complex of symptoms is known as the Plummer-Vinson syndrome. The hair of iron-deficient patients becomes very fine and friable, and the nails break easily. The molecular mechanisms responsible for the alteration of the skin and hair of iron-deficient patients have not been elucidated.

Enzymic changes in tissues other than the erythron and the epithelia have also been reported in iron deficiency; the levels of iron-containing proteins—such as myoglobin, cytochrome, catalase, and others—are reduced. But the decrase in cytochrome activity must affect only the enzyme reserves because it is not accompanied by a corresponding decrease in oxygen uptake.

Other enzymic changes have also been observed in the erythrocyte of the iron-deficient patient, but these changes are difficult to interpret. Increases in acetylcholine esterase and glycerophosphatase, catalase, glutamic oxalacetic transaminase, aldolase, lactic dehydrogenase, and phosphohexose isomerase have decrease reported. These changes in enzyme activities could be only relative changes resulting from the loss of hemoglobin. If such an interpretation of the data is valid, it can be expected that enzyme activity is greater when expressed on the basis of volume, dry weight, or total protein content, but the activity is unchanged when expressed on a cell basis.

This is precisely what has been observed for acetylcholine esterase. Nevertheless, these changes in enzyme activity are of little metabolic consequence because they do not affect the red cell life span as long as the iron deficiency is corrected.

This description of the clinicopathological manifestations of iron deficiency suggests that the disease

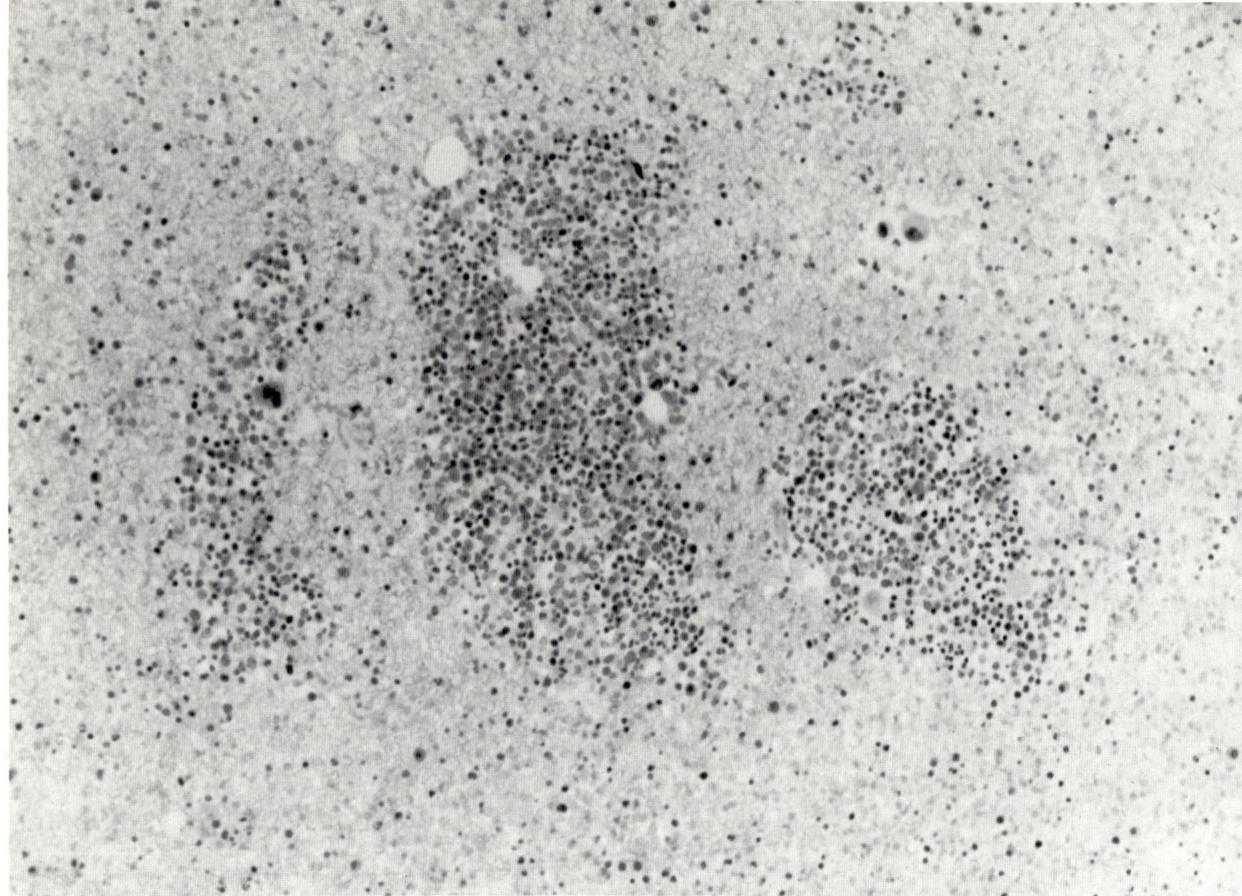

Fig. 6-8. Bone marrow aspirate clot section from an adult patient with iron deficiency anemia. Note hypercellularity and increased erythroid elements. (Maximow stain, ×40; from Dr. George Smith)

is easy to diagnose. Indeed, in most cases of iron deficiency, the severe microcytic hypochromic anemia is rapidly relieved by dietary iron intake. Yet sometimes iron deficiency anemia is difficult to distinguish from the severe hypochromic anemia occurring in thalassemia or in severe infections.

Plasma iron determinations are of little value in these cases because plasma iron is decreased in thalassemia and anemia resulting from infection. In contrast, an increase in the unsaturated iron binding capacity of the plasma is pathognomonic for iron deficiency. The cytological examination of a bone marrow smear stained for iron, however, is the most reliable method of diagnosis. Iron-stain cells are absent from the marrow of iron-deficient patients (see Fig. 6-8).

Although they are not well understood, the changes that take place in cells other than erythrocytes in iron deficiency may be significant in the pathogenesis of iron deficiency. For example, the reticulocyte response to iron administration is delayed in iron deficiency anemia. This delay is believed to result from injuries to the red cell precursors. The defects of the precursors are reflected by lower levels of RNA and DNA synthesis, reduced glycine incorporation in the liver, and reduced iron incorporation in hemoglobin. Thus, iron deficiency anemia appears to be more than a defect in hemoglobin and a disturbance in oxygen transport; it may include some broader cellular disturbance. For example, if iron deficiency interferes with the growth and maturation of the erythroblast, it could also interfere with the growth and maturation of other cells,

Table 6-5. Iron deficiency

1. Low plasma iron level (hypoferremia)	
Normal 110 µg/100 ml	Fe Def.[a] 20 µg/100 ml
2. Increase in transferrin	
Normal 350 µg/100 ml	33% Saturation
Fe Def. 500 µg/100 ml	4% Saturation
3. Decrease in storage iron	
Normal 1250 mg	
Fe Def. 0 mg	
4. Decrease in hemoglobin iron	
Normal 2000 mg	
Fe Def. 600 mg	

[a] Fe Def = Iron Deficiency.

such as the brain cell, and possibly lead to irreversible damage [40]. The pathogenesis of iron deficiency is summarized in Fig. 6-9 and some of the laboratory findings are given in Table 6-5.

Bile Pigments: Metabolism and Jaundice

Hemoglobin Breakdown and Bilirubin Formation

Only about three and a half months after its birth, the red cell is doomed and is phagocytized by the cells of the reticuloendothelial system. The mechanism of the complete degradation of all the components of the red cell is not known, but the fate of the heme of hemoglobin has been investigated extensively. Between 90–100% of the heme moiety of hemoglobin is converted to bile pigment [41–43].

The molecular events that take place during the first step of hemoglobin breakdown are not clear. Two pathways have been postulated. In the first, it is assumed that the iron protoporphyrin complex is first split from globin. Heme is oxidized in the process to hematin, an iron porphyrin derivative in which the iron is oxidized to yield the trivalent hydroxide. The removal of iron from hematin yields protoporphyrin IX. The four pyrrole rings of protophyrin IX are held together by methylene bridges. The oxidation of one of them (the one joining rings x and z) yields a tetrapyrrole chain biliverdin. Bilirubin is formed by the reduction of the central methylene carbon of biliverdin. The pathway that we have just described has been challenged, and it is now postulated that the breakdown of the tetrapyrrole ring takes place in a series of steps, all occurring while the protein is still attached to the tetrapyrrole ring (see Fig. 6-10).

The first steps in hemoglobin breakdown involve the opening of the tetrapyrrole ring and the loss of the iron molecule. The first intermediate is choleglobin, a green iron protein resulting from the split of the α-methylene bridge of heme of hemoglobin. Choleglobin is thus composed of an iron-containing tetrapyrrole chain attached to globin. But the exact structure of the molecule is unknown. Hemoglobin is converted to choleglobin when put in the presence of ascorbic acid and glutathione. This suggests that hemoglobin oxidation may be coupled with the reduction of ascorbic acid and glutathione. Choleglobin is present in normal blood, and its content increases after phenylhydrazine administration.

The degradation of 1 g of hemoglobin yields 30 mg of bilirubin. A normal individual destroys 7.5 g of hemoglobin per day when 1% of his circulating erythrocytes are replaced. At that rate of erythrocyte turnover, the daily bilirubin production equals 250 mg. The maximum capacity of the reticuloendothelial system to convert hemoglobin to bilirubin is five times that amount.

All the bilirubin found in serum does not derive from hemoglobin breakdown—10–30% has a different origin, but the exact source is not clear. It has been postulated to derive from heme synthesized in excess of globin; destruction of the red cells while they are still in the bone marrow; and destruction of other heme protein (myoglobin, catalase, and cytochromes).

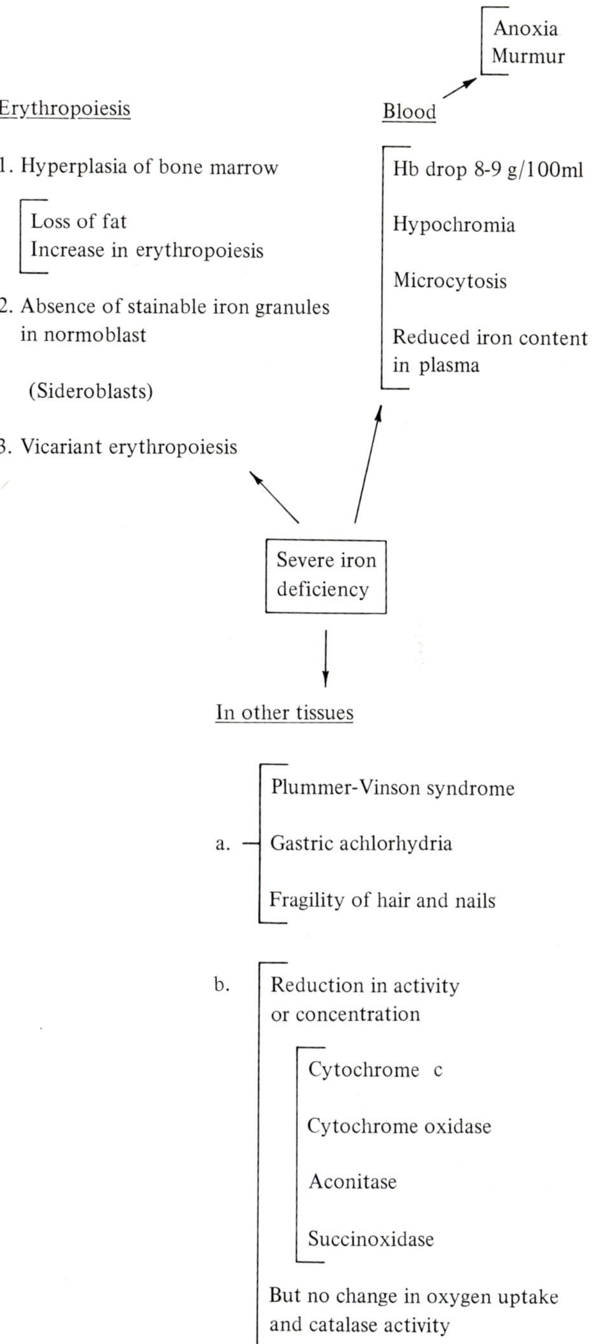

Fig. 6-9. Pathogenesis of iron deficiency

Bilirubin Transport

After it is produced in the reticuloendothelial system, bilirubin has to be transported in the plasma. The

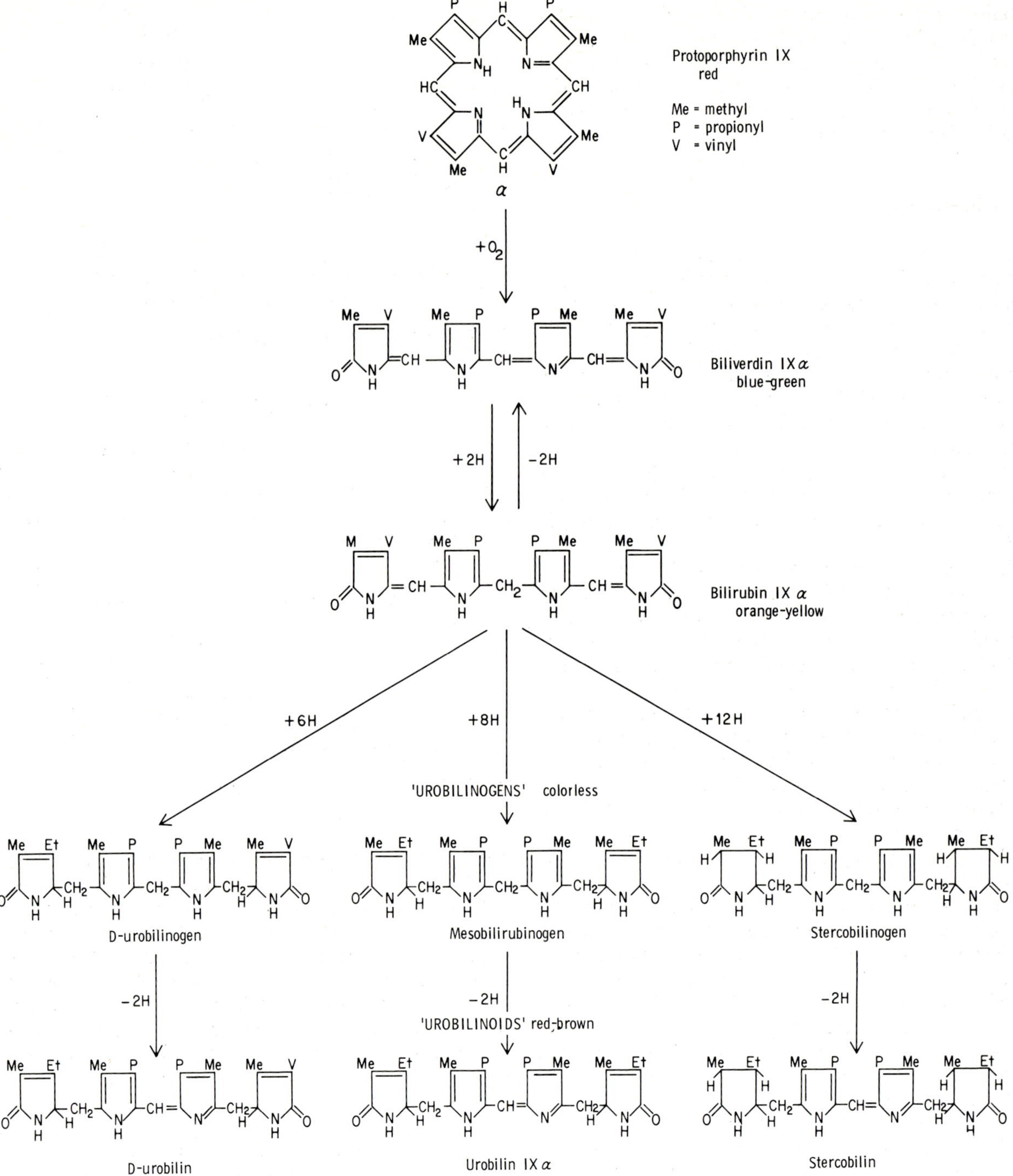

Fig. 6-10. Protoporphyrin breakdown

mechanism by which bilirubin is transferred from the reticuloendothelial cell to the capillary lumen is not known. In the plasma, bilirubin is bound to proteins. Two proteins are involved in complexing bilirubin: serum albumin and α-globulin. The affinity of bilirubin for albumin is much greater than its affinity for α-globulin. Two to three molecules of bilirubin can complex with one molecule of albumin. This complexing between bilirubin and an albumin is necessary for transport because bilirubin is practically insoluble in water at the pH of the blood. The nature of the binding between bilirubin and albumin is not clear, but it resists dialysis under normal conditions. The tetrapyrrole chain can, however, be separated from the protein under the influence of certain organic anions, including drugs such as salicylate and sulfathiazole. The amount of bilirubin normally present in the plasma ranges from 0.5–0.15 mg/100 ml of plasma.

The bilirubin transported in the plasma is eliminated from it by uptake in the liver cell. Again, the mechanism of transfer of bilirubin from plasma to hepatic cell is not known.

Bilirubin Excretion

Bilirubin is not used as such but is excreted by the liver cell. Bilirubin excretion starts with the conversion of bilirubin to bilirubin glucuronide, which was discovered rather recently. It was first observed that bile bilirubin hydrolysis yields two molecules of glucuronic acid per molecule of bilirubin. Bilirubin glucuronide was found in blood serum and was demonstrated to be the compound responsible for the classical van den Bergh reaction.

The formation of the glucuronide involves the propionic side chains of bilirubin, which are esterified to the hydroxyl group of the carbon 1 of glucuronic acid.

For the purpose of testing the bilirubin concentration in the blood, alcohol is usually added to the reaction mixture. Once, by mistake, van den Bergh omitted this step in determining the bilirubin concentration of blood of patients with obstructive jaundice, and this omission led to the discovery of the two types of diagnostic reactions referred to as the direct van den Bergh reaction (not requiring alcohol in the reaction mixture), and the indirect van den Bergh reaction (requiring alcohol). In 1956, it was discovered that indirectly reacting bilirubin is the conjugated product of bilirubin and glucuronide. Both unconjugated and conjugated bilirubin are bound to protein in the plasma.

Bilirubin glucuronide was later obtained by incubating bilirubin with slices, homogenates, or microsomal preparations obtained from liver in the presence of magnesium chloride. The bilirubin glucuronide produced was identified by its ability to react directly with the van den Bergh reagent; by the product of hydrolysis with β-glucuronidase (bilirubin and glucuronic acid); and by its mobility on chromatographic analyis.

Thus, liver microsome preparations undoubtedly catalyze the conversion of bilirubin to bilirubin glucuronide. In that reaction, the donor of the glucuronic acid moiety is UDP-glucuronic acid (see Fig. 6-11), which is derived from glucose through the glucuronic pathway. The uridine diphosphate glucuronic acid used in this reaction is produced by the oxidation of UDP-glucose by a liver supernatant enzyme in the presence of NAD (see Fig. 6-12). The product of that reaction is UDP-glucuronic acid (UDPGA) and NADH. It is not known whether the enzyme involved in the conjugation of bilirubin and glucuronic acid is specific for these two substrates, or if it is an enzyme with broader specificity that also catalyzes the conjugation of other compounds (for example, orthoaminophenol or N-acetylparaminophenol). Borneol inhibits the biosynthesis of bilirubin glucuronide in rat and monkey liver homogenates. In congenital hemolytic jaundice, the ability to conjugate N-acetylparaminophenol, tetrahydrocortisone, and sodium salicylate is reduced, suggesting a competition for a limited amount of enzyme, or substrate, UDPGA. Bilirubin glucuronides are still formed after total hepatectomy and bilateral nephrectomy, indicating that conjugation occurs in organs other than liver and kidney. The mechanism and site of this extrahepatic conjugation are unsettled.

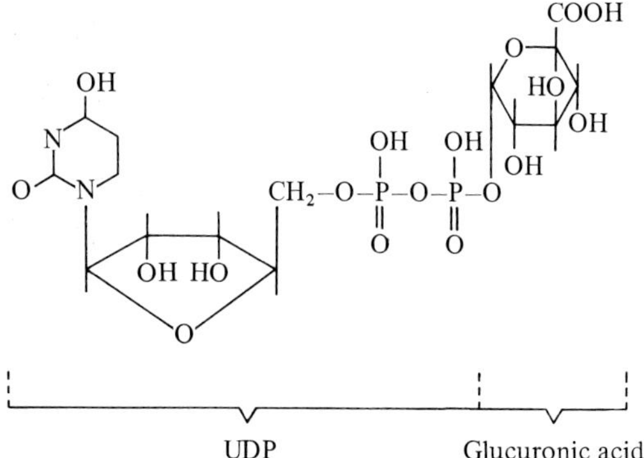

Fig. 6-11. UDP-glucuronic acid (UDPGA)

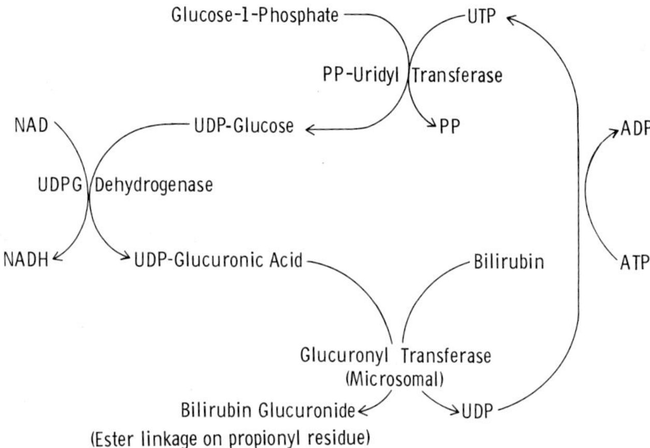

Fig. 6-12. Biosynthesis of bilirubin glucuronide

In liver necrosis, the activity of glucuronyl transferase decreases. This is thought to result from a reduction of microsomes.

Glucuronyl transferase appears in the serum in patients with hepatic necrosis. Thus, even when the liver transferase level is depressed, the serum enzyme can still conjugate bilirubin *in vitro*. In such cases it is possible that the activity of the blood enzyme is the source of conjugated bilirubin found in the serum in patients suffering from hepatic necrosis associated with impaired hepatic glucuronyl transferase activity.

Because the administration of glucuronic acid to newborn children reduces high serum levels of unconjugated bilirubin, it has been proposed that an extrahepatic pathway for the formation of conjugated bilirubin exists. The mechanism of this reaction is not known, and it has been suggested that glucuronic acid induces a shift in the distribution of unconjugated bilirubin from serum to adipose tissue.

Although the conversion of bilirubin to glucuronide is definitely the most important pathway in the metabolism of bile pigment, conversion to sulfate is also observed to a very small extent.

Bilirubin glucuronide synthesized in the hepatic cell is excreted by an unknown mechanism against a concentration gradient into the bile canaliculi, from where it is ultimately poured into the intestine.

Urobilinogens

In the intestine, conjugated bilirubin is split into glucuronic acid and bilirubin, and the bacterial flora of the colon convert the bilirubin molecule into a class of compounds referred to as urobilinogens. The intestinal degradation products of bilirubin include the reduced, colorless urobilinogens and the oxidized pigmented compounds or urobilins. Thus, normally little or no free bilirubin is excreted in the feces. Yet, free bilirubin appears in the stools of patients to whom broad-spectrum antibiotics are administered. The bile pigments are constantly removed from the intestine and returned to the liver (enterohepatic cycle). The reduced bilirubinogen, urobilinogen, mesobilirubinogen, and stercobilinogen are colorless, in contrast to the parent compound bilirubin. After urobilinogen injection in rat, all the reduced pigment is recovered in the bile, and no increase in the amount of bilirubin is observed. Therefore, all reabsorbed urobilinogens probably are re-excreted in the bile. In any event, urobilinogens are normally not found in the urine. Only when excessive amounts of urobilinogens are found in the blood are they excreted in the urine. Blood levels are high when urobilinogen excretion by the hepatic cell is impaired or when the intestinal reabsorption of urobilinogen is accelerated.

Until the advent of isotopes, it was believed that all the urobilinogens in the feces were derived from hemoglobin. Studies on the rate of formation of stercobilin using labeled precursors indicated that such a rate cannot be related to the breakdown of the circulating red cell. It is estimated that 70% of the stercobilin in the feces is derived from circulating red cells, 10–20% is derived from the immature red cells in the bone marrow, and 10% originates from the breakdown of other iron proteins.

The radioactivity of the hemin increases rapidly between the first and the third week after the administration of the isotope. A plateau is then reached and maintained for a period of about 3 months, followed by a slow drop in the radioactivity of the hemin. The radioactive precursor is incorporated into the stercobilin in quite a different pattern; there is a very rapid rise in the radioactivity during the first week, followed first by a quick drop to low values which are maintained for 3 months, then later by a second peak. The late radioactive peak in stercobilins seems easy to explain, since it coincides with the breakdown period of the red cell labeled at the time of the injection. The first radioactive peak in stercobilins is puzzling. It suggests that part of the stercobilin originates from a source of bile pigment with a rapid turnover. The exact source is not known. Such proteins as peroxidases, cytochromes, myoglobin, and catalase obviously are involved, but this does not explain the large amount of label associated with stercobilin in the first week after isotope injection. Part of the stercobilin may be derived from intramedullary breakdown of hemoglobin during red cell maturation. This could explain the high serum bilirubin in some pathological conditions, such as pernicious anemia.

In conclusion, bile pigment metabolism can be divided into at least seven separate steps: (1) hemoglobin breakdown in the reticuloendothelial cell; (2) transfer of bilirubin from the reticuloendothelial to the hepatic cell by way of the blood; (3) conversion of bilirubin to bilirubin glucuronide inside the hepatic cell; (4) excretion of conjugated bile in the bile duct; (5) urobilinogen formation; (6) urobilinogen reabsorption; and (7) re-excretion of urobilinogen in feces and urine (see Fig. 6-13).

Any pathological process that interferes with normal bilirubin production and excretion leads to accumulation of the pigment in blood and tissues, a condition called jaundice.

Jaundice may develop by three different mechanisms: hemolytic jaundice results from the overproduction of bile pigments; obstructive jaundice develops because of a block in the excretion mechanism; and hepatic jaundice is due to impairment of the hepatic secretion of bile.

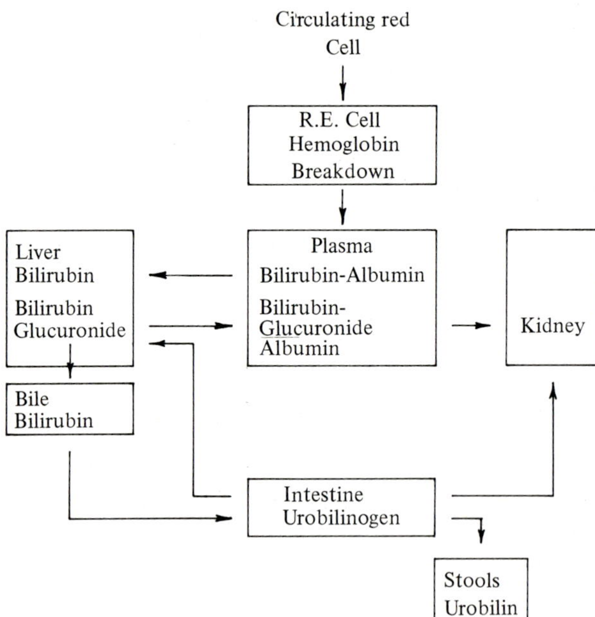

Fig. 6-13. Metabolism of bilirubin

Jaundice

Before describing the special types of jaundice observed classically, it is necessary to review the characteristics common to all forms of jaundice (see Fig. 6-14).

When the amount of bilirubin produced is greater than the amount excreted, bilirubin accumulates in the blood and stains the tissue, leading to yellow discoloration called jaundice. The mechanism by which bilirubin stains tissue is not clear. It is assumed that bilirubin is transferred from plasma proteins to tissue proteins, and, as can be expected, some proteins have a greater affinity for bilirubin than others. For example, elastic tissue avidly picks up bilirubin. Direct bilirubin is more readily fixed by tissues than indirect bilirubin, except in brain of the newborn, where unconjugated bilirubin is more easily picked up by the brain than conjugated bilirubin.

In jaundice, bilirubin may appear in the urine. Two factors seem to govern the renal excretion of bilirubin: the solubility of bilirubin and the renal clearance of bilirubin. Unconjugated bilirubin is not normally excreted by the kidney; only the more soluble conjugated bilirubin is found in the urine. The clearance for the diglucuronide is greater than that for the monoglucuronide. It is not known which part of the nephron excretes bilirubin. Some have postulated that bilirubin is excreted by glomerular filtration, and others have assumed that bilirubin is excreted by the tubules.

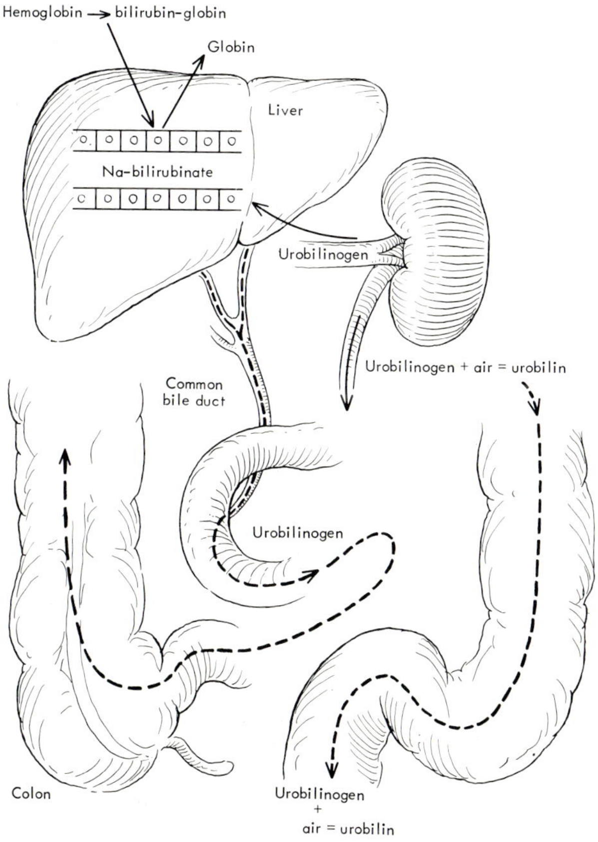

Fig. 6-14. Clinical jaundice

In the adult, the blood-brain barrier prevents the penetration of direct or indirect bilirubin into the cerebrospinal fluid. Ordinarily, only small amounts of bilirubin are found in the cerebrospinal fluid, even in severe jaundice. However, if bilirubinemia develops in the newborn before the blood-brain barrier has matured, high levels of conjugated or unconjugated bilirubin appear in the cerebrospinal fluid. The intraperitoneal injection of bilirubin to newborn rats leads to convulsion and bile staining of the brain. In the newborn infants with high levels of unconjugated bilirubin, a neurological situation known as kernicterus develops. The unconjugated bilirubin is fixed on the brain tissue, particularly in the area of the basal ganglia and also in a few other areas of the brain (see Fig. 6-15). Neurological disorders inevitably follow. The exact pathogenesis of the condition is not clear, but unconjugated bilirubin is known to uncouple oxidative phosphorylation of brain and liver mitochondria and depress oxygen uptake of minced brain tissue. It is not known if conjugated bilirubin affects brain metabolism *in vivo* by similar mechanisms, but the effects of conjugated bilirubin are of only academic interest since high levels of conjugated bilirubin are seldom encountered in the newborn.

Vitamin K analogs and sulfathiazole increase the incidence of and susceptibility to kernicterus in newborns with high levels of unconjugated bilirubin in the plasma. The mechanism by which these compounds act is not known, but it is postulated that they unbind bilirubin and serum albumin.

Neonatal Jaundice

Neonatal jaundice is the consequence of an immature conjugating system in the liver and an accelerated rate of red cell death. Bilirubin conjugation requires UDPGA, which acts as the glucuronic acid donor, and a microsomal enzyme called the transferase. The UDPGA is derived from UDP-glucose, and the conversion of the UDP-glucose to glucuronic acid requires the presence of UDPG dehydrogenase. In newborn children, the activities of both the dehydrogenase and the transferase are very low. In fact, it has been estimated that the activities of these enzymes are so low that the ability of the newborn liver to conjugate bilirubin is only 1–2% that of the normal adult liver. The two enzymes, however, appear rapidly after birth in the newborn liver and reach adult levels of activity within a few weeks after birth. In view of these observations, it is not surprising that all newborns show a certain degree of hemolysis, which reaches a maximum 2 or 3 days after birth and then progressively disappears. But the level of bilirubin in the blood does not correlate with the degree of icterus, and 20% of newborn children never exhibit any clinical signs of jaundice.

Since dehydrogenase and transferase activity is even lower in the fetus, it can be expected that the bilirubin

at its extremities, bone exhibits erythropoietic activity throughout its entire length. For example, the hyperplasia of the marrow in the flat bones of the skull leads to thinning of the cortex of the external tables of the bone, bone distortion, and abnormal calcification processes.

Obstructive Jaundice

In obstructive jaundice, bilirubin glucuronide accumulates in blood. Normally, the path for bilirubin from the hepatic cell to the bile duct is unidirectional. In obstructive jaundice, bilirubin is transferred from the hepatic cell to the blood. This probably results from regurgitation because damage to the hepatic cell would also impair glucuronide conjugation. Various pathways have been postulated for the regurgitation from the hepatic cell into the sinusoid and the lymph spaces, including rupture of the canaliculi and reversal of bilirubin transfer from hepatic cells to sinusoid (see Fig. 6-16). In the beginning of obstructive jaundice, the conjugated bilirubin is a diglucuronide. Later on, the monoglucuronide appears in higher proportion, and finally, unconjugated bilirubin is found in the plasma of these patients. In normal persons, less monoglucuronide than diglucuronide is present in the bile. The fact that in obstructive jaundice higher concentrations of monoglucuronide than of diglucuronide are found suggests that the hepatic function concerned with the conversion of mono- to diglucuronide is impaired. However, the higher amount of monoglucuronide in obstructive jaundice could also result from the contribution of extrahepatic sites of conjugation, or from greater renal clearance for diglucuronide than monoglucuronide.

Hepatic Jaundice

If the damage to the hepatic cell were primarily interference with the mechanisms of bile transport, conjugation, concentration, and excretion, the bilirubin accumulating in the blood would be unconjugated. But since the bilirubin accumulating in the blood in hepatic cellular jaundice is conjugated (bilirubin glucuronide), the main alteration in hepatic jaundice probably involves regurgitation of the bile, either through ruptured bile canaliculi or through some membrane abnormally permeable to the conjugated bile. Even in severe liver disease in which the activity of the glucuronyl serum transferase is reduced, the bilirubin in the plasma is conjugated. This suggests that the enzyme is not rate limiting in bilirubin formation, and that the normal cell contains large excesses of enzyme. Thus, interference with bilirubin excretion is more severe than interference with the activity of the enzymes that conjugate it. Another interpretation of these findings is that the enzyme glucuronyl transferase found in hepatocellular disease is partly responsible for extrahepatic conjugation.

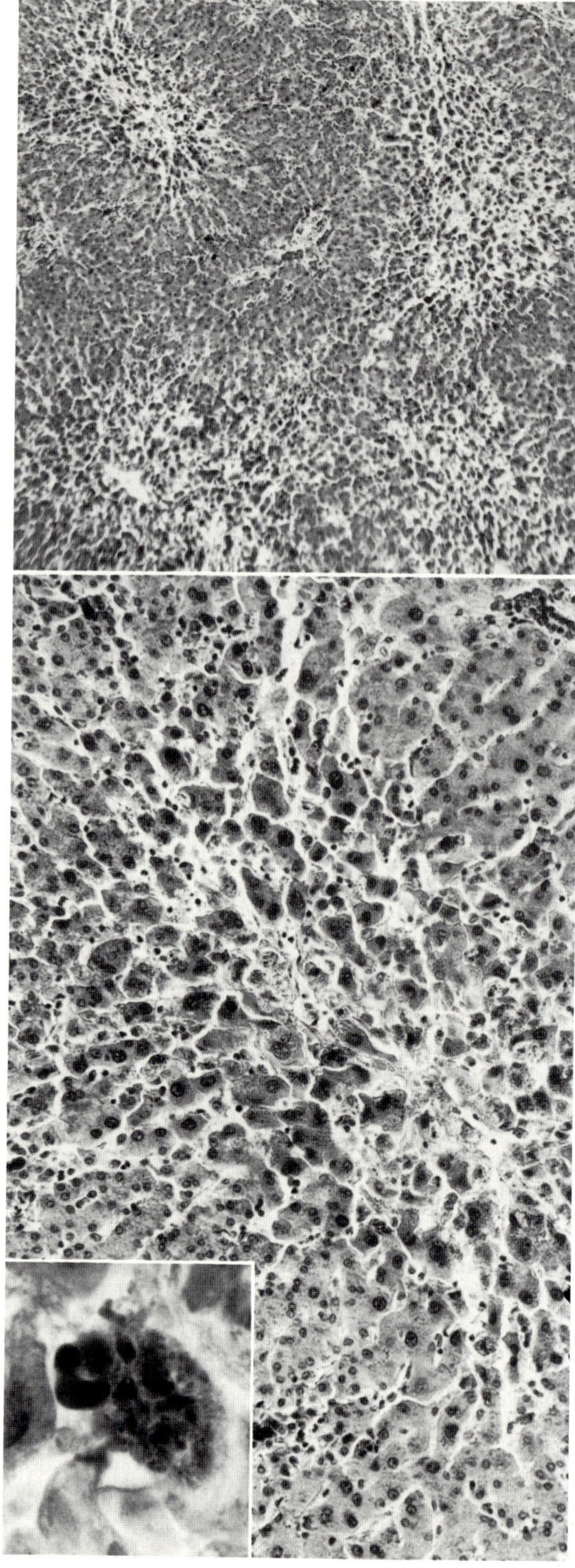

Fig. 6-16. Obstructive jaundice (low, medium, and high-power views) showing bile accumulation in hepatocytes and biliary canals

Cholangiolitic or Intrahepatic Obstructive Jaundice

In a series of hepatic conditions, there is no morphological alteration of either the bile duct or the hepatic cells, but a clinical and biochemical syndrome develops reminiscent of obstructive jaundice. Histologically, bile thrombi accumulate in the bile canaliculi, but the hepatic cell remains normal. No cause for obstruction can be detected, except perhaps for the presence of a few inflammatory cells in the portal tracts. The blood contains conjugated bilirubin, usually the monoglucuronide. That no hepatic damage is involved is indicated by the alkaline phosphatase level of the serum, which is within normal limits; glucuronyl transferase activity in the liver is also normal. Several pathogenic mechanisms have been proposed. One suggests a microobstruction due to compression of the bile canaliculi by inflammatory cells in the portal triad, or as a result of the formation of thrombi of inspissated bile. A second hypothesis postulates increased membrane permeability of the cells of the Herring canal. The third and most plausible hypothesis proposes that the mechanism transporting bile from the hepatic cell is blocked, and the conjugated bile is regurgitated into the blood and lymph. Intrahepatic obstructive jaundice occurs in sheep in South Africa, where it results from the absorption of a toxic grass. The icterogenic compound has been isolated and administered experimentally. These studies demonstrate that the main defect resides in an inability of liver cells to excrete bile.

Pernicious Anemia and Shunt Hyperbilirubinemia

In some diseases characterized by jaundice and the accumulation of unconjugated bilirubin in the blood, the amount of circulating bilirubin exceeds that expected from the rate of cellular death. This may be the case in pernicious anemia and in a disease called shunt hyperbilirubinemia, which developed during the second decade of life in four individuals, three of whom were related. The disease has many of the characteristics of a congenital hemolytic anemia: spherocytosis, reticulocytosis, increased osmotic fragility, and normoblastic hyperplasia of the bone marrow. The response to splenectomy is atypical; splenectomy normalizes the red cells' life span, but the hyperbilirubinemia remains. This observation suggests that the bilirubin originates from sources other than the circulating red cells. Such observations do not imply that the excess bilirubin originates from material other than hemoglobin, because in both pernicious anemia and shunt hyperbilirubinemia at least part of the excess bilirubin could be derived from the death of the red cells during their maturation in the bone marrow.

Gilbert's Disease

In addition to the classical congenital familial nonhemolytic jaundice, there exists another type of nonhemolytic jaundice that occurs either at an early age or in adults. In this disease, unconjugated bilirubin accumulates in the serum (indirect van den Bergh reaction), but the amount of bilirubin glucuronides excreted in the bile is not reduced. The nature of the molecular injury responsible for Gilbert's disease is not clear. The fact that unconjugated bilirubin accumulates suggests interference with the activity of transferase or an increase in glucuronide breakdown by β-glucuronidase. However, bilirubin glucuronide production must be normal since normal amounts are excreted in the bile. Since there is no hemolysis, bilirubin cannot be overproduced at the expense of hemoglobin.

Some investigators have claimed to have demonstrated an interference with transferase, but their results are still incomplete.

Studies of Felsher and Craig suggest an interference with the activity of the uridine diphosphate-glucuronyl transferase, eliminate the participation of β-glucuronidase, but do not exclude the contribution of other unknown factors in the pathogenesis of Gilbert's disease [50].

Identification of the primary injury in Gilbert's disease is complicated by the fact that the disease occurs

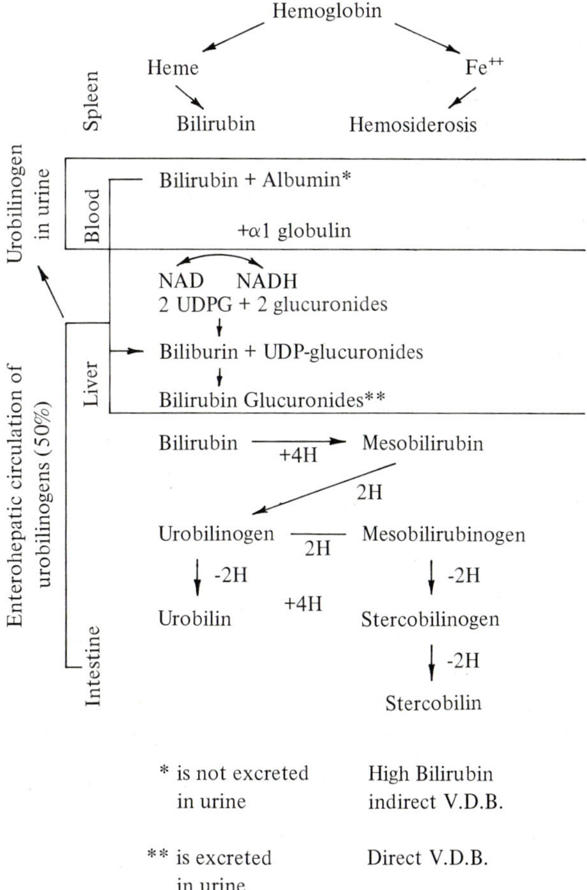

Fig. 6-17. Pathogenesis of hemolytic jaundice

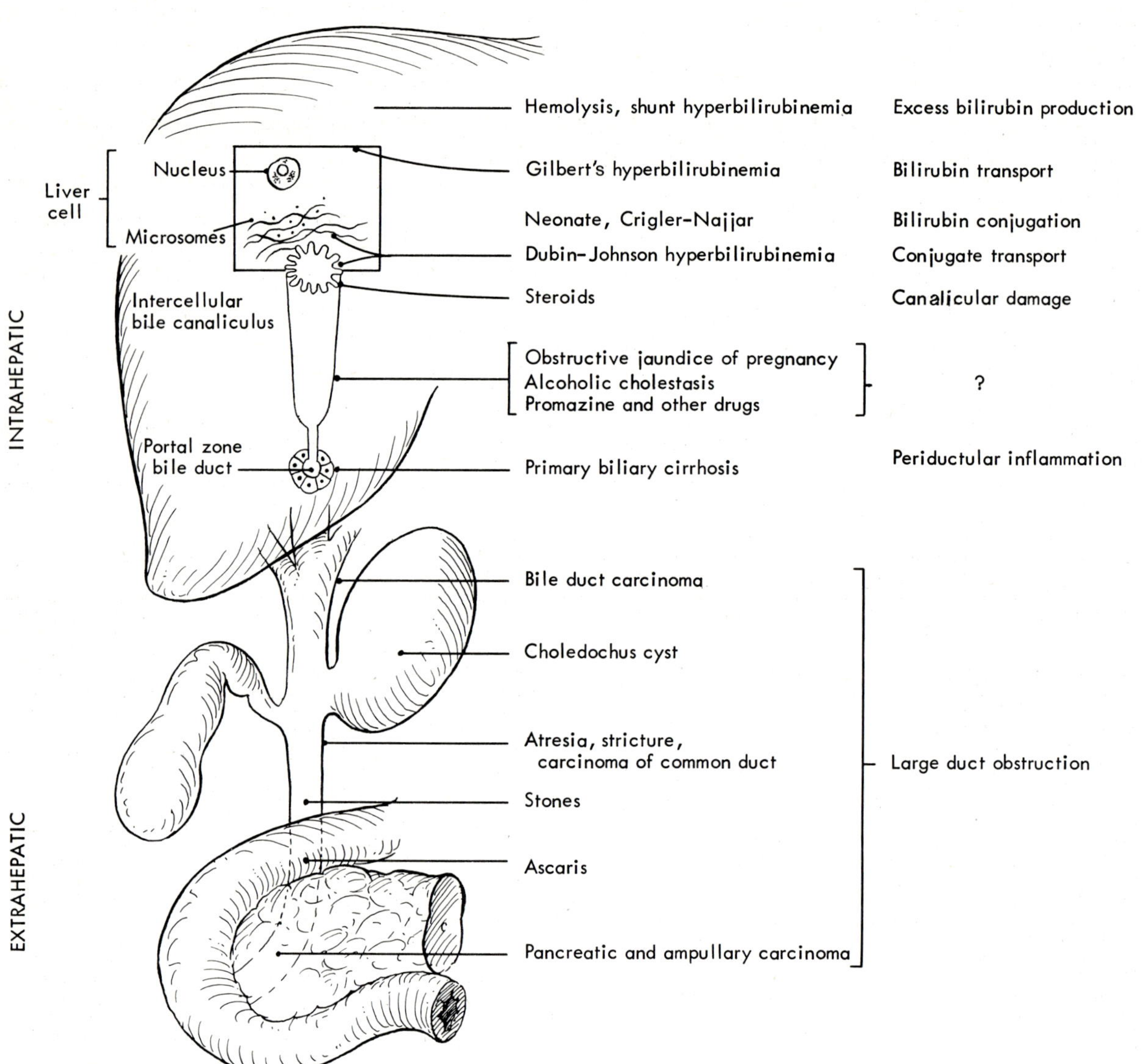

Fig. 6-18. Pathogenesis of nonhemolytic jaundice

years after birth. The enzyme activity is normal in early childhood but drops later. Therefore, if the primary injury is one of the gene itself, the affected gene must control the amount of enzymes synthesized rather than determine the amino acid assemblage in the protein. The delayed appearance of the disease may reflect the development of a repressive mechanism.

Dubin-Johnson Syndrome

A third form of congenital jaundice differs from the preceding one in that the bilirubin that accumulates in the blood is the direct type. Again, the pathogenesis of that disease is not know, but it cannot result from interference with the transferase system since bilirubin glucuronide accumulates in the blood. It has been postulated that the primary injury is an inability of the hepatic cell to excrete the conjugated bile pigment into the lumen of the hepatic canal. This interpretation is of little value since nothing is known of the molecular mechanism involved in that process. The pathogeneses of hemolytic and nonhemolytic jaundice are summarized in Figs. 6-17 and 6-18.

References

1. Gross, F., Naegeli, S.R., Philps, H.D., eds.: Iron metabolism. Int. Ciba Symp. Berlin-Göttingen-Heidelberg-New York: Springer 1964
2. Moore, C.F., Dubach, R.: Metabolism and requirements of iron in human. J. Amer. med. Ass. **162**, 197–204 (1956)
3. Bothwell, T.H., Finch, C.A.: Iron metabolism, 1st ed., p. 128. Boston: Little, Brown and Company 1962

4. Coons, C.M.: Iron metabolism. Annu. Rev. Biochem. **33**, 459–480 (1964)
5. Wheby, M.S., Crosby, W.H.: The gastrointestinal tract and iron absorption. Blood **22**, 416–428 (1963)
6. Bothwell, T.H., Charlton, R.W.: Absorption of iron. Annu. Rev. Med. **21**, 145–156 (1970)
7. Charley, P.J., Sarkar, B., Stitt, C.F., Saltman, P.: Chelation of iron by sugars. Biochim. biophys. Acta (Amst.) **69**, 313–321 (1963)
8. Charley, P.J., Stitt, C.F., Shore, E., Saltman, P.: Studies in the regulation of intestinal iron absorption. J. clin. Med. **61**, 397–410 (1963)
9. Dowdle, E.B., Schacter, D., Schenker, H.: Active transport of Fe59 by everted segments of rat duodenum. Amer. J. Physiol. **198**, 609–613 (1960)
10. Fineberg, R.A., Greenberg, D.M.: Ferritin biosynthesis; acceleration of synthesis by administration of iron. J. biol. Chem. **214**, 97–106 (1955)
11. Loftfield, R.B., Harris, A.: Participation of free amino acids in protein synthesis. J. biol. Chem. **219**, 151–159 (1956)
12. Pullman, T.N., Lavender, A.R., Forland, M.: Synthetic chelating agents in clinical medicine. Annu. Rev. Med. **14**, 175–194 (1963)
13. Westerfeld, W.W.: Effect of metal-binding agents on metalloproteins. Fed. Proc. **20**, 158–176 (1961)
14. Kugelmass, I.: The erythron as a carrier of hemoglobin. In: Biochemistry of blood in health and disease, p. 291–313. Springfield, Illinois: Charles C. Thomas Publisher 1959
15. Harris, J.: The red cell, p. 31–57. Cambridge, Massachussets: Harvard University Press 1963
16. Vuopio, P.: Red cell enzymes in anemia. Scand. J. clin. Lab. Invest. **15**, 1–90 (1963)
17. Mohler, D.N., Majerus, P.W., Minninch, V., Hess, C.E., Garrick, M.D.: Glutathione synthetase deficiency as a cause of hereditary hemolytic disease. New Engl. J. Med. **283**, 1253–1257 (1970)
18. Lajtha, L.G., Oliver, R.: Studies on the kinetics of erythropoieses: A model of the erythron. In: Haemopoiesis: Cell production and its regulation (Wolstenholme, G.E.W., and O'Connor, M., eds.). Proc. Ciba Symp., p. 289–314. Boston: Little, Brown and Company 1960
19. Lajtha, L.G., Gilbert, C.W.: Kinetics of cellular proliferation. Advanc. Biol. Med. Phys. **11**, 1–25 (1967)
20. Carnot, P., Deflandre, C.: Sur l'activité hémopöietique du sérum au cours de la régénération du sang. C. R. Acad. Sci. (Paris) **143**, 384–386 (1906)
21. Erslev, A.J.: Hematology: Control of red cell production. Annu. Rev. Med. **11**, 315–332 (1960)
22. Boggs, D.R.: Homeostatic regulatory mechanisms of hematopoiesis. Annu. Rev. Physiol. **28**, 39–56 (1966)
23. Gordon, A.S.: Humoral influences on blood cell formation and release. In: Haemopoiesis: Cell production and its regulation (Wolstenholme, G.E.W., and O'Connor, M., eds.). Proc. Ciba Symp., p. 325–361. Boston: Little, Brown and Company 1960
24. Linman, J.W., Bethell, F.H.: Factors in the control of haemopoiesis. In: Haemopoiesis: Cell production and its regulation (Wolstenholme, G.E.W., and O'Connor, M., eds.). Proc. Ciba Symp., p. 325–361. Boston: Little, Brown and Company 1960
25. Van Dyke, D.C.: Sources and properties of human urinary erythropoietin, in Haemopoiesis: Cell production and its regulation (Wolstenholme, G.E.W., and O'Connor, M., eds.). Proc. Ciba Symp., p. 397–417. Boston: Little, Brown and Company 1960
26. Stohlman, F., Jr.: The kidney and erythropoiesis. New Engl. J. Med. **279**, 1437–1439 (1968)
27. Beutler, E.: Hematology: Iron metabolism. Annu. Rev. Med. **12**, 195–210 (1961)
28. Gurney, C.W.: Polycythemia vera and some possible pathogenetic mechanisms. Annu. Rev. Med. **16**, 169–186 (1965)
29. Eichhorn, G.L.: Metal chelate compounds in biological systems. Fed. Proc. **20**, 40–51 (1961)
30. Malmström, B.G.: Role of metal-binding in enzymic reactions. Fed. Proc. **20**, 60–67 (1961)
31. Pearson, R.G.: Crystal field theory and the lability of complexes of the first transition series. Fed. Proc. **20**, 14–17 (1961)
32. Isaacson, C., Seftel, H.C., Keeley, K.J., Bothwell, T.H.: Siderosis in the Bantu: The relationship between iron overload and cirrhosis. J. Lab. clin. Med. **58**, 845–853 (1961)
33. Sheldon, J.H.: Haemochromatosis. London: Oxford University Press 1935
34. MacDonald, R.A.: Idiopathic hemochromatosis: Genetic or acquired? Arch. intern. Med. **112**, 184–190 (1963)
35. Block, M.: Hemosiderosis and hemochromatosis. II. In: Iron in clinical medicine (Wallerstein, R.O., and Mettier, S.R., eds.), p. 115–130. Berkeley: University of California Press 1958
36. MacDonald, R.A., Mallory, G.K.: Hemochromatosis and hemosiderosis: A study of 211 autopsied cases. Arch. intern. Med. **105**, 686–700 (1960)
37. Bothwell, T.H., Charlton, R.W., Seftel, H.C.: Oral iron overload. S. Afr. med. J. **39**, 892–900 (1965)
38. Mazur, A., Carleton, A.: Hepatic xanthine oxidase and ferritin iron in the developing rat. Blood **26**, 317–322 (1965)
39. Weintraub, L.R., Conrad, M.E., Crosby, W.H.: The treatment of hemochromatosis by phlebotomy. Med. Clin. N. Amer. **50**, 1579–1590 (1966)
40. Hershko, Ch., Karsai, A., Eylon, L., Izak, G.: The effect of chronic iron deficiency on some biochemical functions of the human hemopoietic tissue. Blood **36**, 321–329 (1970)
41. Fog, J., Jellum, E.: Structure of bilirubin. Nature (Lond.) **198**, 88–89 (1963)
42. Arias, I.M.: Hepatic aspects of bilirubin metabolism, Annu. Rev. Med. **17**, 257–274 (1966)
43. Menken, M., Barrett, P.V.D., Berlin, N.I.: Bilirubin production and excretion. Clinical considerations. Amer. med. Ass. J. **198**, 1273–1276 (1966)
44. Schulman, H.M., Martinez-Medellin, J., Sidloi, R.: The reticulocyte-mediated release of iron and bicarbonate from transferrin: Effect of metabolic inhibitors. Biochim. biophys. Acta (Amst.) **343**, 529–534 (1974)
45. Fielding, J., Speyer, B.E.: Iron transport intermediates in human reticulocytes and the membrane binding site of iron-transferrin. Biochim. biophys. Acta (Amst.) **363**, 387–396 (1974)
46. Brewer, G.J.: 2,3-DPG and erythrocyte oxygen affinity. In: Annual review of medicine (Creger, W.P., Coggins, C.H., and Hancock, E.W., eds.), vol. 25, p. 29–38. Palo Alto, Calif.: Annual Reviews Inc. 1974
47. Arnone, A.: Mechanism of action of hemoglobin. In: Annual review of medicine (Creger, W.P., Coggins, C.H., and Hancock, E.W., eds.), vol. 25, p. 123–130. Palo Alto, Calif.: Annual Reviews Inc. 1974
48. Kubanek, B., Bock, O., Heit, W., Bock, E., Harriss, E.B.: Size and proliferation of stem cell compartments in mice after depression of erythropoiesis. In: Haemopoietic stem cells. Ciba Found. Symp. 13, p. 243–255. Amsterdam: Associated Scientific Publishers 1973
49. Fried, W., Kilbridge, T., Krantz, S., McDonald, T.P., Lange, R.D.: Studies on extrarenal erythropoietin. J. Lab. clin. Med. **73**, 244–248 (1969)
50. Felsher, B.F., Craig, J.R., Carpio, N.: Hepatic bilirubin glucuronidation in Gilbert's syndrome. J. Lab. clin. Med. **81**, 829–837 (1973)
51. Kaplan, S.M., Rothmann, S.A., Gordon, A.S., Rappaport, I.A., Camiscoli, J.F., Peschle, C.: Extrarenal sites of erythrogenin production. Proc. Soc. exp. Biol. (N.Y.) **143**, 310–313 (1973)

Chapter 7

Blood Coagulation

Introduction

Blood coagulation might well be the chemical transformation in the blood known to physicians for the longest time. Still, little is known of the process except for some steps in fibrinogen conversion to fibrin. Therefore, blood coagulation continues to be described in mysterious terms, reminiscent more of a secret code than of chemical nomenclature.

The basic theory of blood coagulation was developed in the beginning of the 20th century and involved three major events: (1) thromboplastin formation, (2) the conversion of prothrombin to thrombin in the presence of calcium, and (3) the transformation of fibrinogen to fibrin. During the last decades, these three steps have remained the cornerstones of all theories on blood coagulation [1, 2]. However, many new factors activating or inhibiting these processes have been discovered, mainly by studying blood coagulation in patients with various types of hemophilia.

Although the biochemical change in the serum during blood coagulation plays a key role, clot formation also involves platelet aggregation and vasoconstriction. The hemostatic response to injury can thus be characterized as follows. Damage to the vascular wall exposes collagen, and tissue thromboplastin is released. The combination of collagen and thromboplastin activates the molecular mechanism of blood coagulation and the aggregation of platelets. Thus, prothrombin is converted to thrombin, and thrombin catalyzes the transformation of fibrinogen to fibrin and fibrin polymerizes Collagen also mediates platelet aggregation, which is followed by release of adenosine diphosphate and formation of a nonadhesive platelet clot. The platelets undergo degradation, and vasoconstrictors are released, contracting the blood vessels. Concentration of coagulating factors and stasis lead to the formation of a red thrombus.

What is remarkable in this homeostatic mechanism is not only that it operates so vigorously and rapidly, but also that it is restricted. A number of factors are responsible for restricting homeostasis to the injured area. Some are molecular and operate *in vitro,* such as: the formation of inhibitors of the coagulating system; the consumption of the various factors involved in clot formation; and the fibrinolytic process of fibrin through the plasminogen-plasmin system. *In vivo,* coagulation is further limited to the injured area by two other mechanisms: the constant dilution of coagulation factors by circulating blood, and clearance of the active substances by the liver.

Distortion of this delicate homeostatic balance may lead to either hemorrhagic disease or thrombi formation. The white thrombi result from interaction between platelets and abnormal vessel walls. Red thrombi are caused by the cooperative effect of hypercoagulability and stasis.

The Coagulation Theory

Thromboplastin is present in tissue juices and in platelets. In experimental studies on blood coagulation, thromboplastin is usually obtained by extracting it with saline from brain or lung. In the blood vessels, the thromboplastin that starts the coagulation process is generated by the disintegration of the blood platelets (which will be discussed in more detail later), but thromboplastin release coincides with the release of vasconstricting substances.

The antihemophilic globulin, the factor absent from the blood in the best-known hemophilia, combines stoichiometrically with a platelet thromboplastic factor, and the two compounds act like precursors, yielding a product itself a precursor. For thromboplastin, the product is prothromboplastin. The platelet thromboplastin component acts like a catalyst; consequently, its concentration does not change during the reaction. The prothromboplastin combines with another compound called the plasma-accelerating globulin to yield a new product, namely, thromboplastin.

Bovine and human antihemophilic globulin have been purified; the molecular weight of both human and bovine AHF is very high (120,000,000). Both the bovine and human proteins are made of several subunits held together by disulfide bonds. Treatment with 2 mercaptoethanol releases the subunits which all migrate as a single band on polyacrylamide gel. In spite of this electrophoretic homogeneity, heterogeneity of the subunit fraction has been demonstrated using ultracentrifugation [40].

The next series of transformations leads to thrombin formation. A precursor of thrombin, the familiar prothrombin whose synthesis in the liver requires vitamin K, combines with thromboplastin stoichiometrically. The reaction requires calcium and a platelet factor referred to as platelet factor 1. The product of the reaction is thrombin; but prothrombin production by the mechanism just described is so slow that the process must be activated for efficient coagulation, and thrombin catalyzes its own production. Thus, thrombin converts a plasma component, factor 1, or the plasma AC globulin into an active catalyst capable of accelerating prothrombin conversion to thrombin.

Thrombin finally catalyzes the only reaction in the cycle that is partially understood—fibrinogen conversion to fibrin. Obviously, these coagulation theories have many variations concerned mainly with the number and specific role of the factors involved. It seems futile to describe these theories in detail since the nature of most of the molecules involved in the reaction is not known.

Each phase of blood coagulation has been considerably oversimplified in the preceding paragraphs to facilitate understanding of the process. But now that the elementary factors involved have been outlined it might be worthwhile to describe some of the complex schemes that have been proposed.

Thromboplastin Formation

Thromboplastin formation is the most mysterious phase of blood coagulation. Compared to the formation of thrombin and fibrin, thromboplastin formation is slow and may therefore be critical in controlling diseases resulting from increased rates of blood coagulation. In his review, Pool [3] distinguishes between the formation of extrinsic and intrinsic thromboplastin activators (see Figs. 7-1 and 7-2).

During the formation of extrinsic thromboplastin, a tissue factor reacts with factors VII and X in the presence of plasma containing calcium and factor V to yield a compound capable of coagulating recalcified plasma. Factor VII regulates the rate while factor X controls the yields of the reaction. Therefore, it is believed that factor X acts as a substrate and factor VII as an enzyme in the formation of extrinsic thromboplastin.

The product of the reaction between the tissue factor, factor X (substrate ?), and factor VII (enzyme ?) reacts practically instantaneously with factor V to yield thromboplastin or the extrinsic prothrombin activator. It is not known whether in the extrinsic reaction factor V, also called proaccelerin, acts itself or only after it is converted to accelerin.

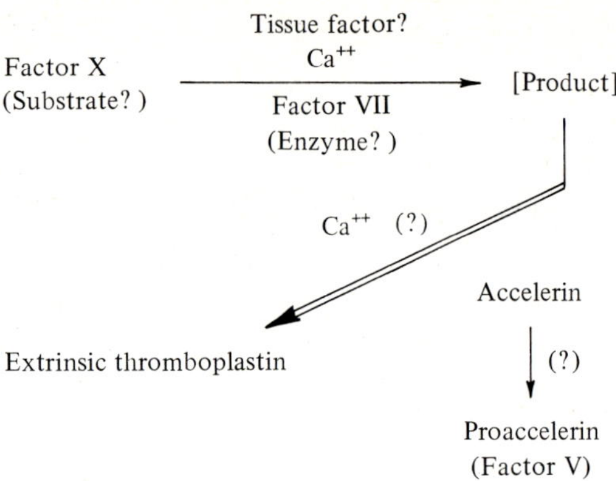

Fig. 7-2. Formation of extrinsic thromboplastin

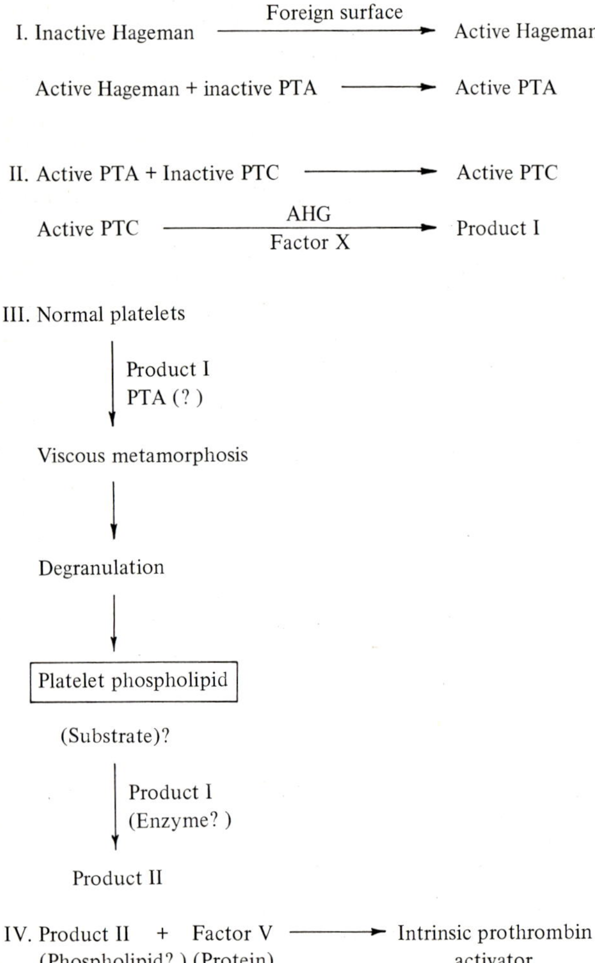

Fig. 7-1. Formation of intrinsic thromboplastin

Factor VII is a protease. It has been purified from bovine plasma. Unlike factor X and prothrombin it circulates in the plasma as an active enzyme. Factor VII requires a tissue factor before it can exert its proteolytic activity on factor X.

Apparently there are two forms of factor VII and one of them is sensitive to DFP. During the reaction with tissue factor, factor VII is hydrolyzed. The peptidase catalyzing the hydrolysis of factor VII is a form of the factor. Nemerson and Esnouf believe that the peptidase activity of the tissue factor is not indispensable for coagulation [41, 42].

The tissue factor is a particle bound protein found in many tissues. It has been purified and the use of peroxide :onjugated antisera has shown that the tissue factor is found in the cell membrane [43].

Factor X, which is the zymogen of the protease responsible for converting prothrombin to thrombin, is made of two polypeptide chains bound by disulfide bridges. Their respective molecular weights as determined by gel filtration are 44,000 and 15,000 daltons, with a total molecular weight of 59,000 daltons for the entire molecule. The amino acid sequence of the heavy chain, which is believed to contain the active site, resembles that of thrombin.

Intrinsic thromboplastin is formed in three steps. The first step involves the plasma thromboplastin component the antihemophilic globulin, the thromboplastin antecedent (PTA), the Hageman factor, and the Stuart-Prower factor. All these factors interact rapidly; they have not been purified, and therefore nothing is known of their molecular structure. With so little information, it would be rather naive to anticipate that the exact mode of interaction of these different factors would be known.

Present working hypotheses suggest that (1) an inactive Hageman factor is converted to an active Hageman factor in the presence of a foreign surface; and

(2) the active Hageman factor reacts with inactive PTA to convert it to active PTA.

In the second step of intrinsic thromboplastin formation, active PTA reacts with inactive plasma thromboplastin component (PTC) to convert it to active PTC. The active PTC reacts with AHG and factor X to yield a compound referred to here as product I. What is substrate and what is enzyme in this reaction are not known. A little is known of product I because it has been possible to purify it chromatographically on DEAE and on continuous flow electrophoresis. Its properties will be discussed below.

The third phase of extrinsic thromboplastin formation seems to be somewhat less complex than the first two steps. This phase involves the interaction of product I and platelet phospholipids. How the phospholipids are released from platelets during the formation of intrinsic thromboplastin is still debated, but two facts are significant: normal platelets have no phospholipid activity, and the development of phospholipid activity is associated with viscous metamorphosis and granular release from platelets. Either product I or PTA has been thought to produce viscous metamorphosis (see below). In the formation of product II, product I is thought to act as an enzyme on the platelet phospholipid to yield product II.

The last step of the formation of intrinsic thromboplastin concerns the conversion of product II to the intrinsic thromboplastin or intrinsic prothrombin activator. The present concept is that product II (a phospholipid) reacts with factor V (a protein) to yield an active lipoprotein. Little is known of product II. Also, it has been claimed to possess esterase activity.

Table 7-1 provides a list of some of the blood coagulating factors and the synonyms used.

Table 7-1

Antihemophilic globulin	AHG; Factor VIII; plasma thromboplastic factor; antihemophilic globulin A; thrombocytolysin; thromboplastinogen; plasmatic cofactor of platelets; coagulase globulin
Plasma thromboplastin component	PTC; factor IX; Christmas factor; antihemophilic globulin B
Plasma thromboplastin antecedent	PTA
Accelerator globulin	Ac-globulin; thrombogen; factor V; labile factor; component A of prothrombin; prothrombin accelerator; proaccelerin; accelerator factor; plasma prothrombin conversion factor; plasmatic cofactor of thromboplastin; prothrombinase
Serum prothrombin conversion factor	SPCA; factor VII; stable component; cofactor V; proconvertin-convertin; prothrombinogen; stable prothrombin conversion factor; cothromboplastin
Thromboplastin	Plasma thromboplastin; tissue thromboplastin; thrombokinase; cytozyme; cephalin

Conversion of Prothrombin to Thrombin

According to Seegers [4], the conversion of prothrombin to thrombin can be catalyzed by a two-enzyme system, thromboplastin and thrombin itself. The compounds participating in this reaction [5–8] are somewhat better known than those involved in thromboplastin formation (prothrombin activator).

Prothrombin, the precursor of thrombin, is present in plasma only in small amounts (0.2% of the protein content of plasma) and can be purified by acid precipitation by adsorption on magnesium hydroxide and fractionation with ammonium sulfate. The purified protein contains glucose (6.5%), neuraminic acid (5%), and unidentified amino sugars. A study of the amino acid composition reveals that the molecule contains 19 different amino acids. Prothrombin has only one N-terminal amino acid (alanine), and no carboxy C-terminal can be detected by hydrolyzing the molecule with carboxypeptidase A or B. Prothrombin is an ellipsoid molecule (119A × 34A) with a molecular weight of 62,700. On electrophoresis of plasma, prothrombin migrates between the α_2- and the β-globulin.

During blood coagulation, prothrombin is activated to yield thrombin. Tissue thromboplastin and calcium are required for the activation. The exact mechanism of activation is not known, and little information is available on the product of the activation—thrombin.

On the basis of his studies of the conversion of prothrombin to thrombin, Seegers has developed what he refers to as the enzymic theory of blood clotting. When prothrombin is incubated in the presence of thromboplastin, autoprothrombin 1-C is obtained. The properties of that compound are still unknown. Now, if Ac-globulin is added to the incubation mixture containing prothrombin and thromboplastin, the site of cleavage of prothrombin by thromboplastin (which acts as a proteolytic enzyme) is modified, and thrombin instead of autoprothrombin 1-C is formed. Autoprothrombin III, a new polypeptide, appears during the reaction. The addition of lipid to the reaction mixture accelerates thrombin formation. From these findings, Seegers has concluded that thromboplastin is a lipoprotein-activated proteolytic enzyme the specificity of which is determined by the Ac-globulin.

It is therefore of interest that prothrombin was found to bind *in vitro* to phospholipid particles, and that the rate of conversion of prothrombin to thrombin was directly related to the binding level.

When bovine prothrombin is incubated in presence of phosphatidylcholine and phosphatidylserine in presence of Ca^{2+} a complex made of equimolecular amounts of each of the phospholipid and the protein is formed [44].

Factor X immobilized on polyacrylamide gel is hydrolyzed by trypsin to yield an active form of factor X(Xa). Xa, also referred to as autoprothrombin, is a protease. It catalyzes the conversion of prothrombin to thrombin, but the reaction is slow in absence of calcium and phospholipids and factor V.

$$\text{Prothrombin} \xrightarrow[\text{Ca}^{++}]{\text{Thromboplastin}} \text{Thrombin} + \text{Autoprothrombin III}$$

AC=Globulin lipid

The difficulties in investigating the conversion of prothrombin to thrombin will be illustrated by describing briefly the work of two laboratories.

Studies of Heldebrandt and his associates in which the activation of prothrombin was investigated using factor Xa yielded four different intermediates. Their molecular weights and their NH_2-terminal amino acids have been determined.

Intermediate	M.W.	NH_2 Terminal
I	51,000	serine
II	41,000	threonine
III	23,000	alanine
IV	14,000	
Prothrombin	70,000	alanine

Intermediates I and III are obtained by treating prothrombin either with Xa or thrombin. Either protease yields intermediates with the same molecular weight and the same amino terminal. Treatment of intermediate I with factor Xa yields intermediates II and IV.

While the treatment of prothrombin with Xa yields intermediates I, II, III, and IV, its treatment with thrombin only yields intermediates I and II. On the basis of these observations a mechanism for the cleavage of prothrombin has been proposed.

In the first step of activation prothrombin is attacked at both its COOH and its NH_2 terminal end to yield intermediates I and III, respectively. This reaction can be catalyzed by either thrombin or factor Xa. In the next step intermediate I is attacked by factor Xa and the cleavage yields intermediate II from its COOH terminal and intermediate IV from its amino terminal region. Intermediate II is believed to be the immediate precursor of thrombin. Its NH_2 terminal end contains the A chain and its COOH terminal segment, the B chain of thrombin. Cleavage of intermediate II yields the active thrombin. In this final step an arginine-isoleucine peptide bond is hydrolyzed at a point joining the COOH terminal amino acid of the A chain and the NH_2 terminal amino acid of the B chain [47–49].

Owen and associates have compared the intermediates obtained after the hydrolysis of prothrombin with factor Xa phospholipids, factor V, Ca^{2+}, and Xa alone. The rapid activation of prothrombin with the complete system yields thrombin and two fragments (1 and 2) that cannot be converted to thrombin. The slow activation by Xa yields 2 intermediates (1 and 2) which can be converted to thrombin. Incubation of prothrombin with thrombin yields intermediate 1 and fragment 1. Intermediate 1 is activated by factor Xa to yield intermediate 2 (mol wt 37,000) and fragment 2 (mol wt 13,000).

$$\underset{48{,}000}{\text{Intermediate 1}} \xrightarrow[\text{Xa}]{} \underset{37{,}000}{\text{Intermediate 2}} + \underset{13{,}000}{\text{Fragment 2}}$$

$$\underset{72{,}000}{\text{Prothrombin}} \xrightarrow[\text{thrombin}]{} \underset{48{,}000}{\text{Intermediate 1}} + \underset{22{,}000}{\text{Fragment 1}}$$

When prothrombin is incubated with Xa phospholipid, factor V, and Ca^{2+} in presence of diisopropylfluorophosphate, an inhibitor of thrombin, a portion of the NH_2 terminal segment of prothrombin is cleaved and has been referred to as fragment F1.2. The polypeptide contains most of the prothrombin sequence extending the thrombin sequence. Fragment 1.2 cannot be cleaved by Xa alone. But when treated with thrombin it yields two fragments: 1 and 2.

Fragment 1 is identical to that obtained after treatment of prothrombin with thrombin and fragment 2 to that obtained after treatment of intermediate 1 with factor Xa [45, 46]. Further studies from the same laboratory revealed that fragment 1 from the NH_2 terminal end binds phospholipids to prothrombin while factor V interacts with fragment 2.

On the basis of these studies the following scheme for prothrombin conversion has been proposed.

$$\text{Prothrombin} \xrightarrow{\text{Factor Xa, Ca}^{2+}} \text{Intermediate 2} + \text{Fragment 1.2}$$

$$\text{Intermediate 2} \xrightarrow{\text{Factor Xa, Ca}^{2+}} \text{Thrombin}$$

Thus prothrombin activation requires the cleavage of two peptide bonds. It can be achieved by factor Xa alone, but the activation is increased fourfold if factor V, phospholipids, and calcium are also present.

Unless rapidly inactivated, thrombin will catalyze the cleavage of fragment 1.2.

$$\text{Fragment 1.2} \xrightarrow{\text{Thrombin}} \text{Fragment 1} + \text{Fragment 2}$$

and split fragment 1 from prothrombin. It thereby removes the site of binding of phospholipid and prevents rapid activation of prothrombin [63–65].

Thrombin has been purified by chromatographic techniques on IRC 50 columns. The amino acid composition of some thrombin preparations has been studied. Thrombin prepared from citrate-activated type I prothrombin lacks tryptophan and half cystine whereas thrombin prepared from extrinsic activation of type I prothrombin contains the 19 amino acids found in prothrombin. One molecule of thrombin contains one isoleucine N-terminal and one aspartic C-terminal [9].

When diisopropylphosphofluoridate-containing radioactive phosphorus is reacted with thrombin, thrombin is inactivated. Thrombin is likely to be a polymer in which the monomer has a molecular weight of approximately 8,000; only the monomeric form is fully active. Thrombin contains three serine residues, but only one of these is labeled by [^{32}P]DIPF, and the amino acid sequence of the serine polypeptide of thrombin is similar to that of chymotrypsin:

Thrombin: Gly–Asp–Ser–Gly–(Gly–AC)

Chymotrypsin: Gly–Asp–Ser–Gly–Glu–Ala–Val

Thrombin attacks synthetic esters, namely tosyl L-arginine methyl ester (TAMe), and when thrombin is treated with DIPF, the loss of esterase activity is proportional to the loss of clotting activity. Studies with synthetic esters have suggested that, as with trypsin, the reaction between thrombin and its substrates occurs in three steps: an enzyme-substrate complex is formed; the acyl portion of the ester is transferred to the enzyme, which becomes acylated while the alcoholic portion of the ester is released; and the enzyme is deacylated and the peptide is transferred to water. The natural substrate of thrombin is fibrinogen. In addition to converting fibrinogen to fibrin, thrombin releases the contractile protein from platelets, activates fibrinase, and may participate in the conversion of prothrombin to thrombin.

Properties of Fibrinogen

Fibrinogen can easily be separated from plasma by salting out because of its low solubility in ammonium sulfate. When fibrinogen is further purified by repeated salting out and chromatographic techniques, a protein is obtained with good antigenic properties. The antigen also precipitates breakdown products of fibrinogens prepared from various animal sources (hog, cow, and horse). All these active antigens have tyrosine as the N-terminal residue. Although the natural enzyme that attacks fibrinogen is thrombin, fibrinogen can also be digested by trypsin. Trypsin digestion of fibrinogen occurs in two steps, a first, rapid step that splits 12 peptide bonds, and a second, slower step attacking 80 other peptide bonds in the fibrinogen molecule [9–17].

The effect of trypsin on fibrinogen is very different from that of thrombin because thrombin hydrolyzes only four arginyl-glycine bonds in the fibrinogen molecule and thus manifests an unusual specificity. It has been postulated that the carbohydrate moiety of fibrin interferes with further attacks of thrombin on the protein.

Fibrinogen is an ellipsoid molecule 600 A long and 50 A wide. Electron microscopic examination of the molecule reveals a unit structure composed of three nodules (65, 50, and 65 A) held together by thin threads (15A thick). The fibrinogen molecule 2700 amino acids) is a dimer, and each half is composed of three polypeptides (α-, β-, and γ-chains) held together by disulfide bridges and hydrogen bonds. There appear to be three interchain bonds, β-α, β-γ, α-γ, and two intrachain bonds are in the α- and γ-chains, respectively.

The N-terminal groups of the three chains of the monomer have been identified. In bovine fibrinogen they are tyrosine, glutamic acid, and threonine. Examination of fibrinogen particles with the electron microscope and comparison of the picture with that obtained with crystalloids of known shapes suggest that the fibrinogen nodules are pentagonal dodecahedrons (see Table 7-2).

Fibrinogen (mol wt 333,000) is a glycoprotein that has the mobility of a globulin. The carbohydrate moiety is composed of a six-oligosaccharide chain of sialic acid, galactose, mannose (linked to asparagine), and glucosamine. Solutions of the purified proteins are highly viscous and birifringent. The sialic acid is likely to occupy a terminal position in the fibrinogen molecule because all the sialic acid can be split by sialidase. The fibrinogen molecule contains polysaccharide chains, and their sequence is postulated sialic acid (1), galactose (1), mannose (2), and hexosamine (2).

Whether the fibrinogen with lower molecular weight (269,000) is derived by chain splitting from fibrinogen of higher molecular weight (320,000) is not certain.

Table 7-2. Amino acid sequence of human fibrinogen

α-chain

	5	10	15	20	25
1	Ala–Asp–Ser–Gly–Glu–Gly–Asp–Phe–Leu–Ala–Glu–Gly–Gly–Gly–Val–Arg–Gly–Pro–Arg–Val–Val–Glu–Arg–His–Gln–Ser–				
27	Ala–Cys–Lys–Asp–Ser–Asp–Trp–Pro–Phe–Cys–Ser–Asp–Glu–Asp–Trp–Asn–Tyr–Lys–Cys–Pro–Ser–Gly–Cys–Arg–Met/				

β-chain

	5	10	15	20
1	Glx[a]–Gly–Val–Asn–Asp–Asn–Glu–Glu–Gly–Phe–Phe–Ser–Ala–Arg–Gly–His–Arg–Pro–Leu–Asp–Lys/			

γ-chain

	5	10	15	20	25
1	Tyr–Val–Ala–Thr–Arg–Asp–Asn–Cys–Cys–Ile–Leu–Asp–Glu–Arg–Phe–Gly–Ser–Tyr–Cys–Pro–Thr–Thr–Cys–Gly–Ile–Ala–				
27	Asp–Phe–Leu–Ser–Thr–Tyr–Gln–Thr–Lys–Val–Asp–Lys–Asp–Leu–Gln–Ser–Leu–Glu–Asp–Ile–Leu–His–Gln–Val–Glu–Asn–				
54	Lys–Thr–Ser–Glu–Val–Lys–Gln–Leu–Ile–Lys–Ala–Ile–Gln–Leu–Thr–Tyr–Asn–Pro–Asp–Glu–Ser–Ser–(Asp, Lys–Pro)–				
79	Met/Leu–Thr–Ile–Gly–Glu–Gly–Gln–Gln–His–His–Leu–Gly–Gly–Ala–Lys–Asp–Val/				

[a] Glx = undefined; may be either Glu or Gln.

In addition to the classical fibrinogen, human plasma contains a fibrinogen with high solubility and lower molecular weight (269,000). This fraction may represent as much as 25% of the total plasma fibrinogen. The high-solubility fibrinogen has mobility and immunological properties similar to those of the regular fibrinogen. Thus, it cannot be separated from regular fibrinogen by electrophoresis or immunoelectrophoresis. Yet the higher solubility fibrinogen clots more slowly than lower solubility fibrinogen. The reasons for these differences are not known. Studies have shown that the two types of fibrinogen molecules do not differ in amino acid terminal groups or tyrosine-tryptophan ratio computed from ultraviolet spectral curves. Greater amounts of sialic acid and hexose were found, however, in the molecule with higher solubility.

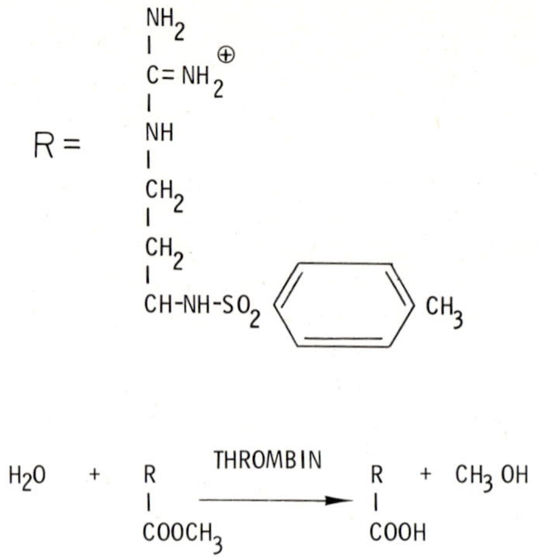

$$R = NH_2-C(=NH_2^{\oplus})-NH-CH_2-CH_2-CH(-NH-SO_2-C_6H_4-CH_3)-$$

$$H_2O + \underset{COOCH_3}{R} \xrightarrow{\text{THROMBIN}} \underset{COOH}{R} + CH_3OH$$

Fig. 7-3. Hydrolysis of tosylarginine methyl ester (TAMe) by thrombin

Conversion of Fibrinogen to Fibrin

Thromboplastin formation and thrombin elaboration are only steps in a cascade of reactions, which to be successful must convert fibrinogen molecules into fibrin. The conversion of fibrinogen into fibrin involves the splitting of peptide bonds, the release of sialic acid, and the rupture of hydrogen bonds.

During its conversion into fibrin, fibrinogen loses some of its negative charges, and the terminal amino groups of the fibrinogen molecules are replaced by glycine residues in the product (fibrin). Two acidic peptides (fibrinopeptide A and B) are released during the hydrolysis of fibrinogen by thrombin. The C-terminal residue of these peptides is arginine. The presence of N-terminal glycine in fibrin and C-terminal arginine in the fibrinopeptide is in keeping with the specificity of thrombin, which acts by splitting arginyl-glycine bonds (see Fig. 7-3). The amino acid sequences of bovine peptide A (18 amino acids) and peptide B (20 amino acids) are listed in Table 7-3. Peptide A contains an unusual biochemical that has not been found in other materials: tyrosine-0-sulfate. The peptide contains 2-N-terminal tyrosine.

The positions of fibrinopeptides A and B in the fibrinogen molecule could be deduced by sophisticated observation of the clotting process. Depending upon the ionic environment, a transparent or a turbid clot forms. Electron microscopic examinations have established that the formation of a transparent clot is associated with end-to-end agglutination of the molecules. In contrast, the formation of a turbid clot results from side-to-side agglutination. The release of fibrinopeptide A modified the dipole movement of fibrinogen in such a way that the product, the fibrin molecules plus peptide B, agglutinates end-to-end. In contrast, the removal of polypeptide B induces side-to-side agglutination.

Thus, the removal of polypeptide A is clearly needed for initial polymerization, but not that of polypeptide B. In fact, polypeptide B is removed much more slowly than polypeptide A.

The rupture of fibrinogen to yield fibrin involves the splitting of four covalent bonds between arginine and glycine, and it also involves the rupture of hydrogen bonds. The breakdown of the peptide bond releases protons that generate heat. Measurement of the heat released during the splitting of the four peptide bonds suggests that only two protons are liberated during the conversion of fibrinogen to fibrin, when in reality four bonds are split. This discrepancy has now been explained. The thermal bombardment result-

Table 7-3. Amino acid sequence of fibrinopeptides

A	
	1 2 3 4 5 6 7 8 9 10 11 12 13 14 15 16 17
Human[a]	Ala–Asp–Ser–Gly–Glu–Gly–Asp–Phe–Leu–Ala–Glu–Gly–Gly–Gly–Val–Arg
Rhesus Monkey[a]	Ala–Asp–Thr–Gly–Glu–Gly–Asp–Phe–Leu–Ala–Glu–Gly–Gly–Gly–Val–Arg
Pig, Boar	Ala–Glu–Val–Gln–Asp–Lys–Gly–Glu–Phe–Leu–Ala–Glu–Gly–Gly–Gly–Val–Arg
B	
	1 2 3 4 5 6 7 8 9 10 11 12 13 14 15 16 17 18 19
Human[b]	Glx[c]–Gly–Val–Asn–Asp–Asn–Glu–Glu–Gly–Phe–Phe–Ser–Ala–Arg
Rhesus Monkey[b]	Asn–Glu–Glu–Ser–Pro–Phe–Ser–Gly–Arg
Pig	Ala–Ile–Asp–Tyr–Asp–Glu–Asp–Glu–Asp–Gly–Arg–Pro–Lys–Val–His–Val–Asp–Ala–Arg

[a] Chimpanzee and gorilla are identical to human.
[b] Glx = undefined; may be either Glu or Gln.

ing from proton release breaks hydrogen bonds (3 or 4), and this process absorbs heat. Therefore, the polypeptide chains of fibrinogen are held together by hydrogen bonds in addition to SH bonds.

During the conversion of fibrinogen to fibrin, 20% of the sialic acid is released in a bound form. Furthermore, preincubation of fibrinogen with neuraminidase enhances the effect of thrombin. Attempts have been made to establish which molecular fraction is susceptible to the action of thrombin. We have already seen that the polysaccharide must be close to the site of action of thrombin on the fibrinogen molecule. The physiological role of the released polypeptide has not been elucidated, but it has been observed that fibrinopeptide B sensitizes smooth muscle to stimulation.

Polymerization of Fibrin

The product of the action of thrombin is fibrin, an unstable monomer that polymerizes in two steps. The first step involves the formation of electrostatic and hydrogen bonds between adjacent monomers. This nonenzymic step is followed by an enzymic step in which calcium is required for the reaction. Thrombin activates a precursor, the fibrin-stabilizing factor, to yield a transpeptidase (or fibrinase) that catalyzes the formation of ε (γ-glutamyl)-lysine cross-links.

While the clot formed when purified thrombin and purified fibrinogen are used is readily soluble in 5 M urea, that formed in the presence of serum is insoluble in urea. This finding indicates that bonds other than hydrogen bonds are formed during clotting in the presence of serum. The hydrogen bonds probably are formed by the sharing of hydrogen by a tyrosine residue of one fibrin molecule with the histidine residue of another fibrin molecule. Two observations support the participation of histidine in clot formation: (1) clots do not form at low pH because at acidic pH histidine grabs the free protons of the buffer and is unable to share the hydrogen of tyrosine; and (2) photooxidation of the histidine residue prevents clot formation.

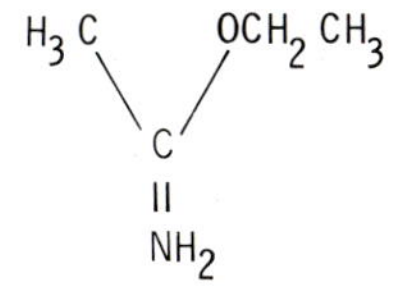

Ethyl acetamidate

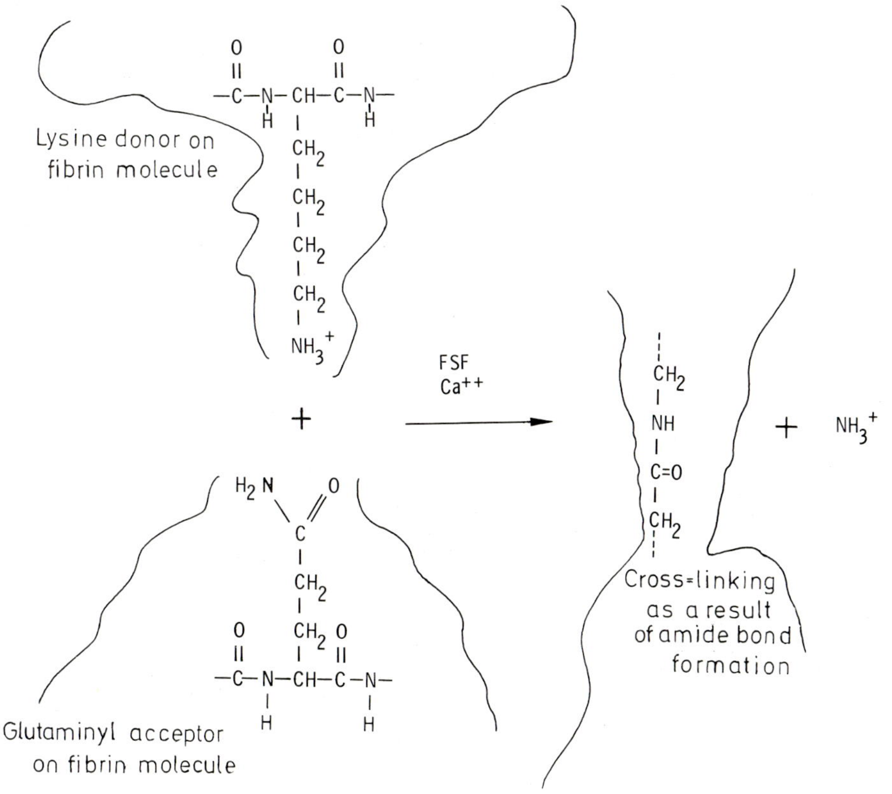

Lysine donor on fibrin molecule

Glutaminyl acceptor on fibrin molecule

FSF Ca^{++}

Cross-linking as a result of amide bond formation

Fig. 7-4. Cross-linking of fibrin

components that are readily precipitated with barium sulfate and then the factor is precipitated by bringing the supernatant to 30 or 40% ammonium sulfate.

The presence of thromboplastin in the blood is illustrated by the ability of the blood to clot in glass vessels without the addition of tissue extracts.

A deficiency in factor XIII of autoimmune origin associated with the administration of isoniazid has been described [50].

Prothrombin Deficiencies

Prothrombin deficiencies may be congenital, deriving from inborn errors of metabolism, or acquired. Congenital hypoprothrombinemia results from an absence of prothrombin A or prothrombin B. This absence leads to a propensity to hemorrhage, particularly in the gums, nose, vagina, rectum, skin, and internal organs. The administration of vitamin K and transfusion may obviate this hemorrhagic tendency [20].

Four siblings have been described with low prothrombin activity (5%), but normal levels of prothrombin when determined immunologically. The defect was shown to be caused by the presence of prothrombin with abnormal electrophoretic mobility referred to as prothrombin Barcelona. The prothrombin is unable to generate more than minute amounts of thrombin [51].

Acquired hypoprothrombinemia may result from a variety of pathological conditions, including vitamin K deficiencies, an inability to synthesize prothrombin, and the administration of antivitamin K compounds. Vitamin K deficiencies develop because of either an inadequate diet or from the absence of synthesis of the vitamin by the flora of the gastrointestinal tract. Such symptoms may have severe consequences when they occur in a newborn who is vitamin K-deficient because the mother absorbs insufficient amounts of vitamin K. In the newborn, such a deficiency is responsible for hemorrhage with the development of bruises.

Vitamin K deficiency may also result from absorption defects in the gastrointestinal tract. Vitamin K is a liposoluble vitamin; thus, whenever there is interference with lipid absorption in the intestine, poor absorption of the vitamin ensues. Two groups of diseases primarily are associated with inadequate lipid absorption—obstructive jaundice and malabsorption conditions in the intestine. Among the latter are included sprue, regional ileitis, intestinal obstruction, intestinal lipodystrophy, short-circuiting anastomosis, hyperperistalsis, chronic dysentery, and ulcerative colitis.

In obstructive jaundice, bile salts are not excreted in the intestine, and therefore their beneficial effects on fat solubilization and lipid absorption through the intestinal epithelium are lost. Consequently, the liposoluble vitamins are lost in the stools. Vitamin K is also synthesized by the bacterial flora of the intestine. When antibiotics are administered to cure intestinal infection by pathogens, they also eliminate the saprophyte flora, thus leading to sterilization of the intestinal tract.

In severe liver disease (postanesthetic, severe cirrhosis, acute yellow atrophy, massive infiltration of the liver with metastatic tumor or leukemic cell, Hodgkin's disease, and massive liver necrosis resulting from chloroform or arsenic administration or from severe traumatic or surgical shock), prothrombin synthesis is blocked as a result of the failing liver metabolism rather than as a consequence of vitamin K deficiency. Therefore, vitamin K administration is ineffective in those cases. In addition, some compounds, like Dicumarol, salicylate, quinine, Tromexan, and phenylindanedione, interfere with prothrombin formation by acting as antivitamin K agents.

Fibrinopenia

At least three types of hereditary diseases have defects resulting in abnormal function of fibrinogen molecules: congenital afibrinogenemia, dysfibrinogenemia, and constitutional fibrinopenia [21–24]. Congenital afibrinogenemia is characterized by bleeding tendencies, but the hemorrhagic crises are much less severe than those that occur in the absence of the antihemophilic globulin. Fibrinogen cannot be detected by biochemical, electrophoretic, or antigenic methods. Since thromboplastin formation or platelet function is not altered in afibrinogenemia, platelet aggregation and viscous metamorphosis are possible, but the clot never evolves into a permanent thrombus because no fibrin mesh is formed. Since the turnover of infused fibrinogen is normal, it is believed that in afibrinogenemia the defect concerns fibrinogen synthesis.

In dysfibrinogenemia, fibrinogen is detectable by electrophoretic and antigenic methods. In fact, there are no obvious physical or immunological abnormalities of fibrinogen found in patients with dysfibrinogenemia and normal fibrinogen. When both types of fibrinogens are submitted to immunoelectrophoresis, reproducible differences in the distribution pattern are observed. Whatever the molecular alteration, blood formation is delayed and is defective when thrombin is added. Interestingly enough, the clotting defect in dysfibrinogenemia is partially restored by the addition of calcium. The role of calcium in this disease is not known.

At least 11 variants of fibrinogen have been described. They differ by their physicochemical properties and functional activities *in vitro*. The functional defects fall in three major categories: (1) unsusceptibility to thrombin with inability to form the monomer; (2) defective polymerization of the monomer; (3) abnormality in the response to factor XIII, the fibrin stabilizing factor [52, 53].

A third form of fibrinogen dyscrasia is of unknown origin. There is a low level of normal fibrinogen; clots form but they are small and soft. The cause of the defect is not known.

Vitamin K Deficiency

In 1929, Dam was studying the role of cholesterol in the diet. For that purpose he prepared a diet extracted with ether. When such a diet was administered to chicks, they developed hemorrhagic diathesis. Dam clearly demonstrated that the hemorrhagic diathesis did not result from the absence of cholesterol, vitamin A, vitamin D, or vitamin E, and he postulated the existence of a new liposoluble vitamin—vitamin K. Dam's finding was confirmed in many laboratories, and the vitamin was concentrated from a number of natural sources.

The factor was called vitamin K, and two quinones were found to be active—vitamin K_1 and vitamin K_2. In addition to the blood coagulation symptoms, vitamin K deficiency led to a decrease in blood levels of prothrombin.

Vitamin K is indispensable for the biosynthesis of a number of blood factors, namely prothrombin, factors VII, IX, and X. In the case of prothrombin, the vitamin is not required for the biosynthesis of the polypeptide chain.

Cattle receiving vitamin K antagonists synthesize a prothrombin precursor with identical amino acid and carbohydrate composition, and immunological properties, but with markedly reduced affinity for calcium. Thus, while normal prothrombin binds four Ca^{2+} ions, the prothrombin elaborated in absence of vitamin K can bind only one Ca^{2+} ion. Normal prothrombin contains an unidentified prosthetic group that is absent from the polypeptide chain in vitamin K deficient prothrombin. A peptide containing the vitamin K moiety has been prepared from normal prothrombin [54–58].

The discovery that the product of proteolytic digestion of prothrombin contains a special dipeptide composed of one serine and one γ-carboxyglutamic acid $(HOOC)_2CH—CH_2—CH(NH_2)—COOH$ has helped to clarify the role of vitamin K in prothrombin activation. The presence of the γ-carboxyglutamic acid was confirmed by titration which established that the prothrombin molecule contained more carboxyl residues than could be predicted on the basis of its amino acid composition and by diborane reduction which reduces an unstable γ carboxyl group to the primary alcohol and allows quantitation of the excess carboxyl groups. The nonthrombin portion of prothrombin contains at least ten residues of γ carboxyglutamic acid and factor X contains fourteen in the light chains. Vitamin K is believed to cause the γ carboxylation, yielding a protein with excess carboxyl groups which is apt to bind the calcium necessary for the conversion of prothrombin to thrombin.

The Role of Platelets in Blood Coagulation

Properties of Platelets

Platelets are generated by the fragmentation of the megakaryocytes in the bone marrow [25–27]. Megakaryocytes are large cells measuring between 35 and 40 μ in diameter, containing a large multilobulated nucleus and abundant eosinophilic cytoplasm. The megakaryocyte probably originates from a primitive cell that undergoes mitosis without cell division. In the rat, it has been estimated that it takes 43–75 hours for megakaryocytes to mature. In humans, megakaryocytes are found only in bone marrow under normal conditions, but they may appear in the bloodstream in severe leukemia or erythremia. The kinetics of platelet production is far from clarified, but 500 million platelets are assumed to be produced each day. Their life span is about 8–9 days, and the total number of the circulating platelets is replaced every third day.

The platelet is an elongated structure approximately 3 μ in diameter. At examination in a darkfield microscope, the platelet has translucent cytoplasm with vacuoles. Romanofsky staining reveals two major components in the platelet; a highly refractile zone containing purple granules, referred to as the granulomere, and a homogeneous pale blue structure called the hyalomere.

The platelets are mainly composed of protein and lipids. Protein represents 60% of the dry weight of platelet. Many of the platelet proteins are enzymes (catalase, hyaluronidase, esterase, β-glucuronidase, nucleotidase, tryptase, tyrosinase, and pyrophosphatase). Many of the enzymes in platelets are hydrolases, which may well reflect platelets' physiological role.

In addition to proteins, platelets contain 15% lipids, of which cephalins constitute 7.5%. Finally, the platelets also contain large amounts of pharmacologically active compounds, namely, serotonin and histamine, and these compounds are released in the blood during platelet disintegration.

If the platelets are lysed experimentally by osmotic shock, two fractions can be recovered, a water-soluble and a liposoluble one. Each of these fractions contains a variety of compounds involved in blood coagulation.

Thus, if the water-soluble fraction is absorbed on tricalcium phosphate and then centrifuged, the supernatant contains a platelet factor referred to as factor I. The tricalcium phosphate pellet contains another factor called factor II, which can be eluted with 0.2% M sodium citrate. Factor II accelerates the conversion of fibrinogen to fibrin by thrombin. Factor III has thromboplastic activity and reverses the anticoagulant effect of heparin, and platelet thromboplastin, or factor IV, reacts with plasma components on active thromboplastin. At least two other factors have been described in platelets—serotonin and fibrinogen that coagulates under the influence of thrombin. Available evidence suggests that fibrinogens in plasma and patelets are identical.

The thromboplastin contained in platelets is readily extracted by incubation of the platelet in saline at 37° for 1 hour. The release is accelerated at low saline concentrations, suggesting that thromboplastin is released by osmotic shock. While 0.4% saline releases all the thromboplastin of normal platelet, distilled water is needed to release the thromboplastin quantitatively in Willebrand-Jergen disease, suggesting that in that disease the platelets are resistant to osmotic shock.

Platelets contain mitochondria, and therefore, it may be anticipated that they have some respiratory activity. Yet the main source of chemical energy in platelets seems to be glycolysis.

The platelets contain a contractile protein (thrombosthenin), which has solubility properties similar to those of the actomyosin group of proteins. The activity of the protein requires the presence of ATP and metallic ions. The protein has been isolated from human platelets, and like actinomycin, the contractile protein possesses ATPase activity. Electron microscopic examination of isolated thrombosthenin revealed a microfibric structure 80–100 A wide with a possible periodic structure [28]. In the presence of ATP, the contractile protein dissociates into an actinlike protein (thrombosthenin A) and a myosinlike (thrombosthenin M) moiety. ATPase activity is nonexistent in thrombosthenin A and weak in thrombosthenin M. When tested alone, neither thrombosthenin A or M is ATP sensitive, but the mixing of A and M thrombosthenin restores ATP sensitivity and ATPase activity.

Platelets have been separated from citrated blood and examined with the electron microscope. Some of the findings confirm the previous observations with the light microscope; others add new detailed information on the platelet ultrastructure. Platelets are surrounded by a single, interrupted membrane 150 A thick; the interruptions give the membrane a "staccato" appearance. Most platelets contain a large number of granulomeres distributed at random throughout the organ, and only a few platelets have a narrow, granule-free peripheral zone referred to as the hyalomere by cytologists. Four types of granules have been recognized: mitochondria, microvesicles, vacuoles, and dense bodies.

Ten minutes after the citrated blood has been supplemented with calcium to restore blood coagulation, three changes develop in the platelets: the platelets acquire bizarre shapes, the membranes lose their staccato appearance, and the granulomeres disappear.

Twenty-six minutes after the blood is recalcified, the platelets have agglutinated in the center of the aggregate. The membranes faint and the granules burst. These changes will be described in more detail later.

The exact physiological role of thrombosthenin is not known, but the contractile protein is assumed to play a role in the retraction of the aggregated platelets and in clot retraction. Because of its ATPase activity, thrombosthenin may split ATP and yield ADP, which is known to stimulate platelet aggregation.

In humans, two populations of platelets have been described: "heavy large" and "lighter small." The two groups can generate ATP, retain cations against concentration gradients, and secure homeostasis, but the heavy large population is more effective than the lighter small in all respects. It is assumed but not established that platelets age, and that the heavy large-type are the younger and the lighter small the older segment of the population [29].

Physicochemical Mechanism of Platelet Agglutination

The sequence of events in homeostasis of larger vessels develops as follows: platelets accumulate at the site of injury (forming a temporary clot), the vascular musculature (in arteries and large veins) constricts, and the platelet undergoes viscous metamorphosis. The mechanisms of aggregation and viscous metamorphosis of platelets will now be briefly reviewed.

One unique property of platelets is their stickiness [30]. Platelets adhere to each other and to the endothelium when discontinuity in the endothelium occurs. Stickiness must be distinguished from viscous metamorphosis and clot formation. Stickiness is a more discrete property of platelets, and it is not unlikely to be important in priming the homeostatic process and the formation of thrombi.

The stickiness of the platelets manifests itself in aggregation of platelets by adhesion to each other and adhesion to surface, for example, glass, collagen, fibers, or endothelium. A number of factors induce platelet aggregation: ADP, thrombi, epinephrine, and serotonin.

When platelet-rich plasma supplemented with anticoagulant is stirred at 37° in the presence of ADP and saline, the platelets clump. The time needed for platelet aggregation is influenced by temperature. The process is optimal at 37°, slowed down at 20°, and nonexistent at 4°. Calcium is indispensable; consequently, EDTA interferes with aggregation. For reasons that are not quite clear, both ATP and AMP influence aggregation. ADP-induced aggregation seems to depend upon the activity of two enzyme systems: glycolysis and an esterase. The addition of iodoacetate to the incubation mixture interferes with aggregation, and aggregation is also blocked by benzoylarginine.

The molecular changes that take place in the ADP-induced aggregation are not known. They are of considerable interest because the elucidation of the mechanism of platelet aggregation might shed some light on the mechanism of cellular cohesion and be relevant to the formation of metastasis. The forces that unite the platelets could be covalent, ionic, or hydrogen bonds; adhesion by modification of surface tension; adhesion under the influence of van der Waals forces; or other similar factors. Little is known of the process. One plausible but obviously oversimplified interpretation is that the sialic group at the surface may be responsible for negative charges that form ionic bonds with divalent calcium ions, which could thus act on ligands between individual platelets. Another hypothesis is that the ADP molecules form hydrogen bonds with the platelet surface. The ADP-loaded platelet would then be held together by the intermediate of calcium ions. But the evidence available suggests that the role of ADP is more than passive. The coenzyme is known to be converted inside the cell and to be metabolized.

Thrombin causes platelet aggregation before it causes clotting; the effect is temperature dependent, but whether or not it requires ADP is not certain. In any event, stimulation of pathways that generate ATP at the expense of ADP interferes with the aggregating effect of thrombin. It has also been proposed that the effect of thrombin is indirect, through the conversion of platelet-absorbed fibrinogen. Thus, thrombin could convert the surface fibrinogen into fibrin, and thereby cause platelet aggregation. This mechanism can not be the only one since platelet aggregation occurs even within fibrinogenic platelets. The mechanism by which serotonin and epinephrine induce platelet aggregation remains obscure.

The circulating platelets are not aggregated, and they do not attach to red cells or to the surface of an intact vascular epithelium, yet when the vascular wall is injured, platelets adhere to the endothelium, forming clumps at the site of injury. Clump formation has been studied in the arteries at the surface of brains. White clumps are formed only if the injury to the artery is relatively severe. Minor injuries are not associated with clump formation unless the vessel is irrigated with ADP, a major injury has been produced at some distance, or the path of the vessel wall is interrupted.

The exposed collagen fiber of the subendothelial layers may form the trap for the circulating platelets. Indeed, when collagen fibers are added to platelet-rich plasma, platelet clumps form at the surface of the fiber. The adhesion requires ADP and is inhibited by EDTA and collagenase. Thus, in vascular injury, the release of an ADP-like substance and the denudation of collagen fiber appear to play a determinant role in platelet adhesion.

During viscous metamorphosis, the platelets lose their individuality and fuse into a large amorphous mass that is structureless on examination with the light microscope. Viscous metamorphosis is accompanied by metabolic alterations in the platelets. Before metamorphosis, the rate of glycolysis and ATP formation are increased. After metamorphosis, ATP is hydrolyzed, possibly by an ATPase released by the platelet, and AMP and ADP appear in the medium.

Such metabolic alterations suggest that the first steps of blood coagulation require energy expenditure. In view of this observation, the fact that some forms of thrombocytopenia are characterized by a depressed rate of glycolysis and reduced ATP formation in platelets acquires particular significance.

Ultrastructural Aspects of Platelet Agglutination

Because of the importance of platelet agglutination in blood coagulation, Brinkhous and his associates have carefully investigated the ultrastructural changes that occur during this process. Brinkhous has studied platelet agglutination in *in vitro* model systems because the condition influencing agglutination can then be controlled better. The observations made after *in vitro* experiments are likely to reflect what happens *in vivo* because the morphological changes that occur *in vitro* mimic the changes *in situ*. Brinkhous described four stages in the platelet agglutination process: preagglutination, agglutination, thrombocytorrhexis, and thrombocytolysis.

As already mentioned, platelets are surrounded by a plasma membrane and contain microvesicles, vacuoles, mitochondria, and some unidentified granules. During the preagglutination period, the intracellular structures migrate toward the center of the platelet and the membrane begins to form pseudopods. These two processes proceed during agglutination. The ultrastructural components conglomerate in a compact mass in the center, and the pseudopods extend further at the periphery. Pseudopods of adjacent platelets become entangled gathering the individual platelets in a composite mass. Thrombocytorrhexis is characterized by the progressive degeneration of the platelet ultrastructure. The unidentified granules disappear first, followed by the mitochondria, the microvesicles, and the vacuoles. The physiological significance of these ultrastructural changes remains to be established. When thrombocytolysis is completed, the platelets have lost their identity and form an amorphous mass studded with remnants of granules or plasmic membranes. Platelet agglutination requires cation and is inhibited by EDTA. It is stimulated by thrombin, ADP, and a thermolabile protein found in plasma referred to as the thrombocyte-agglutinating factor. The mechanism by which these various factors may activate agglutination has been discussed.

The sequence of events associated with intravascular coagulation can be summarized as follows: alteration of the endothelial wall, especially collagen denudation after elimination of the endothelium, is responsible for platelet aggregation with ADP release. The platelets are believed to release ADP even after they have aggregated, stimulating further reversible aggregation. The simple aggregation of the platelets yields the reversible permeable clot, which is converted to an irreversible impermeable clot by thrombin activation through extrinsic or intrinsic thromboplastin formation. Thus, the fibrinogen molecules, which are at the surface of each platelet, are converted to fibrin and entrap the platelets in a permanent blood clot. Moreover, the activation of thrombin degranulates the platelets, releases more ADP and ATP, which is converted to ADP, and aggregates more of the circulating platelets around the clot. Such a process results in the formation of the white thrombus (see below). The size of the white thrombus increases, the focal concentration of the thrombin increases, and circulating fibrinogen is precipitated into a dense fibrin network that entraps circulating white and red blood cells, thus yielding the mixed thrombus (see below).

Thrombocytopenic Purpura

Thrombocytopenia is produced two ways [31]. The low platelet count may result from a general atrophy of the bone marrow and be secondary to aplastic ane-

Table 7-4. Hemorrhagic disease

1. Absence of platelets

Congenital thrombocytopenia:
Passage across placental barrier of maternal antiplatelet antibodies

Essential thrombocytopenia:
Splenic factor inhibits platelet formation in bone marrow by blocking formation of:
- megakaryocytes
- reticuloendothelial cells
- platelets

Symptomatic thrombocytopenia

2. Hemophilia

AHG Deficiency: Classical hemophilia: A hereditary disease transmitted as a sex-linked recessive trait.

PTC (Christmas factor, antihemophilic globulin B) deficiency
Hemophilioid C (Rare hereditary disease, transmitted as a sex-linked recessive trait)

PTA (plasma thromboplastin antecedent) deficiency
Hemophilioid D
(Sex-linked dominant in both sexes)

Ac Globulin (Accelerator globulin) deficiency
Congenital:
- Hemophilioid A transmitted as an autosomal dominant with reduced penetrance

Acquired:
- hepatic disease
- leukemia
- postoperative state
- X-irradiation
- cachexia

SPC (serum prothrombin conversion factor) deficiency
Hemophilioid B
(Autosomal dominant by either sex)

Factor XII deficiency

3. Prothrombin Deficiencies

Congenital	No biosynthesis of prothrombin Abnormal prothrombin
Vitamin K deficiency	
Neonatal	Vitamin K deficiency of mother
Nutritional	Inadequate diet Intestinal sterilization
Gastrointestinal	Inadequate absorption of fat-soluble vitam K Celiac disease Sprue Short-circuiting operation Regional ileitis Intestinal lipodystrophy Bowel obstruction
Hepatic	Failure of prothrombin synthesis in damaged liver acute yellow atrophy Laennec's cirrhosis malignant infiltration phosphorus, arsenic, or chloroform intoxication
Chemical	Anti-vitamin K activity Dicumarol salicylate quinine Tromexan phenylindandione

4. Excessive Anticoagulants

Hyperglobulinemia	The excessive proteins, normal or abnormal, bind to the clotting factor (nonspecific)? infectious diseases (endocarditis) liver disorders cancer pancytopenia agranulocytosis Thromboplastin anticoagulant in chronic glomerulonephritis (specific)? Lupus erythematosus Pemphigus Tuberculosis
Hyperheparinemia	Increase in heparinoid substance leukemia aplastic anemia lymphosarcoma Hodgkin's disease nitrogen mustard and X-irradiation

5. Fibrinogen Deficiencies	*Hereditary* Congenital hypofibrinogenemia Recessive autosomal gene Disafibrinogenemia Fibrinopenia *Symptomatic hypofibrinogenemia* Deficiencies: pellagra scurvy pernicious anemia Bone marrow diseases: myelogenous leukemia polycythemia Liver diseases Infection, scarlet fever, TB, etc.
6. Fibrinolytic Disease	Transfusion reaction Hemolytic crisis Hemorrhagic shock Eclampsia Liver disease Leukemia

mia, or it may be due to selective destruction of platelets and their precursors. Only the second type of thrombocytopenia will be described here. The selective destruction of platelets usually results from immunological disorders. In the adult, antiplatelet immunological factor can be produced in two ways; either the platelet acts as an antigen, and antiplatelet globulins appear in the blood (iodiopathic thrombocytopenia), or drugs combine with the platelet, forming a new antigenic compound (symptomatic thrombocytopenic purpura). The newborn may develop thrombocytopenia due to the appearance in the blood of autoagglutinins that are transmitted from the mother to the fetus through the placenta (neonatal thrombocytopenic purpura).

The pathogenesis of thrombocytopenia will become clearer when the mechanism of production of autoimmune disease is discussed. Only a few salient points are mentioned here. A number of drugs, among them allylisopropylacetyl carbamide (Sedormid), quinine, quinidine, are known to autosensitize platelets. In affected patients, the combination of platelet, drugs, and serum leads to thrombocytopenia; a large number of patients with idiopathic thrombocytopenia have antiplatelet antibodies in their blood. If the blood of such patients is injected to normal individuals, thrombocytopenia occurs.

The immune reaction to platelets is associated with three major alterations of platelet. There is an increased tendency to agglutination, the rate of platelets lysis is accelerated, and the susceptibility of platelets to phagocytosis is increased. Platelet precursors are also altered in thrombocytopenia. The megakaryocytes of the marrow are rare and immature.

Thus, thrombocytopenia results from the alteration of the circulating platelets and their precursors. The platelet number drops from 250,000/mm^3 to 60,000/mm^3. The drop in the platelet count is usually accompanied by hemorrhagic diathesis, taking the form of purpura. Small hemorrhages occur underneath the skin and the mucosa-serosa, especially under those portions of skin or mucosa (stomach) or serosa (visceral pericardium) that cover muscle contracting actively. Occasionally, the hemorrhage may be more extensive and spontaneous or traumatic ecchymosis may develop.

Coagulation tests reveal that in thrombocytic purpura, the bleeding time is prolonged, but the clotting time is normal. None of the soluble factors necessary for blood coagulation is missing. Yet, as may be anticipated from the role of platelets in clot contraction, the clot loses its ability to retract. Repeated hemorrhage may lead to anemia, and a moderate leukocytosis may develop immediately after a hemorrhagic crisis.

The spleen is moderately enlarged in patients with thrombocytopenia, possibly because thrombocytolysis is activated in the spleen. In any event, splenectomy often procures temporary relief or even permanent cure for thrombocytopenia. The various types of thrombocytopenia and other forms of defects in blood coagulation are summarized in Table 7-4.

Vascular Factors in Blood Coagulation

Some investigators, among them A.G. Quick, have suggested that changes in the circulating blood do not constitute the *primum movens* of blood coagulation. In contrast, these authors think that vascular changes control hemorrhage. This concept partly stems from the observation of telangiectasia. The new vessels that are formed in telangectasia do not respond to injury normally. Minimal trauma induces prolonged bleeding, even when all the components necessary for blood coagulation are present in the blood at normal levels.

Hemostasis in small capillaries is particularly intriguing. It can no longer be explained by capillary constriction or platelet formation. No capillary plugs are observed.

Vasoconstriction cannot alone be responsible for hemostasis because in severe thrombocytopenia, vasoconstriction occurs, but hemorrhage is not prevented. It has been proposed that adhesion of the endothelial wall of the capillary causes hemostasis. The mechanism of this adhesive process is not clear; it could result from collapse of the endothelial wall because of destruction of the surrounding connective tissue, compression of the capillary lumen by edema fluid, or collapse of the lumen because of reduced pressure due to activation of precapillary arteriolar sphincters.

The vascular theory of hemostasis postulates a vascular and a hematological step. In the vascular step, the vessel reacts by vasoconstriction, and a "stickiness" develops at the surface of the endothelial walls. This stickiness of the endothelial walls may cause the two opposing walls to adhere to each other and close the vessel. In small capillaries, the occlusion is sufficient to prevent hemorrhage; in larger vessels, occlusion is followed by agglutination of platelets first, and formation of a fibrin clot later. After viscous metamorphosis, a second wave of vasoconstriction occurs, which probably results from serotonin release from platelets. In this second wave, the vasoconstriction is not restricted to the injured vessel but affects the neighboring vessels as well.

Fibrinolysis

After a blood clot has formed, the injured tissue can be restored to normal only if the clot is removed by fibrin digestion. The digestion of fibrin prevents the fibrous organization of the intravascular area [32] in addition to restoring circulation.

Plasma can hydrolyze fibrin under a number of physiological or pathological conditions (stress, shock, burns, inflammation, and even carcinoma of the prostate). The fibrinolytic activity of plasma is usually attributed to the presence of proteolytic activity in the plasma. It is further assumed that plasmin is responsible for all plasmoproteolytic activity because all attempts to separate fibrinolytic from proteolytic activity have failed. These two activities are always recovered in the same fraction whatever method of separation is used (electrophoresis, chromatography, ammonium fractionation, etc.). However, Poly and Kline believe that plasma contains proteolytic enzymes different from plasmin. There seems to be no doubt that the globulin fraction of plasma that possesses fibrinogenase activity affects the physiology and pathology of blood clotting. If a fibrinogenase preparation is injected intracardially in guinea pigs, the blood of these animals losses its ability to coagulate. Clotting ability is restored by the administration of normal blood, but when the administered blood hemolyzes, the clot also lyses—suggesting that fibrinogenase formation is induced during hemolysis.

Attempts to isolate fibrinogen in the blood of guinea pigs early after injection of fibrinogenase failed, yet fibrinogen reappears soon as a result of *de novo* synthesis in liver.

Thus, the main fibrinolytic system of the blood is likely to be the plasmin-plasminogen system. Plasminogen is a globulin found in blood. In the presence of the proper activators, plasminogen is converted into plasmin, a proteolytic enzyme. The blood also contains substances that inhibit plasmin.

Plasminogen has been purified from human blood by classical alcohol fractionation methods, followed by chromatographic analysis on carboxymethyl-cellulose. This preparation contains at least 5–10% impurities, and it is therefore not surprising that immunological studies have demonstrated the presence of two antigens in the chromatographed preparation. But electrophoresis of plasminogen in the Tiselius apparatus yields a single peak. Plasminogen is soluble at an acid (<5) or basic pH (>6), but is only sparingly soluble at the pH of blood.

Plasmin is stable at an acid pH (2.3) but is rapidly autodigested at a neutral pH. The exact molecular weights of plasmin and plasminogen are unknown; they have been estimated to range between 108,000 and 127,000. Both plasminogen and plasmin are suspected to be glycoproteins (containing approximately 1.0–1.5% hexose). The amino acid composition of the proteins has not been established.

The role of fibrinolysis in the restoration of infected areas is best illustrated by the fate of an area repeatedly infected with streptococci. During the first bout of infections, the growing streptococci secrete streptokinase, which activates the fibrinolytic activity of blood and leads to hemorrhage. In the meantime, the organism develops antibodies to streptokinase, and when the infection is repeated, these antibodies prevent plasminogen activation. Therefore, the streptococcal infection is followed by the accumulation of fibrin with the formation of pseudomembranes, as in strep throat.

In a number of diseases, the patient develops multiple coagulation defects. Examination of the blood demonstrates slow formation of a loose and friable clot *in vitro*, even when thrombin is present. This results from increased fibrinolytic activity in plasma. These clinical entities will be further reviewed after the mechanism of plasminogen activation to plasmin is discussed.

Failure to maintain hemostasis even for a short time is bound to be fatal. Consequently, any intravascular fibrinolytic system must include a number of safeguards. The plasminogen-plasmin system is supplemented with a complex molecular machinery that includes activators and inhibitors.

The mechanism by which plasminogen is converted to plasmin in the living organism is not clear, but several so-called kinases of tissue origin have been incriminated. *In vitro*, the conversion of plasminogen to plasmin is achieved by treating the plasma with chloroform or ether or by adding streptokinase. Activators are assumed to convert plasminogen into plasmin by splitting bonds that involve a lysine moiety. This assumption is based on the observation that lysine or lysine analogs (ε-aminocaproic acid) act as competitive inhibitors of the activator. In fact, ε-aminocaproic acid has been used to treat fibrinolytic disorders.

The product of the activation of plasminogen is plasmin, a proteolytic enzyme with optimum activity at a neutral pH (7.5–8.6). Plasmin acts on fibrin and fibrinogen, gelatin, casein, and coagulation factors V and VII. Plasmin also acts on synthetic substrates that are readily attacked by trypsin, but plasmin seems to be different from trypsin.

Ten to fifteen minutes after fibrinogenase injection, the proteolytic activity has completely disappeared, suggesting that the active center of the enzyme is complexed with an inhibitor.

It is not known whether plasmin's digestion of extravascular fibrin results from the proteolytic activity of plasma-plasmin or from tissue plasmins. Kowalski and his associates have demonstrated that human tissue shows fibrinolytic activity after activation with streptokinase. Whether this activity results from the conversion of plasminogen to plasmin is not known.

Whatever role plasmin may play in the digestion of extravascular fibrin, the sequellae of injury clearly vary considerably depending upon the fate fibrin. The resorption of fibrinous exudate prevents organization of the exudate by connective tissue and secures a smoother restoration of the original structure.

Studies of purified plasmin have helped to recognize some of its requirements. The purified enzyme is activated by the addition of nicotinic acid, and the addition of organic phosphate protects the preparation against denaturation. Plasmin can also be protected by arginine-ethylene and lysine-ethylene esters.

The molecule is composed of two polypeptide chains, a heavy A and light B chain. The activation of plasminogen to plasmin involves the cleavage of two peptide bonds, the split of arginyl-valine peptide bond in the COOH terminal portion of plasminogen results in the formation of the two chain plasmin molecule. The other split occurs in the NH_2 terminal portion of the molecule and involves a peptide bond between an unknown amino acid and either an arginine or a lysine residue.

Researchers attempted to identify the active center of plasmin indirectly by kinetic studies of the catalytic properties using N-paratoluene sulfonylarginine methyl ester as substrate. The plotting of the K_m and the maximum velocity with respect to pH yielded a set of curves identical in shape to those obtained with trypsin under similar conditions. The similarities of the kinetics suggest that the active center is identical for both plasmin and trypsin.

The active center is located in the B chain. Its partial sequence has been resolved; the sequence contains a histidine loop, homologous to that found in other proteases [60]. See Chapter 4.

Plasmin activity is eliminated from the blood by autocatalytic digestion, inhibition by antiplasmin agents, and elimination in the urine.

Again, plasmin digests itself at physiological pH, and the proteolytic enzyme activity can be stabilized by adding arginine or lysine ethyl esters to the medium. Antiplasmin factors have been found in at least two different plasmoglobulin components as well as in platelets. Plasmin is inhibited in two stages: in the first, there is immediate reaction between the enzyme and the inhibitor; and in the second, inhibition is much slower. Galovanova has found fibrinogenase activity in normal urine, and it has been assumed that this results from the renal excretion of plasma plasmin independently of plasmin inhibitors.

This description of the plasminogen-plasmin systems illustrates how little is known of this basic physiological system; no substrate, product, activator, or inhibitors have been characterized. The mechanisms triggering activation of plasminogen and inhibition of plasmin are unknown. Yet, the elucidation of some of the steps involved in converting plasminogen to plasmin has helped our understanding of hyperplasminemia.

Hyperproteolytic activity of plasma may result from excessive amounts of plasminogen activator, deficiency in plasmin inhibitors, or appearance in plasma of proteolytic enzymes different from plasmin. Severe anoxia or shock releases large amounts of activators, and the rate of conversion of plasminogen to plasmin exceeds that of inhibition of plasmin activity. Patients with severe liver impairment (liver cirrhosis) may become unable to clear the activator from the blood, but in liver disease the levels of plasmin inhibitor may be low.

An intriguing form of excessive plasmoproteolysis occurs in obstetric patients with abruptio placentae, retained dead fetus, or amniotic fluid embolism. Such incidents are associated with an increased release of fibrin and activation of the plasminogen-plasmin system. Plasmin attacks the fibrinogen and fibrin and releases polypeptide fragments that are resistant to further hydrolysis by plasmin. At least one of the resistant fragments (mol wt 88,000) inhibits the polymerization of fibrin. By changing the equilibrium between fibrinogen and fibrin, the presence of such fragments indirectly interferes with the conversion of fibrinogen to fibrin. But the polymerization inhibitors are eliminated from the blood (half-life 9 hours), and if massive hemorrhage is prevented, recovery is prompt. Lower urinary tract trauma is frequently accompanied by protracted hematuria due to the presence of urokinase in the urine. Urokinase is a thermostabile plasminogen activator that has been purified extensively. The proteolytic activity of urokinase is inhibited by ε-aminocaproic acid, and hematuria can be prevented in patients with urinary trauma by the administration of ε-aminocaproic acid.

Some pathological conditions have been suggested to result from inhibition of fibrinolysis. Lieberman

<u>Increase</u>

Excessive plasminogen activator–

Shock
Liver cirrhosis

Appearance of inhibitors of plasmin–

in Abruptio placentae
Infant death
Amniotic fluid embolism

Appearance in plasma of proteolytic enzymes other than plasmin–
e.g., urokinase

<u>Decrease</u>

Hyalin membrane disease?

Atherosclerosis?

Cerebral and coronary thrombosis?

Fig. 7-5. Pathology of fibrinolysis by plasmin

claims that hyaline membrane formation in the lungs of infants is associated with the presence in plasma of an inhibitor of the plasminogen activator.

Hyperfibrinogenemia and antiplasmin activity have sometimes been incriminated in the development of cerebral and coronary thrombosis. A 28-year-old patient experienced spontaneous thrombotic accidents for 18 years; the condition can be ascribed to increased blood levels of fibrinogen, absence of fibrinolytic activities, and high levels of streptokinase and urokinase inhibitor.

Of course, one of the theories on the pathogenesis of atherosclerosis is that the atherosclerotic plaque starts with the formation of mural thrombi, mainly composed of fibrin. The fibrin layer is soon covered with an endothelial lining and is progressively organized by collagen-producing fibroblasts. The result is a thickened intima that may ultimately undergo fatty degeneration.

In addition to the plasminogen-plasmin system of blood, circulating white cells and the liver reticuloendothelial cells are also involved in scavenging coagulation products from the blood. Neutrophils phagocytize fibrinogen and fibrin, but not fibrin-split products. Reticuloendothelial cells can phagocytize aggregates of fibrin as well as soluble products. The pathological features of fibrinolysis are summarized in Fig. 7-5.

Thrombi and Emboli

If hemorrhagic diseases are dramatic but rare diseases, thrombosis is, with cancer and car accidents, one of the most merciless killers of our time. Thrombosis cripples or kills more often at the time of life when concentrated efforts to serve society reach maximal efficiency [33–35].

There are two major pathological forms of thrombosis, localized or disseminated. Localized thrombi were recognized more than a century ago. Yet the exact definition of the word remains unprecise. The English term is derived from the Greek word $\Theta\rho o\mu\beta o\varsigma$, meaning clot or lump. Pathologists have restricted the original meaning of the word, and thrombus is now used to refer to a clot that forms intravascularly when hemostasis is altered as a result of injury.

Thrombi may form in all branches of the vascular system—heart, arteries, veins, or capillaries. Grossly, thrombi are white, red, or mixed. The most common type of thrombus observed by the pathologist is the mixed type, which develops in a vessel where the blood is circulating. The primary injury is usually some form of endothelial injury, which induces the adherence of platelets, forming a nidus. The nidus grows by apposition of more platelets, and as it grows it slows down the blood flow. This is accompanied by the conversion of fibrinogen to fibrin, and the fibrin entraps erythrocytes and leukocytes in its meshes. This unit lesion (platelet agglutination, fibrin deposition, and entrapping of blood elements) is repeated an indefinite number of times as the thrombus grows (see Fig. 7-6). The alternation of platelets and fibrin deposits gives a striated appearance (lines of Zahn) to the thrombus when it is examined microscopically.

Table 7-5. Major causes of localized thrombosis

Injury to Vascular Walls	
Arteries	Trauma, atherosclerosis
Veins	Varicosities, rheumatic endocarditis, hemorrhoids
Heart	Myocardial infarcts
Poisons	
Endogenous	Eclampsia
Exogenous	Arsenicals, mercurials, potassium chloride, mushroom poisoning, viper venom
Inflammation	
	Thrombophlebitis
	Thromboarteritis
	Periarteritis nodosa
Stasis	
	In auricular appendage, aneurysm of heart and arteries, and varicose veins

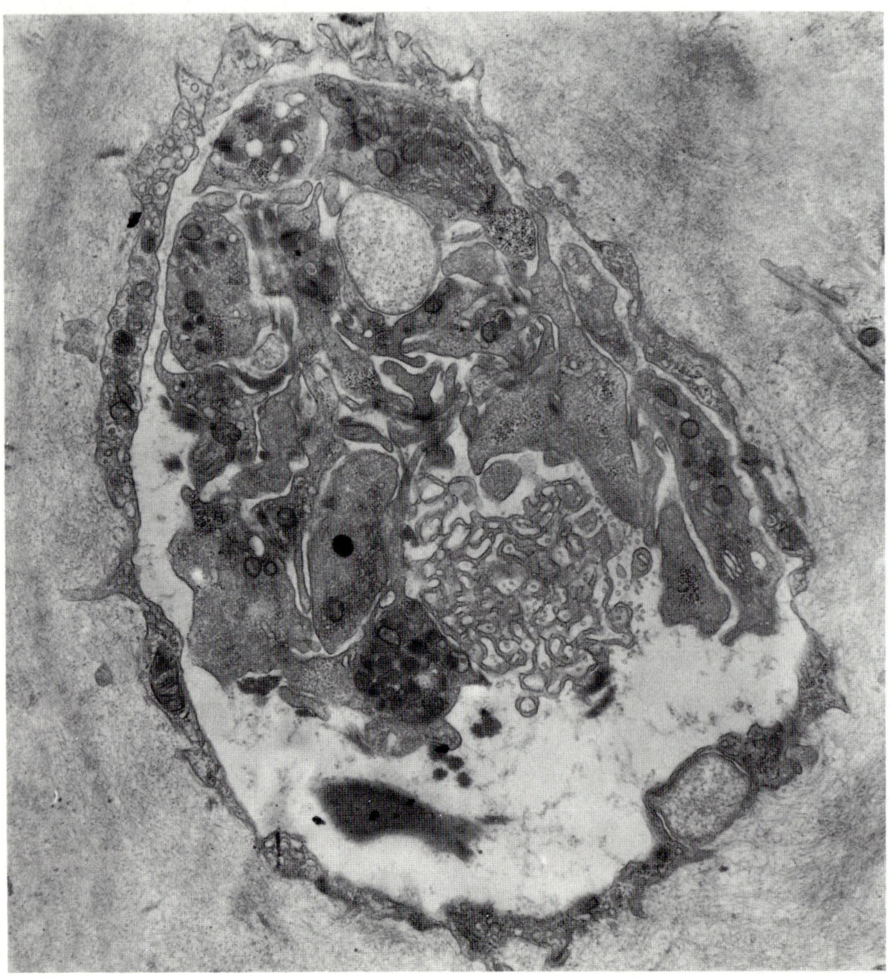

Fig. 7-6. Platelets in thrombosis (×11,000)

White thrombi are formed by platelet agglutination. They develop when extensive fibrin deposition does not occur and at the endothelial surface of the heart valves in rheumatic endocarditis.

In some cases, especially when the flow of the circulating blood is slow, thrombi are composed mainly of red and white cells entrapped in fibrin. These red thrombi are frequently found in thrombotic veins. It has been claimed that such thrombi may form without nidus development and that blood coagulation is primed by the increase of a special factor referred to as the serum thrombotic accelerator.

The injuries that lead to thrombosis are listed in Table 7-5.

On the basis of their location and appearance inside the vascular system, different types of thrombi have been described. The mural thrombus sometimes covers a relatively large surface of the vascular and cardiac wall without altering the lumen or the cavity. Obliterating thrombi occlude the vascular lumen (see Fig. 7-7). Sometimes injuries of the endothelial surface of the valvular structure of the heart lead to the formation of thrombi that are called vegetations. A ball thrombus is a thrombus that comes loose from its mural or valvular attachments and bounces back and forth in the cardiac cavity with the rhythm of the heartbeat.

Obviously, it is important to distinguish postmortem from antemortem thrombi. Postmortem thrombi may be red or yellowish, with the appearance of chicken fat. The clot is moist, resilient, and it can easily be withdrawn from the vascular lumen. In contrast, the antemortem clot is dry, friable, and firm and adheres to the wall. When the clot is extracted, the vascular wall appears rough at the site of thrombus implantation. The thrombus may be autodigested or become organized, infected, or loose.

At the point of its attachment, the thrombus is in direct contact with collagen. From that point, fibroblasts fill the capillaries and invade the fibrin frame of the thrombus and replace it with young fibroblasts and collagen. The capillaries revascularize the obstruction and may ultimately, by coalescing, pierce the obstruction and reestablish circulation, thus forming a canalized thrombus.

When the thrombus contains many leukocytes entrapped in its fibrin meshes, the enzyme released from the dead leukocyte may digest the thrombus. If digestion is complete, the thrombus disappears; if digestion is partial, the thrombus may at autopsy appear like a bag attached to the wall of the vessel filled with yellowish fluid resembling pus.

Bacteria circulating in the blood may colonize the thrombus and attack polymorphonuclears, which will

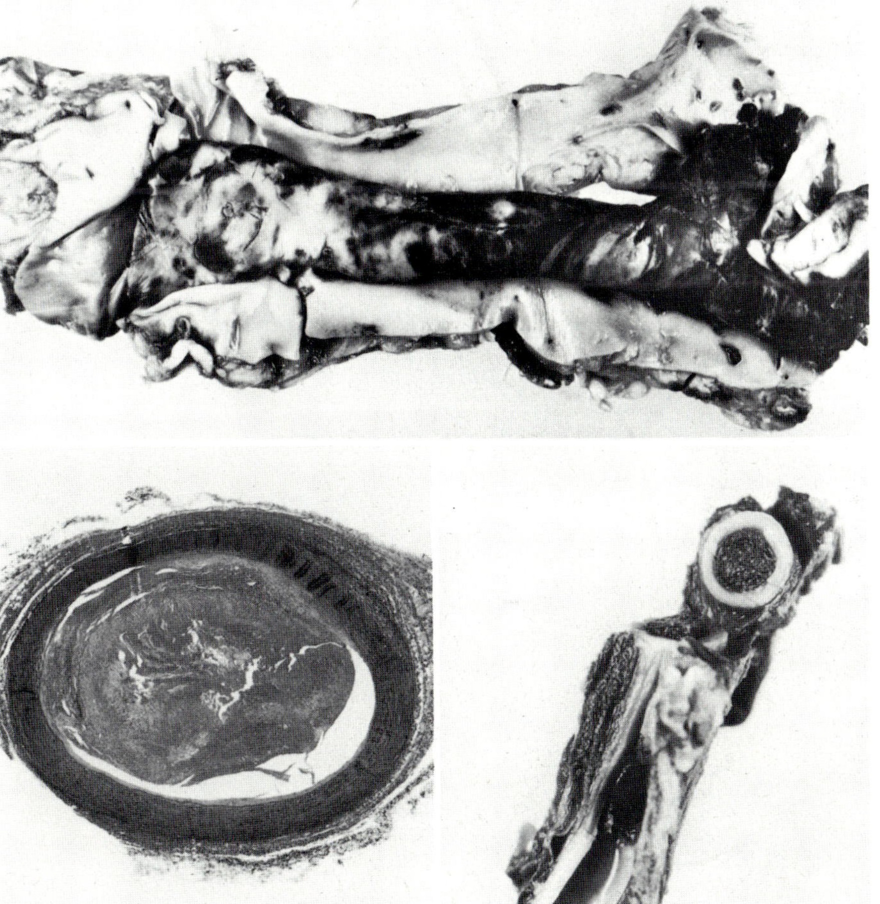

Fig. 7-7. Gross view of thrombosis in aorta and carotid artery (*upper*). Coronary occlusion by thrombosis, gross (*right*) and low power microscopy (*left*)

partially digest the thrombus, releasing it from its point of attachment and converting it into a circulating abcess.

The term embolus comes from the Greek verb meaning to project. Any mass different from circulating blood that is projected into the bloodstream is called an embolus. An embolus can be made of thrombus, fat cells, cancer cells, or even gas (bends).

If an embolus is generated in the peripheral veins, it will go through the right side of the heart and be stopped in the pulmonary artery, causing a death so sudden that changes do not have time to appear in the lung parenchyma. If the embolus is generated in the arteries, the left side of the heart, or the pulmonary vein, it will enter the arterial circulation and ultimately block arteries, such as the renal, cerebral, or splenic. Arterial occlusion causes stasis with further thrombus formation and infarction of the territory vascularized by the occluded artery.

Whenever the foramen ovale is patent, an embolus that originates in the venous system may follow a paradoxical path (retrograde emboli) and enter the arterial system.

The complications of thrombosis may be devastating not only locally but systemically. Venous thrombosis leads to passive hyperemia and necrosis of the

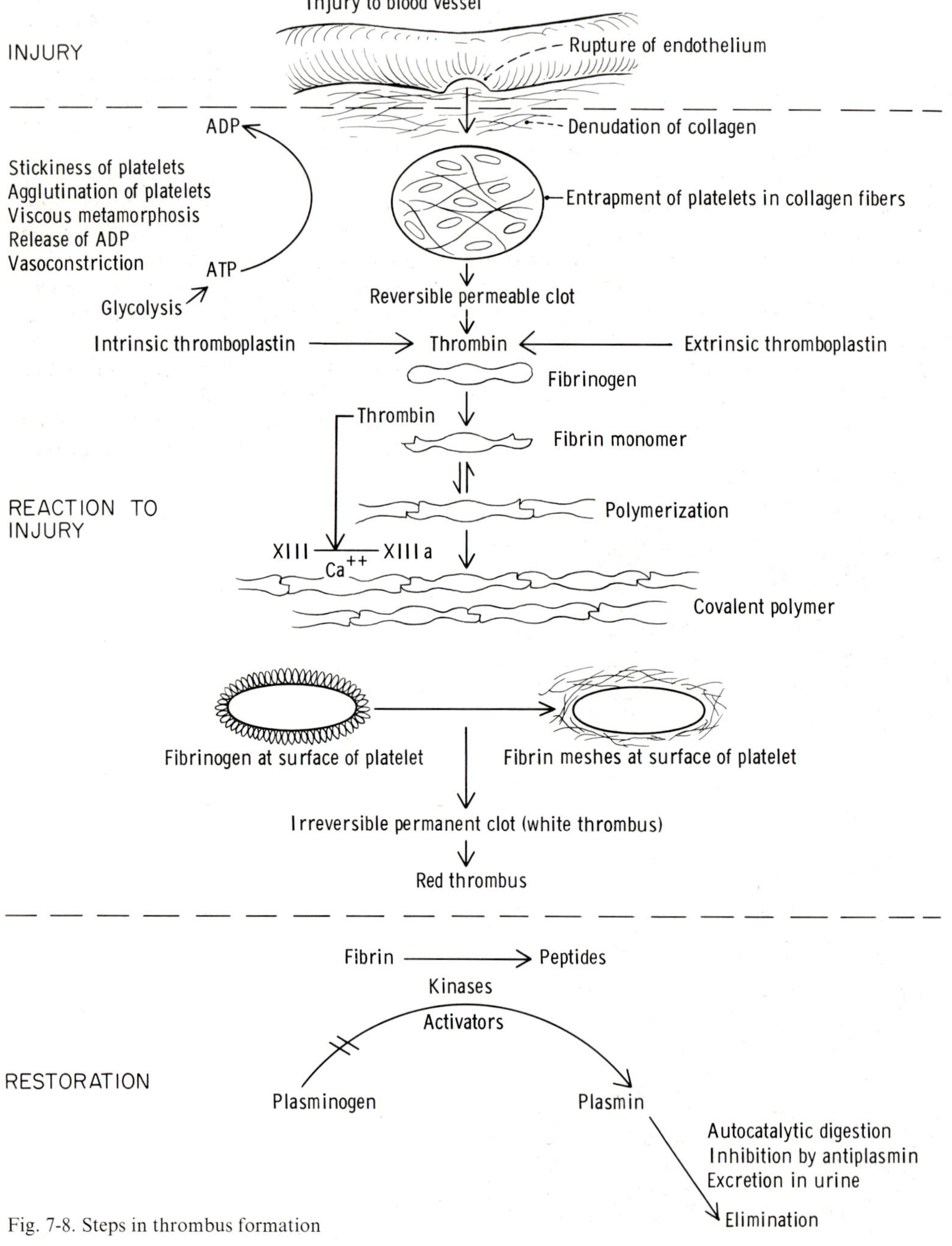

Fig. 7-8. Steps in thrombus formation

anatomical area irrigated by the vein. A typical example is thrombosis of the mesenteric vein. Arterial thrombosis causes infarcts (*e.g.*, myocardial infarcts).

Fragments of the thrombi may become detached, be released in the general circulation, and finally get caught in vessels distant from the thrombotic vessel. As we have seen, these ectopic vascular plugs are called emboli. Venous emboli originate from thrombi that developed in the peripheral veins. Emboli that plug the major pulmonary arteries are often responsible for sudden death. When emboli are found in the peripheral arteries, they usually originate from thrombi in the heart or in an aneurysmal sac. The steps involved with formation of blood clots or thrombi are summarized in Fig. 7-8.

Disseminated Coagulation

In addition to localized thrombi, a number of diseases have been described in recent years in which systemic intravascular clotting occurs. In these diseases widespread intravascular coagulation is associated with the rapid conversion of all the fibrinogen reserves into fibrin (see Table 7-6). When large amounts of fibrin are liberated into the circulation, three courses of evolution are possible: fibrin is temporarily deposited along the vessel wall and no thrombi form; fibrin precipitates massive clotting, particularly in the lung, and death ensues; and a hypercoagulable state associated with systemic capillary thrombosis is followed by a hypocoagulable state associated with hemorrhage [36–39] (see Fig. 7-9).

A detailed description of all the clinical entitites responsible for the development of disseminated intravascular coagulation would, to be sure, provide for an extensive review of morbid anatomy. Only a succinct synthesis of the major clinical and morphological changes are presented here.

Diseases causing disseminated intravascular coagulation have the following points in common. They start with an alteration in the hemostatic mechanism associated with shock, hemorrhagic diathesis, and necrosis. The primary injury in these diseases is acceleration of the clotting process. What triggers the clotting mechanism in each of these cases is not known with certainty, and the pathogenesis of the dissiminated coagulation can only be postulated in each case. These assumptions are based on modern knowledge of blood coagulation. The pathogenesis of the clotting seems obvious in those cases where fibrinolysis inhibition, thrombocythemia, and hypercalcemia occur. Proteolytic enzymes may trigger coagulation by converting prothrombin to thrombin.

Products of hemolyzed red cells and tissue extracts contain thromboplastin. Bacterial endotoxin and antigen-antibody complexes are likely to activate the release of platelet thromboplastin. Viruses or other agents that cause endothelial injury precipitate platelet agglutination at the site of injury. The mechanism triggering disseminated coagulation in hypersensitivity and eclampsia is puzzling, but hypercoagulability has been proposed to result from platelet destruction. The above interpretations are obviously an oversimplified explanation of the pathogenesis of multiple throm-

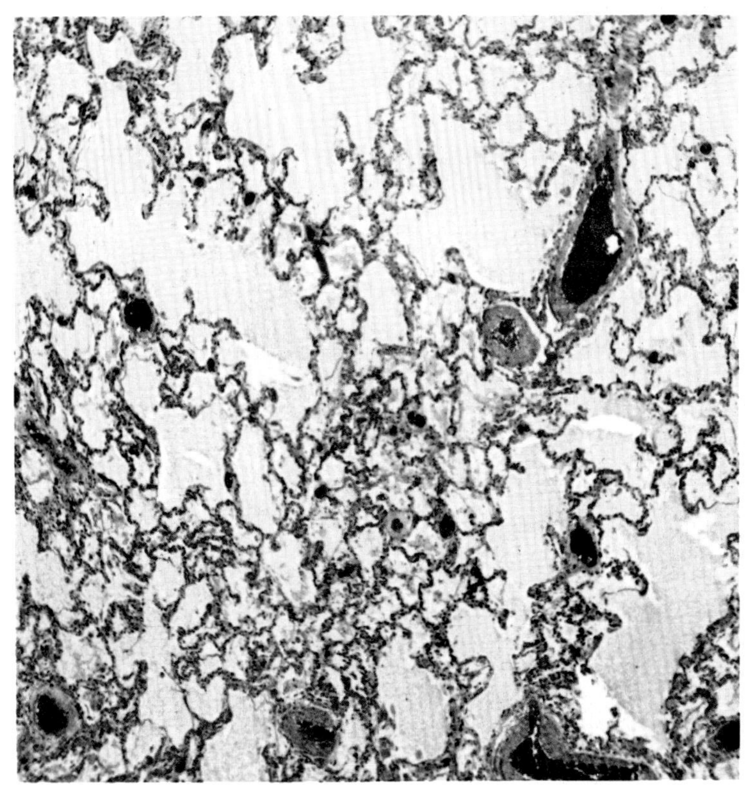

Fig. 7-9. Disseminated intravascular coagulation in lung. Low magnification

Table 7-6. *Conditions that may be associated with disseminated coagulation*

Hemolytic disease	*Introduction of particulate matter into the circulation*
Incompatible blood transfusion reaction	Amniotic fluid embolism
Paroxysmal nocturnal hemoglobinuria	Leischmaniasis
Plasma transfusion reaction	*Anoxia*
Malaria	*Local*
Sickle cell anemia	Intussusception
Favism	Crush syndrome
Paroxysmal cold hemoglobinuria	Epinephrine administration
Release of tissue extracts in circulating blood	Ergot poisoning
Cancer	*General*
Hydatidiform mole	Cardiac arrest
Extracorporeal circulation	*Virus infections causing endothelial damage*
Bacterial toxins	Varicella
Transfusion of bank blood contaminated by bacteria (gram negatives, Pseudomonas, etc.)	Variola
Meningococcemia	Vaccinia
Infected abortions	Rickettsia (Rocky Mountain spotted fever, typhus)
E. coli, Aerobacter aerogenes	*Fibrinolysis inhibition*
Premature rupture of membranes with chorioamnionitis and placentitis (contamination of the amniotic fluid by bacteria from vagina)	Spontaneous
Infantile diarrhea	Injection of ε-aminocaproic acid
Pyoderma gangrenosum (*Pseudonomas aeruginosa*)	*Thrombocythemia*
Cholera	Polycythemia vera
Antigen-antibody complex in nonspecific protein therapy	Myelogenous leukemia
Typhoid vaccine	Hemorrhagic thrombocythemia
Boiled milk	*Miscellaneous*
Diphtheria antitoxin	Giant hemangioma
Introduction of proteolytic enzymes into the circulation	Heat stroke
Acute hemorrhagic pancreatitis	Electrocution
Therapeutic intravenous administration of trypsin	Eclampsia
	Arsphenamine injection
	Acute fatal hyperparathyroidism

bosis, and several different factors may operate synergetically in each instance of disseminated coagulation.

Shock is a constant finding in all cases of disseminated coagulation. In animal experiments, heparin injection prevents clotting and shock, and therefore it is generally accepted that shock is a direct consequence of the disseminated thrombosis. The drop in blood pressure may be slight and reversible, or it may develop into an irreversible state and lead to death. The pathogenesis of shock has been explained as follows. Sudden obstruction of liver and lung capillaries impairs the venous return to the heart, and the cardiac output is reduced.

If the patient survives the shock, he develops hemorrhagic diathesis and necrosis. The results of blood coagulation tests are usually those found in thrombocytopenia (normal clotting time, prolonged bleeding time, poor clot retraction, and positive tourniquet sign). Petechiae and ecchymosis develop under the skin and the mucosa. The severity of the hemorrhagic diathesis is related to the severity of the clotting disturbance. The cause of the diathesis is complex. It involves a sudden depletion of many coagulating factors and accelerated fibrinolysis.

Afibrinogenemia is a prominent alteration in disseminated coagulation. Fibrinogen is low because like other coagulating factors, factors V, VIII, and X have been used in the coagulating process, and also because of an increase in plasmin activity.

The trapping of a large proportion of the platelet population in the disseminated clots leads to a transient thrombocytopenia that is partly responsible for the hemorrhagic diathesis. Fibrinolysis may lead to platelet disintegration in the capillaries, causing ischemia. Such transient ischemia is usually of little consequence. If the multiple thrombi are not resorbed, prolonged ischemia leads to necrosis. The pathological appearance and the clinical symptoms that accompany necrosis depend upon the organs affected.

If the shower of thrombosis in the lung is massive and sudden, the victim may experience chest pain and dyspnea. Sudden interference with blood flow through the lung leads to cor pulmonale and sudden death.

If the thrombi are not excessively numerous and transient, the patient recovers after a period of dyspnea, hyperpnea, and cyanosis. If the fibrinolysis is not intense enough to dissolve the capillary plugs, these pulmonary symptoms will be followed by permanent lung injury. The portion of lung parenchyma ingested by the plugged vessels becomes hemorrhagic and necrotic, and the area of necrosis is later replaced by fibrotic scars. The severity of the pulmonary symptoms is considerably increased if accompanied by chronic pulmonary disease (cyanotic congenital heart disease, primary pulmonary hypertension).

When increased coagulability develops in patients with nonspecific valvular deformities (focal or diffuse), nonbacterial thrombotic endocarditis develops. Nonbacterial can be distinguished from bacterial endocarditis on the basis of negative blood cultures and the presence of multiple emboli in spleen, kidney, and brain.

Massive and transient thrombi in the kidney only induce temporary oliguria and azotemia, but the persistence of renal thrombi leads to tubular necrosis and bilateral cortical necrosis.

Massive thrombosis of the gastrointestinal tract may result from pseudomembranous enteritis. Periportal necrosis or infarcts may be seen; acute liver failure is rare, but more frequently extensive liver necrosis is followed by cirrhosis. Thrombosis in the endocrine organs may lead to fatal consequences; bilateral thrombosis of the adrenal vein is responsible for Addison's disease. The pancreas may exhibit disseminated foci of necrosis or massive hemorrhagic pancreatitis. Portions of the pituitary gland may be destroyed, with the development of various endocrine syndromes.

We have already pointed out that the local pathology of the heart and lungs may be important in determining the intensity of the symptoms. It seems that endothelial or vasomotor changes may determine the localization of the thrombi. The pathogenesis of disseminated coagulation is summarized in Fig. 7-10.

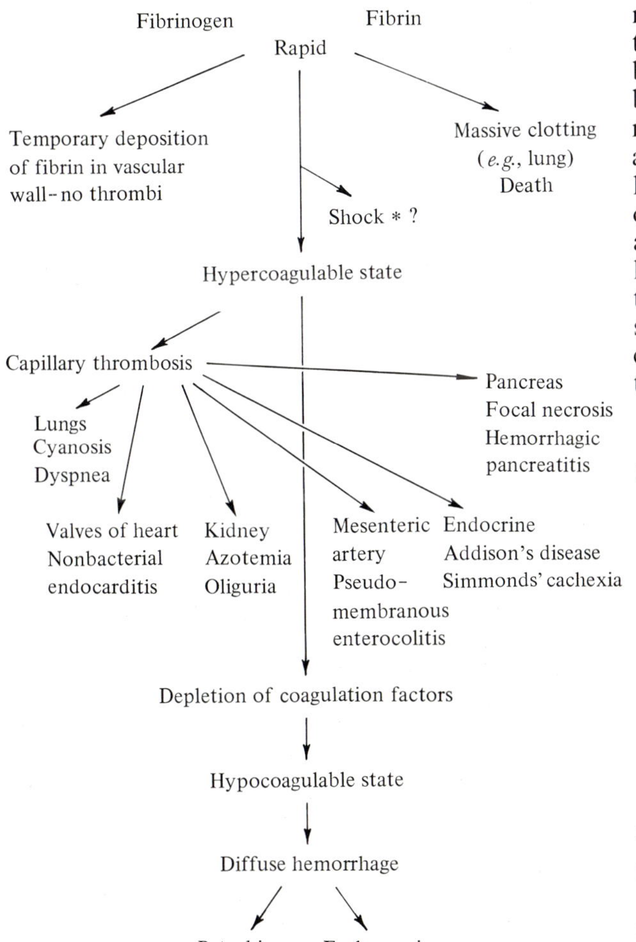

Fig. 7-10. Pathogenesis of disseminated coagulation. * Can be prevented by heparin

Conclusion

The phenomenon of blood coagulation illustrates the ability of the organism to engage and control explosive reactions for the purpose of maintaining a steady state in the face of unexpected environmental threats. The sequence of such reactions is difficult to study because the active compounds are not present in the active form. The reaction is under the surveillance of inhibitors and counterreactions. The products of the various reactions are ephemeral, often produced in small quantities, and sometimes insoluble. Nevertheless, rapid but restrained chain reactions are not unique to the blood coagulation system. They are involved in the release of chemical mediators in inflammation, or after antigen-antibody interaction, in the effectuation of hormone action (*e.g.*, the adenyl cyclase system). Moreover, many a subtle biological event, such as cellular aggregation, initiation of mitosis, or initiation of cellular death, might result from a cascade of rapid reactions as yet unknown. Therefore, a thorough understanding of the molecular events involved in blood coagulation may provide new tools and possibly new thoughts for the study of other biological mechanisms.

References

1. Kugelmass, I.: Biochemistry of blood in health and disease, p. 461–500. Springfield, Illinois: Charles C. Thomas Publisher 1959
2. Esnouf, M., Macfarlane, R.: Enzymology and the blood clotting mechanism. Advanc. Enzymol. **30**, 255–315 (1968)
3. Pool, J.: Thromboplastin formation. Annu. Rev. Med. **15**, 215–232 (1964)
4. Seegers, W.: Enzyme theory of blood clotting. Fed. Proc. **23**, 749–756 (1964)
5. Ferguson, J.: Thromboplastic enzymes. Fed. Proc. **23**, 762–772 (1964)
6. Milstone, J.: Thrombokinase in prime activator of prothrombin: Historical perspectives and present status, Fed. Proc. **23**, 742–748 (1964)
7. Spaet, T.: Nature and action of intermediate product. I. Fed. Proc. **23**, 757–761 (1964)
8. Kline, D.: Blood coagulation: Reactions leading to prothrombin activation. Annu. Rev. Physiol. **27**, 285–306 (1965)
9. Laki, K.: Enzymatic effects of thrombin. Fed. Proc. **24**, 794–799 (1965)
10. Laki, K., Gladner, J.: Chemistry and physiology of the fibrinogen-fibrin transition. Physiol. Rev. **44**, 127–160 (1964)
11. Moore H., Lux, S., Malhotra, O., Bakerman, S., Carter, J.: Isolation and purification of bovine and canine prothrombin, Biochim. biophys. Acta (Amst.) **111**, 174–180 (1965)
12. Mosesson, J.: The preparation of human fibrinogen free of plasminogen. Biochim. biophys. Acta (Amst.) **57**, 204–213 (1962)
13. Lorand, L., Ong, H., Lipinski, B., Rule, N., Downey, J., Jacobsen, A.: Lysine as amine donor in fibrin crosslinking. Biochem. biophys. Res. Commun. **25**, 629–637 (1966)

14. Huseby, R., Murray, M.: Molecular structure of fibrinogen. I. Helical content and the role of the tyrosine moiety in the fibrinogen molecule. Biochim. biophys. Acta (Amst.) **133**, 243–250 (1967)
15. Fuller, G., Doolittle, R.: The formation of crosslinked fibrins: Evidence for the involvement of lysine ε-amino groups. Biochem. biophys. Res. Commun. **25**, 694–700 (1966)
16. Haschemeyer, R., Cynkin, M., Han, L., Trindle, M.: Isolation and amino acid sequences of glycoproteins obtained from bovine fibrinogen. Biochemistry **5**, 3443–3448 (1966)
17. McKee, P., Rogers, L., Marler, E., Hill, R.: The subunit polypeptides of human fibrinogen. Arch. Biochem. Biophys. **116**, 271–279 (1966)
18. Lorand, L.: Physiological roles of fibrinogen and fibrin. Fed. Proc. **24**, 784–793 (1965)
19. Brinkhous, K., Langdell, R., Wagner, R.: Hemostatic disorders; hemophilia and the hemophilioid state. Annu. Rev. Med. **9**, 159–190 (1958)
20. 2299.21. Kattlove, H., Shapiro, S., Spivack, M.: Hereditary prothrombin deficiency. New Engl. J. Med. **282**, 57–61 (1970)
21. Gitlin, D., Borges, W.: Studies on the metabolism of fibrinogen in two patients with congenital afibrinogenemia. Blood **8**, 679–686 (1953)
22. Frick, P., McQuarrie, I.: Congenital afibrinogenemia. Pediatrics **13**, 44–58 (1954)
23. Zervos, N., Vlachos, J., Karpathios, T., Mantas, J.: Giant hemangioma of the spleen with thrombocytopenia and fibrinogen deficiency. Acta paediat. scand. **172**, 206–209 (1967)
24. Hirsch, J., Buchanan, J., DeGruchy, G., Baikie, A.: Hypofibrinogenemia without increased fibrinolysis in leukemia. Lancet **I**, 418–420 (1967)
25. Marcus, A.J.: Platelet function. New Engl. J. Med. **280**, 1213–1220, 1278–1284, 1330–1335 (1969)
26. Bettex-Galland, M., Luscher, E.: Thrombosthenin, the contractile protein from blood platelets and its relation to other contractile proteins. Advanc. Protein Chem. **20**, 1–35 (1965)
27. Mason, R.G., Jr.: Pathogenesis of thrombus: Platelet contribution. Science **138**, 833–838 (1962)
28. Zucker-Franklin, D., Nachman, R.L., Marcus, A.J.: Ultrastructure of thrombosthenin, the contractile protein of human blood platelets. Science **157**, 945–946 (1967)
29. Karpatkin, S.: Human platelet senescence. Annu. Rev. Med. **23**, 101–128 (1972)
30. O'Brien, J.: Platelet stickiness. Annu. Rev. Med. **17**, 275–290 (1966)
31. Baldini, M.: Idiopathic thrombocytopenic purpura. New Engl. J. Med. **274**, 1245–1251, 1302, 1360 (1966)
32. Sherry, S.: Fibrinolysis. Annu. Rev. Med. **19**, 247–268 (1968)
33. Penick, G.: Pathogenesis of thrombosis: Introductory remarks. Fed. Proc. **22**, 1349–1355 (1963)
34. Rodman, N., Mason, R., Brinkhous, K.: Some pathogenic mechanisms of white thrombus formation: Agglutination and self-destruction of the platelet. Fed. Proc. **22**, 1356–1365 (1963)
35. Wessler, S.: Stasis, hypercoagulability and thrombosis. Fed. Proc. **22**, 1366–1370 (1963)
36. McKay, D.G.: Disseminated intravascular coagulation. An intermediary mechanism of disease, New York: Harper & Row Publishers, 1964
37. Penick, G., Roberts, H.: Intravascular clotting: Focal and systemic. Int. Rev. Exp. Path. **3**, 269–328 (1964)
38. Penick, G., Roberts, H., Dejanov, I.: Covert intravascular clotting. Fed. Proc. **24**, 835–839 (1965)
39. Starzl, T., Boehmig, H., Amemiya, H., Wilson, C., Dixon, F., Giles, G., Simpson, K., Halgrimson, C.: Clotting changes including disseminated intravascular coagulation during rapid renal-homograft rejection. New Engl. J. Med. **283**, 383–390 (1970)
40. Legaz, M.E., Schmer, G., Counts, R.B., Davie, E.W.: Isolation and characterization of human factor VIII (antihemophilic factor). J. biol. Chem. **248**, 3946–3955 (1973)
41. Nemerson, Y., Esnouf, M.P.: Activation of a proteolytic system by a membrane lipoprotein: Mechanism of action of tissue factor. Proc. nat. Acad. Sci. (Wash.) **70**, 310–314 (1973)
42. Jesty, J., Nemerson, Y.: Purification of factor VII from bovine plasma; reaction with tissue factor and activation of factor X. J. biol. Chem. **249**, 509–515 (1974)
43. Zeldis, S.M., Nemerson, Y., Pitlick, F.A.: Tissue factor (thromboplastin): Localization to plasma membranes by peroxidase-conjugated antibodies. Science **175**, 766–768 (1972)
44. Gitel, S.N., Owen, W.G., Esmon, C.T., Jackson, C.M.: A polypeptide region of bovine prothrombin specific for binding to phospholipids. Proc. nat. Acad. Sci. (Wash.) **70**, 1344–1348 (1973)
45. Esmon, C.T., Owen, W.G., Jackson, C.M.: The conversion of prothrombin to thrombin; II. Differentiation between thrombin- and factor X_a-catalyzed proteolyses. J. biol. Chem. **249**, 606–611 (1974)
46. Owen, W.G., Esmon, C.T., Jackson, C.M.: The conversion of prothrombin to thrombin; I. Characterization of the reaction products formed during the activation of bovine prothrombin. J. biol. Chem. **249**, 594–605 (1974)
47. Heldebrant, C.M., Mann, K.G.: The activation of prothrombin; I. Isolation and preliminary characterization of intermediates. J. biol. Chem. **248**, 3642–3652 (1973)
48. Heldebrant, C.M., Butkowski, R.J., Bajaj, S.P., Mann, K.G.: The activation of prothrombin, II. Partial reactions, physical and chemical characterization of the intermediates of activation. J. biol. Chem. **248**, 7149–7163 (1973)
49. Nakamura, S., Iwanaga, S., Suzuki, T., Mikuni, Y., Konishi, K.: Amino acid sequence of the peptide released from bovine factor XIII following activation by thrombin. Biochem. biophys. Res. Commun. **58**, 250–256 (1974)
50. Lorand, L., Maldonado, N., Fradera, J., Atencio, A.C., Robertson, B., Urayama, T.: Haemorrhagic syndrome of autoimmune origin with a specific inhibitor against fibrin stabilizing factor (Factor XIII). Brit. J. Haemat. **23**, 17–27 (1972)
51. Benarous, R., Lavergne, J.-M., Labie, D., de Sanchez, J.M., Josso, F.: Isolation and partial characterization of a human prothrombin variant: Prothrombin Barcelona. Biochem. biophys. Res. Commun. **60**, 976–982 (1974)
52. Sherman, L.A., Gaston, L.W., Kaplan, M.E., Spivack, A.R.: Fibrinogen St. Louis: A new inherited fibrinogen variant, coincidentally associated with hemophilia A. J. clin. Invest. **51**, 590–597 (1972)
53. Janssen, C.L., Vreeken, J.: Fibrinogen Amsterdam, another hereditary abnormality of fibrinogen. Brit. J. Haemat. **20**, 287–298 (1971)
54. Nelsestuen, G.L., Suttie, J.W.: The mode of action of vitamin K. Isolation of a peptide containing the vitamin K-dependent portion of prothrombin. Proc. nat. Acad. Sci. (Wash.) **70**, 3366–3370 (1973)
55. Stenflo, J.: Vitamin K and the biosynthesis of prothrombin; III. Structural comparison of an NH_2-terminal fragment from normal and from dicoumarol-induced bovine prothrombin. J. biol. Chem. **248**, 6325–6332 (1973)
56. Stenflo, J.: Vitamin K and the biosynthesis of prothrombin; II. Structural comparison of normal and dicoumarol-induced bovine prothrombin. J. biol. Chem. **247**, 8167–8175 (1972)
57. Suttie, J.W., Grant, G.A., Esmon, C.T., Shah, D.V.: Postribosomal function of vitamin K in prothrombin synthesis. Mayo Clin. Proc. **49**, 933–940 (1974)
58. Nelsestuen, G.L., Zytkovicz, T.H., Howard, J.B.: γ-carboxyglutamic acid; identification and distribution in vitamin K-dependent proteins. Mayo Clin. Proc. **49**, 941–944 (1974)
59. Robbins, K.C., Bernabe, P., Arzadon, L., Summaria, L.: NH_2-terminal sequences of mammalian plasminogens and plasmin S-carboxymethyl heavy (A) and light (B) chain derivatives; a reevaluation of the mechanism of activation of plasminogen. J. biol. Chem. **248**, 7242–7246 (1973)
60. Robbins, K.C., Bernabe, P., Arzadon, L., Summaria, L.: The primary structure of human plasminogen; II. The histidine loop of human plasmin: Light (B) chain active center histidine sequence. J. biol. Chem. **248**, 1631–1633 (1973)
61. Marder, V.J., Budzynski, A.Z., James, H.L.: High molecular weight derivatives of human fibrinogen produced by plasmin; III. Their NH_2-terminal amino acids and comparison with the "NH_2-terminal" disulfide knot. J. biol. Chem. **247**, 4775–4781 (1972)
62. Highsmith, R.F., Rosenberg, R.D.: The inhibition of human plasmin by human antithrombin-heparin cofactor. J. biol. Chem. **249**, 4335–4338 (1974)
63. Esmon, C.T., Jackson, C.M.: The conversion of prothrombin to thrombin. III. The factor X_a-catalyzed activation of prothrombin. J. biol. Chem. **249**, 7782–7790 (1974)
64. Esmon, C.T., Jackson, C.M.: The conversion of prothrombin to thrombin. IV. The function of the fragment 2 region during activation in the presence of factor V. J. biol. Chem. **249**, 7791–7797 (1974)
65. Esmon, C.T., Owen, W.G., Jackson, C.M.: The conversion of prothrombin to thrombin. V. The activation of prothrombin by factor X_a in the presence of phospholipid. J. biol. Chem. **249**, 7798–7807 (1974)

Chapter 8

Hormones

Diseases of the Hypophysis

Anatomy and Histology

The hypophysis is a small, bean-shaped organ (0.6–0.65 g) located in the sella turcica at the base of the skull. On close examination of the gland, an anterior yellowish and a posterior more fibrous and whitish part can be distinguished. The two parts are separated by a pars intermedia, a lamellar structure, grayish and fibrous in appearance. The gland is attached to the floor of the third ventricle by a hollow pedicle, the stalk of the hypophysis.

The anterior lobe of the hypophysis (70–80% of the total weight of the gland) is made of several types of cells that are arranged in long columns forming a complex network. The meshes of the network are filled with small amounts of connective tissue and large capillaries. In some areas of the anterior lobe close to the pars intermedia, these cells are arranged to form pedicles that contain a colloid material.

The cells of the anterior lobe of the hypophysis can be divided into three types according to the staining properties of the cytoplasm; the chromophobes, the eosinophilic, and the basophilic cells. Chromophobe cells contain no granules, but the cytoplasm of the eosinophilic and basophilic cells is packed with large granular elements (eosinophilic or basophilic). Under normal conditions, the eosinophilic and the chromophobe cells are the most numerous. The cells containing basophilic granulations are rare. The relationship between these three types of cells is not clear. It is not known whether the different cells are actually stages of evolution of a single cell type with multiple potentialities or whether they are cells with distinct physiological properties.

The pars intermedia is poor in capillaries and contains small vesicles surrounded by cuboidal cells rich in colloid material, in which degenerated epithelial cells may occasionally be found. The cells forming the lining of the vesicles are of two types: large, elongated cells, which probably originate from the neuroglial or the ependymal cells of the neural part of the hypophysis; and polygonal, clear cells resembling the chromophobe cells of the anterior hypophysis. It is not known whether the clear polygonal cells are identical to or different from the clear polygonal cells in the pars intermedia.

The size of the pars tuberalis of the hypophysis varies considerably depending upon the species. It is large in the cat and dog and small in humans. The pars tuberalis is made of glandular chords that form vesicles containing colloid material. In contrast to the pars intermedia, the pars tuberalis contains many capillaries and neurofibers.

The pars nervosa is made of neuroglial cells and nervous fibers that are most abundant in the stalk of the pituitary. The pars nervosa contains three different types of cells that cannot be distinguished with ordinary staining techniques. The posterior lobe of the hypophysis is connected to the diencephalic nuclei by a band of amyelinic fibers that probably originate from the supraoptic nucleus and cross the tuber cinereum to reach the stalk of the hypophysis. It is not clear whether the posterior hypophysis is connected with other diencephalic nuclei, such as the paraventricular, the tuberian, and the subthalamic bodies.

The hypophysis has an endodermic and ectodermic origin. The first bud of the embryogenesis of the hypophysis originates from an invagination of the epiblast, that epithelium which covers the ceiling of the primitive mouth located in the embryo just anterior to the pharyngeal membrane. The evagination is called Rathke's pocket; its distal portion swells and is soon lined by an epithelium made of cylindrical cells. Later, in embryonic life, the blown-up part of the evagination separates from the pedicle attached to the pharynx.

From the floor of the midbrain, another evagination develops. It takes the form of a hollow diverticulum called the infundibulum. In the fourth week of intrauterine life of the human embryo, the infundibular and entoblastic proliferations join to form the hypophysis. The histogenesis of the hypophysis is usually completed in the ninth month of pregnancy.

The pars anterior, the pars intermediate, and the pars tuberosa are derived from the Rathke pocket. The neural portion of the stalk and the posterior lobe of the pituitary originate from the infundibular invagination.

The histogenesis of the hypophysis explains the existence of the pharyngeal and accessory hypophysis. A small cluster of hypophyseal cells are found in the pharyngeal mucosa. The cellular agglomeration reaches maximum development in young children but disappears as they grow. Even during the peak of its development, the pharyngeal hypophysis never exceeds a few millimeters in diameter.

During its organogenesis, the hypophysis leaves tracks in an area between the pharynx and the hypophysis. Again, small epithelial nodules of hypophyseal cells are inserted within the mucosa. These residues of embryological structures may develop into hypophyseal tumors that are found in the roof of the nasopharynx, in the body of the sphenoid bone, in the capsule of the pituitary gland, within the pars intermedia, and around the infundibulum. Such tumors are most frequently found in the pituitary capsule and the infundibulum.

Growth Hormone

Chemistry

The secretion of a number of hypophyseal hormones was demonstrated with the aid of a variety of techniques (hypophysectomy, injection of hypophyseal extract, and observation of pathological material). The anterior lobe secretes at least four different groups of hormones: somatotropins, gonadotropins, thyrotropins, and adenotropins. Only the role of the somato-

lization from adipose tissue and its accumulation in liver. The mechanisms by which this hypophyseal hormone elicits such alterations in lipid metabolism have, as we shall see later, been partially clarified thanks to the availability of *in vitro* systems. For a long time, these alterations in lipid metabolism were believed to be secondary to changes in nitrogen or carbohydrate metabolism because lipid mobilization coincides with the sparing of carbohydrate, the accumulation of glycogen in liver, and the increase in the rate of protein synthesis.

The significance of the effect of growth hormone on lipogenesis was difficult to evaluate without an adequate *in vitro* test. The use of epididymal fats yielded controversial results, either because of the complex nature of the target or because the hormonal preparation used was contaminated with TSH, which also elicits lipogenesis. Rodwell provided investigators in this field with a marvelous new tool, isolated fat cells that were free of contamination by other types of cells. Furthermore, the isolated cells are equivalent with respect to the diffusibility of the substrates produced. Using this preparation, Fain and his coworker were able to demonstrate that (1) the lipolytic effect of growth hormone is genuine and not due to side effects, presence on cells other than connective tissue cells, or effect of TSH; and (2) the lipolytic effect was inhibited by the administration of actinomycin D or X-irradiation, suggesting that the biosynthesis of new proteins was needed to elicit the lipolytic effect.

In conclusion, the major effect of growth hormone (lipolytic) on lipid metabolism is opposite to that of insulin (stimulates fatty acid deposition). Small doses of growth hormone may elicit an insulinlike effect that is transient, often prolonged by incubation, and replaced by a lipolytic effect.

The effects of growth hormone and insulin on adipose tissue are comparable with respect to glucose utilization but differ with regard to their effects on fatty acid synthesis and glucose use through the hexose monophosphate shunt. Growth hormone and insulin both stimulate glucose utilization by adipose tissue obtained from normal and alloxan-diabetic rats. But although insulin stimulates the synthesis of long-chain fatty acid, or restores to normal the rate of fatty acid synthesis in adipose tissue obtained from alloxan-diabetic rats, growth hormone has no effect on fatty acid synthesis. Whereas insulin stimulates the use of [^{14}C]glucose-1 through the hexose monophosphate shunt preferentially, growth hormone stimulates the use of carbons 1 and 6; in fact, its effect is more marked on carbon 6 than on carbon 1. The measurements of the relative oxidation rates of glucose carbons 1 and 6 suggest that growth hormone stimulates glucose use through the glycolytic pathway preferentially. These observations have led some investigators to conclude that the effect of growth hormone on glucose utilization by adipose tissue does not necessarily result from insulin release.

After growth hormone administration to fasted hypophysectomized rhesus monkeys, the nonesterified free fatty acid fraction of the plasma rises considerably. Although the administration of growth hormone produces secondary effects that suggest that the fatty acid oxidation rate is accelerated (ketosis, decreased respiratory quotient, and reduced lipid stores), an effect of growth hormone on the conversion of long-chain fatty acids, [^{14}C]palmitate to $^{14}CO_2$, has never been conclusively demonstrated *in vivo* or *in vitro*.

Effect on Amino Acids and Protein Synthesis. The administration of growth hormone induces a positive nitrogen balance, which means that less nitrogen is excreted in the urine and the concentration of amino acid nitrogen in the blood is reduced.

The exact site of action of growth hormone in the pathway of protein synthesis is not established, but it has been demonstrated that growth hormone stimulates amino acid transport and incorporation.

The effect of growth hormone on amino acid transport was studied with a nonmetabolizable amino acid—α-aminoisobutyric acid. The administration of growth hormone to female rats stimulates α-aminoisobutyric acid accumulation in kidney, duodenum, and gonads. In contrast, hypophysectomy increases the concentration of [^{14}C]α-aminoisobutyric acid in blood and decreases its concentration in tissues.

The effect of growth hormone on amino acid uptake has been confirmed in *in vitro* experiments. Diaphragms of hypophysectomized rats are unable to accumulate α-aminoisobutyric acid, but the addition of small amounts of growth hormone to the incubation medium normalizes the intracellular α-aminoisobutyric acid concentration.

The stimulation of amino acid uptake is oxygen dependent but is not inhibited by cyanide and dinitrophenol. The effect of growth hormone on α-aminoisobutyric acid has been extended to a number of natural amino acids, and it has been established that growth hormone stimulates the uptake of glycine, alanine, serine, proline, threonine, histidine, tryptophan, glutamine, and asparagine, but is without effect on the uptake of dicarboxylic acids, ethionine, valine, tyrosine, phenylalanine, leucine, lysine, and arginine.

It is not known whether the stimulatory effect of growth hormone on protein synthesis results from the action of the hormone on amino acid uptake or from a direct effect of the hormone on protein synthesis. The administration of growth hormone stimulates the incorporation of [^{14}C]leucine in diaphragm isolated from hypophysectomized rats and of [^{14}C]glycine in bone of hypophysectomized rats. When growth hormone is added to the incubation mixture containing diaphragm or levator muscle of hypophysectomized animals, it stimulates the incorporation of all natural amino acids except dicarboxylic acid. There is no stimulation of [^{3}H]uridine incorporation in the system, suggesting that the amino acid accumulation is not related to increased RNA synthesis.

Studies with cell-free systems have helped to localize the site of action of growth hormone on protein synthesis. The incorporation of [^{14}C]leucine in cell-free systems prepared from hypophysectomized rats is markedly reduced compared to the incorporation of the labeled amino acid into protein of a cell-free system obtained from normal rat.

Treatment of the hypophysectomized animals with growth hormones normalizes the ability of the cell-free system to incorporate amino acids.

In the cell-free systems prepared from hypophysectomized rats, ribosomes lose their capacity to participate in protein synthesis, possibly by dispersion of the polyribosome system into smaller units. Growth hormone is thought to act on protein synthesis in hypophysectomized animals by stimulating amino acid uptake and indirectly increasing the production of messenger RNA.

Effect on Electrolytes. The administration of human growth hormone raises the intracellular levels of electrolytes, leads to a loss of bone calcium, and reduces urinary levels of phosphorus, potassium, and sodium. The increase in intracellular electrolytes may result from an increase in the cellular mass. Mobilization of bone calcium leads to osteoporosis in acromegaly and calciuria. Two explanations offered for this are (1) increased glomerular filtration combined with inhibition of tubular reabsorption, and (2) stimulation of parathyroid secretion.

There is no satisfactory explanation for the reduced levels of phosphorus, potassium, and sodium found in urine of individuals injected with growth hormone. The hormone is believed to stimulate glomerular filtration at the same time that it increases tubular reabsorption. Although the reasons for increased reabsorption of potassium and sodium are obscure, it is possible to speculate on the reasons for phosphorus reabsorption. Increased tubular reabsorption of phosphorus is not accompanied by increased levels of phosphorus in the blood, and therefore phosphorus could be concentrated in the tissue where it stimulates growth.

Like most hormones, growth hormone is believed to bind to specific receptors on the target cells (see section on receptors). Such receptors have been found on human lymphocytes [165].

It is not quite clear how growth hormone exerts its effect at the cellular level. One hypothesis proposes that the major function of the hormone is to stimulate protein synthesis and that this is achieved by the release of insulin-like substances; namely, somatomedin A and B. These substances are found in plasma, but their chemical structure is still unknown. Two somatomedins have been found in plasma; a large (mol wt 44,000) and a small one (mol wt 22,000). Although there is no evidence of conversion of the large into the small somatomedin, it has been proposed that the high is a precursor of the low molecular weight component [166].

Compared to other hormones we know little about the regulation of the secretion of growth hormone (see regulation of secretion of ACTH and TSH). A hypothalamic releasing factor is supposed to exist, but its chemical structure is unknown. A hypothalamic tetradecapeptide, somatostasin, has been discovered which inhibits the release of growth hormone from the hypophysis under physiological (hypoglycemia, exercises, etc.) or pathological (acromegaly) conditions.

Conclusion. Although, as was mentioned previously, it is not surprising that growth hormone manifests its effect in numerous metabolic steps, the information available does not indicate how these various effects relate to growth, which effect is primary, or what common intermediate elicits the multiple responses to the hormone, although somatomedins have been implicated.

The evaluation of the physiological role of growth hormone is further complicated by the fact that it does not seem to be indispensable for growth in all cases. Fetal growth does not seem to be impaired by the absence of growth hormone, and reports have appeared of young hypophysectomized (for tumors of the hypophysis or the hypothalamus) patients who presented normal or superior growth rates with extremely low levels of growth hormone in the plasma (as measured by the radioimmunoassay). The conclusion seems inescapable. Either extremely low levels of growth hormone are compatible with normal growth or other growth factors must be invoked. As Holmes and his associates pointed out, these patients could be missing the growth hormone but could elaborate enough "sulfation factor," which is essential for skeletal growth [9].

Gigantism and Acromegaly

In 1886, Pierre Marie recognized the pituitary origin of acromegaly (from the Greek words ακρος and μεγαλος). Later it was discovered that acromegaly and gigantism result from an adenoma of the eosinophilic cells of the hypophysis, and it was therefore proposed that the eosinophilic cells secrete the growth hormone. This conviction was confirmed by the studies of Smith and Smith, who prepared extracts of two parts of the bovine hypophysis: the lateral region where acidophilic cells are numerous and the median region where acidophilic cells are scarce. The studies clearly demonstrated that extracts obtained from the part of the hypophysis rich in acidophilic cells were much more efficient in stimulating growth in tadpole than extracts of the hypophysis obtained from areas where acidophilic cells were scare.

The effect of eosinophilic cell hyperplasia on growth varies according to age of onset of the adenoma. When the adenoma develops before puberty, it induces a proportionate increase in size of all those organs that are the normal targets for growth hormones (gigantism). After puberty, when the growth of the long bone is completed, only the flat bones and the viscera are stimulated to grow under the influence of the excess growth hormones; therefore, the relative size of the

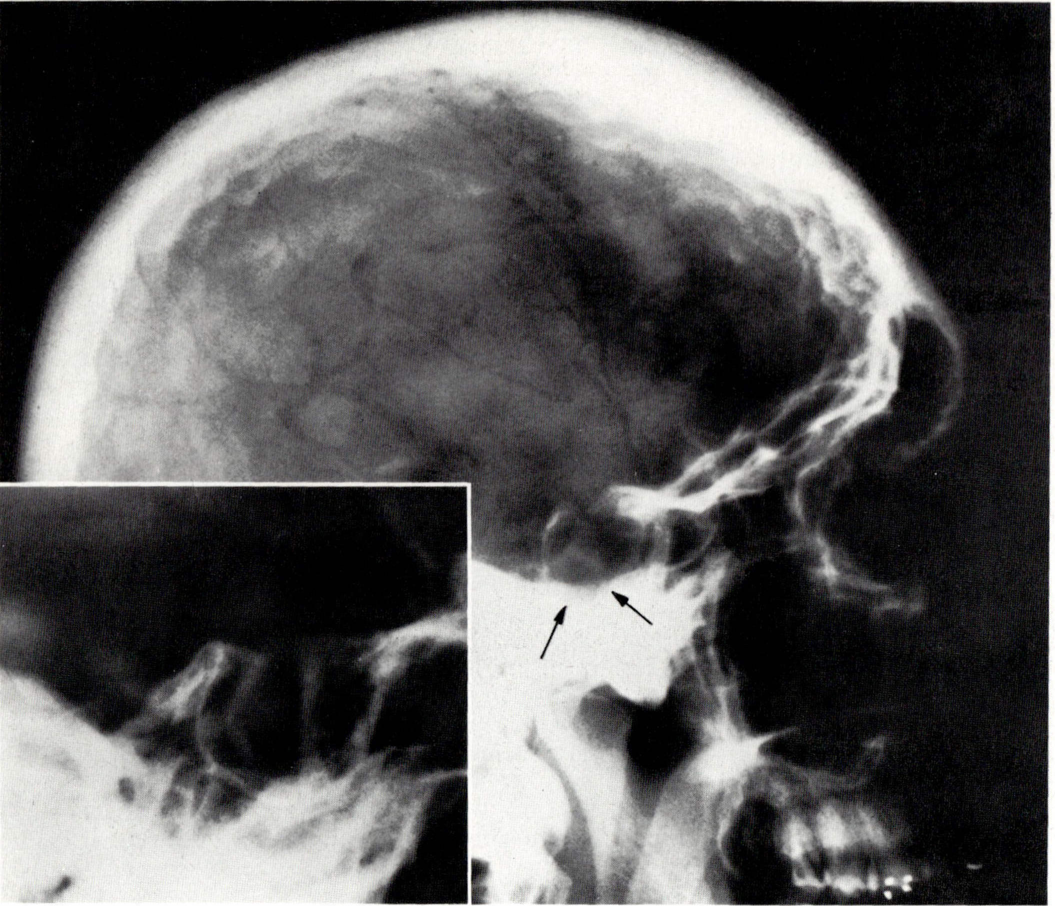

Fig. 8-1. A case of acromegaly with large skull hypertrophy of the cortex, enlargement of the frontal sinuses, and, most relevant, enlargement of the sella turcica (box) as a result of tumoral growth of the hypophysis. (From J. Collins)

bones of the skeleton of the victim is disproportionate (see Fig. 8-1).

Occasionally hyperplasia that starts before puberty persists in the adult and leads to a combination of features typical for gigantism and acromegaly.

Acromegaly and gigantism result from (1) hypersecretion of some hormones of the anterior lobe of the hypophysis (especially the growth hormone), (2) decreased secretion of some other anterior lobe hormones, particularly gonadotropins, and (3) compression of the surrounding tissue at the base of the skull.

The development of radioimmunoassays has permitted measurement of the amount of growth hormone in the blood of acromegalics. The normal blood level may be increased up to 100 times in acromegalics. Patients who develop no further symptoms of acromegaly continue to have excessive amounts of growth hormone in their blood.

In acromegaly there is a tremendous increase in the size of the flat bones of the face, vertebra, phalanges, tarsus, and practically all viscera. Facial bone enlargement involves in particular the malar bone, the bones of the nose, and the lower jaw. The acromegalic has a considerably enlarged face with deep-set eyes, prominent frontal sinuses, a large nose, and a prominent chin. The teeth of the lower mandible are spaced apart, the ears are large, and the lips, tongue, and skin of the nose are very thick (see Fig. 8-2). The vocal cords usually thicken, causing the voice to become deep and rough. The hands and feet are enormous; the hands are often so wide that they appear spadelike. The skin of the scalp is deeply corregated and the hair that grows is thick and coarse. The heart, lung, kidney, spleen, and liver are much larger than normal.

The vertebrae, particularly those of the lumbar spine and thoracic spine, also increase in size. The enlargement involves the anteroposterior axis of the bone more than the lateral axis. Consequently, vertebral enlargement interferes with the patient's flexion, and he has difficulty bending to touch the floor. The increase in skeletal size is not associated with a corresponding increase in muscular strength; and, in spite of his frightening appearance, the acromegalic is often feeble and clumsy.

Skeletal growth in acromegaly cannot simply be explained by an anti-insulinic effect, stimulation of lipolysis, or stimulation of protein synthesis because cartilage formation requires that collagen (conversion of protein to hydroxyproline) and cartilage (synthesis of chondroitin sulfate) be formed. Acromegalic serum

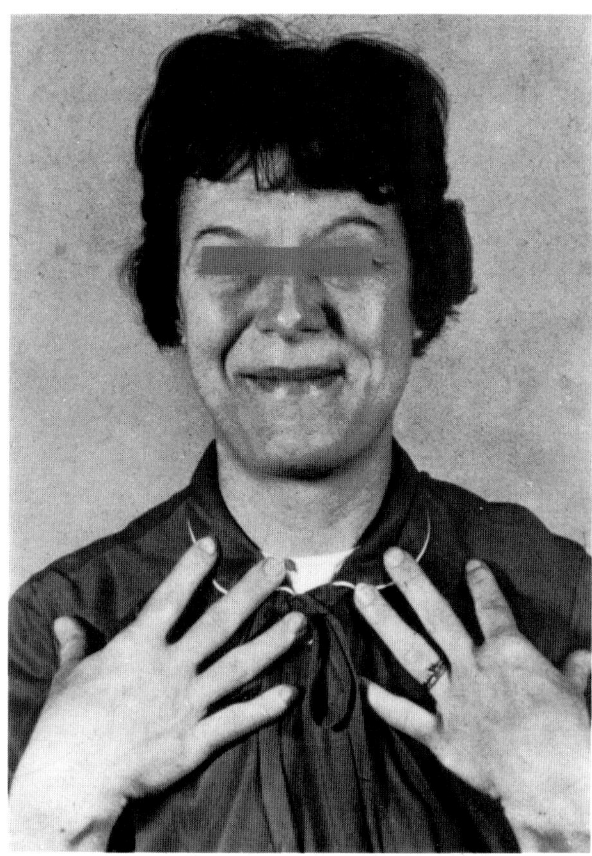

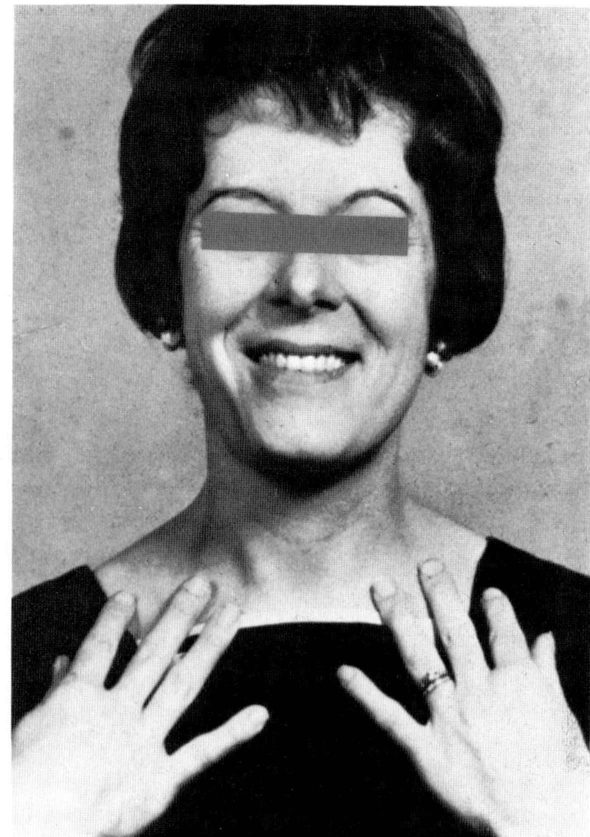

Fig. 8-2. Acromegaly before and after treatment. (From J. Brown)

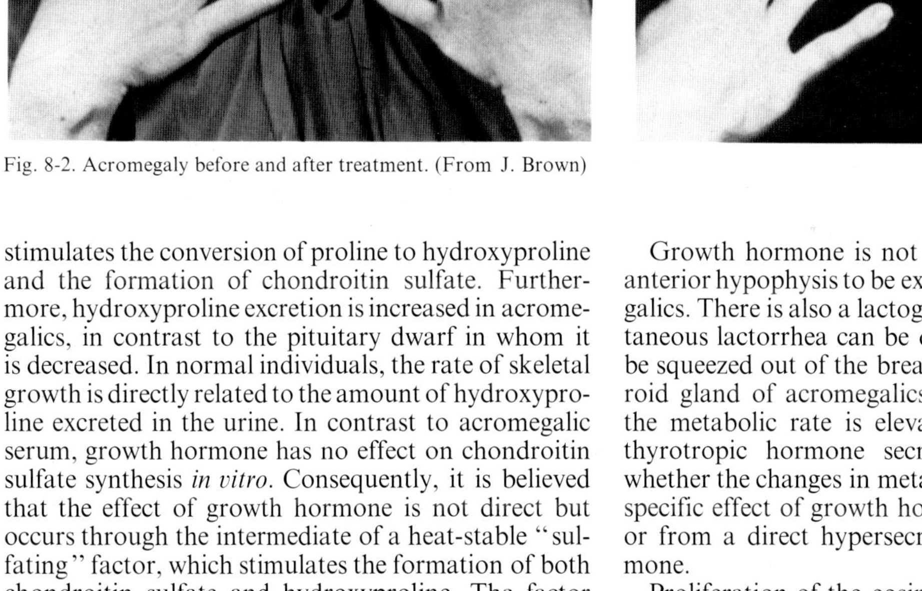

stimulates the conversion of proline to hydroxyproline and the formation of chondroitin sulfate. Furthermore, hydroxyproline excretion is increased in acromegalics, in contrast to the pituitary dwarf in whom it is decreased. In normal individuals, the rate of skeletal growth is directly related to the amount of hydroxyproline excreted in the urine. In contrast to acromegalic serum, growth hormone has no effect on chondroitin sulfate synthesis *in vitro*. Consequently, it is believed that the effect of growth hormone is not direct but occurs through the intermediate of a heat-stable "sulfating" factor, which stimulates the formation of both chondroitin sulfate and hydroxyproline. The factor is believed to disappear after hypophysectomy and reappear after growth hormone administration.

Acromegalics often have increased glucose tolerance, a certain degree of insulin resistance, and a moderate form of diabetes. The pathogenesis of the diabetic changes is not clear. Although the changes are assumed to result from an increase in ACTH secretion, in most cases 17-ketosteroids excretion is not increased. This finding suggests that the adrenocorticotropic hormone is produced in normal amounts in acromegalic patients. Whatever the cause of the anti-insulin effects, experimental evidence suggests that they result from antagonism to the normal action of insulin rather than from a decrease in insulin levels.

Growth hormone is not the only hormone of the anterior hypophysis to be excreted in excess in acromegalics. There is also a lactogenic effect, and if no spontaneous lactorrhea can be detected, milk can usually be squeezed out of the breast of the patient. The thyroid gland of acromegalics is usually enlarged, and the metabolic rate is elevated, suggesting increased thyrotropic hormone secretion. It is not certain whether the changes in metabolism result from an unspecific effect of growth hormone on thyroid growth or from a direct hypersecretion of thyrotropic hormone.

Proliferation of the eosinophilic adenoma leads to compression and atrophy of some of the other cellular components of the hypophysis, particularly the beta cells, with loss of gonadal function in men and amenorrhea in women. Sexual atrophy can be relieved by the administration of estrogens and androgens. In general, sex hormone administration relieves such subjective symptoms as headache, insomnia, irritability, lassitude, numbness, and stinging of the hands. It is not known whether the sex hormones compensate for the reduced gonadostimulin secretion or if they directly counteract the exaggerated somatotropin production.

Some of the symptoms occurring in acromegaly are difficult to explain in the context of our present knowledge of pituitary physiology. For example, inorganic phosphorus is increased and calcium levels are decreased in the sera of acromegalics and children with hyperpituitary conditions. Generalized osteoporosis is common in these patients.

The effect of growth hormone on calcium metabolism has been investigated by somatotropin administration to humans and to hypophysectomized rats. Growth hormone has a calciuric effect and induces a negative calcium balance without affecting the plasma calcium concentration. The mechanism by which this calciuric effect is produced is unknown. Various theories have been proposed, including direct effects of somatotropin on calcium metabolism and an indirect effect resulting from alteration of phosphorus metabolism. A direct effect on calcium metabolism could affect renal calcium transport, calcium absorption in the intestine, or parathyroid function. An effect through alteration of phosphorus metabolism may result from the action of growth hormone on tubular phosphate reabsorption. Whatever the mechanism by which the negative calcium balance is induced after growth hormone is administered, changes in calcium metabolism are reversible and respond to adequate therapy for hypophyseal tumors or to the administration of estrogens and androgens.

Since radioimmunoassays are available, it is now possible to diagnose the early stages of acromegaly and to follow the hormone levels as the disease develops. Furthermore, new types of growth hormone excess have been discovered; a disease has been described in which excess growth hormone is observed in the absence of a pituitary hormone. The disease is believed to result from abnormal stimulation of the hypothalamic secretion.

Hypopituitarism

Hypopituitarism may involve all the hormones of the anterior hypophysis, or it may be restricted to one or a few hormones of the anterior lobe, depending on the patient's age. General hypopituitarism of the anterior lobe results in pituitary dwarfism, infantilism, and Simmonds' cachexia. The restricted forms of hypopituitarism lead to well-known clinical syndromes such as Fröhlich's and the Laurence-Moon-Biedl syndrome.

Simmonds' cachexia is often the consequence of anterior lobe necrosis due to its destruction by shock or septic emboli. The septic emboli may result from ascending pyelonephritis or from pyemia. Simmonds' cachexia is observed more often in women than in men because obstetric shock seems to cause it frequently. But a progressive inflammatory process, the development of a chromophobe adenoma (often associated with compression of the chiasma and seizures), or hemorrhage within the gland should be considered as possible causes of Simmonds' cachexia.

In Simmonds' cachexia, the secretion of all the stimulins elaborated by the anterior pituitary gland is interrupted, growth stops, and secretions from the thyroid, adrenal cortex, and gonads are partially abolished.

The lack of growth hormones results in generalized atrophy of muscle (heart, skeletal muscle) and viscera (liver, kidney, spleen), and at autopsy the most typical finding in patients with Simmonds' cachexia is microsplanchnia. In most cases of Simmonds' disease, the patient becomes dramatically cachectic with a severe negative nitrogen balance.

The exact pathogenesis of the cachexia remains unsettled, but in view of the role of growth hormones in stimulating amino acid uptake and incorporation into protein, cachexia probably results from the lack of growth hormones.

In Simmonds' cachexia, pancreatic and gastric juice secretion is reduced probably as a result of a reduction in the rate of protein synthesis. Without ACTH, the adrenal cortex is unable to secrete those hormones that stimulate gluconeogenesis in liver. As a result, patients with Simmonds' cachexia are severely hypoglycemic and exhibit general weakness, apathy, and unresponsiveness. Insulin administration to patients with generalized anterior hypopituitarism is dangerous because insulin accentuates the hypoglycemia. The adrenal gland and the gonads are unable to elaborate 17-ketosteroids, and the urinary excretion of those compounds is markedly decreased in Simmonds' cachexia. Histologically, the zona reticularis of the adrenal gland is often atrophic.

In Simmonds' cachexia, myxedema develops because of the lack of thyrotropic hormones. The basal metabolism of the victim is low, the patient is oversensitive to cold, and histological examination of the thyroid gland reveals atrophy of the acini with lymphoid infiltration and fibrosis. In Simmonds' cachexia, the primary and secondary sexual organs are markedly atrophied, resulting in lack of libido in men and persistent amenorrhea in women. All patients lose much of their body hair; complete absence of axillary and pubic hair is often observed. Histological examination of the testes demonstrates marked regression of the seminiferous tubes, luteinization of the basal membrane, atrophy of the Leydig cells, and edema of the interstitial tissues. When typical Simmonds' cachexia develops, it is seldom difficult to diagnose, yet it is sometimes necessary to distinguish between generalized anterior hypopituitarism and anorexia nervosa, myxedema, and Addison's disease.

Dwarfism

Dwarfs have been the subjects of novels and fairy tales for centuries. It was only in the early 1900's that the distinction between dwarfs with and without physical abnormalities was made and it was recognized that some attained sexual maturity while others remained infantile.

Pituitary dwarfism may be acquired or hereditary. A dwarf is defined as a person whose stature is 40% lower than normal or three standard deviations to the left of the mean. The causes of dwarfism are numerous. In addition to damage to the hypothalamus and hypophysis one must include other hormonal deficiencies (hypothyroidism, hyperthyroidism, uncontrolled diabetes, Cushing's syndrome and pseudosexual precocity with gonadal tumors). The nonhormonal disorders fall in three major categories: (a) defects in the growth of bones (e.g., achondroplasia), (b) malnutrition (hypocaloric dwarfism, ricketts) and (c) dwarfism associated with organic malfunction resulting from either acquired or hereditary damage to the organ. One distinguishes between renal, hepatic, intestinal, anoxemic dwarfism, etc. For example, renal dwarfism is associated with renal malformation, chronic nephritis, or Franconi's syndrome; hepatic dwarfism with glycogen storage disease; intestinal dwarfism with cystic fibrosis, sprue or megacolon; anoxemic dwarfism with congenital heart disease or chronic anemia. Many of these forms of dwarfism are discussed in other chapters. Acquired pituitary dwarfism is sometimes associated with the development of a craniopharyngioma or another pituitary tumor. Dwarfism results from generalized retardation of skeletal and body growth, and mental deficiency is observed. Pituitary dwarfs often age faster than normal, and their faces may be markedly wrinkled.

Except when the anatomicopathological picture is recognizable (chromophobe adenoma or other types of pituitary tumors), the pathogenesis of the adiposogenital dystrophy is unknown. A congenital alteration of the neurohypophyseal control system is often invoked. The disease is characterized by atrophy of the primary and secondary sex organs, an abnormal distribution of fat with predilection for the lower abdomen, breast, hips, thighs, and mons veneris. Adiposogenitals may be apathetic and intellectually slow, but they are seldom mentally retarded.

The Laurence-Moon-Biedl syndrome is a familial disease characterized by adiposogenital dystrophy, mental deficiency, and numerous other congenital malformations, including skull deformities, atresia of the anus, polydactyly, and retinitis pigmentosa.

In addition to the typical panhypopituitarism resulting from anatomical damage to the anterior lobe of the hypophysis, there are also types of selective hypopituitarism with hypofunction of target organs. Pituitary myxedema, pituitary hypoadrenocorticism, and pituitary hypogonadism can be distinguished. These forms of hypopituitarism in which no hypophyseal lesion is detectable are discussed in more detail in the chapter devoted to the respective target organ. Suffice it to point out here that they are believed to result from abnormal hypothalamic regulation.

A form of idiopathic hypopituitarism resulting in growth retardation with diminished stature, infantilism, and retarded bone age has also been described. This form of idiopathic hypopituitarism should not be confused with a reversible "psychic form" that was described by Powell and his associate. These investigators described a number of patients who showed marked growth retardation with low plasma levels of growth hormone and ACTH. However, all these patients had an important feature in common: they all were deprived of normal affection at home, especially on the part of the father who either drank excessively or had extramarital affairs and was often away from home. In some cases the mother was psychotic, and some of the children had been physically maltreated. The most striking feature was that in all these cases when the child was brought into an emotionally stable environment he grew at a normal rate.

This important clinical observation suggests that the secretion of growth hormone is, like that of gonadotropins, influenced by psychic factors. Well-controlled experimentation or further clinical observations of a similar nature are now needed to substantiate such an interpretation of the results.

Hereditary dwarfism associated with sexual ateliosis is usually transmitted as a hereditary autosomal, sometimes a hereditary dominant, trait. Merimée [167] has classified these dwarfs into two major groups on the basis of determination of growth hormones; one with the other without HGH deficiency. Each group falls into two subgroups: group Ia with insulin hypersensitivity and insulopenia that can be corrected by the administration of growth hormone; and a group Ib with normal insulin sensitivity or hyperinsulinemia. The condition is transferred as an autosomal recessive trait in group Ia and as an autosomal dominant in group Ib. In group IIa and group IIb, the levels of immunoreactive HGH in plasma are normal or high. The groups differ in that the somatomedin is low in the first and normal in the second.

In 1929, Snell discovered a strain of dwarf mice. The dwarfism is the result of a genetic defect that is transmitted as an autosomal recessive mendelian trait. In these animals, the anterior pituitary is unable to elaborate four of the tropins: growth hormone, thyrotropin, prolactin, and possibly corticotropin. However, the mutant secretes gonadotropins. But the basic alteration responsible for this curious disease is not known. Perhaps during histogenesis the anterior hypophysis became incapable of elaborating a precursor cell that would normally differentiate into four different types, each secreting a specific hormone. Another plausible interpretation is that the four polypeptide hormones are secreted in the same cell and that the structural genes for the four polypeptides are, in those mice, linked in a single operon the expression of which is abolished by a mutation in the operator or possibly in regulatory genes [10].

Pygmies are believed to have normal levels of growth hormone in their plasma and elaborate normal levels of somatomedin after the administration of HGH. Although the mechanism of growth retardation in pygmies is not known, it would appear that an inability to respond to somatomedins might be responsible for the condition]167].

In some central African tribes that inhabit the rain forest and rely on hunting for food, all the adults are less than 5 feet tall. They are thin and muscular and weigh an average of 99 pounds. The cause of the dwarfism is not known. They are not likely, as some believed, to be achondroplastic dwarfs since their legs are usually longer than their trunk. Investigators first thought that these individuals had subnormal levels of growth hormone secretion after stimulation. But investigations of growth hormone levels after appropriate stimuli revealed normal responses in the pygmies [11].

Posterior Lobe

In 1895, Oliver and Schafer discovered the effect of pituitary extract on blood pressure. Three years later, they demonstrated that the posterior lobe of the pituitary was the source of these hormones. During the following decade, the milk-releasing, antidiuretic, and oxytocic properties of the posterior extract were discovered [12, 13].

Chemistry of Oxytocin and Vasopressin

When, after numerous unsuccessful attempts, oxytocin and vasopressin were finally isolated, one of the most fascinating chapters of the history of medicine was started. Most of the original work was done in Du Vigneaud's laboratory between 1932 and 1953. It involved oxytocin and vasopressin purification, investigation of their amino acid composition and sequence, synthesis of the polypeptides, and investigation of the relationship between secondary structure and activity of both these hormones.

The hormones were purified by a variety of methods, including countercurrent separation, electrophoresis, and starch column chromatography. Such studies yielded oxytocin and vasopressin preparations with homogenous physical properties. The availability of purified preparations made it possible to investigate the amino acid composition of both these polypeptides. Both are small polypeptides composed of nine amino acids, each of which is present in the polypeptide in equal amounts. Oxytocin and vasopressin have six amino acid residues in common: tyrosine, proline, glutamic acid, aspartic acid, glycine, and cysteine. Both hormones have three amide groups within their formula: glycinamide, glutamine, and asparagine. In both oxytocin and vasopressin, the presence of a 20-member sulfur ring can be demonstrated by oxidation of the disulfide ring with performic acid. (When such a procedure is applied to a linear molecule, it yields two different fragments; if the molecule has a ring structure, only a single compound is formed.) The insulin molecule also contains a 20-atom sulfur ring. The fluorodinitrobenzene reaction followed by hydrolysis led to the release of cysteine and tyrosine residues.

All the pieces of the formulas of both oxytocin and vasopressin fell into place when, in 1953, several laboratories independently published the complete amino acid sequence of both these hormones. It thus became obvious that each of them contained a 20-member disulfide ring and tripeptide side chain. Oxytocins obtained from man, beef, hog, horse, sheep, and chicken all had the same formula and differed from vasopressin mainly by the presence of an isoleucine and a leucine residue instead of phenylalanine and arginine residues. When vasopressins other than the bovine were investigated, it became clear that an arginine and a lysine type of vasopressin existed: the former found in rat, dog, and monkey. Although the monkey and rat have the same amino acid sequence in their vasopressin, man differs from both monkey and rat but has the same vasopressin as steer, hog, and horse. Consequently, the phylogenetic relationships between monkey and man do not necessarily imply a similitude in the protein structure of their hormones (see Table 8-2).

Table 8-2

Cys–Tyr–Phe–Glu(NH_2)–Asp(NH_2)–Cys–Pro–Arg–Gly–(NH_2)

Amino acid sequence of arginine vasopressin.

In addition to publishing the amino acid sequence of these two hormones, Du Vigneaud also described a procedure for their synthesis (which has since been simplified), making it possible for the first time to synthesize analogs of polypeptides and to study a hormone's structure-activity relationships.

The importance of both the 20-member ring and the side chain for biological function has been investigated. These investigations are still in their infancy, and only tentative conclusions can be reached. We shall consider separately the structural changes that inactivate both the oxytocin and the vasopressive effects and those that are necessary to convert oxytocin into vasopressin.

Two types of alteration in functional activity can be introduced into the 20-member ring: those that abolish biological activity and those that are tolerated but modify the intensity of the activity. The activity of the polypeptide hormone can be abolished by replacing the cysteine 1 with mercaptoacetic acid, by introducing an additional amino acid between tyrosine 2 and phenylalanine 3, and by acylation of the amino group of cysteine. These findings underline the importance of the 20-member ring for hormonal activity.

However, the sulfur of cysteine can be replaced by a CH_2 group, and oxytocin activity is maintained. Such an observation conflicts with theories involving the S–S bond in the hormone's mechanism of action.

The presence of glutamine in position 4 and asparagine in position 5 is also important because the replacement by isoglutamine or isoasparagine inactivates the hormone. In fact, isoglutamine-4-oxytocin is an inhibitor of vasopressin.

The basic side chain in position 8 is indispensable. Lengthening of the side chain, replacement of glycine NH_2 by sarcosine, or removal of the N-terminal amino group abolishes the activity of oxytocin (deamido oxytocin). Yet, elimation of the N-terminal amino group does not affect oxytocin, neither does chain branching at the amino group. Replacement of the amino group of the cysteine 1 by β-mercaptopropionic acid, reduction of the sulfhydryl bridge, and substitution of the amino acids in positions 2, 3, and 8 are among the molecular alterations that change the biological activity without abolishing it.

Tyrosine or phenylalanine residues in position 2 are essential for activity. Replacement of the aromatic amino acid by serine inactivates the hormone completely. The phenylalanine derivative is less active than the tyrosine derivative, indicating that the hydroxyl group of tyrosine, although not essential, enhances the hormonal activity.

Dimerization of the oxytocin molecule by incubation in $NaHCO_3$ at pH 8.3 at room temperature yields a biologically inactive compound. When vasopressin is incubated with cysteine or glutathione, the reduction of the disulfide bridge diminishes antidiuretic activity to one-third of its original value. The substitution of the 8-lysine of vasopressin by 1-deamino lysine enhances the antidiuretic activity, but substitution of the 8-lysine by 1-acetyllysine abolishes most of the hormonal activity.

Three alterations of the active polypeptide are known to modify its biological activity by shifting from oxytocic to vasopressic activity. Elimination of the side chain suppresses all vasopressic activity, and the pentopeptide maintains only oxytocic and lactogenic properties. Secondly, substitution of the leucine in position 8 by lysine or arginine enhances vasopressic activity. In contrast, histidine vasopressin has little vasopressic activity. This property thus seems to be a function of the base strength of the amino acid that occupies position 8. The third molecular alteration known to modify activity is the substitution of isoleucine in position 3 of the vasopressin by phenylalanine, which enhances vasopressic activity.

Both oxytocin and vasopressin act on the same target organs. Their action results in ejection of milk by the breast, uterine contraction, increased blood pressure, an antidiuretic effect, and hypertension. It is not known which common denominator in each of the target organs is responsible for its response to the polypeptide hormone. The intensity with which either of these hormones acts on the target organs varies considerably. The sharing of biological properties is not surprising in view of the similarity in the structural formulas of the octapeptides. What is striking is that the simple substitution of isoleucine by phenylalanine and leucine by arginine induces remarkable differences in the activity of both hormones.

Secretion of Posterior Lobe Hormones

The exact site of elaboration of the small polypeptide hormones found in the posterior pituitary is not known. Their secretion is likely to involve a complex of neuroendocrine structures referred to as the neurohypophysis. The neurohypophysis is comprised of at least four different structures: specialized hypothalamic nuclei (supraoptic and paraventricular nuclei), the median eminence, the pituitary stalk, and the posterior lobe of the hypophysis.

The antidiuretic hormone is secreted according to a first-order neuroendocrine mechanism. A central nervous system neuron secretes the active polypeptide, which is transported in the blood to the target organ. This assumption is based on the experiments of Horsley and Clarke, who destroyed the hypothalamic nuclei without affecting the hypophysis itself by approaching the thalamus from above. These experiments demonstrated that destruction of the supraoptic nuclei or of the fiber tracts that lead from these nuclei to the posterior lobe was consistently associated with diabetes insipidus. In contrast, total hypophysectomy does not necessarily induce diabetes insipidus.

Neurosecretory granules have been observed by histochemical methods and electron microscopic examination of the perinuclear region of the neurons of the supraoptic and paraventricular nuclei. Histologically, with the aid of the Gomori stain, the secretory granules can be identified as dark-staining granules. Under the electron microscope, the granules appear as dense vesicles surrounded by a membrane. Secretory granules similar to those in the neuron are recognizable all along the axon in the pituitary stalk. These axons connect the central neuron to the posterior lobe of the hypophysis.

On the basis of morphological observations, it is thought that the active polypeptides are synthesized in the cells of the supraoptic and paraventricular nuclei, and from there they are transported to the posterior pituitary by axoplasmic streaming.

Experiments using tissue slices obtained from the hypothalamus and the neurohypophysis have clearly shown that [^{35}S]cysteine was incorporated in vasopressin extracted from the hypothalamus but not in vasopressin extracted from the neurohypophysis; puromycin interferes with cysteine incorporation in vasopressin.

The role of the pituicyte in the secretory mechanism of the antidiuretic hormone remains unsettled. This cell could act as a storage site for the active polypeptides, which would then be released from the cells of the posterior lobe into the bloodstream as needed. The posterior lobe is more sensitive to changes in the water content of blood because it is not separated from circulating blood by the blood-brain barrier. That the pituicytes affect polypeptide hormone secretion is suggested by the fact that continued stimulation of the neurohypophysis causes the mitotic rate in pituicytes to increase.

At least two lines of indirect evidence indicate that the secretory granules contain the active polypeptide: (1) the colloid material stainable with the Gomori technique contains protein-bound SH or SS groups that are demonstrable histochemically, and (2) the proportion of granules found in the pituicyte changes under physiological conditions known to stimulate or depress ADH secretion. The neural lobe is depleted

centration of the radioisotopes can be detected in the smaller than in the larger follicles. Quantitative radioautography demonstrates that the iodine uptake by the follicles is proportionate to the surface area; consequently, the greater uptake in the smaller follicles appears to result in part from the greater surface volume ratio. The radioisotope is not uniformly distributed throughout the lumen of the follicle, but forms a ring at the periphery of the follicle close to the cell surface. On radioautographs, the ring density is irregular, which suggests that some of these follicular cells participate more actively than others in the iodine pumping.

The thyroid is able to concentrate fantastic quantities of iodine; concentration in the normal thyroid is 20–25 times that in the serum. Although passive diffusion may affect iodine uptake, the concentration process is mainly active. Two different pumping mechanisms have been considered: (1) a trapping mechanism involving specialized protein; and (2) an active transport mechanism involving carriers and energy. In addition to iodine, the thyroid also concentrates other anions—such as fluoroborate and perchlorate—with similar ionic properties, volume, and charges. The uptake of these anions is stimulated by TSH injection. The trapping is thought to be accomplished by a protein with a specific molecular configuration that would chelate iodide and other similar anions, and trapping is believed to require no energy.

The second theory for the mechanism of iodide concentration in the thyroid involves a specific carrier and special energy expenditures. It is based on studies of the effects of two different types of inhibitors—thiocyanate and energy inhibitors—on iodine concentration in the thyroid. Thiocyanate enters the thyroid follicle by simple diffusion, and its concentration within the follicle does not exceed that in the blood. Thiocyanate blocks iodide concentration in the thyroid, preventing the binding of iodide to protein and releasing the iodide already bound to protein. The mechanism by which thiocyanate blocks iodine concentration is not known, but kinetic studies of the effect of thiocyanate indicate that the inhibition is competitive. It has been proposed that thiocyanate competes for iodide with the pumping device and inactivates the pump temporarily by poisoning one of its molecular elements. Of course, nothing is known of the molecule that binds thiocyanate, but such a molecule could be a carrier involved in iodide transport through the cell membrane.

That the active concentration of iodide in the thyroid is energy dependent is suggested by several observations. The active concentration of iodide requires oxygen. It is blocked by metabolic poisons, such as cyanide and azides. There is a correlation between concentration ability and rate of oxidative phosphorylation. Uncouplers such as 2,4-dinitrophenol block iodine uptake in the thyroid.

Iodide Oxidation. Inside the thyroid, iodide is first oxidized to active iodine, which replaces hydrogen atoms occupying various positions in the tyrosine molecule. The mechanism of iodide oxidation is not known, but three different mechanisms have been proposed. The first is an unspecific oxidation mechanism depending only on the high oxidation-reduction potential in the thyroid follicle. The oxidation-reduction potential of the normal thyroid colloid material is high and decreases in hypothyroidism or after antithyroid therapy. The second possibility is that iodide oxidation involves the electron transport chain, in particular the cytochromes. This theory is based on the fact that iodide oxidation can be inhibited by cyanides, azides, carbon monoxide, and sulfides. Thirdly, because catalase interferes with oxidation, theories of a more specific oxidation mechanism have been advanced; and it has been proposed that the oxidation of iodide to iodine requires specific peroxidases. Iodide peroxidase has not been isolated from the thyroid, but a soluble iodide peroxidase has been obtained from the submaxillary gland. The iodide peroxidase is likely to be a protoheme peroxidase.

The source of the hydrogen peroxide needed for the peroxidase reaction is not known. Two different H_2O_2 generators have been proposed. In the first, reduced flavin nucleotide is autoxidized in the presence of O_2, and reduced pyrimidine nucleotide provides the electrons needed to reduce the oxidized flavin nucleotide.

The second pathway for H_2O_2 production involves O_2, NADPH or NADH, and a peroxidase.

I. NAD(P)H ← ┐ +H⁺ ┐ → FMN ← ┐ O_2 → $+H_2O_2$
NAD(P) ← ┘ └ → $FMNH_2$ ← ┘

$$\text{II. NAD(P)} + H^+ + O_2 \xrightarrow[\text{Peroxidase}]{\text{Mn}^{++}} H_2O_2 + \text{NAD(P)}$$

Iodinated Tyrosine Synthesis. The inorganic iodide pumped into the thyroid gland is used to synthesize iodinated tyrosine. The iodinated amino acid is found in thyroglobulin, a protein present in the colloid material of the follicles. The molecular mechanism of the iodination of tyrosine can be compared to that of phenol. During the iodination reaction, two molecular changes occur, one involving the iodide itself, the other the phenol. The iodide undergoes a heterolithic fission and forms an iodide ion and a hypothetical iodonium ion. By losing a proton, the phenol molecule forms an anion, which reacts with the iodonium ion. Enolization of the product of this reaction yields the iodinated phenol. *In vivo* iodination yields both mono- and diiodotyrosine. Monoiodotyrosine forms more rapidly than diiodotyrosine, a finding suggesting that the mechanism of monoiodotyrosine iodinization is different from that involved in diiodotyrosine iodinization.

Various factors are known to influence tyrosine iodination: chronic thyroiditis, cretinism, antithyroid drugs, carcinogens, etc. It is often stated that propylthiouracil blocks tyrosine iodination. This is not exactly the case, because large doses of propylthiouracil allow the formation of at least small amounts of monoiodotyrosine, and small doses of propylthiouracil interfere little with the formation of monoiodinated amino acids. Yet when propylthiouracil is administered, the ratio of monoiodotyrosine to diiodotyrosine is exceptionally large, and reaches values similar to those observed in human thyroid nodules and in thyroids obtained from hypophysectomized rats.

The enzymes involved in tyrosine iodination remain unknown. It has been proposed that in addition to the iodide peroxidase, a specific tyrosine iodinase is also required. Whether this iodinase catalyzes the iodination of both mono- and diiodotyrosine remains to be established.

Inhibitors of tyrosine iodination have been found in supernatants obtained after differential centrifugation of thyroid homogenate. The role of these inhibitors in regulating the rate of iodination and thyroid activity is unknown.

Theoretically, the condensation of mono- and diiodotyrosine could yield four different compounds: 3,3′-diiodothyronine, 3,3′,5′-triiodothyronine, 3,5,3′-triiodothyronine, and thyroxine (see Fig. 8-4). The exact mechanism of the formation of these various iodinated derivatives is not clear, but it has been suggested that they are formed by the condensation of two molecules of iodotyrosines with loss of a side chain. In this reaction, the iodinated tyrosine is converted by the loss of two atoms of hydrogen to the quinoid form, which acts as a free radical and condenses with other molecules of iodinated tyrosine to yield a diphenyl ring. During such a reaction, an oxygen bridge is formed, and one of the side chains of tyrosine is split to yield an α-amino acrylic acid. The α-amino acrylic acid is further metabolized to yield ammonia and pyruvic acid.

The enzymes involved in the condensation of two molecules of iodinated tyrosine have not been identified. The reaction cannot be demonstrated in homogenate, but it has been observed in thyroid slices. In slices the reaction may be accelerated by a number of factors, including the presence of oxygen and iodide. The condensation is also more rapid if either the carboxyl or the amino group of iodotyrosine is engaged in peptide linkages. This observation suggests that the iodotyrosines condense more readily if they are part of a polypeptide chain.

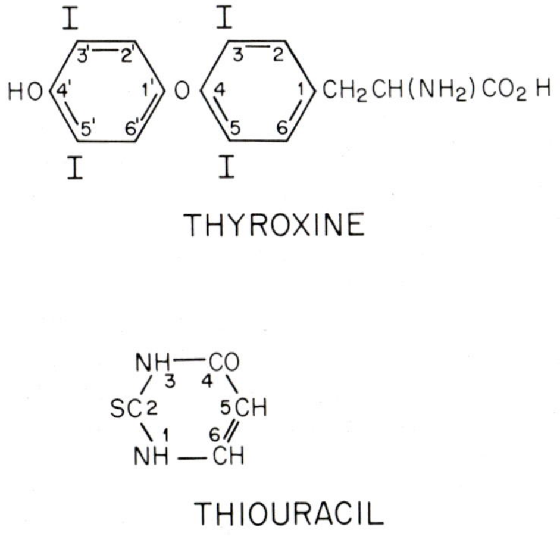

It is not well known at what stage of the thyroglobulin molecule's biosynthesis tyrosine is iodinated. Three proposals have been submitted: (1) iodination of the amino acid before its incorporation into the polypeptide chain (but puromycin does not impair iodination, and no enzymes capable of activating iodotyrosine have been found); (2) iodination of the finished tetrameric globulin; and (3) iodination of the 12 S subunits with concomitant condensation of the unit to yield new protein. However, there seems to be no doubt that thyroglobulin continues to be iodinated even after its excretion into the colloid, indeed the ^{131}I:N ratio increases in the colloid material.

HO–C₆H₂I₂(3′,5′)–O–C₆H₂I₂(3,5)–CH₂CH(NH₂)CO₂H

THYROXINE

NH—CO / SC2 5CH / NH—CH

THIOURACIL

Fig. 8-4. Thyroxine and thiouracil

Thyroglobulin

The various iodinated thyronines are stored in normal thyroid in the form of a large protein molecule, thyroglobulin. Seventy per cent of the colloid material found in the follicles of the thyroid is thyroglobulin [19–22]. By combining salting out and differential centrifugation techniques, the large thyroglobulin molecule was separated from small iodinated protein molecules that usually contaminate the main protein fraction extracted from the colloid material. Relatively little is known of the molecular properties of thyroglobulin. Hog thyroglobulin is a globular protein with a sedimentation constant of approximately 19.2–19.4, and, as may be anticipated, the molecule contains varying amounts of iodinated amino acids. Thyroglobulin is broken into subunits by alkaline pH. The molecule breaks down into two halves at pH values of about 9 and into four elementary subunits when the pH reaches values around 11.7.

In addition to thyroglobulin, other iodinated proteins have been found in the thyroid. Their importance in pathology must not be underestimated. Some iodinated proteins with mobility characteristics identical to those of iodoalbumin escape from the thyroid and are found in the plasma in some pathological conditions. Another iodoprotein is associated with cell particles, and its concentration increases in thyroid tumors.

The mature thyroglobulin molecule is a large molecule, with a molecular weight of 660×10^3 and a sedimentation constant of 19. The molecule is a tetramer that can be broken into its unit component by various chemical procedures.

The monomer has a sedimentation constant of 3 and a molecular weight of 11,500. When the monomer is submitted to electrophoresis on polyacrylamide gel in 8 M urea at pH 10.2, two bands are obtained. But the meaning of this finding is not clear.

Proteolysis with thyroglobulin has yielded two different polypeptides, a smaller one (mol. wt. 1,050) containing mannose and glucosamine, and large one (mol. wt. 3,200) containing sialic acid, fucose, galactose, mannose, and glucosamine.

Thyroglobulin Synthesis. The synthesis and turnover of thyroglobulin have been investigated in Leblond's laboratory with the aid of radioautographic techniques using radioactive leucine. Thyroglobulin synthesis was demonstrated in the cytoplasm of the follicular cells. The new protein accumulates at the apex of the cell and is excreted into the lumen of the follicle. The radioautographic data suggest that all the cells of the follicles participate in protein secretion. Consequently, the absolute amount of the protein accumulating in the lumen of the follicle is directly proportionate to the number of cells present. Leblond's radioautographic studies further indicate that thyroglobulin is iodinated in the lumen of the follicle in an area close to the apex of the follicular cell. Although iodination occurs in the colloid material, it is probably controlled by cellular elements since the absolute amount of iodinated protein formed depends upon the number of cells present.

Studies on the incorporation of labeled amino acids in thyroid slices have provided a more detailed description of the mechanism of synthesis of the hormone. The radioactivity first appears in soluble polypeptides with sedimentation coefficients of 3, 8, or 12. Puromycin or actinomycin blocks the incorporation of the precursor into the soluble polypeptides. The half-life of the messenger RNA for thyroglobulin polypeptide was estimated to be 15–20 hours. Indeed, after inclusion of actinomycin in the incubation mixture, thyroglobulin synthesis continues for several hours. The subunits are transferred from the site of synthesis to an assembly center, in which the subunits are iodinated, carbohydrate units are included in the molecule, and subunits are condensed into a finished protein. Puromycin fails to interfere with the formation of 19S units.

The radioactive proteins that accumulate in the colloid material of the follicles disappear progressively from the lumen; the rate of disappearance is again proportionate to the number of cells present in the follicles. The radioautographic studies have shed no light on the mechanism by which the cell participates in thyroglobulin elimination. Thyroglobulin digestion could occur within the colloid material and result from proteolytic enzyme secretion by the cell, or thyroglobulin hydrolysis could occur within the cell after its active pinocytosis.

Thyroglobulin release. Under normal conditions, thyroglobulin does not leave the thyroid gland. Instead, the active iodinated derivatives, thyroxine and 3,3′,5-triiodothyronine, are released by proteolysis of thyroglobulin. The enzymes involved and the intermediates formed during hydrolysis have not been identified. It is assumed that several polypeptide intermediates appear before the active iodinated compound is completely released.

Like most other organs, the thyroid contains two groups of proteolytic enzymes, some of which have a pH optimum of 3.5 and resemble cathepsin, and others that have a pH optimum around neutrality. Investigators' attempts to purify the hydrolases involved in thyroglobulin breakdown have led to the isolation of two peptidases and two proteases. Thyroglobulin breakdown has been proposed to occur in two steps, one in which the proteases attack the thyroglobulin molecule to yield subunits that are in turn hydrolyzed to yield the active iodinated thyronine or thyroxine under the influence of peptidase.

Deiodination of Mono- and Diiodotyrosine and Diiodothyronine

It has been known for a long time that iodotyrosine and iodothyronine are rapidly deiodinated and that most of the iodine associated with the amino acid appears in the urine in the form of iodide. The dehalogenation was demonstrated in thyroid slices. Studies with thyroid, liver, and kidney homogenate demonstrated that both mono- and diiodotyrosine are deiodinated by thermolabile microsomal enzymes, which require NADPH as coenzyme. Oxygen is not required for the dehalogination reaction, and the reaction is blocked by parahydroxymercuribenzoate but not by cyanide. The same enzyme is capable of deiodinating mono- and diiodotyrosine. When diiodotyrosine is deiodinated, the K_m is much greater (9.2×10^{-7}) than the K_m for the deiodination of monoiodotyrosine (3.7×10^{-7}). The kinetics of the deiodination reaction indicate that the first iodine atoms of the diiodotyrosine molecule are removed at a much slower rate than are the second iodine atoms. The specificity of the dehalogenase may not be restricted to the tyrosine derivative; indeed, an alternative pathway has been observed where the iodinated tyrosine is first converted to pyruvic or lactic acid analogs which are in turn deiodinated.

Thyroxine Metabolism

The exact metabolism of thyroxine is not known, but it has been established that it can be deiodinated, decarboxylated or conjugated to form a glucuronide. When thyroxine is administered, it disappears from the blood within 6 or 7 days after administration, and inorganic iodide accumulates in the urine. Triiodothyronine may very well be the product of thyroxine deiodination. However, when attempts were made to demonstrate the presence of triiodothyronine in the urine of animals injected with labeled thyroxine, the results were controversial. Whereas some failed to observe the appearance of triiodothyronine in urine and bile, others claimed to have identified triiodothyronine in the urine of hepatectomized dogs and thyroidectomized rats.

Three pathways for triiodothyronine degradation have been described: oxidation, deiodination, and conjugation. The products of oxidation to acid have been found in the bile of rats after intraperitoneal administration of triiodothyronine. Triiodothyronine may be deiodinated to yield 3,3′-thyronine. This reaction may be of considerable physiological significance because the dehalogenation of Triac provides a means by which a very active compound can be converted to an inactive substance. Glucuronides and sulfates of triiodothyronine have been found in blood, and it is likely that they are synthesized in the liver.

Although the products of the various degradation reactions are well known, the enzymes involved in the reactions have not been studied except for those involved in the deiodination of tetraiodothyronine and triiodothyronine.

Deiodination of Tetra- and Triiodo-Derivatives

The deiodination of tetra- and triiodo-substrates has been investigated in kidney and liver slices; slices obtained from other tissues have not been used extensively. Using liver slices, investigators demonstrated that both tetra- and triiodothyronine were deiodinated, but the deiodination of triiodothyronine was much slower than that of tetraiodothyronine.

Thyronine is assumed to be the product of the deiodination of tetraiodothyronine. The intermediates formed during the degradation reaction are not known; however, it has been postulated that the acetic analog of tetraiodothyronine, namely, tetraiodothyroacetic acid, or Tetrac, is an intermediate. Albright claims that triiodothyronine is an intermediate in the progressive deiodination of tetraiodothyronine.

Thyroid Hormone in Blood

Mono- and dithyronine never reach the general circulation in appreciable amounts because these compounds are rapidly deiodinated: thyroxine and 3,3′,5-triiodothyronine are found in blood not in a free form, but complexed to proteins. Among the two iodinated thyronines, thyroxine is the principal circulating hormone, and triiodothyronine is present only in small amounts.

The thyroxine in blood is not lost by dialysis and is recovered in the acid precipitate. These findings indicate that the circulating thyroxine is bound to macromolecules. The administration of radiothyroxine followed by paper or starch-gel electrophoresis of the blood proteins has established that circulating thyroxine is bound to albumin, a prealbumin (TBPA), and an α-globulin (TBG).

Serum proteins can bind large amounts of thyroxine. The amount of hormone bound by a specific protein group depends upon the total amount of protein present in the blood and the affinity of the protein for thyroxine. The total binding capacities of TBG, TBPA, and albumin are in reverse order of their affinity for thyroxine (see Table 8-4). Although TBG has the greatest affinity for thyroxine among serum proteins, it has the lowest total binding capacity.

Table 8-4

	Binding affinity[a]	Total binding capacity[b]
TBG	57	25
TBPA	32	256
Albumin	11	1,000

[a] Expressed in per cent of radioactivity recovered in the serum protein fraction after injection of labeled thyroxine (3 μg/100 ml).
[b] Expressed in μg/100 ml determined at saturation concentration of thyroxine.

The binding of thyroxine to serum proteins is unlikely to result from artifacts. Efforts were made to eliminate such a possibility, especially in the binding of thyroxine to prealbumin. The original observation on the binding of thyroxine to prealbumin was made by using Tris maleate buffer for mobility studies on paper electrophoresis. The possibility of an artificial binding between thyroxine and prealbumin was first suspected because thyroxine was not bound to prealbumins when barbital buffer was used. The binding of thyroxine to prealbumin seems to be physiological because it was demonstrated with serum using glycine acetate and veronal buffer. Maybe the most compelling argument in support of a physiological role of TBPA is the observation that the amount of thyroxine bound to TBPA changes (see below).

Attempts were made to isolate the TBG. A preliminary purification from serum by ion-exchange chromatographic analysis on a Dowex-I column equilibrated with acetate buffer (pH 4.5) was followed by repeated ammonium sulfate precipitation, chromatographic analysis on dimethylaminoethyl-cellulose resin, and electrophoresis on cellulose columns. These attempts finally yielded a preparation homogeneous on electrophoresis and in the ultracentrifuge. The amino acid composition and the secondary structure of TBG is not known, and it is still debated whether it is a lipo- or glycoprotein.

The importance of the binding of thyroxine to blood protein must not be underestimated. In many physiological and pathological conditions, thyroxine binding to serum protein is markedly altered. Such observations suggest that binding to serum protein is important in the transport of the hormone from its source to the target organ, and they also provide an important diagnostic tool. Among those conditions that influence the amount of thyroxine bound to serum protein are diseases of the thyroid and shifts in the normal hormonal balance resulting from androgen and estrogen therapy, pregnancy, liver disease, acidosis, certain drugs, hereditary factors, and other causes. The available facts do not always adequately explain these changes in the serum-bound thyroxine, and only tentative interpretations can be given.

In myxedematous patients and thyroidectomized animals, the levels of thyroxine bound to serum protein (PBI) are low; in contrast, they are increased in thyrotoxicosis. Yet in thyrotoxicosis, the elevated PBI is associated with reduced levels of TBPA, increased levels of triiodothyronine, and free thyroxine.

Liver disease may increase or decrease the level of TBG in blood, but it usually doesn't induce symptoms of hypo- or hyperthyroidism. Genetic alterations have been observed that lead to increased or decreased levels of TBG in human euthyroids. Individuals who have increased levels of serum TBG also have increased PBI values (twice normal), and the turnover rate of the PBI is also doubled. A sex-linked dominant hereditary increased TBG capacity has also been described [23].

Respiratory or metabolic acidosis results in low binding capacities of thyroxine by all binding proteins because binding capacity of thyroxine in acidosis may explain the rapid turnover of thyroxine observed in pulmonary emphysema and hyperemia. The effect of drugs on thyroxine binding to serum proteins probably results from competition for the binding site.

A number of pathological conditions (including high fever, cancer, massive infarcts of the brain, heart, and lung) are associated with reduced TBPA but normal TBG capacity.

Free Thyroxine and Active Hormone

The blood, the interstitial fluid, the lymph, and the cerebrospinal fluid contain small amounts of free thyroxine. The role of the free hormone is not clear. The fact that it increases in thyrotoxicosis indicates that it may play a role in physiology.

The chemical structure of the hormone that is active under physiological conditions is still under investigation. Although thyroxine administration undoubtedly corrects the metabolic and clinical alterations that occur after thyroidectomy, and although thyroxine acts on mitochondria *in vitro,* these observations constitute no proof that thyroxine is the compound that is active in normal physiological conditions. The belief that thyroxine is the active hormone was shattered when 3,5′,3′-triiodothyronine was isolated from blood and thyroid and when it was established that triiodothyronine is more potent than thyroxine on a molecular basis. The discovery of active compounds different from thyroxine has stimulated more research on the physiological effects of thyroxine analogs.

History of Thyroxine

The first concrete link between thyroid function and the thyroid gland came from careful observations made at autopsy. In 1874, Gull autopsied two women with myxedema and concluded that thyroid regression was responsible for the disease. Soon after that, myxedema was described in surgically thyroidectomized individuals. About ten years after Gull's discovery, a committee appointed by the Clinical Society of London was requested to study myxedema. The committee concluded that the removal or destruction of the thyroid was responsible for myxedema; and that without the thyroid, the function and structure of almost every organ of the body were altered. In 1891, Murray cured myxedema with glycerol extracts of the thyroid.

A crucial contribution was the crystallization of the active hormone. It was a painstaking task, accomplished by Kendall, who by Christmas day, 1914, had succeeded in preparing a small amount of an active hormone. If one can imagine the thrill experienced by a young man making such a critical discovery, one can conceive of his frustration when for an entire year, Kendall found it impossible to prepare another crystalline sample of his precious hormone. But after 15 months, Kendall's success was complete, and within a few years he had crystallized enough material to elucidate its physiological and some of its chemical properties.

A lecturer in pathological chemistry at the University College Hospital Medical School in London, who was later to become Sir Charles Harrington, elucidated the structure of thyroxine. The degradation of the crystalline compound and the synthesis of an active hormone established that thyroxine was 3,5,3′5′-tetraiodothyronine.

Thyroxine Analogs

After Gross, Pitt-Rivers, and Roche and his coworkers isolated 3,5,3′-triiodothyronine, it was demonstrated that the triiodothyronine is much more active than thyroxine itself. Since, in addition to dehalogenation, thyroxine can also be decarboxylated, deaminated, or conjugated, the physiological activities of the products of these metabolic alterations were also investigated.

Although thyroxine stimulates metamorphosis in the tadpole and oxygen consumption in higher animals, many of the analogs are without effect on oxygen consumption. Because of the dissociation, it proved necessary to investigate the effect of the analog on various physiological reactions including: (1) the metamorphosis of the tadpole; (2) ^{131}I uptake by the thyroid gland; (3) oxygen consumption in rat; and (4) effect on goiter development. The reader is referred to specialized reviews for a comprehensive view of the effect of thyroid analogs, and we shall consider here only the compounds obtained by deiodination in position 5′ and by alteration in the side chain.

Removal of the iodine from position 5′ to produce 3,5,3′-triiodothyronine enhances hormonal activity, but removal of the iodine from position 5 to produce 3,5′,3′-triiodothyronine abolishes hormonal activity almost completely. Modification of the side chain considerably enhances the compound's ability to induce

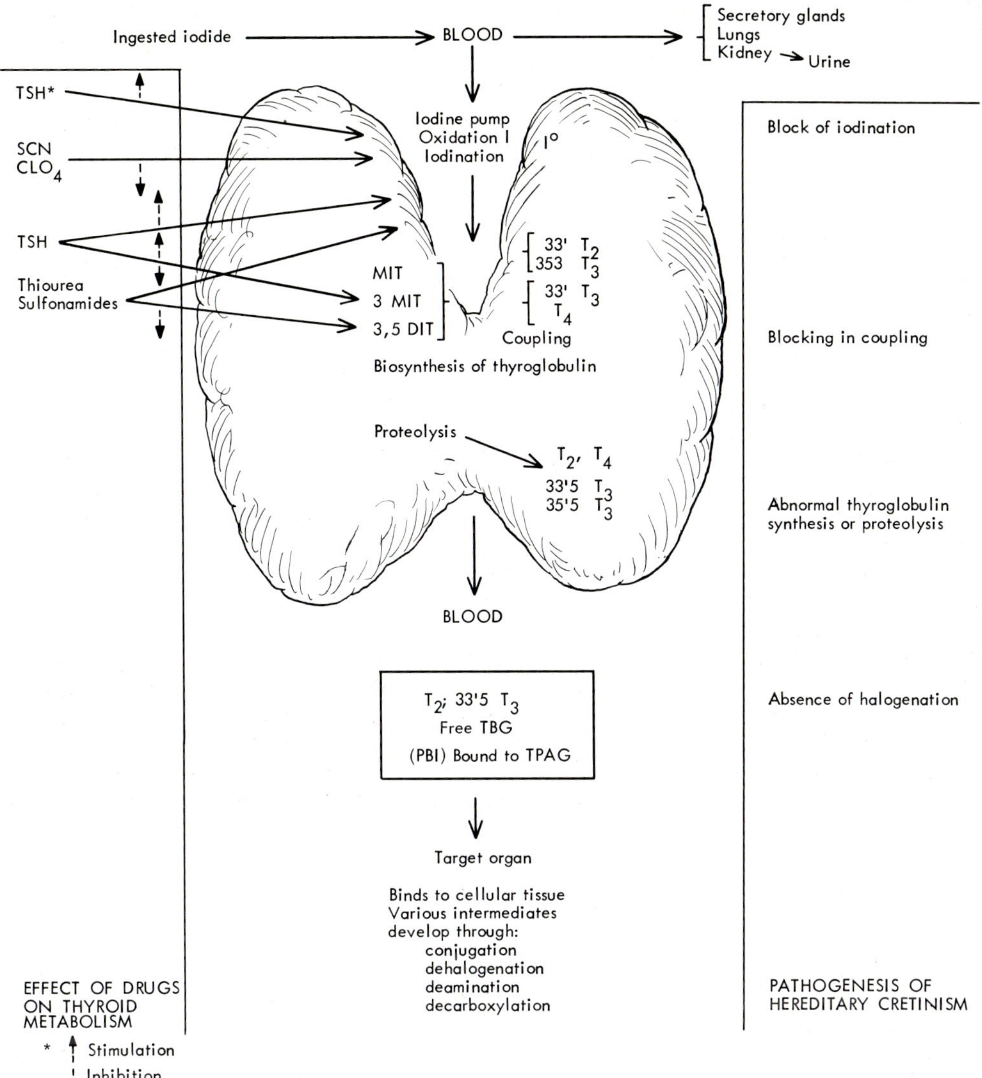

Fig. 8-5. Effects of drugs and inborn errors on the metabolism of thyroid hormones

metamorphosis in the tadpole, but it has little effect on oxygen consumption in high animals.

Studies with thyroxine analog in myxedematous patients have shown that 3,5,3′-triiodothyronine has essentially the same metabolic effects as thyroxine except that effects appear much faster after the administration of 3,5,3′-triiodothyronine (within 24 hours) than after administration of thyroxine (peak activity after 10 days).

A comparison of the affinity of the thyroxine-binding globulin for thyroxine and that of the biologically most effective of the triiodothyronines suggests that the specificity of the binding protein is directed toward the biologically active compound [24].

Studies on myxedematous patients have also suggested that it is possible to dissociate the effect of an analog on the BMR from its effect on blood cholesterol level. Triiodopropionic acid was found to have no effect on BMR although it reduced cholesterol levels markedly. These important observations may lead to the discovery of new pharmacological compounds. The metabolism of thyroid hormone is summarized in Fig. 8-5.

Effect of Thyroid Hormones

The thyroid hormones affect many facets of metabolism [25–32]. The metabolism of nitrogen, carbohydrates, and lipids are all markedly altered in hyper- and hypothyroidism. The biochemical sequence leading to the alteration is not known, yet these metabolic changes are of considerable importance in clinical medicine.

It is somewhat difficult to evaluate the exact effect of thyroxine on nitrogen metabolism because the metabolism is affected by diet and many other factors including the effect of other hormones, such as those of the adrenal cortex and insulin. The increased urinary excretion of nitrogen, metabolism and urea observed in hyperthyroidism suggests that the thyroid hormone does influence nitrogen metabolism by accelerating nitrogen catabolism, particularly of proteins found in the extracellular fluids.

At least two different effects of the thyroid hormones on glucose metabolism can be described: stimulation of glucose absorption through the intestinal mucosa and stimulation of glucose utilization by the cells. The increased intestinal absorption may explain the transient hyperglycemia observed after thyroxine administration, but stimulation of glucose utilization is difficult to reconcile with the tendency toward diabetes in patients with thyrotoxicosis, or with Houssay's classical experiment in which thyroxine administration to pancreatectomized rats accelerated the development of diabetes.

The details of the biochemical mechanism that causes the changes in glucose use and absorption are not known. Studies with perfused livers and liver slices obtained from hyperthyroid animals have shown that liver uses more glucose. The increased glucose utilization is associated with increased activities of glucose-6-phosphatase, glucose-6-phosphate dehydrogenase, and 6-phosphogluconate dehydrogenase. Such findings suggest that the thyroid hormone stimulates gluconeogenesis and the hexose monophosphate shunt in liver.

Thyroid hormones have long been known to affect lipid metabolism. Thyroxine undoubtedly controls cholesterol metabolism; serum cholesterol levels are markedly increased in hypothyroidism and decreased in hyperthyroidism. There are various ways by which thyroxine could cause cholesterol to accumulate in blood: direct stimulation of the pathway involved in cholesterol biosynthesis; block of cholesterol use for further biosynthesis; indirect stimulation of cholesterol synthesis by acceleration of pathways that provide precursors of coenzymes needed for cholesterol synthesis; and indirect stimulation of cholesterol synthesis by blocking pathways that use those precursors involved in cholesterol synthesis. The exact mechanism by which thyroxine induces the accumulation of cholesterol in serum needs to be elucidated. The effect of thyroid hormones on blood cholesterol must be understood because hypothyroidism is known to enhance the development of experimental arteriosclerosis in animals.

The absorption of calcium, sodium, potassium, chloride, magnesium, and iron is thought to be influenced by the thyroid hormone. The results of the studies of the effect of thyroxine on electrolyte metabolism are often contradictory and difficult to interpret; however, it seems accepted that thyroxine affects phosphorus metabolism by increasing the excretion of the metal in both urine and stools.

Thyroxine affects vitamin metabolism by interfering with the requirement of a vitamin and interfering with vitamin biosynthesis. The requirements for thiamine and riboflavin, two vitamins involved in the cell's energy production, are increased in thyrotoxicosis. Thyroxine facilitates nicotinic acid biosynthesis and stimulates the conversion of carotenes to vitamin A.

The blood levels of carotene decrease while those of vitamin A increase in myxedematous patients injected with thyroxine. Whether thyroxine stimulates vitamin A synthesis by increasing carotene absorption or by directly affecting vitamin A synthesis is not known.

The most striking and consistent clinical observations in hyperthyroids are excessive nervousness and a high BMR. In contrast, hypothyroids have a low BMR and are apathetic. Oxygen consumption is high in hyperthyroidism and low in hypothyroidism. After thyroxine administration, oxygen consumption does not increase equally in all tissues of the body. The gonads, the spleen, and the brain do not consume more oxygen under the influence of thyroxine. The effect of thyroxine seems to be determined by sex; the increase in oxygen consumption is greater in men than in women. This effect of the thyroid on basal metabolism is undoubtedly the result of an effect on the mechanism of cellular energy production; and basal metabolism might very well be the hormone's primary target.

Naturally, the respiration rate in tissue obtained from euthyroid, hyperthyroid, and hypothyroid animals has been investigated extensively. Oxygen use is increased in tissues obtained from animals injected with thyroxine and decreased in tissues obtained from hypothyroid animals.

Investigators have for years tried to pinpoint the step in biological oxidations that is affected by thyroxine. In their attempts to solve such a problem, researchers are constantly faced with the necessity of distinguishing between effects that are primary and those that are secondary to adaptation to an increased metabolic rate. This distinction is often difficult to make.

The increases in total number of mitochondria, in the amount of cytochrome c, and in the activity of cytochrome enzymes might be adaptive changes resulting from increased metabolism. These changes are unlikely to determine the rate of oxygen uptake because oxygen use is not controlled by the concentration of the enzymes of the electron transport chain, but rather by the amount of work performed and the amount of ATP used. The concentration of respiratory enzyme even in normal tissues is greater than is needed for maximal respiration rates.

After a number of days, thyroidectomized rats lose approximately half of their electron transport capacity. The injection of thyroxine to these animals restores electron transport capacity within 3 hours, but without stimulating the biosynthesis of the components of the electron transport chain. This finding provides further evidence that the effect of the thyroid hormone is not primarily to stimulate the synthesis of the electron transport chain enzymes but that in some way it must accelerate the use of substrate or secure conservation of energy [33].

Thyroxine induces an increase in the activity of carbonyl aspartate synthetase in tadpole before metamorphosis [34].

Changes in oxidative phosphorylation seem to be more closely related to the primary site of action of the thyroid hormones. Thyroxine uncouples oxidative phosphorylation in mitochondria and causes the energy released during oxidation of the Krebs' cycle intermediates to be less efficiently used for ATP synthesis. A direct effect of thyroxine on oxidative phosphorylation would conveniently explain many metabolic effects of thyroxine, but it would leave unexplained the beneficial effects of small amounts of thyroxine. A direct effect on oxidative phosphorylation is also inconsistent with the fact that uncoupling agents are unable to correct hypothyroidism.

Both these objections have been met by Lardy, who proposed that thyroxine and its analogs act at a specific locus of the oxidative pathway. At low concentrations of the hormone, a rate-limiting step in the chain of reactions that yields chemical energy is uncoupled, and all the steps of the reaction subsequent to the rate-limiting step are accelerated without a marked loss of efficiency in oxidative phosphorylation. In larger doses, the hormone uncouples tightly coupled reactions, and the efficiency of oxidative phosphorylation is lost.

An observation made by Tapley and Lehninger has shed new light on thyroxine's mechanism of action. In contrast to dinitrophenol, thyroxine was found to be without effect on oxidative phosphorylation in digitonin fragments of mitochondria. The difference between the effects of thyroxine on intact mitochondria and digitonin particles can be explained hypothetically: only intact mitochondria convert thyroxine into an active metabolite; and thyroxine does not act at the site of oxidative phosphorylation but interferes with oxidative phosphorylation by an indirect mechanism.

Studies made by Lehninger's group on mitochondria seem to support the second of these possibilities. Isolated mitochondria swell when placed in a medium containing isotonic sucrose, potassium chloride, and buffer. The ability of the mitochondria to swell depends upon the source of the mitochondria. Although liver mitochondria may swell to reach volumes that are four to five times the normal, heart and skeletal mitochondria are slow in swelling and brain mitochondria hardly swell at all. These differences in the ability of mitochondria to swell must be taken into account in interpreting the effect of thyroxine on oxidative phosphorylation. There are two phases in mitochondrial swelling. In the first, the dead space in the mitochondrion rapidly adjusts to the osmotic pressure of the medium. The second phase of mitochondrial swelling is nonosmotic. The swelling is triggered by the addition of inorganic phosphate, reduced glutathione, calcium, or thyroxine to the incubation medium. Anaerobiosis, cyanide, antimycin A, and Amytal block mitochondrial swelling. These observations naturally bring one to conclude that the second phase of mitochondrial swelling is enzymatically controlled and depends upon the amount of adenosine triphosphate present in the medium.

To explain these two stages in mitochondrial swelling, it has been proposed that the mitochondrion can be described as a structure made of two concentric compartments. Without thyroxine or another swelling agent, only the outside compartment is permeable to sucrose, but when thyroxine is added to the medium, the second compartment also becomes permeable to sucrose. During swelling, the NAD bound to the mitochondrial structure is progressively lost, and oxidation is uncoupled from phosphorylation.

Lehninger's studies indicate that thyroxine affects mitochondrial swelling. These studies do not yet exclude the possibility that the hormone acts directly at one or more phosphorylation sites. An effect of thyroxine on the oxidative phosphorylation abilities of mitochondrial fragments obtained by sonication of mitochondria has been demonstrated; thyroxine and triiodothyronine stimulate the oxidation of succinate and NAD and enhance the rate of phosphorylation. In contrast, tetraiodothyroacetic acid and triiodothyroacetic acid uncouple oxidative phosphorylation.

The effect of thyroxine on a number of individual enzyme systems has also been investigated with the hope that the hormone's mechanism of action might be clarified. Studies have been done on four classes of enzymes: phosphatases, dehydrogenases, transhydrogenases, and miscellaneous enzymes.

An accelerated breakdown of high-energy phosphate could readily explain the effect of thyroxine; preincubation of canary liver mitochondria with thyroxine enhances ATP breakdown, and thyroxine increases acyl phosphatase activity in liver and muscle. The principles on which these experiments were conducted undoubtedly were sound, yet the relationship between the effect of thyroxine on high-energy phosphates and the physiological role of the hormone remains unsettled.

If thyroxine stimulates the dehydrogenase that regenerates oxidized pyrimidine nucleotides, then it must stimulate oxygen consumption. Again, it is difficult to decide if the enzymic changes observed are primary effects or adaptive changes. In either case, these studies are important because they provide some clues on the effect of thyroxine and on the control of respiration. Remember that NADPH and NADH are formed in the cell supernatant by the oxidation of glucose through the glycolytic, pentose, and glucuronic pathways. To be reused, the reduced pyridine nucleotides must be reoxidized; this occurs (1) through the mitochondrial electron transport chain; (2) by the transfer of electrons from reduced pyridine to the microsomal cytochrome b system; and (3) by reduction of dihydroxyacetone phosphate to yield diglyceraldehyde. We will briefly discuss here some of the studies on the effect of thyroxine on succinic and α-glycerophosphate dehydrogenase.

Thyroxine stimulates oxygen consumption in fresh heart homogenate incubated with succinate. However,

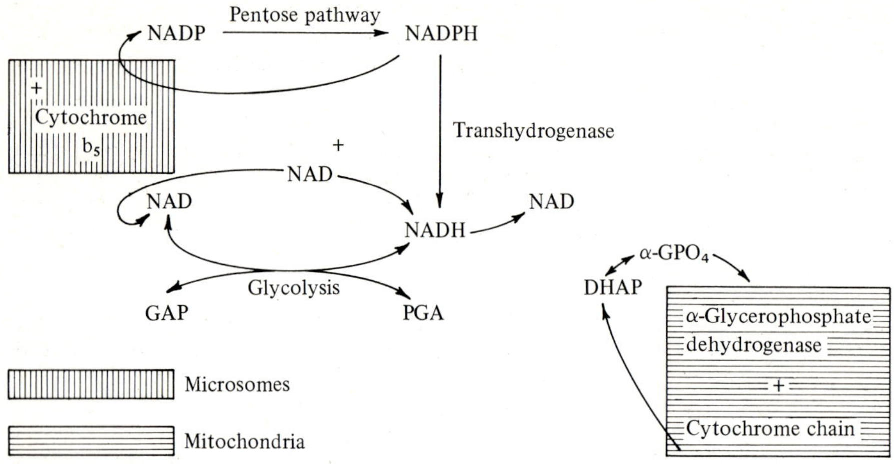

Fig. 8-6. Hypothetical site of action of thyroid hormones. *GAP*: glyceraldehyde-3-phosphate; *PGA*: phosphoglyceric acid; *DHAP*: dihydroxy-acetone phosphate; α-GPO_4: α-glycerophosphate

it is not certain whether this *in vitro* observation is in keeping with the *in vivo* reality.

The administration of thyroxine or triiodothyronine to rats increases five-fold the amount of oxygen consumed by rat mitochondria and markedly stimulates the rate of oxidation of α-glycerophosphate in both liver and kidney mitochondria. The rate of oxidation of α-glycerophosphate returns to normal within ten days after cessation of the thyroid hormone administration. The fact that the increase in α-glycerophosphate oxidation is blocked by cyanide suggests that the oxidation of glycerophosphate elicited by thyroid hormone is linked to the oxidation-reduction of cytochrome oxidase. Specific assay of α-glycerophosphate dehydrogenase demonstrates a marked increase in the activity of that enzyme. If thyroxine and ethionine are administered simultaneously, α-glycerophosphate dehydrogenase doesn't increase; thus, the increase in dehydrogenase activity must result from new enzyme synthesis. These results suggest that outside of the mitochondria, the oxidation of NADH is coupled to the reduction of dihydroxyacetone phosphate to yield α-glycerophosphate, and that α-glycerophosphate is reoxidized inside the mitochondria by α-glycerophosphate dehydrogenase, yielding dihydroxyacetone PO_4, which diffuses out of mitochondria to be reduced again. Thus, α-glycerophosphate acts as a shuttle for the transfer of hydrogen from the cytosol to the mitochondria.

Thyroid hormone administration markedly increases the activity of microsomal NADPH cytochrome c reductase. This again facilitates the oxidation of extramitochondrial reduced pyridine nucleotides. These results are summarized in Fig. 8-6.

The administration of thyroxine was also found to inhibit transhydrogenase, an enzyme that catalyzes the reversible conversion of NADH to NADPH.

$$\text{NAD} + \text{NADPH} \underset{\text{Transhydrogenase}}{\rightleftarrows} \text{NADH} + \text{NADP}$$

When transhydrogenase activity is blocked, NADPH accumulates, and either the processes requiring NADP will be blocked, or some pathway for NADPH reduction must be accelerated.

It was mentioned that the administration of thyroid hormones increases the activity of NADPH cytochrome c reductase, a process that accelerates NADPH oxidation through a pathway not coupled with phosphorylation.

If NADPH is oxidized through the microsomal cytochrome system, then the pathway reducing NADP should be stimulated and the rate of glucose oxidation through the hexose monophosphate shunt should be accelerated in those tissues where the shunt operates. However, studies on the effect of thyroxine on glucose oxidation through the shunt pathways have yielded controversial results.

Most of the reported metabolic effects of the thyroid hormone do not satisfactorily explain the hormone's mode of action either because the effects occur too late after the administration of the hormone or because serious discrepancies between *in vivo* and *in vitro* observations exist. Therefore, the effect of the hormone is reinvestigated each time a new biochemical pathway is elucidated. The biosynthetic pathways for proteins have not escaped this intrusion by the thyroid endocrinologist. Researchers have known for a long time that human myxedema is associated with a slowing down of the anabolic processes, including reduction in the rate of protein synthesis. Anabolic pathways of hypothyroids are normalized by thyroid hormone administration. Similarly, the rate of protein synthesis as measured by amino acid incorporation is reduced in thyroidectomized rats, and the ratio of DNA to RNA is decreased. Thyroxine administration restores to normal amino acid incorporation and the RNA: DNA ratio.

In an attempt to pinpoint the site of action of the thyroid hormone in the pathway for protein synthesis, incorporation of labeled amino acid in cell-free systems of thyroidectomized animals was compared to that in similar systems of thyroidectomized animals injected with thyroxine. Thyroidectomy slowed amino acid incorporation, and injection of thyroxine restored it to normal. The step in protein synthesis that is stimulated by thyroxine was believed to be the transfer of amino acyl tRNA to ribosomes. *In vivo*, the administration of thyroxine to thyroidectomized animals also stimulates the incorporation of precursor in rapidly labeled nuclear RNA and increases RNA polymerase activity.

A controversy has arisen about the participation of the mitochondria in thyroxine's stimulation of protein synthesis in cell-free systems. Whereas some investigators claimed to have observed the effect in microsomal preparations completely devoid of mitochondria, others claim that the presence of mitochondria in the cell-free preparation is indispensable to stimulation. The mitochondria do not act simply by generating ATP because their replacement by a classical ATP-generating system does not contribute to the stimulation of protein synthesis in microsomes of thyroidectomized animals injected with thyroxine.

It has also been claimed that only mitochondrial RNA is responsible for the stimulatory effect of the hormone on protein synthesis. Concentrations of thyroxine 25×10^{-5} to 10^{-6} molar times smaller than those used to uncouple oxidative phosphorylation stimulated the incorporation of amino acids in liver mitochondria obtained from intact rats. The effect occurs within minutes after incubation of the system with the hormone.

Appealing as it is, the hypothesis that the thyroid hormone exerts its entire metabolic effect by stimulation of protein synthesis in mitochondria or in the endoplasmic recticulum is unlikely; metabolic effects of the thyroid hormone continue to be expressed when actinomycin D is administered simultaneously [35].

The thyroid gland may be damaged in many ways. A comprehensive description of the anatomical pathology of the thyroid will not be attempted, but since the pathogenesis of hyper- and hypothyroidism cannot be properly understood without an adequate morphological description of the lesion, the major forms of thyroid injuries are outlined (see Table 8-5). The reader is referred to specialized books for further information on the anatomical pathology of the thyroid.

Alterations of the thyroid either leave the hormone balance intact (as in some deficiency goiters or in benign or malignant tumors) or hyper- or hypothyroidism will follow damage to the thyroid or to organs controlling thyroid function.

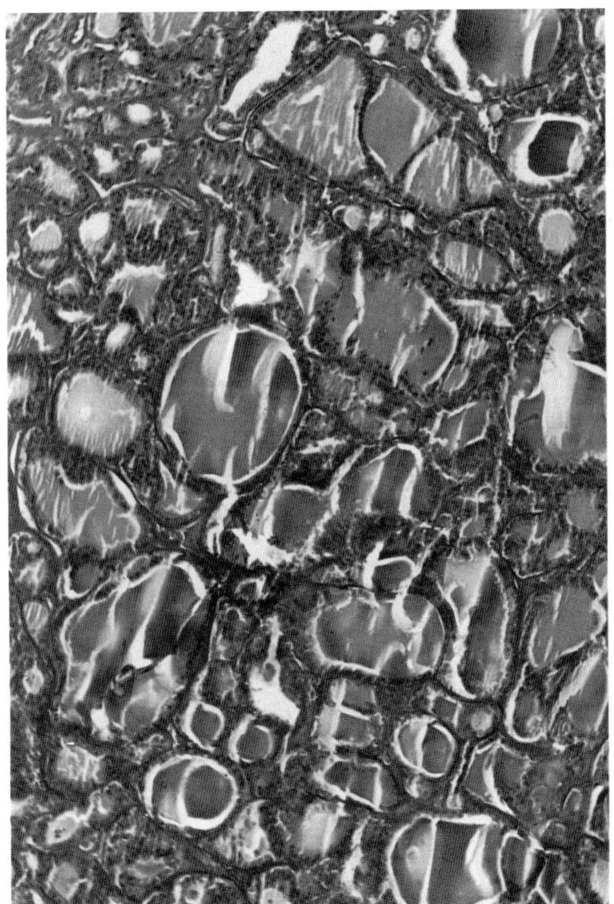

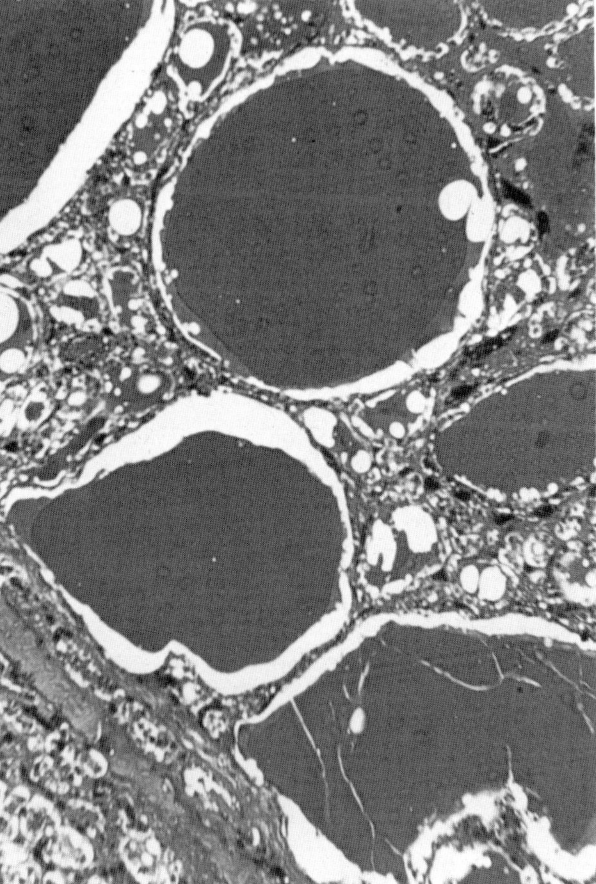

Fig. 8-7. Histological appearance of normal and colloid goiter

Table 8-5. Pathology of the Thyroid

I. Deficiency: Iodine

Absolute: Goitrogenic regions
Relative: Puberty, adolescence, pregnancy, lactation; it usually leads to colloid goiter, sometimes to nodular goiter

II. Toxic agents:

Thiocyanate (therapy for hypertension, cabbage)
Thiouracil (cabbage)
L5-vinyl, 2-thioxazolidone (rutabaga, turnip, cabbage)

III. Degeneration:

Amyloidosis rare; it may be primary or secondary to pulmonary tuberculosis
Hemosiderin: in hemochromatosis (hemosiderin)

IV. Inflammation:

Acute: diffuse thyroiditis; rare:
septicemia or pyemia;
Localized: abscess;
Chronic: nonspecific;
Specific:

1. *Tuberculosis* associated with miliary tuberculosis rarely localized
2. *Syphilis* gumma
3. *Granulomatous:*
   Viral disease?
   Autophagia of the colloid?
   Occurs between 30 and 50 years of age
   Gross: Moderate enlargement, tenderness
   On section: firm yellowish disseminated spots
   Microscopic: Giant cell granuloma lymph, plasma, epitheloic, and giant cells
4. *Lymphocytic thyroiditis*
   Proliferation of lymphoid tissue in the thyroid gland
   Gross: Diffusely enlarged gland; grayish-white, firm and rubbery on section
   Microscopic: The acinar structure is replaced by lymphoid center
5. Hashimoto's disease:
   Lymphoid proliferation in the thyroid associated with degenerative process in the glandular epithelium
   Cause unknown; it has been suggested that it results from an autoimmune reaction
   Gross: Moderate uniform enlargement, whitish, firm, rubbery consistency, irregular fibrosis, no lesions in surroinding tissues
   Microscopic: Small follicles containing little colloid, lined with large cell with an eosinophilic cytoplasm (oxyphilic, Askanazy cell), moderate fibrosis, occurs frequently in women 45–60 years of age
6. Riedel's struma:
   Rare disease most frequently found in women
   Gross: Replacement of the glandular parenchyma by dense fibrous connective tissue; grossly the thyroid is moderately and often unilaterally enlarged, it may be adherent to the surrounding tissue, but the striking characteristic is its hardness
   Etiology: Evolutionary stage of Hashimoto?
   Primary injury due to toxins

V. Benign defects:

1. *Diffuse colloid goiter:*
   Physiological hyperplasia in response to iodine deficiency
   Gross: Uniform enlargement of the gland; on section, surface is amber color, translucent, and honeycombed; (fibrous tracts separate parenchymal nodules) occasional hemorrhage and cysts
   Microscopic: Enlargement of acini with flattened epithelial lining
   Abundant colloid
   Small foci of epithelial hyperplasia usually associated with euthyroidism (see Fig. 8-7)
2. Nodular goiter:
   Results from repeated hyperplasia alternating with hypoplasia
   Gross: Irregular enlargement of the thyroid gland
   The cut surface presents meaty and firm area of hyperplasia separated by a trabecular pattern of fibrous tissue and alternating with translucent foci rich in colloid and zones of necrosis, hemorrhage, cystic degeneration, calcification, and hyalinization; it may be associated with eu-, hyper-, or hypothyroid (see Fig. 8-8)
3. Exophthalmic goiter:
   Associated with hyperthyroidism
   Gross: The gland undergoes a variable degress of enlargement ranging from none to very marked on section the gland has a homogenously meaty, finely lobulated appearance; firm in consistency
   Microscopic: Proliferation of the epithelial cells
   The cells are tall, columnar cells with a distinct Golgi apparatus, mitochondria, and large basal nuclei that often exhibit mitosis
   The acinar lining projects papillary proliferations into the enlarged acinar lumina; the colloid has almost completely disappeared
   A moderate lymphocytic hyperplasia accompanies the epithelial changes
4. Adenoma:
   Difference from goiter:
   Goiter presents diffuse swelling or multiple nodules
   Origin:
   Fetal rests
   Excess TSH
   Adenoma: Localized encapsulated tumor

Types:	
Colloid	Colloid-containing follicles with cuboidal or flat cell lining
Fetal	Minute follicles with small lumina lined by small cuboidal cells
Hürthle cell	Cords of large polygonal or cuboidal cells with eosinophilic granular cytoplasm and small hyperchromic nuclei
Papillary	Delicate papillary proliferation of the acinar lining within the lumina of acini

Types:	
Papillary	Malignant version of papillary adenoma: Differential diagnosis between adenoma an carcinoma sometimes difficult; ramified papillary projection with central vasculoconjunctive core and mono- or multilayered epithelium whose cells may be more or less differentiated Poor ^{131}I Uptake (slow growth)
Follicular	Follicles with more or less abundant colloid, with more or less well-differentiated lining Invades surroinding tissues Metastasizes in bone Good ^{131}I Uptake
Undifferentiated	Few or no follicles Trabeculae of more or less anaplastic cells growth, invasion, and rapid metastasis No or poor ^{131}I uptake

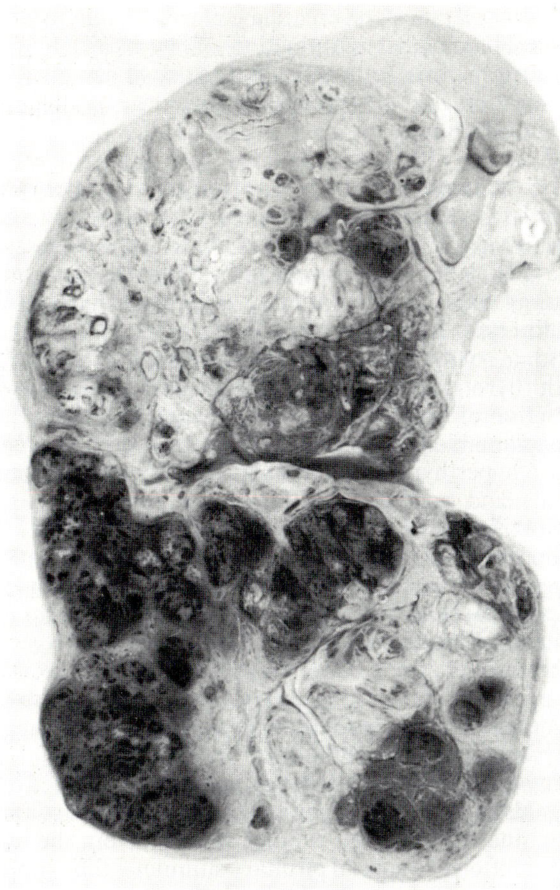

Fig. 8-8. Adenomatous thyroid

Hyperthyroidism

Three famous physicians have contributed to our early knowledge of hyperthyroidism. Parry, an Englishman, described the disease but left it to his son to publish, in 1825, the first description of hyperthyroidism. In 1835, Graves, an Irish physician, attracted attention to the relationship between some symptoms now known to be associated with hyperthyroidism and the thyroid gland. Von Basedow in 1840 meticulously described the complete clinical symptomatology of hyperthyroidism [36].

Typical hyperthyroidism, or Graves' disease, is characterized by a number of symptoms, the most typical of which are exophthalmia, tachycardia, and thyroid enlargement. The clinicopathological alterations in hyperthyrodism will first be described and then the pathogenesis of the disease will be explained.

Pathology and Symptoms

Hyperthyroidism may develop suddenly or progressively. Children or young adults are seldom affected; the disease is seen most frequently between the third and fifth decade. Women are more frequently affected than men. In typical Graves' disease, the thyroid gland is uniformly enlarged like a tumor in the neck that can be seen and palpated when the patient is swallowing. The gland has a normal consistency at first but later becomes firm. The blood supply in the gland may be so increased that a murmur can be identified in morderate cases and a thrill heard in severe cases.

On section of the gland obtained after surgery or at autopsy, the surface has lost the follicular appearance typical of the normal thyroid; instead it has acquired a homogeneously meaty consistency. Histologically, there is marked proliferation of the acinar cells, which become tall, columnar cells with a distinct Golgi apparatus and numerous mitochondria. The cells often contain a dividing nucleus. The acinar lining projects richly ramified papillary proliferations within the lumen of the acini. The colloid material, which in the normal thyroid fills all acini, is practically absent in Graves' disease.

In addition to the morphological changes related to the glandular function, a moderate lymphocytic hyperplasia accompanies the epithelial changes. The symptoms that accompany thyroid hyperfunction include: (1) alterations of thyroid hormone metabolism; (2) alterations in target organs; and (3) changes of unknown origin.

The metabolism of the thyroid gland can be investigated clinically by a number of different techniques such as studies on the concentration of iodine in blood, iodine excretion, thyroidal uptake of radioactive iodine, and determination of iodine in blood. The concentration of total iodine in blood is increased in hyperthyroid individuals. The increase is mainly in protein-bound iodine as it is determined in alcohol or zinc hydroxide precipitates.

In spite of the increased PBI observed in thyrotoxicosis, the concentrations of TBG are not altered in hyperthyroidism, but the binding capacity of TBPA seems to be reduced.

More iodine is present in the urine (five times the normal) and feces of hyperthyroid patients than of normal individuals. Yet, if radioactive iodine is administered to hyperthyroid individuals, the urinary exretion of the hormone is less than in normal persons; freshly administered iodine is apparently retained in the body. The reduced excretion of radioactive iodine probably results from greater uptake of radioiodine in the thyroid; this increase in thyroidal uptake of radioiodine is consistently observed in patients with hyperthyroidism after oral or intravenous administration of radioiodine. Thyroid slices obtained from hyperthyroid patients clear the iodine from the medium at a much faster rate that slices obtained from normal individuals.

Dialysis of blood through cellophane followed by chromatographic analysis of the dialysate has permitted the demonstration of small amounts of thyroxine in blood (approximately 6.5 μg of iodine per 100 ml). In thyrotoxicosis, the concentration of free thyroxine in blood increases to as high as five times normal. When [^{131}I]thyroxine is administered intravenously to

be simple to pinpoint its cellular site of action. TSH stimulates the ^{32}P uptake by phospholipids in thyroid slices. ^{32}P is probably transferred to ATP and in turn to L-α-glycerophosphate, an intermediate between ATP and phosphatidic acid. Thus, when ^{32}P- labeled glycerophosphate is incubated with thyroid slices, incubated in a hypotonic medium for the purpose of increasing the membrane permeability to the phosphorylated derivative, ^{32}P is incorporated in phosphatidic acid, phosphatidylinositol, but not in phosphatidylcholine.

However, when [$^{32}P-2^{3}H$] glycerophosphate was used as a precursor, the ratio of $^{32}P:^{3}H$ in the various phospholipids differed from that initially present in the glycerophosphate. This observation suggests that in addition to direct conversion of glycerophosphate to phosphatidic acid, other still unidentified reactions are also involved [200].

In the hypophysis, TSH is probably elaborated in those cells of the anterior hypophysis known as basophils. Partial thyroidectomy and goitrogenic substances induce an enlargement and an increase in number of those cells.

The transplantation of homologous or autologous adenohypophyseal tissue in the anterior chamber of the eye of hypophysectomized rats and mice is associated with normal ^{131}I uptake and normal blood ratios of thyroxine versus thyrotropin. But the transplant cannot maintain the weight of the thyroid within normal limits. Histologically, the thyroids of hypophysectomized mice carrying a hypophyseal transplant were intermediate between those of normal and hypophysectomized mice. To explain these discrepancies in the effect of thyroid function and thyroid weight, Greer has proposed that two factors control the performance of the thyroid gland: one is exclusively hypophyseal and regulates iodine uptake, and the other is hypothalamic and regulates thyroid growth.

Hypothalamic Control of TSH Secretion

The evidence for neurological control of THS secretion comes from delicate microsurgery and electrostimulation and microinjections of the brain. The section of the pituitary stalk in rat and rabbit diminishes the increase in thyroid weight following thiouracil administration, reduces iodine uptake and release in the thyroid, and produces histological signs of hypopituitarism in the thyroid.

Destruction of the cells in the anterior portion of the hypothalamus, probably near the midline between the median eminence and the optic chiasma, decreases thyroid function. Yet the thyroids of animals in which such hypothalamic lesions have been produced respond normally to TSH stimulation, and it can therefore be concluded that the hypothalamic injury interferes with thyroid function by blocking TSH secretion in the anterior pituitary. The hypothalamic centers controlling TSH, ACTH, or gonadotropin secretion are different because lesions that interfere with the secretion of one of these stimulins do not necessarily interfere with the secretion of the others.

The location of the thyrotropic center within the anterior hypothalamus varies depending upon the animal. In dog and hamster, the anterior portion of the median eminence must be in the injured area to inhibit TSH secretion. In guinea pigs, the lesion must include the paraventricular nuclei, and in rat it must include regions of anterior hypothalamus posterior and ventral to the paraventricular nuclei.

These findings on the role of the hypothalamus in controlling TSH secretion were confirmed by selective electrostimulation of the anterior hypothalamus, which increases the activity of the thyroid gland.

Microinjections of amounts of thyroxine too small to have a systemic effect on the adenohypophysis or the anterior hypothalamus reduce TSH secretions. The effect is immediate when thyroxine is injected in the hypophysis. It is delayed when thyroxine is injected in the hypothalamus, but the minimal dose necessary to inhibit TSH secretion is the same whether thyroxine is injected in the hypothalamus or the hypophysis.

The anterior hypothalamus secretes a neurohormone (TRH), that is transported from the brain center to the hypophysis by way of the hypophyseal-portal system.

The thyroid releasing hormone binds to receptors in the pituitary and like many other hormones activates adenyl cyclase and increases the intracellular levels of cyclic AMP.

Insofar as TRH has been purified, the effects of the purified hormone were tested on the hypophysis *in vitro*. TRH stimulates the synthesis and the release of TSH in mouse pituitary. The effect of TRH differs from those of histamine and stress, which increase TRH levels in the pituitaries without provoking TRH release.

The incubation of pituitary glands with both TRH and thyroxine results in the block of the TRH effect (synthesis and release of TSH) by thyroxine. The mechanism by which thyroxine blocks the TRH effect is most intriguing because the block is relieved by addition of inhibitors of protein synthesis (actinomycin and cycloheximide).

In summary, thyrotropin secretion is feedback regulated by circulating levels of T_3 and possibly T_4. Increased levels of these compounds in the blood result in a decreased level of thyrotropin. The thyroid hormones block the action of TRH on the pituicyte by stimulating the formation of a protein which masks the effect for TRH without interfering with the elaboration of TSH. The molecular mechanism by which the newly synthesized protein interferes with TRH activity is not known.

The structure of TRH is L-pyroglutamyl-L-histidyl-L-prolineamide, a very small peptide. Of course, it is not excluded that *in vivo* the releasing factor is a larger polypeptide. But the active polypeptide has been synthesized, and the synthetic product has properties identical to those of the natural.

Inasmuch as TRH is believed to exert its hormonal function after binding to the plasma membrane of the pituitary cell, which secretes thyrotropin, the molecular configuration of the hormone is of interest because such configuration will influence the binding properties. Molecular biologists believe that the molecule is bent to form a hairpin turn, which is then stabilized by two specific hydrogen bonds [41].

It is appropriate to devote a few lines here to the releasing factors. We encounter them each time we review the mechanism of hypophyseal hormone secretion. And yet only the structures of TRH and LRH seem to be known with some degree of certainty. All the others have been suspected to exist since 1930, and some have been partially purified.

These releasing factors have been prepared from the hypothalamus, but that does not exclude the possibility that they may also be found in other parts of the nervous system.

A sensitive and specific radioimmunoassay for TRH has established that the hypothalamus contained only 31.2% of the total brain content of TRH and that appreciable amounts of the hormone were found in all areas of the brain except the cerebellum [170–171].

The search for releasing factors of pituitary hormones started with an attempt to purify the corticotropin releasing factor. Although these attempts were unsuccessful, they led to the discovery of a thyrotropin releasing factor (TRF) which was ultimately purified in mg amounts from 500 tons of sheep brains, containing 50 tons of hypothalamic fragments. An active tripeptide was isolated with the following amino acid composition; glutamic acid, histidine and proline, but with no detectable NH_2 terminal. The amino acid sequence of the tripeptide was determined by synthesis and the active compound was shown to be glu-his-pro-NH_2, in which the C-terminal fraction is amidated, as is the case for many other polypeptide hormones (gastrin, vasopressin, secretin, ACTH) [172–173].

The discovery of the existence and the determination of the structure of the thyroid-releasing hormone have led to important clinical applications with respect to the diagnosis of thyroid diseases. Only a few examples will be presented.

After injection of 200 μg of TRH intravenously, the TSH levels in the blood rise in normal individuals to a peak 20 minutes after injection.

In patients with hyperthyroidism, whether it is due to nodular hyperplasia or Graves' disease, the response to the injection of the TRH does not elicit an increase in TSH secretion because in such patients the high levels of circulating T_3 and T_4 inhibit the hypophyseal response.

In primary hypothyroidism the injection of TRH causes an excessive release of TSH. In contrast, in secondary hypothyroidism the hypothyroid patient does not respond to TRH by excessive secretion of TSH [176]. Luteinizing hormone releasing factor (LRF), follicle stimulating hormone releasing factor (FRF), corticotropin releasing factor (CRF), growth hormone releasing factor (GRF), prolactin releasing factor (PRF), and melanocyte stimulating releasing factor (MRF) have been described. All these factors are believed to be polypeptides.

The biosynthesis and breakdown mechanism of these releasing factors are unknown. Their mechanism of action is also obscure. Their releasing effect is not inhibited by protein inhibitors; therefore, it is not likely that they act by stimulating the biosynthesis of the active hypophyseal hormone. Calcium and sodium are needed for their activity. The coming years will probably yield crucial information on the regulation of hormonal release.

In summary, the hypothalamus secretes small polypeptides, 3 to 13 amino acids long, which are released in the hypophyseal portal system. When they reach the hypophysis they specifically bind to receptors in the target cells where they probably activate adenyl cyclase and modify calcium permeability. Ultimately a specific hypophyseal tropin is released that activates other endocrine glands, the thyroid, the gonads, the adrenal, etc. The binding site of the releasing hormone is the target of the negative feedback exerted by the various hormones such as T_3, T_4, etc. on the hypophysis. The feedback inhibition involves protein synthesis, but the kind of protein that is synthesized has not been identified. When the releasing hormone has done its work, it is released in the blood, degraded by the plasma and eliminated in the urine. The hormones are target specific, but not species specific. Those hormones whose amino acid sequence is known can be synthesized. Because of their target specificity, these hormones may provide new diagnostic or therapeutic tools [172, 174, 175].

In addition to a neurohumoral and the hypophyseal control centers, the thyroid is itself capable of self-regulation. At least three different mechanisms of autoregulation have been described. First, Halmi [42] has proposed that iodine in the thyroid inhibits the iodine pump and regulates the iodine content in the gland, and that this effect is independent of TSH secretion. It occurs without TSH as well as with constant amounts of TSH. Second, low-iodine diets increase the ratio of MIT to DIT. This condition could lead to an increased ratio of triiodothyronine versus tetraiodothyronine. Third, the amount of organic iodine released by the thyroid is proportionate to the amount stored in the gland. After propylthiouracil administration, the amount of hormone released gradually decreases as the stores are depleted. The gradual decrease of the released hormone is not affected by TSH.

Pathogenesis of Graves' Disease

There are two aspects to the pathogenesis of Graves' disease; the type of hormonal defect and the cause of the hormonal defect.

If studies of regulation of thyroid hormone secretion have not revealed the ultimate pathogenic mechanism of Graves' disease, they allow meaningful speculation. The primary insult could be at the level of the neurohypophysis, the hypophysis, the thyroid or the target organ. Among the possible causes are (1) increased secretion of TRH with resultant increased secretion of TSH, (2) increased sensitivity of hypophyseal receptors to TRH, (3) increased levels of TSH either because of increased production of the hormone in the hypophysis or as a result of a decreased rate of breakdown of the circulating hormone, (4) disruption of the regulatory feedback mechanism between hypophysis and thyroid, either because the two glands secrete their product autonomously or because the regulatory mechanism of the hypophysis is set too high and responds only to high levels of thyroxine, (5) abnormalities of the receptor sites for TSH (however, experiments in which the binding of labeled TSH to thyroid membranes was measured revealed that the binding capacity of membranes prepared from normal and Graves' diseased thyroid were the same), (6) hypersecretion of thyroxine, (7) hypersentivity of the target tissues to thyroid hormones. The pathogenic mechanism of Graves' disease is further complicated by the finding that in addition to TSH, the hypophysis also secretes other thyrotropins with chemical and immunological properties different from those of TSH. Increased levels of LATS were believed to be responsible for all the symptoms of Graves' disease, including the ocular symptoms. Measurements of LATS activity in hyperthyroids have yielded inconsistent results, possibly because of the inadequacies of the biological testing method available.

Neonatal hyperthyroidism has, however, provided new arguments in favor of a pathogenic role of LATS in hyperthyroidism. Some children are born with hyperthyroidism, including exophthalmos. The symptoms are transient and subside within a month after birth. The severity and duration of the symptoms are related to the levels of LATS in the mother's and infant's blood.

LATS is believed to have a half-life of 20–30 days, a finding in keeping with the duration of the symptoms in neonatal hyperthyroidism. Available evidence suggests that LATS is a protein with a sedimentation constant of 7 and electrophoretic properties similar to those of globulin. LATS appears to be an immunoglobulin of the IgG type and possess the properties of an antibody.

Even if the hormonal defect in Graves' disease were identified, the cause of the abnormality would still be unknown. Available evidence points toward an autoimmune mechanism in which the normal tolerance for thyroid tissue breaks down.

Some of the observations that have led to these conclusions are (1) the finding of antibodies to either thyroglobulin or thyroidal microsomal antigens, (2) the deposition of IgG, IgM, IgE and complement along the basement membrane of diffuse thyroid goiters, (3) the delayed cell mediated immunity to thyroid antigens, (4) the presence of the LATS immunoglobulins in the serum, (5) the presence in blood of immunoglobulins which bind to TSH receptors, (6) the increased proportion of circulating lymphocytes forming nonimmune rosettes in patients with Graves' disease, (7) the frequent enlargement of the thymus in patients with Graves' disease [177].

Hypothyroidism

Hypothyroidism occurs in children (cretinism) and in adults (myxedema; see Fig. 8-10). Myxedema may be primary and result from direct injury to the thyroid gland, or be secondary and due to interference with thyrotropin secretion. In contrast to primary myxedema, which is resistant to TSH injection, secondary myxedema can be corrected by thyrotropin administration.

The pathogenesis of primary myxedema is not always easy to trace except when hypothyroidism follows thyroidectomy or massive necrosis of the thyroid. Autoimmune reactions probably are at least in some cases responsible for thyroid destruction. In myxedema, the thyroid is usually small and hard. Histologically, the typical follicular pattern has disappeared and is replaced by fibrosis. The thyroid is goitrous in a few cases, and presents the histological characteristics of hyperplasia.

Hypothyroids have a low basal metabolism (30–40% below the normal values) and gain weight readily. They are mentally slow and usually quite apathetic.

The changes in iodine metabolism are quite typical. Iodine uptake by the thyroid is low, and the levels of protein-bound iodine are markedly reduced in the blood. Except in special cases (see below), the binding properties of TBG and TBPA in blood are not changed. Determination of free thyroxine concentrations in blood of hypothyroids has yielded values lower than 20% of normal (1.5 µg of free thyroxine iodine per 100 ml). If these symptoms are readily explainable on the basis of modern knowledge of thyroid physiology and biochemistry, the causes of the

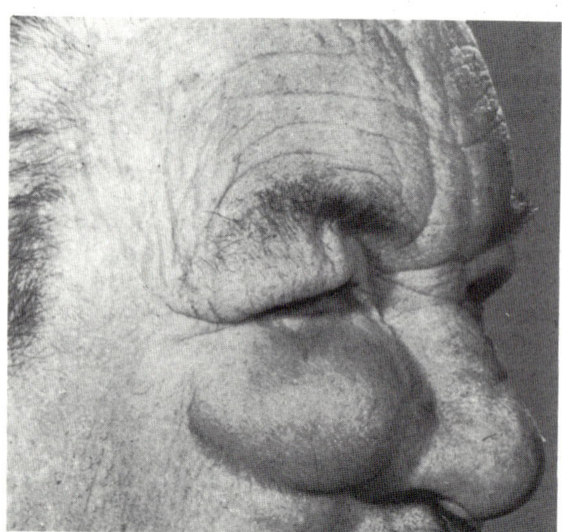

Fig. 8-10. Myxedema. (From J. Brown)

high blood cholesterol levels and the deposition of mucopolysaccharides in connective tissue of hypothyroids remain mysterious.

In contrast to hypothyroids who have low serum cholesterol levels (120–125 mg/100 ml), patients with myxedema have blood cholesterol levels ranging between 260 and 400 mg/100 ml. The interrelationship between the thyroid hormone and cholesterol metabolism needs to be clarified further. The increased level of serum cholesterol observed in hypothyroids could result either from increased synthesis or from a decreased rate of cholesterol breakdown. Studies of cholesterol and fatty acid synthesis in hyper- and hypothyroid rats have indicated that the absolute levels of the synthesis of both fatty acid and cholesterol increase with the increase in oxygen consumption. Weiss and Marx have shown that in rats thyroid hormone accelerates the conversion of labeled cholesterol to acidic products and the excretion of the labeled products.

Such observations indicate that the absolute rate of cholesterol synthesis is increased in hyperthyroidism and decreased in hypothyroidism. These findings can be correlated with the alteration in the levels of activity of the enzymes of the hexose monophosphate shunt (an important source of NADPH). The activities of the enzymes of the hexose monophosphate pathway are increased in hyper- and decreased in hypothyroidism. In spite of the considerable interest that such findings generate, their correct interpretation will have to wait until the mechanisms controlling the rate of cholesterol synthesis are adequately understood.

It seems safe to conclude that in hypothyroids the increase in serum cholesterol levels results from a shift in the rate of use versus the rate of synthesis, and that, in spite of low rates of synthesis, cholesterol increases in serum because of the low rate of use.

The effect of thyroid hormones on fatty acid metabolism probably resembles the effect of the hormone on cholesterol metabolism. Fatty acid synthesis increases in hyperthyroidism and decreases in hypothyroidism. In contrast, the rate of oxidation of butyrate is accelerated in hyperthyroidism and decreased in hypxthyroidism. The levels of serum phospholipids are increased in hypothyroidism. The mechanism of this metabolic alteration is not known.

Patients with myxedema often present carotenemia, which may give the skin a yellowish color.

The effect of thyroid hormones on steroid metabolism is not restricted to cholesterol; the metabolism of steroid hormones is also affected. An adequate understanding of the effects of thyroid hormones on steroid hormones requires a thorough knowledge of the intermediary metabolism of these hormones (see chapter on steroid hormones).

In the dermal tissues of hyperthyroid individuals, the accumulation of a mucinous substance, water accumulation, and mast cells are seen. In myxedema, a mucinous substance that reduces Fehling's solution after hydrochloric acid hydrolysis and stains metachromatically with toluidine blue accumulates in the subcutaneous tissue. The mucinous compound is a mucoprotein containing hyaluronic and chondroitin sulfuric acids. The accumulation of water probably results from the great capacities of hyaluronic acid to bind with water. The biochemical mechanisms responsible for the polysaccharide accumulation have not been elucidated.

In well-developed hypothyroidism, well-granulated mast cells are found in the connective tissue. After the patient is treated with thyroxine, the granules of the mast cells disappear and small degranulated cells take their place.

The changes in the subcutaneous tissue are associated with alterations of the epithelial layer of the skin. Prominent among them are hyperkeratosis and atrophy of the subcutaneous glands and the pilary bodies. The skin of myxedematous individuals is cold, scaly, and dry, and the hair of the scalp and all other parts of the body falls out. The nails are brittle and atrophic. Subcutaneous tissue thickening erases the fine features of the facies, thus all patients with myxedema have a coarse appearance. The subcutaneous tissue throughout the body is swollen, making the condition resemble generalized edema. The edema is nonpitting at the start, but in long-standing myxedema, pitting edema of the ankles may develop.

The loose connective tissue underlying the mucosa of the mouth, nose, larynx, bronchi, and intestine undergoes changes similar to those observed in the skin. Accumulation of mucoprotein in the subendothelial layer of the heart capillaries has been described. These alterations are associated with permeability changes, which might well explain the pericardial effusion common in myxedema. Mucoprotein accumulation in the media of the aorta has been reported to induce aneurysmal swelling of the vessels. Thickening of the vocal chords is responsible for the hoarseness of patients with myxedema.

Mucopolysaccharide infiltration is not restricted to the subcutaneous, submucosal, or subendothelial connective tissues, but it also involves muscle fiber. The accumulation of the mucinous substance in the intestinal wall leads to atrophy of the musculature with constipation, megacolon, and sometimes ileus paralyticus. Mucoprotein accumulation in myocardial muscle fibers weakens the heart, leading to dilation of all the heart chambers.

A microcytic, hypochromic, or macrocytic anemia often accompanies myxedemia. The mechanism of production of the anemia remains unknown.

Hypothyroidism seems also to affect calcium metabolism. Although the serum levels of calcium and phosphate are unaltered in myxedema, less calcium than normal is excreted. The calcium is retained on the bone matrix, and roentgenographic examination of the bone may reveal a greater density of texture.

Cretinism

In the newborn, hypothyroidism leads to cretinism. The two types of cretinism are endemic, resulting from iodine deficiency, and sporadic, which results from inborn errors of metabolism. In iodine-deficient areas, women with goiters, which may or may not be hypothyroid, give birth to goitrous hypothyroid children.

A number of inborn errors have been described in patients with sporadic cretinism: (1) a defect in the iodine-trapping mechanism; (2) an inability to convert inorganic iodide to iodine; (3) a lack of thyroid peroxidase [178]; (4) an inability to couple iodotyrosine to form iodothyronines and thyroglobulins; (5) a lack of dehalogenase; (6) an interference with thyroglobulin metabolism; and (7) a defect in thyrotropin secretion [179–180].

A patient was described who was unable to concentrate iodide in saliva, gastric juice, or thyroid slices; in that case the metabolic block obviously involved the iodine-pumping mechanism.

Patients with a defect in the conversion of inorganic to organic iodide are of two types. The first group of patients present only congenital cretinism; the second group has, in addition, goiter and congenital nerve deafness.

All cretins have an organification defect. They take up iodide in the thyroid much more rapidly than normal, but as soon as the peak of uptake is reached, the labeled iodine is lost from the thyroid. These observations indicate that the iodine pump functions normally in these patients, but that the thyroid is unable to retain the halogens. This conclusions finds further substantiation in the effect of perchlorate and thiocyanate on iodine metabolism. The administration of these two compounds to patients with the organification defect produces a precipitous release of iodine from the thyroid. Organically bound iodine is absent from the thyroid surgically removed from these patients. Moreover, organic iodides are not formed when particulate fractions of the thyroids are incubated with ^{131}I and an H_2O_2 generating system. Peroxidase has been proposed as the enzyme lacking in this form of hypothyroidism.

In patients with the coupling defect, iodine uptake by the thyroid is high, but the release of the radioactive halogen from the gland follows an almost normal pattern. Chemical analysis of thyroid biopsies of such cretins yields high levels of MIT and DIT. In contrast, thyroxine is virtually absent from the gland.

The critical observation in patients with the deiodinase defect is that the administred DIT is quantitatively excreted in the urine. Slices of the thyroid obtained from such patients are unable to dehalogenate ^{131}I-labeled diiodotyrosine.

The mechanism by which the absence of the dehalogenase leads to low thyroxine levels and cretinism is not clear. Two different theories have been proposed. The first postulates the existence of an additional defect; namely, an inability to couple iodotyrosine to form T_3 and T_4. The second proposes that the absence of dehalogenase leads to a glandular hyperfunction in which hormone precursors are released before they can be used for thyroglobulin biosynthesis. It has now been established that the dehalogenase defect results from the absence of a single autosomal recessive gene.

The fourth type of inborn error of thyroid hormone metabolism is characterized by the presence of a butanol-insoluble iodinated component in the serum. When this component was further purified and submitted to the action of proteolytic enzymes, iodinated tyrosines and thyronines were released in the digest. Except that these abnormal iodinated proteins have been identified, the molecular defect responsible for this form of cretinism is not understood. It is often assumed that the defect results from the absence of one or more enzymes involved in thyroglobulin degradation.

In addition to the symptoms already described in myxedema, cretins also present retardation of growth and of mental and sexual development. They are classified as imbeciles or idiots.

The alterations that determine the most obvious symptoms are those of the skeleton, especially the skull, vertebral bodies, and long bone. At birth the base of the skull is short, the cartilaginous junction between the pre- and postsphenoid bones remains uncalcified, the frontal sutures are wide, and the posterior fontanelles are large. The centers of ossification do not appear in the long bones, and epiphyseal growth is virtually nonexistent. These alterations of skeletal growth lead to typical dwarfism. Cretins have a large head compared with the size of the rest of the body and a short forehead. Their neck, hands, and feet are short. The trunk is longer than the legs. In addition to dwarfism, myxedema gives a quite typical appearance to the facies, which resembles that of adult myxedema.

Some hereditary diseases affecting thyroid metabolism manifest themselves only at a later age. Two examples will be given.

An acquired form of hypothyroidism resulting from the presence of an abnormal thyroid peroxidase has been described in a 16 year old girl. The peroxidase is inactive unless heme is added to the incubation mixture. It is believed that the hereditary defect results in the elaboration of an apoenzyme molecule which is incapable of properly binding the prosthetic group.

Hypothyroidism developed in a man of 19 as a result of a deficiency in secretion of TRH [180].

The Adrenal Cortex—Function and Diseases

Anatomical Considerations

The adrenal gland is composed of the cortex and medulla. These two parts of the human adrenal have a different embryonic origin; the cortex originates from the mesoderm, and the medulla is ectodermic. Early in embryonic life, the proliferation of the coelomic mesothelium forms an interrenal cluster of cells. These changes occur in the long axis of the embryo, in the area between the root of the mesentery and the prominence of the mesonephros. The three stages of embryonic development of the adrenal are: (1) the independent development of the mesodermic and the ectodermic buds, (2) a period during which the ectodermic and mesodermic cell clusters are in close contact, and (3) a stage when the proliferating ectodermic cells infiltrate the mesodermic structures. The anatomical relationship between cortex and medulla varies depending upon the animal species. In selachians, the cortex and the medulla are separate; they are close together in teleosts; and they are intertwined in batrachians, reptiles, and birds. Thus, the three stages of embryonic development are in some way reminiscent of philogenetic development. Therefore, the embryologic development of the adrenal has often been used to support the old theory stating that ontogenesis recapitulates philogenesis.

The medulla originates from the ectoderm. Primitive ectodermal cells called sympathogonia migrate down the neural crest and infiltrate the primitive mesodermal cell clusters, which are at the origin of the cortex. In the medulla, these primitive sympathogonia later differentitate into neuroblasts, ganglion cells, or chromaffin cells, depending on their location. The adrenal gland develops early in human embryonic life. Adrenals are visible in the human embryo 21 mm long.

The adrenal of the newborn is relatively large. At birth it weighs one-third of the total weight of the kidney, but the gland involutes very quickly during the first year. The involution mainly results from the disappearance of a fraction of the cortex adjacent to the medulla.

The cortex of the adrenal is composed of three cellular layers: glomerulosa, fascicularis, and reticularis. The glomerulosa is made of small, narrow cells tightly packed together to form clusters or columns. In some species, this part of the adrenal resembles the renal glomeruli histologically. The fascicularis is made of radially arranged columns of cells. The reticularis is composed of a network of richly interlaced epithelial cells forming meshes occupied by large capillary sinuses. The cells of the glomerulosa contain only small and few vacuoles. Because the cells of the fascicularis are large and stuffed with lipid inclusions, they are sometimes referred to as spongiocytes.

The adrenals have two types of circulatory systems. The first is made of superficial arterial branches that ramify into numerous small sinusoid-type capillaries. The system vascularizes the cortex of the adrenal. A second circulatory system is composed of perforating arteries that procede straight into the medulla of the adrenal, where they divide into tiny capillaries. The venous blood derived from the medulla and the cortex is collected into separate circulatory systems, and the blood running in these two systems is mixed only in the larger veins. Except for those nerves controlling the vascular wall, there seems to be no direct innervation of the adrenal gland. Such a finding suggests that adrenal secretion is entirely under hormonal control.

Histochemical techniques have been used to identify the adrenal hormones. Such investigations include the use of lipid staining for cholesterol and neutral fat, and staining of aldehyde groups with Schiff's sulfur leukofuchsin or with the phenylhydrazine method described by Bennett. The use of basophilic stains has indicated that in the embryo, ribonucleic acid is abundant in the cells of all layers of the adrenal cortex. In contrast, marked basophilia is found only in the glomerulosa of the adult. Brownish pigmented granules have been found in the reticularis. Because these inclusins are often found in association with pyknotic nuclei, it has been postulated that the brownish granule results from cellular degeneration [43].

Histophysiology

An important contribution of the histochemical studies to adrenal physiology was in the unravelling of the relationship between the presence of lipids and adrenocortical secretion. When adrenocortical secretion is stimulated, the cells, which at first are rich in lipoid material, are rapidly depleted. This depletion involves not only those staining with Sudan (total fat) but also lipids, which are birifringent in polaroid light. In addition, components that react with Feulgen or phenylhydrazine are reduced. These findings suggest that the adrenal cells filled with lipids are those that are rich in adrenal secretion.

Two theories have been proposed to explain the relationship between the three different types of adrenal cells. They are referred to as a unicist and a pluralist theory. The unicist theory proposes that the cells of the glomerulosa are stem cells from which the cells in all other layers originiate. Through progressive differentiation of the glomerulosa, cells transform into cells typical of the fascicularis and the reticularis, and the cells of the reticularis are those undergoing progressive cellular degeneration. Da Costa objected to this theory on the grounds that mitosis is no more common in the glomerulosa than in the reticularis, and in adrenal development the cells typical of the reticularis (according to Da Costa) are really the first to appear. Da Costa has proposed his own version of the unicist theory. He says that the reticularis is the stem cell from which all other cell types are derived.

Greep and Deane have criticized both forms of the unicist theory on the basis that: cell division and degradation are by no means the appanage of either the glomerulosa or the reticularis, and culture of glomerulosa cells never leads to transformation of a glomerulosa cell into a reticularis cell [44]. In addition to these negative criticisms of the unicist theory, Greep and Deane present an argument in favor of the pluralist theory. Careful correlation of histological observations and physiological studies of the adrenal cortex indicate that each cell type of the adrenal has a specific function or role. The glomerulosa cell secretes aldosterone; this secretion does not occur in other cells of the adrenal cortex and is independent of ACTH control.

In contrast, the secretion of glucocorticoid hormone probably occurs in the fascicularis and is under the direct influence of ACTH. The reticularis secretes sex hormones, and their secretion is influenced by gonadotropin.

Many observations have been made which support this theory, including the existence of the so-called x-zone of the internal part of the cortex. The x-zone is a cellular layer the importance of which changes considerably depending upon the sex of the animal. During the first pregnancy of female mice, a zone of dark cells containing few lipid inclusions has been observed. After pregnancy, the zone of darker cells involutes rapidly. A similar x-zone is observed in males but only after birth. The zone involutes rapidly within a few weeks after birth, and the regression can be retarded by castration or accelerated by the administration of testosterone or progesterone. The influences of sex hormones on the x-zone in the reticularis suggest that the reticularis affects sex hormone secretion. This assumption is further supported by the fact that in tumors producing active steroids, it is precisely in that zone that the major cellular alterations occur.

The effect of hypophysectomy on the adrenal cortex has provided an important argument in favor of the pluralist theory. Hypophysectomy does not cause death, probably because this intervention does not alter aldosterone secretion. Yet, the metabolism of protein, lipids, and carbohydrates is considerably modified in hypophysectomized animals. Such modification indicates that glucocorticoid secretion is affected by hypophysectomy. Histological examination of adrenals of hypophysectomized animals reveals that the median and internal parts of the cortex are atrophied, while the glomerulosa remains intact or may even be hypertrophied. In contrast, the injection of hypophyseal extract causes hypertrophy of the fascicularis.

Adrenocortical Hormones

Once attention had been attracted to the role of the adrenal in physiology by Addison's observations on the disease that now bears his name, considerable efforts were made to concentrate active adrenal extract. In 1930, an adrenal extract was prepared that was capable of keeping patients with Addison's disease alive. Within the decade, 28 pure crystalline compounds were isolated from the adrenal cortex. Reichstein [45] prepared deoxycorticosterone by partial synthesis. One of the most striking effects of adrenocorticoid extracts was their ability to endow those injected with great vigor. These properties apparently prompted the Nazi government to purchase tons of adrenals from Argentina. The adrenal extracts were allegedly injected to invigorate Luftwaffe pilots. This threatening rumor encouraged the United States government to prepare pure adrenal hormones. The magnitude of such a task can be appreciated if it is kept in mind that the adrenals of approximately 10,000 head of cattle are needed to prepare not much more than 10 mg of any of the active glucocorticoids.

These war efforts led to the discovery of two major groups of corticoid hormones, the glucocorticoid and the mineralocorticoid hormones. The most active glucocorticoids are cortisone and cortisol. Both these hormones have anti-insulin effects but are without marked effect on sodium retention. Quite to the contrary, they may stimulate sodium and chloride excretion. Aldosterone has only mild corticoid effects, but it has a marked ability to influence the kidney to retain sodium. 17α-Hydroxydeoxycorticosterone, 19-hydroxydeoxycorticosterone, 6B-hydroxycorticosterone, and deoxycorticosterone have a mild mineralocorticoid effect. Corticosterone, 11-dehydrocorticosterone, and 19-hydroxycorticosterone exert physiological effects intermediate between those of the glucocorticoids and the mineralocorticoids.

Steroid Hormone Biosynthesis

Two observations suggest that cholesterol is a precursor of steroid hormones. First, cholesterol concentrations are high in the adrenal, and the cholesterol content of the endocrine gland decreases when the cortex is stimulated by ACTH. Second, by using a deuterium label, Block demonstrated that cholesterol is converted into pregnanediol, a compound readily isolated from urine of pregnant women [46, 47].

Although cholesterol may serve as a precursor of steroid hormones, investigators still debate whether or not cholesterol is a required or the only intermediate. Other compounds formed from squalene may serve as precursors of steroid hormones. The slow transfer of radioactivity from cholesterol to testosterone in testes slices obtained from animals stimulated by chorionic hormones indicates that intermediates other than cholesterol participate in steroid hormone biosynthesis. Whereas chorionic hormones markedly stimulate the incorporation of acetate into cholesterol, the specific activity of testosterone is not markedly increased.

The conversion of cholesterol to corticosteroids was further confirmed by *in vitro* experiments. The first reaction in this biochemical transformation is NADP and ATP dependent. It results in the splitting of the cholesterol side chain to yield pregnenolone and isocaproic acid (see Fig. 8-11). Pregnenolone is a 21-carbon compound with a β-carboxyl group in position 3 and an unsaturated bond between carbons 5 and 6. The exact mechanism of the splitting reaction is not clear, although it would appear that 20α-hydroxycholesterol and 22α-hydroxycholesterol can serve as intermediates. Their role has been questioned [192].

It has been proposed that the monohydroxylate is further hydroxylated to yield a dihydroxylate. A desmolase capable of attacking either 20- or 22-hydroxy derivatives yields, by splitting the side chain, pregnenolone and isocaproic aldehyde, which is converted to isocaproic acid. It may be significant that the carbon-to-carbon split that is catalyzed by desmolase is sensitive to ACTH, LSH, HCG, and angiotonin II (which

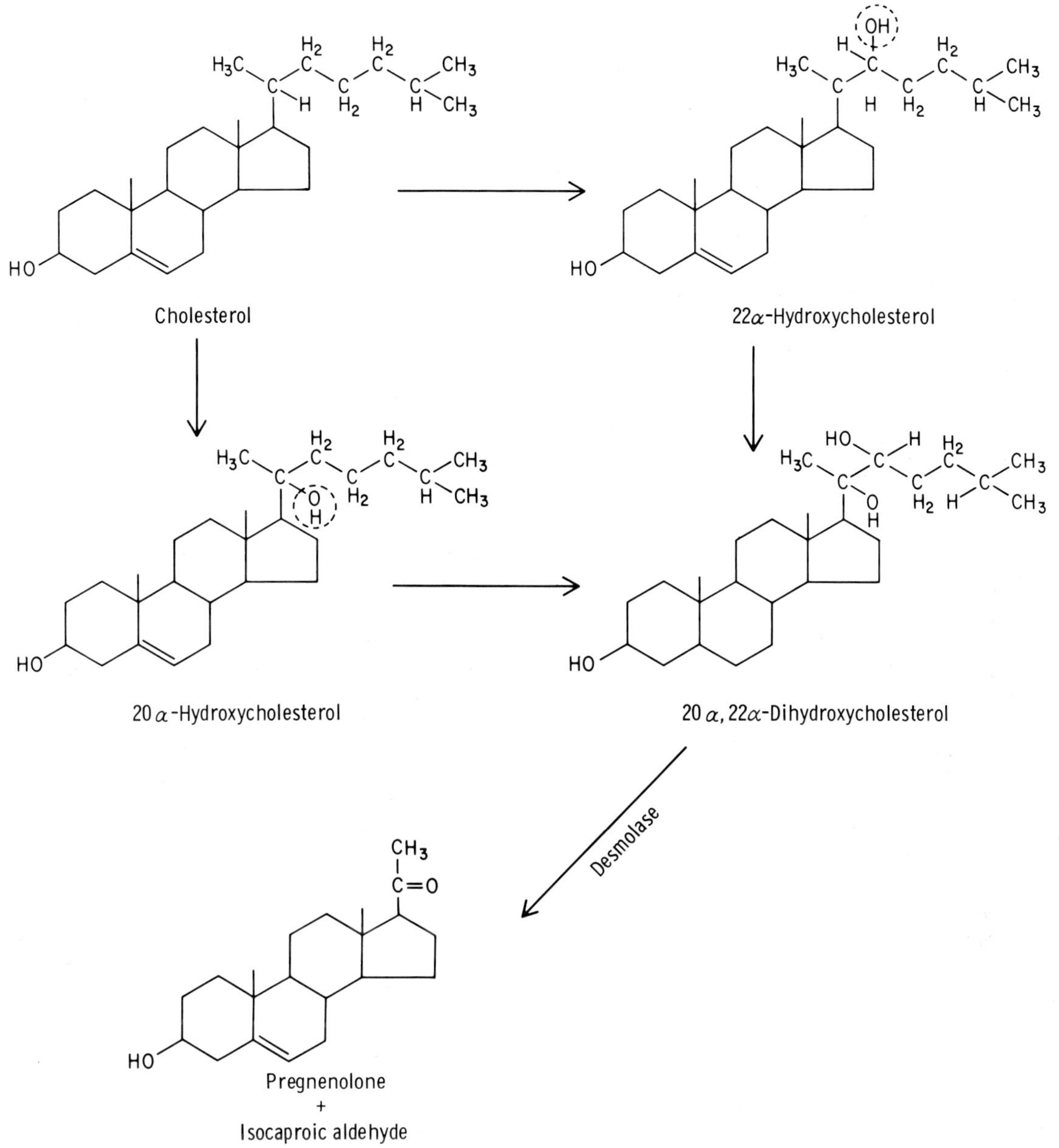

Cholesterol

22α-Hydroxycholesterol

20 α-Hydroxycholesterol

20 α, 22α-Dihydroxycholesterol

Desmolase

Pregnenolone
+
Isocaproic aldehyde

Fig. 8-11. Conversion of cholesterol to pregnenolone

stimulates the formation of aldosterone). The availability of pregnenolone appears to constitute the rate-limiting step in steroid hormone biosynthesis. The reaction occurs in the mitochondria in the adrenal, and the effect of ACTH seems to be elicited through the action of cyclic AMP.

However, it seems unlikely that the conversion of cholesterol to pregnenolone would be the only ACTH-sensitive step. If it were, glucocorticoid steroidogenesis and also mineralosteroidogenesis in hypophysectomized animals should be blocked.

Enzymes catalyzing the conversion of cholesterol to pregnanediol have been found in the adrenals, gonads, and placenta, but not in liver.

Although large amounts of cholesterol accumulate in the adrenal cortex and in the corpus luteum, almost no cholesterol is found in the testes and follicles. To explain these differences, it has been proposed that the precursor is stored in the adrenal and the corpus luteum because the need for glucocorticoids and progesterone is usually more acute than the need for testosterone and androgens.

In the next step of steroid biosynthesis, the 3β-hydroxyl group of pregnenolone is oxidized in the presence of NADP, yielding NADPH and the ketone derivative: progesterone (see Fig. 8-12). The dehydrogenation is followed by an isomerization involving the transfer of the unsaturated double bond from its position between carbons 5 and 6 to a new position between carbons 4 and 5. The enzyme that catalyzes the oxidative reaction is the $\Delta 5,3\beta$-hydroxysteroid dehydrogenase. The $\Delta 5,3\beta$-hydroxysteroid dehydrogenase is found in all tissues synthesizing steroid hormones; it is tightly bound to the cell fraction preparation called microsomes. It is specific for $\Delta 5,3\beta$-hydroxyl and will not react with the $\Delta 5,3\beta$-hydroxyl. The enzyme is inactive if ring D is eliminated, but a double bond is not absolutely necessary. The double bond migrates spontaneously at a slow rate after β-dehydrogenation with a purified dehydrogenase preparation.

The migration is accelerated in homogenates in which an isomerase has been demonstrated. An iso-

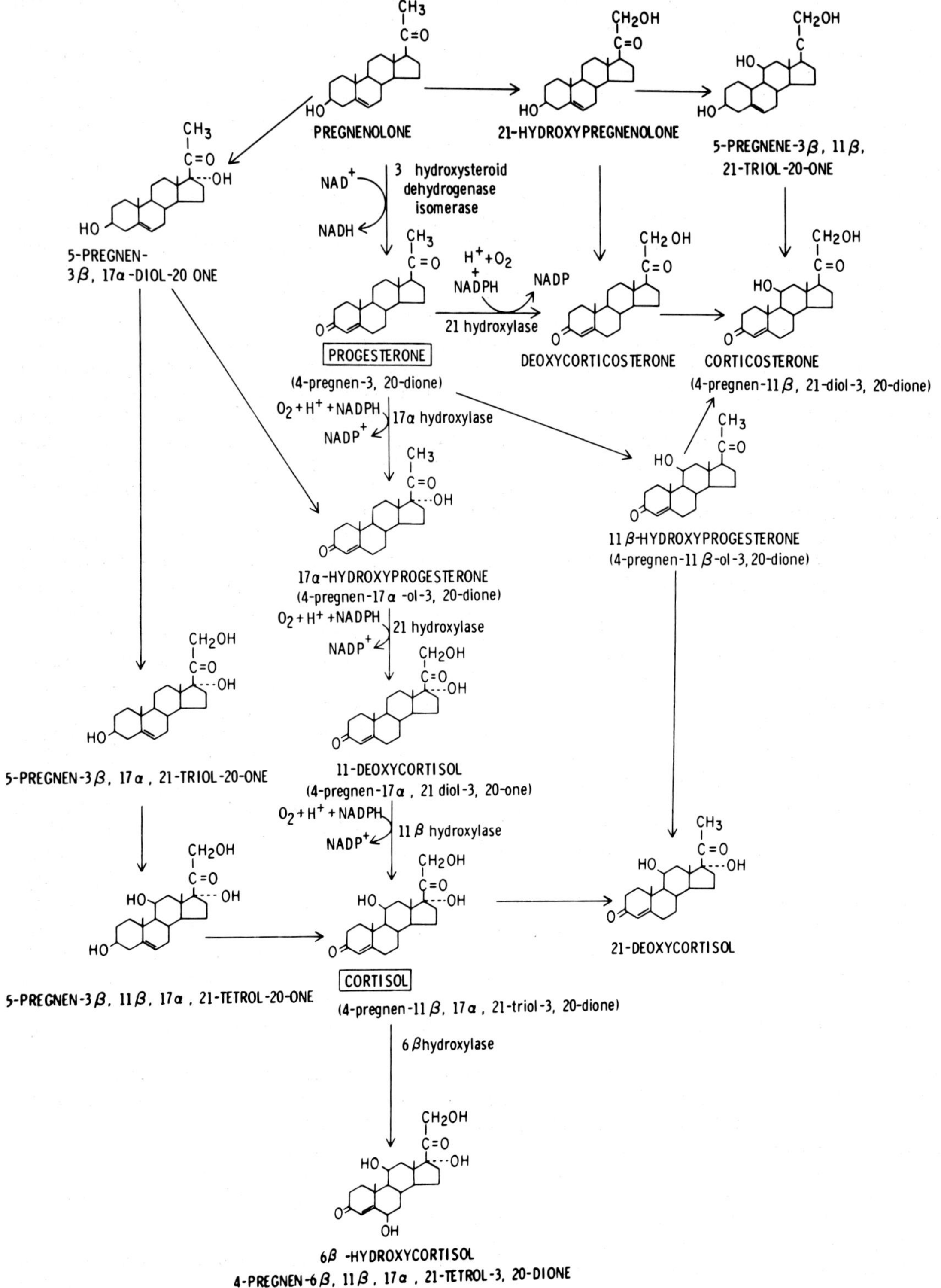

Fig. 8-12. Biosynthesis of progesterone and cortisol

merase catalyzing the migration of a Δ^5 bond to the Δ^4 position of the 3-ketosteroid has also been found in human blood, liver supernatant fluid, and pseudomonas grown in the presence of testosterone. The enzyme has been partially purified from bacterial extract. The isomerization is practically irreversible, and at the end of the incubation only the Δ^4 analog is found. No cofactor requirements are known for isomerase activity. Studies of the isomerase reaction in labeled water have demonstrated that there is no exchange of the hydrogen molecule of the steroid with that of water. This fact indicates that the bond is transferred within the molecule.

The conversion of progesterone into hydroxycorticosteroids involves three separate hydroxylations. It is not known in what sequence these hydroxylations occur, but the following sequence has been postulated: (1) hydroxylations in position 17 yielding the 17α-hydroxyprogesterone; (2) hydroxylation in position 11, yielding the 11,17-dihydroxyprogesterone; and (3) a hydroxylation in position 21, yielding cortisone. When hydroxylation first involves the carbons in positions 21 and 11, hydroxylation in position 17 is impossible. This observation suggests that steroids further hydroxylated in positions 21 or 11 cannot attach themselves to the enzyme hydroxylating the position 17. Yet a compound already hydroxylated in position 21 can react with the 11-hydroxylase, and thus deoxycorticosterone may be converted to corticosterone; in contrast, the compound hydroxylated in position 11 cannot be converted to either 17- or 21-hydroxylated steroids.

The 11-hydroxylase has been solubilized, and its properties and mode of action have been partially elucidated. It is a metal enzyme markedly inhibited by diethylthiocarbamate. The hydroxylase is different from cytochrome or from cytochrome oxidase because azides or cyanides have no effect on the hydroxylation.

The hydroxylation reaction consists in a steric exchange between a hydrogen and a hydroxyl. It involves the hydrogen of NADPH and atmospheric oxygen, but there is no exchange with the hydrogen atoms of water, and hydrogen peroxide is not a likely intermediate. The hydroxylation system is probably composed of three different enzyme molecules because *in vitro* three heat-labile nondiffusable compounds must be combined to yield full activity. It has been suggested that while one enzyme forms the hydroxyl, a second activates the substrate and a third transfers the electron from NADPH to form H_2O. The enzyme transferring the electron is probably an iron porphyrin of a type similar to cytochrome. Kimura *et al.* have shown that the hydroxylation of 11β-, 18 steroid and the cholesterol side chain cleaving hydroxylation reactions are carried out in mitochondria through a special electron transport chain located in the inner mitochondrial membrane, which includes adrenodoxin reductase, a flavoprotein, adrenodoxin, an iron sulfur protein and cytochrome (P_{450}). The electron transport chain is found in the adrenal, the testis and the ovary. The adrenodoxin reductase has been purified. It has a molecular weight of 54,000 with one molecule of FAD per molecule of protein. Adrenodoxin (mol wt 12,500) contains two atoms of iron and sulfur per mole of protein [192]. Specific hydroxylases have been found in the adrenal; those catalyzing the conversion of progesterone to the 11-hydroxyl derivative are mitochondrial enzymes, and those catalyzing the 17- and 21-hydroxylation are microsomal enzymes.

Studies in which adrenal glands were perfused with steroids of known structure have established that 21-hydroxypregnenolone and deoxycorticosterone could serve as corticosterone precursors. 21-Deoxycortisol and 11β-hydroxyprogesterone are converted to cortisone and corticosterone, in contrast to 17α-hydroxyprogesterone, which can be converted only to cortisol. On the basis of such studies, a number of alternative pathways for glucocorticoid steroidogenesis have been described. The conversion of progesterone to corticoid hormone is shown in Fig. 8-12.

17-Hydroxyprogesterone is not only a precursor of corticoid hormones, but also of androgens and estrogens (see Fig. 8-13). By splitting the side chain, which is attached to the 17th carbon, 17-hydroxyprogesterone is converted to a 19-carbon ketone. Acetic acid and androstendione are formed in the process. The androstene-3,17-dione in testes or adrenals may be converted to testosterone by reduction of the ketone group of position 19.

The enzyme involved in that reaction is a hydroxylase;* however, it does not catalyze a direct reaction between substrate and active OH. The reaction occurs in two steps. A first enzyme weakens the carbon-to-carbon bond between carbon 17 and the side chain by introduction of the 17-hydroxyl group. A second enzymic attack splits the side chain. The overall reaction requires NADPH and oxygen.

Several other pathways have been described for testosterone biosynthesis. One involves dehydroepiandrosterone. This pathway has been found in both adrenal and testes. Pregnenolone is hydroxylated in position 17 to yield 17α-hydroxypregnenolone, which after splitting of the side chain is converted to dehydroepiandrosterone. Dehydrogenation of the 3-hydroxyl and isomerization of the Δ5-bond yields androst-4-ene-3,17-dione, the immediate precursor of testosterone. Thus, when the pathways involving progesterone and that involving dehydroepinandrosterone are compared, it is clear that a major difference resides in the sequence of oxidation of the 3 hydroxyl versus the side chain splitting. In the progesterone pathway, oxidation precedes the desmolase reaction, whereas in the second pathway the side chain splitting precedes the dehydrogenation.

It has been suggested that dehydroepiandrosterone can be formed directly from cholesterol, through either a 17α,20α-dihydroxycholesterol or 17α,20α,22-trihydroxycholesterol.

* The difference between the mode of action of the splitting enzymes and other hydroxylases resides in the fact that with the splitting enzymes, there is no direct exchange between the hydrogen atom attached to the carbon and an activated hydroxyl group.

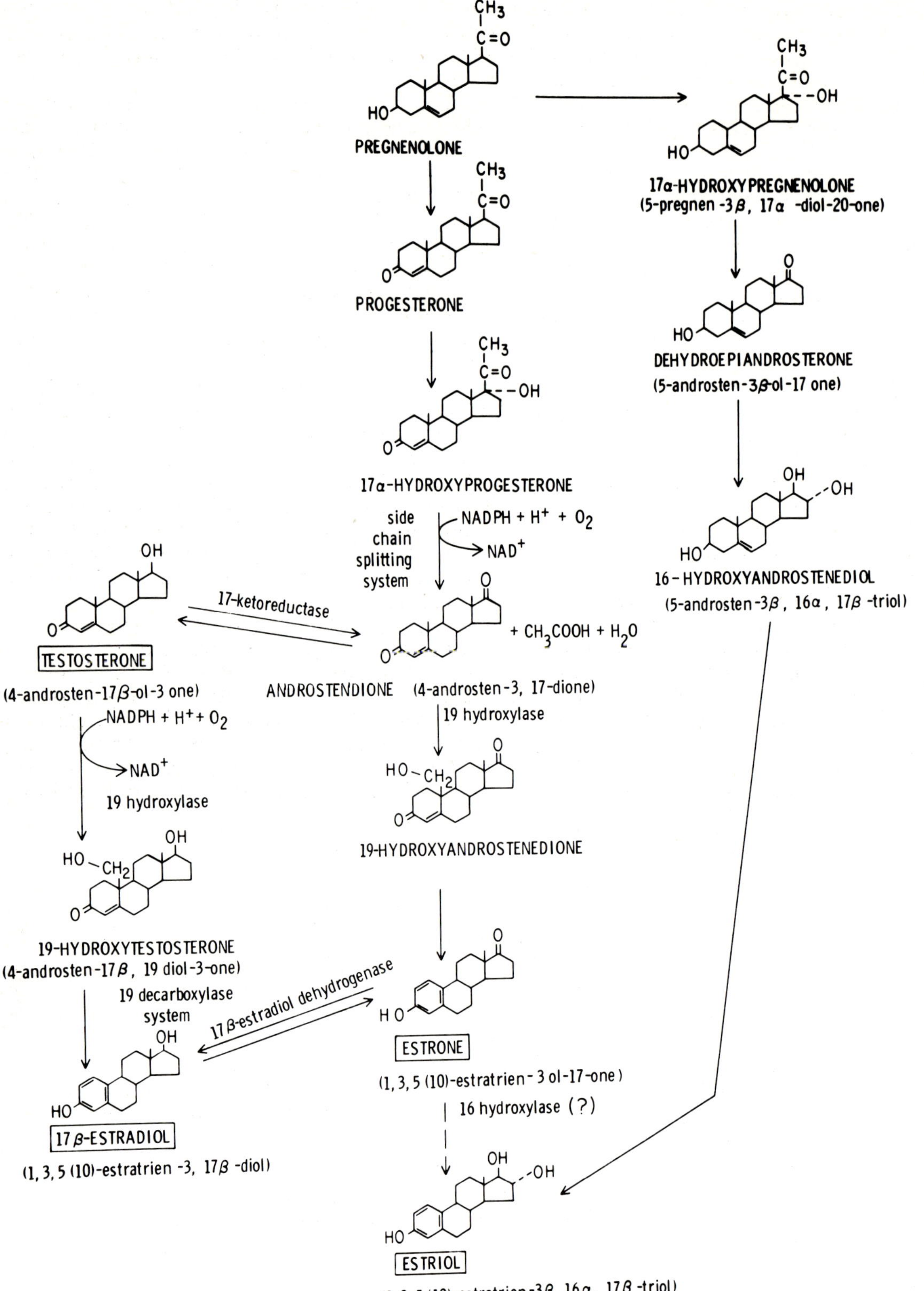

Fig. 8-13. Biosynthesis of sex hormones

The bulk of dehydroxyepiandrosterone is converted to sulfate and excreted. However, the conjugation reaction is reversible, and consequently labeled androsterone is in the urine after injection of labeled dehydroxyepiandrosterone sulfate. Adrenal tissues can convert dehydroepiandrosterone to androstenedione or testosterone.

Biosynthesis without intermediate formation of androstenedione has been described using homogenates of human polycystic ovaries. Testosterone acetate is then assumed to be an intermediate.

In another pathway, 17α,20α-dihydroxycholesterol or 17α,20α-trihydroxycholesterol is an intermediate between cholesterol and dehydroepiandrosterone. Androgens are also assumed to be synthesized in liver and nongonadal tissue from 21-hydroxylated derivatives such as 3β,17α,21-trihydroxy-5α-pregnane-11,20-dione and 11-deoxycorticosol, which by splitting of the side chain yield dehydroepiandrosterone or androst-4-ene- 3,17-dione.

Cortisol and 21-deoxycortisol can by splitting of the side chain yield 11β-hydroxyandrost-4-ene-3,17-dione, which may yield either adrenosterone or 11β-hydroxytestosterone. The biosynthesis of testosterone is summarized in Fig. 8-14.

There are several pathways for estrogen biosynthesis. Estradiol-17β is synthesized from testosterone and 19-hydroxytestosterone is an intermediate. Another pathway involves 19-hydroxytestosterone, which is in turn converted to 19-oxotestosterone. 19-Oxotestosterone may be either converted directly to estradiol-17β and acetate, or it may yield 19-carboxytestosterone, which by decarboxylation yields 19-entostestosterone, an immediate precursor of estradiol-17β. A pathway involving the conversion of testosterone to 17β-hydroxyandrosta-1,4-dien-3-one, 17β-hydroxy-19-carboxyandrosta-1,4-dien-3-one has been described.

Estradiol is readily converted to estrone because of the presence of a dehydrogenase specific for a steroid with a benzene ring, the 17β-estradiol dehydrogenase. The enzyme requires NAD as a coenzyme like other dehydrogenases; it contains sulfhydryl groups and is inactivated by copper, mercury, iron, and zinc. The aromatization of the A ring of the steroid molecule after hydroxylation of the carbon 19 of testosterone to yield 17β-estradiol is an important reaction. The mechanism of that reaction is not known. The enzyme system acting as catalyst requires NADPH and atmospheric oxygen, and the catalyst has been prepared from placenta microsomes. The biosynthesis of estrogens from testosterone is reviewed in Fig. 8-15.

On the basis of *in vitro* studies using adrenal glomerulosa tissues two pathways have been proposed for aldosterone biosynthesis. One starts with progesterone as a precursor and involves 11-deoxycorticosterone and possibly 18-hydroxycorticosterone and corticosterone as intermediates; and in the other pathway, corticosterone is bypassed, and the 18-hydroxylation precedes the 11-hydroxylations.

All tissues synthesizing steroid hormones can convert acetate to cholesterol and can form pregnenolone and progesterone. Furthermore, since they all originate from the germinal ridge, it is not surprising that they all have the enzyme activities necessary for most hydroxylations and for splitting the side chain, aromatization, etc. There are only two known exceptions to this generalization: the absence of 11-hydroxy-

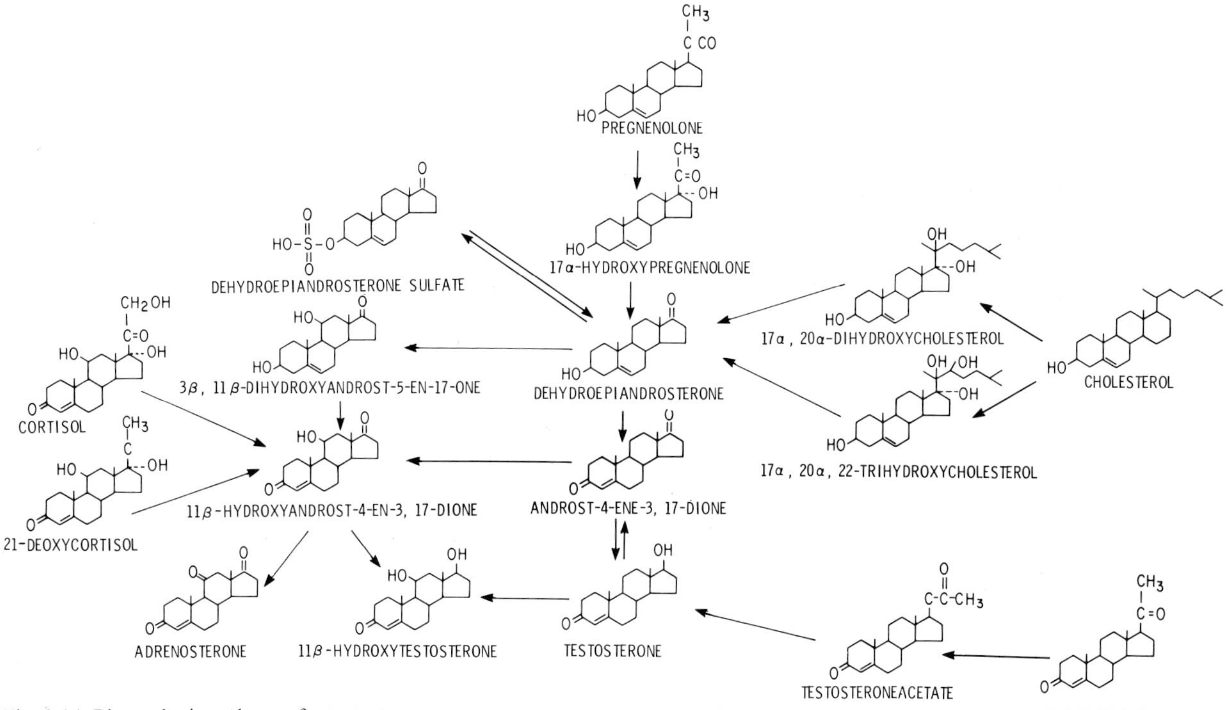

Fig. 8-14. Biosynthetic pathways for testosterone

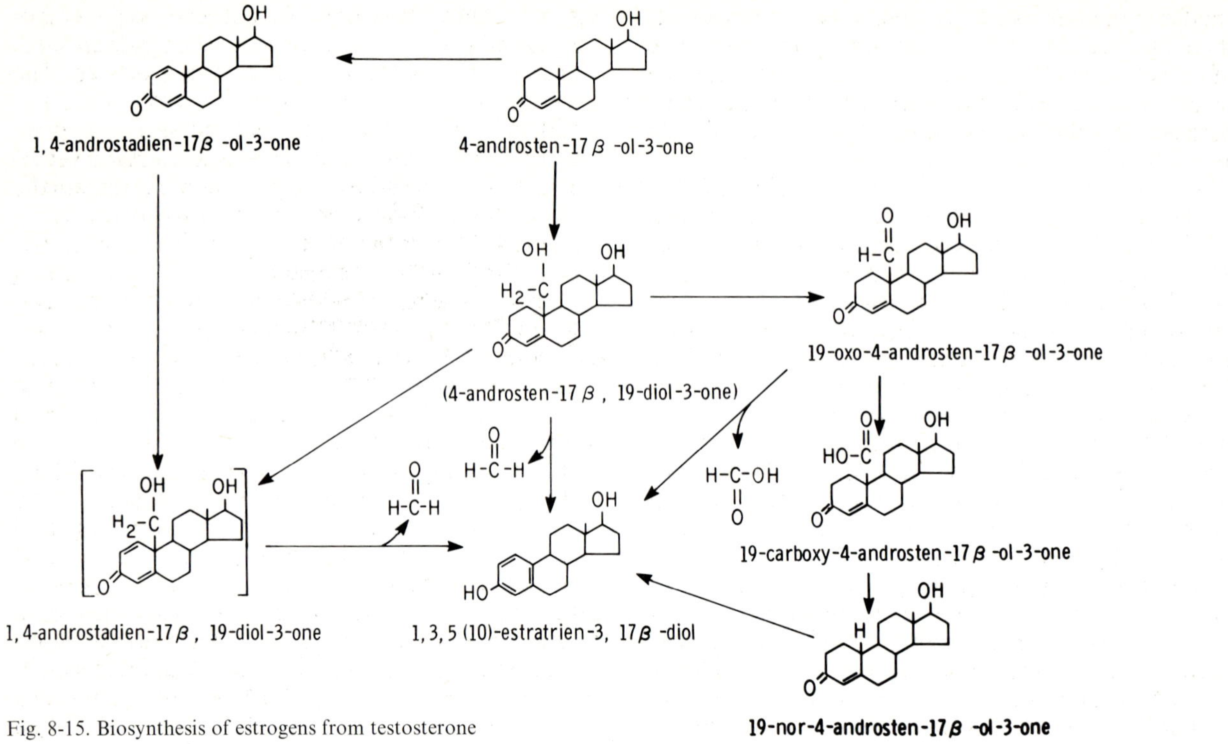

1,4-androstadien-17β-ol-3-one

4-androsten-17β-ol-3-one

(4-androsten-17β, 19-diol-3-one)

19-oxo-4-androsten-17β-ol-3-one

19-carboxy-4-androsten-17β-ol-3-one

1,4-androstadien-17β, 19-diol-3-one

1,3,5 (10)-estratrien-3, 17β-diol

19-nor-4-androsten-17β-ol-3-one

Fig. 8-15. Biosynthesis of estrogens from testosterone

lases in gonads, and the incapacity of the adrenal cortex to convert androstenedione to testosterone. Thus, the difference between the various steroid-synthesizing organs is quantitative rather than qualitative.

The sex steroid hormones are produced by the mature cells of the gonad. Estrogens are produced mainly by the ovary during maturation of the follicles and by the placenta. Progesterone is secreted during the elaboration of the corpus luteum and by the placenta. The interstitial cell of the testicle is the main source of testosterone.

Inactivation of Steroid Hormones. The active steroid hormones can be inactivated in various ways. Most inactivation reactions occur in kidney or liver. In liver, the inactive product is excreted either into the bile and carried from there to the feces, or into the general circulation and from there to the urine. In kidney, the inactivated steroids are excreted directly in urine.

Steroid hormones can be inactivated by conjugation and degradation. Conjugation involves the formation of either glucuronides or aryl sulfates. Steroids are conjugated with glucuronic acid in the presence of a liver microsomal enzyme and UDP-glucuronic acid.

Ten per cent of the urinary sulfur is in the form of organic sulfate. Organic sulfates are formed in liver by the conjugation of sulfates with aromatic compounds, steroids, or other hydroxyl compounds. Inorganic sulfate is first activated in the presence of ATP and a specific enzyme, then the inorganic sulfate is converted to adenosine-3′-phosphate-5′-phosphosulfate with pyrophosphate as a by-product of the reaction. The active sulfate is then transferred to the hydroxyl group of the steroid.

Enzymes, glucuronidases, and aryl sulfatases capable of splitting the conjugated products (glucuronates or sulfates) are found in a variety of animal tissues. These are acid hydrolases assumed to be associated with the lysosomal fraction.

The reactions involved in steroid degradation are: (1) reduction of the Δ^4 double bond; (2) dehydrogenation of the hydroxyl in positions 3, 6, 11, 16, 17, and 20; (3) hydroxylation of carbons 6, 7, 16, and 18; and (4) splitting of the side chain of the steroids of a 21-carbon steroid to yield a C_{19} compound.

The enzymes involved in reducing the 4,5 double bond of the A ring of the steroid hormone are called Δ^4-reductases. These reductases have been found in liver, but they have not been solubilized or purified. The reductases are substrate specific and require NAD as a coenzyme. The specificity of the enzymes also extends to the product, and one can distinguish between Δ^4,5α-reductases and Δ^5,5β-reductases. Both types of reductases have been found in liver. Whereas the α-reductase is associated with the microsomal fraction, the β-reductase is found in liver supernatant.

The activities of the reductases vary depending upon the physiological and the pathological conditions. In the reduction of the double bond of the A ring of cortisone in liver cirrhosis is inhibited, but the reduction of the double bond of other steroids remains unimpaired. In other liver diseases or in hypothyroidism, reduction of the double bond of the A ring of cortisol and 11-deoxycortisol is inhibited. Δ^4,α-reductases are found only in liver of male rats, in association with smaller amounts of Δ^5,α-reductases. Consequently, there is a definite sex distribution of the two types of reductases, yet this difference in distribution seems to be independent of the secretion of sex hormones because the differences occur before puberty [48, 49].

The second type of reaction that inactivates some of the active steroids involves enzymes that catalyze the keto group reduction to yield a hydroxyl group. The enzymes catalyzing such reactions are called dehydrogenases when the reaction is freely reversible and reductases when it is irreversible. None of these enzymes has been purified extensively, and the intracellular distribution is not always known. These enzymes have been found in microsomes and in the supernatant fluid. This group of enzymes can be classified on the basis of their specificity for the substrate. Two properties of the substrate would determine specificity: the position of the hydroxyl group of the steroid and the steric configuration of the molecule.

Many hydroxysteroid dehydrogenases have been discovered. These enzymes are important not only because they regulate the amount of active steroids found in a given organ, but also because they influence the nature and the amounts of steroid derivatives excreted in the urine. 3-Ketosteroids are reduced to yield 3-hydroxysteroids under the catalytic influence of a 3-hydroxysteroid dehydrogenase, which has been found in microsomes and the soluble fraction of rat liver. The 3-hydroxysteroid dehydrogenases have a broad specificity, and their activity is not restricted by the structural changes occurring outside the ring. They require either NAD or NADP as coenzyme. The product of the reaction has the 3α-steroid configuration. Small amounts of 3β-hydroxysteroids can be formed. It is not known whether formation of the 3β-hydroxysteroid results from the action of a specific 3β-hydroxysteroid dehydrogenase or from the unspecific action of the 17β-hydroxysteroid dehydrogenase on the 3-ketosteroid [50].

A 6ε-hydroxysteroid dehydrogenase catalyzing the conversion of 6-ketoestradiol to 6ε-hydroxyestradiol has been found in rat liver.

11-Hydroxysteroid dehydrogenases are widely distributed among rat tissue. They have been found in liver, kidney, placenta, and peripheral tissue. Prolonged treatment of mice with cortisol increases the ability of the thymus but not that of spleen and lymph nodes to convert cortisol to cortisone. The 11-hydroxyketosteroid dehydrogenase requires NAD or NADP as coenzyme.

A 17-hydroxysteroid dehydrogenase acting on testosterone has been found in kidney and liver. A 16-hydroxysteroid dehydrogenase, which converts estrone to estradiol, was found in liver and placenta. A 20-hydroxysteroid dehydrogenase acting on cortisone has been found mainly in liver and kidney. The activity of the enzyme is reduced after castration in males, but not in females. Wiest has partially purified a 20α-hydroxysteroid dehydrogenase acting on progesterone and 17-hydroxyprogesterone. The purified enzyme requires NADPH. This observation excludes the possibility that the hydroxysteroid dehydrogenases act as transhydrogenases, as had been proposed. The activity of the 20α-hydroxysteroid dehydrogenases is modified by the animal's hormonal balance. The presence of progesterone inhibits and a higher level of estrogen or follicle-stimulating hormones increases the activity of the 20α-hydroxysteroid dehydrogenase.

A number of hydroxylases are capable of inactivating active steroids. These hydroxylases can be classified by the site of action on the steroid ring, and one can distinguish among 6β-, 7β-, 16β-, and 18β-hydroxylases. The 6β-hydroxylases act on progesterone, cortisol, estrone, estradiol, and estriol. The activity of these enzymes explains the presence of these hydroxylated compounds in the urine. 16α-Hydroxylases are suspected of acting on a number of substrates. Many of the products of these reactions have been found in normal urine. Some of the products are increased under pathological conditions; *e.g.*, 3α,16α-dihydroxy-5β-pregnan-20-none is found in the urine of patients with the adrenogenital syndrome. The existence of an adrenal cortex carcinoma is associated with the excretion in the urine of 3β,16α-dihydroxy-5-androstene-17-one, 3β,7α,16α-trihydroxy-5-androsten-17-one and 3β,16α-dihydroxy-5-androstene-7,17-dione. The urine of pregnant women contains 18-dihydroxyestrone, suggesting that an 18-hydroxylase exists.

Metabolic Effects of Corticoid Hormones

Effect on Protein Metabolism. Corticoid hormones affect various steps of protein metabolism: amino acid penetration in the cells, intracellular biosynthesis of amino acids from small precursors, protein synthesis, and protein catabolism. In discussing the effect of corticoid hormones on protein synthesis, it is necessary to distinguish between the effects of the glucocorticoid on muscle and liver. The injection of C11-oxygenated corticosteroid increases the excretion of urinary nitrogen, with loss of tissue nitrogen (*e.g.*, in heart and kidney) [51].

The negative nitrogen balance probably results from a dual effect of the hormonal acceleration of protein catabolism and a decrease in the rate of protein synthesis (at least in the peripheral tissue).

Studies of the incorporation of labeled amino acid and labeled amino acid precursors (Krebs cycle intermediate or bicarbonate) in diaphragm have dissociated the effect of cortisone on amino acid penetration from its effect on amino acid incorporation and protein synthesis. The results demonstrated that adrenalectomy is without effect on the amino acid uptake by isolated diaphragm. In contrast, corticosteroids inhibit amino acid uptake by the isolated diaphragm. Adrenalectomy enhances and cortisone depresses the incorporation of ^{14}C-labeled amino acid, glucose, bicarbonate, and dicarboxylic acid into protein of the diaphragm. It has also been found that corticosteroids decrease amino acid incorporation into proteins *in vitro* in kidney, skin, small intestine, and epididymal tissue preparations.

Although cortisone interferes with protein synthesis in peripheral tissue,* the hormone stimulates protein synthesis in liver. The increased rate of protein synthesis results from an effect of the hormone on the penetration of the amino acid through the cell membrane. The effects of the steroid hormones on the amino acid penetration were demonstrated with the aid of α-aminoisobutyric acid, a nonmetabolizable amino acid the intracellular penetration of which is regulated by mechanisms similar to those controlling glycine penetration. The penetration of α-aminoisobutyric acid is competitively inhibited by valine. Two hours after the injection of cortisol, the aminoisobutyric acid content of the liver is increased 70%. The increase in amino acid uptake may explain the increase of protein synthesis in liver observed early after cortisone injection. In that connection, it is of interest that an accelerated turnover of serum albumin has been reported early after the injection of cortisone. Since the concentrations in the levels of the blood protein are not changed, it seems that increased synthesis only compensates for increased catabolism.

Studies on cell-free systems of liver are still in a preliminary stage and have yielded results difficult to reconcile with the notion that cortisone stimulates protein synthesis. These studies suggest that adrenalectomy increases and large doses of cortisone decrease amino acid incorporation into proteins. Studies in which the Zamecnik-Keller systems were used suggest that the hormonal imbalance affects the microsomes rather than the supernatant fluid of the liver.

Effect on Gluconeogenesis. Cortisone accelerates gluconeogenesis. The increased amounts of glucose that are made available explain the hyperglycemia, the increase in blood levels of lactic acid, the modified glucose tolerance curve, and the accumulation of glycogen in the liver that are observed after corticoid hormone injection. The greater sensitivity to insulin, the amelioration of diabetes, the fasting hypoglycemia, and the depletion of muscle glycogen observed after adrenalectomy or in Addison's disease are also compatible with an effect of corticosteroids on gluconeogenesis [52–54].

At least four different mechanisms may contribute to explaining the stimulation of gluconeogenesis by cortisone: (1) gluconeogenesis is secondary to the effect of cortisone on protein or amino acid metabolism; (2) cortisone interferes with glucose usage; (3) cortisone interferes with glycogen synthesis; and (4) corticosteroid directly stimulates gluconeogenesis.

Cortisone could increase the free amino acid pool and thereby make the amino acids available for gluconeogenesis through reversing the glycolytic pathway. The effects of cortisone on transaminase and glutamic dehydrogenase are consistent with such a theory. Adrenalectomy reduces the level of transaminase activity, and the enzyme activity is restored to normal by the injection of cortisone or hydrocorticosterone. Corticosterone and hydrocorticosterone slightly stimulate the activity of purified glutamic dehydrogenase. The enhancement of the activity of both these enzymes could increase the amount of amino acids that are converted into Krebs cycle substrates, yet the physiological significance of the effect of cortisone on these two enzymes remains to be established.

That the effect of corticosteroids on gluconeogenesis results from an indirect effect on protein metabolism is unlikely. Sometimes the effect of cortisone on protein and carbohydrate metabolism is not quantitatively related. When corticosteroid hormones are administered to fed animals, considerable changes in glycogen deposition are induced, but there are no changes in the rate of protein catabolism.

When corticosteroids are administered to animals starved for 48–72 hours, a paradoxical effect is observed. In spite of an increase in carbohydrate storage in the liver, muscle, and body fluid, the new glucose made available by the hormone is not used for oxidation. This observation suggests that corticoid hormones depress glucose utilization. Such findings are difficult to reconcile with the effects of cortisone and adrenalectomy on phosphorylase, phosphoglucomutase, and phosphohexose isomerase activity. The activities of these three enzymes are depressed by adrenalectomy, but are normalized by cortisone administration.

Cortisone could interfere with glucose usage by reducing hexokinase activity, and glucosteroid could interfere with glucose phosphorylation by blocking ATP synthesis. Cortisone inhibits oxygen uptake by rat mitochondria, and indirect evidence suggests that cortisone interferes with cytochrome oxidase activity. No effect of cortisone on the succinate dehydrogenase or the NADH cytochrome c reductase systems has been observed. If an effect of cortisone on ATP synthesis could be substantiated, such an effect could readily explain the interference of cortisone with fatty acid and protein synthesis. In keeping with this possibility is the observation that progesterone and deoxycorticosterone, like dinitrophenol, swell mitochondria, activate latent ATPase, and uncouple oxidative phosphorylation. But the physiological significance of the effect of cortisone on ATP synthesis remains to be established.

Cortisol injections have been reported to increase glucose-6-phosphatase and fructose-1,6-diphosphatase activity. When ethionine is administered in addition to cortisone, phosphatase activity is not increased. This finding suggests that the increase in enzyme activity results from *de novo* synthesis of an adaptive enzyme in response to an increase of substrate.

Ashmore has proposed that the primary effect of corticoid hormones is on glycogen synthesis from glucose-6-phosphate, rather than on glucose usage

* Among the unsolved problems raised by the effect of cortisone is its site of action. It is not clear whether or not cortisone acts on peripheral tissue *in vivo*. Corticoid hormones have little or no effect on glucose usage or protein metabolism in eviscerated animals.

through the Embden-Meyerhof pathway. This assumption is based on the analysis of the specific activities of glycolytic and Krebs cycle intermediates derived from labeled glucose injected in normal and cortisol-treated animals.

Recent developments in the study of the mechanism controlling gluconeogenesis (see Chapter 16) have inspired those investigators interested in the mechanism of action of cortocosteroids. Studies of the effect of corticosteroids on gluconeogenesis have been performed in liver and kidney of adrenalectomized animals. The results obtained after the injection of corticosteroid are compared to those obtained in simply adrenalectomized animals. In these studies, the incorporation of various types of potential precursors into glucose is followed. But these studies have yielded conflicting results, and therefore definite interpretation of the effect of corticosteroid on the enzymes involved in gluconeogenesis is not available.

For example, gluconeogenesis from lactate, pyruvate, fructose, and alanine in perfused livers from starved adrenalectomized rats was compared with gluconeogenesis in nonadrenalectomized rats. In two laboratories in which gluconeogenesis from pyruvate and lactate was investigated, opposite results were obtained. One laboratory claimed that adrenalectomy decreased gluconeogenesis, while the other could not demonstrate such an effect.

The ability of kidney slices of adrenalectomized animals to generate glucose from pyruvate, succinate, and glutamate is lower than the glucogenic activity of kidney obtained from nonadrenalectomized animals. Corticosteroid injection to adrenalectomized animals before the kidney slices are collected restores glucogenic activities to normal.

Although a conclusion on the site of action of glucocorticosteroids in gluconeogenesis must remain tentative, data available indicate that the effect of the hormone is localized between pyruvate and triose phosphate. Therefore, it is relevant that the administration of corticosteroids *in vivo* increased the activity of phosphoenolpyruvic and possibly of pyruvic carboxylase.

The effects of glucocorticosteroids on the carboxykinase is entirely abolished by actinomycin D, yet actinomycin D only partially abolishes gluconeogenesis. Consequently, factors other than *de novo* synthesis of gluconeogenic enzymes seem to be involved in regulating gluconeogenesis. A group of German investigators believe that the increased gluconeogenesis that follows glucocorticosteroid administration results in part from activation of pyruvate carboxylase by increased levels of acetyl-CoA and from increased levels of glucose-6-phosphate. Although the mechanism of this increase is not clear (inhibition of phosphofructokinase by citrate has been eliminated), this finding probably reflects a kind of negative control of gluconeogenesis through interference with glycolysis or with glycogen synthesis.

The same group of investigators also established that alanine is an inhibitor of pyruvic kinase; therefore, they proposed that a second type of negative control could be the result of the effect of corticosteroids on the amino acid mobilization.

Effect on Lipid Metabolism. Our knowledge of the effect of cortisone on lipid metabolism is still fragmentary. Interpretation of the results is complicated by the fact that the effect of the hormone seems to vary depending upon the source of the adipose tissue. Adrenalectomy stimulates and corticoid injections decrease lipogenesis in the adipose tissue of the mesentery. The decreased lipogenesis induced by corticosteroids is accompanied by release of free fatty acids. When corticosterone and hydrocortisone are added to epididymal adipose tissue incubated *in vitro,* the hormones fail to stimulate lipogenesis from [^{14}C]pyruvate, but they accelerate fatty acid release, and the lipolytic effect is completetly blocked by actinomycin D. Consequently, one effect of glucocorticoids on some of the adipose tissues seems to be to accentuate lipid catabolism.

Clinical examination of patients injected with corticoid hormones or afflicted with corticosteroid hypersecretion reveals that the effect of corticosteroids on lipid metabolism cannot be simple. Excess corticosteroids are responsible for abnormal lipid distribution in some parts of the adipose tissue (cheek, subclavicular, and thoracic pads), which explains the development of the moon-shaped face and the buffalo hump so typical of hypercorticosteroidism.

Effect on Lymphopoietic Tissue. The changes in the hematopoietic tissue are the most striking morphological changes induced by glucocorticoid administration. The changes include destruction of the lymphoid tissue under acute administration of the hormone and stimulation of hematopoietic activity when the hormone's activity is consistently increased, as in Cushing's disease.

Modern descriptions of the effects of corticoid hormones on the lymphoid system can be related to old observations made by pathologists, particularly by Dustin. Dustin observed that toxin or trauma induced the appearance of numerous pyknoses in thymus, lymph nodes, and Peyer's patches of the intestinal mucosa. Later it became clear that these effects were observed not only after exposure to toxins or trauma, but following all forms of stress. Seleye demonstrated that the pyknoses were part of a general reaction of the hypophyseal adrenal system to nonspecific stimuli, including exposure to cold, extensive physical fatigue, and other factors.

A role of the adrenal in this reaction was suspected because it was found that stress is associated with depletion of the lipoid material of the adrenal cortex. Furthermore, such changes were not observed if preliminary hypophysectomy had been performed. Similar injuries could, however, be induced by ACTH or glucocorticoid hormone injection; yet, the administration of ACTH to adrenalectomized animals has no effect on the lymphoid tissue.

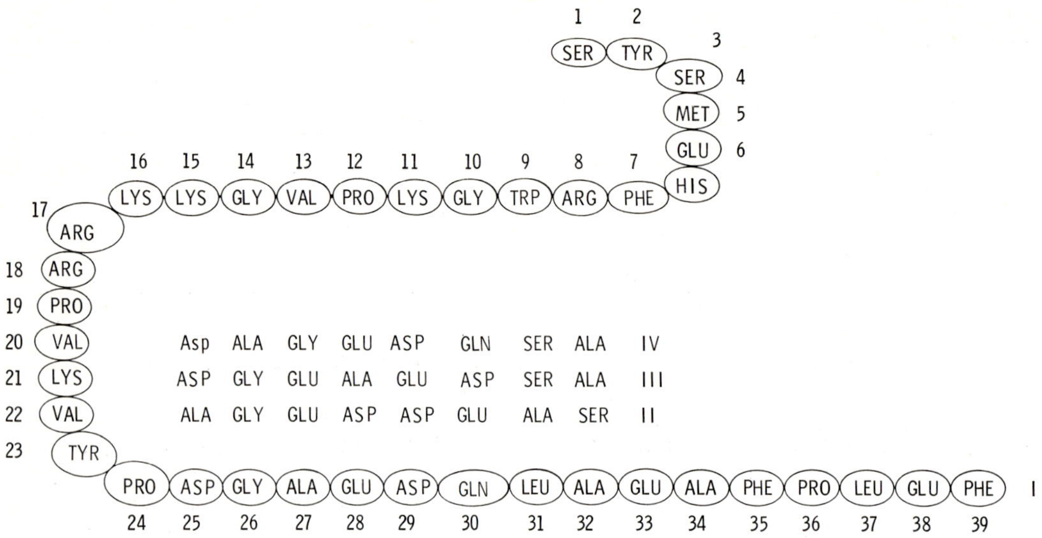

Fig. 8-16. A 13-amino acid sequence from N-terminal is common to α-MSH and ACTH, with the exception that the NH_2-terminal serine is acetylated in α-MSH. *I*: porcine ACTH; *II*: sheep ACTH; *III*: ox ACTH; *IV*: human ACTH

the molecule play any biological role at all? Comparison of the amino acid sequence and activity might be interesting in that respect. We have seen that there are only minor differences between the amino acid sequences of pig, sheep, and human hormone. They involve the C-terminal peptide in an area of the molecule between residues 25 and 33, a fraction of the molecule not essential to the mere manifestation of activity. Nevertheless, there is considerable difference in the activities of the quantitative aspects of the preparations. For example, the activity of the human hormone is only 25% of that of pig hormone,* and about 15% of that of sheep hormone. The reasons for these differences are not clear. The amino acid sequence of the C-terminal chain, although not essential to activity, may be important in protecting the rest of the molecule or in inactivating part of the molecule.**

Two corticotropins have been described, corticotropin A and corticotropin B. At first they were thought to differ by their amino acid sequence, namely, with respect to residues 25–28. Thus, while the sequence (starting with 25) was asparagine, glycine, alanine, and glutamine in corticotropin B; it was assumed to be glycine, alanine, glutamine, and asparagine in corticotropin A. It now seems that the sequence described for corticotropin B is the only correct one.

Yet two different hormonal preparations can be obtained from pig hypophyseal extract (A_1 and A_2); they are identical with respect to their amino acid composition, but they differ in their chromatographical properties. The A_2 polypetide has a free carboxyl on the glutamic acid of residue 30, whereas A_1 has a glutamine residue in that position. Naturally, such a difference would cause A_2 to migrate more rapidly toward the anode on starch gel electrophoresis. The relationship between corticotropin A_1 and A_2 and corticotropin A and B has now been clarified. Corticotropin A is a mixture of A_1 and A_2 but contains primarily A_1, and corticotropin B is a mixture of A_1 and A_2 but is primarily composed of A_2.

A discussion of the amino acid sequence of corticotropin should be accompanied by a review of the molecular structure of the melanocyte-stimulating hormones, for there are indeed many analogies between the structures of the melanocyte-stimulating hormones and the corticotropins. Many hormonal factors may directly or indirectly influence skin pigmentation. However, three observations made in frogs and tadpoles are pertinent to our modern understanding of the control of pigmentation. In frogs, hypophysectomy and injection of pineal gland extract lighten the skin color, and injection of hypophyseal extract increases pigmentation. The agent responsible for these effects has been prepared from both the hypophysis and the pineal gland.

The melanocyte-stimulating hormones are relatively small polypeptide chains found in the neurohypophysis. The two groups (α and β) of melanocyte-stimulating hormones differ essentially by the length of the polypeptide chain. MSH has been prepared from pig, beef, horse, monkey, and human pituitary. Fractionation of the extract produces the three active polypeptides: α-MSH, β-MSH, and corticotropin. Both α- and β-MSH are found in all animals. There is no species variation in the amino acid sequence of the α-MSH. In contrast, a small species variation exists in the structure of β-MSH as indicated in Table 8-6. All known MSH's including corticotropin have a common denominator: a heptapeptide composed of methionine,

* The amino acid sequence of the corticotropins extracted from pig has been somewhat confusing.

** It is interesting to note that 5 of the 8 amino acids involved in the polypeptide fraction between residues 5 and 33 are acidic.

Table 8-6

α-MSH Pig, Beef, Horse, Monkey	Ac–Ser–Tyr–Ser–Met–Glu–His–Phe–Arg–Try–Gly–Lys–Pro–Val–NH_2
β-MSH	
Pig	Asp–Glu–Gly–Pro–Tyr–Lys–Met–Glu–His–Phe–Arg–Try–Gly–Ser–Pro–Pro–Lys–Asp
Beef	Asp–Ser–Gly–Pro–Tyr–Lys–Met–Glu–His–Phe–Arg–Try–Gly–Ser–Pro–Pro–Lys–Asp
Horse	Asp–Glu–Gly–Pro–Tyr–Lys–Met–Glu–His–Phe–Arg–Try–Gly–Ser–Pro–Arg–Lys–Asp
Monkey	Asp–Glu–Gly–Pro–Tyr–Arg–Met–Glu–His–Phe–Arg–Try–Gly–Ser–Pro–Pro–Lys–Asp
Human	Ala–Glu–Lys–Lys–Asp–Glu–Gly–Pro–Tyr–Arg–Met–Glu–His–Phe–Arg–Try–Gly–Ser–Pro–Pro–Lys–Asp

glutamic acid, histidine, phenylalanine, arginine, threonine, and glycine, α-MSH has been extracted from pig, beef, horse, and monkey, but not from humans. It is composed of 3 amino acids, 3 preceding and 3 following the heptapeptide. The sequence is identical to that of the 13 N-terminus polypeptide or corticotropin with two main additions: acetylation of the N-terminus of serine and the formation of the carboxamide on the carboxyl group of the terminal valine.

Species differences in the β-MSH involve the number or sequence of the amino acids of the polypeptidic chain that precedes the heptapeptide. The chain that follows the heptapeptide is the same in all species. The sequence of the carboxy-terminal polypeptide chain is serine, proline, proline-lysine, and aspartic, except in horse where one of the proline residues is replaced by an arginine residue. The human β-MSH is the largest active polypeptide; an octapeptide precedes the heptapeptide and forms the N-terminal polypeptide chain. The composition of the octapeptide is alanine, glutamine, lysine, asparagine, glutamine, glycine, proline, tyrosine, and arginine. Monkey β-MSH is identical to human β-MSH except for the absence of the first 3 amino acids of the N-terminus octapeptide. Thus, a pentapeptide—asparagine, glutamine, glycine, proline, tyrosine, and arginine—precedes the central heptapeptide. Pig, beef, and horse MSH are identical to monkey MSH except that lysine is substituted for the arginine residue.

ACTH is secreted in the anterior lobe of the hypophysis and, although it is not clear which cells secrete ACTH, the basophils are considered the source of ACTH because the cytoplasm of these cells is degranulated in stress, and basophilic tumors of the hypophysis develop in Cushing's disease. These arguments are not accepted by all; it has been suggested that the hyperplasia of the cells of the anterior hypophysis that develops after adrenalectomy involves either chromophobe or acidophilic cells. The treatment of a Cushing's adenoma preparation with fluorescent anti-ACTH serum yielded fluorescence in the chromophobic cells. In adrenalectomized rats injected with tritiated glycine, the isotopes accumulate in the chromophobic cells; therefore, it has been concluded that these cells synthesize ACTH.

The administration of agents which block corticosteroid synthesis and stimulate ACTH secretion leads to the hyperplasia of a cell type resembling the chromophobe cells. The hyperplastic cell contains small acidophilic granules.

Control of ACTH Secretion. While the neurohypophysis is technically part of the brain, the anterior lobe of the hypophysis is disconnected from the central nervous system except for the existence of a common circulatory system, the portal system of the hypophysis. These differences between the anatomical structures of the anterior and posterior lobes are at the origin of the differences between the mechanisms controlling ACTH secretion. The secretion of ACTH is probably not triggered by any neurosecretory mechanism, like that triggering vasopressin secretion.

In fact, ACTH secretion is regulated by a neurohormonal mechanism and a humoral feedback mechanism. The neurohormonal control of ACTH secretion involves what has been referred to as a third-order neuroendocrine mechanism. Three steps are involved in this control. A stimulus (stress, for example) brings the central nervous system to secrete a hormone (CRF), which acts on another endocrine structure in the anterior lobe of the hypophysis where it induces the secretion of second hormone (ACTH), which ultimately acts on the target endocrine gland—the adrenal cortex [40, 56–60].

The mechanism controlling ACTH secretion cannot be understood without first reviewing the anatomy of the connections between the hypothalamus and hypophysis.

Two major structures form the walls of the third ventricle: the thalamus and the hypothalamus. The hypothalamus is separated from the thalamus by a deep groove called the hypothalamic sulcus. The hypothalamic region is an anatomical region delineated by the mammary body rostrally, the chiasma caudally, and the divergent optical tract laterally. The hypothalamus can be subdivided into preoptic, supraoptic, tuberal, and mammary regions.

Several types of nuclei have been described in the various thalamic regions (see Fig. 8-17). The hypothalamic nuclei are coordinating centers for neurological circuits that control eating, sleeping, mating, body temperature, and endocrine regulation. It is not always known which specific neurological structure controls these activities, but the supraoptic and the paraventricular nuclei are known to control diuresis.

Whereas the hypothalamus receives its blood supply from the circle of Willis, the hypophysis is vascularized

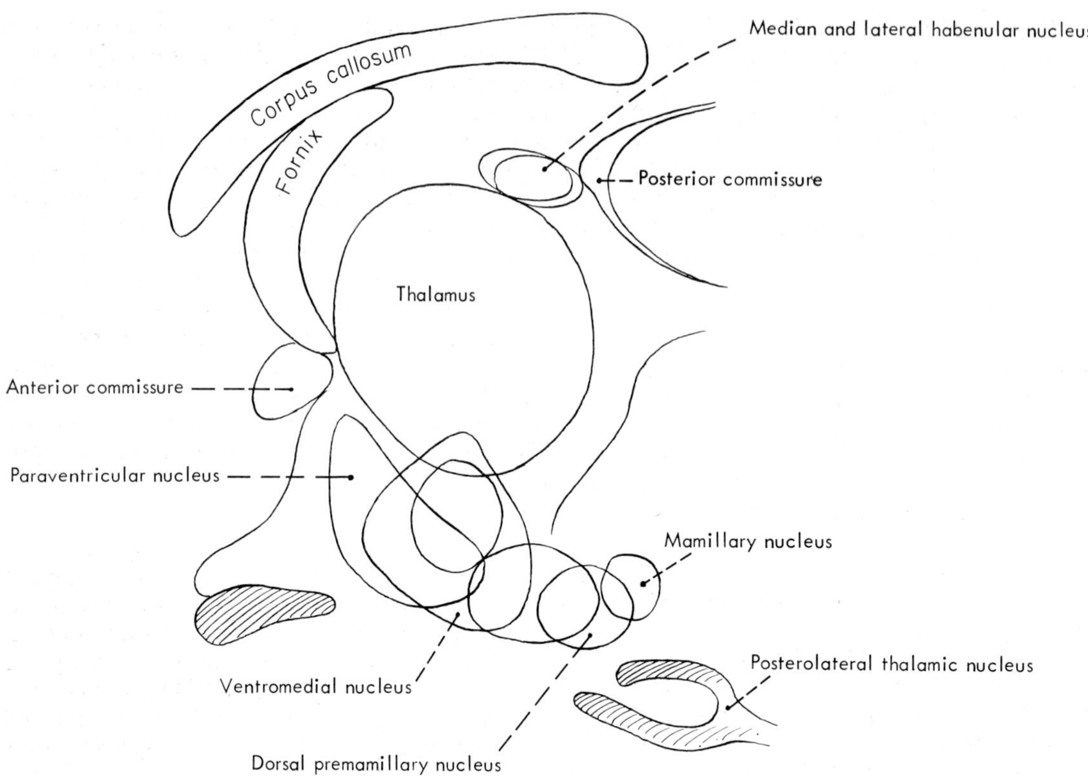

Fig. 8-17. Anatomy of hypothalamic nuclei

by a completely independent arteriolar system. The arteriolar blood supply is derived from two superior and two inferior hypophyseal arteries. The superior hypophyseal artery originates from and courses in a direction perpendicular to the stalk. At some distance from the stalk of the hypophysis, the superior hypophyseal artery bifurcates to yield a horizontal branch that extends toward the stalk and a long descending branch that parallels the stalk to the anterior lobe of the hypophysis. The descending branch crosses the anterior lobe to yield terminal branches. The terminal branches anastomose with the terminal branches of the inferior hypophyseal artery, which is itself a branch of the intracavernous portion of the internal carotoid. The system of hypophyseal arteries terminates into two separate capillary beds. The primary plexus is derivied chiefly from the superior hypophyseal artery. The blood of the primary capillary bed drains down through the long and short channels of the hypophyseal portal system into a secondary capillary bed, which forms the sinusoid of the adenohypophysis.

From this description of the hypophyseal vasculature, it is obvious that the anterior lobe receives no direct arterial blood but receives all its blood through the hypophyseal venous system.

The portal vessels of the hypophysis play a considerable role in the physiology of the hypophysis. They constitute an important link between the hypothalamus and the anterior lobe of the hypophysis. When the hypophysis is disconnected from its stalk and its normal vascular system and is transplanted at a distance from its original anatomical location, little glandular function can be detected; but if the gland is grafted in the median eminence or in an area where the portal system can regenerate and revascularize the graft, then the endocrine functions return as vascular regeneration progresses.

Direct evidence for the role of the portal system in ACTH secretion has been obtained by collecting blood from these vessels and demonstrating the presence of ACTH-releasing activity in the plasma.

The nervous fibers of the tubular hypophyseal tract connect the hypothalamus with the adenohypophysis. The anatomical details of this connection are still debated. Fibers have been described that originate in the tubular infundibular nucleus, course into the infundibulum and the upper part of the infundibular stem, and terminate at the wall of the capillary loops of the primary portal plexus.

It is postulated that the releasing hormones are formed in the hypothalamic nucleus to be carried through the nervous fibers via the capillaries of the portal space, the blood of which transports them to the adenohypophysis.

Neurohumoral releasing factors stimulate the anterior pituitary cell to secrete specific tropins (LSH, FSH, TSH, and ACTH). The evidence for the hypothalamic control of ACTH release by the adenohypophysis was obtained by electrostimulation of hypothalamic nuclei, destruction of the median eminence, sectioning of the stalk, and transplantation of the pituitary at a distance from its normal anatomical location.

When hypophyses were transplanted in the kidneys of hypophysectomized animals, they were found to contain ACTH but were unable to release it. However, when the hypophysis is transplanted in the median eminence, the levels of plasma gonadotropins are maintained close to normal. Excision of most of the brain, except for the hypothalamus, has led to the conclusion that an intact median eminence and pituitary stalk complex is indispensable for normal activation of the adrenal by the pituitary.

ACTH Release. A battery of techniques is available for testing ACTH release; for example, measurements of ascorbic acid released from the adrenal, determination of adrenal weights, and estimation of lymphopenia and eosinophilia. The pitfalls of these methods have been reviewed in detail in specialized articles, but it seems safe to conclude that the release of the stimulins of the anterior pituitary is under hypothalamic control. The hypothalamus functions as a synaptic region between the anterior pituitary and those parts of the central nervous system that respond to stress. It is not known whether the median eminence secretes a hormone that stimulates the anterior lobe of the hypophysis to secrete ACTH. In fact, little is known of the molecular structure of the hormone that is elaborated in the central nervous system and carried to the anterior lobe of the hypophysis. Vasopressin was first thought to be responsible for stimulating ACTH secretion. This theory has now been abandoned, and it is now proposed that a specific polypeptide hormone, the corticotropin-releasing factor (CRF), is responsible for ACTH secretion. This releasing factor is thought to be elaborated in the neurons of the median eminence and then transferred by a neurosecretory mechanism along nervous fibers that connect the median eminence to the capillaries of the primary plexus of the portal vessel. But the possibility that the corticotropin-releasing factor is transferred to the posterior lobe by a neurosecretory mechanism and then from the posterior to the anterior lobe by anastomosis between posterior and anterior lobe arteries has not been excluded.

By applying the now classical techniques (ion-exchange chromatography, countercurrent distribution, zone electrophoresis) for protein fractionation to the separation of ACTH-releasing factors, two releasing factors have been obtained: α-CFR and β-CFR.

α-CRF is a mixture of two different polypeptides, α_1-CRF and α_2-CRF. α_2-CRF has the same amino acid sequence as α-MSH except for the N-terminal serine. The structure of α_1-CRF is still unknown.

β-CRF has been separated from α-MSH, and a partial amino acid sequence has been reported: acetyl-Ser-Tyr-Cys-Phe-His-(Asn, Gln)-Cys-(Pro, Val)-Lys-Gly-NH_2.

There seems to be no doubt that the stimulation of the hypothalamic nuclei induces the biosynthesis of releasing factor because secretion of ACTH by the adenohypophysis is blocked by the administration of puromycin and actinomycin. Almost nothing is known of the release of CRF itself. Implantation of crystalline atropine in the anterior hypothalamus markedly inhibits the response of the adrenal cortex to stress. On the basis of this information, hypothalamic control of pituitary corticotropin release has been proposed to in some way involve acetylcholine. Histochemical observations made on the hypothalamus are at least compatible with this conclusion. Enzymatic histochemistry established that acetylcholinesterase and monoamine oxidase are present in the hypothalamus. Moreover, the presence of norepinephrine and other biologically active monoamines has been demonstrated in the hypothalamus by fluorescence methods. All these findings suggest that cholinergic and monoaminogernic transmission may take place in the hypothalamus.

Effect of the Hypophysis on Adrenal Weight. In addition to a neurohormonal mechanism controlling the secretion of steroid hormones by the adrenals, it has also been proposed that the pituitary possesses an inherent ability to stimulate steroid secretion by the adrenal cortex. The ability of the pituitary to maintain adrenal weight and function has been investigated either by section of the pituitary stalk or by heterotopic transplantation of the endocrine gland in hypophysectomized animals. Section of the pituitary stalk in the monkey with insertion of plates between the cut ends and the stalk to prevent portal revascularization has yielded controversial results. Although some investigators found that this procedure interferes with steroidogenesis, others failed to observe cortical involution. Studies with heterotopic transplants have not yielded conclusive information on the control of steroidogenesis by ACTH. Pituitaries implanted under the renal capsule of hypophysectomized rats failed to maintain adrenal weight and function, whereas pituitaries implantated in the anterior chamber of the eye of hypophysectomized rats restored adrenal weight and function.

There seems to be no question that ACTH increases protein synthesis and turnover of rapidly labeled RNA in the nucleus. Separation of the new proteins formed on polyacrylamide gel indicates that the increase in protein synthesis induced by ACTH is selective. The molecular mechanism by which the rate of protein synthesis is increased is believed to involve an increase in amino acyl transferase and an enhancement of polysome activities.

A single injection of ACTH stimulates adrenal growth, but a second injection has no such effect. In view of these results, it has been proposed that ACTH has a dual effect on adrenal growth, a stimulatory effect that assures a growth increase, and a longer lasting inhibitory effect on DNA, RNA, and protein synthesis [61].

Humoral Feedback Mechanism. The most compelling evidence in favor of a humoral feedback mechanism comes from an observation made in adrenalectomized animals. Although ACTH levels in the blood of normal animals are almost impossible to determine, the ACTH levels of the blood are easily measured in adrenalectomized rats because the hormone is secreted in large amounts after adrenalectomy. Glucocorticoid injection interferes with ACTH secretion. It is not known whether the feedback mechanism involves the hypothalamic nuclei. The exogenous administration and endogenous liberation of ACTH increase corticosteroid production in the adrenal cortex and the steroid level in the adrenal blood flow.

Metabolic Effects of ACTH. Numerous metabolic changes in response to ACTH have been described in the adrenal cortex. They include an increase in basic and acid phosphatase activities, an increase in the CoA content of the cortex, increased RNA content and turnover, and increased rates of protein synthesis.

An important contribution to determining the mode of action of ACTH is the observation that adenosine-3′,5′-monophosphate stimulates corticogenesis. ACTH was also found to induce adenosine-3′,5′-monophosphate accumulation within the adrenal cortex. On the basis of these findings, Haynes and Berthet assumed that ACTH stimulates the formation of adenosine-3′,5′-monophosphate cyclic adenylate, which activates phosphorylase. Phosphorylase hydrolyzes glycogen to yield glucose-1-phosphate. Glucose-1-phosphate is converted to glucose-6-phosphate by phosphoglucomutase. In the adrenal, the glucose-6-phosphate is used preferentially for oxidation through the hexose monophosphate shunt. Oxidation of the carbon 1 of glucose generates NADPH, a coenzyme necessary for steroid synthesis. Although the role of cyclic AMP in steroidogenesis is well established the remainder of the hypothesis has found no supportive evidence.

Another hypothesis describing the mode of action of ACTH has recently gained some support. It proposes that ACTH modifies cellular permeability through the intermediate of adenosine-3′,5′-monophosphate. The increased penetration of xylose and α-aminoisobutyric acid under the influence of ACTH has been established. It also has been proposed that ACTH stimulates the formation of cyclic adenylate, which in turn induces the synthesis of new mitochondrial proteins. The appearance of the new protein is associated with the development of activity of the enzyme involved in the conversion of cholesterol to pregnenolone.

We have seen already that the conversion of cholesterol to pregnenolone involves 20α-hydroxy- and 20α, 22-dihydroxycholesterol as intermediates, and that the side chain cleavage of the dihydroxy compound completes the reaction. One or both hydroxylations are believed to be rate-limiting reactions for steroidogenesis that are under the influence of ACTH. ACTH is believed to stimulate the reaction through adenosine cyclic phosphate. It is not known how the cyclic AMP acts. Although stimulation of NADPH and effects on mitochondrial permeability cannot be excluded, it is remarkable that the adenosine-3′,5′-cyclic phosphate was found to stimulate directly steroidogenesis that had been in sonically disrupted adrenal mitochondria [62, 63].

Elevated concentrations of 3′,5′-cyclic AMP appear within 1 minute after ACTH is added to adrenal slices. The increases in the cyclic nucleotide precede steroidogenesis and persist throughout the period of accelerated steroid secretion. Increased steroidogenesis can also be produced by adding 3′,5′-cyclic AMP without ACTH. For these reasons, 3′,5′-cyclic AMP is believed to be the intracellular mediator of adrenocorticotropic hormone action on the adrenal. In this stimulation of 5′-nucleotides by ACTH, the following sequence of events seems to take place: an equilibrium is established between the hormone levels in the medium and the amount of hormone bound to binding sites on the cell surface (receptor). After binding of ACTH, adenyl cyclase is activated. The enzyme converts ATP to 3′,5′-cyclic AMP, but the activity of the enzyme degrading cyclic AMP is not changed [64].

The steroidogenic effects of ACTH can be completely suppressed by puromycin and cycloheximide administration, suggesting that new protein synthesis is needed to stimulate steroidogenesis. Nevertheless, studies with actinomycin suggest that the templates involved in dictating the proteins needed for acute steroidogenesis are stable.

In addition to stimulating corticosteroidogenesis in the adrenals, ACTH also stimulates lipogenesis in adipose tissue and the release of ascorbic acid from the adrenals.

The reasons for the presence of ascorbic acid in the adrenals are not clear, but the adrenals have long been known to contain ascorbic acid; in fact, the vitamin was first extracted from the adrenal.

Under the influence of ACTH, free fatty acids and glycerol concentrations increase in adipose tissue. ACTH stimulates lipolysis in adipose tissue by activating a hormone-sensitive lipase. However, ACTH does not act directly on the lipase. There are at least two other intermediate messengers: adenylate cyclase and cyclic adenylate. The molecular mechanism of action of ACTH on lipolysis will be discussed in more detail in the section devoted to adipose tissue metabolism.

The physiological significance of the adipolytic effect of ACTH is not clear. Although the lipolytic effect of the hormone is readily demonstrated *in vitro,* an effect *in vivo* proved difficult to substantiate unless large doses of the hormone were administered. Investigators have postulated that the effect *in vivo* may exist at physiological doses but that it is small and of short duration. Furthermore, the lytic effect of physiological doses of ACTH is soon masked by the insulinlike effect that follows the hormone's administration.

Adrenalectomy stimulates urea formation. The mechanism by which this effect occurs is not known, but it may result from an effect of ACTH on amino acid transport. ACTH is transported in the blood attached to a protein, most likely a euglobulin. It is not known how ACTH is inactivated. Proteases attacking ACTH have been found in blood, but there is evidence that ACTH is destroyed in the adrenal gland itself. Studies with labeled ACTH have been inconclusive because they were not able to distinguish between inactivation resulting from binding and that resulting from protein hydrolysis.

Among the corticosteroids affecting electrolyte metabolism, aldosterone is by far the most active. The maintenance of adequate sodium stores is essential to normal homeostasis. Sodium homeostatis is secured by reabsorption of sodium from the glomerular filtrate through the tubules. The main function of aldosterone is to stimulate this reabsorption in the distal tubules. Thus, sodium excretion is strictly proportionate to glomerular filtration of sodium in adrenalectomized animals or in patients with Addison's disease. The increased sodium excretion is probably responsible for the increased potassium retention resulting from potassium-sodium ion exchange in the distal tubules. Without aldosterone, sodium is lost in the urine and increased amounts of sodium are also found in saliva, sweat, and feces. The mechanism controlling aldosterone secretion and the mode of action of the mineralocorticoids will be discussed in more detail in the chapter devoted to electrolyte metabolism.

In conclusion, stress or other stimuli release CRF from the hypothalamus. CRF probably binds to receptors in special hypophyseal cells and causes the release of ACTH which binds to the membrane of the cells of the inner portion of the adrenal stimulating adenyl cyclase activity and raising the intracellular concentrations of cyclic AMP with resultant corticosteroidogenesis.

Cushing's Syndrome

Hypersecretion of adrenal glucocorticosteroids leads to a syndrome described by Cushing in 1932. Hyperadrenalism may be of hypophyseal or adrenal origin. Hypersecretion of hypophyseal origin results from the elaboration of excessive amounts of ACTH. When of adrenal origin, hypersecretion may or may not be associated with pathological changes of the adrenal. Three types of adrenal changes have been described in association with Cushing's syndrome: adenocarcinoma, adenomas, and adrenal hyperplasia (see Fig. 8-18). In 70% of the patients with Cushing's disease, no tumor is found; only a widening of the fasciculаris and the reticularis is observed [65–70].

The disease affects both sexes, but the incidence is five times greater in men than in women. Cushing's syndrome may develop at all ages, but it is most frequently observed between 30 and 40 years of age. At autopsy, the few pathological changes observed are basophilic adenoma or hyalinization of the basophilic cells of the hypophysis, selective obesity, osteoporosis, and cardiomegaly. The basophilic adenoma of the hypophysis is usually so small that it seldom induces changes in the sella. Basophilic adenoma was first thought to be the cause of Cushing's disease, but it is now clear that Cushing's disease is associated with basophilic adenoma (15–45%) in only a small proportion of cases. The polygonal basophilic cells are large and well delineated, but irregular in size. They may or may not be affected by characteristic hyalin changes that are most typically observed in Cushing's disease. In 1935, Crooke first described hyalinzation of the basophils in Cushing's disease. The hyalin material at first accumulates at the periphery of the nucleus

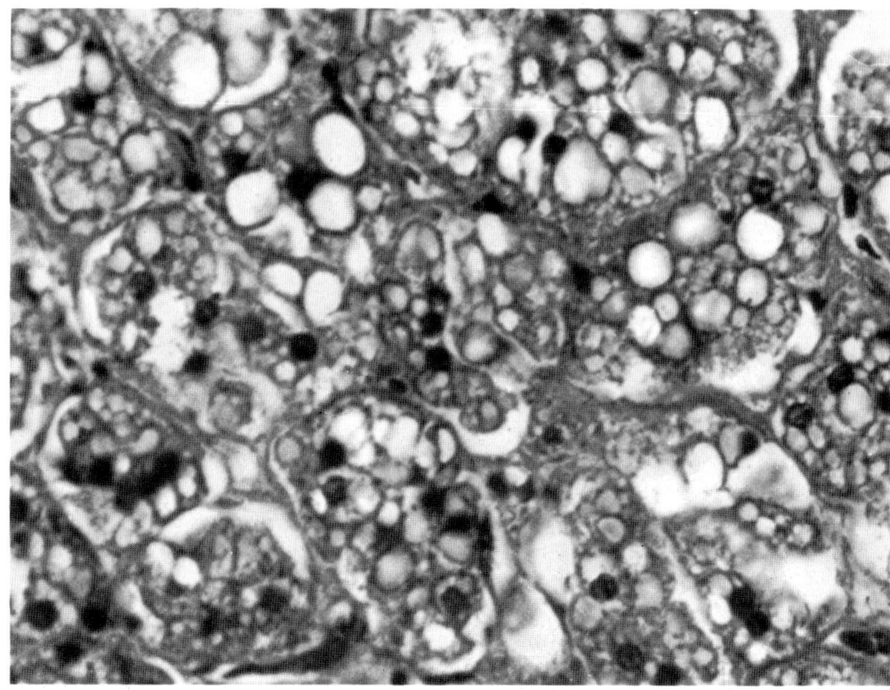

Fig. 8-18. Hormones showing adrenal hyperplasia with aldosteronism

and spreads toward the periphery of the cytoplasm. The significance of hyalinization of the basophil is not clear, but it is linked to the secretion of excessive amounts of ACTH because it also occurs after therapeutic administration of ACTH.

The accumulation of fat is highly selective in Cushing's disease; the obesity spares the extremities and involves instead the subclavicular region, where it forms what is called the buffalo hump. The facial changes are typical in patients affected with Cushing's syndrome (see Fig. 8-19). Wrinkles disappear, and the face becomes round. The lips form what has been described as a fish mouth, and the plump appearance of the face has brought clinicians to describe these patients as moon-faced.

Patients with Cushing's disease develop marked osteoporosis which affects the skull, the vertebrae, and other bones. The spotty loss of calcium in the flat bones of the skull leads to the appearance of the moth-eaten skull and the softening of the vertebrae may be so marked that the intervertebral disc burrows into the body of the vertebra. The pathogenesis of osteoporesis in Cushing's disease is far from elucidated. Osteoporosis can result from decreased osteoblastic activity, increased osteoclastic activity, a negative calcium balance, or the elaboration of an abnormal matrix.

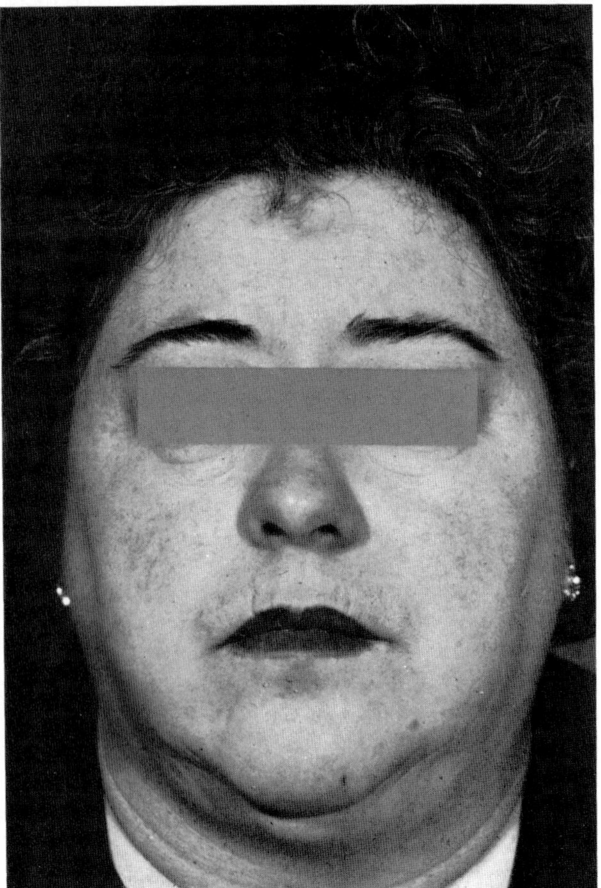

Fig. 8-19. Cushing's syndrome. (From J. Brown)

Histological evidence suggests that the excess glucocorticoid osteoporosis results more from interference with osteoblastic than from an increase of osteoclastic activity.

The negative nitrogen balance in hyperglucocorticosteroid activity indicates that the osteoporosis may result from increased breakdown of the matrix, but no conclusive evidence of this has been obtained. Reduced calcium absorption has been reported in animals and in man after injection of glucocorticoid. Whether osteoporosis results from antivitamin D action is not clear. The parathyroid appears normal; calciuria occurs and calcium is deposited in the renal medulla. However, hypercalciuria is produced only by administration of large doses of glucocorticosteroid or is observed in severe Cushing's disease.

The kidneys of patients with Cushing's disease may show broadening of the capillary wall of the glomeruli and nephrocalcinosis.

The fact that the injection of ACTH or deoxycorticosterone can induce hypertension in experimental animals suggests that the hypertension observed in patients with Cushing's disease may well be of adrenocortical origin. The effect of deoxycorticosterone on blood pressure is usually considered to be only "permissive," which means that it is not the only factor involved in determining blood pressure changes. Thus, deoxycorticosterone administration to adrenalectomized rats will only partially restore blood pressure to normal. The blood pressure of rats made hypertensive experimentally drops after adrenalectomy. The administration of deoxycorticosterone to animals induces nephrosclerosis and increased sodium retention through enhanced tubular reabsorption. Both these effects are likely to contribute to the development of hypertension in Cushing's disease. The role of anephrotensin, a substance that accumulates in rats injected with deoxycorticosterone, has not been investigated in humans.

The hypertension in Cushing's disease first leads to myocardial hypertrophy; later, congestive heart failure develops with dilation of the heart chamber. Cardiomegaly develops as a result of hypertrophy and dilation of both chambers of the heart.

The blood of patients with Cushing's disease shows changes typical of hypercorticosteroidism and diabetes. The level of 17-hydroxycorticosteroids is increased in the urine. This is not pathognomonic of Cushing's disease, and such an increase must be differentiated from that attributable to stress or liver insufficiency. On the other hand, the levels of blood 17-corticosteroids are usually not markedly increased in Cushing's disease, and they barely exceed the normal values.

Cortisol production varies with age, sex, and disease. Daily cortisol production is slightly lower in women than in men. The levels of cortisol production decrease with age somewhat faster in men than in women. Cortisol production is lower than normal in Addison's disease, cirrhosis, hypopituitarism, and hypothyroidism. It is elevated in hyperthyroidism and obesity (50%), but the largest increases are observed in Cushing's disease, in which the levels of cortisol production may be 250% of normal.

Patients with Cushing's disease have moderate, insulin-resistant diabetes with hyperglycemia and glucosuria. The diabetes of patients with Cushing's syndrome is seldom accompanied by the development of ketosis.

The skin of the abdomen is usually thin, and the rupture of the subcutaneous tissue leads after scarring to the formation of purple striae, which resemble those observed in pregnancy or after rapidly increasing obesity.

It is sometimes possible to differentiate Cushing's syndrome resulting from adrenal tumor and that due to adrenal hyperplasia. ACTH injection in patients with Cushing's syndrome resulting from hyperplasia increases cortisone levels, but ACTH is without effect in patients with Cushing's syndrome resulting from adrenal tumors.

The panoply of symptoms described above have all been described in cases of Cushing's disease. However, the full panoply is not found in all cases. Correct diagnosis requires judicious analysis of the symptoms. Thus, among patients with Cushing's disease, obesity is the most common symptom and is present in 88% of the patients; however, only 38% have the typical abnormal fat distribution (truncal obesity) described above. Hypertension is found in 74% of the cases.

In conclusion, only obesity and hypertension are consistent findings. Among the clinical findings, the most reliable seems to be that of free cortisol excreted in the urine. The results of the administration of dexamethasone (which suppresses urinary 17-ketogenetic steroid excretion) may be suggestive, but not always conclusive. Measurements of 17-ketosteroid excretion or of 11-hydroxycorticosteroid excretion are of little or limited value.

Whereas the pathologist can weave, from the experiences plucked during many years in different countries, a large canvas describing all the events that occur in Cushing's disease, the diagnostician is somewhat like the archaeologist faced with a few broken pieces. On the basis of these scarce elements, he must quickly make a diagnosis and choose a course of therapy. There is thus a big gap between the amount of information that one ought to collect on the basis of his knowledge of pathogenesis and that which would be useful in making a diagnosis. However, this does not imply that a thorough knowledge of pathogenesis is not indispensable for correct diagnosis. The more facts that have been established, the greater the chance of finding a combination fitting with a pattern typical for a given disease. Computers may help in studying both pathogenesis and diagnosis: in pathogenesis by building models, in diagnosis by identifying patterns. The responsibility of the new generation of physicians is to make medicine a mathematical science as accurate as possible and to be aware of all its limitations.

Sex Hormones

Histogenesis of the Germinal Gland

The germinal gland develops in the coelomic epithelium by proliferation of the germinal epithelium of Waldeyer. This thickening of the coelomic epithelium occurs in the middle of the body of Wolf. Primary gonocytes, large clear cells with round nuclei, appear. The primary gonocytes are associated with smaller cells of mesothelial origin. Both the gonocytes and the mesothelial cells proliferate actively producing a bud called the sexual bud, which penetrates deep into the mesenchymal tissue. In the male, these sexual cords produce the seminiferous tubules. At a later stage of embryonic life, interstitial cells develop in between the seminiferous tubules. These interstitial cells exist only in the primordium of the male genital glands and are absent from the female genital glands. Their presence in histological preparations can be used to distinguish between a male and a female fetus.

In females there are two successive proliferations of the sexual cords. The first proliferation produces buds resembling those that develop in males and yield rudimentary seminiferous tubules but no interstitial cells. This first attempt to produce a germinal gland is abortive, and the rudiments become the medullary cords. The second proliferation of the germinal epithelium leads to the formation of the cord of Pflüger. Inside these tubules are gonocytes that do not proliferate actively but increase in size, and even in the embryo they resemble ovocytes.

Hence, during fetal development, the female gonads contain all the germinal cells that will be available during life. During the development of the female genital gland, there is sometimes a third proliferation of small tubules referred to as epithelial invagination. Usually these structures do not contain gonocytes. They may be the starting point of some tumoral processes.

The adult ovary has a superficial epithelium resting on a thick layer of connective tissue that contains follicles at various stages of development. This cortical zone of the ovary surrounds a fibrous medullary part that contains no follicles but contains arterioles, venules, lymphatics, and the cellular cords which appeared during the first proliferation of the germinal epithelium.

The follicle undergoes various stages of development. The primary follicle contains an ovocyte in the center surrounded by a single layer of epithelial cells and a thin fibrous layer. Later, the epithelial cells proliferate actively, generating several layers surrounding the ocyte. This new follicular epithelium is referred to as the granulosa. The granulosa, which at first forms a compact multilayer envelope surrounding the oocyte, ultimately develops cavities containing poorly stainable fluid. At this stage, the follicle is surrounded by a thick layer of connective tissue called the theca follicularis. The cavity increases in size and fluid accumulates. In women, the mature follicle is 1–1.5 cm in diameter and appears like a translucent vesicle bulging at the surface of the ovary.

The central cavity of the follicle is called the antrum, and the liquid it contains is referred to as liquor follicularis. The mesenchymal cells of the theca are voluminous and are loaded with lipid granules. The theca is delimited by a thin membrane, the membrana propria, which is composed of follicular cells. During the maturation of the follicle, the follicular cells proliferate into a budding mass bulging inside of the lumen of the follicle and surrounding the oocyte. During ovulation, the follicular membrane that surrounds the oocyte ruptures and releases the liquor follicularis, dragging the oocyte with it.

Three or four hours after the follicular rupture, the follicle is replaced by the corpus luteum, a small structure measuring 1–2 cm in diameter. The corpus luteum is yellowish because the cells contain large amounts of lipid, and it often contains a central clot due to hemorrhage. The corpus luteum disappears after 2 weeks in the absence of pregnancy, but it remains well developed throughout pregnancy.

Details of follicular formation and ovulation remain unsettled. There is some indication that the liquor is formed by filtration of plasma rather than as a result of venous or lymphatic obstruction. Increased hydrostatic pressure in the follicle is unlikely to cause its rupture. Rupture seems to result from focal alteration of the site of the stigma. Electron microscopic examination of the stimaga reveals collagen digestion and cell separation.

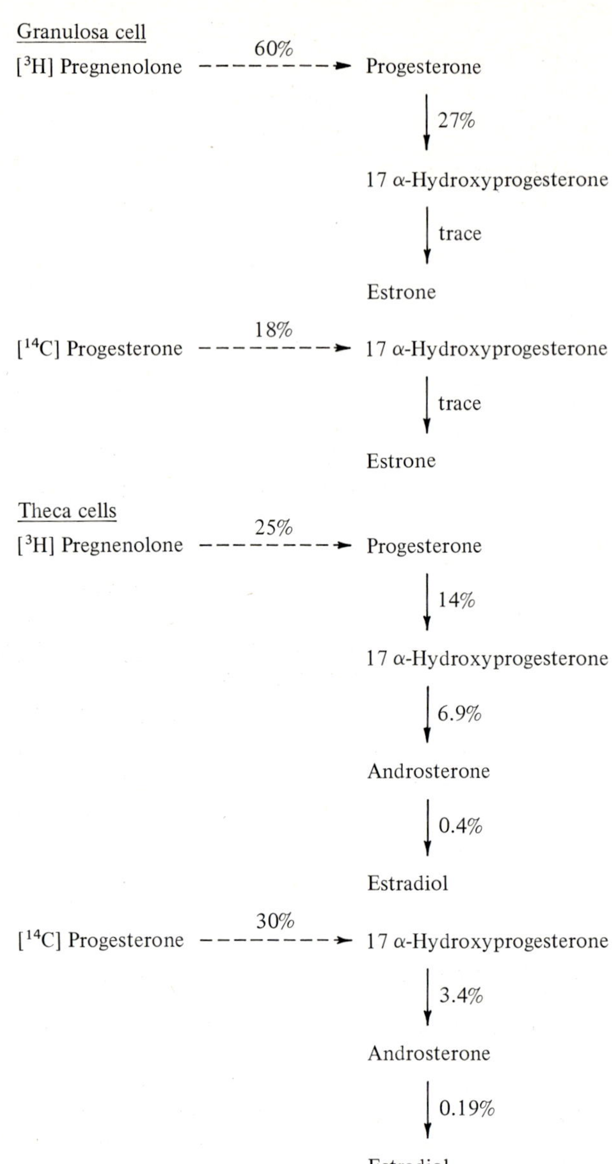

Fig. 8-20. Distribution of steroid enzymes in various tissues

Estrogen and Progesterone in Ovaries

A look at the metabolic map reveals that there are two major pathways for estrogen synthesis, one involves progesterone as an intermediate, and the other bypasses the Δ^4 isomerization. The relative activities of the Δ^4 pathway (progesterone) and of the Δ^5 pathway (pregnenolone) can be compared by studying the fate of labeled progesterone with that of labeled pregnenolone [71].

In a typical experiment, a female is injected with FSH to stimulate the formation of granulosa cells, then ovarian slices or homogenates are incubated with [^{3}H]pregnenolone and [^{14}C]progesterone-4. After incubation, various metabolites are isolated. Similar experiments have been performed using theca cells (see Fig. 8-20). These data clearly indicate that the granulosa cells convert pregnenolone to progesterone using the Δ^4 pathway. In contrast, the theca cells convert little pregnenolone to progesterone, and the major pathway in the theca cells is the Δ^5 pathway. Thus, there is no isomerization of pregnenolone to progesterone, but estrogens are synthesized by direct aromatization of dehydroepiandrosterone. Therefore, two cell types are side by side in the same organ, and each cell type differs quantitatively by the activity of some key enzyme. As a result, one pathway predominates in one type of cell, while another pathway predominates in the other type. This is an interesting observation of differentiation at the molecular level. These observations are summarized in Fig. 8-21.

The factors that determine the expression of one pathway over another are not known. Activation or inactivation of enzymes may result from derepression or repression of the genome or modification of the rate of enzyme synthesis by regulation of transcription or by regulation of positive or negative feedback mechanisms.

The exact role of hormones in these metabolic shifts is not clear. However, animal experiments clearly demonstrate that follicles induced by FSH are unable to convert pregnenolone to progesterone, whereas the conversion takes place clearly if LH is also administered.

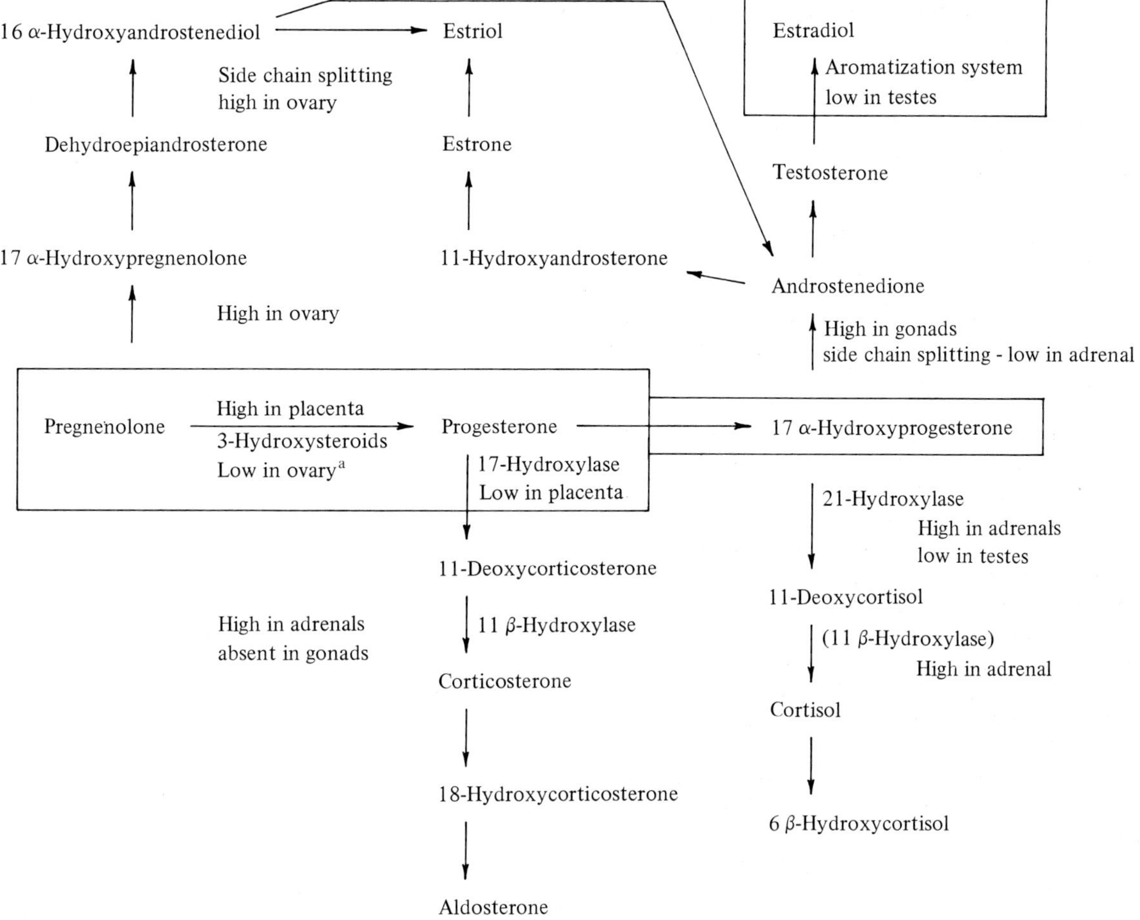

Fig. 8-21. Distribution of steroid enzymes in various tissues. [a] But increases where corpus luteum is formed

In the ovary, at least one instance of feedback inhibition has been demonstrated. As in the adrenal, the 17α-hydroxylation of progesterone is catalyzed in the ovary by a $NADPH_2$-dependent microsomal enzyme. There is evidence that progesterone accumulation inhibits the 17α-hydroxylation.

Neither ovaries made pseudopregnant by injection of pregnant mare serum or human chorionatrophic gonadotropin nor the ovaries of hypophysectomized mice elaborate estrogen unless LH is administered.

Menstrual Cycle and Effects of Female Sex Hormones

The alternate ovarian secretion of estrogen and progesterone determines the development of the menstrual cycle (see Fig. 8-22). The frequency and duration of the cycle vary depending upon the animal. The cycle occurs only once a year in the deer and the bat, twice in the dog and the lamb. The duration of the cycle is 4–5 days in the mouse, 2 weeks in the guinea pig, 3 weeks in the cow, 21 days in the *Macaca rhesus* and 28 days in women.

The cycle starts with the secretion of FSH by the hypophysis. The hormone stimulates the development of one of the many immature follicles in the cortex of the ovary into a mature graafian follicle. Studies done in the mare show that the follicular fluids contain most of the metabolic intermediates between cholesterol and estradiol-17β. The mucosa of the uterus is markedly modified under the influence of this steroid. The cells at the bottom of the glandular structures in the mucosa of the body of the uterus divide rapidly and cover the surface of the uterine cavity with a proliferating glandular epithelium. Mitosis is frequent in the glandular epithelium. The connective tissue that separates elongated glands with narrow lumina also proliferates, and the number of capillaries in the stroma increases.

Between the 10th and the 16th day after the onset of menstruation, ovulation is assumed to occur. The follicle ruptures as described above, and the thickened endometrium elaborates glycogen and secretes a thick mucoid material. If fecundation does not occur, the arteries that supplied the endometrium with blood shrink, ischemia develops, and the endometrium involutes. The involution is followed by opening of the constricted arteries and menstrual hemorrhage.

Even before the vaginal discharge is completed, the torn glands of the endometrium regenerate a new surface epithelium, which covers the myometrium.

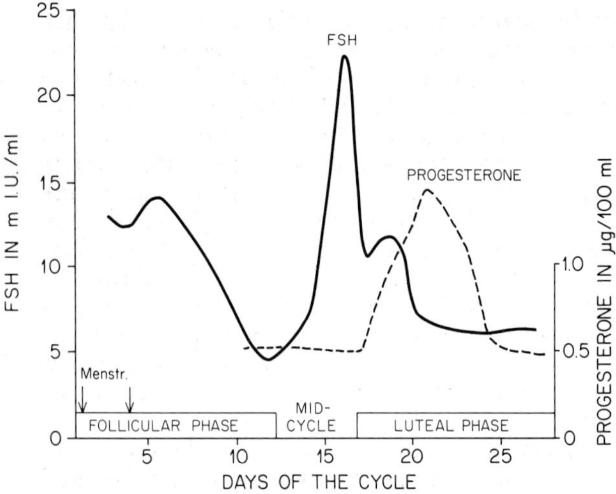

Fig. 8-22. Levels of FSH and progesterone in plasma during menstrual cycle

After fertilization, the human ovum, which is already segmented, burrows in the endometrium, accompanied by the outer cells of the blastocyst, which participates in forming the placenta; the uterine mucosa at that time presents an edematous stroma with glands rich in glycogen, lipoid material, and mucoid secretion.

The predominant steroid in the fluid of the hemorrhagic follicle of the mare is progesterone. After studying the histological changes in the ovary of the mare and making a biochemical study of hormones in the follicular liquor, R.V. Short has developed a theory that accounts for steroid synthesis in the ovary. Short refers to his theory as the two-cell-type theory. The theory proposes that the cells of the theca interna synthesize estrogens, while progesterone is synthesized by the granulosa cells. In the beginning, the cells of the theca interna contain all the enzymes necessary for 17β-estradiol synthesis from cholesterol. In contrast, the granulosa cells contain little 17-hydroxylase and 17-desmolase, and therefore these cells can synthesize only progesterone. Early in the cycle, the cells of the theca interna in the intact follicle are richly vascularized and synthesize large amounts of estrogens. These estrogens are liberated in the follicular liquor, and from there pass into the general circulation. In contrast, the granulosa cells are poorly vascularized, and their products are not secreted actively. When the follicle ruptures, the capillaries invade the granulosa layer, and progesterone secretion is activated. Hypophyseal secretion is also believed to control the histological changes in the ovary during the cycle.

Electron microscopists have described subcellular changes in the granulosa cells as they become luteinized: the rod-type of mitochondria change into spherical-type mitochondria by developing tubular, agranular endoplasmic reticulum. Although it seems logical to associate these changes with modifications of the bioenergentic pathways and with steroidogenesis, the real meaning of the cellular alterations remains obscure.

In addition to affecting the endometrial mucosa, estrogens induce changes in the mucosa of the vagina and of the cervical glands. The vaginal epithelium proliferates during the follicular phase, and glycogen accumulates in the cells of the superficial layer. Estrogens also stimulate the cervical glands to elaborate mucus. The growth of the breast at puberty is likely to be influenced by both estrogens and progesterone. Estrogens probably stimulate the proliferation of ductile and stromal tissue, while progesterone influences acinar tissue growth. During pregnancy, the combined action of progesterone and estrogens, in addition to stimulating growth and maintaining the decidua, also influences the vascularization of the vulva, the vagina, the cervix, and the myometrium.

Hormones in Pregnancy

The complex shift in hormonal balance occurring during pregnancy will not be discussed here except to point out that an important role of the placenta is to secrete a variety of hormones. It has been suggested that the placenta secretes chorionic gonadotropin, corticotropin, somatotropin, luteotropin, vasopressin, relaxin, estrogen, androgens, and progesterone.

Although the elaboration of chorionic gonadotropin, estrogens, and progesterone seems well established, it is not known whether the placenta secretes other hormones. The mechanism controlling placental secretion has not been established, but placental secretions are adjusted by an autoregulatory mechanism and fetal circulation.

The reasons for massive hormonal secretion during pregnancy are not known. It has been proposed that estrogens stimulate rapid uterine growth, that chorionic gonadotropins maintain the luteinic functions, and that progesterone inhibits premature uterine contractions. One of the difficulties encountered in interpreting the role of the endocrine secretions in the placenta results from considerable species differences, and it seems necessary to interpret data in humans only on the basis of investigations of the hormonal secretion of human or primate placentas.

Effects of Male Sex Hormones

The secretions of the sex glands have intrigued mankind since the beginning of civilization, but it was a 72-year-old man, Brown Sequard, who demonstrated the therapeutic value of gonadal extract. Brown Sequard injected himself with testicular extract and claimed to have recovered physical vigor and intellectual alertness.

The effect of the male hormone is too well known to be described here in great detail. The observations that Pezard made in 1911 illustrate the effect of the male hormone most dramatically. This biologist castrated roosters, and soon these fierce animals lost their aggressive character and became unable to awaken the sleeping countryside with their resounding chant. The crest, a badge of masculinity, disappeared, and the king of the poultry yard became a modest capon. Yet the injection of male hormones soon restored the flamboyant combs and the martial tunes.

In man, androgens stimulate the growth of the external genitalia, the prostate, the seminal glands; and the hair of the face, pubis, axilla, and chest. In addition, male sex hormone affects the retention of nitrogen, potassium, phosphorus, sodium, and chloride. The male sex hormone is secreted in the testicle by clusters of large polygonal cells, called the Leydig cells, located between the seminiferous tubules. It is not established whether these cells originate from the germinal epithelium or from mesenchymal cells. Interstitial cells are abundant in the human fetus, but they regress and reappear only at puberty. Thus, a fetal interstitial system must be distinguished from that of an adult. (But the testes of the impuberal animal should not be considered a static organ; a study on beef testicles shows that many qualitative and quantitative changes occur between birth and puberty.) The cytoplasm of the Leydig cells contains lipofuscin granules and refractile crystalloids called the crystalloids of Reinke. Testosterone is also made in the adrenals, probably in the x-zone of the cortex.

A number of steroids possess androgenic activites; they have a keto or hydroxyl group on the carbon 17. Testosterone is the prototype of androgenic hormones. Numerous attempts have been made to alter the molecular structure of testosterone for the purpose of investigating the relationship between structure and activity. For further information, the reader is referred to specialized textbooks.

Mechanisms of Action of Steroid Hormones

The sex hormones appear to have one property in common—they stimulate the proliferation of the primary or secondary sex organs. Naturally, investigators have been intrigued by the mechanism involved in stimulating growth by these rather simple molecules [49, 51, 72–79].

The effect of steroid hormones on protein molecules *in vitro* has been investigated extensively. The results can be summarized as follows: Steroid hormones can modify enzyme activities acting directly as coenzymes or by combining with a part of the molecule distant from the active center, and thereby modifying the configuration of the molecule. The binding of the steroid hormones with noncatalytic proteins may also change the molecular configuration and alter cellular permeability or modify the cell's normal metabolic compartmentalization. Investigator have proposed that steroid hormones act as coenzymes in the transhydrogenase reaction. This property will be discussed in more detail when we review the effect of steroid hormones on transhydrogenases.

The effect of steroids on the molecular structure of glutamic dehydrogenase is intriguing. Glutamic dehydrogenase has a molecular weight of 1,000,000 when the protein is concentrated, but upon dilution, the molecule breaks down into four subunits, each with a molecular weight of 250,000. The glutamic dehydrogenase molecule can be broken down even at high concentration of the enzyme by adding steroids (estrogenic, androgenic, and progestational) to the medium. ADP prevents the steroid-induced disaggragation, and kinetic studies indicate that the activity of the enzyme can be correlated with the ability of the steroid to promote reversible dissociation. Glutamic dehydrogenase dehydrogenates alanine in addition to acting on glutamic acid. In the presence of steroids, although the glutamic dehydrogenase activity is reduced, the alanine dehydrogenase activity is increased.

Some steroids inhibit purified aldehyde dehydrogenase without splitting the enzyme in subunits, and it has been proposed that the effect of steroids on aldehyde dehydrogenase is produced by inducing minor changes in the enzyme molecule's configuration. Similar effects have been observed with pyruvic kinase, which is inhibited by diethylstilbestrol and estrogenic steroids.

That steroids can modify the protein configuration is suggested by the fact that when testosterone is bound to serum albumin, it induces an increase in the helical content of the protein.

The effects of steroids on protein molecules cannot now be correlated with their physiological activity. However, steroid inhibition of glucose-6-phosphate dehydrogenase is more directly relevant to the physiological role of the hormone. It has been proposed that inhibition of the rate of NADP reduction by the glucose-6-phosphate dehydrogenase of the adrenal may play a role in regulating hormone production in these tissues. We have already mentioned that the ultimate effect of most sex steroids is to stimulate cellular proliferation. This can be achieved by modifying cellular permeability, stimulating bioenergetic pathways, and directly stimulating biosynthetic pathways.

When injected to intact or ovariectomized rats or rabbits, estrogens stimulate the uptake of α-aminoisobutyric acid, and the free amino acid levels in the uterus are increased. Similarly, the addition of diethylstilbestrol or estradiol disulfate to an Ehrlich ascites cell *in vitro* stimulates the amino acid uptake by the cells. These findings are interpreted by postulating that the hormones affect amino acid transport. The hormone could, however, stimulate the penetration of the amino acid by affecting the usage of amino acids through metabolic pathways rather than by directly affecting the transport mechanism.

A new avenue of investigation was opened when it was observed that 17β-estradiol stimulates NAD reduction in the presence of human placenta homogenates and isocitrate. This was a puzzling observation because most of the isocitrate dehydrogenase found in a soluble fraction requires NADP as a coenzyme. Williams and Ashman demonstrated that the stimulation of NAD reduction resulted from an effect of the steroid on a transhydrogenase, which would catalyze the conversion of NADPH to NADH. The transhydrogenase was thought to be identical to the 17β-hydroxysteroid dehydrogenase. Some investigators claim to have separated the transhydrogenase from the steroid dehydrogenase by partial purification. It is not clear how the effect of the steroid on the transhydrogenase reaction is transcribed into a metabolic effect.

Steroids could of course stimulate NADPH formation and accelerate electron transport and ATP formation. The primary effect of steroids on cellular proliferation would thus result from an increase in the amount of chemical energy available to the cell. NADPH could not itself be used for generating energy because the oxidation of NADPH via cytochrome is not coupled with oxidative phosphorylation. Since these original observations on the effect of steroids on transhydrogenase, the stimulation of other transhydrogenation systems has been reported: androstene-3,17-dione stimulates purified 3α-hydroxysteroid dehydrogenase. Androgens stimulate the transhydrogenases in rat prostate, and an estradiol-sensitive transhydrogenase has been found in cancerous and noncancerous rats' breasts.

If steroids acted primarily on transhydrogenase, the effect of the hormone would manifest itself only on the transhydrogenase of target organs and would not affect transhydrogenase of organs that are not under the control of steroid hormones. Steroids do not activate transhydrogenase obtained from various mitochondrial sources or purified transhydrogenase obtained from *Pseudonomas fluorescens*. In contrast, an estrogen-sensitive transhydrogenase has been obtained from rat liver mitochondria.

In addition to steroids, pharmacological compounds such as serotonin and epinephrine activate transhydrogenases. A particulate pyrimidine nucleotide transhydrogenase was obtained from anterior beef pituitary, which is stimulated by epinephrine, estradiol, and serotonin.

It was suggested above that steroid hormones act by stimulating bioenergetic pathways, and that as a result of this stimulation, the biosynthesis of phospholipids, proteins, and nucleic acids is activated. After estrogen administration there is a rapid uptake of precursors: ^{32}P in phospholipids, glycine or other amino acids into proteins, and serine and formate into the purine and pyrimidines of the nucleic acid.

Nevertheless, the sequence of events in the uteri of estrogen-stimulated animals argues against a primary effect of the hormone on the bioenergetic pathway because the changes in the increase in the rate of biosynthesis—especially that of proteins—usually precede the changes in the rate of respiration. An alternative hypothesis suggests that the primary effect of the steroid hormone is on protein biosynthesis.

Estrogen stimulates the penetration of amino acid into the cell. This could result from a modification in the cell permeability or from an acceleration of amino acid usage. McCorquodale and Mueller demonstrated that the administration of estradiol to ovariectomized rats stimulates amino acid activation in the uterus homogenate [80]. But not all enzymes involved in amino acid activation were affected; only seven of them were stimulated. The findings of Mueller and his associates were confirmed in another laboratory using ventral rat prostate. In contrast, no effect of estrogens on amino acid activation could be observed in liver.

Arguing that the results obtained on homogenized tissues and on purified enzymes might provide a distorted view of the role of steroids on cellular metabolism, Chayen et al. have made a study of some steroid effects using what they call the method of cellular biochemistry. The details of the methodology will not be reviewed. The analyses are made on thin tissue slices (20 μ) that are sectioned with a cryostat after cooling the tissue at −70° C in *n*-hexane for one minute. Apparently cooling under such conditions does not lead to the formation of water crystals that disrupt intracellular structures. When the slices are incubated in aqueous solution they may lose up to 70% of their protein nitrogen. This can be avoided by adding polyvinyl alcohol, a polypeptide derived from degraded collagen, or Ficoll. Methods have been developed to measure the formation of NADH and NADPH, the permeability of the lysosomal membranes, and electron transport in mitochondria. For details see [183]. By adding an intermediate hydrogen carrier, phenazine methosulfate, one can measure the amount of NADPH generated in either the glucose-6-PO_4 dehydrogenase or in the phosphogluconate reaction.

$$\text{Glucose 6 } PO_4 + \xrightarrow[\text{NADP dehydrogenase}]{\text{Glucose 6 } PO_4} \text{6-phosphogluconate} + \text{NADH} + H^+$$

$$\text{Neotetrazolium} + \xrightarrow[\text{NADPH tissue bound diaphorase}]{-2H} \text{NADP} + \text{colored formazan}$$

The use of neotetrazolium salts in absence of phenazine methosulfate provides a quantitative measurement of NADPH produced. The stain can either be eluted or measured on sections by microdensitometry.

In the experiments the total NADH produced by a given enzyme is measured in presence of phenazine methosulfate; and the hydrogen transferred from NADPH to neotetrazolium is measured in absence of phenazine methosulfate. Two different types of NADPH hydrogen can be distinguished—a type I produced in presence of neotetrazolium in absence of phenazine methosulfate, and a type II which is the difference between the total NADH produced in presence of phenazine methosulfate and type I.

Although not absolutely compelling, available evidence suggests that type I hydrogen is channeled through the microsomal electron transport chain including the p_{450} cytochrome for hydroxylation and that type II is channeled into biosynthetic pathways. When the effects of steroids were determined on these *in vitro* slices, it was observed that the hormone inhibits NADH production by glucose-6-phosphate dehydrogenase, but is without effect on that produced by 6-phosphogluconate dehydrogenase; and that steroids fall in five classes with respect to their effect on type I and type II hydrogen. (1) Neither type is affected. (2) The total level of NADPH produced is unchanged, but the type II is increased at the expense of type I. (3) Both type I and type II are inhibited. (4) Only type II is inhibited. (5) Only type I is inhibited.

A study between structural relationship and the effect on type I and type II hydrogen pathways suggested that: (1) Pregnenolone; dehydroepiandrosterone; 5β-pregnane-3α,20α-diol, 5α-pregnane-3β, 20α-diol; and 3α-actiocholanolone, 17α-hydroxypregnenedione inhibit type I pathway and, therefore, their effect falls in class 3 or 5. All these steroids have an α or a β hydroxy group on a normal steroid A ring. (2) Cortisone, corticosterone, hydrocortisone and prednisolone, compounds with an α or β hydroxy on C_{11}, inhibit type II pathway and therefore fall into classes 3 and 4. (3) 11-Deoxy-17-hydroxycorticosterone, testosterone, progesterone stimulate pathway 2 at the expense of pathway 1 (class 2). All these steroids have a C_{4-5} double bond, a 3-α keto group, but no substituent at C_{11}. As we have seen the steroids with the latter type of substituent depress type II.

Although it may be too early to feel absolutely secure about the functional structural correlations, these studies establish that steroids, depending upon their structure, can guide the NADPH hydrogen generated by the hexose monophosphate shunt into either the hydroxylation pathway through the cytochrome p_{450} electron transport chain or into biosynthetic pathways.

Because structural relationships are maintained in the type of slice used in these experiments, it is possible to show interaction between various cell compartments that cannot be demonstrated convincingly in homogenates. For example, it is possible to show a synergistic effect between oxidations in the hexose monophosphate shunt and oxidation in mitochondria (glutamate or succinate dehydrogenation). Progesterone and pregnenolone, but not cortisone, inhibit the increased succinate dehydrogenase activity observed in this preparation when the hexose monophosphate shunt is stimulated. The effect of steroid hormones on lysosomal permeability will be discussed in the chapter on inflammation.

A theory proposing that the effect of steroids is to stimulate amino acid activation and, as a result, protein synthesis is not without objections. If the primary effect of steroid hormones is to activate amino acids, the rate of synthesis of all proteins should be increased in the target organ. In contrast, steroids stimulate the biosynthesis of highly specific proteins, and therefore it seems more logical to assume that the steroids act at a step of the biosynthetic pathway that determines the specificity of the proteins synthesized. Such a mechanism could involve transcription of messenger RNA on DNA template. The rate of incorporation of labeled nucleotide triphosphate into the rapidly labeled nuclear RNA by the methods described by Weiss is markedly decreased in castrated animals. These changes are accompanied by changes in the activity of the RNA polymerase.

Mueller and his associates [79] have observed that estrogen injection stimulates RNA synthesis in the uterus of castrated rats and increases the RNA polymerase activity. The stimulation of the RNA polymerase activity precedes the acceleration in protein synthesis induced by estrogen and is blocked by actinomycin D. These experiments of Mueller and his associates are at the origin of the "hormone gene" theory, which proposes that the hormone acts on the genome. In a first step, the hormone stimulates the biosynthesis of a specific messenger RNA, which in turn stimulates the formation of specific proteins. The new proteins are needed for the synthesis of new breeds of messenger RNA. The new messenger RNA dictates the biosynthesis of a number of new proteins that amplify the effect of the hormone.

Hamilton [77] expanded Mueller's model thanks to the development of adequate techniques for separating nuclear from cytoplasmic uterine fractions. The sequence of events in the uteri of ovariectomized rats after the administration of estradiol-17β can be summarized as follows: The injected hormones quickly pass the cellular and nuclear membranes, and within minutes after injection a small but consistent fraction of the hormone binds to the nucleus, where it is believed to stimulate the biosynthesis of nuclear RNA. It is not clear which type of RNA is stimulated first, but Hamilton believes that it is chromosomal and ribosomal RNA. In any event, new messenger must be formed at some time in the sequence because the population of polyribosomes increases in the cytoplasm.

The hormone gene theory has been extended to other steroid hormones (ecdysone, aldosterone, corticosterone). The effect of ecdysone, a steroid hormone secreted by the thoracic glands of insects, is of particular interest. The mating and metamorphosis of the insect are partly regulated by the ecdysone, the structure of which is still unknown. Within 15–30 minutes after injection of the hormone, the giant chromosomes of the salivary glands of these insects develop typical "puffs" (site of DNA-dependent RNA synthesis). The site of the puffing is specific, and the degree of puffing is proportional to the amount of hormone injected.

In spite of its appeal, the hormone gene theory leaves many questions unanswered. First, it remains to be established that the effects on the genome and protein synthesis are primary and not secondary to subtle changes in other cellular mechanisms. Second, the selectivity of the hormone to specific target cells, specific chromosomes, and specific sites within the chromosome must be elucidated. And, third, there is no information on the molecular interaction between the genome and the hormone.

From this discussion on the hormones' mode of action, it appears that a characteristic property of the hormones is their specificity for the target organ. What is the combination of such specificity? Is it selective permeability, specific binding, or a unique molecular arrangement that permits the hormone to act only on the target organs? Studies on the binding of hormones to the target organs have yielded new clues as to the specificity of their mode of action [81–83].

Jensen and Jacobson [84] were among the first to demonstrate that tritiated estradiol binds selectively to the uterus and the vagina. The significance of this mode of binding is still not clear. It is not known whether the purpose of the binding is to activate, inactivate, or store active hormone. In fact, it is not even certain that the binding is physiologically significant. In any event, studies from Gorski's laboratory [81] have established that after injection of estradiol-17β or after incubation of the hormone with target organs, a large amount of the hormone is bound to a protein with a sedimentation constant of 9.5 and a molecular weight of 200,000. Tissue fractionation studies of the uterus revealed that 30% of the hormone is associated with the cytosol, whereas more than 50% is in the particulate fraction (nucleus and myofibrils). The protein does not bind to nonestrogenic steroid and is not found in nontarget tissues. The 9.5 S receptor described by Gorski's group is believed to participate in the initial uptake of estradiol. Some aspects of the mode of action of estrogens are represented schematically in Fig. 8-23.

The existence of cytosol and nuclear receptors has been broadly confirmed. They have been found in several laboratories and in different target tissues.

In the nucleus the estrogens bind to chromatin from which they can be dissociated in the form of a 5S hormone-receptor complex. While the cytosol receptor is found in uteri of animals untreated with estrogens, pretreatment with estrogens is needed for the appearance of a 5 S complex in the nucleus. Moreover, when isolated nuclei are incubated with estradiol, the 5 S complex will appear only if cytosol is present in the incubation mixture. These experiments and others

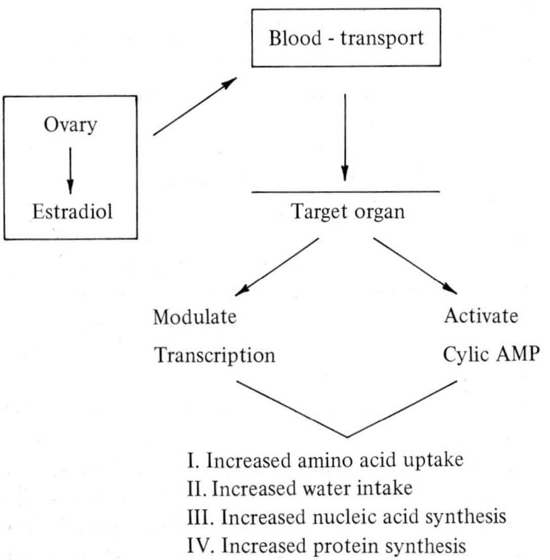

Fig. 8-23. Hypothetical mode of action of estrogens

demonstrate the cytosol dependence of nuclear binding.

Experiments with progestins have yielded results similar to those observed with estrogens and we have already mentioned that similar findings were made with glucocorticosteroids.

All steroid hormones (estrogen, progestins, glucocorticosteroids and aldosterone) are now believed to exert their specific physiological action in a three step sequence: binding to a specific cytosol receptor, transfer from cytosol to a specific nuclear receptor associated with chromatin, and modulation of gene expression in the target cell [184].

The Hypophysis and Sex Hormones

The role of the hypophysis in the development of male or female sexual characteristics became obvious when hypophysectomies could readily be performed [85–87]. When hypophysectomy is done before puberty, the testicles, the uterus, and the ovaries fail to develop normally. When hypophysectomy is performed in adults, the male and female genital organs atrophy. This harmful situation is reversed by injection of hypophyseal extracts. Clever experiments of Zondeck and Asheim in Germany and Smith and Ingell in the United States demonstrated that the implantation of fragments of the anterior hypophysis in muscles of hypophysectomized animals restores genital organ development. Attempts were then made to extract the active principles from the anterior lobe of the hypophysis. Alkaline extracts were obtained that stimulate the formation of the corpus luteum and acid extracts that stimulate the development of the follicles. It was observed that gonadotropins were also present in human blood and urine, particularly during pregnancy, menopause, after ovariectomy, and in some patients in the tumors of the gonads or of the adrenal.

FSH has been prepared from sheep and swine. The two preparations differ considerably in their molecular weights (67,000 and 29,000) and isoelectric points (4.5 and 4.8). Both hormones are glycoproteins, and the sheep preparation contains 1.5% of carbohydrates, calculated as hexosamine. The original molecule can be hydrolyzed extensively with crystalline pepsin without complete loss of activity. This suggests that the preparation is composed on an artificial mixture of proteins, or that it consists of a large protein molecule with a relatively small active center. The sheep preparation is rather stable in solution; at pH 7–8 it withstands heat at 75° C for one-half hour. Like growth hormones, the gonadotropins may show some species specificity. Indeed, whereas monkey gonadotropin produces ovulation in monkeys regularly, the effect of sheep gonadotropin on monkeys is inconsistent.

The luteinizing hormone has been prepared from the same sources. Again, the molecular weight varies from sheep (40,000) to swine (100,000); but in this case the sheep hormone has the lowest molecular weight. LSH is also a glycoprotein, and hexosamine and mannose have been found in its molecule. The isoelectric point of LSH is 4.6 for the sheep hormone, and 7.5 for the swine hormone.

The human gonadotropin, discovered in urine by Zondeck and Asheim, induces ovulation in the human in the presence of adequate FSH stimulation. The trophoblastic hormonal activity is now known to be the result of a single compound related to LSH biologically and immunologically. However, human LSH and human gonadotropic hormones are probably not identical proteins.

ICSH is a gonadotropin that stimulates the growth of the ventral prostate in hypophysectomized rats. It has been prepared as a homogeneous protein with a molecular weight of 30,000 and an isoelectric point of 7.3.

All the gonadotropins are glycopeptide hormones and are composed of two subunits, α and β. Much of their amino acid sequence is known. Treatment of LSH with diluted acid dissociates the polypeptide into two unidentical subunits which can be separated by a variety of methods including countercurrent distribution and chromatography. Neither subunit is active by itself but when α and β subunits are recombined 80% of the original activity is recovered. The α subunit of human LSH contains 89 residues. The carbohydrate moiety is attached to asparagine residues 49 and 75. The exact composition of the carbohydrate moiety is unknown but it includes hexose, fucose, hexosamine, and sialic acid. The addition of sialic acid is believed to occur after the polypeptide is synthesized. The presence of sialic acid on the glycopeptide is critical to its half life. There is a close correlation between the level of desialylation and the clearance of the hormone from the plasma.

The β subunit of human LSH is composed of 115 amino acid residues including 12 cysteine residues with a glycine C terminus and a serine N terminus. The carbohydrate moiety is believed to be linked to asparagine 13 and 30.

Human FSH has also been divided into α and β subunits, but little is known about them.

The amino acid sequence of the α and β subunits of human chorionic gonadotropin is known. The α subunit contains 92, the β subunit 147 amino acid residues. Except for deletion of the N terminus in human LSH-α, the α subunits of human LSH and HCG are identical. The β subunits from HCG and human LSH differ by the deletion of 30 amino acid residues at the C terminus in LSH-β and by the carbohydrate placement. While human LSH-β has one site of carbohydrate attachment, HCG-β has five.

The tertiary structure and, in particular, the location of disulfide bridges in the gonadotropins are incompletely known [185–187].

In women, the known effects of FSH are the stimulation of follicular growth, maturation of the oocytes, and ovulation. LSH stimulates the transformation of the granulosa cells and the thecal cell into lutein cells of the corpus luteum. In women, LSH facilitates the ripening of the follicle, the rupture of the follicle, and the development of the corpus luteum. In men, it stimulates testosterone secretion by the Leydig cells. The administration of pure LSH does not have any effect on the target organ in the absence of FSH, which suggests that the response to one of the gonadotropins depends on the simultaneous presence of the two hormones.

FSH and LSH stimulate spermatogenesis and the descent of the testes. In contrast, prostate and seminal vesicular growth depend only on LSH and ICSH. FSH appears in the urine of the castrates.

The secretion of sex hormones clearly changes with age and sex. The menstrual cycle appears regularly in women until interrupted by pregnancy and menopause. Patterns of sex hormone secretion change in both men and women at puberty. To secure the undisturbed sequential reappearance of the different sex hormones, a sophisticated control mechanism is needed. Sex hormone secretion is regulated by the target organ, the hypophysis, and the hypothalamus.

The gonads secrete steroid hormone, the hypophysis elaborates gonadotropins, and the hypothalamus elaborates factors capable of releasing the tropins. Three types of stimuli are believed to act on the hypothalamic centers that regulate gonadotropin secretion: (1) Chemical (steroids and drugs); (2) sensory (sight, touch, etc.); (3) stress by release of cholinergic and adrenergic agents.

Feedback Regulation

A direct feedback loop operates between the hypophysis and the gonads. Such a feedback mechanism was suspected because of the structural changes that took place in the hypophysis and the target organs under various experimental or clinical endocrinological combinations. Examples include the changes in the anterior hypophysis after castration, and the change observed in ovaries transplanted in the spleen of castrated rodents.

The basophil cells of the delta series change whenever sexual function ceases as a result of aging or castration. The number of the basophils, which are believed to store gonadotropins, increases in the anterior pituitary, and the cells are bloated with numerous small vacuoles. However, these so-called castration cells are seen only in animals and not in man.

The sex hormones of the gonads interfere with gonadotropin production in the hypophysis. Thus, although FSH stimulates estrogen secretion in the ovaries, the increased estrogen levels in the blood that follow depress FSH secretion by the hypophysis. This ovarian hypophyseal balance can readily be disturbed by implantation of an ovary in the spleen of castrated men or women. Under those circumstances, estrogens never reach the hypophysis, FSH secretion is uninterrupted, and the implanted ovary hypertrophies. Under appropriate conditions, the implanted ovary may develop cancer. A similar mechanism has been invoked to explain the mechanism of development of cancers of the ovary after bilateral irradiation of these organs.

More recently, it has been possible to directly measure the changes in gonadotropin levels in blood of experimental animals under various endocrinological conditions. The plasma levels of FSH and LSH increase within a week and reach a plateau within a month after ovariectomy.

If estradiol is injected starting on the day of castration, LSH rises only slightly. Progesterone injections have no effect on the increases in plasma levels of LSH in castrated women, but the combination of progesterone and estrogen abolishes the rise in LSH levels completely. To eradicate the rise in plasma FSH observed after castration, doses of estrogen much larger than those required to abolish increased LSH levels are needed, a finding which suggests that LSH and FSH are two different hormones.

Fluctuation in the plasma levels of LSH have been detected during the cycle in rats just before ovulation. Plasma LSH levels in the rats increase to values almost as high as those found in ovariectomized women. At the same time, the hypophyseal LSH drops and the maturing follicle secretes estrogens and progesterone. In this situation the ovarian hormones exert a positive feedback effect on the hypophysis by stimulating LSH secretion.

Endocrinologists have attempted to determine at which level of the regulatory circuit the steroid hormones exert their feedback effect. Two lines of investigations have been used (1) implantation of steroids in hypothalamus or hypophysis, and (2) administration of releasing factors, such as LSH-RF and FSH-RF, in ovariectomized animals. If the feedback affects the hypophysis rather than the hypothalamus, then releasing factors should not induce LSH and FSH secretion by the hypophysis. Nevertheless, these hormones are secreted in ovariectomized animals injected with releasing factors. To conclude from the implantation of steroid and injection of releasing factors that the site of the feedback effect has been identified would be unfair. One preferred hypothesis proposes that the feedback loop affects both the hypophysis and the hypothalamus, but that the hypothalamus is more sensitive than the hypophysis.

Changes in the sensitivity of the hypothalamic receptor to the feedback impulse help to explain the hormonal changes that take place at puberty and the absence of cyclic secretion of LSH in males. Before puberty, the hypothalamus is very sensitive to feedback inhibition by even small amounts of steroids, and the undeveloped gonads secrete enough steroids to inhibit the hypothalamic neurosecretion. At puberty, hypothalamic sensitivity to steroids changes rapidly but gradually, and much larger amounts are needed to inhibit the hypothalamus. Only time will tell whether this appealing hypothesis will be substantiated by further investigation. Alternative explanations for the development of puberty are that at the outset of puberty the hypophyseal receptors become more sensitive to the releasing factors, or that the gonadal receptors become more sensitive to the tropins. However, such interpretations conflict with some of the transplantation experiments. Newborn hypophyses transplanted under the hypothalamic region of adult animals respond like adult hypophyses. Similarly, the implantation of young ovaries in adult castrated women causes the ovaries to mature. And in any event, it would appear that the cause of puberty coincides with a resetting of the sensitivity of one or more receptors along the feedback circuit that links gonads, hypophysis, and hypothalamus. Such a resetting must result from very specific changes in the molecular structure of the receptor in the sensitive cells. Unfortunately we cannot even begin to guess what these molecular changes might be.

Whereas LSH secretion is cyclic in women, in men gonadotropin secretion is constant. Again, it has been proposed that the machinery that secures the smooth cyclic increases and decreases in LSH secretion is located in the hypothalamus rather than the hypophysis. A single injection of testosterone to a newborn rat will disrupt the clock that regulates the cycle. It is not likely that the ovary or the hypophysis of the "androgenized" women is disrupted because if these organs are implanted in other castrated and hypophysectomized animals, the animals respond cyclically. By exclusion, the hypothalamus is again incriminated. Studies of adrenal stimulation and focal destruction of the hypothalamus have provided conclusive evidence for hypothalamic participation in the control of gonadotropin secretion. In rats and rabbits, ovulation is evoked by stimulation of the tuberal region. Precocious puberty has been induced in man and rat as a result of hypothalamic lesions. Lesions of the median eminence abolish the feedback response of plasma FSH on LSH levels after ovariectomy. The hypothalamus probably affects hypophyseal secretion by elaborating releasing factors, at least two of which have been described. One stimulates the secretion of FSH; the other that of LSH. These releasing factors are polypeptide hormones. The LSH releasing hormone is a decapeptide (see section on releasing hormones).

Pathogenesis of Diseases of the Sex Glands

Considering the relative simplicity of the molecular structure of steroid hormones and the huge amount of work devoted to studying these hormones' mechanism of action, one would hope to be able to present a lucid explanation of the mechanism of action. Yet, we are still far from such a proposition, and unless a dramatic breakthrough occurs within a short time, the solution is not within our reach. The student should therefore not anticipate a description of the pathogenesis of the diseases of the sex glands in molecular terms.

The biosynthesis of sex hormones is catalyzed by enzymes formed in the adrenal and the gonads. The genetic information that determines the biosynthesis of these enzymes is stored in the DNA of the gametes. These sex chromosomes participate in determining the elaboration of the enzyme mosaic necessary for the development of male and female characteristics. In addition to the acquired sexual anomalies, hereditary anomalies of the sexual organs also exist. These hereditary anomalies result from chromosomal anomalies.

Hereditary Diseases of the Sex Glands

Before they differentiate into a mature oocyte or spermatocyte, the primitive germinal cells contain a diploid contingent of chromosomes. As the gonia develop into mature germ cells, two main events occur: a reductive kinesis or meiosis, and differentiation. During meiosis, the nucleus divides twice and the chromosome once. In the first or leptotene period of meiotic division, the chromosomes become apparent but are not longitudinally divided. The leptotene period corresponds to the prophase of a mitotic division. Next is the zygotene, a period during which the homologous chromosomes pair. Usually the pairing begins at the centromere, and it involves strictly homologous regions of the two homologous chromosomes. During the pachytene, the total number of chromosomes is reduced by half. The individual chromosome, called the pachytene bivalent, is made of two homologous chromosomes—one paternal and one maternal. It is twice as thick as normal and contains two centromeres.

Each chromosome of the bivalent divides into two chromatids, and the bivalent is now composed of four chromatids. But, in contrast to chromatid separation in mitotic prophase, these chromatids are held together in some points of their structure. These points of attachment are referred to as chiasmata. At the chiasmata, two of the four chromatids form an x. The chiasmata provide for genetic crossover and probably result from the breaking followed by the refusion of the chromatids.

During the meiotic division, the chromosomes of the bivalent are separated by a process resembling classical mitosis, and two haploid cells are formed. Yet the chromosomes that separate during the first anaphase are not identical to the maternal and paternal chromosomes that paired during the zygotene phase. The chromosomes that separate during meiosis are new combinations of segments of the paternal and maternal chromosomes due to the interchange of segments by chiasmata formation.

The main difference between the second meiotic division and the classical mitosis resides in the fact that the second meiotic division involves a haploid number of chromosomes.

In conclusion, during meiosis, the paternal and maternal chromosomes are briefly paired to be separated again during the first meiotic division. Any failure of pairing or separation leads to an abnormal distribution of chromosomes in the gamete.

Klinefelter's Syndrome. In trisomy, there is a diploid set of homologous chromosomes plus one extra chromosome. Geneticists have classified the chromosomes according to their size and shape, and each chromosome is referred to by a number. The most classical example of trisomy is mongolism, which is believed to result from nondisjunction in the mother. In mongolism, the oocyte contains two 21 chromosomes. (The type of trisomy is referred to by the number of the chromosome that is trisomic, usually using the new international standard nomenclature; for example, in mongolism the type of trisomy is trisomy 21). After fertilization, each cell of the embryo contains three chromosomes, 21, two derived from the oocyte and one from the spermatocyte.

Other congenital anomalies have been associated with chromosomes 13, 15, and 18. Trisomy X has been observed in mentally defective women who are fertile and capable of producing normal offspring.

Trisomy XXY leads to the development of Klinefelter's syndrome in males (see Fig. 8-24).

The presence of the Y chromosome determines the differentiation of the primitive embryonic gonad into testes. Its presence, in addition to two X chromosomes (47 chromosomes instead of 46), is responsible for the development of sexual ambivalence (Klinefelter's syndrome). But we shall see that the syndrome also occurs in patients with different karyotypes. These phenotypically male individuals have small testicles with agenesis and sometimes hyalinization of the seminiferous tubules. In contrast, foci of hyperplasia of the Leydig cells are found.

The victims of this chromosomal anomaly are often mentally retarded and sterile. The levels of urinary gonadotropins are usually high in patients with Klinefelter's syndrome, but those of 17-ketosteroids are decreased. Examination of sex chromatin is always positive. The syndrome is not rare—it has been reported in 1 of 400 newborns. Although in the typical Klinefelter syndrome, the karyotype is XXY, the syndrome has also been observed in patients with 48 chromosomes (48XXYY or 48XXXY) and 49 chromosomes (49XXXYY or 49XXXXY).

As in many chromosome anomalies, patients with Klinefelter's syndrome present mosaicism, or the presence of cells with different karyotypes in the same individual. Since all cells of the body descend from the fertilized egg, it can be expected that in the normal person all cells will have the same karyotype. This is not the case in patients with chromosomal disorders. In these patients, some cells may have a normal karyotype of the male or female, while others have an abnormal karyotype. Twenty percent of patients with Klinefelter's syndrome exhibit mosaicism.

The mechanism of production of chromosomal aberrations is discussed in another chapter. Suffice it to point out that failure of a pair of chromosomes to separate during meiosis generally leads to a trisomic zygote (XXX or XXY) if the ovum carries two X chromosomes, and a monosomic zygote if the ovum carries no X chromosome.

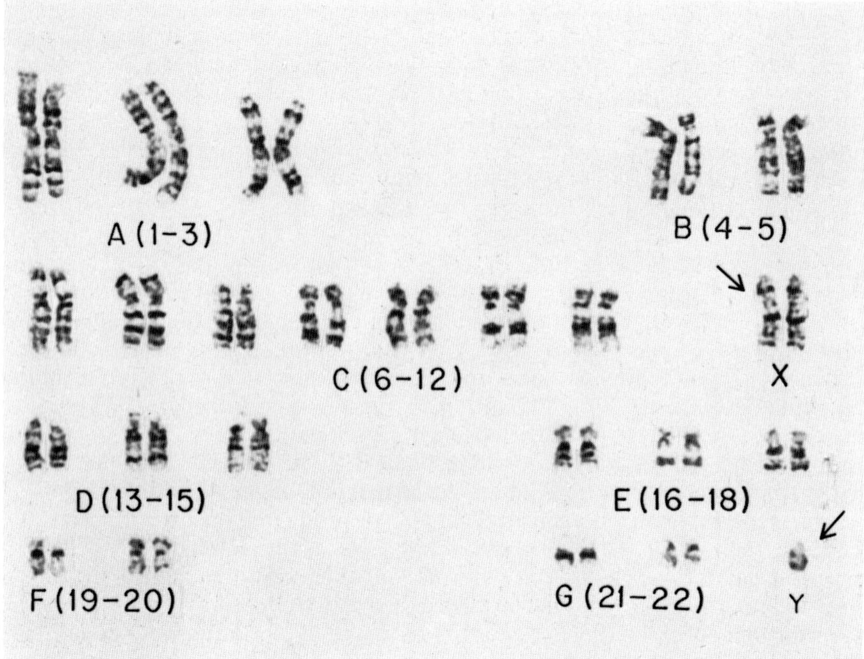

Fig. 8-24. Giemsa-banded karyotype with 47 chromosomes and 3 sex chromosomes, the usual pattern observed in the Klinefelter's syndrome. (From R. Sparkes)

Turner's Syndrome. In Turner's syndrome (XO; 45 chromosomes), there is a congenital absence of the gonads, and the ovaries exist only in the form of slender streaks of connective tissue. The patients affected by the Turner syndrome are phenotypically females and have a vagina and uterus. However, at puberty they do not menstruate (amenorrhea), and their breasts never develop. Needless to say they are sterile, in contrast to XO mice, which can procreate. Without female hormones negative feedback is abolished, and, as a result, these patients excrete large amounts of gonadotropins in the urine [88, 89].

Since Turner's syndrome results from the absence of the Y chromosome (see Fig. 8-25), it may be relevant that XO mice have been produced by irradiating mice immediately after fertilization, which suggests that the Y chromosome is extremely sensitive to radiation at this time. When dysfunction occurs after fertilization, such mosaicisms as XX/XO and XXX/XO have been observed. Such karyotypes may be associated with the clinical manifestations of Turner's syndrome.

XXX Type of Trisomy. Patients have been observed with duplicate sex chromatin bodies in the cells of the buccal mucosa, with resistant amenorrhea, infantile external genitalia, underdeveloped breasts, and high levels of urinary gonadotropins. Chromosome studies on cultured bone marrow cells demonstrated the presence of 47 chromosomes with the XXX type of trisomy.

XYY Type of Trisomy. A form of trisomy in which one X and two Y chromosomes are present has been described. The individuals carrying this genotype are usually tall, and it has been claimed that they might be unusually aggressive and tend to engage in criminal and violent activities. Therefore, a large number of the carriers of such phenotypes are believed to be found in jail or in psychiatric hospitals.

Whether such suppositions are correct remains to be seen; indeed, the chromosomal anomaly occurs frequently (1 in 700 people). Most available surveys have been made on inmates, and no survey of the general population is available. Unlike XYY mice, these individuals are fertile and procreate normal offspring. The extra Y chromosome seems to be lost during meiosis. A similar connection of the chromosome pattern occurs in fertile XXX females in which the cells of the offspring contain either XX or XY patterns.

Other Sex Chromosomal Anomalies. A number of chromosomal anomalies other than XO lead to sterility and amenorrhea in women. They include deletion of the short or long arm of the X chromosome and deletion of the short arm with duplication of the long arm ring formation. Deletion of the long arm has no effect on stature, and neither does it lead to the development of the other congenital anomalies often associated with Turner's syndrome, but it does cause amenorrhea and sterility. In contrast, deletion of the short arm leads to typical Turner's syndrome. Thus, although both arms of the chromosome are needed for normal development of the ovary, only the short arm is needed for normal development of stature.

Ring X chromosomes are likely to develop as a result of breaks of the short and long arms of the X chromosome, and, as may be anticipated from this pathogenic mechanism, considerable variation in symptomatology occurs in patients carrying such karyotypes.

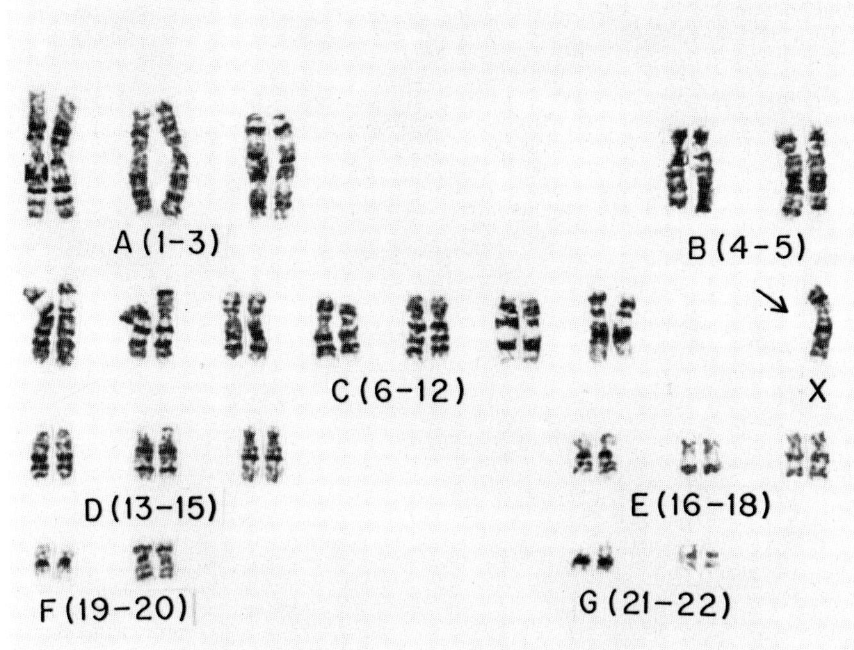

Fig. 8-25. Giemsa-banded karyotype demonstrating 45 chromosomes with only one sex chromosome, an X, the usual pattern in the Turner syndrome. (From R. Sparkes)

Hermaphroditism. True hermaphroditism is a rare condition in which the patient carries both ovarian and testicular elements. (Pseudohermaphroditism is more common; in this condition, the patient has either a male or female gonad, but the exterior genitalia resemble those of the opposite sex. Pseudohermaphroditism may result from three defects: lack of androgen, inability of androgens to act on target organ, and absence of regression of müllerian ducts [188].) Usually both the male and female tissues are found in a single organ, the ovotestis, which contains rudimentary follicles and seminiferous tubules. Spermatogenesis is nonexistent. A few rare cases have been reported in which a testis is found on one side and an ovary on the other. The disease is associated with either normal chromosome patterns or with mosaicism. Some types of mosaicism that have been reported in hermaphrodism are XYY/XO and XYY/XY/XO.

These considerations on the anomalies of the sex chromosomes and their clinical manifestations raise the important question of the role sex chromosomes play in gene expression. We have already seen how differences occur depending upon whether the short or the long arm of the X chromosome is missing.

A puzzling problem is that of the presence of two X chromosomes in the female. According to gene theory, duplication of the chromosome material in the genotype should lead to quantitative duplication of gene expression in the phenotype. The expression in the phenotype of a number of enzymes, blood coagulation proteins, etc. is known to be linked to the presence of X chromosomes. Yet the activity or the concentration of these proteins is the same in male and female. Therefore, Mary Lyon has proposed that one of the X chromosomes is repressed [90]. Repression is believed to occur early in embryonic life. The morphological counterpart of the repressed chromosome is the chromatin body seen in female cells. Of further interest is the fact that the XXX karyotype has two Barr bodies and two X chromosomes that duplicate late.

In support of Mary Lyon's hypothesis is the fact that the number of chromatin bodies found in individuals is equal to the number of X chromosomes in the karyotype minus one. However, the second X chromosome cannot be considered to be totally unnecessary. If it were, the XO karyotype would not lead to the clinical symptomatology found in Turner's syndrome, nor would an XXY karyotype result in a phenotype different from XY. It has been proposed that the second X chromosome might affect the expression of the genotype in the phenotype simply by its presence. The number of ridges in the fingerprint decreases as the number of genetically inactive X chromosome increases. The Y chromosome unquestionably determines the male sex characteristics, and the male sex characteristics develop in the phenotype even if the karyotypes contain as many as three X chromosomes. Whether the Y chromosome determines the appearance of the phenotype because it stores specific genes or simply because of its presence in the karyotype has not been established.

Inborn Errors of Corticosteroid Metabolism

Four inborn errors of corticosteroid metabolism have been described: deficiencies in hydroxysteroid dehydrogenase, and in 12-, 21-, or 11-hydroxylase [89, 90].

In the 3β-hydroxysteroid dehydrogenase deficiency, the conversion of pregnenolone Δ^5 or pregnenolone derivatives to progesterone is blocked. Such a block precludes the formation of glucocortico- and mineralocorticosteroids. Consequently, most individuals with complete dehydrogenase deficiency die in early infancy. A small number of cases have been reported in which the enzymic block is incomplete, and enough cortisol is secreted to secure survival.

The exact site of the enzymic block has been identified by biochemical and histochemical analyses of the hydroxysteroid dehydrogenases in adrenal tissues of afflicted infants. The block is responsible for a somewhat pathognomonic urinary excretory pattern. There is an accumulation in the urine of Δ^5-3β-hydroxysteroids (pregnenolone, 17-hydroxypregnenolone, and pregnanetriol), 17-ketosteroids with a Δ^5 structure, and dehydroepiandrosterone; small amounts of Δ^5 21-hydroxysteroid derivatives have also been found in the urine of these patients, indicating that 21-hydroxylation is not impaired. Consistently absent are pregnanetriol and 11-ketopregnanetriol, two compounds that accumulate in the urine of patients with 21-hydroxylase deficiency.

The accumulation of 17-ketosteroids with androgen activity is responsible for partial virilization of the female, leading to pseudohermaphroditism. In males, the disease is not associated with sexual precocity, but with incomplete development of the genital organ manifested by hypospadias and cryptorchidism.

In the classical adrenogenital syndrome, the block involves the 21-hydroxylase. The absence of enzyme activity has been established by determination of 21-hydroxylase in the adrenals of patients afflicted with adrenogenital syndrome. Although homogenates of normal adrenals convert 17α-hydroxyprogesterone to the 21-derivative, no such conversion occurs in the diseased adrenal. It has also been established that 17α-hydroxyprogesterone accumulates in patients with the adrenogenital syndrome. Isotope detection studies have indicated that the diseased adrenal secretes 80 times as much 17α-hydroxyprogesterone than the normal adrenal. Furthermore, twice the normal amount of 17α-hydroxyprogesterone is converted to etiocholanolone and androsterone in the diseased adrenal. At least 20 different steroids are found in excess of normal in patients with adrenogenital syndromes. Of particular significance are pregnanetriol, 17 α-hydroxypregnenolone, 11-ketopregnanetriol, and 17-ketosteroids. Pregnanetriol has been shown to be derived by reduction of 17 α-hydroxyprogesterone. The metabolite is excreted in normal individuals but only in small amounts. Cortisol administration reduces pregnanetriol excretion (see Fig. 8-26).

The levels of 17-ketosteroids (sign of androgenic activity) are increased in patients with virilizing tumors. The 17-ketosteroids fall into three major groups: those that are not substituted in carbon 11, 11-deoxy-, and 11-oxyketosteroids. In normal individuals, the ratio of 11-hydroxy to 11-oxysteroids is 3. In patients with virilizing adrenal tumors, the ratio drops to 1, indicating a relative increase of 11-oxysteroids.

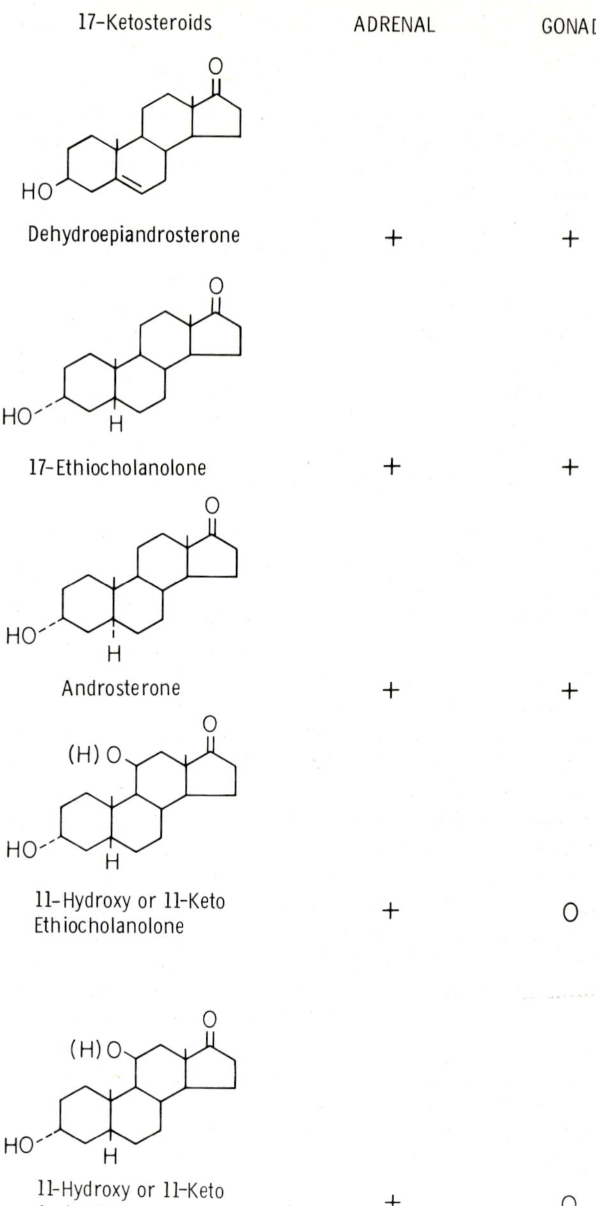

Fig. 8-26. Sources of principal urinary metabolites

The 17-ketosteroids increased in the urine include dehydroepiandrosterone, androsterone, etiocholanolone, and 11-hydroxy- and 11-oxy-derivatives. Thus, it would appear that as a result of the block in 21-hydroxylation, the rate of conversion of 17-hydroxyprogesterone to carbon 19 compounds is increased. These results indicated that the adrenals of these patients are capable of 11-hydroxylations in spite of the fact that 21-hydroxylation does not take place. As a result of 11-hydroxylation, 11-ketosteroids, and especially 11-ketopregnaetriol, are excreted in excessive amounts in the urine of patients with adrenogenital syndrome. Thus, 17-hydroxyprogesterone is directly hydroxylated in position 11 of the steroid molecule to yield 21-deoxycortisol, which is then converted to 3α-, 17α-, 20α-trihydroxy-5β-pregnan-11-one. The detection of this compound is of considerable diagnostic value because it is found only in syndromes associated with adrenal hyperplasia (adrenogenital syndrome, Cushing's disease). It is not found in patients with tumors of the adrenal.

Side chain splitting in the absence of or after hydroxylation in position 11 results in an overall increase in 17-ketosteroids, some of which are biologically active (androstenedione, 11β-hydroxyandrostenedione, dehydroepiandrosterone) and are responsible for the virilizing symptoms. Whether or not testosterone is formed in excess in patients with the adrenogenital syndrome is not certain. High levels of testosterone glucuronide have been found in some patients with adrenal hyperplasia.

In the absence of cortisone, the negative feedback on the anterior hypophysis does not operate. Consequently, the hypophysis is induced to secrete ACTH. The high level of ACTH stimulates proliferation of the adrenal cortex, and androgens are secreted in excess. In the female, cortical hyperplasia at birth leads to enlargement of the clitoris, and possibly to a urogenital sinus and an imperforated vagina. Thus, the newborn female is often taken for a male. The victim is a pseudohermaphrodite. Precocious growth of sexual hair, irreducible amenorrhea, and ovarian atrophy will occur in these patients.

In the male, the penis is enlarged and may reach adult sizes at a very early age. The scrotum, the prostate, and the seminal vesicles develop precociously. The hair of the beard, the axillary regions, and the pubis grows in infancy. The general body growth of the child is accelerated, yet the adult male pseudohermaphrodite is usually shorter than normal because the epiphysis closes prematurely in patients with excess androgens. 17-Ketosteroid secretion in the urine is markedly elevated, but it can be normalized rapidly by the administration of small doses of cortisone.

In most forms of the adrenogenital syndrome, the block in the 21-hydroxylase activity is not absolutely complete, and enough aldosterone can be synthesized to maintain the sodium balance in the blood. Nevertheless, in some cases the block is complete; the patient loses salt rapidly and experiences an Addison-like crisis.

Although there is a good correlation between the severity of salt loss and the degree of impairment of 21-hydroxylase activity, the salt loss effect has sometimes been claimed to result from the elaboration of salt-losing hormones rather than from interference with aldosterone synthesis. 3α,16α-Dihydroxy-5α(or 5β)-pregnan-20-one has been incriminated, but the salt-losing effects of these compounds have not been established in humans.

In another form of the adrenogenital syndrome, 11β-hydroxylase is missing. The disease results in progressive hyperpigmentation, virilism, and hypertension. The urinary pattern is somewhat specific. There is an increase in 17-ketosteroids, but none of the 17-ketosteroids is a hydroxy-α-oxy derivative. In contrast, large amounts of tetrahydro-11-deoxycortisol and tetrahydrodeoxycorticosterone are found in the urine of these patients, in addition to dehydroepiandrosterone, androsterone, and etiocholanolone. Pregnanetriol is scarce in the urine compared to the amount observed in the 21-hydroxylation defects. The accumulating precursors are deoxycortisol and its derivatives; deoxycorticosterone is believed to be responsible for hypertension.

A case with a deficiency in 17α-hydroxylase has been described. In such a case, the biosynthesis of either 17-pregnenolone (precursors of 17-ketosteroid) or 17-progesterone (precursors of cortisol) is impossible. Consequently, there is a decrease in the biosynthesis of androgens, estrogens, and cortisol.

Pathways not involving 17α-hydroxylation are not altered, and there is no decrease in deoxycorticosterone, which derives from the 21-hydroxylation of progesterone, and corticosterone, which is formed by the 11β-hydroxylation of deoxycorticosterone.

In such a case of 17-hydroxylase deficiency, the levels of these compounds in tissues and plasma are increased 10–40 times above normal. However, aldosterone production decreases in these patients, and this decrease is believed to result from secondary inhibition of aldosterone synthesis by other mineralocorticoids, deoxycorticosterone and corticosterone. In any event, the disease is associated with hypogonadism and excess mineralocorticoid activity.

The plasma renin levels are undetectable, and there is increased urinary potassium excretion with hypokalemia. (Low renin levels and high urinary potassium with hypokalemia are also observed in primary hyperaldosteronism.) The patient also sufferes from severe hypertension. Low levels of estrogen and cortisol are associated with increased plasma levels of LH and ACTH.

Assuming that the decreased β-hydroxydehydrogenase and 11-, 17-, and 22-hydroxylase activities result from the decreased production of molecules responsible for catalyzing each of the reactions, rather than from the elaboration of inhibitors or from the development of increased activities of antagonistic enzymes (e.g., 21-dehydroxylases), several different macromolecular mechanisms can be invoked to explain the changes in enzyme activities.

A point mutation of a structural gene resulting in the elaboration of a catalytic molecule with activity lower than normal could explain the various degrees of deficiencies. Of course, retaining of some residual activity could also result from the existence of smaller quantities of isozymes.

By analogy with what is observed in bacteria, an alteration of a regulatory gene would be likely to be associated with multiple-enzyme defects. Cases of combined 21-hydroxylase and hydroxysteroid dehydrogenase deficiencies have been described. In considering combined deficiencies, one must keep in mind the possibility of a single mutation affecting a gene that is responsible for elaborating a cofactor indispensable to the activity of several enzymes.

The development of the various adrenal enzymes in the fetal adrenal may also be of considerable significance in interpreting the lack of activity observed after birth. β-Hydroxysteroid dehydrogenase is low in fetal liver; thus, the Δ^5 pathway prevails (pregnenolone—dehydroepiandrosterone), and cortisol synthesis is low. The relatively large adrenals found *in utero* may result from compensatory hyperplasia in response to low cortisol secretion. In early embryonic life, 21- and 11-hydroxylase activities are low, but they increase with the age of the fetus.

Thus adrenal gland maturation, expressed in limited molecular terms, corresponds at least in part with the progressive increase in preexisting enzyme activities. What determines this increase? Plausible answers to this question include: elimination of inhibition, reduction of protein catabolism, elaboration of new cofactors, *de novo* synthesis of an isozyme, or acceleration of the biosynthesis of the preexisting enzyme. If the increased enzyme activity results from acceleration of the catalytic protein's biosynthesis, what triggers the acceleration, when does it trigger it, and does regulation involve transcription or translation?

Adrenal Carcinoma

The cortisol-androgen ratio of normal adrenals is 40:1; in adrenal carcinoma, this ratio is 1:3. Consequently, in adrenal adenomas or carcinomas, virilization rather than excessive glucocorticoid secretion occurs.

Virilizing Adrenal Adenoma

A small virilizing adrenal adenoma producing testosterone and responsive to gonadotropin has been described [189].

Stein-Leventhal Syndrome

The ovaries synthesize androgens. The levels of 17-ketosteroids in the urine were found to be decreased in ovariectomized individuals. Dehydroxyandrosterone has been found in ovarian blood, and the plasma levels of testosterone are higher in the normal than in ovariectomized females. When ovarian tissue from patients with Stein-Leventhal syndrome is incubated *in vitro,* testosterone and other androgens are formed. The three pathways for androgen synthesis in the ovary are outlined in Fig. 8-14.

Either benign or malignant ovarian tumors may be associated with virilization. Virilization has been observed in patients with adrenal rest tumors, Leydig cell tumors, granulosa cell tumors, Brenner tumors, and arrhenoblastomas. In patients with hirsutism associated with amenorrhea and sterility, the androstenedione and testosterone levels may be increased five times, while the urinary levels of dehydroxyandrosterone may be increased ten times.

The polycystic ovary is a disease of particular interest with respect to sex hormone biosynthesis. The polycystic ovary was described in 1935 by Stein and Leventhal in patients with obesity, moderate hirsutism, and amenorrhea. The large ovary usually is discovered on palpation, and surgery reveals large, pale polycystic ovaries. The cysts are surrounded by a fibrous capsule with unusual activity of the theca interna. Wedge resection of the ovary restores menstruation and relieves sterility. These ovaries produce excessive amounts of testosterone or other androgenic steroids but estrogen levels are normal. This explains the absence of flushes and the abnormal proliferation of the endometrium. The increase in the level of androgenic steroids is believed to result from a block in either the 3-β-ol-dehydrogenase or in the step involved with the oxygenation of 11-hydroxytestosterone to yield estradiol-17-β. A block prevents the conversion of pregnenolone to progesterone in the dehydrogenase, and pregnenolone is preferentially metabolized to 17α-hydroxypregnenolone and dehydroxyandrosterone.

Thus, in the Stein-Leventhal syndrome, the Δ^5 pathway has preponderance over the Δ^4 pathway, yielding precursors of testosterone. Without 11-hydroxytestosterone oxidation, testosterone (a precursor of estrogen) is not converted to estrogen. The enzyme defects just described are probably restricted to only part of the organ because, as already mentioned, normal levels of estrogens are found in patients with Stein-Leventhal syndrome, and wedge section cures the disease [91].

Whether this molecular interpretation of Stein-Leventhal syndrome will survive the tide of further publications remains to be seen. In any event, the exact pathogenesis of the disease is not clear, but it has been proposed that the disease results from an aberration of the hypothalamic control of gonadotropin release with excess and uncyclic secretion of luteinizing hormone.

Steroid Metabolism in Liver Disease

Since most steroid hormones are degraded in the liver one may anticipate that the pattern of steroid hormones in blood and urine will be altered in severe liver disease, such as cirrhosis of the liver. This is the case, but the interpretation of the observation is complicated by the existence of the enterohepatic circulation of steroid hormones.

In liver cirrhosis in man the plasma levels of: androstenedione and estradiol are increased, testosterone are decreased, and the sex hormone binding proteins are increased, with larger amounts of bound testosterone and estradiol.

There is also an increased level of urinary estrogen. The peripheral conversions of androstenedione to estrone and of testosterone to estradiol are increased and the conversion of estrone to its water soluble and protein bound metabolites is decreased with increased concentrations of estrone in liver. Finally, there is a decrease in 2-hydroxylation, 16α-hydroxyestrone reduction, and 16α-hydroxylation. Estrogen metabolite conjugation is normal in liver cirrhosis. Some of the key metabolic conversions of steroid hormones relevant to liver disease are illustrated below. All estrogen metabolites are conjugated or sulfated. They are excreted in the bile and 80% of the excreted metabolites are reabsorbed after deconjugation or desulfatation by the intestinal microflora. During reabsorption by the intestinal cell the steroid metabolites are again conjugated with glucuronic acid yielding among other metabolites estriol-3-glucuronide which is not further metabolized, but excreted directly into the urine.

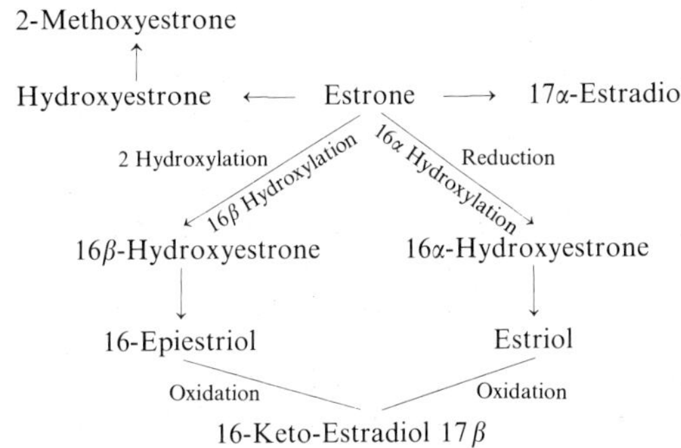

Conversion of androstenedione to estrone and of testosterone to estradiol by peripheral tissue (fat cells) rather than liver probably explains the increased urinary excretion of estrogen in patients with liver cirrhosis.

The increased levels of estradiol stimulate the biosynthesis of the steroid hormone binding protein. The reduction of estriol formation in liver cirrhosis is likely to result in part from cholestasis which interrupts the enterohepatic circulation and in more severe cases from poor uptake of estrogens by the hepatic cells, thus preventing the conversion of estrogen to estriol.

Cholestatic conditions, without impairment of other liver functions, inhibit the reduction of 16α-hydroxyestrone to estriol and the conversion of estrone to 16α-hydroxyestrone and estriol [190].

Diabetes of Pancreatic Origin, the Mode of Action of Insulin, and Hypoglycemia

Introduction

In diabetes mellitus, glucose accumulates in the blood and is spilled in the urine as a result of acquired or hereditary insulin deficiency. The disease has been known since antiquity. It was described in the Ebers papyrus. The word diabetes is derived from the Greek verb *διαβαινειν*, meaning to go through. The name diabetes was selected because of the polyuria that frequently afflicts diabetics.

The physicians of the Middle Ages and the Renaissance, in spite of the crude diagnostic methods available, could recognize the pathognomonic symptom of diabetes—glucosuria. With magnificent detachment from his fellow man's natural repulsion, Thomas Willis remarked that "the urine of the sick is wonderfully sweet and has a honeyed taste." In 1775, the urine of diabetics was evaporated and the presence of sugar was established. In 1815, Chevreul established that the sugar excreted in diabetic urine was glucose.

The first recorded observation relating diabetes to the pathology of the pancreas came from Thomas Cooley who, in 1788, described diabetes in patients with chronic pancreatitis and severe pancreatic lithiasis. The classical experiments of Minskowski and von Mering in Strasbourg demonstrated that diabetes developed after pancreatectomy and conclusively established the role of the pancreas in diabetes.

After the histological studies of Laguesse revealed the distinction between two major secretory systems in the pancreas—namely, the islets and the exocrine pancreas, Banting and Best demonstrated that insulin is secreted by the islet cell. The critical experiment was the ligation of the pancreatic duct, which destroys the exocrine pancreas but leaves the endocrine pancreas intact. Banting and Best showed that such a procedure does not induce diabetes.

Factors Influencing the Onset of Diabetes

A number of factors affect the incidence and establishment of diabetes. The mechanism by which these factors precipitate the appearance of the disease is often unclear. The role of heredity in diabetes is now well established. Observations made on siblings and twins in diabetic families have led to the conclusion that diabetes is transmitted by an autosomal recessive gene. Yet diabetes becomes manifest only in late adolescence and adulthood. Therefore, new physiological or pathological conditions must appear during development to precipitate the expression of the genotype into the phenotype.

Physiological conditions that play such a role are stress, pregnancy, and puberty; pathological conditions are obesity, trauma, and hemochromatosis. Participation of these various factors in the onset of diabetes will be briefly discussed in the following paragraphs.

Diabetes is eight times more common among persons whose weight is 25% above normal, and the severity of the diabetes correlates with degree of obesity. The effect of overfeeding can be demonstrated experimentally. Whereas subtotal pancreatectomy in animals is compatible with normal glucose metabolism, overfeeding of partially pancreatectomized animals induces hyperglycemia and glucosuria. The mechanism by which obesity leads to diabetes is not clear. It is assumed that the increase in total metabolic activity exerts a strain on the insulin resources and is thereby responsible for the appearance of diabetes among predisposed individuals.

Clinicians are often inclined to implicate specific dietary habits in the onset of diabetes. However, it appears that none of the ordinary foodstuffs—carbohydrates, proteins, and lipids—has such a specific effect, and overindulgence *per se* facilitates the onset of diabetes. The total caloric intake is more important in triggering diabetes than any specific type of food.

The first manifestations of diabetes often appear during a period of stress or after trauma. The mechanism by which the symptoms are precipitated by trauma is not known, but the relationship between diabetes and trauma is of medicolegal importance, for example, in cases of fractures of the skull, which can lead to the onset of diabetes.

Many diseases have been associated with diabetes. In some the association is not justified, and in others a definite cause-effect relationship exists. Whereas there is no reason to assume, as was often done in the past, that syphilis is responsible for the appearance of diabetes, diabetes is almost a constant finding in hemochromatosis. Insulin deficiency and the diabetes that it causes can result from surgical pancreatectomy or from massive destruction of the endocrine pancreas, which occurs in severe pancreatitis, pancreatic lithiasis, or carcinoma of the pancreas. Hyperglycemia and glucosuria have been observed in 25% of the patients with carcinoma of the pancreas.

In few diseases are the biochemical, pathological, and clinical observations quite as profuse as they are in diabetes. Nevertheless, the pathogenesis of diabetes remains one of the most challenging problems of pathology. Although the disease is characterized by a rather typical clinical symptomatology, the metabolic alterations that have been described are extremely complex, and the morphological changes are often inconsistent.

Clinicopathological Correlation

In this review of diabetes, the clinical and the pathological manifestations in diabetes will be described, and then the pathogenesis of each symptom will be traced.

The mechanism of action of insulin is discussed below. Insulin deficiency interferes with the adequate utilization of glucose. The sugar accumulates in the blood, and hyperglycemia, glucosuria, dehydration, increased diuresis, loss of electrolytes, and polydypsia ensue. By a mechanism that may or may not be related to the primary defect in carbohydrate metabolism observed in diabetes, fatty acid metabolism is also affected. A critical phenomenon is the accumulation of 2-carbon compounds that condense to form 4-carbon compounds, generating ketosis and acidosis. These symptoms will now be reviewed individually.

Hyperglycemia is a constant finding in severe diabetes. The severity of hyperglycemia increases with the severity of diabetes. Although hyperglycemia is transitory in mild diabetes and develops only in the morning, in severe diabetes it is continuous (with an increase in blood glucose levels in the morning). The lower daytime level of hyperglycemia probably results from higher glucose oxidation rates during the day because of greater muscular activity.

The changes observed in the glucose tolerance curves of diabetic patients are related to hyperglycemia. Whereas in normal individuals the administration of glucose is followed by a rapid rise of the blood-glucose content to levels slightly above normal followed by a rapid drop to normal values, in diabetics the blood glucose level rises slowly and reaches a plateau high above the normal values.

When the amount of glucose entering the glomerular filtrate is greater than 225 mg/minute, glucose is spilled in the urine.

Glucose is lost through the glomeruli when 100 mg of glucose per ml is in the blood; large amounts of glucose (up to 200 g/day in the younger person with severe, uncontrolled diabetes) can be lost in the urine of diabetes. In uncontrolled diabetics, up to 50–100 g of glucose is lost a day. Because so much glucose is lost in the urine rather than reabsorbed in the renal tubule, maximum reabsorption of water is also prevented, and polyuria, dehydration, and polydypsia follow. Because much of the glucose spilled in the urine is derived from glycogen, fat, and protein storage in the body, severe tissue wasting occurs, and in untreated diabetics the glycogen reserves of the liver and muscles are exhausted.

At least two symptoms are related to the increased protein breakdown observed in diabetes: (1) the increased amount of nitrogen in the urine, especially during periods of acidosis (when total nitrogen excretion per day reaches 15–25 g); and (2) a drop in protein content of the body in general, and a reduced concentration in serum albumin in particular.

By mechanisms that will be discussed later, the alteration of fat metabolism observed in diabetes leads to the formation of ketone bodies; about half of the keto acids are excreted unchanged. The high concentration of keto acids in diabetic urine explains the low pH (4.6 instead of 6) and the need for large amounts of 0.1 normal alkali (25 milliequivalents) to titrate the urine. The other half of the keto acids are neutralized, mainly at the expense of ammonia formed in the kidney from glutamine or amino acids. Consequently, the amount of ammonia ions in diabetic urine may be as much as 10 times normal.

Sodium, potassium, and calcium ions may also be used to neutralize the keto acids; therefore, in addition to the loss of electrolytes resulting from polyuria, electrolytes are also lost in the process of keto acid neutralization. If acidosis develops, the electrolyte loss will be further increased by repeated vomiting.

Since insulin is elaborated in β cells of the islets of the pancreas, the fact that lesions of the islets occur is important to our understanding of the pathogenesis of diabetes [93–95]. Yet it is not known whether the primary injury in diabetes involves the β cells. In alloxan-diabetic rats, the β cells of the pancreas first

advocated as a mode of treatment for diabetic retinitis that is resistant to more conservative therapy. The mode of participation of the hypophysis in the pathogenesis of diabetic retinitis remains unknown.

On gross examination, the kidneys of diabetics usually appear normal, but after careful histological studies, a typical glomerulonephrosclerosis is often found. The kidney lesion first described by Kimmelstiel and Wilson is characterized by the accumulation of hyalin nodules in the glomerular capillaries. At first, hyalinization is restricted to part of the glomerulus, but later it may involve its entire mass. Hyalinization of the afferent and efferent arterioles [98–100] is often associated with glomerular hyalinization.

Light microscopy (see Fig. 8-28) reveals the formation of a focal lesion or a diffused thickening of the capillary wall. The hyalin masses are acidophilic, except for some flattened nuclei that may be in the areas of hyalinization. Electron micrographs of the injured glomeruli demonstrate that the primary injury is a thickening and infolding of the capillary membrane associated with the formation of hyalin material either within the cytoplasm of the endothelial cells or between the endothelial cells. Some investigators believe that the hyalin deposition starts in the cell, but others think that it begins in the extracellular space. Electron microscopic studies (see Fig. 8-29) of kidney biopsies obtained from diabetic patients have shown that the hyalin substance is initially observed near the primary branches of the afferent arterioles and later in the peripheral arterioles, obliterating the glomeruli completely. Examination of such biopsy material has established that there is no correlation between the incidence and severity of Kimmelstiel-Wilson disease and the severity of diabetes; yet the incidence of glomerulosclerosis is closely correlated with the incidence of retinopathy.

Retinitis and Kimmelstiel-Wilson disease are probably two manifestations of the generalized hyalin degeneration of small vessels. Hyalin degeneration similar to that observed in the retina and the kidney has been described in the vessels of the mammary glands of diabetic patients. The pathogenesis of human glomerulonephrosclerosis is known. Similar hyalin degeneration has been induced in rabbit glomeruli by cortisone injection. The appearance of the hyalin material in the capillaries coincides with the development of diabetes. The experiments suggested that the hyalin masses resulted from capillary thrombosis, with globulin rather than fibrin as the principal component. Thrombosis is then followed by the focal infarction of part of the glomeruli.

Lendrum [101] believes that in the diabetic kidney there is an abnormal permeability of the vascular lining leading to "plasmatic vasculosis," or the escape of plasma from the capillaries of the glomeruli. In addition to hyalin deposition, fibrin deposition has been described. Fibrin is found in the Bowman's capsules or in the basal membranes of the convoluted tubules, where it forms crescentiform homogeneous masses called fibrin caps. Such fibrous accumulations are not characteristics only of diabetes but are also found in lupus erythematosus and glomerulonephritis. They are easily recognized on electron micrographs because of their density. Although in many cases Kimmelstiel-Wilson disease is symptomless, occasionally glomerulosclerosis leads to the development of a nephrotic syndrome with albuminuria and uremia.

Kimmelstiel-Wilson disease should not be confused with the Armanni-Ebstein diabetic nephropathy, which is characterized by the accumulation of glycogen in the epithelial cells of the renal tubule. The Armanni-Ebstein syndrome is pathognomonic of diabetes and is related to the severity of the disease. It has become rare since the discovery of insulin. The principal site of glycogen accumulation is in the proximal tubules, and the injury seems to affect preferentially and practically exclusively those nephrons that penetrate the corticomedullar junction.

There are few studies on the biochemistry of the glomerular membrane in diabetes. Except for a significant decrease in cystine and sialic acid, the glomerular membrane of the diabetic is not abnormal [193].

A lead that has often been followed in studying the pathogenesis of the vascular manifestations of diabetes is measurement of the serum glycoprotein levels. Glycoproteins are carbohydrates containing proteins in which the sugars or their derivatives are inserted either at one end or within the polypeptide chain. The binding between sugar and polypeptide is firm and usually covalent. Glycoproteins can be distinguished on the basis of their mobility on electrophoresis or with respect to their carbohydrate content. Although some glycoproteins exhibit the mobility of globulins, others move like albumins upon electrophoresis. "Glycoid" glycoproteins have a low carbohydrate content (0.5–4%) in contrast to "mucoid" glycoproteins, which have carbohydrate content greater than 4%.

Two observations have provided a rational basis for the study of serum glycoprotein levels in vascular diseases associated with diabetes. The incorporation of sugar into glycoprotein is insulin independent, and therefore in diabetes the utilization of glucose could be deviated from insulin-dependent into insulin-independent pathways. Glycoproteins have been detected at the early stages of the thickening of the basement membrane of the glomeruli in diabetic glomerulosclerosis, as well as during capillary thickening in the retina, ciliary body, and muscle in diabetes.

An overall increase in serum glycoproteins has been reported in diabetics. The rise of glycoproteins of the globulin type is greater than that of the albumin type. These changes occur whether or not diabetes is associated with any vascular alteration. In contrast, in patients with arteriosclerosis without diabetes, the increase in serum glycoproteins is less marked than in diabetics, and it seems to involve only one group of glycoproteins the α_2-glycoglobulins.

Another somewhat far-fetched theory suggests that growth hormone stimulates lipase activity with increased plasma levels of free fatty acids. In turn, increased fatty acid oxidation yields acetyl CoA which

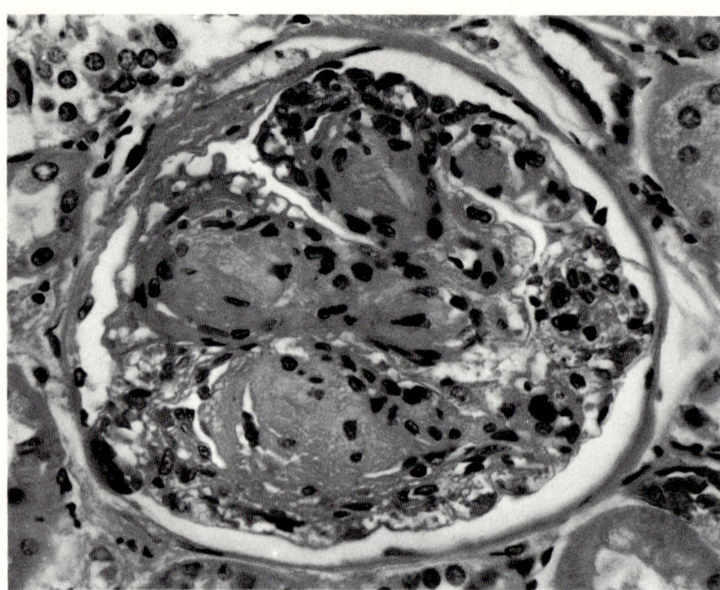

Fig. 8-28. Nodular diabetic glomerulosclerosis. Deposits form in the mesangial regions of glomerular lobules as described by Kimmelstiel and Wilson. Hematoxylin and eosin stain (supplied by Harrison Latta)

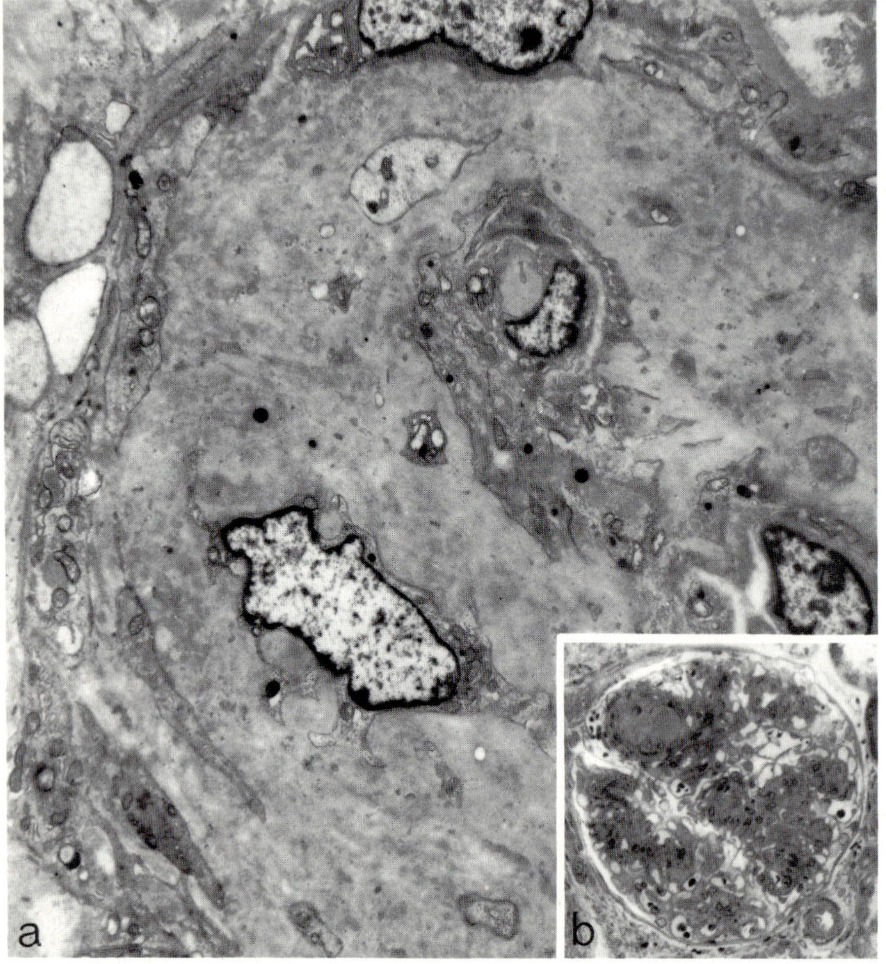

Fig. 8-29a and b. Electron micrograph showing diabetic glomerulonephrosis (a ×3,750, b ×150; from Dr. Zamboni)

is converted to citrate. Citrate inhibits phosphofructokinase thereby blocking the utilization of glucose in the glycolytic pathway. As a result fructose-6-phosphate, glucose-6-phosphate, and glucose accumulate. The excess of hexoses leads through the aldo reductase reaction to the formation of sorbitol which is believed to be the agent causing diabetic angiopathy. Most tissues are impermeable to sorbitol, but the polyol enters the cells of the capillaries and may directly or indirectly be responsible for cataracts, neuropathy, etc. [205].

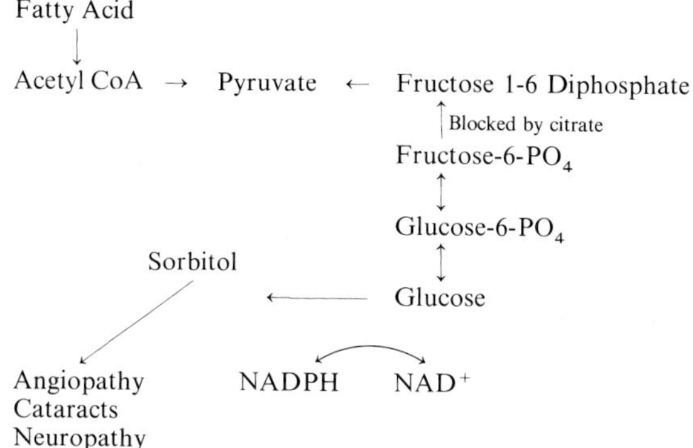

As a result of vascular or neurogenic alteration, amyotrophies and osteopathies have been observed in diabetes. The muscle atrophy involves the individual fibers and is usually not associated with an inflammatory reaction. The electron microscope may reveal a marked increase in the granular osmiophilic material between myofibrils. Frequently, the basement membrane of the capillaries nourishing the muscle is greatly thickened. Abnormalities of motor end plates resulting in a soap bubble appearance have been described in diabetes.

Osteopathy of various degrees of severity including osteoporosis, juxta-articular osteopathy, bone defects, foci of osteolysis, periosteal reaction, and osteosclerosis of the shaft have been described, primarily in the foot of the diabetic but occasionally elsewhere. Three types of mechanisms have been invoked to explain the bone lesions: (1) ischemia, although it has been argued that this is unlikely because osteolysis does not occur in ischemic bone—indeed, calcification takes place in bone; (2) infarct of the small arteries without reduction in the blood circulation; and (3) lesions of the nerve.

It is generally accepted that the incidence of infection among diabetics is increased. Tuberculosis of the lungs, pyelonephritis, and carbuncles frequently occur. In fact, it is advisable to examine the urine for the presence of sugar in patients with repeated occurrence of carbuncles. In spite of researchers' attempts to explain the increased susceptibility of diabetics to infection, a conclusive answer to that problem is not yet available. It seems that the participation of immunological factors can be excluded. Some workers have suggested that the exudative cellular reaction is impaired in patients with poorly controlled diabetes. A number of different factors have been invoked to explain the relatively high incidence of pyelonephritis among diabetics: the existence of urinary tract infection resulting from vascular disease, the increased susceptibility to all forms of infection, the effect of glucosuria on bacterial growth in urine, and the large number of catheterizations necessary for treatment of diabetic complications.

About half of the untreated or long-standing diabetics develop peripheral neuritis. The symptoms are diffuse aches and pains without the signs of peripheral nerve involvement, a typical bilateral sciatica with abolition of the achillian reflexes, severe ataxia with intentional tremor in the leg reminiscent of tabes dorsalis (pseudotabes peripherica), and severe neuritis of a single nerve, for example, the ulnar or anterior tibial nerve resulting in paralysis and sensory loss.

The pathogenesis of the diabetic neuropathy is not clear, although metabolic alterations are frequently implicated. It has also been suggested that the neuropathy results from alteration of the vasculature of the intra- and perineural arterioles. Dolman's careful histological examination seems to have eliminated such a possibility. Dolman investigated the histological changes that occur in the nervous system in patients with diabetic neuropathy and demonstrated the existence of degeneration in the dorsal columns and patchy demyelinization of the sciatic, femoral, and anterior and posterior tibial nerves [102].

Because cholesterol and phospholipid synthesis is likely to be necessary for the maintenance of the integrity of the structural lipoproteins in nervous tissue, especially in the Schwann cells and the oligodendrocytes, and also because cholesterol and fatty acid synthesis is decreased in livers of diabetics, cholesterol and fatty acid synthesis has been measured in the central and peripheral nervous tissue of animals made diabetic with alloxan. The synthesis of both compounds is decreased in diabetic animals in the central but not in the peripheral nervous system. Whether the extension of such studies will someday lead to the elucidation of the pathogenesis of diabetic neuropathy remains to be seen.

The increased need for insulin in pregnant diabetics is well known. In diabetics or prediabetics (persons who have an abnormal glucose tolerance curve), a transitory hyperglycemia develops during pregnancy as a result of the stimulation of gluconeogenesis by the corticoid hormones. Hyperglycemia may be responsible for miscarriage or for the delivery of a large infant presenting so-called pseudoerythroblastosis, cardiomegaly, and hypoglycemia. This combination of symptoms develops in all the children of the same mother and becomes more severe with successive pregnancies. That the corticosteroid hormones might be involved in the pathogenesis of this injury is suggested by the fact that the injection of cortisone (2–5 mg per kg of body weight per day) to pregnant rabbits causes abortion. If lower amounts of the hormones are injected, pregnancy is not interrupted, but the newborn rabbits are large and both the placenta and the fetus contain large amounts of glycogen. These effects of diabetes or prediabetes on pregnancy are of great clinicial importance because surveys have demonstrated that patients with abnormal carbohydrate tolerance tests can be protected against the deleterious effects of the transitory hyperglycemia by insulin administration.

Causes of Insulin Deficiency in Diabetes

Although all the symptoms of diabetes have been known for a long time, their pathogenesis is not clear, and the cause of the insulin deficiency and the nature of the primary metabolic alteration in diabetes remain unknown.

Several lines of evidence show that insulin is elaborated in the β cells of the pancreas. First of all, the islets, rather than the exocrine pancreas, clearly are involved in insulin secretion because pancreatic duct ligation leads to atrophy of the exocrine gland without inducing diabetes.

The β cells are considered to be the source of insulin because these cells are the first to be altered in diabetes produced by alloxan or after prolonged insulin administration. Furthermore, the β cells are also altered when insulin secretion is stimulated, for example, after partial hepatectomy, glucose administration and administration of diabetogenic hormones or of various sulfa drugs. In all of these cases, the β cells undergo a variety of morphological alterations, including degranulation, vacuolization, increased rate of mitosis, and enlargement of the Golgi apparatus and nucleus.

In view of the role played by the islets in insulin secretion, the causes of the lack of insulin are obvious when the β cells are eliminated, such as in surgical elimination of the pancreas or in chronic pancreatitis. The lack of insulin is much more difficult to explain in idiopathic diabetes. At least four different theories have been proposed. The first assumes that the hereditary defect in diabetes affects the pancreatic β cells and interferes with the ability to produce insulin. We have already seen that the morphological changes in the diabetic pancreas are hyalinization, fibrosis, and hydropic degeneration. But it is not clear whether these injuries constitute the primary cause or are the consequence of the lack of insulin. A second possibility is that the primary lesion in diabetes does not concern insulin, but is related to a metabolic defect that induces greater needs for insulin. Again, this hypothesis would seem to be easy to test by determining the amount of insulin in the blood of normal and diabetic patients. However, the quantitative determination of the insulin content of blood is difficult because of the presence in blood of factors that either destroy or antagonize insulin.

Insulin production could be normal in diabetes, but the insulin could be inactive because of the presence of antagonizing factors in blood or other tissues. Insulin can be inactivated several ways—by specific inhibitors, by specific antibodies, or by catabolic enzymes. Furthermore, diabetic patients, especially those with acidosis, carry insulin antagonists of an unknown nature. The presence of these antagonists is not related to previous insulin injections.

An inhibitor bound to albumin (sinalbumin) has been repeatedly described. It is thought by some that sinalbumin is formed by the binding of B chains (resulting from insulin degradation) to albumin [103].

Immunological Properties and Breakdown of Insulin

Since insulin is a protein, it can be expected to elicit immunological reactions. In considering the immunological properties of insulin, immune reactions against foreign antigens must be distinguished from autoimmune reactions.

Insulin antibodies have been separated by paper electrophoresis from the blood of insulin-treated patients. These antibodies act on insulin labeled with iodine 131, and a cross reaction occurs between the antibodies and insulin obtained from humans, pork, and beef. Insulin antibodies appear in the blood after bovine or foreign insulin administration. The amount of antibody is proportional to the amount of insulin injected. The antibody insulinic complex has little insulinlike activity. Insulin antibody develops in those patients who receive large doses of insulin, and it is therefore impossible to judge whether the insulin resistance results from antibody formation or from the appearance of agents with autoimmune activity.

In untreated diabetics, fibrinoid necrosis may be found in the vascular wall. Moreover, fluorescent-tagged insulin locates in the pancreas and the kidney, suggesting that insulin antibodies might be found at these sites. On the basis of such observation, some researchers have suggested that various forms of diabetes result from autoimmune reactions to insulin. Whether this is true remains to be established.

Liver and kidney primarily, but other tissues as well, contain a rather specific enzyme glutathione-insulin transhydrogenase, which in the presence of glutathione catalyzes the cleavage of the α from the β chain as a result of the reduction of the S-S bond that holds the two chains together. Proteolytic enzymes, then, degrade most of the reduced chain, except perhaps for a small amount of the β chain, which may escape into the blood and bind to other molecules.

The presence of a specific insulinase in the pancreas has often been claimed, but the preparations are too impure to permit definite conclusions. The insulinase preparation now available may well be a mixture of several proteolytic enzymes. The role of these enzymes, if any, in the pathogenesis of diabetes remains to be established.

The lack of insulin activity observed in diabetes could result from elaboration of an abnormal insulin molecule by the β cells of the pancreas. Whether the amino acid composition of the insulin obtained from normal and diabetic pancreata is different remains to be demonstrated.

Circulating antibodies to pancreatic islet cells were found in some patients with diabetes mellitus suggesting that autoimmunity may play a role in the pathogenesis of at least some form of diabetes. The antibody is of the IgG class and can fix complement.

Factors Controlling Glucose Levels in Blood

The mechanism of action of insulin will be easier to understand if the factors controlling glucose levels in the blood are reviewed first. Blood glucose originates from two main sources: digestion or conversion of the carbohydrates contained in foods, and generation of glucose in liver and kidney from glucose-6-phosphate. The amount of glucose in the blood depends on the utilization of the metabolite by several major routes: glycogen synthesis, gluconeogenesis, and glucose oxidation (glycolysis, the hexose monophosphate shunt, the glucuronic pathway). In spite of the complexity of the routes available to glucose metabolism, the amount of glucose in the blood is maintained within narrow limits by several hormones that regulate glucose utilization and gluconeogenesis.

An ordinary diet contains mainly two types of digestible carbohydrates: polysaccharides and disaccharides. The polysaccharides include large amounts of starch and small amounts of glycogen, amylose, and amylopectin; the disaccharides are sucrose and lactose. The poly- and disaccharides cannot be absorbed through the intestinal wall unless they are broken down to the monosaccharide by the action of enzymes in the salivary glands, the pancreas, and the intestinal wall; these enzymes are α-amylase of the pancreas and the salivary glands, and maltase, the oligo-1,6-glucosidase, and sucrase in the intestinal wall.

The amylases in animal tissue are α-amylases, or enzymes that catalyze the hydrolysis of the α-1,4-glucosidic linkage of the polysaccharide, leading to a reduction in viscosity and a decrease in molecular weight. The reducing hemiacetyl group that appears during hydrolysis of the polysaccharide is in the α-optical configuration and mutarotates downward. The pancreatic and the salivary gland amylases cannot be distinguished from each other. These enzymes have an average molecular weight of 50,000; they are slightly water-soluble, acidic proteins, containing 1 g atom of calcium per mole. The calcium is essential for activity and also stabilizes the secondary structure of the enzyme, probably enabling the amylases to perform their physiological function in systems rich in proteases.

Trypsin does not digest the α-amylase molecule unless the metal ion has been removed. The enzyme-substrate complex (amylase and starch) probably tightens the configuration of the amylase molecule, thereby further strengthening the binding of the metal. These observations are of interest because of the close association between amylase and proteolytic enzymes. In the pancreas, amylase is synthesized in the endoplasmic reticulum and is then transferred, probably through the Golgi apparatus, to the zymogen granules, which find their way to the intestine. In the zymogen granules, amylase is associated with several proteolytic enzymes. Most of the proteolytic enzymes are inactive within the zymogen granules, but when they have reached the intestinal lumen they are activated. At that point starch is still being digested; therefore, the amylases must be protected against the action of the proteolytic enzymes. The binding of calcium to the enzyme probably is responsible for such protection.

The Michaelis-Menten theory of enzyme action is based on the assumption that enzymic reactions occur in two steps. The first is the formation of an enzyme-substrate complex, the second is the action of the enzyme on the substrate. In the second step of the amylase reaction, the glucosidic bond is split. Although variations in pH or in the concentration of activators (chloride or monovalent ions) have no effect on the formation of the enzyme-substrate complex, these changes in the environment affect the splitting of the glucosidic bond. With the aid of H_2O_{18}, workers have found that it is the glycosidic C_1 carbon-oxygen bond that is split.

Maltases* catalyze the hydrolysis of the maltose obtained by polysaccharide digestion. Maltases have been found in the intestine and in the liver. The liver enzyme has been partially purified. Little is known of the mechanism of action of maltases, but it is assumed that a histidine residue in the polypeptide chain functions catalytically. In pig intestine, maltase is practically absent immediately after birth, and appears only after 25 days. Oligo-1,6-glucosidase is an enzyme catalyzing a low molecular weight, α-1,6-linked glucose moiety. The enzyme is without effect on dextran, glycogen, or limit dextrin. Isomalto-triose is hydrolyzed more rapidly than isomaltose at equimolecular concentration.

A lactase that specifically hydrolyzes the disaccharide lactose has also been found in the intestine. Little is known about these enzymes. Lactases are β-galactosidases. In animals they have been found in the intestine and in liver. The enzymes hydrolyze the lactose molecule to yield the two sugars, glucose and galactose.

Sucrase (also called saccharase and invertase) hydrolyzes sucrose. In contrast to the abundant information available on the yeast invertase, little is known of the mammalian sucrases. The significance of invertase and lactase in absorption is illustrated by inborn errors of metabolism in which these two enzymes are absent in the intestinal secretion. (The fact that lactase deficiency does not interfere with growth indicates that the galactose needed for biosynthesis of brain lipids or lens proteins can be synthesized endogenously in amounts sufficient to fulfill the metabolic requirements.) In that case, there is an intolerance to lactose or sucrose with no increase in blood glucose levels or without an increase in the levels of disaccharides

* The maltase reaction can be reversed. ^{14}C-labeled glucose incubated in the presence of a maltase preparation obtained from the intestinal mucosa can be incorporated into maltose.

in the blood, but with the appearance of watery stools of low pH containing lactic and other volatile acids, resulting from microbial fermentations.

Although a small fraction of the carbohydrates are digested in the upper digestive tract, most are digested in the duodenum and the jejunum.

The intestinal absorption of monosaccharides has been investigated in a number of laboratories, but much of our knowledge of that process stems from work done by Crane and his associates. Sugars are absorbed in the upper part of the intestine, and the ability of the intestinal mucosa to absorb sugars decreases continually in those portions of the intestine located downward from the pylora.

Using the everted sac technique, Crane determined the requirements for absorption of a sugar. First, absorption does not occur by simple diffusion because the rate of absorption is independent of the concentration within the intestinal lumen since some sugars that diffuse rapidly, like pentoses, are in fact absorbed more slowly than others, like glucose.

The absorption of sugars depends on a source of energy and on electrolyte movement. Agents uncoupling oxidative phosphorylation, such as dinitrophenol, salicylate, and triethylene sulfate inhibit the active transport of glucose. But the inhibition of glucose transport by DNP can be corrected by sodium fluoride and iodoacetate, which interfere with glycolysis. It is not known in what form the sugar is transported through the intestinal membrane, but the amount of phosphorylated organic compounds increases as glucose is transported.

In 1958, Riklis and Quastel [104] observed that sodium ion is necessary for sugar transport in the intestine. This finding was confirmed not only in intestine but also in kidney and other cells. Moreover, ouabain, an inhibitor of sodium transport, blocks sugar concentration by epithelial cells. On the other hand, sodium and water absorption is increased in rat intestine when sugar is actively transported.

When the effect of ions on the activity of sucrase was studied *in vitro,* it became obvious that the ion requirement for sugar transport through the brush border and that of sucrase were the same. The molecular requirement for the transport of the hexoses is extremely low; for example, only the hydroxyl group in position 2 seems essential. Yet a specific carrier site seems to exist because the sugars compete with each other for transport, suggesting that all the sugars transported by the active mechanism use a common pathway. The minimal requirements for the transport of a hexose sugar are a D-pyranose structure, a methyl or substituted methyl group on carbon 5, and a hydroxyl group in the glucose configuration at carbon 2.

The significance of these findings will become clearer when we discuss models for sugar transport in another chapter. Suffice it to point out now that such observations were at the origin of the theory of the "sodium-glucose" carrier complex.

Whatever the chemical forms of the absorbed material may be, the absorbed sugars are released in the blood after hydrolysis to the free hexose (glucose, galactose, and fructose). The blood fructose levels are somewhat lower than would be expected from the amounts absorbed, probably because part of the absorbed fructose is converted to glucose within the intestinal wall. This conversion is assumed to involve the breakdown of the fructose molecule to yield trioses. The remainder of the blood fructose is converted to glucose mainly in the liver, where it is used for glycogen synthesis. Galactose is converted to glucose and glycogen only in liver.

Glucose must be converted to glucose-6-phosphate for further metabolic transformation. Glucose-6-phosphate may then be used for glycogen synthesis or oxidized through glycolysis, the hexose monophosphate shunt, and the glucuronic pathway.

It was established long ago that in addition to glycogen, stored fats and proteins may be transformed into glucose. Triglycerides are hydrolyzed to yield glycerol, which can be used through the glycolytic pathway, and fatty acids. Fatty acids are transported in blood in the form of the nonesterified fatty acids and oxidized in liver to yield acetyl CoA, which can be used for gluconeogenesis through the reversal of the Krebs cycle and the glycolytic pathway.

Proteins are hydrolyzed to amino acids, which enter the Krebs cycle after transportation in the blood and transamination in the liver. The reversal of the Krebs cycle and the Embden-Meyerhof pathway converts the Krebs cycle substrates into glucose. The conversion of glucose-6-phosphate to glucose is catalyzed by a microsomal enzyme, glucose-6-phosphatase. The mechanisms controlling gluconeogenesis are discussed in more detail in Chapter 16.

The blood glucose formed through gluconeogenesis can be used by the peripheral tissue for glycogen synthesis or for glycolysis, or it may reach the kidney where it is filtered through the glomeruli and reabsorbed in the tubules. Normally, all the glucose excreted in the glomerular filtrate is reabsorbed, but once the amount of glucose contained in the glomerular filtrate reaches 220 mg/minute or more, reabsorption is incomplete, and part of the blood glucose is spilled in the urine.

A variety of hormones influence these different pathways. Their action will be described in detail in other parts of the book. Only a panoramic view of the hormonal control of glucose metabolism will be presented here. Insulin facilitates glucose penetration or usage by the peripheral tissue. In the liver, insulin is thought to facilitate both glucose utilization and output. Epinephrine accelerates the hydrolysis of glycogen to yield glucose-1-phosphate. The hormones of the adrenal cortex stimulate gluconeogenesis from lipids. Growth hormones probably stimulate the adrenal cortex and thereby stimulate gluconeogenesis indirectly. A pituitary inhibitor blocks hexokinase and prevents glucose usage by peripheral tissue. The factors involved in hormonal control of glucose metabolism are summarized in Fig. 8-30.

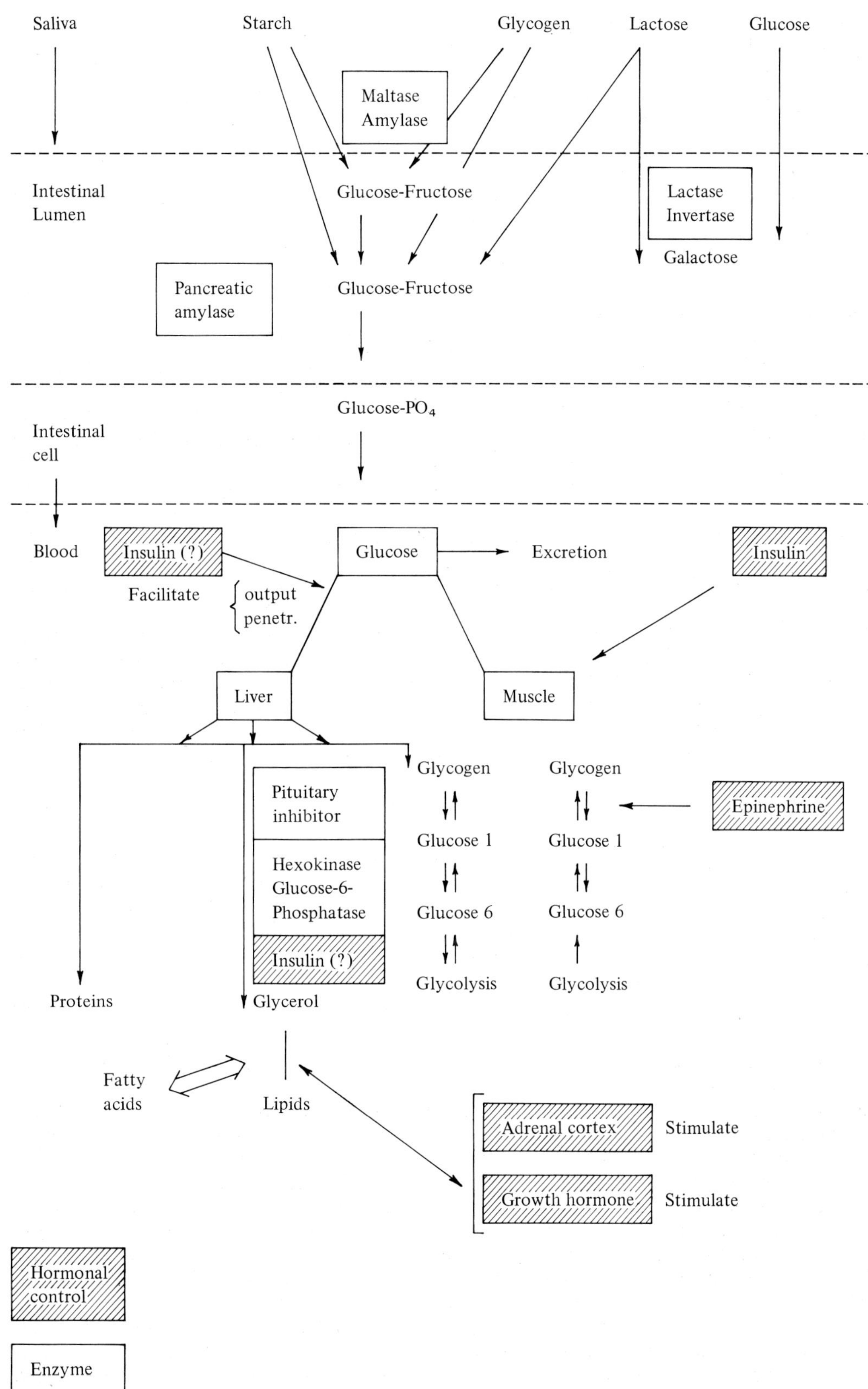

Fig. 8-30. Hormonal control of glucose metabolism (except for gluconeogenesis)

Insulin Secretion

Morphology

The histophysiology of the islets of Langerhans has been reviewed by Lacy [95]. The islets contain three types of cells: (1) the α cell, which secretes glucagon (20–30% of the overall population); (2) the β cell, which secretes insulin (60–70% of the cell population); and (3) the D cell for which no secretory function has been described. Except for the presence of specific granules, the ultrastructural details of each cell type do not differ from those of any other cell that actively synthesizes protein.

A remarkably clean preparation of secretory granules of the islets of Langerhans of rat pancreas was isolated in Lacy's laboratory, starting with islet isolated after treatment with collagenase. The granules were then separated from the crude homogenate, by differential centrifugation in discontinuous sucrose gradient. Their ultramicroscopic appearance in the pellet was the same as in the section and they contained clucagon and insulin [193].

The granules which contain the respective hormones are elaborated in the endoplasmic reticulum. Saccular cytoplasmic dilatations, which at first are granulated, become smooth and are soon filled with electron-dense material. The granule, surrounded by smooth membranes, migrates toward the periphery of the cell where it is ejected in a process involving microvilli formation. When the sac contacts the outer layer of the cell membrane, the two membranes fuse and the smooth sac membrane ruptures, ejecting its granular content into the extracellular space. This process leaves behind depressions of the membrane that are thought to be at the origin of the microvilli formation. This mechanism of granular ejection is activated by those substances that stimulate insulin secretion.

Factors Affecting Insulin Secretion

Insulin secretion seems to be regulated in part by a simple feedback loop [105]. Carbohydrate releases the hormone; high levels of insulin in the blood block insulin secretion in the pancreas. Superimposed on this simple feedback loop are complex interactions between the β cells and a number of metabolites and hormones. As a result, it is difficult to describe what regulates insulin secretion.

In isolated perfused pancreas, the amount of insulin released is proportional to the amount of glucose in the perfusate. The effect of glucose is to stimulate the secretion of stored insulin in the β granule and to accelerate new insulin formation.

The response of the pancreas to glucose is biphasic; a rapid release and then a sustained one. The first results from the release of stored insulin, the second from the elaboration in the islet cell of mRNA coding for proinsulin which is converted to insulin (see below). In releasing insulin stores the glucose binds to glucoreceptors located at the surface of the cell membrane and triggers calcium entry but does not seem to activate adenyl cyclase. The binding to membranes and the calcium flux cause alteration in the conformation of microtubules with displacement of the granules. More insulin is released on oral than intravenous glucose administration probably because glucagon, secretin, pancreozymin, and gastrin release insulin by increasing the sensitivity of the β cell to glucose [197].

Mannose, although less effective than glucose, also stimulates insulin secretion. Fructose, which is not metabolized in the β cell, has little or no effect on insulin secretion. Galactose, xylose, and L-arabinose are ineffective. The glucose effect depends on an intact secretory mechanism and glucose metabolism.

If the integrity of the membranes of either the β granule or the β cell is altered—for example, by removing calcium from the medium—the effect of glucose on insulin secretion is lost. Inhibitors of glucose metabolism such as deoxyglucose, mannoketoheptose, or glucosamine, block the glucose effect. But such analogs do not disrupt the secretory chain because insulin release can still be stimulated with tolbutamide. Other metabolites that stimulate insulin release are citrate, ketones, and leucine. The effect of citrate is not constant. It was observed to stimulate insulin secretion in duck but not in rabbit pancreas. Whether the effect of citrate is an indication that ATP is needed for insulin release remains to be seen.

Insulin release by ketone bodies would provide a most elegant feedback mechanism by which ketosis could normalize its own level by stimulating glucose usage. Yet, although an insulin-releasing effect has been claimed when β-hydrobutyric acid and acetoacetate are administered *in vivo*, *in vitro* ketone bodies inhibit insulin release.

Leucine induces hypoglycemia and a rise in immunologically active insulin by a mechanism that is still unknown. Whether other amino acids exhibit similar properties is not certain.

The effective level of insulin in tissues is influenced by a number of hormones: ACTH, cortisone, glucagon, epinephrine, and growth hormone. The fact that glucagon administered *in vivo* or added to pancreas *in vitro* induces the release of immunologically active insulin suggests that a feedback loop exists, linking α and β cells. The detailed circuit inserted between the terminals of the loop is not known.

Glucagon may exert its effect on the β cells by activating adenyl cyclase. Epinephrine also releases insulin in the pancreas by activating adenyl cyclase.

That growth hormone, ACTH, and cortisone stimulate the secretory activity of the β cell is well established, yet a direct effect of these hormones on the β cells has not been demonstrated.

Only the effect of growth hormone and glucagon will be discussed. A number of observations demonstrate that growth hormone might play a role in the pathogenesis of diabetes. The injection of growth hormone or of human chorionic somatotropin in animals and man impairs insulin secretion in the cell. Plasma growth hormone is elevated in individuals with juvenile diabetes or in those whose insulin hypoglycemia has been induced by insulin injection.

It appears that it is not growth hormone itself, but a degraded polypeptide that is responsible for the diabetogenic effect. Peptides have been prepared from bovine pituitaries, ones that inhibit fatty acid synthesis from acetate; another that is frankly diabetogenic. These findings together with earlier clinical observations have led to a new working hypothesis for the pathogenesis of diabetes.

Years ago a diabetic patient with marked diabetic retinitis developed pituitary insufficiency and, surprisingly, the retinitis disappeared. This finding led to the performance of hypophysectomies as a treatment of retinitis and although the therapy was not consistently effective it did help some patients, usually stabilizing the lesion for a period of one or two years. Diabetic dwarfs with hereditary growth hormone defect are free of angiopathy. To explain these facts it is proposed that the genetic defect in diabetes results in the selective breakdown of growth hormone molecules (mol wt 20,000) into active fragments; one that causes impaired glucose uptake by tissues and inhibition of the release of insulin after glucose stimulation, and another that stimulates collagen synthesis resulting in thickening of the capillaries [194, 195].

In the 1920's when pure preparations of insulin were difficult to obtain, some preparations for intravenous administration were shown to contain a hyperglycemic factor that was later named glucagon. The factor was purified from insulin preparations and crystallized in 1953. Glucagon is a straight chain polypeptide hormone composed of 29 amino acids and devoid of disulfide bonds. It is elaborated in the alpha cells of the islet of the pancreas and is extractable from pancreas after beta cell destruction.

A substance similar to glucagon is also secreted by the gastrointestinal tract. It has immunological and biological properties related to those of glucagon, but it is probably not identical. It is assumed to be secreted by the argyrophilic cells of the intestine.

Intravenous injections of glucagon raise blood sugar for 20 to 30 minutes by stimulating glycogenolysis in the liver as a result of conversion of inactive phosphorylase B into active phosphorylase A.

Decreases in blood sugar stimulate glucagon secretion by the pancreas. Yet, hyperglycemia caused by glucagon stimulates insulin secretion. In addition to causing hyperglycemia, glucagon stimulates the liberation of free fatty acids from adipose tissue, in contrast to insulin which promotes the storage of triglycerides. The hyperglycemic and lipolytic actions of glucagon are mediated by cyclic AMP which converts inactive phosphorylase and inactive lipase to the active enzyme via phosphorylation of serine residues through protein kinases.

A bihormonal pathogenic mechanism involving both insulin deficiency and hyperglucagonemia as a result of alteration of both the α and β cells of the pancreas has been proposed. The following arguments have been marshalled in favor of such a theory. All known forms of hyperglycemia are associated with hyperglucagonemia [206]. Even in the absence of insulin, glucagon is needed to induce hyperglycemia. Moreover, after total pancreatectomy, hyperglucagonemia has been detected probably as a result of secretion of glucagon by the gastric fundus or the duodenum. The suppression of glucagon secretion with somatostatin prevents hyperglycemia even in pancreatectomized animals [205].

Without insulin there is a reduction of glucose uptake by muscle and fat cells and a reduced level of production of glucose by liver causing prolonged postprandial hyperglycemia. The hyperglucagonemia normally overcomes the block of endogenous glucose in liver by insulin and, as a result, in the absence of insulin excessive amounts of endogenous glucose are produced and released in the extracellular fluid, thus accentuating the hyperglycemia. Similarly, a correlation between ketosis and hyperglucagonemia has been reported. This suggests that glucagon may also contribute to the pathogenesis of ketosis by stimulating the formation of ketone bodies in the liver. Thus, the lack of insulin would cause lipolysis with release of free fatty acids whose converstion to ketone bodies is accelerated as a result of hyperglucagonemia.

In livers perfused with a nonrecirculating medium containing amino acid concentrations equal to those found in plasma of normal rats, glucagon produces changes in intracellular and extracellular concentration of amino acids. The changes include: intracellular utilization of glycine, alanine, glutamine, and phenylalanine; intracellular production of glycine, isoleucine, and valine; and inward transport of lysine. In other experiments the uptake of non-metabolized amino acid (alpha-aminoisobutyrate and 1-aminocyclopentane carboxylic acid) was increased by glucagon.

The changes in amino acid metabolism caused by glucagon resemble those found in diabetic animals and it has been proposed that many of the changes in hepatic metabolism observed in diabetes and fasting result from the action of glucagon and catecholamines acting unopposed by insulin. Several workers have reported that glycine, alanine, and other glucogenic amino acids are decreased in the liver and plasma after fasting and in diabetes. Perfused livers from alloxan diabetic rats put out increased amounts of valine, isoleucine, and leucine. These three amino acids are also increased in the sera of diabetic humans and of alloxan diabetic rats and dogs. The change of serum amino acids in diabetes results, at least in part, from alterations in hepatic amino acid metabolism mediated by a relative excess of glucagon and catecholamine in blood [208].

In 1946, Loubatiere described the hypoglycemic effect of sulfonylureas. These compounds now are important in the treatment of diabetes. The two major categories of hypoglycemic agents are the sulfonylureas (tolbutamide, chlorpropamide, acetohexamide, and tolazamide), and the biguanides.

After tolbutamide administration, the granules in the pancreas disappear, the amount of extractable insulin is decreased. These effects have been detected when fragments of dog pancreas are incubated with tolbutamide *in vitro*.

Thus, tolbutamide undoubtedly stimulates insulin secretion, but whether it does so by a mechanism similar to that of glucose stimulation, or whether it acts through the intermediate action of other active molecules is in question. In any event, the biosynthetic activity of the β cell is stimulated by tolbutamide as a result of a direct action of the drug or because of a compensatory mechanism. Under the effect of tolbutamide, the nucleus of the β cell enlarges. New DNA and RNA are synthesized. Whether the increased metabolism of the nucleic acid is associated with new insulin synthesis is not known.

The mechanism of action of the hypoglycemic drugs is not clear, but tolbutamide and chlorpropamide have been found to inhibit cyclic AMP phosphodiesterase and thereby increase the intracellular concentrations of cyclic AMP.

Biguanides do not release insulin but act by stimulating glucose usage by peripheral tissues. Biguanides act on mitochondria *in vitro* and either uncouple oxidative phosphorylation or inhibit oxidative enzymes.

Structure and Synthesis of Insulin

A molecule of insulin is made up of two polypeptide chains: an A chain composed of 21 amino acids, with asparagine as carboxy terminal and glycine as amino terminal, and a B chain composed of 30 amino acids with phenylalanine as amino terminal and alanine as carboxy terminal. The two chains are held together by two S-S bridges linking the residues A-7 and B-7, and residues A-20 and B-29. The insulin molecule contains a third S-S bridge within the A chain, linking the residues A-6 to A-11. The exact structural requirements for hormonal activity are not known, but the alanine terminal is not essential for activity whereas the asparagine is. In the β chain, residues 1–6 and 23–30 can be eliminated without losing hormonal activity. It is not clear whether the residues 1–5 of the α chain are required for insulin activity. Acetylation of the ε-amino group, sulfation of the three serines and one threonine, and iodination of the tyrosine are of little consequence with respect to insulin activity (see Table 8-7).

Three mechanisms have been proposed for the molecular assemblage of insulin: (1) separate biosynthesis of A and B polypeptides with subsequent formation of the S-S bridge; (2) assembly of small polypeptide units, some of which contain the cysteine residues to provide the disulfide bridge; and (3) elaboration of a single "precursor" polypeptide chain, which contains both the A and B chain and a connecting unit [106].

Evidence assembled in human islet tumors, in rat islet cells, and in fetal bovine slices has established the existence of a proinsulin, a precursor of insulin. The amino acid sequences of porcine and bovine proinsulin are known: NH_2B chain Arg-Arg-C peptide Lys-Arg-A chain COOH (see Fig. 8-31). The C peptide forms the connecting unit in human proinsulin and is composed of 31 amino acids. Proinsulin is synthesized in the endoplasmic reticulum, transferred to the cisterna, and from there to the Golgi. Where the proinsulin is converted to insulin is unknown, but it has been suggested that the precursor is broken down into separate chains in the Golgi apparatus. After proteolysis of proinsulin, both proinsulin and the C peptide are stored in the zymogen granule.

Specific immunoassays for proinsulin have established that the cellular pool of proinsulin is small compared to that of insulin.

Little is known of proinsulin metabolism. Conversion of the precursor to the active hormone requires proteolysis, and two proteolytic steps are thought to be needed.

The enzymes causing the conversion of proinsulin to insulin are found in secretory granules. Two enzymes are involved; an endopeptidase with trypsin-

Table 8-7. The structure of ox insulin

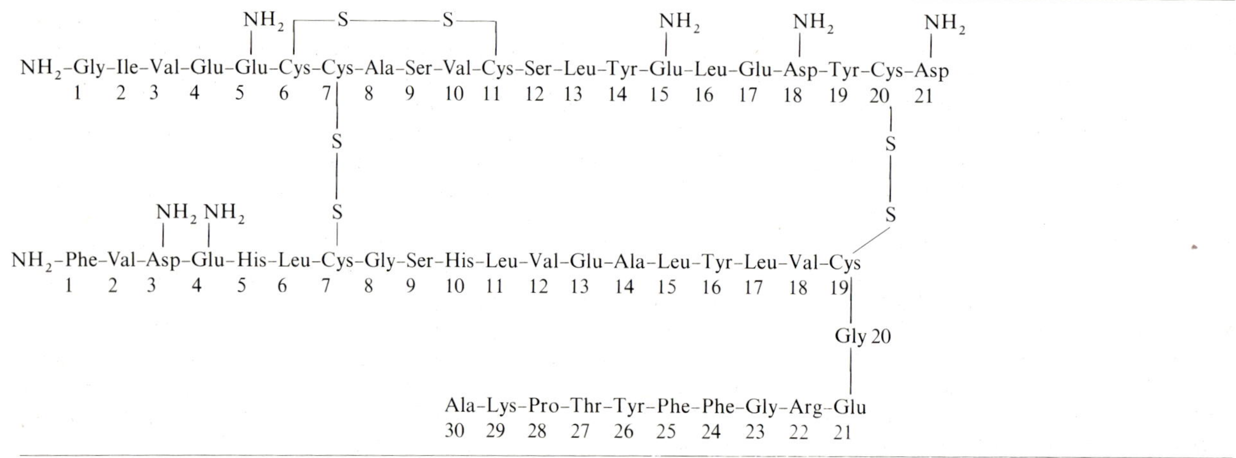

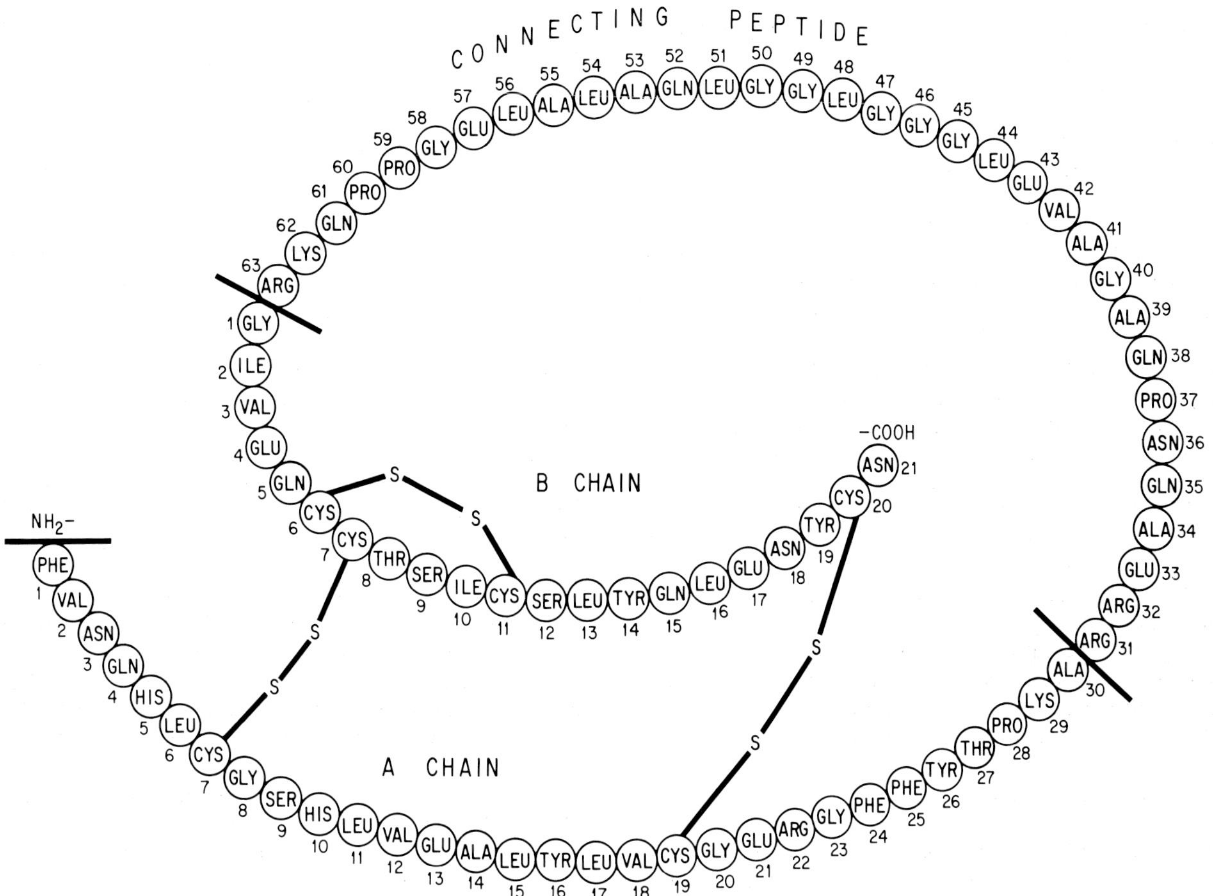

Fig. 8-31. Proposed covalent structure of porcine proinsulin

like specificity cleaves the C terminus side at a point where two basic residues link insulin A and B chains to the connecting polypeptide chain, an exopeptidase with substrate specificity similar to that of carboxypeptidase B removes the C terminal residues from insulin on the C peptide.

When the β cell granules are excreted by exocytosis they contain not only insulin, but also equimolecular concentrations of C peptide. It is not known whether the C peptide has any special physiological function. The rate of mutation acceptance in its amino acid sequence is much higher than that for insulin suggesting that the main role of the polypeptide is in determining the tertiary structure of proinsulin.

In retrospect, biosynthesis of a single polypeptide chain precursor for insulin is not surprising, because it brings the biosynthesis of insulin in line with that of other proteins, such as ribonuclease, containing sulfhydryl bridges. Here the amino acid sequence of the single polypeptide chain determines the tertiary structure, and this induces molecular folding until a minimal energy state is reached. The folding brings about the spontaneous formation of the S-S bond.

The significance of these studies on insulin biosynthesis to the pathogenesis of diabetes is considerable because the findings add a new dimension to the many pathogenic possibilities. Hereditary diabetes could conceivably result either from a mutation in the connecting unit, which interferes with the cleavage, of proinsulin or from the deletion of enzymes indispensable for cleavage. Acquired diabetes might result from distortions in the regulation of proinsulin synthesis or cleavage. Clearly, accurate diagnosis of diabetes may in the future require, in addition to diagnostic procedures already in use, determination of proinsulin and evaluation of the rate of proinsulin breakdown.

Proinsulin levels are abnormal in obesity, chronic renal failure, and severe hypokalemia. Determination of proinsulin has been particularly useful in the diagnosis of β cell tumors.

The conversion of proinsulin to insulin raises a number of intriguing questions. First, is this phenomenon unique? Second, why does the structural gene code for an inactive protein? The phenomenon is certainly not unique although all details of structure and conversion of the prohormone into hormone are not known. Prohormones of vasopressin, parathormone, gastrin, glucagon, MSH β, and ACTH have been known to exist. In the case of insulinoma and parathormone adenoma the tumors may secrete large amounts of the extended hormones that appear in the blood and may serve as a diagnostic tool [196].

Extended Polypeptide Hormones

Parathormone

Extended peptide linked through a basic residue to the N terminus. The N terminal extension has the following amino acid sequence: Lys-Ser-Val-Cys-Arg

Gastrin

Extended peptide linked to the N terminal residue; lysine is the linking amino acid

Glucagon

8 additional residues attach to the carboxy terminal end

MSH β

90 residues which contain the MSH β sequence with amino acids 41 and 58 attached to the remaining peptide chain by basic residues

ACTH

Suspected, but structure unknown

Vasopressin

Structure unknown but suspected because of the delay of incorporation of amino acids in vasopressin and the persistence of appearance of vasopressin after blocking protein synthesis

In 1965, insulin was being synthesized organically in Germany and Pennsylvania [108–110], and this work caused intense excitement. Both laboratories built the two polypeptide chains (A and B), amino acid by amino acid, and finally succeeded in bringing together the two oxidation chains of cystine to obtain an active product, but in very low yields (0.2–1%). At about the same time, Chinese investigators announced organic synthesis with a 50% yield. This feat was accomplished by oxidation of the two separate chains at pH 10.6. Whether or not their biological analysis permitted them to reach such a conclusion remains to be confirmed. In any event, the fact that the hormone could be synthesized by organic methods suggests that the structure of insulin first described by Sanger must be correct.

Insulin Transport and Breakdown

The exact mechanism of insulin transport in plasma is not known. Insulinlike material has been found in plasma. At least part of the ILA (insulinlike activity) of plasma is believed to be bound to protein. The bound ILA can be separated by chromatography. The chromatographed material is active on adipose tissue but not on diaphragm, suggesting that adipose tissue, not diaphragm, contains agents capable of separating the ILA from protein. In addition to bound ILA, plasma contains "free" ILA not absorbable on Dowex columns and active on both adipose tissue and diaphragm. Not all ILA is typical insulin because only a fraction of the plasma ILA is inactivated by insulin antibodies. Therefore, antibody suppressible and nonsuppressible ILA are referred to as typical and atypical insulin, respectively. It has been proposed that the atypical insulin is insulin made somewhere other than the pancreas.

Some of the enzymes suspected to be involved in insulin breakdown have already been discussed. Whatever the mechanism of insulin disappearance may be, the insulin elaborated in the pancreas is rapidly removed from the blood. Some is believed to be destroyed in the liver (50%); some is excreted in the urine (1–5%); and the remainder is used in the peripheral tissue. The rate of insulin breakdown may be of considerable importance in regulating its mode of action. Thus, it is relevant that the role of insulin breakdown varies depending upon the nutritional or hormonal status. Breakdown is increased in obesity and decreased in prolonged starvation [111]. Pregnancy accelerates insulin breakdown, but hypophysectomy and thyroidectomy slow it down.

Metabolic Effects of Insulin: Carbohydrate Metabolism

Whatever the primary molecular alteration in diabetes might be, major disturbances in carbohydrate metabolism must occur; therefore these changes will be considered first [112–115]. Studies of the metabolic alteration in diabetes and the biochemical effects of insulin are so closely interrelated that they cannot be discussed separately. Consequently, the remaining section of this chapter will deal with (1) the alterations of carbohydrate metabolism under the influence of insulin and as a result of diabetes; (2) insulin's primary site of action; and (3) the carbohydrate alteration of lipid and protein metabolism under the influence of insulin and diabetes.

Although insulin and diabetes are likely to affect carbohydrate metabolism in most tissues of the body, this discussion will center on their effect in muscle and liver. Fundamental differences between carbohydrate metabolism in muscle (no gluconeogenesis) and in liver (active gluconeogenesis) necessitate discussing the effect of insulin and diabetes on these organs separately.

Muscle

Experiments on insulin's mode of action on muscle metabolism have been carried out *in vivo* and *in vitro*. The *in vivo* studies include investigations on the conversion of [^{14}C]glucose to CO_2 or glycogen. (The balance studies done before isotopes were made available do not need to be reviewed here because the data so obtained would add little to the observations made with ^{14}C-labeled precursors.)

The *in vivo* determination of glucose oxidation by analysis of the respiratory CO_2 derived from radioactive glucose does not distinguish between the effect of insulin on glycolysis and its effect on the tricarboxylic acid cycle. Such *in vivo* experiments cannot yield information on the effect of insulin on specific enzymes, but they provide information on the effect of insulin on the overall glucose metabolism. These experiments have led to the conclusions that insulin accelerates glucose oxidation in muscle; glucose oxidation is considerably reduced in both diabetic and pancreatectomized rats; and insulin stimulates glycogen synthesis in muscle of normal and eviscerated animals.

Studies of the effect of muscle *in vitro* have included a variety of preparations, namely, heart-lung preparations, perfused limbs, perfused heart, and diaphragm. All these biological systems differ by the relative amount of muscle tissue in the preparation.

In considering the effect of insulin or diabetes on carbohydrate metabolism of muscle preparation *in vitro*, the effect of insulin on the conversion of glucose to CO_2 and the fate of the various possible pathways for glucose—glycolysis, the tricarboxylic acid cycle, the hexose monophosphate shunt, and glycogen synthesis—will be examined separately.

The observations made *in vitro* have shed some light on numerous problems related to the mode of action of insulin and the pathogenesis of diabetes. Among such problems are the primary site of action of insulin, the intracellular localization of the glycolytic cycle, and the pathway of glycogen synthesis in muscle.

In a series of experiments done in Hastings' laboratory, glucose utilization, glucose dissimilation to CO_2, and glycogen synthesis were measured *in vitro* in rat diaphragm obtained from normal and diabetic animals which received various hormonal treatments, including adrenalectomy and hypophysectomy. Studies were done with and without insulin and findings were that (1) in nondiabetic animals, without insulin, the amount of glucose used and radioactive CO_2 produced is increased by adrenalectomy, hypophysectomy, and combined adrenalectomy and hypophysectomy; (2) glucose utilization and dissimilation are not affected by hypophysectomy or adrenalectomy in diabetic animals in the absence of insulin; (3) glycogen synthesis is depressed in the diaphragms of diabetic animals; (4) glycogen synthesis is depressed in diaphragms obtained from adrenalectomized animals; (5) glycogen synthesis is increased in diaphragms obtained from nondiabetic hypophysectomized animals, or from nondiabetic animals that have been hypophysectomized and adrenalectomized; and (6) glycogen synthesis is increased even in those diaphragms obtained from diabetic animals that have been hypophysectomized and adrenalectomized.

When insulin is administered to these various groups of experimental animals, glycogen synthesis is increased in all cases, and the dissimilation of glucose to CO_2 is increased in all cases except in diaphragms obtained from diabetic animals not hypophysectomized or adrenalectomized. The reasons for these differences in the effect of insulin on glucose usage and its effect on glycogen synthesis still remain unexplained.

Villee and his associates [116, 117] also studied pyruvic acid dissimilation and usage as well as lactic acid accumulation in the presence and absence of insulin. Many of these experiments were performed in animals submitted to different hormonal treatments. Whereas in normal muscle insulin increases the amount of pyruvate formed from glucose, insulin has no effect on the amount of pyruvate utilized by muscle in normal or adrenalectomized animals.

The amount of pyruvate dissimilated in diabetic muscle is smaller than that in normal muscle, but pyruvate dissimilation can be normalized by the addition of insulin *in vitro*. This suggests that there is more than one metabolic defect in diabetes, and that in addition to interference with glucose usage, a block in the oxidation pathway of pyruvate may also exist.

Insulin stimulates glycogen synthesis in muscle incubated *in vitro*. This stimulation results from the increased incorporation of glucose into glycogen rather than from a reversal of the steps of the glycolytic pathway. While the total glycogen synthesized is increased in muscle under the influence of insulin, the proportion of glycogen synthesized from pyruvate is reduced. However, the absolute amount of glycogen synthesized from pyruvate in the presence and absence of insulin is unaltered.

These experiments indicate that insulin facilitates glucose usage for glycolysis and glycogen synthesis, but does not affect the usage of pyruvate for glycogen or CO_2 formation. Yet, in diabetes, in addition to a block in the utilization of glucose for glycolysis or glycogen synthesis, there is also a block in the dissimilation of pyruvate to CO_2.

The results obtained in these studies on the usage of glucose and pyruvate have been analyzed to pinpoint insulin's site of action. The pathways for glycogen synthesis utilizing glucose and pyruvate as starting substrate differ in the following way: to be used glucose must be phosphorylated to glucose-6-phosphate through the hexokinase reaction; in contrast, that reaction is bypassed when pyruvate is used as substrate. Since insulin affects glucose but not pyruvate utilization, it has been concluded that insulin must affect the hexokinase reaction; yet, these results can also be explained by saying that insulin stimulates glucose penetration in the cell (see below).

Villee and his associates claim that experiments in which glucose and pyruvate were added simultaneously to the medium provided additional evidence for the hexokinase theory. When both the substrates are added, the amount of CO_2 produced from glucose is decreased, while the amount of glycogen synthesized from glucose is increased. In addition, the total amount of glycogen synthesized when the two substrates are added simultaneously is greater than the sum of the amounts of glycogen synthesized from either glucose or pyruvate. Such experiments indicate that the mutase is perfectly capable of catalyzing the utilization of the substrate formed from both glucose (through the hexokinase reaction) and pyruvate (through the reversal of glycolysis). Consequently the rate-limiting step in glycogen synthesis cannot be at the site of the mutase reaction, and again by exclusion, the rate-limiting step must involve the hexokinase reaction. The findings at most provided only indirect evidence for a direct effect of insulin on hexokinase. In fact, these experiments permit only the conclusion that the rate-limiting step in glucose usage for glycogen synthesis precedes the mutase reaction.

Shaw and Stadie [118, 119] have reported interesting experiments relating to insulin's mode of action. Their results suggest that two glycolytic pathways operate in the cell. One is insulin sensitive, the other insulin insensitive; one is located inside of the cell, the other at the cell surface. These conclusions were reached (1) from studies on insulin's effect on glucose incorporation into lactic acid and glycogen in rat diaphragm; (2) by investigating the exchange between intermediates in the washed diaphragm and those in the medium; and (3) by direct determination of the specific activities of some of the intermediates in the washed diaphragm. The first experiments were done with diaphragms equilibrated aerobically in the presence of a phosphate saline medium of pH 7.4. These studies demonstrated that insulin consistently stimulates the incorporation of [^{14}C]glucose into glucose-6-phosphate, glucose-1-phosphate, and glycogen. There is no exchange between the glucose-6-phosphate, glucose-1,6-phosphate, and fructose-1,6-phosphate of the medium and the same intermediates in the diaphragm, and the added intermediates are not incorporated into glycogen.

Insulin is without effect on the incorporation of glucose into lactic acid when glucose-6-phosphate and fructose-1,6-diphosphate are added to the medium in the presence of glucose. The intermediates of the medium exchange with those in the diaphragm, and the added intermediates are incorporated into glycogen.

These studies on the incorporation of intermediates into glycogen of washed diaphragms revealed the existence of two different pathways for the usage of the intermediates. The first is geared toward glycogen synthesis and is insulin sensitive. This pathway is probably intracellular since the product of the reaction is found inside the cell.

Studies of the incorporation of precursors into lactic acid demonstrated a second pathway in which the intermediates were readily exchanged with the intermediates added to the medium; therefore, the second pathway might be located at the surface of the diaphragm.

These experiments are remarkable examples of the use of biochemical techniques for elucidating the intracellular division into compartments, and they shed new light on insulin's mechanism of action. Both the intracellular and the extracellular pathways require hexokinase for activity; and if insulin acts on hexokinase, why would it stimulate one pathway but not the other?

Much of the metabolism of carbohydrates, lipids, and proteins is performed in liver. Consequently, marked changes in liver metabolism are to be expected in diabetes and after insulin administration. Nevertheless, it is still debated whether the effect of diabetes and insulin is primary on liver or secondary to changes observed in peripheral tissue. Diabetes and insulin affect several aspects of liver metabolism, including glucose utilization, glucose output, protein synthesis, fatty acid synthesis, and ketone body synthesis and formation.

Before discussing the effect of diabetes and insulin on carbohydrate metabolism in liver, it is necessary to point out some of the difficulties encountered in studying glucose metabolism in liver. Studies on the effect of glucose utilization in liver are complicated by the fact that the liver is capable of gluconeogenesis, in which glucose-6-phosphatase catalyzes the conversion of glucose-6-phosphate to glucose. Moreover, in many early experiments on the effect of insulin on carbohydrate metabolism in liver, crude preparations of insulin were used, and such preparations were often contaminated by glucagon (a hyperglycemic factor).

Liver

Although insulin probably affects glucose transport in muscle (see below), in liver a primary effect of insulin on hexokinase could not be excluded. Studies performed on dogs with complete end-to-side portocaval shunt, on dog liver perfused *in situ*, and on liver slices suggested that insulin had no effect on glucose uptake in liver.

In contrast, studies by Chaikoff and his associates [120] on the uptake of glucose and [^{14}C]fructose in diabetic animals indicated that at least one of the metabolic blocks observed in diabetes involved glucose phosphorylation. Many of the observations made in Chaikoff's laboratory were confirmed by Hastings, who established that the phosphorylation of glucose was four times greater in hepatic tissues of normal animals than in those of diabetic animals.

The administration of insulin to the intact animal normalized the ability of liver slices to phosphorylate glucose, yet the effect of insulin in restoring the metabolic distortion was not immediate, and only after 20–24 hours of insulin administration was phosphorylation normal. Hastings and his group also pointed out that although the administration of insulin *in vivo* may restore the ability of diabetic liver slices to phosphorylate glucose, insulin does not correct all the metabolic alterations associated with diabetes.

Considerable liver hypertrophy persisted even after 48 hours of insulin administration. These studies on the effect of insulin on liver metabolism were followed by an investigation of the time that is needed to restore to normal, by insulin administration, several of the metabolic alterations associated with diabetes. The effect of prolonged insulin administration in diabetic animals on several multiple-enzyme systems and on single metabolic steps in liver were investigated. These studies included measurements of CO_2 production from glucose and fructose, fatty acid synthesis at the expense of glucose and fructose, glucose-6-phosphate formation, and glucose-6-phosphatase activity. These studies indicated that in the reversal of these impaired reactions, there is at first an exaggerated compensation which is followed by a slow return to normal. During the period of overcompensation, glycogen synthesis in diabetic rats injected with insulin is eight times greater than normal. The authors did not explain this effect of insulin on glycogen synthesis, but it is known that the increased glycogen synthesis occurs in spite of persistent high glucose-6-phosphatase activity. The increase in glycogen synthesis is thought to be caused by the increase in food uptake that occurs during the first 48 hours after insulin administration.

After 2 weeks of administration of small doses of insulin, all the metabolic alterations associated with diabetes in both liver and diaphragm are reversed. Since prolonged insulin administration corrected all the metabolic alterations in alloxan-diabetic rats, insulin-treated, alloxan-diabetic rats are useful for studying the acute effect of insulin withdrawal.

The formation of glucose-6-phosphate in diaphragm and liver slices obtained from insulin-treated, alloxan-diabetic rats is markedly reduced 10–14 hours after insulin withdrawal. Glucose phosphorylation can be restored to normal within 4 hours after resuming administration of the hormone. These experiments suggested that insulin affected the utilization of glucose in both liver and muscle by stimulating glucose penetration in the cell or by activating the enzymes that phosphorylate glucose to glucose-6-phosphate, or in liver by inhibiting glucose-6-phosphatase.

An effect on glucose-6-phosphatase activity in liver is unlikely because glucose usage is increased in both muscle and liver, and because changes in glucose-6-phosphatase under the influence of insulin are slow in liver. These studies on insulin withdrawal in insulin-treated, alloxan-diabetic rats suggested that the mode of action of insulin is similar in liver and muscle, but they did not indicate whether its primary action is on the penetration of glucose into the cell or on hexokinase activity.

A major difference between carbohydrate metabolism in muscle and liver is that liver is capable of gluconeogenesis. A direct effect of insulin on glucose output by liver could explain the effect of insulin on glucose levels in the blood and on hyperglycemia in diabetics. The glucose output of liver was investigated by perfusion of liver with a solution containing labeled glucose. The specific activity of the glucose was determined before and after the perfusate had passed through the liver. Since only the liver can introduce glucose into the circulation, the dilution of the specific activity of the radioisotopes probably resulted from glucose output through the liver. The glucose-specific activity of the perfusate dropped progressively because of dilution by unlabeled glucose produced by the liver. When insulin was added to the perfusate, the specific activity of the glucose in the perfusate no longer dropped consistently but was maintained at constant levels. In spite of some positive reports in the literature, a consistent effect of insulin on glucose output was not demonstrated in every laboratory that investigated the problem.

Although a single injection of insulin has no significant effect on glucose output by the liver, repeated injections seem to slightly reduce hepatic glucose output. In any event, this effect of insulin on glucose output is not likely to explain the hyperglycemia observed in diabetes.

Hastings and his associates studied the effect of insulin on glucose uptake and output in liver slices. The reduced glucose uptake and output of liver slices obtained from diabetic animals could not be brought to normal by adding insulin *in vitro,* yet liver slices of diabetic animals treated with insulin have normal glucose uptake and output.

Glucose metabolism was restored to normal within 6 hours after insulin injection. Adrenalectomy and hypophysectomy also restored glucose output in slices obtained from livers of diabetic animals, but these hormonal treatments did not affect glucose uptake. It is therefore interesting that glucose-6-phosphatase activity increases markedly in diabetic rats. Furthermore, insulin administration *in vivo* restores glucose-6-phosphatase activity to normal, a typical delayed effect of insulin occurring only after prolonged administration of the hormone.

Developments in our understanding of the mechanism controlling gluconeogenesis illuminated the earlier experiments on glucose output in perfused livers. The regulation of gluconeogenesis is discussed elsewhere, only the effect of pancreatic hormones on gluconeogenesis will be considered here.

Glucagon stimulates gluconeogenesis and insulin slows it down. In perfused liver glucagon stimulates gluconeogenesis from pyruvate, lactate, and oxaloacetate (but not from alanine). Crossover studies have established that glucagon acts on gluconeogenesis by activating pyruvic carboxylase.

Studies made in Williamson's laboratory [121] conclusively proved that gluconeogenesis is controlled by the rate of fatty acid oxidation; oleic acid was found to double the glucose production by the perfused livers from lactate, pyruvate, or alanine. When lactose was used as a precursor and decanoyl carnite (an inhibitor of polymityl carnitaic transferase) was added to the perfusion medium, the stimulatory effect of oleic acid was completely abolished.

The mechanism activating the pacesetter enzyme is not known. Two possibilities are a direct effect on glycogen, possibly through cyclic AMP, or an indirect effect resulting from the action of glucagon on lipolysis. In this mechanism, it is proposed that the free fatty acids are oxidized in liver, and the acetyl CoA acts as an activator of pyruvic carboxylase.

Williamson studied the effect of glucagon on animals treated with anti-insulin serum (AIS). He found that whereas the effects of glucagon were of short duration in normal animals, they were of long duration in animals injected with AIS. Insulin administration restores the effect of glucagon to a transient one.

Therefore, it appears that glucagon stimulates insulin release as a result of the hyperglycemia that it produces, and that insulin counteracts the effects of glucagon, possibly by stimulating glucose utilization in the liver. Moreover, AIS administration produces effects resembling those of glucagon, and insulin administration relieves all the symptoms. Thus, researchers have concluded that the effects of an acute lack of insulin—hyperglycemia and others—result from glucagon release. Then glucagon and insulin levels in the blood are inversely related.

In diabetes, insulin decreases and glucagon increases [122], and therefore the gluconeogenic effect of glucagon probably contributes to hyperglycemia.

Thus although many, if not all, acute effects of insulin deprivation may result from glucagon-induced hyperglycemia, long-range effects are probably easiest to explain by modification in the rates of enzyme synthesis. For example, alloxan diabetes is ultimately associated with increased activity of the enzymes that determine gluconeogenesis and decreased activity of the enzyme specific for glycolysis. Insulin restores the normal balance. Studies involving the administration of inhibitors of protein synthesis have established the long-range effects of alloxan diabetes and insulin. Whether one can claim that insulin acts as an inducer of key enzymes in glycolysis and as a repressor of key enzymes in gluconeogenesis remains to be seen. Such an assumption requires reevaluation of the operon concept as it applies to mammals by the nature of inducers and repressors in mammalian tissue.

In alloxan-diabetic rats, the amount of glucose oxidized through the hexose monophosphate shunt is below normal, and the activities of at least two important enzymes of the shunt pathway, glucose-6-phosphate and 6-phosphogluconate dehydrogenase, are much lower in alloxan-diabetic animals than in normal animals.

Effect of Insulin on Glucose Uptake in Brain, Lens, and Tumor Cells

In the above paragraphs the effects of insulin on glucose transport in the muscle cell were considered. Obviously, the interpretation of the effect of insulin on liver is complicated by the fact that liver can generate intracellular glucose. We have already alluded to sugar transport in the intestine that is insulin-independent; insulin's action on sugar penetration in the brain, the lens, and the tumor cells will be briefly considered.

Glucose easily penetrates the blood-brain barrier of nephrectomized and eviscerated animals, but its distribution in the brain is not affected by insulin. Fructose and ribose do not pass the brain-blood barrier at all. Yet fructose is a good substrate for brain hexokinase and is metabolized in brain homogenates. From the observations made *in vitro,* fructose seems to be a good substrate for brain metabolism *in vivo.* But such an interpretation is false since direct *in vivo* experiments have shown that fructose cannot enter the brain.

Ross [125, 126] studied the transport of nonelectrolytes through the ciliary body and demonstrated that the permeability of a number of substances was related to their solubility in lipids. The membrane is most permeable to those compounds most soluble in lipids. However, there are exceptions—namely, glucose, galactose, and xylose, which are transported through the barrier presented by the ciliary body at a much greater rate than all lipid-soluble substances studied. Ross's experiments suggest that the ciliary body has a highly selective membrane; glucose and galactose uptake can be increased by insulin injection, but xylose uptake is unaffected by insulin. Ross extended his studies to isolated lenses and demonstrated that insulin stimulates glucose uptake considerably in lenses obtained from normal animals, but has only a small effect on lenses obtained from diabetic animals.

Cori and his associates [127] studied the kinetics of sugar penetration in tumor cells and found them to obey the Michaelis-Menten equation, a finding which excludes the possibility of simple diffusion. When the Michaelis constant obtained for glucose penetration in the tumor cell was compared to that for purified hexokinase, both constants were found to be the same. These findings suggest that there is no barrier to glucose penetration at the level of the cell membrane, and that the hexokinase concentration is rate limiting for the utilization of glucose.

In conclusion, there seems to be no doubt that insulin can modify glucose transport through cell membranes. Yet, from the example given, cells obviously vary in permeability to glucose. Some permit free entry, but others are impermeable to glucose in the absence of insulin. Muscle cells, fibroblasts, and fat cells do not permit free entry of glucose. Insulin is required to activate a specific transport mechanism at low or normal concentration of glucose in the blood. But glucose will enter these cells in the absence of insulin if the glucose concentration in the blood is excessive.

Site of Action of Insulin on Carbohydrate Metabolism

Although many metabolic effects of insulin on carbohydrate metabolism have been described in detail, discussion of the primary site of action of the hormone has often been omitted.

The effects of insulin on the utilization of substrate in glycolysis and on glycogen synthesis in muscle and liver indicated that insulin acted before these reactions occurred and must therefore affect either glucose transport or glucose phosphorylation. The early hypotheses on the site of action of insulin can be categorized as those that hold that insulin acts by facilitating glucose transport through the cellular membrane, and those that propose that insulin activates the enzyme that catalyzes glucose phosphorylation.

The modern view proposes that insulin binds to specific receptors in all tissues where it exerts an effect. When it affects glucose transport, the formation of an insulin-receptor complex stimulates a carrier protein. When it stimulates macromolecular synthesis, insulin may either enhance ionic flow and precursor uptake or exert its effect through a second messenger.

A complete review of the arguments for and against both hypotheses is beyond the scope of this book; however, a brief outline of the work in that field illustrates the difficulties encountered when a hormone's mechanisms of action are investigated.

Transport Theory

The transport theory stems in great part from the work of Levine and his associates [123]. Insulin's effect on the usage of a variety of compounds was investigated by measuring their disappearance in the blood of eviscerated and nephrectomized animals. Since eviscerated animals were used, the animals could be kept alive only by infusing glucose, a procedure which complicates the interpretation of the results; however, the effect of insulin on blood concentrations of galactose was studied in the absence of glucose by performing artificial respiration during the test period. Insulin was found to promote the disappearance from the blood of some sugars but not of others, and the effect on the disappearance of the sugar was shown to be unrelated to the rate at which the sugar is metabolized.

Insulin induces a rapid drop of the blood concentration of glucose and galactose, but whereas the former is actively metabolized, the latter is not because muscle exhibits no appreciable galactokinase activity. When galactose is administered to eviscerated and nephrectomized animals, all the galactose can be recovered from the carcass in a free form. In contrast to galactose and glucose, other metabolizable hexoses or pentoses, such as fructose, mannose, and rhamnose, do not respond to the action of insulin in eviscerated and nephrectomized animals. Two nonmetabolizable pentoses—arabinose and xylose—are readily eliminated from the extracellular space under the influence of insulin.

Examination of the formulas of the insulin-sensitive sugars reveals that all have a similar configuration in carbons 1, 2, and 3. These considerations brought Levine to conclude that sugar transport does not occur by simple diffusion but through a specific insulin-sensitive transfer mechanism.

Large concentrations of one insulin-sensitive sugar prevent the usage of other sensitive sugars. Levine and his associates observed that large concentrations of galactose prevent glucose oxidation by peripheral tissue. The competition between galactose and glucose for the insulin-sensitive system suggests that both sugars are transported by the same mechanism.

Insulin has no effect on the removal of urea, creatine, and sucrose from the blood of eviscerated and nephrectomized animals. This observation suggests that insulin's effect is specific for sugars.

The work of Levine and his associates has been confirmed and extended in several laboratories using different methods on a variety of biological preparations. Drury and Wick confirmed that galactose competes with glucose for the insulin-sensitive system in the eviscerated and nephrectomized rabbits, but in contrast to Levine's observations, Drury and Wick found that muscle is not completely unable to use galactose. Using the radioactive galactose, Drury and Wick showed that small amounts of radioactivity were recovered in the respiratory CO_2 of eviscerated animals.

A competition between galactose and glucose [124] for insulin was demonstrated in the perfused heart and glucose oxidation was found to be reduced in the presence of galactose. A similar competition was demonstrated between glucose and fructose. Glucose drastically reduces the intracellular concentration and oxidation of fructose. In contrast, fructose has little influence on the penetration and utilization of glucose. A similar reciprocal influence of glucose on fructose was demonstrated on isolated diaphragms.

The minimum concentration of insulin required to promote hexose uptake was found to be the same for glucose and galactose. Similarly, the optimum concentration of insulin (that for which any further increase in the concentration of the hormone is not accompanied by an increase in sugar uptake) was found to be the same for both sugars.

Many aspects of insulin's mode of action were further studied in Parks' laboratory. Parks and his associates measured the effect of insulin on nephrectomized, eviscerated animals and on the distribution of infused sugars by determining the ratio between the free sugar in the tissues and the concentration in the medium. The results confirm previous observations indicating that insulin increases the intracellular concentration of galactose, glucose, and xylose—three sugars appreciably metabolized by muscle. In addition, the effect on the intracellular concentration of the sugar was greater in heart and diaphragm than in gastrocnemius muscle. This was thought to result from the greater vascularization and activity in heart and diaphragm.

The net increase of intracellular glucose in muscle constitutes further evidence for the transport theory, because in muscle it cannot be argued that the increase in intracellular glucose results from gluconeogenesis. Moreover, in isolated diaphragm incubated in the presence of insulin and radioactive glucose, the specific activity of the intracellular glucose is the same as that of the extracellular glucose. This finding suggests that the sugar originates from the extracellular space. Similar results were obtained on perfused heart and on isolated diaphragm incubated with insulin at low temperature, under which conditions glucose usage is delayed.

The experiments described above illustrate the effect of insulin on muscle permeability to hexoses. These observations show that muscle membrane must be included among the sites of action of insulin. Insulin promotes the penetration of sugars within the muscle cell either by modifying the permeability of the membrane or by causing the sugars to combine with a specific carrier capable of transferring the sugars from the outside to the inside.

A variety of arguments support the carrier theory. First, as noted previously, the effect of insulin is specific. Insulin stimulates the intracellular penetration of a restricted number of sugars. Among the factors determining penetration is the configuration in carbons 1, 2, and 3. Alteration of the configuration of only one of these carbons is sufficient to interfere with the sugar's penetration. Wick and Drury have shown that sorbitol, which has a structure similar to that of glucose except in carbon 1, is insulin insensitive.

In conclusion, some of the most important observations made on insulin's mechanism of action in muscle can be summarized as follows: the intracellular concentration of glucose is increased after insulin administration; insulin stimulates the transport of metabolizable and nonmetabolizable sugars with an identical configuration in carbons 1, 2, and 3; and different sugars compete for the site of action of insulin. On the basis of such observations, it has been suggested that insulin promotes sugar transport through the muscle cell membrane by activation of a specific transfer system rather than by an increase in membrane permeability.

The available evidence suggests that insulin facilitates the penetration of glucose into muscle intracellular space, and that in diabetes the penetration of sugar is impaired. These observations raise new problems with respect to insulin's mode of action. Does insulin promote the penetration of sugars within the cell, either by simply modifying the permeability of the membrane or by facilitating the combination of the sugar with a specific carrier that takes the sugar inside the cell?

The carrier theory is not without fault. Some sugars with configurations in carbons 1, 2, and 3 different from that of glucose were found to be insulin sensitive. The stimulation of muscular activity has an effect similar to that of insulin on glucose transport (this finding suggests that some reaction involved in glucose oxidation is the rate-limiting step controlling the penetration of glucose in the cell). Cortisone, which is known to induce insulin resistance, has no effect on the insulin sensitivity of galactose. And, finally, some of the arguments supporting the carrier theory can also be invoked by proponents of other theories.

Effect of Insulin on Individual Enzymes and Multiple-Enzyme Systems: Hexokinase Theory

The hexokinase theory originally rested on four groups of experimental observations: (1) insulin's ability to relieve the inhibition of crude hexokinase activity caused by extracts of the anterior hypophysis or the adrenal cortex; (2) indirect evidence that insulin activates hexokinase *in vivo;* (3) the existence of hexokinase inhibitors in the blood of diabetic animals; and (4) the absence of an effect of insulin on pathways that bypass the hexokinase reaction. In discussing the effects of insulin on hexokinase, we must again distinguish between peripheral tissue and liver.

It should be made clear from the start that a direct effect of insulin on glucose phosphorylation has never been satisfactorily demonstrated in muscle or brain. (In 1946, Berger and his associates discovered that the addition of insulin stabilizes hexokinase activity. This was, however, an unspecific effect that could also be produced by many other proteins.) Consequently, all arguments in support of an effect of insulin on hexokinase in those tissues are indirect. Brief analysis of the following results shows that much of this work on the action of insulin on hexokinase needs confirmation.

The hexokinase theory originated from experiments by Colowick, Cori, and Slein [128]. The combined effects of three hormones—an anterior pituitary extract, an adrenocortical extract, and insulin—were studied on crude or purified hexokinase preparations obtained from kidney, brain, liver, and heart of normal and alloxan-diabetic rats. We shall first present the results obtained on normal rats, and then those obtained on diabetic rats.

In the normal rat, a labile anterior pituitary extract inhibited crude hexokinase in brain, liver, kidney, and heart. The inhibition was counteracted by insulin. When, in addition to the anterior pituitary extract, an adrenocorical extract was put in the presence of crude muscle hexokinase preparations, the inhibitory effect was greater than that seen with the anterior pituitary extract alone. The adrenocortical extract alone had no effect on hexokinase prepared from normal muscles.

From these results it was concluded that in normal individuals insulin's effect is antagonized by an anterior hypophyseal extract, and that the effect of the hypophysis is mediated through an adrenocortical extract.

In alloxan-diabetic animals, the following observations were made. The hexokinase activity of several tissue extracts was very low; hexokinase activity was not directly stimulated by insulin; hexokinase activity was inhibited by an anterior pituitary extract, except in brain preparation; the inhibitory effect of the anterior pituitary extract was counteracted by insulin; contrary to what was observed when muscles obtained from normal animals were used, the adrenocortical extract inhibited alloxan-diabetic muscle hexokinase in the absence of the anterior pituitary extract; and the effect of the adrenocortical extract could be relieved by insulin.

Because of the significance of such results, several laboratories tried to repeat this work, and, although it could be demonstrated that an anterior pituitary extract inhibits hexokinase, it soon became clear that this effect on hexokinase and the diabetogenic effect of the pituitary hormone are unrelated.

When pituitary extracts with a powerful diabetogenic effect were used, the inhibition of hexokinase and the relief of the inhibitory effect by insulin were found to be very irregular. Furthermore, many of the results claimed by Cori and his associates in diabetic animals could not be confirmed. No difference between the hexokinase activity in normal and diabetic rats was found in two laboratories that investigated this problem. No effect of an adrenocortical extract could be demonstrated on the hexokinase activity of muscle extracts from either normal or diabetic animals, and the combination of insulin and adrenocortical extract had no effect on the hexokinase activity of normal or diabetic muscle extracts. Some of the experiments described above were performed on a large number of animals and included extensive statistical analysis of the results, leaving no doubt as to the validity of the conclusions. The inconsistency of the effect of inhibitors and the changes of hexokinase activity in diabetic animals ruin one of the main arguments in favor of the hexokinase theory. Extensive research is needed to establish these observations without doubt.

Studying the effect of diabetic plasma on hexokinase activity of rat brain, Weil-Malherbe [129] observed an inhibition of enzyme activity in half of the cases, but insulin injection successfully eliminated the inhibitory effect. Similarly, Bornstein and Park [130] observed the inhibition of glucose uptake in rat diaphragm incubated in the presence of sera obtained from alloxan-diabetic rats. The inhibitory properties disappeared either by adding insulin to the incubation medium or by simultaneous hypophysectomy and adrenalectomy of the diabetic animals.

In view of more recent data that have been accumulated on the effects of alloxan and insulin on gluconeogenic and glycolytic enzymes, these long-range effects appear to result from modification of the pattern of enzyme synthesis. Despite the fact that growth hormones and cortisone had no effect on glucose uptake *in vitro*, injection of the hormones into hypophysectomized and adrenalectomized diabetic rats restored the inhibitory properties in the serum.

Bornstein also showed that the inhibitory factor is a lipoprotein present in the sera of diabetic rats and diabetic humans. The factor inhibits glucose uptake and also prevents incorporation of [^{14}C]glycine into glutathione and proteins. The first effect is reversible by the addition of insulin to the medium. The second is only partially reversible. Although the lipoprotein fraction obtained from normal serum had no effect on glucose uptake, the lipoprotein obtained from humans during the hypoglycemic period that follows insulin administration inhibits glucose uptake in a manner similar to that in the diabetic serum.

In conclusion, although blood seemed to contain an insulin antagonist—possibly of pituitary origin—the appearance of which could well have been controlled by the adrenals, there is not much evidence (except for Weil-Malherbe's experiments on brain) that insulin acts directly on hexokinase. Moreover, the antagonistic effect on insulin and the inhibitory effect on hexokinase could be two unrelated phenomena.

The stimulation of brain hexokinase activity by hypophysectomy and adrenalectomy does not necessarily result from the absence of a hypophyseal or adrenal hormone. In these experiments on the effect of hypophysectomy and adrenalectomy on hexokinase activity, the enzyme assays were done several days after hypophysectomy, and the changes in hexokinase activity could be secondary to hyperglycemia. Hypophysectomy and adrenalectomy reduce the amount of insulin in the blood and lead to hyperglycemia, which may in turn cause increased hexokinase activity.

Attempts to provide further evidence on the role of insulin on the hexokinase reaction include studies on insulin's effect on certain pathways of carbohydrate metabolism that bypass the hexokinase reaction. For example (1) absence of an effect of insulin on the usage of a sugar such as fructose, which is metabolized after phosphorylation by enzymes other than hexokinase; (2) the fact that fructose oxidation is unaltered in alloxan-diabetic animals; and (3) the effect of insulin on glycogen biosynthesis in diaphragm or liver slices from [^{14}C]glucose and labeled pyruvate.

When eviscerated rats are injected with [^{14}C]glucose, 90% of the sugar is recovered in the carcass 1 hour after injection. If insulin is injected simultaneously with the glucose, the amount of sugar recovered from the carcass drops to 2%. These results demonstrate that in the presence of insulin, glucose is very actively used. After 1 hour, 63% of fructose is recovered, and its utilization is not affected by insulin. Two assumptions were made to account for the differences between glucose and fructose metabolism. First, the hexokinase that phosphorylates glucose is different from the enzyme that phosphorylates fructose (fructokinase). Second, insulin acts only on hexokinase. This interpretation is not entirely warranted, and the results are somewhat controversial because although some investigators found that fructose is poorly used by muscle tissues, others have shown a definite effect of insulin on fructose usage by muscle. The low rate of glucose utilization described in this experiment is also surprising and does not correlate with results obtained in other laboratories.

The results of at least three other groups of experiments were considered to provide further evidence for an effect of insulin on peripheral hexokinase.

These observations are that glycolysis or the tricarboxylic acid cycle is unimpaired in alloxan-diabetic rats, the rate of conversion of fructose to glucose is unaltered in alloxan-diabetic rats, and alloxan diabetes does not interfere with the activity of fructokinase, mutase, isomerase, or glucose-6-phosphatase. On the basis of these findings, it was deduced by exclusion that the metabolic block in alloxan-diabetic rats must occur above the metabolite fructose-6-phosphate, and consequently involves glucose phosphorylation or penetration in the cell.

If insulin acted on hexokinase, one might expect that the biosynthesis of glycogen through pathways that bypass the hexokinase reaction would not be impaired in alloxan-diabetic rats and would not be influenced by insulin. Hastings and his associates [131] observed this when they compared the biosynthesis of glycogen from pyruvate and [^{14}C]glucose. These investigators discovered that although alloxan diabetes does not interfere with the incorporation of pyruvate into glycogen, diabetes markedly reduces the incorporation of [^{14}C]glucose. Similarly, insulin has no effect on the biosynthesis of glycogen from fructose, but it considerably enhances the biosynthesis of glycogen from glucose.

In conclusion, studies on the effects of insulin on hexokinase in muscle have yielded either inconclusive results or have provided evidence that may as well be interpreted to indicate that insulin's primary action is on glucose transport rather than on phosphorylation.

In liver the situation is different; studies of the effect of insulin on glucose phosphorylation have revealed the existence of two hexokinases—the classical hexokinase and glucokinase. Of these two, only the glucokinase is decreased in alloxan diabetes and responds to insulin injection (see section on metabolic regulation in Chapter 16).

In spite of the large amount of work that has been done in many different laboratories all over the world, the information on carbohydrate metabolism in diabetic muscle and liver cannot be organized in an unshakable pathogenic theory. The facts can be summarized, but interpretation must be kept open.

In muscle, diabetes causes reduced glucose uptake, a reduction of glucose dissimilation when glucose or pyruvate is used as substrate, and reduced glycogen synthesis from glucose. Insulin administration restores glucose penetration, glucose dissimilation, and glycogen synthesis when glucose is used as a substrate. In contrast, when fructose and pyruvate are used as precursors, no restorative effect is observed.

These findings suggest that in the diabetic muscle the lack of insulin interferes with glucose penetration.

In contrast, in liver, insulin has no effect on glucose penetration, but it may decrease glucose output, stimulate glucokinase, glucose dissimilation, and glycogen synthesis.

The diabetic liver exhibits decreased glucokinase activity, interference with dissimilation of glucose and possibly of pyruvic acid, and stimulation of glucagon secretion. The administration of insulin to diabetics corrects liver metabolic defects slowly and seldom completely.

Effect of Insulin on Pathways Other Than Carbohydrate Metabolism

ATP Synthesis. Inasmuch as the first steps in glucose metabolism start with the phosphorylation of hexoses in the presence of ATP, a direct effect of insulin on the formation of the high-energy phosphates could explain the effect of the hormone on glucose utilization, and an impairment of the rate of ATP synthesis could be responsible for the biochemical lesions of diabetes. Without ATP, glucose cannot enter the glycolytic cycle. Insulin could affect phosphorus metabolism by increasing the PO ratio, stimulating the rate of oxidative phosphorylation, and improving the effectiveness of ATP usage for energy-requiring reactions. The ^{32}P turnover of various phosphate compounds of liver decreases in diabetes and increases after insulin administration. Whether this indicates a primary effect of insulin on the usage of the phosphorylated compound or an effect secondary to the action of insulin on glucose utilization is not known. The ATP concentration in livers of alloxan-diabetic rats is low and can be normalized by insulin administration. A direct effect of insulin on the mitochondrial oxidative phosphorylation has not yet been demonstrated.

The effect of insulin on phosphorus metabolism in muscle was studied in three ways: the incorporation of radioactive phosphorus in various phosphorylated compounds, balance studies in isolated diaphragms in which inorganic total phosphorus and labile phosphorus were measured as well as the rate of glycogen synthesis, and *in vitro* phosphorylation studies of creatine.

In vivo incorporation studies with ^{32}P demonstrate that insulin increases ^{32}P incorporation in various phosphorylated compounds, such as glucose-6-phosphate, ATP, and phosphocreatine. The effect of insulin on the incorporation varies with the nature of the final product. In diabetic animals, turnover of the labile phosphorus of ATP and of the phosphorus of glucose-1-phosphate is decreased. This indicates a slower rate of phosphorylation of ATP and hexoses.

Protein Synthesis. Two observations illustrate the role of insulin in protein synthesis: The restoration of normal levels of protein synthesis in diabetics following insulin administration, and insulin's permissive role in stimulating anabolic processes, triggered by sex hormones, gonadotropins and corticosteroids.

In many patients with diabetes there is a negative nitrogen balance and reduced level of protein synthesis, best illustrated in diabetic children who fail to grow. Injection of insulin restores the nitrogen balance to normal. Muscle excised from diabetic animals has a reduced rate of protein synthesis. The addition of insulin *in vitro* restores the anabolic process to normal. The administration of glucocorticoid increases activity of tyrosine amino transferase, the injection of insulin further enhances the enzyme activity.

Adding insulin to media in which diaphragms are incubated with labeled amino acid stimulates the incorporation of the amino acid into protein. The injection of insulin to normal rats increases the incorporation of labeled amino acids into liver microsomes, and insulin stimulates the incorporation of [^{14}C]glycine and [^{14}C]phenylalanine into the protein of liver slices of alloxan-diabetic rats. The stimulation of amino acid incorporation into protein under the effect of insulin is independent of the effect of the hormone on glucose penetration in the cells, since it occurs when glucose is absent from the medium. The effect of insulin on protein synthesis raises the question of insulin's site of action in the sequence of steps that lead to the elaboration of the polypeptide chains including: amino acid transport, transcription or translation.

A potential effect of insulin on transcription is suggested by the effect of actinomycin D on protein synthesis in adipose tissue and liver.

Adipose tissue obtained from diabetic animals shows a decrease of the incorporation of labeled amino acid into protein and in the activity of a number of enzymes: hexokinase II, glucose-6-PO_4 dehydrogenase, phosphofructokinase and pyruvate kinase. Amino acid incorporation and enzyme activities return to normal after injection of insulin. The effect of insulin is, at least in the case of hexokinase II and glucose-6-PO_4 dehydrogenase, impaired by the injection of actinomycin D indicating that insulin stimulates *de novo* protein synthesis by triggering the biosynthesis of new messenger RNA.

Adipose tissue obtained from fasted animals is low in hexokinase II activity. The activity of the enzyme is restored to normal by incubating the tissue with insulin. Labeled histidine is incorporated in the hexokinase. The incorporation of histidine in the enzyme and the increase of activity are blocked by actinomycin D.

In liver insulin stimulates the activity of enzymes that catalyze the anabolic events which lead to storage of glycogen and lipids and depresses catabolic enzymes and gluconeogenesis. Typical of the effect of insulin is its action on glucokinase. Glucokinase activity drops in diabetic rats; restoration of enzyme activity results from *de novo* enzyme synthesis rather than reduced breakdown as shown by the incorporation of labeled precursors. Actinomycin D blocks incorporation of label and development of enzyme activity. The half life of labeled glucokinase is not affected by insulin. To explain this highly specific effect of insulin on stimulation or depression of enzyme synthesis, it has been proposed that insulin acts either directly or through a second messenger at the gene level.

In liver glucocorticoids induce the synthesis of tyrosine aminotransferase. This activity of the enzyme is further increased if insulin is added to the incubation medium, but while actinomycin blocks the enzyme induction by glucocorticoids it is without effect on the stimulation of activity elicited by insulin and therefore, insulin is also believed to act at the level of translation.

Effect of Insulin and Diabetes on Adipose Tissue. For a long time, the role of adipose tissue in metabolism was underestimated. It is now clear that the adipose tissue cell is more than a site of lipid storage. It is also an actively metabolizing cell—performing lipid synthesis and breakdown—with well-developed bioenergetic and biosynthetic pathways. The adipose tissue is formed of closely packed cells filled with lipid. The cells of the adipose tissue are separated by thin layers of collagen fibers, which may or may not contain lymphocytes and plasma cells. The cell of the adipose tissue is large and is filled with lipid droplets with very thin layers of cytoplasm and a small flattened nucleus that is usually pushed out at the periphery of the cell. The fat cells are organized into small groups of cells called lobules. Each lobule has its own connective tissue framework and a network of small capillaries.

Many of the properties of adipose tissue have been discussed elsewhere in this book, and only a brief summary of the metabolic pathways involved in lipogenesis and lipolysis will be presented here. Adipose tissue, like muscle, contains a complete glycolytic pathway; but adipose tissue lacks glucose-6-phosphatase, fructose-1,6-diphosphatase, and glycerokinase. Thus, the pathways for glycolysis in adipose tissue resemble those of muscle rather than liver. Adipose tissue contains no glycerokinase and therefore cannot use the glycerol that is released by the hydrolysis of triglycerides. The L-glycerol-1-phosphate necessary for lipid synthesis must be provided from the oxidation of glucose through glycolysis. Adipose tissue has an active hexose monophosphate shunt, which is stimulated during lipogenesis. Although some assume that an increase in the amount of glucose oxidized through the hexose monophosphate shunt is responsible for the stimulation of lipogenesis, others have proposed that the stimulation of the monophosphate shunt is secondary to increased fatty acid synthesis.

To explain the interaction between fatty acid synthesis and the activity of the hexose monophosphate shunt, researchers have postulated the following sequence of events. When the rate of fatty acid synthesis is increased, the levels of available NADPH are reduced, and the NADPH concentration is increased. The shift in the NADPH/NADP ratio in favor of the oxidized coenzyme stimulates the activity of the hexose monophosphate shunt.

Adipose tissue synthesizes triglycerides at the expense of circulating triglycerides. Lipoprotein lipase digests the triglycerides to yield free fatty acid and glycerol. By a mechanism that is still not elucidated, adipose tissue cells take up the fatty acids and convert them into triglycerides in the presence of α-glycerophosphate, provided through the glycolytic pathway. In adipose tissue, triglycerides are broken down under the influence of a specific lipase to yield glycerol and nonesterified fatty acids, namely, oleic and palmitic acid. The free fatty acids are released from the adipose tissue cell and transported in the serum bound to albumin. The levels of circulating nonesterified fatty acids are extremely low, but their turnover is rapid. Adipose tissue metabolism is controlled by a number of hormones. Only the effects of insulin and glucagon will be discussed here. Insulin stimulates fatty acid uptake and triglyceride synthesis in the adipose tissue *in vivo* and *in vitro*, probably by facilitating the utilization of glucose. The activation of lipid metabolism in adipose tissue under the influence of insulin is accompanied by invagination of the plasma membrane with formation of pinocytic vacuoles.

The mechanism of stimulation of lipogenesis under the influence of insulin is not clear. Theories that have been proposed include a stimulation of α-glycerophosphate synthesis through the Embden-Meyerhof pathway, and a stimulation of NADPH formation. The latter concept is in keeping with the effect of insulin on the utilization of glucose through the hexose monophosphate shunt.

In adipose tissues, insulin accelerates the dissimilation of glucose to CO_2 through the Embden-Meyerhof pathway and the hexose monophosphate shunt and increases its utilization for glycogen and fatty acid synthesis. Insulin is without effect on fatty acid uptake and lipogenesis when glucose is absent from the medium. And the studies of Fain and Loken [132] have established that the antilipolytic effect of insulin is blocked by trypsin. Trypsin does not affect its inhibition through other metabolic interferences. A protein factor, possibly a receptor, probably is needed for insulin's action on adipose tissue.

Insulin deficiency impairs several steps in lipogenesis as a result of: decreased penetration of glucose and utilization in glycolytic pathway with reduced formation of α-glycerophosphate, decreased pyruvate dehydrogenase and acetyl CoA carboxylase activities with reduced fatty acid chain initiation, decreased fatty acid synthetase activity.

Insulin deficiency also stimulates lipolysis resulting in the breakdown of triglycerides with release of free fatty acid, mainly palmityl CoA. The free fatty acids are carried to the liver where palmityl CoA inhibits fatty acid synthesis by interfering with several metabolic steps including: citrate extrusion from mitochondria, initiation and elongation of fatty acids with increased acetyl CoA in mitochondria resulting in conversion of pyruvate to oxaloacetate, which is in turn metabolized to malate. The mitochondrial membrane is permeable to malate, which passes to the cytosol where it is converted to phosphoenolpyruvate which enters the glycolytic pathway in reverse and generates new glucose.

In diabetes, glucose uptake by adipose tissue is reduced, and the dissimilation of glucose through the glycolytic and the pentose pathways, glycogen synthesis, and lipogenesis are impaired. The amount of lipid synthesized from glucose, acetate, and pyruvate is reduced in the adipose tissue of diabetic animals. In contrast, the release of nonesterified fatty acids is accelerated in diabetic animals, an observation of considerable significance for the production of ketosis.

Regulation of Insulin Secretion

Regulation of insulin secretion is often described as a simple feedback loop between glucose and insulin. However, insulin is released by a number of biochemicals, and several hormones directly or indirectly influence insulin release. Moreover, the mode of release varies with the type of stimulus, and acute and delayed responses can be distinguished [133].

Some improvement in understanding of the regulation of insulin secretion was achieved when radioimmunoassays became available for insulin determination.

The basic level of insulin in the blood varies from subject to subject. For example, the basic level is usually higher than normal in obese individuals. The intravenous administration of a single injection of glucose causes an immediate insulin release (with a 2–5-minute peak). The response is acute and makes stored insulin available. Thus, a short feedback loop exists between glucose and insulin stores.

Chronic glucose infusion releases already synthesized, stored insulin and it also induces *de novo* synthesis. After approximately 45 minutes of glucose infusion, the machinery for insulin synthesis is activated.

Not much more is known about this induction of insulin, except that a number of other metabolites—such as fructose, ribose, and xylitol—may, like glucose, elicit delayed insulin responses. Moreover, deoxyglucose and mannoheptulose, which block glucose metabolism, interfere with the insulin response to glucose. On the basis of these findings, it has been proposed that the inducer of the delayed insulin response is not glucose, but one of its metabolites.

A short feedback loop also links insulin stores to fatty acid. Insulin inhibits free fatty acid mobilization and elevation of fatty acid levels in plasma, stimulates insulin release.

One injection or infusion of amino acid stimulates the acute or the delayed response to insulin. The ingestion of food rich in protein gives an intermediate response.

The interpretation of the effects of other hormones on insulin secretion is complicated by the fact that these hormones may directly or indirectly influence carbohydrate metabolism.

Glucagon is believed to stimulate insulin secretion in man. Secretin, pancreozymin, and gastrin have all been found to stimulate insulin release. Acetylcholine and carbamylcholine stimulate insulin secretion *in vitro;* in contrast, epinephrine and norepinephrine inhibit it. Thus, in pheochromocytoma, acute insulin responses are inhibited.

Insulin Binding

If insulin acts by facilitating glucose transport through the cell membrane, the mode of action of insulin must be studied at a level of integration other than the one at which a hormone acts directly on an enzyme.

Working on the assumption that cellular integrity is a prerequisite of insulin activity, Stetten and his associates investigated the binding of insulin to a variety of tissues, including muscle, adipose tissue, and mammary gland. In their first experiment, these investigators incubated one hemidiaphragm in the presence of insulin and the other in the absence of insulin under various conditions of temperature, pH, and insulin concentration. They then measured the insulin effect by determining the increase in glycogen synthesis in the insulinized muscle as compared to that in the noninsulinized muscle. The result demonstrated that insulin binds to the muscle rapidly. With physiological concentration of insulin, the binding is independent of pH, requires a high temperature coefficient, and is unaffected by prolonged washing. The binding also responds to hormonal action, as can be expected from the effect of hormones in intact animals; for example, hypophysectomy increases the rate of binding, whereas diabetes decreases it.

All these findings suggested that insulin binding is not an artifact but reflects a true physicochemical combination between muscle cells and insulin. This conclusion found further support from experiments made by Stetten with radioactive insulin (^{35}S- and ^{131}I-labeled). The use of labeled insulin permitted Stetten to demonstrate that the insulin concentration is much higher within the tissues than in the medium, and the amount of insulin bound per unit mass was found to be the same when the hormone was injected as when it was added to the incubation mixture. After measuring the amount of bound insulin, the investigators could express the metabolic activity due to insulin—namely, glycogen synthesis per unit mass of bound insulin. These studies established that hormonal activity is proportional to the amount of hormone bound to the muscle cell. Hormonal activity was expressed per unit mass for the first time in these experiments.

Insulin binds to epididymal fat as well as to diaphragm. When iodinated insulin is used, most of the insulin can be liberated by adding SH reagents to the incubation mixture. Insulin binding to muscle cells can be prevented by including N-ethylmaleimide in the incubation medium. SH reagents, such as PCMB and iodoacetate, may in a limited way mimic the action of insulin, possibly by combining with the same structural components of the cell as insulin does. PCMB and iodoacetate promote sugar transport into the rat diaphragm.

These preliminary studies on the binding of insulin led to the discovery of insulin receptors.

The binding of insulin to the cell membrane is not random, but takes place at specific sites called receptors. For example, the adipose cell is believed to contain 10,000 overt binding sites and many more occult binding sites. The receptors are located at the periphery of the cell membrane. Experiments in which insulin is bound to plastic beads have demonstrated that the hormone is capable of exerting its metabolic effects on glucose penetration, amino acid uptake, protein synthesis, DNA and RNA synthesis, and lipogenesis.

Similarly, when "ghosts" of lipid cells are inverted, insulin does not bind to the membrane, but when noninverted ghosts are incubated with insulin, the hormone binds and is found, after inversion, inside of the ghost.

Although the details of the chemical composition of the insulin receptors are not known, treatment of membranes with various enzymes has revealed that the receptor is a glycoprotein. A drastic digestion by trypsin markedly reduces the binding of insulin to the membrane by destroying the receptor sites, milder digestion only reduces the affinity of the binding without receptor destruction.

Treatment with phospholipase, instead of reducing the number of binding sites, increases them considerably indicating that a large number of occult binding sites apparently specific for insulin exist. The biological significance of these occult sites remains unknown.

A combined treatment with neuraminidase and β galactosidase abolishes binding of insulin with, of course, elimination of all its physiological effects. Treatment with neuraminidase alone does not prevent the binding, but interferes with the cell's response to insulin as if the presence of sialic acid were indispensable to communicate the message imparted from the formation of the insulin receptor complex to the various cell constituents responsible for the insulin response.

Partial purification of the receptor on affinity chromatography columns made of insulin-agarose have yielded a glycoprotein with a molecular weight of approximately 300,000.

It is not likely that the binding is caused by the formation of covalent bonds as it was believed at first. Weaker types of bonds are believed to be involved since cold bound insulin can readily be replaced by labeled insulin.

The binding of insulin to the receptor is clearly linked to the many metabolic functions of the hormone. Therefore, one may ask what is the line of communication between the receptor hormone complex on the membrane, the change in permeability for glucose, the stimulation of protein synthesis in the endoplasmic reticulum, the sites of synthesis of glycogen and lipids.

Two observations have given some clues as to the mode of transfer of information from membrane to cytoplasm. Insulin inhibits adenylate cyclase and thereby reduces the concentration of cyclic AMP in the cells, but stimulates guanylate cyclase and the synthesis of cyclic GMP, an inhibitor of adenylate cyclase.

The mechanism of inactivation of adenylate cyclase is not known. It may be a direct influence of the insulin receptor complex on the conformation of the membrane enzyme, or it may result from increased levels of cyclic GMP. It is also possible that at least some of the metabolic effects of insulin, for example, glucose transport, are mediated through conformational changes of the membrane. The exact molecular mechanisms by which such changes take place remain to be described.

The role of insulin receptors in stimulation of growth is puzzling. Substances like lectins and concanavalin A, which are capable of eliciting an insulin effect, also stimulate mitosis. Insulin stimulates mitosis in some tissue, for example, the mammary gland. Whether the lectins bind to insulin receptors to stimulate mitosis is not known. The fact is that lectins are capable of stimulating mitosis in cells devoid of insulin receptors such as lymphocytes, possibly because the lymphocyte contains partial insulin receptors. But it cannot be excluded that lectins act at other sites of the membrane at one or more steps in the chain that leads from insulin binding to mitosis. The interpretation of the effect of lectins on mitosis is further complicated by the fact that transformed lymphocytes and lymphocytes obtained from patients with leukemia contain complete insulin receptors.

The role of insulin receptors in insulin resistant diabetes is unknown. In experimental animals the number and, as far as it is known, the structure of the receptor are unchanged in insulin resistance induced by starvation, prednisone or streptozotocin. A decrease of insulin binding to liver cell membranes has been described in obese mice with hypoglycemia, a recessively inherited trait, probably as a result of a decrease of total number of receptors [198].

Ketosis

The liver is exceptionally well equipped to oxidize fatty acids. The ketone bodies (acetone, hydroxybutyrate, and aceto acetic acid) are among the products of fatty acid oxidation.

In contrast to peripheral tissue, which can readily oxidize ketone bodies, the liver is poorly equipped for ketone body utilization. Consequently, whenever the production of ketone bodies in liver exceeds the ability of the peripheral tissue to use them, ketosis takes place [134, 135].

Excess ketone bodies are produced when carbohydrates are not available as a source of energy. In humans, starvation and diabetes are frequently associated with ketosis. In animals, ketosis is often observed in pregnant or lactating ruminants. This discussion will center primarily on the pathogenesis of ketosis in diabetes [136, 137], but first the biochemistry of ketosis will be briefly reviewed.

Ketone Body Metabolism

The β-oxidation of fatty acids involves the conversion of fatty acids to the acyl-CoA derivative, which is in turn reduced and converted to β-hydroxyacyl CoA. After reduction to the β-ketoacyl CoA by the hydroxyl CoA dehydrogenase, the acyl derivative is split in the presence of thiolase to yield acyl-CoA and an acyl derivative of the original fatty acid shorter by two carbons. When even-numbered fatty acids are oxidized, 4-carbon compounds, β-hydroxybutyryl CoA, and acetoacyl CoA are formed. All the reactions of β-oxidation of fatty acids are readily reversible, except for the thiolase reaction, the equilibrium of which lies in the direction of the cleavage.

Acetoacetyl CoA could be the stump of the successive β-oxidation of even-numbered fatty acids, or it could result from the condensation of two molecules of acetyl CoA. In the last step of even-numbered, straight-chain fatty acid oxidation, the split involves a reaction between the acetoacetic enzyme complex and reduced CoA. The product of the reaction is acetyl CoA and a complex between the enzyme and acetic acid.

Acetoacetic acid can be formed from acetyl CoA by two different pathways. The first involves the formation of β-hydroxy-β-methylglutaryl CoA, and a cleavage of that compound to yield acetoacetic acid and acetyl CoA. The second is a simple deacylation of acetoacetyl CoA in the presence of water and acetoacetyl CoA deacylase. The hydroxymethylglutarate CoA condensing enzyme is an SH enzyme that is found in mitochondria and microsomes. The enzyme catalyzing the cleavage of hydroxymethylglutarate CoA is also an SH enzyme, but it is found only in mitochondria. The differences in the intracellular distribution of these two enzymes may be of considerable significance. Hydroxymethylglutarate CoA is a precursor of cholesterol (hydroxymethylglutarate reductase reduces hydroxymethylglutarate CoA to yield mevalonic acid), and in the absence of the cleavage enzyme, it yields cholesterol in microsomes, whereas in mitochondria it is split to acetoacetic acid and acetyl CoA.

Two observations suggest that ketosis in diabetes results from the combination of the accelerated hydroxymethylglutarate shunt and the increased rate of deacylation. Mitochondrial deacylation is the major pathway for converting acetoacetyl CoA to acetoacetic acid in both normal and alloxan-diabetic livers, and a marked increase in activity of acetoacetyl CoA deacylase and a moderate increase in the hydroxymethylglutarate cleavage enzymes is observed in alloxan-diabetic rats.

The activity of hydroxymethylglutarate CoA reductase, which produces mevalonic acid—a precursor of cholesterol—was unchanged in diabetic rats. The observations made on diabetic rats contrast with those made in fasted animals in which ketosis is likely to result from activation of the hydroxymethylglutarate CoA shunt pathway, probably due to decreased activity of the hydroxymethylglutarate CoA reductase. In the presence of NADH and a specific mitochondrial dehydrogenase, acetoacetic acid is reduced to yield β-hydroxybutyric acid, one of the ketone bodies that is excreted in the urine in ketosis. In fact, D-β-hydroxybutyric acid represents 50–75% of the blood content of ketone bodies. Therefore, hydroxybutyric acid metabolism assumes a particular importance.

The conversion of acetoacetic acid to hydroxybutyric acid restitutes the 4-carbon compounds to the general metabolism because the liver contains enzymes that will activate β-hydroxybutyric acid to yield β-hydroxybutyric CoA, whereas acetoacetic acid cannot be acylated by the liver.

A specific dehydrogenase catalyzes the conversion of β-hydroxybutyric CoA to acetoacetyl CoA, which can then be split under the catalytic effect of the thiolase, and the product of the reaction can be further oxidized through the Krebs cycle.

An NADPH-dependent acetoacetyl CoA reductase catalyzes the conversion of acetoacetyl CoA to D-β-hydroxybutyryl CoA. This reaction is the reversal of the β-hydroxybutyryl CoA dehydrogenase reaction. The D-β-hydroxybutyryl CoA dehydrogenase and the reductase have been separated from the L-β-hydroxyacetyl CoA dehydrogenase. Liver mitochondria contain a racemase that catalyzes the conversion of D-β-hydroxybutyryl CoA to L-β-hydroxybutyryl CoA in the presence of NAD. The physiological role of that enzyme remains to be established. Mitochondria also contain a dehydrogenase, L-β-hydroxybutyryl dehydrogenase, which catalyzes the conversion and reduction of acetoacetyl CoA in the presence of NAD.

In view of the complexity of ketone body metabolism, it is not surprising that the mechanism of ketone body production in human diabetes or in pregnant and lactating ruminants is far from understood. Yet the progress of the last decades has permitted the proposals of a number of hypotheses to explain the pathogenesis of ketosis.

The mechanism by which ketosis is produced is unclear. At least five mechanisms can be postulated: (1) reduced usage of the acetyl CoA produced during normal fatty acid oxidation; (2) limited availability of oxaloacetate; (3) acceleration of the rate of fatty acid oxidation; (4) reduced levels of NADPH and (5) a retarded utilization of ketone bodies by peripheral tissues.

The information concerning the last theory is controversial. Whereas Stadie claims that insulin does not affect ketone body usage, the work of several laboratories indicates that insulin facilitates the utilization of ketone bodies in skeletal muscle or in diaphragm of diabetic rats and controlled diabetic rats. (Acetone can be metabolized to acetate, formate, and 3-carbon glycolytic intermediates in rats. Such transformations occur slowly, even in intact rats, and are unaltered by alloxan diabetes.)

In view of the fact that a diet rich in carbohydrates has antiketogenic properties, it was assumed that fats are burned in the fire of the carbohydrate, and that the accumulation of acetyl CoA results from insufficient production of oxaloacetate. Because of these low levels of oxaloacetate, citric acid condensation from oxaloacetate and 2-carbon compounds is not possible.

Since pyruvic acid carboxylation via the malic enzyme is the main source of oxaloacetate, slow glycolysis may result in deceleration of the Krebs cycle. The slowing down of glycolysis may result from reduced enzyme activity or from insufficient amounts of substrate; the former possibility has been eliminated by the experiments of Chaikoff and his group. These investigators injected lactate, pyruvate, and acetate, and measured their conversion to CO_2. They observed that CO_2 formation was the same in diabetics as in nondiabetics. Insufficiency of Krebs cycle substrate is also unlikely, because CO_2 production, which is derived mainly from the tricarboxylic acid cycle, is unimpaired in diabetes. In addition to being used for citric acid formation, acetyl CoA is also a key building block for fatty acids.

The rate of fatty acid synthesis is reduced in diabetes, and, as a result, more acetyl CoA may be available for other metabolic reactions, such as the formation of ketone bodies.

The utilization of [^{14}C]glucose and [^{14}C]fructose for fatty acid synthesis is reduced in liver slices obtained from diabetic rats. The block in fatty acid synthesis was assumed to result from reduced NADPH levels, a consequence of reduced utilization of glucose-6-phosphate through the hexose monophosphate shunt. The reduction in the oxidation of the carbon 1 of glucose could result from poor penetration or from deficient phosphorylation of glucose in the cell, or from increased activity of glucose-6-phosphatase in liver. Relevant to this theory is the fact that acetic acid incorporation into fatty acid of slices of diabetic liver can be stimulated by adding NAD or NADP to the medium, and the stimulation by NADP is much greater than that by NAD.

However, these observations provide only indirect evidence for a role of NADPH in the development of ketosis in diabetes. When attempts were made to measure directly the ratio of NAD to NADH in tissues of diabetic and normal animals, no significant differences were found. This old theory must be evaluated in the light of new information on fatty acid synthesis. Wakil [140] has clearly established that the carboxylation of acetyl CoA is the rate-limiting reaction in fatty acid synthesis. This is a reaction in which acetyl CoA is converted to malonyl CoA in the presence of acetyl CoA carboxylase, a biotin enzyme.

This reaction is stimulated by adding the tricarboxylic acid cycle intermediates, and citric acid is particularly effective in that respect. Thus, is interesting that in starved diabetic rats acetyl CoA carboxylase activity is reduced. Furthermore, Frohman and his associate have found that the concentration of the Krebs cycle intermediate is greatly reduced in liver of diabetic rats [141].

Another mechanism by which fatty acid synthesis could be inhibited in diabetic rats is a feedback inhibition of the acetyl CoA carboxylase by the longer chain length fatty acid. Fatty acid synthesis in crude liver and avocado homogenates can be blocked by adding free fatty acid or albumin-bound fatty acid to the media. But the acyl-CoA derivative of the free fatty acid rather than the free fatty acid appears to be the inhibitor. The acetyl CoA accumulation could result from the inhibition of citrate synthetase by palmityl CoA described by Lynnen. In any case diabetic ketosis is unlikely to result from retarded fatty acid synthesis because a similar and sometimes greater retardation of fatty acid synthesis is observed in starvation, and ketone body accumulation in starvation does not compare to such accumulation in diabetes.

Chaikoff and his associates [138, 139] have studied simultaneously the breakdown of [^{14}C]palmitate to acetyl CoA and the reusage of the labeled activated acetate, and they demonstrated that the reduced synthesis accounts for only a small fraction of the accumulated 4-carbon compounds.

Since retarded usage does not explain the accumulation of ketone bodies, theories invoking accelerated production have been proposed. The exact mechanism by which the acceleration of fatty acid oxidation is triggered is not clear, but without insulin in diabetic animals the blood levels of nonesterified fatty acid increase. In addition, nonesterified fatty acids released in the blood are oxidized in the liver; and Chaikoff and his associates have demonstrated that [^{14}C]palmitate is oxidized twice as fast in diabetic as in normal liver.

To explain how increased fatty acid oxidation might lead to ketosis, investigators have proposed that in the face of accelerated fatty acid oxidation, the Krebs cycle becomes more truly rate limiting and 2-carbon compounds accumulate.

One theory on the mechanism of ketosis production proposes that the amount of oxaloacetic acid produced in diabetes is too small to permit the condensation with acetyl CoA to form the 6-carbon Krebs cycle substrates. Such a theory seems to imply that the oxidation rate of the Krebs cycle substrates is reduced in diabetes. But we have seen that early evidence on the oxidation of Krebs cycle substrates in diabetes is controversial. Indeed, the dissimilation of [^{14}C]acetate to $^{14}CO_2$ is as rapid in diabetic rats as in nondiabetic rats. Yet different results were obtained in diabetic dogs in which the breakdown of [^{14}C]acetate to $^{14}CO_2$ is impaired.

In human diabetes, there is some indication that insulin stimulates the utilization of pyruvate through the tricarboxylic acid cycle, whereas in diabetes above

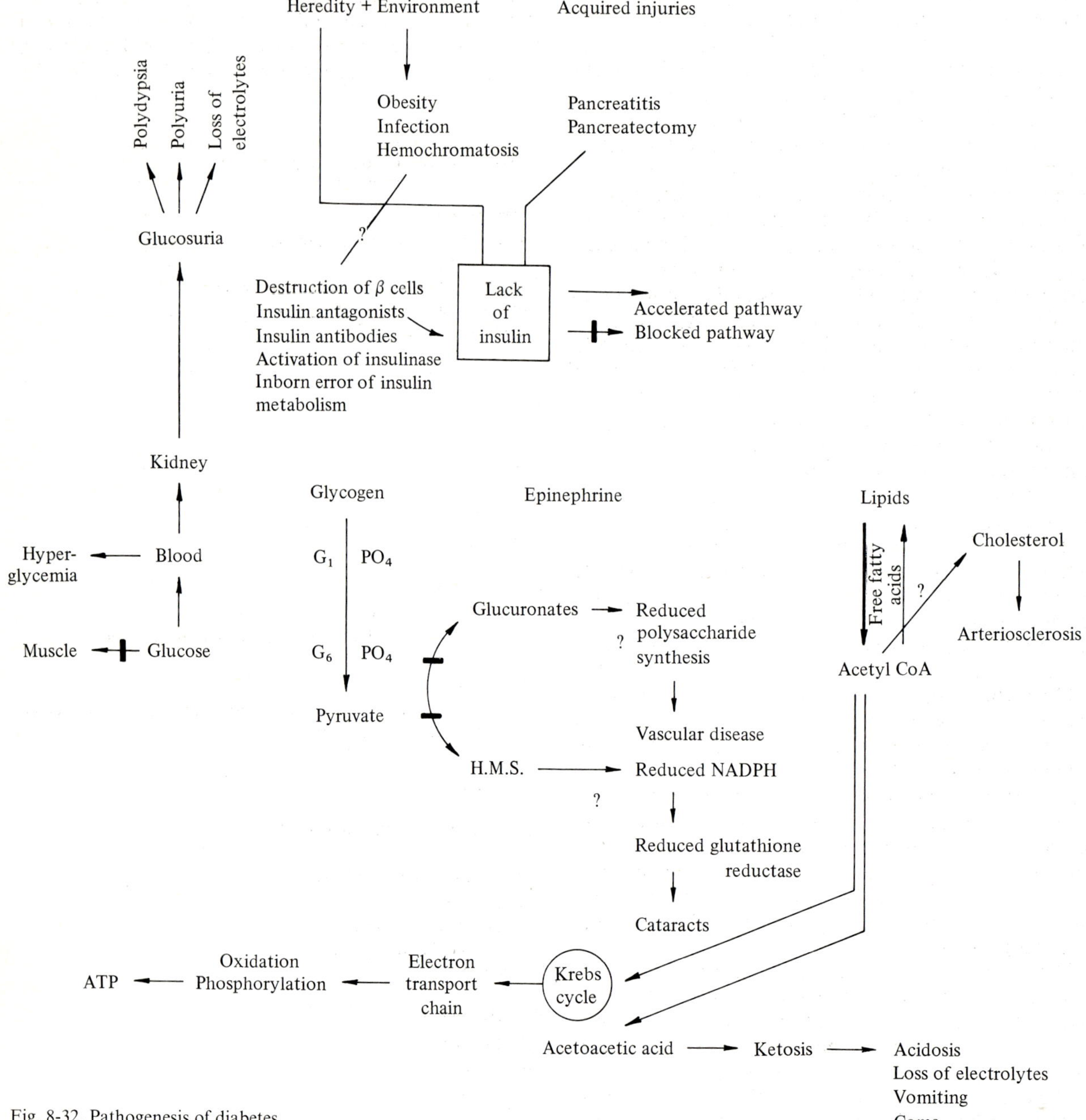

Fig. 8-32. Pathogenesis of diabetes

normal amounts of pyruvate are detracted from the tricarboxylic acid cycle to be used for glucose formation through the reversal of the glycolytic pathway. But the administration of a Krebs cycle substrate, such as malate or succinate, has no effect on ketonuria.

However, the role of oxaloacetate in the pathogenesis of ketosis has become clearer with our knowledge of the mechanism regulating gluconeogenesis and the function of ketone bodies as respiratory fuel.

Only in the 1960's was it established that peripheral tissue, such as heart and kidney, could actively metabolize ketone bodies to use as respiratory fuel. In fact, glucose and fatty acids complement each other as respiratory fuels. Consequently, when glucose is low (starvation or ketosis in cattle) or unavailable (diabetes), glucagon is secreted, inducing lipogenesis in adipose tissue with liberation of free fatty acids.

The oxidation rate of fatty acids appears to be proportional to the concentration in plasma, and, as mentioned above, major products are ketone bodies, which can serve as respiratory fuel. In moderate forms of ketosis, glucagon secretion is antagonized by that of insulin; consequently ketone bodies do not accumulate because ç their utilization by peripheral tissues is accelerated and their genesis interrupted. In several forms of ketosis, homeostasis breaks down. In diabetics, the lipolytic effect of glucagon is not compensated for, and the increase in free fatty acids and ketone bodies in plasma is unchecked. Moreover, utilization of ketone bodies by peripheral tissue could be reduced.

An important consequence of this distortion of homeostasis is that the more severe the gluconeogenesis, the more severe the ketosis. One substrate—oxaloacetate—is the pivot that regulates the coincidence of the two processes. Oxaloacetate is at the crossroads between glycolysis, gluconeogenesis, and the Krebs cycle.

It is formed from the carboxylation of pyruvate and, to a lesser degree, from the deamination of aspartate or through the metabolic conversion of other ketogenic amino acids through a pathway involving transamination and successive oxidation from succinate to fumarate and malate.

The carboxylation of pyruvate condenses the tricarbon compound with a molecule of acetyl CoA; consequently, oxaloacetate formation drives acetyl CoA out of the metabolic traffic.

In gluconeogenesis, phosphopyruvate carboxylase permits the surmount of a thermodynamic hurdle and determines the conversion from oxaloacetate back to pyruvate, which can now enter the gluconeogenic stream, provided, of course, that the common steps for glycolysis and gluconeogenesis are placed on the gluconeogenic track.

The preponderance of gluconeogenesis in diabetes is achieved by increasing the amount of phosphopyruvic carboxylase. Oxaloacetate is consequently withdrawn from the oxidative pathways, and, as a result, the combustion of acetyl CoA is impaired.

The overproduction of nonesterified free fatty acid in diabetes has also been suggested as the primary injury in the development of ketosis; free fatty acids have been assumed to accumulate in liver cells and uncouple oxidative phosphorylation. This assumption is based on an observation made by Pressman and Lardy indicating that free fatty acids uncouple oxidative phosphorylation. Without insulin, the nonesterified free fatty acids are released from the adipose tissue, and their serum concentration is increased. The increase in serum concentration of nonesterified free fatty acids leads to increased uptake of the free fatty acid by the liver cell. The excess intracellular free fatty acids uncouple oxidative phosphorylation. Loss of energy resulting from uncoupling leads to accelerated fatty acid oxidation with an increased production of 2-carbon compounds. If the activity of the Krebs cycle is assumed to be rate limiting for acetyl CoA oxidation, there will be an excess of 2-carbon compounds, which are at the source of the ketone bodies found in ketosis. The pathogenesis of diabetes is summarized in Fig. 8-32.

Table 8-8. Classification of Hypoglycemia

Excess Insulin
Islet tumors—benign or malignant
Islet hyperplasia
Tumors secreting insulin or insulinlike substances
Administration of exogenous insulin
Infants of diabetic mothers
Insulin Release
Leucine
Some drugs
Glucagon Deficiency
Lack of Glucose
Starvation
Rapid Glucose Intake
Rapid glucose intake with insulin response
Rapid absorption of glucose in postgastrectomy patients
Defective Gluconeogenesis
Generalized hepatic dysfunction (hepatitis, cirrhosis, cancer, metastasis, chronic passive congestion etc.)
Prematurity
Drugs: alcohol
Automonic Nervous System Disorder
Miscellaneous Endocrine Disorders
Hypopituitarism
Adrenocortical insufficiency
Adrenal medulla unresponsiveness
Myxedema
Miscellaneous Hereditary Disease
Galactosemia
Glycogen storage disease
Maple syrup disease
Fructosuria

Hypoglycemia

Hypoglycemia, low levels of sugar in the blood, is associated with a number of diseases; a classification is presented in Table 8-8 [142, 143].

Clinicopathological Correlation

Symptoms appear only when the level of glucose in the blood drops to 40 mg/100 ml. The pathogenesis of the symptoms involves an epinephrine response and cerebral dysfunction due to lack of glucose. After a rapid fall of blood glucose, the patient becomes nervous and sweaty; his heart palpitates, his face flushes, and he feels hungry. As the drop in blood glucose continues, cerebral manifestations begin. They usually consist of headache, visual disturbance, thick speech, bizarre behavior related to fasting and physical activity, and twitching; but they may develop into transitory hemiplegia and seizures.

The cause of the condition is usually obvious if one remembers the mechanism controlling glucose levels in the blood.

Pancreatic Hypoglycemic Tumors

The pathogenesis of many of these hypoglycemic conditions has been discussed in other chapters of this book. Only the special cases—pancreatic and nonpancreatic tumors, leucine initiation of hypoglycemia, and some hypoglycemic agents—will be considered here.

Two major forms of islet tumors are adenomatosis of the islet and generalized hyperplasia of the islet. Both conditions may be associated with hyperinsulinism. Islet cell adenomas are often small (see Fig. 8-33), well-localized tumors; occasionally they are multiple. Some tumors grow into such large masses that they cause pressure syndromes. Pancreatic hyperinsulinic tumors are rarely malignant. The therapy is obvious—excision of the tumor. These tumors are often difficult to diagnose. Danish clinicians have improved chances of accurate localization by the use of arteriography.

Still, in most cases, the surgeon must perform a blind partial pancreatectomy. Fortunately, such a procedure is successful more often than would be expected on the basis of chance because most functional islet tumors are in the tail or body of the pancreas. Since diabetes seldom develops if approximately one-third of the pancreas is left in place, a blind partial pancreatectomy including the tail and an adjacent portion of the body may include the islet tumor but will not result in hypoinsulinism. If, on histological examination, generalized hyperplasia rather than an isolated tumor is found, total pancreatectomy may be indicated.

Histologically, islet tumors are formed from chords of islet cells. The benign tumors are well encapsulated. The appearance of the individual cells in the malignant tumor is not very different from that in the benign tumor, but there is no fibrous capsule in the former. In contrast, surrounding tissues are invaded, and metastasis may develop. Occasionally, benign or malignant tumors of the islet cells may appear in ectopic pancreatic formation.

The major morbid changes observed after hypoglycemia occur in the brain. Glucose is the brain's primary source of food; the brain's respiratory quotient is 1, and it uses 100 g of glucose a day. Histological changes in the brain of hypoglycemic patients resemble those observed in hypoxia.

Nonpancreatic Hypoglycemic Tumors

A number of nonpancreatic tumors induce hypoglycemia. These tumors may be located anywhere in the

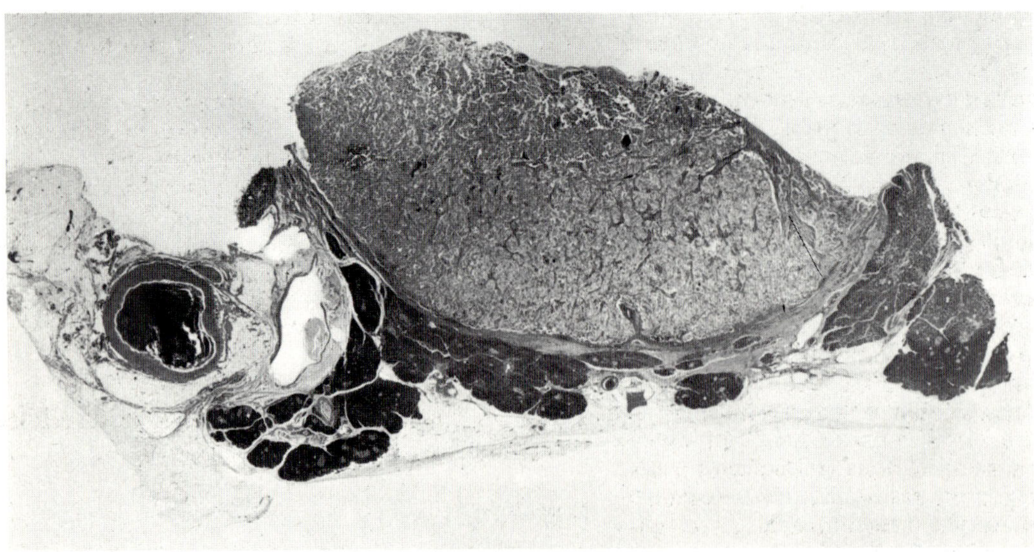

Fig. 8-33. Low-power microscopic view of an islet cell tumor. The tumor is separated from the normal pancreas by a thin capsule

body, but they are usually of mesenchymal origin. Some nonpancreatic hypoglycemic tumors are fibrosarcomas, mesotheliomas, neurofibromas, neurofibrosarcomas, rhabdomyosarcomas, and lymphosarcomas. Sometimes hypoglycemia develops with a spindle cell carcinoma. While the pathogenesis of the hypoglycemia is unclear, it seems that the insulin level in the blood is not increased, and the glucose levels of the arteries and veins supplying the tumor do not differ significantly.

Hypoglycemia usually appears to result from excessive usage of glucose by the tumor, possibly because of the presence inside the tumor of insulin or insulinlike substances.

In some cases it has been proposed that the tumor elaborates substances that block hepatic glucose release and adipose tissue lipolysis. Although this type of tumor is usually incurable, excision of the tumor usually eliminates the hypoglycemia.

Various Hypoglycemic Agents

Administration of leucine to adults or young children has no or little effect on insulin release, but in some children leucine injection causes hypoglycemia and increased levels of circulating insulin.

These patients have convulsions at an early age, abnormal EEG patterns, and glucose levels of 30 and 31 mg/100 ml after oral administration of leucine. Daily administration of diazoxide may satisfactorily control the hypoglycemia.

A number of chemicals or drugs produce hypoglycemia. Alcohol is believed to exert its hypoglycemic effect by decreasing hepatic release of glucose. Propranolol blocks the release of insulin in normal humans whose pancreas has been stimulated with isoproteranol. Propranolol should be used with discretion in patients susceptible to hypoglycemia.

A number of hypoglycemic agents interfere with gluconeogenesis. Tryptophan acts indirectly via a metabolite, quinaldinic acid, which prevents the conversion of oxaloacetate to phosphoenolpyruvate. The hypoglycemic action of alanine is probably related to the metabolite quinaldinic acid. Other hypoglycemic agents, such as 5-methoxyindole-2-carboxylic acid, act differently from quinaldinic acid in that they are glycogenolytic, increase the plasma concentration of nonesterified fatty acid, and decrease blood levels of ketone bodies. Incubation of mitochondria with these two hypoglycemic agents leads to interference with the oxidation and carboxylation of pyruvate. Adding octanoate to the incubation mixture restores the processes to normal; moreover, when octanoate is included in the perfusion medium of liver, it prevents the alteration and restores normal gluconeogenesis. Therefore, investigators believe that the quinaldate and the 5-methoxyindole-2-carboxylate act on gluconeogenesis by interfering with mitochondrial pyruvate metabolism [144].

Hypoglycemia in the Newborn

In the newborn infant there is a very dramatic switch from dependency on the mother to self reliance for sources of energy. In general the newborn is well prepared for this change because the fetus uses most of the glucose provided by the mother to build up glycogen and lipid stores.

The glycogen stores in the fetus at term are twice those in the adult in the liver and ten times those in adult muscle. After birth the liver stores of glycogen are almost exhausted within hours because of conversion to glucose. The muscle stores take longer to be depleted.

The lipid stores are also mobilized after birth. Within minutes the blood levels of free fatty acid increase markedly as a result of lipolysis. The fatty acid oxidation yields acetyl CoA which is converted to ketone bodies which can readily be used as a source of energy by the brain. Of course with normal feeding the muscle and liver stores are soon restored.

Thus, in the newborn glycogenolysis, gluconeogenesis and lipolysis are placed in gear at a rate far greater than in the adult. These various pathways involve batteries of enzymes with multifacet regulations of their catalytic properties. Therefore it is not surprising that the delicate mechanism that has evolved in the newborn to provide glucose is easily distorted and that the newborn is particularly susceptible to hypoglycemia.

The causes of hypoglycemia in the newborn fall in two categories: idiopathic defects in the production of chemical sources of energy and hyperinsulinemia.

Defective levels of glucose may result from malnutrition of the mother, hypoxia, prematurity or enzyme defects impairing glycogenolysis, gluconeogenesis or lipolysis.

The hypoglycemia of prematurity results in part from immatiurity of the enzyme machinery needed to mobilize or build up the stores and in part from the large size of the brain relative to liver. The brain/liver mass ratio exceeds what it is at term, therefore, the liver cannot provide the chemical fuel needed by the brain.

Hyperinsulinemia occur in insulinomas, erythroblastosis fetalis and diabetes. The causes of hyperinsulinemia in erythroblastosis fetalis are unknown.

The fetus depends upon the mother for its supply of glucose, the glucose levels in fetal blood are increased when the mother is diabetic causing stimulation of insulin secretion by the fetus with hyperplasia of the islet and hyperinsulinemia in the fetus with increased storage of glycogen and fat.

Common Denominators in Hormone Action: Receptors and Second Messengers

Receptors

The preceding paragraphs introduced a number of circulating chemicals of various compositions, including complex proteins, which elicit specific metabolic changes in target organs. In their journey through the body, these substances encounter millions of cells that remain indifferent to their presence, but a few cells establish an intimate relationship with the hormone, which ultimately reprograms the metabolism of these cells. What determines such specificity and controls when cell and hormone interact, what molecular mechanisms are activated for the purpose of reprogramming cellular function? Although a detailed description of the events involved in the implementation of hormone action is still not within our reach, the mysteries of endocrinology are slowly giving place to concrete molecular descriptions.

In the discussion of diabetes, the specificity of insulin binding to target organs was mentioned. The mode of action of estrogens illustrates that specific cytoplasmic receptors have been isolated from various target tissues. From these findings and others, the concept of receptors emerged. Thus, the specificity of hormone action is now believed to be determined by the presence in the target cell of special molecules with high affinity for the hormone. The receptor can be located anywhere in the cell, provided, of course, that the cell membrane is permeable to the hormone. With large hormonal molecules (such as insulin), this arrangement is unlikely, but smaller molecules (such as corticosteroids) can penetrate the membrane and reach the cytoplasm. These small molecules may bind to a protein receptor, and under the influence of cytoplasmic protease become attached to nuclear receptors [145].

Insulin stimulates the utilization of glucose and the formation of glycerol, and fatty acid in adipose cells. The hormone binds to the adipose cell membrane through a specific receptor, which is located on the exterior surface of the cell membranes. Insulin bound to saccharose beads can bind to the surface and also elicits the hormonal effects in the target cell. The mechanism by which binding of the hormone to the receptor activates glucose transport is not known, but there is no indication that other intermediates, such as cyclic AMP, are involved. The receptor appears to contain several sulfhydryl groups essential to binding a polypeptide portion with rapid turnover.

The glucagon receptor is a glycoprotein, but in contrast to what is observed with insulin, treatment with phospholipase inhibits hormone receptors binding rather than activating it.

Epinephrine and norepinephrine receptors have been detected on the membrane of the target cells. They are much more numerous than insulin receptors and less specific in that stereoisomers show similar binding properties, but elicit different physiological responses.

Acetylcholine receptors are found on the muscle plates and have been isolated from cells of electric organs. They are certainly lipoproteins and possibly glycolipoproteins. Their concentration on the surface of the target organ is much greater than that for insulin. These differences in concentration of receptor sites are believed to have physiological meaning.

The intracellular type of receptors has been best studied with glucocorticoids, especially dexamethasone, which *in vivo* and *in vitro* affects various target tissue. One effect of dexamethasone is the induction of aminotransferase in hepatoma cells (see section on gene expression). Tompkins and his associates have established that the steroid enters the cells and binds to specific cytoplasmic proteins. The bound complex is then activated by proteases and becomes attached to the nucleus. This is rapidly followed by elaboration of the messenger RNA for the transaminase. Cytoplasmic receptors for corticosteroids are lacking in fetal tissues, and the absence of such receptors is believed to be responsible for the inability of fetal tissue to respond to corticosteroid induction. Corticosteroid receptors have been found in lung, the complete differentiation of which depends upon the availability of corticosteroids, and in the hypophysis, in which ACTH secretion is feedback regulated by corticosteroids [146].

Clearly, all the recently discovered binding sites for hormones cannot be reviewed here [147–158].

Clear-cut situations in which disease results from the absence of receptors have not been convincingly established, except perhaps in the obese diabetic rat in which insulin resistance apparently results from the inability to bind the hormone. We have already mentioned that the parathormone receptor defect might be responsible for pseudohypoparathyroidism. A similar mechanism has been proposed for inherited vasopressin-resistant diabetes insipidus.

Cyclic AMP

Introduction

With the possible exception of prostaglandins, few discoveries in the last decennium have been as far reaching and have had as unpredictable results as the discovery of cyclic AMP (see Fig. 8-34). The word cyclic AMP and the name Sutherland are almost synonymous because the pioneering studies and the original working hypothesis emanated from Sutherland's laboratory [159–164].

Hormones are chemicals elaborated in a source organ and transported to the target organ where they exert specific effects. The simplest conceivable

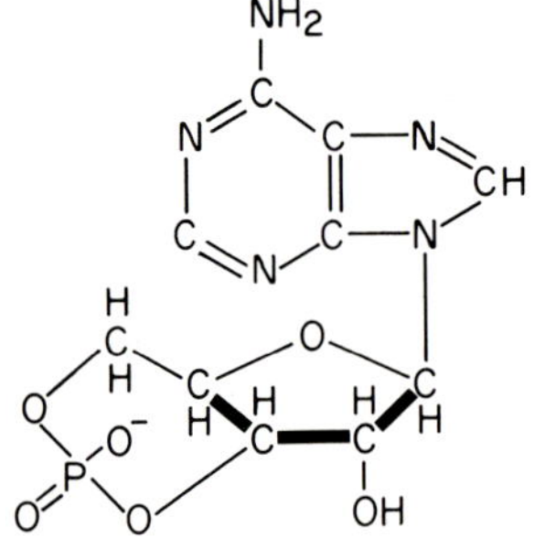

Fig. 8-34. Formula of cyclic AMP

mechanism for hormone action is the application of the stimulus followed by the immediate release of a preexisting hormone, and transport of the hormone to the target organ, which is directly triggered. Although such a mechanism may function in some cases, on the basis of studies to be described, Sutherland proposed that many hormones elicit their effects through a chemical intermediate that activates whatever events are triggered by the hormone in the target cell. Such intermediates are called second messengers. Although cAMP is the prototype of second messengers, other known chemicals are also believed to act as such—namely, cyclic GMP and prostaglandins—and other second messengers probably remain to be discovered.

The significance of the function of cAMP is best understood by reviewing the original experiments. Around 1940, some of the unanswered questions in biochemistry concerned glycogen metabolism, its synthesis and breakdown (see Chapter 1). The study of glycogen breakdown to glucose-1-phosphate by epinephrine and glucagon (the hyperglycemic factor) brought what seemed to be a simple metabolic conversion into the broad scope of endocrinology.

Sutherland was among those who attempted to dissect the molecular mechanism of action of epinephrine and glucagon. He discovered that these hormones catalyzed a kinase that converted an inactive into an active phosphorylase, and with his colleagues he demonstrated the conversion in homogenates. That was, in fact, the first demonstration of a hormonal effect in a cell-free system. To improve on the system, the investigators tested the hormone on a high-speed supernatant known to contain the phosphorylase, but they found it to be ineffective. Ingenious recombination experiments established that the particulate fraction (nuclei, mitochondria, ER, cell membranes, etc.) produced a thermostable factor, which when added to the high-speed supernatant mimicked the hormonal effect. The thermostable compound was identified and is now known as cAMP. It was soon established that cAMP is formed from ATP through the catalytic action of an enzyme, adenyl cyclase, which was later established to be located in the cell membrane. Moreover, the cell was found to contain a phosphodiesterase that readily converts cAMP to AMP. As work accumulated, what started as an unexpected finding expanded into an elegant generalization for hormone action, and more and more hormones (today more than a dozen) were found to act through cAMP.

A number of criteria were defined to establish whether a hormonal effect is mediated through cAMP. First, the concentration of cAMP must rise before hormonal effects become apparent. Second, the response of the target tissue must be proportionate to the amount and the biological activity of the primary messenger. Third, antagonists to a hormone that raises cAMP should reduce the levels of cAMP. Fourth, inhibitors of those enzymes that interfere with cAMP breakdown should enhance the hormonal effect.

In addition to the study of glycogenolysis, a classical example of the mediation of hormone action by cAMP, other studies include epinephrine stimulation of heart contraction and the induction of lipolysis by a variety of hormones. Epinephrine stimulates cardiac contraction almost instantaneously, the stimulation is inhibited by adrenergic blocking agents, and cAMP mimics the effect of epinephrine. When epinephrine is added to the isolated heart, the levels of cAMP are considerably increased within minutes, and when contraction is blocked, the levels of cAMP are reduced in proportion to the effectiveness of the blocking agent.

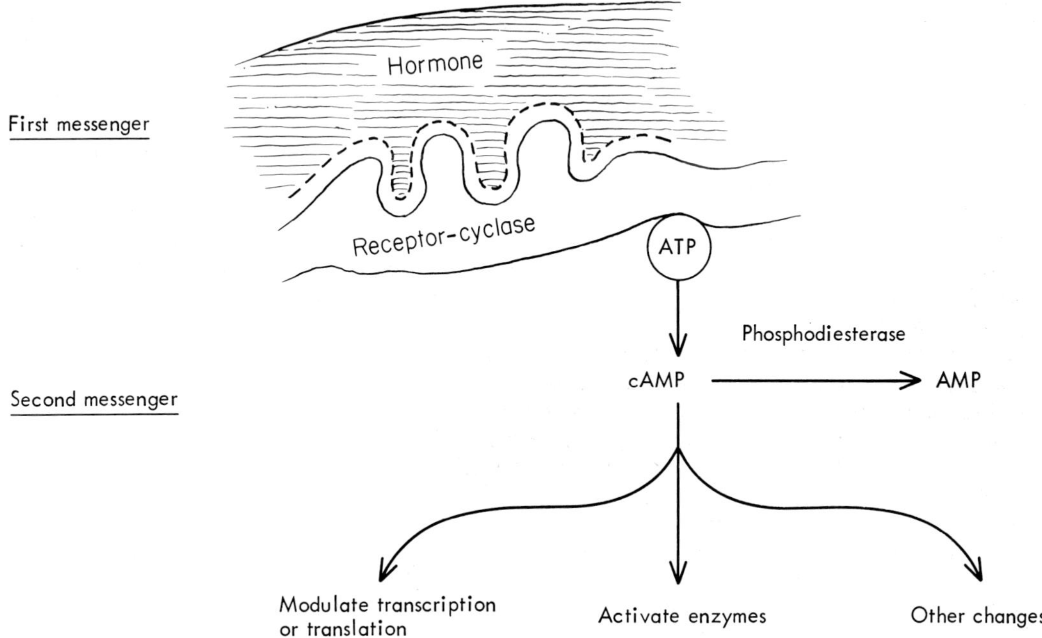

Fig. 8-35. Mode of action of cyclic AMP

At least seven hormones, including ACTH, epinephrine, and glucagon, attack the adipose tissue and lead to the breakdown of triglycerides to yield free glycerol and fatty acids. After each hormone has attached to its own specific receptor, they all act through a common mechanism—the stimulation of adenyl cyclase—with proportionate elaboration of cAMP. The cAMP activates the lipase, possibly by phosphorylation of the enzyme. Insulin and certain prostaglandins inhibit lipolysis and decrease the level of cAMP in the cell. In contrast, theophylline and caffeine inhibit the phosphodiesterase and thereby increase lipolysis in the fat cells (see Fig. 8-35).

Cyclic AMP Metabolism

Adenyl Cyclase. Adenyl cyclase seems to be ubiquitous and has been found in bacteria, higher plants, and mammals. It is particulate, and therefore, must be attached to the membranes of the endoplasmic reticulum, the mitochondria, or the plasma membrane. At least in some cases, like the fat cell, adenyl cyclase has been shown to be located in the plasma membrane. The enzyme has been postulated to exist in various isozymic forms. However, the evidence for this is unconvincing, and it is generally accepted that when different hormones exert their effects on a cell by raising the levels of cAMP, they do so by activation of the same cyclase.

In the case of glucagon an activating protein called the transducer is believed to be intercalated between the adenyl cyclase and the receptor. The transducer is activated by GTP.

Glucagon + receptor
↓ GTP
Activation of transducer
↓
Activation of adenyl cyclase
↓ cAMP
Activation of protein kinase
↓
Phosphorylase kinase
↓
Activate glycogen phosphorylase

The mechanism of action of adenyl cyclase stimulators or inhibitors is unclear except for fluoride, which stimulates purified preparations of the enzyme. All other known inhibitors or stimulators are inactive on the purified enzyme, but they can modulate its activity when the enzyme is in the particulate form. Therefore, investigators have proposed that binding to specific membrane receptors modifies the configuration of membrane-bound adenyl cyclase. The existence of receptors specific for the various lipolytic hormones has been demonstrated on the fat cell membrane. All such receptors (except those for epinephrine) are destroyed by treatment with trypsin, which suggests that they are proteins. The specificity of the mode of action of hormones whose action is mediated through cAMP has been explained by proposing that the adenyl cyclase is composed of two segments—a specific hormonal receptor and a common catalytic unit. Binding of the hormone to the receptor activates the catalytic unit, and AMP is converted to cAMP. The phenotype of the target cell then determines which events are put in gear by the cAMP.

cAMP Phosphodiesterase. cAMP phosphodiesterases have been found in bacteria and mammals, in mammals they have been purified from heart, brain, liver. The heart and liver enzymes are found in the cytosol, the brain enzyme is found partly in the cytosol and partly in the microsomes.

Little is known about the regulation of cAMP phosphodiesterase activity except that most mammalian enzymes are activated by imidazoles and inhibited by methyl xanthines. An effect of insulin on the diesterase has been claimed but not confirmed, and insulin is generally believed to interfere with lipolysis by inhibiting the cyclase.

Mode of Action of Cyclic AMP

Because many hormones that act through cAMP are also responsible for elaborating new enzyme molecules that can be inhibited by anctinomycin, researchers have suggested that cAMP affects some cells by modulating transcription. Convincing evidence for a role of cAMP in regulating transcription comes from experiments in bacteria relating to β-galactosidase induction (see Chapter 2), in which exogenous cAMP overcomes catabolite repression in *E. coli.*

In mammalian cells, direct evidence for cAMP's role in modulating transcription is unavailable, and the proposition that cyclic AMP affects transcription by phosphorylating histone or nonhistone nuclear proteins is premature since such an effect has not been convincingly established, and it is not known whether such phosphorylation is involved in transcription. Similarly, an effect of cAMP on translation in eukaryotes is questionable. Both inhibitory and stimulatory effects have been claimed. All effects of cAMP in cells are more likely to be traced to direct action on enzymes; namely, the conversion of an inactive to an active form, or *vice versa.*

The best known example for the mode of action of cAMP on metabolic regulation is its effect on glycogen metabolism (Chapter 1). Cyclic AMP exerts its glycogenolytic effect by simultaneously stimulating glycogen catabolism and inhibiting its anabolism. This is achieved by converting an inactive phosphorylase to an active phosphorylase, and an active glycogen synthetase to an inactive glycogen synthetase. In either case, cAMP activates a protein kinase, which converts a dephosphorylated protein to a phosphorylated protein. The phosphorylated protein is active in phosphorylase, inactive in glycogen synthetase.

Cyclic AMP stimulates glycolysis in the liver fluke *Fasciola hepatica* and in mammalian heart muscles.

Such an effect appears to be mediated by regulating the activity of the rate-limiting enzyme of glycolysis, phosphofructokinase. Cyclic AMP affects such activity by relieving the inhibition by ATP of effective levels of the substrate fructose-1,6-phosphate and by converting an inactive form of phosphofructokinase to an active form through subunit aggregation.

Hormones that stimulate steroidogenesis in the adrenal, the testis, and the corpus luteum—such as ACTH, luteinizing hormone, and prostaglandins (E_1, E_2, and E_1-α)—all stimulate the adenyl cyclase of the target cell, and their effects be mimicked by cAMP.

In the acrenal, a hormone's effect on steroidogenesis concerns at least three steps in steroid metabolism: (1) a protein synthesis-independent stimulation of the conversion of cholesterol ester to cholesterol; (2) an actinomycin-, cycloheximide—, and puromycin-sensitive stimulation of the conversion of cholesterol to pregnenolone; and (3) an activation of C-11 β-hydroxylase, the enzyme that converts 11-deoxycorticosterone to corticosterone.

Cyclic AMP activation plays a double role: amplification and regulation.

The enzyme substrate relationship is far from a monogamous one in which one molecule of enzyme binds and converts one molecule of substrate, but rather a more or less fast changing polygamous one in which the enzyme reacts with a large number of molecules of substrate in a unit of time (enzyme turnover). Let's assume that the turnover of enzyme I is 100, and that the product of enzyme I is enzyme II with a turnover of 50, and that the product of enzyme II is enzyme III with a turnover of 200. Because of the cascade of reactions that precede the formation of enzyme III, the number of molecules produced by enzyme III will be $100 \times 50 \times 200$ rather than the 200 that would have been made by a reaction scheme involving only enzyme III. It has been estimated that because of the amplification caused by the cascade of reaction, a single molecule of glucagon can release 10^3 molecules of glucose from glycogen.

Cyclic AMP modulates a great number of complex metabolic reactions, gluconeogenesis, lipogenesis, mitosis, etc. One may wonder how one single compound can be responsible for such a variety of biochemical changes in the cell. Specificity is, in part, determined by the hormonal receptor at the surface of the cell membrane and by the enzymic mosaic of the target cell. But are there other common denominators to the action of cAMP in all cells? Protein kinases have been proposed as the common target of cAMP. Little is known about the molecular properties, specificities, and effectors of protein kinase.

In general protein kinases are composed of two subunits: an active and an inhibitory one. The activation of the enzyme by cAMP results from the combination of the nucleotide with the inhibitor. Protein phosphorylation by protein kinase results from the transfer of one phosphate group of ATP to a serine residue in the enzyme sequence. Surely some protein kinases are activated by cAMP which bind to the inhibitory portion of the molecule, but it is not certain that all protein kinases are cAMP dependent. Whenever the concentrations of cAMP are no longer large enough to complex with the inhibitory subunit, the activated protein kinases are deactivated by phosphoprotein phosphatases.

Every cell contains protein kinases. Sometimes several different protein kinases are found in the same cells. Although it has not been established that all protein kinases are activated by cAMP, it is certain that many are. Neither is there much information on the specificity of protein kinase. It would seem that in most cases any protein kinase can phosphorylate any protein in spite of the molecular diversity of the kinase. Among the proteins that can be phosphorylated are plasma membrane, microtubules, microsomes, nucleic acid proteins, and histones.

A good illustration of the regulatory properties of cAMP occurs in liver stimulated by glucagon. cAMP activates a protein kinase which phosphorylates both phosphorylase and glycogen synthetase, but while phosphorylase is activated, glycogen synthetase is inhibited by the phosphorylation. Consequently, cAMP at the same time stimulates glycogen breakdown and inhibits glycogen synthesis.

Similarly, a complex sequence of steps leads to the synthesis of glycogen (see metabolic regulation, Chapter 16). A protein kinase transforms active glycogen synthetase (I form) into an inactive phosphorylated form (D form). There are also inactive and active forms of the protein kinase; the conversion to the active enzyme is cAMP dependent. It is believed that insulin reduces the level of cAMP and thereby inhibits the protein kinase and the glycogen synthetase [200].

The events triggered by hormones and sometimes by cAMP are unidirectional (for example, the stimulation of steroidogenesis in the adrenal) or bidirectional (for example, stimulation of mitosis with low doses of cAMP, and inhibition of mitosis at high doses of cAMP; see Chapter 16). A new interpretation of the regulation of the bidirectional events rests on the discovery of cyclic GMP. Again, the stimulation of myocardial contraction by epinephrine or isoproteranol induces a proportionate increase in cAMP levels, and a decrease in cGMP also takes place. The acetylcholine-induced decrease in myocardial contractility is associated with a twofold increase in cGMP and only a small or delayed decrease in cAMP.

Cyclic GMP was found in urine in 1963, but its metabolic role and origin were unknown until it was discovered that cGMP is present in mammalian tissues including brain, liver and kidney. A guanylate cyclase which synthesizes cGMP was discovered in most mammalian tissues. In fact, the enzyme is found in other organisms such as insects. In contrast to adenylate cyclase, guanylate cyclase is found not only in the cell membrane, but in the cytoplasm. Hormones that stimulate adenyl cyclase do not stimulate guanylate cyclase. Cellular concentrations of cGMP are 100 times lower than that of cAMP. When a search was made for a role of cGMP in cellular metabolism, it was found that under most circumstances the cGMP acted antagonistically to cAMP and thereby served as a regulatory counterpart of cAMP.

The first example came from the study of acetylcholine on the myocardium. Acetylcholine is an α-adrenergic agent. In myocardium, it decreases the rate and force of contraction, in contrast to epinephrine (β-adrenergic agent), which increases these. The administration of acetylcholine increases the intracellular concentrations of cGMP in heart 300%.

Several other hormones increase the intracellular levels of cGMP including polypeptide hormones, oxitocin, insulin, calcitonin, angiotensin and cholecystokinin, α-adrenergic agents, serotonin, histamine, epinephrine and prostaglandin $F_2\alpha$.

While cAMP hyperpolarizes the postsynaptic membrane, cGMP depolarizes it. At least some of the hormonal action can be mimicked provided that low doses of cGMP are used. Thus while 10^{-10} M cGMP elicits acetylcholinelike effects on heart cells in culture, 10^{-10} M cGMP stimulates DNA synthesis in thymic lymphocytes, 10^{-9} enhances chemotaxis toward human lymphocytes.

An example of the antagonistic roles of cAMP and cGMP relates to their effect on cell division. Cyclic AMP and agents that stimulate cAMP formation inhibit the sequence of reactions that stimulate cell division in rats, mice and human lymphocytes grown in culture in presence of phytohemagglutinin. In contrast, in inhibited cells stimulated to divide by the addition of fresh serum, cGMP is increased and cAMP is decreased. In endometrial cells stimulated by estradiol, cGMP is increased. cGMP has been shown to be able to substitute for the cocarcinogen in two stage carcinogenesis induced by benzenepyrene. The levels of cGMP are double in skin cells of psoriatic patients compared to those of skin cells in normal individuals [201].

Calcium and Hormone Action

As pointed out by Rasmussen [201, 203] the stimulation of cyclic AMP is unidirectional. We know how it is produced, but we do not know how it is turned off after the cell response. We have seen many examples in which the ultimate product of a biological stimulus is inhibited by the initial stimulus thus closing what is called a feedback loop (see Inborn Errors—Hormones). Rasmussen believes that if the actions of calcium and cAMP are coupled, such a feedback loop can be achieved. Calcium concentrations in the extracellular fluid, the cytosol and the mitochondria are 10^{-3}, 10^{-7}, 10^{-2} molar respectively, thus the calcium pool in the cytosol is much lower than that of the extracellular pool and that of the mitochondria is greater than that of the extracellular fluid and cytosol. Most of the calcium in mitochondria is found in the form of a calcium PO_4 salt, but the concentration of ionic calcium is still 10^{-4}. To build up such great stores of calcium energy is needed, and one can only assume that such stores of calcium play an important role in cellular metabolism. The work of Rasmussen and his associates suggests that the flow of calcium from various cellular compartments contributes to the convergance of the hormonal message.

There are many examples in mammalian cells in which the effectuation of hormone action is associated with variations in calcium concentration in the different cellular compartments. Such hormonal actions include stimulation of corticosteroid secretion in the adrenal, the effect of parathyroid hormone on renal tubules, reabsorption and release of calcium from the bone, the effect of melanocyte hormone on melanophores, the stimulation of smooth muscle contraction by acetylcholine and its retraction by epinephrine, the effect of epinephrine on heart contraction, etc. The role of calcium in the sequence of steps following the hormonal stimulus is often difficult to interpret.

Rasmussen and his associates have, however, clarified the role of calcium fluxes in the secretion of the salivary gland of the blowfly. Such flies possess, in their abdomen, a noninnervated gland which secretes isotonic KCl for the purpose of dissolving the food taken in by the insects. Serotonin release stimulates secretions as well as an increase in cAMP concentrations. In absence of calcium, serotonin induces the formation of cAMP, but there is no secretion. Addition of calcium to a gland primed with serotonin or the simultaneous addition of calcium and cAMP restores the secretion process. These and other observations clearly indicate that cAMP and calcium act synergistically.

Studies of calcium movement under the influence of serotonin and cAMP revealed that serotonin stimulates both calcium uptake and calcium loss from the secretory cell. cAMP does also stimulate calcium efflux. In addition to stimulating K^+ secretion, cAMP is also believed to increase the levels of calcium of the cytosol by releasing calcium from the mitochondrial stores and to modulate the permeability of the cell membrane to chloride ions.

These various findings, seemingly puzzling and sometimes contradictory, have been organized into a plausible working hypothesis by Rasmussen and his associates. The cycle starts with the application of the stimulus, serotonin, which activates adenylcyclase and increases cAMP.

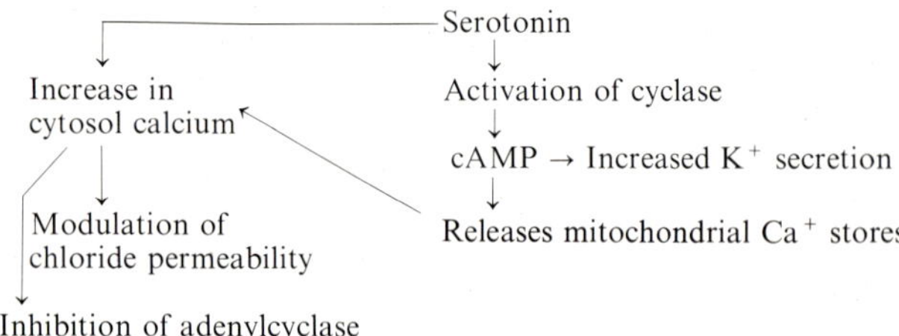

At the same time serotonin increases the permeability to calcium. Since the extracellular calcium concentrations are much higher than that of the cytosol, the net result is a flow of calcium from the extracellular fluid into the cytosol. cAMP stimulates K^+ secretion and releases the stores of calcium from mitochondria thereby contributing the high calcium levels in the cytosol.

The high cytosol concentration of calcium stimulates chloride secretion and inhibits adenylcyclase, thus closing the feedback loop. Only three examples of the role of calcium as an endocrine messenger will be given.

The administration of glucose leads to the release of insulin from pancreatic β-cells. The release can be mimicked by the administration of dibutyric cAMP and by theophylline, which interferes with cAMP breakdown. Calcium deprivation *in vivo* or *in vitro* interferes with insulin release. Again, these findings suggest a synergic effect of calcium and cAMP in the release of insulin by high blood glucose concentrations. The evidence suggests that (1) glucose increases the levels of calcium in the cytosol, either by blocking cytosol efflux or by stimulating cytosol influx. (2) cAMP releases calcium from the mitochondrial stores and thereby increases the calcium levels of the cytosol; (3) the insulin release is not due to the formation of cAMP, but to these two independent modes of increasing calcium concentration in the cytosol.

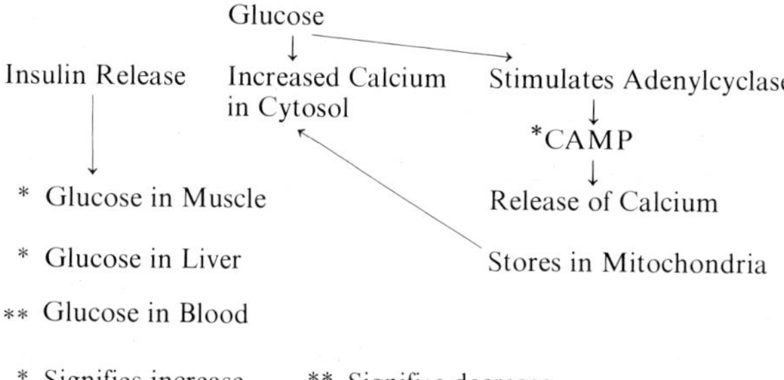

Hormones secreted in the hypothalamus, which releases pituitary hormones, are believed to bind to specific receptors of the membranes of the pituicyte. This action is dependent upon the presence of Ca^{++}. Therefore, it has been proposed that the hormone depolarizes the cell membrane increasing the permeability to calcium which would, in turn, release the hormone. Whatever the role of calcium, it is known in the case of thyroid releasing hormone that it stimulates adenylcyclase and cAMP.

During neurogenic stimulation of muscle, cAMP activates phosphorylase, thus converting glycogen to glucose-1-PO_4, the source of energy for muscle contraction. Phosphorylase is, however, also activated by calcium and it is therefore significant that neurogenic stimulation of muscle also induces a flow of calcium inside the muscle cells. Cooperative action between cAMP and calcium seems to be needed to produce the hormonal effect.

Conclusion

The integration of optimal metabolic activity in the body depends upon the elaboration of humoral messengers including hormones. The mechanisms that determine the appearance of a hormone in the blood are varied. In the simplest form, a single metabolite (*e.g.*, glucose) induces the blood level of the hormone needed for its utilization (*e.g.*, insulin). In other situations a complex neuroendocrine machinery is put in gear before the desired metabolic change in the target cell takes place. Thus, a stimulus of the central cortex brings about the formation of releasing factors, which in turn make trophins available. The released trophins (*e.g.*, ACTH) reach an endocrine source (*e.g.*, the adrenal), where they stimulate the elaboration of specific hormones (*e.g.*, corticosteroids). These hormones in turn bring about the expected metabolic changes in the target organ (see Fig. 8-36).

A central problem in endocrinology is that of recognizing the target organ by a specific humoral factor. The recognition is molecular as it is for enzymes and substrate, antigen and antibody. The target cells contain specific molecules, or receptors, with special affinity for the hormone in question. The receptor may be at the surface of the cell, and then the binding of the hormone must be coupled with changes in membrane permeability (*e.g.*, for glucose in the case of insulin) or with activation of enzymes capable of elaborating second messengers (*e.g.*, adenyl cyclase). The receptors may be intracellular, thus the hormone may directly or after metabolic conversion affect a unit of specificity (DNA, RNA, proteins) or a catalytic unit (substrate, enzyme, product).

Clearly, new unifying concepts have emerged in hormone action, but much more research is needed before each molecular step linking stimulation to effect for individual hormone can be described, and some of the most subtle hormonal effects may remain to be discovered.

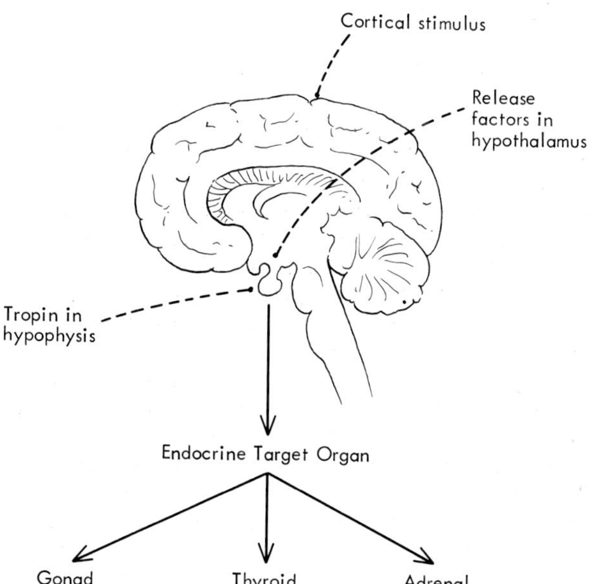

Fig. 8-36. Schematic representation of hormonal action

References

1. Knobil, E., Hotchkiss, J.: Growth hormone. Annu. Rev. Physiol. **26**, 47–74 (1964)
2. Daughaday, W.H., Parker, M.L.: Human pituitary growth hormone. Annu. Rev. Med. **16**, 47–66 (1965)
3. Armstrong, D.T.: Reproduction. Annu. Rev. Physiol. **32**, 439–470 (1970)
4. Behrens, O.K., Grinnan, E.L.: Polypeptide hormones. Annu. Rev. Biochem. **38**, 83–112 (1969)
5. Blizzard, R.M. (ed.): Human pituitary growth hormone, Report, 54th Ross Conf. Pediatric Res., Ross Laboratories, Columbus, Ohio (1966)
6. Hanson, L.A., Roos, P., Rymo, L.: Heterogeneity of human growth hormone preparations by immuno-gel filtration and gel filtration electrophoresis. Nature (Lond.) **212**, 948–949 (1966)
7. Li, C.H., Liu, W.K., Dixon, J.S.: Human pituitary growth hormone. XII. The amino acid sequence of the hormone. J. Amer. chem. Soc. **88**, 2050–2051 (1966)
8. Schwyzer, R.: Chemistry and metabolic action of nonsteroid hormones. Annu. Rev. Biochem. **33**, 259–286 (1964)
9. Holmes, L.B., Frantz, A.G., Rabkin, M.T., Soeldner, J.S., Crawford, J.D.: Normal growth with subnormal growth-hormone levels. New Engl. J. Med. **279**, 559–566 (1968)
10. Lewis, U.J., Cheever, E.V., Van der Laan, W.P.: Studies on the growth hormone of normal and dwarf mice. Endocrinology **76**, 210–215 (1965)
11. Rimoin, D.L., Merimée, T.J., Rabinowitz, D., McKusick, V.A., Cavalli-Sforza, L.L.: Growth hormone in African pygmies. Lancet **II**, 523–526 (1967)
12. Kleeman, C.R., Cutler, R.E.: The neurohypophysis. Annu. Rev. Physiol. **25**, 385–432 (1963)
13. Share, L.: Vasopressin; its bioassay and the physiological control of its release. Amer. J. Med. **42**, 701–712 (1967)
14. Wollman, S.H.: Concentration and organic binding of radioiodine by the thyroid gland. Ann. N.Y. Acad. Sci. **86**, 354–361 (1960)
15. Pitt-Rivers, R.: Some factors that affect thyroid hormone synthesis. Ann. N. Y. Acad. Sci. **86**, 362–372 (1960)
16. Rall, J.E., Robbins, J., Edelhoch, H.: Iodoproteins in the thyroid. Ann. N. Y. Acad. Sci. **86**, 373–399 (1960)
17. Stanbury, J.B.: Deiodination of the iodinated amino acids. Ann. N. Y. Acad. Sci. **86**, 417–439 (1960)
18. Degroot, L.J.: Stimulation and inhibition of thyroid iodinating enzyme systems. Biochim. biophys. Acta (Amst.) **136**, 364–374 (1967)
19. Fraser, R.: Endocrinology: The thyroid. Annu. Rev. Med. **11**, 171–186 (1960)
20. Solomon, D.H., Dowling, J.T.: The thyroid. Annu. Rev. Physiol. **22**, 615–650 (1960)
21. Ingbar, S.H., Galton, V.A.: Thyroid. Annu. Rev. Physiol. **25**, 361–384 (1963)
22. Rosenberg, I.N., Bastomsky, C.H.: The thyroid. Annu. Rev. Physiol. **27**, 71–106 (1965)
23. Hodgson, S.F., Wahner, H.W.: Hereditary increased thyroxine-binding-globulin capacity. Mayo Clin. Proc. **47**, 720 (1972)
24. Schussler, G.C.: Thyroxine-binding globulin: Specificity for the hormonally active conformation of triiodothyronine. Science **178**, 172 (1972)
25. Lehninger, A.L.: Thyroxine and the swelling and contraction cycle in mitochondria. Ann. N. Y. Acad. Sci. **86**, 484–493 (1960)
26. Bronk, J.R.: The influence of thyroxine and related compounds on oxidative rate and efficiency of phosphorylation in liver mitochondria and submitochondrial particles. Ann. N. Y. Acad. Sci. **86**, 494–505 (1960)
27. Lardy, H.A., Lee, Y.P., Takemori, A.: Enzyme responses to thyroid hormones. Ann. N. Y. Acad. Sci. **86**, 506–511 (1960)
28. Money, W.L., Kumaoka, S., Rawson, R.W., Kroc, R.L.: Comparative effects of thyroxine analogues in experimental animals. Ann. N. Y. Acad. Sci. **86**, 512–544 (1960)
29. Tata, J.R.: Basal metabolic rate and thyroid hormones. Advanc. Metab. Disord. **1**, 153–184 (1964)
30. Sokoloff, L., Roberts, P.A., Januska, M.M., Kline, J.: Mechanisms of stimulation of protein synthesis by thyroid hormones *in vivo*. Proc. nat. Acad. Sci. (Wash.) **60**, 652–659 (1968)
31. Vecchio, G., Salvatore, M., Salvatore, G.: Biosynthesis of thyroglobulin *in vivo*: Formation and polymerization of subunits in the rat and guinea pig. Biochem. biophys. Res. Commun. **25**, 402–408 (1966)
32. Pastan, I.: Biochemistry of the nitrogen-containing hormones. Annu. Rev. Biochem. **35**, 369–404 (1966)
33. Bronk, J.R.: Thyroid hormone: Effects on electron transport. Science **153**, 638–639 (1966)
34. Cohen, P.P.: Biochemical aspects of metamorphosis: Transition from ammonotelism to ureatelism. Harvey Lect. Series **60**, 119–154 (1964–1965)
35. Werner, S.C., Nauman, J.A.: The thyroid. Annu. Rev. Physiol. **30**, 213–244 (1968)
36. Pitt-Rivers, R., Tata, J.R.: The chemistry of thyroid diseases. Springfield, Ill.: Charles C. Thomas Publisher 1960
37. Bogdanove, E.M.: Regulation of TSH secretion. Fed. Proc. **21**, 623–627 (1962)
38. Nadler, N.J.: Synthesis and release of thyroid hormones. Fed. Proc. **21**, 628–629 (1962)
39. Guillemin, R.: The adenohypophysis and its hypothalamic control. Annu. Rev. Physiol. **29**, 313–348 (1967)
40. Burgus, R., Guillemin, R.: Hypothalamic releasing factors. Annu. Rev. Biochem. **39**, 499–526 (1970)
41. Blagdon, D.E., Rivier, J., Goodman, M.: Proposed tertiary structure for the hypothalamic thyrotropin-releasing factor. Proc. nat. Acad. Sci. (Wash.) **70**, 1166 (1973)
42. Halmi, N.S.: Thyroidal iodide transport. Vitam. and Horm. **19**, 133–163 (1961)
43. Moon, H.D. (ed.): The adrenal cortex. New York: Paul B. Hoeber 1961
44. Greep, R.O., Deane, H.W.: The cytology and cytochemistry of the adrenal cortex. Ann. N. Y. Acad. Sci. **50**, 596 (1969)
45. Reichstein, T., Shoppee, C.W.: The hormones of the adrenal cortex. Vitam. and Horm. **1**, 346–413 (1943)
46. Dorfman, R.I., Ungar, F.: Metabolism of steroid hormones. New York: Academic Press 1965
47. Sih, C.J., Whitlock, H.W., Jr.: Biochemistry of steroids. Annu. Rev. Biochem. **37**, 661–694 (1968)
48. Samuels, L.T., Eik-Nes, K.B.: Metabolism of steroid hormones. In: Metabolic pathways (Greenberg, D., ed.) 3rd ed., vol. II, p. 169. New York: Academic Press 1967
49. Tomkins, G.M., Maxwell, E.S.: Some aspects of steroid hormone action. Annu. Rev. Biochem. **32**, 677–708 (1963)
50. Engel, L.L., Langer, L.J.: Biochemistry of steroid hormones. Annu. Rev. Biochem. **30**, 499–524 (1961)
51. Litwack, G., Kritchevsky, D. (eds.): Actions of hormones on molecular processes. New York: John Wiley&Sons 1964
52. Seubert, W., Henning, H.V., Schoner, W., L'äge, M.: Effects of cortisol on the levels of metabolites and enzymes controlling glucose production from pyruvate. Advanc. Enzyme Regul. **6**, 153–187 (1968)
53. Söling, H.D., Kaplan, J., Erbstoeszer, M., Pitot, H.C.: The role of hormones in glucose repression in rat liver. Advanc. Enzyme Regul. **7**, 171–182 (1969)
54. Scrutton, M.C., Utter, M.F.: The regulation of glycolysis and gluconeogenesis in animal tissues. Annu. Rev. Biochem. **37**, 249–302 (1968)
55. Van Lancker, J.V.: Hydrolases and cellular death. In: Metabolic conjugation and metabolic hydrolyses (Fishman, W.H., ed.), vol. 1, p. 356–418. New York: Academic Press 1970
56. Szentágothai, J., Flerkó, B., Mess, B., Halász, B.: Hypothalamic control of the anterior pituitary; an experimental-morphological study. Budapest: Publishing House of Hungarian Acad. Sci. 1962
57. Randle, P.J.: Endocrine control of metabolism. Annu. Rev. Physiol. **25**, 291–324 (1963)
58. Venning, E.H.: Adenohypophysis and adrenal cortex. Annu. Rev. Physiol. **27**, 107–132 (1965)
59. McCann, S.M., Dhariwal, P.S., Porter, J.C.: Regulation of the adenohypophysis. Annu. Rev. Physiol. **30**, 589–640 (1968)
60. Reichlin, S.: Functions of the median-eminence gland. New Engl. J. Med. **275**, 600–607 (1966)
61. Bransome, E.D., Jr.: Stimulation and inhibition of adrenal RNA synthesis by ACTH. Curr. Mod. Biol. **1**, 21–23 (1967)
62. Ney, R.L., Davis, W.W., Garren, L.D.: Heterogeneity of template RNA in adrenal glands [rat]. Science **153**, 896–197 (1966)
63. Roberts, S., McCune, R.W., Creange, J.E., Young, P.L.: Adenosine 3′,5′-cyclic phosphate: Stimulation of steroidogenesis in sonically disrupted adrenal mitochondria. Science **158**, 372–374 (1967)
64. Taunton, O.D., Roth, J., Pastan, I.: Studies on the adrenocorticotropic hormone-activated adenyl cyclase of a functional adrenal tumor. J. biol. Chem. **244**, 247–253 (1969)
65. Ross, E.J., Marshall-Jones, P., Friedman, M.: Cushing's syndrome: Diagnostic criteria. Quart. J. Med. **35**, 149–192 (1966)
66. Bondy, P.K.: Metabolism: The adrenal cortex. Yb. Med. 469–492 (1967–1968)
67. Espiner, E.A.: Urinary cortisol excretion in stress situations and in patients with Cushing's syndrome. J. Endocr. **35**, 29–44 (1966)
68. Nelson, D.H., Sprunt, J.G., Mims, R.B.: Plasma ACTH determinations in 58 patients before or after adrenalectomy for Cushing's syndrome. J. clin. Endocr. **26**, 722–728 (1966)
69. Jones, I.C., Bellamy, D.: Hormonal mechanisms in the homeostatic regulation of the vertebrate body with special reference to the adrenal cortex. Soc. exp. Biol. Symp. (Great Britain) XVIII, 195–236 (1964)
70. Crane, M.G., Harris, J.J.: Desoxycorticosterone secretion rates in hyperadrenocorticism. J. clin. Endocr. **26**, 1135–1143 (1966)
71. Richardson, G.S.: Ovarian physiology (continued). New Engl. J. Med. **274**, 1008–1015 (1966)
72. Kowal, J.: Metabolic events associated with steroid biosynthesis in adrenal tissue cultures. Trans. N. Y. Acad Sci. **31**, 359–378 (1969)
73. Samuels, L.T.: Metabolism of the steroid hormones. Prog. Chem. Fats Other Lipids **3**, 395 (1955)
74. Kupperman, H.S.: The endocrine status of the transsexual patient. Trans. N. Y. Acad. Sci. **29**, 434–439 (1967)
75. Valadares, J.R.E., Singhal, R.L., Parulekar, M.R.: 17 β-estradiol: Inducer of uterine hexokinase. Science **159**, 990–991 (1968)
76. Ui, H., Mueller, G.C.: The role of RNA synthesis in early estrogen action. Proc. nat. Acad. Sci. (Wash.) **50**, 256–260 (1963)
77. Hamilton, T.H.: Control by estrogen of genetic transcription and translation: Binding to chromatin and stimulation of nucleolar RNA synthesis

are primary events in the early estrogen action. Science **161**, 649–661 (1968)
78. Wicks, W.D., Kenney, F.T.: RNA synthesis in rat seminal vesicles: Stimulation by testosterone. Science **144**, 1346–1347 (1964)
79. Mueller, G.C., Herranen, A.M., Jervell, K.F.: Studies on the mechanism of action of estrogens. Recent Progr. Hormone Res. **14**, 95–129 (1958)
80. McCorquodale, D.J., Veach, E.G., Mueller, G.C.: The incorporation *in vitro* of labeled amino acids into the proteins of normal and regenerating rat liver. Biochim. biophys. Acta (Amst.) **46**, 335–343 (1961)
81. Gorski, J., Toft, D., Shyamala, G., Smith, D., Notides, A.: Hormone receptors: Studies on the interaction of estrogen with the uterus. Recent Progr. Hormone Res. **24**, 45–80 (1968)
82. Shyamala, G., Gorski, J.: Estrogen receptors in the rat uterus: Studies on the interaction of cytosol and nuclear binding sites. J. biol. Chem. **244**, 1097–1103 (1969)
83. Jensen, E.V., Hurst, D.J., DeSombre, E.R., Jungblut, P.W.: Sulfhydryl groups and estradiol-receptor interaction. Science **158**, 385–387 (1967)
84. Jensen, E.V., Jacobson, H.I.: Basic guides to the mechanism of estrogen action. Recent Progr. Hormone Res. **18**, 387–414 (1962)
85. Cross, B.: The hypothalamus in mammalian homeostasis. Soc. exp. Biol. Symp. (Great Britain) XVIII, 157–193 (1964)
86. Jones, G.S.: Induction of ovulation. Annu. Rev. Med. **19**, 351–372 (1968)
87. Everett, J.W.: Central neural control of reproductive functions of the adenohypophysis. Physiol. Rev. **44**, 373–431 (1964)
88. Federman, D.D.: Disorders of sexual development. New Engl. J. Med. **277**, 351–360 (1967)
89. Richardson, G.S.: Ovarian physiology (continued). New Engl. J. Med. **274**, 1121–1134 (1966)
90. Hirschhorn, K.: Chromosomes: Growth and development: Chromosomal influences on sexual differentiation. In: Environmental influences on genetic expression: Biological and behavioral aspects of sexual differentiation (Kretchmer, N., and Walcher, D.N., eds.), No. 2, p. 83–90. Washington, D.C.: Superintendent of Documents 1969
91. Segre, E.J., Klaiber, E.L., Lobotsky, J., Lloyd, C.W.: Hirsutism and virilizing syndromes. Annu. Rev. Med. **15**, 315–324 (1964)
92. Kendall, J.W., Sloop, P.R., Jr.: Dexamethasone-suppressible adrenocortical tumor. New Engl. J. Med. **279**, 532–535 (1968)
93. Lazarus, S.S., Volk, B.W.: The pancreas in human and experimental diabetes. New York: Grune & Stratton 1962
94. Wells, L.J., Lazarow, A.: Organ cultures of pancreases of fetuses from diabetic rats: Effects of high-glucose media upon the granulation of the beta cells and upon the insulin content of the media. Diabetes **16**, 846–851 (1967)
95. Lacy, P.E.: Pathology of the islets of Langerhans. Path. Annu. **1**, 352–370 (1966)
96. Bloodworth, J.M.B., Jr.: Histochemistry and electron microscopy of diabetic retinopathy. In: Small blood vessel involvement in diabetes mellitus (Siperstein, M.D., Colwell, S.R., and Meyer, K., eds.). Proc. Symp. p. 81–87. Washington, D.C.: Amer. Inst. Biol. Sci. 1964
97. Israel, M.: Diabetic retinopathy. Roslyn Heights, New York: Vascular Research Foundation 1968
98. Churg, J., Dachs, S.: Diabetic renal disease: arteriosclerosis and glomerulosclerosis. Path. Annu. **1**, 148–171 (1966)
99. Gullick, H.D.: Carcinoma of the pancreas; a review and critical study of 100 cases. Medicine (Baltimore) **38**, 47–84 (1959)
100. Farquhar, M.G., Hopper, J., Jr., Moon, H.D.: Diabetic glomerulosclerosis: Electron and light microscopic studies. Amer. J. Path. **35**, 721–753 (1959)
101. Lendrum, A.C.: The hypertensive diabetic kidney as a model of the so-called collagen diseases. Canad. med. Ass. J. **88**, 442–452 (1963)
102. Dolman, C.L.: The morbid anatomy of diabetic neuropathy. Neurology (Minneap.) **13**, 135–142 (1963)
103. Vallance-Owen, J.: Insulin antagonists and inhibitors. Advanc. Metab. Disord. **1**, 191–215 (1964)
104. Riklis, E., Quastel, J.H.: Effects of metabolic inhibitors on potassium-stimulated glucose absorption by isolated surviving guinea pig intestine. Canad. J. Biochem. **36**, 363–367 (1958)
105. Levine, R., Mahler, R.: Production, secretion, and availability of insulin. Annu. Rev. Med. **15**, 413–432 (1964)
106. Porte, D., Jr., Bagdade, J.D.: Human insulin secretion: An integraded approach. In: Annual review of medicine (DeGraff, A.C., and Creger, W.P., eds.), vol. 21, p. 219 –240. Palo Alto, Calif.: Annual Reviews Inc. 1971
107. Rubenstein, A.H., Steiner, D.F.: Proinsulin. In: Annual review of medicine (DeGraff, A.C., and Creger, W.P., eds.), vol. 22, p. 1–18. Palo Alto, Calif.: Annual Reviews Inc. 1971
108. Katsoyannis, P.G.: Synthetic insulins. Recent Progr. Hormone Res. **23**, 505–563 (1967)
109. Steiner, D.F.: Evidence for a precursor in the biosynthesis of insulin. Trans. N. Y. Acad. Sci. **30**, 60–68 (1967–1968)
110. Steiner, D.F., Cunningham, D., Spigelman, L., Aten, B.: Insulin biosynthesis: Evidence for a precursor. Science **157**, 697–700 (1967)
111. Kreisberg, R.A., Boshell, B.R., DiPlacido, J., Roddam, R.F.: Insulin secretion in obesity. New Engl. J. Med. **276**, 314–319 (1967)
112. Stadie, W.C.: Current concepts of the action of insulin. Physiol. Rev. **34**, 52–100 (1954)
113. Rieser, P.: Insulin, membranes and metabolism. Baltimore: Williams & Wilkins 1967
114. Krahl, M.E.: The action of insulin on cells. New York: Academic Press 1961
115. Levine, R., Haft, D.E.: Carbohydrate homeostasis (Part II). New Engl. J. Med. **283**, 237–246 (1970)
116. Villee, C.A., White, V.K., Hastings, A.B.: Metabolism of C^{14}-labeled glucose and pyruvate by rat diaphragm muscle *in vitro*. J. biol. Chem. **195**, 287–297 (1952)
117. Villee, C.A., Hastings, A.B.: The metabolism of C^{14}-labeled glucose by the rat diaphragm *in vitro*. J. biol. Chem. **179**, 673–687 (1949)
118. Shaw, W.N., Stadie, W.C.: Coexistence of insulin-responsive and insulin-non-responsive glycolytic systems in rat diaphragm. J. biol. Chem. **227**, 115–123 (1957)
119. Shaw, W.N., Stadie, W.C.: Two identical Embden-Meyerhof enzyme systems in normal rat diaphragms differing in cytological location and response to insulin. J. biol. Chem. **234**, 2491–2496 (1959)
120. Chaikoff, I.L.: Metabolic blocks in carbohydrate metabolism in diabetes. Harvey Lect. Series **47**, 99–125 (1951–1952)
121. Williamson, J.R., Kreisberg, R.A., Felts, P.W.: Mechanism for the stimulation of gluconeogenesis by fatty acids in perfused rat liver. Proc. nat. Acad. Sci. (Wash.) **56**, 247–254 (1966)
122. Randle, P.J., Garland, P.B., Hales, C.N., Newsholme, E.A., Denton, R.M., Pogson, C.I.: Interactions of metabolism and the physiological role of insulin. Recent Progr. Hormone Res. **22**, 1–48 (1966)
123. Levine, R., Goldstein, M.S.: On the mechanism of action of insulin. Recent Progr. Hormone Res. **11**, 343–380 (1955)
124. Fisher, R.B.: Insulin and the transport of sugars. In: The mechanism of action of insulin (Young, F.G., Broom, W.A., and Wolff, F.W., eds.), p. 35–46. Oxford: Blackwell Scientific Publications 1960
125. Ross, E.J.: Insulin and permeability of cell membranes to glucose. Nature (Lond.) **171**, 125 (1953)
126. Ross, E.J.: Influence of insulin on permeability of blood-aqueous barrier to glucose. J. Physiol. (Lond.) **116**, 414–423 (1952)
127. Cori, C.F.: Enzymatic reactions in carbohydrate metabolism. Harvey Lect. Series **41**, 253–272 (1945–1946)
128. Colowick, S.P., Cori, C.F., Slein, M.W.: The effect of adrenal cortex and anterior pituitary extracts and insulin on the hexokinase reaction. J. biol. Chem. **168**, 583–596 (1947)
129. Weil-Malherbe, H.: An inhibitor of hexokinase in the plasma of diabetics. Nature (Lond.) **165**, 155 (1950)
130. Bornstein, J., Park, C.R.: Inhibition of glucose uptake by the serum of diabetic rats. J. biol. Chem. **205**, 503–511 (1953)
131. Hastings, A.B., Renold, A.E., Teng, C.T.: Effects of ions and hormones on carbohydrate metabolism. Recent Progr. Hormone Res. **11**, 381–400 (1955)
132. Fain, J.N., Loken, S.C.: Response of trypsin-treated brown and white fat cells to hormones. Preferential inhibition of insulin action. J. biol. Chem. **244**, 3500–3506 (1969)
133. Porte, D., Jr., Bagdade, J.D.: Human insulin secretion: An integrated approach. Annu. Rev. Med. **21**, 219–240 (1970)
134. Bressler, R.: The biochemistry of ketosis. Ann. N. Y. Acad. Sci. **104**, 735–752 (1963)
135. Krebs, H.A.: The regulation of the release of ketone bodies by the liver. Advanc. Enzyme Regul. **4**, 339–353 (1966)
136. Shreeve, W.W.: Diabetic ketosis. Ann. N. Y. Acad. Sci. **104**, 772–786 (1963)
137. Stadie, W.C.: Ketogenesis. Diabetes **7**, 173–179 (1958)
138. Lossow, W.J., Brown, G.W., Jr., Chaikoff, I.L.: The action of insulin in sparing fatty acid oxidation: A study with palmitic acid-1-C^{14} and octanoate-1-C^{14}. J. biol. Chem. **220**, 839 (1956)
139. Felts, J.M., Doell, R.G., Chaikoff, I.L.: The effect of insulin on the pathways of conversion of glucose to fatty acids in the liver. J. biol. Chem. **219**, 473–478 (1956)
140. Wakil, S.J.: Mechanism of fatty acid synthesis. J. Lipid Res. **2**, 1–24 (1961)
141. Frohman, C.E., Orten, J.M.: Tracer studies of the acids of the tricarboxylic acid cycle. I. The fate of labeled acetate in the livers of normal and diabetic rats. J. biol. Chem. **216**, 795–799 (1955)
142. Cornblath, M.: Familial carbohydrate intolerance and hypoglycemia. Annu. Rev. Med. **17**, 161–178 (1966)
143. Marble, A.: Oral hypoglycemic agents. Annu. Rev. Med. **12**, 135–150 (1961)
144. Hanson, R.L., Ray, P.D., Walter, P., Lardy, H.A.: Mode of action of hypoglycemic agents. I. Inhibition of gluconeogenesis by quinaldic acid and 5-methoxyindole-2-carboxylic acid. J. biol. Chem. **244**, 4351–4359 (1969)
145. Lefkowitz, R.J.: Isolated hormone receptors, physiologic and clinical implications. New Engl. J. Med. **288**, 1061–1066 (1973)
146. Cuatrecasas, P.: Insulin receptor of liver and fat cell membranes. Fed. Proc. **32**, 1838–1846 (1973)
147. Cuatrecasas, P.: Interaction of concanavalin A and wheat germ agglutinin with the insulin receptor of fat cells and liver. J. biol. Chem. **248**, 3528–3534 (1973)
148. Karlin, A.: Molecular interactions of the acetylcholine receptor. Fed. Proc. **32**, 1847–1853 (1973)
149. Dufau, M.L., Charreau, E.H., Catt, K.J.: Characteristics of a soluble gonadotropin receptor from the rat testis. J. biol. Chem. **248**, 6973–6982 (1973)

150. Cuatrecasas, P.: Properties of the insulin receptor isolated from liver and fat cell membranes. J. biol. Chem. **247**, 1980–1991 (1972)
151. Milgrom, E., Thi, L., Atger, J., Baulieu, E.E.: Mechanisms regulating the concentration and the conformation of progesterone receptor(s) in the uterus. J. biol. Chem. **248**, 6366–6374 (1973)
152. Samuels, H.H., Tsai, J.S.: Thyroid hormone action in cell culture: Demonstration of nuclear receptors in intact cells and isolated nuclei. Proc. nat. Acad. Sci. (Wash.) **70**, 3488–3492 (1973)
153. Marx, S.J., Woodward, C., Aurbach, G.D., Glossmann, H., Keutmann, H.T.: Renal receptors for calcitonin binding and degradation of hormone. J. biol. Chem. **248**, 4797–4802 (1973)
154. Selinger, Z., Batzri, S., Eimerl, S., Schramm, M.: Calcium and energy requirements for K^+ release mediated by the epinephrine α-receptor in rat parotid slices. J. biol. Chem. **248**, 361–368 (1973)
155. Batzri, S., Selinger, Z., Schramm, M., Robinovitch, M.R.: Potassium release mediated by the epinephrine α-receptor in rat parotid slices; properties and relation to enzyme secretion. J. biol. Chem. **248**, 361–368 (1973)
156. Gospodarowicz, D.: Properties of the luteinizing hormone receptor of isolated bovine corpus luteum plasma membranes. J. biol. Chem. **248**, 5042–5049 (1973)
157. Chen, T.C., DeLuca, H.F.: Receptors of 1,25-dihydroxycholecalciferol in rat intestine. J. biol. Chem. **248**, 4890–4895 (1973)
158. Sica, V., Parikh, I., Nola, E., Puca, G.A., Cuatrecasas, P.: Affinity chromatography and the purification of estrogen receptors. J. biol. Chem. **248**, 6543–6558 (1973)
159. Sutherland, E.W.: Studies on the mechanism of hormone action. Science **177**, 401–408 (1972)
160. Jost, J.P., Rickenberg, H.V.: Cyclic AMP. Annu. Rev. Biochem. **40**, 741–774 (1971)
161. Bitensky, M.W., Gorman, R.E.: Chemical mediation of hormone action. Annu. Rev. Med. **23**, 263–284 (1972)
162. Kolata, G.B.: Cyclic GMP: Cellular regulatory agent? Science **182**, 149–151 (1973)
163. Liddle, G.W., Hardman, J.G.: Cyclic adenosine monophosphate as a mediator of hormone action. New Engl. J. Med. **285**, 560–566 (1971)
164. Pastan, I.: Cyclic AMP. Sci. Amer. **227**, 97–105 (1972)
165. Lesniak, M.A., Gorden, P., Roth, J., Gavin, J.R., III: Binding of ^{125}I-human growth hormone to specific receptors in human cultured lymphocytes. Characterization of the interaction and a sensitive radioreceptor assay. J. biol. Chem. **249**, 1661–1667 (1974)
166. Hall, K., Luft, R.: Growth hormone and somatomedin. In: Advances in metabolic disorders (Levine, R., and Luft, R., eds.), vol. 7, p. 1–36. New York: Academic Press 1974
167. Merimée, T.J.: Isolated growth hormone deficiency and related disorders. In: Annual review of medicine (Creger, W.P., Coggins, C.H., and Hancock, E.W., eds.), vol. 25, p. 137–142. Palo Alto, Calif.: Annual Reviews Inc. 1974
168. Dousa, T.P.: Cellular action of antidiuretic hormone in nephrogenic diabetes insipidus. Mayo Clin. Proc. **49**, 188–199 (1974)
169. Breslow, E.: The neurophysins. In: Advances in enzymology (Meister, A., ed.), vol. **40**, p. 271–333. New York: John Wiley & Sons 1974
170. Winokur, A., Utiger, R.D.: Thyrotropin-releasing hormone: Regional distribution in rat brain. Science **185**, 265–267 (1974)
171. Brownstein, M.J., Palkovits, M., Saavedra, J.M., Bassiri, R.M., Utiger, R.D.: Thyrotropin-releasing hormone in specific nuclei of rat brain. Science **185**, 267–269 (1974)
172. Guillemin, R.: Hypothalamic hormones: Releasing and inhibiting factors. Hosp. Pract. **8**, 111–120 (1973)
173. Blackwell, R.E., Guillemin, R.: Hypothalamic control of adenohypophysial secretions. Annu. Rev. Physiol. **35**, 357–390 (1973)
174. Besser, G.M., Mortimer, C.H.: Hypothalamic regulatory hormones: A review. J. clin. Path. **27**, 173–184 (1974)
175. Wilber, J.F.: Thyrotropin releasing hormone: Secretion and actions. In: Annual review of medicine (Creger, W.P., Coggins, C.H., and Hancock, E.W., eds.), vol. 24, p. 353–364. Palo Alto, Calif.: Annual Reviews Inc. 1973
176. Shishiba, Y., Shimizu, T., Yoshimura, S., Shizume, K.: Direct evidence for human thyroidal stimulation by LATS-protector. J. clin. Endocr. **36**, 517–521 (1973)
177. Valenta, L.J., Bode, H., Vickery, A.L., Caulfield, J.B., Maloof, F.: Lack of thyroid peroxidase activity as the cause of congenital goitrous hypothyroidism. J. clin. Endocr. **36**, 830–844 (1973)
178. Fleischer, N., Lorente, M., Kirkland, J., Kirkland, R.: Synthetic thyrotropin releasing factor as a test of pituitary thyrotropin reserve. J. clin. Endocr. **34**, 617–624 (1972)
179. Miyai, K., Azukizawa, M., Kumahara, Y.: Familial isolated thyrotropin deficiency with cretinism. New Engl. J. Med. **285**, 1043–1048 (1971)
180. Krüskemper, H.L., Beisenherz, W., Herrmann, J., Kley, H.K.: Hypothyreose mit isoliertem Mangel an thyreotropin-releasing-hormon (TRH). Dtsch. med. Wschr. **97**, 76–81 (1972)
181. Baxter, J.D., Harris, A.W., Tomkins, G.M., Cohn, M.: Glucocorticoid receptors in lymphoma cells in culture: Relationship to glucocorticoid killing activity. Science **171**, 189–191 (1971)
182. Rinehart, J.J., Sagone, A.L., Balcerzak, S.P., Ackerman, G.A., LoBuglio, A.F.: Effects of corticosteroid therapy on human monocyte function, New Engl. J. Med. **292**, 236–241 (1975)
183. Chayen, J., Bitensky, L., Butcher, R.G., Altman, F.P.: Cellular biochemical assessment of steroid activity. In: Advances in steroid biochemistry and pharmacology (Briggs, M.H., and Christie, G.A., eds.), vol. 4, p. 1–60. London: Academic Press 1974
184. Jensen, E.V., DeSombre, E.R.: Mechanism of action of the female sex hormones. In: Annual review of biochemistry (Snell, E.E., Boyer, P.D., Meister, A., and Sinsheimer, R.L., eds.), vol. 41, p. 203–230. Palo Alto, Calif.: Annual Reviews Inc. 1972
185. Vaitukaitis, J.L., Ross, G.T.: Recent advances in evaluation of gonadotropic hormones. In: Annual review of medicine (Creger, W.P., Coggins, C.H., and Hancock, E.W., eds.), vol. 24, p. 295–302. Palo Alto, Calif.: Annual Reviews Inc. 1973
186. Blackwell, R.E., Guillemin, R.: Hypothalamic control of adenohypophysial secretions. In: Annual review of physiology (Comroe, J.H., Jr., Edelman, I.S., and Sonnenschein, R.R., eds.), vol. 35, p. 357–390. Palo Alto, Calif.: Annual Reviews Inc. 1973
187. Tager, H.S., Steiner, D.F.: Peptide hormones. In: Annual review of biochemistry (Snell, E.E., Boyer, P.D., Meister, A., and Richardson, C.C., eds), vol. 43, p. 509–538. Palo Alto, Calif.: Annual Reviews Inc. 1974
188. Wilson, J.D., Harrod, M.J., Goldstein, J.L., Hemsell, D.L., MacDonald, P.C.: Familial incomplete male pseudohermaphroditism, type 1. Evidence for androgen resistance and variable clinical manifestations in a family with the Reifenstein syndrome. New Engl. J. Med. **290**, 1097–1103 (1974)
189. Werk, E.E., Jr., Sholiton, L.J., Kalejs, L.: Testosterone-secreting adrenal adenoma under gonadotropin control. New Engl. J. Med. **289**, 767–770 (1973)
190. Adlercreutz, H.: Hepatic metabolism of estrogens in health and disease, New Engl. J. Med. **290**, 1081–1083 (1974)
191. Hochberg, R.B., McDonald, P.D., Feldman, M., Lieberman, S.: Studies on the biosynthetic conversion of cholesterol into pregnenolone. Side chain cleavage of some 20-φ-tolyl analogs of cholesterol and 20α-hydroxycholesterol. J. biol. Chem. **249**, 1277–1285 (1974)
192. Kimura, T., Nakamura, S., Huang, J.J., Chu, J-W, Wang, H-P, Tsernoglou, D.: Electron transport system for adrenocortical mitochondrial steroid hydroxylation reactions: The mechanism of the hydroxylation reactions and properties of the flavoprotein-iron-sulfur protein complex. Ann. N. Y. Acad. Sci. **212**, 94–106 (1973)
193. Kefalides, N.A.: Biochemical properties of human glomerular basement membrane in normal and diabetic kidneys. J. clin. Invest. **53**, 403–407 (1974)
194. Krahl, M.E.: Endocrine function of the pancreas. In: Annual review of physiology (Comroe, J.H., Jr., Sonnenschein, R.R., Zierler, K.L., eds.), vol. 36, p. 331–360. Palo Alto, Calif.: Annual Reviews Inc. 1974
195. Cahill, G.F., Jr., Soeldner, J.S.: Diabetes, glucagon and growth hormone. New Engl. J. Med. **291**, 577–578 (1974)
196. Steiner, D.F., Kemmler, W., Tager, H.S., Peterson, J.D.: Proteolytic processing in the biosynthesis of insulin and other proteins. Fed. Proc. **33**, 2105–2115 (1974)
197. Hellman, B., Idahl, L-Å, Lernmark, Å., Täljedal, I.-B.: The pancreatic β-cell recognition of insulin secretagogues: Does cyclic AMP mediate the effect of glucose? Proc. nat. Acad. Sci. (Wash.) **71**, 3405–3409 (1974)
198. Senior, B.: Neonatal hypoglycemia. New Engl. J. Med. **289**, 790–793 (1973)
199. Schneider, P.B.: A site of action of thyrotropin. Stimulation of the conversion of glycerophosphate to phosphatidic acid in bovine thyroid slices. J. biol. Chem. **247**, 7910–7914 (1972)
200. Goldberg, N.D.: Cyclic nucleotides and cell function. Hosp. Pract. **9**, 127–142 (1974)
201. Rasmussen, H.: Cell communication, calcium ion and cyclic adenosine monophosphate. Science **170**, 404–412 (1970)
202. Rasmussen, H., Goodman, D.B.P., Tenenhouse, A.: The role of cyclic AMP and calcium in cell activation. Critical Reviews Biochem. **1**, 95–148 (1972)
203. Rasmussen, H.: Ions as "second messengers". Hosp. Pract. **9**, 99–107 (1974)
204. Beisswenger, P.J., Spiro, R.G.: Studies on the human glomerular basement membrane. Composition, nature of the carbohydrate units and chemical changes in diabetes mellitus. Diabetes **22**, 180–193 (1973)
205. Beaumont, P., Schofield, P.J., Hollows, F.C.: Growth hormone, sorbitol and diabetic capillary disease. Lancet **I**, 579–581 (1971)
206. Unger, R.H.: The essential role of glucagon in the pathogenesis of diabetes mellitus. Lancet **I**, 14–16 (1975)
207. Dobbs, R., Sakurai, H., Sasaki, H., Faloona, G.: Glucagon: Role in the hyperglycemia of diabetes mellitus. Science **187**, 544–547 (1975)
208. Mallette, L.E., Exton, J.H., Park, C.R.: Effects of glucagon on amino acid transport and utilization in the perfused rat liver. J. biol. Chem. **244**, 5724–5728 (1969)
209. MacCuish, A.C., Irvine, W.J., Barnes, E.W., Duncan, L.J.P.: Antibodies to pancreatic islet cells in insulin-dependent diabetics with coexistent autoimmune disease. Lancet **I**, 7869 (1974)
210. Hamlym, A.M., Morris, J.W., Lunzer, M.R., Puritz, H.: Portal hypertension with varices in unusual sites. Lancet **II**, 1531 (1974)
211. Bottazzo, G.F., Florin-Christensen, A., Doniach, D.: Islet-cell antibodies in diabetes mellitus with autoimmune polyendocrine deficiencies. Lancet **II**, 1279–1283 (1974)
212. Schrier, R.W., Berl, T.: Nonosmolar factors affecting renal water excretion (Part I and II). New Engl. J. Med. **292**, 81–88, 141–145 (1975)

Chapter 9

Alterations of Body Fluids and Electrolyte Metabolism

The body fluids do not constitute a homogeneous solution of electrolytes. Body fluids are classically categorized as intracellular and extracellular. The extracellular compartment is further divided into an intra- and extravascular compartment. The intravascular fluid constitutes the blood plasma, but even blood plasma does not form a homogeneous compartment. The composition of plasma varies with anatomical location and physiological conditions. The electrolyte compositions of plasma obtained from venous and arterial blood differ, and there are diurnal variations in the electrolyte concentrations of the plasma.

The relative proportion of extra- and intracellular fluid has been determined by the dilution technique. Chemicals of known distribution in the body fluids are administered, and the volume of a given fluid compartment can be measured by determining the concentration of these chemicals. For example, Evans' blue, bromosulftalein, iodinated albumin, and iodine 131 are used to determine plasma volume. Urea, thiourea, deuterium, and tritiated water are useful for determining the total body fluids. Thiocyanate, iodide, sulfate, mannitol, sucrose, raffinose, chloride 36, chloride 38, and bromide 32 are used to determine extracellular space.

The division of the body fluid into intra- and extracellular systems is somewhat arbitrary because none of the above methods gives identical results, and thus only an approximate distribution of water and electrolytes is available.

There are only minor differences between the electrolyte composition of plasma and the rest of the extracellular fluid. The main difference between these two fluid systems resides in the higher protein concentration in plasma. Consequently, electrolyte shifts between the two body systems follow the laws of Donnan. Thus, the concentration of cations in the various interstitial compartments will indirectly be regulated by the amount of protein in those fluid systems, but in general, the plasma concentration of anions is higher than that of cations.

The electrolyte concentrations of intra- and extracellular fluids differ in many ways. Among the most significant are the low sodium and chloride concentration and the high potassium concentration in intracellular fluid. These differences in electrolyte composition are not explained, but they do not result from the permeability of the cell membrane to sodium because radioactive sodium is rapidly distributed between extra- and intracellular fluid. Yet after radioactive sodium is injected, the sodium concentrations of the respective fluid systems are not altered.

Inasmuch as the cell membrane is freely permeable to water, the osmotic pressure of the various compartments of the body fluids is usually assumed to be the same (see above), and osmotic pressure is a constant of the living organism.

The electrolyte concentrations in the various body fluids also remain remarkably constant. Several organs regulate electrolyte concentrations: intestine, lung, sweat glands, and kidney. In mammals, the kidney is the most important organ in controlling electrolyte metabolism. The osmolarity of the medium is naturally a function of such cations as sodium in the extracellular fluid and magnesium and potassium in the intracellular fluids. Anions such as chloride and bicarbonate, which are mainly found in the extracellular fluid, also affect the osmolarity in the body fluids. Among the compounds determining osmotic pressure, sodium and chloride ions are of primordial importance because they are present in high concentrations. In contrast, the concentrations of other solutes are too low to be significant in determining osmotic pressure. The protein molecule is too big for proteins to participate significantly in determining osmotic pressure. Hence, shifts in the concentrations of sodium and chloride ions regulate the osmotic pressure of the intra- and extracellular fluids.

The pH of the body fluids and their electrolyte concentrations are closely interrelated. The mechanism controlling this relationship is complex. Many aspects of electrolyte metabolism will be discussed in the following pages. A brief review of kidney function will facilitate the understanding of electrolyte metabolism.

Anatomy of the Kidney

Gross examination of a frontal section of the kidney reveals two components, a dark medulla and a yellowish cortex. The medulla has 10–12 dark triangular malpighian pyramids with their apex directed toward the center and their base directed outward. The summit of the pyramid is rounded and contains a small hole called the foramina papillaria. Smaller pyramids project outward from the bases within the cortex, forming the slender pyramids of Ferrein.

The cortex forms a narrow peripheral band of yellow tissue spotted with red dots or streaks. The band sends projections inward between the pyramids, forming the columnae renalis, or the columns of Bertin.

A renal lobe is composed of a pyramid and the cortical substance that surrounds it. In some animals, like the pig, these functional lobules are separated anatomically by a fissure. In man, this lobulation exists only during fetal life.

Careful microdissection of the kidney reveals its unit structure—the nephron, which is composed of the glomerulus and the renal tubules. The malpighian glomerulus, with a diameter ranging between 120 and 250 μ, is essentially a bundle of intertwined and coiled capillaries. One of the anatomical particularities of this bundle of capillaries is that it is included between two arteries, the vas afferens glomeruli and the vas efferens glomeruli. The vas afferens, a branch of the interlobular arteries, ramifies into tiny and richly coiled capillaries. After ramification, these capillaries all converge to form the vas efferens. When it emerges from the glomerulus, the adventitia of the vas efferens is thickened by a layer of granular cells outside the muscularis of the arteriole. This cellular cuff probably is important in regulating blood flow through the glomeruli. Both the vas efferens and the vas afferens emerge close to each other at the same pole of the glomerulus, called the vascular pole.

Two structures can be recognized at the immediate periphery of the glomerulus: a thin hyaline membrane, called the basal membrane, and an epithelial layer of flattened cells. Where the epithelial layer reaches the vascular pole, the lining projects outward and forms a peripheral epithelial membrane (Bowman's capsule) resting on a basal membrane, which is in continuity with the interstitial tissue. The epithelium of Bowman's capsule touches the epithelium of the proximal convoluted tubule.

The renal tubule is composed of a central loop extending between two coiled tubules, the proximal and distal convoluted tubules. The proximal convoluted tubule is 15 μ in diameter and 14 mm long. It is lined with cuboidal cells with a typical brush border. The proximal tubule is in continuity with the loop of Henle, which is formed of a narrow descending branch that penetrates deep in the medulla, bends to project outward and vertically toward the glomerulus, to continue in an ascending branch that is wider than the descending branch of Henle's loop. Close to the glomerulus the ascending branch meets the distal convoluted tubules. The distal convoluted tubules are lined by clear cells without a brush border, and they open in collecting tubes that are lined respectively with cuboidal and cylindrical cells. An interesting morphological characteristic of the cells of the collecting tubules is the presence of an exoplasmic thickening, which probably plays a role in preventing the passage of the urinary components from the lumen into the cell.

As it penetrates the hilum of the kidney the renal artery divides into several branches, one for each lobule. These arteriae interlobularis travel along the borderline that separates two lobules. The arteries penetrate in the central part of the kidney and reach the cortex, where they form arcs from which small branches arise that provide the afferent branches of the glomeruli. The afferent branches ramify extensively to form the coiled capillaries of the glomeruli.

The vas efferens emerges from the glomerulus and its blood is collected into vessels that run along the branches of the collecting tubes and the loop of Henle. The collecting tubules descend with the descending branch of the loop of Henle, bend in the depth of the medulla, and are directed outward toward the cortex. The anatomy of these vessels is of the greatest importance to kidney function; indeed, the anatomical relationship between the loop of Henle and the arteriolae rectae determines the concentration of urine in the collecting tubules.

The arteriolae rectae divide in small capillaries in the depth of the medulla. The blood of these capillaries is collected in a subcapsular venous network called the venulae stellatae, from which the venae radiata emerge and the vena interlobularis reassemble to form arciform trunks at the bases of the pyramid.

The main function of the kidney is urine formation. Urine is composed of water and solutes normally found in the blood.

Osmotic Pressure

In the seventeenth century, a learned French priest, L'Abbé Nollet, amazed and amused the French intelligentsia of the salons with his experiments. One of the most popular experiments was to place a few spoons of sugar in a dried pig bladder that was then plunged into a bucket of water. To the surprise of the audience, the bladder swelled more and more until it finally burst [1, 2].

Similar observations of the botanist Pfeffer, who had described the shrinking of plant cells placed in a concentrated medium, attracted the attention of the chemist Van't Hoff, who was working on the kinetic theory of gases. The results of both experiments were explained by proposing that water molecules move from the more concentrated solution toward the more dilute one. This movement, however, could be demonstrated only if the two solutions were separated by a membrane permeable to the solvent but impermeable to the solute. Membranes with such restricted selective permeabilities were called semipermeable.

The force that "pulls" the solvent through the membrane toward the more concentrated solution is called osmotic pressure. It can be exactly balanced by applying hydrostatic pressure to the membrane. Consequently, when π, osmotic pressure $=$ p, hydrostatic pressure, fluid will not move.

Van't Hoff developed a mechanistic theory of osmotic pressure, which most of us learned in school. His theory was based on the broad generalizations (1) that the osmotic pressure was directly proportional to the concentration of solute; and (2) that the osmotic pressure was directly proportional to the absolute temperature, T^0. His assumptions permitted the formulation of an equation for osmotic pressure

$$\pi V = mRT$$

where m is the molar concentration, T the absolute temperature, and R the gas constant. Van't Hoff's equation was reminiscent of the equation developed for gases. (In fact, for very dilute solutions, it was found that R has a value close to the gas constant.)

The logical conclusion is that the osmotic pressure is equal to the pressure that the molecules would exert if they were a perfect gas and occupied a volume equal to that of the solution.

The Van't Hoff equation, although adequate at very low concentrations of solute, is not applicable to relatively concentrated solutions, such as those found in biological material.

Van't Hoff proposed that the osmotic pressure resulted from the pressure exerted by the individual molecule of solute on the semipermeable membrane and assumed that the pressure exerted by the molecules of solvent was equal on both sides. But the number of molecules of solvent is not the same on both sides; it is smaller on the side containing the solute. Furthermore, it is the solvent and not the solute that moves from one side of the membrane to the other. It would seem more logical, therefore, to assume that the osmotic pressure results from the difference between the number of molecules of solvent at both sides of the membrane.

We have already mentioned that the Van't Hoff theory proved to be extremely useful as a first approximation, especially in the case of very dilute solutions; but the relationship $\pi V = mRT$ did not stand when more concentrated solutions were used.

To express osmotic pressure in an equation that would better reflect reality, physicists abandoned the kinetic treatment of the problem of osmotic pressure and resorted to a thermodynamic approach. Thus, it was assumed that the fluid tends to flow from a higher energy state toward a lower energy state.

The energy levels, or chemical potential of the solvent, can be modified by adding a solute or by applying pressure. Solvents in dilute solution have a higher chemical potential than solvents in a concentrated solution. Solutions under high hydrostatic pressure have a higher chemical potential than solvents under lower hydrostatic pressure.

We will not attempt to derive the equation that expresses the chemical potential of the solution in function of pressure and concentration of solute.

$$\Delta u_1 = PV_1$$

The chemical potential is directly proportional to the increase in pressure and the partial molar volume

$$\Delta M_2 = RT \ln \frac{n^1}{n^1 + n^2}$$

n^1 = g/molecule of solvent
n^2 = g/molecule of solute
g/molecule = number of molecule divided by Avogadro's number.

At equilibrium,

$$PV_1 = RT \frac{n^1}{n^1 + n^2}$$

From these two equations, a third can be derived that applies to dilute solutions,

$$\pi = RTm$$

where m is the molar concentration of solvent. The formula is reminiscent, of course, of the Van't Hoff formula. The originality lies in the way it was derived, and because of the way it was derived, the formula indicates the limitations of the equation.

To derive the formula, it was necessary to assume that all the molecules of the solute were identical in mass, size, and intermolecular force. This is a form of perfection that is never achieved in solutions, not even in so-called ideal solutions, which are very dilute. Consequently, the formula is not expected to express changes in osmotic pressure accurately. The gap between ideality expressed by the formula and the reality of the measurements can be filled in by introducing an empirical coefficient in the formula. Thus,

$$\pi = \phi RTm$$

in which ϕ is the molar osmotic coefficient. ϕ is valid only for a specific solute dissolved in a specific solvent. It expresses the deviation of the mixture from ideality. In an ideal solution $\phi = 1$.

The significance of ϕ will become clearer when we attempt to identify the role played by protein in determining osmotic pressure. In cells, the osmotic pressure results mainly from the pressure of proteins. The application of the principle of thermodynamics to the calculation of the entropy of mixing of proteins will illustrate the importance of ϕ. In calculating the entropy of mixing of an ordinary solute, one assumes that the volume of a molecule of solute is equal to that of a molecule of solvent.

In the case of proteins, the volume of the protein ($\bar{V}$) molecule is obviously much greater than that of the solvent ($\bar{V}_2$). If we assume that $\bar{V} = \chi \bar{V}_2$, and if there are N_1 sites, each of which is occupied by a molecule of the solvent, there are χN_2 sites for the protein molecules, and the total number of sites available is $N_1 + N_2$.

Obviously, the neighbor macromolecule is excluded from the area occupied by another macromolecule. In fact, the so-called excluded volume is much greater than the volume of the macromolecule. Indeed there is a zone around the macromolecule in which it is impossible to locate the center of another macromolecule.

For a sphere, the "excluded volume" has a radius twice that of the spherical molecule. Consequently, if

$$\tfrac{4}{3}\pi \Sigma^3 = \chi \text{ sites}$$
$$\tfrac{4}{3}\pi (2\Sigma^3) = 8\chi \text{ sites}$$

So if the excluded area is $M_1 — \chi M_2 — 8\chi$ for the second molecule that enters the volume containing the mixture, the N_2 molecule will be excluded from a volume $N_1 — N_2 — 8\chi N_2$. This type of computation permits investigators to develop equations for the entropy of mixing:

$$\Delta S = R \frac{V_2}{\chi}(1 + 4V_2 + 21.3\,V_2^2 \ldots)$$

for the chemical potential:

$$\Delta \mu_1^3 = RT \frac{V^2}{\chi}(1 + 4V_2 + 21.3\,V_2^2 \ldots)$$

for osmotic pressure:

$$\pi = RTC\,(1 + 4\,V_2 + 21.3\,V_2^2 \ldots)$$

and, finally, for the molar osmotic coefficient:

$$\phi' = 1 + 4\,V_2 + 21.3\,V_2^2 \ldots$$

With this type of formula, the osmotic coefficient can be calculated for protein solutions, which are often very close to the empirical value.

The osmotic effect of protein is also affected by hydration and ion binding. When a protein is sedimented in the ultracentrifuge, a certain amount of water and salt moves with the protein. The amount of bound water is referred to as the "selective solvation." The selective solvation decreases the osmotic effect of proteins. In contrast, the presence of bound ions in the protein increases the osmotic effect of the protein.

The osmotic pressure inside cells is difficult to predict on the basis of the formulas derived above because the permeability properties of the cell membrane are complex and it is not known how many macromolecules, particularly proteins, are in solution.

The cell membrane is not a true semipermeable membrane—that is to say, permeable to water and impermeable to solute. The cell membrane is permeable to ions: more to anions than to cations. This is particularly true under nonphysiological conditions, such as anoxia or prolonged exposure to cold. Nonelectrolytes like urea and glucose traverse the cell membrane either by a simple diffusion process or by active transport.

Although morphologically most cellular membranes look alike, great differences are found in the osmotic permeability of cells. These differences probably result from differences in cell size and the existence of water-filled pores in the membrane. The permeability of the cell to water is not only a function of the membrane and the difference between osmotic pressures inside and outside the cell, but it is restricted by the rate of diffusion inside the cell cytoplasm.

Some of the molecules that pass the membrane are converted to macromolecules inside the cell (purine, pyrimidine precursors, amino acids). Others are catabolized and completely oxidized to CO_2 and water as a result of membrane transfer and cellular metabolism. The molecular population inside of the cell changes constantly. The very structural properties of the cell (existence of endoplasmic reticulum, mitochondria, etc.) suggest that at least a fraction of the intracellular proteins are in a solid state and do not participate in the osmotic effect exerted by proteins. Biophysicists have not agreed on how much of the protein population is in the solid state and how much is in true solution.

In spite of all these reservations, the cell membrane, for all practical purposes, can be considered to be an effective semipermeable membrane separating two osmotic media that are maintained in equilibrium. This implies that the intracellular osmotic pressure is isotonic to that of the body fluids. Such a concept has occasionally been challenged; it has been claimed that the intracellular osmotic pressure is greater than that of the body fluid. The maintenance of intracellular hypertonicity would require, of course, that water is actively eliminated from the cells.

An empirical formula is used to express the osmotic pressure inside of cells.

$$\pi(V - b) = \pi^0(V_0 - b)$$

π^0 is isotonic external osmotic pressure, V_0 represents the original cell volume, and b is the osmotic inactive volume, or the difference between the total volume of the cell and the water volume in the cell; b is equal to the sum of the volume of the solute and that of undissolved material in the cell.

In this equation, $(V_0 - b)$ should be equal to the isotonic water content Wm. When b or Wm was measured directly, it was found that the isotonic water content was consistently greater than $(V_0 - b)$.

$$Wm > V_0 - b \quad \text{or} \quad \frac{V_0 - 1}{Wm} = R < 1.$$

It can be demonstrated that

$$R = 1 - \frac{\Delta\phi\,\pi_0}{\Delta\phi_0\,\pi}.$$

From the equation on osmotic pressure in cells,

$$\pi(V - b) = \pi_0(V_0 - b)$$

$$\frac{V - b}{V_0 - b} = \frac{\pi_0}{\pi}$$

The change in volume should bear a linear relationship to the reciprocal of the osmotic pressure. This is the case for erythrocytes. When erythrocytes are placed in solutions of decreasing osmolarity, they increase in volume, and the increase in the hematocrit readings is proportional to the reciprocal of the osmolarity. Now the concentration of hemoglobin in the red cells is known, and so is the osmotic coefficient of hemoglobin at a given concentration. Consequently, R can be calculated.

In erythrocytes, the values for R are close to 1 $(R = 1)$. This means that the osmotic effect of the erythrocyte can be explained entirely in terms of the physico-chemical properties of hemoglobin, and the erythrocytes may be considered to be perfect osmometers.

Urine Formation

Urine formation involves glomerular filtration, reabsorption, and tubular secretion. The capillaries of the glomeruli are inserted between two arteries. The hydrostatic pressure and the permeability of the glomerular capillaries are greater than in ordinary capillaries. Because of these properties, the glomerulus is well suited to operate as an ultrafilter that lets all the components of plasma pass through the membrane except the plasma proteins. (Some physiologists believe that even a small amount of albumin passes through the glomerulus, and that the albumins are reabsorbed in the tubules.) The glomerular property that determines ultrafiltration is obviously the permeability of the capillary membrane. This permeability is assumed to result from the existence of pores, the size of which would permit the passage of small solutes but would exclude the extrusion of large molecules such as protein. In contrast, polysaccharides of the molecular size of inulin and dextran pass freely through the glomerulus. (However, no morphological evidence has ever unquestionably established the existence of pores that connect the plasma membrane to the glomerular fluid.)

Filtration of plasma through the glomeruli occurs according to the following formula:

$$GFR = Kp(Pb - Pc) - \pi b$$

Kp is a constant that is related to the permeability and the area occupied by the glomeruli in the kidney. Pb is the hydrostatic pressure of the blood in the capillaries, a positive factor in ultrafiltration. Pc is the hydrostatic pressure of the fluid in Bowman's capsule, a force opposing ultrafiltration through the capillary wall. And πb is the osmotic pressure resulting from protein concentration in the plasma (25–30 ml Hg).

It is not known whether all the glomeruli participate in glomerular filtration, and it has been suggested that the GFR can be changed by modifying the number of glomeruli that participate in the process.

The rate at which urine is filtered through the glomeruli can be measured by determining the concentration in the urine of substances, such as mannitol and inulin, that are filtered through the glomeruli without being reabsorbed in the tubules. Such studies revealed that the glomerular filtrate remains constant at 130 cc/minute in spite of the variations in systemic arterial pressure. This observation indicated that the kidney can regulate its own vascular pressure. Not all investigators accept the existence of an autoregulatory mechanism. The disagreement stems from contradictory results obtained on isolated perfused rat kidney. Some investigators find that arterial pressure and renal blood flow are strictly related; others fail to observe such a relationship. Differences in the technical approaches can probably explain these discrepancies.

Yet such contradictory results make it somewhat difficult to decide which of these observations made *in vitro* represents the true *in vivo* situation. Two different theories remain. One proposes that the rise in systemic blood pressure is associated with a constriction of the afferent arterioles sufficient to maintain the capillary pressure, the renal blood flow, and the glomerular filtration constant. In contrast, a drop in systemic arteriolar blood pressure is accompanied by a dilatation of the afferent arteriole. Another theory proposes that the autoregulation of the glomerular filtration results from the changes in interstitial tissue pressure, which are due to changes in the systemic blood pressure. When the systemic blood pressure rises, the hydrostatic pressure of the interstitial tissue rises, and the afferent arterioles are compressed. The available evidence suggests that the autoregulation of glomerular filtration is of myogenic origin. Thureau has proposed that renin release regulates the tone of the afferent arterioles.

Autoregulation is not observed only in the kidney—it has recently been described in muscle. In perfused muscle, a sudden increase in the pressure of the perfusate induces an initial vascular dilatation, which is followed by a return to the normal pressure in the muscle even when the pressure of the perfusate continues to be elevated.

Before we leave the subject of glomerular filtration, it should be added that the glomerulus is not totally impermeable to proteins. Physiological studies have shown that the permeability of the glomerulus to proteins is related to the dimension of the molecules and thus, in most cases, to their molecular weight. The smaller the molecular weight, the easier the molecule traverses the glomerular filter. The threshold seems to be around 70,000; any molecule with a higher molecular weight does not pass the glomerular filter. However, even smaller molecules do not traverse the glomerular filter at the same rate.

Studies with ferritin suggested that the principal glomerular filtration barrier is the basal membrane. Further investigation using peroxidases of various sizes were done in Karnofsky's laboratory. Peroxidases can easily be detected by testing for their activity after passage through the membrane with 3′-diaminobenzidine in the presence of H_2O_2. The reaction yields an insoluble, brown, electron-opaque product at the site of enzyme activity. Peroxidases from different sources vary in size; for example, the molecular weight of horseradish peroxidase is 40,000, while that of human myeloperoxidase is between 160,000 and 80,000. As a result, differential permeability can be investigated.

Karnofsky's experiments indicated that in addition to the basal membrane, a second glomerular filtration barrier, the epithelial slits, acts as a fine filter. The latter filter ultimately determines the nature of the protein content of the glomerular filtrate. It has also been proposed that the glomerular epithelium is capable of retrieving the proteins that might have leaked through the basal membrane. Indeed, there seems to be epithelial uptake of ferritin; however, peroxidases (horseradish peroxidase or human myeloperoxidase) are not phagocytized by the epithelium. Uptake of both horseradish and human bone marrow peroxidase is prominent in mesangial cells, indicating that the mesangial cells are active in removing proteins that have leaked through the basement membrane; but neither endothelial nor epithelial pinocytosis seems to play a major role in the passage of the peroxidases through the glomerular wall.

Convoluted Tubule, Loop of Henle, and Urine Formation

Although the glomerular filtrate is identical to plasma except for the absence of protein in the filtrate, the final composition of the urine is very different from that of plasma. Urine may be diluted or concentrated. Various segments of the nephron participate in different ways in urine formation.

To understand the mechanism of urine concentration, one must retrace the fate of the fluid in the various segments of the kidney. In the glomerulus, the membranes of the Bowman's capsule cells allow passage of all plasma components except protein. The ultrafiltrate is markedly reduced in volume as it passes through the proximal tubule. In fact, only 20% of the original volume reaches the distal portion of the proximal convoluted tube. The volume of the ultrafiltrate is reduced due to passive water reabsorption. "Passive" means that no known molecular mechanism exists for the transport of water from the lumen of the proximal tubule to the interstitial tissue. However, the movement of water follows that of sodium. In the proximal tubule, sodium is excreted actively into the interstitial tissue, and as a result, the osmotic pressure of the interstitial tissues increases. This draws water from the lumen of the tubule into the interstitial environment of the medulla because the tubule is highly permeable to water.

As the urine leaves the proximal tubule to enter the loop of Henle, it also leaves the medulla to enter the cortex. The importance of these territorial changes will become more obvious later [3].

In the loop of Henle, the fate of urine in each limb must be distinguished. The major difference between the two limbs results from their difference in permeability to water. The descending limb is highly permeable to water. Consequently, the osmotic pressure is the same in the descending loop as in the surrounding interstitial tissue. In the ascending limb, the high water permeability is lost, but the membrane continues to remove sodium from the urine and to transfer sodium into the interstitial tissue. Consequently, the urine of the ascending loop is more dilute than that of the surrounding medium. Thus, diluted urine leaves the loop of Henle in the medulla to reach the distal convoluted tubule in the cortex. Depending upon the water content and osmotic pressure of the blood, dilute or concentrated urine will be excreted. The difference between the formation of the two types of urine depends upon the removal of water.

Sodium removal continues in the collecting tubules as well as in the convoluted tube. Consequently, in the absence of water reabsorption, dilute urine is excreted. Water reabsorption under the influence of the antidiuretic hormone results in the formation of concentrated urine.

The kidney is unexpectedly efficient in concentrating urine. The reason for this efficiency resides in the kidney's anatomical organization. In 1909, as a result of an extensive study in comparative physiology and microanatomy, Peter concluded that the length of the loop of Henle correlated with maximum urine concentration. In 1942, Kuhn and Ryffel enounced a theory for the changes of concentration that take place as a solution moves in opposite directions in two parallel tubes connected by a hairpin that is bent and separated by a semipermeable membrane. Such a system generates a countercurrent multiplier. The detailed mathematical development of the theory will not be discussed here [4, 5].

The concept can more easily be understood by examining an analogous situation—heat transfer in solution. If the solution flows in a longitudinal tube and the heating element is placed in the center of the tube, only the outflowing water will be heated, and the maximum temperature cannot exceed that reached by the delivery of the heat at the point of contact between the flowing solution and the heating element (see Fig. 9-1).

Consider a tube of the same length and caliber that is bent in the middle. The two ends are parallel and in close contact, and the wall of the tube readily conducts heat as if it were made of metal, for example. The heating element is placed at the hairpin bend. A solution is introduced at one end. The temperature of the solution is measured at various distances from the origin of the tube and plotted graphically. In such a tube, the maximum temperature of the solution is much greater than that in the straight tube because the incoming water is heated by outgoing water. The plot result is a hairpin-shaped curve (see Fig. 9-2).

Consider a hairpin tube with a semipermeable membrane that is filled with a solution of electrolyte. If the hairpin bend is dipped in the container filled with a solution with an electrolyte concentration greater than that of the solution in the tube, the concentration in the ascending loop of the tube will be increased as it was for T°. The concentration is greatest at the loop and lowest at the outflowing edge in the heated tube.

In the kidney, there is an increasing concentration of gradient in electrolytes in the interstitial tissue as one moves from cortex deeper into the medulla. This

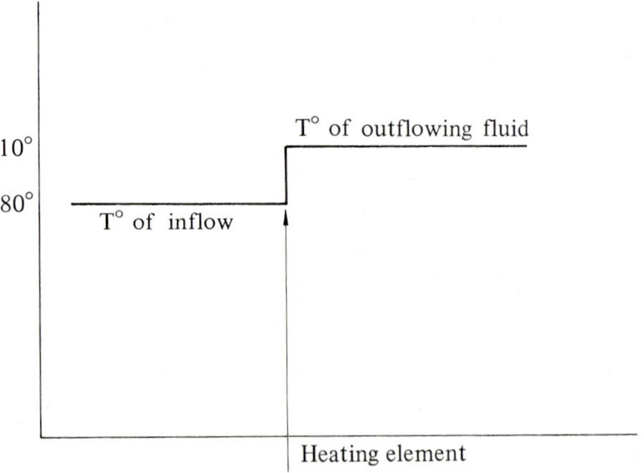

Fig. 9-1. Temperature curve in longitudinal tube

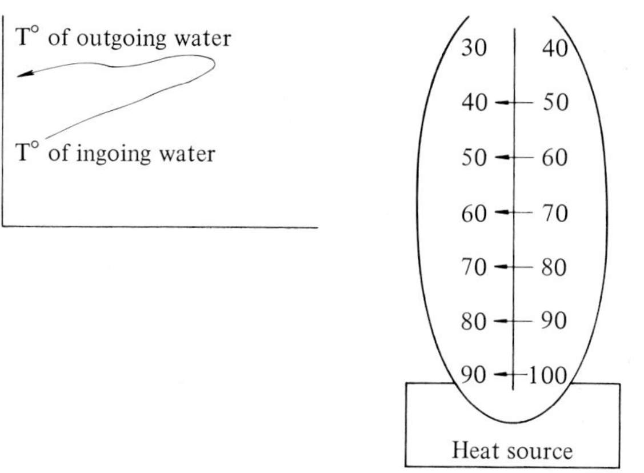

Fig. 9-2. Temperature curve in bent tube (*left*); temperature distribution in bent tube (*right*)

results from the removal of electrolytes from the urine in the ascending tubule and from the special vascular arrangement, which permits the concentration of the electrolytes in the medulla to remain high.

Sodium and chloride, but not water, are actively removed from the ascending limb, transferred to the interstitial fluid, and from there sent to the descending limb. Thus, the isotonic urine formed in the proximal convoluted tube becomes more and more concentrated as it descends, reaches a maximum concentration at the bend, and becomes hypotonic (by active removal of electrolytes) in the ascending limb.

The capillary loop follows the tubes, but neither limb of the capillary tube reabsorbs or excretes sodium actively. Consequently, the concentration gradient in the capillaries is the same as in the interstitial fluid. Therefore, in this system outflowing blood loses its solute to the inflowing blood by countercurrent exchange, and the solutes are continuously recirculated in the hypertonic area.

The water that is removed from the urine does not dilute the interstitial fluid because the hypotonic urine formed in the ascending branch is concentrated by water loss only when it reaches the cortex, where it is separated from the hyperosmotic medulla. Moreover, blood flow in the cortex is so great that the electrolyte concentration of the interstitial tissue is not modified by water withdrawal under the effect of ADH.

This isotonic urine leaves the convoluted to enter the collecting tubule in the medulla, which is hyperosmotic, and reaches an osmotic pressure equal to that of the surrounding medium. More water is eliminated from the tubule in the presence of ADH.

In conclusion, the loops of Henle and the vasa recta form a countercurrent system capable of developing a high concentration of solutes in the medulla. The hyperosmotic force of the medulla is then used to extract water from the fluid contained in the collecting ducts.

This is obviously an oversimplified presentation of the mechanism of urine concentration, and the reader is referred to reviews and specialized articles for further details [3, 4, 67, 69].

Reabsorption and Secretion in the Kidney

During the passage through the tubules, the glomerular filtrate is not only concentrated, but its composition is considerably modified by reabsorption and secretion in the tubular cells. Several forms of tubular reabsorption are discussed in the paragraphs on sodium, potassium, chloride, and water metabolism. It appears that the mechanisms of electrolyte and water absorption vary greatly. Absorption can be passive, as is the case for chloride, or active and influenced by hormones, as is the case for sodium and water (in at least some portions of the nephron) [6].

Among the organic compounds that are normally or pathologically eliminated through the urine, glucose, uric acid, the amino acids and urea are of particular significance. The glucose present in the glomerular filtrate is quantitatively reabsorbed in the proximal tubules, and this reabsorption process is postulated to involve a specific "carrier" system.

During renal transport, glucose is successively phosphorylated and dephosphorylated.* The concentration of the carrier in the cells is limited, and when the glucose levels in the blood exceed the renal threshold for glucose, glucosuria occurs, as in diabetes mellitus. Occasionally, the tubular cells are deficient in their carrier system and glucose is spilled in the urine. The intrinsic deficiency in the carrier with glucosuria is termed renal glucosuria. Glucosuria may also result from a reduction of the total number of tubular cells because of renal injury; however, in most renal injuries, the glomerular filtration rate is reduced enough so that all the glucose that reaches the proximal tubule is adequately reabsorbed. The glycoside phlorhizin inhibits glucose reabsorption without interfering with the metabolism of the renal cells.

The molecular structure of the carrier involved in glucose transport is unknown. Mutarotase has been postulated to be located in the luminar pole of the tubular cell, where it converts glucose to a metabolite more readily absorbed than glucose. The absorption of the metabolite generates a concentration gradient that favors diffusion of glucose into the bloodstream. The mutarotase theory still rests on shakey grounds. Compounds similar to glucose (L-deoxyglucose) but containing no mutarotable groups are absorbed in the tubules; and although mutarotase activity is found in kidney, the intracellular location of mutarotase remains to be established.

The uric acid present in the glomerular filtrate is reabsorbed in the proximal tubule, and the uric acid

* The exact mechanism of transport is not known, and other transport systems have been described in other cells. In the red cells it has been proposed that glucose transport involves the formation of a dimer, and one molecule acts as a carrier for the other. In the intestinal cell, glucose is assumed to form a complex with sodium. The complex passes the membrane and then the sodium is removed by active transport.

found in the urine is there because of active secretion in the lower parts of the nephron. The molecular mechanism of uric acid secretion is unknown. It is unlikely that gout affects the uric acid secretion; in contrast, in inherited uric acid lithiasis, a disease not accompanied by hyperuricemia, a defect in proximal reabsorption and distal secretion has been assumed. Salicylate, chlorothiazide, lactate, acetazolamide, and pyrazinamide interfere with uric acid excretion, possibly by blocking secretion. Uric acid excretion is reduced only when small doses of the compounds are used, the administration of large doses enhances uric acid excretion.

Many aspects of amino acid excretion are discussed in the sections devoted to amino acidurias. To review, the amino acids in the glomerular filtrate are reabsorbed in the proximal segments of the renal tubule. For some of these amino acids—glycine, L-alanine, L-glutamic, L-lysine, L-arginine—the reabsorption mechanism is active. The basic amino acids arginine, lysine, and ornithine seem to share a common transport mechanism. Aspartic acid competitively inhibits the transport of glutamic acid, suggesting that a single transport mechanism is involved in transporting acid amino acids. The reabsorption of the amino acids is not exclusively active, and passive diffusion also occurs. There is no evidence that amino acids are secreted in the tubular content.

Hemoglobin is reabsorbed in the proximal tubules and in the distal tubule at the site of maximum sodium reabsorption. The reabsorption of electrolytes, organic acids, and organic bases is discussed in the sections devoted to electrolyte metabolism and acid-base regulation. The fate of urea in the kidney is discussed later [7].

Water Metabolism

The body water has a dual source, ingestion and metabolism (oxidation of fats, carbohydrates, and proteins). The amount of water that is contributed by metabolism (300 ml per day) is far from negligible. Normally, water intake is regulated by the sensation of thirst, a sensation triggered by changes in the osmotic pressure of the blood. A nervous center in the brain in the floor of the fourth ventricle responds to osmotic pressure differences between the cells of the nervous center and those of the circulating fluid. An increase in the osmotic pressure of the circulating fluid induces the sensation of thirst. The body water is excreted through the urine, feces, sweat, and respiration. Under abnormal conditions, large amounts of water can be lost by diarrhea, excessive sweating, hyperpnea, or repeated vomiting. Under normal conditions, the kidney regulates the amount of water in the body fluids. The regulation is achieved both by water excretion and by reabsorption of the water in the glomerular filtrate. The amount of water that passes through the kidney is considerable, and 90–160 ml of water passes through the glomeruli every minute. Of that amount, 80–85% of the water is reabsorbed and the remainder is excreted.

Reabsorption of water through the renal tubule controls water metabolism. This process adjusts itself to changes in the water intake. For example, in beer-drinking individuals, large amounts of water are excreted in the urine. In contrast, water reabsorption is increased in dehydrated patients.

As mentioned before, the antidiuretic hormone is important in water reabsorption (the hormone's molecular properties and mechanism of action are discussed in the section on the hormones of the posterior lobe). In spite of the importance of the antidiuretic hormone in regulating water metabolism, the control it provides is not indispensable for water reabsorption. Twenty per cent of the water present in the glomerular filtrate can be reabsorbed even in absence of antidiuretic hormones. Consequently, in diabetes insipidus, no more than 80% of the water passing through the glomerulus is excreted.

The tubular reabsorption of water that occurs in the absence of antidiuretic hormones explains some of the findings that are made in water intoxication, which is observed in psychiatric patients who drink excessively or in surgical patients to whom excessive amounts of fluid are administered. In those patients, 20% of the water intake is reabsorbed, and the excess water accumulates in the body fluids leading to edema. Edema of the brain is responsible for the headaches, vomiting, and stupor observed in water intoxication. Even under conditions of water deprivation, some water is, in healthy individuals, excreted in the urine. The excreted water is then of metabolic origin, and the amount excreted is regulated by the amounts of solutes present in the glomerular filtrate that are not reabsorbed in the tubules.

In patients who excrete excessive amounts of solutes (ingestion of large amounts of salt, diabetes, Addison's disease) diuresis will increase. In severe dehydration, the regulatory mechanism tends to maintain the osmolarity of the blood constant by eliminating some of the solutes, particularly sodium and chloride. Consequently, even in severe dehydration, the osmolarity of the blood is not considerably increased, and it is therefore dangerous to administer pure water to victims of severe dehydration. Among the victims of concentration camps of World War II, such a procedure led to even further loss of sodium and to the development of severe cramps in the limbs.

Importance of Water in Biology

The experiments of the 17th century Flemish physician Van Helmont probably first illustrated the importance of water in biology. The experiment may have lacked the precision of those of his contemporary Galileo, but it had such a romantic flare that it deserves to be related. Van Helmont planted a small willow (5 pounds) in a large container filled with 200 pounds of desiccated dirt. He covered the receptacle with a perforated top and sprinkled the experimental vegetable with rain water for 5 years. Although this seems to be an enslaving enterprise, considering the frequency of rainfall in Flanders, it is unlikely that Van Helmont needed to sprinkle the precious plant more than a few times a year. After 5 years, in the fall when the closely watched willow had shed its leaves, Van Helmont carefully weighed the tree and the dirt, which was again desiccated. Only 2 ounces of dirt was gone, probably lost in manipulation. Yet the tree had gained 162 pounds, not counting the leaves that were shed for 5 years. There was no doubt in Van Helmont's mind that water had provided the building blocks for the tree.

Since then we have learned that water constitutes 50–80% of living plants and animals. However, biological water should not be regarded as a space filler. Such a concept would be erroneous because many of the special properties of living matter result from the peculiar properties of water. To think of macromolecules (such as glycogen, protein, and nucleic acids) and water as two separate entities is to ignore the close interaction between water and macromolecules. Moreover, as a solvent water plays an important part in distributing ions inside and outside the cell.

Water Structure

The hydrogen molecule has a 1s electron. The oxygen molecule has two electrons in the 1s orbital, two in the 2s orbital, and four in the 2p orbital (see Fig. 9-3).

When hydrogen and oxygen combine to form water, a nonconjugated polyatomic molecule is formed.

Thus the hydrogen molecule contains an unpaired electron in 1s, and the oxygen atom contains unpaired electrons in 2py and 2px. In forming the water molecule, two hybrid 3p (linear) orbitals are formed, yielding two covalent H—O bonds. The length of the O—H bond is 0.99 A. (The radii of the atomic orbitals are 0.53 A for the 1s orbital of hydrogen, 0.45 A for the 2p orbital of oxygen.)

In the water molecule, the 1s orbitals of each of the hydrogen atoms overlap with the px and py orbitals of the oxygen. In the oxygen molecule, the angle between these two orbitals is 90°. Spectroscopic measurements indicate that in the water molecule the angle between the orbitals is 104°. The angle widens because of a slight hybridization of the sp orbitals on the oxygen molecule and the polarization of the molecule. In the water molecule, the electron cloud is not symmetrically distributed, but the distribution of the

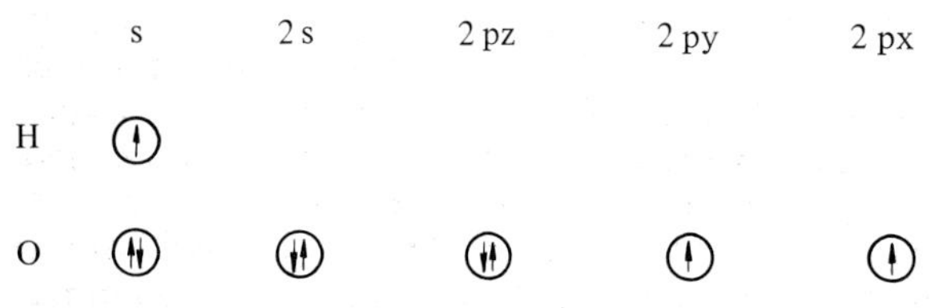

Fig. 9-3. Water structure

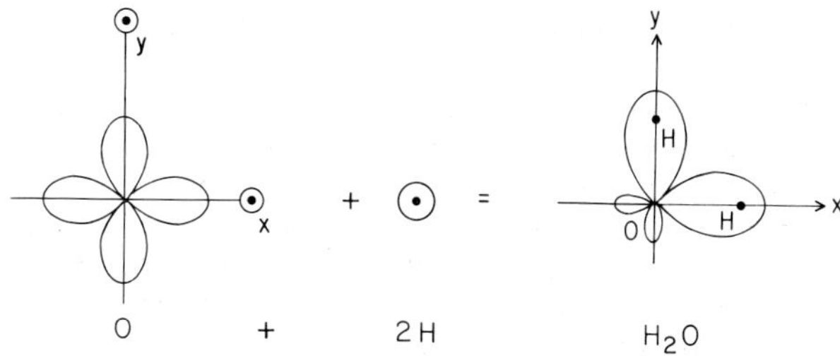

Fig. 9-4. Molecular orbitals in hydrogen atoms

charges is such that after combining to form water, the oxygen atom becomes strongly negative and the hydrogen atom strongly positive. The repulsion of the two (highly positive) hydrogen atoms tends to widen the angle between the bonds (see Fig. 9-4).

As a result of the asymmetric distribution of the charges, the water molecule has a negative and a positive pole, like a magnet. Consequently, when placed in a magnetic field, the water molecule can be expected to orient itself the same way the needle of a compass orients itself in the magnetic field of the earth.

The dipole moment of the molecule is the force that makes the molecule turn, orient itself, and acquire polarity in a given field. In other words, the dipole moment is a measure of the degree of polarity. The value of the dipole moment is expressed in Debye units (electrostatic unit of charge $\times$ cm $\times 10^{-18}$). The dipole moment of the water molecule is comparatively high: 1.81 Debye units. The high dipole moment of water explains its tendency to form hydrogen bonds. In a hydrogen bond, one atom of hydrogen binds two other atoms. This type of bonding occurs with highly electronegative atoms (F, O, N). The bond may bind different molecules or may bind a hydrogen to an electronegative atom within a given molecule. The bond is thus electrostatic in nature. The binding energy of hydrogen bonds is relatively low.

C—H	87
O—H	110
N—H	84
S—H	86
O—H...H	5–10
	Cal/mole

Water's high dipole moment also explains the high dielectric constant of water. The mode in which a high dipole moment influences the properties of the water molecule can best be illustrated by comparing H_2O to H_2S:

Hydrogen bonds in water

	Melting point	Boiling point
H_2O	0	100
H_2S	−85.5	−60

In each case, the values for water are much higher than those for H_2S; although one might have expected some similarity between the properties of water and H_2S since a similar type of bonding is involved in forming both types of molecules. Because the boiling point usually increases with molecular weight the boiling point of water should be lower than that of H_2S.

The bonding in H_2S is similar to the bonding of water except that the former involves the 3p electrons of sulfur rather than the 2p electrons of oxygen. As a result, the length of the S—H bond is greater than that of the O—H bond, thus the polarity of the S—H bond is smaller. Consequently, the repulsion between the H atoms is much less in H_2S than in H_2O. The difference in polarity explains the difference between the properties of water and H_2S.

	Length (A)	Polarity (D)[a]	Bond energy (Cal)
O—H	0.96	1.5	110
S—H	1.35	0.7	86

[a] Debye units.

Water molecules are so polar that the negative pole of the molecule (O) with its lone electron pair is attracted by the positive pole of another molecule (H) 176° away, forming a hydrogen bond between the two. As a result, the water molecules are not isolated but form chains of molecules. Thus, in reality water behaves as a polymer (H_2O) n and this is probably why its boiling point is unexpectedly high.

Hydrogen bond formation is best understood by looking at the geometrical structure of ice crystals. Ice is a complete polymer. Each molecule of HOH is bonded to its neighboring molecules by hydrogen bonds. The result is a hexagonal lattice. The packing of the molecules is loose, and a hole (water cage) is in the center of each hexagon.

In forming the polymer, one atom of hydrogen is bound to two atoms of oxygen, and one atom of oxygen is bound to four atoms of hydrogen. As a result of this intensive hydrogen bonding, the molecules are less packed in the water crystal than in the liquid state, a property which explains why water, in contrast to other liquids, expands when it freezes and why ice floats at the surface of liquid water. (The density of water is 1000; that of ice is 0.9998. Thus, the density increases by 9% when ice melts.) Liquid water retains some of the properties of the crystal. Measurements of the heat of ice fusion indicate that only 15% of the hydrogen bonds are disrupted when ice melts.

The biological role of water would be easier to interpret if the arrangement of the molecules in liquid were known. In spite of much investigation, water's molecular structure remains uncertain, and our knowledge of the biological role of water gives us only a glimpse of what is to be learned in the future. Thus, among some unoriented molecules, many water molecules seem to be arranged in a crystalline fashion. In liquid water, oriented molecules are in "clusters," and the molecules of the clusters are most likely held together by hydrogen bonds. Let us assume that a cluster is formed by the molecules A, B, C, D, E, etc. held together by hydrogen bonds 1, 2, 3, 4, etc. Hydrogen bonding between A and B will cause an electronic perturbance in B. This electronic perturbance will in turn influence the hydrogen bonding between B and C. As a result, the bonding energy is increased. Thus, because of the electronic perturbation in atom B, bond 2 is stronger than bond 1, bond 3 is stronger than 2, 4 is stronger than 3, and so on. However, thermodynamic calculations suggest that the gain in energy resulting from the perturbation of the electronic layers is very small.

When a molecule of water is bound to another molecule of water by a single hydrogen bond, the energy of the bond is 4.5 Cal. In 4-bound molecules, the average energy per hydrogen bond is 6 Cal. The change in energy mainly involves the proton involved in bonding and hardly affects the lone electron pair of the oxygen.

Opinions vary with respect to the stability of the bonds involved in cluster formation. Some investigators feel that the bond is stable once it is formed, and although it may be bent or stretched until it breaks, once broken, it is not readily restored. Others believe that the electronic perturbation generated by hydrogen bonding is responsible for cooperative resonance between a large number of molecules. A water cluster in which these electronic perturbations are constantly passing back and forth from one molecule to another is sometimes referred to as a flickering iceberg.

Several groups of investigators have attempted to calculate the size of such a cluster. Some models propose that the cluster is formed by 90 molecules at 0°. When the temperature of water increases, the size of the cluster decreases and drops to 28 at 60°. Other models have proposed clusters containing as many as 600 molecules.

The geometrical distribution of the water molecules in liquid water is still debated [8–14]. Both hexagonal and pentagonal arrangements have been proposed. The Danford-Levy model postulates that the "ice cluster" of liquid water has a hexagonal lattice as in ice, but the lattice is expanded and the water cages are filled with isolated water molecules. Such an expanded lattice may be expected to present many more protionic defects than "ice crystals."

The isolated H_2O molecules could be bound to the lattice by hydrogen bonds. The bonding between isolated water molecules and the "icelike lattice" is likely to occur at points where the crystalline structure presents protionic defects (empty or doubly occupied bonding sites).

The Danford-Levy model, which seemed to satisfy the results obtained by X-ray crystallography, has been rejected by Frank on thermodynamic grounds. The Danford-Levy model seems to require the formation of hydrogen bonds that are so long (2.94 A) that the bond energy would not be sufficient to hold the molecules together.

A pentagonal model has been proposed. The model is made up of 15 atoms: 10 oxygen, 5 hydrogen. The 10 oxygen atoms are placed in two parallel pentagons, each atom occupying the summit of the angles of a regular pentagon. The two pentagons are held together by bonds to hydrogen atoms. The pentagon layers can extend on their sides into icelike hexagonal structures. This combination of hexagon and pentagon has the advantage that all the hydrogen bonds have normal length and consequently are not under stress as in the Danford-Levy model. Furthermore, a model including both pentagonal and hexagonal configurations constitutes an excellent intermediate structure between the typical ice structure (hexagonal) and the structure taken by water when bound to proteins (which is believed to be pentagonal).

This brief description of some structural models of the water molecules illustrates the complexity of the problem. Kavanau [13] has remarked that there are at least four ice structures, and he predicts that clusters with different structures and of different sizes will be found in biological water.

Despite the fact that the exact structure of biological water is unknown, some important generalizations can be made, which are relevant to the biological role of water. First, compared with the ice crystal, the "ice-cluster" of water is small. Second, although even the ice crystal presents protionic defects, these defects are much more common in the water cluster. And, finally, within the water cage, the cluster contains interstitial water molecules bound by hydrogen bonds to an ice-like structure.

As we shall see, present knowledge of water structure can help explain many aspects of a number of biological events—for example, membrane transport, charge transfer, protein hydration, and the mechanism of action of anesthetics.

Hydration and Water Organization

Anhydrous copper sulfate is white, but as the dry powder is hydrated, blue crystals are formed by electrostatic binding of the water molecules to the ions. When the blue hydrate is heated, the water is progressively released, and it was calculated that five molecules of water were bound to one molecule of copper sulfate.

Now, if an anhydrous salt, such as NaCl, is dissolved in water, there are two steps in the process of solution: (1) the dissociation of the crystal lattice holding the NaCl molecules together, and (2) hydration of the ions. The first of these processes requires energy; consequently, energy is absorbed during ionization. The second step results in energy release.

The "hydration energy" for a given ion is the algebraic sum of the amount of energy needed to pull the water molecules off their crystalline structure and the amount of energy required to bind the water to the metallic ion. The term "bound water" is misleading if it is understood that the water molecule is permanently attached to the ion. In reality, the duration of the attachment is very short (10^{-6} second). This is still far longer than the time required for the molecule to go through a complete vibration (10^{-13} second). Therefore, the water molecules of the solvent (icelike structure) must be constantly exchanged with those bound to the ion. Thus, the intermediate zone contains water molecules in transit from one form of "organized" water to another form of "organized" water.

Understandably, the structural organization of the water molecules in the vicinity of the ion varies as the distance from the nucleus of the ion increases. The water molecules close to the ion are rigidly bound to the ion and consequently are in a high state of organization. The water molecules at some distance from the ion's nucleus are in their normal semicrystalline cluster-type of organization. In the area between the bound water molecules and the normal, icelike water is an intermediate zone in which the water molecules are schizophrenic; some participate in an icelike structure and some are bonded to the crystal.

Within the transit zone, the water molecules are partially disorganized. Thus, adding ions to a solvent disorganizes part of the water molecules, which explains the lower heat capacity of the solution. (Less heat is required to increase the temperature of a solution of NaCl than is required to increase the temperature of pure water.) Quite to the contrary, adding nonpolar compounds to pure water increases its heat capacity. This finding suggests that the nonpolar compounds have an organizing effect on the water molecules, and indeed, a combination of physicochemical studies has led to the proposition that nonpolar compounds organize the water molecules into pentagonal lattices. Five molecules of water combine to form a pentagonal plane, the side of which may combine with other water molecules to form a dodecahedron.

The pentagonal arrangement forms water cages, in which the nonpolar molecules can be lodged conveniently. These hydrates are usually stable and have a melting point far above that of water (30° or higher). Many compounds (including inert gases) with widely different chemical properties form such hydrates. Significantly, the hydration energy is the same for all these polar hydrates, suggesting that the hydration energy is needed to organize the solvent rather than the solute.

Water Structure and Proton Transfer

In a previous section, the significance of the hydration shell surrounding the apatite crystals in mineralization and demineralization was discussed. Structured water could be important in proton transfer, and thus in oxidation-reduction. Whether water plays such a role *in vivo* is not certain, but when radioactive Fe^{3+} is placed in solution containing cold Fe^{2+}, radioactive Fe^{2+} is recovered. Consequently, an electron must have been transferred from Fe^{3+} to Fe^{2+}. The charge of both ions is too large to permit a direct transfer from the valence orbital of Fe^{3+} to that of Fe^{2+}. The forces repelling these two ions would keep them so far apart that an overlapping of their orbitals is prohibited.

To explain the conversion of the Fe^{3+} to Fe^{2+}, investigators have postulated that the two hydrated ions are linked by a structured chain of water molecules. The assumption is that one of the electrons of the Fe^{2+} valence shell enters the valence shell of an oxygen atom of a neighboring water molecule. The water molecule releases a hydrogen-free radical (one hydrogen atom and one electron), which is transferred from water molecule to water molecule until it reaches a water molecule close to $^{*}Fe^{+3}$, at which point the electron is transferred to $^{*}Fe^{+3}$, yielding Fe^{+2}. H^{+} (protons) and H^{-} (hydrate ions) may be transferred the same way.

Proteins and Water Structure

Of considerable importance are the effects of proteins on water structure. Proteins contain three different types of "groups" with respect to their effect on water structure: (1) the hydrate group which hydrates as ions would, (2) the nonpolar groups, and (3) those with surfaces with sites amenable to form hydrogen bonds.

Many amino acids contain nonpolar side chains that act as "organizers" of the water molecules, for example, the CH_3 of alanine, the $\overset{CH}{H_3C\diagup\ \diagdown CH_3}$ of valine, the $CH_2{-}CH(HC)(CH_2)$ of leucine, and the CH_2 of phenylalanine.

Moreover, in a protein sequence, some of these nonpolar groups project outside of the main chain. They are frequently repeated, often in close proximity to each other. As a result, their effect on the solute is coordinated. Together they build a sheet of water molecules that are restricted in their movements. The protein and the organized water are so intimately interrelated that removal of the water may lead to the collapse of the protein structure and denaturation. The sheet of water molecules may also conceal some of the normal reactivity of the naked protein chain. For example, in some proteins the presence of some SH groups can be revealed only after denaturation.

In addition to amino acids with polar side chains (glutamic, aspartic, lysine, arginine, histidine) and amino acids with nonpolar side chains (alanine, valine, leucine, isoleucine, phenylalanine), protein chains also contain amino acids whose side chains can form hydrogen bonds with a suitable donor (serine, threonine, tyrosine, tryptophan, asparagine, glutamine). Many of the side chains form hydrogen bonds between amino acids of the same chain. Such bonding contributes to the formation of a tightly woven molecule with a rigid tertiary structure. Yet, in some proteins—for example, collagen molecules—some of the group remain free to form hydrogen bonds with water molecules. A good understanding of the relationship between solute and collagen fiber can come only from a detailed description of the collagen molecule. Suffice it now to remark that according to the Rich-Crick model for the collagen fiber, N-H and C=O groups protrude from the main chain. Thus, the molecule is spiked, with many sites favoring hydrogen bonding with water molecules.

In view of their influence on water structure, the distributions of charges on cellular and subcellular membranes acquire particular significance. For example, Kavanau has speculated that proton transfer would be hindered in membrane zones in which fixed charges are abundant, whereas proton transfer would be facilitated when the membrane structure includes methyl or methylene groups. Whether this is the case remains to be established; but interpretations of membrane transport in molecular terms will have to take into account the effect of the membrane's molecular structure on water structure.

Hypothetical Mode of Action of Anesthetics and Water Structure

The molecular mechanism of action of anesthetics has defied pharmacologists and biochemists for more than five decades. To be sure, it is puzzling that compounds as different from each other as rare gases and chloroform manifest the same biological activity. Most anesthetics are liposoluble, and the effectiveness of an anesthetic correlates with its solubility in lipids.

However, such a correlation does not explain how consciousness is lost. In 1961, Pauling* proposed what he called a molecular theory of anesthesia. The basic argument in Pauling's theory is that all anesthetics have only one property in common: they take part in the formation of clathrate crystals. In fact, in the case of such inert compounds as xenon, the formation of clathrate is the only possible property that could be invoked to explain the anesthetic effect. The principle of the effect of the anesthetic gas is as follows. The gas and the water molecules are organized into a crystalline form having water cages, in which the gas molecules are lodged as guests. The presence of the gas stabilizes the hydrate. The melting temperature of the hydrate is a measure of its stability. The structures of many such hydrates have been investigated.

Although only the xenon hydrate will be described, all these hydrates have common features. The oxygen atoms of a water molecule are surrounded by four hydrogen atoms. One oxygen forms covalent bonds with two hydrogens of its own molecule, and hydrogen bonds with hydrogens of two adjacent water mole-

* The same Pauling who wrote *The Nature of the Chemical Bond* discovered the hydrogen bond, discovered the electrophoretic alteration in sickle cell anemia, and first proposed plausible molecular models for protein structure.

cules. The oxygen is in the center, and the four hydrogens are at the corners of a regular tetrahedron. Consequently, if four straight lines were traced linking the oxygen to the hydrogen, the lines would be separated by an angle of 108°. Twenty of such molecules are organized into 12 side structures (dodecahedrons), or geometrical figures limited by 12 regular pentagonal faces.

It is impossible to fill all the spaces in these dodecahedrons, and openings are left between them. Some of the openings are pentagonal, some are hexagonal (formed by 6 dodecahedrons).

Sodium Metabolism

The main source of the sodium in the body is the sodium chloride used for cooking or preserving foods. The high intake of sodium chloride does not result from physiological needs, but from the presence of salt in food. Salt is tasty because man's tongue and the chordae tympanae contain sensory fibers that recognize the taste of salt. In contrast, man possesses no nervous structure capable of tasting distilled water, and one may well wonder if it was the Creator's intention to have us drink this impalatable liquid.

The existence of a salt tax* at some time in the history of practically every country of the world (except the United States and Sweden) emphasizes the importance of salt in gastronomic customs.

The word salary originated from the Latin (*salus*) and was introduced because part of the earnings of some of the functionaries of the Roman Republic** were payed with cakes of salt, which were brought to Rome through the famous *Via Salaria*. In spite of the extensive use of salt in most cultures, it is unlikely that primitive man used salt, and even in recent history, some cultural groups (North American Indians, Numidian nomads, and Bedouins) do not add salt to their diet.

The ingested sodium is absorbed in the intestine through a passive process that occurs in the direction of the concentration gradient. When the sodium concentration in the intestinal lumen is high, the sodium diffuses from the intestinal lumen into the extracellular fluid until the sodium concentration in the extracellular fluid equals that of the intestinal lumen. When the sodium concentration in the intestinal lumen is low (as in severe diarrhea), the sodium moves out of the intracellular fluid into the intestine, and severe sodium loss may occur.

The normal human body contains 2–3 equivalents of sodium (130–140 mEg/liter or 300–335 mg/100 ml). The body sodium is distributed into intracellular and extracellular pools. The sodium contained in these pools exchanges readily with the exogenous sodium. Isotope studies demonstrate that the intracellular sodium is rapidly exchanged with the extracellular sodium. In addition to the two large sodium pools, a smaller sodium pool is located in the skeleton. The sodium in bone is only slowly exchanged with exogenous sodium. The presence of sodium in the bone has been discussed in the chapter on calcification.

* The most famous of such taxes was the gabelle, which was introduced early in the history of France and survived until the French Revolution. To assure that each of their subjects would contribute their fair share to the internal revenue, the king of France imposed the purchase of a certain quantity of salt each year.

** This custom apparently still exists in Ethiopia and Tibet.

Sodium Pump

The transport of sodium against a concentration gradient is a phenomenon common to all cells that has been investigated in typical physiological models. The most popular model is the giant axon of the squid and the salt excretory gland of the albatros.

Electrodes can be introduced in the giant axon of the squid. The sodium concentration inside the axon is low, and that of potassium is high. At the outside, the potassium concentration is low, and that of sodium is high.

The extracellular sodium diffuses freely within the cell until a certain intracellular concentration is reached (about 20% that of potassium). Any excess of sodium is immediately eliminated from the intracellular fluid against a high concentration gradient (the sodium concentration of the extracellular fluid is 5–10 times that of the intracellular fluid) by an active process called the sodium pump.

Although all the details of the operation of the sodium pump are not clear, it is known that energy is used in the process. The process is slowed down at low temperatures during anoxemia and by metabolic inhibitors. If erythrocytes are kept at 0°, potassium is rapidly lost from the erythrocyte into the medium, and sodium enters the cell. A similar potassium loss may be observed in hypoxia. Oxygen is consumed while sodium is excreted against the concentration gradient. Blocking of bioenergetic pathways with antimetabolites inactivates the sodium pump.

The salt gland of the albatros pumps sodium against a concentration gradient, and the gland may achieve intracellular sodium concentrations as high as 1 M. The operation of the gland is controlled by cholinergic nerve branches, and acetylcholine stimulates the gland. The center of the stimulation is in the brain and is under osmoregulatory control. When a hypertonic solution is injected, the gland secretes sodium. If in contrast the gland is incubated in water, it loses sodium and there is a sharp drop in phosphatidic acid, and if ^{32}P and acetylcholine are added to the incubated mixture, ^{32}P is rapidly incorporated into the phosphatidic acid.

Tissue fractionation studies of the albatros' salt gland have demonstratey that the phosphatidic acid concentration is 10 times greater in the microsomes than in other cell fractions. Electron microscopic examination demonstrates a dense network of smooth endoplasmic reticulum membranes that are connected with the cell membrane.

Studies on brain have led to new interpretation of the relationship between phosphatidic acid and the sodium pump. The incorporation of ^{32}P into the endoplasmic reticulum of brain is stimulated by acetylcholine, ATP, and diglyceride kinase. Sodium is an absolute requirement for this effect. The addition of sodium to the medium stimulates the incorporation of ^{32}P into phosphatidic acid, in contrast to potassium, which has no effect.

On the basis of such studies, Hokin proposed a cyclic type of reaction in which sodium combines with phosphatidic acid to yield sodium phosphatidate. Sodium phosphatidate is assumed to be formed at the inner side of the membrane.

In conclusion, the active transport of sodium and potassium is essential to muscular excitability, absorption in kidney and intestine, and maintenance of the volume of all cells. Moreover, the active transport of sugar and of amino acids is determined by the distribution of potassium and ions in the cell. The energy source for sodium transport is ATP. Although it is believed that in most cells a large amount of the energy generated serves to activate the sodium pump, there are few instances where we know the stoichiochemical relationship between molecular transport and ATP breakdown. In the red cell, the energy generated by the hydrolysis of one ATP molecule to ADP serves to pump three Na^{++} outward and 1 K^+ inward.

The sodium pump is located in the cell membrane. The details of its molecular structure are unknown but an enzyme discovered by Skou in 1957 is very likely to be an essential part of it.

Sodium-Potassium-Dependent ATPase

An ATPase has been found in association with cell membranes. The enzyme is stimulated by sodium and inhibited by cardiac glycosides (ouabain and digitalis). If potassium is present, the stimulation by sodium is enhanced. The enzyme originally described in crab nerve and erythrocyte ghost was later found in a number of tissues—kidney, ciliary body, brain, giant axon, ascites cell, skin cell, lens, leukocyte, and skeletal and heart muscle.

The ATPase has been associated with electrolyte flow. It is believed to catalyze the outward movement of sodium and the inward movement of potassium. Studies with erythrocyte ghost suggest that the enzyme activity directs the electrolyte flow. In erythrocytes, sodium is extracted and potassium enters the cell. Digitalis inhibits the enzyme only at the outside, and sodium stimulates it only at the inside.

Under special conditions using erythrocyte ghosts, the activity of the enzyme and, for that matter, of the sodium pump can be reversed.

The simple situation of the toad bladder better illustrates the polarization of the electrolyte flow. In the toad bladder, the enzyme is at the serosal surface of the mucosa, where sodium is extruded in the plasma and potassium penetrates the cell. This is also the part of the cell membrane that is sensitive to ouabain inhibition. Consequently, the serosal site of the cell membrane is freely permeable to potassium but not to sodium. In contrast, the epithelial site of the membrane was found to be permeable to sodium but not to potassium. As a result of the differences of the epithelial and serosal sites of the toad bladder membranes, the flow of sodium is unidirectional. The sodium that diffuses passively at the epithelial site is extruded actively at the serosal site, and the potassium that freely enters the cell at the serosal site can leak back into the plasma.

The enzyme is in the intact cells or even in tissue fragments bound to microsomal and plasma membranes. The purified enzyme is composed of two polypeptides with a molecular weight of 84,000 to 100,000 daltons for the larger. The smaller is a glycoprotein with an estimated molecular weight of 55,000 daltons. The chemical composition of both polypeptides varies with the source. The large portion is the one which becomes phosphorylated during the reaction. The role of the glycoprotein is not clear. It has not been possible to restore activity by reassociating the two polypeptides.

Among the requirements for enzyme activity are ATP, Mg^{++}, Na^+, K^+ and lipids. The requirement for ATP is not absolute. The hydrolysis of ITP, GTP, and CTP can replace that of ATP, but these nucleotides are much less effective than ATP. The hydrolysis of UTP is ineffective in the sodium-potassium—dependent ATPase (Na K ATPase).

Na^+ and K^+ are required for enzyme activity, but an excess of either of these ions will inhibit the effect of the other. Therefore optimal activity is only achieved with a ratio of Na^+:K^+ of 10:1 and 5:1. While the requirement for Na^+ is absolute, Rb^+, NH_4^+, Cs^+ and Li^+ can replace potassium, but these ions are much less effective than K^+. Optimal activity of the enzyme is achieved with concentrations of Mg^{++} that equal that of the substrate. Other divalent cations Mn^{++} and Co^{++} are ten times less effective than Mg^{++}. Whether Ca^{++} affects the activity of the Na K ATPase remains to be seen.

The enzyme is inactivated by detergent or phospholipase and reduction in activity correlates with the amount of lipids removed. Moreover, activity can be restored by adding lipids to the delipidized enzyme. The chemical composition and the molecular interaction between lipid and protein *in vivo* are unknown.

The function of the Na K ATPase is a complex one. It must (1) pick up one Na^+ in the inside of the cell and transfer it to the extracellular fluid against a concentration gradient, (2) pick up a K^+ in the extracellular fluid and then transfer it inside the cell. Obviously this driving of ions through the cell membrane against concentration gradients requires energy and the ion transport must therefore be closely coupled to the hydrolysis of ATP to ADP.

It is not known whether there are two different binding sites, one for Na^+ and one for K^+, or a single site for both cations.

Innumerable studies of the catalytic and kinetic properties of the enzyme indicate that the reaction takes place in a sequence of steps. In step one the enzyme binds to ATP to yield phosphorylated enzyme and ADP,

$$E + ATP \xrightleftharpoons{Mg^{+} + Na^{+}} E{=}P + ADP.$$

The molecular details of the binding of ATP to the protein are not known except that: binding is inhibited by K^+ and ouabain, the inhibition of K^+ is reversed by Na^+, and the 6-amino group of the purine ring is indispensable for the reaction. Hydrolysis of the phosphorylated enzyme at various pH's indicates that it contains an acylphosphate forming either L-glutamyl or an L-aspartyl-β-phosphate residue.

In the second step of the reaction the phosphorylated enzyme undergoes conformational changes leading to a different molecular form of the phosphorylated enzyme. The potential significance of these conformational changes will be discussed later.

$$E_1{-}P \rightleftharpoons E_2{-}P$$

The conformation change requires higher concentrations of Mg^{++} than the phosphorylation of E and the conversion of E_1-P to E_2-P is almost irreversible. It is believed that Ca^{++}, which inhibits the Na K ATPase, can stimulate phosphorylation but inhibits the conformation change.

In the third step the newly conformed phosphorylated enzyme binds K^+ to yield a K-E_2-P complex.

$$E_2{-}P + K^+ \rightleftharpoons K{-}E_2{-}P$$

In the fourth step the potassium bound E_2-P complex is in presence of water dephosphorylated by a K^+ dependent phosphatase. In red cells there is a correlation between the activity of the K^+ activated phosphatase and the Na K ATPase. After dephosphorylation the K^+ remains attached to the E_2 conformation. When it is released the E_2 enzyme is reconverted to the E_1 type in presence of Na^+.

$$KE_2 \rightarrow E_2 + K^+ \qquad E_2 \xrightleftharpoons{Na^+} E_1$$

Except for the fact that the interaction of cardiac glycoside with the Na K ATPase is very complex, the mode of inhibition of the enzyme by these drugs is not understood.

In absence of detailed knowledge of the structure of Na K ATPase it is impossible to provide accurate models of the ionic transport system. Several have been proposed, only one will be described.

At the start of the reaction the conformation of the protein is such that the binding site of Na^+ faces the inside of the cell, Na^+ activates phosphorylation of the protein which then undergoes a conformation change with the following consequence: the binding site of Na^+ rotates from the inside to the outside bringing the Na^+ in contact with the extracellular fluid and the affinity for Na^+ binding is reduced, while that for K^+ is increased. When the enzyme is dephosphorylated by the K^+ dependent phosphatase, it returns to its original conformation bringing the K^+ binding site inside of the cell and potassium is released [70–72].

Evidence that the Na K ATPase plays an important role in sodium diuresis and urine concentration has been obtained in experiments in which digoxin is injected directly into a dog's renal artery. Such an experiment induced unilateral sodium excretion as a result of the specific inhibition of ATPase in renal cortex (proximal and distal tubules) and medulla (loops of Henle). Inhibition of the ATPase prevented sodium reabsorption with natriuresis and impairment of urine concentration [15–17].

However, sodium reabsorption and Na K ATPase activity in the kidney are not closely correlated since the activity of the Na K ATPase is greater in the medulla than in the cortex and the bulk of sodium reabsorption is thought to occur in the cortex (proximal and distal tubules). No satisfactory explanation has been proposed for these differences.

Na K ATPase is likely to be present in all cells where it maintains the electrolytic balance, but it is of particular significance in the kidney. Although in addition to Na^+ the kidney also reabsorbs glucose, potassium, phosphate and amino acid, on a stoichiometric basis Na^+ reabsorption accounts for 70% of renal transport.

In the kidney the Na K ATPase is present in both cortex and medulla, but its activity is maximal in the outer medulla where it is responsible for 40% of sodium reabsorption through sodium potassium exchange. The source of chemical energy is both aerobic and anaerobic.

Inhibition of the Na K ATPase with ouabain does not completely prevent sodium reabsorption. Sixty percent of the sodium continues to be reabsorbed suggesting that other mechanisms of sodium transport must exist, for example, an Na/Cl/H_2O transport [73].

For further details on the properties of the Na^+ K^+ ATPase, the reader is referred to the recent issue of the Annals of the New York Academy of Sciences [84].

Sodium Reabsorption in the Kidney

Most of the sodium in the glomerular filtrate is reabsorbed in the renal tubules. Thus, while 20,000 mEq of sodium appears daily in the glomerular filtrate of an adult male, only 200 mE is excreted.

Sodium is reabsorbed in the proximal and the distal tubules. A major fraction of filtered sodium and water is reabsorbed in the proximal tubule. Reabsorption is isotonic, and consequently the sodium concentration in the proximal tubule is not markedly changed. However, sodium reabsorption in the proximal tubule is not passive. A number of experiments have suggested the existence of an active process. For example, if an electric potential is generated within the lumen of the tube and interferes with sodium movement, sodium is reabsorbed anyway.

The mechanism regulating sodium reabsorption in the proximal tubules is still debated. Ten to fifteen per cent of the filtered sodium load is reabsorbed in the distal tubule. This reabsorption is no longer isotonic but occurs against a concentration gradient (from low concentration in urine toward high concentration in interstitial tissue). Moreover, only part of the sodium reabsorption is accompanied by cation absorption. The electrolytic balance is maintained in part by replacing sodium with potassium, hydrogen, and ammonium ions.

The potassium ions are provided from the intracellular pool through the intermediate of the extracellular fluid. The hydrogen ions are provided by the ionization of H_2CO_3, a compound synthesized at the expense of blood CO_2 by a zinc protein, carbonic anhydrase. The distal tubule cell seems to possess a single ion-exchange system of an unknown nature, which exchanges either hydrogen or potassium for sodium. This explains the competition between hydrogen and sodium in the exchange process. Nevertheless, a certain degree of sodium is reabsorbed independently of the ion-exchange mechanism, and sodium is reabsorbed even when large amounts of carbonic anhydrase inhibitors are present.

Although sodium-potassium exchange in the medulla may derive its source of energy from anaerobic glycolysis, the fact that the exchange is affected by cyanide, antimycin A, and 2,4-dinitrophenol suggests that the aerobic pathway provides, under aerobic conditions, part of the energy necessary for the sodium-potassium exchange. Researchers have proposed that coenzyme Q is directly involved in the exchange, but this has not been established.

Investigations of the sodium-potassium exchange made on tissue slices of renal cortex and renal medulla revealed that oxygen consumption was related to potassium uptake. In addition to a basal rate of oxygen consumption, an additional oxygen uptake was observed which could be correlated with the cation exchange. This observation led to the conclusion that oxygen uptake could be regulated by the concentration of cations in the cell.

Regulation of Sodium Reabsorption

Inasmuch as sodium intake may vary considerably, the maintenance of sodium homeostasis requires a broadly adjustable mechanism of renal sodium excretion and reabsorption. Regulatory mechanisms for sodium reabsorption in the kidney include: (1) rapid adjustment of the levels of tubular reabsorption through the glomerular filtration rate; (2) secretion of aldosterone, which stimulates sodium reabsorption; (3) the possible elaboration of a natriuretic hormone that stimulates secretion.

The details of the sequence of molecular events responsible for adjusting renal sodium excretion are far from completely known. Moreover, although there is much evidence that these three regulatory mechanisms interact in healthy and diseased individuals, the degree and the mechanism of these interactions are not at all clear.

Most of the sodium reabsorbed in the distal portion of the nephron is regulated by aldosterone. In contrast, only 50% of the sodium reabsorbed in the proximal tubule appears to be aldosterone controlled. The mechanism regulating aldosterone-independent reabsorption is unknown, but it has been suggested that in the normal individual sodium secretion is adjusted to sodium intake, and that sodium secretion is regulated primarily by proximal reabsorption followed by further fine adjustment of reabsorption in the distal tubule under the influence of aldosterone. Clinical and experimental observations (in animals with sodium lack or depletion) have provided evidence for the existence of factors other than aldosterone regulating sodium metabolism. A few of these observations will be described.

All patients with congestive heart failure do not become edematous. Several investigators obtained evidence that explains why edema develops. They observed that patients with congestive heart failure often have reduced renal blood flow and glomerular filtration rate. If the electrolyte and water balance of the body fluid is to be maintained in the presence of reduced filtration rates, tubular reabsorption must also be reduced. Reduced glomerular filtration with normal tubular reabsorption must lead to water and salt retention.

The existence of a feedback loop between glomerular filtration rate and tubular reabsorption has been established. Glomerular filtration and sodium reabsorption in the proximal tubules increase and decrease together. The following type of experiment best illustrates this relationship. Saline is administered to a dog to expand the extracellular fluid volume and induce sodium excretion. When a balloon catheter introduced in the femoral artery of the dog is inflated to reduce arterial pressure from 170 to 90 mm Hg, filtration drops 10%, but reabsorption drops 50%. The adjustment of reabsorption to filtration is not affected by decerebration, adrenalectomy, hepatectomy, or renal denervation.

In conclusion, changes in glomerular filtration are within seconds followed by the adjustment in the rates of tubular reabsorption. The nature of the signal between glomerulus and tubule remains unknown, but it almost certainly involves no nerves or known neurogenic factors.

Aldosterone

Sodium is excreted in the renal tubules, and this is partially regulated by aldosterone. Aldosterone is elaborated in the glomerulosa cells of the adrenal cortex (see Fig. 9-5). The biosynthesis of aldosterone is discussed in Chapter 8. One of the most dramatic obser-

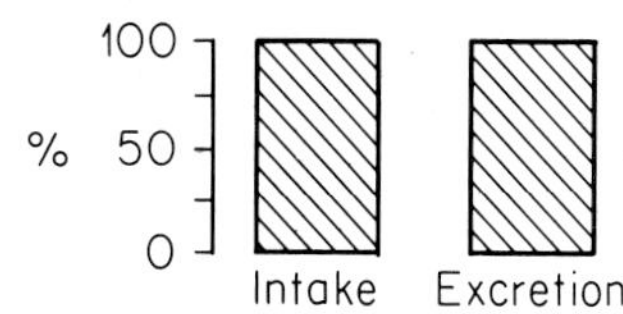

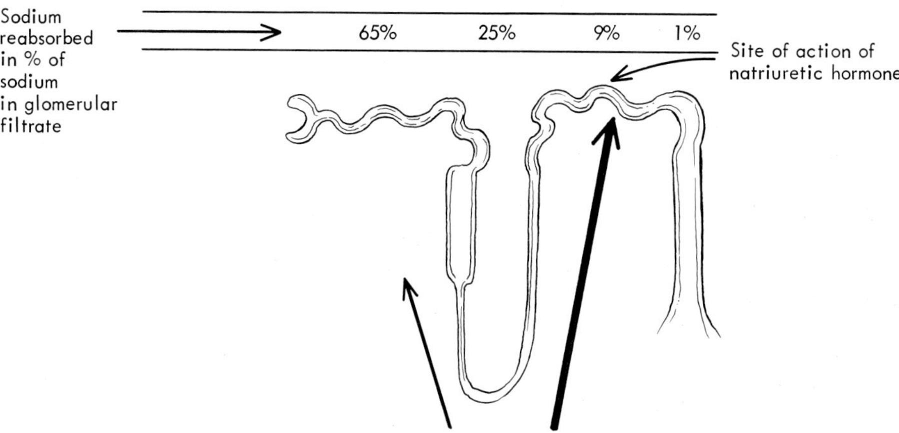

Fig. 9-5. Sodium balance

vations providing evidence for aldosterone secretion by the glomerulosa was obtained at autopsy. A patient who had Cushing's syndrome developed fatal pancreatitis after adrenal exploration. At autopsy, one of the adrenals was hypoplastic, and the other was completely necrotic except for the glomerulosa. In the hypoplastic adrenal, 36% of the steroid hormone secreted was aldosterone. In the necrotic adrenal, 95% of the steroid extracted was aldosterone [18–22].

Experimentally or clinically, aldosterone secretion can be initiated by a number of different stimuli, including sodium deprivation, potassium deprivation, contraction of the body fluid volume, and renin release. Although several of these individual stimuli may affect the adrenal gland through common pathways, the point where they cross and the extent of the overlap are not always clear. Therefore, the effects of each stimulus will be described separately.

Sodium deprivation increases aldosterone secretion, probably by modifying the volume of the body fluids. This conclusion was reached because of observations that: (1) the administration of water (with Pitressin to prevent diuresis) depresses aldosterone secretion in spite of the hyponatremia; (2) if the body fluids are expanded by administering normal saline, aldosterone excretion is stimulated although the sodium levels of the blood are not modified; and (3) dehydration leads to aldosterone secretion, although the reduction in body fluid volume resulting from dehydration is associated with hypernatremia. From these observations, changes in body fluid volume appear to affect aldosterone secretion, and it has been stated that the expansion of the body fluids reduces aldosterone secretion. This conclusion, which is based on experimental observation, needs to be reconciled with clinical observations.

Aldosterone secretion is increased in patients with cirrhosis and cardiac failure, diseases associated with an increase in body fluid volume. To interpret the paradoxical observations made in experimental animals and in patients, investigators have proposed that aldosterone secretion is controlled by selective variations within the total fluid volume, rather than from changes in the total body fluid volume.

From what has been presented, the effect of sodium on aldosterone secretion seems to depend not on the sodium concentration, but on changes in blood volume that accompany the changes in sodium concentrations. Potassium also affects aldosterone secretion, and this effect does not appear to result from changes in the body fluid volume. Potassium deprivation increases aldosterone secretion.

If aldosterone secretion is controlled by changes in the body fluid volume, one could anticipate that the changes in volume are perceived by receptor centers and transferred neurologically to the central nervous system. Constriction of the carotid, which normally affects aldosterone secretion, is without effect if the vessel has been deprived of its normal innervation. Constriction of the inferior vena cava leads to hypersecretion of aldosterone. This effect is not altered by vagotomy; therefore, the vagus nerve is not necessary for transfer of the impulse that originates in the receptor center.

Not much is known of the anatomical location of the receptor center, the structure and path of the afferent and efferent neurons, or the molecular mechanism by which the receptor regulates the changes in volume.

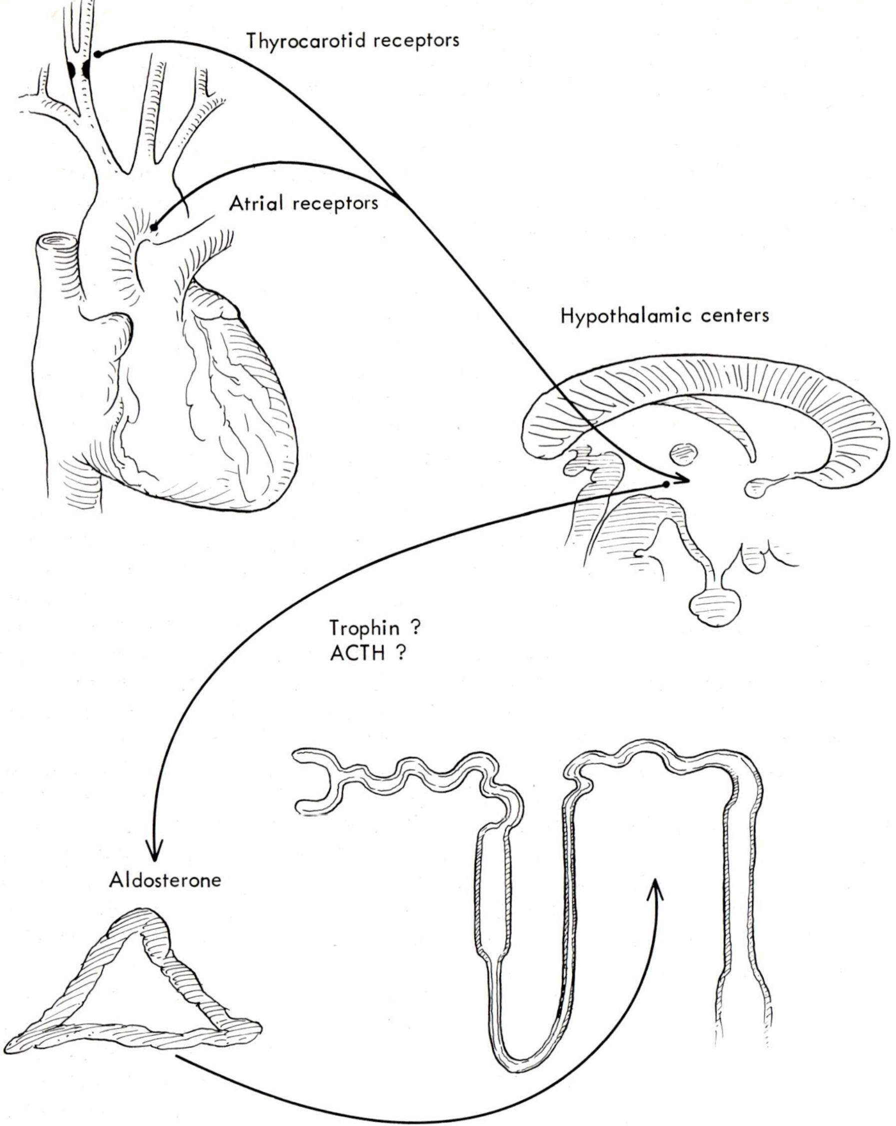

Fig. 9-6. Mechanisms by which arterial receptors regulate aldosterone secretion

The location of receptors in various areas of the body has been suspected; for example, a stretch receptor is believed to be in the right atrium. At least two other sets of receptors located in the arterial system have been identified and found to be sensitive to change in blood volume. Once sensitized, these receptors send a message to the kidney parenchyma, which responds by modulating sodium secretion. These centers are located in the carotid arteries near the origin of the thyroid arteries and in the juxtaglomerular apparatus. Constriction of the carotid low in the neck stimulates the carotid receptor, which then sends impulses to a diencephalic center; once activated, this center sends a message to the adrenal instructing it to secrete aldosterone. The nature of the mode of transfer of the message is not clear. Although ACTH secretion by the hypophysis has been suspected, other tropins are also believed to exist.

That the central nervous system affects aldosterone secretion is not questioned. The removal of the diencephalon and the telencephalon induces a drop in aldosterone secretion within 4 hours; in contrast, the removal of the cerebral cortex has no effect on aldosterone secretion.

Ventral hypothalamic lesions reduce both aldosterone and cortisol secretion. However, aldosterone secretion is independent of cortisol secretion because (1) sodium depletion induces aldosterone secretion without affecting cortisol secretion; (2) chronic cortisone administration decreases cortisol but not aldosterone secretion; and (3) lesions of the cat's midbrain that involve the pons have no effect on cortisol secretion but stimulate aldosterone secretion (see Fig. 9-6).

Hypophysectomy consistently reduces aldosterone secretion. A "glomerulotrophin" stimulates aldosterone secretion, and although ACTH may influence aldosterone secretion, it is unlikely to be the principal trigger in the mineralocorticoid secretion. A diencephalon extract has been obtained which contains no ACTH but stimulates aldosterone secretion. Urinary lipoproteic factors that stimulate aldosterone secretion have been obtained. The molecular structure of the glomerulotrophin is unknown, and its site of secretion has not been established. Structures in the mesencephalon, diencephalon, hypophysis, pineal gland, and kidney have all been invoked as the site of elaboration of the glomerulotrophin.

These observations on the effect of body fluid volumes and electrolyte concentrations on aldosterone secretion may have some bearing on the pathogenesis of classical clinical changes. For example, in renal tubular acidosis there is an obligatory sodium loss. The plasma potassium levels are decreased, and aldosterone secretion is stimulated. The administration of bicarbonate, a procedure which usually leads to potassium diuresis, induces potassium retention in these cases. To explain such intricate interactions between hormonal secretion, body fluid volume, and ion concentration, one could assume that the sodium loss, which is obligatory in renal acidosis, leads to reduction in the body fluid volume followed by a loss of potassium. The administration of bicarbonate, which prevents sodium loss and thereby influences the body fluid volume, reduces aldosterone secretion and prevents the loss of potassium ions.

In orthostatic hypotension sodium is lost and aldosterone secretion is decreased, in spite of the observation that ACTH administration stimulates glucocorticoid secretion in patients with orthostatic hypotension as well as in normal individuals. Orthostatic hypotension presumably affects aldosterone secretion, but it is not known in what way.

Juxtaglomerular Apparatus

Before discussing the role of angiotensin in regulating aldosterone secretion, the origin of renin, the enzyme that produces angiotensin I, must be reviewed.

Renin is made in the juxtaglomerular apparatus, a specialized system in the kidney at the vascular pole of the glomerulus, or more precisely in the wall of the afferent arteriole of the glomerulus [23–27]. At the point where the vessel enters the glomerulus, the smooth muscle cells of the media are replaced by large clear cells. The juxtaglomerular cells form the sheet of cells surrounding the arteriolar intima just before the arteriole folds into the capillary bundle that forms the glomerulus. The juxtaglomerular cells should not be confused with the nearby cells of the macula densa, which is an epithelial structure that is part of the ascending limb, rather than a vascular structure like the juxtaglomerular apparatus. When the ascending limb enters the cortex, the loop of Henle forms through the macula a sort of connection between the afferent arteriole, its juxtaglomerular apparatus, and the lumen of the loop. The contact is established through an eliptic cluster of tall, thin, tubular cells. In addition to the juxtaglomerular cells and the cells of the macula densa, the juxtaglomerular region contains a third kind of cell—the lace cells—located in a triangular area delimited at the afferent and efferent arterioles and the macula densa (ascending branch of the loop of Henle). Together the macular, juxtaglomerular, and lace cells constitute the juxtaglomerular apparatus, or the juxtaglomerular complex.

The juxtaglomerular cells are characterized by the presence of several pale granules containing lipids, proteins, and carbohydrates. Electron microscopic examination of the juxtaglomerular cells reveals that in addition to Golgi, mitochondria, ribosomes, and smooth endoplasmic reticulum, these cells contain osmophilic granules surrounded by a single membrane. The granules are likely to contain two hormones, erythropoietin and renin. The role of erythropoietin and the evidence for its concentration in the juxtaglomerular cells are discussed in Chapter 15. Renin converts angiotensinogen to angiotensin. On the basis of observations demonstrating that the degree of granulation of the juxtaglomerular cells parallels the degree of hypertension, Goormachtig postulated that the juxtaglomerular cells secrete a hypertensive substance. Furthermore, exogenous administration of renin degranulates the juxtaglomerular cells. Some investigators believe that renin is made in the macula densa, and a portion of the renin stock could be synthesized in the macula and stored in the juxtaglomerular cells. Indeed, studies on isolated kidney cells with fluorescent antibodies suggest that the juxtaglomerular cell is a site of renin storage. Light microscopic and electron microscopic hyperplasias of the juxtaglomerular apparatus are shown in Fig. 9-7 and 9-8.

The mechanism controlling the rate of renin synthesis and release by juxtaglomerular cells is not known. Except for the fact that both intrarenal and extrarenal factors are involved, the stimuli for renin secretion remain unknown. Two major hypotheses regarding the intrarenal factors have been proposed: (1) the cells of the juxtaglomerular apparatus function as baroreceptors; and (2) the cells of the macula densa function as osmoreceptors [27]. The juxtaglomerular cells are believed to be exquisitely sensitive baroreceptors capable of detecting intraluminal pressure changes. Renin is released whenever the pressure decreases. The location of these cells at the distal portion of the afferent arteriole makes them uniquely suited to function as monitors. Changes in osmolarity, sodium concentration, and amount of sodium delivered per unit time have also been incriminated in triggering renin release on the basis of micropuncture and microdiffusion studies near the juxtaglomerular apparatus.

Changes in blood volume are among the most important external factors that influence renin secretion. For example, reduction of the blood volume through moderate hemorrhage results in increased renin secretion, which can be blocked by anesthetizing the renal nerve. Similarly, ventriculosternal perfusion with

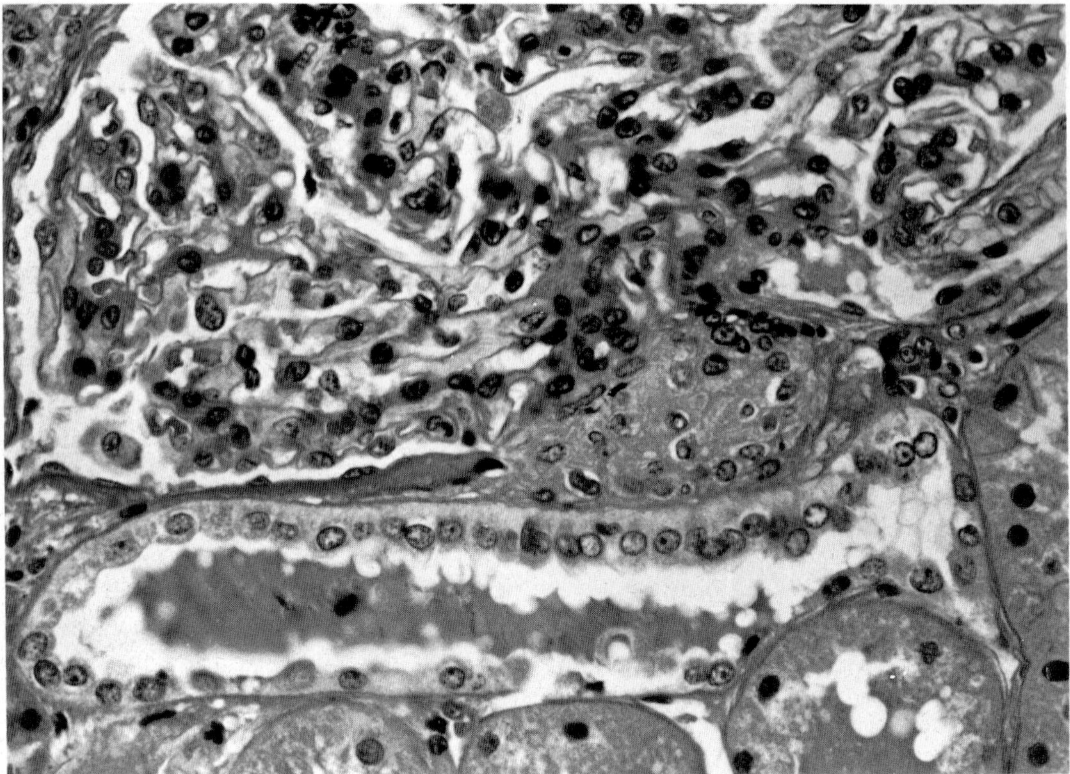

Fig. 9-7. Hyperplastic juxtaglomerular apparatus in renovascular hypertension. The juxtaglomerular cell mass (polar cushion) is at the hilus of the glomerulus, next to the macula densa and between the afferent arteriole (which enters the glomerulus here on the right side) and the efferent arteriole (which is not in the plane of this section). Instead of its normal 6 to 8 cells, this polar cushion contains over 30 cells. Hematoxylin and eosin stain. ×520. (Supplied by Harrison Latta)

hyposmotic, low-sodium solution stimulation of pressor areas in the mesencephalon and medulla oblongata stimulates renin secretion, and this is inhibited by denervation. These findings clearly suggest that changes in blood volume are in some way communicated to the central nervous system, which in turn translates them into nervous impulses directed to the kidney where they elicit renin secretion.

Mills [28] has reviewed the intricacies of the regulation of renin release. Molecular details are not available.

Renin-Angiotensin System

The renin substrate is a large glycoprotein with a molecular weight of approximately 60,000, synthesized in liver and released in the extracellular fluid and the plasma. The attack of renin on its substrate occurs primarily in the plasma and secondarily in the blood vessel wall and the nephron.

Renin has been purified to electrophoretic homogeneity from hog kidneys [86]; but nevertheless, little is known about it except for the facts that it has a molecular weight of approximately 40,000 and that it splits a leucyl-leucyl bond that links the N terminal decapeptide angiotensin I to the bulk of the protein.

Angiotensin I is an inactive prohormone which is converted to the active octapeptide angiotensin II by a dipeptidyl carboxypeptidase, an enzyme believed to be identical to that which inactivates bradykinin.

The angiotensin converting enzyme (peptidyldipeptide hydrolase or kinase II) has been purified from hog kidney and plasma and from human kidney and lung. The hog kidney enzyme is a glycoprotein forming a single polypeptide chain with a molecular weight of 125,000 [85].

The enzyme is found in plasma and most tissue, but the vascular bed of the lung is exceptionally rich in the enzyme and most of the conversion of the decapeptide into the octapeptide occurs in lung.

Asp–Arg–Val–Tyr–Ile–His–Pro–Phe–His–Leu–Leu–Val–Tyr–Ser

Renin (between Leu–Leu)

Carboxypeptidase (between Phe–His)

The hectapeptide acts at at least three different sites: cardiovascular, nervous and adrenal. Angiotensin II constricts the arterioles, and this effect is believed to be mediated through the sympathetic nervous system. Angiotensin II acts on the cells of the zona glomerulosa of the adrenal stimulating the secretion of aldosterone and of the adrenal medulla releasing epinephrine. The major function of angiotensin II is to maintain circulatory homeostasis. Although the hormone is not needed in animals or men with normal salt intake,

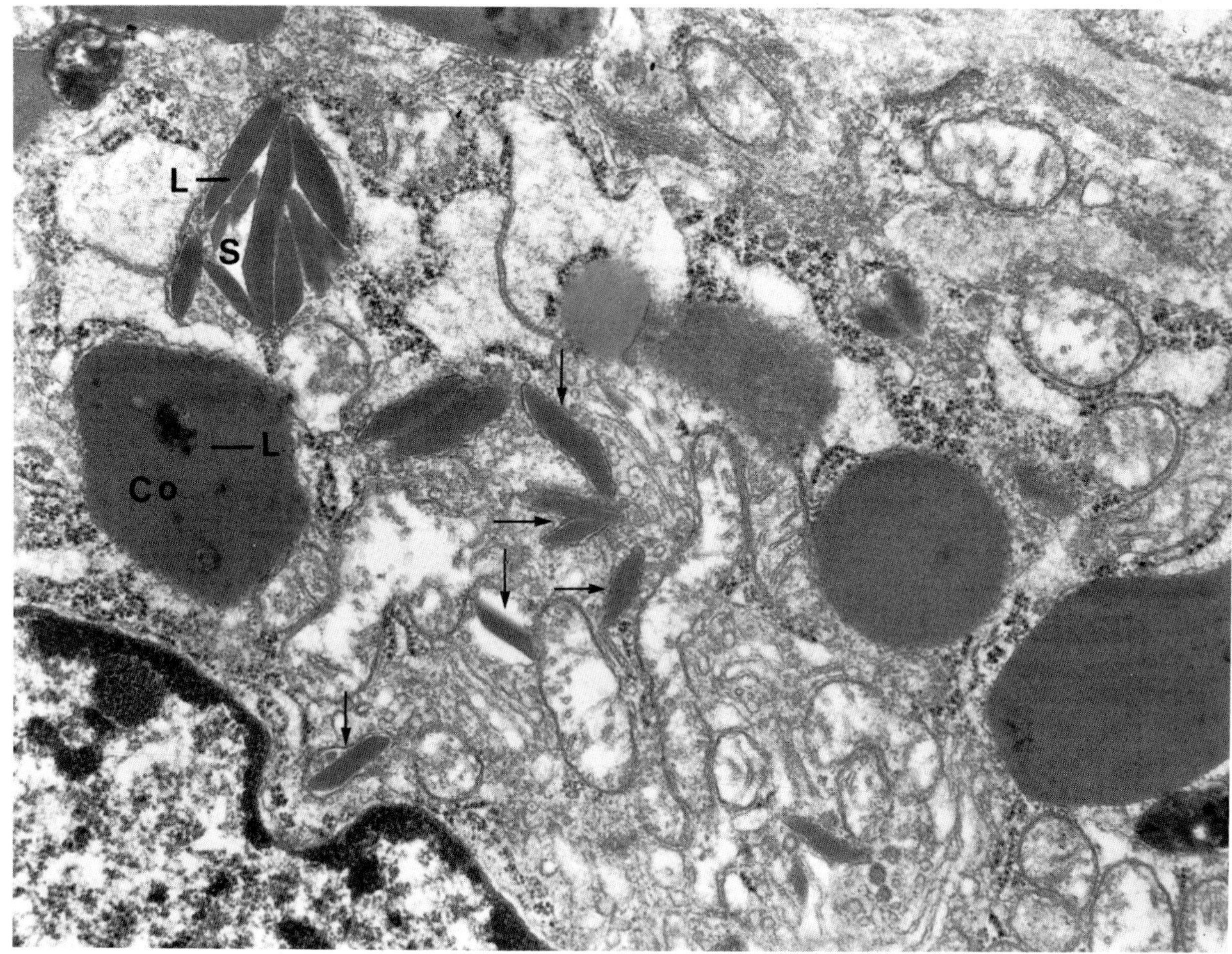

Fig. 9-8. Portion of the cytoplasm of a granular cell from the juxtaglomerular apparatus of a patient with renovascular hypertension. Rhomboid-shaped protogranules (*arrows*) are in the cisternae of the Golgi apparatus. A sac containing multiple protogranules and vesicles is in the upper left (*S*). Below it is a conglomerate (*Co*). Notes lines of fusion (*L*) of the protogranules in the conglomerate and in the upper protogranules in the upper left of the sac. A mature granule is at the lower right. ×30,000. (From Dr. Barajas)

it is required to maintain normal blood pressure when blood volume depletion is caused, for example, by salt depletion, administration of diuretics, or thoracic caval constriction. Thus, if under the conditions just listed one administers agents blocking the action of angiotensin II (antibodies or competitive peptides), the blood pressure drops and renin levels rise.

This finding is in keeping with observations made on the isolated perfused kidney in which vasoconstriction decreases while vasodilation increases renin output. The mode of action of angiotensin II at the cellular level is unknown except for the fact that it binds to several cell structures including mitochondria, nuclei and microsomes and that it stimulates cyclic AMP [74–75].

Mode of Action of Aldosterone

We have already seen that several steroid hormones secreted by the adrenal cortex influence mineral metabolism. Aldosterone is the prototype of these mineralocorticoids.

Aldosterone maintains the sodium levels of the body by stimulating the reabsorption of sodium in the renal tubules. In adrenalectomized dogs and in patients with Addison's disease, sodium excretion is proportional, but not equal, to glomerular filtration. Aldosterone acts in the distal tubule at the site of sodium reabsorption and potassium excretion, but much of the information on aldosterone's mode of action and on renal tubular cells has been obtained indirectly by studying the effect of the hormone on simpler systems. Aldosterone stimulates active sodium transport across the bladder and skin of toads. This effect of aldosterone is manifest only after a lag period.

During the lag period the hormone is bound, and activity in the bioenergetic pathway is increased. Whether aldosterone acts by stimulating the bioenergetic pathway or by modifying the permeability of the membrane to sodium remains to be established. Aldosterone was first believed to affect only the distal

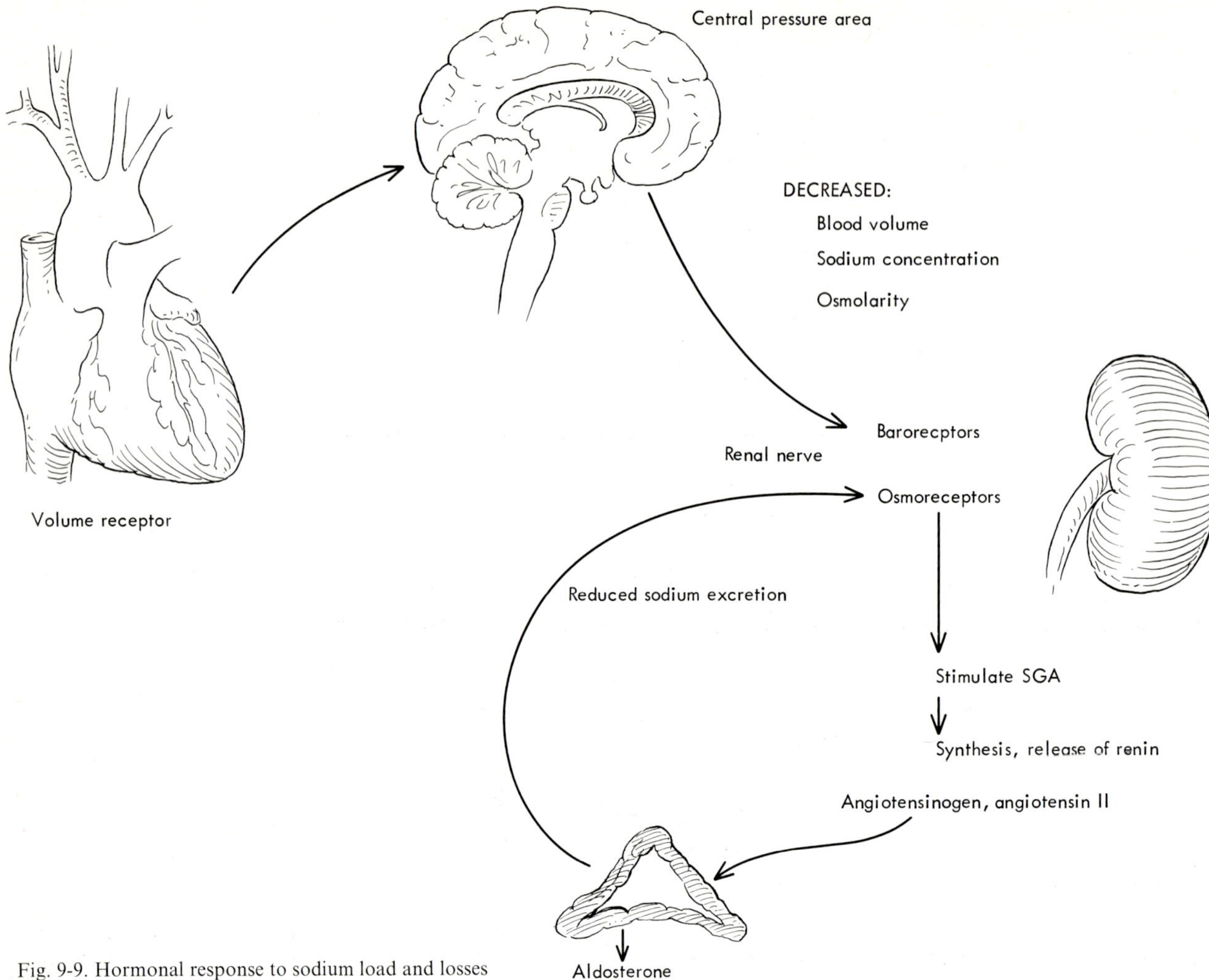

Fig. 9-9. Hormonal response to sodium load and losses

tubule. However, more precise micropuncture experiments have demonstrated an effect of aldosterone on the proximal tubule.

Edelman and Fimognari [20] have reviewed published studies on aldosterone's mode of action and have proposed a working hypothesis in which nuclear RNA synthesis and protein synthesis are involved. It is postulated that in the nucleus aldosterone induces the synthesis of new messenger RNA, which in the cytoplasm stimulates the biosynthesis of an enzyme.

Other investigators have also shown that actinomycin interferes with the action of aldosterone *in vivo*, and that aldosterone binds to a nuclear receptor believed to be a protein. Binding between receptor and aldosterone requires the presence of SH groups on the protein. The role played by the complex of aldosterone and protein in gene expression is still not clear, but the complex of hormone and protein is believed to induce the biosynthesis of a new protein. The most popular hypothesis proposes that the new protein is in some way involved in NADH electron transport and coupled to phosphorylation of ADP. It is not certain whether the protein produces NADH or whether it facilitates NADH oxidation. The ATP produced in the oxidation-phosphorylation process could then be used as fuel for the sodium pump.

Although it is not possible to discuss here the pros and cons of other theories on aldosterone's action, the possibility that the induced protein is either a sodium permease or a protein that stimulates the sodium pump directly has not been excluded.

Factors Other Than Aldosterone That Influence Sodium Excretion

At least three other groups of factors believed to regulate sodium secretion are neurogenic factors, switches from salt-conserving to salt-wasting nephrons, and natriuretic hormones. When dietary sodium is low, renal excretion decreases within 4 days to levels approaching zero. Reduced renal excretion results in part from aldosterone production. However, a decrease in urinary sodium has been observed with normal levels of aldosterone secretion. Consequently,

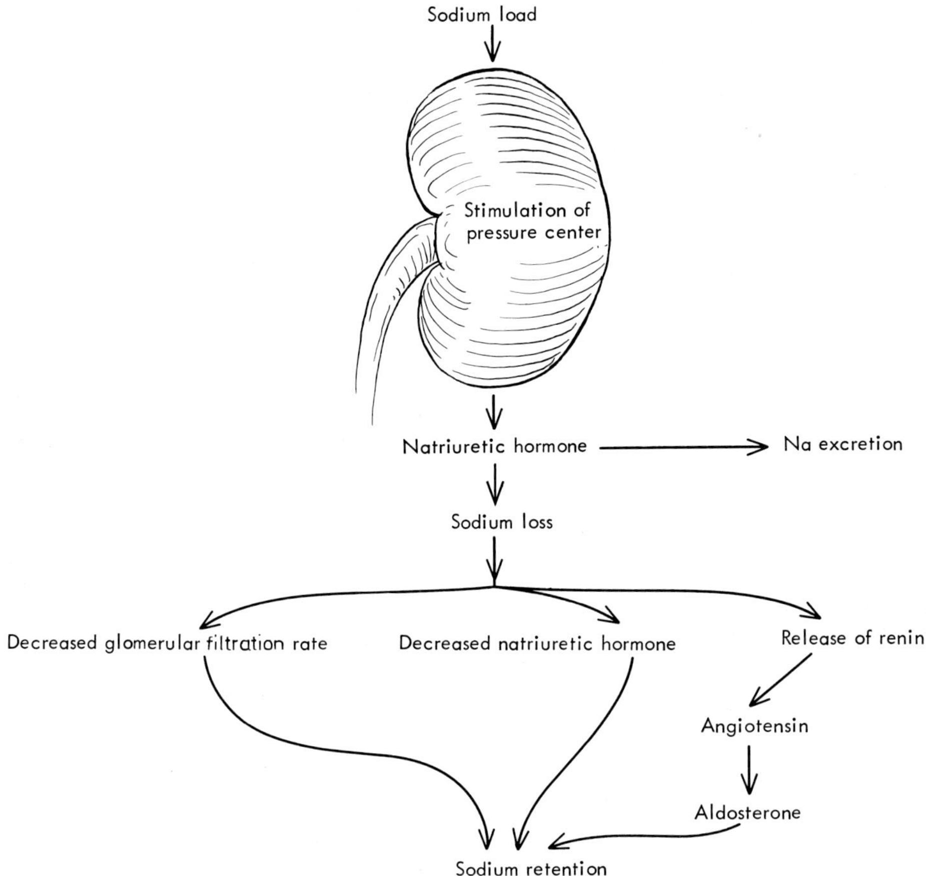

Fig. 9-10. Hypothetical role of natriuretic hormone

other mechanisms must be involved. Two findings suggest that the autonomic nervous system might regulate sodium excretion: first, the impaired ability of patients with insufficiency of the autonomic nervous system to conserve sodium; second, the facilitation of sodium excretion induced by sympathetic blockage in normal subjects given saline infusions or treated with sodium-retaining steroid [29]. Studies with agents that block adrenergic secretion have shown that adrenergic blockage impairs sodium conservation during sodium deprivation by altering tubular sodium reabsorption.

When rats are loaded with salt, the type of nephrons participating in filtration seems to shift. Often salt-loading filtration doubles in superficial cortical nephrons, while it decreases in the juxtaglomerular segments. Thus, the responsibility for reabsorption is transferred from long salt-conserving nephrons to short salt-wasting nephrons. This would permit adjustment of total reabsorption without modifying the reabsorption capacity of individual nephrons. Although a shift in the type of nephrons may occur, more recent evidence suggests that it is not sufficient to explain the adjustment to salt reabsorption that takes place during saline infusion.

Wardner established that animals and humans infused with salt excrete sodium in their urine, even if they are injected with large amounts of retaining steroid or when glomerular filtration is markedly reduced. The inescapable conclusion is that glomerular filtration rates or aldosterone levels in the blood do not alone regulate sodium reabsorption.

A number of hypotheses have been proposed that attempt to link sodium load, expansion of the vasculature and blood dilution to the reduced sodium reabsorption. Interpretations involving both hormonal and intrarenal mechanisms have been advocated. None of the hypotheses has satisfactorily explained the reduced sodium absorption following a sodium load.

In reviewing the various hypotheses about regulation of sodium metabolism, Mills [28] proposed that renal sodium excretion depends on a pressure-sensitive mechanism near the juxtaglomerular cells that releases an active natriuretic hormone promptly when the site is stimulated by an increase in pressure. Conversely, sodium retention is believed to be caused by a decrease in this substance, by a fall in glomerular filtration rate, and by the release of renin to increased aldosterone production [29, 30].

Although many laboratories have attempted to demonstrate the existence of a natriuretic hormone, the most conclusive evidence in support of its existence was obtained in an isolated perfused dog kidney with blood of "expanded dogs." A decrease in tubular sodium reabsorption was demonstrated [31], even when renal blood flow was reduced and in the absence of renal vasodilation.

Nevertheless, claims of partial purification of the natriuretic hormone are not definitive. Moreover, well-known hormones, such as calcitonin and α-NSH, could also be responsible for the natriuretic effect. It is questionable that either of these hormones plays a considerable role in regulating sodium metabolism. Hormone responses to sodium loads and losses are summarized in Fig. 9-9, and the hypothetical role of the natriuretic hormone is illustrated in Fig. 9-10.

Diseases Associated with Disequilibrium of Sodium Metabolism

Primary Aldosteronism

The discovery of aldosterone and of hyperaldosteronism are linked to observations made on individuals exposed to excessive heat. The invasion of North Africa in World War II by the forces of the Axis brought a large number of allied troops to live in a hot and wet climate. In face of this necessity the U. S. Army initiated a project aimed at the study of adaptation to heat and humidity. The study revealed that exposure to heat is associated with increased urinary nitrogen, decreased urinary sodium and decreased excretion of sodium in sweat. After a few days of exposure to heat the homeostatic regulatory mechanisms of the body take over. The negative nitrogen balance decreases and the sweat gland retention of sodium is compensated by increased sodium excretion in the urine. When deoxycorticosteroids were administered experimentally similar observations were made: transient hyponatruria without change in the sodium excretion in sweat glands. These observations were ultimately responsible for the discovery of aldosterone [76]. When in 1954 J.W. Conn saw a patient with episodic weaknes, hypernatremia, hypokalemia, and alkalosis, he suspected overproduction of aldosterone.

In patients with primary aldosteronism, the aldosterone levels in the blood and the amount of aldosterone excreted in the urine are markedly increased. Primary aldosteronism may or may not be associated with the development of a tumor of the adrenal cortex. When a tumor exists, it is usually small and weighs less than 6 g. Chemical analysis of the adrenal indicates that the cortex is rich in aldosterone, probably as a result of hypersecretion. Eighty-five per cent of the patients have a single or multiple adenoma of the adrenal cortex. Bilateral hyperplasia of the adrenals is found in 10% of the patients.

Primary aldosteronism is more common in women (2.5 times) than in men. It develops in adults between the ages of 30 and 50 years, and may or may not be associated with hypertension. In the classical form the symptoms include hypernatremia (see Fig. 9-11), hypokalemia, and alkalosis. The levels of free aldosterone are increased in the urine, but the levels of 17-ketosteroids and 17-hydroxysteroids are usually within normal range. The disease leads to muscular and renal dysfunction in about 60% of patients, who develop facial paresthesia. Muscular dystrophy with periodic paralysis occurs in approximately 30% of the victims, and 10% develop tetany with Chvostek and Trousseau signs. Almost all patients are polydypsic and polyuric, excrete urine with low specific gravity, and are unre-

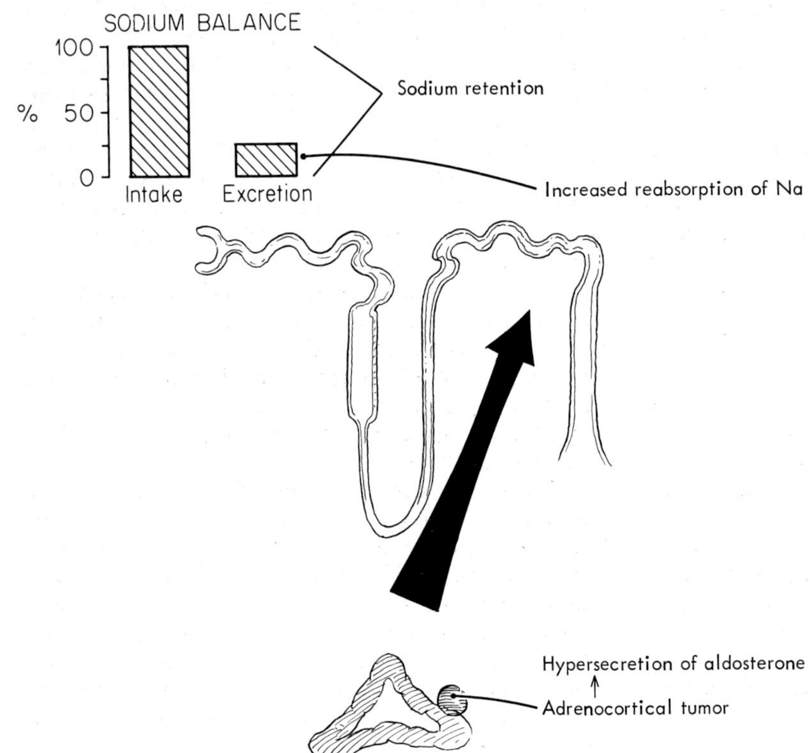

Fig. 9-11. Pathogenesis of hypernatremia in primary aldosteronism

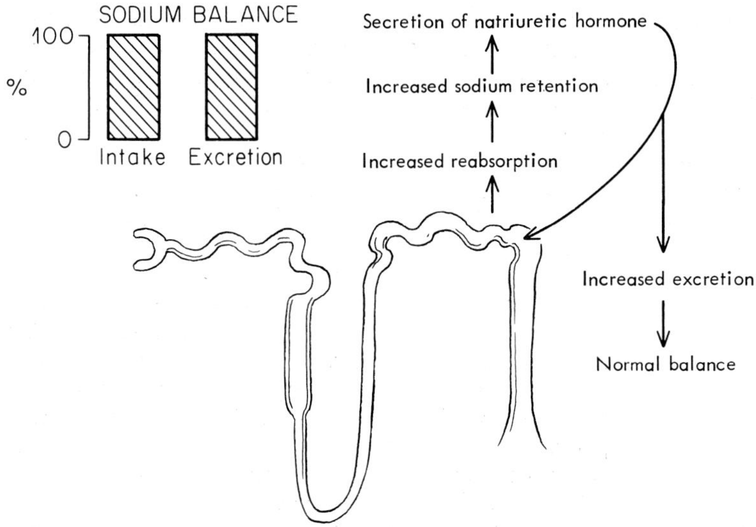

Fig. 9-12. Adaptation to hypernatremia in hyperaldosteronism

sponsive to dehydration and to vasopressin administration. Often patients have alkaline urine and mild albuminuria.

Hypokalemia is associated with the excretion of large amounts of potassium in the urine and with the development of an electrocardiographic picture typical for patients with low potassium levels. The prognosis varies with the patient's age; whereas young patients seem to respond well to bilateral adrenalectomy, older ones are often not improved by surgical intervention.

Hypernatremia is not an absolute finding in primary hyperaldosteronism, and a mechanism of renal adaptation has been proposed. Thus, in progressive aldosteronism, salt retention activates a nonaldosteronic mechanism of sodium regulation, which increases sodium excretion in the distal tubules without affecting the proximal tubule, thus reestablishing the balance between sodium intake and excretion. Whether the postulated natriuretic hormone [32] is involved remains to be seen (see Fig. 9-12).

Conn has remarked that all forms of primary aldosteronism are not associated with hypokalemia. Normokalemic hyperaldosteronism can be distinguished from secondary aldosteronism by low levels of renin in the plasma. These patients usually have smaller tumors (3 to 16 mm, compared to 8 to 50 mm in patients with hypokalemia). The syndrome is believed to result from slow development of aldosterone tumors. At the beginning the patients have a relatively mild overproduction of aldosterone and hypertension with normokalemia. The pathogenesis of the condition is assumed to be the following: the tumor grows, there is a low grade of sodium retention with expansion of the extracellular and intravascular fluid volume and plasma renin is decreased. The decrease in renin and angiotensin II leads to a decrease of aldosterone secretion. It takes months or years for the extracellular fluid to slowly expand enough to prevent further escape of sodium and hypertension sets in as a result of hypervolemia.

Secondary Aldosteronism

Secondary aldosteronism results in high blood levels of aldosterone in the absence of a primary adrenal defect. Secondary aldosteronism develops when one of the stimuli of aldosterone secretion is overactive. Consequently, at least three types of conditions can lead to secondary aldosteronism: excess of renin, ACTH, or potassium. The condition encountered most often clinically is that in which renin is overproduced because of increased secretion by the juxtaglomerular apparatus or, more rarely, because of overproduction of precursors or extrarenal secretion (for example, by tumors). Renal secretion is stimulated when: (1) arterial blood flow to the kidney is reduced (stimulation of baroreceptors), either as a result of the reduction in effective blood volume or renal artery obstruction; or (2) osmoreceptors are stimulated as a result of hyponatremia. The arterial blood volume can be reduced by factors that decrease the blood volume (reduced water, sodium, or both), passage of fluid from the blood vessels to the extracellular compartment, and pregnancy (see Table 9-1).

When hyperaldosteronemia results from renal artery obstruction it is often curable, as is the hypertension it causes. The vasoconstriction associated with hypertension is believed to cause focal ischemia, which results in hypersecretion of renin. In fact, hyperaldosteronemia disappears with the therapeutic correction of the hypertension.

A form of hyperaldosteronemia may result from a unique renal lesion of the juxtaglomerular apparatus. The symptoms usually consist of: (1) delayed development of the bone; (2) hypokalemia with alkalosis and excessive urinary potassium loss (the hypokalemia is refractory to potassium administration); (3) increased aldosterone levels in blood despite a large sodium intake; and (4) normal blood pressure even though levels of an "angiotensinlike" substance in blood are increased.

Table 9-1. Conditions that lead to secondary aldosteronism

Stimulation of the renin-angiotensin system

I. Excess angiotensin secretion
 A. Stimulation of baroreceptors
  a) Reduced arterial blood flow
   Hemorrhage
   Vomiting
   Diuretic therapy
   Low salt intake
   Salt wasting (adrenal insufficiency)
  b) Loss of fluid to the extracellular space (edema)
   Hepatic cirrhosis with ascites[a]
   Nephrotic syndrome
   Congestive heart failure[b]
   Idiopathic edema
  c) Pregnancy[c]
 B. Obstruction to renal blood flow
  a) Renal artery obstruction
   Arteriosclerosis
   Anatomical defect
   Compression by tumors
  b) Renal ischemia
   Shock
   Accelerated or malignant hypertension

II. Stimulation of osmoreceptor
 A. Hyponatremia

III. Hyperproduction of renin
 A. Overproduction of renin precursor in liver as a result of estrogen intake
 B. Hyperplasia of juxtaglomerular apparatus
 C. Renin-producing renal tumor

Excess ACTH

I. Exogenous administration

II. Adrenal hyperplasia associated with 17-hydroxylase or 21-hydroxylase deficiency

Excess potassium

[a] In hepatic cirrhosis, hyperaldosteronemia can result in part from ineffective aldosterone metabolism by the hepatic cell.
[b] Most patients with cardiac edema do not have increased aldosterone levels.
[c] In pregnancy, the effective blood flow is thought to be reduced because of general venous dilatation and renal salt loss due to high levels of progesterone. Despite the high aldosterone level in pregnancy, the hormone appears to have no metabolic effect.

The following pathogenesis has been proposed for this form of hyperaldosteronemia. A low threshold of the vascular wall has been postulated to be responsible for the compensatory renin secretion. The increased secretion of renin stimulates aldosterone secretion by the adrenal cortex. As a result of the hyperaldosteronemia, sodium is retained and potassium lost.

This form of aldosteronism must be distinguished from primary aldosteronism due to adrenal hyperplasia and ordinary secondary aldosteronism. In primary aldosteronism, sodium deprivation results in a reduction of sodium in blood and potassium in urine, potassium retention is facilitated, and, as a result, hypokalemia is corrected. In the juxtaglomerular disease, urinary potassium loss persists in spite of sodium deprivation. The secondary aldosteronism differs from the primary aldosteronism by the existence of hypertension or edema in the latter.

Secondary aldosteronism has been described in a few patients with various types of renin generating tumors, juxtaglomerular tumors, Wilm's tumors, neuroblastoma and even a case of lung tumor. These are usually young patients with severe hypertension, high plasma renin, secondary hyperaldosteronism with hypokalemia. This form of hypertension can be distinguished from secondary aldosteronism resulting from abnormalities of the renal vasculature by arteriography which reveals a vascular pattern suggesting a tumor or a filling defect rather than arteriostenosis [77].

Most clinical cases of secondary aldosteronism are associated with hypertension or edema. Of course, in hypertension, Starling's law still applies; in edema, it does not apply. In either case, persistent hyperaldosteronemia results in sodium retention, which expands the volume of the body fluid by promoting water retention. In hypertension, the expansion concerns mainly the blood volume, and hypertension is aggravated. In edema, the expansion mainly affects the extracellular fluid, and edema is exacerbated. Persistent hyperaldosteronemia in both hypertension and edema leads to hypokalemia with muscle weakness and even paralysis, polyuria, and metabolic alkalosis [33] (see Fig. 9-13).

Periodic Paralysis

Periodic paralysis is a rare hereditary disease observed in patients of practically any age (the reported cases include patients from 1 to 56 years old). The disease is transmitted by an autosomal recessive gene and is somewhat more common among women than men. Patients with periodic paralysis have attacks of flaccid paralysis, preceded by hypersecretion of aldosterone. Usually there are no changes in adrenocortical histology.

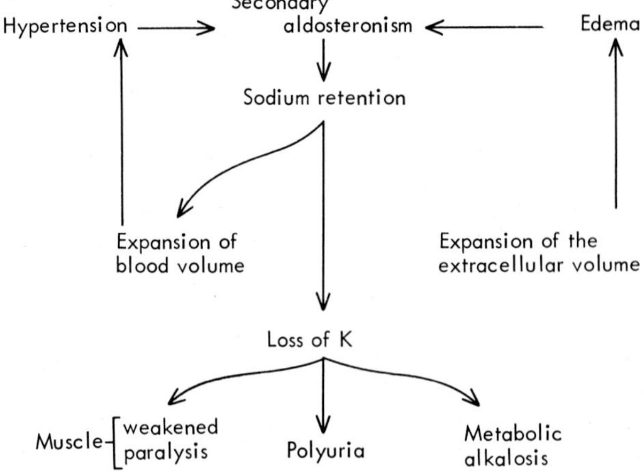

Fig. 9-13. Pathogenesis of secondary aldosteronism

The paralysis affects the striated muscles primarily, excluding the respiratory muscles and those of the face. The muscles of the lower limbs are attacked first, then those of the arms, the trunk, and the neck. Physical signs are non existent between the crises of paralysis, but the tendon reflexes and muscle excitability to electrical stimulation are often abolished during a crisis. Occasionally, the myocardium is also paralyzed, and the cardiomegaly, arrhythmias, bradycardia, and hypertension observed in patients with periodic paralysis may result from cardiac muscle paralysis. The gastrointestinal symptoms (vomiting, nausea, and flatulence) in these patients are sometimes attributed to involvement of the smooth muscle of the gastrointestinal tract.

The crucial biochemical finding was obtained by studies of muscle biopsies. The potassium concentration inside the muscle cells was found to be doubled during the attack. The intramuscular potassium concentration returns to normal rapidly when the attack is over. In contrast, the sodium concentration in the muscle cells is unchanged, and the serum potassium concentration and the amount of potassium excreted in the urine are decreased during the crisis. The biochemical findings suggest that periodic paralysis is associated with a shift in the distribution of potassium among the body fluids. The pathogenesis of the disease remains obscure. It could involve a primary defect in the muscle membrane or an excessive stimulus of the mechanism controlling aldosterone secretion.

Histologically, separation and rarefaction of myofibrils occur with vacuolization of the muscle fiber. Ultrastructural investigations have shown the vacuoles to be membrane-bound vesicles, some of which represent dilations of the sarcoplasmic reticulum. Although detailed descriptions of the various ultrastructural types of vacuoles that occur in periodic paralysis are available, the reasons for, and the mechanisms of, formation of these vacuoles remain obscure [34].

Addison's Disease

Addison's disease was described in 1855. In his original observation, Addison remarked that the disease is characterized by the now classical triad of symptoms—electrolyte imbalance, hypertension, and hyperpigmentation. Addison's disease is a rare disease that affects both sexes with the same frequency. Before the corticosteroid hormones were discovered, the disease was inevitably fatal.

The adrenal insufficiency that develops in Addison's disease results from the extensive yet not total destruction of the cortex of both adrenal glands. Various pathological conditions that may be responsible for this destruction include bilateral tuberculosis (68%), destructive atrophy (19%), amyloid degeneration (2%), and neoplasm (1%). In the remaining 10% of cases, Addison's disease results from a number of miscellaneous degenerative processes (venous thrombosis, arterial emboli, syphilis, and pressure atrophy). The physician should beware of the sudden withdrawal of intensive steroid therapy, since this could lead to iatrogenically induced adrenal insufficiency.

Adrenal tuberculosis is usually of the fibrocaseous type and affects both cortex and medulla. Although it is not known whether the adrenal infection is primary, adrenal tuberculosis is often associated with tuberculosis of the urogenital tract. The tuberculosis usually starts in one gland and progressively invades the contralateral gland. Adrenal insufficiency develops only when 90% of the cortex of both adrenal glands is destroyed.

In destructive atrophy of the adrenal, the normal cellular structure of the cortex is progressively replaced by lymphocytes, plasmocytes, and monocytes. The thymus is often enlarged, and many organs are infiltrated by lymphocytes. Destructive atrophy of the adrenal cortex could result from an autoimmune process.

The other changes found at autopsy in Addison's disease depend upon the course of the disease. For example, Addison's disease may be associated with generalized tuberculosis or generalized amyloidosis. Secondary changes, such as localized atrophy of the gland's basophilic cells, have been described in the hypophysis of patients with Addison's disease.

The heart of patients with Addison's disease is usually small, and it often contains brown pigment. Histological examination reveals focal degeneration of the myocardium. Usually there are no changes in the sex organs of victims of Addison's disease, but occasionally fibrosis of the testicles and ovaries has been observed.

In Addison's disease, secretion of all corticoid hormones formed in the adrenal cortex is markedly reduced. The lack of aldosterone is responsible for the major electrolyte imbalance—a massive loss of sodium through the kidney, salivary glands, sweat glands, and the gastrointestinal tract. The sodium loss is accompanied by a loss of water, and dehydration and its symptoms develop. The reduction in the water of the extracellular space leads to reduced renal blood flow and reduced glomerular filtration. The changes in renal physiology may lead to azotemia.

The sodium loss is associated with chloride and potassium retention; and the blood of addisonians contains increased levels of potassium and chloride and decreased sodium. The sodium loss combined with the poor circulation may be responsible for the severe, frequent muscle cramps often experienced by patients with Addison's disease.

If cortisone and hydrocortisone are not administered, the glycogen storages of the liver are rapidly depleted, and severe hypoglycemia results. The hypoglycemia may not be present at all times but occur only as a result of starvation or administration of a low-carbohydrate diet. Obviously, this hypoglycemic crisis cannot be relieved by epinephrine injection, and patients with Addison's disease are oversensitive to insulin.

A drop in the blood levels and urine content of the 17-ketosteroids may be anticipated in Addison's

The obvious alternative to the sorption theory is the membrane theory, which proposes that the specific molecular and structural properties of the cell membrane determine the discrimination between potassium and sodium. Studies of the kinetics of penetration of the potassium ion in the cell are consistent with the fact that the passage of the cell membrane is the rate-limiting step for intracellular potassium accumulation.

In 1867, Traube proposed that the selectivity of the membrane resulted from the presence of pores at the membrane's surface. Later, Conway proposed that the membrane is a lipoproteic sieve with its pores filled with water. The assumption was then made that the diameter of the pore is intermediate between that of hydrated sodium and hydrated potassium ions. As a result, potassium ions can pass the barrier but sodium ions are stopped. Again, such models are oversimplifications. According to the theory, the passage of a cation is determined by its mobility in a given field and by the size of the hydrated ion. The velocities of rubidium, cesium, and potassium under a gradient of 1 volt/cm are almost identical. In addition, the diameter of potassium is assumed to be equal to that of cesium and rubidium. Why should the cell membrane, then, be less permeable to cesium and rubidium than to potassium?

Two mechanisms have been proposed to explain these discrepancies. One is that the number of hydration molecules influences permeability; the other is that permeability to a cation is an all-or-nothing effect, and that the cation penetrates the membrane only if the diameter of the pore resembles that of the ion.

Refer to specialized texts for the evaluation of the validity of these explanations. The determination of the diffusion rate of tritiated water and the measurement of osmotic flow are consistent with the existence of pores. Assuming that cell membranes are semipermeable, one can estimate the average size of the pores by comparing the theoretical osmotic pressure inside the cell with the measured values. If the semipermeable membrane is continuous, the calculated values should closely match the measured values. In contrast, if the membrane is leaky because of the pores, the osmotic pressure inside the cell will be lower than that calculated. Such measurements have permitted the calculation of the average pore size of the membrane surrounding human erythrocytes: 3.5–4.2 A. An average pore size of 4 A would be consistent with a theory proposing that the selective permeability results from the size of the pores in the membrane. But, in view of the difficulties encountered in measuring intracellular osmotic pressure, one may question the validity of measuring pore sizes in cells other than erythrocytes. In conclusion, although the exact mechanism of the selectivity is not clear, there seems to be little doubt that the cell membrane is responsible for the selectivity.

Because of the selectivity of the cell membrane, the distribution of the cations inside and outside the membrane is unequal. Yet this distribution takes place according to rigid laws, those predicted by Gibbs and discovered by Donnan. Consider the practical importance of the unequal distribution of the ions with respect to cellular electricity. When electrodes are placed at the surface of a muscle fiber and inside the fiber, a voltmeter connecting the two electrodes will register a difference in potential of 70 mV. Similarly, a difference in potential can be detected between the surface and the inside of nerve axons. In fact, a difference of potential exists between the surface and the inside of every cell in the body. Thus, everything happens as if the cell membrane were polarized with negative charges inside and positive charges outside. The importance of this polarization stems from the fact that the nervous impulse is associated with a depolarization of the membrane and a reversal of the distribution of the charges on both sides of the membrane. What is the molecular basis for polarization and depolarization? The selective concentration of potassium inside the membrane is responsible for polarization, and the influx of sodium and the efflux of potassium are responsible for depolarization.

As pointed out already, since the inside of the cell is separated from the surrounding medium by a membrane with selective permeability, the potassium and chloride ions on both sides of the membrane are distributed according to Donnan distribution. The Donnan law states that the charges on both sides of the membrane are to be neutralized, and that the product of the diffusible ions on one side of the membrane equals the product of the diffusible ions on the other. In any cell, the diffusible cations (only potassium, since the membrane is virtually impermeable to sodium) traverse the membrane in order to neutralize the intracellular fixed negative charges (organic phosphates and carboxyl groups).

Thus, the diffusible ions are distributed as follows: $K_2Cl_2 = K_LCl_L$, according to the Donnan equation. The potential that develops at the surface of the cell as a result of the uneven distribution of the cations can be calculated with the aid of the Nernst equation:

$$E = \frac{RT}{nF} \ln \frac{[K+]i}{[K+]o}$$

R = gas constant
T = absolute temperature
n = valency of ion
F = Faraday

If the calculated potential equals the actual potential in the cell, then the membrane potential is in fact a potassium potential. There is enough agreement between the calculated and the measured values to conclude that this is probably true. Yet, some have objected that the actual and theoretical values correlate only within narrow limits of potassium concentration. For example, although there is a good correlation at concentration of K equal to 0.5 mM, at concentration of 2.5 mM, the actual intracellular potential, the calculated values are 8 mV higher than the measured potential.

Whether these discrepancies are due to a lower activity coefficient of potassium inside of the cell as compared with the activity coefficient outside of the cell, or whether they are due to leakage of potassium outside and penetration of sodium inside the cell is not certain. Better agreement between theoretical and actual values can sometimes be obtained by calculating the potential by the Goldman equation rather than the Nernst equation. The Goldman equation does not express the membrane potential in function of the intracellular movements of potassium, but it takes into consideration the participation of other cations, particularly sodium. Thus, in Goldman's equation the potential is expressed as a function of the relative permeabilities and the electrochemical gradients of each ion.

$$E = \frac{RT}{F} \ln \frac{Pk[\mathrm{K}]\, L + PKa[\mathrm{Na}]\, i + Pcl\,[\mathrm{Cl}]\, o}{PK[\mathrm{K}]\, o + PKa[\mathrm{Na}]\, o + Pcl[\mathrm{Cl}]\, i} \cdots$$

In the resting cell, the membrane potential is negative with respect to the interior of the cell. During excitation, the resting potential is reversed, and the membrane becomes positive with respect to the interior of the cell. The reversal of membrane potential is followed by rapid restoration to normal. The changes in the electrical potential are all explicable in terms of electrolytes flowing in and out of the cell. As mentioned previously, most of these studies were done on the giant axon of the squid. Immediately after excitation, the permeability properties of the membrane are markedly modified. The mechanism that made the membrane permeable to potassium is blocked, and the membrane becomes permeable to sodium, which pours inside of the axon thereby upsetting the resting cationic distribution and reversing the membrane potential. In a fraction of a second, the membrane's permeability to sodium is lost and that to potassium is restored. To reestablish the resting potential, the cell must eliminate the accumulated sodium; this is accomplished by activating the sodium pump.

It is generally assumed that the inward movement of potassium is passive and is in some way coupled to the expulsion of sodium ions. However, active potassium transport has been observed in the red cell, and tumor cells appear to lose their permeability to potassium and require an active transport mechanism to maintain the intracellular potassium concentration.

This sequence of events has been followed by studying the effect of changes of the electrolyte composition of the bathing medium on the action potential, and by following the inward and outward flow of labeled electrolytes under the influence of nerve excitation. For example, the amplitude of the action potential is reduced when the sodium level in the bathing medium is low, and radioactive sodium was found to enter the axon after excitation. The propagation of the impulses along the axon is readily understood if it is kept in mind that electric currents generate ion movements and that, conversely, ion movements generate electric current. The current generated by the inward movement of sodium and the outward movement of potassium in a small portion of the axon induces ion movement in the immediately adjacent fraction of the axon, thus propagating the initial potential disturbance.

At this point, the Nernst potential is mainly a sodium potential, and it depends upon the distribution of the sodium ions on both sides of the membrane.

$$E = \frac{RT}{nF} \ln \frac{[\mathrm{Na}]\, i}{[\mathrm{Na}]\, o}$$

Potassium in Cellular Metabolism

Whereas sodium participates in metabolism mainly by its cationic properties, potassium is more directly involved in metabolism. Potassium stimulates the activity of a specific enzyme—pyruvic kinase—and is required for the phosphorylation of fructose-1-phosphate to fructose-1,6-diphosphate. Similarly, potassium stimulates acetyl kinase activity. Many alterations in the bioenergetic pathways of the cell are accompanied by changes in the intracellular concentration of potassium. After insulin administration, some of the potassium of the extracellular fluid is transferred inside the cells. During oxidative phosphorylation, potassium accumulates inside the mitochondria, and dinitrophenol uncouples the ion penetration and the oxidation.

A direct role of potassium in protein synthesis has not been established, but at least two observations suggest that potassium may in some way influence that metabolic pathway: during protein synthesis potassium accumulates inside the cell; and the hypokalemia observed in surgical patients is often accompanied by a negative nitrogen balance. In fact, it has been claimed that low dietary potassium levels may interfere with growth, yet the mechanism of such an interference is unknown.

Protein breakdown is accompanied by a loss of intracellular potassium, and starvation is associated with hyperkalemia. Hyperkalemia may seriously complicate uremia observed when severe renal failure occurs, and then the excess potassium should be eliminated by extracorporeal dialysis.

Ringler found that isolated muscle would not contract when placed in distilled water, but would readily do so when placed in tap water. The potassium content of muscle is low in myasthenia and muscular dystrophy; it is high in myasthenia gravis.

Low potassium levels were found in the right ventricle in patients with backward heart failure; hyperkalemia interferes with the normal response of the cardiac muscle to digitalis. In contrast, potassium depletion is sometimes associated with blocking myocardial contraction and hypersensitivity to digitalis. The mechanism by which potassium influences muscle contraction is not known. Investigators have proposed that the myocardial fiber is surrounded by a polarized membrane with potassium ions at the inside and sodium ions at the outside. The role of potassium in nervous excitation has already been discussed.

The potassium content of the intracellular fluid is difficult to determine. The ratio of the potassium concentration in the erythrocyte to that in plasma is generally assumed to reflect the ratio in intra- and extracellular fluids in general. In the cell, potassium is usually at the inner side of the cell membrane.

Chloride Metabolism

Most of the chloride in the body fluid is in the extracellular fluid, and only a small amount of the chloride ions are in the intracellular fluid. (There are 95–110 mEq of chloride per liter of extracellular fluid, and only one mEq of chloride per liter of intracellular fluid.) In contrast, in the intracellular fluid, mainly protein and phosphate anions are present. In the extracellular fluid, the concentration of bicarbonate is second to that of chloride.

Chloride ions in the body are from the same source as sodium chloride. Thus, the source of sodium differs from that of chloride since sodium may be ingested in the form of nonchloride salts. Researchers have proposed that this difference in the sources of sodium and chloride is responsible for the relatively high sodium concentrations in the blood as compared to blood chloride concentrations.

The mechanism of chloride excretion has not been completely elucidated; however, it is known that it is similar to that of sodium excretion in the distal tubules. In the distal tubules, the regulatory mechanism controlling sodium conservation does not affect chloride. Chloride reabsorption is not directly influenced by aldosterone or by carbonic anhydrase.

Chloride reabsorption appears to depend on the amount of reabsorbed sodium and on the amount of bicarbonate needed to maintain a constant blood pH. Although normally sodium reabsorption is accompanied by chloride reabsorption in impending acidosis, carbonic anhydrase forms H_2CO_3 at the expense of water and CO_2. As a result, the ratio of sodium to chloride may vary considerably, and these changes are due mainly to changes in chloride concentration. For example, the ingestion of sodium bicarbonate or ammonium chloride alters the sodium-chloride ratio (increasing it in the former, decreasing it in the latter).

However, the changes in ratio are of short duration when the kidneys are intact. In acidosis (diabetic or respiratory) not complicated by kidney disease, the plasma chloride levels drop. In nephrosclerosis or nephrocalcinosis, the chloride concentration may be high and that of sodium low in the blood as a result of the distal tubules' inability to reabsorb sodium. In contrast, the injury of the distal tubules does not affect chloride absorption. Large amounts of chloride are present in the stomach fluids.

The blood chloride levels are reduced in some forms of alkalosis—for example, when chronic vomiting occurs—as in pyloric stenosis or other types of obstructions.

Acids and Bases

The term acid is derived from the Greek οζυς (sour), and originally a substance that tasted sour when in solution in water was called acid. Vinegar (οζος) a product of wine fermentation (*vinaigre* in French) is probably the oldest manufactured acid. Vinegar is simply an impure and dilute solution of acetic acid. In fact, vinegar was used for technical purposes in ancient Greece and possibly in Egypt. In 400 BC, Theophrastus proposed a formula for white lead pigment that called for mixing vinegar with metallic lead.

The ancient Greeks knew that if the fermentation of wine is not stopped at the proper moment, the tasteful alcoholic beverage turns into a sour mixture good for only medicinal purposes or preserving vegetables. In Mesopotamia, a medicinal vinegar prepared from a root was believed to remove poisons from the body. In other potions, vinegar was often used as a drug solvent. The Bible often refers to vinegar, and we know from Ruth 2:14 that Jews used to dip bread in diluted vinegar.

Although muriatic acid (hydrochloric acid) and sulfuric acid were known for a long time mainly because they were used to separate silver from gold, the notion of acid as a class originated only in the 16th century.

Alkalies* have been used since antiquity for cleaning, religious and medicinal purposes, and for glass making. The Egyptians used to harvest natron, a crude form of sodium bicarbonate. Natron was found in the Nile valley, mainly in three areas, the most famous being the Wadi Natrum. These were areas that were flooded by the Nile at the time of the *Crues*. When the water withdrew, it left a salty deposit called natron. Natron was used in religious services (burned with incense as a symbol of purity), in food preservation, and in glass industry.

In other civilizations (Mesopotamian, Greek), plant ash (potash) was used as a source of alkali. In contrast to natron (as may be expected), these plant ashes were rich in potassium. Mesopotamian sources reveal that plant ashes were used for glass making, and the Bible reports the use of borite ashes by the Old Testament fullers (Malachi 3:2). Ammonia obtained from stale urine was used in the textile industry in the Middle Ages. This historical introduction emphasizes the fact that many of the acids and alkalies used in the prescientific age were of biological origin.

The concept of acids and bases cannot be dissociated from the studies of Cavendish, who showed that water is made of two gases—hydrogen and oxygen—in the

* The ancient word for burned is *kaldi*, from which the Arab word *algali* seems to be derived.

proportions two to one. Lavoisier, who developed the theory of combustion and stated that oxygen was responsible for the peculiar properties of acid, also contributed to our knowledge of acids and bases. In Lavoisier's time, acids were thought to resemble little balls covered with minute spikes that hurt the tongue. Lavoisier's contemporaries also thought that during neutralization acids entered tiny holes at the surface of a base.

In the late 18th century, a medical student named Berzelius was assigned a thesis on the action of electricity on chemicals in solution. The discovery of potassium by Davy possibly resulted from Berzelius' endeavors. Davy also demonstrated the structure of Berzelius' muriatic acid (HCl) using electrolytic techniques. During his career, Berzelius discovered many new elements and measured their atomic weight. Yet the effect of electricity on chemicals in solution remained unexplained until Svante Arrhenius, who later collaborated with van't Hoff and Ostwald, tackled that problem. In his thesis for a doctorate in philosophy, Arrhenius established and later explained the relationship between the ease of passage of electricity in solution and the concentration of the solute. After a cold reception in Sweden and in Germany, the thesis reached Willhaus Ostwald in Riga. Later Arrhenius became aware of van't Hoff's theory of solution and considered it an important argument supporting his own theory of dissociation. The triumvirate is responsible for the ion concept.

Arrhenius' theory is simple. Acids are composed of a base and a hydrogen ion. The base is either a metal or a hydroxyl group. In solution, acid and base partly dissociate and yield, respectively, free H and OH. During neutralization, H and OH combine to yield water. The hydrogen ions do not exist as such but are hydrated to yield a hydronium ion H_3O_2. Thus, when one speaks of hydrogen ion concentration, he means hydronium ion concentration. The limitation of Arrhenius' theory is obvious. It is restricted to compounds that dissociate to yield H or OH ions and to water solvent. A more general theory of the acid-base concept was developed independently by Brøunsted and Lowry in 1923. (It is said that Brøunsted derived his theory from a thesis of a young student, Eric Warburg, who later became a professor of medicine at the University of Copenhagen.) Brøunsted's theory, a generalization of Arrhenius' theory, proposes that the acid is a compound prone to yield protons, whereas the base is one prone to accept protons.

At about the time that the electronic theory of valence developed, Lewis proposed the electronic theory for acidity. In Lewis' view, an acid is any molecule—radical or ion—in which the electron shell is incomplete and is therefore capable of accepting one or more electron pairs. In contrast, a base is a compound capable of donating electrons.

Usanovitch has proposed a definition of acid and base that takes all the acid-base theories into account.

When salts are placed in solution, they are ionized to yield cations and anions. If the solution is then traversed by an electric current by dipping an electrode into it, the anions concentrate at the negative pole and the cations at the positive pole of the battery. Now, if 1 atomgram of salt is placed on the cathode, to discharge that amount of material from the cathode one would need a quantity of electricity equal to 1 Faraday (96,540 coulomb).

The difference between a strong electrolyte or acid and a weak electrolyte or acid is that the strong is completely dissociated in solution, whereas the weak is only partly dissociated. Consequently, the properties of the dilute solution of strong acids are equivalent to the sum of the properties of the dissociated ions.

Arrhenius believed that the decrease in conductivity observed with increased concentration resulted from decreased dissocation. In 1923, Debby and Hückel were able to establish that the decrease in conductivity was not due to reduced dissociation but from interionic reaction.

Polybasic acids like HP_3O_4 dissociate in stages—the first atom of hydrogen ionizes easily, but HP_2O_4 is a weak acid and ionizes slowly, and HPO_4 is a very weak acid.

Water, H_2O, is partially dissociated to yield H^+, OH^-. At 25°, 1 liter of water contains 10^{-7} ion-grams of hydrogen. According to mass action, the equation for the dissociation of water is

$$\frac{[H^+][OH^-]}{[OH_2]} = k/[H^+][OH] = kH_2O$$

The change in the concentration of H_2O is negligible, and consequently kH_2O may itself be considered a constant k'.

$$[H^+][OH^-] = k' = 10^{-14}$$

$$[H] \quad = \frac{10^{-4}}{OH}$$

To facilitate the notation of hydrogen ion concentration, Sorenson proposed the use of $pH = \log H^+$ at neutrality [40–43]. Difficulty with that notation is that it does not distinguish between hydrogen ion activity and concentration.

$$[H^+]^2 = 10^{-14}$$

$$[H] \quad = 10^{-7}$$

$$pH \quad = \log 10^{-7} = 7$$

Blood pH, Acidosis, and Alkalosis

The concentration of H^+ ions in the blood is maintained strictly constant. The mean pH of the blood is 7.398, and changes in a few hundredths of a pH unit are pathological. Extreme changes in pH lead to death by acidosis (pH 6.8) or alkalosis (pH 7.6).

The regulation of the acid-base balance* involves the transfer of H^+ ions** to yield either new base or new acid. When the pH of the blood drops below normal values, the total concentration of free proton donors (acids) increases whereas that of free proton acceptors (bases) decreases. In the body fluids Na^+, K^+, Ca^{++}, and Mg^{++} act as bases by combining with Cl^-, HCO_3^-, PO_4^-, and SO_4^- proteinate, and organic acids.

The constancy of the blood pH is continuously challenged by the introduction of metabolic products in the bloodstream. The ultimate products of metabolism contain acids and bases that must be neutralized if the blood pH is to remain constant.

The daily acid production exceeds the available basic intake and production by 100 mEq. Carbohydrate and fat metabolism produces CO_2 in normal individuals, and that of proteins and phospholipids produces sulfuric acid, phosphoric acid, and organic acids. These acids are eliminated in the lung and in the kidney via the blood. These acids must be neutralized while they are transported from their site of formation to the site of elimination. The buffer system in the red cell, the serum, and the tissues neutralizes the acids.

Buffers are mixtures of poorly ionized acids (weak acids) and their highly ionized salt. In the presence of such mixtures, the addition of strong acids or strong bases has little effect on the pH.

Plasma contains at least three major buffer systems: $BHCO_3/H_2CO_3$, B_2HPO_4/BH_2PO_4, B proteinate/protein. The bicarbonate carbonic acid buffer is the most important of the three because of its relatively high concentration. Phosphate buffers are of little value because of their low concentration. Proteins contain imidazole groups (pK 7.6), ammonium groups of lysine, guanine groups of arginine (p$K>7.0$), carboxyl groups of aspartic and glutamic acid (p$K<4.0$). It is obvious from the pK values that only the imidazole group can function efficiently as a buffer system at the pH of blood.

Cell pH

In the red cell, hemoglobin plays a considerable role in adjusting pH. Hemoglobin contains 8% histidine and 35 imidazole groups, making the protein molecule an effective buffer at blood pH.

The reversible conversion of reduced hemoglobin to oxyhemoglobin also helps to maintain a constant blood pH. Oxyhemoglobin is a stronger acid than reduced hemoglobin, and more alkalies are neutralized by oxyhemoglobin than by reduced hemoglobin. When the CO_2 diffuses from tissues into blood, the oxygen leaves the blood for the tissues, and the oxyhemoglobin is converted to reduced hemoglobin. The strongest acid is replaced by the weakest, and consequently less base is needed to neutralize the CO_2 and maintain the pH of blood at 7.4. The reverse occurs in the lung where the CO_2 is eliminated and oxygen is taken up to yield oxyhemoglobin.

Hemoglobin also directly neutralizes CO_2 by the formation of carbaminohemoglobin ($CO_2 \cdot Hb$). In $CO_2 \cdot Hb$, the free amino group of the protein combines with CO_2 to form $^-$NHOOCH groups.

The exact pH of cells is unknown, as is, of course, the exact pH of the intracellular organelles. Histochemical studies suggest that the intracellular pH may vary considerably and is much lower than that of body fluids (7.0). Phosphate and proteins are the most important intracellular buffers.

Bicarbonate Ion Concentration and Blood pH

Among all the buffer systems discussed above, bicarbonate is particularly important because of its ability to capture protons (strong base) and also because the CO_2 produced is constantly eliminated through the lung and kidney. Consequently, changes in blood pH can readily be adjusted by modifying the rate of CO_2 elimination in the lungs or kidneys.

The relationship between blood pH, bicarbonate ion concentration, and partial pressure of CO_2 is expressed in the Henderson-Hasselbalch equation:

$$\mathrm{pH} = 6.11 + \log \frac{[\mathrm{HCO_3}]}{P_{CO_2} \times 0.03}$$

P_{CO_2} partial pressure of CO_2 in alveolar air; $[HCO_3^-]$, concentration in plasma.

In this equation, the values of the numerator are determined by metabolic activities and renal excretion. The values in the denominator are determined by the respiration rate.

The kidney regulates the acid-base balance by two major functions: acidification of the urine and excretion of ammonia [44–46].

Acidification of the Urine

The acidification of the urine involves the excretion of H^+ from the tubular cell into the tubular fluid. Modern concepts on the excretion of H^+ are based

---

* Any substance that releases free protons in solution is an acid; any substance that captures free protons is a base. For example, HCl is an acid that releases free H^+ in solution, the corresponding base is Cl

$$HCl \leftrightharpoons H^+ + Cl^-.$$

In bicarbonate, H_2CO_3 is the acid and HCO_3^- is the base

$$H_2CO_3 \leftrightharpoons H^+ + HCO_3^-.$$

The symbol H^- does not refer to hydrogen ions but to a proton or a hydrogen nucleus carrying one positive charge and deprived of an electron shell. Protons readily combine with NH_3^- and OH^- OH^- groups.

** X-Ray studies, magnetic resonance studies, and studies of the electric properties of water solutions suggest that protons are not free in water but exist as a complex with a molecule of water to form hydronium or oxonium (H_3O^+).

on three major postulates: (1) the secreted H^+ is exchanged for Na^+ in the tubular fluid;* (2) K^+ can compete with H^+ for exchange with Na^+; and (3) the source of H^+ in the cell is carbonic acid, which is formed by the hydration of the carbon dioxide that diffuses into the tubular cell or is formed there as an ultimate product of metabolism.

CO_2 can be hydrated spontaneously, but in the tubular cell the conversion of CO_2 into H_2CO_3 is catalyzed by an enzyme called carbonic anhydrase. Carbonic anhydrase activities are greater in the cortical portion of the kidney than in the medulla. This is in agreement with micropuncture and stop-flow studies, which suggest that the urine is acidified in the distal part of the nephron, possibly in the collecting ducts. Yet the proximal tubule may also participate in hydrogen secretion.

Carbonic anhydrase is assumed to be located at the surface of the membrane of the tubular cells [47, 48]. Carbonic anhydrase is a small zinc protein found in many animal tissues, but its concentration is highest in the kidney tubules' cells, the erythrocyte, and some cells of the gastric mucosa. The substrates of the carbonic anhydrase reaction are carbon dioxide and water; the product is carbonic acid. The enzyme has been purified from erythrocytes, and its molecular weight is about 30,000. The purified enzyme preparation contains 0.21% zinc, probably 1 atom of zinc per molecule of enzyme. The zinc is tightly bound to the enzyme molecule and cannot be removed by dialysis or electrodialysis. The presence of zinc in the molecule is essential to activity because when zinc is removed from the molecule by extended incubation with 1–10 phenanthroline, the enzyme's activity reflects the zinc content of the preparation.

The zinc-free molecule can be reactivated by adding other metals, such as calcium, and inhibitors of carbonic anhydrase, such as metal poisons or sulfonamides, which appear to interfere with enzyme activity by reacting with the zinc.

Human erythrocyte carbonic anhydrase is a single polypeptide chain with 259 amino acid residues [78].

In the tubular fluid, the proton generated by the hydration of CO_2 may react with HCO_3^-, $NaHPO_4^-$, and Cl^- to yield H_2CO_3, NaH_2PO_4, and HCl, respectively.

Table 9-2 Electrolyte movement

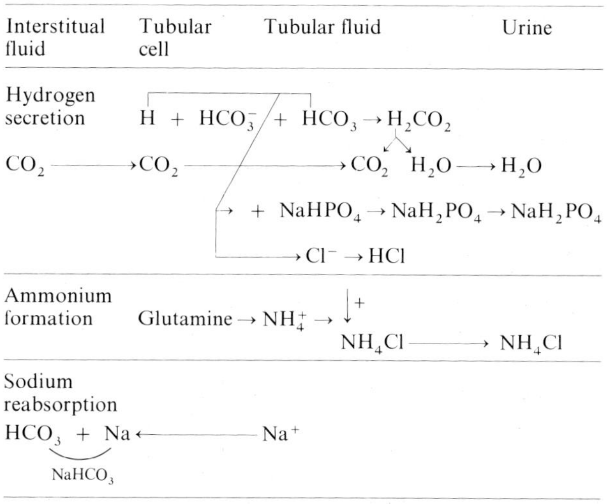

Interstitial fluid	Tubular cell	Tubular fluid	Urine
Hydrogen secretion	$H + HCO_3^-$	$+ HCO_3 \rightarrow H_2CO_2$	
$CO_2 \longrightarrow$	$CO_2 \longrightarrow$	CO_2 $H_2O \longrightarrow$	H_2O
	$\rightarrow$	$+ NaHPO_4 \rightarrow NaH_2PO_4 \rightarrow$	NaH_2PO_4
	$\longrightarrow$	$Cl^- \rightarrow HCl$	
Ammonium formation	Glutamine $\rightarrow NH_4^+ \rightarrow$	$\downarrow +$ $NH_4Cl \longrightarrow$	NH_4Cl
Sodium reabsorption $HCO_3 + Na \leftarrow$ ($NaHCO_3$)	$\longleftarrow$	Na^+	

The exact tubular site of bicarbonate reabsorption is not known, but stop-flow and micropuncture experiments suggest that it coincides with the site of sodium absorption. Although the mechanism of bicarbonate reabsorption is unknown, no extrarenal hormone appears to control it.

Bicarbonate Absorption

It would seem from the previous discussion that all the bicarbonate reabsorbed depends upon the excretion of H^+ and, consequently, depends upon carbonic anhydrase activity. Yet experiments in which carbonic anhydrase inhibitors were used suggest that 75% of the filtered bicarbonate is absorbed by mechanisms independent of carbonic anhydrase.

Secretion of Ammonia

An important function of the kidney in regulating the acid-base balance is ammonium secretion. Ammonia probably forms in the collecting tubes. The tubule cells form ammonia by deaminating glutamine. The ammonia reacts with protons to yield ammonium ions. It is not known whether these ammonium ions are formed within the tubular cell or within the tubular fluid. In any event, the trapping of protons by ammonia and the excretion of the ammonium ion into the tubular fluid confer buffering properties to the tubular fluids, reduce the difference between the pH of the tubular and the intracellular fluid, and further facilitate the excretion of hydrogen ions.

Although glutamine is generally said to be the only source of ammonia, researchers have suggested that several other amino acids, among them asparagine and glycine, can act as ammonia precursors. However, experiments by Van Slyke and his associates seem to leave no doubt that glutamine is the single major source of ammonia in the kidney.

Glutaminase splits the amide group of glutamine to yield ammonia and glutamate. The enzyme is present in most tissues, but highest activities are found in the kidney—especially the cortex. Although the kidney enzyme has been highly purified, the purification has failed to yield a crystalline preparation or even a preparation that yields single boundaries after ultracentrifugation or electrophoresis. The intracellular distribution of the enzyme is not established with certainty, but glutaminase has been claimed to be a mitochondrial enzyme.

* A passive hydrogen transfer from the tubular cell into the tubular fluid is unlikely to occur because it would require, according to calculations, a potential gradient of 180 mv.

Respiration and Blood pH

The rate and the volume of respiration regulate the pulmonary CO_2 output. The pons and the medullary portion of the brain contain a respiratory center that is sensitive to changes in partial pressure of CO_2 and changes in pH. The reabsorption of bicarbonate by the renal tubules is related to the partial pressure of CO_2 in arterial blood. Consequently, there is a constant feedback of information between the blood pH, the partial pressure of CO_2 in the alveolae, and the rate and volume of CO_2 expired. The relationship between blood and respiratory function is discussed in more detail in the section devoted to respiratory acidosis.

Acidosis and Alkalosis

From the above succinct review of the mechanism regulating blood pH, it is clear that acidosis or alkalosis may result from metabolic, renal, or pulmonary dysfunctions. Depending upon the severity of the dysfunction, the electrolyte imbalance will not be associated with severe changes in blood pH (compensated acidosis or alkalosis), or the blood pH may reach values incompatible with survival.

Metabolic Acidosis and Alkalosis

Metabolic acidosis may have at least four origins: loss of cations, retention of anions, electrolyte shifts that favor hydrogen accumulation in the ECF, and overproduction of organic acids.

Adrenal insufficiency is responsible for renal sodium loss and failure of urine acidification in the tubules. Similarly, the reabsorption of sodium and the excretion of H^+ are impaired in severe kidney disease.

Since potassium can replace H^+ in the tubular exchange of sodium, pathological conditions that lead to potassium retention (hyperkalemia) also lead to acidosis. Hyperkalemia may result from excess potassium intake, low potassium excretion, or severe shifts in the distribution of body electrolytes.

Excessive amounts of K^+ have been administered in treating hypokalemia, especially in patients with glomerulonephritis or diabetic nephropathy.

In the absence of aldosterone (Addison's disease), sodium is lost but potassium is retained. In sodium depletion, no sodium is available in the tubular fluid for exchange with potassium, and hyperkalemia and acidosis result.

Severe glomerulonephritis or terminal stages of chronic nephritis, pyelonephritis, or nephrosclerosis with oliguria or anuria may lead to hyperkalemia. The amount of potassium present in the ECF is so small compared to the amount in tissues that extrusion of potassium from a relatively small number of cells into the ECF may lead to hyperkalemia. This occurs in hypothermia or after muscular trauma, myocardial infarction, or severe burns.

In renal disease (nephrosclerosis, nephrocalcinosis, acetazolamide therapy) in which sodium is lost without a corresponding loss of chloride, chloride is retained and metabolic acidosis develops. Overproduction of lactic acid leading to acidosis is observed in hypoxia, hemorrhagic shock, and physical exertion.

Excessive amounts of keto acids are formed in diabetes and starvation. The molecular mechanism of formation of ketone bodies was discussed in the section on ketosis.

β-Hydroxybutyric and acetoacetic acids are strong acids that require neutralization by cations when they are excreted in the urine. Sodium is excreted and the lungs compensate for the acidosis by hyperventilation; as for the kidneys, they form more ammonium ions for the purpose of neutralizing the keto acids. When the compensation fails, the pH drops—possibly to values as low as 7, nausea is replaced by vomiting with severe dehydration, and drowsiness develops into stupor and finally coma. Surprisingly little is known of the metabolic changes that take place in the kidney while metabolic acidosis is established.

Acidosis can readily be produced by ammonium chloride. When an animal is given ammonium chloride, all the ammonia is neutralized and converted to urea, as if the animal had been given a solution of strong acid.

Some of the metabolic changes in the kidney in such experiments are known. Sodium is temporarily depleted, and then ammonium excretion is markedly increased. The increased excretion of ammonium ions is accompanied by *de novo* synthesis of glutaminase. Although the adaptation just described can be expected, other metabolic changes observed in the kidney of animals made acidotic experimentally are somewhat more difficult to understand. They include increased activity of the enzymes of the hexose monophosphate shunt, increased rates of gluconeogenesis and lipogenesis, and compensatory hypertrophy of the kidney, somewhat reminiscent of that observed after unilateral nephrectomy [48].

Metabolic alkalosis develops whenever the bicarbonate concentration is increased without a corresponding increase in the partial pressure of CO_2. Clinical metabolic alkalosis develops after excessive bicarbonate administration, persistent vomiting, diuretic and steroid therapy, and potassium loss.

The ingestion of excessive amounts of bicarbonate usually does not affect the body fluid's electrolyte balance because of the kidney's ability to excrete $NaHCO_3$. Consequently, the ingestion of bicarbonate is accompanied by increased potassium secretion. The ingestion of $NaHCO_3$ results in excretion of both $NaHCO_3$ and $KHCO_3$. Alkalosis develops if the potassium intake is inadequate (if potassium intake is inadequate, the body tends to conserve it). Therefore, instead of being exchanged for K^+, the sodium that is reabsorbed in the renal tubule is exchanged for H^+; as a result, reabsorption of sodium bicar-

Table 9-3. Causes of potassium depletion

Reduced intake	
Parenteral feeding without the addition of K^+	
Excessive losses	
In feces:	Diarrhea
In urine:	Administration of mercurial diuretics or steroid hormones
	Cushing's syndrome
	Primary aldosteronism

bonate is enhanced, aggravating the alkalosis. Potassium depletion may result from reduced intake or excessive losses (Table 9-3).

The gastric juice contains high H^+ concentrations (up to 10 mEq/100 ml). Consequently, repeated vomiting results in marked loss of H^+ with concomitant alkalosis. If food and drink can be taken in soon after the episode of vomiting, the compensatory excretory mechanisms of the kidney will prevent serious alkalosis. However, if the disease that caused the vomiting persists and all food and drink are regurgitated, more H^+ is lost and the alkalosis will persist. In fact, under those circumstances the kidney's adjustments to the loss of electrolytes in the vomitus further exacerbate the alkalosis.

To understand the pathogenetic mechanism of progressive alkalosis in the case of intractable vomiting, one must keep in mind: (1) that the vomitus contains, in addition to H^+, considerable amounts of Na^+, K^+, and Cl^-, and (2) that ion excretion or reabsorption results not from the passage of an individual ion through the membranes, but from an exchange of cations for anions or anion-cation complex, and *vice versa*. In the extracellular fluid, the positive charges of the sodium cation (Na^+) are balanced mainly by the negative charges of Cl^- and HCO_3^-. If the charges are to remain balanced, sodium reabsorption must be accompanied by the reabsorption of an equivalent amount of Cl^-, or each Na^+ reabsorbed must be exchanged for K^+ or H^+. In intractable vomiting, the first adaptive step involves increased excretion of H^-CO_3 in the urine as an attempt to compensate for the alkalosis generated by the loss of H^+. The H^-CO_3 cannot be excreted alone, it must be accompanied by equivalent amounts of Na^+ and K^+.

Obviously, patients who vomit are dehydrated, and as a result their ECF volume is reduced. Moreover, the increased K^+ and Na^+ excretion further reduce the ECF volume. The kidney reacts to this reduced ECF volume by increasing sodium reabsorption by inhibiting the natriuretic factor or by stimulating aldosterone secretion, or both.

Since the chloride content is low in these patients, chloride will not compensate the sodium reabsorption electrostatically. In contrast, the urinary H^-CO_3 levels are high; therefore, one would expect these anions to be driven to be reabsorbed with the Na^+, thus aggravating the alkalosis. However, mainly an exchange function takes place in hypochloremia. (This is because of the differences in the reabsorption ability of the tubule's membrane. Iron chloride is readily reabsorbed, but H^-CO_3 is not and therefore, the sodium is exchanged for K^+ and H^+, increasing the hypokalemia and the alkalosis.) In some cases aciduria may develop in the face of alkalosis.

Inasmuch as this paradoxical aciduria results from sodium and potassium depletion, it is not surprising that the urine may again become alkaline if sodium and potassium are administered. Moreover, ammonium chloride administration to such patients may result in retention of ammonium with loss of bicarbonate in the urine, leading to acidosis. From the pathogenetic mechanism discussed, it would seem that chloride administration in the form of sodium or potassium chloride would alleviate alkalosis. In fact, the management of such patients usually involves controlled rehydration, sodium chloride administration, and compensation for the potassium depletion [49, 50].

Tubular Acidosis

We have seen that bicarbonate is reabsorbed primarily in the proximal tubule and that the urine is acidified and ammonia excreted in the distal tubule. Any inherited or acquired injury that interferes with bicarbonate reabsorption or urine acidification will lead to acidosis. (Although interference with ammonia excretion could, in theory, also lead to acidosis, it seldom does.)

Tubular acidosis is primary (an isolated inherited metabolic disorder) or secondary (a metabolic disorder associated with inherited or acquired lesions). It is characterized by the inability to maintain normal blood pH as a result of interference with bicarbonate reabsorption or a defect in renal hydrogen excretion. By definition, glomerular filtration is not altered in renal tubular acidosis.

Primary distal tubular acidosis (Albright's disease) is an inherited disease believed to be transferred by an autosomal dominant gene with low penetrance in males. The exact molecular mechanism responsible for the injury remains unknown, but a metabolic distortion in the tubular cell is suspected. In primary distal tubular acidosis, the distal tubule cannot maintain an adequate pH gradient between blood and tubular fluid. As a result, chronic acidosis develops, causing anorexia, lethargy, and dyspnea. Because of the decrease in distal sodium-hydrogen exchange, sodium is lost in the urine in the form of sodium bicarbonate. The reduced level of $NaHCO_3$ in serum is associated with a shrinkage of the extracellular fluid volume. The drop in sodium is not coupled with a reduction in the blood chloride levels. The absence of increased chloride excretion coupled with a reduced extracellular fluid volume is responsible for the hyperchloremia observed in patients with Albright's disease.

The increased sodium-potassium exchange—which is a consequence of (1) the inability to exchange sodium for hydrogen, and (2) the secondary aldosteronism that develops as a result of the decreased ECF volume—leads to hypokalemia, which is believed to be partly responsible for the inability to concentrate urine.

In an attempt to compensate for the acidosis, bone salts are mobilized. Unless the patient is treated with alkalies, bone salt mobilization results in osteomalacia. Calcium is excreted in the urine, and the combination of high calcium levels, high pH, and low citrate concentration in the tubular fluid facilitates precipitation of calcium salts, thus leading to nephrocalcinosis. Secondary hyperparathyroidism develops as a consequence of the negative calcium balance. The inability to concentrate urine and the accompanying polyuria result in part from the hypokalemia and in part from the hypercalciuria.

Starvation, malnutrition, vitamin D deficiency, intoxication, hormonal imbalances, and renal injury can cause secondary renal tubular acidosis.

Proximal Tubular Acidosis

In normal individuals all the bicarbonate that passes the glomerular filter is reabsorbed in the proximal tubule, up to a maximum concentration of bicarbonate in the blood (26 mM/liter in adults, 22 mM/liter in infants). If serum levels exceed these thresholds, bicarbonate is excreted in the urine.

In patients with proximal tubular acidosis, the threshold is lowered and bicarbonate appears in the urine with relatively low levels of bicarbonate in the serum. The loss of bicarbonate in the urine leads to acidosis by reducing the alkaline reserve and by inhibiting distal hydrogen secretion. Thus, the distal tubules are flooded with urine rich in bicarbonate, which impairs hydrogen secretion. At least three mechanisms for bicarbonate reabsorption have been described in the proximal convoluted tubule.

Reabsorption of most of the bicarbonate is coupled with hydrogen ion excretion. Thus, when $NaHCO_3$ is filtered, the salt is dissociated in the urine to Na^+ and HCO_3^-. The sodium is reabsorbed, and in the presence of H^+ ions, which are generated in the tubular cells, HCO_3 and carbonic anhydrase, found at the surface of the tubular cells, yield CO_2 and H_2O.

$$HCO_3^- + H^+ \rightarrow CO_2 + H_2O$$

CO_2 diffuses freely into the tubular cell, where it is hydrated again by carbonic anhydrase to yield HCO_3^- and H^+.

$$CO_2 + H_2O \rightarrow H_2CO_3 \rightarrow H^+ + HCO_3^-$$

Thus, the H^+ that is needed to neutralize HCO_3 of the tubular fluid is generated.

Part of the HCO_3^- reabsorption is the consequence of the direct (not involving carbonic anhydrase) reaction of CO_2 with H_2O to yield H_2CO_3. This mechanism depends upon the pCO_2, and it explains why the threshold of bicarbonate reabsorption varies with the level of pCO_2, in blood (see Pulmonary Acidosis).

It is also believed that a fraction of HCO_3^- is reabsorbed independently of any form of H^+ excretion, whether it be carbonic anhydrase dependent or independent. Thus, proximal tubular acidosis could result from low carbonic anhydrase activity, ineffective diffusion of hydrogen ions, or a defect in the H^+-independent reabsorption of bicarbonate. It is not known which distortion is responsible for proximal renal tubular acidosis.

Patients with primary proximal tubular acidosis usually have stable serum HCO_3^- levels in spite of the reduced acid excretion. The compensatory mechanism is not known, but reduced endogenous acid production can probably be excluded. A more likely adaptation is the formation of additional bases at the expense of bone $CaCO_3$. Some patients with primary proximal acidosis present skeletal demineralization, reduced levels of CO_3^- in the hydroxyapatite crystals, and increased calcium losses in the feces. Patients with proximal renal acidosis are usually symptomless, and except in cases of acid overproduction, the prognosis is usually favorable.

Two hereditary diseases, Fanconi's syndrome and Lowe's syndrome, are associated with tubular acidosis. The Fanconi syndrome is a multiple proximal tubular defect including interference with amino acid, glucose, and phosphate reabsorption. In most cases there is also acidosis that is proximal in origin, but the mechanism of the acidosis is unknown. Lowe's syndrome, or oculocerebrorenal syndrome, is transmitted in a sex-linked recessive fashion. Affected patients have the Fanconi syndrome and also mental retardation, cataracts, glaucoma, and hypotonia.

Lung Function, Gas Exchange, and Pulmonary Acidosis

An adequate understanding of the acid-base balance in the blood demands an elementary knowledge of the mechanism of respiration. In respiration, oxygen is used in metabolism and carbon dioxide and pairs of hydrogen atoms are produced. Thus, respiration refers to the consumption of oxygen and the production of CO_2.

Respiration occurs at various levels of organization: mitochondria, intact cell, the lungs, and the entire organism. Here we are mainly concerned with gas exchange across the alveolar membrane, which occurs during breathing.

Breathing is a mechanical phenomenon in which rigid (the thoracic wall), contractile (muscle), and elastic (pulmonary alveoli) structures operate in a bellows-like maneuver to take in oxygen-rich air and expel air rich in CO_2. These gas exchanges (which lead to changes in gas content) occur at the boundary separating the external from the internal milieu—namely, the alveolar wall.

The nose, mouth, larynx, trachea, bronchi, and bronchioles are elastic and contractile pipelines communicating the deep-seated alveoli with the atmosphere.

The Lung

The alveolar wall is the elementary structure at which gas is exchanged. Each alveolus is a polyhedral structure with one missing side. The walls of the alveoli have been assumed to be perforated by pores. The missing side permits a communication between alveolar air and the terminal branches of the bronchi. The pores connect alveolar cavities. In pneumonia, fibrin may fill these pores, facilitating transfer of bacteria from one alveolus to the other.

The wall of the alveolar cavity is composed of a layer of alveolar cells, a basal membrane, and a layer of capillaries. The alveolar cavity is covered with a lining of flat epithelial cells, with a nucleus, cytoplasm and all cellular organelles. Generations of histologists have debated the properties of the alveolar cells. Electron microscopists have now established that the epithelial lining is continuous, except for the presence of pores that go through the septa.

Although embryogenesis suggests that the alveolar wall is of entodermal origin, this was often doubted, mainly because some properties reminiscent of macrophages, which are of mesodermal origin, were attributed to the alveolar cells. Indeed, histologists taught that the alveolar cells of the lung were part of the reticuloendothelial system and were capable of phagocytosis. Modern concepts propose that it is not the epithelial cell *per se* that performs the reticuloendothelial functions. Instead, macrophages, which are present in large amounts in the alveolar septa, are brought into the alveolar lumen to phagocytize red cells, carbon, etc.

The alveolar linings are separated by conjuntivovascular septae. These septae contain large amounts of reticulin, conjunctive, and elastic fibers. The walls of the alveoli contain no smooth muscle fibers; these are found only around the terminal bronchioles at the junction of bronchiole and the alveolar sacs. An extremely rich lacework of capillaries traverses the septae, uniting the terminal branches of the pulmonary arteries with those of the pulmonary veins. The branches of the pulmonary arteries follow bronchioles, bronchi, and veins on their way to the pleura.

The human lungs contain close to a billion alveoli. The total surface of the alveolar walls is 90 square meters, or approximately the surface of a square with a 30-foot side. The surface of the capillary network is 1.5 times that of the alveolar lining, or 140 square meters. Consequently, the amount of blood passing through the pulmonary network per minute is considerable—24 liters. (In view of the enormous pool of blood that passes through the lungs, it is not surprising that slowing of the circulation causes blood in the pulmonary vasculature to stagnate, leading to passive congestion. In fact, up to 20% of the total blood volume can accumulate in the lungs.) Each liter of blood can dissolve 162 cm^3 of oxygen per minute, so 3.5 liters of oxygen is dissolved per minute by the blood passing through the pulmonary network. These facilities for gas exchange far exceed the physiological needs at rest, when only 0.1 of the respiratory capacity of each lung is used.

Breathing results from the combined activities of several muscles: the diaphragm, the intercostal muscles (internal and external), the abdominal scalenus, the sternocleidomastoid, and the intrinsic muscles of the larynx. Primarily the diaphragm and the intercostal muscles are used during quiet breathing. The diaphragm rises 1.5 cm in expiration during quiet breathing and 7 cm in expiration during deep breathing. The active inspiration movements of the lungs result from intercostal muscle contraction. These muscles lead to a downward movement of the diaphragm and an outward movement of the chest, movements that generate the positive pressure in the virtual space that separates parietal from visceral pleura.

To understand the changes in interpleural pressure, one must consider the fetus. In the fetus, the airless lung fills the entire thoracic cavity, and the visceral and parietal pleura are in close apposition. In the first inspiration, the thoracic cavity, visceral pleura, interpleural space, and parietal pleura are all pulled outward. The alveoli open and are filled with air. The lung is thus inflated by expanding the thoracic cage and the diaphragm. The distension of the lungs is opposed by its elasticity. These conflicting forces generate a pressure that is negative with respect to the atmospheric pressure. During infancy, the volume of the thoracic cage increases more rapidly than that of the lung, there is a greater tendency for the lung to distend, and, as before, the distension is opposed by the inherent elasticity of the lung, thus increasing the negative pressure.

The negative pressure that exists in the pleural cavity is best illustrated by describing two types of cadavers. If a body is frozen and the thoracic cavity is opened, all the viscera of the thorax fill the thoracic cavity entirely. In contrast, when the thoracic cavity is opened under ordinary autopsy conditions, a large amount of free space is available. The heart and lungs have collapsed. Thus, one may conclude that in the living, a force is pulling on the walls of the lungs and the heart, stretching them out. If the diameter of the thoracic cavity is further widened, this force, the negative pressure in the interpleural space, is further increased. Consequently, the lungs passively expand and more air rushes into the alveolar sacs.

The anatomical structure of the rib cage facilitates these changes in the dimensions of the thoracic cavity. Each rib constitutes a semicircle joining the spine with the sternum. However, the half circle is not on a single horizontal plane. Two major slopes can be distinguished: (1) a posterior slope that brings the ribs downward and outward; (2) an anterior slope also oriented downward and outward. Thus, by raising the ribs horizontal and bringing down the diaphragm, the thoracic cavity can be markedly expanded. The lungs can be distended with air, and, because of the elasticity of their framework, they can also collapse. A normal lung resembles a resilient sponge, with its cavities filled with air. Resilience results from distensibility and elasticity. If distensibility is decreased, greater efforts must be made at inspiration. If elasticity is impaired, expiration can no longer be passive.

In expiration, the elasticity of the thoracic wall and, if necessary, the contraction of the diaphragm and the abdominal muscles, expel the inspired air. This process of ventilation continues uninterrupted until death. The movements are repeated rhythmically; the rhythm is altered only if the physiological conditions are markedly modified or if pathological conditions interfere.

Obviously, the mechanism controlling breathing is of capital importance. The rhythmic regulation of breathing is controlled through a complex neurological arrangement, the center of which is in the upper medulla and the pons.

Respiratory Centers

In contrast to heart muscles, which possess inherent rhythm, the respiratory muscles have no inherent rhythm and lose their ability to contract when their nervous connections are severed. Considering the complex patterns of breathing that develop under different physiological or pathological conditions, one must inescapably conclude that the nervous mechanism controlling breathing is extremely complex. Legallois and Flourens were responsible for pioneer findings in this area. On the basis of transections of the central nervous system and topical destructions, they developed the concept of a bilateral "vital node" located on the midline close to the obex.

The respiration rate is altered by changes in the chemical composition of the blood, and also as a result of emotional stress or voluntary efforts. Passing from quiet respiration to dyspnea affects different nerves, such as the intercostal nerve. Breathing patterns are controlled by a complex network of neurons connected to efferent fibers: chemoreceptors, stretch receptors, pressure receptors. The efferent fibers send impulses to the motor nuclei in the spinal cord, and from there the impulse is transferred to the respiratory muscles.

The primary center, located in the medulla, is made up of a complex of interconnected neurons that are part of the reticular system. Although the neurons are not condensed to form typical nuclei, it is possible to distinguish, on the basis of stimulation by microelectrodes, neurons that generate impulses responsible for inspiration from neurons generating impulses that lead to expiration. Therefore, we speak of an inspiration and an expiration center. Researchers believe that these centers are widely interconnected, and that inspiratory neurons are intermingled with expiratory neurons.

A complete transection of the nervous axis above the medulla is compatible with life. However, the pattern of breathing is abnormal. After transection between the medulla and the pons, the quiet pattern of breathing is replaced by gaps occurring at wide intervals. The primary respiratory center is thought to constitute the minimum neuron arrangement compatible with breathing. The fine adjustment of the breathing pattern is controlled in other respiratory centers, such as the pneumotaxic and apneustic centers.

The pneumotaxic center is in the upper pons. From it, neuron impulses are generated, periodically inhibiting the inspiratory center. Consequently, the continued excitation of the inspiratory center is converted into a rhythmic discharge: inspiration followed by rest (expiration).

If both the pneumotaxic center and the vagal fibers are transected, prolonged inspiratory apneusis develops. An apneustic center seems to be located in the lower pons, between the inspiratory and pneumotaxic centers. The discharge of the apneustic center stimulates the inspiratory center. Under normal conditions, the apneustic center is periodically restrained by vagal reflexes and pneumotaxic impulses, when uncontrolled apneustic breathing takes place.

From this brief discussion, it seems obvious that the rhythmicity of respiration must reside in some specific properties of the neurons of the inspiratory center. The exact mechanism of rhythmicity is not yet known, but three theories have been proposed: (1) spontaneous rhythmic discharge of the inspiratory center neuron; (2) continued discharge of the neurons of the respiratory center with periodic inhibition from other centers; (3) low excitability of the neurons of the inspiratory center with periodic excitation by discharges generated in other neurons.

It is not possible to discuss here the different hypotheses proposed to explain the rhythmicity of respiration, and only the more generally accepted views are presented.

In certain cold-blooded species, the brain stem continues to discharge impulses to the respiratory muscles after all afferent paths have been disconnected. This spontaneous activity of the respiratory neurons is controlled by the combined activation of the inspiratory and expiratory centers.

The primary control resides in the inspiratory center, and the expiratory center acts in a secondary fashion by intermittently interrupting the respiratory discharge. The excited inspiratory center is believed to send impulses to the pneumotaxic center, which stimulates neurons in the expiratory center to discharge impulses that inhibit the inspiratory center.

In conclusion, investigators believe that in the intact animal an autonomic rhythm in the medullary center is enhanced by impulses from the pneumotaxic center and from the vagus (as a result of inflation). These impulses stimulate the expiratory center, which in turn inhibits the inspiratory center.

The breathing pattern changes without stimulation of the airway reflexes and without changes in the chemical composition of the blood. Under those conditions, the increase in respiration can be closely correlated with patterns obtained on electroencephalograms. In these special circumstances, the increased respiration appears to be associated with increased activity of the cortex. The increased cortical activity sends impulses to the respiratory center through connections with the reticular activating system, which has been shown to generate a positive feedback impulse that increases the discharge of the inspiratory neuron. In contrast, the reticular system inhibits those neurons that exert a negative feedback on the inspiratory center. As a result, inspiratory movements are not only facilitated, but their length and amplitude are increased, providing increased ventilation.

This increased ventilation in response to peripheral excitation is believed to be only part of a general stimulation of the reticular system, a sort of alerting reaction that prepares the individual for increased muscular activity.

The neurons of the medullary reticular formation appear to be in some way connected with neurons in the hypothalamus. Indeed, hypothalamic depression may lead to decreased bicarbonate and electrically invoked inspiratory responses [51].

Receptors

The respiratory center can be thought of as a black box, the output of which is transmitted to motor neurons, from which the fibers of the phrenic nerve and the intercostal nerves originate. The input to the center is generated in various types of receptors: stretch receptors, pressure receptors, and chemoreceptors. The sum of these incoming impulses generates a discharge that is then transferred to the motor neurons. Excitation of at least 12 respiratory reflexes begins in different receptors in the viscera. These impulses are generated in the receptors and are transferred to the central nervous system through the vagus.

One of the most studied of these reflexes is the Hering-Breuer reflex. Inflation of the lung decreases the frequency of the inspiration; conversely, deflation of the lung increases respiratory activity. This reflex is lost if the vagus nerve is cut or if the vagus fibers are cooled to 8° C or 10° C. The impulses generated by the sensitive vagal fiber inhibit the nervous impulses that generate muscle contraction. As a result, the electromyographic pattern of the isolated muscles is abolished when the Hering-Breuer reflex is stimulated. The exact locations of the receptors are not known, but they are assumed to be in the bronchi or the bronchioles because there is no correlation between the electrical activity of the efferent fibers and the pressure that stretches alveoli.

The stimuli generated by these impulses facilitate the termination of inspiration by maintaining a tonic inhibition of the inspiration center during the entire breathing cycle, and thereby regulating the work involved in breathing, such as to secure minimal energy expenditure.

The deflation effect is transmitted by a series of impulses different from those responsible for the inflation effect. Indeed, if the vagus nerves are cooled below 8° C, response to deflation continues. Little is known of the mechanism and the physiological significance of these reflexes.

Until the discovery of chemoreceptors, the Hering-Breuer reflex was the only pulmonary reflex known. Therefore, it was assumed to be important in the autoregulation of respiration. Although the vagus impulses contribute to the rhythmicity of breathing, it is clear that respiration is controlled mainly by the chemical changes that take place in the blood.

Chemoreceptors

The purpose of ventilation is to secure oxygen intake and CO_2 output. Adequate regulation of these gas exchanges is vital for the survival of every cell. Thus, the ventilation rate would be expected to be regulated according to the needs for oxygen or by the excess in CO_2. Consequently, the concentration of the gases would exert a feedback control over the activity of the respiratory center. Such a mechanism would require sensitive receptors in the peripheral or central nervous system. At first it was thought that chemoreceptors were in the alveolar wall and that the respiratory center responded to changes in gas composition in the alveoli. Other investigators suspected that the chemoreceptors were centrally located. Only when Heymans discovered the carotid body was the concept of peripheral chemoreceptors accepted. It is now clear that both peripheral and central chemoreceptors exist.

The two major peripheral chemoreceptors are the carotid and the aortic bodies. The central chemoreceptors are probably localized close to the respiratory center in the medulla.

Between 1927 and 1930, Heymans discovered the chemoreceptor reflexes in the carotid and aortic bodies. He showed that when stimulated, these receptors excite the respiratory center.

The carotid bodies have been known since 1743. They are small, pinkish neurovascular structures 1–2 mm in diameter; they are located at the posterior aspect of the common carotid, where it divides into the internal and external carotids. Carotid bodies are vascularized by the occipital branch of the external carotid. The efferent nerve fibers of the carotid body converge into the carotid branch of the ninth cranial nerve, or glossopharyngeal nerve.

The histological structure of the carotid bodies is very typical. The chemoreceptor cell contains unstable granules that disappear when the nerve endings of the carotid bodies are stimulated. The mechanism by which the chemical stimulus is converted to an electrical impulse is not known. Aortic chemoreceptors are found between the aortic arch and the pulmonary artery on the dorsal aspect of the pulmonary artery.

The arterial blood supply of the carotid body is brought by the occipital and ascending pharyngeal arteries, the venous blood drains into the internal jugular veins.

Histologically, three structures can be distinguished in the carotid bodies: epithelioid cells, vascular sinuses, and nerve endings. The epithelioid cells are assembled in cords in close contact with the vascular sinuses and with nerve endings. This rich innervation of the epithelioid cells is derived from branches of the carotid sinus nerves, which arise from the glossopharyngeal trunk. The carotid body is activated by severe anoxemia, hypercapnia, and lowered pH.

The aortic body's structure resembles that of the carotid body. The aortic body is at the root of the right subclavian artery and scattered around the transverse part of the aortic arch.

Assuming that the epithelioid cells are the actual chemoreceptors, the histological arrangement assures maximum efficiency by securing an abundant blood supply through the sinuses and a concentrated array of afferent and efferent nerve endings.

It was first shown that increased concentrations of hydrogen ions in the cerebrospinal fluid stimulated ventilation. At first investigators believed that a hydrogen ion directly affected the respiratory center, but it was then demonstrated that the center responding to the hydrogen ion was located in the lateral recesses of the 4th ventricle, or in the adjacent subarachnoid space near the roots of the 8th and 10th cranial nerves. These intracranial receptors now appear to be the site of the respiratory response to inhaled CO_2. The responsive area is located at lateral surface of the ventricle and bound medially by the paramedial tract, laterally by the roots of the 8th, 9th, 10th, and 11th cranial nerves, rostrally by the pons, and it extends caudally about 6 mm.

If one plots on the same graph the respiratory activity in liters per minute versus the percentage of CO_2 in the inspired air, or the percentage of oxygen in the inspired air, the two curves have completely different shapes. The oxygen curve remains flat until the percentage of oxygen in the inspired air drops to values as low as 10%; only then does respiration increase rapidly. With these low levels of oxygen in the inspired air, the arterial oxygen saturation is less than 90%, and the alveolar partial pressure of oxygen is less than 60 mm Hg. Such a degree of anoxemia leads to loss of visual acuity and deterioration of mental activities. Obviously, a control mechanism with a setting that responds only when the organism is at the verge of catastrophe does not constitute an adequate mechanism controlling the respiration rate. Yet, low levels of oxygen undoubtedly can stimulate respiratory activity, mainly by increasing the depth of breathing.

The exact mechanism by which this occurs was not clear until the carotid and the arotic bodies were discovered by Heymans. Until then, anoxemia was assumed to affect the rate of breathing by modifying the blood's pH rather than through a direct action of oxygen. Yet hyperpnea following anoxemia had been observed in the presence of hypocapnia and alkalosis. Sophisticated explanations of this paradox were proposed, including the elaboration of unidentified blood-borne chemicals.

Under normal conditions of aeration, the CO_2 content of the inspired air is negligible, but if it rises artificially or under experimental conditions, an increase to 1% leads to an increase in ventilation. When the CO_2 levels reach 4% in the inspired air, the ventilation rate is doubled, and from then on the rate of ventilation continues to increase proportionally to the increase in CO_2 in the inspired air until the CO_2 proportion reaches 10%.

Two nerve mechanisms are activated by modifying the CO_2 levels. One is the aortic and carotid bodies, which respond to CO_2 as chemoreceptors and generate a reflex impulse that stimulates respiration; the other is a central effect on either the respiratory center or another center the location of which is not known. It is also not certain whether CO_2 *per se* stimulates the central nervous center or whether the increase in hydrogen ions, which is associated with the increased CO_2 levels, is responsible for stimulating the central nervous center. Although the response to CO_2 clearly involves central nervous mechanisms, it is not known to what extent these two mechanisms participate in the physiological response to CO_2. Investigators have proposed that the chemoreceptors constitute a second line of defense that secures pulmonary ventilation when the sensitivity of the respiratory centers is dulled—for example, under the influence of an anesthetic. These conclusions are in conflict with those of Heymans and his associates who withdrew the cerebrospinal fluid from anesthetized dogs and irrigated the cysterna with saline. Such a procedure does not modify the respiratory response in these dogs when the partial pressure of CO_2 of the inspired air is high. These results suggest that the central mechanism is not essential for triggering the respiratory response.

In view of the effect of the partial pressure changes of CO_2 on the rate of ventilation, it was questioned whether or not the hydrogen ion *per se* had any effect on chemoreceptors or on the respiratory centers. In metabolic acidosis, pulmonary ventilation is increased despite normal arterial partial pressures of CO_2. Since under those conditions there is no indication of anoxemia, it seems that the hydrogen ion must elicit the respiratory response. We have seen already how modifications of the cerebrospinal fluid's acidity stimulate respiration and how the response from specialized centers in the nervous system is elicited by increases in the hydrogen ion concentration. The question one can raise is whether the effects of the increase of the partial pressure of CO_2 and an increase of the hydrogen ion concentration may be adequate. Gray believes that they are; however, his views have been challenged by others. It seems well established that under a certain partial pressure, CO_2 levels have little effect on the respiratory response elicited by lack of oxygen.

The decrease in oxygen concentration leads to an increase in the rate of ventilation. However, the medulla is not sensitive to changes in oxygen tension in the brain, nor does the respiratory response to lack of oxygen take place if the carotid and aortic bodies are eliminated. Consequently, it is now accepted that the respiratory response to lack of oxygen results primarily from the firing of chemoreceptors in the carotid and aortic bodies.

Pulmonary Acidosis

When pulmonary ventilation is reduced, the partial pressure of CO_2 increases in the blood. The excess of CO_2 is caused by the fact that the diffusion of CO_2 through the alveolar wall is much slower than that of oxygen.* Inasmuch as the CO_2 is rapidly hydrated to yield H_2CO_3, which is in turn dissociated to yield $H^+ + HCO_3^-$, the absolute concentration of hydrogen ions is increased in blood, and acidosis develops in hypoventilation.

Both acute and chronic lung disease may lead to acidosis, but the pulmonary reserves are considerable in normal individuals. The arterial tension in oxygen and CO_2 remains constant even after heavy exercise, when the oxygen intake may be 10–15 times that of the resting values. Only when the lungs have been damaged extensively or when the bellows capacity is seriously impaired do changes in the blood gases take place. Most forms of severe pulmonary (pneumonia, pulmonary edema) and pleural (pleuritis, emphysema, pneumothorax) diseases are associated with pulmonary acidosis. Depression of the respiratory center in poliomyelitis and surgical anesthesia also leads to hypoventilation and acidosis. In surgical anesthesia, the hypoventilation may be acute and kill the patient unexpectedly.

Attempts of the body to compensate for pulmonary acidosis include: first, exchange of plasma chloride with HCO_3 of the red cells. This exchange increases the bicarbonate levels in plasma and restores the bicarbonate-carbonic acid ratio to normal, although the levels of both these compounds are elevated. Second, a rise in the concentration and stimulation of pulmonary ventilation, thereby reducing the P_{CO_2} as a result of the rise of the H^+ concentration. Third, acceleration of Na^+ reabsorption in the form of bicarbonate and an increase in the formation of NH_3 by renal tubules**. In acute hypercapnia, blood $[H^+]$ increases as a direct linear function of P_{CO_2} as expressed by the equation:

$$[H^+]\ \text{nEq/L} + 0.76\ P_{CO_2}\ \text{mm Hg} + 9.3.$$

* When the alveolar gas exchanges are severely inadequate, the first manifestation is usually a drop in the oxygen tension of the blood. In most patients this is followed by increased CO_2 tension. Only rarely are the oxygen tension of the blood decreased and normal CO_2 levels maintained, except in perfusion or diffusion disturbances. In patients with chronic bronchitis, emphysema, and tuberculosis, the CO_2 drops to 20–30 mm Hg and the P_{CO_2} to 80–90 mm Hg.

** Biochemists have investigated the mechanism by which ammonia formation is stimulated by the renal tubules in pulmonary acidosis. Acidotic dogs show no increase in glutaminase, glutamic dehydrogenase, or glutamic oxaloacetic dehydrogenase; in contrast, the activities of these enzymes are markedly increased in kidneys of rats made acidotic. Nothing is known about the enzyme changes that occur in man.

Yet, renal acid excretion does not increase significantly and serum HCO_3^- is increased only slightly. In acute hypercapnia, the blood buffers,* principally hemoglobin, protect against a rise in cellular pH with only a limited attempt to directly neutralize the acid by mobilizing the alkaline reserve.

If acute hypercapnia does not constitute a threat to normal individuals with good blood circulation and normal renal function, in a surgical patient in shock with reduced renal blood flow, hypoventilation may mean death as a result of rapidly worsening acidosis. When blood pH falls to low values, the neuromuscular impulses that control heart beats are distorted, and ventricular fibrillation may develop.

In chronic hypercapnia, the full compensatory mechanism is placed in gear. The central events are increased acid secretion by the kidney (ammonium plus titrable acid) and increased reabsorption of bicarbonate (HCO_3^-). Indeed, the maximum renal reabsorption of bicarbonate rises with the increase in P_{CO_2} levels in blood.

Chronic pulmonary failure may be further complicated by metabolic disturbances tending to metabolic alkalosis or metabolic acidosis. The mechanism leading to alkalosis is not always clear, but among the factors that may influence it are the loss of hydrogen and Cl ions, because of vomiting or because of selective Cl and potassium depletion as a result of undernourishment, and prolonged treatment with diuretics. It is usually assumed that severe respiratory acidosis is always accompanied by metabolic acidosis. This reasoning is based on the fact that when the same CO_2 tensions are achieved in the blood *in vivo* and *in vitro*,** the plasma concentration of bicarbonate for identical pH's is lower *in vivo* than *in vitro*. In reality, this bicarbonate deficit seems to result because: (1) the buffer curve of the blood CO_2 has a lower slope *in vivo* than *in vitro*; and (2) hyperventilation *in vivo* leads to lactic acid accumulation in he blood.

Metabolic acidosis is, however, a rare complication of chronic pulmonary disease, except in cases of acute hypoxemia, which may lead to transient lactic acid accumulation in the blood.

Since the renal generation of HCO_3 is coupled with Cl elimination, the total concentration of anions in the plasma is unchanged in respiratory acidosis, and electron neutrality is maintained.

The body's potassium levels are believed to be reduced in respiratory acidosis. Whether this results from adaptation to high partial pressure of CO_2 or is due to abnormal loss or insufficient intake is not clear.

If the lung condition that leads to the respiratory acidosis is corrected, the gas exchanges in the lung may be normalized before the adaptation mechanism in the kidney has been interrupted. Consequently, the kidney will continue to excrete hydrogen ions even

* The major buffer of the extracellular fluid is the carbonate-bicarbonate system (H_2CO_3/Na HCO_3), and its buffering capacity cannot absorb the large changes in $[H^+]$ observed in acute hypercapnia.

** Perfused lungs.

though the blood CO_2 levels are normal. As a result, bicarbonate is generated, plasma Cl concentration remains low, and hypokalemia occurs. In other words, a condition resembling uncompensated metabolic alkalosis follows. In such cases, the urinary hydrogen ion excretion can be reduced by administering KCl. Thus, in some cases of recovery from respiratory acidosis, abolition of the renal compensatory mechanism is not in phase with the restoration of the gas exchanges and the decline in the partial pressure of CO_2 in the blood.

Respiratory Alkalosis

In cases of hyperventilation of the lung, respiratory alkalosis results from a decrease in the level of CO_2 in the blood. Hypoxemia leads to a transient accumulation of lactic acid. If the conditions responsible for alkalosis are maintained for a prolonged period, the kidney reacts to the transient acidosis by decreasing the excretion of hydrogen ions and increasing that of HCO_3. However, studies in an individual exposed to high altitudes suggest that this renal adaptation mechanism is very slow.

Edema

Edema results from the abnormal accumulation of fluid in the interstitial tissue. Edema may be localized, resulting from local changes in vascular permeability or hydrostatic pressure. Systemic edema is associated with changes in protein or electrolyte content of the body fluids. (The causes of allergic and inflammatory edema are discussed in separate sections.) Obstruction of the lumina of the veins or lymphatics induces changes in capillary hydrostatic pressure or prevents lymphatic drainage [52].

Pulmonary Edema

Pulmonary edema is caused by the transfer of fluid from the lumina of the alveolar capillaries into the alveolar cavities. Theoretically, such a transfer results from a change in capillary permeability, an increase in the intracapillary hydrostatic pressure, a decrease of the osmotic forces that maintain the fluid inside the capillary lumen, or suction of the fluid from the capillaries into the alveoli as a result of a high negative pressure [53]. Although each of these mechanisms can produce pulmonary edema, individual cases of pulmonary edema often result from a combination of pathogenetic mechanisms.

Changes in capillary permeability may be caused by direct damage to the capillary wall, as when toxins are inhaled (phosgene, ozone, etc.), or as a result of the release of vasoactive substances, such as kinin or histamine (see section on inflammation). The cause of the release of the vasoactive substances may be local, as in a pulmonary infection, or systemic, such as the release of a bacterial toxin, an infection, or an immune disease. Increased capillary hydrostatic pressure may be caused by cardiogenic left ventricular failure, mitral stenosis, congenital or acquired pulmonary venous occlusion, and the administration of excessive amounts of fluids. Avoidance of increased fluid overload is of particular significance when the osmotic pressure of plasma is reduced because of low levels of plasma proteins.

Lymphatic obstruction, resulting from lymphangitis as in silicosis or carcinomatosis, may be the source of pulmonary edema, especially if the patient has elevated left side pressure. Removing large amounts of fluid from the pleural cavity may generate such strong negative pressures that the collapsed lung will suddenly respond, causing fluid to be extruded from the capillaries into the alveolar walls.

Pulmonary edema is also associated with central nervous system lesions and with exposure to high altitudes; the pathogenesis of this edema is not known. Pathologists know that cerebral damage is often accompanied by pulmonary edema. Pulmonary edema can be produced experimentally by occluding the carotid arteries, destroying the hypothalamus, or injecting fibrinogen-fibrin in the cysterna. In rats, vagotomy interferes with the development of pulmonary edema of central origin. In contrast, destruction of the vagal nucleus and vagotomy induce pulmonary edema in guinea pigs. These puzzling observations have led to the theory that the nervous system affects the pathogenesis of pulmonary edema.

Borison and Kovack placed guinea pigs in parabiosis and produced pulmonary edema by vagotomy in one of the animals. As a result of this experiment, pulmonary edema developed in both animals, suggesting that a humoral factor may also be involved in the pathogenesis of pulmonary edema.

Anoxemia may induce pulmonary edema. In fact, pulmonary edema occurs after prolonged exercise among people who live at high altitudes. It is also observed among those who return to high altitudes after a 1- or 2-week stay at sea level. In such cases, it has been suggested that pulmonary edema results from vasoconstriction of the venules distal to the capillaries.

Even the pathogenesis of pulmonary edema observed in heart disease has been questioned. Indeed, the old hemodynamic theory, which proposes that edema results from left ventricular insufficiency, is no longer in vogue. It was believed that the decreased output in the left side of the heart and the systemic circulation was compensated for by an increased output of the right ventricle and increased hydrostatic pressure in the lung. The theory was supported by experiments in which the aorta was completely occluded or the left ventricle crushed. However, these extreme procedures can hardly be equated with even the worst pathological conditions. Moreover, pulmonary edema often occurs in patients with intact left ventricles.

Table 9-4. Pathogenesis of edema

Local:

Allergic reaction

Inflammatory reaction

Venous obstruction:
- Thrombosis
- Varicose veins
- Compression by tumor or aneurysm

Lymphatic obstruction:
- Phlegmasia alba dolens
- Carcinomatous spread
- Infection by microfilaria *Wuchereria bancrofti*
- Congenital malformation of lymph vessels (Milroy's disease)

Pulmonary edema:
- Anoxemia
- Inhalation of toxins
- Inflammation
- Chronic heart failure
- Acute obstruction of pulmonary veins
- Cerebral trauma

Systemic:

Water retention:
- Renal insufficiency
- Congestive heart failure
- Hyperproduction of ADH

Sodium retention:
- Overconsumption
- Reduced filtration
  - Acute glomerulonephritis
  - Congestive heart failure
- Increased resorption
  - Hyperaldosteronemia

Hypoproteinemia:
- Idiopathic cyclic
- Loss:
  - Nephrotic syndrome
- Deficiency:
  - Starvation
- Deficient production:
  - Liver disease (cirrhosis)

In conclusion, the clinical and experimental observations on pulmonary edema suggest that its pathogenesis is rather complex and may involve humoral, hemodynamic, and neurogenic factors (see Table 9-4). How these factors interact remains to be established.

The gross appearance of the lungs in pulmonary edema is typical. At the opening of the thoracic cavity, the lungs bulge. They are swollen, pale, and loaded with fluid. Their consistency is firm, and the organs are more than twice their normal weight. On sectioning, clear or bloody fluid mixed with air oozes out of the alveoli.

Systemic Edema

Systemic edema may be the consequence of water retention, salt retention, or hypoproteinemia. When sodium is retained, the osmotic pressure of the body fluids increases; as a result, the secretion of the antidiuretic hormone is stimulated, and water is retained until iso-osmolarity is reestablished and water accumulates in the extracellular fluid. A drop in the plasma protein levels reduces the blood's osmotic pressure, and iso-osmolarity is established by water migration from the vascular compartment to the intracellular compartment, where the protein concentration is unaltered. Except for a moderate hypokalemia of unknown origin, the electrolyte content of the blood is not altered in edema resulting from hypoproteinemia.

Overhydration does not occur with normally functioning kidneys. It is rare even with chronic kidney disease. In acute kidney disease, overhydration may be preceded by hypernatremia. This is suggested by the fact that in the early stages of the edema, the plasma sodium levels are low.

A damaged kidney cannot eliminate all the water that is ingested and produced through metabolism. Water retention has a triple effect—it turns off ADH secretion, reduces the osmolarity of the plasma and the extracellular fluid, and increases the plasma volume. The increase in plasma volume triggers the volume receptors and stimulates diuresis. If water diuresis does not compensate for the water intake and production, the hypotonicity is corrected by sodium retention, and the vicious circle is closed.

In congestive heart failure, the renal venous pressure increases. As a result, glomerular filtration is reduced. Water is retained, at least in the beginning of the congestive failure. Serum sodium is moderately decreased, and sometimes potassium is markedly decreased. Although overhydration, salt retention, and hypoproteinemia may trigger edema, often as the disease progresses more than one of these factors participate in the development of edema. Congestive heart failure may lead to overhydration, hyponatremia, and liver damage with hypoproteinemia. Nephrotic patients are sensitive to overhydration. In nephrosis and liver cirrhosis, the loss of protein leads to hypoproteinemia. Secondary aldosteronism develops by a mechanism that is not clear.

Now that we have provided a panoramic outline of the pathogenesis of edema, it may be worthwhile to discuss the mechanism of production of edema in cardiac failure and in cases of cyclic idiopathic edema.

Salt and Water Retention in Cardiac Failure

The hemodynamic failure associated with chronic heart disease leads to several homeostatic distortions that act in concert to produce edema. They are reduced glomerular filtration, increased aldosterone levels, and disturbance of the adaption to salt retention [54–57].

When the pumping action of the myocardium fails, renal blood flow and the glomerular filtration rates are reduced. The reduced GFR results in increased water and salt retention and in stimulation of aldosterone via the renin-angiotensin system.

Aldosterone secretion is increased in cardiac failure. Feces of cardiac patients are abnormally low in sodium and high in potassium. This finding is associated with plasma levels of aldosterone that are 25–50 times normal (4–5 mg). That the increased aldosterone levels contribute to the edema is suggested by the fact that the severity of the failure correlates with the plasma levels of aldosterone. The causes of this hyperaldosteronemia are probably varied. Among them must be included the stimulation of secretion through activation of the juxtaglomerular system, which is induced to secrete renin. Renin attacks the angiotensinogen of the plasma, yielding angiotensin I, which in the lung is converted to angiotensin II. The angiotensin II reaches the adrenal, where it stimulates aldosterone secretion. (This function is independent of increased peripheral arteriolar resistance that is elicited by angiotensin.) The increased secretion is probably associated with decreased aldosterone catabolism because of reduced blood flow in the liver, the organ in which most if not all aldosterone is metabolized.

When a normal human ingests an average amount of sodium for several days and aldosterone is administered, reduced sodium excretion may be expected. Surprisingly, after a few days (usually 5) of aldosterone administration, sodium excretion returns to normal. Plasma sodium levels increase (from 142 mEq/liter to 152 mEq/liter) and potassium excretion increases while plasma potassium levels decrease. Thus, the normal individual "escapes" the consequence of sustained high plasma levels of aldosterone. This escape mechanism is lost in cardiac patients, who maintain normal plasma and urinary levels of potassium, but retain sodium. In that respect cardiac patients resemble dogs in whom a large arteriovenous fistula has been installed. These animals are not in cardiac failure, but after the sustained administration of large doses of aldosterone, the electrolyte content of plasma and urine resembles that of cardiac patients because the dogs lose their ability to "escape" sustained hyperaldosteronemia.

These results are interpreted by assuming the existence of an extra-adrenal sodium-retaining factor that stimulates sodium reabsorption or prevents sodium excretion. In any event, abnormal levels of this factor are believed to contribute to the pathogenesis of edema in cardiac failure.

Cyclic Idiopathic Edema

Some women become edematous periodically and irregularly. The edema involves the face, hands, and feet and is a nonpitting type. Usually it is only loosely connected with menstrual periods, but seems to be associated with activity. The more active the woman, the more extensive the edema. Puffiness of the face and sometimes interference with the flexion of fingers occur. In fact, some women may gain 2–4 pounds a day and might find it necessary to change dress size during the day. The patient usually drinks large quantities of water, but urinates little despite a frequent sense of bladder fullness [58].

A number of pathogenetic mechanisms have been proposed, none of which is satisfactory. Although there is a vague relationship between edema and menstruation, the levels of estrogens and progesterone are never increased in these patients. Aldosterone is often elevated, but this is believed to be a consequence of the edema rather than its cause. There is no clear evidence of a failure in the escape mechanism. Perhaps the most plausible pathogenetic interpretation is that the disease results from a failure of capillary permeability.

Dehydration

Pure Water Loss or Deprivation

In spite of the negative water balance, the victim continues to excrete 300 ml of water (basal diuresis) in the urine, and to lose 400–800 ml of water through respiration and sweat. Only 300 ml of water is produced metabolically; therefore, once a water balance is negative, the drop in total water content of the body falls rapidly until death ensues.

The reduced water content leads to hyperosmolarity, which affects renal function two ways. It stimulates water resorption or ADH secretion and increases electrolyte excretion. Depending upon the water intake, this latter adaptation may maintain homeostasis, but body water content continues to drop. In addition to its normal load of electrolytes, the kidney is forced to excrete an amount of electrolyte sufficient to restore normal osmolarity. The increased rate of electrolyte excretion has consequences that are important for the progress and therapy of the dehydration syndrome. First, the amount of electrolytes excreted by the kidney exceeds the concentration power, and, despite dehydration, more water is lost in the urine to dilute the electrolytes. Second, the total concentration of electrolytes in the blood drops below normal values. Consequently, if plain water is administered to dehydration victims, hyposmolarity ensues. The hyposmolarity blocks ADH secretion and, as a result, water diuresis follows and the patient loses the benefit

Table 9-5. Pathogenesis of dehydration

Water deprivation	Castaways, desert walkers Senile patients Esophageal stricture
Loss of water:	
In sweat	Excessive heat
In urine	Diabetes insipidus Glucosuria (osmotic diuresis)
In feces	Diarrhea
Loss of salt:	
Excessive salt intake	Polyuric states
Excessive urinary salt excretion	Nephritis Addison's disease
Administration of diuretics	
Loss of body fluids (external or internal):	
Hemorrhage	
Burns	
Ascites	
Fistulas	

of the plain water therapy. In cases of dehydration, it is wise to administer a saline or Ringer's solution, at least in the beginning of therapy. Table 9-5 lists the principal causes of dehydration.

Urea Metabolism and Uremia

Amino Acid and Protein Metabolism in Liver

In mammals, urea is formed in the liver from the waste products of the organism's nitrogenous compounds. Most of the urea is excreted through the kidney. In renal insufficiency, uremia frequently develops.

Urea is synthesized from the ammonia derived from protein, polypeptide, and amino acid catabolism. Consequently, the amount of urea in the blood is partly determined by protein and amino acid metabolism. Therefore, in considering urea production, the source of the ammonia and the biosynthesis of urea must be discussed. The ammonia is derived mainly from amino acid breakdown. In liver, the amino acid pool is derived from the breakdown products of ingested and endogenous proteins and from amino acids synthesized *de novo*.

Many aspects of the usage of dietary protein were discussed in the chapter on nutrition. After the protein is digested by the proteolytic enzymes of the gastrointestinal tract, the free amino acids are transported inside the cells of the intestinal mucosa and from there released into the portal circulation. The portal blood carries the free amino acids in close contact with the hepatic cells. The amino acids are transported through the cell membrane inside the liver cell.

Amino Acid Transport

The molecular mechanism of amino acid transport through cell membranes is not clear; it seems to require energy and may be hormonally controlled. To study amino acid transport in the intestines, investigators introduce the amino acid solution inside an intestinal loop and determine the rate of the appearance of the amino acid in the solution bathing the serosal surface. Some investigators study amino acid transport using everted intestinal loops with the serosa inside.

Intestinal transport studies with a variety of amino acids have demonstrated that amino acids (except dicarboxylic acid) are transported across the intestinal wall against a concentration gradient, and chemical energy is necessary for the transport. Adding cyanide or dinitrophenol to the incubation mixture inhibits amino acid transport. The active transport is specific for the L isomer; the D isomer may be transported across the mucosa through passive diffusion. Although different carrier systems seem to exist, a given carrier system probably is shared by several amino acids. This explains the competition between amino acids for transport, and the simultaneous malabsorption of several amino acids in inborn errors of metabolism that involve the carrier system. For example, ornithine and lysine are poorly absorbed in cystinuria, and all neutral amino acids except proline and hydroxyproline are poorly absorbed in Hartnup disease.

Although pyridoxal phosphate is required for amino acid transport, its exact role is not known. Several investigators have attempted to define the molecular requirement for amino acid transport. Some of the most detailed studies were done with histidine. The studies suggest that the transport mechanism reacts with at least three groupings on the asymmetric carbon—the amino group, the carboxyl group, and the hydrogen of the carbon atom.

Sodium activates amino acid uptake; despite much research, the mechanism of this activation remains unknown. At least two mechanisms have been proposed; sodium stimulates the production of energy needed for transport in an unspecific fashion. There is coupled binding of sodium and amino acid at the carrier site, and the movement of the sodium down its concentration gradient provides the energy needed for transport.

The Liver

Once inside the liver cell, the dietary amino acid becomes part of the cell's amino acid pool. The breakdown of endogenous proteins in liver and other organs also contributes to that pool.

The liver is a large organ located in the peritoneal cavity immediately underneath the left and central portions of the diaphragm. It is composed of at least three functional systems—the hepatic cell, the reticuloendothelial cell, and the biliary tract—a supporting stroma, and a complex vasculature. Two types of efferent circulation in the liver—the portal and arterial systems—are formed by branches of the hepatic artery. The functional, vascular, and structural components of the liver are histologically integrated to form units called the hepatic lobule, which is centered around a branch of the portal vein. The large polygonal hepatic cells form cords that radiate perpendicularly to the long axis of the vein. The cords are separated by a capillary network called the sinusoids, which are lined with small, star-shaped cells. On section, these cells often appear as small, triangularly shaped cells with a fusiform nucleus often located excentrically. The Kupffer's cells exhibit typical reticuloendothelial function. At the periphery of the lobules are Kiernan's spaces, triangular spaces composed of connective tissue and containing a classical triad: an artery, a vein, and a biliary canal.

On gross examination of a liver section, one recognizes red dots corresponding to the central vein, surrounded by brown polygonal zones made of hepatic cells and alternating with gray areas of the portal spaces. This one-dimensional description of the hepatic lobule does not adequately describe the functional unit, and anatomists and physiologists have coordinated their efforts to reconstitute the functional unit in three dimensions. Refer to specialized texts for a review of the functional anatomy of the liver. It suffices to point out that serious distortion of the liver architecture impairs liver function and also may modify vascular pressure and lead to portal hypertension (see section on lithiasis).

The metabolic functions of hepatic cells are numerous. In addition to containing a wide range of bioenergetic pathways (glycolysis, hexose monophosphate shunt, glucuronic pathway, Krebs cycle, oxidative phosphorylation, etc.), biosynthetic pathways (fatty acid, protein, lipid, amino acid, vitamins, etc.), and storage pathways (glycogen and lipid synthesis), the liver exhibits a number of special functions: gluconeogenesis, urea synthesis, synthesis of a large number of coagulating factors and plasma proteins, detoxification, and others.

Because it is such a vital organ, the liver has unusual reserve capacities. Consequently, normal function is maintained even in the face of extensive damage. For example, when large portions of the liver are replaced by invading cells, the liver functions normally. Because of the liver's reserve capacity, there can be no extremely sensitive tests for liver function, and when changes in liver function are detected, severe and sometimes irreversible damage has already occurred. The liver tests include evaluation of secretion, evaluation of specific protein synthesis, and measurements of enzymes released in the blood.

Protein Turnover in Liver

The turnover of endogenous proteins is a problem of such importance in body chemistry that it is worthwhile to consider it here. For the sake of clarity, this brief discussion of protein turnover is restricted to nonregenerating liver. In nonregenerating liver, protein turnover is essentially restricted to cytoplasmic proteins. Abundant evidence indicates that the constitutive and catalytic proteins of liver are constantly renewed. When the specific activity of liver proteins is measured at various times after injecting a precursor ([^{15}N]glycine, [^{36}S]methionine, or [^{14}C]leucine), the plot of the specific activity versus time can be divided into two theoretical phases. (Practically all the data are complicated by the reusage of the labeled amino acids freed during protein catabolism.) When the labeled precursor is in the blood, the specific activity rises steadily until a peak or a plateau is reached. This indicates that new proteins are synthesized using the labeled compound as the precursor.

In the second phase, the specific activity of the protein decreases as a result of dilution of the labeled protein with newly synthesized unlabeled proteins. Studies in which [^{14}C]leucine was injected in mice and the specific activity of the liver protein was compared with that of the free leucine in the tissues yielded turnover rates ranging form 6–124 hours. These studies suggested that the turnover rate varies considerably depending upon the individual protein.

During these radioisotope studies, the total protein content of liver remains unchanged; therefore, the stock of newly synthesized protein must be balanced by the catabolism or excretion of an equivalent amount of protein. Remarkable experiments of Shoenheimer and his associates [59] using ^{15}N-labeled amino acid as precursor of the total protein contained in the organism have clearly established that proteins are continuously broken down, and that the endogenous amino acid pool is freely mixed with the exogenous amino acid pool.

Starvation is a good example of discrete turnover taking place. Glucose is the major source of aerobic and anaerobic energy. In starvation the major source of glucose is glycogen. Once the glycogen is exhausted, the proteins provide the only source of gluconeogenic substrate. Although lactate and pyruvate, which are glucose-derived substrates, can be reconverted into glucose by the reversal of the glycolytic pathway, these substrates are not very effective in gluconeogenesis because of their scarcity. Moreover, the enzymes capable of converting even chain fatty acids are missing in mammals while odd chain fatty acids, which could be used in gluconeogenesis, constitute only 5% of the fatty acid contingent. Thus proteins from muscle are the major source of glucose in prolonged starvation. The proteins are broken down in a discrete fashion that remains unknown, but certainly docs not involve the release of lysosomal enzymes. The free amino acids are reutilized for protein synthesis except for alanine, which constitutes only 5% of the muscle amino acid component, and to a small extent, glutamine. Both these amino acids are transported from muscle to liver where they are used in the process of gluconeogenesis, and one speaks of a glucose alanine cycle, involving the conversion of alanine into pyruvate through the catalytic action of the proper transaminase [79].

Some physiological or pathological stimuli induce the liver cell to synthesize or break down some protein selectively. Even during starvation the content of all liver protein does not drop simultaneously. For example, while the activities of catalase, xanthine oxidase, alkaline phosphatase, and acid phosphatase drop at various rates as starvation progresses, that of glucose-6-phosphatase increases. Hydrocortisone and tryptophan administration induces a massive increase in tryptophan peroxidase activity. In either case, at least part of the increase in enzyme activity results from *de novo* enzyme synthesis. If tryptophan administration is interrupted, the activity of the peroxidase returns to normal. During the induction, turnover rates of other proteins do not change.

A helpful distinction is between discrete turnover of one or a few selective proteins and bulk turnover of large segments of the cytoplasm.

Schimke studied the fate of arginase in liver when the protein content of a diet is shifted from low to high and *vice versa* by following the incorporation of [^{3}H]lysine in the enzyme. In changes from low- to high-protein diets, enzyme synthesis accelerates and the rate of breakdown decreases. The reverse occurs in changes from high- to low-protein diets.

After the induction of endoplasmic reticulum membranes and enzymes, enzymes and membranes disappear when the inductive stimulus is withdrawn. The enzyme disappears faster than the membrane, but focal cytoplasmic degradation does not develop, suggesting that the lysosomal hydrolases are not likely to be involved in discrete turnover of the endoplasmic reticulum [80].

Svoboda and his associates [81] investigated the disappearance of microbodies after their multiplication under the influence of various chemical stimuli and concluded that engulfment in autodigestive vacuoles could only, in part, explain their dissolution. In conclusion, there is ample evidence that the liver catabolizes all or most of its protein, and that under the influence of special stimuli it catabolizes some proteins selectively.

The mechanism by which proteins are catabolized in liver is not elucidated. Like most other organs, the liver contains an impressive arsenal of proteolytic enzymes. Some act optimally at alkaline pH, others at acid pH. The activity of the alkaline proteases is usually recovered in the supernatant fraction after differential centrifugation of the liver homogenate. The activity of the acid protease is associated with all cell fractions, but the highest specific activity is found in the lysosomal pellet. Whether the alkaline or the acid proteases attack the endogenous liver proteins is not known.

The alkaline proteases are active at pH 7 and are not strictly localized. Therefore, they could be mobilized any time and anywhere in the cell at physiological pH to destroy selective proteins. But how do they select the protein to be destroyed? How are they kept harmless between digestions? The acid proteases are found in association with other acid hydrolases in granules that can be separated by differential centrifugation. The origin of these granules is still debated.

It is not known whether the granules are active vesicles that are transported to the site of excretion to digest the protein (primary lysosomes), or if they are the product of focal cytoplasmic degradation. The granules have been observed in viable cells, and therefore they probably represent areas where the cytoplasm turns over. Even if it were established that primary lysosomes exist in liver, how do they select a specific protein for digestion? Or do they release their enzyme contingent to destroy an entire section of cytoplasm simply because it contains an undesirable amount of a given protein? Much research is needed before the mechanism of intracellular protein digestion can be described.

Some Aspects of Amino Acid Metabolism

Part of the amino acid pool of the liver is derived from *de novo* synthesis. Nutritional experiments and studies of the incorporation of labeled precursors ([^{14}C]sucrose, CO_2C_{14}, [^{14}C]glucose) have permitted the grouping of the 21 amino acids in the body into three major categories: essential, nonessential, and quasi-essential amino acids. Humans cannot synthesize any of the eight essential amino acids. Humans make all nonessential amino acids from simple carbon and nitrogen sources. The quasi-essential amino acids are partially synthesized using the skeleton (phenylalanine from tyrosine) or a fraction of an essential amino acid (cysteine and cystine from methionine) as a precursor. The biosynthetic pathways of the individual amino acids have been reviewed in the sections on vitamins and inborn errors of metabolism.

Amino acids can be excreted in the urine or used in metabolism. Some amino acids are excreted in the urine free of the conjugates, others are conjugated. Among the conjugates are hippurates, taurocholate, glycocholate, acetylated derivatives, and phenylacetylglutamine.

Hippuric Acid Synthesis

When benzoate is administered to mammals, it is excreted as benzoylglycine, a normal constituent of urine in man.

The conjugation of glycine with benzoic acid to yield hippurate requires energy and in liver slices is inhibited by cyanide. The reaction occurs in three steps: (1) the acylation of benzoic acid in the presence of ATP to yield benzoyl AMP and pyrophosphate; (2) the conversion of benzoyl AMP to the coenzyme A derivative; and (3) in the presence of a mitochondrial enzyme, the exchange of the coenzyme A moiety of benzoyl AMP with glycine to yield hippurate and coenzyme A. Glycine-N-acylase is the enzyme that catalyzes the transfer. It has been purified from beef liver mitochondria.

Benzoate + ATP → Benzoyl AMP + PP
Benzoyl AMP + CoA → Benzoyl CoA + AMP
Benzoyl CoA + Glycine → Hippuric Acid

Cholate Synthesis

The bile acid, cholate, which is formed in liver, is converted to cholyl CoA in the presence of a microsomal enzyme, ATP and CoA. A supernatant enzyme catalyzes the coupling of taurine or glycine with cholyl CoA, and CoA is released in the reaction.

Amino Acid Acetylation

Acetyl coenzyme A is the acyl donor in the acetylation of N-acetylglutamate and N-acetylaspartate.

L-aspartate + Acetyl CoA → N-acetyl-L-Aspartate
Aspartic acid N-acetylase + CoASH

The acetylase has been partially purified from brain; it exhibits a high degree of specificity for both acetyl CoA and aspartic acid.

Phenylacetate, after conversion to the coenzyme A (in the presence of ATP) derivative (phenylacetyl CoA), is coupled with glutamine to yield phenylacetylglutamine. The product of the reaction is found in normal urine. The enzyme catalyzing the reaction is present in liver and kidney. The reaction is assumed to develop in three steps. First, the enzyme reacts with phenylacetic acid and ATP to yield enzyme-bound phenylacetyl adenylate and pyrophosphate. Second, the enzyme-bound phenylacetyl adenylate reacts with reduced coenzyme A to yield free phenylacetyl-S-CoA, AMP, and the free enzyme. Third, phenylacetyl CoA reacts with glutamine and a phenylacetylglutamine synthetase to yield the final product.

The amino acids found in the liver amino acid pool can be degraded or used for biosynthesis (protein, coenzymes, glutathione, glutamine, purines, and pyrimidines).

Glycine, glutamine, and aspartic acid are involved in purine nucleotide biosynthesis. Aspartic acid and glycine play a role in the biosynthesis of pyrimidine nucleotides. Protein biosynthesis has already been discussed. Glutathione is a tripeptide composed of glutamine, cysteine, and glycine; its biosynthesis involves the formation of two peptide bonds. The tripeptide has been synthesized *in vitro* in pigeon liver homogenate. The reaction requires ATP (2 moles per mole of glutathione) and two enzymes.

$$\text{L-glutamic acid} + \text{L-cysteine} \xrightarrow{\text{Glutamylcysteine synthetase}} \text{L-glutamylcysteine} + \text{ADP} + H_3PO_4$$

$$\text{L-glutamylcysteine} \underset{\text{ATP} \curvearrowright \text{ADP}}{\overset{\text{Mg} \quad \text{ATP} \curvearrowright \text{ADP}}{\xrightleftharpoons[\quad]{\text{Tripeptide synthetase}}}} \text{Glutathione}$$

Glutamine is synthesized in pigeon liver homogenates from glutamic acid and ammonia. The reaction requires ATP and a specific enzyme, glutamine synthetase. ADP is formed and inorganic phosphate is released in the reaction. The mechanism of the reaction is similar to that involved in glutathione biosynthesis.

Glutamine synthetase was found in the microsomes and the supernatant of rat brain and liver. The microsomal enzyme is weakly bound to the endoplasmic reticulum. The sheep enzyme has been purified and has a molecular weight of 450,000. Glutamine is bound to the enzyme in at least four bindings sites. The postulated mechanism of the enzymic reaction involves: (1)

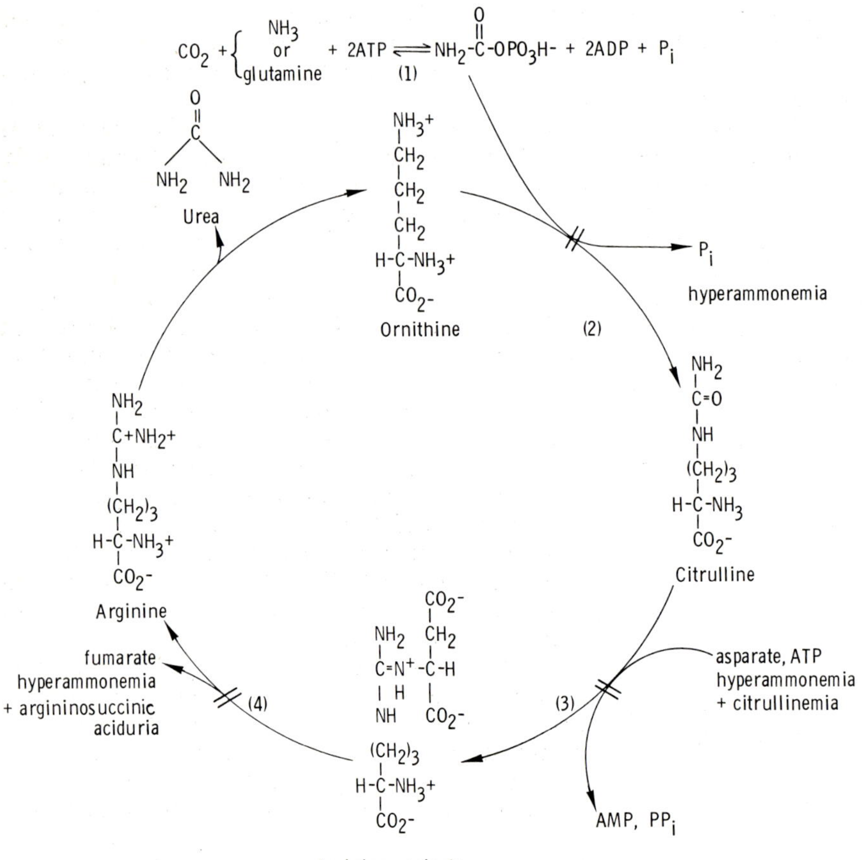

Fig. 9-14. Inborn errors of metabolism in urea cycle (see p. 591)

binding of the enzyme and ATP; (2) binding of the enzyme and glutamate; (3) phosphorylation of glutamate at the enzyme surface to yield glutamyl phosphate; and (4) the reaction of glutamyl phosphate with ammonia and glutamine whereby inorganic phosphate and ADP are released.

In considering amino acid catabolism, one must distinguish the catabolism of the carbon chain from that of the nitrogen moiety. The breakdown of the carbon chain of the amino acids yields carbon units that can be used in carbohydrate metabolism, acetate metabolism, or the metabolism of single carbon units. The fate of the carbon units of the individual amino acids has been discussed in other sections of this book, and only a synopsis of the results will be presented here. The carbon skeletons of isoleucine, phenylalanine, threonine, tryptophan, valine, histidine, alanine, arginine, aspartic acid, glycine, proline, glutamic acid, and hydroxyproline are ultimately converted to pyruvic acid.

The product of amino acid catabolism may be used in the Krebs cycle to generate ATP, or it may be converted to glucose and glycogen through the reversal of the Krebs cycle and the glycolytic pathway. The gluconeogenetic properties of amino acids should not be underestimated by the physician, because of the sparing effect on proteins created by adding carbohydrates to the diet. (The control of gluconeogenesis will be discussed in Chap. 6.)

The biochemical mechanism by which carbohydrates spare proteins is not known. Carbohydrate breakdown yields ATP, NADH, and amino acid precursors—all compounds needed for amino acid or protein synthesis, and each of them could alone be responsible for the sparing effect. To assume that in the presence of excess carbohydrates the cell selects carbohydrates rather than protein as an energy source is obviously an oversimplified interpretation of the sparing effect. Yet, glucose and fructose administration reverses urea formation and glycine catabolism.

The carbon skeletons of isoleucine, leucine, phenylalanine, and alanine can yield acetoacetate, and these amino acids may therefore be ketogenic.

Methionine, tryptophan, glycine, and serine yield carbon-l units that may be used in transmethylation or in reactions involving folic acid and its derivatives. Carbon-l metabolism is discussed in the section on folic acid.

Transmethylation

The reactions involved in transmethylation are discussed in Chapters 3 and 10. Only the overall equation is shown here.

$$\text{L-methionine} + \text{ATP} \rightarrow \text{S-adenosylmethionine} + P_i + PP$$
$$\downarrow$$
$$\text{Homocysteine} + \text{Adenosine}$$

Urea Synthesis

One of the terminal products of nitrogen metabolism is NH_2. Ammonia is a rather reactive ion that attacks other ions and molecules unless it is rapidly eliminated or modified. Most animals convert ammonia into less toxic molecules such as urea and uric acid.

In invertebrates, half of the excretory nitrogen is in the form of ammonia. The other half is excreted in the form of a number nitrogen-containing organic molecules: urea (0–20%), uric acid (0–50%), amino acid or creatinine (3–30%). Insects differ from other invertebrates in that a large amount of their excretory nitrogen is in the form of uric acid (50%–80%). Fish can excrete ammonia, whereas the teleosts excrete most nitrogen in the form of ammonia. Elasmobranchii excrete only small amounts of their metabolized nitrogen in the form of ammonia, but eliminate most in the form of uric acid.

The terrestrial vertebrates are uricotelic, excreting mainly uric acid, or ureotelic, excreting mainly urea. With the exception of reptiles, birds, and the dalmatian dog, most terrestrial vertebrates are ureotelic.

In addition to being a small neutral molecule, urea has a considerable advantage over uric acid as a terminal product of nitrogen metabolism in that it is much more soluble. As has been the case for many other pathways in intermediary metabolism, the understanding of ureas biosynthesis began with the availability of isotopic nitrogen and carbon. The administration of [^{15}N]ammonium salt and ^{14}C-labeled bicarbonate salt demonstrated that urea nitrogen derives from ammonia and urea carbon from respiratory CO_2.

Krebs and Henseleit [60] performed key experiments in 1932 by studying the deamination of amino acids, all in *in vitro* systems. While substituting one amino acid after another for ammonia for the production of urea, they discovered that arginine stimulated urea formation.

It had been known since 1904 that arginase splits arginine to yield ornithine and urea, but Krebs and Henseleit's results suggested that the stimulatory effect of arginine on urea formation could not be explained simply by the conversion of substrate to product. The result indicated that the arginine or a derivative was constantly recycled, and thus was made available again and again for the biosynthesis of urea. Thus, in these experiments careful quantitative analysis of the concentrations of arginine and ornithine showed that the sum of these concentrations did not decrease; in contrast, the ammonia disappeared at a rate equivalent to that of urea formation.

Additional *in vitro* experiments yielded the clues needed to identify the intermediate of the "urea cycle" when it was found that ornithine and citrulline could effectively replace arginine in stimulating urea formation. Armed with this evidence, Krebs proposed a mechanism of urea formation in which arginase catalyzes the hydrolytic cleavage of one molecule of urea per molecule of argine to yield one molecule of ornithine. In the presence of ammonia and CO_2, ornithine is converted to citrulline, and one molecule of water is formed in the process in the presence of another molecule of ammonia. The citrulline restores arginine and another molecule of water. The arginine is thus again made available for hydrolysis to yield urea and ornithine. In the process, one molecule of urea and one molecule of water are generated; two NH_5 and one CO_2 molecules are consumed. This was essentially the level of our understanding of the urea cycle in the beginning of 1950.

As knowledge accumulated, the original simple scheme mushroomed into a complex metabolic map that sometimes defies the best of biochemists. The easiest way to understand the sequence of steps involved in the urea cycle is to review the key steps of the formation of citrulline, arginine, and urea one by one.

The cycle starts with carbamyl phosphate formation (this reaction was discussed in the section on pyrimidine biosynthesis). Carbamyl phosphate synthetase catalyzes the condensation of "active CO_2" with NH_4^+ to yield carbamyl phosphate, a precursor of pyrimidines and urea.

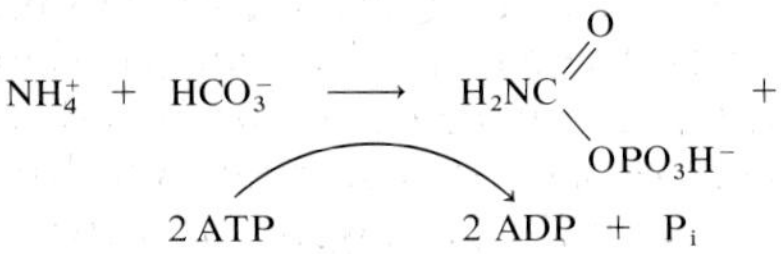

$$NH_4^+ + HCO_3^- \longrightarrow H_2NC(=O)OPO_3H^- + H^+$$

2 ATP → 2 ADP + P_i

Ornithine carbamyl transferase catalyzes the irreversible transfer of one molecule of carbamyl phosphate on the terminal amino group of a molecule of ornithine to yield citrulline and P_i.

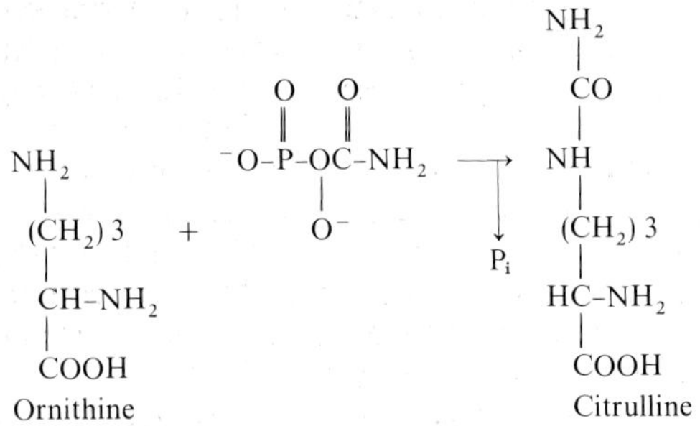

$$NH_2\text{-}(CH_2)_3\text{-}CH(NH_2)\text{-}COOH + {}^-O\text{-}P(=O)(O^-)\text{-}OC(=O)\text{-}NH_2 \longrightarrow NH_2\text{-}CO\text{-}NH\text{-}(CH_2)_3\text{-}HC(NH_2)\text{-}COOH + P_i$$

Ornithine Citrulline

Two enzymes are needed to convert citrulline to arginine—argininosuccinate synthetase and arginine synthetase. The first enzyme catalyzes the condensation of citrulline and L-aspartic acid in the presence of ATP and Mg^{++}. A molecule of water is lost in the reaction, and a C=N bond is formed between the -CO-carbon of citrulline and the NH_2 nitrogen of aspartic acid.

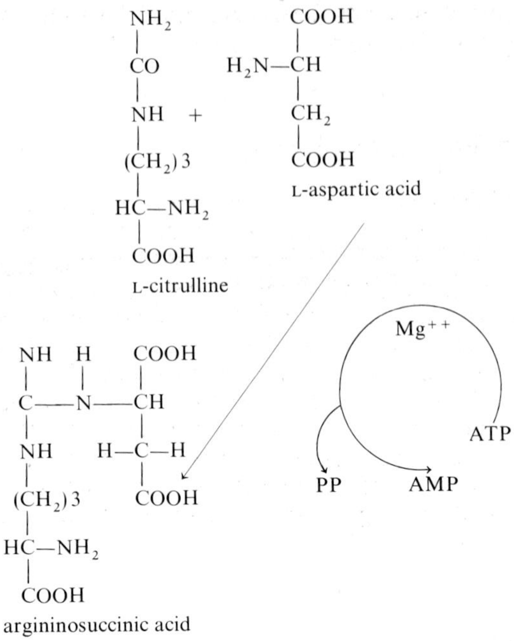

$$NH_2\text{-}CO\text{-}NH\text{-}(CH_2)_3\text{-}HC(NH_2)\text{-}COOH + H_2N\text{-}CH(COOH)\text{-}CH_2\text{-}COOH$$

L-citrulline L-aspartic acid

Mg++, ATP → PP, AMP

$$\downarrow$$

$$C(=NH)(NH\text{-}(CH_2)_3\text{-}HC(NH_2)\text{-}COOH)\text{-}NH\text{-}CH(COOH)\text{-}CH_2\text{-}COOH$$

argininosuccinic acid

Arginine synthetase yields one molecule of fumaric acid and one molecule of L-arginine from one molecule of argininosuccinic acid.

$$\text{argininosuccinic acid} \longrightarrow HOOC\text{-}CH{=}CH\text{-}COOH + NH_2\text{-}C(=NH)\text{-}NH\text{-}(CH_2)_3\text{-}HC(NH_2)\text{-}COOH$$

argininosuccinic acid fumaric acid arginine

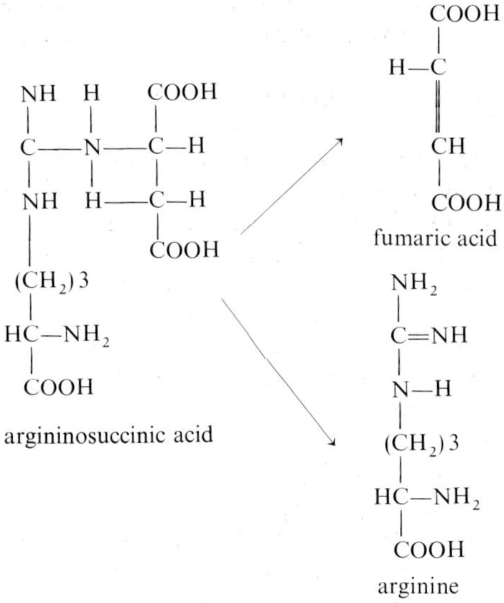

Available evidence suggests that the step catalyzed by arginine synthetase constitutes the rate-limiting step in the overall functioning of the urea cycle. As pointed out above, arginine catalyzes the conversion of arginine to ornithine.

Certain species of tadpoles excrete ammonia during the premetamorphic phase and urea after the onset of metamorphosis. In those animals, two of the urea cycle enzymes are present in small amounts during the premetamorphic phase—carbamyl phosphate synthetase and arginine synthetase. Metamorphosis can be induced by the administration of thyroxine, and the onset of metamorphosis is associated with the biosynthesis of the two missing enzymes [61].

Inborn Errors of the Urea Cycle

Three inborn errors of metabolism involving enzymes of the urea cycle have been demonstrated. These diseases are observed in children less than a year old. They all cause hyperammonemia and have a characteristic neurological and gastrointestinal symptomatology: nausea, vomiting, agitation, convulsions, stupor, coma, and general mental retardation (see Fig. 9-14).

Liver biopsies have revealed three different groups of patients: those with low ornithine carbamyl transferase levels and hyperammonemia, those with low arginosuccinic acid synthetase and citrullinemia, and those with low argininosuccininase and argininosuccinicaciduria [62].

Uremia

Since the biosynthesis of urea takes place in the liver, in severe liver damage one may expect increases in blood ammonia and amino acid levels. Although such increases may affect the pathogenesis of hepatic coma, direct measurement of serum ammonia or amino acid levels of patients with liver disease is of little diagnostic value.

Urea is a waste product, so after it is produced by the liver the kidneys must excrete it. Whenever there is impairment of urine excretion or formation, urea will accumulate in the patient's blood, leading to uremia. The blood nitrogen levels are of considerable importance in medicine, not only because their determination is invaluable for the diagnosis and prognosis of renal insufficiency, but also because they indicate the presence of the uremic syndrome.

In addition to electrolyte imbalances, uremic patients present neuropsychiatric, cardiovascular, hematopoietic, and metabolic symptoms. Although this combination of symptoms may be typical enough for diagnosis, the pathogenesis of uremia is not well understood. A patient developing advanced chronic uremia first experiences anorexia, general wasting, progressive anemia, and dyspnea. As the condition worsens, the patient develops nausea, headaches, hiccups, vomiting, severe diarrhea, muscular twitching, widespread bone demineralization, hemorrhagic diathesis, and paresthesia. It is not certain whether a common demoninator can explain all these symptoms (uremia toxin), or whether each of the symptoms results from a unique mechanism. Much of what is known of the pathogenesis of uremia's symptomatology is speculative or based on contradictory information.

The gastrointestinal symptoms (nausea, vomiting, and possibly diarrhea) are often attributed to the elaboration of ammonia by the gastrointestinal flora. Normally this ammonia would be absorbed in the blood and converted to urea in the liver. A drastically restricted protein intake causes marked improvement in the gastrointestinal symptoms of uremia.

Congestive heart failure often occurs in conjunction with uremia. A number of factors—such as hypertension, arteriosclerosis, and anemia—may contribute to the heart failure, but more specific to the uremic syndrome is the increased intravascular volume and plasma volume that appear in uremics. Whether blood urea accumulation causes metabolic alterations of the heart muscles and results in myocardial malfunction remains to be established.

An interesting complication of uremia is the development of so-called uremic pericarditis. In the early stage of pericarditis, urea precipitates at the surface of the pericardium. This precipitation results in the development of a fibrinous exudate, which covers both the parietal and visceral aspects of the pericardium, giving the typical appearance of the bread and butter pericardium. Despite this extensive fibrinous exudate, constrictive pericarditis is a rare complication of uremic pericarditis.

Anemia, usually normochromic and normocytic in nature, is a frequent complication of chronic renal failure. However, there is no clear-cut correlation between the degree of anemia and the blood levels of urea nitrogen. At least two mechanisms have been proposed to explain the pathogenesis of the anemia: complicating renal failure and a defect in erythropoietin production. Since the kidney produces erythropoietin, severe chronic kidney disease might be expected to reduce the amount of the hormone produced. However, observations made in bilaterally nephrectomized animals and patients reveal that although erythropoietic activity is reduced after bilateral nephrectomy, some erythropoietin continues to be produced; therefore, the kidneys appear nonessential for erythropoietin production. Bone marrow of bilaterally nephrectomized patients responds normally to hypoxia. It has also been proposed that the red cell is modified due to alteration of the membranes or because of a distortion in intracellular metabolism. However, convincing evidence for any pathogenetic mechanism for anemia associated with uremia is not available.

The hemorrhagic diathesis observed in patients with uremia probably results from a defect in platelet function because of platelets' inability to provide thromboplastic substance or platelet factor IV.

As already mentioned, irritability, inability to concentrate, and borderline mental distortions are observed in uremic patients. These symptoms may develop into severe disorientation, delusions, and psychosis as the disease evolves. Dialysis does not always restore the psychiatric balance in uremic patients. Some patients have been observed to deteriorate mentally even when the blood urea content is reduced. Two explanations have been proposed for this discrepancy. One mechanism proposes that urea is removed more rapidly from extra- than intracellular fluid; as a result, water is pumped in the intracellular space, causing cerebral edema. Another mechanism suggests that there is a temporary acid-base disequilibrium after dialysis. Thus, dialysis corrects acidosis and thereby increases the bicarbonate concentration in the blood. However, as the bicarbonate levels increase in the serum, they increase even more rapidly in the cell. The intracellular increase in bicarbonate exceeds the ability of bicarbonate to diffuse through the cell membrane, and respiratory alkalosis results.

A number of neurological symptoms are also observed, including muscle cramps, muscular hyperirritability, decreased nerve conduction, and defective sensory perception. The pathogenesis of these distortions is not known. A number of metabolic changes in the blood take place in uremic patients. They include increased levels of α-amino nitrogen (lysine, tyrosine, tryptophan, glycine, and glutamine), decreased plasma concentration of dietary essential amino acids (valine, isoleucine, tryptophan, methionine, leucine, and phenylalanine), moderate hyperuricemia, distortion of the acid-base balance, and other changes.

Marked hypoglycemia is unusual in uremia, but the glucose tolerance of these patients is often decreased. In fact, experiments of Perkoff and associates [63] have shown that the administration of a urea load to a normal patient also leads to a decrease in glucose tolerance. This effect of uremia is not clear; it could involve an alteration of the mechanism of utilization of glucose through interfering with the glucose metabolic pathways or by modifying the permeability of the peripheral tissue to glucose. An effect on peripheral tissue could result from the anti-insulin effect, or uremia could modify the sensitivity of peripheral tissue to insulin.

Uremia Toxin

Clinicians and investigators have often attempted to explain the complex pathogenesis of the uremic syndrome by the development of a single toxin. A typical experiment suggesting the existence of a uremic toxin is that of John Merrill [64], who dialyzed patients against a vat containing urea at the same concentration as that in the blood. Although the urea concentration had not been changed in the blood, the patient felt much better as a result of the dialysis. It was therefore concluded that factors other than urea were involved in uremia. It is often suspected that the derivatives of guanidine—especially methylguanidine and guaninidoacetic and guaninidosuccinic acids—are at least partly responsible for the uremic symptomatology. The role of the guanidine derivatives has been implicated, particularly in the pathogenesis of the neuropsychiatric and platelet defects [65, 66].

Bricker [82] has taken another view of the pathogenesis of uremia. Rather than postulating that the systemic symptoms result from the elaboration of a toxin, he believes that they are secondary effects resulting from adaptation of individual surviving nephrons. The adaptation is for the purpose of maintaining kidney function.

Normal human kidneys contain approximately 2,000,000 nephrons. The anatomical consequence of chronic renal disease is a progressive destruction of nephrons. As a result there is a quantic loss of function. The smallest number of nephrons needed to sustain life is 22,000.

One can distinguish three stages in progressive chronic renal disease. In the earliest stage, the reserve nephrons are lost without functional alteration; next the nephrons are destroyed with adaptation of remaining healthy nephrons; and finally the remaining intact nephrons are insufficient to sustain life.

If a normal person takes in 7 g of sodium (or 120 mEq per min per day), he excretes 0.5% of the sodium filtered in the glomeruli. Now if a patient has chronic renal disease, the total number of nephrons destroyed precipitates a marked drop in GFR and in Na^+ filtered. Yet, the patient is still capable of maintaining the external sodium balance.

The individual nephrons may have a potential capacity of 2, 10 or even 60 nephrons depending upon the chemicals that they must deal with. This change in the level of excretion is believed to be exerted by specific hormones. Adaptation thus consists in shifting from reabsorption to excretion and hormones are responsible for the shift. The price for adaptation would then be an imbalance of hormonal activity which, in turn, is responsible for at least some of the symptoms of uremia.

An example of this theory is the secondary hyperparathyroidism observed in uremia. With average phosphate intake and normal glomerular filtration, a small portion of phosphate is excreted and most is reabsorbed to maintain the external phosphate balance. In uremia, because of nephron destruction, only a small fraction of the filtered phosphate is reabsorbed and most is excreted. As a result, PTH secretion is increased leading to secondary hyperparathyroidism.

Lithiasis

Introduction

The excretory products of the kidneys and the excretory glands may be eliminated directly through a specific canal into an environment in direct contact with the outside—such as the skin, the mouth, the salivary gland, the intestine, or the pancreatic duct—or they may be collected into reservoirs, such as the gallbladder and the urinary bladder, before they are drained to the outside. Under normal conditions all excretory products remain in solution. Damage to the epithelium of the excretory canals or the reservoirs or changes in the composition of the secretions (because of abnormal intake or metabolic alterations) may lead to the precipitation of the metabolites in solution, causing obstructive masses called stones to be formed.

Anatomical and Geographic Distribution

Renal Calculi

Stones may form in all parts of the urinary excretory system, from the proximal tubules to the bladder and the urethra. Urinary stones are formed by the precipitation of organic or inorganic compounds normally in solution in the urine. Precipitation results from alterations of the urinary composition or the structure of the linings of the excretory canals. The gross appearance of stones is determined by their location in the urinary tract and by their chemical composition.

Pathologists distinguish uric acid, calcium oxalate, ammonium magnesium phosphate, cysteine, and other types of stones. Reference to a specific mineral or biochemical compound concerns only that compound present in the largest amount, because in addition to the major compound, the stone contains other chemicals; the nature of these chemicals will be discussed later.

Stones can be distinguished by their color, shape, surface appearance, consistency, and the appearance of the cut section.

Although most of this discussion concerns stones originating in the kidney, lithiasis may be primary to the bladder or the ureters. Whereas phosphate and oxalate* stones develop more frequently in the kidney, bladder stones are usually of the urate type. The geographical distribution of urinary lithiasis varies considerably. The disease is common in central Russia, China, Canton, and India. However, in India and China bladder stones are the type seen most often, whereas renal stones are more common in America.

* The oxalate stone is often rough and spiny; consequently, it causes irritation and hemorrhage, which are responsible for the stone's dark brown color.

In general, renal lithiasis is rare in early age, but its incidence increases to reach a maximum in the 6th and 7th decades. It is slightly more common in men than in women, and in 30% of the cases it is bilateral.

Lonsdale [87, 88] has reviewed the epidemiology of urinary lithiasis and concludes that bladder stones have become rare in adult man. The bladder stones Lonsdale found were urate stones and were believed to result from faulty diet. Bladder stones, composed of urate and oxalates, are seen in young people, rarely in the West, but commonly in India, Turkey, and Thailand. In contrast, kidney stones are a relatively frequent ailment in the West (200,000 new cases a year in the United States). Kidney stones are usually made of calcium oxalate, calcium phosphate, or $MgNH_4$ phosphates. The incidence seems to be highest among those with sedentary professions, and renal lithiasis is said to constitute an occupational hazard among airplane pilots.

Even within the United States, the distribution of renal lithiasis varies from place to place. The incidence of stones is approximately 20% in Tampa and 29% in Miami, but it is only 4% in Cleveland and 2.4% in New York. The environmental factors that influence this variation in incidence are not known.

Studies in a London hospital indicate that approximately 55% of the stones are found in the kidney, 20% in the ureteral tract, 22% in the bladder, and less than 3% in the urethra.

Naturally, the composition of the stone influences its distribution in the urinary tract. The smoothness of both the uric acid and the cysteine stones facilitate their passage through the ureter, so ureteral stones of this type are rarely found. When ureteral stones develop, they occur in the pelvic portion of the organ. Examples of kidney stones are shown in Figs. 9-15 and 9-16.

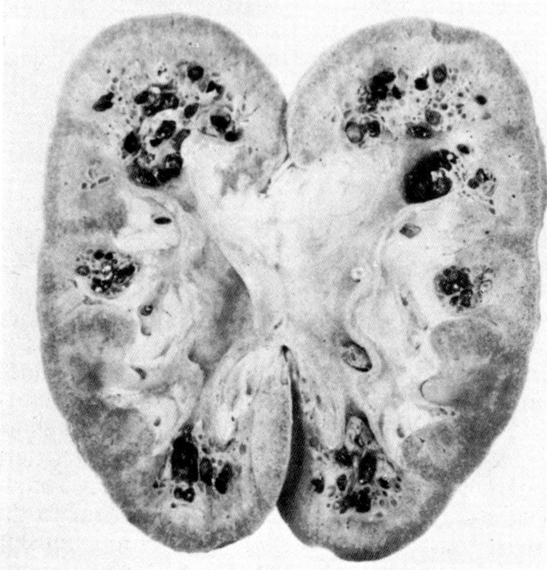

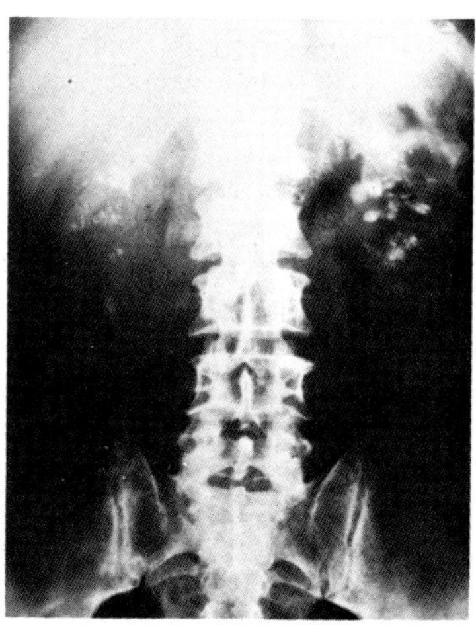

Fig. 9-15. Kidney stones, gross pathology and X-ray

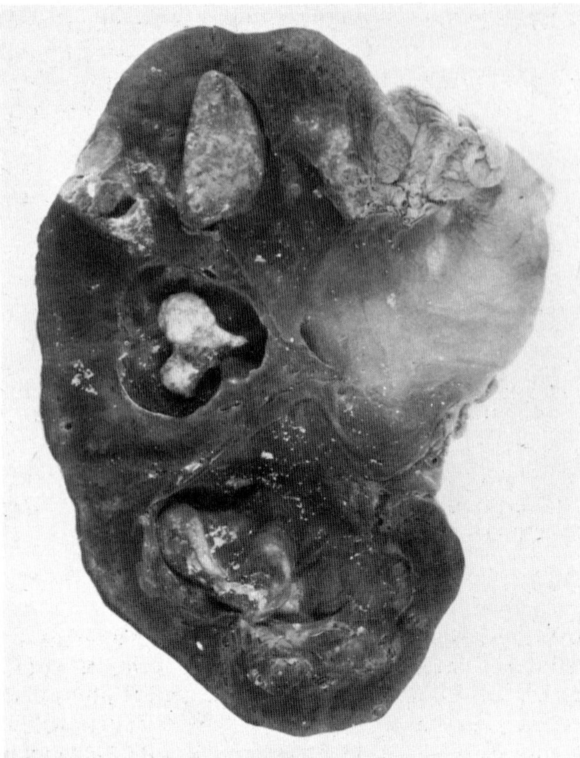

Fig. 9-16. Kidney stones

Pathogenesis of Urinary Lithiasis

At least three properties of urine influence the formation of stones: the concentration of the major stone component, urinary pH, and the presence or appearance of compounds that facilitate the precipitation of a given substance apt to form stones [89].

Many conditions result in the increased concentration of special solutes in the urine, and many of these have been or will be described in more detail in other chapters. For example, primary or secondary hyperparathyroidism releases calcium from bone and leads to its precipitation in kidneys. In some inborn errors of metabolism, metabolites accumulate in the blood, are excreted in the kidney, and under the appropriate conditions are precipitated in the excretory system. Gout, cysteinuria, and oxaluria are among those diseases that cause renal lithiasis.

The urine pH is important for the development of stones. Calcium stones develop preferentially at an alkaline pH and uric acid stones at acid pH. Among other factors that modulate renal pH, infection is critical. For example, Proteus species split urea to yield ammonia; therefore, infection by this type of bacteria is often followed by stone formation.

Changes in the concentration of solutes, other than major precipitating components, that are present in the urine are also significant. It has been established that high concentrations of sodium chloride interfere with the formation of the phosphatic urelites in sheep.

Colloid materials are among the urinary components present in small amounts that seem to play a key role in calculi formation. Urinary protein may be most critical in that respect. On the basis of their physical, chemical, and immunological properties, urinary proteins can be grouped into two major categories: those similar and those different from proteins in serum. Thus, electrophoresis of urine yields a number of peaks with mobilities similar to those of the α-, β-, and γ-globulins of serum.

Normal individuals excrete 31 mg of albumin in their urine daily. The studies of the physical and chemical properties of the urinary albumin have established that the urinary protein is identical to the blood protein. Both serum and urinary albumin have the same mobilities on free-starch electrophoresis and immuno-electrophoresis. The sedimentation constants and the diffusion rates of the two proteins are similar, and their molecular weight must be of the order of 73,000. The two proteins react identically antigenically.

However, in spite of some gross similarities, the protein compositions of urine and serum differ markedly. The ratio of albumin to globulin, which varies from 1.3 to 3 in serum, is 0.5 in urine. A number of proteins normally found in serum are absent from urine (fibrinogen, ceruloplasmin, α_1, α-, and β-lipoproteins); a prealbumin, an α_1-, α_2-, and a γ-globulin have been found in urine, but not in serum.

Renal lithiasis can be produced experimentally by a variety of means: hyperparathyroidism, excessive administration of vitamin A, or a deficiency in vitamin A, etc. The development of what has been called stone crisis is associated with histological modification, usually in the proximal convoluted tubules. The proximal tubule may or may not be enlarged, but almost always its cytoplasm contains increasing amounts of a substance that is both PAS and von Kossa positive, suggesting the presence of polysaccharides and phosphates. These changes have been observed in a variety of animals, rodents, cattle, and even humans, and they are present in numerous genitourinary diseases including infection, cancer, and others. Nevertheless, the significance of these changes is unclear, and it is not certain that they are related to stone formation. The possibility that the substances that accumulate in the renal tubules are not of tubular origin but rather are reabsorbed from urine by pinocytosis has not been excluded.

In addition to the precipitated mineral or organic compound, calculi usually contain a matrix of proteic nature. The composition of this matrix has been studied by immunological techniques based on the development of specific antibodies giving precipitin reactions in the presence of the appropriate antigen. On the basis of these studies, investigators have divided proteins found in calculi into two major groups: antigens that are similar to and antigens that are different from those found in serum. Thus, all stones contain proteins normally found in serum or urine.

In addition, a mucoprotein is found in all calculi, and it is immunologically more prominent than all other proteins found in stones. Therefore, this mucoprotein is called matrix substance A. The matrix protein yields a single precipitin line of Ouchterlony preparations. It has been concentrated and examined by electrophoresis and ultracentrifugation. In the ultracentrifuge, it moves as a single boundary, but its spreading suggests polydispersity. The molecular weight has been calculated to be of the order of 30,000–40,000 daltons. At pH 8.6 the matrix A substance has a mobility intermediate between that of serum α_2 und β components, and its isoelectric point is estimated to be around pH 4.5. Its amino acid composition is known. The protein contains 18 different amino acids. Glucosamine, small amounts of galactosamine, mannose, and traces of fucose provide the carbohydrate moiety.

A substance with similar immunological properties has been found in the kidney, where it is more abundant in the cortex than in the medulla. According to Boyce and King [90, 91], the mucoprotein is consistently found in the kidneys of patients with recently formed stones, but is rarely present in patients with other kidney diseases in which stones are not formed (*e.g.*, renal carcinoma and severe renal arteriosclerosis). The mucoprotein has not been detected in normal kidneys or in patients with pyelonephritis, and the compound cannot be immunologically detected in blood serum, human saliva, or bone matrix. The only other tissue where this matrix substance has been detected is the intestine. (Boyce and King think that this is not a coincidence and that the reason for its presence in kidney and intestine is that both these organs are involved in calcium absorption.) But most important to the pathogenesis of lithiasis are the mucopolysaccharides found in urine.

Urinary Mucoproteins and Lithiasis

Human and animal urine contains variable amounts of low molecular weight, alcohol-soluble proteins that contain carbohydrates (up to 40%). Changes in the mucoprotein composition of urine could precipitate stone formation as a result of either qualitative or quantitative alterations. Therefore, it is significant that the concentration of total nondialyzable urinary mucoproteins is considerably increased in calculus disease (5 times).

Special efforts were made to specify and identify some of the urinary mucoproteins; in humans a uromucoid referred to as Tams protein and in bovine species one called bovine urinary mucoprotein have been specially studied. The origin of those uromucoids is not clear; at first they were believed to be derived from the transitional epithelium of the bladder, but later they were suspected to have a renal origin. The Tams protein has an average molecular weight of approximately 7×10^6. However, it occurs in fragments of various sizes. The typical protein is 25,000 A long and 100 A wide and has repeated beading approximately every 110 A. The Tams mucoprotein contains 21% polysaccharide and 9% sialic acid. It is assumed that there are two polysaccharide prosthetic groups per bead and that each prosthetic group contains four repeating structures composed of eight hexosamine, glucosamine, and galactosamine, six galactose, three mannose, one fucose, and four sialic acid residues. The total molecular weight of the polysaccharide prosthetic group is 12,000. The protein is insoluble in salt and precipitates in threadlike structures in concentrations of 0.58 M sodium chloride.

A similar threadlike protein has been found in bovine urine. It is approximately 150 A wide, the beading occurs every 200 A, and it may be 25,000–40,000 A long. The bovine urinary protein contains 4% sialic acid, 6% hexose, and 5% hexosamine. It is not known if glycoproteins with similar properties are responsible for the PAS-positive reaction observed in the renal tubules of sheep receiving an experimental diet. The use of fluorescent antibodies could help in answering that question. The role of the mucoproteins in forming stone matrices is unknown.

Comparative Pathology of Urolithiasis

In the field of lithiasis, as in many other areas of pathology, the study of animal diseases has provided valuable information on pathogenesis. Lithiasis is common among cattle, particularly beef. In the northwestern plains of North America, 4% of the beef living on the range have urolithiasis with clinical symptoms. The stones are formed mainly of silicon dioxide (SiO_2) (which is found in high concentration in forage) and a protein matrix. The mechanism of formation of these stones has been reviewed [92].

Experimental urolithiasis is produced in beef by feeding them a diet supplemented with K_2HPO_4. PAS stains demonstrate that glycoproteins accumulate in the cells of the proximal and distal convoluted and collecting tubules. The proximal convoluted tubules seem to be more affected than the other segments of the nephron. PAS-positive casts are also found in the urine. Moreover, examination of the urine by ultrafiltration, electrophoresis, and chromatography on DEAE-cellulose columns demonstrated the presence of a mucoprotein similar to that found in human stones.

The administration of a high-phosphate diet to castrated male sheep produced phosphatic calculosis. Triple phosphate crystals appear in the urine as early as 24 hours after the diet is instituted. Precipitation occurs when the urine pH reaches values of 8.2–8.5. The concentrations of magnesium, ammonium, and P_i and the overall ionic strength modulate the precipitation. The triple phosphate crystals are agglomerated to form small, round bodies 3–7 μ in diameter called microcalculi. Later, microcalculi are packed together into an organic envelope.

In the kidney, PAS-positive* granules appear in the

* A positive PAS stain suggests the presence of neutral glycoproteins; however, these staining techniques are not rigidly specific, and it cannot be excluded that the component detected may be an acidic carbohydrate protein, such as chondromucoid or collagen, that has lost its metachromatic properties because of depolymerization or loss of free reactive groups.

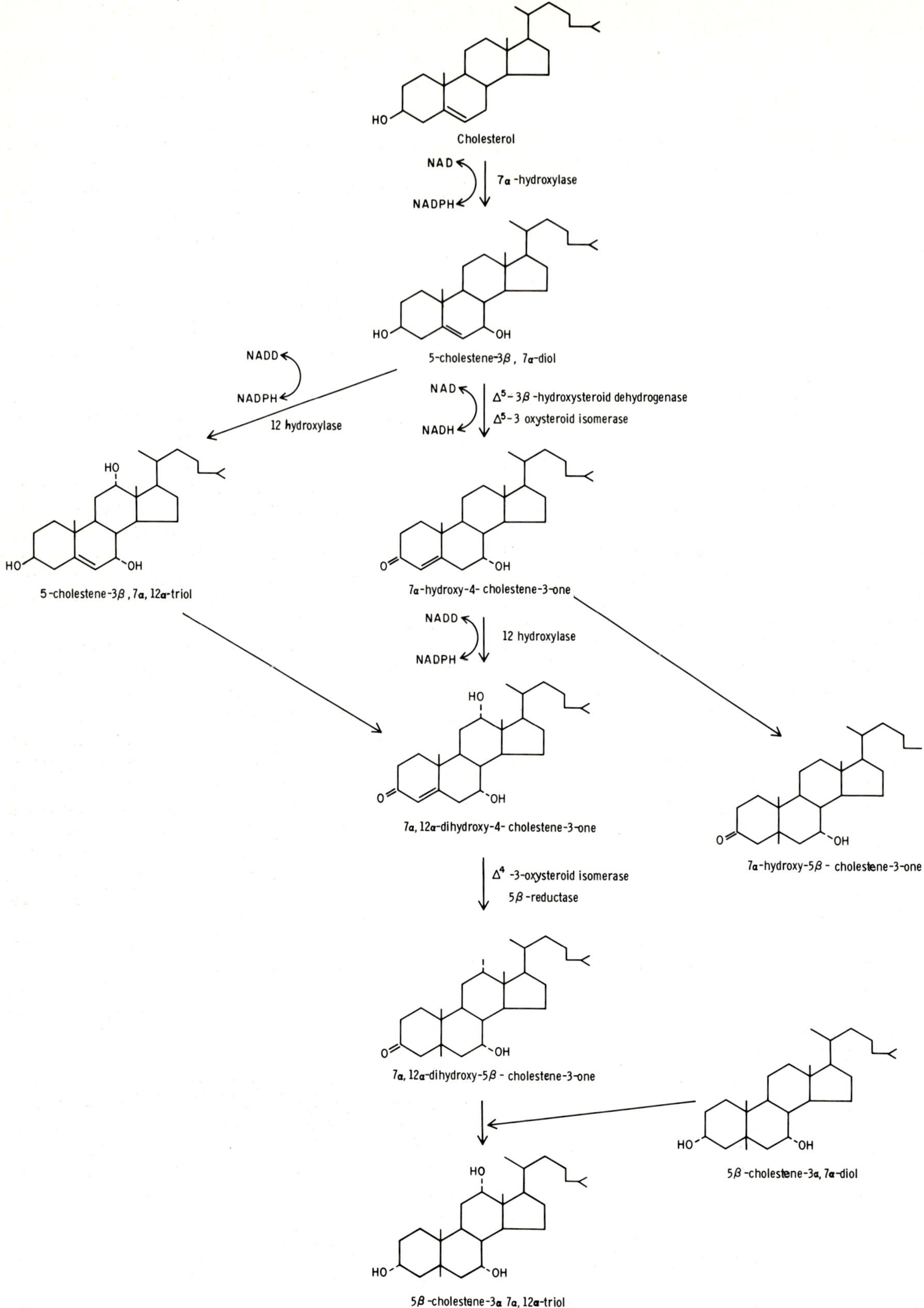

Fig. 9-18. Metabolism of bile acids

the bile salt levels in the blood. Similarly, severe liver disease which interferes with the reabsorption of bile salts will change the serum distribution. Before discussing such alterations, we will review briefly the anatomy of bile ducts and gallbladder [95–98].

The intrahepatic bile ducts converge to form larger channels that ultimately meet to form the large right and left hepatic ducts, each irrigating half of the liver. The hepatic ducts combine to form the common duct, which is soon joined by a smaller duct emanating from the gallbladder, the cystic duct. The common duct runs behind the first portion of the duodenum and the head of the pancreas and enters the posterior aspect of the second portion of the duodenum, forming an elevation at the surface of the mucosa and piercing it to yield what is referred to as the sphincter of Oddi.

The gallbladder is a pear-shaped organ in the lower right portion of the liver. The base of the organ, the fundus, is covered with peritoneum and emerges below the lower border of the liver in direct contact with the abdominal wall. The body narrows into an undulated stalk forming the cystic duct, which joins the hepatic duct.

Cholecystokinin relaxes the sphincter of Oddi and contracts the gallbladder, and as a result bile is secreted in the intestine. When the sphincter closes again, the bile flows into the duct and refluxes into the cystic duct and the gallbladder. The pressure of bile in the gallbladder is usually the same as or slightly lower than that in the duct because the gallbladder simultaneously relaxes and withdraws water from the bile as it is contained in the reservoir.

Biliary fistulas in rats or humans increase the level of bile acid production markedly, primarily as a result of increasing activity of cholesterol 7α-hydroxylase and hydroxymethylglutamyl CoA reductase. In contrast, bile acids repress the activities of these two enzymes.

Cholestyramine is an ion-exchange resin that traps the bile salts and thereby secures their elimination in the intestine. The bile salts are continuously excreted in the intestine; bile acid formation from cholesterol increases, and serum cholesterol levels drop. Existing evidence indicates that cholestyramine administration and ileal bypass are the most effective measures available to reduce serum cholesterol in hypercholesterolemic patients.

The concentration of bile salts in feces, urine, and bile by necessity varies under some pathological conditions. Two groups of organic diseases affect bile salt distribution: those of the intestine and liver.

When bacteria invade the upper intestine, as in an intestinal blind loop, they deconjugate bile. The deconjugated bile is not absorbed, and instead of being absorbed normally, lipids are excreted in the feces (steatorrhea).

Bile acids and lipids are lost in the feces when a large portion of the intestine is bypassed because disease has caused obstruction (for example, in obstruction by cancer), or because the intestinal segment has been bypassed surgically to limit lipid reabsorption, as in severe cases of obesity.

Depending upon their severity, liver diseases may impair bile salt reabsorption, interfere with bile salt conjugation, or both. Detailed discussion of the changes in bile salt concentration in feces, urine, and bile is not necessary; it suffices to remark that the concentration of biliary bile salts in serum depends on the rate of production and on intestinal and hepatic reabsorption. The ability of the liver cells to reabsorb bile salts is believed to be the most critical factor; therefore, measurements of the bile salt levels in serum can be helpful for diagnosing certain liver diseases. The concentration of primary bile salts in the serum is significant, whereas that of secondary bile salts is too low to influence distribution changes markedly.

In severe liver injury, the level of bile salts is increased in the serum. The amount of chenodeoxycholate increases over that of cholate, so the cholate-chenodeoxycholate ratio is greater than 1.

Gallstones have plagued men throughout history and have spared neither prince nor pauper. They afflict 10% of American men and 20% of American women. In the year that President Johnson had gallstones removed, 5,000,000 entered hospitals for similar purposes, sometimes cheerfully and sometimes miserably sick [99–104].

Grossly, there are two major types of gallstones: pure cholesterol or cholesterol hydrate stones and mixed stones. The first type are yellowish, friable, and primarily composed of radiating cholesterol crystals. The mixed stones include various calcium salts such as carbonates, bilirubinates, and phosphates. Often the major components, cholesterol and pigmented salt, are arranged in successive, concentric layers. An example of gallstone is shown in Fig. 9-19.

As is often the case for biological products, the term "pure" is a misnomer; no stones are made of pure cholesterol or concentric layers of pure cholesterol and calcium bilirubinate. Stones often contain other components such as iron, proteins, and mucopolysaccharides. In fact, even the purest cholesterol stones seem to contain small amounts of protein or polysaccharide at their center, a finding that is in keeping with the concept that gallstones, like kidney

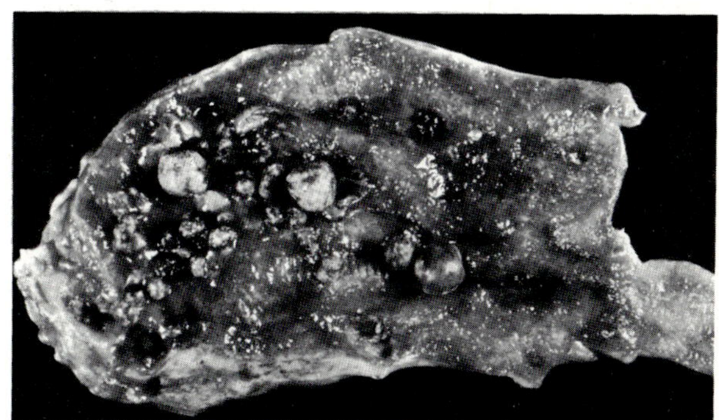

Fig. 9-19. Gallstone

stones, may be started by the precipitation of a hypersaturated solution on a nidus. In gallstones, such nidi might include ingested foreign bodies (vegetable fibers, bone fragments, and even pins), surgically introduced artifacts (such as sutures), or a clump of desquamated epithelial cells. In typhoid fever, live and dead bacteria may conglomerate in the gallbladder and form a nidus for a gallstone.

Stones do not always develop into a single rounded or oval mass. Often smaller stones form which may be multifaceted because of friction against each other, or may conglomerate into an irregular mass. Sometimes the stones are so small that the bile appears as if sand had been dropped in it. In fact, under some circumstances, the separation of the liquid and solid phases is not clear cut, and the bile is viscous and mushy.

If one examines the pathogenic mechanism of gallstone formation, one thing is certain—the solute that is precipitated must exist in the form of a supersaturated solution. However, this does not necessarily imply that all supersaturated solutions precipitate to form stones because such states may be transient and small crystals might go back into solution. Whether a protein or mucoid nidus is needed to precipitate a supersaturated solute is not known. If a nidus is needed, it must remain in the gallbladder, thus stasis will play a role in the pathogenesis. Therefore, among other mechanisms by which bile stasis may influence the pathogenesis of gallstones, the maintenance of the potential nidus is one.

The pathogenesis of gallstones can essentially be reduced to a matter of solubility of the various solutes that constitute the stone—cholesterol in cholesterol stones and cholesterol, calcium carbonate, phosphate, or bilirubinate in mixed stones.

The pathogenesis of cholelithiasis has been investigated most extensively for cholesterol stones. Clearly, when everything else is constant, the solubility of cholesterol depends upon the relative concentrations of bile acids, phospholipids, and lecithin. This relationship can be expressed by plotting the percentage of cholesterol, lecithin, and bile salts on the left, right, and lower sides of an isosceles triangle. The surface delineated by the triangle can then be divided into various zones of cholesterol solubility. In zone one, delineated by line A, cholesterol is soluble, and line A thus joins the points at which equilibrium solubility is achieved for cholesterol for various relative concentrations of the two other components. Zone two is a zone of apparent solubility, the upper limit of which separates it from a zone in which cholesterol inescapably is precipitated (see Fig. 9-20).

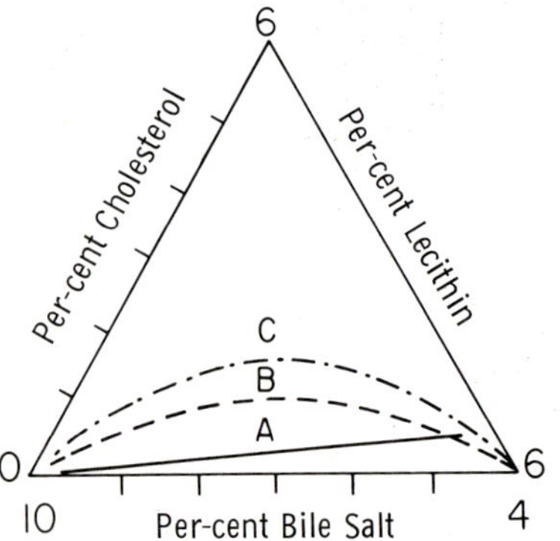

Fig. 9-20. Triangular phase diagram of cholesterol solubility

However, the role of cholesterol solubility in the pathogenesis of gallstones cannot be appreciated unless one keeps in mind that the relative concentrations of cholesterol, bile salts, and phospholipids vary during the day. A number of factors, including exogenous cholesterol intake and rates of cholesterol and phospholipid synthesis,* cholesterol breakdown, bile acid formation, and bile acid reabsorption, vary during the day. As a result, the relative concentration of the triad changes constantly so that the bile cholesterol may switch from solubility to supersaturation with metastability, or even to a state in which the cholesterol is precipitated. If the newly developed crystals find a heterogenous nidus or a pure cholesterol nidus and the bile remains supersaturated, the precipitated cholesterol will not dissolve and stones will form.

The mechanism of supersaturated bile formation may involve increased cholesterol excretion or decreased levels of bile acids and phospholipids. In patients with gallstones, lithogenic bile is produced by the liver, in some cases as a result of high cholesterol, in others, like in Indian women, because of a combination of high cholesterol production and a relatively low production of bile salts [106]. Cholesterol synthesis is markedly accelerated in cholestasis because the feedback control of cholesterol synthesis noted in normal liver tissue is not operative; consequently, elevated hepatic cholesterol synthesis may be at least partly responsible for the hypercholesterolemia observed in cholestasis [107]. In those cases in which the lithogenic bile results from high cholesterol levels, it was observed that the cholesterol secretion by the hepatic cells returns to normal after cholecystectomy. Such a finding suggests that in some way the gallbladder influences bile secretion in the liver and induces that organ to secrete lithogenic bile [108, 111].

Conclusion

Mammalian organisms, including humans, have learned to deal with solutes by maintaining them in various levels of solubility. Thus, calcium crystallizes to form calcium apatite in a very organized fashion to build bones. Lipid molecules of the cell membranes exist in a fluid crystal structure and play a key role in their properties.

* Little attention has been payed to the metabolism of the lecithin that appears in bile. However, it has been established that the administration of labeled precursors of phospholipids brings about the appearance of labeled phospholipids in the bile [105].

Those compounds that need to be excreted either for function or for elimination are usually kept under conditions of optimal solubility. But when these dynamic states are drastically altered, the compounds precipitate and form stones that obstruct either reservoir or excretory canals. The common denominator in all these events is the need for a supersaturated solution. Perhaps other factors of unknown nature are also needed to precipitate renal or cholecystic lithiasis. Thus, careful investigation of the dynamic changes in the fluid composition helps to prevent stone formation and in some cases may point the way to effective medical dissolution of preexisting stones. In the case of lithiasis, as in many other diseases, new diagnostic procedures and therapies were conceived from a daring marriage of physics and pathology.

Cholestasis

Etymologically, cholestasis means stoppage of bile flow. Although there is some confusion in the literature as to the definition of cholestasis, Popper and Schaffner [109] distinguish two types: mechanical and intrahepatic. Mechanical cholestasis results from a block between the bifurcation of the hepatic duct and the ampulla of Vater. In hepatic cholestasis, the patient presents clinical, laboratory, and histological symptoms identical to those observed in extrahepatic cholestasis, but there are no radiological or clinical signs of extrahepatic biliary obstruction. The pathogenesis of intrahepatic cholestasis will be easier to understand if we briefly outline the anatomical and physiological basis of bile secretion.

Because the lobule concept could explain only part of the pathology seen in liver (for example, in some cases only part of a lobule is affected; in others, two adjacent lobules are affected), pathologists and anatomists attempted to identify a smaller unit that could better account for liver pathology: the acinus.

The acinus is that portion of liver supplied by a primary afferent branch of the portal vein and hepatic artery. If we look at a bell-shaped hepatic lobule hanging from and traversed by a branch of the hepatic vein, the portal space runs on its side containing the artery, the vein, and the bile duct (see Figs. 9-21 and 9-22). At some point the vein and the arteriole yield three branches, two lateral and one apical, that are the primary afferent branches. They leave the connective tissue of the portal tract to penetrate in the substance of the lobule, where they further ramify into smaller branches that enter the sinusoid of the lobule. The territory irrigated by these primary afferent arterioles, the acinus, forms a somewhat irregular ovoid mass of tissue inserted between two central veins. The tissue mass is composed mainly of hepatic cells and sinusoids lined by Kupffer's cells. The central zone of the acinus (zone 1) receives the fresh blood supply and may therefore be the first to be injured by toxins in blood. The peripheral zone (zone 3) receives a blood supply partially exhausted in its oxygen and nutrients and is therefore more susceptible to anoxemia.

The hepatic cells form partially anastomosed cords one cell thick that radiate from the periphery of the

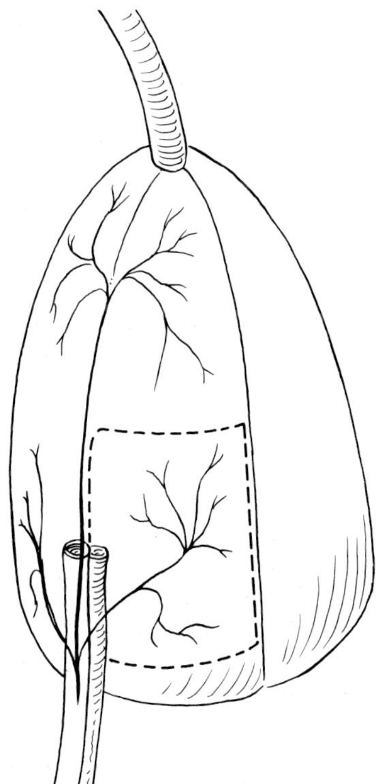

Fig. 9-21. Hepatic lobules—longitudinal view around the central vein. Modified after L.G. Whitley from "A Companion to Medical Studies". Blackwell Scientific Publications, Oxford and Edinburgh

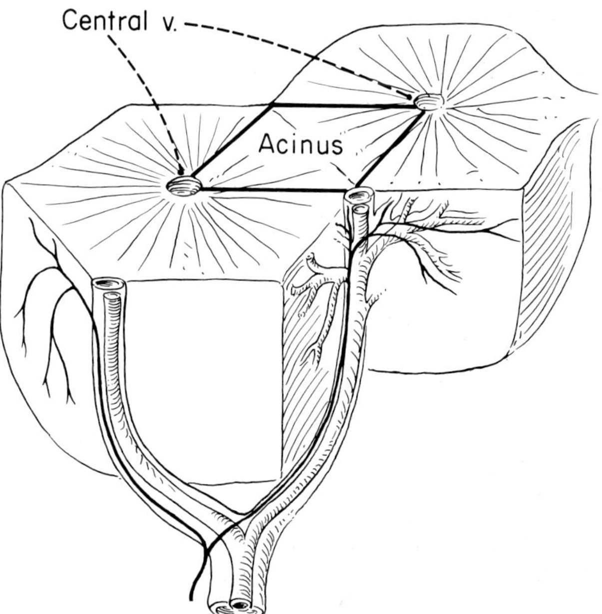

Fig. 9-22. Hepatic acinus. Modified after L.G. Whitley from "A Companion to Medical Studies". Blackwell Scientific Publications, Oxford and Edinburgh

lobule toward the central vein. The cords are separated by capillaries lined by the phagocytic Kupffer's cells. Between the hepatic cells and the endothelium of the sinusoid is the space of Disse containing fluid collagen and reticulum fibers. Two portions of the surface of each hepatic cell are in contact with sinusoids, and the remaining surface is in contact with several other liver cells. The membrane develops numerous microvilli that protrude in the space of Disse.

The smaller bile canaliculi are formed by the hepatic cell itself. The lateral portion of the cell presents a small groove which when combined with similar grooves from a number of other hepatic cells forms a channel in which the bile is excreted. The canaliculi converge toward the portal tracts and open in the bile ductules.

The hepatocytes are held together by junctional complexes; a polysaccharide, the glycocalyx, covers the luminal surface of the bile canaliculi.

Because there is no evidence of damage to the bile ducts, it was at first assumed that intrahepatic jaundice was caused by kinking of the connection between hepatic cells and the bile ducts, the cholangioles. Inflammation and fibrosis were believed to cause the kinking. This pathogenic interpretation became untenable when jaundice and cholestasis were observed with drugs such as steroids or chlorpromazine without inflammation.

Bile salts are the determinant component in bile secretion, and they regulate the secretion of cholesterol, bilirubin, and phospholipids. In the hepatic cell and in the bile, at the proper bile salt concentration, micelles are formed from bile salts, phospholipids, cholesterol, and bilirubin. It is believed that in cholestasis there is interference with micelle secretion, possibly as a result of disturbances in the bile salt concentration. This is followed by alterations of canalicular microvilli, enlargement of the Golgi, and proliferation of the endoplasmic reticulum.

An unsolved problem is the cause of cellular damage in biliary obstruction. The accumulation of bile acids is suspected. In man the normal cholic acid level is approximately 30 m moles/g; this level is increased 7 times in biliary obstruction. Yet the extent of histological injury does not parallel the cholic acid level. In contrast, it parallels that of chenodeoxycholic acid. Although the levels of chenodeoxycholic acid are low at the onset of biliary obstruction, they increase gradually and may after weeks reach levels between 70 and 400 ng per g of liver. Chenodeoxycholic acid, but not the trihydroxylated cholic acid, is toxic for microsomes *in vitro*. Although it is likely that *in vivo* chenodeoxycholic acid is responsible for cellular damage, it is not known whether this results from damage to microsomes or cell membranes [110].

References

1. Dick, D.A.T.: Cell water. London: Butterworth and Col., Ltd. 1966
2. Potts, W.T.W.: Osmotic and ionic regulation. Annu. Rev. Physiol. **30**, 73–104 (1968)
3. Gottschalk, C.W.: Osmotic concentrations and dilution of the urine. Amer. J. Med. **36**, 670–685 (1964)
4. Barker, J.N.: The renal countercurrent concentrating mechanism. Med. Clin. N. Amer. **47**, 873–886 (1963)
5. Barajas, L.: Renin secretion: An anatomical basis for tubular control. Science **172**, 485–487 (1971)
6. Kotyk, A., Janaček, K.: Cell membrane transport; principles and techniques. In: The kidney, p. 433–444. New York: Plenum Press 1970
7. Ullrich, K.J., Marsh, D.J.: Kidney, water and electrolyte metabolism. Annu. Rev. Physiol. **25**, 91–142 (1963)
8. Bland, J.H.: Introductory remarks: Some usual, unusual properties of liquids in general and water in particular. Fed. Proc. **25**, 951–953 (1966)
9. Scatchard, G.: Water: a review. Fed. Proc. **25**, 954–957 (1966)
10. Berendsen, H.J.C., Migchelsen, C.: Hydration structure of collagen and influence of salts. Fed. Proc. **25**, 998–1002 (1966)
11. Berendsen, H.J.C.: Water structure in biological systems. Fed. Proc. **25**, 971–976 (1966)
12. Catchpool, J.F.: Effect of anesthetic agents on water structure. Fed. Proc. **25**, 979–985 (1966)
13. Kavanau, J.L.: Remarks on water structure. Fed. Proc. **25**, 977–978 (1966)
14. Klotz, I.M.: Water. In: Horizons in biochemistry (Kasha, M., and Pullman, B., eds.), p. 523–550. New York: Academic Press 1962
15. Skou, J.C.: The influence of some cations on an adenosine triphosphatase from peripheral nerves. Biochim. biophys. Acta (Amst.) **23**, 394–401 (1957)
16. Martinez-Maldonado, M., Allen, J.C., Eknoyan, G., Suki, W., Schwartz, A.: Renal concentrating mechanism: Possible role for sodium-potassium activated adenosine triphosphatase. Science **165**, 807–808 (1969)
17. Katz, A.I., Epstein, F.H.: Physiologic role of sodium-potassium-activated adenosine triphosphatase in the transport of cations across biologic membranes. New Engl. J. Med. **278**, 253–261 (1968)
18. Bransome, E.D., Jr.: Adrenal cortex. Annu. Rev. Physiol. **30**, 171–212 (1968)
19. Fanestil, D.D.: Mechanism of action of aldosterone. Annu. Rev. Med. **20**, 223–232 (1969)
20. Edelman, I.S., Fimognari, G.M.: On the biochemical mechanism of action of aldosterone. Recent Progr. Hormone Res. **24**, 1–44 (1968)
21. Ehrlich, E.N.: Aldosterone, the adrenal cortex, and hypertension. Annu. Rev. Med. **19**, 373–398 (1968)
22. Forster, R.P.: Kidney, water and electrolytes. Annu. Rev. Physiol. **27**, 183–232 (1965)
23. Hartroft, P.M.: The juxtaglomerular complex. Annu. Rev. Med. **17**, 113–122 (1966)
24. Barajas, L., Latta, H.: Structure of the juxtaglomerular apparatus. Circulat. Res. **21**, Suppl. II, 15–28 (1967)
25. Laragh, J.H.: Renin, angiotensin, aldosterone, and hormonal regulation of arterial pressure and salt balance; introductory remarks. Fed. Proc. **26**, 39–41 (1967)
26. Tobian, L.: Renin release and its role in renal function and the control of salt balance and arterial pressure. Fed. Proc. **26**, 48–54 (1967)
27. Vander, A.J.: Control of renin release. Physiol. Rev. **47**, 359–382 (1967)
28. Mills, I.H.: Renal regulation of sodium excretion. Annu. Rev. Med. **21**, 75–98 (1970)
29. Gill, J.R., Bartter, F.C.: Adrenergic nervous system in sodium metabolism. II. Effects of guanethidine on the renal response to sodium deprivation in normal man. New Engl. J. Med. **275**, 1466–1471 (1966)
30. Paglia, D.E., Holland, P., Baughan, M.A., Valentine, W.N.: Occurrence of defective hexosephosphate isomerization in human erythrocytes and leukocytes. New Engl. J. Med. **280**, 66–71 (1969)
31. Lee, M.R.: Renin-secreting kidney tumours, a rare but remediable cause of serious hypertension. Lancet **1971II**, 254–255
32. Knox, F.G.: Role of the proximal tubule in the regulation of urinary sodium excretion. Mayo Clin. Proc. **48**, 565–573 (1973)
33. Kaplan, N.M.: Secondary aldosteronism: With observations on the definition of hypokalemia. Amer. J. clin. Path. **54**, 316–323 (1970)
34. Engel, A.G.: Evolution and content of vacuoles in primary hypokalemic periodic paralysis. Mayo Clin. Proc. **45**, 774–814 (1970)
35. Zarkowsky, H.S., Oski, F.A., Sha'afi, R., Shohet, S.B., Nathan, D.G.: Congenital hemolytic anemia with high sodium, low potassium red cells. I. Studies of membrane permeability. New Engl. J. Med. **278**, 573–581 (1968)
36. Wakim, K.G.: Predominance of hyponatremia over hypo-osmolarity in simulation of the dialysis disequilibrium syndrome. Mayo Clin. Proc. **44**, 433–460 (1969)
37. Kernan, R.P.: Cell K. London: Butterworth & Co., Ltd. 1965
38. Josephson, B.: Chemistry and therapy of electrolyte disorders. In: Potassium metabolism, p. 51–67. Springfield, Ill.: Charles C. Thomas Publisher 1960
39. Diamond, J.M., Wright, E.M.: Biological membranes: the physical basis of ion and non-electrolyte selectivity. Annu. Rev. Physiol. **31**, 581–646 (1969)
40. Bates, R.G.: Acids, bases and buffers. Ann. N.Y. Acad. Sci. **133**, 25–33 (1966)
41. Siggaard-Andersen, O.: Titratable acid or base of body fluids. Ann. N.Y. Acad. Sci. **133**, 41–58 (1966)
42. Astrup, P., Engle, K., Jorgensen, K., Siggaard-Andersen, O.: Definitions and terminology in blood acid-base chemistry. Ann. N.Y. Acad. Sci. **133**, 59–65 (1966)

43. Butler, T.C., Waddell, W.J., Poole, D.T.: The pH of intracellular water. Ann. N.Y. Acad. Sci. **133**, 73–77 (1966)
44. Elkinton, J.R.: Acid-base disturbances in renal disease. Ann. N.Y. Acad. Sci. **133**, 195–209 (1966)
45. Bittar, E.E.: Cell pH. London: Butterworth & Co., Ltd. 1963
46. Lotspeich, W.D.: Metabolic aspects of acid-base change. Science **155**, 1066–1075 (1967)
47. Maren, T.H.: Carbonic anhydrase: Chemistry, physiology and inhibition. Physiol. Rev. **47**, 595–781 (1967)
48. Enns, T.: Facilitation by carbonic anhydrase of carbon dioxide transport. Science **155**, 44–47 (1967)
49. Giebisch, G.: Coupled ion and fluid transport in the kidney. New Engl. J.Med. **287**, 913–919 (1972)
50. Schwartz, W.B., van Ypersele de Strihou, Kassirer, J.P.: Role of anions in metabolic alkalosis and potassium deficiency. New Engl. J. Med. **279**, 630–639 (1968)
51. Redgate, E.S.: Hypothalamic influence on respiration. Ann. N.Y. Acad. Sci. **109**, 606–618 (1963)
52. Gill, J.R., Jr.: Edema. Annu. Rev. Med. **21**, 269–280 (1970)
53. Robin, E.D., Cross, C.E., Zelis, R.: Pulmonary edema. Part II. New Engl. J. Med. **288**, 292–304 (1973)
54. Johnston, C.I., Davis, J.O., Howards, S.S., Wright, F.S.: Cross-circulation experiments on the mechanism of the natriuresis during saline loading in the dog. Circulat. Res. **20**, 1–10 (1967)
55. Urquhart, J., Davis, J.O., Higgins, T.J., Jr.: Stimulation of spontaneous secondary hyperaldosteronism by intravenous infusion of angiotensin II in dogs with an arteriovenous fistula. J. clin. Invest. **43**, 1355–1366 (1964)
56. Davis, J.: The physiology of congestive heart failure. In: Handbook of physiology: Circulation (Hamilton, W.F., and Dow, P., eds.), p. 2071–2122. Washington, D.C.: Amer. Physiol. Soc. 1965
57. Davis, J.O.: The mechanisms of salt and water retention in cardiac failure. Hosp. Prac. **5**, 63–76 (1970)
58. Thorn, G.W.: Approach to the patient with "idiopathic edema" or "periodic swelling." J. Amer. med. Ass. **206**, 333–338 (1968)
59. Barnes, F.W., Jr., Schoenheimer, R.: On biological synthesis of purines and pyrimidines. J. biol. Chem. **151**, 123–139 (1943)
60. Krebs, H.A., Henseleit, K.: Untersuchungen über die Harnstoffbildung im Tierkörper. Klin. Wschr. **11**, 757–759 (1932)
61. Rubin, H.: The behavior of cells before and after virus-induced malignant transformation. Harvey Lect. Series **61**, 118–143 (1965–66)
62. Berger, R., Broyer, M.: Anomalies héréditaires du cycle de l'urée. Presse méd. **76**, 1183–1184 (1968)
63. Perkoff, G.T., Thomas, C.L., Newton, J.D., Sellman, J.C., Tyler, F.H.: Mechanism of impaired glucose tolerance in uremia and experimental hyperazotemia. Diabetes **7**, 375–383 (1958)
64. Merrill, J.P., Hampers, C.L.: Uremia. I. New Engl. J. Med. **282**, 953–961 (1970)
65. Hampers, C.L., Soeldner, J.S., Doak, P.B., Merrill, J.P.: Effect of chronic renal failure and hemodialysis on carbohydrate metabolism. J. clin. Invest. **45**, 1719–1731 (1966)
66. Herschman, J.M., Givens, J.R., Cassidy, C.E., Astwood, E.B.: Long-term outcome of hyperthyroidism treated with antithyroid drugs. J. clin. Endocr. **26**, 803–807 (1966)
67. Renkin, E.M., Robinson, R.R.: Glomerular filtration. New Engl. J. Med. **290**, 785–792 (1974)
68. Wright, F.S.: Intrarenal regulation of glomerular filtration rate. New Engl. J. Med. **291**, 135–141 (1974)
69. Morel, F., de Rouffignac, C.: Kidney. In: Annual review of physiology (Comroe, J.H., Jr., Edelman, I.S., and Sonnenschein, R.R., eds.), vol. 35, p. 17–54. Palo Alto, Calif.: Annual Reviews Inc. 1973
70. Dahl, J.L., Hokin, L.E.: The sodium-potassium adenosinetriphosphatase. In: Annual review of biochemistry (Snell, E.E., Boyer, P.D., Meister, A., and Richardson, C.C., eds.), vol. 43, p. 327–356. Palo Alto, Calif.: Annual Reviews Inc. 1974
71. Baker, P.F.: The sodium pump in animal tissues and its role in the control of cellular metabolism and function. In: Metabolic pathways (Hokin, L.E., ed.), third ed., vol. VI: Metabolic transport, p. 243–268. New York: Academic Press 1972
72. Caldwell, P.C.: Membranes and ion transport (Bittar, E.E., ed.), vol. I, p. 433. New York: Wiley-Interscience 1970
73. Lewy, P.R., Quintanilla, A., Levin, N.W., Kessler, R.H.: Renal energy metabolism and sodium reabsorption. In: Annual review of medicine (Creger, W.P., Coggins, C.H., and Hancock, E.W., eds.), vol. 24, p. 365–384. Palo Alto, Calif.: Annual Reviews Inc. 1973
74. Oparil, S., Haber, E.: The renin-angiotensin system, I. New Engl. J. Med. **291**, 389–401 (1974)
75. Peart, W.S.: Renin-angiotensin system. New Engl. J. Med. **292**, 302–306 (1975)
76. Conn, J.W.: Primary aldosteronism and primary reninism. Hosp. Pract. **9**, 131–140 (1974)
77. Eddy, R.L., Sanchez, S.A.: Renin-secreting renal neoplasm and hypertension with hypokalemia. Ann. intern. Med. **75**, 725–729 (1971)
78. Lin, K-T D., Deutsch, H.F.: Human carbonic anhydrases. XII. The complete primary structure of the C isozyme. J. biol. Chem. **249**, 2329–2337 (1974)
79. Felig, P., Wahren, J.: Protein turnover and amino acid metabolism in the regulation of gluconeogenesis. Fed. Proc. **33**, 1092–1097 (1974)
80. Orrenius, S., Ericsson, J.L.E.: Enzyme-membrane relationship in phenobarbital induction of synthesis of drug-metabolizing enzyme system and proliferation of endoplasmic membranes. J. Cell Biol. **28**, 181–198 (1966)
81. Svoboda, D.J., Reddy, J.K.: Some biologic properties of microbodies (peroxisomes). In: Pathobiology annual (Ioachim, H.L., ed.), vol. 4, p. 1–32. New Yorl: Appleton-Century-Crofts 1974
82. Bricker, N.S.: Adaptations in chronic uremia: Pathophysiologic "tradeoffs". Hosp. Pract. **9**, 119–126 (1974)
83. Kelch, R.P., Kaplan, S.L., Biglieri, E.G., Daniels, G.H.: Hereditary adrenocortical unresponsiveness to adrenocorticotropic hormone. J. Pediat. **81**, 726–736 (1972)
84. Askari, A. (ed.): Properties and functions of $(Na^+ + K^+)$-activated adenosinetriphosphatase. N.Y. Acad. Sci. **242**, 1–741 (1974)
85. Oshima, G., Gecse, A., Erdos, E.G.: Angiotensin I-converting enzyme of the kidney cortex. Biochim. biophys. Acta (Amst.) **350**, 26–37 (1974)
86. Murakami, K., Inagami, T.: Isolation of pure and stable renin from hog kidney. Biochem. biophys. Res. Commun. **62**, 757–763 (1975)
87. Lonsdale, K.: Human stones. Science **159**, 1199–1207 (1968)
88. Lonsdale, K.: Human stones. Sci. Amer. **219**, 104–111 (1968)
89. Williams, H.E.: Nephrolithiasis. New Engl. J. Med. **290**, 33–38 (1974)
90. Boyce, W.H., King, J.S., Jr.: Present concepts concerning the origin of matrix and stones. Ann. N.Y. Acad. Sci. **104**, 563–578 (1963)
91. King, J.S., Jr., Boyce, W.H.: Immunological studies on serum and urinary proteins in urolith matrix in man, II. Metabolic aspects of urolithiasis in man and animals. Ann. N.Y. Acad. Sci. **104**, 579–591 (1963)
92. Keeler, R.F.: Silicon metabolism and silicon-protein matrix interrelationship in bovine urolithiasis. Ann. N.Y. Acad. Sci. **104**, 592–611 (1963)
93. Kritchevsky, D., Nair, P.P.: Chemistry of the bile acids. In: The bile acids (Nair, P.P., and Kritchevsky, D., eds.), vol. 1, p. 1–10. New York: Plenum Press 1971
94. Danielsson, H.: Mechanisms of bile acid biosynthesis. In: The bile acids (Nair, P.P., and Kritchevsky, D., eds.), vol. 2, p. 1–32. New York: Plenum Press 1973
95. Miettinen, T.A.: Clinocal implications of bile acid metabolism in man. In: The bile acids (Nair, P.P., and Kritchevsky, D., eds.), vol. 2, p. 191–247. New York: Plenum Press 1973
96. Bekersky, I., Mosbach, E.H.: Effects of hormones on bile acid metabolism. In: The bile acids (Nair, P.P., and Kritchevsky, D., eds.), vol. 2, p. 249–257. New York: Plenum Press 1973
97. Lack, L., Weiner, I.M.: Bile salt transport systems. In: The bile acids (Nair, P.P., and Kritchevsky, D., eds.), vol. 2, p. 33–54. New York: Plenum Press 1973
98. Carey, J.B., Jr.: Bile salt metabolism in man. In: The bile acids (Nair, P.P., and Kritchevsky, D., eds.), vol. 2, p. 55–82. New York: Plenum Press 1973
99. Palmer, R.H.: Experimental cholelithiasis. In: The bile acids (Nair, P.P., and Kritchevsky, D., eds.), vol. 2, p. 153–190. New York: Plenum Press 1973
100. Javitt, N.B., McSherry, C.K.: Pathogenesis of cholesterol gallstones. Hosp. Prac. **8**, 39–48 (1973)
101. Grundy, S.M.: Cholesterol—Bile acid interactions in gallstone pathogenesis. Hosp. Prac. **8**, 57–65 (1973)
102. Evans, D.F., Cussler, E.L.: Physicochemical considerations in gallstone pathogenesis. Hosp. Prac. **9**, 133–140 (1974)
103. Gunn, A., Keddie, N.: Some clinical observations on patients with gallstones. Lancet **1972II**, 239–241
104. King, J.E., Schoenfield, L.J.: Lithocholic acid, cholestasis, and liver disease. Mayo Clin. Proc. **47**, 725–730 (1972)
105. Tompkins, R.K., King, W., III.: Investigations of the enterobiliary metabolism of lecithin. Surgery **75**, 243–252 (1974)
106. Grundy, S.M., Metzger, A.L., Adler, R.D.: Mechanisms of lithogenic bile formation in American Indian women with cholesterol gallstones. J. clin. Invest. **51**, 3026–3043 (1972)
107. Smallwood, R.A., Jablonski, P., Watts, J. McK.: Intermittent secretion of abnormal bile in patients with cholesterol gall stones. Brit. med. J. **4**, 263–266 (1972)
108. Simmons, F., Ross, A.P.J., Bouchier, I.A.D.: Alterations in hepatic bile composition after cholecystectomy. Gastroenterology **63**, 466–471 (1972)
109. Popper, H., Schaffner, F.: Pathophysiology of cholestasis. Hum. Path. **1**, 1–24 (1970)
110. Greim, H., Czygan, P., Popper, H.: Mechanism of necrosis induced by hepatotoxic bile acids. In: IV. Workshop on experimental liver injury. "Pathogenesis and mechanisms of liver cell necrosis", Freiburg, November 9 and 10, 1974 (Abstract)
111. Shaffer, E.A., Braasch, J.W., Small, D.M.: Bile composition at and after surgery in normal persons and patients with gallstones. Influence of cholecystectomy. New Engl. J. Med. **287**, 1317–1322 (1972)

Subject Index

Related Titles

The Dynamic Structure of Cell Membranes

Editors: D.F. Hölzl Wallach, H. Fischer (22. Colloquium der Gesellschaft für Biologische Chemie, 15.–17. April 1971 in Mosbach/Baden)

This book comprises tutorial reviews of the most exciting areas of membrane biology by noted worldwide authorities. Its diverse, interdisciplinary material is knit together in a comprehensive round-table discussion.

Protein-Calorie Malnutrition

A Nestlé Foundation Symposium. Editor: A.v. Muralt

The symposium on protein-calorie malnutrition, which gathered experts from 12 countries, discussed, among other subjects mainly the possibility of a biochemical assessment of human adaptation to a low protein intake. It became clear that there is no single and unique method, but that the approach has to be made with a "battery" of biochemical methods. A new trend developed at this symposium: the introduction of highly refined biochemical methods in the study of world problems of nutrition.

Connective Tissues

Biochemistry and Pathophysiology. Editors: R. Fricke, F. Hartmann. Editorial Board: E. Buddecke, R. Fricke, F. Hartmann, H. Muir, K. Kühn

These symposium proceedings present current research on connective tissues and cover the latest information on metabolism and catabolism in their various components, including the findings of biochemists, immunobiologists, and pathophysiologists. Each chapter is virtually a self-contained review of one particular area of the field.

Journals

Berichte Biochemie und Biologie

Biochemistry – Biology

Referierendes Organ der Deutschen Botanischen Gesellschaft und der Zoologischen Gesellschaft
Managing Editors: G. Czihak, R. Lehmann, G. Schoser, E. Schwartz, J. Schwoerbel, L. Träger, M. Wiedemann, I. Ziegler

Berichte Physiologie, physiologische Chemie und Pharmakologie

Physiology – Physiological Chemistry – Pharmacology

Edited by K. Lang
Managing Editors: K. Lang in cooperation with A. Fricker

European Journal of Biochemistry

Published on behalf of the Federation of European Biochemical Societies (FEBS)

Editor-in-Chief: C. Liébecq

Histochemistry

Managing Editor: T.H. Schiebler

Molecular and General Genetics

An International Journal

Managing Editors: G. Melchers, H. Stube

Springer-Verlag
Berlin Heidelberg New York